SYSTEM IDENTIFICATION
(SYSID'03)

A Proceedings volume from the 13th IFAC Symposium on System Identification, Rotterdam, The Netherlands, 27 – 29 August 2003

Edited by

P.M.J. Van den HOF
Delft Center for Systems and Control,
Delft University of Technology,
Delft, The Netherlands

B. WAHLBERG
Royal Institute of Technology,
Stockholm, Sweden

S. WEILAND
Department of Electrical Engineering
Eindhoven University of Technology,
Eindhoven, The Netherlands

(In four volumes)

Volume 4

Published for the

INTERNATIONAL FEDERATION OF AUTOMATIC CONTROL

by

ELSEVIER LTD

ELSEVIER Ltd
The Boulevard, Langford Lane
Kidlington, Oxford OX5 1GB, UK

Elsevier Internet Homepage
http://www.elsevier.com

Consult the Elsevier Homepage for full catalogue information on all books, journals and electronic products and services.

IFAC Publications Internet Homepage
http://www.elsevier.com/locate/ifac

Consult the IFAC Publications Homepage for full details on the preparation of IFAC meeting papers, published/forthcoming IFAC books, and information about the IFAC Journals and affiliated journals.

First edition 2004

Library of Congress Cataloging in Publication Data

A catalogue record for this book is available from the Library of Congress

British Library Cataloguing in Publication Data

A catalogue record for this book is available from the British Library

ISBN 0-08-043709 5
ISSN 1474-6670

Printed and bound in the United Kingdom
Transferred to Digital Print 2010

To Contact the Publisher

Elsevier welcomes enquiries concerning publishing proposals: books, journal special issues, conference proceedings, etc. All formats and media can be considered. Should you have a publishing proposal you wish to discuss, please contact, without obligation, the publisher responsible for Elsevier's industrial and control engineering publishing programme:

Christopher Greenwell
Publishing Editor
Elsevier Ltd
The Boulevard, Langford Lane
Kidlington, Oxford
OX5 1GB, UK

Phone: +44 1865 843230
Fax: +44 1865 843920
E.mail: c.greenwell@elsevier.com

General enquiries, including placing orders, should be directed to Elsevier's Regional Sales Offices – please access the Elsevier homepage for full contact details (homepage details at the top of this page).

13th IFAC SYMPOSIUM ON SYSTEM IDENTIFICATION (SYSID 2003)

Sponsored by
International Federation of Automatic Control (IFAC)
IFAC Technical Committees on:
- Modeling, Identification and Signal Processing (MISP)
- Adaptive Control and Tuning (ACT)

Co-sponsored by
IEEE Control Systems Society
Division of Automatic Control (MRBT) of the Royal Institution of Engineers in The Netherlands (KIVI)
The Netherlands Organisation of Scientific Research (NWO)
Royal Netherlands Academy of Arts and Sciences (KNAW)
Dutch Institute of Systems and Control (DISC)
Faculty of Applied Sciences, Delft University of Technology (TUD)
Delft Center for Systems and Control (TUD)
Department of Electrical Engineering, Eindhoven University of Technology, The Netherlands (TU/e)
Stichting Meten en Regelen ER-THE, The Netherlands

Organizing Committee
P.M.J. Van den Hof – Delft University of Technology, Delft, The Netherlands
B. Wahlberg – Royal Institute of Technology, Stockholm, Sweden
S. Weiland – Eindhoven University of Technology, Eindhoven, The Netherlands

IPC Task Force
M. Deistler
M. Gevers
L. Ljung
M. Morari
J. Schoukens
P.M.J. Van den Hof
M. Viberg
B. Wahlberg

International Programme Committee (IPC)
P.M.J. Van den Hof; The Netherlands (Co-Chair)
B. Wahlberg, Sweden (Co-Chair)

P. Albertos; Spain
B. Anderson; Australia
E. Bai; USA
M. Basseville; France
R. Bitmead; USA
S. Bittanti* **; Italy
M. Blanke; Denmark
J. Bokor; Hungary
M. Campi; Italy
H.F. Chen; P.R. China
J. Chen; USA
R. de Callafon; USA
M. Deistler; Austria
B. de Moor; Belgium
J.J. Fuchs; France
K. Godfrey; UK

G. Goodwin; Australia
M. Gevers; Belgium
P. Guillaume; Belgium
L. Guo; P.R. China
H. Hjalmarsson; Sweden
H. Kimura; Japan
R. Kosut; USA
V. Krishnamurty; Australia
K. Kumamaru; Japan
I. Landau; France
J.H. Lee; USA
L. Ljung; Sweden
P. Mäkilä; Finland
T. McKelvey; Sweden
M. Milanese; Italy
M. Morari; Switzerland
B. Ninness; Australia
R. Ortega**; France
G. Picci; Italy
R. Pintelon; Belgium
B. Polyak; Russia
P. Regalia; France
D. Rivera; USA
W. Scherrer; Austria
J. Schoukens; Belgium
R. Schumann; Germany
R. Smith*; USA
T. Söderström*; Sweden
T. Sugie; Japan
R. Tempo; Italy
J. van Schuppen; The Netherlands
M. Verhaegen; The Netherlands
S. Veres; UK
M. Viberg; Sweden
A. Vicino; Italy
E. Walter; France

** Appointed by IFAC Technical Committee MISP*
*** Appointed by IFAC Technical Committee ACT*

National Organizing Committee (NOC)

P.M.J. Van den Hof (Finances, contacts NMO, Public Relations)
S. Weiland (PC Secretariat, Paper handling, Website)
A.C.P.M. Backx (Industrial participation, Sponsors)
M.H.G. Verhaegen (Publications)
Y. Zhu (Exhibitions)
T. Van der Weiden (Local arrangements, Technical and Social Events)

PREFACE

These Proceedings contain all the technical material presented at the 13th IFAC Symposium on System Identification (SYSID 2003), held in the Conference Center "De Doelen", Rotterdam, The Netherlands from 27 – 29 August 2003.

The SYSID symposium is organized every three years and is among the most successful symposia organized by IFAC. This has been the first SYSID symposium in the 3rd millennium and the second SYSID symposium to take place in The Netherlands, following The Hague symposium in 1973.

Being the only worldwide symposium that is fully directed towards system identification, it is the ideal opportunity for researchers and industrial engineers from very many disciplines to present and discuss the developments, the results and the future challenges in all aspects of modelling dynamical systems on the basis of experimental data.

The symposium covered all major aspects of system identification, experimental modelling, signal processing and adaptive control from theoretical and methodological developments to practical applications in a wide range of application areas. For the 13th edition of this symposium, the International Program Committee has taken steps to position SYSID 2003 as a meeting place where scientists and engineers from several research communities can meet to discuss issues related to these areas.

A total of 350 delegates from 40 different countries attended the conference. 100 of the participants were PhD students, showing that system identification is a very vital field of research. Out of a total of 422 papers that were submitted to SYSID 2003, the IPC selected 333 papers and these were incorporated in the final program. The selection was based on two referee reports per paper. The final program of the symposium was composed of 3 plenary papers, 6 semi-plenary papers, 232 papers in oral sessions, 82 posters and 10 software demonstrations. The Preprints of this Symposium appeared on CD-ROM and were distributed among the participants of the symposium. The Proceedings of SYSID-2003 contain 321 papers.

We hope that you, as reader or as researcher in the area of System Identification, will find the contents of these Proceedings useful and informative for your professional work.

We would like to thank all members of the International Program Committee (IPC), members of the IPC Taskforce and members of the National Organizing Committee for their work in the organization of this symposium and in the preparation of these Proceedings. We would also like to thank many friends and colleagues for their help and support in many practical matters related to SYSID 2003.

The editors,

Paul Van den Hof
Bo Wahlberg
Siep Weiland.

CONTENTS

VOLUME 1

PLENARY PAPER
FROM EXPERIMENTS TO CLOSED-LOOP CONTROL

From Experiments to Closed Loop Control — 1
H. HJALMARSSON

IDENTIFICATION FOR CONTROL

Exploratory Modelling for Controller Optimization — 15
S.M. VERES

Connecting PE Identification and Robust Control Theory: The Multiple-Input Single-Output Case. Part I: Uncertainty Region Validation — 21
X. BOMBOIS, P. DATE

Connecting PE Identification and Robust Control Theory: The Multiple-Input Single-Output Case. Part II: Controller Validation — 27
X. BOMBOIS, P. DATE

Relation Between Uncertainty Structures in Identification for Robust Control — 33
S.G. DOUMA, P.M.J. van DEN HOF

Strong Robustness Measures for Sets of Linear SISO Systems — 39
M. CADIC, S. WEILAND, J.W. POLDERMAN

Using a Sufficient Condition to Analyze the Interplay Between Identification and Control — 45
H. HJALMARSSON, H. JANSSON

NONLINEAR IDENTIFICATION

Structure Selection with ANOVA: Local Linear Models — 51
I. LIND, L. LJUNG

On Identification of Hammerstein Systems Using Excitation with a Finite Number of Levels — 57
T. MCKELVEY, C. HANNER

Fast Approximate Identification of Nonlinear Systems — 61
J. SCHOUKENS, J. NEMETH, P. CRAMA, Y. ROLAIN, R. PINTELON

Gaussian Processes Framework for Validation of Linear and Nonlinear Models — 67
A. LUNDGREN, J. SJÖBERG

Functional Analytic Framework for Model Selection — 73
M. SUGIYAMA

Robust Complexity Criteria for Nonlinear Regression in NARX Models — 79
J. de BRABANTER, K. PELCKMANS, J.A.K. SUYKENS, B. de MOOR, J. VANDEWALLE

IDENTIFICATION OF MIMO COMMUNICATION CHANNELS

Analysis of MIMO Channel Measurements — 85
G. Del GALDO, M. HENNHÖFER, M. HAARDT

Performance Evaluation of MIMO Channel Prediction Algorithms Using Measurements — 91
T. SVANTESSON, J.W. WALLACE

High-Resolution Channel Parameter Estimation for Communication Systems Equipped with Antenna Arrays 97
B.H. FLEURY, X. YIN, P. JOURDAN, A. STUCKI

Analysis of Spectral-Based Localization of Spatially Distributed Sources 103
M. TAPIO, M. VIBERG

Ray Tracing Interpretation of Multiple-Input Multiple-Output Wireless Systems 109
P.F. DRIESSEN

Computationally Efficient Blind MMSE Receivers for Long Code WCMDA Using Time-Varying Systems Theory 115
A.-J. van der VEEN, L. TONG

ESTIMATION IN PHYSICAL AND MEDICAL SYSTEMS

Maximum Likelihood Identification of Quantum Systems for Control Design 121
R.L. KOSUT, H. RABITZ, I.A. WALMSLEY

Maximum Likelihood Estimation of Signal Amplitude and Noise Variance from Complex Valued Data 127
A.J. den DEKKER, J. SIJBERS

Reliable Nonlinear Identification in Medical Applications 133
L.Y. WANG, G.G. YIN, H. WANG

Pattern Recognition of EEG Signals During Right and Left Motor Imagery 139
K. INOUE, G. PFURTSCHELLER, C. NEUPER, K. KUMAMARU

From Dynamic Metabolic Modeling to Unstructured Model Identification of Complex Biosystems 145
J.E. HAAG, A.VANDE WOUWER, P. BOGAERTS

Flow Controlled Non-Invasive Ventilation Considering Mask Leakage and Spontaneous Breathing 151
F. DIETZ, A. SCHLOßER, D. ABEL

STOCHASTIC SYSTEMS

Estimation and Identification of Non-Stationary Functional Series TARMA Models 157
A.G. POULIMENOS, S.D. FASSOIS

Modelling Multivariate Pollutant Time Series with Wavelet Functions 163
G. NUNNARI, D. LONGO

Estimating the Lyapunov Exponents of Chaotic Time Series Based on Polynomial Modelling 169
M. ATAEI, A. KHAKI-SEDIGH, B. LOHMANN

Sampling Density Design for Particle Filters 175
M. ŠIMANDL, O. STRAKA

Diffusive Representation of *N*-Th Order Fractional Brownian Motion 181
J. SEMBIRING, K. SOEMINTAPOERA, T. KOBAYASHI, K. AKIZUKI

APPLICATIONS OF SYSTEM IDENTIFICATION

Multi-Channel Active Noise Control for Uncertain Secondary Channels 187
Y. OHTA, H. OHMORI, A. SANO

Channel Estimation and Coupling Wave Cancellation in OFDM Relay Station 193
L. SUN, A. SANO

Application of System Identification for the Prediction of Avalanche Hazard 199
J. MILEK, B. BRABEC

Models for Incoming Calls Forecasting in a Customer Attention Center 205
M.R. ARAHAL, P.P. FERNANDO, E.F. CAMACHO

Modeling the Relationships Between the Users DB and the Web-Log File of a Large Virtual Community 211
S.M. SAVARESI, S. GARATTI, S. BITTANTI

FINANCIAL ECONOMETRICS

A Short Introduction to Time-Varying Volatility in Financial Time Series 217
B. HANZON

Forecasting Emerging Equity Market Volatility Using Nonlinear GARCH Models 221
D. van DIJK

Stochastic Properties of Multivariate Time Series Equations with Emphasis on ARCH 227
A. RAHBEK

A Rational Probability Density Approach to Stochastic Volatility Estimation 233
B. HANZON

SEMI-PLENARY
SNIPPETS OF IDENTIFICATION THEORY IN COMPUTER VISION

Snippets of System Identification in Computer Vision 237
S. SOATTO, A. CHIUSO

SEMI-PLENARY
INTERVAL ANALYSIS FOR GUARANTEED NONLINEAR PARAMETER ESTIMATION

Interval Analysis for Guaranteed Nonlinear Parameter Estimation 249
E. WALTER, M. KIEFFER

IDENTIFICATION IN AUTOMOTIVE SYSTEMS

Online Detection of Tyre Pressure Deflation in Passenger Cars 261
J. SHAH, M. BÖRNER, R. ISERMANN, Y.G. SRINIVASA

A Subspace-Based Identification Approach for the Analysis of Road Vehicles Yaw Dynamics Around Steering-Pad Conditions 267
S.M. SAVARESI, E. SILANI, S. BITTANTI, F. FARACHI

Identification and Fault Detection of an Active Vehicle Suspension 273
D. FISCHER, M. ZIMMER, R. ISERMANN

Non-Adaptive Neural Automotive Sideslip Virtual Sensor 279
M. BATTIPEDE, D. DANESIN, P. KRIEF, G. SASSI, M. VELARDOCCHIA

Parametric Identification of the Car Dynamics 285
G. VENTURE, M. GAUTIER, W. KHALIL, P. BODSON

Simulating Energy Consumption of Auxiliary Units in Heavy Vehicles 291
N. PETTERSSON, K.H. JOHANSSON

SENSOR IDENTIFICATION AND MONITORING

Prior Characterization of the Performance of Software Sensors 297
I. BRAEMS, M. KIEFFER, E. WALTER

Model Based Source Localisation by Distributed Sensors for Point Sources and Diffusion 303
J. MATTHES, L. GRÖLL

Continuous-Time Model Identification by Using Adaptive Observer 309
K. IKEDA, Y. MOGAMI, T. SHIMOMURA

Optimal Filtering of Nonlinear Systems Based on Pseudo Gaussian Densities 315
U.D. HANEBECK

A Total Least Squares Approach to Sensor Characterisation 321
P.C.F. HUNG, S. MCLOONE, G. IRWIN, R. KEE

IDENTIFICATION OF NONLINEAR SYSTEMS I

Estimation and Validation of Semi-Parametric Dynamic Nonlinear Models 327
Y. ROLAIN, W. van MOER, J. SCHOUKENS

Nonlinear System Modelling Using the RBF Neural Network-Based Regressive Model 333
H. PENG, T. OZAKI, Y. TOYODA, K. NAKANO

Modeling and Linearization of Nonlinear Dynamic Systems 339
J.G. NÉMETH, J. SCHOUKENS

Linear Parameter Estimation and Predictive Constrained Control of Wiener/Hammerstein Systems 345
K.J. LATAWIEC, C. MARCIAK, R. ROJEK, G.H.C. OLIVEIRA

Identification of Wiener Systems Using Reduced Complexity Volterra Models 351
R. HACIOĞLU, G.A. WILLIAMSON

Structure Selection for Polynomial NARX Models Based on Simulation Error Minimization 357
L. PIRODDI, W. SPINELLI

MECHANICAL AND AEROSPACE APPLICATIONS

Nonlinear Identification of a Two Link Robotic System Using Dynamic Neural Networks 363
S. TORRES, V.M. BECERRA

Neural Network System Identification for a Low Pressure Non-Linear Dynamical Subsystem Onboard the Alicia II Climbing Robot 369
D. LONGO, G. MUSCATO, G. NUNNARI

Measurement of Young's Modulus Via Modal Analysis Experiments: A System Identification Approach 375
R. PINTELON, P. GUILLAUME, K. de BELDER, Y. ROLAIN

A Novel Algorithm for Fully Autonomous Star Identification 381
S. BITTANTI, E. de MARCHI, M. GIRANZANI, M. LOVERA, B. LÜBKE-OSSENBECK, E. SILANI

Fast Model Updates and Simulation for Efficient Flight Control Software Design 387
H. FRIEHMELT, D. ROHLF

CLOSED-LOOP IDENTIFICATION

Continuous-Time Identification of First-Order Plus Dead-Time Models from Step Response in Closed Loop 393
F.S. COELHO, P.R. BARROS

Identification of Simple Continuous-Time Models from Relay Feedback 399
G.H.M. de ARRUDA, P.R. BARROS

Continuous-Time Model Identification of Systems Operating in Closed-Loop 405
M. GILSON, H. GARNIER

Multivariable Closed-Loop System Identification of Plants Under Model Predictive Control 411
E. de KLERK, I.K. CRAIG

Dead Time Measurement of Closed Loop System by Wavelet 417
T. TABARU, S. SHIN

Closed Loop Identification Method Using a Subspace Approach 423
M. POULIQUEN, M. M'SAAD

INDUSTRIAL APPLICATION OF IDENTIFICATION

Model Identification of a Multivariable Industrial Furnace 429
M. BARRERAS, M. GARCÍA-SANZ

Extended Fuzzy GK Clustering with Application to Identification of an Automatic Voltage Regulation Loop Dynamics 435
L. REN, G.W. IRWIN

On Simplified Modelling Approaches to SMB Processes 441
V. GROSFILS, C. LEVRIE, M. KINNAERT, A.VANDE WOUWER

Optimal Filtering for Bilinear Systems and its Application to Terpolymerization Process State Identification 447
M. BASIN, M.A. ALCORTA-GARCIA

Neural Prediction of Cylinder Air Mass for AFR Control in SI Engine 453
G. BLOCH, Y. CHAMAILLARD, G. MILLERIOUX, P. HIGELIN

Contribution to Identification of Thermo-Mechanic Interaction at Vibrating Rubber-Like Materials 459
L. PEŠEK, L. PŮST, F. VANĚK

PROCESS CONTROL SYSTEMS

Identification of a High Efficiency Boiler by Support Vector Machines Without Bias Term 465
M. VOGT, K. SPREITZER, V. KECMAN

Implementing GA-Based Predictive Controller for on-line Control of a Process Mini-Plant 471
Y.Y. NAZARUDDIN, F. MAULANA

Long-Range Optimal Model and Multi-Step-Ahead Prediction Identification for Predictive Control 477
R. HABER, U. SCHMITZ, R. BARS

Predictive Control of Flow Quantity and Sloshing-Suppression During Back-Tilting of a Ladle for Batch-Type Casting Pouring Processes 483
K. TERASHIMA, K. YANO, M. KANEKO

CLOSED LOOP AND PERFORMANCE ISSUES

Optimal Prefiltering in Iterative Feedback Tuning 489
R. HILDEBRAND, A. LECCHINI, G. SOLARI, M. GEVERS

Identification of Performance Limitations in Control Using General SISO Models 495
J. MÅRTENSSON, H. HJALMARSSON

Control Loop Performance Monitoring by CUSUM Algorithms for Local Linear Hypotheses 501
M. KINNAERT, R. HANUS, C. PARLOIR

Model Approximation of Plant and Noise Dynamics on the Basis of Closed-Loop Data 507
J. ZENG, R.A. de CALLAFON

IV Methods for Closed-Loop System Identification 513
M. GILSON, P. van den HOF

Coprime Factor Perturbation Models for Closed-Loop Model Validation Techniques 519
M. CROWDER, R.A. de CALLAFON

REPRODUCING KERNELS I

An Introduction to Reproducing Kernel Hilbert Spaces and Why they are so Useful 525
G. WAHBA

An Introduction to Smoothing Spline ANOVA Models in RKHS, with Examples in Geographical Data, Medicine, Atmospheric Sciences and Machine Learning 531
G. WAHBA

Robust Design with Nonparametric Models: Prediction of Second-Order Characteristics of Process Variability by Kriging 537
L. PRONZATO, É. THIERRY

Geostatistical Models and Kriging 543
H. WACKERNAGEL

Hilbert Space Embeddings in Dynamical Systems 549
A.J. SMOLA, S.V.N. VISHWANATHAN

Bayesian Input Selection for Nonlinear Regression with LS-SVMs 555
T. van GESTEL, M. ESPINOZA, J.A.K. SUYKENS, C. BRASSEUR, B. de MOOR

VOLUME 2

BLIND ESTIMATION AND EQUALIZATION

Blind Turbo Equalization Using the Constant Modulus Algorithm 561
P.A. REGALIA

A New Method for Channel Estimation and Data Detection in the Context of Turbo Equalisation 567
S. PERREAU, G. GORLIER

On the Applicability to Correlated Sources of a Blind Channel Equalization Method Robust to Order Overestimation 573
R. LÓPEZ-VALCARCE

Blind Estimation with Signal Scrambling 579
H. XU, X. SONG, S. DASGUPTA

Blind Channel Shorteners 585
C.R. JOHNSON Jr., R.K. MARTIN, J.M. WALSH, A.G. KLEIN, C.E. ORLICKI, T. LIN

Multiple Antenna System Equalization Using Semi-Blind Subspace Identification Methods 591
C. ZHANG, R.R. BITMEAD

CONTINUOUS TIME IDENTIFICATION

The Identification of Continuous-Time Linear and Nonlinear Models: A Tutorial with Environmental Applications 597
P.C. YOUNG, H. GARNIER, A. JARVIS

Continuous-Time System Identification of A Food Extruder: Experiment Design and Data Analysis 609
L. WANG, P.J. GAWTHROP, C. CHESSARI, T. PODSIADLY

Identification of Continuous Time Models Using Discrete Time Data 615
N.R. KRISTENSEN, H. MADSEN, S.B. JØRGENSEN

On Possibilities for Estimating Continuous-Time ARMA Parameters 621
E.K. LARSSON, M. MOSSBERG

On the Interpretation of a Continuous–Time Model Identification Method in Terms of Regularization 627
S. MOUSSAOUI, D. BRIE, A. RICHARD

INPUT DESIGN

A Survey of Readily Accessible Perturbation Signals 633
K.R. GODFREY, A.H. TAN, H.A. BARKER

Multiple Input Design for Real-Time Parameter Estimation in the Frequency Domain 639
E.A. MORELLI

Minimizing the Worst-Case ν-Gap by Optimal Input Design 645
R. HILDEBRAND, M. GEVERS

Identification of Resonant Systems Using Periodic Multiplicative Reference Signals 651
W.J. DUNSTAN, R.R. BITMEAD

Aircraft Parameter Estimation by Using the Optimal Input Design and Linear Matrix Inequalities 657
C. JAUBERTHIE, L. DENIS-VIDAL, G. JOLY-BLANCHARD

The Performance of Multilevel Perturbation Signals for Nonlinear System Identification 663
H.A. BARKER, A.H. TAN, K.R. GODFREY

IDENTIFICATION FOR FLIGHT TEST EXPLORATION

Applying System Identification to Assess the Vibro-Acoustic Behaviour of Airplanes 669
B. PEETERS, R. RUOTOLO, A. VECCHIO, H. van der AUWERAER

Subspace Identification Combined with New Mode Selection Techniques for Modal Analysis of an Airplane 675
I. GOETHALS, B. de MOOR

Flight Flutter Analysis Using Frequency-Domain System Identification Techniques 681
P. GUILLAUME, P. VERBOVEN, B. CAUBERGHE

Real-Time Modal Analysis and its Application for Flutter Testing 687
T. UHL, M. BOGACZ

Statistical Approach to Flutter Monitoring 693
L. MEVEL, M. BASSEVILLE, A. BENVENISTE

Reliable System Identification for Large Flexible Space Structures 699
V. BABUŠKA, S.L. LACY, R.S. ERWIN, A.M. MELIN

IDENTIFIABILITY

Identifiability Analysis of a Class of Systems Described By Convolution Equations 705
L. BELKOURA

Identification of Fully Parameterized Linear and Nonlinear State-Space Systems by Projected Gradient Search 711
V. VERDULT, N. BERGBOER, M. VERHAEGEN

A Differential Geometric Viewpoint on Local Identifiability and Identification Part I: Theory 717
B. EITZINGER, K. SCHLACHER

A Differential Geometric Viewpoint on Local Identifiability and Identification Part II: Application 723
B. EITZINGER, K. SCHLACHER

Identifiability of Nonlinear Homogeneous Polynomial Systems 729
R. PEETERS, B. HANZON

PLENARY PAPER
SYSTEM IDENTIFICATION FOR STRUCTURAL DYNAMICS AND VIBROACOUSTICS DESIGN ENGINEERING

System Identification for Structural Dynamics and Vibroacoustics Design Engineering 735
H. van der AUWERAER

SELECTED TOPICS IN IDENTIFICATION

A Personal View on the Development of System Identification 747
M. GEVERS

System Identification Via a Computational Bayesian Approach 759
B. NINNESS, S. HENRIKSEN

A New Information Theoretic Approach to Order Estimation Problem 765
S. BEHESHTI, M.A. DAHLEH

Conditions for Local Convergence of Maximum Likelihood Estimation for Armax Models 771
G.C. GOODWIN, J.C. AGÜERO, R.E. SKELTON

A Nonparametric Approach to Model Selection 777
M. BEKARA, A.-K. SEGOUANE, F. GILLES

REPRODUCING KERNELS II

An Introduction to Learning with Reproducing Kernel Hilbert Spaces 783
M. PONTIL

Sparse Gaussian Processes: Inference, Subspace Identification and Model Selection 789
L. CSATÓ, M. OPPER

Sparse Kernel Methods 795
S.R. GUNN

A Generalised LS–SVM 801
J. VALYON, G. HORVÁTH

Adaptive Kernel Methods 807
A. KUH

Subspace Regression in Reproducing Kernel Hilbert Space 813
L. HOEGAERTS, J.A.K. SUYKENS, J. VANDEWALLE, B. de MOOR

IDENTIFICATION OF NONLINEAR BLOCK MODELS

Frequency Domain Identification of Wiener Models 819
E.-W. BAI

Non-Parametric Identification of Non-Linearity in Hammerstein Systems 825
W. GREBLICKI, P. ŚLIWIŃSKI

Generation of Enhanced Initial Estimates for Wiener Systems and Hammerstein Systems 831
P. CRAMA, J. SCHOUKENS, R. PINTELON

User Choices and Model Validation in System Identification Using Nonlinear Wiener Models 837
T. WIGREN

Approximation of Feasible Parameter Set in Worst Case Identification of Block-Oriented Nonlinear Models 843
L. GIARRÉ, G. ZAPPA

Parameters Set Evaluation of Wiener Models from Data with Bounded Output Errors 849
V. CERONE, M. MILANESE, D. REGRUTO

NEW RESULTS IN SUBSPACE IDENTIFICATION

Constructing the State of Random Processes with Feedback 855
A. CHIUSO, G. PICCI

Closed-Loop Subspace Identification with Innovation Estimation 861
S.J. QIN, L. LJUNG

A Frequency Domain Subspace Algorithm for Mixed Causal, Anti-Causal LTI Systems 867
R. FRAANJE, M. VERHAEGEN, V. VERDULT, R. PINTELON

A Stochastic Realization in a Hilbert Space Based on "LQ Decomposition" with Application to Subspace Identification 873
H. TANAKA, T. KATAYAMA

Subspace-Based Identification Methods Using Schur Complement Approach 879
Y. TAKEI, H. NANTO, S. KANAE, Z.-J. YANG, K. WADA

Recursive Subspace Identification for Continuous-/Discrete-Time Stochastic Systems 885
A. OHSUMI, Y. MATSUÜRA, K. KAMEYAMA

IDENTIFICATION FOR PROCESS CONTROL: INPUT DESIGN

"Plant-Friendly" System Identification: A Challenge for the Process Industries 891
D.E. RIVERA, H. LEE, M.W. BRAUN, H.D. MITTELMANN

Multi-Objective Input Signal Design for Plant-Friendly Identification 897
S. NARASIMHAN, R. SRINIVASAN, R. RENGASWAMY

Control-Relevant Design of Periodic Test Input Signals for Iterative Open-Loop Identification of Multivariable FIR Systems 903
J.H. LEE

Constrained Signal Design Using Approximate Prior Models with Application to the Tennessee Eastman Process 909
T. LI, C. GEORGAKIS

Constrained Minimum Crest Factor Multisine Signals for "Plant-Friendly" Identification of Highly Interactive Systems 915
H. LEE, D.E. RIVERA, H.D. MITTELMANN

IDENTIFICATION OF MECHANICAL SYSTEMS

Online Identification of a Robot Using Batch Adaptive Control 921
B. BUKKEMS, D. KOSTIĆ, B. de JAGER, M. STEINBUCH

Dynamic Identification of a Compactor Using Splines Data Processing 927
C.-E. LEMAIRE, P.-O. VANDANJON, M. GAUTIER

Non-Stationary Mechanical Vibration Modeling and Analysis Via Functional Series TARMA Models 933
A.G. POULIMENOS, S.D. FASSOIS

Globally Convergent Adaptive Tracking of Angular Velocity with Inertia Identification and Adaptive Linearization 939
A.K. SANYAL, M. CHELLAPPA, J.L. VALK, J. AHMED, J. SHEN, D.S. BERNSTEIN

On Vision-Based Kinematic Calibration of n-Leg Parallel Mechanisms 945
P. RENAUD, N. ANDREFF, G. GOGU, P. MARTINET

A Geometric Approach to Motion Tracking in Manifolds 951
J.G. SILVA, J.S. MARQUES, J.M. LEMOS

SOFTWARE SESSION I

Version 6 of the System Identification Toolbox 957
L. LJUNG

Process Identification, Controller Tuning and Control Circuit Simulation Using MS Excel 963
H.M. SCHAEDEL

Developments for the MATLAB CONTSID Toolbox 969
H. GARNIER, M. GILSON, E. HUSELSTEIN

detectNARMAX: A Graphical User Interface for Structure Detection of NARMAX Models Using The Bootstrap Method 975
E. SHAFAI, M. BIANCHI, H.P. GEERING

SEMI-PLENARY
DATA-BASED METHODS IN PROCESS CONTROL

Data-Based Methods for Process Analysis, Monitoring and Control 981
J.F. MacGREGOR

SEMI-PLENARY
SUBSPACE ALGORITHMS

Subspace Algorithms 993
D. BAUER

FILTERING AND ESTIMATION

Optimal Filtering for Linear Systems with Multiple Delays in Observations 1005
M. BASIN, R. MARTINEZ-ZUNIGA

The Information Analysis in Joint Problem of Continuous-Discrete Filtering and Generalized Extrapolation 1011
N.S. DYOMIN, I.E. SAFRONOVA, S.V. ROZHKOVA

Guaranteed Ellipsoidal State Estimation for Uncertain MIMO Models 1017
B.T. POLYAK, S.A. NAZIN, C. DURIEU, É. WALTER

Regularized Robust Estimators for Time Varying Uncertain Discrete-Time Systems 1023
A. SUBRAMANIAN, A.H. SAYED

Minimax L_2-E_2 FIR Filters for Deterministic Continuous-Time State Space Signal Models 1029
S.H. HAN, W.H. KWON

Numerically Reliable H_∞ – Synthesis of Estimators Based on J – Lossless Factorisations 1035
P. SUCHOMSKI

DIAGNOSIS, DETECTION AND TRACKING

Statistical Analysis of Subspace-Based Method for Direction Estimation Without Eigendecomposition 1041
J. XIN, A. SANO

Fault Detection of Non-Linear Systems Based on Multi-Form Quasi-Armax Modeling and its Application to the Ship Benchmark 1047
K. KUMAMARU, K. INOUE, Y. HOSOYAMADA, T. SÖDERSTRÖM

A Comparison of Two Methods for Stochastic Fault Detection: The Parity Space Approach and Principal Components Analysis 1053
A. HAGENBLAD, F. GUSTAFSSON, I. KLEIN

Identification of Object's Movement Models in a Radar Tracking Filter 1059
M. SANKOWSKI, Z. KOWALCZUK

Estimation and Tracking of Quasi-Periodically Varying Processes 1065
M. NIEDŹWIECKI, P. KACZMAREK

VOLUME 3

IDENTIFICATION OF NONLINEAR SYSTEMS II

A Pruning Method for the Identification of Polynomial NARMAX Models 1071
L. PIRODDI, W. SPINELLI

Generalized Orthonormal Basis Selection for Expanding Quadratic Volterra Filters 1077
A.Y. KIBANGOU, G. FAVIER, M.M. HASSANI

A Localised Forgetting Method for On-Line Adaptation of Gaussion RBFN Models 1083
D.L. YU, J.B. GOMM, D.W. YU, D. WILLIAMS

Subspace Identification of Switching Model 1089
K.M. PEKPE, K. GASSO, G. MOUROT, J. RAGOT

Application-Oriented Neural Modelling 1095
K. LI, G. IRWIN

IDENTIFICATION METHODS

Closed-Form Frequency Estimation Using Second-Order Notch Filters 1101
S.M. SAVARESI, S. BITTANTI, H.C. SO

L_1 Prediction Error System Identification: A Modified AIC Rule 1107
J.C. CARMONA, M. OULADSINE, M. EL ADEL

On Parameter Estimation of ARMAX Model Via BCLS Method 1113
L.-J. JIA, S. KANAE, Z.-J. YANG, K. WADA

Estimation in the Presence of Interferences 1119
J.J. FUCHS

Autoregressive Spectral Analysis with Randomly Missing Data 1125
P.M.T. BROERSEN, S. de WAELE, R. BOS

Estimating Unknown Probability Density Functions for Random Parameters of Stochastic ARMAX Systems 1131
H. WANG, Y. WANG

CONTROLLER TUNING AND IDENTIFICATION

Iterative Controller Tuning by Minimization of a Generalized Decorrelation Criterion 1137
L. MIŠKOVIĆ, A. KARIMI, D. BONVIN

Subspace Identification Based PID Control Tuning 1143
A. SANCHEZ, M.R. KATEBI, M.A. JOHNSON

Evolutionary Tuning of PID Parameters 1149
T. YAMAMOTO

Adaptive, Cautious, Predictive Control with Gaussian Process Priors 1155
R. MURRAY-SMITH, D. SBARBARO, C.E. RASMUSSEN, A. GIRARD

Controller Design for Systems Suffering Nonlinear Distortions 1161
M. SOLOMOU, D. REES, N. CHIRAS

How the Output Saturation of a Regulator Influences the Reachable Performance and Robustness Measures 1167
L. KEVICZKY, C. BÁNYÁSZ

APPLICATIONS OF IDENTIFICATION

Random Loading Identification of a Plastic Glass Cantilever Beam 1173
D. LI, X. GUO, H. LI

On Sequential Identification of a Diffusion Type Process with Memory 1179
U. KÜCHLER, V. VASIL'IEV

Incremental Identification of Transport Coefficients in Distributed Systems 1185
A. BARDOW, W. MARQUARDT

On the Structure of Static Balanced Flow Systems 1191
E. WEYER, A. GLEIß, M. DEISTLER, K. GRUBER, T. MATYUS

Endogeneity and Identification in Transportation Systems: Econometric Relationships to Partial Observability 1197
N.K. JUVVA, V.N. SHANKAR, S. CHAYANAN

Tool for Equal Opportunity Evaluation in Dynamic Organizations 1203
P. ALBERTOS, I. BENÍTEZ, J.L. DÍEZ, J.A. LACORT

BIOENGINEERING SYSTEMS

Linearization in the Parameters Via Differential Algebra Techniques 1209
M.P. SACCOMANI

A Penalty Function Approach to HIV/AIDS Model Parameter Estimation 1215
R. FILTER, X. XIA

Sensitivity Analysis and Parameter Identification of Wastewater Treatment System Based on Activated Sludge Models 1221
J. SATO, H. OHMORI

A Methodology for Nonlinear System Identification Using Volterra Series. Application to an Anaerobic Digestor 1227
G. BIBES, P. COIRAULT, R. OUVRARD, J.P. STEYER

Some Relations of Sensitivity Functions in Bio-Reactor Models 1233
J.A.R. PÉREZ, J.L.N. HERRERO

An Experimental Object-Oriented Modelling of an Hydraulic Valley 1239
T. BASTOGNE, A. LIBAUX

PARTICLE FILTERS

Particle Filters for System Identification with Application to Chaos Prediction 1245
F. GUSTAFSSON, P. HRILJAC

Particle Filters for System Identification of State-Space Models Linear in Either Parameters or States 1251
T. SCHÖN, F. GUSTAFSSON

Fault Detection, Isolation and Diagnosis with Particle Filters for Nonlinear Stochastic Systems 1257
V. KADIRKAMANATHAN, P. LI

Monte Carlo Mixture Kalman Filter and its Application to Space-Time Inversion 1263
T. HIGUCHI, J. FUKUDA

A Particle Implementation of the Recursive MLE for Partially Observed Diffusions 1269
A. GUYADER, F. LE GLAND, N. OUDJANE

Online Sampling for Parameter Estimation in General State Space Models 1275
C. ANDRIEU, A. DOUCET, V.B. TADIĆ

WIENER HAMMERSTEIN MODELS

Nonlinear Structure Identification with Application to Wiener-Hammerstein Systems 1281
D.J. LEITH, W.E. LEITHEAD, R. MURRAY-SMITH

Identification of a Wiener System with Some General Discontinuous Nonlinearities 1285
F. GUO, G. BRETTHAUER

Nonlinear Model Identification Using Working Point Variables 1291
Y. ZHU

Identification of Wiener-Hammerstein Models with Cubic Nonlinearity Using LIFRED 1297
A.H. TAN, K.R. GODFREY

Performance Investigation of SLICOT Wiener Systems Identification Toolbox 1303
V. SIMA

IDENTIFICATION USING BASIS FUNCTIONS

Rational Bases Generated by Blaschke Product Systems 1309
F. SCHIPP, J. BOKOR

More on Sparse Representations in Arbitrary Bases 1315
J.J. FUCHS

On Spectral Analysis Using Models with Pre-Specified Zeros 1321
B. WAHLBERG

Identification of Rational Spectral Densities Using Orthonormal Basis Functions 1327
A. BLOMQVIST, G. FANIZZA

Orthonormal Basis Functions for Modeling Continuous-Time Fractional Systems 1333
M. AOUN, R. MALTI, F. LEVRON, A. OUSTALOUP

Adaptive Laguerre Time Scaling Factor in Predictive Control 1339
M. EL ADEL, M. OULADSINE, J.C. CARMONA

SUBSPACE IDENTIFICATION AND APPLICATIONS

Identification of MIMO State Space Models for Helicopter Dynamics 1345
M. LOVERA

Estimation of Damped and Undamped Sinusoids with Application to Analysis of Electromagnetic FDTD Simulation Data 1351
T. McKELVEY, T. RYLANDER, M. VIBERG

Application of a Recursive Subspace Identification Algorithm to Change Detection 1357
H. OKU

Subspace-Based Modal Identification and Monitoring of Large Structures: A Scilab Toolbox 1363
L. MEVEL, M. GOURSAT, M. BASSEVILLE, A. BENVENISTE

Identifying Positive Real Models in Subspace Identification by Using Regularization 1369
I. GOETHALS, T. van GESTEL, J. SUYKENS, P. van DOOREN, B. de MOOR

Modeling Human Gaits with Subtleties 1375
A. BISSACCO, P. SAISAN, S. SOATTO

IDENTIFICATION IN LARGE SCALE SYSTEMS

Reduction of Large-Scale Groundwater Flow Models Via the Galerkin Projection 1381
P.T.M. VERMEULEN, A.W. HEEMINK, C.B.M TE STROET

Model Reduction for Large-Scale Linear Applications 1387
K. WILLCOX, A. MEGRETSKI

Reduced Order Modeling of an Industrial Feeder Model 1393
P. ASTRID, S. WEILAND, A. TWERDA

INDUSTRIAL APPLICATIONS OF IDENTIFICATION

Identification of the Topology of a Power System Network 1399
Y. HASSAINE, E. WALTER, M. DANCRE, B. DELOURME, P. PANCIATICI

LPV Identification of a Diesel Engine Torque Model 1405
X. WEI, L. DEL RE

Identification and Control of a PV-Supplied Separately Excited DC Motor Using Universal Learning Networks 1411
A. HUSSEIN, K. HIRASAWA, J. HU

Validation of Stability for an Induction Machine Drive Using Experiments 1417
H. MOSSKULL, B. WAHLBERG, J. GALIC

Automatic Steering Control System Design Utilizing a Visual Feedback Approach - System Identification and Control Experiments with a Radio-Controlled Car 1423
S. ADACHI, T. FUJIHIRA, Y. FUJIWARA

Application of RBF-Type ARX Modeling and Control to Gas Turbine Combined Cycle SCR Systems 1429
Y. TOYODA, H. PENG, T. OZAKI, K. NAKANO, H. SHIOYA

SOFTWARE SESSION II

Automatic Time Series Identification Spectral Analysis with MATLAB Toolbox ARMASA 1435
P.M.T. BROERSEN

MULTI-EDIP – An Interactive Software Package for Process Identification 1441
J. KASPRZYK

KALMTOOL for Use with MATLAB 1447
M. NØRGAARD, N.K. POULSEN, O. RAVN

The $ADAPT_X$ Software for Automated and Real-Time Multivariable System Identification 1453
W.E. LARIMORE

Frequency Domain System Identification Toolbox for MATLAB: Automatic Processing – from Data to Models 1459
I. KOLLÁR, R. PINTELON, Y. ROLAIN, J. SCHOUKENS, G. SIMON

PLENARY PAPER
PREDICTION ALGORITHMS: COMPLEXITY, CONCENTRATION AND CONVEXITY

Prediction Algorithms: Complexity, Concentration and Convexity 1465
P.L. BARTLETT

IDENTIFICATION AND PHYSICAL MODELING

Grey–Box Model Calibrator and Validator 1477
T. BOHLIN, A.J. ISAKSSON

Initialization of Physical Parameter Estimates 1483
P.A. PARRILO, L. LJUNG

Parameter Estimation in Linear Differential-Algebraic Equations 1489
M. GERDIN, T. GLAD, L. LJUNG

Model Validation in Non-Linear Continuous-Discrete Grey-Box Models 1495
J. HOLST, E. LINDSTRÖM, H. MADSEN, H.A. NIELSEN

Identification of Mechanical Parameters in Drive Train Systems 1501
A.J. ISAKSSON, R. LINDKVIST, X. ZHANG, M. NORDIN, M. TALLFORS

Identification and Model Predictive Control of a pH Neutralization Process Based on Linear and Wiener Models 1507
J.C. GÓMEZ, A. JUTAN

IDENTIFICATION OF NONLINEAR SYSTEMS

Local Modelling of Nonlinear Dynamic Systems Using Direct Weight Optimization 1513
J. ROLL, A. NAZIN, L. LJUNG

Optimality in SM Identification of Nonlinear Systems 1519
M. MILANESE, C. NOVARA

A Suboptimal Bootstrap Method for Structure Detection of Nonlinear Output-Error Models 1525
S.L. KUKREJA

Identification of Nonlinear Parametrically Varying Models Using Separable Least Squares 1531
F. PREVIDI, M. LOVERA

Modeling and Identification of Rate-Independent Hysteresis Using a Semilinear Duhem Model 1537
J. OH, D.S. BERNSTEIN

Least Squares Harmonic Signal Analysis Using Periodic Orbits of ODEs 1543
T. WIGREN, E. ABD-ELRADY, T. SÖDERSTRÖM

VOLUME 4

EDUCATION AND TRAINING

Educational Aspects of Identification Software user Interfaces 1549
L. LJUNG

An Identification Course on the Web: Rationale, Realization and Students' Evaluation 1555
R. GUIDORZI, I. PAGANI, R. DIVERSI

Control Related Topics in Identification - Closed Loop Experiments and Identification for Control 1561
R.R. BITMEAD, R.A. de CALLAFON

Teaching Semiphysical Modeling to Chemical Engineering Students Using a Brine-Water Mixing Tank Experiment 1567
D.E. RIVERA

Estimating Parameters in a Lumped Parameter System with First Principle Modeling and Dynamic Experiments 1573
R.A. de CALLAFON

RECURSIVE AND SUBSPACE IDENTIFICATION

Recursive Subspace Identification Based on Projector Tracking 1579
M. LOVERA

Subspace Identification and ARX Modeling 1585
M. JANSSON

Parallel QR Implementation of Subspace Identification with Parsimonious Models 1591
S.J. QIN, L. LJUNG

A New Recursive Method for Subspace Identification of Noisy Systems: EIVPM 1597
G. MERCÈRE, S. LECOEUCHE, C. VASSEUR

Canonical Correlation Partial Least Squares 1603
U. KRUGER, S.J. QIN

Frequency-Domain System Identification Techniques for Experimental and Operational Modal Analysis 1609
P. GUILLAUME, P. VERBOVEN, B. CAUBERGHE, S. VANLANDUIT, E. PARLOO, G. de SITTER

PROCESS CONTROL: THEORY

Data-Driven Modeling of Nonlinear and Time-Varying Processes 1615
D. BONNÉ, S.B. JØRGENSEN

PID Parameter Cycling to Tune Industrial Controllers: a New Model-Free Approach 1621
J. CROWE, M.A. JOHNSON, M.J. GRIMBLE

Stepwise Refinement of Sparse Grids in Data Mining Applications 1627
M. BRENDEL, W. MARQUARDT

Iterative Identification for Control and Robust Performance of Bioreactor 1633
K. BØJSTRUP, H.H. NIEMANN, N.K. POULSEN, S.B. JØRGENSEN

Modified Subspace Identification Method for Building a Long-Range Prediction Model for Inferential Control 1639
Y. PAN, J.H. LEE

Identification and Model Predictive Control of an Industrial Glass-Feeder 1645
L. HUISMAN, S. WEILAND

APPLICATION OF SYSTEM IDENTIFICATION

Computationally Efficient Estimation of Wave Propagation Functions of Viscoelastic Materials 1651
K. MAHATA, T. SÖDERSTRÖM, L. HILLSTRÖM

Identification of Underlying Intensity Processes of Interference Patterns 1657
L. NÁDAI, J. BOKOR, A EDELMAYER

Fractional Multimodels - Application to Heat Transfer Modeling 1663
R. MALTI, M. AOUN, J.-L. BATTAGLIA, A. OUSTALOUP, K. MADANI

A Recursive Algorithm for Estimating Parameters in a One Dimensional Diffusion System 1669
B. BHIKKAJI, T. SÖDERSTRÖM, K. MAHATA

Regularization Method in Infrared Image Processing 1675
S. DATCU, L. IBOS, Y. CANDAU, S. MATTEÏ, N. RAMDANI

Filtering of Stochastic Volatility Model 1681
S. AIHARA, A. BAGCHI

OPTIMAL FILTERING

State Estimation for Nonlinear Continuous Systems in a Bounded-Error Context 1687
T. RAÏSSI, N. RAMDANI, Y. CANDAU

Multigrid Design in Point-Mass Approach to Nonlinear State Estimation 1693
M. ŠIMANDL, J. KRÁLOVEC

An Efficient Nonlinear Adaptive Observer with Global Convergence 1699
Q. ZHANG, A. XU, G. BESANÇON

Adaptive Observer for Discrete Time Linear Time Varying Systems 1705
A. GUYADER, Q. ZHANG

Linear Dynamic Filtering with Noisy Input and Output 1711
I. MARKOVSKY, B. de MOOR

The *p*-Norm Generalization of the LMS Algorithm for Adaptive Filtering 1717
J. KIVINEN, M.K. WARMUTH, B. HASSIBI

SEMI-PLENARY
IDENTIFICATION OF LINEAR SYSTEMS WITH NONLINEAR DISTORTIONS

Identification of Linear Systems with Nonlinear Distortions 1723
J. SCHOUKENS, R. PINTELON, T. DOBROWIECKI, Y. ROLAIN

SEMI-PLENARY
SOME PROBLEMS IN STATISTICAL INFERENCE FOLLOWING MODEL SELECTION

Some Problems in Statistical Inference following Model Selection 1735
B.M. PÖTSCHER

USER CHOICES IN SUBSPACE IDENTIFICATION

Choosing Integer Parameters in Subspace Methods: A Survey on Asymptotic Results 1741
D. BAUER

Asymptotic Variances of Subspace Identification by Data Orthogonalization and Model Decoupling 1747
A. CHIUSO, G. PICCI

A Finite Sample Comparison of Automatic Model Selection Methods 1753
D. BAUER, S. de WAELE

On the Number of Rows and Columns in Subspace Identification Methods 1759
B.L.R. de MOOR

Aspects and Experiences of User Choices in Subspace Identification Methods 1765
L. LJUNG

Inferring Multivariable Delay and Seasonal Structure for Subspace Modeling 1771
W.E. LARIMORE

IDENTIFICATION OF STATIC AND DYNAMICAL NONLINEAR SYSTEMS

Mathematical Results Concerning Kernel Techniques 1777
R. SCHABACK

Multi-Output Suppport Vector Regression 1783
E. VAZQUEZ, E. WALTER

Set Membership Identification of Piecewise Affine Models 1789
A. BEMPORAD, A. GARULLI, S. PAOLETTI, A. VICINO

Piecewise-Linear Output-Error Models 1795
F. ROSENQVIST, A. KARLSTRÖM

CMAC with Linear Functional Weights 1801
Q. GAN, E. ROSALES

Optimal Expansions of Discrete-Time Volterra Models Using Laguerre Functions 1807
R.J.G.B. CAMPELLO, G. FAVIER, W.C. AMARAL

IDENTIFICATION AND MODEL VALIDATION

Quantification of the Variance of Estimated Transfer Functions in the Presence of Undermodeling 1813
R. HILDEBRAND, M. GEVERS

Reliable Parameter Estimation in Presence of Uncertain Variables that are not Estimated 1819
I. BRAEMS, L. JAULIN, M. KIEFFER, N. RAMDANI, E. WALTER

Validation Test Based Parameter Uncertainty Versus Analysis-Based Confidence Bounds 1825
S.G. DOUMA, X.J.A. BOMBOIS, P.M.J. van den HOF

Empirical Estimation of Parameter Distributions in System Identification 1831
W.J. DUNSTAN, R.R. BITMEAD

Uncertainty of Transfer Function Modeling Using Prior Estimated Noise Models 1837
R. PINTELON, J. SCHOUKENS, Y. ROLAIN

The Size of the Membership-Set in a Probabilistic Framework 1843
H. AKÇAY

MODEL APPROXIMATION

Connections Between L_2-Model Reduction and Balanced Truncation 1849
W. SCHERRER, F. TJÄRNSTRÖM

Recursive Exact H∞ Identification from Impulse Response Measurements 1855
O. KANEKO, P. RAPISARDA

Properties of Optimal Solutions in L_1 Identification Problem 1861
M. NAMVAR, A. BESANÇON-VODA

Optimal Approximation and Model Quality Estimation for Nonlinear Systems 1867
P.M. MÄKILÄ

Linear Models of Nonlinear FIR Systems with Gaussian Inputs 1873
M. ENQVIST, L. LJUNG

An Algebraic Method for System Reduction of Stationary Gaussian Systems 1879
D. JIBETEAN, J.H. van SCHUPPEN

PARAMETER ESTIMATION AND CONVERGENCE

Separable Least Squares Data Driven Local Coordinates 1885
T. RIBARITS, M. DEISTLER, B. HANZON

Optimal Yule Walker Method for Pole Estimation of ARMA Signals 1891
M. JANSSON, P. STOICA

Initializing Parameter Estimation Algorithms Under Scarce Measurements 1897
P. ALBERTOS, R. SANCHIS, I. PEÑARROCHA

Robust Parameter Estimation for Uncertain Gross-Error Models 1903
K. UOSAKI, K. SAITO, T. HATANAKA

Limit Covariance of Estimation Error for Quasistationary Functions 1909
A.E. BARABANOV

IDENTIFICATION OF HYDROLOGIC SYSTEMS

Structural Identification of Multivariate Neural Networks for Rainfall Runoff Modelling 1915
G. CORANI, G. GUARISO, S. CASTELLI

Parameter and State Regularization for Prediction of Distributed Hydrologic Systems 1921
E.E. van LOON, K.J. KEESMAN

Time–Delay Estimation of a Managed River Reach from Supervisory Data 1927
M. THOMASSIN, T. BASTOGNE, A. RICHARD, A. LIBAUX

Geohydrological Application of a Nonlinear Physically Based Time Series Model 1933
W.L. BERENDRECHT, A.W. HEEMINK, F.C. van GEER, J.C. GEHRELS

On Physical and Data Driven Modelling of Irrigation Channels 1939
S.K. OOI, M.P.M. KRUTZEN, E. WEYER

Identification and On-Line Estimation of the Unsaturated Hydraulic Conductivity in Presence of Forced Air Convection Based on a Distributed-Parameter Model 1945
O. SCHOEFS, D. DOCHAIN, R. CHAPUIS, R. SAMSON, M. PERRIER

ERRORS IN VARIABLE IDENTIFICATION

Confidence Regions for Non-Parametric Errors-In-Variables Estimates 1951
W.P. HEATH

A New Criterion in EIV Identification and Filtering Applications 1957
R. DIVERSI, R. GUIDORZI, U. SOVERINI

Strongly Consistent Parameter Estimate for Error-In-Variables Model 1963
H.-F. CHEN

Ellipsoid Set Refinement by Simultaneous Use of Multiple Hyperplane Cuts 1969
D. JOACHIM, J.R. DELLER Jr.

Identification Methods in a Unified Framework 1975
I. VAJK

Author Index 1981

IFAC
Publications
www.elsevier.com/locate/ifac

EDUCATIONAL ASPECTS OF IDENTIFICATION SOFTWARE USER INTERFACES

Lennart Ljung *

* *Division of Automatic Control, Linköping University,
SE-58183, Linköping, Sweden, email:* `ljung@isy.liu.se`

Abstract: Apparently many users of identification techniques learn the topic only via use of commercial software. This may or may not include the software manual. This means that the software user interface – graphical or not – plays a major role in teaching identification theory and methodology to large number of users. This contribution deals with such educational / pedagogical aspects software user interfaces.
In particular we focus on issues to hide certain design variables as defaults, and what can be done in case no defaults are obvious. Other questions are how to force the user to appreciate and understand the quality of an identified model, and to know what optional design choices and methods that are available, in particular if there is no Graphical User Interface (GUI).

1. INTRODUCTION

It is illuminating to compare sales figures of text books on System Identification, such as (Ljung 1999), (Ljung and Glad 1994), with figures of circulation of commercial software packages on Identification, such as (Ljung 2000). They show that most users of the software have not read (or at least not bought) comprehensive treatments of the topic. Combine that with the well-known fact that most software users do not open the manual, and we realize that a large number of users actually learn system identification from just running software packages.

This means that the software user interface (syntax, help texts, graphical aids, etc) plays a major educational role in teaching System Identification. It is the purpose of the current contribution to discuss such issues. Compare also with other identification software packages like (Van Overschee *et al.* 1994), (Larimore 1997), (Landau 1990), (Garnier and Mensler 2000), and (Kollar 1994).

These packages have some features in common, but have also chosen different paths to syntax and GUI-features. Not much has been written about the educational aspects of the different choices, and it would be interesting to see more research around this topic.

2. WHAT ARE THE SALIENT FEATURES OF THE IDENTIFICATION PROCESS?

What are the most important steps/items in the identification process that any user must understand? We would like to point the the following things:

- The role of subjective judgments vs automatic procedures.
- The concept of *model validation.*
- Data requirements for reasonable results.

2.1 *Subjective and Objective Decisions*

System Identification is as much an art as a science. Models of physical processes are built and estimated for a certain purpose. It could be control design, decision making, simulation, monitoring, and several other things. Depending on the model's purpose, different aspects of the model may be of different importance. For control design, for example, the low frequency properties of the system may be estimated with low accuracy (since the control system typically will have sensitivity function that is small at low frequencies). On the other hand, the properties around the intended cross over frequency need to be estimated with much higher accuracy.

The model quality can therefore not be summarized in a single number, but the user has to evaluate the model properties in a subjective way, taking the intended model use into account. This means, unfortunately, that there cannot be a fully automated route to the ideal model, but the user will have to interfere with subjective decisions. This is perhaps the single most important feature of the process, and it requires the user not only to know what properties to study, but also to know what measures can be taken to construct models with improved desired properties.

Unfortunately the list of model structures, possibilities, options and design variables is very long and clearly overwhelming to the newcomer. This is the nature of the pedagogical challenge in identification software: To make clear what are the choices, at the same time as not choking the user.

2.2 *Model Validation*

It is necessary to understand that a model is not necessarily "good" just because it is "best" within a certain chosen class of models. It is necessary to twist and turn the model to look at it from different angles to evaluate if it might be good enough for its purpose. The different techniques for this are more or less sophisticated. While a software package must contain a multitude of such options, it is important to "force" or "encourage" the inexperienced user to look at least at plots that compare model outputs with measured outputs. Such comparisons are very intuitive, immediate and instructive. They do not require any statistical understanding, except for the fact that they should be carried out on a fresh *validation* data set that was not used to estimate the model.

A special challenge of such comparisons is to make clear what the difference between "simulated" are "predicted" outputs are. For example, when building a neural network dynamical model, such as NARX, the easy and natural comparison will be between 1-step ahead model predictions and measured outputs – it is much more complicated to compute the simulated model output in typical Neural Network estimation packages. Now, a (fast sampled) model with excellent one step ahead predictions could very well be useless, but this could be a subtle message to convey.

2.3 *Data Quality*

One of the most common problems when inexperienced users apply identification software is that the data is not sufficiently exciting for the chosen model structure. There is both a technical difficulty and an information-theoretical issue involved in this.

It is not so difficult to use common sense to understand and explain that complicated models cannot be built from noisy, single step responses: The system must simple have had occasion to show its dynamical properties in an unambiguous way, to allow for a reasonable model. This information issue revolves around the concept of *persistence of excitation*, and there should be an early-on indication to the user how complex models can be reliably estimated from the data at hand.

There is a technical side that is more treacherous: It is quite common to see applications where the data is a single step or impulse response, where the step/impulse occurs in the first few data points. Such data could very well be used for reasonable modeling if the signal–to–noise ratio is good. However, many estimation techniques (like estimating ARX-model and subspace methods) effectively shift the data (up to a number of samples that is about the model order) in order to avoid estimation data points prior to the start of the data collection. This makes the data look like a response to a constant input, and no reasonable models can be built. It is my experience that this problem is the single most common reason for complaints about the identification process from first-time users.

3. DEALING WITH THE WORK FLOW

Identifying models from real data is as much an art as a science. The identification process involves a very large number of choices:

- Character of model structure (State-space/ ARX/ Output Error etc)
- Size of model structure (Model orders and delays)
- Choice of method (Subspace/Instrumental Variables/ Prediction error Method ...)
- Options in the method (For example: should initial filter conditions be estimated or not)
- Algorithmic aspects in the method (For example: auxiliary orders in subspace methods, parameters association with function minimization for prediction error methods)
- Pretreatment of data (Detrending/ Prefiltering/ Decimation/ etc)
- Validation process

It is indeed a challenge to make the inexperienced user aware of available options and the choices that have to be made, without leaving him/her in total confusion. Actually, to avoid confusion one simply has to enforce certain choices and hide options for a user who does not need or want to know about them.

This problem of how to guide the user with a gentle and insightful hand through the identification marsh has to be solved in different ways in a GUI (Graphical User Interface) and in a non-GUI environment.

In a GUI-environment one may hint the possibility of e.g. prefiltering, by, say, having a pop-up menu `focus`. Sooner or later the user will check the tooltip of this menu, investigate its options, and at least intuitively understand that the model fit can focused to different frequency ranges.

It is much more difficult in a non-GUI environment to inform that such an option actually exists. It is then very helpful to have data and models represented in an object-oriented way. This will, e.g., allow suitably tailored descriptions of models and data without any specific commands. In MATLAB this is accomplished by the function `display` that is called in the appropriate object method directory for any object. An average user will also know that `get` and `set` for any object will give a succinct summary of options, without having to read help texts.

General GUI aspects are discussed in Section 5 below. In a non-GUI environment, the problem has been dealt with by a general command `advice` in (Ljung 2003*a*). This is further discussed in Section 4.

4. THE ADVICE FEATURE

The function `advice` in (Ljung 2003*a*) takes any data object or any model object as an argument and delivers a text that comments the objects and gives some hints how to proceed.

This is accomplished by a hidden property (called `utility`) of these objects, that is a structure with several fields. The MATLAB functions `assignin` and `inputname` allows you to update a workspace variable (in this case its `utility`-property) even if is not the output of a function called. This makes it possible to adjoin to a model or a data set a "diary" of what has been done, and thus give relevant advice of further things that can be tested.

For data objects, `advice` tests things like

- excitation
- feedback in data
- if a step or impulse occurs "too early" in the set (cf the discussion in Section 2.3.)
- possible detrending
- points to the possibility to assign input intersample behavior
- etc

For model objects `advice`

- points out that `compare` and `resid` should be run if this has not been done. It also checks whether this was done on a separate validation data set.
- comments on what has been seen in the residual analysis test `resid`. Based on χ^2-test it gives advice whether higher order dynamics and noise models are required. Furthermore tests are performed to check whether feedback might be present in the (validation) data.
- checks if the model order can be reduced.
- allows comparisons of several models, `advice(m1,m2,m3)` and points out which one should be preferred.
- etc

The function is best illustrated by an example of how it works: (See also (Ljung 2003*b*))

```
>>load iddata1
>>advice(z1)

All your inputs have been denoted as
'zero order hold' ('zoh'), i.e.
they are assumed to be piecewise
constant over the sampling interval.
If the input is a sampled continuous
signal and you plan to build or
convert to time continuous models,
it is recommended to mark the
Intersample property as 'First
order hold': Data.int = 'foh' or
Data.int = {'foh','foh', ...}
for multiinput signals.
You may mix 'zoh' with 'foh' for
the different inputs in the latter
case.

All inputs and outputs are not zero
mean. It it generally recommended to
remove the means by DAT = DETREND(DAT),
except in the following two cases:

1. The signals are measured relative
   to a level that corresponds to a
   physical equlibrium. This could e.g.
   be the case if step responses
   are recorded from an equlibrium point.
2. There is an integrator in the
   system, and the input and output
   levels are essential to describe
   the effect of the integration.
```

```
There is no significant indication of
feedback in the data.

The input is persistantly exciting of
order more than 50. This means that you
will not encounter identifiability
problems if estimating models of order
lower  than so.

>>m=arx(z1,[1 1 1]);
>>advice(m)

You should run a comparison test

  COMPARE(VDATA,m)

where VDATA preferably is a different
data set from the estimation data z1.
You can then run ADVICE(m) again.

You should run a residual test

  RESID(VDATA,m)

where VDATA preferably is a different
data set from the estimation data z1.
You can then run ADVICE(m) again.

>>compare(z1,m);
>>resid(z1,m);
>>advice(m)

There is a very strong indication that
the dynamics of the model is
not adequately described.
A first general advice is to run
RESID(VDATA,m,'FR') to check in which
frequency ranges the model error is
present. If the model error is unacceptable,
you will have to increase the model order.

In particular you should pay attention to
lags 2  3  6 from input 1.
Modify KU and the orders NB so that
these lags are included in the model.


There is a very strong indication that
the residuls are not white. To get a
good noise model you need to increase
the orders associated with the noise
parameters, or just increase
the order of a state-space model.
```

5. GUI ISSUES

In several ways, a graphical user interface (GUI) makes it easier to make the software accessible to inexperienced users. The GUI can display choices in various menus that point to what possibilities are available in a certain situation. More advanced design variables can be hidden deeper into nested dialogs etc.

There are two basic issues or compromises in designing a GUI. The first one concerns the trade off between

- Immediate appeal and understanding for the first time user
- Ease and convenience of use for the power user in "model production mode"

Both these properties are of course desirable, but they are partly in conflict with each other. Many tests are required to find a good balance.

The second important question when designing a GUI for system identification is whether it should be

- *process-oriented* or
- *result-oriented.*

A process-oriented GUI gives a flowchart of the identification process: first select data, then pre-process data, then select a model structure, then estimate a model, then validate the estimated model, etc. One works, so to speak "from left to right" in this flowchart. This no doubt has advantages for the first time user, but may turn out to be tiresome as one produces models *en masse* that have to be compared with each other.

In (Ljung 2003*a*) a result-oriented GUI is chosen. The focus is to keep an explicit view of the modified data sets and the different models that have been produces during an identification session. This is essential for the experienced user, for whom the biggest problem may be to keep track of what he/she has done. For the inexperienced user, a visually prominent board with empty slots for several models also carries the important message that the identification task is not finished just because you have constructed one model. On the other hand such a user needs more help in getting started.

Figure 1 shows the result-oriented main `Summary Board` of the GUI in (Ljung 2003*a*). It contains all derived data sets and all estimated models. It also has the main menus for selecting ways of modifying (preprocessing) data and for estimating models of various structures.

6. HOW TO HIDE MANY NOT-SO-IMPORTANT CHOICES?

The identification process is characterized by a myriad of choices, parameters and design variables. Just listing the choices will be utterly confusing to the newcomer. The obvious way to solve

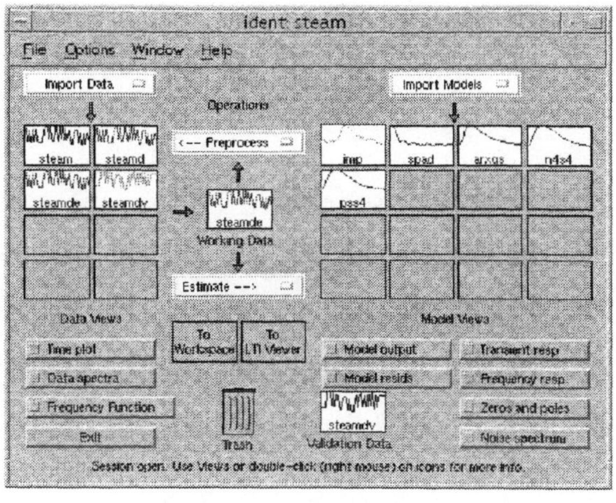

Fig. 1. The main board of the GUI of the System Identification Toolbox

this is to allow a syntax that does not require all choices be specified but uses defaults in most cases. This is standard practice, and is no problem is case typical defaults are data-independent. Such examples are `MaxIter` for the maximum number of iterations to perform when searching for the minimum, or `Tolerance` for the tolerance to decide whet to stop iterations.

Another case is when the desired value of a design variable indeed depends on data properties. An example of this is `InitialState`: Should the initial state (filter initializations for the predictor filters) be estimated, or set to zero? The answer to this question is data/system dependent: If the true predictor filter is slowly decaying and/or the data record is short, then it may be very important to estimate the initial conditions. This fact is not so easy to explain or make visible to a new user. The solution in (Ljung 2003*a*) is to use the default value `auto`, which means that an automatic choice is made in the course of the calculations. Then the model property `EstimationInfo` will, after the model is estimated, contain a field that gives the value that was chosen, `Estimate`, `Zero` or `BackCast`.

Another example of this kind concerns the important choice of various horizons and weights in so called subspace identification methods. The theory of the influence of these options is not well understood even among researchers, so it would be futile to force the user to make the choices. Again the value `auto` for the corresponding properties is the solution, with an option for expert users to try explicit values. The choice `auto` also gives the further advantage that the underlying defaults can be changed as research progresses.

7. CONCLUSIONS

In this paper we have pointed to some issues in software interfaces that will guide the inexperienced user through the complex process of practical system identification. For many users that may be the only effective training in the theory and practice of identification. The educational/pedagogical aspects of the interface are thus most essential. Not much has been written about this, and more research around these questions would be welcome.

8. REFERENCES

Garnier, H. and M. Mensler (2000). The CONTSID toolbox: A Matlab toolbox for CONtinuous-Time System IDentification. In: *Proc. 12th IFAC Symposium on Identification*. Santa Barbara (USA).

Kollar, I. (1994). *Frequency Domain System Identification Toolbox for use with MATLAB*. The MathWorks Inc.. Natick, MA, USA.

Landau, I. D. (1990). *System Identificaiton and Control Design Using P.I.M. + Software*. Prentice Hall. Engelwood Cliffs.

Larimore, W. E. (1997). *Adapt$_X$ Automated System Identification Software, Users Manual*. Adaptics, Inc.. 1717 Briar Ridge Road, McLean, VA.

Ljung, L. (1999). *System Identification - Theory for the User*. 2nd ed.. Prentice-Hall. Upper Saddle River, N.J.

Ljung, L. (2000). *System Identification Toolbox for use with* MATLAB*. Version 5.*. 5th ed.. The MathWorks, Inc. Natick, MA.

Ljung, L. (2003*a*). *System Identification Toolbox for use with* MATLAB*. Version 6.*. 6th ed.. The MathWorks, Inc. Natick, MA.

Ljung, L. (2003*b*). Version 6 of the system identification toolbox. In: *Proc. of the IFAC Conference on System Identification, SYSID'03* (P. van den Hof, Ed.). Rotterdam, The Netherlands.

Ljung, L. and T. Glad (1994). *Modeling of Dynamic Systems*. Prentice Hall. Englewood Cliffs.

Van Overschee, P., B. De Moor, H. Aling, R. L. Kosut and S. Boyd (1994). *Xmath interactive system identification module*. Integrated Systems, Inc.. Santa Clara, CA.

IFAC
Publications
www.elsevier.com/locate/ifac

AN IDENTIFICATION COURSE ON THE WEB: RATIONALE, REALIZATION AND STUDENTS' EVALUATION

Roberto Guidorzi*, Ilaria Pagani* and Roberto Diversi**

**CITAM, University of Bologna, Italy*
***DEIS, University of Bologna, Italy*

Abstract: The role of net-based tools in both traditional and remote learning is continuously increasing and will eventually define new integrated environments. This paper reports the opinions of the students that have followed a course on dynamic system identification based on hypertexts, asynchronous tutoring and platform-independent virtual laboratories. *Copyright © 2003 IFAC*

Keywords: identification, e-learning, web-distributed courses, distance learning, open learning.

1. INTRODUCTION

E-learning is a label sometimes confused with specific technologies and, in fact, the very term is somehow misleading since learning is a complex process centered not on technologies but on man. Technologies, however, play a fundamental role in storing, retrieving and transmitting information as well as in establishing communication channels linking persons located in different places. A role of paramount importance has been and is, in fact, played by that great technological advancement constituted by the introduction of mobile types print, dating back to 1500. All present educational systems are based on books and on synchronous environment in which teachers and students are located in the same room at the same time. The purpose of the simultaneous presence of teachers and students is to establish a high level of vertical and transversal interaction that, as is well known, characterizes high level of teaching and learning.

Present-day information and net-based communication technologies allow, however, to create learning communities free from time and space limits associated to traditional teaching environments. Despite the miming of traditional synchronous environments that has been, sometimes, performed by means of video conference tools, the great potentiality of e-learning is clearly based on its possibilities of establishing high levels in learning inside communities that adopt asynchronous interaction channels. E-learning must thus be considered not as a surrogate of traditional learning for remote students (distance learning) but as a new powerful way to create effective learning communities in which distance can be measured more properly in Kbit/s rather than in kilometers.

The course considered in this paper concerns the identification of dynamical systems i.e. the modeling of real processes on the basis of measures performed on their inputs and outputs.

The content is organized as follows. Section 2 describes the rationale of the course while Section 3 contains some realization details. Section 4 reports some evaluations expressed by the students. Short concluding remarks are finally reported in Section 5.

2. RATIONALE OF THE COURSE

The design goals have been:

• *Economic efficiency*. Specialistic courses are often followed by a reduced number of students inside single faculties or single universities and are thus penalized by modest economic efficiencies. These courses have however almost always strong connections with research and their elimination would lead to a dispersion of cultural resources not compatible with maintaining high qualitative standards. The Web distribution channel that has been adopted in this project is transversal with respect to faculties and universities and allows to reach a critical mass of students, sufficient to conjugate economic efficiency with quality and a large variety of educational choices.

• *Transferring the control on learning times from teachers to students.* A peculiar aspect of the channel that has been adopted concerns its asynchronousness. This breaks the dogma on the unity of time and space at the basis of traditional teaching and the consequences are not irrelevant: the temporal control on learning is transferred to students that can thus achieve a more flexible compatibility with their social environment (family and work constraints, etc.).

• *Remote tutoring.* This project does *not* concern self-learning in the traditional sense (autonomous acquiring of knowledge and self-evaluation); the tutor plays a role of paramount importance in evaluating with continuity the level of the knowledge of the learner and in guiding him to the final exam which has an absolutely traditional structure but should lose every aleatory aspect since mere legal validation of an evaluation already performed *in itinere*.

• *Effective laboratory activity performed remotely.* Specialistic courses cannot be effective without laboratory activities where the students can experience situations concerning their subsequent professional activities. Laboratory practice can play a remarkable role both in the learning phase (situated learning) and in the evaluation phase. The tools that have been developed in this project allow an effective remote laboratory practice on every hardware and software platform.

• *The use of the course also outside the academic environment.* The flexibility of the channel used for its distribution does not confine the use of the course to interfaculty or interuniversity environments. It allows, on the contrary, extending the learning offer to external contexts (industry, institutions, lifelong learning etc.). These possibilities can contribute to improve the global economic efficiency and also stimulate a better tailoring of the contents to real needs.

• *The development of a modular structure allowing extracting specific paths useful from diploma to doctorate levels.* The project relies on four modules allocated at two levels that allow the extraction of partial courses that can be inserted also at different moments of the same curriculum without any dispersion of credits.

3. REALIZATION

The course leads its students to proficiency in using identification techniques, i.e. the techniques that allow the construction of mathematical models for dynamic processes on the basis of measures, affected by errors, performed on these processes. Every aspect of reality is constituted by dynamic processes. Possible applications extend from biology to macro economy, from finance to ecology, control of industrial processes (chemical, electrical, petro-chemical, mechanical), meteorology, medicine etc.

These contents allow the implementation of learning paths based on four modules corresponding to traditional courses with lengths between 30 and 90 hours and educational target spacing from Diploma to Doctorate levels (Figure 1).

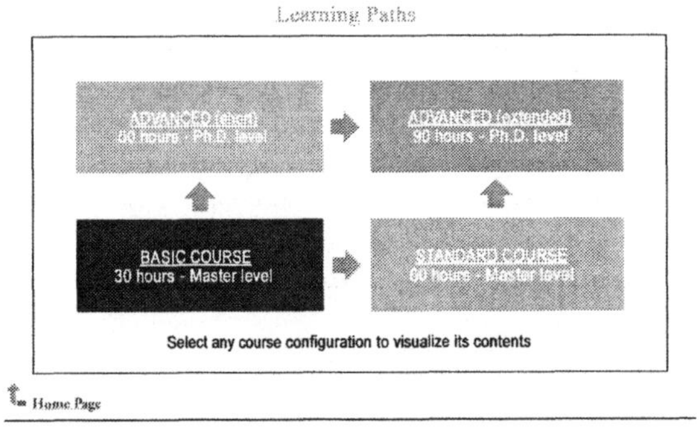

Figure 1 - Modules composing the course

The levels of interaction implemented in the course concern:

• The interaction between learner and contents (hypertexts and video clips);
• The interaction between learner and tutor;
• The interaction between learner and real problems (problem solving).

The hypertexts have been composed using TeX and Adobe PDF (Portable Document Format) technology. The well-known potentialities of TeX have thus been joined with the operative flexibility and platform independence of PDF. The exclusive use of vector fonts assures a high level of quality even on low-resolution output devices.

The well-known effectiveness of replies to FAQ (Frequently Asked Questions) offers another possibility for integrating the contents of the course. The replies inserted on the server are given in the form of short video clips. No part of the contents is, on the contrary, exposed by means of video clips

A high degree of flexibility in the distribution of a course can be achieved avoiding the use of channels requiring a synchronous communication between tutors and students (video, audio etc.). The communication between tutors and students that has been implemented in the course takes advantage of the asynchronous possibilities offered by Web servers; among other advantages also the real-time reception control.

The target of transferring professional know-how to students can be achieved only allowing the application of their notions to real situations. The data base of the course contains many data sets belonging to different areas ranging from finance to ecology, from macro economy to meteorology, chemical, petrochemical and electrical processes, mining etc.

The identification laboratory has been designed to offer the possibility of applying to real processes the identification methodologies described in the course. It has been entirely developed in Java and is based on efficient implementations of identification algorithms. It allows the students to perform locally, and in a way completely independent from their hardware and software platform, modeling experiences working on the data sets contained in the process data base.

4. USERS' EVALUATION

Part of the information collected from the users after completing the learning path and sustaining the final exam is described by the percent response to the questions that follow; for a more complete picture see Guidorzi and Pagani (2002).

The graphical and computational tools available inside the course have been developed using the JAVA language. Which are, in your opinion, the more interesting characteristics of JAVA?

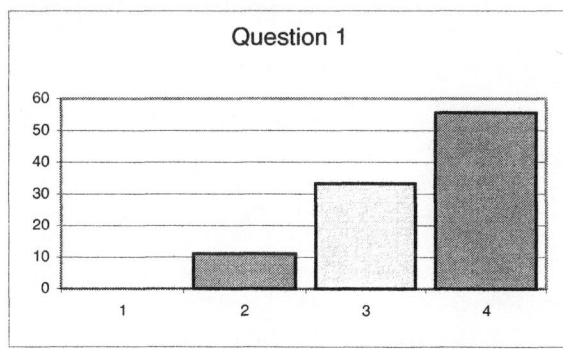

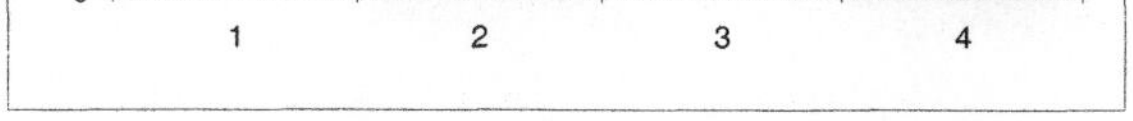

1 – The possibility of realizing software modules (Applets) independent from the user's hardware. 2 – The possibility of realizing software modules (Applets) independent from the user's operating system. 3 – The possibility for the user to download not only data but also specific computing environments. 4 – The possibility of developing software running on the client instead of the server.

What's your technical evaluation of the tutor-student interaction?

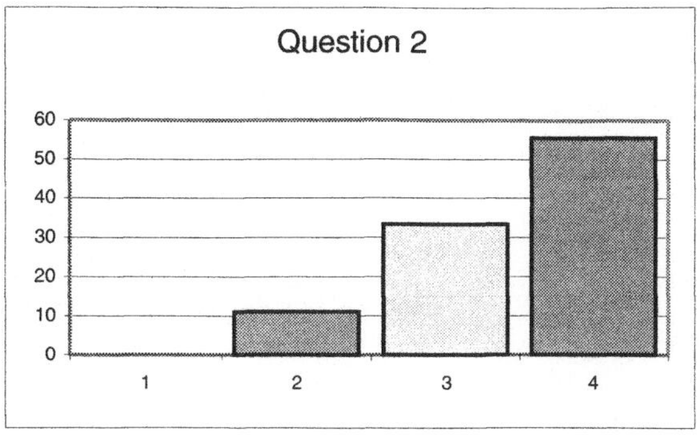

1 – Not adequate. 2 – Adequate. 3 – Efficient. 4 – Very efficient .

One of the key features of the course project concerns a tutoring that follows every student from the beginning of the learning path to the final examination. How do you compare this approach with the on-request assistance present in traditional courses?

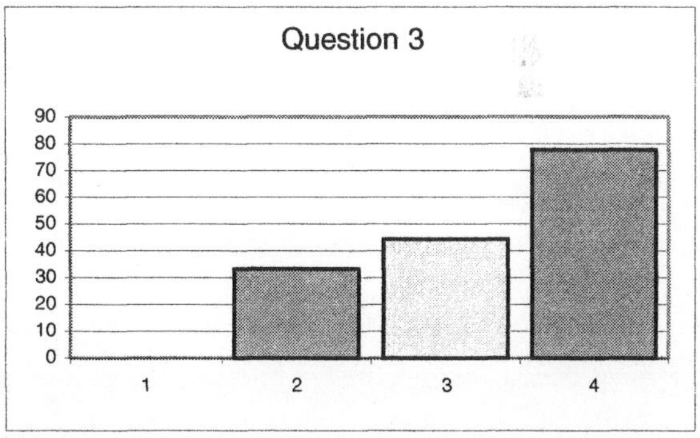

1 – Without real advantages. 2 – Effective. 3 – Very effective. 4 – Capable of eliminating the uncertainties associated with traditional final exams.

What's your evaluation of your tutoring experience?

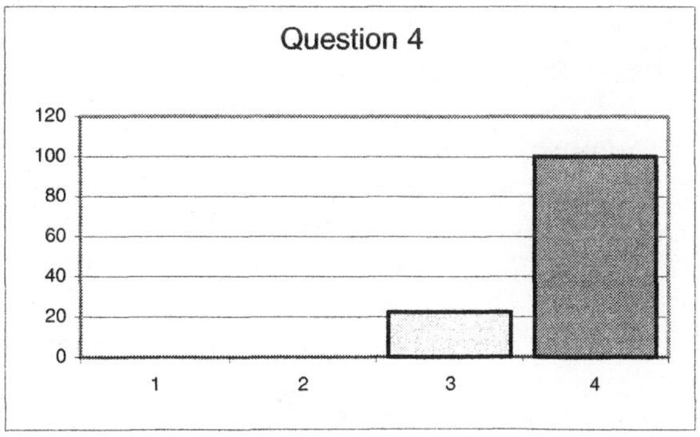

1 – I have not used this service. 2 – I have used this service but its relevance has been quite modest. 3 – Tutoring helped me on a psychological basis. 4 – Tutoring has had a positive effect in evaluating the learning process.

Evaluate, giving a mark between 1 and 5, the following characteristics of the process database available on the course server:

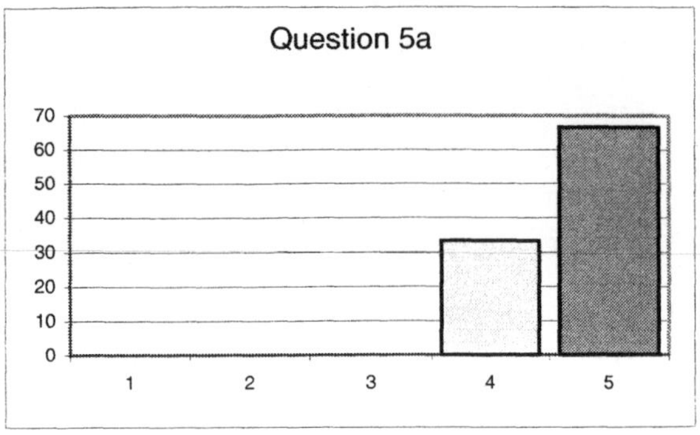

Significance of the available processes.

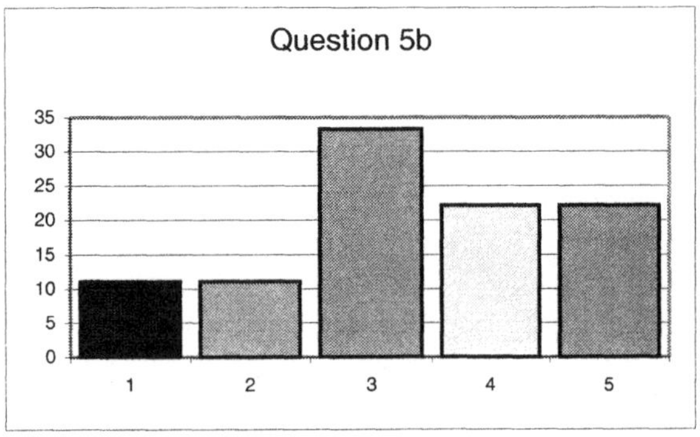

Process information adequacy.

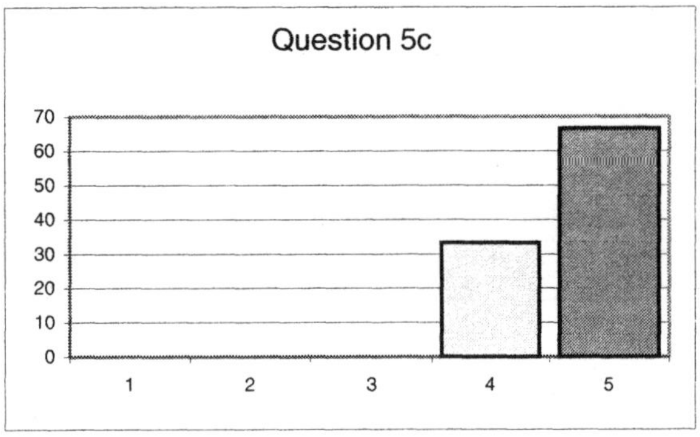

Easy access to data.

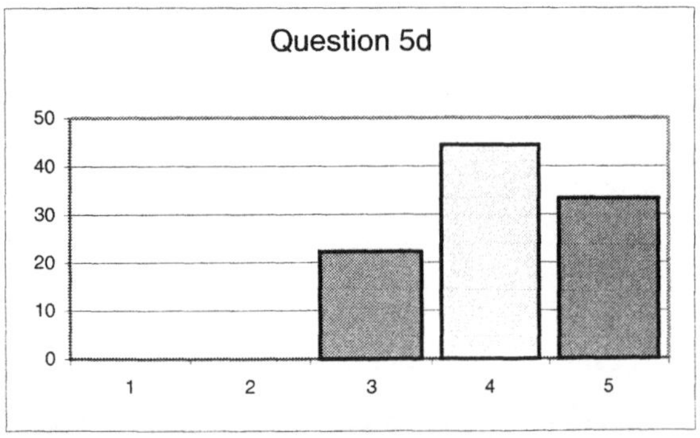

Adequacy of visualization facilities.

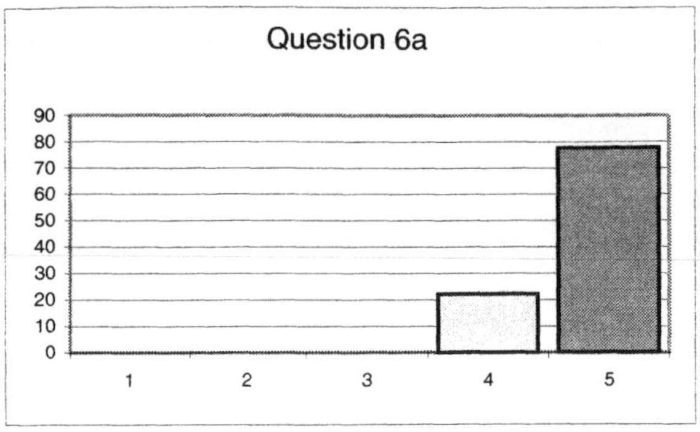

Easy use.

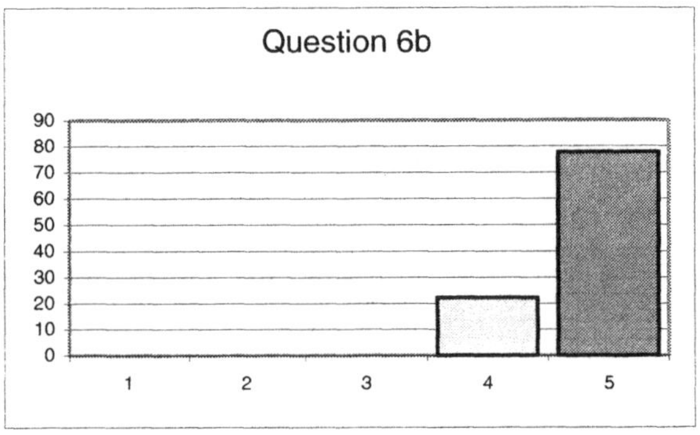

User interface.

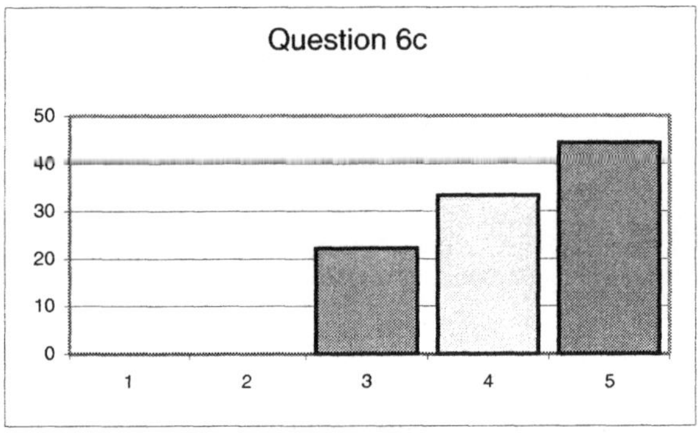

Interaction with the process data base.

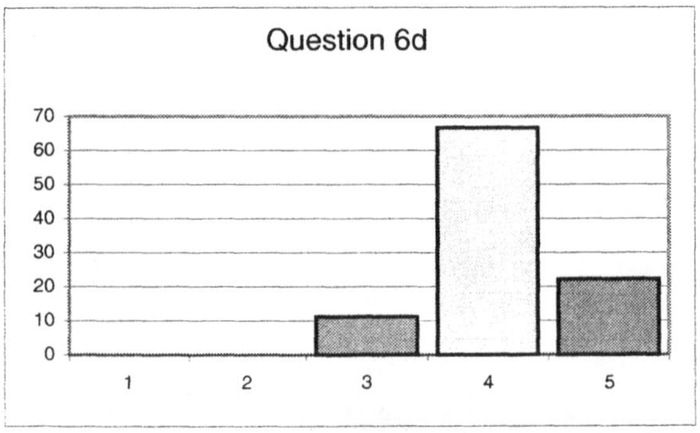

Efficiency of the implemented algorithms

Evaluate, giving a mark between 1 and 5, the following characteristics of the virtual identification laboratories:

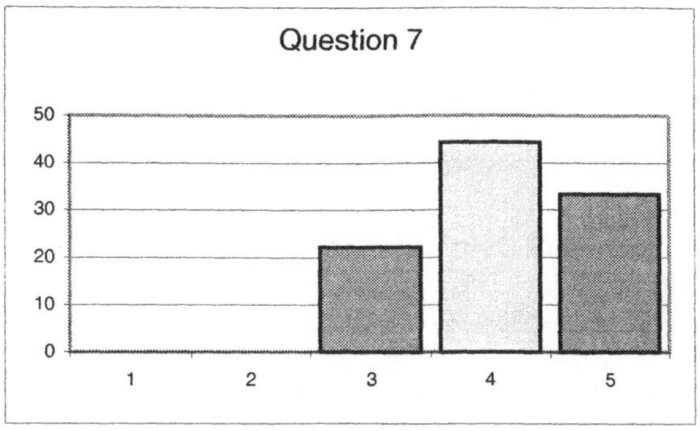

Robustness of the algorithms

Which role plays the use of virtual laboratories operating on real processes in the comprehension of the theoretical aspects of identification?

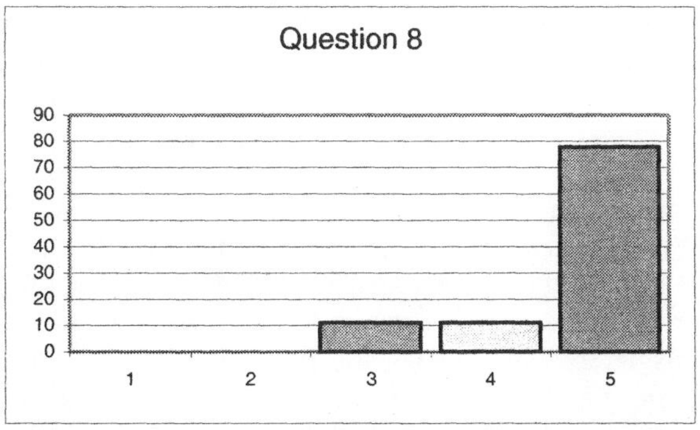

1 – Inessential. 2 – Modest. 3 – Useful. 4 – Very useful. 5 – Essential.

How do you evaluate the professional capabilities deriving from working with real data in the operative contexts available in the virtual laboratories?

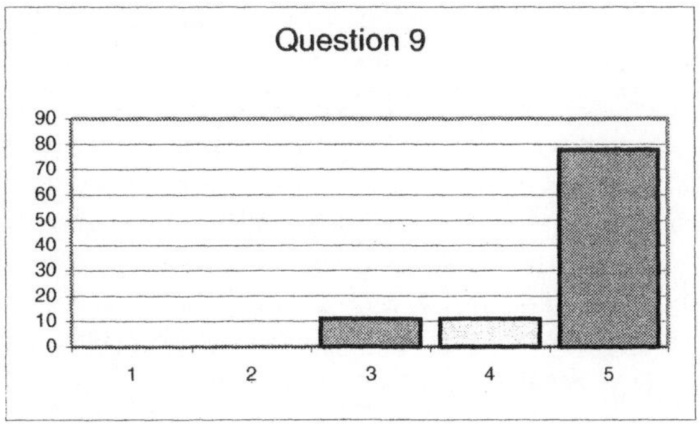

1 – Inessential. 2 – Modest. 3 – Useful. 4 – Very useful. 5 – Essential.

How do you compare the contents of this course with those of other specialistic courses?

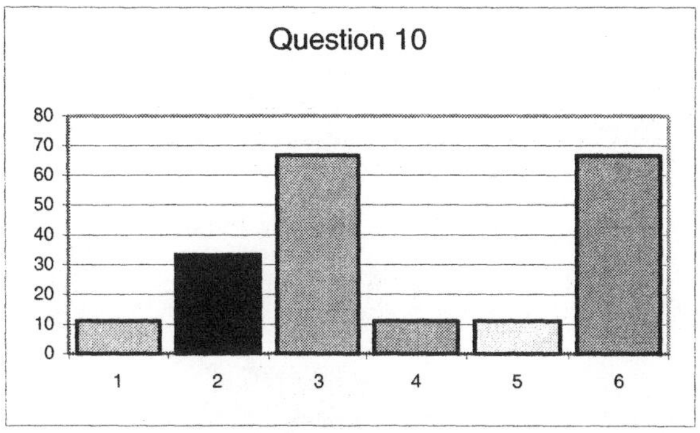

1 – Poorer. 2 – At the same level. 3 – More application-oriented. 4 – Less application-oriented. 5 – Less profession-oriented. 6 – More profession-oriented.

How do you compare the difficulties that you have met in this course with those associated with other specialistic courses?

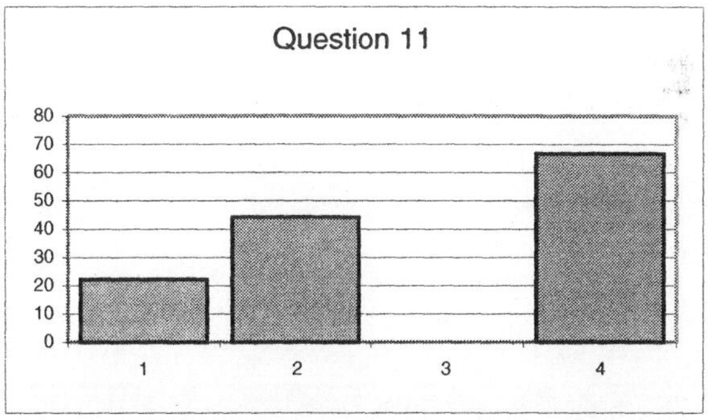

1 – No significant differences. 2 – The learning path has been easier because of the available tools. 3 – The learning path has been less easy. 4 – The learning path has been easier because of tutoring.

How do you evaluate the absence of traditional lectures in this course?

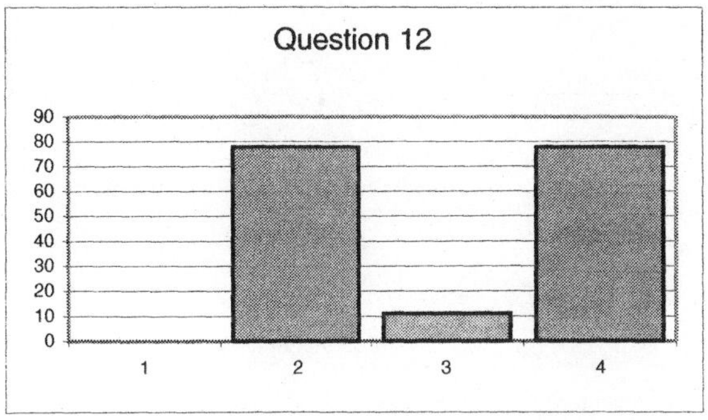

1 – A feature not properly balanced by other features. 2 – A feature suitably balanced by tutoring. 3 – A feature with a negative influence on the whole learning path. 4 – A feature without negative influence on the whole learning path.

A feature of paramount importance of this course is its asynchronous design that does not rely on the unity of place and time associated with traditional teaching environments. How do you evaluate this feature?

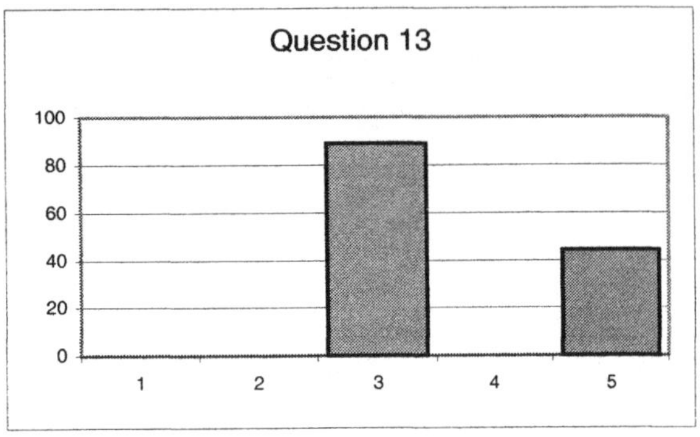

1 – Inessential. 2 – Very relevant but negative. 3 – Very relevant and positive because of the associated flexibility. 4 – Relevant only for limited categories like working students and handicapped persons. 5 – Of general interest and more relevant for specific categories like previous ones.

Would you choose other courses based on the same architecture of this course?

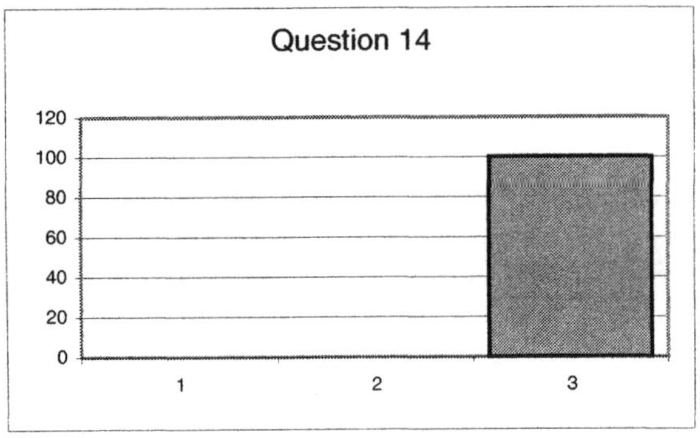

1 – I don't know. 2 – No. 3 – Yes.

Defining as "merit factor" of this course the ratio between the acquired knowledge and the necessary effort and time when compared with other courses, your evaluation is (0.5=double effort, 1=same effort, 2=half effort):

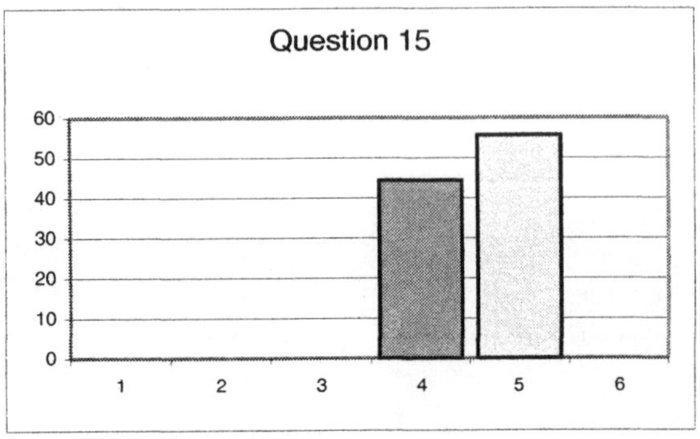

1 – 0.5. 2 – 0.75. 3 – 1. 4 – 1.25. 5 – 1.5. 6 –2.

5. CONCLUDING REMARKS

This paper has presented the more relevant opinions of the students of an identification course preliminarly described by Guidorzi *et al.* (2000).

A first consideration that can be deduced from the opinions formulated by the students concerns their positive evaluation of the role of the asynchronous tutoring implemented in the course. This service has been used by all students who evaluate its role as very effective and capable of eliminating the uncertainties associated with final exams. It is interesting to note also that the presence of a tutoring service has been considered as sufficient to balance the absence of presential lectures.

A very positive evaluation has been formulated also for the availability of virtual laboratories and of a data base of real processes to be used in identification experiments. This is a key feature of the course and the development of platform-independent virtual laboratories implementing high-efficiency algorithms suitable also for on-line identification has required a substantial amount of work. These laboratories have been evaluated as easy to use, endowed with efficient and robust algorithms and properly interfaced with the process database. It is interesting to note the high appreciation for the role of laboratory activities in the comprehension of the various aspects of identification theory on one side and in acquiring professional abilities in the field of identification on the other.

The comparison with other specialistic courses sees this course as more application-oriented and richer of profession-oriented contents.

The global appreciation for this specific course is confirmed by the fact that all students declare that they would select it again.

Finally, the "merit factor" of the course, defined as the ratio between the effort necessary to acquire a certain amount of knowledge and the time necessary to achieve this goal has been evaluated as remarkably higher (between 25 and 50%) than in traditional courses; this evaluation is probably associated with the asynchronous fruition of the course.

REFERENCES

Guidorzi, R.P., U. Soverini, P. Castaldi and R. Diversi (2000). A modular approach in designing an environment for teaching system identification. *Proc. of SYSID 2000,* Santa Barbara, California.

Guidorzi, R.P. and I. Pagani (2002). Evaluation of experience with a web based multi-campus e-learning course at the University of Bologna. *Proc. of 2002 annual EDEN Conference*, Granada, Spain.

IFAC Publications
www.elsevier.com/locate/ifac

CONTROL RELATED TOPICS IN IDENTIFICATION - CLOSED LOOP EXPERIMENTS AND IDENTIFICATION FOR CONTROL

R.R. Bitmead *,[1] **R.A. de Callafon** *

* *University of California, San Diego*
Dept. of Mechanical and Aerospace Engineering
9500 Gilman Drive
La Jolla, CA 92093-0411, U.S.A

Abstract: This paper summarizes the organization of a graduate course on control related topics in identification. The course is taught at the department of Mechanical and Aerospace Engineering at the University of California, San Diego and analyzes the problems of estimation on closed-loop data and control relevant approximation of systems. As part of the course, several case studies are reviewed and include the well documented examples of a sugar cane crushing mill and the identification of the marginally stable electromechanical system found in a CD-ROM player. *Copyright © 2003 IFAC*

Keywords: identification; education; closed-loop; electromechanical systems; process control

1. INTRODUCTION

Experimental data and system identification techniques can be used to estimate dynamical models that are useful in control applications. Models useful for control design are typically of low order and capture the essential closed-loop dynamical behavior needed for control design purposes. Instead of optimizing models for standard prediction or simulation objectives, control relevant models and control related identification are optimized for closed-loop control objectives.

The theory of Prediction Error methods can be extended to include the estimation of linear dynamical models from time domain observations obtained under closed-loop or feedback controlled conditions. Items such as bias, variance, experiment design, and closed-loop optimality are of concern during the experiments and estimation of control-relevant models. The expertise in the area on control relevant and closed-loop system identification can be combined in a graduate course on system identification.

The graduate course *MAE283B Approximate Identification and Control* is a second course in System Identification offered in the Mechanical & Aerospace Engineering Department at the University of California, San Diego. Its place in the course sequence follows *MAE283A Parametric Identification: theory and methods* and, for many students, *MAE284 Robust and Multivarable Control*. The prescribed text is Lennart Ljung's textbook (Ljung 1999), although some material is also taken from the first edition (Ljung 1987). Next to this standard material, various research papers are included in the course and the recently published text book on Iterative Identification and Control (Albertos and Sala 2002) which contains many theoretical results and relevant applications for this course.

The primary focus of the material in this course is to provide students with understanding of the interconnections between modeling and control with a concentration on approximate models derived and validated using experimental data. Recently emerging

[1] This work was supported the US National Science Foundation under grant number ECS-0200449

techniques of iterative (closed-loop) system identification and (model-based) control design were a feature using sophisticated practical examples as a central pedagogical tool. Because of the emphasis on approximation in modeling, the thrust of the material is towards understanding compromises (bias and variance versus excitation) of model fitting versus questions of consistency or asymptotic normality.

2. PRESENTATION OF BIAS ANALYSIS

2.1 Frequency domain expressions

Although prediction errors methods provide a framework for model estimation, challenging problems in the field of system identification lie in the area of approximation of complex systems for control design purposes. Models intended for control design, may require a good approximation of the critical closed-loop behavior of the system to design reliable robust controllers. The objective of finding approximate models becomes even more challenging when closed-loop observations need to be used for identification purposes.

The frequency-domain formulation of Linear System Identification in a quasi-stationary setting is used. For the filtered prediction error

$$\varepsilon_f(t,\theta) = L(q)\varepsilon(t,\theta);\ \varepsilon(t,\theta) := y(t) - y(t|t-1,\theta)$$

where $y(t)$ is a measured output signal and $y(t|t-1,\theta)$ denotes the one-step ahead prediction, the quasi-stationary setting allows variance properties of the prediction error to be represented in the frequency domain

$$\bar{\mathrm{E}}\{\varepsilon_f^2(t,\theta)\} = \frac{1}{2\pi}\int_{-\pi}^{\pi} \Phi(\omega,\theta)\,\mathrm{d}\omega \qquad (1)$$

where the frequency domain integral expression relies on the asymptotic expressions for the variance of the prediction error and application of Parseval's formula.

2.2 Open-loop analysis

To present the bias results by means of the standard integral expressions, first a slightly modified version of the standard and intuitively appealing open-loop bias expression from (Ljung 1987) is presented. In the standard open-loop framework the data generating system is represented by

$$y(t) = G_0(q)u(t) + v(t),\ v(t) = H_0(q)e(t)$$

where, for the moment, the additive noise $v(t) = H_0(q)e(t)$ on the output $y(t)$ is assumed to be uncorrelated with the input $u(t)$. The prediction error $\varepsilon(t,\theta)$ can be written as

$$\begin{aligned}\varepsilon(t,\theta) = H(q,\theta)^{-1}((G_0(q) - G(q,\theta))u(t)+ \\ (H_0(q) - H(q,\theta))e(t)) + e(t)\end{aligned} \qquad (2)$$

where $e(t)$ is a white noise with variance λ_0. As both $H_0(q)$ and $H(q,\theta)$ are monic noise filters and $e(t)$ is a white noise, $\bar{\mathrm{E}}\{e(t)\tilde{e}(t)\} = 0$ where $\tilde{e}(t) := (H_0(q) - H(q,\theta))e(t)$. Due to the open-loop experiments, $\bar{\mathrm{E}}\{e(t)u(t-\tau)^T\} = 0\ \forall\tau$ and as a result, the spectrum $\Phi(\omega,\theta)$ of the (filtered) prediction error in (2) is given by

$$\begin{aligned}|G_0 - G_\theta|^2\Phi_u\frac{|L|^2}{|H_\theta|^2} + \\ |H_0 - H_\theta|^2\lambda_0\frac{|L|^2}{|H_\theta|^2} + \lambda_0\end{aligned} \qquad (3)$$

where the arguments of $e^{j\omega}$ and θ have been dropped for brevity and clarity of the bias formula. The result in (3) can be used to explain the tradeoff in approximate open-loop identification:

- Optimization of θ aims at 'whitening' the prediction error, as $G(q,\theta) = G_0(q)$ and $H(q,\theta) = H_0(q)$ (consistent estimation) makes $\Phi(\omega,\theta) = \lambda_0$.
- Choice of a fixed (stable and stably invertible) noise filter $H_\star(q)$ is equivalent to the choice of a prediction error filter $L(q) = H_\star^{-1}(q)$.
- In case joint parameters occur in the parametrization of $G(q,\theta)$ and $H(q,\theta)$, there is a tradeoff between modeling H_0 versus G_0. It can be seen from (3) that this tradeoff is highly determined by the signal to noise ratio $\Phi_u(\omega)/\lambda_0$.

For control applications, the modeling and approximation of G_0 is often considered of more importance than the noise dynamics H_0. Certainly, this is true for stability as the estimated (low order) model $G(q,\hat{\theta})$ is used for control design purposes and closed-loop stability solely depends on the properties of G_0. The tradeoff between modeling H_0 versus G_0 in approximate identification can be eliminated by either choosing a fixed noise filter (Output Error model structure) or an independent parametrization (Box-Jenkins model structure). Both come with the price of an unavoidable non-linear optimization, but allow an explicit bias tuning.

2.3 Closed-loop analysis

One of the problems in approximate closed-loop identification of plant and noise dynamics on the basis of closed-loop data, is the correlation of the noise $e(t)$ with the signals $\{u(t),y(t)\}$ in the closed-loop. Due to the noise correlation, an approximate identification method that directly uses the input and output of the plant G_0 and ignores the feedback, will lead to biased approximation results for the system dynamics.

The spectrum $\Phi(\omega,\theta)$ of the (filtered) prediction error in (2) in case $u(t)$ is correlated with $e(t)$ has been summarized in (Ljung 1999) as follows

$$\frac{L}{|H_\theta|^2}\begin{bmatrix}\tilde{G} & \tilde{H}\end{bmatrix}\begin{bmatrix}\Phi_u & \Phi_{ue} \\ \Phi_{eu} & \lambda_0\end{bmatrix}\begin{bmatrix}\tilde{G}^* \\ \tilde{H}^*\end{bmatrix} + \lambda_0$$

where the arguments of $e^{j\omega}$ and θ have been dropped for brevity and $\tilde{G} := (G_0 - G_\theta)$, $\tilde{H} := (H_0 - H_\theta)$

and Φ_{ue} indicates the correlation between u and e. By using Schur's complement

$$\begin{bmatrix} \Phi_u & \Phi_{ue} \\ \Phi_{eu} & \lambda_0 \end{bmatrix} = \begin{bmatrix} I & 0 \\ \frac{\Phi_{eu}}{\Phi_u} & I \end{bmatrix} \begin{bmatrix} \Phi_u & 0 \\ 0 & \lambda_0 - \frac{|\Phi_{eu}|^2}{\Phi_u} \end{bmatrix} \begin{bmatrix} I & \frac{\Phi_{ue}}{\Phi_u} \\ 0 & I \end{bmatrix}$$

with respect to Φ_u and definition of

$$B(e^{j\omega}, \theta) = \frac{(H_0(e^{j\omega}) - H(e^{j\omega}, \theta))\Phi_{ue}(\omega)}{\Phi_u(\omega)}$$

allows the spectrum $\Phi(\omega, \theta)$ of the (filtered) prediction error in (2) to be written as

$$\begin{aligned} &|G_0 + B_\theta - G_\theta|^2 \Phi_u \frac{|L|^2}{|H_\theta|^2} + \\ &|H_0 - H_\theta|^2 \left(\lambda_0 - \frac{|\Phi_{ue}|^2}{\Phi_u} \right) \frac{|L|^2}{|H_\theta|^2} + \lambda_0 \end{aligned} \quad (4)$$

and illustrates the bias effect in case $\Phi_{ue}(\omega) \neq 0$:

- Optimization of θ still aims at 'whitening' the prediction error, as $H_\theta = H_0$ yields $B_\theta = 0$ and $G_\theta = G_0$ (consistent estimation) makes $\Phi(\omega, \theta) = \lambda_0$.
- In an approximate identification where $H_\theta \neq H_0$, the model G_θ will approximate $G_0 + B_\theta$ even in the case where there exists a parameter θ for which $G_\theta = G_0$. B_θ clearly indicates an undesirable bias for the plant model G_θ.
- Due to the ratio of L and H_θ in (4), the choice of a fixed (stable and stably invertible) noise filter $H_\star$ remains equivalent to the choice of a prediction error filter $L = H_\star^{-1}$.

In case joint parameters occur in the parametrization of $G(q, \theta)$ and $H(q, \theta)$, there is again a tradeoff between modeling H_0 versus $G_0 + B_\theta$ in approximate identification. Unfortunately, the tradeoff cannot be eliminated by choosing a fixed noise filter (Output Error model structure) or an independent parametrization (Box-Jenkins model structure) to focus on the approximate identification of G_0 only. The choice of a fixed noise filter $H_\star \neq H_0$ would be unable to eliminate the bias term

$$B_\star = \frac{(H_0 - H_\star)\Phi_{ue}}{\Phi_u}$$

resulting in a biased estimation of the plant dynamics G_0 that highly depends on the chosen noise filter $H_\star$ and the (unknown) noise dynamics Φ_{ue}.

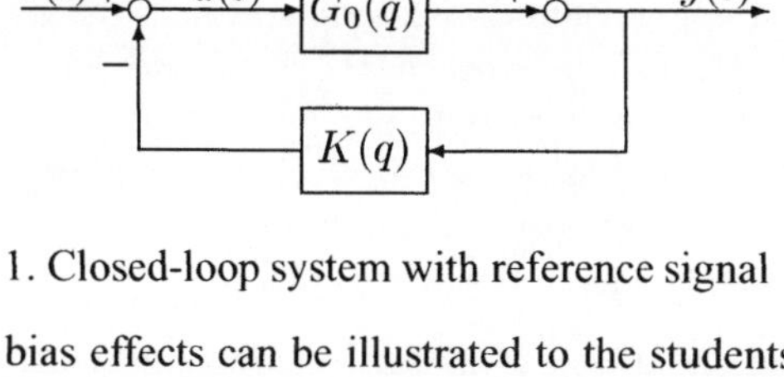

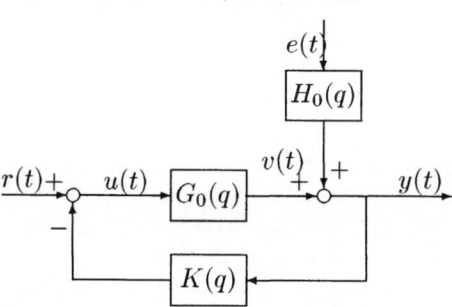

Fig. 1. Closed-loop system with reference signal

The bias effects can be illustrated to the students, by analyzing the correlation between input $u(t)$ and noise $e(t)$ for a closed-loop system as indicated in Figure 1. With Figure 1, the data coming from the plant $G_0(q)$ and subjected to external reference signal $r(t)$ and additive noise $H_0(q)e(t)$ operating under closed-loop condition can be described as follows:

$$\begin{aligned} y(t) &= G_0(q)S_{in}(q)r(t) + S_{in}(q)H_0(q)e(t) \\ u(t) &= S_{in}(q)r(t) - K(q)S_{in}H_0(q)e(t) \end{aligned} \quad (5)$$

where $S_{in}(q)$ is the input sensitivity function defined by

$$S_{in}(q) = \frac{1}{1 + K(q)G_0(q)}$$

With the reference signal $r(t)$ uncorrelated with the noise $e(t)$

$$\begin{aligned} B_\theta &= (H_0 - H_\theta)\frac{\Phi_{ue}}{\Phi_e}\frac{\Phi_e}{\Phi_u} \\ &= (H_\theta - H_0)KS_{in}H_o\frac{\lambda_0}{\Phi_u} \end{aligned}$$

which clearly indicates the (implicit) character of the bias B_θ in case of closed-loop experiments. The bias B_θ depends on the noise model H_θ being estimated, the way in which the noise is present on the input ($KS_{in}H_0$) and the signal to noise ratio Φ_u/λ_0.

The effect of the bias can also be illustrated for the extreme situation where no reference $r(t)$ is present on the closed-loop system to provide sufficient excitation. For that purpose, an alternative formulation is used: substitution of (5) in the formulation of the prediction error (2) yields the following prediction error

$$\begin{aligned} &\varepsilon(t, \theta) = H_\theta^{-1}S_{in}(G_0 - G_\theta)r(t) + \\ &H_\theta^{-1}S_{in}((I + G_\theta K)H_0 - H_\theta)e(t) + S_{in}e(t) \end{aligned} \quad (6)$$

With at least one step delay in the product $G_\theta K$, both $(1+G_\theta K)H_0(q)$ and H_θ are monic filters. As a result, $\bar{\mathrm{E}}\{e(t)\tilde{e}(t)\} = 0$, as $e(t)$ is a white noise signal, where $\tilde{e}(t) := ((1 + G_\theta K)H_0 - H_\theta)e(t)$. With the reference signal $r(t)$ uncorrelated with $e(t)$ we also find $\bar{\mathrm{E}}\{e(t)r(t - \tau)^T\} = 0 \ \forall \tau$ and the spectrum $\Phi(\omega, \theta)$ of the (filtered) prediction error in (2) can be written as

$$\begin{aligned} &|G_0 - G_\theta|^2 \Phi_r \frac{|S_{in}|^2|L|^2}{|H_\theta|^2} + \\ &|(1 + G_\theta K)H_0 - H_\theta|^2 \lambda_0 \frac{|S_{in}|^2|L|^2}{|H_\theta|^2} + |S_{in}|^2 \lambda_0 \end{aligned} \quad (7)$$

which also gives clear insight in the bias effects. In case of lack of external excitation of the closed-loop system it can be seen that:

- The estimation of models G_θ and H_θ is done such that $H_0 - H_\theta + H_o G_\theta K$ is minimized. As a result, the actual plant G_0 does not play a role in the estimation of G_θ.
- In case a fixed noise model $H_\star$ is chosen (OE model structure), the optimal model G_θ is given by $G_\theta = -K^{-1}$, and we would be estimating the inverse of the controller.

The analysis presented above demonstrates clearly to the students that the lack of consistency and a tunable bias expression for the estimation of the model G_θ and H_θ in closed-loop identification.

3. PRESENTATION OF VARIANCE RESULTS

To complete the analysis and provide students with the concepts on the tradeoff between bias and variance in typical System Identification problems, the standard variance result in (Ljung 1999) are presented. The results are presented without going into the details of the technical conditions involved with the formulation of the variance results.

Consider estimate $\hat{\theta}_N$ determined by minimizing the Least Squares criterion

$$V_N = \frac{1}{2N}\sum_{t=1}^{N}\varepsilon^2(t,\theta)$$

and define

$$\psi(t,\theta_\star) = \left.\frac{\partial\epsilon(t,\theta)}{\partial\theta}\right|_{\theta=\theta_\star}$$

then subject to the technical conditions:

- model structure is linear and uniformly stable
- data are generated by stable linear filtering of quasistationary signals with finite moments of $4+\delta$
- $\theta_N \to \theta_\star$ w.p.1 as $N \to \infty$ for bounded $\theta_\star$
- $\dfrac{\partial^2 V_\theta}{\partial\theta^2} > 0$
- $\sqrt{N}\mathrm{E}\left\{\frac{1}{N}\sum_{t=1}^{N}\psi(t,\theta_\star)\epsilon(t,\theta_\star) - \bar{\mathrm{E}}\psi(t,\theta_\star)\epsilon(t,\theta_\star)\right\} \to 0$ as $N \to \infty$

the following asymptotic variance expression of the parameter estimate $\hat{\theta}_N$ can be given (Ljung 1999):

$$\sqrt{N}(\hat{\theta}_N - \theta_\star) \sim N(0, P_\theta)$$

$$P_\theta = [V''(\theta_\star)]^{-1}Q[V''(\theta_\star)]^{-1} \qquad (8)$$

$$Q = \lim_{N\to\infty} N\bar{\mathrm{E}}\left\{[V'(\theta_\star,N)][V'(\theta_\star,N)]^T\right\}$$

The background and technical implications behind the variance expression formula are emphasized by mentioning that the result deals with the asymptotic normality and variance of the parameter error. Furthermore it must be stressed that it is possible to extend the parameter variance expression to the frequency domain in a direct way.

$$\mathrm{Cov}\begin{bmatrix}\hat{G}_N(e^{j\omega})\\ \hat{H}_N(e^{j\omega})\end{bmatrix} \sim \frac{n}{N}\begin{bmatrix}\Phi_u(\omega) & \Phi_{eu}(\omega)\\ \Phi_{ue}(\omega) & \lambda_0\end{bmatrix}^{-1} \qquad (9)$$

The frequency domain expression of (9) treats the transfer function estimation error distribution. The limitations of the variance expression are pointed out to the students by mentioning that the frequency domain variance expression in (9):

- is asymptotically valid only as $N \to \infty$
- relies on weak convergence — no large deviations
- fundamentally assumes (G_0, H_0) has a parametrization with $\theta = \theta_0$ and $\theta_N \to \theta_0$
- indicates that the covariance increases with the number of parameters n, and decreases with the number of data points N
- depends in a sensible way on noise to signal ratio
- the exact calculation depends on the criterion used and provides the measure of asymptotic efficiency of the estimator.

Although limited in application, the variance expression results in (8) and (9) and the bias expressions in (4) and (7) give insight in the variance and bias tradeoff in experiment design and system identification. More detailed information on (closed-loop) variance expressions are summarized in the paper by Gevers *et al.* (2001) that is presented to the students during the course. It is illustrated that the spectrum of the reference or input signal can be used to influence bias and variance aspects during experiment design. For closed-loop experiments, the controller K also provides a valuable tool to influence the signal properties.

Post processing of the data (after the actual experiments) can be done by the choice of a suitable data filter L and the choice of the model class in the identification of plant model G_θ and noise model H_θ. Important design variables are the model structure with the number of parameters and the possibility to use an independent parametrization of plant and noise dynamics. But it must be stressed that identification on the basis of closed-loop data requires special attention: standard open-loop identification techniques that directly use the input u and output y signals of the plant is bound to give biased result. What remains is the motivation to perform closed-loop experiments, despite the pitfalls of biased estimation.

4. IDENTIFICATION AND CONTROL

4.1 Why closed-loop experiments?

To argue for the role of experiments measured in closed-loop, several basic results are mentioned to the students that motivate the usefulness of closed-loop experiments. One of the first steps towards the interaction between identification and control has been made in Åström and Wittenmark (1971) and Gevers and Ljung (1986). In Gevers and Ljung (1986) it is mentioned from a variance point of view that optimal models can be found via a Prediction Error estimation method that uses closed-loop experiments and appropriate data filters. As such, the usefulness of closed-loop experiments as opposed to open-loop experiments to model a plant was shown to be fruitful. Unfortunately, the desired closed-loop experiments and the appropriate data filter contains knowledge of the controller yet to be designed.

Argumentation of optimal models for control design from a bias point of view is built on the observation that an approximate identification of a model is allowed, as long as the approximate model G_θ takes into account its intended use – the design of a high performing controller K for the actual plant G_0. In case a norm function is used to characterize the performance of a feedback system, the performance of a controller K applied to the actual plant G_0 can be delineated by $\|J(G_0, K)\|$. Even if a controller K is available, the performance $\|J(G_0, K)\|$ cannot be evaluated precisely, as the plant G_0 is unknown. From a bias point of view, the role of the model G_θ can be seen as providing upper and lower bounds for $\|J(G_0, K)\|$ via triangular inequalities

$$\left| \|J(G_\theta, K)\| - \|J(G_0, K) - J(G_\theta, K)\| \right| \leq \|J(G_0, K)\|$$
$$\|J(G_0, K)\| \leq \|J(G_\theta, K)\| + \|J(G_0, K) - J(G_\theta, K)\|$$

that were presented in Schrama (1992). For a given controller K, the minimization of $\|J(G_0, K) - J(G_\theta, K)\|$ provides a tight upper and lower bound for $\|J(G_0, K)\|$ and constitutes a so-called control relevant identification problem (Van den Hof and Schrama 1995). In this identification problem, a model G_θ is found by minimizing the difference between closed-loop performance criteria. Obviously, closed-loop experiments are required to solve such an identification problem.

4.2 Dealing with closed-loop experiments

Now that the setting and motivation for closed-loop experiments has been established, the methodologies for dealing with closed-loop data are presented. The direct approach consists of applying a standard open-loop prediction error method directly to the input $u(t)$ and output $y(t)$ signals, ignoring any possible feedback and the reference signal $r(t)$. From the analysis in Section 2.3 it is obvious that this approach leads to estimation results in case of approximate identification for which the bias cannot be tuned explicitly.

Following the analysis of identification on the basis of closed-loop data, possible solutions and methods to the closed-loop identification problem are presented in the course. The closed-loop identification methods are presented by providing a short overview of the method and a copy of the papers that summarize the details of the methodology. It is beyond the scope of this paper to present the methods detail, but it can be mentioned here that in the presentation of these methods, a distinction is made between the following approaches:

- The first class of methods presented in the course is based on a reparametrization of the closed-loop identification problem.
- The second class of methods presented to deal with closed-loop data are two-stage methods, where the estimation of approximate models on the basis of closed-loop data is solved in two estimation steps.

For the first class of methods, reparametrization is done by using the knowledge of the controller and parametrizing the closed-loop transfer function in terms of the controller and the open-loop model to be estimated. Methodologies that are reviewed in the course are the indirect estimation method, tailor-made parametrization (Landau and Karimi 1997) and recursive estimation methods (de Bruyne *et al.* 1999).

In the prediction error framework, the reparametrization of the closed-loop transfer function involves the minimization of a closed-loop prediction error

$$\varepsilon_{cl}(t, \theta) = y(t) - P(q, \theta) r(t)$$

where $P(q, \theta)$ can be parametrized using a customized parametrization

$$P(q, \theta) = \frac{G(q, \theta)}{1 + K(q) G(q, \theta)} \tag{10}$$

using the explicit knowledge of the controller $K(q)$. Alternatively, $P(q, \theta)$ can be freely parametrized and the knowledge of the controller is used to recompute the model $G(q, \theta)$ via

$$G(q, \theta) = \frac{P(q, \theta)}{1 - P(q, \theta) K(q)} \tag{11}$$

In the customized parametrization (10) the order of the model $G(q, \theta)$ can be controlled. The computation of the model $G(q, \theta)$ via (11) in general increases the model order due to the free parametrization of the closed-loop transfer function $P(q, \theta)$.

For the second class or the two-stage methods, the first step is used to create a (noise free) instrument that will used in the second step to recast the closed-loop identification problem into an open-loop one. The instrument in the first step of the two-stage methods is typically a filtered closed-loop signal. With the closed-loop data given in (5), an estimate of the (input) sensitivity function S_{in} can be obtained by minimizing the closed-loop prediction error

$$\varepsilon_1(t, \theta) = u(t) - S(q, \theta) r(t)$$

in the first step of the method, where $S(q, \theta)$ is used to model the (input) sensitivity function $S_{in}(q)$ as accurately as possible. The estimated model $\hat{S}(q) = S(q, \hat{\theta})$ can be used to create a filtered input signal

$$\hat{u}(t) := \hat{S}(q) r(t)$$

that will be uncorrelated with the noise $e(t)$ present on the (unfiltered) input signal $u(t)$. The (noise free) instrument can be used to perform an equivalent open-loop identification problem by minimizing the prediction error

$$\varepsilon_2(t, \theta) = y(t) - G(q, \theta) \hat{u}(t)$$

in the second step of the method. Closed-loop identification method that fall under this category are presented in the course and include the instrumental variable method (Ljung 1999), the two-stage method (Van den Hof and Schrama 1993) and the coprime factor based methods (Van den Hof *et al.* 1995, Anderson 1998).

4.3 Link between identification and control

Estimating approximate models suitable for control requires closed-loop experiments to approximate the closed-loop and control-relevant aspects of the system. This is illustrated by evaluating the bias expressions associated to the different the closed-loop identification methods. It is shown that for most of the methods presented in the course, the closed-loop prediction error exhibits a spectrum $\Phi(\omega, \theta)$ that is given by

$$|G_0 - G_\theta|^2 \Phi_r \frac{|S_{in}|^2 |L|^2}{|H_\star|^2} \qquad (12)$$

for a prediction error model estimation with a fixed noise filter $H_\star$. It can be observed that (12) is equal to the first term in (7) and the closed-loop identification methods have eliminated the bias effects due to the closed-loop noise. This allows for an explicit tuning of the bias expression of the model G_θ.

The merits of approximate identification and the use of closed-loop data in estimating approximate models can be found in the triangular inequalities that represent the interaction between model based control and identification of models for control. The idea of alternately minimizing $||J(G_\theta, K) - J(G_0, K)||$ in via system identification and $||J(G_\theta, K)||$ via a control design problem forms a basis for many of the iterative schemes or control relevant identification approaches listed in the literature (Van den Hof and Schrama 1995). In such an iterative scheme, the control relevant identification of a (nominal) model G_θ and the design of a model-based controlled K are applied iteratively with the aim to minimize the overall performance $||J(G_0, K)||$ of the feedback controlled plant G_0.

4.4 Case studies

To illustrate the work that has been done in the field of control relevant identification and model based control design, a short overview of iterative methods is presented at the end of the course. The iterative identification and control methods are presented by means of applications and case studies that illustrate the effectiveness of closed-loop identification methods and model-based control design to obtain high performance feedback control systems.

The case studies that are presented at the end of the course are the "control relevant identification and servo design for a compact disc player" and the "iterative identification and control: a sugar cane crushing mill" that both can be found in Albertos and Sala (2002). Both case studies illustrate the use of closed-loop experiments and model-based control design to enhance the performance of a feedback controlled system. The case studies are used to demonstrate in concrete form the principles of closed-loop identification presented during the course.

5. SUMMARY

This paper shows the organization of a course that focuses on estimation techniques for closed-loop or feedback controlled systems. The course gives both a theoretical and practical introduction to closed-loop identification methods and control relevant experimentally based modeling. Instead of focusing on models that are optimized for standard prediction or simulation objectives, models are optimized for closed-loop control objectives. As part of the course, two case studies are reviewed: a sugar cane crushing mill and the identification of a marginally stable electromechanical system.

REFERENCES

Albertos, P. and A. Sala (2002). *Iterative Identification and Control.* Springer Verlag, London, UK.

Anderson, B.D.O. (1998). From Youla-Kucera to identification, adaptive and non-linear control. *Automatica* **34**, 1485–1506.

Åström, K.J. and B. Wittenmark (1971). Problems of identification and control. *Journal of Mathematical Analysis and Applications* **34**, 90–113.

de Bruyne, F., B.D.O. Anderson, M. Gevers and N. Linard (1999). Gradient expressions for a closed-loop identification scheme with a tailor-made parametrization. *Automatica* **35**(11), 1867–1871.

Gevers, M., L. Ljung and P.M.J. Van Den Hof (2001). Asymptotic variance expressions for closed-loop identification. *Automatica* **73**(5), 781–786.

Gevers, M.R. and L. Ljung (1986). Optimal experiment design with respect to the intended model application. *Automatica* **22**, 543–554.

Landau, I. and A. Karimi (1997). An output error recursive algorithm for unbiased identification in closed-loop. *Automatica* **33**, 933–938.

Ljung, L. (1987). *System Identification: Theory for the User.* Prentice-Hall, Englewood Cliffs, New Jersey, USA.

Ljung, L. (1999). *System Identification: Theory for the User (second edition).* Prentice-Hall, Englewood Cliffs, New Jersey, USA.

Schrama, R.J.P. (1992). Accurate identification for control design: the necessity of an iterative scheme. *IEEE Trans. on Automatic Control* **37**, 991–994.

Van den Hof, P.M.J. and R.J.P. Schrama (1993). An indirect method for transfer function estimation from closed loop data. *Automatica* **29**, 1523–1527.

Van den Hof, P.M.J. and R.J.P. Schrama (1995). Identification and control - closed loop issues. *Automatica* **31**, 1751–1770.

Van den Hof, P.M.J., R.J.P. Schrama, R.A. de Callafon and O.H. Bosgra (1995). Identification of normalised coprime plant factors from closed-loop experimental data. *European Journal of Control* **1**(1), 62–74.

IFAC Publications
www.elsevier.com/locate/ifac

TEACHING SEMIPHYSICAL MODELING TO CHEMICAL ENGINEERING STUDENTS USING A BRINE-WATER MIXING TANK EXPERIMENT

D. E. Rivera [1]

Control Systems Engineering Laboratory
Department of Chemical and Materials Engineering
Arizona State University, Tempe, Arizona 85287

Abstract: The Chemical Engineering program at Arizona State offers an integrated series of core courses that teach students how conservation and accounting principles can be applied to describe engineering phenomena across disciplines. A brine-water mixing tank experiment was introduced in the third course in the series (ECE 394C: Understanding Engineering Systems Via Conservation) as a capstone modeling project for the recitation portion of the course. The experiment provides students with "hands-on" experience on a real-life system incorporating process, electrical, and mechanical components, as well as real-time data acquisition and control. A major feature of the brine-water tank project is that students apply a comprehensive system identification procedure relying on semiphysical (a.k.a. "grey box") models to complement their understanding of first-principles modeling. This paper describes the brine-water tank experiment, presents the formulation of the semiphysical parameter estimation problem, and describes the comprehensive procedure that students undertake to go from process data to validated plant models. *Copyright © 2003 IFAC*

Keywords: system identification education, semiphysical modeling

1. INTRODUCTION

ECE 394 Systems, Understanding Engineering Systems via Conservation, is the third in an experimental core curriculum developed at Texas A&M which has been part of the Chemical Engineering curriculum since the fall of 1992. Students traditionally take ECE 394 Systems in the spring semester of their junior year. This four credit hour course is structured with three lecture hours a week and one weekly 2-hour recitation. The course stresses the broad-based use of accounting and conservation principles to model systems involving process, electrical, and mechanical components (separately and in combination). Another principal course objective is the use of computer-based tools to model engineering systems of practical interest.

[1] to whom all correspondence should be addressed; phone:(480) 965-9476, email:daniel.rivera@asu.edu

In ECE 394 Systems, students are confronted with the "reality" of engineering systems from the very first lecture. Students are made aware that real systems are:

- dynamic/unsteady-state ("steady-state is a figment of the imagination"),
- nonlinear,
- multivariable (i.e., possess multiple inputs and outputs),
- uncertain (i.e., models of real systems lack accuracy),
- stochastic (i.e., real systems are subject to random behavior, and as such cannot be always described by deterministic models). Precision errors will always be present in models.

Students are also presented in the first lecture (and frequently reminded thereafter) of the saying attributed to famous statistician Professor G.E.P. Box of the University of Wisconsin, *"all models are wrong, but some models are useful*." Students work in recitation as part of three-person teams. Two individual reports and three group presentations are required as part of the modeling project.

The brine-water tank experiment (Figure 1) is used in ECE 394 Systems as an ongoing project to bring students to reconcile the abstraction of mathematical modeling with the realities of a practical system. The main objective of this experiment is to develop, via first principles and semiphysical modeling techniques, *useful* mathematical models of the tank behavior displaying good predictive ability. Specifically, the students are asked to model the dynamic response of salt concentration in the outlet stream (c) and level in the tank (h) to changes in the inlet brine flowrate (q_c), the fresh water flowrate (q_w), and outlet flow (q_F). The tank is interfaced to a industrial-scale real-time computing platform, namely a Honeywell TotalPlant Solution System (previously known as the TDC3000, Figure 3). The engineer is capable of adjusting all three tank flows via the TDC3000 regulatory control points FIC100, FIC101, and LIC100 (see Figure 1). The experiment also requires students to generate a suitable calibration between the signal generated from an on-line conductivity sensor and the salt concentration (in g/ℓ) for the outlet stream in the tank.

The paper is organized as follows. Section 2 presents a brief description of the experimental apparatus. Section 3 discusses the first-principles model for the tank and the corresponding derivation of a semiphysical model for this system. Section 4 describes the steps involved in developing a comprehensive semiphysical modeling procedure, beginning from experimental design and concluding with model validation. Section 5 presents a summary and conclusions.

2. EXPERIMENTAL DESCRIPTION

Figure 1 shows both the process and the instrumentation used in this experiment. The flow of tap water to the process is regulated by measuring the flow with an orifice meter and changing the valve position on the water line according to an algorithm in a regulatory control point in the TDC3000. This control loop is assigned the tagname FIC100. Similarly, the flow of a concentrated salt solution is controlled with loop FIC101. The level in the tank is measured with a differential pressure cell (d/p) with one leg connected to the bottom of the tank and the other leg open to the atmosphere. The regulatory control point LIC100 compares this level with a desired level and manipulates the flow through the drain line. The salt concentration leaving and entering the tank is measured with conductivity cells and read into the system via analog input points CI100 and CI102, respectively. The conductivity measurements are displayed as the PVs (Process Values) of CI100 and CI102. By setting the appropriate instrument range limit parameters in the system (PVEUHI and PVEULO) the students are able to implement a linear correlation relating the raw 4-20ma signal from the conductivity cells to a sensible value for concentration in units of g/ℓ. CIC100 is a regulatory control point used in a subsequent course (ChE 461 Introduction to Process Control) (Rivera *et al.*, 1996) which adjusts the salt inlet flowrate setpoint (FIC101.SP) to keep exit stream salt concentration at setpoint (CI100.PV); students are asked to leave this point on MANUAL throughout the experiment.

3. BRINE-WATER TANK MODELING

3.1 *First-principles modeling*

During lecture and through homework assignments, students use MATLAB with SIMULINK to develop a first-principles dynamical model describing the effect of the various system inputs on the level and salt concentration. The principles of conservation of total mass and accounting of the salt species in the tank are used to derive this model. The level dynamics of the system are described by a differential equation arising from the conservation of total mass in the system

$$\frac{dh}{dt} = \frac{1}{A}(q_w - q_F + \frac{\rho_c}{\rho}q_c) \quad (1)$$

while the dynamics of salt in the tank are modeled by accounting for this species in the system:

$$\frac{dc}{dt} = \frac{q_c}{V}(c_c - \frac{\rho_c}{\rho}c) - \frac{q_w c}{V}, \quad V = hA \quad (2)$$

A, the crossectional area of the tank, ρ_c and ρ, the inlet brine and inlet water/outlet stream densities (respectively), and c_c, the inlet brine concentration, are constant-valued parameters in the model. An example of the SIMULINK window built by students is shown in Figure 2. Furthermore, Matlab with SIMULINK can be used compare the results of the first-principles nonlinear model with the responses obtained from its linearized equivalent at an operating condition; this enables students to evaluate the modeling errors associated with linearization.

3.2 *Semiphysical modeling*

The derivation of the semiphysical model follows along the line of the analysis presented in Lindskog (1996). Assuming constant volume in the tank (as the result of tight level control in the system) and constant densities for all streams, the first-principles model per Equations 1 and 2 reduces to:

$$\frac{dc}{dt} = \frac{q_c\, c_c}{V} - \frac{(q_c + q_w)\, c}{V} \quad (3)$$

Using a forward-difference approximation on the derivative (for a sampling time T) leads to

$$\frac{c(t+1)-c(t)}{T} = \frac{q_c(t)\,c_c(t)}{V} - \frac{(q_c(t)+q_w(t))\,c(t)}{V} \quad (4)$$

which solving for $c(t+1)$ yields

$$c(t+1) = c(t) + \frac{q_c(t)\,c_c(t)\,T}{V} - \frac{(q_c(t)+q_w(t))\,c(t)\,T}{V} \quad (5)$$

Rearranging and consolidating terms leads to the semiphysical structure

$$c(k) = c(k-1) + \theta_1 q_c(k-1)c_c(k-1) + \theta_2 q_c(k-1)c(k-1) + \theta_3 q_w(k-1)c(k-1) \quad (6)$$

Estimates of θ_1, θ_2, and θ_3 can be obtained from the first-principles model

$$\theta_1 = \frac{T}{V} \qquad \theta_2 = -\frac{T}{V} \qquad \theta_3 = -\frac{T}{V} \quad (7)$$

or alternatively, they can be estimated from plant data by recognizing that θ_1, θ_2, and θ_3 are linear in the parameters and hence linear regression can be readily applied.

4. A COMPREHENSIVE SEMIPHYSICAL MODELING EXPERIENCE

Having recognized that parameter estimation in semiphysical modeling constitutes a regression problem, students are then asked to perform a series of tasks that comprise a comprehensive identification procedure. These include:

(1) *Experimental Design.* Students are asked to use the first-principles MATLAB/SIMULINK model to design an informative experiment on the system. The design consists of a series of step changes of varying magnitude and duration that are intended to highlight the nonlinear behavior of the system and take into account the dominant time dynamics. The experiment must not exceed a 2 hr time period (the length of a recitation session) and must avoid taking the sensors and actuators past their limits. Figure 4 shows a TDC3000 data screen for a typical experimental run designed by the students. Various experimental runs are performed during the course of two weeks in the semester, and these are used to serve as estimation and validation data sets for the ensuing parameter estimation problem.

(2) *Model structure selection and parameter estimation.* Students are then asked to develop a Matlab program that uses regression analysis to estimate parameters of the semiphysical model. In addition to the three-parameter model structure shown in Equation 6, the program must also estimate parameters for the following difference equation model structures:

Four parameter model (Version A):

$$c(k) = \theta_4 c(k-1) + \theta_1 q_c(k-1)c_c(k-1) + \theta_2 q_c(k-1)c(k-1) + \theta_3 q_w(k-1)c(k-1) \quad (8)$$

Four parameter model (Version B):

$$c(k) = c(k-1) + \theta_1 q_c(k-1)c_c(k-1) + \theta_2 q_c(k-1)c(k-1) + \theta_3 q_w(k-1)c(k-1) + \theta_4 \quad (9)$$

Five parameter model:

$$c(k) = \theta_4 c(k-1) + \theta_1 q_c(k-1)c_c(k-1) + \theta_2 q_c(k-1)c(k-1) + \theta_3 q_w(k-1)c(k-1) + \theta_5 \quad (10)$$

The "four parameter" and "five parameter" models have more degrees of freedom and therefore allow greater flexibility in improving the goodness-of-fit as compared to the "three-parameter difference equation.

(3) *Model Validation.* Ultimately, the goal of model validation is to determine the model structure and parameters leading to predictions that are both physically meaningful and result in lower errors when compared on a validation data set (i.e., a data set other than the one used for estimation). The semiphysical model estimates are compared against each other and against the responses obtained from the first-principles model (in both continuous-time and difference equation form). In addition, students are asked to compute, display, and plot the maximum and Root-Mean-Square (RMS) errors for both the estimation and crossvalidation data sets. The RMS and maximum errors are determined on the basis of the residual time series

$$e_{resid}(k) = c(k) - \hat{c}(k) \qquad k = 1, \cdots, N \quad (11)$$

which is the difference between the measured concentration $(c(k))$ and that estimated from a model $(\hat{c}(k))$. N is the total number of observations in the data set. The RMS error is computed as

$$RMSerr = \left(\frac{1}{N}\sum_{k=1}^{N} e^2_{resid}(k)\right)^{1/2} \quad (12)$$

while the maximum error consists of the largest absolute magnitude in the residuals, determined by

$$MAXerr = \max_{k} |e_{resid}(k)| \qquad (13)$$

$$k = 1, \cdots, N$$

(4) *Reflection.* Determining which model (semiphysical or first-principles) is "best" is not enough. Students are asked to examine their experience with the system and list all possible sources of error and prioritize them in order of importance. The inquisitive student will recognize problems related with the calibration of measurements, the relative effect of the simplifying assumptions, and similar circumstances. Ultimately, the students realize the importance of semiphysical modeling and of working with data as a valuable tool in modeling.

An illustration of the various steps with some representative test data sets is shown in Figures 5 through 12. These plots are generated using the Matlab/SIMULINK files developed by the students throughout the course of the semester. Estimation data (consisting in this case of one step change each for the inlet brine and fresh water flows) is shown in Figures 5 and 6. The relative agreement between the first principles and 3-parameter semiphysical model results can be seen in Figure 5. Simulation results that include the two four-parameter models and the five-parameter model are shown in Figure 7. All semiphysical models closely agree, and as reflected in the RMS values (Figure 8), increasing the number of parameters yields improved goodness of fit in the estimation data set. Parameter estimates are presented on the Matlab command window and compared to first-principles coefficients; Figure 9 shows the values obtained for the four-parameter model (Version A). Simulation results on the validation data set (Figure 10) indicate that this model has the best predictive ability over all the evaluated models. This is reflected in both a better visual fit in the simulation as well as superior RMS and MAX errors (Figures 11 and 12, respectively).

5. SUMMARY AND CONCLUSIONS

The brine-water mixing tank is a relatively simple experiment that, while originating from the field of chemical engineering, can be readily taught to students across disciplines. The experiment described in this paper exposes students to significant concepts in modeling, identification, and numerical computing in a challenging experimental and real-time information setting. Semiphysical modeling is introduced in a meaningful way while demanding only a modest mathematical background from students: knowledge of differential equations, basic numerical methods, and regression analysis. Copies of the Matlab/SIMULINK files implementing this procedure (as well as some sample data files) can be obtained by request from the author at (`daniel.rivera@asu.edu`).

6. REFERENCES

Lindskog, P. (1996). Methods, Algorithms, and Tools for System Identification Based on Prior Knowledge. PhD thesis. Linköping University, Sweden. Dept. of Electrical Engineering.

Rivera, D.E., K.S. Jun, V.E. Sater and M.K. Shetty, "Teaching Process Dynamics and Control Using an Industrial-Scale Real-Time Computing Environment," *Computer Applications in Engineering Education*, Computer-Aided Chemical Engineering Education Special Issue, Vol 4., No. 3, pp 191-205, 1996.

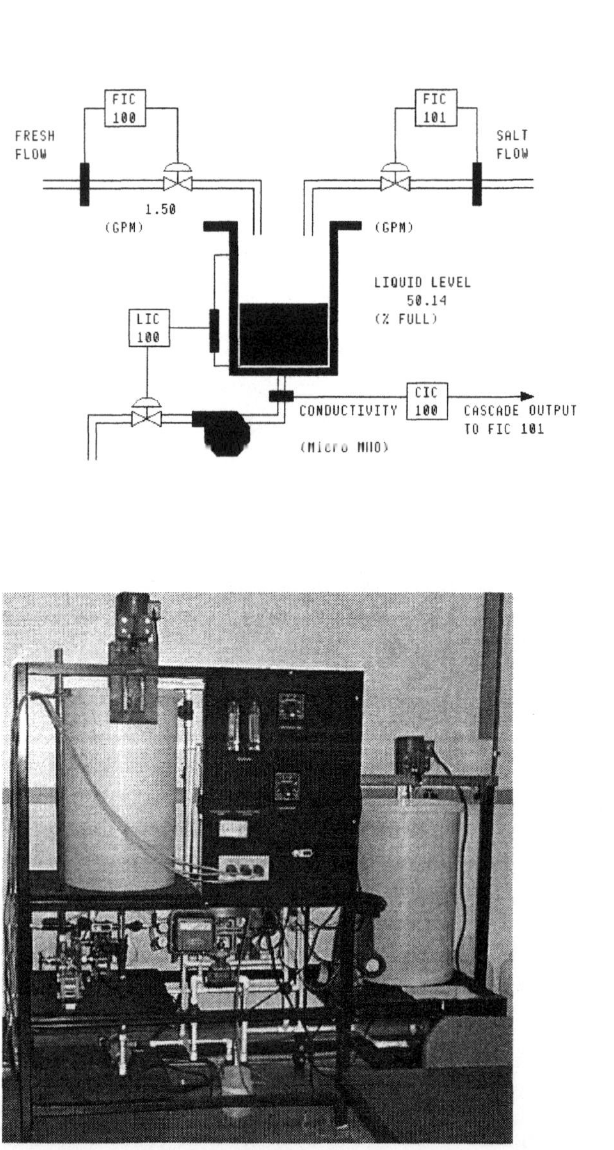

Fig. 1. Brine-water mixing tank schematic (top) and photograph (bottom).

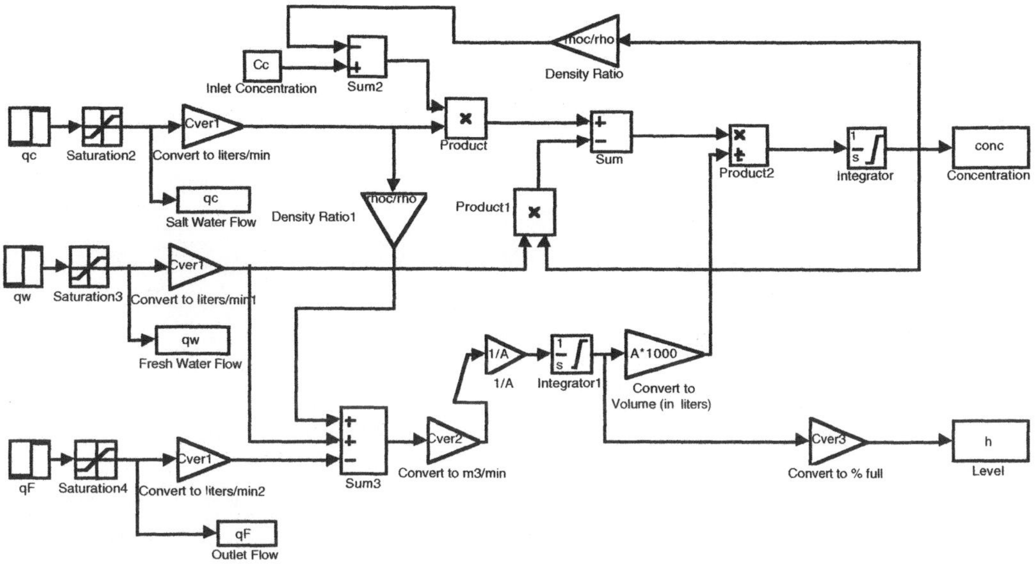

Fig. 2. SIMULINK window for Brine-Water Mixing Tank First-Principles Model

Fig. 3. Representative cluster of Universal and Global User Stations for ASU's TotalPlant Solution System.

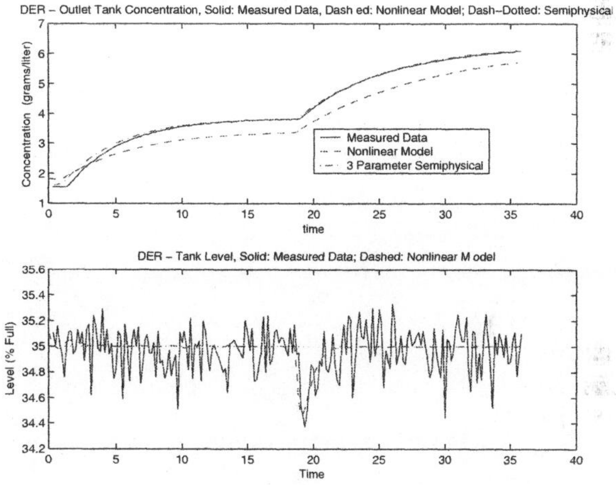

Fig. 5. Output time series for the estimation data set.

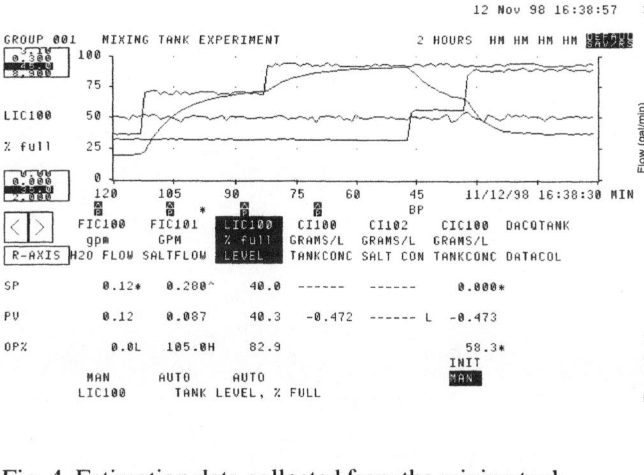

Fig. 4. Estimation data collected from the mixing tank, shown on a Honeywell TPS group display

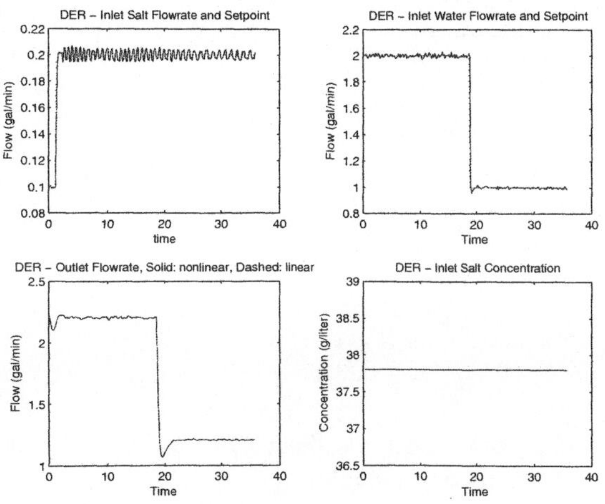

Fig. 6. Input time series for the estimation data set.

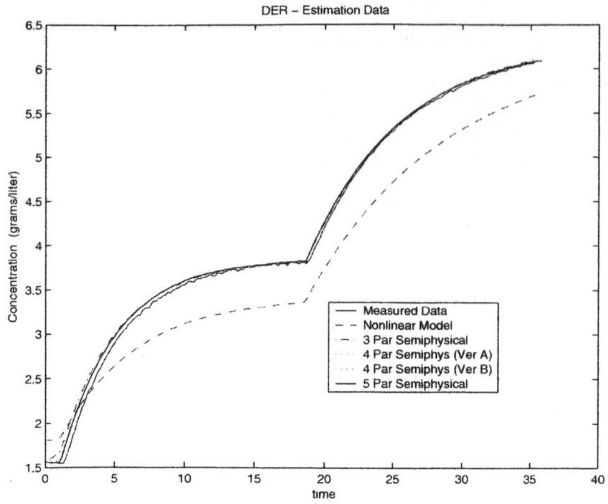

Fig. 7. Simulation results on the estimation data set for the first-principles and semiphysical models.

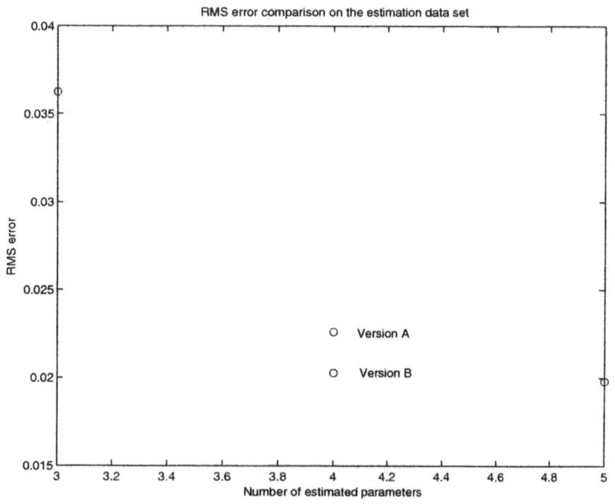

Fig. 8. RMS error comparison on the estimation data set.

```
Four Parameter Model - Ver A
Estimated Theta1  = 0.019993
First principles, Theta1 = 0.013557

Estimated Theta2  = 0.047278
First principles, Theta2 = -0.014069

Estimated Theta3  = -0.015284
First principles, Theta3 = -0.013557

Estimated Theta4  = 0.98189
First principles, Theta4 = 1

Estimation Data Results
RMS error = 0.022619
MAX error = 0.056326

Validation Data Results
RMS error = 0.22734
MAX error = 0.46546
```

Fig. 9. Parameter estimates for the four-parameter semiphysical model (Version A), compared with coefficients obtained from first-principles.

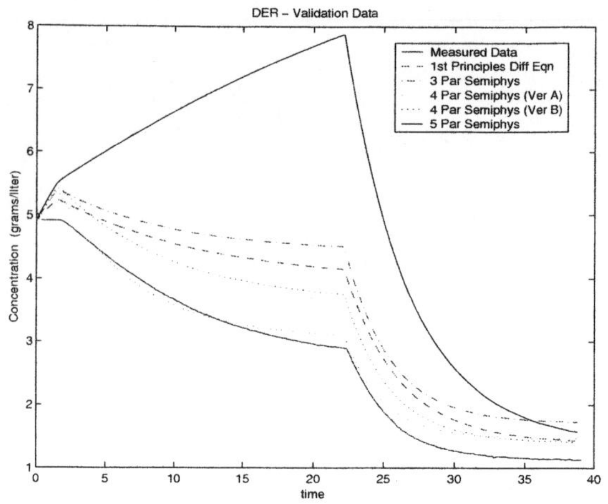

Fig. 10. Simulation results for the validation data set.

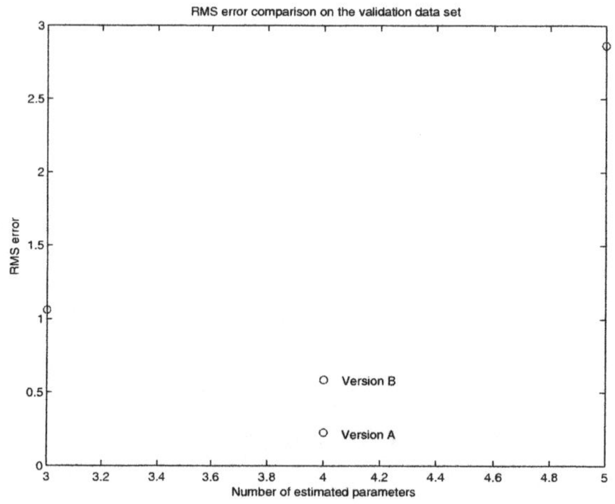

Fig. 11. RMS error comparison on the validation data set.

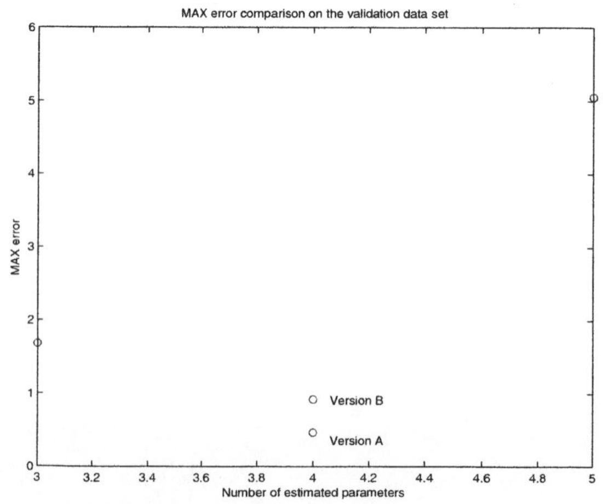

Fig. 12. MAX error comparison on the validation data set.

IFAC Publications
www.elsevier.com/locate/ifac

ESTIMATING PARAMETERS IN A LUMPED PARAMETER SYSTEM WITH FIRST PRINCIPLE MODELING AND DYNAMIC EXPERIMENTS

R.A. de Callafon *

* *University of California, San Diego*
Dept. of Mechanical and Aerospace Engineering
9500 Gilman Drive
La Jolla, CA 92093-0411, U.S.A

Abstract: Commercially available mechanical systems are available to teach and demonstrate the principles behind dynamics and control. A single system can be used for basic dynamic analysis in an undergraduate class to teaching and applying sophisticated identification techniques in a graduate class. In this paper it shown how a commercial system is used at the undergraduate level to estimate lumped parameter coefficients using multiple step responses and first principle modeling. At the graduate level, the same commercial system is used to teach concepts of system identification for the estimation of models for a multi-degree of freedom mechanical system. *Copyright © 2003 IFAC*

Keywords: identification; education; lumped parameter models

1. INTRODUCTION

Dynamic models are important to illustrate the main concepts in dynamic system analysis and linear control. With a mathematical description of a linear dynamic system, either in the form of a differential equation, a transfer function or a state space model, main concepts such as dynamic response, stability and feedback control can be taught and demonstrated (Dorf and Bishop 2000, Stefani *et al.* 2001, Franklin *et al.* 2001). Consequently, derivation of a dynamic model should be an integral part of a course on dynamic system analysis and control system design.

A fundamental step in constructing a dynamic model is based on first principle modeling. Especially in undergraduate engineering education where students develop a background in analyzing equations of motion, thermodynamics and circuit theory, dynamic models are based on governing equations obtained from the various disciplines (Bryson 1994, Morari and Zafiriou 1997, Franklin *et al.* 2001). With applications and students coming from various disciplines, a challenge in teaching control system design is to demonstrate that dynamic models arising from different disciplines have a similar dynamic structure and can be subjected to the same dynamic system analysis needed in control system design.

Important components in dynamic system analysis are the transient response and the frequency response of a model. Time and frequency domain analysis illustrate the main dynamic concepts of a model, but also illuminate the similarities between dynamic models from various disciplines. Unfortunately, in most undergraduate courses on dynamic system analysis and control system design, time and frequency domain analysis is used only to illustrate the dynamic behavior of a system. It is beneficial to include a reversal of this information and use dynamic responses to *derive* the dynamic behavior of the system.

Estimation of models on the basis of data can be integrated in undergraduate education by providing dynamic analysis of time and frequency response data with the purpose of characterizing dynamic model properties. In this paper it is shown how this can be done for a flexible mechanical system on the basis of step response experiments.

2. ESTIMATION OF MODELS FROM DATA

Estimating models from data in general requires a substantial background in the field of system identification. This is one of the primary reason not to include data based modeling techniques in basic undergraduate courses on dynamic system analysis and control system design. However, without a background in system identification, estimation of models from data can still be done by using relatively simple experiments. Such experiments may include step responses that provide an intuitive understanding of the dynamic behavior. Experiments of this nature are particular useful in a laboratory environment where students are asked to develop models for control system design on the basis of measured data. When combined with first principle modeling, data based modeling illustrates the possibility to obtain parameters of a system from dynamic experiments.

In this paper it is illustrated how a commercially available mechanical lumped parameter system is used to teach the basic concepts of model parameter estimation from experimental data. By combining first principle modeling and standard vibration analysis, mass, damping and spring parameters are estimated on the basis of simple step experiments that are carried out in an undergraduate laboratory course. At the graduate level, the same commercial system is used to teach concepts of system identification by estimating of models for a multi-degree of freedom mechanical system.

The objective of this paper is to illustrate the use of a commercially available mechanical lumped parameter system to estimate system parameters using multiple step responses and first principle modeling. The application of simple step experiments in a laboratory environment enables students to derive models from data, whereas the lumped parameter system can also be used to derive a model from first principles. The experiment is currently used in an undergraduate laboratory course on control system design at the Mechanical and Aerospace Engineering department at the University of California, San Diego.

3. LABORATORY EXPERIMENT

3.1 Mechanical system

The rectilinear (and torsional) system used for the laboratory experiments discussed in this paper are mechanical systems with multiple mass, spring and damper element with one-dimensional motion. The systems consist of mass, spring and damper components and a picture of the rectilinear system is depicted in Figure 1. The systems are commercially produced by Educational Control Products (ECPsystems.com) and are equipped with a hardware interface and a user-friendly software environment for data acquisition and controller implementation purposes.

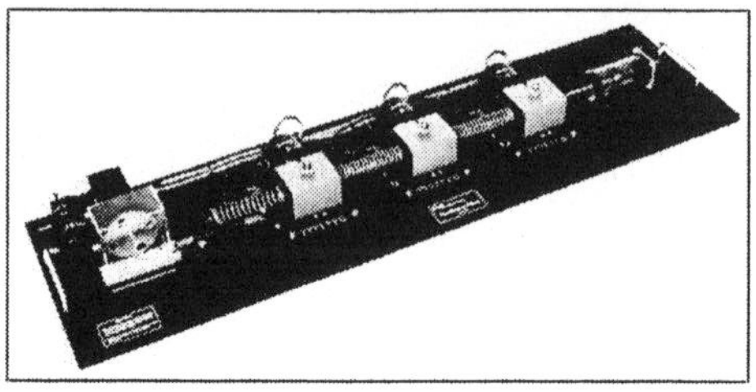

Fig. 1. Rectilinear 3 mass system used for dynamic experiments

The mechanical system depicted in Figure 1 consists of several carts with adjustable weights connected by spring elements. Optional air-restriction dampening devices can be added to increase the damping coefficients of the overall mechanical system. To simplify the analysis, the mechanical system in Figure 1 is equipped with only 2 charts connected via a spring element. As a result, a 2 mass mechanical system is created where each mass or inertia has a positioning freedom. A schematic diagram of the 2 mass system is depicted in Figure 2.

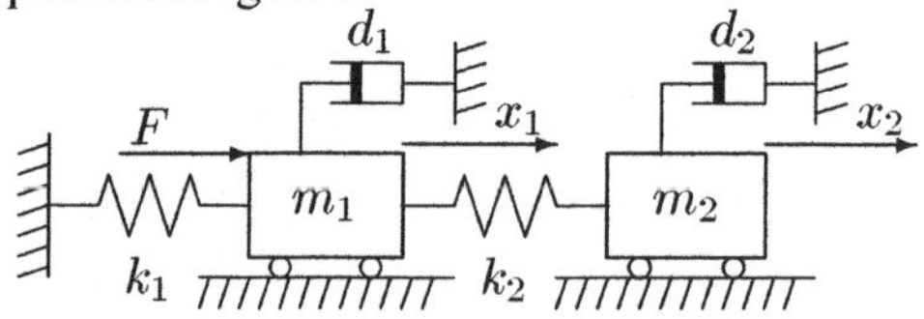

Fig. 2. Schematic view of 2 mass rectilinear system

A force F can be applied to the first mass m_1 via a linear DC-motor. The purpose of the laboratory experiment is to model and control the vibrations of the 2 mass system. The DC-motor is chosen such that its dynamics is negligible compared to the dynamics of the mechanical vibrations of the system. A dynamic model is required to develop a controller to reduce the residual vibrations and to change the position x_1 of a mass/inertia m_1 as fast as possible by means of a controlled force F.

3.2 First principle modeling

Using standard analysis based on 2nd Newton's law, the equations of motion for the lumped parameter system in Figure 2 can be derived. The equations of motion are given by

$$m_1\ddot{x}_1 = -k_1x_1 - d_1\dot{x}_1 - k_2(x_1 - x_2) + F \quad (1)$$

$$m_2\ddot{x}_2 = k_2(x_1 - x_2) - d_2\dot{x}_2 \quad (2)$$

where m_1, m_2 represent the mass or inertia, while x_1, x_2 represent displacement of the masses. The coefficients d_1, d_2 and k_1, k_2 represent respectively damping and stiffness parameters of the mechanical system and F denotes the applied control force.

In order to find the dynamical relation between control force F and the displacement x_1 of the first mass m_1, x_2, $\dot{x}_2$ and $\ddot{x}_2$ are eliminated from (1) and (2). By means of a Laplace transformation, the equation of motions (1) and (2) for this 2 mass system can be written in the matrix representation

$$T(s)\begin{bmatrix} x_1(s) \\ x_2(s) \end{bmatrix} = \begin{bmatrix} F(s) \\ 0 \end{bmatrix}$$

with

$$T(s) = \begin{bmatrix} m_1 s^2 + d_1 s + (k_1 + k_2) & -k_2 \\ -k_2 & m_2 s^2 + d_2 s + k_2 \end{bmatrix} \quad (3)$$

From (3) it can be seen that the computation of a solution of x_1 (and x_2) involves the inversion of a matrix $T(s)$. For a 2 mass model, $T(s)$ is a 2×2 matrix and the computation of the inverse only requires the computation of the determinant of $T(s)$. With the determinant $d(s)$ of $T(s)$ given by

$$(m_1 s^2 + d_1 s + k_1 + k_2)(m_2 s^2 + d_2 s + k_2) - k_2^2$$

it can be concluded that $T^{-1}(s)$ is given by

$$\frac{1}{d(s)}\begin{bmatrix} m_2 s^2 + d_2 s + k_2 & k_2 \\ k_2 & m_1 s^2 + d_1 s + (k_1 + k_2) \end{bmatrix}.$$

With this analysis it follows that the resulting transfer function $G(s)$, that relates the control effort $F(s)$ to the position $x_1(s)$, is given by

$$G(s) = \frac{b_2 s^2 + b_1 s + b_0}{a_4 s^4 + a_3 s^3 + a_2 s^2 + a_1 s + a_0} \quad (4)$$

where the coefficients are

$$\begin{aligned} & & a_4 &= m_1 m_2 \\ b_2 &= m_2 & a_3 &= (m_1 d_2 + m_2 d_1) \\ b_1 &= d_2 & a_2 &= (k_2 m_1 + (k_1 + k_2) m_2 + d_1 d_2) \\ b_0 &= k_2 & a_1 &= ((k_1 + k_2) d_2 + k_2 d_1 \\ & & a_0 &= k_1 k_2 \end{aligned}$$

For the accurate prediction of the flexibilities in the 2 mass lumped parameter model, the mass m_1, m_2, the stiffness k_1, k_2 and the damping d_1, d_2 have to be determined. Estimation of these parameters on the basis of (dynamic) laboratory experiments provides valuable insight in the basic concepts of model parameter estimation in an undergraduate course.

4. EXPERIMENTS FOR PARAMETER ESTIMATION

4.1 Single mass experiments

To facilitate the estimation of the parameters of the 2 mass mechanical system, experiments with only a single mass system are used. Performing the experiments on a single mass is possible, due to the nature of the lumped parameter system. By either physically disconnecting the masses or restricting the displacement of one of the masses, a single mass system is obtained.

The rationale behind the usage of single mass experiments is the ability to isolate the various resonance modes in the lumped parameter system. For a standard 2nd order mass/spring/damper system, the relation between force input F and displacement y is given by the transfer function

$$G(s) = \frac{1}{ms^2 + ds + k} = C \cdot \frac{\omega_n^2}{s^2 + 2\beta\omega_n s + \omega_n^2} \quad (5)$$

where

$$\begin{aligned} C &= \frac{1}{k} && \text{steady state gain} \\ \omega_n &= \sqrt{\frac{k}{m}} && \text{undamped resonance frequency} \\ \beta &= \frac{d}{2\sqrt{km}} && \text{damping ratio} \end{aligned}$$

The relationship between mass m, damping d,stiffness k, the (undamped) resonance frequency ω_n and the damping ratio β is taught at the undergraduate level in standard vibration analysis. This knowledge can be exploited to estimate the model parameters by a sequence of well-planned experiments, where in each experiment only one degree of freedom of the mechanical system is analyzed.

Following the schematic diagram in depicted in Figure 2, the following three experimental conditions can be constructed:

- Disconnect mass m_1 and m_2 by removing spring k_2, resulting in a single mass system with mass m_1, stiffness k_1 and damping d_1.
- Connect mass m_1 and m_2 by spring k_2, but restrict motion of mass m_2. This results in a single mass system with a mass m_1 and stiffness $k_1 + k_2$.
- Restrict motion of mass m_1. This results in a single mass system with mass m_2, stiffness k_2 and damping d_2. It should be noted that no control force F can be applied to mass m_2 in this configuration.

The first experiment can be used to gather information about the the parameters m_1, d_1 and k_1 by observing the steady stage gain $C = 1/k_1$, the undamped resonance frequency $w_n = \sqrt{k_1/m_1}$ and the damping ratio $\beta = \frac{d_1}{2\sqrt{k_1 m_1}}$. As no control force can be applied to m_2, the second experiment is used to estimate the sum of the stiffness from which k_2 can be computed by using the knowledge of k_1 obtained from the first experiment. With the knowledge of the stiffness k_2, the last experiment is used to estimate the mass m_2 and damping d_2 parameters by again observing the undamped resonance frequency $w_n = \sqrt{k_2/m_2}$ and the damping ratio $\beta = \frac{d_2}{2\sqrt{k_2 m_2}}$.

4.2 Step response experiments

The dynamic behavior of the lumped 2 mass system in Figure 2 can be determined by performing relatively simple single mass experiments. The parameters estimated in each experiment are combined to form the complete model of the mechanical system.

For the estimation of the parameters C, ω_n and β in each experiment based on a single mass system, a step experiment will be used. The response to a step input signal can be computed analytically for a 2nd order system and gives rise to a straightforward estimation of the parameters of a single mass system from the observed data.

The displacement $y(t)$ due to a step input $F(t) = U,\ t \geq 0$ is given by

$$\begin{aligned} y(t) &= \mathcal{L}^{-1}\left\{C \cdot \frac{\omega_n^2}{s^2+2\beta\omega_n s+\omega_n^2} \cdot \frac{U}{s}\right\} \\ &= CU\left[1 - \mathrm{e}^{-\beta\omega_n t}(\cos\omega_d t + \phi\sin\omega_d t)\right] \end{aligned} \tag{6}$$

where

$$\omega_d = \omega_n\sqrt{1-\beta^2}$$
$$\phi = \frac{\beta}{\sqrt{1-\beta^2}}$$

The step response of a single mass system with damping $0 < \beta < 1$ is an exponentially decaying sinusoidal function. As a result, the damped resonance frequency ω_d, the damping β and the static gain or DC-gain $\frac{1}{k}$ can be estimated from an observed step response.

5. ESTIMATION OF MODEL PARAMETERS

5.1 Direct estimation

For the estimation of the parameter, consider the step response depicted in Figure 3. From the observed step response the following parameters can be estimated.

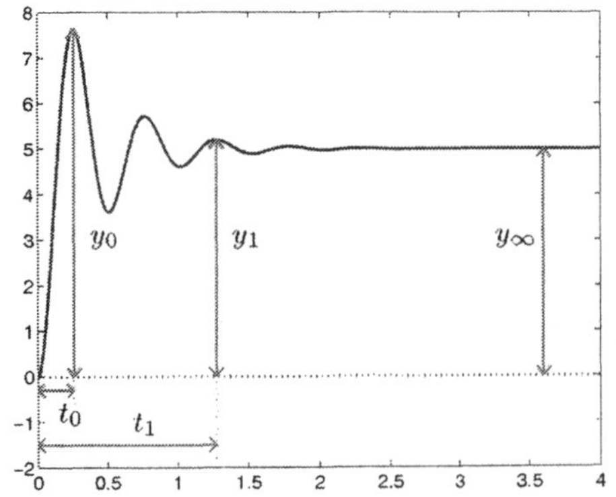

Fig. 3. Typical step response of single mass system

- The steady state behavior is given by

$$\lim_{t\to\infty} y(t) := y_\infty = CU$$

and with U known as the step size of the input signal

$$\hat{C} = \frac{y_\infty}{U} \tag{7}$$

is an estimation of the parameter C.

- From the oscillation in the step response $y(t)$, the damped resonance frequency ω_d can be estimated. For that purpose, consider the time measurements t_0 and t_1 to distinguish two subsequent maximum values of oscillations in the output $y(t)$, then

$$\hat{\omega}_d = 2\pi\frac{n}{t_1 - t_0} \tag{8}$$

gives an estimate for the damped resonance frequency ω_d of the single mass system, where n is the number of oscillations between the two subsequent maximum values of output $y(t)$.

- From the decay of the oscillation in the step response $y(t)$, the damping coefficient β can be estimated. For that purpose, consider the difference of the steady state value y_∞ with two subsequent maximum values y_0 and y_1 of oscillations in the output $y(t)$. With the analytic solution of the step response given in (6) it can be verified that

$$\widehat{\beta\omega_n} = \frac{1}{t_1 - t_0}\ln\left(\frac{y_0 - y_\infty}{y_1 - y_\infty}\right) \tag{9}$$

is an estimate for the product of the damping ration β and the (undamped) resonance frequency ω_n.

Combining the estimates in (8) and (9) yields

$$\hat{\omega}_n = \sqrt{\hat{\omega}_d^2 + (\widehat{\beta\omega_n})^2}$$
$$\hat{\beta} = \frac{\widehat{\beta\omega_n}}{\omega_n}$$

and gives estimates for the undamped resonance frequency ω_n and the damping ratio β. With the values of the estimates $\hat{C}$, $\hat{\omega}_n$ and $\hat{\beta}$, the values of the mass m, damping d and stiffness k in (5) (up to a scaling constant) are computed via

$$\begin{aligned} \hat{k} &= \frac{1}{\hat{C}} && \text{(stiffness constant)} \\ \hat{m} &= \frac{1}{\hat{C}\,\hat{\omega}_n^2} && \text{(mass/inertia)} \\ \hat{d} &= \frac{2\hat{\beta}}{\hat{C}\,\hat{\omega}_n} && \text{(damping constant)} \end{aligned} \tag{10}$$

which concludes the estimation of the system parameters from a single step experiment. The parameter estimates in (10) are unbiased estimates of the system parameters, provided step experiment are used. To improve the variance properties of the estimates in case of noise experiments, averaging of the estimates over multiple step experiments can be performed. The simple step experiments provide means to estimate the unknown parameters in the 2 mass system. The parameters are estimated by a sequence of step experiments that only require the knowledge of the dynamic behavior of a standard 2nd order system.

5.2 Estimation via step response realization

It can be observed that the estimation of the parameters in (7)-(9) is based on only three discrete data points: y_0, y_1 and y_∞ with the corresponding time elements t_0, t_1 and t_∞. With the help of the (continuous time) analytic solution of the step response in (6), an explicit expression for the parameters estimates of the single mass/spring/damper can be obtained. Although the parameter estimation gives insight in determining system parameters from dynamic experiments, the parameter estimation is highly influenced by noise. An extra level of complexity can be added to the estimation by including all the data points of the step response in the parameter estimation.

One possible solution would be to pose an optimization using a parametrized continuous model (Ljung and Glad 1994, Ljung 1999) to find the optimal values for the mass m, stiffness k and damping d parameters. Although this is a viable solution, non-linear optimization techniques would be required to solve the parameter estimation technique. An alternative would be to construct a model using realization techniques (Ho and Kalman 1966, Kung 1978), which only requires standard matrix manipulations to estimate a model (Vandewalle and de Moor 1991). Matrix manipulations such as singular value decompositions are taught at the undergraduate level. Therefore, realization based techniques can be easily adapted to the discrete time step response measurements obtained from the laboratory experiments.

For the estimation via the step response realization, consider the step response data $y(t)$. Since $y(t)$ is (discrete time) step response data,

$$y(t) = \sum_{\tau=0}^{t} g(\tau)$$

where $g(\tau)$ indicates the (unmeasured) discrete time impulse response data. By forming the matrix

$$S := \begin{bmatrix} y(1) & y(2) & \cdots & y(n) \\ y(2) & y(3) & \cdots & y(n+1) \\ \vdots & \vdots & \vdots & \vdots \\ y(n) & y(n+1) & \cdots & y(2n-1) \end{bmatrix} - \begin{bmatrix} y(0) & y(0) & \cdots & y(0) \\ y(1) & y(1) & \cdots & y(1) \\ \vdots & \vdots & \vdots & \vdots \\ y(n-1) & y(n-1) & \cdots & y(n-1) \end{bmatrix}$$

from the step response measurement $y(t)$, $t = 1, \ldots, 2n-1$, it can be observed that

$$S = HI_u \tag{11}$$

where H is the standard impulse response based Hankel matrix

$$H = \begin{bmatrix} g(1) & g(2) & \cdots & g(n) \\ g(2) & g(3) & \cdots & g(n+1) \\ \vdots & \vdots & \vdots & \vdots \\ g(n) & g(n+1) & \cdots & g(2n-1) \end{bmatrix}$$

and I_u is an upper triangular matrix with ones on the diagonal and in the upper triangular part. Due to the structure of H and I_u in (11) it can be verified that $rank(S) = rank(H)$. With $rank(S) = r$, a decomposition

$$\begin{gathered} S = S_1 S_2 \\ S_1^T, S_2 \in R^{r\times n},\ rank(S_1^T) = rank(S_2) = r \end{gathered} \tag{12}$$

can be computed via a singular value decomposition of the matrix S similar to the realization algorithm of Kung (1978).

For a discrete time system the impulse response $g(\tau)$ satisfies $g(\tau) = CA^{\tau-1}B$, where (A, B, C) denote the matrices of a state space realization of the system. With the elements of $g(\tau)$ present in H and the structure of I_u in (11) it can also be verified that

$$\bar{S} = S_1 A S_2$$

where

$$\bar{S} := \begin{bmatrix} y(2) & y(3) & \cdots & y(n+1) \\ y(3) & y(4) & \cdots & y(n+2) \\ \vdots & \vdots & \vdots & \vdots \\ y(n+1) & y(n+2) & \cdots & y(2n) \end{bmatrix} - \begin{bmatrix} y(1) & y(1) & \cdots & y(1) \\ y(2) & y(2) & \cdots & y(2) \\ \vdots & \vdots & \vdots & \vdots \\ y(n) & y(n) & \cdots & y(n) \end{bmatrix}$$

As a result, the A-matrix of the state space realization can be computed by

$$A = \bar{S}_1 \bar{S} \bar{S}_2 \tag{13}$$

where $\bar{S}_1$ and $\bar{S}_2$ indicate respectively the left and right inverse of the matrices S_1 and S_2 in the decomposition (12).

The step response realization technique can be applied to any step response measurement to find a state space model based on standard realization techniques (Kung 1978). For the step response experiment of a single mass system considered in Section 4 of this paper, the rank of S in the decomposition (12) will be 2. As a result, a 2×2 matrix A will be estimated on the basis of the step response experiments from which the (continuous time) system parameter m, k and d have to be computed.

For a lightly damped single mass system, the eigenvalues of the matrix A will appear in a complex conjugate pair. Since the matrix A is the discrete time equivalent of the continuous time system, the eigenvalue $\lambda(A)$ can be used to estimate the equivalent continuous time natural frequency ω_n and damping ratio β via

$$\begin{aligned} \hat{\omega}_n &= \left| \frac{\ln \lambda(A)}{\Delta T} \right| \\ \hat{\beta} &= -\cos \tan^{-1} \frac{imag \ln \lambda(A)}{real \ln \lambda(A)} \end{aligned} \tag{14}$$

by assuming a zero order old discrete time equivalent model with a sampling time of ΔT seconds. With an estimate of $\bar{C}$ based on the steady state value of the step response $y(t)$, the estimates in (14) can be used to compute the estimate of the mass m, spring k and damping d parameters in (10).

6. ILLUSTRATION OF PARAMETER ESTIMATION

Consider the measured step response of a one of the single mass experiments depicted in Figure 4. The data is obtained from the rectilinear ECP system by applying a 0.8 Volt step input on the DC-motor and measuring the position of the mass in encoder counts with a sampling time of 9 msec.

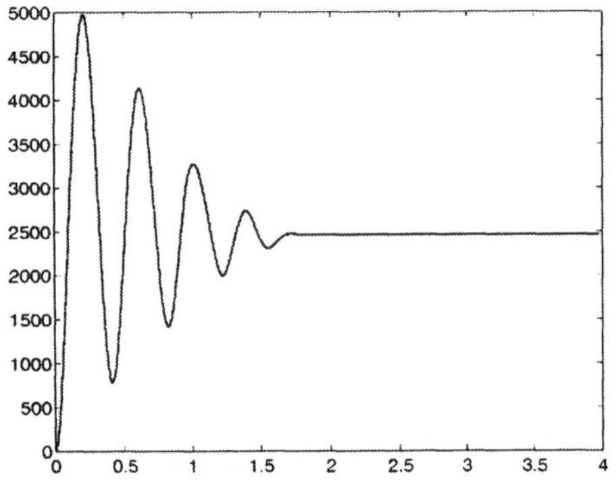

Fig. 4. Measured step response of a single mass using the ECP rectilinear system

Using the first and third peak in the step response oscillations for estimation purposes one finds

$$\hat{\omega}_d = = 2\pi\frac{2}{1.05-0.21} \approx 4.76\pi$$
$$\hat{\beta\omega_n} = \frac{1}{0.84}\ln\left(\frac{4975-2464}{3276-2464}\right) \approx 1.34$$

giving the estimated values:

$$\hat{\omega}_n = \sqrt{\hat{\omega}_d^2 + (\hat{\beta\omega_n})^2} \approx 4.78\pi$$
$$\hat{\beta} = \frac{\hat{\beta\omega_n}}{\hat{\omega}_n} \approx 0.0895 \qquad (15)$$

Estimation of a 2×2 state space matrix A using the step response realization yields a step response matrix S with 2 singular values significantly larger than the remaining singular values. Computation of a rank 2 decomposition S_1 and S_2 via a singular value decomposition gives a A-matrix via (13) that is given by

$$A = \begin{bmatrix} 0.9744 & -0.0710 \\ 0.2428 & 0.9900 \end{bmatrix}$$

Computation of the complex conjugate eigenvalue pair $\lambda(A)$ and the equivalent continuous time natural frequency ω_n and damping ratio β via (14) yields the estimates

$$\hat{\omega}_n \approx 4.7014\pi$$
$$\hat{\beta} \approx 0.0686 \qquad (16)$$

The differences between (15) and (16) can be attributed to the difference in information used from the data to estimate the parameters. In case of noise free data obtained from an actual 2nd order system, both estimation results would be similar. However, due to the friction present in the mechanical system, the damping is not consistent throughout the experiment.

7. CONCLUSIONS

Without a comprehensive background in system identification, estimation of lumped parameters in a mechanical system from experimental data can be done by using relatively simple experiments. Such experiments may include step responses that provide an intuitive understanding of the dynamic behavior. In this paper it shown how a commercial educational lumped parameter system with two degrees of freedom is used at the undergraduate level to estimate mechanical parameters using multiple step responses and first principle modeling. By a series of well thought experiments, students find the parameters in a first principle model and compare these results to simulations based on the first principle model.

REFERENCES

Bryson, A. (1994). *Control of Spacecraft and Aircraft.* Princeton University Press.

Dorf, R.C. and R.H. Bishop (2000). *Modern Control Systems.* Pearson Education.

Franklin, G.F., J.D. Powel and A. Emami-Naeini (2001). *Feedback Control of Dynamic Systems.* Prentice Hall, Upper Saddle River, NJ, USA.

Ho, B.L. and R.E. Kalman (1966). Effective construction of linear state-variable models from input/output functions. *Regelungstechnik* **14**, 545–548.

Kung, S.Y. (1978). A new identification and model reduction algorithm via singular value decomposition. In: *Proc. 12th Asilomar Conference on Circuits, Systems and Computers.* Pacific Grove, USA. pp. 705–714.

Ljung, L. (1999). *System Identification: Theory for the User (second edition).* Prentice-Hall, Englewood Cliffs, New Jersey, USA.

Ljung, L. and T. Glad (1994). *Modeling of Dynamic Systems.* Prentice-Hall, Englewood Cliffs, New Jersey, USA.

Morari, M. and E. Zafiriou (1997). *Robust process Control.* Prentice Hall.

Stefani, R.T., G. Hostetter, B. Shahian and C.J. Savant (2001). *Design of Feedback Control Systems.* Oxford University Press.

Vandewalle, J. and B. de Moor (1991). On the use of the singular value decomposition in identification and signal processing. *Numerical Linear Algebra, Digital Signal Processing and Parallel Algorithms* pp. 321–360.

IFAC Publications
www.elsevier.com/locate/ifac

RECURSIVE SUBSPACE IDENTIFICATION BASED ON PROJECTOR TRACKING

Marco Lovera

Dipartimento di Elettronica e Informazione, Politecnico di Milano,
Piazza Leonardo da Vinci 32, 20133 Milano, Italy,
Tel. +39-02-23993592, Fax +39-02-23993412,
E-mail: lovera@elet.polimi.it

Abstract: The problem of MIMO state space recursive identification is considered and analyzed using subspace model identification (SMI) techniques. In this paper the use of projection tracking techniques for the update of the observability subspace is proposed: existing results are used for the output error case and a novel instrumental variable (IV) projection tracking approach is proposed, to accommodate for arbitrary correlation of the disturbances. Simulation results show the performance achievable with the given algorithms. *Copyright © 2003 IFAC*

Keywords: Subspace methods, Recursive algorithms, Identification algorithms, State space models, Tracking systems.

1. INTRODUCTION

The derivation of efficient recursive subspace model identification (SMI) algorithms has been an active area of research in the last few years; in particular, the application of recursive subspace model identification (RSMI) has been proposed in fault detection (see (Lovera *et al.*, 1997; Oku *et al.*, 1997)) and adaptive control (Hale and Qin, 2002) problems.

Various solutions to the RSMI problem have been proposed in the literature (see, e.g., (Verhaegen and Deprettere, 1991; Lovera *et al.*, 2000; Oku and Kimura, 2002)), with different characteristics in terms of computational load, tracking performance etc. The challenging problem with RSMI is that the SVD is computationally burdensome to update and an unbiased updating scheme must be worked out in the presence of both measurement and process noise. Consequently, all of the above cited RSMI algorithms apply certain updating techniques that avoid direct application of the SVD while preserving the unbiasedness of the original batch algorithms.

In this paper the focus will be on the recursive version of SMI algorithms of the MOESP class. The first such algorithm (Verhaegen and Deprettere, 1991) had the drawback of requiring that the disturbances acting on the system output be spatially and temporally white. On the other hand, the algorithms of (Lovera *et al.*, 2000) could deal with more general types of perturbations thanks to the use of instrumental variables. In particular, the algorithms proposed in the cited paper have been developed by keeping in mind very different goals defined in a number of previous contributions. In (Gustafsson, 1997) the focus was on computational efficiency; this has led to the so-called subtraction approach for the update of the data matrices and to the idea of applying subspace tracking algorithms to the RSMI problem. In (Lovera and Verhaegen, 1998) the main point was the estimation accuracy: to this purpose, recursive versions of the so-called PI and PO MOESP SMI schemes were derived.

The aim of this work is to present some recent developments in the RSMI class of algorithms, obtained by means of a combination of the approaches studied in (Lovera *et al.*, 2000) with some new results in the field of array signal processing. More precisely, the use of the projector tracking scheme developed in (Utschick, 2002) is proposed for the update of the observability subspace of the model, and an instrumental variable (IV) version of the tracking scheme is proposed in order to take into account the presence of process noise in the problem formulation. The main advantage of the projector approach with respect to the PAST subspace tracking technique and its IV variants used in (Lovera *et al.*, 2000) is that no approximations are introduced in the formulation of the tracking problem. Also, note that the algorithms discussed herein operate on the data almost only by means of unitary operations, which makes them particularly attractive from a numerical point of view. Finally, note that the problem of replacing the projection approximation subspace tracking algorithm in RSMI has been also recently studied in (Mercere *et al.*, 2003), using a different approach.

2. OVERVIEW OF THE MOESP ALGORITHMS.

The MOESP family of subspace identification algorithms consists of a number of variants, corresponding to different types of perturbations that can be tolerated on the recorded input and output samples, the class of systems and the nature of the input signal. Consider the finite dimensional, linear time-invariant (LTI) state space model:

$$\begin{aligned} x_{t+1} &= Ax_t + B\tilde{u}_t + f_t & u_t &= \tilde{u}_t + w_t \\ \tilde{y}_t &= Cx_t + D\tilde{u}_t & y_t &= \tilde{y}_t + v_t \end{aligned} \quad (1)$$

where $x_t \in \mathbb{R}^n$, $u_t \in \mathbb{R}^m$, $y_t \in \mathbb{R}^p$ and the triples $\{f_t, v_t, w_t\}$ are additive perturbations to be defined in more detail for the different variants individually, the key problem is the consistent estimation of the column space of the extended observability matrix Γ_i, defined as:

$$\Gamma_i^T = \left[C^T \; (CA)^T \; (CA^2)^T \; \ldots \; (CA^{i-1})^T\right]$$

from measured input-output (i-o) samples $\{u_t, y_t\}$. From the estimate of the column space of Γ_i, one can derive an estimate of the matrices A and C (up to a similarity transformation) in a straightforward way by exploiting the shift invariance of the column space of Γ_i. As for the estimation of B and D, it can be based on the minimization of the simulation error over the identification data set, thus leading to a conventional linear least squares problem which can also be recursively updated as described in (Lovera *et al.*, 2000).

Define the following block-Hankel matrix

$$Y_{t,i,j} = \begin{bmatrix} y_t & y_{t+1} & \cdots & y_{t+j-1} \\ y_{t+1} & y_{t+2} & \cdots & y_{t+j} \\ \vdots & \vdots & \ddots & \vdots \\ y_{t+i-1} & y_{t+i} & \cdots & y_{t+i+j-2} \end{bmatrix}$$

and similarly, construct $U_{t,i,j}$ from input samples. Considering the special case of absence of the perturbation terms f_t, v_t, w_t, the data equation is compactly denoted as:

$$Y_{t,i,j} = \Gamma X_{t,j} + HU_{t,i,j} \quad (2)$$

where H is given by

$$H = \begin{bmatrix} D & 0 & \cdots & 0 \\ CB & D & \cdots & 0 \\ CAB & CB & \cdots & 0 \\ \vdots & \vdots & \vdots & \vdots \\ CA^{i-2}B & CA^{i-3}B & \cdots & D \end{bmatrix}. \quad (3)$$

and

$$X_{t,j} = \left[x_t \; x_{t+1} \; \cdots \; x_{t+j-1}\right].$$

A summary of the considered MOESP identification algorithms is given in the following. The consistency result implicitly assumes appropriate persistency of excitation conditions (see the cited papers for details).

The OM scheme (Verhaegen and Dewilde, 1992): This scheme considers the RQ factorization of the compound matrix:

$$\begin{bmatrix} U_{t,i,j} \\ Y_{t,i,j} \end{bmatrix} = \begin{bmatrix} R_{11}(\bar{t}) & 0 \\ R_{21}(\bar{t}) & R_{22}(\bar{t}) \end{bmatrix} \begin{bmatrix} Q_1(\bar{t}) \\ Q_2(\bar{t}) \end{bmatrix} \quad (4)$$

with the time index $\bar{t}$ equal to $t+i+j-2$. Then a consistent estimate of the column space of Γ is provided via an SVD of the matrix $R_{22}(\bar{t})$ under the same assumptions on the perturbations as stated for the EM scheme.

The PI/PO (Verhaegen, 1994) schemes: These schemes consider the following RQ factorization:

$$\begin{bmatrix} U_{t+i,i,j} \\ W_{t,i,j} \\ Y_{t+i,i,j} \end{bmatrix} = \begin{bmatrix} R_{11}(\bar{t}) & 0 & 0 \\ R_{21}(t) & R_{22}(t) & 0 \\ R_{31}(\bar{t}) & R_{32}(\bar{t}) & R_{33}(\bar{t}) \end{bmatrix} \begin{bmatrix} Q_1(\bar{t}) \\ Q_2(\bar{t}) \\ Q_3(\bar{t}) \end{bmatrix} \quad (5)$$

with $\bar{t}$ equal to $t+2i+j-1$. Here $W_{t,i,j} = U_{t,i,j}$ in the PI scheme, $W_{t,i,j} = [U_{t,i,j}^T \; Y_{t,i,j}^T]^T$ in the PO scheme. Then a consistent estimate of the column space of Γ is provided via an SVD of the matrix $R_{32}(\bar{t})$ under the assumptions that $f_t \equiv 0, w_t \equiv 0$ and v_t is an ergodic sequence satisfying $E[u_t v_s^T] = 0 \quad \forall t,s$ (PI scheme) $w_t \equiv 0$ and f_t, v_t are ergodic sequences satisfying

$$E[\begin{bmatrix} f_t \\ v_t \end{bmatrix} [f_s^T \; v_s^T]] = \begin{bmatrix} Q & S \\ S^T & R \end{bmatrix} \delta_{s,t}$$

with $\delta_{s,t}$ denoting the Kronecker delta function, which are furthermore independent from the input u_t (PO scheme).

3. RECURSIVE UPDATE OF THE OBSERVABILITY SUBSPACE.

The first step in the development of a recursive updating scheme for Γ_i is to derive a low rank update/downdate of the matrix from which the column space of Γ_i is estimated. This will be outlined in the subsequent sections. In particular, in Sections 3.1 and 3.2 the updating schemes of (Lovera and Verhaegen, 1998) and (Lovera *et al.*, 2000) respectively will be presented.

3.1 *The OM update*

The basic idea of this approach is to update directly the QR factorization (4). For, consider the explicit form of the data equation, namely:

$$Y_{t+i,i,j+1} = \Gamma_i \left[X_{t+i,j} \; x_{t+i+j}\right] + HU_{t+i,i,j+1} + \left[E_{t+i,i,j} \; \phi_{E_f}\right]$$

Then we consider the following factorization, as in the OM scheme described in Section 2:

$$\begin{bmatrix} U_{t+i,i,j} \\ Y_{t+i,i,j} \end{bmatrix} = \begin{bmatrix} R_{11}(\bar{t}) & 0 \\ R_{21}(\bar{t}) & R_{22}(\bar{t}) \end{bmatrix} \begin{bmatrix} Q_1(\bar{t}) \\ Q_2(\bar{t}) \end{bmatrix}$$

When a new data point is obtained, the decomposition must be updated as:

$$\begin{bmatrix} U_{t+i,i,j+1} \\ Y_{t+i,i,j+1} \end{bmatrix} = \begin{bmatrix} R_{11}(\bar{t}) & 0 & \phi_{U_f} \\ R_{21}(\bar{t}) & R_{22}(\bar{t}) & \phi_{Y_f} \end{bmatrix} \begin{bmatrix} Q_1(\bar{t}) & 0 \\ Q_2(\bar{t}) & 0 \\ 0 & 1 \end{bmatrix}$$

where, in particular

$$R_{21}(\bar{t}) = \Gamma X_{t+i,j} Q_1(\bar{t})^T + HR_{11}(\bar{t}) + E_{t+i,i,j} Q_1(\bar{t})^T$$

A sequence of Givens rotations is used to annihilate the vector ϕ_{U_f} and bring back the R factor to block lower triangular form. Using the fact that the input u_t is independent from the perturbations v_t and f_t and letting

$$R_{11}(\bar{t}+1) = R_{11}(\bar{t})P_{11}(\bar{t}+1) + \phi_{U_f} P_{21}^T(\bar{t}+1) \quad (6)$$

$$R_{21}(\bar{t}+1) = R_{21}(\bar{t})P_{11}(\bar{t}+1) + \phi_{Y_f} P_{21}^T(\bar{t}+1) \quad (7)$$

we obtain for the R factor:

$$\begin{bmatrix} R_{11}(\bar{t}+1) & 0 & 0 \\ R_{21}(\bar{t}+1) & R_{22}(\bar{t}) & \bar{\phi}_{Y_f} \end{bmatrix}. \quad (8)$$

Then, it holds that

$$R_{22}(\bar{t}+1)R_{22}^T(\bar{t}+1) = R_{22}(\bar{t})R_{22}^T(\bar{t}) + \bar{\phi}_{Y_f}\bar{\phi}_{Y_f}^T. \quad (9)$$

However, by feeding the vector $\bar{\phi}_{Y_f}(\bar{t}+1)$ into an SVD updating scheme we would typically obtain a biased estimate of the column space of Γ_i. This is due to the rather restrictive consistency assumptions of the OM scheme. A possible rescue to this problem is the introduction of instrumental variables. Therefore, for the update of the approximation of the column space of Γ_i, an IV-based SVD update technique must be used.

3.2 *The PI/PO update*

The main idea in this algorithm is to update the more complex RQ factorization of the PI/PO schemes by means of Givens rotations. In this section, the application of this approach to the PI version of MOESP is proposed (see (Lovera *et al.*, 2000) for details).

Consider the PI MOESP described in Section 2; at the generic time instant $\bar{t}$, the RQ factorization of the data matrix is given by (5) where in this case $\bar{t} = t + j + 2i - 1$. Assume now that the following new set of i-o data vectors become available:

$$\phi_{U_f}(\bar{t}+1) = \left[u_{t+i+j}^T \; \cdots \; u_{t+2i+j-1}^T\right]^T \quad (10)$$

$$\phi_{Y_f}(\bar{t}+1) = \left[y_{t+i+j}^T \; \cdots \; y_{t+2i+j-1}^T\right]^T \quad (11)$$

and define the vector $\phi_{U_p}(\bar{t}+1)$ as,

$$\phi_{U_p}(\bar{t}+1) = \left[u_{t+j}^T \; \cdots \; u_{t+i+j-1}^T\right]^T \quad (12)$$

Then, a new column is added to the data matrices and the decomposition must be written as:

$$\begin{bmatrix} R_{11}(\bar{t}) & 0 & 0 & \phi_{U_f} \\ R_{21}(\bar{t}) & R_{22}(\bar{t}) & 0 & \phi_{U_p} \\ R_{31}(\bar{t}) & R_{32}(t) & R_{33}(\bar{t}) & \phi_{Y_f} \end{bmatrix} \begin{bmatrix} Q_1(\bar{t}) & 0 \\ Q_2(\bar{t}) & 0 \\ Q_3(\bar{t}) & 0 \\ 0 & 1 \end{bmatrix} \quad (13)$$

Givens rotations are then used twice to update the factorization. They are first applied in order to zero out the elements of vector ϕ_{U_f}, bringing the R factor to the form

$$\begin{bmatrix} R_{11}(\bar{t}+1) & 0 & 0 & 0 \\ R_{21}(\bar{t}+1) & R_{22}(\bar{t}) & 0 & \bar{\phi}_{U_p} \\ R_{31}(\bar{t}+1) & R_{32}(\bar{t}) & R_{33}(\bar{t}) & \bar{\phi}_{Y_f} \end{bmatrix}. \quad (14)$$

Subsequently, the elements of $\bar{\phi}_{U_p}$ are zeroed in a similar way, to give:

$$\begin{bmatrix} R_{11}(\bar{t}+1) & 0 & 0 & 0 \\ R_{21}(\bar{t}+1) & R_{22}(\bar{t}+1) & 0 & 0 \\ R_{31}(\bar{t}+1) & R_{32}(\bar{t}+1) & R_{33}(\bar{t}) & \bar{\bar{\phi}}_{Y_f} \end{bmatrix} \quad (15)$$

Then it is easy to show that the "square" of block $R_{32}(\bar{t}+1)$, can be written as:

$$R_{32}(\bar{t}+1)R_{32}(\bar{t}+1)^T = R_{32}(\bar{t})R_{32}(\bar{t})^T + \bar{\phi}_{Y_f}\bar{\phi}_{Y_f}^T - \bar{\bar{\phi}}_{Y_f}\bar{\bar{\phi}}_{Y_f}^T \quad (16)$$

Thus, in this case the subspace estimate at time $\bar{t}+1$ is related to the one at time $\bar{t}$ via the combination of an update and a downdate.

Remark 1. In (Gustafsson, 1997) an updating scheme based on a "subtraction" approach was proposed. First of all, formulate a column-by-column version of (2):

$$Y_{t,i} = \Gamma x_t + HU_{t,i} + E_{t,i} \quad (17)$$

where $Y_{t,i} = Y_{t,i,1}$. Here, $E_{t,i}$ denotes a vector that consists of both the measurement noise and the filtered process noise. Then, introduce the modified output signal $Z_{t,i} \overset{\text{def}}{=} Y_{t,i} - HU_{t,i}$ and note that $Z_{t,i} = \Gamma x_t + \bar{E}_{t,i}$. This update is clearly the simplest among the proposed methods, however it requires the knowledge of previous estimates of B and D. On the other hand, the PI/PO scheme, which is possibly the most accurate one, suffers from the highest computational cost. Therefore, the OM scheme seems to provide an adequate compromise solution.

Remark 2. A forgetting scheme can be easily included in the above recursions for the update of the RQ factorisations.

4. PROJECTOR BASED PAST AND IV-PAST.

The most important step in RSMI identification is the update of the SVD of the observability subspace. In (Gustafsson, 1997; Lovera and Verhaegen, 1998; Lovera *et al.*, 2000), it was proposed to exploit the close relationship between array signal processing and SMI to derive efficient update schemes for the SVD. More precisely, the PAST algorithm (Yang, 1995) and its instrumental variables modification IV-PAST (Gustafsson, 1998) were applied and modified to derive an efficient partial SVD update in the different SMI schemes considered in Section 2. PAST and IV-PAST, as will be briefly summarised in the following, provide RLS-like algorithms for the update of the signal subspace of a covariance matrix. The recent literature on subspace tracking, however, has shown that it might be useful, under many respects, to try and update the projection matrix associated with the observability subspace, rather than an orthonormal basis for the subspace itself. This has led to the development of a projector tracking version of PAST (see (Utschick, 2002)), and similar, geometrically inspired gradient based approaches to projection tracking (see (Fuhrmann, 1997; Srivastava and Fuhrmann, 1997)). In this Section an overview of PAST and its projection based version will be proposed, and a IV projection tracking algorithm will be derived, along the lines of (Gustafsson, 1998).

4.1 *Projector based PAST scheme.*

Consider a random vector $x \in \mathbb{R}^m$, and study the following *unconstrained* criterion:

$$V(W) = E\left\|x - WW^T x\right\|^2 \tag{18}$$

with a matrix argument $W \in \mathbb{R}^{m\times n}$, $m > n$, that without loss of generality is assumed to have full rank (=n). Let the eigenvalue decomposition of $R_x = E[xx^T]$ be given as

$$R_x = Q\Lambda Q^H \tag{19}$$

with $Q = [q_1, \ldots, q_m]$, $\Lambda = \text{diag}(\lambda_1, \ldots, \lambda_m)$. Furthermore, assume that R_x is positive definite. The eigenvalues are ordered as $\lambda_1 \geq \lambda_2 \geq \ldots \geq \lambda_m$. Then, the following theorem holds ((Yang, 1995)):

Theorem 1. The global minimum of $V(W)$ is attained if and only if $W = Q_n T$ where Q_n contains the n dominating eigenvectors of R_x. Here T is an arbitrary unitary matrix. Furthermore, all other stationary points are saddle points.

This Theorem allows one to reformulate the problem of computing the n dominating eigenvectors of a positive definite matrix into an unconstrained optimization problem. In particular, if the expectation operator in (18) is replaced with an exponentially weighted finite summation, and the so-called *projection approximation* is introduced, an efficient RLS-like algorithm for the update of the signal subspace Q_n of R_x is easily derived.

As illustrated in (Utschick, 2002), a different approach to the solution of the optimisation problem associated with the cost function (18) can be obtained by considering as independent variable the projection matrix $P = WW^T$ rather than the orthonormal basis W. In particular, it is easy to see that optimising $V(W)$ in (18) with respect to W is equivalent to optimising

$$\tilde{V}(P) = E\left\|x - Px\right\|^2 = tr(R_x) - tr(R_x P), \tag{20}$$

with respect to P, with the constraints $P = P^T = P^2$ and subject to the rank constraint rank$(P) = n$. Tracking of a real, $m \times m$ projection matrix P of rank n can be carried out by means of the update rule

$$P(\phi,t) = G(\phi,t)P(t-1)G(\phi,t)^T \tag{21}$$

where

$$G(\phi,t) = \prod_{k=m-1}^{1} \prod_{l=k+1}^{m} G^{kl}(\phi_{kl}), \tag{22}$$

$G^{kl}(\phi_{kl})$ being the Givens rotor (see (Golub and Van Loan, 1989; Murnaghan, 1938)) with characteristic submatrix

$$G^{kl}_{\begin{bmatrix} kk & kl \\ lk & ll \end{bmatrix}} = \begin{bmatrix} \cos(\phi_{kl}) & -\sin(\phi_{kl}) \\ \sin(\phi_{kl}) & \cos(\phi_{kl}) \end{bmatrix}, \tag{23}$$

and $\phi = \{\phi_{kl}\}$. At each time step, the elements of ϕ are determined according to the gradient rule

$$\phi_{kl} = -\mu \frac{\partial \tilde{V}(\phi)}{\partial \phi_{kl}}, \tag{24}$$

where μ is the gain of the gradient iteration. The gradient of $\tilde{V}(P(\phi))$ is given by (Utschick, 2002)

$$\frac{\partial \tilde{V}(P(\phi))}{\partial \phi_{kl}} = \sum_{i=1}^{m} (P_{l,i}\hat{R}_{x\,i,k} - P_{k,i}\hat{R}_{x\,i,l}), \tag{25}$$

and $P_{i,j}$, $\hat{R}_{x\,i,j}$ are the (i,j) elements of P and $\hat{R}_x$, respectively.

4.2 *The projector based IV-PAST scheme.*

An IV generalization of PAST has been proposed in (Gustafsson, 1998), in order to deal with situations in which the measurements of x are affected by measurement noise with arbitrary and unknown covariance matrix. In this scenario it is assumed that the cross correlation matrix $R_{x\xi}$ associated with x and an instrumental variable vector [1] $\xi \in \mathbb{R}^{l\times 1}$ has a low rank $(= n)$ structure:

$$E[x\xi^T] = \Gamma\Psi \tag{26}$$

[1] A typical choice for the IV vector is a (possibly filtered) sequence of past input samples.

where $\Gamma \in \mathbb{R}^{m\times n}$, $\Psi \in \mathbb{R}^{n\times l}$ both have full rank n $(m,l \geq n)$. Then, introduce the criterion

$$V(W(t)) = \left\| R_{x\xi}(t) - W(t)W^T(t)R_{x\xi}(t) \right\|_F^2. \quad (27)$$

From (Gustafsson, 1998) we have the following result:

Theorem 2. Let $R_{x\xi}(t)$ have the SVD

$$R_{x\xi}(t) = Q\Sigma V^T \quad (28)$$

Then the global minimum of $V(W(t))$ is obtained if and only if $W(t) = Q_n T$ where Q_n contains the n dominating left singular vectors of $R_{x\xi}(t)$ and T is an arbitrary unitary matrix. All other stationary points are saddle points.

Applying the projection approximation approach, a recursive algorithm can be derived (see (Gustafsson, 1998)), however, note that the IV-PAST cost function (27) can be written in terms of the projection matrix $P = WW^T$ as

$$\tilde{V}(P) = \left\| R_{x\xi} - PR_{x\xi} \right\|_F^2 =$$
$$= tr\left[(R_{x\xi} - PR_{x\xi})(R_{x\xi} - PR_{x\xi})^T \right] \quad (29)$$

so that letting $R = R_{x\xi}R_{x\xi}^T$

$$\tilde{V}(P) = tr(R) - tr(RP), \quad (30)$$

which has the same structure as the cost function of the PAST problem (see equation (20)). Therefore, a projection tracking approach can be followed also in the IV case, along the lines of the previous subsection. More precisely, taking into account that $P = WW^T$, the recursion for the update of the signal subspace at time t can be outlined as follows.

(1) Use the new I/O data to update the RQ factorisation as in Sections 3.1-3.2 or the subtraction as in Remark 1 (possibly using a forgetting mechanism).
(2) From a suitable choice of IVs, compute $\hat{R}_{x\xi} = x\xi^T$ and $\hat{R} = \hat{R}_{x\xi}\hat{R}_{x\xi}^T$.
(3) On the basis of $\hat{R}$ and the previous estimate $\hat{P}(t-1)$ compute ϕ_{kl} according to

$$\phi_{kl} = -2\mu \sum_{i=1}^{m} (P_{l,i}\hat{R}_{x\xi\ i,k} - P_{k,i}\hat{R}_{x\xi\ i,l}). \quad (31)$$

(4) Update the subspace estimate $\hat{W}(t)$ according to

$$\hat{W}(t) = G(\phi,t)\hat{W}(t-1), \quad (32)$$

with $G(\phi,t)$ as defined in equation (22).
(5) Compute matrices A and C, and (possibly) initialise the recursion for B and D.

Remark 3. As pointed out in (Utschick, 2002) in the framework of direction of arrival problems for the approach to projector PAST, the algorithms lends itself naturally to a modification for blockwise processing of data, by suitably choosing how many data are used at each time step in the estimation of R_x. The same holds for the projector version of IV-PAST proposed herein, in which a blockwise estimate of $R_{x\xi}$ can be used..

Remark 4. It should be mentioned that the presented recursive schemes can be extended to the PO_EIV errors in variables algorithm and to the estimation of the linear part of Wiener type nonlinear models, along the lines of (Lovera *et al.*, 2000).

5. SIMULATION EXAMPLES

The recursive projection tracking algorithms proposed in this paper have been used in a series of simulated experiments. All the simulations presented in this section have been performed using Matlab; the recursive algorithms have been initialised using batch implementations of OM and PI MOESP. For the sake of conciseness only a SISO example will be presented in this Section. A first order, linear discrete time system of the form

$$x(t+1) = a(t)x(t) + u(t) + w(t) \quad (33)$$
$$y(t) = x(t) + v(t) \quad (34)$$

has been considered, where

$$a(t) = \begin{cases} 0.5 & t < 100 \\ 0.7 & t > 100 \end{cases} \quad (35)$$

and the problem of recursive estimation of a has been studied, with u, w and v realisations of white Gaussian noise, with standard deviation 1, 0.05 and 0.05 respectively. The algorithms illustrated in Sections 4 and 4.2 have been applied, using the OM scheme for the update of the RQ factorisation (without forgetting) and past input samples as IVs.

The results obtained in the estimation of a are illustrated in Figures 1 and 2 for each of the two algorithms. In each case, the time histories of the estimate of a obtained for different values of the gain μ are shown. As can be seen, low values of μ provide accurate steady state performance at the expense of a slower tracking response, while increasing μ leads to a better tracking capability.

6. CONCLUDING REMARKS

The problem of recursive identification in the framework of subspace methods has been considered and a novel recursive subspace identification algorithm has been derived, based on a projection tracking approach. For the recursive version of OM-MOESP, the results of (Utschick, 2002) could be applied, while in order to derive recursive versions of the IV-based subspace algorithms, the projection tracking idea has been extended to the IV-PAST cost function.

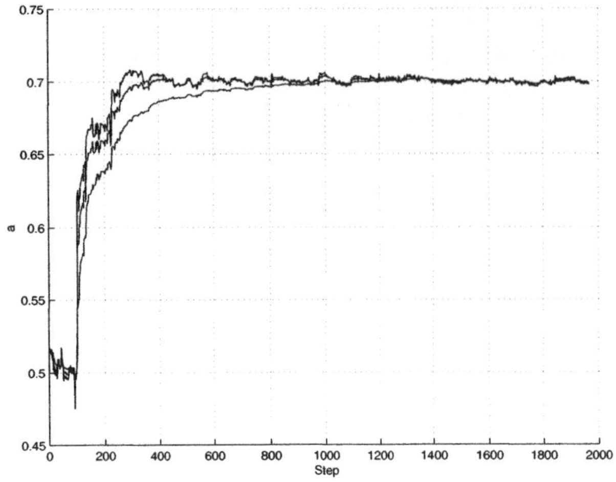

Fig. 1. Estimated eigenvalue: case with measurement noise only ($\mu = 0.1, 0.2, 0.3$).

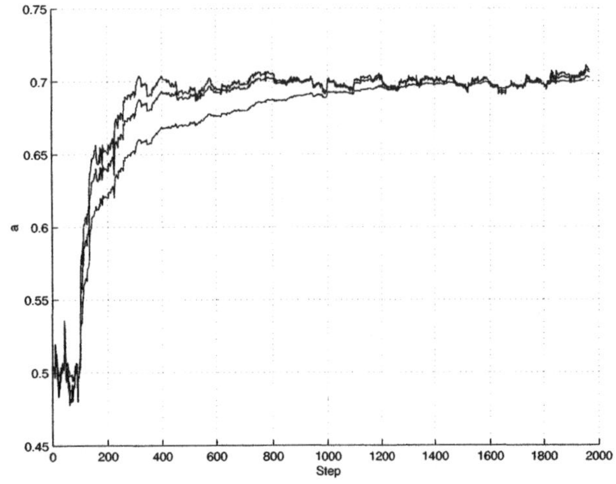

Fig. 2. Estimated eigenvalue: case with process and measurement noise ($\mu = 0.1, 0.2, 0.3$).

7. ACKNOWLEDGEMENTS

Paper supported by the MIUR project "Innovative Techniques for Identification and Adaptive Control of Industrial systems".

8. REFERENCES

Fuhrmann, D. (1997). A geometric approach to subspace tracking. In: *Proc. of the 31st Asilomar Conference on Signals, Systems and Computers*. Vol. 1. pp. 783–787.

Golub, G. and C. Van Loan (1989). *Matrix Computations*. John Hopkins University Press.

Gustafsson, T. (1997). Recursive system identification using instrumental variable subspace tracking. In: *Proc. of the 11th IFAC Symposium on System Identification, Fukuoka, Japan*.

Gustafsson, T. (1998). Instrumental variable subspace tracking using projection approximation. *IEEE Transactions on Signal Processing* **46**(3), 669–681.

Hale, E. and J. Qin (2002). Subspace model predictive control and a case study. In: *Proc. of the American Control Conference, Anchorage, Alaska*.

Lovera, M. and M. Verhaegen (1998). Recursive subspace identification of linear and non-linear Wiener type models. In: *1998 IEEE Conference on Control Applications, Trieste, Italy*.

Lovera, M., T. Gustafsson and M. Verhaegen (2000). Recursive subspace identification of linear and non-linear Wiener state space models. *Automatica* **36**(11), 1639–1650.

Lovera, M., T. Parisini and M. Verhaegen (1997). Fault detection: a subspace identification approach. In: *Proc. of the 39th Conference on Decision and Control, Orlando, Florida*.

Mercere, G., S. Lecoeuche and C. Vasseur (2003). A new recursive method for subspace identification of noisy systems: EIVPM. In: *13th IFAC Symposium on System Identification, Rotterdam, The Netherlands*.

Murnaghan, F. D. (1938). *The theory of group representations*. Dover Publications, Inc.

Oku, H. and H. Kimura (2002). Recursive 4SID algorithms using gradient type subspace tracking. *Automatica* **38**(6), 1035–1043.

Oku, H., G. Nijsse, M. Verhaegen and V. Verdult (1997). Change detection in the dynamics with recursive subspace identification. In: *Proc. of the 39th Conference on Decision and Control, Orlando, Florida*.

Srivastava, A. and D. Fuhrmann (1997). Gradient flows on projection matrices for subspace estimation. In: *Proc. of the 31st Asilomar Conference on Signals, Systems and Computers*. Vol. 2. pp. 1317–1321.

Utschick, W. (2002). Tracking of signal subspace projectors. *IEEE Transactions on Signal Processing* **50**(4), 769–778.

Verhaegen, M. (1994). Identification of the deterministic part of mimo state space models given in innovations form from input-output data. *Automatica* **30**(1), 61–74.

Verhaegen, M. and E. Deprettere (1991). A fast, recursive MIMO state space model identification algorithm. In: *Proc. of the 1991 Conference on Decision and Control*. pp. 1349–1354.

Verhaegen, M. and P. Dewilde (1992). Subspace model identification, part 1: output error state space model identification class of algorithms. *International Journal of Control* **56**(5), 1187–1210.

Yang, B. (1995). Projection approximation subspace tracking. *IEEE Transactions on Signal Processing* **43**(1), 95–107.

IFAC Publications
www.elsevier.com/locate/ifac

SUBSPACE IDENTIFICATION AND ARX MODELING

Magnus Jansson *,[1]

* *Royal Inst. of Technology (KTH), Stockholm, Sweden*

Abstract: In this paper we present a new identification method that points at the close relationship between high order ARX modeling and subspace identification. A high order ARX model is utilized to obtain initial estimates of certain Markov parameters. These parameters are then used to restructure the data model used for subspace identification to facilitate the estimation of the state sequence. Based on the estimated state sequence, the system parameters are estimated by linear regression. The method is shown to be competitive to existing subspace methods by a simulation example. The method can also be used, without modification, on closed loop data in contrast to most previously published subspace identification methods. *Copyright © 2003 IFAC*

Keywords: multivariable, identification

1. INTRODUCTION

The subspace methods for system identification estimates linear state-space models directly from time-discrete observations. The ideas behind these methods go back to classical state-space realization theory. The main observation used in the more recent algorithms (Van Overschee and De Moor, 1996; Verhaegen, 1994; Larimore, 1983; Peternell *et al.*, 1996) is that, under the assumption that there exists a true underlying finite order linear time invariant system, an estimate of the observability matrix or the state-trajectory can be obtained from the singular value decomposition of a certain data cross correlation matrix. In contrast to the more classical approaches of maximum likelihood or prediction error methods (PEM) (Ljung, 1987), the subspace methods avoid the use of (multivariable) canonical forms and the system order is the only structural parameter needed. The subspace methods are also computationally attractive since only standard matrix operations are utilized to calculate the estimates and they do not use iterative optimization techniques as, in general, are needed for the computation of the PEM estimates.

The "standard" subspace methods referred to above have problems when data are collected in closed loop. The estimates will then in general be biased since the feedback introduces a correlation between the input and the noise. Assuming a delay in the feedback loop, PEM can provide consistent estimates even on closed loop data (Forssell and Ljung, 1999). (Ljung and McKelvey, 1996) used this fact and proposed to utilize a high order ARX model to circumvent the problem of subspace methods for closed loop data. They used the ARX model to build a bank of predictors from which the state sequence can be estimated similarly to the standard subspace methods. Some other subspace methods for closed loop data have also appeared (see, e.g., (Gustafsson, 2001) and the references therein). However, no method seems to perform satisfactorily in all cases. Similar to PEM it would be desirable with a subspace method that works satisfactorily regardless of whether the data are collected in open or closed loop. This was also the main motivation behind the work presented here.

The method of this paper is inspired by the ideas in (Ljung and McKelvey, 1996) and is closely related to the canonical correlation analysis (CCA) method

[1] Corresponding author: Magnus Jansson, Dept. of Signals, Sensors and Systems, Signal Processing, Royal Inst. of Technology (KTH), SE-100 44 Stockholm, Sweden. Email: magnusj@s3.kth.se, fax +46 8 7907260.

(Peternell *et al.*, 1996; Larimore, 1983). Similar to (Ljung and McKelvey, 1996) we use in a first initial step an ARX model to try to avoid the problem with closed loop data. However, we use the ARX model in a completely different way. Rather than using predictors constructed from the ARX model, we use a certain set of the estimated Markov parameters obtained from the ARX model. These Markov parameters are used to enforce more structure into the data model used by the subspace identification approaches (similar ideas have also appeared in, e.g., (Peternell *et al.*, 1996; Chui and Maciejowski, 1998)).

2. PROBLEM FORMULATION

Consider a linear time-invariant system which can be described in innovations form by the state-space equations:

$$\begin{cases} \mathbf{x}(t+1) = \mathbf{A}\mathbf{x}(t) + \mathbf{B}\mathbf{u}(t) + \mathbf{K}\mathbf{e}(t) \\ \mathbf{y}(t) = \mathbf{C}\mathbf{x}(t) + \mathbf{D}\mathbf{u}(t) + \mathbf{e}(t) \end{cases} \tag{1}$$

where $\mathbf{u}(t) \in \mathbb{R}^{n_u}$ and $\mathbf{y}(t) \in \mathbb{R}^{n_y}$ denote the input and output signals, respectively. $\mathbf{x}(t) \in \mathbb{R}^n$ is the state-vector and n is the system order. We assume that the system is minimal in the sense that the system cannot be described by a state-space model of order less than n. $\mathbf{e}(t) \in \mathbb{R}^{n_y}$ denotes the zero mean white innovation process.

In the model (1), the matrices $\mathbf{A} \in \mathbb{R}^{n\times n}$, $\mathbf{B} \in \mathbb{R}^{n\times n_u}$, $\mathbf{C} \in \mathbb{R}^{n_y\times n}$, $\mathbf{D} \in \mathbb{R}^{n_y\times n_u}$, $\mathbf{K} \in \mathbb{R}^{n\times n_y}$ and the covariance matrix of the innovations are unknown and need to be estimated from observations of $\mathbf{u}(t)$ and $\mathbf{y}(t)$ for $t = 1, 2, \ldots, N$.

3. ESTIMATION METHOD

In the following we will present the ideas behind the new method for estimating $\{\mathbf{A}, \mathbf{B}, \mathbf{C}, \mathbf{D}, \mathbf{K}\}$ of this paper. Before doing that we will need to introduce some notation often used in the subspace identification literature. We will also review the CCA (canonical correlation analysis) subspace identification method of (Peternell *et al.*, 1996) (see also (Larimore, 1983)) since our proposed method borrows ideas from CCA.

Let us first introduce some preliminary notation. Define the vectors of stacked inputs, outputs and innovations as

$$\mathbf{y}_f(t) = \left[\mathbf{y}^T(t)\ \mathbf{y}^T(t+1)\ \ldots\ \mathbf{y}^T(t+f-1)\right]^T$$
$$\mathbf{u}_f(t) = \left[\mathbf{u}^T(t)\ \mathbf{u}^T(t+1)\ \ldots\ \mathbf{u}^T(t+f-1)\right]^T$$
$$\mathbf{e}_f(t) = \left[\mathbf{e}^T(t)\ \mathbf{e}^T(t+1)\ \ldots\ \mathbf{e}^T(t+f-1)\right]^T$$

where $f > n$ is an integer chosen by the user. By employing (1), we can establish the relation

$$\mathbf{y}_f(t) = \Gamma\mathbf{x}(t) + \Phi\mathbf{u}_f(t) + \Psi\mathbf{e}_f(t) \tag{2}$$

where

$$\Gamma = \begin{bmatrix} \mathbf{C} \\ \mathbf{CA} \\ \vdots \\ \mathbf{CA}^{f-1} \end{bmatrix}$$

$$\Phi = \begin{bmatrix} \mathbf{D} & \mathbf{0} & \ldots & \mathbf{0} \\ \mathbf{CB} & \mathbf{D} & & \\ \mathbf{CAB} & \mathbf{CB} & \ddots & \vdots \\ \vdots & & \ddots & \mathbf{0} \\ \mathbf{CA}^{f-2}\mathbf{B} & & \mathbf{CB} & \mathbf{D} \end{bmatrix}$$

$$\Psi = \begin{bmatrix} \mathbf{I} & \mathbf{0} & \ldots & \mathbf{0} \\ \mathbf{CK} & \mathbf{I} & & \\ \mathbf{CAK} & \mathbf{CK} & \ddots & \vdots \\ \vdots & & \ddots & \mathbf{0} \\ \mathbf{CA}^{f-2}\mathbf{K} & & \mathbf{CK} & \mathbf{I} \end{bmatrix}.$$

Define

$$\tilde{\mathbf{A}} = (\mathbf{A} - \mathbf{KC})$$
$$\tilde{\mathbf{B}} = (\mathbf{B} - \mathbf{KD}).$$

It is well known that we can rewrite (1) as follows

$$\begin{cases} \mathbf{x}(t+1) = \tilde{\mathbf{A}}\mathbf{x}(t) + \tilde{\mathbf{B}}\mathbf{u}(t) + \mathbf{K}\mathbf{y}(t) \\ \mathbf{y}(t) = \mathbf{C}\mathbf{x}(t) + \mathbf{D}\mathbf{u}(t) + \mathbf{e}(t) \end{cases} \tag{3}$$

From (3) it is also clear that

$$\mathbf{x}(t) = \sum_{k=0}^{p-1} \tilde{\mathbf{A}}^k[\mathbf{K}\mathbf{y}(t-k-1)\ \tilde{\mathbf{B}}\mathbf{u}(t-k-1)] + \tilde{\mathbf{A}}^p\mathbf{x}(t-p)$$

Assuming $\tilde{\mathbf{A}}$ is stable, this implies that the state can at least in principle be estimated by a linear combination of past inputs and outputs by choosing p "large enough." This is used in for example the CCA subspace method (see (Peternell *et al.*, 1996; Larimore, 1983)). Let us replace the state in (2) by the estimate

$$\hat{\mathbf{x}}(t) = \mathscr{K}\mathbf{p}(t), \tag{4}$$

where $\mathscr{K}$ is a matrix of unknown coefficients and $\mathbf{p}(t)$ a vector containing delayed inputs and outputs p steps back:

$$\mathbf{p}(t) = \left[\mathbf{y}^T(t-1)\ \mathbf{y}^T(t-2)\ \ldots\ \mathbf{y}^T(t-p)\right.$$
$$\left.\mathbf{u}^T(t-1)\ \mathbf{u}^T(t-2)\ \ldots\ \mathbf{u}^T(t-p)\right]^T$$

The first step of the CCA subspace method is to estimate the coefficient matrices $\Gamma\mathscr{K}$ and Φ by an "unstructured least squares linear regression approach" in the equation

$$\mathbf{y}_f(t) \approx \Gamma\mathscr{K}\mathbf{p}(t) + \Phi\mathbf{u}_f(t) + \Psi\mathbf{e}_f(t)$$

That is, $\mathbf{y}_f(t)$ is regressed on $\mathbf{p}(t)$ and $\mathbf{u}_f(t)$ to form unstructured estimates $\widehat{\Gamma\mathscr{K}}$ and $\hat{\Phi}$ of $\Gamma\mathscr{K}$ and Φ, respectively. The above step is where the CCA estimates typically will be biased for closed loop data due to the correlation between $\mathbf{u}_f(t)$ and $\mathbf{e}_f(t)$. The next step of

CCA is to utilize $\hat{\Phi}$ in an attempt to remove the effects of the future inputs on the future outputs by forming

$$\mathbf{z}(t) \triangleq \mathbf{y}_f(t) - \hat{\Phi}\mathbf{u}_f \approx \Gamma\mathscr{K}\mathbf{p}(t) + \Psi\mathbf{e}_f(t). \quad (5)$$

This equation can be viewed as a low rank linear regression problem in $\Gamma\mathscr{K}$. The main interest in CCA lies in the estimation of $\mathscr{K}$ so that the state can be subsequently estimated as in (4). This can be done by performing a canonical correlation analysis on $\mathbf{z}(t)$ and $\mathbf{p}(t)$ as follows: Let

$$\mathbf{M} = (\hat{\mathbf{R}}_{\mathbf{zz}})^{-1/2}(\hat{\mathbf{R}}_{\mathbf{zp}})(\hat{\mathbf{R}}_{\mathbf{pp}})^{-1/2}$$

where we have defined the sample correlation matrix between two signals $\mathbf{z}(t)$ and $\mathbf{p}(t)$ as

$$\hat{\mathbf{R}}_{\mathbf{zp}} = \frac{1}{N}\sum_{t=1}^{N}\mathbf{z}(t)\mathbf{p}^T(t).$$

Next, compute the singular value decomposition

$$\mathbf{USV}^T = \mathbf{M}$$

where $\mathbf{U}$ and $\mathbf{V}$ are orthonormal matrices of left and right singular vectors, respectively, and $\mathbf{S}$ is a diagonal matrix containing the singular values in nonincreasing order along the diagonal. Let $\mathbf{V}_n$ denote the matrix consisting of the n first columns of $\mathbf{V}$; that is, the matrix made of the "dominating" right singular vectors. The CCA estimate of $\mathscr{K}$ is then given by

$$\hat{\mathscr{K}} = \mathbf{V}_n^T(\hat{\mathbf{R}}_{\mathbf{pp}})^{-1/2}$$

and the corresponding estimated state sequence is

$$\hat{\mathbf{x}}(t) = \mathbf{V}_n^T(\hat{\mathbf{R}}_{\mathbf{pp}})^{-1/2}\mathbf{p}(t). \quad (6)$$

The system matrices can now be estimated by linear regression in the state space model equations (1) by replacing the true state with the estimate (6). $\mathbf{C}$ and $\mathbf{D}$ are obtained by regressing $\mathbf{y}(t)$ on $\hat{\mathbf{x}}(t)$ and $\mathbf{u}(t)$. An estimate of the innovation sequence is then obtained as the residual of that regression. Finally, the matrices $\mathbf{A},\mathbf{B}$ and $\mathbf{K}$ are estimated by regressing $\hat{\mathbf{x}}(t+1)$ on $\hat{\mathbf{x}}(t)$, $\mathbf{u}(t)$, and the estimated innovation sequence. This concludes the review of the CCA identification method (Peternell *et al.*, 1996; Larimore, 1983).

Let us now turn to the idea of the method of this paper. First we will use (3) to rewrite (2) on the following alternative form

$$\mathbf{y}_f(t) = \tilde{\Gamma}\mathbf{x}(t) + \tilde{\Phi}\mathbf{u}_f(t) + \tilde{\Psi}\mathbf{y}_f(t) + \mathbf{e}_f(t) \quad (7)$$

where

$$\tilde{\Gamma} = \begin{bmatrix} \mathbf{C} \\ \mathbf{C}\tilde{\mathbf{A}} \\ \vdots \\ \mathbf{C}\tilde{\mathbf{A}}^{f-1} \end{bmatrix} \quad (8)$$

$$\tilde{\Phi} = \begin{bmatrix} \mathbf{D} & \mathbf{0} & \dots & \mathbf{0} \\ \mathbf{C}\tilde{\mathbf{B}} & \mathbf{D} & & \\ \mathbf{C}\tilde{\mathbf{A}}\tilde{\mathbf{B}} & \mathbf{C}\tilde{\mathbf{B}} & \ddots & \vdots \\ \vdots & & \ddots & \mathbf{0} \\ \mathbf{C}\tilde{\mathbf{A}}^{f-2}\tilde{\mathbf{B}} & & \mathbf{C}\tilde{\mathbf{B}} & \mathbf{D} \end{bmatrix} \quad (9)$$

$$\tilde{\Psi} = \begin{bmatrix} \mathbf{0} & \mathbf{0} & \dots & \mathbf{0} \\ \mathbf{CK} & \mathbf{0} & & \\ \mathbf{C}\tilde{\mathbf{A}}\mathbf{K} & \mathbf{CK} & \ddots & \vdots \\ \vdots & & \ddots & \mathbf{0} \\ \mathbf{C}\tilde{\mathbf{A}}^{f-2}\mathbf{K} & & \mathbf{CK} & \mathbf{0} \end{bmatrix}. \quad (10)$$

An important observation is that (7) is nothing but the stacked outputs of an ARX model, which in general is of infinite order. However, again assuming $\tilde{\mathbf{A}}$ to be stable we can approximate this by truncating the ARX model similar to what is done in the CCA method. The main idea of our method (that we will call SSARX in the following) is to first estimate a "high order" ARX model to get (unstructured) estimates of the impulse response coefficients $\mathbf{D}$, $\mathbf{C}\tilde{\mathbf{A}}^k\tilde{\mathbf{B}}$ and $\mathbf{C}\tilde{\mathbf{A}}^k\mathbf{K}$ for $k = 0, 1, \dots, f-2$. If we use these estimated impulse response coefficients to form estimates $\hat{\tilde{\Phi}}$ and $\hat{\tilde{\Psi}}$ of (9) and (10), we can write

$$\mathbf{y}_f(t) - \hat{\tilde{\Phi}}\mathbf{u}_f(t) - \hat{\tilde{\Psi}}\mathbf{y}_f(t) \approx \tilde{\Gamma}\mathscr{K}\mathbf{p}(t) + \mathbf{e}_f(t). \quad (11)$$

Similar to CCA (cf. (5)), we view this last relation as a low rank linear regression in $\tilde{\Gamma}\mathscr{K}$. The remaining steps are completely analog to the CCA estimation procedure as outlined above. We finally note that $\hat{\tilde{\Phi}}$ and $\hat{\tilde{\Psi}}$ should be consistent estimates if we allow the order of the ARX model to increase. This holds also for closed loop data. Clearly the formulation of the low rank linear regression model as in (11) is what differs SSARX from CCA.

4. SIMULATION EXAMPLE

To indicate the potential of the SSARX method suggested in this paper we consider the same example as was used in (Ljung and McKelvey, 1996). This example considers both open and closed loop identification of the following system

$$y(t) = \frac{0.21q^{-1} + 0.07q^{-2}}{1 - 0.6q^{-1} + 0.8q^{-2}}u(t) + \frac{1}{1 - 0.98q^{-1}}e(t)$$

where $e(t)$ is zero mean white noise with variance 4. In the open loop situation the input $u(t)$ is zero mean white noise with unit variance. In the closed loop case the input is given by

$$u(t) = r(t) - y(t)$$

where the external input $r(t)$ is white noise with unit variance.

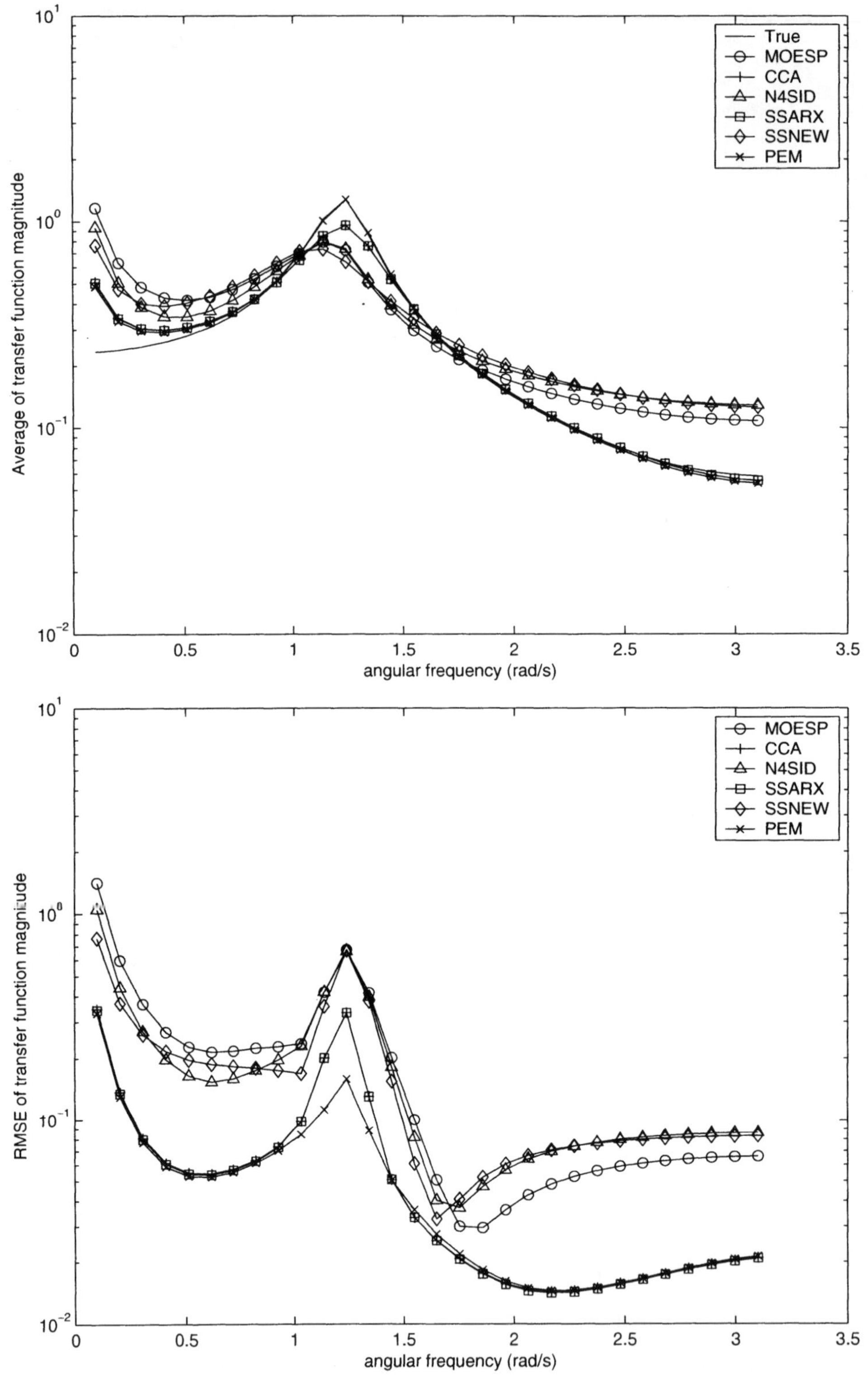

Fig. 1. The top graph shows the average of the estimated magnitudes of the transfer function from $u(t)$ to $y(t)$ for the open loop case. The bottom graph shows the corresponding RMS errors of the estimated magnitudes. $f = p = 10$ for all methods.

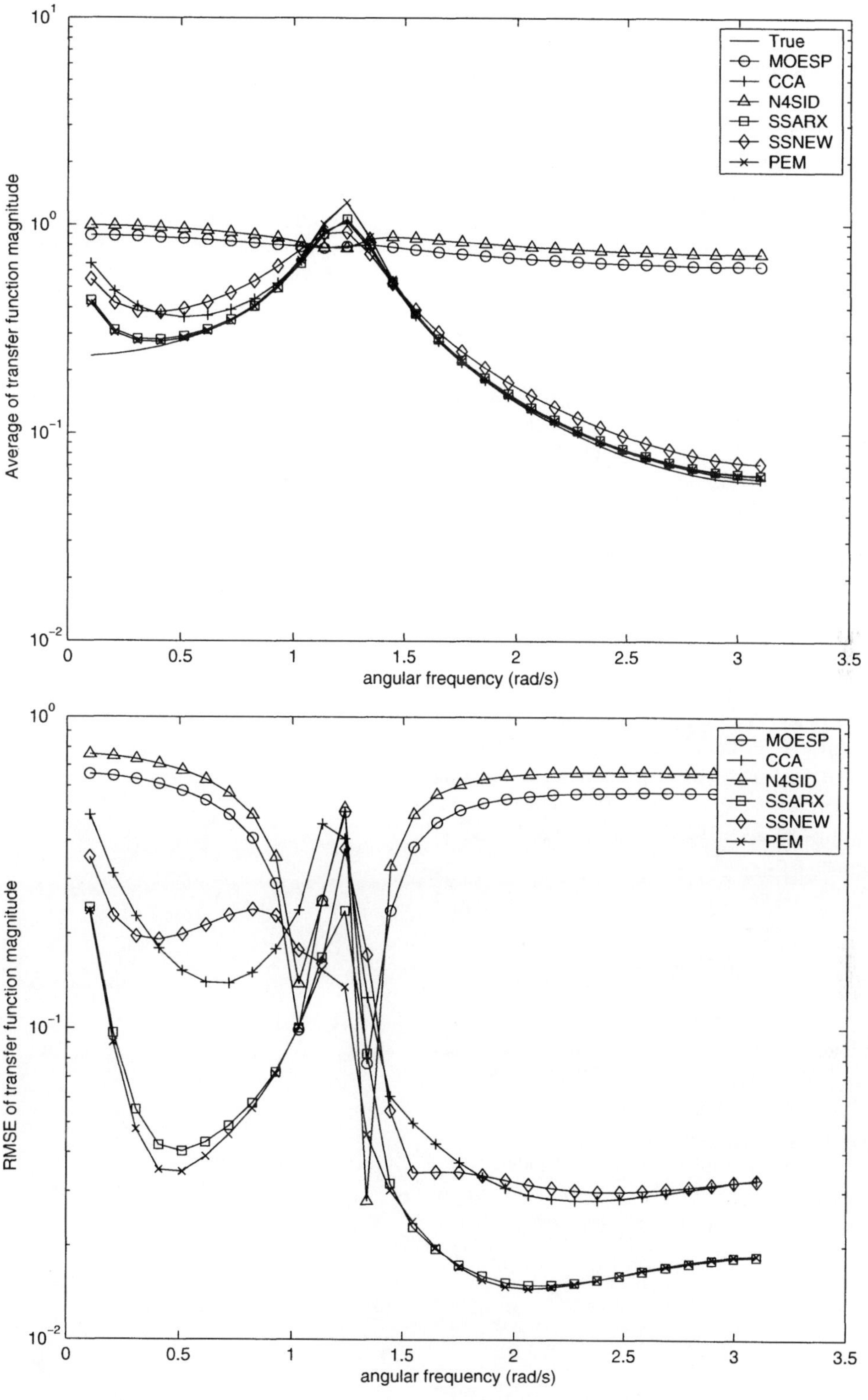

Fig. 2. The top graph shows the average of the estimated magnitudes of the transfer function from $u(t)$ to $y(t)$ for the closed loop case. The bottom graph shows the corresponding RMS errors of the estimated magnitudes. $f = p = 10$ for all methods.

For both the open and closed loop cases we use 50 independent Monte Carlo simulations. In each simulation run we estimate third order state space models by different subspace identification methods based on $N = 3000$ samples of $u(t)$ and $y(t)$. (We use a larger data set than what was used in (Ljung and McKelvey, 1996) simply to get more reasonably accurate models.) The results are presented in Fig. 1 and Fig. 2. The figures show the average as well as the root mean square error (RMSE) of the estimated magnitude of the transfer function from $u(t)$ to $y(t)$. The methods included in the comparison are: MOESP (Verhaegen, 1994), CCA (Larimore, 1983; Peternell *et al.*, 1996), N4SID (the so called "robust" version in (Van Overschee and De Moor, 1996), SSNEW (the method proposed in (Ljung and McKelvey, 1996)), SSARX of this paper and the prediction error method (Ljung, 1987). We remark that for all methods we constrain $\mathbf{D}$ to be zero. This is essential especially for SSARX and SSNEW to give consistent estimates in the closed loop case. The similarity of CCA and SSARX is clearly visible in the open loop case (see Fig. 1) whereas the other methods perform significantly worse for this example. Among the considered subspace methods, only SSNEW and SSARX are in general consistent in the closed loop case, which is demonstrated in Fig. 2. However, at least for this example, the performance of SSARX appears to be superior to that of SSNEW, and in fact very close to the performance of PEM.

5. CONCLUSIONS

In this paper we have presented a new subspace identification method (SSARX). The method combines ideas from ARX modeling with the CCA subspace method and is able to handle data collected both in open and closed loop. The method was shown to outperform many existing subspace methods in two presented simulation examples. In fact, the accuracy of SSARX in these examples was very close to that of PEM.

6. REFERENCES

Chui, N.L.C: and J.M. Maciejowski (1998). Subspace identification – a markov parameter approach. Technical Report CUED/F-INFENG/TR.337. Cambridge University.

Forssell, U. and L. Ljung (1999). Closed-loop identification revisited. *Automatica* **35**, 1215–1241.

Gustafsson, T. (2001). Subspace identification using instrumental variable techniques. *Automatica* **37**, 2005–2010.

Larimore, W. E. (1983). "System Identification, Reduced-Order Filtering and Modeling via Canonical Variate Analysis". In: *Proc. ACC*. pp. 445–451.

Ljung, L. (1987). *System Identification: Theory for the User*. Prentice-Hall. Englewood Cliffs, NJ. 2nd ed. 1999.

Ljung, L. and T. McKelvey (1996). Subspace identification from closed loop data. *Signal Processing* **52**(2), 209–216.

Peternell, K., W. Scherrer and M. Deistler (1996). "Statistical Analysis of Novel Subspace Identification Methods". *Signal Processing* **52**(2), 161–177.

Van Overschee, P. and B. De Moor (1996). *Subspace Identification for Linear Systems: Theory–Implementation–Applications*. Kluwer Academic Publishers.

Verhaegen, M. (1994). "Identification of the Deterministic Part of MIMO State Space Models given in Innovations Form from Input-Output Data". *Automatica* **30**(1), 61–74.

IFAC Publications
www.elsevier.com/locate/ifac

PARALLEL QR IMPLEMENTATION OF SUBSPACE IDENTIFICATION WITH PARSIMONIOUS MODELS

S. Joe Qin * and **Lennart Ljung** **

* *Department of Chemical Engineering*
The University of Texas at Austin
Austin, TX 78712
e-mail:qin@che.utexas.edu.
** *Department of Electrical Engineering*
Linkoping University
Linkoping, Sweden

Abstract:
In this paper we reveal that the typical subspace identification algorithms use non-parsimonious model formulations, with extra terms in the model that appear to be non-causal. These terms are the causes for inflated variance in the estimates and partially responsible for the loss of closed-loop identifiability. We then propose a parallel parsimonious formulation of a new subspace identification algorithm and demonstrate the effectiveness of the proposed algorithm via simulation. *Copyright © 2003 IFAC*

Keywords: subspace identification; parsimonious formulation; parallel projections

1. INTRODUCTION

Subspace identification methods (SIM) are attractive not only because of their numerical simplicity and stability, but also for their state space form that is very convenient for estimation, filtering, prediction, and control. A few drawbacks, however, have been experienced with SIMs:

(1) The estimation accuracy is in general not as good as the prediction error methods (PEM), represented by large variance.
(2) The application of SIMs to closed-loop data is still a challenge, even though the data satisfy identifiability conditions for traditional methods such as PEMs.
(3) The estimation of B and D is more problematic than that of A and C, which is reflected in the poor estimation of zeros and steady state gains.

In this paper, we are concerned with the reasons why subspace identification approaches exhibit these drawbacks and propose parsimonious SIMs for open-loop applications. First of all, we start with the analysis of existing subspace formulation using the linear regression formulation (Jansson and Wahlberg, 1998; Knudsen, 2001). From this analysis we reveal that the typical SIM algorithms actually use non-parsimonious model formulation, with extra terms in the model that appear to be non-causal. These terms, although conveniently included for performing subspace projections, are the causes for inflated variance in the estimates and partially responsible for the loss of closed-loop identifiability.

2. ANALYSIS OF SUBSPACE MODEL FORMULATION

Parsimoniousness is a general rule is regression analysis and system identification. The typical subspace identification models, however, are not parsimonious and even non-causal as will be revealed in this section. We begin with an innovation model formulation,

$$x_{k+1} = Ax_k + Bu_k + Ke_k \quad (1a)$$
$$y_k = Cx_k + Du_k + e_k \quad (1b)$$

where $y_k \in R^{n_y}$, $x_k \in R^n$, $u_k \in R^{n_u}$, and $e_k \in R^{n_y}$ are the system output, state, input, and innovation, respectively. A,B,C,D and K are system matrices with appropriate dimensions.

An extended state space model can be formulated as

$$Y_f = \Gamma_f X_k + H_f U_f + G_f E_f \quad (2a)$$
$$Y_p = \Gamma_p X_{k-p} + H_p U_p + G_p E_p \quad (2b)$$

where the extended observability matrix

$$\Gamma_f = \begin{bmatrix} C \\ CA \\ \vdots \\ CA^{f-1} \end{bmatrix} \quad (3)$$

and the Toeplitz matrices are

$$H_f = \begin{bmatrix} D & 0 & \cdots & 0 \\ CB & D & \cdots & 0 \\ \vdots & \vdots & \ddots & \vdots \\ CA^{f-2}B & CA^{f-3}B & \cdots & D \end{bmatrix} \quad (4a)$$

$$G_f = \begin{bmatrix} I & 0 & \cdots & 0 \\ CK & I & \cdots & 0 \\ \vdots & \vdots & \ddots & \vdots \\ CA^{f-2}K & CA^{f-3}K & \cdots & I \end{bmatrix} \quad (4b)$$

The input and output data are arranged in the following Hankel form:

$$U_f = \begin{bmatrix} u_k & u_{k+1} & \cdots & u_{k+N-1} \\ u_{k+1} & u_{k+2} & \cdots & u_{k+N} \\ \vdots & \vdots & \ddots & \vdots \\ u_{k+f-1} & u_{k+f} & \cdots & u_{k+f+N-2} \end{bmatrix} \quad (5a)$$

$$\triangleq [u_f(k) \cdots u_f(k+N-1)] \quad (5b)$$

$$U_p = \begin{bmatrix} u_{k-p} & u_{k-p+1} & \cdots & u_{k-p+N-1} \\ u_{k-p+1} & u_{k-p+2} & \cdots & u_{k-p+N} \\ \vdots & \vdots & \ddots & \vdots \\ u_{k-1} & u_k & \cdots & u_{k+N-2} \end{bmatrix} \quad (5c)$$

$$\triangleq \begin{bmatrix} u_p(k-p) & \cdots & u_p(k-p+N-1) \end{bmatrix} \quad (5d)$$

Similar formulations are made for Y_f, Y_p, E_f, and E_p. Subspace identification methods minimize the following objective function (Overschee and Moor, 1996),

$$[\, \hat{L}^1 \; \hat{L}^2 \; \hat{L}^3 \,] = \arg\min\{ \left\| Y_f - L^1 Y_p - L^2 U_p - L^3 U_f \right\|_F^2 \} \quad (6)$$
$$= \arg\min\{ \sum_{j=0}^{N-1} \left\| y_f(k+j) - \begin{bmatrix} L^1 & L^2 & L^3 \end{bmatrix} \begin{bmatrix} y_p(k-p+j) \\ u_p(k-p+j) \\ u_f(k+j) \end{bmatrix} \right\|^2 \}$$

Denoting

$$L^1 = \begin{bmatrix} L^1_{11} & L^1_{12} & \cdots & L^1_{1p} \\ L^1_{21} & L^1_{22} & \cdots & L^1_{2p} \\ \vdots & & \ddots & \\ L^1_{f1} & L^1_{f1} & & L^1_{fp} \end{bmatrix} \triangleq \begin{bmatrix} L^1_1 \\ L^1_2 \\ \vdots \\ L^1_f \end{bmatrix} \quad (7a)$$

$$L^2 = \begin{bmatrix} L^2_{11} & L^2_{12} & \cdots & L^2_{1p} \\ L^2_{21} & L^2_{22} & \cdots & L^2_{2p} \\ \vdots & & \ddots & \\ L^2_{f1} & L^2_{f1} & & L^2_{fp} \end{bmatrix} \triangleq \begin{bmatrix} L^2_1 \\ L^2_2 \\ \vdots \\ L^2_f \end{bmatrix} \quad (7b)$$

$$L^3 = \begin{bmatrix} L^3_{11} & L^3_{12} & \cdots & L^3_{1f} \\ L^3_{21} & L^3_{22} & \cdots & L^3_{2f} \\ \vdots & & \ddots & \\ L^3_{f1} & L^3_{f1} & & L^3_{ff} \end{bmatrix} \triangleq \begin{bmatrix} L^3_1 \\ L^3_2 \\ \vdots \\ L^3_f \end{bmatrix} \quad (7c)$$

the above problem is equivalent to f separate subproblems:

$$\begin{bmatrix} \hat{L}^1_i & \hat{L}^2_i & \hat{L}^3_i \end{bmatrix} =$$
$$\arg\min\{ \sum_{j=0}^{N-1} \left\| y(k+j+i-1) - \begin{bmatrix} L^1_i & L^2_i & L^3_i \end{bmatrix} \begin{bmatrix} y_p(k-p+j) \\ u_p(k-p+j) \\ u_f(k+j) \end{bmatrix} \right\|^2 \quad (8)$$
for $i = 1, 2, \ldots, f$

For the moment consider the first subproblem, that is, $i = 1$. In this case the problem implies that the following model is specified:

$$y(k) = \begin{bmatrix} L^1_1 & L^2_1 & L^3_1 \end{bmatrix} \begin{bmatrix} y_p(k-p) \\ u_p(k-p) \\ u_f(k) \end{bmatrix} + v(k)$$
$$= \begin{bmatrix} L^1_1 & L^2_1 \end{bmatrix} \begin{bmatrix} y_p(k-p) \\ u_p(k-p) \end{bmatrix}$$
$$+ L^3_{11} u(k) + \sum_{j=2}^{f} L^3_{1j} u(k+j-1) + v(k) \quad (9)$$

Note that the third term on the RHS of the above equation is non-causal and unnecessary. Therefore, we can make the following statements about the typical SIM formulation in general.

(1) The model format used in SIM during the projection step is non-causal. This would result in non-causal models in the projection step. Although the non-causal terms are ignored at the step to estimate B, D, all the model parameters estimate have inflated variance due to the fact that extra and unnecessary terms are included in the model, making the model nonparsimonious.
(2) Because of the extra terms that turn out to be 'future' inputs, SIMs in general have

problems with closed-loop data using direct identification methods. Most SIMs usually project out U_f as follows:

$$Y_f/\Pi_{U_f}^{\perp} = \Gamma_f X_k/\Pi_{U_f}^{\perp} + G_f E_f/\Pi_{U_f}^{\perp} \quad (10)$$

where $\Pi_{U_f}^{\perp} = I - U_f^T(U_f U_f^T)^{-1}U_f$. Because of the non-causal terms in the model, $\frac{1}{N}E_f U_f^T \neq 0$ as $N \to \infty$ for closed-loop data. As a consequence, many SIMs fail to work on closed loop data, except for a few SIM algorithms that avoid this projection (Chou and Verhaegen, 1997; Wang and Qin, 2001; Wang and Qin, 2002).

(3) Because U_f contains extra rows due to the extra terms, the projection in Eq. 10 tends to reduce the information content unncessarily even for open-loop data, leading to inefficient use of the data.

(4) These non-causal terms will have negligible coefficients only when the number of data is very large and process is very well excited. For limited number of samples or non-white input signals, SIM algorithms tend to have large estimation errors.

To avoid these problems the SIM model must not include these non-causal terms. We propose a parallel QR implementation of a parsimonious subspace identification method (PARSIM) which removes these non-causal terms by enforcing triangular structure of the Toeplitz matrix H_f at every step of the SIM procedure. The parallel PARSIM (PARSIM-P) method involves a bank of least squares problems in parallel. This idea was presented in (Qin *et al.*, 2002), In this paper, the bank of least squares problems is implemented via QR factorization. Optimal weighting is derived for this PARSIM-P method. An optimal estimate of the B, D matrices is given using the Kalman filter structure. Numerical simulation is used to demonstrate the effectiveness of the proposed method.

3. PARALLEL PARSIM METHOD

The key idea in the proposed method is to exclude those non-causal terms of U_f. To accomplish this we partition the extended state space model row-wise as follows:

$$Y_f = \begin{bmatrix} Y_{f1} \\ Y_{f2} \\ \vdots \\ Y_{ff} \end{bmatrix}; \; Y_i \triangleq \begin{bmatrix} Y_{f1} \\ Y_{f2} \\ \vdots \\ Y_{fi} \end{bmatrix}; i = 1, 2, \ldots, f \quad (11)$$

Partition U_f and E_f in a similar way to define U_{fi}, U_i, E_{fi}, and E_i, respectively, for $i = 1, 2, \ldots, f$. Denote further

$$\Gamma_f = \begin{bmatrix} \Gamma_{f1} \\ \Gamma_{f2} \\ \vdots \\ \Gamma_{ff} \end{bmatrix} \quad (12a)$$

$$H_{fi} \triangleq \begin{bmatrix} CA^{i-2}B & \cdots & CB & D \end{bmatrix} \quad (12b)$$

$$\triangleq \begin{bmatrix} H_{i-1} & \cdots & H_1 & H_0 \end{bmatrix} \quad (12c)$$

$$G_{fi} \triangleq \begin{bmatrix} CA^{i-2}K & \cdots & CK & I \end{bmatrix} \quad (12d)$$

$$\triangleq \begin{bmatrix} G_{i-1} & \cdots & G_1 & G_0 \end{bmatrix} \quad (12e)$$

$$\forall i = 1, 2, \cdots, f$$

where H_i and G_i are the Markov parameters for the deterministic input and innovation sequence, respectively. We have the following partitioned equations:

$$Y_{fi} = \Gamma_{fi}X_k + H_{fi}U_i + G_{fi}E_i \qquad \forall i = 1, 2, \cdots, f \quad (13)$$

Note that each of the above equation is guaranteed causal.

3.1 *Parallel Estimation of* Γ_{fi} *and* H_{fi}

By eliminating $e(k)$ in the innovation model through iteration, it is straightforward to derive the following relation (Knudsen, 2001),

$$X_k = L_z Z_p + A_K^p X_{k-p} \quad (14)$$

where

$$L_z \triangleq \begin{bmatrix} \Delta_p(A_K, K) & \Delta_p(A_K, B_K) \end{bmatrix} \quad (15a)$$

$$\Delta_p(A, B) \triangleq \begin{bmatrix} A^{p-1}B & \cdots & AB & B \end{bmatrix} \quad (15b)$$

$$A_K \triangleq A - KC \quad (15c)$$

$$B_K \triangleq B - KD \quad (15d)$$

Substituting this equation into Eq. 13, we obtain

$$Y_{fi} = \Gamma_{fi}L_z Z_p + \Gamma_{fi}A_K^p X_{k-p} + H_{fi}U_i + G_{fi}E_i \qquad \forall i = 1, 2, \cdots, f \quad (16)$$

Since the second term in the RHS of Eq. 16 tends to zero as p tends to infinity, we have the following least squares estimates:

$$\begin{bmatrix} \hat{\Gamma}_{fi}L_z & \hat{H}_{fi} \end{bmatrix} = Y_{fi} \begin{bmatrix} Z_p \\ U_i \end{bmatrix}^+ \qquad \forall i = 1, 2, \cdots, f \quad (17)$$

Augmenting all estimates $\hat{\Gamma}_{fi}L_z, \forall i = 1, 2, \cdots, f$, we have

$$\begin{bmatrix} \hat{\Gamma}_{f1}L_z \\ \hat{\Gamma}_{f2}L_z \\ \vdots \\ \hat{\Gamma}_{ff}L_z \end{bmatrix} = \hat{\Gamma}_f L_z \quad (18)$$

Now we have the following parallel PARSIM algorithm to estimate Γ_{fi} and H_{fi}, for $i = 1, 2, \cdots, f$.

[Algorithm 1] Parallel PARSIM (PARSIM-P)

(1) Perform the following least squares estimates,

$$\left[\hat{\Gamma}_{fi}L_z \ \hat{H}_{fi}\right] = Y_{fi}\begin{bmatrix} Z_p \\ U_i \end{bmatrix}^+ \quad (19)$$
$$\forall i = 1, 2, \cdots, f$$

(2) Perform SVD for the following weighted matrix

$$W_1(\hat{\Gamma}_f L_z)W_2 = USV^T \quad (20)$$

where W_1 is nonsingular and $L_z W_2$ does not lose rank. We choose

$$\hat{\Gamma}_f = W_1^{-1}US^{1/2} \quad (21)$$

which will yield the estimate for A and C (Verhaegen, 1994).

(3) The estimates for $H_{i-1}, \forall i = 1, 2, \cdots, f$ can be calculated by averaging repeated estimates

$$\hat{H}_{i-1} = \frac{1}{f-i+1}\sum_{j=i}^{f} \hat{H}_{fj}(:,(j-i)lf+1:(j-i+1)lf)$$
$$\forall i = 1, 2, \cdots, f$$
$$(22)$$

which can be used to estimate B and D.

[Theorem 1]

Algorithm 1 gives *consistent* estimates for Γ_f and $H_{i-1}, \forall i = 1, 2, \cdots, f$ under the following conditions:

(1) The past horizon $p \to \infty$.
(2) The input $u(k)$ and innovation sequence $e(k)$ are uncorrelated, i.e.,
$\frac{1}{N}E_iU_i^T \to 0$ as $N \to \infty$.
(3) $\{C, A\}$ is observable and $\{A, \begin{bmatrix} B & K \end{bmatrix}\}$ reachable.

[Proof] See Appendix A.

[Remark 1] For finite past horizon p the algorithm is biased, but the bias decays to zero exponentially with p. If p is too large in practice, however, large variance is expected for the estimates. Therefore, it is necessary in practice to use a finite p for the best trade-off. Cross-validation can be used to select an optimal p.

[Remark 2] Because of Condition 2 in Theorem 1 where the $E_iU_i^T$ term requires no correlation between future u_k and past e_k, this PARSIM-P algorithm is biased for direct closed loop identification.

3.2 *Optimal Weighting and QR Implementation*

3.2.1. *Optimal Weighting* In the conventional SIM formulation under open-loop conditions,

$$E_f\Pi_{U_f}^{\perp} \to E_f \text{ as } N \to \infty \quad (23)$$

since E_f is uncorrelated with U_f. Therefore Eq. 10 becomes:

$$Y_f\Pi_{U_f}^{\perp} = \Gamma_f X_k\Pi_{U_f}^{\perp} + G_fE_f \quad (24)$$

Post-multiplying $\frac{1}{N}Z_p^T$ to eliminate the noise term,

$$Y_f\Pi_{U_f}^{\perp}Z_p^T = \Gamma_f X_k\Pi_{U_f}^{\perp}Z_p^T \quad (25)$$

Van Overshee and De Moor (1995) show that all SIM methods do SVD on the following weighted matrix:

$$W_rY_f\Pi_{U_f}^{\perp}Z_p^TW_c = W_r\Gamma_f X_k\Pi_{U_f}^{\perp}Z_p^TW_c \quad (26)$$

where W_r and W_c are the row and column weighting matrices, respectively. In CVA $W_r = (Y_f\Pi_{U_f}^{\perp}Y_f^T)^{-1/2}$ which basically normalizes the output variables to achieve a maximum likelihood type of weighting. Gustafsson (Gustafsson, 2002) shows that an approximately optimal weighting for W_c is

$$W_c = (Z_pZ_p^T - Z_pU_f^T(U_fU_f^T)U_fZ_p^T)^{-1/2}$$
$$= (Z_p\Pi_{U_f}^{\perp}Z_p^T)^{-1/2} \quad (27)$$

which is used in CVA and MOESP. Substituting Eq. 27 into Eq. 26, and replacing X_k with L_zZ_p as an instrumental variable, we obtain,

$$W_rY_f\Pi_{U_f}^{\perp}Z_p^TW_c = W_r\Gamma_f L_zZ_p\Pi_{U_f}^{\perp}Z_p^T(Z_p\Pi_{U_f}^{T}Z_p^T)^{-1/2}$$
$$= W_r\Gamma_f L_z(Z_p\Pi_{U_f}^{\perp}Z_p^T)^{1/2} \quad (28)$$

Comparing Eq. 28 with Eq.20, the equivalent weightings for the PARSIM-P algorithm is

$$W_1 = W_r = (Y_f\Pi_{U_p}^{\perp}Y_f^T)^{-1/2} \quad (29a)$$
$$W_2 = (Z_p\Pi_{U_f}^{\perp}Z_p^T)^{1/2} \quad (29b)$$

Jansson and Wahlberg (1996) show that the row-weighting W_1 (or W_r) has no influence on the asymptotic accuracy of the estimate.

3.2.2. *QR Implementation for Γ_f* To implement the PARSIM-P algorithm efficiently, we perform the following QR decomposition

$$\begin{bmatrix} \begin{bmatrix} Z_p \\ U_i \end{bmatrix} \\ Y_{fi} \end{bmatrix} = \begin{bmatrix} R_{11,i} & \\ R_{21,i} & R_{22,i} \end{bmatrix}\begin{bmatrix} Q_{1,i} \\ Q_{2,i} \end{bmatrix} \quad (30)$$

Using this relation in Eq. 19, we obtain

$$\left[\hat{\Gamma}_{fi}L_z \ \hat{H}_{fi}\right] = R_{21,i}R_{11,i}^+ \quad (31)$$

Notice that the QR decomposition in Eq. 30 can be implemented recursively. Given that the ith QR decomposition is performed, the next QR decomposition can be performed on

$$\begin{bmatrix} Z_p \\ U_{i+1} \\ Y_{f,i+1} \end{bmatrix} = \begin{bmatrix} Z_p \\ U_i \\ U_{f,i+1} \\ Y_{f,i+1} \end{bmatrix} = \begin{bmatrix} R_{11,i}Q_{1,i} \\ U_{f,i+1} \\ Y_{f,i+1} \end{bmatrix} \quad (32)$$

which is in QR decomposition form except for the last two rows.

3.2.3. *QR Implementation for K* Once $\hat{\Gamma}_f$ is known in Step 2 of Algorithm 1, the Kalman filter gain K can be estimated similar to Wang and Qin (Wang and Qin, 2002).

With large p, substituting Eq. 14 into Eq. 2 leads to:

$$Y_f = \Gamma_f L_z Z_p + H_f U_f + G_f E_f \quad (33)$$

Therefore,

$$Y_f \Pi^{\perp}_{\begin{bmatrix} Z_p \\ U_f \end{bmatrix}} = G_f E_f \Pi^{\perp}_{\begin{bmatrix} Z_p \\ U_f \end{bmatrix}} = G_f E_f \quad (34)$$

since E_f is not correlated with Z_p and U_f in open-loop.

Performing QR decomposition,

$$\begin{bmatrix} Z_p \\ U_f \\ Y_f \end{bmatrix} = \begin{bmatrix} R_{11} & & \\ R_{21} & R_{22} & \\ R_{31} & R_{32} & R_{33} \end{bmatrix} \begin{bmatrix} Q_1 \\ Q_2 \\ Q_3 \end{bmatrix} \quad (35)$$

then

$$R_{33} Q_3 = G_f E_f \quad (36)$$

Denoting $e_k = F e_k^*$ such that $cov(e_k^*) = I$,

$$G_f E_f = (G_f \otimes F) E_f^* \triangleq G_f^* E_f^* \quad (37)$$

where $\otimes$ denotes Kronecker product and

$$G_f^* = \begin{bmatrix} F & 0 & \cdots & 0 \\ CKF & F & \cdots & 0 \\ \vdots & \vdots & \ddots & \vdots \\ CA^{f-2}KF & CA^{f-3}KF & \cdots & F \end{bmatrix} \in \Re^{m.f \times m.f} \quad (38)$$

From Eqs. 36 and 37 and using the fact that Q_3 is an orthonormal matrix, we choose

$$\hat{E}_f^* = Q_3 \quad (39a)$$
$$\hat{G}_f^* = R_{33} \quad (39b)$$

Therefore,

$$\hat{F} = R_{33}(1:m, 1:m) \quad (40)$$

and K can be calculated from G_f^* using Γ_f.

3.2.4. *Optimal Estimation for B and D* With A and C estimates, Section 10.6 in (Ljung, 1999) gives an effective approach to estimate B and D with an output error formulation. Here we give a modified approach to estimating B, D optimally using A, C, K and F for the general innovation form.

From the innovation form of the system we have:

$$x_{k+1} = A_K\, x_k + B_K\, u_k + K\, y_k \quad (41)$$

The process output can be represented as

$$y_k = C(qI - A_K)^{-1} x_0 + [C(qI - A_K)^{-1} B_K + D] u_k + C(qI - A_K)^{-1} K y_k + e_k \quad (42)$$

or:

$$[I - C(qI - A_K)^{-1} K] y_k = C(qI - A_K)^{-1} x_0 + [C(qI - A_K)^{-1} B_K + D] u_k + e_k \quad (43)$$

using $e_k = F e_k^*$ where e_k^* has an identity covariance matrix, and defining

$$\tilde{y}_k = F^{-1}[I - C(qI - A_K)^{-1} K] y_k \quad (44a)$$
$$G(q) = F^{-1} C(qI - A_K)^{-1} \quad (44b)$$
$$D^* = F^{-1} D \quad (44c)$$

we obtain,

$$\begin{aligned} \tilde{y}_k &= G(q) B_K u_k + D^* u_k + G(q) x_0 \delta_k + e_k^* \\ &= G(q) \otimes u_k^T\, vec(B_K) + I_m \otimes u_k^T\, vec(D^*) \\ &\quad + G(q) x_0 \delta_k + e_k^* \end{aligned} \quad (45)$$

where $vec(B_K)$ and $vec(D^*)$ are vectorized B_K and D^* matrices along the rows. δ_k is the Kronecker delta function. Now $vec(B_K)$, $vec(D^*)$ and x_0 can be estimated using least squares from the above equation. The B, D matrices can be backed out as:

$$\hat{D} = F \hat{D}^* \quad (46a)$$
$$\hat{B} = \hat{B}_K + K \hat{D} \quad (46b)$$

3.2.5. *PARSIM-P with QR Implementation* We summarize the above procedure into following algorithm that implements PARSIM-P with QR decomposition.

[Algorithm 2] PARSIM-P with QR decomposition

(1) Calculate $\Gamma_{fi} L_z$ as in Eq. 31 via a bank of QR decompositions for $i = 1, 2, \cdots, f$. Form $\Gamma_f L_z$ and calculate A and C estimates from SVD of $W_1 \hat{\Gamma}_f L_z W_2$ as in Algorithm 1.

(2) Find the Kalman filter gain K and matrix F using A and C estimates.
(3) Calculate B, D estimates from A, C, K and F estimates.

Using these three steps, it is straightforward to show that the estimates are consistent if the prior steps are consistent.

4. SIMULATION RESULTS

The counter example proposed in (Jansson and Wahlberg, 1998) is used here to test the effectiveness of the proposed parallel parsimonious method. The case of $K = [-0.21 \quad -0.559]^T$ is used here. For comparison any standard SIM algorithms can be used, but we choose the MOESP algorithm in (Verhaegen, 1994). Figure 1 shows the poles and zero estimates using the PARSIM-P algorithm and MOESP. In this simulation, we choose N=2000, f=3, p=5. It can be seen from the results that the PARSIM-P gives better estimates of the poles than the MOESP algorithm, which knows the benefit of the parsimonious formulation. The zero estimates from the PARSIM-P is much better than the MOESP estimates.

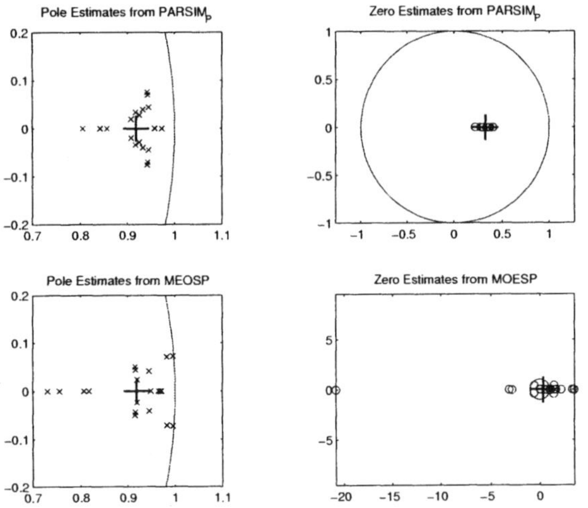

Fig. 1. Poles and zero estimates for the counter example.

5. CONCLUSIONS

The conventional subspace identification models are non-parsimonious with extra non-causal input terms. The extra terms in the models are responsible for large variance in SIM estimates and partially responsible for the loss of closed-loop identifiability. The proposed parallel subspace identification algorithm overcomes these problems.

ACKNOWLEDGMENTS

Financial support from National Science Foundation under CTS-9985074 and a Faculty Research Assignment grant from University of Texas is gratefully acknowledged.

6. REFERENCES

Chou, C.T. and Michel Verhaegen (1997). Subspace algorithms for the identification of multivariable dynamic errors-in-variables models. *Automatica* **33**(10), 1857–1869.

Gustafsson, Tony (2002). Subspace-based system identification: weighting and pre-filtering of instruments. *Automatica* **38**, 433–443.

Jansson, Magnus and Bo Wahlberg (1998). On consistency of subspace methods for system identification. *Automatica* **34**(12), 1507–1519.

Knudsen, Torben (2001). Consistency analysis of subspace identification methods based on a linear regression approach. *Automatica* **37**, 81–89.

Ljung, L. (1999). *System Identification: Theory for the User.* Prentice-Hall, Inc.. Englewood Cliffs, New Jersey.

Overschee, Peter Van and Bart De Moor (1996). *Subspace Identification for Linear Systems.* Kluwer Academic Publishers.

Qin, S.J., J. Wang and L. Ljung (2002). Subspace identification methods using parsimonious model formulation. In: *AIChE Annual Meeting.* Indianapolis, IN.

Verhaegen, Michel (1994). Identification of the deterministic part of MIMO state space models given in innovations form from input-output data. *Automatica* **30**(1), 61–74.

Wang, Jin and S. Joe Qin (2001). Principal component analysis for errors-in-variables subspace identification. In: *Proc. of the 40th IEEE Conf. On Decision and Control.* Orlando, FL. pp. 3936–3941.

Wang, Jin and S. Joe Qin (2002). A new subspace identification approach based on principal component analysis. *J. Proc. Cont.* **12**, 841–855.

APPENDIX A

Condition 1 implies the initial state has zero effect on the estimate since A_K is always stable. From Condition 2 we have $\frac{1}{N}E_i Z_p^T \rightarrow 0$ as $N \rightarrow \infty$, that is, the effect of noise is zero under condition 2. It is straightforward to show that Condition 3 is need for L_z to have full row rank and Γ_f full column rank.

IFAC Publications
www.elsevier.com/locate/ifac

A NEW RECURSIVE METHOD FOR SUBSPACE IDENTIFICATION OF NOISY SYSTEMS : EIVPM

Guillaume Mercère [*,**] **Stéphane Lecoeuche** [*,***]
Christian Vasseur [*]

[*] *Laboratoire I^3D - Bâtiment P2, USTL - 59655 Villeneuve d'Ascq - France*
[**] *EIPC, Campus de la Malassise - BP39 - 62967 Longuenesse Cedex -France - gmercere@eipc.fr*
[***] *Ecole des Mines de Douai - Rue Charles Bourseul - BP 838 - 59508 Douai Cedex -France*

Abstract: In this article, a new recursive identification method based on subspace algorithms is proposed. This method is directly inspired by the Propagator Method used in sensor array signal processing to estimate directions of arrival (DOA) of waves impinging an antenna array. Particularly, a new quadratic criterion and a recursive formulation of the estimation of the subspace spanned by the observability matrix are presented. The problem of process and measurement noises is solved by introducing an instrumental variable within the minimized criterion. *Copyright © 2003 IFAC*

Keywords: Identification algorithms; Recursive estimation; State-space models; Subspace methods; Subspace tracking; Quadratic criterion

1. INTRODUCTION

Subspace state-space system identification (4SID) methods have been particularly developed during the last two decades (Moonen *et al.*, 1989; Verhaegen, 1994; Van Overschee and De Moor, 1996). These techniques are well adapted to identify a state-space representation of a multivariable time-invariant system from measured I/O data. Contrary to PEM (Ljung, 1999), they don't require any nonlinear regression. They only need few structural parameters and are based on robust numerical tools such as QR factorization and Singular Value Decomposition (SVD). However, these tools, which are appropriate for off-line identification, are difficult to implement on-line due to their computational complexity. In fact, in many on-line identification scenarios, it is desirable to update the model as time goes on with the minimal computational cost. Recent researches on recursive algorithms for subspace identification have been developed to avoid the application of the SVD.

The most convincing on-line results (Gustafsson, 1997; Lovera, 1998; Oku and Kimura, 2002) have been obtained by introducing some techniques used in signal processing to find the directions of arrivals (DOA). One goal of this field of research is the conception of low computational cost adaptative algorithms for the location of moving sources. The method mainly used in recursive subspace identification is the Yang's algorithm PAST (Projection Approximation Subspace Tracking) (Yang, 1995). In this case, a minimization problem is considered with respect to a fourth-order cost function so as to retrieve the signal subspace from I/O data. A projection approximation is introduced in order to reduce the minimization task to an exponentially weighted least-square problem (Yang, 1995). Thus, recursive least-squares (RLS) methods can be used to find the signal subspace.

Unfortunately, these recursive methods are based on an approximation. In order to overcome this imprecision, a new quadratic cost function is proposed in this paper. The minimization of this second-order criterion leads to the subspace spanned by the observability matrix (i.e. the signal subspace). This new algorithm is inspired from the Propagator Method (PM) (Munier and Delisle, 1991). This technique proposes the decomposition of the observability matrix into two parts which can be easily extracted from the data. Thus, the properties of the linear propagator operator are adapted to estimate the signal subspace.

In their original form, the linear methods such as PAST or PM need strong hypotheses on the disturbances acting on the studied process. In fact, it must be assumed that the noise-covariance matrix is proportional to the identity. During any problem of identification, the reachable I/O data are most of the time disturbed at the same moment by measurement and process noises. The obtaining of reliable estimations from such data can only be realized if the effects of the noise terms are removed. The proposed solution to treat this problem is to introduce an instrumental variable in the minimized criterion in order to eliminate the effect of perturbations without losing the informations contained in the I/O data.

2. BACKGROUND OF THE SUBSPACE RECURSIVE IDENTIFICATION TECHNIQUES

2.1 Problem formulation and notations

Consider a nth order causal linear time-invariant state-space model with l undisturbed outputs and m undisturbed inputs, collected respectively in $\tilde{\mathbf{y}}(t)$ and $\tilde{\mathbf{u}}(t)$:

$$\begin{cases} \mathbf{x}(t+1) = \mathbf{A}\mathbf{x}(t) + \mathbf{B}\tilde{\mathbf{u}}(t) + \mathbf{w}(t) \\ \tilde{\mathbf{y}}(t) = \mathbf{C}\mathbf{x}(t) + \mathbf{D}\tilde{\mathbf{u}}(t) \end{cases} \tag{1}$$

where $\mathbf{w}(t) \in \mathbb{R}^n$ is the process noise. The measured input and output signals are modeled as:

$$\begin{aligned} \mathbf{u}(t) &= \tilde{\mathbf{u}}(t) + \mathbf{f}(t) \\ \mathbf{y}(t) &= \tilde{\mathbf{y}}(t) + \mathbf{v}(t) \end{aligned} \tag{2}$$

with $\mathbf{f}(t) \in \mathbb{R}^m$ and $\mathbf{v}(t) \in \mathbb{R}^l$ the measurement noises. All these three noises are assumed to be zero-mean white noise and statistically independent of the past noise-free input $\tilde{\mathbf{u}}$. Furthermore, the measurement noises $\mathbf{f}(t)$ and $\mathbf{v}(t)$ are assumed to be independent of the state $\mathbf{x}(t)$.

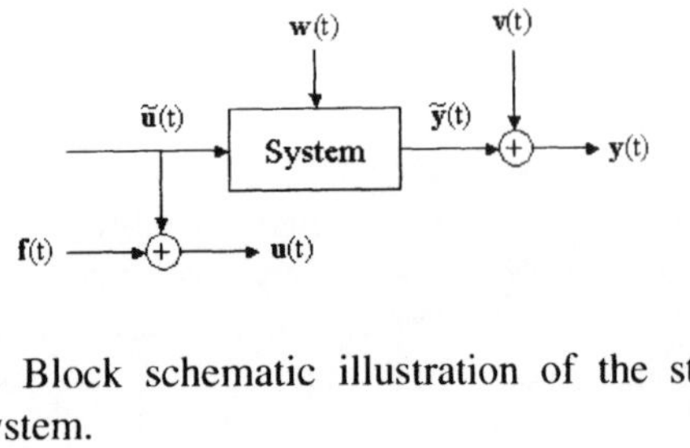

Fig. 1. Block schematic illustration of the studied system.

The problem is to estimate *recursively* a state-space realization from the updates of the disturbed I/O data $\mathbf{u}(t)$ and $\mathbf{y}(t)$. For that purpose, introduce the stacked $l\alpha \times 1$ output vector:

$$\mathbf{y}_\alpha(t) = [\mathbf{y}^T(t) \cdots \mathbf{y}^T(t+\alpha-1)]^T \tag{3}$$

where $\alpha > n$ is a user-defined integer. From the state-space representation (1) and (2), it is easy to show recursively that the output vector verifies (the stacked input and noise vectors are defined in the same way than $\mathbf{y}_\alpha(t)$):

$$\begin{aligned} \mathbf{y}_\alpha(t) = &\boldsymbol{\Gamma}_\alpha \mathbf{x}(t) + \boldsymbol{\Phi}_\alpha \mathbf{u}_\alpha(t) \\ &\underbrace{-\boldsymbol{\Phi}_\alpha \mathbf{f}_\alpha(t) + \boldsymbol{\Psi}_\alpha \mathbf{w}_\alpha(t) + \mathbf{v}_\alpha(t)}_{\mathbf{b}_\alpha(t)} \end{aligned} \tag{4}$$

where $\boldsymbol{\Gamma}_\alpha \in \mathbb{R}^{l\alpha \times n}$ is the observability matrix of the system:

$$\boldsymbol{\Gamma}_\alpha = \left[\mathbf{C}^T (\mathbf{CA})^T \cdots (\mathbf{CA}^{\alpha-1})^T\right]^T \tag{5}$$

$\boldsymbol{\Phi}_\alpha \in \mathbb{R}^{l\alpha \times m}$ and $\boldsymbol{\Psi}_\alpha \in \mathbb{R}^{l\alpha \times n}$ are two lower triangular Toeplitz matrices. Furthermore, introduce a notation useful in the following sections :

$$\mathbf{z}_\alpha(t) = \mathbf{y}_\alpha(t) - \boldsymbol{\Phi}_\alpha \mathbf{u}_\alpha(t) = \boldsymbol{\Gamma}_\alpha \mathbf{x}(t) + \mathbf{b}_\alpha(t) \tag{6}$$

The first stage of subspace methods is typically the extraction of the observability matrix from I/O data. Indeed, from this matrix, it is possible to extract the state-space matrices of the system (Van Overschee and De Moor, 1996). In relation (4), the extended observability matrix spans a n-dimensional signal subspace which focuses on the informations about the direct relation between the output and state vectors. The various subspace techniques differ in the way the observability matrix is estimated. The extraction of a low dimensional subspace from a large space is, most of the time, performed by an SVD. This fundamental step is the main problem of recursive 4SID formulation. In fact, the SVD is computationally burdensome to update in an adaptative algorithm. The first researches on this topic suggested to perform the SVD on compressed I/O data matrices in order to decrease the computational load (Verhaegen and Deprettere, 1991; Cho *et al.*, 1994). However, it was always necessary to apply the SVD at every update. An efficient method to avoid this mathematical tool in system identification was firstly proposed in (Gustafsson, 1997). Under the assumption that the order of the system is a priori known, T. Gustafsson suggested to update an estimate of the extended observability matrix thanks to the use of an algorithm named PAST.

2.2 Review of PAST

PAST (Projection Approximation Subspace Tracking) was originally introduced into array signal processing by Yang (Yang, 1995). In order to correctly understand the principle of this method, it is useful to present the context of the sensor array signal processing. The sensor array signal model can be written as:

$$\mathbf{z}(t) = \boldsymbol{\Gamma}(\theta)\mathbf{s}(t) + \mathbf{b}(t) \tag{7}$$

where $\mathbf{z}(t) \in \mathbb{C}^{M\times 1}$ is the output of the M sensors of the antenna at time t, $\mathbf{\Gamma}(\theta) \in \mathbb{C}^{M\times n}$ the steering matrix for a direction of arrival vector θ, $\mathbf{s}(t) \in \mathbb{C}^{n\times 1}$ the vector of the n ($n < M$) random impinging waves and $\mathbf{b}(t) \in \mathbb{C}^{M\times 1}$ the measurement noise. B. Yang proposed an unconstrained cost function to estimate the range of $\mathbf{\Gamma}(\theta)$ ($(.)^H$ denotes the Hermitian transposition, $\|.\|$ the Euclidean vector norm and $E[.]$ the expectation operator):

$$J(\mathbf{W}) = E\|\mathbf{z} - \mathbf{W}\mathbf{W}^H\mathbf{z}\|^2 \tag{8}$$

and proved that it has neither local minimum nor maximum except for a unique global minimum which corresponds to the signal subspace. Thus, the minimisation of (8) provides an expression of $\mathbf{\Gamma}(\theta)$ in a particular basis. In order to update this matrix recursively, it was proposed to replace the expectation operator in (8) with an exponentially weighted sum:

$$J(\mathbf{W}(t)) = \sum_{i=1}^{t} \lambda^{t-i}\|\mathbf{z}(i) - \mathbf{W}(t)\mathbf{W}^H(t)\mathbf{z}(i)\|^2 \tag{9}$$

where λ is a forgetting factor ($0 < \lambda \leq 1$) ensuring that past data are downweighted (tracking capability). The key idea of PAST is to approximate $\mathbf{W}(t)^H\mathbf{z}(i)$ in (9) with:

$$\mathbf{h}(i) = \mathbf{W}^H(i-1)\mathbf{z}(i) \tag{10}$$

This so-called projection approximation lets the criterion (9) be quadratic in $\mathbf{W}(t)$:

$$\bar{J}(\mathbf{W}(t)) = \sum_{i=1}^{t} \lambda^{t-i}\|\mathbf{z}(i) - \mathbf{W}(t)\mathbf{h}(i)\|^2 \tag{11}$$

Assuming that the covariance of the measurement noise $\mathbf{b}(t)$ is proportional to the identity matrix, it is also possible to estimate the signal subspace by using efficient recursive tools like RLS algorithms.

T. Gustafsson (1997) was the first to use this criterion in order to estimate the subspace spanned by the observability matrix. In fact, assuming that an estimate $\hat{\mathbf{\Phi}}_\alpha(t-1)$ of the Toeplitz matrix $\mathbf{\Phi}_\alpha$ can be constructed, in the same way than (4), the following approximation can be considered:

$$\check{\mathbf{z}}_\alpha(t) = \mathbf{y}_\alpha(t) - \hat{\mathbf{\Phi}}_\alpha(t-1)\mathbf{u}_\alpha(t) \simeq \mathbf{\Gamma}_\alpha\mathbf{x}(t) + \mathbf{b}_\alpha(t) \tag{12}$$

Then, the terms of (12) can easily be connected with those of (7) (cf. tab. 1). Thus, assuming that the noise-covariance of $\mathbf{b}_\alpha$ is proportional to the identity matrix (e.g. $\mathbf{f}(t) = \mathbf{w}(t) = 0$ and $\mathrm{Cov}(\mathbf{v}) = \sigma^2\mathbf{I}$), it is possible to apply the PAST criterion to (12) so as to estimate the observability matrix.

Table 1. Relations between signal processing and subspace identification

$\mathbf{z}_\alpha(t) \in \mathbb{R}^{l\alpha\times 1}$	$\mathbf{z}(t) \in \mathbb{C}^{M\times 1}$
$\mathbf{\Gamma}_\alpha \in \mathbb{R}^{l\alpha\times n}$	$\mathbf{\Gamma}(\theta) \in \mathbb{C}^{M\times n}$
$\mathbf{x}(t) \in \mathbb{R}^{n\times 1}$	$\mathbf{s}(t) \in \mathbb{C}^{n\times 1}$
$\mathbf{b}_\alpha(t) \in \mathbb{R}^{l\alpha\times 1}$	$\mathbf{b}(t) \in \mathbb{C}^{M\times 1}$

With the PAST criterion, the signal subspace is derived by minimizing the modified cost function $\bar{J}(\mathbf{W})$ instead of $J(\mathbf{W})$. Hence, the estimated column-subspace is slightly different from the one reachable with the original cost function. Theoretically, the columns of $\mathbf{W}$ minimizing the criterion $J(\mathbf{W})$ are orthonormal. Even if this property is not necessary to extract the state-space matrices, the minimization of $\bar{J}(\mathbf{W})$ leads to a matrix having columns that are not exactly orthonormal. This property evolves during the recursive minimization since, under some conditions, the minimizer of $\bar{J}(\mathbf{W})$ converges to a matrix with orthonormal columns (Yang, 1995). This evolution can be interpreted as a slow change of basis which does not guarantee that $\mathbf{\Gamma}_\alpha(t)$ and $\mathbf{\Gamma}_\alpha(t-1)$ are in the same state-space basis. This corruption could pose a problem during the extraction of the estimates $\hat{\mathbf{A}}$, $\hat{\mathbf{B}}$, $\hat{\mathbf{C}}$ and $\hat{\mathbf{D}}$. In order to avoid these difficulties due to the use of an approximate algorithm, a new quadratic criterion is proposed in the next section.

3. A NEW QUADRATIC SUBSPACE TRACKING CRITERION

The new proposed cost function is inspired by a signal processing method for source bearing estimation. This technique, named the Propagator Method (PM) (Munier and Delisle, 1991), is used to find the direction of arrival without requiring the eigen-decomposition of the cross-spectral matrix of the received signals (Marcos *et al.*, 1995).

3.1 The Propagator Method

The propagator is a linear operator which provides the decomposition of the observation space into a noise subspace and a source subspace. In order to well understand the key point of this method, consider again the sensor array signal model (7). Assuming that the steering matrix $\mathbf{\Gamma}$ is of full rank, n rows of $\mathbf{\Gamma}$ are linearly independent, the others being expressed as a linear combination of these n rows. Under this hypothesis and after a reorganisation of the sensors outputs so that the first n rows of $\mathbf{\Gamma}$ are linearly independent, it is possible to partition the steering matrix according to:

$$\mathbf{\Gamma} = \begin{bmatrix} \mathbf{\Gamma}_1 \\ \mathbf{\Gamma}_2 \end{bmatrix} \begin{matrix} \}\in \mathbb{C}^{n\times n} \\ \}\in \mathbb{C}^{M-n\times n} \end{matrix} \tag{13}$$

The propagator is the unique linear operator $\mathbf{P} \in \mathbb{C}^{n\times M-n}$ defined as follows:

$$\mathbf{\Gamma}_2 = \mathbf{P}^H\mathbf{\Gamma}_1 \tag{14}$$

From the works of S. Marcos et al. (1995) on the source bearing estimation, on the basis of the similarity between (6) and (7) (cf. tab. 1) and under the hypothesis that $\{\mathbf{A}, \mathbf{C}\}$ is observable, the extended observability matrix can be decomposed in the following way:

$$\mathbf{\Gamma}_\alpha = \begin{bmatrix} \mathbf{\Gamma}_{\alpha_1} \\ \mathbf{\Gamma}_{\alpha_2} \end{bmatrix} \begin{matrix} \}\in \mathbb{R}^{n\times n} \\ \}\in \mathbb{R}^{l\alpha-n\times n} \end{matrix} \tag{15}$$

where $\mathbf{\Gamma}_{\alpha_1}$ is the block of the n independent rows and $\mathbf{\Gamma}_{\alpha_2}$ the matrix of the $l\alpha - n$ others. Thus, there is a unique $\mathbf{P} \in \mathbb{R}^{n \times l\alpha - n}$ such as:

$$\mathbf{\Gamma}_{\alpha_2} = \mathbf{P}^T \mathbf{\Gamma}_{\alpha_1} \tag{16}$$

It is also easy to verify that:

$$\mathbf{\Gamma}_{\alpha} = \begin{bmatrix} \mathbf{\Gamma}_{\alpha_1} \\ \mathbf{\Gamma}_{\alpha_2} \end{bmatrix} = \begin{bmatrix} \mathbf{I}_n \\ \mathbf{P}^T \end{bmatrix} \mathbf{\Gamma}_{\alpha_1} = \mathbf{Q}_s \mathbf{\Gamma}_{\alpha_1} \tag{17}$$

This means that the columns of $\mathbf{\Gamma}_{\alpha}$ are linear combinations of the columns of $\mathbf{Q}_s$. Thus:

$$\text{span}\{\mathbf{Q}_s\} = \text{span}\{\mathbf{\Gamma}_{\alpha}\} \tag{18}$$

The new quadratic criterion proposed in this paper is based on this property (18). Indeed, this equality implies, from the knowledge of the propagator $\mathbf{P}$, the ability to find an expression of the observability matrix in a particular basis. Thus, assuming that the order n is known, an estimation of the subspace spanned by the observability matrix is available by estimating $\mathbf{P}$. For that purpose, consider the equation (6). After an initial reorganisation such that the first n rows of $\mathbf{\Gamma}_{\alpha}$ are linearly independent, the following partition can be introduced:

$$\mathbf{z}_{\alpha}(t) = \mathbf{\Gamma}_{\alpha}\mathbf{x}(t) + \mathbf{b}_{\alpha}(t) = \begin{bmatrix} \mathbf{z}_{\alpha_1}(t) \\ \mathbf{z}_{\alpha_2}(t) \end{bmatrix} \begin{matrix} \} \in \mathbb{R}^{n \times 1} \\ \} \in \mathbb{R}^{l\alpha - n \times 1} \end{matrix} \tag{19}$$

In the no-noise case, it is easy to show that:

$$\mathbf{z}_{\alpha_2} = \mathbf{P}^T \mathbf{z}_{\alpha_1} \tag{20}$$

In the presence of noise, this relation holds no longer. An estimation of $\mathbf{P}$ can however be obtained by minimizing the following cost function:

$$J(\hat{\mathbf{P}}) = E\|\mathbf{z}_{\alpha_2} - \hat{\mathbf{P}}^T \mathbf{z}_{\alpha_1}\|^2 \tag{21}$$

the unicity of $\hat{\mathbf{P}}$ being ensured thanks to the convexity of this criterion.

The criterion (21) is, by definition, quadratic and reduces the determination of the range of $\mathbf{\Gamma}_{\alpha}$ to the estimation of an $l\alpha - n \times n$ matrix. Its minimization is possible by using techniques such as RLS or TLS algorithms (Ljung, 1999). However, these techniques give biased estimates when the noise $\mathbf{b}_{\alpha}(t)$ is not white. In order to obtain an unbiased estimation of the signal subspace, an instrumental variable is introduced in the previous criterion so as to be applicable even if colored disturbances act on the system, as done by Gustafsson (1997).

3.2 *Instrumental variable tracking*

Since the covariance matrix of the noise $\mathbf{b}_{\alpha}$ is rarely proportional to the identity matrix, it is necessary to modify the criterion (21) so as to be applicable with colored disturbances. This correction is realized by introducing an instrumental variable $\boldsymbol{\xi}(t) \in \mathbb{R}^{\gamma \times 1}$ ($\gamma \geq n$) in (21), assumed not be correlated with the noise but enough correlated with $\mathbf{x}(t)$:

$$J_{IV}(\hat{\mathbf{P}}) = E\|\mathbf{z}_{\alpha_2}\boldsymbol{\xi}^T - \hat{\mathbf{P}}^T \mathbf{z}_{\alpha_1}\boldsymbol{\xi}^T\|^2 \tag{22}$$

Since only a finite number of data are accessible in practice, replacing the expectation operator with a finite exponentially weighted sum, the previous criterion becomes:

$$J_{IV}(\hat{\mathbf{P}}(t)) = \sum_{i=1}^{t} \lambda^{t-i} \|\mathbf{z}_{\alpha_2}(i)\boldsymbol{\xi}^T(i) - \hat{\mathbf{P}}^T(t)\mathbf{z}_{\alpha_1}(i)\boldsymbol{\xi}^T(i)\|^2 \tag{23}$$

Assuming an instrumental variable can be constructed, the minimization of (23) can be realized by two different manners according to the number of instruments in $\boldsymbol{\xi}$. Indeed, if $\gamma = n$, using the matrix inversion lemma, it is possible to find a first IVPM algorithm:

$$\mathbf{K}(t) = \frac{\boldsymbol{\xi}^T(t)\mathbf{R}(t-1)}{\lambda + \boldsymbol{\xi}^T(t)\mathbf{R}(t-1)\mathbf{z}_{\alpha_1}(t)} \tag{24a}$$

$$\mathbf{P}^T(t) = \mathbf{P}^T(t-1) + [\mathbf{z}_{\alpha_2}(t) - \mathbf{P}^T(t-1)\mathbf{z}_{\alpha_1}(t)]\mathbf{K}(t) \tag{24b}$$

$$\mathbf{R}(t) = \frac{1}{\lambda}[\mathbf{R}(t-1) - \mathbf{R}(t-1)\mathbf{z}_{\alpha_1}(t)\mathbf{K}(t)] \tag{24c}$$

where $\mathbf{R}(t) = \{E[\mathbf{z}_{\alpha_1}(t)\boldsymbol{\xi}^T(t)]\}^{-1} = \mathbf{C}^{-1}_{\mathbf{z}_{\alpha_1}\boldsymbol{\xi}}(t)$.

In (Söderström and Stoica, 1989), it was proved that the accuracy of the estimate obtained from an instrumental variable method increases with the number of instruments. It would be interesting to improve the efficiency of the previous algorithm by increasing the number of used instruments (i.e. $\gamma > n$). In that case, the minimization of the criterion (23) appeals to a technique named the Extended Instrumental Method (Friedlander, 1984). The application of such a technique gives the recursive updating formulae of appendix A, named EIVPM, the main step being:

$$\mathbf{P}^T(t) = \mathbf{P}^T(t-1) + (\mathbf{g}(t) - \mathbf{P}^T(t-1)\mathbf{\Psi}(t))\mathbf{K}(t) \tag{25}$$

The complexity of this new algorithm is comparable to the one of PAST or its by-products.

Remark 1. From the second iteration, the updating of the estimated subspace has always the same writing:

$$\hat{\mathbf{\Gamma}}_{\alpha} = \begin{bmatrix} \mathbf{I}_n \\ \hat{\mathbf{P}}^T \end{bmatrix} \tag{26}$$

This means that, after a short transient period, the recursive estimation is made in the same state-space basis. This property is an important asset for the extraction of the state-space matrices.

4. RECURSIVE IDENTIFICATION SCENARIO AND INSTRUMENTAL VARIABLE CHOICE

In this section are presented the main stages of the recursive update of the state-space matrices. First of all, assume that an estimate of the system matrices is available at time $t-1$. It is then possible to build an estimate of the matrix $\mathbf{\Phi}_{\alpha}(t-1)$ by noticing, for example, that the first block column of $\hat{\mathbf{\Phi}}_{\alpha}(t-1)$ expresses itself simply according to $\hat{\mathbf{B}}(t-1)$, $\hat{\mathbf{D}}(t-1)$ and $\hat{\mathbf{\Gamma}}_{\alpha}(t-1)$, the other blocks being simple partitions of this one. From this estimate, the approximation $\check{\mathbf{z}}_\alpha(t)$ (cf. eq. (12)) can be constructed.

The following step is the setting up of the instrumental variable. Since the available data are temporal data, the easiest way to get a good instrumental variable is to use delayed I/O data. As far as figure 1 is concerned, the instruments must be correlated with present and future data $\tilde{\mathbf{y}}_\alpha$ and/or $\tilde{\mathbf{u}}_\alpha$ but uncorrelated with $\mathbf{w}_\alpha$, $\mathbf{f}_\alpha$ and $\mathbf{v}_\alpha$. Introduce the following notation:

$$\mathbf{u}_\beta(t) = \left[\mathbf{u}^T(t-\beta-\mu) \ \cdots \ \mathbf{u}^T(t-1-\mu) \right] \quad (27)$$

where β and μ are user-defined integers. This variable, as $\mathbf{y}_\beta$, represents the available past I/O data from which the instrument vector can be constructed. These data are not noise-free. It is then necessary to study the correlation between the past and future reachable signals to focus on the informative data uncorrelated with the disturbances. By considering table 2, the

Table 2. Correlation between the available I/O data.

	$\tilde{\mathbf{u}}_\beta$	$\mathbf{f}_\beta$	$\mathbf{w}_\beta$	$\tilde{\mathbf{y}}_\beta$	$\mathbf{v}_\beta$
$\tilde{\mathbf{u}}_\alpha$	0	0	0	0	0
$\mathbf{f}_\alpha$	0	0	0	0	0
$\mathbf{w}_\alpha$	0	0	0	0	0
$\tilde{\mathbf{y}}_\alpha$	✓	0	✓	✓	0
$\mathbf{v}_\alpha$	0	0	0	0	0

only past data uncorrelated with present and future disturbances but correlated with the informative signal $\tilde{\mathbf{y}}_\alpha$ is the past measured input vector $\mathbf{u}_\beta \equiv \{\tilde{\mathbf{u}}_\beta, \mathbf{f}_\beta\}$. Thus, the choice of the instrumental variable is:

$$\boldsymbol{\xi}(t) = \mathbf{u}_\beta(t) \quad (28)$$

where β is chosen so that $\gamma = m\beta > n$ and μ so that the temporal correlation length of the instrumental variable is larger than that of the noise.

By feeding the EIVPM algorithm (cf. Appendix A) with $\check{\mathbf{z}}_\alpha(t)$ and $\boldsymbol{\xi}(t)$, a new estimation of the observability matrix is reached. Then, the last stage consists in the extraction of system-matrices. Firstly, the estimation of $\hat{\mathbf{A}}(t)$ and $\hat{\mathbf{C}}(t)$ is made as in the non-recursive case thanks to the shift-invariance structure of $\boldsymbol{\Gamma}_\alpha$. Given the estimates $\hat{\mathbf{A}}(t)$ and $\hat{\mathbf{C}}(t)$, the matrices $\hat{\mathbf{B}}(t)$ and $\hat{\mathbf{D}}(t)$ can be obtained from a linear regression problem ((Van Overschee and De Moor, 1996) chap. 4), the second part of the regressor being calculable recursively (e.g. (Lovera, 1998)).

5. EXAMPLES

In this section, the results of a numerical simulation are presented to illustrate the performances of the new recursive algorithm EIVPM compared with the extended instrumental version of the PAST method named EIVPAST (Gustafsson, 1997).

Consider the following fourth-order MIMO system found in the appendix diskette of (Van Overschee and De Moor, 1996) (`sta_demo.m`):

$$\mathbf{x}_{k+1} = \begin{bmatrix} 0.603 & 0.603 & 0 & 0 \\ -0.603 & 0.603 & 0 & 0 \\ 0 & 0 & -0.603 & -0.603 \\ 0 & 0 & 0.603 & -0.603 \end{bmatrix} \mathbf{x}_k + \begin{bmatrix} 1.1650 & -0.6965 \\ 0.6268 & 1.6961 \\ 0.0751 & 0.0591 \\ 0.3516 & 1.7971 \end{bmatrix} \tilde{\mathbf{u}}_k + \begin{bmatrix} 0.1242 & -0.0895 \\ -0.0828 & -0.0128 \\ 0.0390 & -0.0968 \\ -0.0225 & 0.1459 \end{bmatrix} \mathbf{e}_k \quad (29)$$

$$\tilde{\mathbf{y}}_k = \begin{bmatrix} 0.2641 & -1.4462 & 1.2460 & 0.5774 \\ 0.8717 & -0.7012 & -0.6390 & -0.3600 \end{bmatrix} \mathbf{x}_k + \begin{bmatrix} -0.1356 & -1.2704 \\ -1.3493 & 0.9846 \end{bmatrix} \tilde{\mathbf{u}}_k \quad (30)$$

$$\mathrm{Cov}(\mathbf{e}) = \begin{bmatrix} 0.0176 & -0.0267 \\ -0.0267 & 0.0497 \end{bmatrix} \quad (31)$$

To be in the case of figure 1, measurement noises are added to the undisturbed I/O data. Thus, the input $\tilde{\mathbf{u}}$, a Gaussian white noise sequence with variance 1, is contaminated with a zero-mean white noise sequence $\mathbf{f}$ with variance 0.64. The output $\tilde{\mathbf{y}}$ is also disturbed by a Gaussian white measurement noise such as the signal-to-noise ratio equals $16dB$. Now, let the initial system matrices be randomly generated under the constraint that the absolute value of the maximum eigenvalue of $\hat{\mathbf{A}}(0)$ is less than 1 (stability requirement). To reduce the effect of this random initialization, consider the following forgetting factor:

$$\lambda(t) = \min\{\lambda_0 \lambda(t-1) + 1 - \lambda_0, \lambda_{final}\} \quad (32)$$

For this example, $\lambda_0 = \lambda(t=0) = 0.99$ and $\lambda_{final} = 0.999$. Furthermore, the following user-defined quantities are applied: $\alpha = \beta = 6$ and $\mu = 1$.

In figure 2, the eigenvalues trajectories obtained respectively with the new EIVPM algorithm and EIVPAST are presented. In both cases, the eigenvalue estimations are close to the true values.

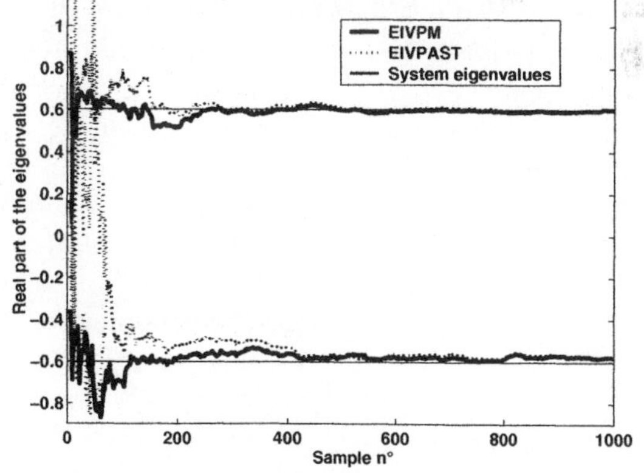

Fig. 2. Estimated real part of the eigenvalues using the new EIVPM algorithm and EIVPAST.

Even if both results seems to be equally accurate (both methods present the same behaviour after 500 loops), the speed of convergence of EIVPM is higher than the one of EIVPAST. This asset is confirmed in figure 3 since, for each eigenvalue, the normalized mean quadratic error for EIVPM is globally smaller than the one supplied by EIVPAST.

The figure 4 emphasizes the real benefit of EIVPM in term of accuracy, the initialization step being aside. By avoiding to take account of the 100 first iterations, the normalized mean quadratic error for both methods

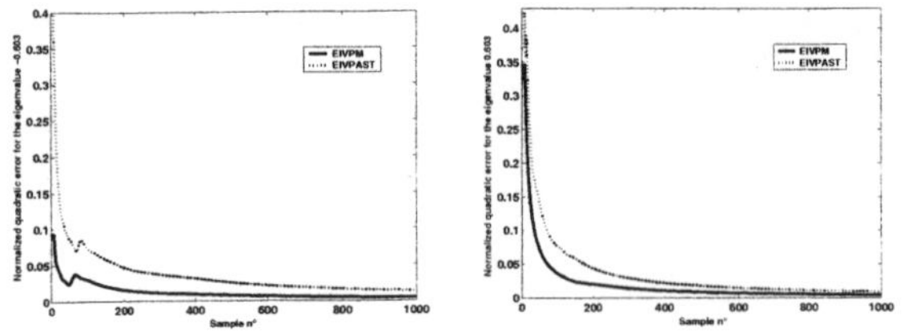

Fig. 3. Normalized mean quadratic error of both eigenvalues.

is plotted again. The mean quadratic error of EIVPM is constantly smaller than with EIVPAST (it can be selectively up to eight times smaller). Thus, EIVPM is more accurate both during and after the initialization stage.

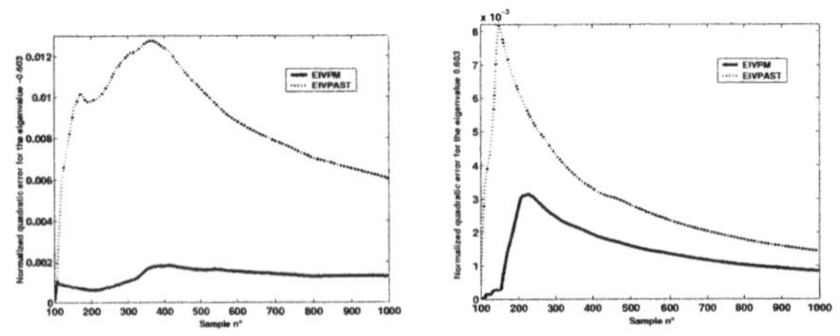

Fig. 4. Normalized mean quadratic error of both eigenvalues aside from the initialization step.

6. CONCLUSION

In this paper, the problem of recursive identification of a state-space system with process and measurement noises is considered. In order to update the model online with the minimal computational cost, a new recursive algorithm is presented. Under the assumption that the state order is a priori known, a new quadratic criterion has been proposed to estimate recursively the subspace spanned by the observability matrix. The use of an instrumental technique has allowed to annihilate the effects of the disturbances during the recursive subspace identification. The performances of the EIVPM algorithm have been compared with those of EIVPAST on a simulation example. The good results of EIVPM in terms of accuracy and speed of convergence have been underlined.

REFERENCES

Cho, Y. M., G. Xu and T. Kailath (1994). Fast recursive identification of state space models via exploitation displacement structure. *Automatica* **30**(1), 45–60.

Friedlander, B. (1984). The overdetermined recursive instrumental variable method. *IEEE Transactions on Automatic Control* **4**, 353–356.

Gustafsson, T. (1997). Recursive system identification using instrumental variable subspace tracking. In: *the* 11th *IFAC Symposium on System Identification*. Fukuoka, Japan.

Ljung, L. (1999). *System identification. Theory for the user*. 2nd ed.. PTR Prentice Hall Information and System Sciences Series. T. Kailath, Series Editor.

Lovera, M. (1998). Subspace identification methods: theory and applications. PhD thesis. Politecnico di Milano.

Marcos, S., A. Marsal and M. Benidir (1995). The propagator method for source bearing estimation. *Signal Processing* **42**(2), 121–138.

Moonen, M., B. De Moor, L. Vandenberghe and J. Vandewalle (1989). On and off line identification of linear state space models. *International Journal of Control* **49**(1), 219–232.

Munier, J. and G. Y. Delisle (1991). Spatial analysis using new properties of the cross spectral matrix. *IEEE Transactions on Signal Processing* **39**(3), 746–749.

Oku, H. and H. Kimura (2002). Recursive 4SID algorithms using gradient type subspace tracking. *Automatica* **38**(6), 1035–1043.

Söderström, T. and P. Stoica (1989). *System identification*. Prentice Hall International Series in Systems and Control Engineering. New York.

Van Overschee, P. and B. De Moor (1996). *Subspace identification for linear systems. Theory, implementation, applications*. Kluwer Academic Publishers.

Verhaegen, M. (1994). Identification of the deterministic part of mimo state space models given in innovations form from input output data. *Automatica* **30**, 61–74.

Verhaegen, M. and E. Deprettere (1991). A fast, recursive mimo state space model identification algorithm. In: *the* 30th *Conference on Decision and Control*. pp. 1349–1354.

Yang, B. (1995). Projection approximation subspace tracking. *IEEE Transactions on Signal Processing* **43**(1), 95–107.

Appendix A. THE EIVPM ALGORITHM

$$\mathbf{P}^T(t) = \mathbf{P}^T(t-1) + (\mathbf{g}(t) - \mathbf{P}^T(t-1)\mathbf{\Psi}(t))\mathbf{K}(t)$$

$$\mathbf{g}(t) = \begin{bmatrix} \hat{\mathbf{C}}_{\mathbf{z}_{\alpha_2}\xi}(t-1)\boldsymbol{\xi}(t) & \mathbf{z}_{\alpha_2}(t) \end{bmatrix}$$

$$\mathbf{\Lambda}(t) = \begin{bmatrix} -\boldsymbol{\xi}^T(t)\boldsymbol{\xi}(t) & \lambda \\ \lambda & 0 \end{bmatrix}$$

$$\mathbf{q}(t) = \hat{\mathbf{C}}_{\mathbf{z}_{\alpha_1}\xi}(t-1)\boldsymbol{\xi}(t) \quad \mathbf{\Psi}(t) = \begin{bmatrix} \mathbf{q}(\mathbf{t}) & \mathbf{z}_{\alpha_1}(t) \end{bmatrix}$$

$$\mathbf{K}(t) = (\mathbf{\Lambda}(t) + \mathbf{\Psi}^T(t)\mathbf{M}(t-1)\mathbf{\Psi}(t))^{-1}\mathbf{\Psi}^T(t)\mathbf{M}(t-1)$$

$$\hat{\mathbf{C}}_{\mathbf{z}_{\alpha_1}\xi}(t) = \lambda\hat{\mathbf{C}}_{\mathbf{z}_{\alpha_1}\xi}(t-1) + \mathbf{z}_{\alpha_1}(t)\boldsymbol{\xi}^T(t)$$

$$\hat{\mathbf{C}}_{\mathbf{z}_{\alpha_2}\xi}(t) = \lambda\hat{\mathbf{C}}_{\mathbf{z}_{\alpha_2}\xi}(t-1) + \mathbf{z}_{\alpha_2}(t)\boldsymbol{\xi}^T(t)$$

$$\mathbf{M}(t) = \frac{1}{\lambda^2}(\mathbf{M}(t-1) - \mathbf{M}(t-1)\mathbf{\Psi}(t)\mathbf{K}(t))$$

$$\text{with } \mathbf{M}(t) = \left(\hat{\mathbf{C}}_{\mathbf{z}_{\alpha_1}\xi}(t)\hat{\mathbf{C}}^T_{\mathbf{z}_{\alpha_1}\xi}(t)\right)^{-1}$$

IFAC
Publications
www.elsevier.com/locate/ifac

CANONICAL CORRELATION PARTIAL LEAST SQUARES

Uwe Kruger * **and S. Joe Qin** **

* *Intelligent Systems and Control Research Group, Queen's University of Belfast, BT9 5AH, U.K.*
** *Department of Chemical Engineering, The University of Texas at Austin, TX78712, U.S.A.*

Abstract: In this paper, a novel algorithm, referred to as Canonical Correlation Partial Least Squares or CCPLS is introduced. The core algorithm is based on the deflation procedure of partial least squares and the cost function of Canonical Correlation Analysis for determining linear combinations of two sets of variables.

Keywords: Correlation Coefficients, Covariance, Parameter Identification, Identification Algorithms, Least Squares Estimation, Linear Systems

1. INTRODUCTION

The theory of Canonical Correlation Analysis (CCA) was developed by Hotelling in the mid 1930s (Hotelling, 1935; Hotelling, 1936). CCA is designed to extract linear combinations between two sets of variables that have maximum correlation. In fact, CCA is based on a cost function that is subject to a number of orthogonal constraints.

The principles of Partial Least Squares (PLS) were pioneered by Wold in the mid 1960s (Wold, 1966*a*; Wold, 1966*b*). On the basis of this early work, further developments resulted in the date of birth of PLS in 1977 (Wold, 1982).

In this paper, the integration of the deflation procedure of the PLS algorithm into the CCA algorithm, is proposed. This integration gives rise to a new algorithm, termed Canonical Correlation Partial Least Squares or CCPLS. More precisely, the cost function for determining the score vectors are provided by the CCA algorithm and the deflation procedure is delivered by the PLS algorithm. It is demonstrated that some of the constraints that guarantee the orthogonality properties of the CCA algorithm are omitted and the PLS deflation procedure is considered instead.

The next section provides a brief overview of the CCA and the PLS algorithms. Section 3 describes a preliminary CCPLS algorithm obtained by combining the features of the CCA and the PLS algorithms. Then, the properties of CCPLS are shown in Section 4. By blending these properties into the preliminary CCPLS algorithm, a more advanced CCPLS algorithm is developed in Section 5 and conclusions concerning this paper are given in Section 6.

2. CCA AND PLS ALGORITHMS

Both CCA and PLS are designed to analyse the relationships between two sets of variables. In this paper, the variable sets relate to the predictor and the response variables of a process. The observations of each variable are stored in matrices, where each column refers to a particular process variable and each row refers to a particular instance in time. The predictor and response matrices are denoted by $\mathbf{X}$ and $\mathbf{Y}$ respectively:

$$\mathbf{X} = \left[\mathbf{x}_1 \; \mathbf{x}_2 \; \cdots \; \mathbf{x}_M \right] \in \mathbb{R}^{K \times M} \qquad (1)$$
$$\mathbf{Y} = \left[\mathbf{y}_1 \; \mathbf{y}_2 \; \cdots \; \mathbf{y}_N \right] \in \mathbb{R}^{K \times N},$$

where $\mathbf{x}_1, \ldots, \mathbf{x}_M$ and $\mathbf{y}_1, \ldots, \mathbf{y}_N$ are column vectors of $\mathbf{X}$ and $\mathbf{Y}$ respectively, K represents the number of observations and M and N are the number of predictor and response variables, respectively. Typically, $\mathbf{X}$ and $\mathbf{Y}$ are mean centered and appropriately scaled prior to the applications of CCA and PLS. It is assumed throughout this paper that $K \gg M, N$ and that the ranks of $\mathbf{X}$ and $\mathbf{Y}$ is M and N respectively. It should be noted that these assumptions are more for the convenience of the presentation rather than imposing restrictions onto the presented methods.

2.1 The CCA Algorithm

Originally, CCA is designed to extract linear combinations between two sets of variables, i.e. the predictor and the response variables, that have maximum correlation:

$$\mathbf{t}_k = \mathbf{X}\mathbf{w}_k, \ \mathbf{u}_k = \mathbf{Y}\mathbf{v}_k. \qquad (2)$$

The variables $\mathbf{t}_k$ and $\mathbf{u}_k$ and the coefficients $\mathbf{w}_k$ and $\mathbf{v}_k$ are termed the kth canonical variates and k^{th} canonical coefficients respectively. The correlation coefficient between $\mathbf{t}_k$ and $\mathbf{u}_k$ is defined as the canonical correlation coefficient r_k.

In order to compare CCA with PLS, the kth canonical coefficients are referred to as weight vectors, i.e. the kth w- and v-weight vector, and the canonical variates are termed score vectors. The kth pair of score vectors is determined to maximize the cost function $J_k^{(\mathbf{t},\mathbf{u})}$:

$$J_k^{(\mathbf{t},\mathbf{u})} = \mathbf{t}_k^T \mathbf{u}_k, \qquad (3)$$

subject to the following constraints:

$$\begin{aligned} G_k^{(\mathbf{t})} &= \|\mathbf{t}_k\|_2^2 - 1 = 0 \qquad (4) \\ G_k^{(\mathbf{u})} &= \|\mathbf{u}_k\|_2^2 - 1 = 0 \\ \mathbf{G}_k^{(\mathbf{t})} &= \begin{bmatrix} \mathbf{t}_1^T & \mathbf{u}_1^T \\ \mathbf{t}_2^T & \mathbf{u}_2^T \\ \vdots & \vdots \\ \mathbf{t}_{k-1}^T & \mathbf{u}_{k-1}^T \end{bmatrix} \mathbf{t}_k = \mathbf{0} \\ \mathbf{G}_k^{(\mathbf{u})} &= \begin{bmatrix} \mathbf{u}_1^T \\ \mathbf{u}_2^T \\ \vdots \\ \mathbf{u}_{k-1}^T \end{bmatrix} \mathbf{u}_k = \mathbf{0}, \end{aligned}$$

where $\|\circ\|_2^2$ denotes the squared length of a vector and $G_k^{(\circ)}$ and $\mathbf{G}_k^{(\circ)}$ represent constraints. Since each score vector has been set to unit length, the correlation coefficients, r_k, are equal to $\mathbf{t}_k^T \mathbf{u}_k$. With respect to the above cost function, the first pair of weight vectors, $\mathbf{w}_1$ and $\mathbf{v}_1$, is given by the eigenvectors associated with the largest eigenvalue of the following matrix expressions:

$$\begin{aligned} \text{for } \mathbf{w}_1 &: \left[\mathbf{X}^T\mathbf{X}\right]^{-1} \mathbf{X}^T\mathbf{Y} \left[\mathbf{Y}^T\mathbf{Y}\right]^{-1} \mathbf{Y}^T\mathbf{X} \qquad (5) \\ \text{for } \mathbf{v}_1 &: \left[\mathbf{Y}^T\mathbf{Y}\right]^{-1} \mathbf{Y}^T\mathbf{X} \left[\mathbf{X}^T\mathbf{X}\right]^{-1} \mathbf{X}^T\mathbf{Y}, \end{aligned}$$

2.2 The PLS Algorithm

PLS determines weight and score vectors by maximising the cost function:

$$J_k^{(\mathbf{t},\mathbf{u})} = \mathbf{t}_k^T \mathbf{u}_k = \mathbf{w}_k^T \mathbf{X}^T \mathbf{Y} \mathbf{v}_k, \qquad (6)$$

subject to the following constraints:

$$\begin{aligned} G_k^{(\mathbf{w})} &= \|\mathbf{w}_k\|_2^2 - 1 = 0 \qquad (7) \\ G_k^{(\mathbf{v})} &= \|\mathbf{v}_k\|_2^2 - 1 = 0. \end{aligned}$$

In contrast to CCA, which incorporates constraints for determining subsequent score vectors, PLS uses a deflation procedure. The contribution of the kth pair of score vectors towards the predictor and response matrices is subtracted (deflated) from these matrices, (Geladi and Kowalski, 1986):

$$\begin{aligned} \mathbf{X}_{k+1} &= \mathbf{X}_k - \mathbf{t}_k \mathbf{p}_k^T \qquad (8) \\ \mathbf{Y}_{k+1} &= \mathbf{Y}_k - \mathbf{t}_k \mathbf{q}_k^T, \end{aligned}$$

where $\mathbf{X}_k$, $\mathbf{Y}_k$ and $\mathbf{X}_{k+1}$, $\mathbf{Y}_{k+1}$ are the predictor and response matrices after $k-1$ and k deflation steps, respectively. $\mathbf{p}_k$ and $\mathbf{q}_k$ are loading vectors that represent the contribution of the t-score vector towards the predictor and the response variables respectively. They are determined as follows:

$$\begin{aligned} \mathbf{p}_k &= \frac{\mathbf{X}_k^T \mathbf{t}_k}{\mathbf{t}_k^T \mathbf{t}_k} \qquad (9) \\ \mathbf{q}_k &= \frac{\mathbf{Y}_k^T \mathbf{t}_k}{\mathbf{t}_k^T \mathbf{t}_k}. \end{aligned}$$

A parametric regression matrix between the predictor and the response variables can be obtained as follows:

$$\mathbf{B}_{PLS} = \mathbf{W} \left[\mathbf{P}^T \mathbf{W}\right]^{-1} \mathbf{Q}^T, \qquad (10)$$

where $\mathbf{B}_{PLS}$ represents the regression matrix and $\mathbf{W}$, $\mathbf{P}$ and $\mathbf{Q}$ are matrices where the w-weight and the p- and q-loading vectors are stored as column vectors in successive order.

3. A PRELIMINARY CCPLS ALGORITHM

Whilst the CCA algorithm calculates each pair of score variables to be maximally correlated, the PLS algorithm determines pairs of score variables that have maximum covariance (Höskuldsson, 1988). In order to determine subsequent weight and score variables, PLS utilises a deflation procedure and CCA relies on the incorporation of additional constraints.

Given the above discussion, it is possible to replace the constraints of the CCA algorithm by the deflation procedure of the PLS algorithm. Furthermore the PLS cost function, for determining the weight and score variables or vectors, can be replaced by that of the CCA algorithm. This gives rise to a novel algorithm, referred to as CCPLS. A preliminary CCPLS algorithm is graphically illustrated in Fig. 1.

PLS	CCA	CCPLS
1) Determine k^{th} pair of weight vectors		
$\mathbf{w}_k^T\mathbf{X}_k^T\mathbf{Y}_k\mathbf{v}_k$ $\|\mathbf{w}_k\|_2^2-1=0$ $\|\mathbf{v}_k\|_2^2-1=0$	$\mathbf{w}_k^T\mathbf{X}^T\mathbf{Y}\mathbf{v}_k$ $\|\mathbf{X}\mathbf{w}_k\|_2^2-1=0$ $\|\mathbf{Y}\mathbf{v}_k\|_2^2-1=0$ $\mathbf{W}_{k-1}^T\mathbf{X}^T\mathbf{X}\mathbf{w}_k=0$ $\mathbf{V}_{k-1}^T\mathbf{Y}^T\mathbf{Y}\mathbf{v}_k=0$ $\mathbf{V}_{k-1}^T\mathbf{Y}^T\mathbf{X}\mathbf{w}_k=0$	$\mathbf{w}_k^T\mathbf{X}_k^T\mathbf{Y}_k\mathbf{v}_k$ $\|\mathbf{X}_k\mathbf{w}_k\|_2^2-1=0$ $\|\mathbf{Y}_k\mathbf{v}_k\|_2^2-1=0$
2) Determine k^{th} pair of score vectors		
$\mathbf{t}_k=\mathbf{X}_k\mathbf{w}_k$ $\mathbf{u}_k=\mathbf{Y}_k\mathbf{v}_k$	$\mathbf{t}_k=\mathbf{X}\mathbf{w}_k$ $\mathbf{u}_k=\mathbf{Y}\mathbf{v}_k$	$\mathbf{t}_k=\mathbf{X}_k\mathbf{w}_k$ $\mathbf{u}_k=\mathbf{Y}_k\mathbf{v}_k$
3) Incorporate k^{th} LVs to determine $(k+1)^{th}$ LVs		
$\mathbf{p}_k=\frac{\mathbf{X}_k^T\mathbf{t}_k}{\mathbf{t}_k^T\mathbf{t}_k}$ $\mathbf{q}_k=\frac{\mathbf{Y}_k^T\mathbf{t}_k}{\mathbf{t}_k^T\mathbf{t}_k}$ $\mathbf{X}_{k+1}=\mathbf{X}_k-\mathbf{t}_k\mathbf{p}_k^T$ $\mathbf{Y}_{k+1}=\mathbf{Y}_k-\mathbf{t}_k\mathbf{q}_k^T$	$\mathbf{W}_k=[\mathbf{W}_{k-1}\ \ \mathbf{w}_k]$ $\mathbf{V}_k=[\mathbf{V}_{k-1}\ \ \mathbf{v}_k]$	$\mathbf{p}_k=\mathbf{X}_k^T\mathbf{t}_k$ $\mathbf{q}_k=\mathbf{Y}_k^T\mathbf{t}_k$ $\mathbf{X}_{k+1}=\mathbf{X}_k-\mathbf{t}_k\mathbf{p}_k^T$ $\mathbf{Y}_{k+1}=\mathbf{Y}_k-\mathbf{t}_k\mathbf{q}_k^T$

Fig. 1. Relationship between PLS, CCA and CCPLS

In comparison to CCA, CCPLS does not require constraints to incorporated. In comparison to PLS, CCPLS computes score variables that have maximum correlation instead of score variables that have maximum covariance.

The steps of the preliminary CCPLS algorithm for determining the kth pair of weight, score and loading vectors are presented in Table 1.

The first step of this algorithm is to determine the weight vectors $\mathbf{w}_k$ and $\mathbf{v}_k$ as the dominant eigenvectors of the matrices $\left[\mathbf{X}_k^T\mathbf{X}_k\right]^\dagger\mathbf{X}_k^T\mathbf{Y}_k\left[\mathbf{Y}_k^T\mathbf{Y}_k\right]^\dagger\mathbf{Y}_k^T\mathbf{X}_k$ and $\left[\mathbf{Y}_k^T\mathbf{Y}_k\right]^\dagger\mathbf{Y}_k^T\mathbf{X}_k\left[\mathbf{X}_k^T\mathbf{X}_k\right]^\dagger\mathbf{X}_k^T\mathbf{Y}_k$, respectively. Since this represents the maximum of the CCA cost function for the first pair of weight vectors, $\mathbf{w}_k$ and $\mathbf{v}_k$, given in Equations (6), this step is inherited from CCA. The determination of the weight vectors is followed by the calculation of the score vectors, $\mathbf{t}_k$ and $\mathbf{u}_k$. This is also the case for both the CCA and the standard PLS algorithms.

The third step is to scale the score vectors to unit length, which is required for CCA, prior to the determination of the regression coefficient (PLS) or the correlation coefficient (CCA). Then the loading vectors $\mathbf{p}_k$ and $\mathbf{q}_k$ are computed, which is also required for the PLS deflation procedure.

The predictor and the response matrices are then deflated. At this point, the CCPLS algorithm can either be terminated or the $(k+1)^{th}$ pair of weight, score and loading vectors can be computed. After this iterative algorithm is terminated, the CCPLS regression matrix is calculated.

4. PROPERTIES OF THE CCPLS ALGORITHM

This section presents the results from an analysis of the properties of CCPLS in relation to CCA and PLS. Abbreviating LT as a lower triangular matrix, SY as a symmetric matrix and DI as a diagonal matrix, a summary of the geometric properties of the weight, score and loading vectors are given in Table 2.

Table 2. Properties of the Weight-, Score- and Loading Vectors

	CCA	PLS	CCPLS
$\mathbf{W}^T\mathbf{W}$	SY	DI	SY
$\mathbf{V}^T\mathbf{V}$	SY	SY	SY
$\mathbf{W}^T\mathbf{P}$	/	LT	DI
$\mathbf{V}^T\mathbf{Q}$	/	SY	DI
$\mathbf{P}^T\mathbf{P}$	/	SY	SY
$\mathbf{Q}^T\mathbf{Q}$	/	SY	SY
$\mathbf{T}^T\mathbf{T}$	DI	DI	DI
$\mathbf{U}^T\mathbf{U}$	DI	SY	DI
$\mathbf{T}^T\mathbf{U}$	DI	LT	DI

In addition, the deflation of the matrix product $\left[\mathbf{X}_k^T\mathbf{X}_k\right]^\dagger\mathbf{X}_k^T\mathbf{Y}_k\left[\mathbf{Y}_k^T\mathbf{Y}_k\right]^\dagger\mathbf{Y}_k^T\mathbf{X}_k$ is equivalent to $\left[\mathbf{X}^T\mathbf{X}\right]^\dagger\mathbf{X}^T\mathbf{Y}\left[\mathbf{Y}^T\mathbf{Y}\right]^\dagger\mathbf{Y}^T\mathbf{X}_k$, which is proven in Appendix A.

5. CCPLS ALGORITHM

Given the properties of the preliminary CCPLS algorithm, an algorithm that is similar in approach to the NIPALS algorithm (Geladi and Kowalski, 1986) can be established. The resultant algorithm, termed as the CCPLS algorithm, is able to determine the score, weight and loading vectors for the kth LV using one iteration procedure.

The CCPLS algorithm can be initiated by computing $\mathbf{\Sigma}_{XX}^{-1}=\left[\mathbf{X}^T\mathbf{X}\right]^{-1}$ and $\mathbf{\Sigma}_{YY}^{-1}=\left[\mathbf{Y}^T\mathbf{Y}\right]^{-1}$ as well as selecting the initial t-score vector as some column of $\mathbf{X}$. The ith iteration step can the be carried out by calculating the q-loading vector

$$\mathbf{q}_k^{(i)}=\mathbf{Y}_k^T\mathbf{t}_k^{(i)}, \tag{11}$$

the v-weight vector

$$\mathbf{v}_k^{(i)}=\mathbf{\Sigma}_{YY}^{-1}\mathbf{q}_k^{(i)} \tag{12}$$

$$\mathbf{v}_k^{(i)}=\frac{\mathbf{v}_k^{(i)}}{\left\|\mathbf{Y}\mathbf{v}_k^{(i)}\right\|_2},$$

Table 1. The Preliminary CCPLS Algorithm

Step	Preliminary CCPLS Algorithm	Contributing Algorithm
1	Determine $\mathbf{w}_k^*$ and $\mathbf{v}_k^*$ as the dominant eigenvectors of the matrices $\left[\mathbf{X}_k^T\mathbf{X}_k\right]^\dagger \mathbf{X}_k^T\mathbf{Y}_k \left[\mathbf{Y}_k^T\mathbf{Y}_k\right]^\dagger \mathbf{Y}_k^T\mathbf{X}_k$ and $\left[\mathbf{Y}_k^T\mathbf{Y}_k\right]^\dagger \mathbf{Y}_k^T\mathbf{X}_k \left[\mathbf{X}_k^T\mathbf{X}_k\right]^\dagger \mathbf{X}_k^T\mathbf{Y}_k$, respectively	CCA
2	Calculate the score vectors $\mathbf{t}_k^* = \mathbf{X}_k\mathbf{w}_k^*$ and $\mathbf{u}_k^* = \mathbf{Y}_k\mathbf{v}_k^*$	CCA/PLS
3	Scale $\mathbf{t}_k^*$ and $\mathbf{u}_k^*$ to unit length $\mathbf{t}_k = \frac{\mathbf{t}_k^*}{\left\Vert\mathbf{t}_k^*\right\Vert_2}$ and $\mathbf{u}_k = \frac{\mathbf{u}_k^*}{\left\Vert\mathbf{u}_k^*\right\Vert_2}$ and rescale $\mathbf{w}_k^*$ and $\mathbf{v}_k^*$ $\mathbf{w}_k = \frac{\mathbf{w}_k^*}{\left\Vert\mathbf{t}_k^*\right\Vert_2}$ and $\mathbf{v}_k = \frac{\mathbf{v}_k^*}{\left\Vert\mathbf{u}_k^*\right\Vert_2}$	CCA
4	Determine regression/correlation coefficient $r_k = \mathbf{t}_k^T\mathbf{u}_k$	CCA/PLS
5	Calculate loading vectors $\mathbf{p}_k = \mathbf{X}_k^T\mathbf{t}_k$ and $\mathbf{q}_k = \mathbf{Y}_k^T\mathbf{t}_k$	PLS
6	Deflate predictor and response matrix $\mathbf{X}_{k+1} = \mathbf{X}_k - \mathbf{t}_k\mathbf{p}_k^T$ and $\mathbf{Y}_{k+1} = \mathbf{Y}_k - \mathbf{t}_k\mathbf{q}_k^T$ Go to step 1 or terminate algorithm and go to step 7	PLS
7	Compute regression matrix	PLS

the u-score

$$\mathbf{u}_k^{(i)} = \mathbf{Y}\mathbf{v}_k^{(i)}, \tag{13}$$

the p-loading vector,

$$\mathbf{p}_k^{(i)} = \mathbf{X}^T\mathbf{u}_k^{(i)},$$

the w-weight vector,

$$\mathbf{w}_k^{(i)} = \mathbf{\Sigma}_{XX}^{-1}\mathbf{p}_k^{(i)} \tag{14}$$

$$\mathbf{w}_k^{(i)} = \frac{\mathbf{w}_k^{(i)}}{\left\Vert\mathbf{X}\mathbf{w}_k^{(i)}\right\Vert_2},$$

and the $(i+1)^{th}$ t-score vector

$$\mathbf{t}_k^{(i+1)} = \mathbf{X}\mathbf{w}_k^{(i)} \tag{15}$$

in successive order. The number of iterations is determined by checking two subsequently obtained t-score vectors, i.e.:

$$\left\Vert\mathbf{t}_k^{(i+1)} - \mathbf{t}_k^{(i)}\right\Vert_2^2 < \epsilon, \tag{16}$$

where ϵ is a given threshold, e.g. 10^{-10}. On convergence, the correlation coefficient is then given by $r_k = \mathbf{t}_k^T\mathbf{u}_k$. Before determining subsequent weight, score and loading vectors, the response matrix can be deflated:

$$\mathbf{Y}_{k+1} = \mathbf{Y}_k - \mathbf{t}_k\mathbf{q}_k^T \tag{17}$$

Note that the inverse of the cross product matrices, $\left[\mathbf{Y}^T\mathbf{Y}\right]^{-1}$ and $\left[\mathbf{X}^T\mathbf{X}\right]^{-1}$, remain unchanged. The complete CCPLS algorithm is presented in Table 3.

If enough LVs have been iteratively calculated, the regression matrix between the predictor and response variables can be calculated. Since the p-loading and w-weight vectors are mutually orthonormal, the calculation of the regression matrix, $\mathbf{B}_{CCPLS}$ reduces to $\mathbf{W}\mathbf{Q}^T$ in relation to the calculation of the PLS regression matrix in Equation (10).

It is proven in Appendix B that the CCPLS and the preliminary CCPLS algorithms are equivalent.

6. CONCLUSIONS

In this paper, a new algorithm, denoted as Canonical Correlation Partial Least Squares or CCPLS, was introduced. CCPLS was evolved by placing the deflation procedure of Partial Least Squares (PLS) into the Canonical Correlation Analysis (CCA) and by placing the cost function for determining the canonical variables of CCA into PLS.

The CCPLS algorithm is based on a deflation procedure and hence, requires loading vectors to be determined. It was shown that the weight and loading vectors are mutually orthonormal for the predictor and the response variables. Furthermore, the score vectors are also mutually orthogonal.

Appendix A. DEFLATION OF $\left[\mathbf{X}_K^T\mathbf{X}_K\right]^\dagger \mathbf{X}_K^T\mathbf{Y}_K \left[\mathbf{Y}_K^T\mathbf{Y}_K\right]^\dagger \mathbf{Y}_K^T\mathbf{X}_K$

The proof that $\left[\mathbf{X}_k^T\mathbf{X}_k\right]^\dagger \mathbf{X}_k^T\mathbf{Y}_k \left[\mathbf{Y}_k^T\mathbf{Y}_k\right]^\dagger \mathbf{Y}_k^T\mathbf{X}_k$ is equal to $\left[\mathbf{X}^T\mathbf{X}\right]^\dagger \mathbf{X}^T\mathbf{Y} \left[\mathbf{Y}^T\mathbf{Y}\right]^\dagger \mathbf{Y}^T\mathbf{X}_k$ is given in 3 steps, which are shown below.

Step 1 It can be shown that (i) the matrix product $\mathbf{X}_k^T\mathbf{Y}_k$ can be written as $\mathbf{X}^T\mathbf{Y}_k$ or $\mathbf{X}_k^T\mathbf{Y}$ and that (ii) the constraints that force the score vectors to be of unit length, i.e. $\Vert\mathbf{X}_k\mathbf{w}_k\Vert_2 - 1 = 0$ and $\Vert\mathbf{Y}_k\mathbf{v}_k\Vert_2 - 1 - 0$, are equal to $\Vert\mathbf{X}\mathbf{w}_k\Vert_2 - 1 = 0$ and $\Vert\mathbf{Y}\mathbf{v}_k\Vert_2 - 1 - 0$, respectively.

Table 3. Steps of the CCPLS Algorithm

Step	Description	Equation
1	Determine matrix products	$\mathbf{\Sigma}_{XX} = \mathbf{X}^T\mathbf{X}, \mathbf{\Sigma}_{XX}^{-1}, \mathbf{\Sigma}_{YY} = \mathbf{Y}^T\mathbf{Y}, \mathbf{\Sigma}_{YY}^{-1}$
2	Set and scale initial t-score vector	$\mathbf{t}_k^{(0)} =$ some column of $\mathbf{X}, \mathbf{t}_k^{(0)} = \frac{\mathbf{t}_k^{(0)}}{\left\Vert \mathbf{t}_k^{(0)} \right\Vert_2}$
3	Determine q-loading vector	$\mathbf{q}_k^{(i)} = \mathbf{Y}_k^T\mathbf{t}_k^{(i)}$
4	Compute v-weight vector	$\mathbf{v}_k^{(i)} = \mathbf{\Sigma}_{YY}^{-1}\mathbf{q}_k^{(i)}$
5	Scale v-weight vector	$\mathbf{v}_k^{(i)} = \frac{\mathbf{v}_k^{(i)}}{\left\Vert \mathbf{Y}\mathbf{v}_k^{(i)} \right\Vert_2}$
6	Calculate u-score vector	$\mathbf{u}_k^{(i)} = \mathbf{Y}\mathbf{v}_k^{(i)}$
7	Determine p-loading vector	$\mathbf{p}_k^{(i)} = \mathbf{X}^T\mathbf{u}_k^{(i)}$
8	Calculate w-weight vector	$\mathbf{w}_k^{(i)} = \mathbf{\Sigma}_{XX}^{-1}\mathbf{p}_k^{(i)}$
9	Scale w-weight vector	$\mathbf{w}_k^{(i)} = \frac{\mathbf{w}_k^{(i)}}{\left\Vert \mathbf{X}\mathbf{w}_k^{(i)} \right\Vert_2}$
10	Determine t-score vector	$\mathbf{t}_k^{(i+1)} = \mathbf{X}_k\mathbf{w}_k^{(i)}$
11	Check for convergence	$\left\Vert \mathbf{t}_k^{(i+1)} - \mathbf{t}_k^{(i)} \right\Vert_2 \geq \epsilon$ goto step 3 $(i = i+1)$ $\left\Vert \mathbf{t}_k^{(i+1)} - \mathbf{t}_k^{(i)} \right\Vert_2 < \epsilon$ goto step 12
12	Compute k^{th} correlation coefficient	$r_k = \mathbf{t}_k^T\mathbf{u}_k$
13	Deflate predictor matrix	$\mathbf{X}_{k+1} = \mathbf{X}_k - \mathbf{t}_k\mathbf{p}_k^T$
14	Compute $(k+1)^{th}$ LV (goto step 2), or compute regression matrices	$\mathbf{B}_{CCPLS} = \mathbf{W}_R\mathbf{Q}_R^T$

(i) $\mathbf{X}_k^T\mathbf{Y}_k$=$\mathbf{X}^T\mathbf{Y}_k$=$\mathbf{X}_k^T\mathbf{Y}$: The deflation of the predictor and response matrices can be carried out as follows:

$$\mathbf{X}_k = \left[\mathbf{I}_K - \mathbf{t}_{k-1}\mathbf{t}_{k-1}^T\right]\mathbf{X}_{k-1} \qquad \text{(A.1)}$$
$$\mathbf{Y}_k = \left[\mathbf{I}_K - \mathbf{t}_{k-1}\mathbf{t}_{k-1}^T\right]\mathbf{Y}_{k-1}$$

It is shown in Table 1 that t-score vectors are mutually orthonormal. This implies that by incorporating $(k-1)$ deflation steps, the predictor and the response matrices can be written as:

$$\mathbf{X}_k = \left[\mathbf{I}_K - \mathbf{T}\mathbf{T}^T\right]\mathbf{X} \qquad \text{(A.2)}$$
$$\mathbf{Y}_k = \left[\mathbf{I}_K - \mathbf{T}\mathbf{T}^T\right]\mathbf{Y}$$

Therefore,

$$\mathbf{X}_k^T\mathbf{Y}_k = \mathbf{X}^T\mathbf{Y}_k = \mathbf{X}_k^T\mathbf{Y} \qquad \text{(A.3)}$$

It can be shown in the same way that $\mathbf{X}_k^T\mathbf{X}_k = \mathbf{X}_k^T\mathbf{X} = \mathbf{X}^T\mathbf{X}_k$ and that $\mathbf{X}_k^T\mathbf{Y}_k = \mathbf{Y}_k^T\mathbf{Y} = \mathbf{Y}^T\mathbf{Y}_k$.

(ii) $\mathbf{X}_k\mathbf{w}_k = \mathbf{X}\mathbf{w}_k$ and $\mathbf{Y}_k\mathbf{v}_k = \mathbf{Y}\mathbf{v}_k$: Knowing $\mathbf{X}_k^T\mathbf{X}_k = \mathbf{X}_k^T\mathbf{X} = \mathbf{X}^T\mathbf{X}_k$, gives rise to:

$$\mathbf{w}_k^T\mathbf{X}_k^T\mathbf{X}\mathbf{w}_k = \mathbf{w}_k^T[\mathbf{X} - \mathbf{T}\mathbf{P}^T]^T\mathbf{X}\mathbf{w}_k. \qquad \text{(A.4)}$$

Given that the w-weight and p-loading vectors are mutually orthonormal, the above equation reduces to:

$$\mathbf{w}_k^T\mathbf{X}_k^T\mathbf{X}\mathbf{w}_k = \mathbf{w}_k^T\mathbf{X}^T\mathbf{X}\mathbf{w}_k. \qquad \text{(A.5)}$$

Similarly, knowing $\mathbf{Y}_k^T\mathbf{Y}_k = \mathbf{Y}_k^T\mathbf{Y} = \mathbf{Y}^T\mathbf{Y}_k$ and that the v-weight and the q-loading vectors are mutually orthonormal, it can be shown that:

$$\mathbf{v}_k^T\mathbf{Y}_k^T\mathbf{Y}_k\mathbf{v}_k = \mathbf{v}_k^T\mathbf{Y}^T\mathbf{Y}\mathbf{v}_k \qquad \text{(A.6)}$$

Step 2 Using the results of the first step, it is now shown that the matrix product $\left[\mathbf{X}_k^T\mathbf{X}_k\right]^\dagger \mathbf{X}_k^T\mathbf{Y}_k\left[\mathbf{Y}_k^T\mathbf{Y}_k\right]^\dagger\mathbf{Y}_k^T\mathbf{X}_k$ can be simplified to $\left[\mathbf{X}^T\mathbf{X}\right]^{-1}\mathbf{X}_k^T\mathbf{Y}\left[\mathbf{Y}^T\mathbf{Y}\right]^{-1}\mathbf{Y}^T\mathbf{X}_k$. The cost function for determining the weight and score vectors, $J_k^{(\mathbf{w},\mathbf{v})}$, is as follows:

$$J_k^{(\mathbf{w},\mathbf{v})} = \mathbf{w}_k^T\mathbf{X}^T\mathbf{Y}_k\mathbf{v}_k \qquad \text{(A.7)}$$
$$G_k^{(\mathbf{w})} = \mathbf{w}_k^T\mathbf{X}^T\mathbf{X}\mathbf{w}_k - 1 = 0$$
$$G_k^{(\mathbf{v})} = \mathbf{v}_k^T\mathbf{Y}^T\mathbf{Y}\mathbf{v}_k - 1 = 0.$$

The solution of the above optimisation problem is given by:

$$\frac{\partial J_k^{(\mathbf{w},\mathbf{v})}}{\partial \mathbf{w}_k} - \lambda_k^{(1)}\frac{\partial G_k^{(\mathbf{w})}}{\partial \mathbf{w}_k} = \mathbf{0} \qquad \text{(A.8)}$$
$$\frac{\partial J_k^{(\mathbf{w},\mathbf{v})}}{\partial \mathbf{v}_k} - \lambda_k^{(2)}\frac{\partial G_k^{(\mathbf{v})}}{\partial \mathbf{v}_k} = \mathbf{0},$$

which leads to:

$$\mathbf{X}^T\mathbf{Y}_k\mathbf{v}_k - 2\lambda_k^{(1)}\mathbf{X}^T\mathbf{X}\mathbf{w}_k = \mathbf{0} \qquad \text{(A.9)}$$
$$\mathbf{Y}_k^T\mathbf{X}\mathbf{w}_k - 2\lambda_k^{(2)}\mathbf{Y}^T\mathbf{Y}\mathbf{v}_k = \mathbf{0},$$

where $\lambda_k^{(1)}$ and $\lambda_k^{(2)}$ are Lagrangian multiplier. By using the above equations, the kth w-weight vectors is equal to the largest eigenvalue of the following matrix expression:

$$\left[\left[\mathbf{X}^T\mathbf{X}\right]^{-1}\mathbf{X}_k^T\mathbf{Y}\left[\mathbf{Y}^T\mathbf{Y}\right]^{-1} \mathbf{Y}^T\mathbf{X}_k - 4\lambda_k^{(1)}\lambda_k^{(2)}\mathbf{I}_N\right]\mathbf{w}_k = \mathbf{0}. \qquad \text{(A.10)}$$

Consequently, only the cross product matrix $\mathbf{X}^T\mathbf{Y}_k$ needs to be deflated.

Step 3 Incorporating the deflation procedure into the matrix product $\mathbf{X}_k^T\mathbf{Y}\left[\mathbf{Y}^T\mathbf{Y}\right]^{-1}\mathbf{Y}^T\mathbf{X}_k$ gives rise to:

$$\mathbf{X}^T\left[\mathbf{I}_K - \mathbf{T}\mathbf{T}^T\right]\mathbf{Y}\left[\mathbf{Y}^T\mathbf{Y}\right]^{-1}\mathbf{Y}^T\left[\mathbf{I}_K - \mathbf{T}\mathbf{T}^T\right]\mathbf{X}. \quad (A.11)$$

Abbreviating $\mathbf{T}\mathbf{T}^T = \mathbf{\Theta}$ and $\mathbf{Y}\left[\mathbf{Y}^T\mathbf{Y}\right]^{-1}\mathbf{Y}^T = \mathbf{\Psi}$, the above expression can be simplified as follows:

$$\begin{aligned}\mathbf{X}^T\left[\mathbf{I}_K - \mathbf{\Theta}\right]\mathbf{\Psi}\left[\mathbf{I}_K - \mathbf{\Theta}\right]\mathbf{X} &= \mathbf{X}_k^T\left[\mathbf{\Psi} - \mathbf{\Psi}\mathbf{\Theta}\right]\mathbf{X} \\ \mathbf{X}^T\left[\mathbf{I}_K - \mathbf{\Theta}\right]\mathbf{\Psi}\left[\mathbf{I}_K - \mathbf{\Theta}\right]\mathbf{X} &= \mathbf{X}^T\left[\mathbf{\Psi} - \mathbf{\Theta}\mathbf{\Psi}\right]\mathbf{X}_k.\end{aligned} \quad (A.12)$$

Given that $\mathbf{\Psi}\mathbf{X}_k = \widehat{\mathbf{X}}_k$, the above equation can be written as: $\widehat{\mathbf{X}}_k^T\mathbf{X} - \widehat{\mathbf{X}}_k^T\mathbf{\Theta}\mathbf{X} = \mathbf{X}^T\widehat{\mathbf{X}}_k - \mathbf{X}^T\mathbf{\Theta}\widehat{\mathbf{X}}_k$. In fact, the matrix products $\widehat{\mathbf{X}}_k^T\mathbf{\Theta}\mathbf{X}$ and $\mathbf{X}^T\mathbf{\Theta}\widehat{\mathbf{X}}_k$ are equal to zero which is proven below. Note that $\widehat{\mathbf{X}}_k^T\mathbf{\Theta}\mathbf{X}$ is the transpose matrix to $\mathbf{X}^T\mathbf{\Theta}\widehat{\mathbf{X}}_k$ and hence, only the case of $\mathbf{X}^T\mathbf{\Theta}\widehat{\mathbf{X}}_k = \mathbf{0}$ is considered here:

$$\mathbf{X}^T\mathbf{\Theta}\widehat{\mathbf{X}}_k = \mathbf{P}_M\mathbf{T}_M^T\mathbf{\Theta}\mathbf{\Psi}\left[\mathbf{I}_K - \mathbf{\Theta}\right]\mathbf{T}_M\mathbf{P}_M^T \quad (A.13)$$

Note that the subscripts on $\mathbf{T}$ and $\mathbf{P}$ indicate how many column vectors vectors are stored in these matrices. It is shown that $\mathbf{T}_{k-1}^T\mathbf{\Psi}\mathbf{T}_R = \mathbf{T}_{k-1}^T\mathbf{\Psi}\mathbf{T}_{k-1}$

$$\begin{aligned}\mathbf{T}_{k-1}^T\mathbf{\Psi}\mathbf{T}_R &= \mathbf{W}_{k-1}^T\mathbf{X}^T\mathbf{\Psi}\mathbf{X}\mathbf{W}_R \quad (A.14)\\ &= \mathbf{W}_{k-1}^T\mathbf{X}^T\mathbf{X}\mathbf{W}_R\mathbf{R}_R^2 \\ &= \mathbf{T}_{k-1}^T\mathbf{T}_R\mathbf{R}_R^2 \\ &= \left[\mathbf{T}_{k-1}^T\mathbf{\Psi}\mathbf{T}_{k-1}\ \mathbf{0}\right],\end{aligned}$$

where $\mathbf{R}_R^2$ is a diagonal matrix storing the squared values of the correlation coefficients, r_k^2, and therefore,

$$\begin{aligned}\mathbf{X}^T\mathbf{\Theta}\mathbf{\Psi}\mathbf{X}_k = &\mathbf{P}_{k-1}\mathbf{T}_{k-1}^T\mathbf{\Psi}\mathbf{T}_{k-1}\mathbf{P}_{k-1}^T \quad (A.15)\\ &-\mathbf{P}_{k-1}\mathbf{T}_{k-1}^T\mathbf{\Psi}\mathbf{T}_{k-1}\mathbf{P}_{k-1}^T = \mathbf{0}.\end{aligned}$$

The above equation outlines that $\mathbf{X}_k^T\mathbf{Y}\left[\mathbf{Y}^T\mathbf{Y}\right]^{-1}\mathbf{Y}_k^T\mathbf{X}_k$ is equal to $\mathbf{X}^T\mathbf{Y}\left[\mathbf{Y}^T\mathbf{Y}\right]^{-1}\mathbf{Y}^T\mathbf{X}_k$

Appendix B. EQUALITY OF BOTH CCPLS ALGORITHMS

By substituting Equations (11) to (13) into Equation (14), the calculation of the w-weight vector is as follows:

$$\mathbf{w}_k^{(i)} = \mathbf{\Sigma}_{XX}^{-1}\mathbf{X}^T\mathbf{Y}\mathbf{\Sigma}_{YY}^{-1}\mathbf{Y}_k^T\mathbf{t}_k^{(i)}. \quad (B.1)$$

Now, substituting Equation (17) into Equation (B.1) leads to:

$$\mathbf{w}_k^{(i+1)} = \mathbf{\Sigma}_{XX}^{-1}\mathbf{X}^T\mathbf{Y}_k\mathbf{\Sigma}_{YY}^{-1}\mathbf{Y}_k^T\mathbf{X}\mathbf{w}_k^{(i)}. \quad (B.2)$$

Therefore, the w-weight vector obtained by the CCPLS algorithm, is an eigenvector of the same matrix as the one used in the preliminary CCPLS algorithm.

By substituting Equations (11), (13), (14) and (15) into Equation (12) gives rise to:

$$\mathbf{v}_k^{(i+1)} = \mathbf{\Sigma}_{YY}^{-1}\mathbf{Y}_k^T\mathbf{X}\mathbf{\Sigma}_{XX}^{-1}\mathbf{X}^T\mathbf{Y}\mathbf{v}_k^{(i)}, \quad (B.3)$$

which shows that both algorithms produce the same v-weight vector.

It is now shown that the vectors $\mathbf{q}_k = \mathbf{\Sigma}_{YY}\mathbf{v}_k$ and $\mathbf{p}_k = \mathbf{X}^T\mathbf{u}_k$ point to the same direction as the vectors $\mathbf{q}_k = \mathbf{Y}^T\mathbf{X}^T\mathbf{w}_k$ and $\mathbf{p}_k = \mathbf{\Sigma}_{XX}\mathbf{w}_k$, respectively. By analysing Equation (A.9), it can be seen that:

$$\begin{aligned}\mathbf{X}^T\mathbf{Y}_k\mathbf{v}_k &\propto \mathbf{X}^T\mathbf{X}\mathbf{w}_k =: \mathbf{p}_k \quad (B.4)\\ \mathbf{Y}_k^T\mathbf{X}\mathbf{w}_k &\propto \mathbf{Y}^T\mathbf{Y}\mathbf{v}_k =: \mathbf{q}_k\end{aligned}$$

REFERENCES

Geladi, P. and B. R. Kowalski (1986). Partial least squares: A tutorial. *Analytica Chimica Acta* **185**, 1–17.

Höskuldsson, A. (1988). Pls regression methods. *Journal of Chemometrics* **2**, 211–228.

Hotelling, H. (1935). The most predictable criterion. *J. Educ. Psychol.* **26**, 139–142.

Hotelling, H. (1936). Relations between two sets of variables. *Biometrica* **28**, 321–377.

Wold, H. (1966*a*). In: *Research Papers in Statistics* (F. David, Ed.). Wiley. New York.

Wold, H. (1966*b*). Estimation of principal components and related models by iterative least squares. In: *Multivariate Analysis* (P. Krishnaiah, Ed.). pp. 391–420. Academic Press. New York.

Wold, H. (1982). In: *A Second Generation of Multivariate Analysis* (C. Fornell, Ed.). Preager. New York.

IFAC Publications
www.elsevier.com/locate/ifac

FREQUENCY-DOMAIN SYSTEM IDENTIFICATION TECHNIQUES FOR EXPERIMENTAL AND OPERATIONAL MODAL ANALYSIS

Patrick Guillaume, Peter Verboven, Bart Cauberghe, Steve Vanlanduit, Eli Parloo, Gert De Sitter

Vrije Universiteit Brussel (VUB)
Department of Mechanical Engineering
Acoustics & Vibration Research Group
Peinlaan 2, B-1050 Brussel, BELGIUM

Abstract: In this paper frequency-domain estimators will be presented for application in the field of modal analysis. In "Experimental Modal Analysis" system identification is used to model mechanical systems with a few inputs and hundreds of outputs. This requires adapted estimators designed to handle large amount of data in a reasonable amount of time. Next, attention will be paid to "Operational Modal Analysis" and it will be shown how the modal parameters can be estimated from output-only data. Finally, a combined experimental/operational identification approach will be introduced.

Keywords: Modal analysis, vibration, parameter estimation, multivariable, aerospace systems, automotive.

1. INTRODUCTION

It is well known that (mechanical) structures can resonate, i.e. that small forces can result in important deformation, and possibly, damage can be induced in the structure. Resonant vibration is mainly caused by an interaction between the inertial and elastic properties of the materials within a structure. Resonance is often the cause of, or at least a contributing factor to many of the vibration and noise related problems that occur in structures and operating machinery. To better understand any structural vibration problem, the resonant frequencies of a structure need to be identified and quantified.
The application of system identification to vibrating structures resulted some 20 years ago in a new research discipline in mechanical engineering known as "Experimental Modal Analysis" (EMA).
Today, modal analysis has become a widespread means of finding the modes of vibration of a machine or structure. In every development of a new or improved mechanical product, structural dynamics testing on product prototypes is used to assess its real dynamic behaviour.
Typical for modal analysis is the very large number of outputs. The number of inputs (excitation points) is typically in the order of 1 to 10, while the number of outputs (response measurements) can reach more than 1000 points when using optical measurement equipment such as for instance a scanning laser Doppler vibrometer.
This huge amount of data requires dedicated algorithms that balance between accuracy and memory/computing needs.
In Section 2 a 'dedicated' frequency-domain least-squares estimator will be presented. Based on these results, it is possible to implement more sophisticated identification methods such as for instance the frequency-domain Maximum Likelihood (ML) estimator.
In Section 3 attention will be paid to "Operational Modal Analysis" and it will be shown how the modal parameters can be estimated from output-only data. Finally, in Section 4, a combined experimental-operational identification approach will be introduced and the conclusions will be drawn.

2. FREQUENCY-DOMAIN MODAL PARAMETER ESTIMATION

2.1 Multivariable Transfer Function Modelling

Many multivariable transfer-function models are available (Kalaith, 1980). In this paper, a common-denominator transfer function model will be considered.

The relationship between output o ($o = 1,\ldots,N_o$) and input i ($i = 1,\ldots,N_i$) can be modelled in the frequency domain as

$$\hat{H}_k(\omega) = \frac{N_k(\omega)}{d(\omega)} \tag{1}$$

for $k = 1,\ldots,N_oN_i$ (where $k = (o-1)N_i + i$). The numerator polynomial between output o and input i equals

$$N_k(\omega) = \sum_{j=0}^{n} \Omega_j(\omega) B_{kj} \tag{2}$$

while

$$d(\omega) = \sum_{j=0}^{n} \Omega_j(\omega) A_j \tag{3}$$

is the common-denominator polynomial. The real-valued coefficients A_j and B_{kj} are the parameters to be estimated.
All these coefficients are grouped together in one column vector $\boldsymbol{\theta} = [\boldsymbol{\beta}_1^T,\ldots,\boldsymbol{\beta}_{N_oN_i}^T,\boldsymbol{\alpha}^T]^T$ with

$$\boldsymbol{\beta}_k = \begin{Bmatrix} B_{k0} \\ B_{k1} \\ \vdots \\ B_{kn} \end{Bmatrix}, \quad \boldsymbol{\alpha} = \begin{Bmatrix} A_0 \\ A_1 \\ \vdots \\ A_n \end{Bmatrix} \tag{4}$$

Several choices are possible for the polynomial basis functions $\Omega_j(\omega)$. For a discrete-time domain model, the functions $\Omega_j(\omega)$ are usually given by $\Omega_j(\omega) = \exp(-\mathrm{i}\omega T_s \cdot j)$ (with T_s the sampling period) while for a continuous-time domain model $\Omega_j(\omega) = (\mathrm{i}\omega)^j$. The bad numerical conditioning of the continuous-time domain approach can be improved by using for instance orthogonal Forsythe polynomials (at the expense of an increase of the computation time).

2.2 Modal Parameter Estimation Based on Frequency Response Measurements

In modal analysis, measurements of Frequency Response Functions (FRFs) are commonly used. Most often, modal tests take place in laboratory conditions, and so, the applied forces can be measured together with the response of the structure (e.g., accelerations). Starting from the measured FRFs, $H_k(\omega_f)$ (with $k = 1,\ldots,N_oN_i$ and $f = 1,\ldots,N_f$), estimates of the transfer-function coefficients can be obtained by minimizing the following nonlinear least-squares (NLS) cost function with respect to the parameter vector $\boldsymbol{\theta}$

$$\ell_{\mathrm{NLS}}(\boldsymbol{\theta}) = \sum_{k=1}^{N_oN_i} \sum_{f=1}^{N_f} \left| \varepsilon_k^{\mathrm{NLS}}(\omega_f, \boldsymbol{\theta}) \right|^2 \tag{5}$$

where the (weighted) NLS equation error, $\varepsilon_k^{\mathrm{NLS}}(\omega_f, \boldsymbol{\theta})$, is defined as

$$\varepsilon_k^{\mathrm{NLS}}(\omega_f, \boldsymbol{\theta}) = W_k(\omega_f) \left(\frac{N_k(\omega_f, \boldsymbol{\beta}_k)}{d(\omega_f, \boldsymbol{\alpha})} - H_k(\omega_f) \right) \tag{6}$$

with $W_k(\omega_f)$ an arbitrary weighting functions. The quality of the estimate can often be further improved by using an adequate weighting function such as

$$W_k(\omega_f) = \frac{1}{\sqrt{\mathrm{var}\{H_k(\omega_f)\}}} \tag{7}$$

By doing so, the quality of the measured FRFs can be into account: FRF measurements with a small variance, $\mathrm{var}\{H_k(\omega_f)\}$, have an important contribution to the cost function while noisy FRF measurements are penalized.

To solve this NLS problem, good starting values are required. Starting values can be obtained via (sub-optimal) linear least-squares estimators.

2.3 The Least-Squares Complex Frequency-domain (LSCF) Approach

Note that the nonlinear least-squares problem can be approximated by a (sub-optimal) linear least-squares one. Indeed, by multiplying $\varepsilon_k^{\mathrm{NLS}}(\omega_f, \boldsymbol{\theta})$ with $d(\omega, \boldsymbol{\alpha})$, one obtains an equation error that is linear in the parameters

$$\begin{aligned} \varepsilon_k^{\mathrm{LS}}(\omega_f, \boldsymbol{\theta}) &= d(\omega_f, \boldsymbol{\alpha}) \cdot \varepsilon_k^{\mathrm{NLS}}(\omega_f, \boldsymbol{\theta}) \\ &= W_k(\omega_f)\left(N_k(\omega_f, \boldsymbol{\beta}_k) - d(\omega_f, \boldsymbol{\alpha}) H_k(\omega)\right) \\ &= W_k(\omega_f) \sum_{j=0}^{n} \left(\Omega_j(\omega_f) B_{kj} - \Omega_j(\omega_f) A_j H_k(\omega_f) \right) \end{aligned} \tag{8}$$

Because the equations (8), for $f = 1,\ldots,N_f$, are "linear-in-the-parameters", they can be reformulated in matrix notations as

$$\boldsymbol{\varepsilon}_k^{\mathrm{LS}}(\boldsymbol{\theta}) = \begin{Bmatrix} \varepsilon_k^{\mathrm{LS}}(\omega_1, \boldsymbol{\theta}) \\ \vdots \\ \varepsilon_k^{\mathrm{LS}}(\omega_{N_f}, \boldsymbol{\theta}) \end{Bmatrix} = \begin{bmatrix} \mathbf{X}_k & \mathbf{Y}_k \end{bmatrix} \cdot \begin{Bmatrix} \boldsymbol{\beta}_k \\ \boldsymbol{\alpha} \end{Bmatrix} = \mathbf{J}_k \cdot \begin{Bmatrix} \boldsymbol{\beta}_k \\ \boldsymbol{\alpha} \end{Bmatrix} \tag{9}$$

with

$$\mathbf{X}_k = \begin{bmatrix} W_k(\omega_1)[\Omega_0(\omega_1), \Omega_1(\omega_1), \ldots, \Omega_n(\omega_1)] \\ \vdots \\ W_k(\omega_{N_f})[\Omega_0(\omega_{N_f}), \Omega_1(\omega_{N_f}), \ldots, \Omega_n(\omega_{N_f})] \end{bmatrix} \tag{10}$$

$$\mathbf{Y}_k = \begin{bmatrix} -W_k(\omega_1) H_k(\omega_1)[\Omega_0(\omega_1), \Omega_1(\omega_1), \ldots, \Omega_n(\omega_1)] \\ \vdots \\ -W_k(\omega_{N_f}) H_k(\omega_{N_f})[\Omega_0(\omega_{N_f}), \Omega_1(\omega_{N_f}), \ldots, \Omega_n(\omega_{N_f})] \end{bmatrix} \tag{11}$$

The (weighted) linear least-squares problem is found by minimizing

$$\begin{aligned}\ell_{\mathrm{LS}}(\boldsymbol{\theta}) &= \sum_{k=1}^{N_o N_i} \sum_{f=1}^{N_f} \left|\varepsilon_k^{\mathrm{LS}}(\omega_f, \boldsymbol{\theta})\right|^2 \\ &= \sum_{k=1}^{N_o N_i} \mathrm{Re}\left(\left(\boldsymbol{\varepsilon}_k^{\mathrm{LS}}(\boldsymbol{\theta})\right)^H \cdot \boldsymbol{\varepsilon}_k^{\mathrm{LS}}(\boldsymbol{\theta})\right) \\ &= \sum_{k=1}^{N_o N_i} \left(\begin{bmatrix} \boldsymbol{\beta}_k^T & \boldsymbol{\alpha}^T \end{bmatrix} \cdot \begin{bmatrix} \mathbf{R}_k & \mathbf{S}_k \\ \mathbf{S}_k^T & \mathbf{T}_k \end{bmatrix} \cdot \begin{Bmatrix} \boldsymbol{\beta}_k \\ \boldsymbol{\alpha} \end{Bmatrix} \right)\end{aligned} \tag{12}$$

with $\mathbf{R}_k = \mathrm{Re}(\mathbf{X}_k^H \mathbf{X}_k)$, $\mathbf{S}_k = \mathrm{Re}(\mathbf{X}_k^H \mathbf{Y}_k)$, and $\mathbf{T}_k = \mathrm{Re}(\mathbf{Y}_k^H \mathbf{Y}_k)$. Note that cost function (12) is equivalent to

$$\ell_{\mathrm{LS}}(\boldsymbol{\theta}) = \boldsymbol{\theta}^T \cdot \mathrm{Re}(\mathbf{J}^H \mathbf{J}) \cdot \boldsymbol{\theta} \tag{13}$$

with $\mathbf{J}$ the so-called Jacobian matrix, which is given by

$$\mathbf{J} = \begin{bmatrix} \mathbf{X}_1 & 0 & \cdots & 0 & \mathbf{Y}_1 \\ 0 & \mathbf{X}_2 & & 0 & \mathbf{Y}_2 \\ \vdots & & \ddots & & \vdots \\ 0 & 0 & & \mathbf{X}_{N_o N_i} & \mathbf{Y}_{N_o N_i} \end{bmatrix} \tag{14}$$

The Jacobian matrix $\mathbf{J}$ of this least-squares problem has $N_f N_o N_i$ rows and $(n+1)(N_o N_i+1)$ columns (with $N_f >> n$, where n is the order of the polynomials).

In the minimum of the cost function the derivatives of (12) with respect to the unknown coefficients $\boldsymbol{\beta}_k$ and $\boldsymbol{\alpha}$ have to be zero

$$\frac{\partial}{\partial \boldsymbol{\beta}_k} \ell_{\mathrm{LS}}(\boldsymbol{\theta}) = 2\left(\mathbf{R}_k \boldsymbol{\beta}_k + \mathbf{S}_k \boldsymbol{\alpha}\right) = \mathbf{0}, \quad k = 1, \ldots, N_o N_i \tag{15}$$

$$\frac{\partial}{\partial \boldsymbol{\alpha}} \ell_{\mathrm{LS}}(\boldsymbol{\theta}) = 2\left[\sum_{k=1}^{N_o N_i} \left(\mathbf{S}_k^T \boldsymbol{\beta}_k + \mathbf{T}_k \boldsymbol{\alpha}\right)\right] = \mathbf{0} \tag{16}$$

Substitution of (15), $\boldsymbol{\beta}_k = -\mathbf{R}_k^{-1} \mathbf{S}_k \cdot \boldsymbol{\alpha}$, in (16) yields

$$\left[2 \sum_{k=1}^{N_o N_i} \left(\mathbf{T}_k - \mathbf{S}_k^T \mathbf{R}_k^{-1} \mathbf{S}_k\right)\right] \cdot \boldsymbol{\alpha} = \mathbf{M} \cdot \boldsymbol{\alpha} = \mathbf{0} \tag{17}$$

with $\mathbf{M} = 2 \sum_{k=1}^{N_o N_i} \left(\mathbf{T}_k - \mathbf{S}_k^T \mathbf{R}_k^{-1} \mathbf{S}_k\right)$.

Equations (15) and (16) are the so-called normal equations, which are usually formulated as

$$2 \begin{bmatrix} \mathbf{R}_1 & 0 & \cdots & \mathbf{S}_1 \\ 0 & \mathbf{R}_2 & & \mathbf{S}_2 \\ \vdots & & \ddots & \vdots \\ \mathbf{S}_1^t & \mathbf{S}_2^t & \cdots & \sum_{k=1}^{N_o N_i} \mathbf{T}_k \end{bmatrix} \cdot \begin{Bmatrix} \boldsymbol{\beta}_1 \\ \boldsymbol{\beta}_2 \\ \vdots \\ \boldsymbol{\beta}_{N_o N_i} \\ \boldsymbol{\alpha} \end{Bmatrix} = 2\,\mathrm{Re}(\mathbf{J}^H \mathbf{J}) \cdot \boldsymbol{\theta} = \mathbf{0} \tag{18}$$

The size of the square matrix $\mathbf{M}$ in the "reduced" normal equations (17) is $n+1$, and thus much smaller than the size of $\mathrm{Re}(\mathbf{J}^H \mathbf{J})$ in (18).

2.4 *Fast Implementation of the Reduced Normal Equations*

Examining the matrices $\mathbf{R}_k = \mathrm{Re}(\mathbf{X}_k^H \mathbf{X}_k)$, $\mathbf{S}_k = \mathrm{Re}(\mathbf{X}_k^H \mathbf{Y}_k)$, and $\mathbf{T}_k = \mathrm{Re}(\mathbf{Y}_k^H \mathbf{Y}_k)$ in more details reveals that the entries of these matrices equal

$$\begin{aligned} R_k(r,s) &= \mathrm{Re}\left[\sum_{f=1}^{N_f} \left(\left|W_k(\omega_f)\right|^2 \cdot \Omega_{r-1}^H(\omega_f) \Omega_{s-1}(\omega_f) \right)\right] \\ S_k(r,s) &= -\mathrm{Re}\left[\sum_{f=1}^{N_f} \left(\left|W_k(\omega_f)\right|^2 H_k(\omega_f) \cdot \Omega_{r-1}^H(\omega_f) \Omega_{s-1}(\omega_f) \right)\right] \\ T_k(r,s) &= \mathrm{Re}\left[\sum_{f=1}^{N_f} \left(\left|W_k(\omega_f) H_k(\omega_f)\right|^2 \cdot \Omega_{r-1}^H(\omega_f) \Omega_{s-1}(\omega_f) \right)\right] \end{aligned} \tag{19}$$

If a discrete time-domain model is used, i.e. $\Omega_j(\omega_f) = \exp(-\mathrm{i}\omega_f T_s \cdot j)$, and if the frequencies are uniformly distributed (i.e. $\omega_f = f \cdot \Delta\omega$, $f = 1, \ldots, N_f$, with $\Delta\omega = 2\pi / N T_s$), then, the above summations can be rewritten as

$$\begin{aligned} R_k(r,s) &= \mathrm{Re}\left[\sum_{f=1}^{N_f} \left(\left|W_k(\omega_f)\right|^2 \cdot e^{\mathrm{i}2\pi(r-s)f/N} \right)\right] \\ S_k(r,s) &= -\mathrm{Re}\left[\sum_{f=1}^{N_f} \left(\left|W_k(\omega_f)\right|^2 H_k(\omega_f) \cdot e^{\mathrm{i}2\pi(r-s)f/N} \right)\right] \\ T_k(r,s) &= \mathrm{Re}\left[\sum_{f=1}^{N_f} \left(\left|W_k(\omega_f) H_k(\omega_f)\right|^2 \cdot e^{\mathrm{i}2\pi(r-s)f/N} \right)\right] \end{aligned} \tag{20}$$

One can readily verify that the above matrices have a Toeplitz structure, i.e. entry (r,s) only depends on $(r-s)$.

For instance

$$\mathbf{S}_k = \begin{bmatrix} s_k(0) & s_k(-1) & \ddots & s_k(-n) \\ s_k(1) & s_k(0) & \ddots & \ddots \\ \ddots & \ddots & \ddots & s_k(-1) \\ s_k(n) & \ddots & s_k(1) & s_k(0) \end{bmatrix} \tag{21}$$

with $s_k(r) = -\mathrm{Re}\left[\sum_{f=1}^{N_f} \left(\left|W_k(\omega_f)\right|^2 H_k(\omega_f) \cdot e^{\mathrm{i}2\pi r f/N} \right)\right]$

For symmetric Toeplitz matrices (such as $\mathbf{R}_k$ and $\mathbf{T}_k$) only the first column has to be computed. Moreover, these entries can be computed in a time-efficient way by means of the Fast Fourier Transform (FFT) algorithm.

2.5 *Solving the Reduced Normal Equations*

To remove the parameter redundancy of transfer function model (1) (and to avoid the trivial solution with all coefficient equal to zero), a constraint has to be imposed on the coefficients. This can be done, for instance, by imposing that one of the coefficients is equal to a non-zero constant value. Assume, for instance, that the last coefficient of α is constrained to 1 (i.e. $\alpha(n+1)=1$). In that case, the "reduced" normal equations become

$$\mathbf{A}\cdot\mathbf{x}=\mathbf{b} \tag{22}$$

with

$$\begin{aligned}\mathbf{A}&=\mathbf{M}(1:n,1:n)\\ \mathbf{b}&=-\mathbf{M}(1:n,n+1)\end{aligned} \tag{23}$$

The least-squares estimate of α is given by

$$\hat{\boldsymbol{\alpha}}_{\mathrm{LS}}=\begin{Bmatrix}\mathbf{x}\\1\end{Bmatrix} \tag{24}$$

with $\mathbf{x}=\mathbf{A}^{-1}\cdot\mathbf{b}$. Once $\hat{\boldsymbol{\alpha}}_{\mathrm{LS}}$ is known, $\boldsymbol{\beta}_k=-\mathbf{R}_k^{-1}\mathbf{S}_k\cdot\boldsymbol{\alpha}$ (see (15)) can be used to derive all $\boldsymbol{\beta}_{\mathrm{LS},k}$ coefficients. This approach, which takes into account the structure of the normal equations, is much faster than solving (18) directly (approximately $N_o^2N_i^2$ times faster).

2.6 *Mode Selection Using a Stabilization Chart*

In modal analysis, a stabilization chart is an important tool that is often used to assist the user in separating the physical system poles from mathematical ones. A stabilization chart is obtained by repeating the analysis for increasing model order n. For each model order, the poles are calculated from the estimated denominator coefficients. The stable poles (i.e. the poles with a negative real part in the Laplace domain) are then presented graphically in ascending model order in a so-called "stabilization chart" (see Fig. 1). On the vertical axis the model order is given; the horizontal axis represents the damped natural frequency of the estimated stable poles. The symbols in Figure 6 are used to denote how well poles are stabilizing. E.g., the symbol 's' means that the variation over consecutive model orders of the damped natural frequency is smaller than 1% while the damping ratio varies with less than 5%. Estimated poles corresponding to physically relevant system modes tend to appear for each estimation order at nearly identical locations, while the so-called mathematical poles, i.e. poles resulting from the mathematical solution of the normal equations but meaningless with respect to the physical interpretation, tend to jump around. These mathematical poles are mainly due to the presence of disturbing noise on the measurements.

The LSCE estimator (Least Squares Complex Exponential) is probably the most frequently used technique in industry. The LSCE estimator is a time-domain technique (AR model) that makes use of impulse response functions to derive the modal poles and participation factors. In Fig. 1 the stabilization chart of the LSCE estimator is compared with the frequency-domain least squares estimator. It turns out that in many applications, the frequency-domain estimator is able to generate quite clear stabilization charts compared to the LSCE approach (Guillaume, *et al.*, 1998; Van der Auweraer, *et al.*, 2001).

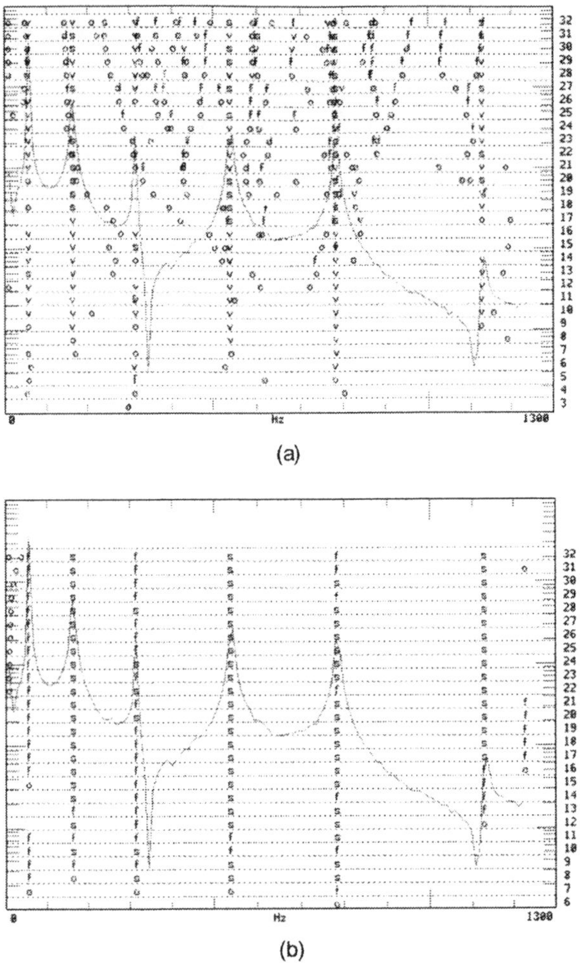

Fig. 1. Stabilization chart obtained with (a) a time-domain estimator (LSCE) and (b) the frequency-domain least-squares estimator (LSCF).

2.7 *The Maximum Likelihood (ML) Approach*

Least-squares techniques are often used in modal analysis because they are able to handle large amounts of data in a reasonable amount of time. However, when the accuracy of a least-squares approach is not good enough, more sophisticated estimators have to be used. In this section a 'dedicated' multivariable implementation of the frequency-domain maximum likelihood estimator will be presented (see Guillaume, *et al.*, 1998).

Assuming the FRFs to be uncorrelated, the (negative) log-likelihood function reduces to

$$\ell_{\mathrm{ML}}(\boldsymbol{\theta})=\sum_{k=1}^{N_oN_i}\sum_{f=1}^{N_f}\frac{\left|\dfrac{N_k(\omega_f,\boldsymbol{\beta}_k)}{d(\omega_f,\boldsymbol{\alpha})}-H_k(\omega_f)\right|^2}{\operatorname{var}\{H_k(\omega_f)\}} \tag{25}$$

Note that the maximum likelihood (ML) problem is equivalent to a weighted nonlinear least-squares one with equation error (see (6))

$$\varepsilon_k^{\mathrm{ML}}(\omega_f,\boldsymbol{\theta})=\frac{1}{\sqrt{\operatorname{var}\{H_k(\omega_f)\}}}\left(\frac{N_k(\omega_f,\boldsymbol{\beta}_k)}{d(\omega_f,\boldsymbol{\alpha})}-H_k(\omega_f)\right) \tag{26}$$

The maximum likelihood estimator takes the quality of the measured FRFs into account: FRF measurements with a small variance, $\mathrm{var}\{H_k(\omega_f)\}$, have an important contribution to the cost function (25) while noisy FRF measurements are penalized.

The maximum likelihood estimate $\hat{\boldsymbol{\theta}}_{\mathrm{ML}}$ is obtained by minimizing (25). This can be done by means of a Gauss-Newton optimization algorithm, which takes advantage of the quadratic form of the cost function (25). The Gauss-Newton iterations are given by

$$\begin{aligned} &\text{(a) solve } \mathrm{Re}(\mathbf{J}_p^H \mathbf{J}_p)\boldsymbol{\delta}_p = -\mathrm{Re}(\mathbf{J}_p^H \mathbf{r}_p) \text{ for } \boldsymbol{\delta}_p \\ &\text{(b) set } \boldsymbol{\theta}_{p+1} = \boldsymbol{\theta}_p + \boldsymbol{\delta}_p \end{aligned} \tag{27}$$

with $\mathbf{r}_p = \mathbf{r}(\boldsymbol{\theta}_p)$, $\mathbf{J}_p = \partial \mathbf{r}(\boldsymbol{\theta})/\partial\boldsymbol{\theta}|_{\boldsymbol{\theta}_p}$ and

$$\mathbf{r}(\boldsymbol{\theta}) = \begin{Bmatrix} \varepsilon_1^{\mathrm{ML}}(\omega_1, \boldsymbol{\theta}) \\ \vdots \\ \varepsilon_1^{\mathrm{ML}}(\omega_{N_f}, \boldsymbol{\theta}) \\ \varepsilon_2^{\mathrm{ML}}(\omega_1, \boldsymbol{\theta}) \\ \vdots \\ \varepsilon_{N_o N_i}^{\mathrm{ML}}(\omega_{N_f}, \boldsymbol{\theta}) \end{Bmatrix} \tag{28}$$

The Jacobian matrix $\mathbf{J}_p$ has the same structure as the matrix $\mathbf{J}$ given in (14). Also here it is possible to form the normal equations (i.e. $\mathrm{Re}(\mathbf{J}_p^H \mathbf{J}_p)$ and $\mathrm{Re}(\mathbf{J}_p^H \mathbf{r}_p)$) in a time-efficient way.

3. FREQUENCY-DOMAIN TECHNIQUES FOR OPERATIONAL MODAL ANALYSIS

Cases exist where it is rather difficult to apply an artificial force and where one has to rely upon available ambient excitation sources. In such cases, it is practically impossible to measure this ambient excitation, and consequently, the responses are the only signals that can be measured. Traditionally, one assumes that the outputs are the results of a stochastic process with white noise sources as inputs. The need to perform output-only modal analysis probably emerged first in civil engineering, where it is very difficult and expensive to excite constructions like bridges and buildings by an artificial excitation that exceeds the natural vibrations due to traffic and wind. Also in mechanical engineering, operational modal analysis proved to be useful: for instance to obtain modal parameters of a car during road testing or an airplane in flight conditions. Often operational test conditions differ from laboratory test conditions, because during the in-operation tests the real loading conditions on the structure are present (suspension pre-strains of a car on the road, aero-elastic interactions, ...). For example, an airplane during flight conditions has a (totally) different dynamical behaviour than an airplane tested in laboratory conditions. Another benefit of output-only modal analysis is the fact that a linear modal of the system is obtained around the real working point of operation.

The relation between the measured outputs (response signals) and the unknown inputs (forces) is given by a transfer function relation in the frequency domain

$$\mathbf{X}(\omega_f) = \mathbf{H}(\omega_f)\mathbf{F}(\omega_f) \tag{29}$$

with $\mathbf{X}$ and $\mathbf{F}$ respectively the spectrum of the output and input signals; $\mathbf{H}$ represents the transfer function matrix. The covariance of the output spectra is given by

$$\begin{aligned} \mathbf{C}_{\mathbf{X}}(\omega_f) &= E\{\mathbf{X}(\omega_f)\mathbf{X}^H(\omega_f)\} \\ &= \mathbf{H}(\omega_f)E\{\mathbf{F}(\omega_f)\mathbf{F}^H(\omega_f)\}\mathbf{H}^H(\omega_f) \end{aligned} \tag{30}$$

with E the expected value. Inserting the modal model and assuming that the unknown forces can be represented by white noise ($E\{\mathbf{F}(\omega_f)\mathbf{F}^H(\omega_f)\} = \mathbf{C}_{\mathbf{F}}$ ($\mathbf{C}_{\mathbf{F}}$ is independent of the frequency), one can show that

$$\mathbf{C}_{\mathbf{X}}(\omega_f) = \sum_{m=1}^{N_m} \left(\frac{\boldsymbol{\psi}_r \mathbf{Q}_r^T}{j\omega_f - \lambda_r} + \frac{\boldsymbol{\psi}_r^* \mathbf{Q}_r^H}{j\omega_f - \lambda_r^*} + \frac{\mathbf{Q}_r \boldsymbol{\psi}_r^T}{-j\omega_f - \lambda_r} + \frac{\mathbf{Q}_r^* \boldsymbol{\psi}_r^H}{-j\omega_f - \lambda_r^*} \right) \tag{31}$$

with $\mathbf{Q}_r$ a vector playing the same role as a modal participating vector, but without any physical meaning. If λ_r is a physical pole then $\boldsymbol{\psi}_r$ will correspond with a mode shape vector of dimension N_o (number of output signals).

Beside the quadrate symmetry of the poles, (31) does not differ from a modal model. So, one concludes that all algorithms developed to estimate modal parameters from FRF data can still be applied with this approach.

4. COMBINED FREQUENCY-DOMAIN IDENTIFICATION TECHNIQUES

In some in-operational testing applications, exciters are used to inject more energy in the system. This is for instance done during flight flutter tests, where artificial forces are applied on the wings of the airplane using special equipment.
In that case, input signals are available and it is again possible to use classical EMA identification techniques to estimate the modal parameters from the input/output (or FRF) measurements.
However, by doing so, the effect on the natural forces will be treated as ambient "noise". Traditional EMA techniques will remove this "noise" by averaging the measurements. This is in contradiction with the output-only approach were the modal parameters are estimated using response data due to the ambient excitation only. Clearly, the ambient noise is not just "noise" but it contains useful information about the system.

To make an optimal use of the operational data, both measured (artificial) and unmeasured (natural) forces should be taken into account. By doing so, all the available information in the measured data can be optimally used.

One can write that output o consists of two contributions: a forced part due to the applied (and measured) forced and a second unknown part due to the ambient excitation, which is modelled by means of stationary white noise sources $E_o(\omega_f)$,

$$X_o(\omega_f) = \underbrace{\frac{\mathbf{N}_o(\omega_f)}{d(\omega_f)}\mathbf{F}(\omega_f)}_{\text{FORCED}} + \underbrace{\frac{M_o(\omega_f)}{d(\omega_f)}E_o(\omega_f)}_{\text{AMBIENT}} \quad (32)$$

Note that both (forced and ambient) transfer function models have the same poles (common denominator). Assuming the natural forces $E_o(\omega_f)$ to be normally distributed, the transfer functions occurring in (32) can be identified by minimizing (see, Cauberghe, *et al.*)

$$\ell_{\text{CLSF}}(\boldsymbol{\theta}) = \sum_{o=1}^{N_o}\sum_{f=1}^{N_f} \frac{\left|d(\omega_f,\boldsymbol{\alpha})X_o(\omega_f) - \mathbf{N}_o(\omega_f,\boldsymbol{\beta}_o)\mathbf{F}(\omega_f)\right|^2}{\left|M_o(\omega_f,\boldsymbol{\delta}_o)\right|^2} \quad (33)$$

The main idea of the combined approach is illustrated in Fig. 2. The third mode (around 4.7 Hz) is not well excited by the applied force (dotted green line). However, it is well excited by the ambient forces (dashed red line). Traditional FRF-based identification techniques, which try to eliminate the contribution of the process noise, will not be able to well estimate this mode. For this mode, an output-only approach would even be better. The combined experimental/operational approach uses the information available in the forced as well as ambient contribution resulting in improved estimates of all modes.

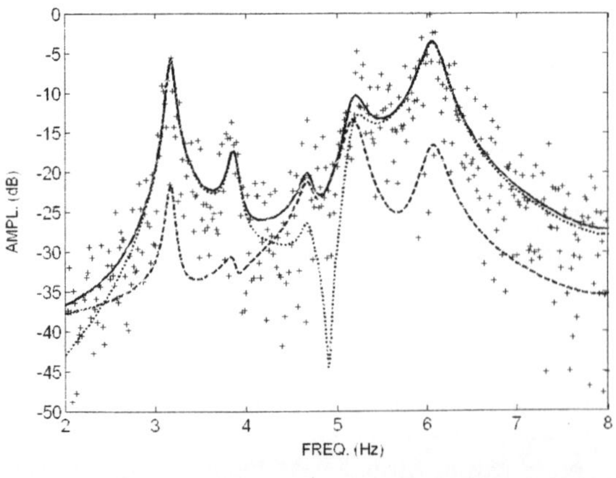

Fig. 2. Combined experimental/operational approach. Crosses: Measured auto-power spectrum of the response. Solid line: estimated auto-power spectrum. Dotted line: estimated forced contribution. Dashed line: estimated ambient contribution.

5. CONCLUSIONS

In this paper dedicated frequency-domain estimators for experimental modal analysis have been described. Attention has been paid to operational modal analysis and the idea of a combined experimental/operational identification approach has been introduced.

REFERENCES

Cauberghe B., P. Guillaume, P. Verboven, and E. Parloo. Identification of modal parameters including unmeasured forces and transient effects. *Journal of Sound and Vibration*, accepted for publication.

Guillaume P., P. Verboven and S. Vanlanduit (1998), Frequency-Domain Maximum Likelihood Identification of Modal Paramaters with Confidence Intervals. *Proceedings of the International Conference on Noise and Vibration Engineering (ISMA-23)*, Leuven (Belgium), September 16-18, pp. 359-366.

Kailath, T. (1980). *Linear Systems*. Prentice Hall.

Pintelon R., P. Guillaume, Y. Rolain, J. Schoukens and H. Van hamme (1994). Parametric Identification of Transfer Functions in the Frequency Domain - a Survey. *IEEE Trans. Autom. Control*, vol. AC-39, no. 11, pp. 2245-2260.

Pintelon R. and J. Schoukens (2001). *System Identification: A Frequency Domain Approach*. IEEE Press and John Wiley & Sons.

Van der Auweraer H., P. Guillaume, P. Verboven and S. Vanlanduit (2001). Application of a Fast-Stabilizing Frequency Domain Parameter Estimation Method. *ASME Journal of Dynamic Systems, Measurement and Control*, vol. 123, no. 4, pp. 651-658.

Verboven P. (2002). *Frequency Domain System Identification for Modal Analysis*, PhD Thesis, Vrije Universiteit Brussel (VUB).

IFAC Publications
www.elsevier.com/locate/ifac

DATA-DRIVEN MODELING OF NONLINEAR AND TIME-VARYING PROCESSES

Dennis Bonné * **Sten Bay Jørgensen** *

* *Department of Chemical Engineering, Technical University of Denmark, DK-2800 Lyngby, Denmark, Email: (db or sbj)@kt.dtu.dk, Fax: +45 4593 2906*

Abstract: In this contribution, it is proposed using a novel interpretation of generalized ridge regression to identify a 1-dimensional grid of interdependent linear models from operation data. Such 1-dimensional model grids can be used to model repeated finite horizon, nonlinear and non-stationary process operations. These finite horizon process operations include chemical process operations such as start-ups, grade transitions, shut-downs, and of course batch, semi-batch and periodic processes. Explicitly, it is proposed to identify sets of interdependent linear models using modified generalized ridge regression/Tikhonov regularization that penalizes weighted discrepancies between one linear model and the models in its neighborhood. Penalizing weighted discrepancies between neighboring linear models induce both model interdependency and designed model properties.

Keywords: Parameter estimation, Non-linear systems, Time-varying systems

1. INTRODUCTION

Production strategies such as products-on-demand and first-to-marked have impelled the need for flexible and specialized production methods during the last decade. In the chemical production industry, the products-on-demand strategy necessitates frequent grade-transitions in plants that are otherwise operated continuously at steady state. The products-on-demand strategy may also imply the traditional continuous type process equipment is replaced with flexible batch type process equipment, which facilitates production of more diverse products. Furthermore, during design and scale-up most products are produced in batch type operations. This means that time-to-marked can be reduced if the desired production capacity can be achieved with multiple (pilot scale) batch type production units. Such batch type operations are however nonlinear and non-stationary. Hence, the traditional steady state considerations and linear process approximations do not apply these batch type operations.

In section 2 it is thus proposed to model batch type processes with a time-varying grid of linear models. Identification of these model grids is addressed in section 3 where it is proposed to use model property based regularization to overcome excessive variance. The methods proposed in sections 2 and 3 are applied to an industrial case study in section 4 where it is demonstrated how models may be obtained both time and cost efficiently from historical process data. Finally conclusions are given.

2. TIME-VARYING MODELS

In chemical process plants several batch type operations are repeated as part their continuous operation. These batch type operations are typically characterized by finite horizons and nonlinear and non-stationary dynamic behavior.

Most often the nonlinear dynamics of infinite horizon, stationary process operations can be approximated with a moderate set of local linear time-invariant models, each of which describes a characteristic region in the operation window. These regions described by local models, will often be characterized by a set of active constraints. For non-stationary process operations however, the set of active constraints will change with time. In fact, to operate finite horizon processes in an optimal fashion, a specific sequence of constraints is tracked during operation. This means that local approximations of characteristic regions are not sufficient to describe the process operation. The transitions between these locally approximated characteristic regions are also needed to provide a complete description of non-stationary process operation.

The periodicity of repeated finite horizon process operations however, make it possible to model the evolution from each sample point to the next in an operation with one *grid-point* LTI model. In this fashion, both the time variation within the characteristic regions and the transitions between these may be approximated with a grid of grid-point models. Thus, such model sets give a complete description of non-stationary process operations. The finite horizon means that the model set will be finite. The periodic way in which the same process operation is repeated means that several measurements from the individual sample points are available for identification. That is, the time evolution of a process variable is sampled at specific sample points during the finite horizon operation, and as the finite horizon operation is repeated, several measurements are collected from every sample point. With multiple data points from one specific sample point, a grid-point model can be identified for this sample point. Explicitly, in addition to the time dimension, data from finite horizon process operations also evolve in an operation index dimension. This additional dimension in data from repeated finite horizon process operations is illustrated in Figure 1.

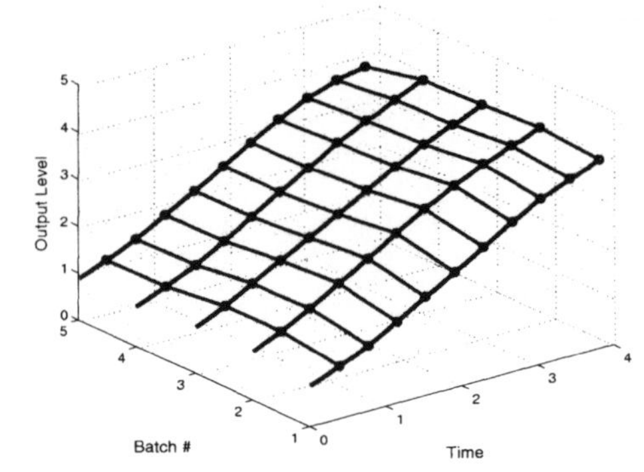

Fig. 1. The three dimensions of data from repeated finite horizon process operations, i.e., output level, time, and operation/batch index. The circles represent the sample points. It is demonstrated how measurements from the same sample point can be collected from consecutive operations, thus facilitating identification of a grid-point model for each sample point.

2.1 Model Parameterization

Given the discussion above, repeated finite horizon process operations are modeled with sets of dynamic grid-point LTI models. Such a set of grid-point LTI models could also be referred to as a Linear Time Varying (LTV) model. These grid-point LTI models can be parameterized in a number of ways – e.g. as Output Error (OE) models, AutoRegressive models with eXogenous inputs (ARX), State Space (SS) models, etc. In the present contribution the ARX model parameterization was chosen. This choice of parameterization offers a relatively good multivariable system description with a moderate number of model parameters.

As repeated finite horizon process operations progresses, different inputs and outputs may be used depending on the current phase of the operation and hence in order to model repeated finite horizon process operations it is convenient to define the following variables and references for each time step t: Input variable $u_t \in \mathbb{R}^{n_u(t)}$ with reference $\bar{u}_t \in \mathbb{R}^{n_u(t)}$, output variable $y_t \in \mathbb{R}^{n_y(t)}$ with reference $\bar{y}_t \in \mathbb{R}^{n_y(t)}$, and disturbance variable $w_t \in \mathbb{R}^{n_y(t)}$. Using an ARX model parameterization, the output deviation $\bar{y}_t - y_t$ at time t may be given as a weighted sum of $n_A(t)$ past output deviations and $n_B(t)$ past input deviations

$$
\begin{aligned}
\bar{y}_t - y_t = & - a_{t,t-1}\Big(\bar{y}_{t-1} - y_{t-1}\Big) - \dots \\
& - a_{t,t-n_A(t)}\Big(\bar{y}_{t-n_A(t)} - y_{t-n_A(t)}\Big) \\
& + b_{t,t-1}\Big(\bar{u}_{t-1} - u_{t-1}\Big) + \dots \\
& + b_{t,t-n_B(t)}\Big(\bar{u}_{t-n_B(t)} - u_{t-n_B(t)}\Big) \\
& + w_t
\end{aligned} \tag{1}
$$

where $n_A(t), n_B(t) \in [1, \dots, t]$ are the grid-point ARX model orders and $a_{i,j} \in \mathbb{R}^{n_y(i),n_y(j)}$ and $b_{i,j} \in \mathbb{R}^{n_y(i),n_u(j)}$ are the grid-point ARX model parameter matrices. Note, as the grid points are modeled with individual grid-point models, the sample points t do not have to be equidistantly spaced in time. Let N be the process operation length (or number of samples) and define the input $\boldsymbol{u}$, output $\boldsymbol{y}$, shifted output $\boldsymbol{y}^0$, and disturbance $\boldsymbol{w}$ profiles as

$$\boldsymbol{u} = \begin{bmatrix} u_0' \; u_1' \; \dots \; u_{N-1}' \end{bmatrix}'$$
$$\boldsymbol{y} = \begin{bmatrix} y_1' \; y_2' \; \dots \; y_N' \end{bmatrix}' \tag{2}$$
$$\boldsymbol{y}^0 = \begin{bmatrix} y_0' \; y_1' \; \dots \; y_{N-1}' \end{bmatrix}'$$
$$\boldsymbol{w} = \begin{bmatrix} w_1' \; w_2' \; \dots \; w_N' \end{bmatrix}'$$

Note, not all initial conditions y_0 are measurable and/or physically meaningful — e.g. off-gas measurements. Thus the ARX model set may be expressed in matrix form

$$\bar{\boldsymbol{y}} - \boldsymbol{y} = -\boldsymbol{A}(\bar{\boldsymbol{y}}^0 - \boldsymbol{y}^0) + \boldsymbol{B}(\bar{\boldsymbol{u}} - \boldsymbol{u}) + \boldsymbol{w} \tag{3}$$

where $\boldsymbol{A}, \boldsymbol{B}$ are structured lower block triangular matrices. To exemplify, if it is assumed that $n_A(t) = n_A$ and $i - n_A = 0$, then $\boldsymbol{A}$ has the following structure

$$\boldsymbol{A} = \begin{bmatrix} a_{1,0} & & & & \\ \vdots & \ddots & & & \\ a_{i,0} & & a_{i,i-1} & & \\ & \ddots & \vdots & \ddots & \\ & & a_{N,i-1} & \cdots & a_{N,N-1} \end{bmatrix}$$

The profile $\boldsymbol{w}$ is a sequence of disturbance terms caused by bias in the reference input profile $\bar{\boldsymbol{u}}$, the effect of process upsets, and the modeling errors from linear approximations. This means that the disturbance $\boldsymbol{w}$ contains contributions from both repeated disturbances, such as recipe/input bias, model bias, and erroneous sensor readings, as well as from random disturbances, which occur with no correlation from operation to operation. It thus seems resoanble to model the disturbance profile $\boldsymbol{w}$ with a random walk model with respect to the operation index k

$$\boldsymbol{w}_k = \boldsymbol{w}_{k-1} + \boldsymbol{v}_k \tag{4}$$

where $\boldsymbol{v}_k = [v_{k,1}', v_{k,2}', \dots, v_{k,N}']', v_{k,t} \in \mathbb{R}^{n_y(t)}$, represents a sequence of non-repeated disturbances that are assumed to be zero-mean, independent and identically distributed. Considering the difference between two successive operations

$$\begin{aligned} \Delta \boldsymbol{y}_k &= \boldsymbol{y}_k - \boldsymbol{y}_{k-1} \\ &= \boldsymbol{A}(\boldsymbol{y}_k^0 - \boldsymbol{y}_{k-1}^0) - \boldsymbol{B}(\boldsymbol{u}_k - \boldsymbol{u}_{k-1}) \\ &\quad + \boldsymbol{w}_k - \boldsymbol{w}_{k-1} \\ &= \boldsymbol{A}\Delta \boldsymbol{y}_k^0 - \boldsymbol{B}\Delta \boldsymbol{u}_k + \boldsymbol{v}_k \end{aligned} \tag{5}$$

an ARX model (5) that is independent of the reference profiles $(\bar{\boldsymbol{y}}, \bar{\boldsymbol{u}})$ and repeated disturbances has been obtained. With such an ARX model the path is prepared for multivariable, model-based monitoring, control, optimization, and of course simulation.

3. MODEL IDENTIFICATION

With the ARX model (5) derived above, the parameterization of the model is in place, however the model orders and the model parameters still need to be determined from process data. One major drawback of the proposed parameterization is the immense dimensionality of the resulting set of models — in practice this immense dimensionality will render any standard Least Squares (LS) identification problem ill-conditioned. It turns out however, that as the grid-point models are progressively constrained by the smoothness of the model grid, the conditioning of the identification problem improves.

3.1 Data Pretreatment

In industry, the process variables $\tilde{z}_{k,\tilde{t}}(p) \in \mathbb{R}$ are most often logged individually at times $\tilde{T}(k, \tilde{t}, p)$, giving $N_{\tilde{z}}(k,p) + 1$ observations of variable p in operation k. What is needed however, is up to $N + 1$ noise free observations of the variables at times $T(t)$ in the N_B process repetitions available for identification. These noise free or expected observations can be estimated using local polynomial regression and If the profile of process variable p in operation k is defined as $\tilde{\boldsymbol{z}}_{k,p}$, then the estimation problem can be given explicitly (Hastie *et al.*, 2001) as

$$\hat{\bar{z}}_{k,t}(p) = \boldsymbol{s}_{k,p,t}\tilde{\boldsymbol{z}}_{k,p} \tag{6}$$

where $\boldsymbol{s}_{k,p,t}$ is a smoothing vector. If it is further assumed that process variable p will be used throughout the operation, then the estimated profile of variable p in operation k is given as

$$\begin{aligned} \hat{\bar{\boldsymbol{z}}}_k(p) &= \begin{bmatrix} \boldsymbol{s}_{k,p,0}' \; \boldsymbol{s}_{k,p,1}' \; \dots \; \boldsymbol{s}_{k,p,N}' \end{bmatrix}' \tilde{\boldsymbol{z}}_k(p) \\ &= \boldsymbol{S}_{k,p}\tilde{\boldsymbol{z}}_k(p) \end{aligned} \tag{7}$$

Let the true observation $\bar{\boldsymbol{z}}_k \in \mathbb{R}^{(n_y(t)+n_u(t))}$ be given as

$$\bar{\boldsymbol{z}}_k = \hat{\bar{\boldsymbol{z}}}_k + \boldsymbol{\omega}_k \tag{8}$$

where $\hat{\bar{\boldsymbol{z}}}_k$ is the estimated observation and $\boldsymbol{\omega}_k$ is a sequence of estimation errors. The estimation error $\boldsymbol{\omega}_k$ will consist of both systematic errors such as the height of a characteristic peek being underestimated due to excessive smoothing and/or *trimming the hills and filling the valleys* due to too low local regression order, and random errors. Thus the estimation error $\boldsymbol{\omega}_k$ is modeled with a random walk with respect to the operation index k

$$\boldsymbol{\omega}_k = \boldsymbol{\omega}_{k-1} + \boldsymbol{\nu}_k \tag{9}$$

where $\boldsymbol{\nu}_k$ represents a sequence of non-repeated estimation errors that are assumed to be zero-mean. Consider the expected difference between two successive operations, then

$$E\{\Delta\bar{\boldsymbol{z}}_k\} = \hat{\bar{\boldsymbol{z}}}_k - \hat{\bar{\boldsymbol{z}}}_{k-1} + E\{\boldsymbol{\nu}_k\} = \Delta\hat{\bar{\boldsymbol{z}}}_k \tag{10}$$

is given as the difference between their respective estimates. The expected output and input difference profiles which are all contained in $\Delta\tilde{z}_k$, are thus given as

$$\begin{aligned} E\{\Delta y_k\} &= \Delta\hat{y}_k, \quad E\{\Delta y_k^0\} = \Delta\hat{y}_k^0 \\ E\{\Delta u_k\} &= \Delta\hat{u}_k \end{aligned} \tag{11}$$

3.2 Parameter Estimation

Several suggestions to how (sets of) LTI or (periodic) LTV models should be identified from data can be found in literature. All these authors employ some or other coefficient shrinkage or subspace method to improve the conditioning of the identification problem and hence lower the variance of the model parameter estimates. Simoglou *et al.* (2002) suggested estimating a set of independent, overlapping local LTI SS models using Canonical Variant Analysis (CVA). Instead the present contribution proposes estimating a grid/set of interdependent grid-point LTI ARX models using a novel interpretation of generalized ridge regression.

The ARX model (5) can be formulated as linear regression

$$\Delta y_k = \Delta x_k \theta + v_k \tag{12}$$

where $\Delta x_k = \Delta x_k(\Delta y_k^0, \Delta u_k)$ is a structured regressor matrix with past outputs and inputs and $\theta = \theta(A, B)$ is a column parameter vector with the model parameters from the ARX model. Taking the expectation of the linear regression (12) and recalling (11) we find that

$$\Delta\hat{y}_k = E\{\Delta x_k\}\theta + E\{v_k\} = \Delta\hat{x}_k\theta \tag{13}$$

with $\Delta\hat{x}_k = \Delta x_k(\Delta\hat{y}_k^0, \Delta\hat{u}_k)$. This means that, if the process variable estimation error model (9) is a valid approximation, then estimation of model parameters from the pretreated data will give unbiased model parameter estimates. Although unbiased, model parameter estimates based on data from a single operation would have excessive variance. Thus to lower the variance of model parameter estimates, all available data should be used for the model parameter estimation

$$\begin{aligned} Y &= \left[\Delta\hat{y}'_1 \; \Delta\hat{y}'_2 \; \dots \; \Delta\hat{y}'_{N_B}\right]' \\ &= \left[\Delta\hat{x}'_1 \; \Delta\hat{x}'_2 \; \dots \; \Delta\hat{x}'_{N_B}\right]' \theta = X\theta \end{aligned} \tag{14}$$

The linear system (14) would however, most likely still be rank-deficient and solving it in a Least Squares (LS) sense would still produce estimates with low bias, but excessive variance. Such excessive model parameter variance would despite the low bias, yield poor model predictions (Larimore, 1996). Hence, to improve the predictive capabilities of an estimated model the variance of the estimated model parameters must be further reduced.

A possible approach to reducing the variance of model parameter estimates is to enforce that the estimated model possesses some desired model properties. One such model property could be that neighboring grid-point models are analogous in the sense that they exhibit similar behavior. In fact, without this property, the model would be a *set* of independent models and not a *grid* of interdependent models. Enforcing model properties however, inevitably introduce bias into the model parameter estimates. There will thus be a trade-off between the bias and variance of the model parameter estimates and this trade-off will determine the predictive capabilities of estimated models. A parameter estimation method that could incorporate model properties into LS estimates is generalized ridge regression, which also is referred to as Tikhonov regularization. We thus propose to estimate the model parameters by solving the extended LS problem

$$\begin{aligned} \hat{\theta}_\Lambda &= \arg\min_{\theta}[(Y - X\theta)'(Y - X\theta) \\ &\qquad + (\Lambda L\theta)'(\Lambda L\theta)] \\ &= \left(X'X + L'\Lambda^2 L\right)^{-1} X'Y \end{aligned} \tag{15}$$

where the penalty $\Lambda L\theta$ is a column vector of weighted differences between parameters in neighboring grid-point models. E.g. one element in $\Lambda L\theta$ could be the difference between the lag one autoregression of output j onto output i (on the 0th subdiagonal in A) at times t and $t+1$; $\lambda_{t,i,j}^{A,D_0}(a_{t,t-1}(i,j) - a_{t+1,t}(i,j))$ and another element could be the difference between the lag one regression of input j onto output i at time t and the lag two regression of input j onto output i at time $t+1$ (in column j of the tth block column in B); $\lambda_{t,i,j}^{B,C_j}(b_{t,t-1}(i,j) - b_{t+1,t-1}(i,j))$. In this fashion, the structured penalty matrix L maps the parameter vector θ into the desired parameter differences and the diagonal regularization matrix Λ weights the parameter differences. The estimated parameter vector $\hat{\theta}_\Lambda$ is a function of the regularization matrix Λ, which determines the shrinkage and hence the trade-off between bias and variance. This means that the regularization matrix Λ can be used to tune the predictive capabilities of the model estimate. Through the particular choice of penalty matrix L, the regularization matrix Λ also determines the interdependency between the grid-point models in the model grid.

3.3 Model Orders and Regularization Weights

Several methods for choosing (optimal) regularization weights can be found in literature (Hansen, 1996), but all of these consider either scalar regularization weights or diagonal penalties. In the present work it is proposed to simply select a regularization matrix from a finite set $\Lambda \in \Omega_\Lambda$, that yield near minimum mean squared prediction error, when the estimated model is cross-validated through *pure-simulation.* That is, given the *pure-simulation* prediction error profile $\boldsymbol{\zeta}_{\boldsymbol{\Lambda},k}$ from cross-validation operation k

$$\boldsymbol{\zeta}_{\boldsymbol{\Lambda},k} = \Delta\hat{\boldsymbol{y}}_k^{val} - \boldsymbol{H}(\hat{\boldsymbol{\theta}}_{\boldsymbol{\Lambda}})\Delta\hat{y}_{k,0}^{val} + \boldsymbol{G}(\hat{\boldsymbol{\theta}}_{\boldsymbol{\Lambda}})\Delta\hat{\boldsymbol{u}}_k^{val} \quad (16)$$

the regularization matrix $\boldsymbol{\Lambda}$ is the solution to the discrete optimization problem

$$\begin{aligned} \boldsymbol{\Lambda} = \arg\min_{\boldsymbol{\Lambda}} & \left[\gamma_{\boldsymbol{\Lambda}} = \sum_{k=1}^{N_B^{val}} \boldsymbol{\zeta}'_{\boldsymbol{\Lambda},k}\boldsymbol{\zeta}_{\boldsymbol{\Lambda},k} \right] \\ s.t. \quad & \boldsymbol{\Lambda} \in \boldsymbol{\Omega}_{\boldsymbol{\Lambda}} \\ & (\boldsymbol{X}'\boldsymbol{X} + \boldsymbol{L}'\boldsymbol{\Lambda}^2\boldsymbol{L}) \text{ nonsingular} \end{aligned} \quad (17)$$

where N_B^{val} is the number of difference profiles available for cross-validation. In this fashion, the computational burden of solving (17) is determined by the number of elements in the finite set $\boldsymbol{\Omega}_{\boldsymbol{\Lambda}}$.

Thus far only estimation of a specific ARX parameterization, i.e., an ARX model with model orders $n_A(t)$ and $n_B(t)$ for $t = 1, \ldots, N$ has been considered. These model orders are however unknown and will also have to be identified from data. This means that in addition to the regularization weighting matrix $\boldsymbol{\Lambda}$ also the model orders can be used to tune the predictive capabilities of the model estimate. Traditionally, ARX model orders are selected based on minimization of measures such as Final Prediction Error (FPE) or Akaike's Information Criterion (AIC) both of which are proportional to the optimal value of the LS objective being minimized as part of the identification, to prevent modeling noise/disturbance characteristics, i.e., overfit. Overfit is however also prevented if the ARX model orders are selected based on minimization of the mean squared prediction errors from cross-validation of the estimated models. This means that the ARX model orders can be selected based on minimization of

$$\begin{aligned} \begin{bmatrix} \{n_A(t)\}_{t=1}^N \\ \{n_B(t)\}_{t=1}^N \end{bmatrix} = & \min_{\substack{\{n_A(t)\}_{t=1}^N \\ \{n_B(t)\}_{t=1}^N}} [\gamma_{\boldsymbol{\Lambda}}] \\ & s.t. \quad \boldsymbol{\Lambda} \text{ given by (17)} \end{aligned} \quad (18)$$

If the ARX model orders are assumed constant throughout the operation $n_A = n_A(t)$ and $n_B = n_B(t)$ for $t = 1, \ldots, N$, then (18) simplifies to

$$\begin{aligned} (n_A, n_B) = & \min_{n_A, n_B} [\gamma_{\boldsymbol{\Lambda}}] \\ & s.t. \quad \boldsymbol{\Lambda} \text{ given by (17)} \end{aligned} \quad (19)$$

4. APPLICATION

To demonstrate the capability of the proposed data-driven models, an industrial, production scale *Bacillus protease* fermentation has been modeled from historical data (Novozymes A/S). The modeling objective was prediction of the on-line measured variables used to supervise the fermentation as well as the product activity (PA) which is measured sparsely off-line. That is, the objective was to obtain a model that can predict the course and outcome of a batch given the batch recipe and its initial conditions. For this *Bacillus protease* fermentation the batch recipe consists (essentially) of reference profiles for two substrate/nutrient feeds (S_1 and S_2), an alkaline feed (S_3), pressure (P), and temperature (T) — i.e., these reference profiles are the inputs of the model. For most of these inputs the current practice is however, that the reference profile is either not logged or not manipulated from batch to batch. As a temporary workaround, perfect control was assumed and the reference profiles were replaced by the realized profiles. As is common practice in supervision of fermenters, Dissolved Oxygen (DO), Carbon dioxide Evolution Rate (CER), Oxygen Uptake Rate (OUR), and pH were chosen as process indicators. Along with the product activity these process indicators makeup the outputs of the model. All inputs and outputs are used throughout the batch ($n_u(t) = n_u = 5$ and $n_y(t) = n_y = 5$ for $t = 0, \ldots, N-1$ and $t = 1, \ldots, N$, respectively).

The historical on-line data was smoothened and re-sampled to 30 minutes intervals using local constant regression with a tri-cube Kernel and between 6 – 73 nearest neighbors. The historical product activity data was sub-sampled to 30 minutes intervals with linear interpolation. Before identification the batch difference profiles were normalized. The model was identified using data from 29 batches and cross-validated using data from 9 batches.

In the model parameter estimation, both the time evolution of the model parameters (e.g. $(a_{t,t-1}(i,j) - a_{t+1,t}(i,j))$) and the non-smoothness of the step responses of the estimated model (e.g. $(b_{t,t-1}(i,j) - b_{t+1,t-1}(i,j))$) were penalized. To ease the computational burden of selecting optimal ARX model orders, the ARX model orders were assumed constant throughout the batch ($n_A(t) = n_A$ and $n_B(t) = n_B$ for $t = 1, \ldots, N$). The identified regularization weights ranged between 0 and 8.39E5. The identified ARX model orders are given in Tables 1 and 2.

The quality of the identified ARX model was quantified through the mean of the AIC and FPE from the 29 identification batches and the mean of the mean prediction error norm (FIT = $\sqrt{\gamma_{\boldsymbol{\Lambda}}}/(Nn_y)$) from the 9 cross-validation batches. These quality measures are listed in Table 3. One of the cross-validation batches is given in Fig. 2.

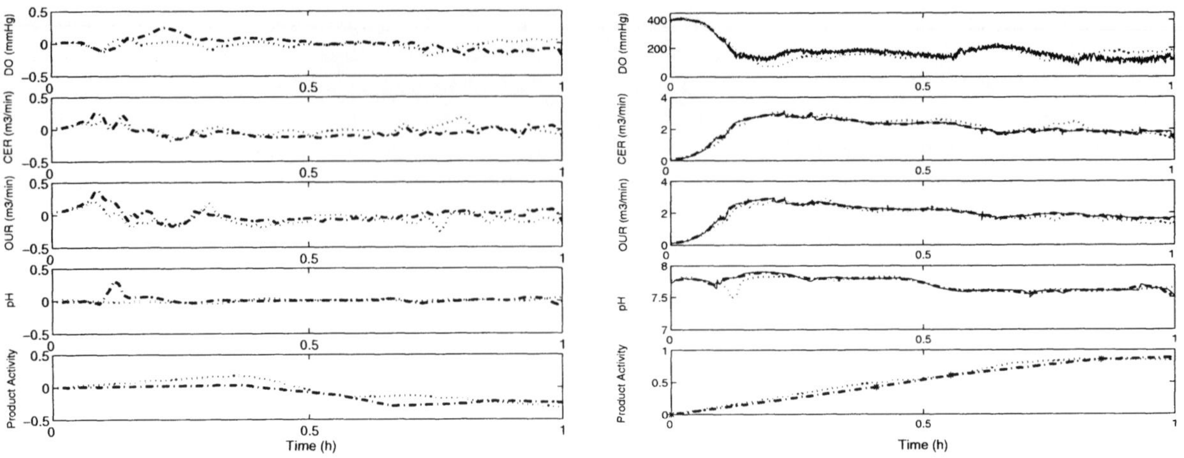

Fig. 2. Example of cross-validation of an industrial *Bacillus protease* fermentation model. The five outputs, Dissolved Oxygen (DO), Carbon dioxide Evolution Rate (CER), Oxygen Uptake Rate (OUR), pH, and product activity, are predicted given information about their initial conditions and the batch recipe (assuming perfect control) — i.e., *pure-simulation* prediction. In the left figure the difference profiles are given and in the right figure the profiles themselves are given. The solid lines (or '∗' for the product activity) are the historical measurements as logged (only shown on the right figure), the dash-dotted lines are the pretreated data (as the data from which the model was identified, but not used in the identification), and the dotted lines are the model predictions.

5. CONCLUSION

In the present paper it is proposed to model finite horizon, time-varying and nonlinear process operations with 1-dimensional grids of interdependent ARX models. Such model grids can be used for both off- and on-line monitoring, prediction, control, and optimization applications. It is further proposed that these model grids are obtained time and cost efficiently through identification from historical process data using ridge regression. By identifying all the ARX models in a model grid simultaneously, the interdependency of the ARX models can be used to reduce the variance of their estimates and thereby improve the predictive capabilities of the estimated model grid. The proposed data-driven modeling scheme has been demonstrated through modeling of an industrial fermentation process.

Table 1. Identified AutoRegressive (***A***) *Bacillus protease* fermentation model orders.

	DO	OUR	CER	pH	PA
DO	2	0	0	0	0
CER	0	2	1	0	1
OUR	1	0	9	3	1
pH	0	0	0	2	1
PA	0	0	6	1	2

Table 2. Identified eXogenous input (***B***) *Bacillus protease* fermentation model orders.

	S_1	S_2	T	P	S_3
DO	2	2	2	6	2
CER	2	2	2	2	1
OUR	2	0	0	4	7
pH	1	2	3	2	1
PA	0	20	12	7	2

REFERENCES

Hansen, Per Christian (1996). *Rank-Deficient and Discrete Ill-Posed Problems.* Polyteknisk Forlag. Lyngby.

Hastie, Travor, Robert Tibshirani and Jerome Friedman (2001). *The Elements of Statistical Learning.* Springer. New York.

Larimore, Wallace E. (1996). Statistical optimality and canonical variate analysis system identification. *Signal Processing* **52**(2), 131–144.

Simoglou, A., E. B. Martin and A. J. Morris (2002). Statistical performance monitoring of dynamic multivariate processes using state space modelling. *Computers and Chemical Engineering* **26**(6), 909 – 920.

Table 3. Quantified quality of the identified *Bacillus protease* fermentation model. Note, the model quality is quantified in terms of normalized variables.

Measure	Value
Mean AIC	6.74
Mean FPE	2687
Mean FIT	0.123

IFAC
Publications
www.elsevier.com/locate/ifac

PID PARAMETER CYCLING TO TUNE INDUSTRIAL CONTROLLERS: A NEW MODEL-FREE APPROACH

James Crowe **Michael A. Johnson** **Michael J. Grimble**

Industrial Control Centre, University of Strathclyde,
Glasgow, Scotland, U. K.
Author for Correspondence Email: m.johnson@eee.strath.ac.uk

Abstract: A brief review of the background to the method of Iterative Feedback Tuning is presented. This introduces the idea of optimising a cost function to produce restricted structure controllers or to benchmark controllers for performance assessment. The new method of controller parameter cycling is introduced to achieve the same objectives as the IFT method. An advantage of the new method is the ease with which Hessian information is obtained to enhance the optimisation procedure. A multivariable example demonstrating the controller parameter cycling method is given. *Copyright © 2003 IFAC*

Keywords: model-free, iterative feedback tuning, optimal control, restricted structure control, process control.

1. INTRODUCTION

Today in the Process Industries, the first choice for a control algorithm is still usually PID although this situation is changing slowly (Blevins *et al*, 2002). The main reason for this lies in the fact that most industry based control practitioners have an understanding of how the closed loop performance of particular processes will be affected by changes in the PID controller parameters. Where there is a gap in the process knowledge, rudimentary parametric and non-parametric data may be determined by performing step tests, by the application of a relay experiment (Astrom and Hagglund, 1985; Yu, 1999) or using a phase locked loop test (Crowe and Johnson, 1998; 1999). If the control practitioner has such parametric or non-parametric data then rule based methods can be used to determine a PID controller (O'Dwyer, 2001a; 2001b).

Process operators are often under economic pressure to maximise the production from industrial processes and to reduce the cost of production. One means of helping to achieve these goals is to benchmark the existing controller against some form of performance index value (Harris, 1989; Thornhill *et al*, 1999). This type of approach entails the acceptance of a simple mathematical index as capturing desired control performance. For example, the Harris (1989) method uses the minimum variance criterion as a benchmark index function for controller performance assessment. It is a short step from minimum variance indices to LQ and LQG cost functions as the benchmark criterion. The use of LQ and LQG cost functions has lead to recent research focussing on (a) more extended benchmarking formulations and (b) restricted structure control problems.

A key problem with both LQ/LQG benchmarking and restricted structure areas has been a requirement for an explicit parametric process model on which to base the necessary computations. In quite a different approach, particularly to the solution of restricted structure controller problems, Hjalmarsson and colleagues have performed cost function optimisation using only input and output data from the process without the intermediate step of having to produce an explicit process model (Hjalmarsson *et al*, 1994; 1998; Lequin *et al*, 1999). This technique is called Iterative Feedback Tuning (IFT). A frequency domain version is given by Kammer *et al* (2000). This avoidance of an *explicit* identified process model in the method has often led this approach to be termed a *model-free* approach.

The key step in Iterative Feedback Tuning is the computation of the cost function gradient so that the cost function can be optimised with respect to the controller parameters. Hjalmarsson and colleagues would also like to compute second order information residing in the cost function Hessian. Unfortunately, the Hjalmarsson IFT method is rather cumbersome at extracting the desired second order information, and this difficulty was one of the motivating reasons for looking again at the IFT method. Hence, this paper also follows the theme of optimising a cost function using an *implicit* model or *model-free* approach. However instead of using special system inputs to compute gradients, controller parameter cycling is used to find the same quantities, and also produce the second order Hessian information in much more direct manner. The theory of the new method used to minimise a LQ cost function online with respect to the parameters of a restricted structure controller is given in Section 2 of the paper. In Section 3 some implementation issues are addressed, and in section 4, an application example is given for the design of a restricted structure decentralised (PID) controller for a (2×2) multivariable process. Conclusions and references close the paper.

2. ITERATIVE COST MINIMIZATION

The seminal papers for the Iterative Feedback Tuning method due to Hjalrmarsson *et al* (1994; 1998) use a system description involving a stochastic process output disturbance, a two-degrees of freedom control law, and a stochastic optimisation approach In this paper a deterministic LQ optimal control problem is employed.

2.1 A deterministic LQ optimal control problem.

The system is assumed to be multi-input, multi-output with a one degree of freedom control law as is shown in Figure 1.

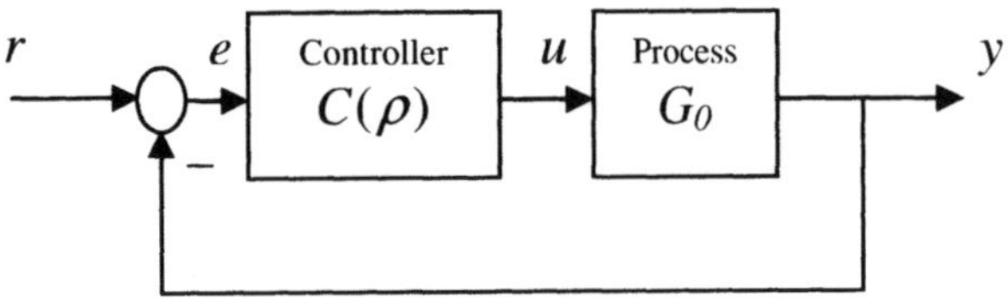

Figure 1 Unity feedback control loop

The controller $C(\rho)$ is given in terms of the controller parameter vector, $\rho\in\Re^{n_c}$, where n_c is the number of controller parameters. The notation for the time signals *e, u,* and *y* are given as, $e_t(\rho)$, $u_t(\rho)$ and $y_t(\rho)$. The subscript *t* denotes the dependence of these signals on time, and the dependence on the controller parameter is given using vector, ρ.

IFT Optimisation Problem Using the control signals of Figure 1, the LQ cost function to be minimised is given by

$$J(\rho)=\frac{1}{T_f}\int_0^{T_f}\begin{pmatrix}(L_e e_t(\rho))^T(L_e e_t(\rho))\\ +\lambda(L_u u_t(\rho))^T(L_u u_t(\rho))\end{pmatrix}dt \qquad (1)$$

Where, $e_t\in\Re^r, u_t\in\Re^m, \lambda\in\Re^+, \rho\in\Re^{n_c}$ and n_c is the number of controller parameters. The time domain operators L_e and L_u are used to weight the error and control signal contributions to the cost function. The relative contribution between the error and control signals is adjusted by means of λ.

The Generic Optimisation Problem

$$\min_{w.r.t.\ \rho\in\Re^{n_c}} J(\rho)$$

with (i) $\rho>0\in\Re^{n_c}$, (ii) $C(\rho)$ closed loop stabilising.

This is seen to be a fixed structure or restricted structure LQ optimal control problem (Grimble and Johnson, 1988). Incorporating a limit process $T_f\to\infty$ yields the steady state optimisation problem. The condition $\rho>0$ ensures that the PID controller parameters are positive.

IFT Numerical Optimisation
The optimisation framework follows the simple iterative gradient algorithm

Algorithm 1 Basic IFT Optimisation
Step 1 Initialisation
Choose cost weighting, λ, costing time interval, T_f
Choose convergence tolerance, ε
Set loop counter, $k=0$
Choose initial controller parameter vector, $\rho(k)$
Step 2 Gradient Calculation
Calculate gradient, $\frac{\partial J}{\partial\rho}(k)$. If $\left\|\frac{\partial J}{\partial\rho}(k)\right\|<\varepsilon$ then stop
Step 3 Update Calculation
Select the update parameters, γ_k and R_k
Compute $\rho(k+1)=\rho(k)-\gamma_k R_k^{-1}\frac{\partial J}{\partial\rho}(k)$
Set $k=k+1$ and go to Step 2
Algorithm End

Remarks
(a) The selection of γ_k can be a fixed step or line search step.
(b) Setting $R_k=I$ gives a steepest descent optimisation routine.
(c) Setting $R_k=H(\rho(k))$ where *H* is the Hessian matrix produces a Newton iteration for the optimisation, where, $H_{ij}=\frac{\partial^2 J}{\rho_i\rho_j}$.

Mahathanakiet *et al* (2002) investigated the IFT method on a simple deterministic LQ cost function for a wastewater treatment plant. Leibniz's Theorem for the Differentiation of an Integral (Abramowitz and Stegun, 1972) was used to derive the gradient function for the simplified problem as,

$$\frac{\partial J}{\partial\rho}=\frac{1}{T_f}\int_0^{T_f}\{e_t(\rho)\frac{\partial e_t(\rho)}{\partial\rho}+\lambda u_t(\rho)\frac{\partial u_t(\rho)}{\partial\rho}\}dt \qquad (2)$$

The gradient was then constructed using system responses and special input signals. In this paper, a quite different approach is taken to the gradient and Hessian generation step.

2.2 Generating the Gradient and Hessian

A classical gradient computation would be based on perturbing the gain vector from $\rho(k)$ to $\rho(k)+\Delta\rho$ and calculating the respective cost function, $J(\rho(k)),\ J(\rho(k)+\Delta\rho)$ so that numerical differences could be used to calculate an expression for the gradient. Further numerical perturbations can be used to calculate the Hessian information. It would then be possible to use a steepest descent or Newton method to calculate updated controller parameters. This method suffers from the problems of large numbers of gain perturbations and system response generations to calculate a gradient and a large number of gradient iterations to attain the (local) minimum. In the following a method of perturbing the gain is given such that estimates of both the gradient and the Hessian are obtained. This allows improved numerical routines, over steepest descent, to be used to reduce the number of iterations required to achieve convergence to the minimum of the cost function.

Assume that the controller parameters are perturbed by $\Delta\rho(t_\Delta)\in\Re^{n_c}$, then Taylor's Theorem gives,

$$J(\rho+\Delta\rho(t_\Delta))=J(\rho)+\Delta\rho(t_\Delta)^T\frac{\partial J}{\partial\rho}+R_2(\xi) \tag{3}$$

$$J(\rho+\Delta\rho(t_\Delta))=J(\rho)+\Delta\rho(t_\Delta)^T\frac{\partial J}{\partial\rho}+\frac{1}{2}\Delta\rho(t_\Delta)^T H(\rho)\Delta\rho(t_\Delta)+R_3(\xi) \tag{4}$$

$\frac{\partial J}{\partial\rho}\in\Re^{n_c}$, $R_2(\xi)=\frac{1}{2}\Delta\rho(t_\Delta)^T H(\xi)\Delta\rho(t_\Delta)$,

$R_3(\xi)$ is a third order residual,

$H_{ij}=\frac{\partial^2 J}{\partial\rho_i\partial\rho_j}$, $H\in\Re^{n_c\times n_c}$

and $\rho_i<\xi_i<\rho_i+\Delta\rho_i(t_\Delta)$.

Using these Taylor expansions, gradient and Hessian extraction is given by two new results. Sinusoidal function orthogonality plays a key role in the proof of the propositions. For the gradient result it is useful to define a set of integers which will define multiples of a fundamental gain perturbation frequency, ω_0. Introduce a set of n_c integers, denoted $\Im_G(n_c)$ with $\Im_G(n_c)=\{n_i,\ n_i\neq n_j,\ i,j\in[1,..n_c]\}$.

Proposition 1 Gradient extraction
Consider a controller gain vector perturbation, $\Delta\rho(t_\Delta)\in\Re^{n_c}$, where $\Delta\rho_i(t_\Delta)=\delta_i\sin n_i\omega_0 t_\Delta$, $\delta_i\in\Re,\ n_i\in\Im_G(n_c); i=1,..,n_c$.

Set $T_0=2\pi/\omega_0$, then for $i=1,..,n_c$,

$$\int_0^{T_0}J(\rho+\Delta\rho(t_\Delta))\sin n_i\omega_0 t_\Delta dt_\Delta=\left(\frac{\delta_i\pi}{\omega_0}\right)\frac{\partial J}{\partial\rho_i}+O(\delta_{\max}^2) \tag{5}$$

where $\delta_{\max}=\max\{\delta_i, i=1,..n_c\}$ **********************

The result requires a suitable selection of the perturbation integration period, T_0, then the fundamental frequency is calculated as $\omega_0=2\pi/T_0$. The proposition also requires the integers $n_i,\ i=1,..,n_c$ to satisfy $n_i\in\Im_G(n_c)$.

Sinusoidal function orthogonality also plays a key role in the result for the extraction of the Hessian information. The Hessian is symmetric and has $n_c(n_c-1)/2$ unknown elements. For an orthogonality based on a sine perturbation of the gain parameters and a cosine extraction of the Hessian, introduce a set of $n_c(n_c-1)/2$ integers, denoted $\Im_H^{s\sim c}(n_c)$. Given the set of gain excitation multiples, $\Im_G(n_c)$ then set $\Im_H^{s\sim c}(n_c)$ comprises integers, n_{ij} such that (i) $n_{ij}=n_i+n_j$, (ii) $n_{ij}\neq n_{i_1j_1}$, (iii) $n_{ij}\neq n_{i_1}-n_{j_1}$ for all pairs $(i,j),(i_1,j_1)$ where $i=1,..n_c; j=i,..,n_c$ and $i_1=1,..n_c; j_1=i,..,n_c$.

Proposition 2 Hessian extraction
Consider a sinusoidal controller gain vector perturbation, where $\Delta\rho_i(t_\Delta)=\delta_i\sin n_i\omega_0 t_\Delta$ with $\delta_i\in\Re,\ n_i\in\Im_G(n_c); i=1,..,n_c$.

Set $T_0=2\pi/\omega_0$, then for set $\Im_H^{s\sim c}(n_c)$,

$$\int_0^{T_0}J(\rho+\Delta\rho(t_\Delta))\cos n_{ij}\omega_0 t_\Delta dt_\Delta=\left(-\frac{\delta_i\delta_j\pi f(i,j)}{4\omega_0}\right)\frac{\partial^2 J}{\partial\rho_i\partial\rho_j}+O(\delta_{\max}^3) \tag{6}$$

$f(i,j)=\{1, i=j;\ or\ \ 2, otherwise\}$, and

$\delta_{\max}=\max\{\delta_i, i=1,..n_c\}$ ***************************

The proofs of the propositions define the sets $\Im_G(n_c),\ \Im_H^{s\sim c}(n_c)$, and it is useful to note that other variants of the propositions are possible (Crowe, 2003). For example, cosine perturbation of the gain elements followed by cosine extraction of the Hessian elements can be constructed. In this case, the integer set denoted $\Im_H^{c\sim c}(n_c)$ which specifies the frequencies used by the cosine Hessian extraction functions has to be defined. Thus, $\Im_H^{c\sim c}(n_c)$ has $n_c(n_c-1)/2$ integers n_{ij} such that,

(i) $n_{ij}\notin\Im_G(n_c)$ (ii) $n_{ij}=n_i+n_j$,

(iii) $n_{ij} \neq n_{i_1 j_1}$, (iv) $n_{ij} \neq n_{i_1} - n_{j_1}$ for all pairs (i, j), (i_1, j_1) where $i = 1,...n_c; j = i,..,n_c$ and $i_1 = 1,...n_c; j_1 = i_1,..,n_c$.

Apart from these theoretical issues, other numerical considerations are necessary to convert the theory into working algorithms as outlined in the next section.

3. IMPLEMENTATION ISSUES

3.1 Numerical Selections for the Algorithm

Selection of T_f, T_0, ω_0 In the calculation of the cost function it is not practical to allow the online experiment to continue for an infinitely long time. Thus, the integration period is fixed to $T_f = 5 \times \tau(\text{dominant})$, where it is assumed that the system has settled to within 1% of the steady state value after this time has elapsed, and then set $T_0 = T_f$. This gives the frequency of the basic perturbation frequency as, $\omega_0 = 2\pi / T_0$.

Selection of Gain Perturbation amplitudes The selection of the gain perturbation amplitudes $\delta_i \in \Re$ will be problem dependent. However there are several considerations here, firstly the perturbation should not cause closed loop instability, and secondly, the size can be made to minimise the disturbance to the system outputs. This latter issue will be of particular concern to production personnel when conducting online experiments. Finally, accuracy of gradient and Hessian extraction as given in the propositions is directly linked to the perturbation size.

Constructing sets $\Im_G(n_c)$, $\Im_H^{s\sim c}(n_c)$, $\Im_H^{c\sim c}(n_c)$ A frequency multiple of ω_0 must be assigned to each gain parameter perturbation, $\Delta\rho_i(t_\Delta)$. The choice of frequency multiples is dependent on the choice of whether a sine or cosine controller gain perturbation function is chosen. If only the gradient is to be extracted, then the rules for the construction of $\Im_G(n_c)$ are straightforward. If, however the Hessian is also to be extracted then the construction of the pair $\Im_G(n_c)$, $\Im_H^{s\sim c}(n_c)$ or the pair $\Im_G(n_c)$, $\Im_H^{c\sim c}(n_c)$ is a little more involved. Table 1 shows feasible frequency multiples for the case of a single PID controller, with three gains.

Thus, for a sine perturbation and a cosine extraction function for the Hessian, the duplication of a frequency between the $\Im_G(n_c)$ and the set $\Im_H^{s\sim c}(n_c)$ is permissible (see for example in Table 1, that $n_2 = n_{11} = 4$ in $\Im_H^{s\sim c}(n_c)$).

Table 1 Feasible frequency multiples

$\Im_G(n_c)$	$n_1 = 2$	$n_2 = 4$	$n_3 = 5$
$\Im_H^{s\sim c}(n_c)$	$n_{11} = 4$	$n_{12} = 6$	$n_{13} = 7$
	--	$n_{22} = 8$	$n_{23} = 9$
	--	--	$n_{33} = 10$
$\Im_G(n_c)$	$n_1 = 1$	$n_2 = 5$	$n_3 = 8$
$\Im_H^{c\sim c}(n_c)$	$n_{11} = 2$	$n_{12} = 6$	$n_{13} = 9$
	--	$n_{22} = 10$	$n_{23} = 13$
	--	--	$n_{33} = 16$

Selection of Integration formula The time scales over which the cost function is calculated and the parameter perturbation signals evolve are separate. The cost function is calculated in real time, whereas the parameter perturbation signal uses the time-domain, t_Δ. Standard numerical integration formulas are used to compute the two extraction integrals,

$$\int_0^{T_0} J(\rho + \Delta\rho(t_\Delta)) \sin n_i \omega_0 t_\Delta dt_\Delta$$

and $$\int_0^{T_0} J(\rho + \Delta\rho(t_\Delta)) \cos n_{ij} \omega_0 t_\Delta dt_\Delta .$$

These use the discrete time where $t_\Delta = kT$, $k = 0,1,2,\cdots,N$, and T is the integration step size or sampling interval. Clearly, step size, T has to be chosen to achieve maximal accuracy from the integration method and provide sufficient resolution for the maximum frequency of the perturbation and extraction signals.

3.2 The controller parameter cycling algorithm

To create a new controller parameter cycling method, the optimisation Algorithm 1 is used with the estimates of the cost function gradient and Hessian from the theory of Propositions 1 and 2. In an application, the controller structure will be fixed *a priori*, and can be used to define the integer sets, $\Im_G(n_c)$, $\Im_H^{s\sim c}(n_c)$ or $\Im_G(n_c)$, $\Im_H^{c\sim c}(n_c)$.

The new algorithm finds estimates of both the gradient and the Hessian and uses these in a Newton algorithm to determine the next set of controller parameters whilst minimising the cost function. It is known that Newton type algorithms are not globally convergent and if the algorithm is initialised outside the region of convergence, the parameter updates are chosen to move in the direction of the convergence region. Similarly if the Hessian estimate is negative definite then this cannot be used and additional steps are used for the parameter updates prior to the Hessian estimate becoming positive definite. This has been discussed in Ljung (1987) and involves replacing the Hessian by, $H_{LM} = \frac{\partial J}{\partial \rho}\frac{\partial J}{\partial \rho}^T + \alpha I$

where $\alpha \in \Re^+$.

This is known as the Levenberg-Marquardt procedure. When the Hessian becomes positive definite, the Newton method reverts to using the Hessian estimate in the parameter updates. The algorithm is given as follows.

Algorithm 2 Optimisation by Controller Parameter Cycling

Step 1 Initialisation, $\rho \in \Re^{n_c}$

Choose cost weighting, λ, cost time interval, T_f

Set $T_0 = T_f$, compute $\omega_0 = 2\pi / T_0$

Choose perturbation sizes, $\{\delta_i,\ i = 1,..,n_c\}$

Find sets, $\Im_G(n_c)$, $\Im_H^{s\sim c}(n_c)$, $\Im_G(n_c)$, $\Im_H^{c\sim c}(n_c)$

Determine N, set $T = T_0 / N$

Choose convergence tolerance, ε

Set loop counter, $k = 0$, choose $\rho(k)$

Step 2 Gradient and Hessian Calculation

Calculate gradient, $\frac{\partial J}{\partial \rho}(k)$ (Proposition 1)

And Hessian, $H_{ij}(k) = \frac{\partial^2 J}{\partial \rho_i \partial \rho_j}$, $H \in \Re^{n_c \times n_c}$

(Proposition 2)

If $\left\| \frac{\partial J}{\partial \rho}(k) \right\| < \varepsilon$ and $H(k) > 0 \in \Re^{n_c \times n_c}$ then stop

Step 3 Update Calculation

Select or calculate the update step size, γ_k.

If $H(k) > 0 \in \Re^{n_c \times n_c}$ compute

$$\rho(k+1) = \rho(k) - \gamma_k H(k)^{-1} \frac{\partial J}{\partial \rho}(k)$$

Else compute

$$\rho(k+1) = \rho(k) - \left[\left[\frac{\partial J}{\partial \rho}(k) \right] \left[\frac{\partial J}{\partial \rho}(k) \right]^T + \alpha_k I \right]^{-1} \frac{\partial J}{\partial \rho}(k)$$

Set $k = k + 1$ and go to Step 2

Algorithm End

The implementation of Proposition 1 and 2 as numerical procedures involves exploiting the symmetry properties of the integrals to be evaluated and careful selection of the integration step size. Details of these issues which reduce the computational burden are reported by Crowe (2003).

4. APPLICATION RESULTS

An application of the method to a two input, two output system controlled by a decentralised PID controller system is reported below.

4.1 Multivariable Process - Algorithm Setup

The system to be controlled has the transfer function matrix (Zhuang and Atherton, 1994),

$$G(s) = \frac{1}{d(s)} \begin{bmatrix} 1.5s + 1 & 0.15s + 0.2 \\ 0.45s + 0.6 & 0.96s + 0.8 \end{bmatrix}$$

$$d(s) = 2s^4 + 8s^3 + 10.5s^2 + 5.5s + 1$$

This system is considered to be typical of those found in process industries. The PID controllers to be tuned are of the parallel type and are given as,

$$K_{ii}(s) = K_{Pii}(s) + \frac{K_{Iii}(s)}{s} + K_{Dii}s, i = 1,2$$

Thus, $n_c = 6$, and the fixed structure controllers are to be tuned such that the LQ cost function to be minimised is given by

$$J = \int_0^{T_f} \left(e(t)^T e(t) + u(t)^T \Lambda u(t) \right) dt$$

In this case, unit weightings are given to the error terms, and equal weighting is given to each of the control terms using, $\Lambda = diag\{0.01, 0.01\}$. The initial PID controller was derived using a relay experiment and Ziegler-Nichols rules. Throughout the tuning procedure the controller parameter perturbation size was, $\delta_i = \delta = 0.001,\ i = 1,...,6$. The cost function integration period was chosen to be $T_f = 20$, hence, $T_0 = T_f = 20$, and $\omega_0 = 0.1\pi$ (rad.s^{-1}). The control perturbation frequencies were chosen via the integer set, $\Im_G(n_c) = \{1, 4, 5, 16, 19, 20\}$.

4.2 Algorithm Results

Figures 2 and 3 below show the 2-norm of the cost function gradient and the cost function values for the algorithm iterations.

For the first three iterations of the algorithm the Hessian estimate was negative definite and the Hessian was replaced by the Levenberg-Marquardt procedure with, $\alpha I = 0.01 I_6$. At the fourth iteration it was found that the algorithm returned negative values for the updated integral gain parameters for both controllers. Consequently the algorithm was adjusted using $\alpha I = 0.1 I_6$. After two further iterations the Hessian estimate was positive definite and was used in the Newton update. Since the Hessian estimate was corrupted by noise its use had to be conservative. To do this the controller parameter update size was limited using, $\gamma_k \in (0,1)$. For the next two iterations, $\gamma_k = 0.1$ and the values of the 2-norm of the cost function gradient and cost function value decreased. For iterations 10 and 11, $\gamma_k = 1$. After iteration 11, the cost function value and gradient 2-norm increased. The step size was reset at $\gamma_k = 0.1$ and a new iteration 11 performed where the cost function value and gradient 2-norm decreased; the step size $\gamma_k = 0.1$ was retained for the remaining iterations. From Figure 2 and 3 the evolution of the cost function gradient 2-norm shows that the minimum of the cost function is found after approximately 15 iterations of the algorithm when the cost function gradient 2-norm is $O(10^{-3})$. Thus, the controller parameters at iteration 15 give the minimum value of the cost function. In order to ensure that further iterations of the algorithm do attain a minimum, a large number of additional iterations were performed. It can be seen from Figures 2 and 3 that the cost function values tend to an asymptotic value and that the cost function gradient 2-norm limits at zero.

5. CONCLUSIONS

The method of Iterative Feedback Tuning has certain practical shortcomings; a particular one is that the generation of Hessian information is not a simple operation. This paper reported a new model-free procedure where the theory for generating second order information is not particularly difficult. The new method is termed controller parameter cycling.

The paper also reported the development of a first version of a numerical algorithm for the new procedure. Whilst the procedure was successful in tuning a multivariable decentralised PID controller, certain issues remain to be investigated further:

(a) Experience with several examples has shown that many of the cost functions of fixed structure controllers are often very flat and it is necessary to ensure that sensible parameter updates are used in the routine. In the current version of the algorithm, the Levenberg-Marquardt procedure is used to try to ensure that successive parameter updates continue in the cost minimising direction; it would be useful to examine other procedures.

(b) The implementation of the gradient and Hessian extraction involves a heavy computational burden in terms of system runs and minimising iteration steps. Symmetry can be exploited to reduce the computational load and an intelligent strategy can be used to ensure the algorithm uses computed data to maximum effect in the sequence of iteration steps. These topics are currently under research.

REFERENCES

Abramowitz M and I. A. Stegun (Eds.), 1972, Handbook of Mathematical Functions, Dover Publications, ISBN 0-486-61272-4

Astrom K. J and T. Hagglund, 1985, Method and an Apparatus in Tuning a PID Regulator, US Patent.

Blevins T. L., G. K. McMillan, W. K. Wojsznis and M. W. Brown, 2002, Advanced Control Unleashed, ISA Press,

Crowe J, 2003, PhD Thesis (In Preparation), University of Strathclyde, Glasgow, UK.

Crowe J. and Johnson M.A., 1998, New Approaches To Non-Parametric Identification For Control Applications, Preprints IFAC ASCSP Workshop, pp 309-314, Glasgow, Scotland U. K., 26-28, Aug.

Crowe J. and Johnson M.A., 1999, A New Non-Parametric Identification Procedure For Online Controller Tuning, Procs. ACC, pp. 3337-3341, San Diego, U. S. A.,

Grimble M. J. and M. A. Johnson, 1988, Optimal Control and Stochastic Estimation, Volume 1, John Wiley, UK.

Harris T. J., 1989, Assessment of control loop performance, Can. J. Chem. Eng., Vol 67, pp 856-861

Hjalrmarsson H., S. Gunnarsson, and M. Gevers, 1994, A Convergent Iterative Restricted Complexity Control Design Scheme, Procs CDC, Florida, USA, December.

Hjalrmarsson H., S. Gunnarsson, and M. Gevers, 1998, Iterative Feedback Tuning: Theory and Applications, IEEE CSS Magazine, August, pp 26-41.

Kammer L. C., R. R. Bitmead and P. L. Bartlett, 2000, Direct iterative tuning via spectral analysis, Automatica, Vol 36, pp 1301-1307

Lequin O., M. Gevers and L. Tiest, 1999, Optimising The Settle-time With Iterative Feedback Tuning, IFAC Triennial, Beijing, China, Paper I-36-08-3, pp 433-437.

Mahathanakiet K., M. A. Johnson, A. Sanchez and M. Wade, 2002, Iterative Feedback Tuning and an Application to Wastewater Treatment Plant, Procs Asian Control Conference, pp 256-261, Singapore, September.

O'Dwyer A, 2001a,b A summary of PI and PID controller tuning rules for processes with time delay, Part 1: PI controller tuning rules, pp 175-180; Part 2: PID controller tuning rules, pp 242-247; IFAC Workshop on Digital Control, Terrasse, Spain, 5-7 April,

Thornhill, N. F., M. Oettinger and P. Fedenczuk, 1999, Refinery-wide control loop performance assessment, J. Process Control, Vol. 9, pp 109-1214.

Yu C. C., 1999, Autotuning of PID controllers: Relay Feedback Approach, Springer Verlag London.

Zhuang M., and D. P. Atherton, 1994, PID Controller Design for a TITO System, IEE Proc.- CTA, Vol. 141, No. 2, March 1994.

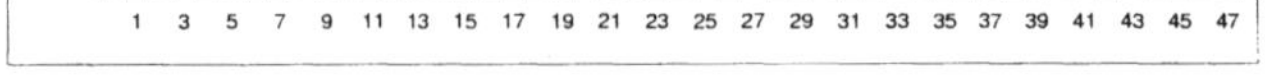

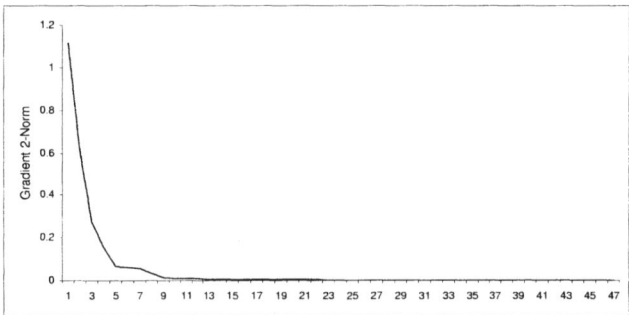

Figure 2 Evolution of ||Gradient||.

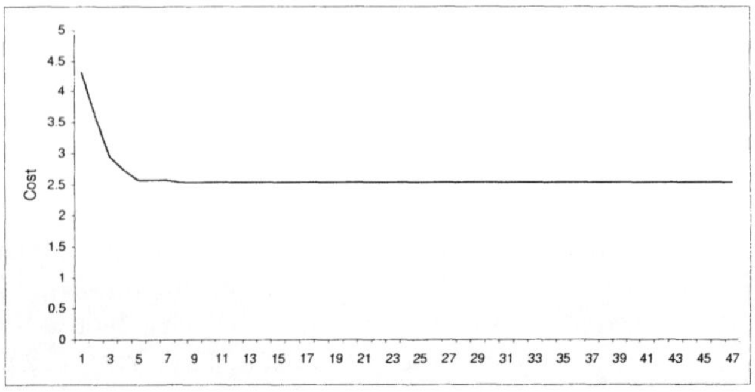

Figure 3 Evolution of Cost Function Value

IFAC
Publications
www.elsevier.com/locate/ifac

STEPWISE REFINEMENT OF SPARSE GRIDS IN DATA MINING APPLICATIONS

Marc Brendel, Wolfgang Marquardt [1]

Lehrstuhl für Prozesstechnik, RWTH Aachen University, D-52064 Aachen, Germany

Abstract: A stepwise refinement approach for sparse grids is proposed to recover an unknown function from a set of corrupted data. The work is based on the approach of Garcke *et al.* (2001), which is extended to cover non-uniform discretizations in each dimension. Starting from an initially coarse discretization, the multivariate grid is gradually refined using sensitivity analysis. An appropriate level of refinement is identified by the *L-curve* criterion or split-sample validation, capable to resolve functional details while suppressing measurement noise. It is shown that the algorithm allows reliable identification of the best possible functional approximation. *Copyright © 2003 IFAC*

Keywords: Sparse grids, functional approximation, data mining, adaptive refinement

1. INTRODUCTION

In recent years, sparse grid methods are attracting increasing interest for the solution of partial differential equations (PDEs), integral equations, numerical integration, function interpolation and approximation. They promise a significant reduction in problem size with approximation accuracies which come close to those of full grid methods. The sparse grid approach stems from the work of Yserentant (1986) on multi-level splitting of finite-element spaces. It has been mainly used for the solution of PDEs thereafter. Griebel *et al.* (1992) proposed a grid combination technique, where the problem is solved on a set of full subgrids, giving rise to parallelization on today's supercomputers.

Recently, Garcke *et al.* (2001) proposed an approach to sparse grid functional approximation in a data mining context. In engineering sciences such functional approximations are essential in applications such as identification of the input/output behavior of a stationary process or the construction of dynamic hybrid models, but may also find use for structure selection of nonlinear ARMAX models. Here, sparse grid methods promise a significant reduction of problem size associated with large, high-dimensional data sets originating e.g from modern high-resolution measurement techniques.

Garcke *et al.* (2001) use multidimensional sparse grids with uniform discretization in each dimension for function reconstruction from a set of data corrupted by measurement noise. To obtain stable solutions for this ill-posed problem, Tikhonov regularization has been applied in addition to the regularization properties inherent to problem discretization (Kirsch, 1996). Cross validation has been used to identify appropriate regularizations. However, for function approximation in engineering sciences, in some cases different resolutions in the various dimensions may reduce computational complexity significantly.

In this work, an extension of the approach of Garcke *et al.* (2001) is presented, covering non-

[1] Corresponding author. E-mail: marquardt@lfpt.rwth-aachen.de

uniform discretizations in each dimension. As a first step, a restriction to the regularization inherent to discretization is made. Starting on an initially coarse multivariate grid, the dimension in which refinement is most promising is defined by a sensitivity analysis in each step. The algorithm results in a hierarchically refined sequence of sparse grids, from which an appropriate refinement in terms of generalization is identified using the *L-curve* criterion or split-sample validation. It is shown that the algorithm allows reliable identification of the best possible functional approximation from data, capable to resolve functional details while suppressing measurement noise.

2. DIMENSIONALLY UNIFORM SPARSE GRIDS FOR FUNCTION APPROXIMATION

In the following, the work of Garcke *et al.* (2001) on functional approximation is briefly summarized.

Consider the set of data

$$S = \{(\mathbf{x}_i, y_i) \in \mathbb{R}^d \times \mathbb{R}\}_{i=1}^M \tag{1}$$

obtained by sampling an unknown function f of some function space V defined over $\mathbb{R}^d$. The sampling process is disturbed by noise of unknown magnitude. The objective is now to recover the unknown function f from the given data as good as possible. This problem is ill-posed in the sense of Hadamard (Engl *et al.*, 1996). A general formulation of the approximation problem is

$$\min_{f \in V} \frac{1}{M} \sum_{i=1}^{M} (f(\mathbf{x}_i) - y_i)^2 + \lambda \Phi(f), \tag{2}$$

with the first term, $(f(\mathbf{x}_i) - y_i)^2$, measuring the approximation error and the second term, $\Phi(f)$, representing a regularization term to enforce smoothness of f. λ is a regularization parameter to adjust the trade-off between data and regularization error.

If the problem is restricted to a finite dimensional subspace $V_N \subset V$, the function f is replaced by

$$f_N(\mathbf{x}) = \sum_{j=1}^{N} \theta_j \varphi_j(\mathbf{x}) \tag{3}$$

where the Ansatz functions $\{\varphi_j\}_{j=1}^N$ should span V_N and preferably form a basis for V_N. The coefficients $\{\theta_j\}_{j=1}^N$ denote the degrees of freedom.

Projection of the solution into the finite dimensional subspace V_N has inherent regularizing properties (see e.g. Binder *et al.* (2002)). Based on this fact, this work will restrict attention to regularization inherent to discretization. Equation (2) then reads as

$$\min_{f_N \in V_N} \frac{1}{M} \sum_{i=1}^{M} (f_N(\mathbf{x}_i) - y_i)^2. \tag{4}$$

The projection to V_N, i.e. the function $f_N(\mathbf{x})$, can be determined using a conventional finite element discretization on an equidistant grid Ω_n with mesh size $h_n = 2^{-n}$ for each coordinate direction. To cope with the curse of dimensionality inferred from multivariate data sets S, a sparse grid discretization technique is applied instead, where only $O(h_n^{-1}(\log(h_n^{-1}))^{d-1})$ grid points are used for discretization instead of $O(h_n^{-d})$ for a full grid (Bungartz and Griebel, 1999). The sparse grid solution is shown to be nearly as accurate as the solution on a full grid. The advantage of the sparse grid technique compared to full grid discretization becomes obvious especially for high dimensionality.

Griebel *et al.* (1992) proposed a technique to compose sparse grid approximations from a sequence of full subgrids $\{\Omega_\mathbf{l}, \mathbf{l} \in \mathbb{N}^d\}$ with multi-index $\mathbf{l} = (l_1, ..., l_d) \in \mathbb{N}^d$. These subgrids $\Omega_\mathbf{l}$ have mesh sizes $h_\mathbf{l} = (h_{l_1}, ..., h_{l_d}) = (2^{-l_1}, ..., 2^{-l_d})$. A grid generally has different mesh sizes in different coordinate directions but is equidistant within each direction. The grid points contained in such a grid are $\mathbf{x}_{\mathbf{l},\mathbf{j}} = (x_{l_1,j_1}, ..., x_{l_d,j_d})$ with $x_{l_t,j_t} = j_t 2^{-l_t}, j_t = 0, ..., 2^{l_t}$. Here, the vector of independent variables $\mathbf{x}$ is restricted to $\mathbf{x} \in \Psi = [0,1]^d$ without loss of generality.

On each grid $\Omega_\mathbf{l}$ the space $V_\mathbf{l} = \text{span}\{\phi_{\mathbf{l},\mathbf{j}}, j_t = 0, ..., 2^{l_t}, t = 1, ..., d\}$ is spanned by the d-dimensional piecewise d-linear hat functions $\phi_{\mathbf{l},\mathbf{j}}(\mathbf{x})$. The one-dimensional functions $\phi_{l_t,j_t}(x_t)$ are created from the one-dimensional mother function $\phi(x)$ by dyadic dilation and translation:

$$\phi_{\mathbf{l},\mathbf{j}}(\mathbf{x}) := \prod_{t=1}^{d} \phi_{l_t,j_t}(x_t), \tag{5}$$

$$\phi_{l_t,j_t}(x_t) := \phi\left(\frac{x_t - j_t h_{l_t}}{h_{l_t}}\right), \tag{6}$$

$$\phi(x) := \begin{cases} 1 - |x| & \text{if } x \in]\text{-}1,1[\\ 0 & \text{otherwise} \end{cases}. \tag{7}$$

A subgrid solution of (4) with $f_\mathbf{l} \in V_\mathbf{l}$ is then

$$f_\mathbf{l}(\mathbf{x}) = \sum_{\mathbf{j}} \theta_{\mathbf{l},\mathbf{j}} \phi_{\mathbf{l},\mathbf{j}}(\mathbf{x}) \tag{8}$$

with the parameters $\boldsymbol{\theta}$ obtained from $\mathbf{J}^T\boldsymbol{\theta} = \mathbf{y}$, where $\mathbf{J}$ is a rectangular $N \times M$ matrix with entries $J_{j,i} = \phi_j(\mathbf{x}_i), i = 1, .., M; j = 1, .., N$. The vector $\mathbf{y}$ contains the data and has length M. The unknown vector $\boldsymbol{\theta}$ contains the degrees of freedom θ_j and has length $N = \prod_{t=1}^{d}(2^{l_t} + 1)$.

The approximation $f_n^{(c)}$ on sparse grid $\Omega_n^{(c)}$ is combined from full subgrid approximations $f_\mathbf{l}$ by

$$f_n^{(c)}(\mathbf{x}) := \sum_{q=0}^{d-1}(-1)^q \binom{d-1}{q} \times \sum_{|\mathbf{l}|_1=n+(d-1)-q} f_{\mathbf{l}}(\mathbf{x}), \tag{9}$$

where n is the level of approximation, $f_{\mathbf{l}}$ is the full grid approximation of S on $\Omega_{\mathbf{l}}$ and index (c) denotes the composition from subgrids. The second term comprises full grids with all index permutations of $\mathbf{l}$ satisfying $|\mathbf{l}|_1 = n + (d-1) - q$, $q = 0, ..., d-1, l_t > 0$ with $|\mathbf{l}|_1 := \sum_{t=1}^{d} l_t$.

3. STEPWISE GRID REFINEMENT

3.1 Non-uniform mesh sizes

To account for grids, which do not have uniform mesh sizes in all dimensions, the grid combination equation (9) is extended to non-uniform sparse grids $\Omega_{\mathbf{k}}^{(c)}$ with level index $\mathbf{k} = (k_1, ..., k_d) \in \mathbb{N}^d$:

$$f_{\mathbf{k}}^{(c)}(\mathbf{x}) := \sum_{q=0}^{d-1}(-1)^q \binom{d-1}{q} \times \sum_{|\mathbf{l}|_1=n+(d-1)-q} f_{\mathbf{l}+\mathbf{l}^\star}(\mathbf{x}) \tag{10}$$

where $\mathbf{l}^\star = (k_1 - n, ..., k_d - n)$, $n = \min(\mathbf{k})$ characterizes the extension of the largest possible uniform sparse grid $\Omega_n^{(c)}$ to the non-uniform sparse grid $\Omega_{\mathbf{k}}^{(c)}$. Exemplarily, the non-uniform sparse grid $\Omega_{(5,3)}^{(c)}$ is constructed as $\Omega_{(5,3)}^{(c)} = \Omega_{(5,1)} \oplus \Omega_{(4,2)} \oplus \Omega_{(3,3)} \ominus \Omega_{(4,1)} \ominus \Omega_{(3,2)}$.

3.2 Generic refinement algorithm

The refinement algorithm starts with the coarse grid of level $\mathbf{k}_1 = (1, ..., 1) \in \mathbb{N}^d$, which is gradually refined thereafter. The following refinement algorithm is proposed:

(1) calculate solution $f_{\mathbf{k}}^{(c)}$ on current grid $\Omega_{\mathbf{k}}^{(c)}$ of level $\mathbf{k} = (k_1, ..., k_d) \in \mathbb{N}^d$,
(2) create extended grids in each dimension by bisection,
(3) for each extended grid, calculate sensitivity of residual w.r.t. the set of parameters,
(4) select the grid showing highest sensitivity as next step.

The algorithmic steps are explained in more detail subsequently.

Solution on current grid $\Omega_{\mathbf{k}}^{(c)}$ The solution $f_{\mathbf{k}}^{(c)}$ on sparse grid $\Omega_{\mathbf{k}}^{(c)}$ is calculated from equation (10), composing the solution from the set of full subgrids $\Omega_{\mathbf{l}+\mathbf{l}^\star}$.

Grid extension Bisect current grid of level $\mathbf{k}$ in each dimension (e.g. $\mathbf{k}_1^{\text{ext}} = (k_1+1, k_2)$, $\mathbf{k}_2^{\text{ext}} = (k_1, k_2+1)$ in 2 dimensions). The augmented sets of parameters $\boldsymbol{\theta}_{\mathbf{k}_t^{\text{ext}}}, t = 1, ..., d$ are derived by interpolation of $\boldsymbol{\theta}_{\mathbf{k}}$, leaving $f_{\mathbf{k}_t^{\text{ext}}}^{(c)}(\mathbf{x})$ identical to $f_{\mathbf{k}}^{(c)}(\mathbf{x})$. The set of parameters $\boldsymbol{\theta}_{\mathbf{k}}$ of $f_{\mathbf{k}}^{(c)}(\mathbf{x})$ is composed from the parameter sets $\boldsymbol{\theta}_g$, $g = 1..s$ of $f_{\mathbf{l}_g+\mathbf{l}^\star}(\mathbf{x})$ as $\boldsymbol{\theta}_{\mathbf{k}} = (\boldsymbol{\theta}_1^T, ..., \boldsymbol{\theta}_s^T)^T$, where s denotes the number of subgrids needed for composition of sparse grid $\Omega_{\mathbf{k}}^{(c)}$.

Sensitivity calculation Sensitivity calculation is performed on each $\Omega_{\mathbf{k}_t^{\text{ext}}}^{(c)}$, $t = 1, ..., d$, where indices ext and t are dropped subsequently for easier reading. The squared residual $\xi_{\mathbf{k}}$ of sparse grid approximation $f_{\mathbf{k}}^{(c)}(\mathbf{x})$ and data $\mathbf{y}$ is

$$\xi_{\mathbf{k}} = \frac{1}{2}\left|\left|f_{\mathbf{k}}^{(c)}(\mathbf{x}) - \mathbf{y}\right|\right|_2^2. \tag{11}$$

Differentiating $\xi_{\mathbf{k}}$ w.r.t. the set of parameters $\boldsymbol{\theta}_{\mathbf{k}}$ of $f_{\mathbf{k}}^{(c)}$ results in

$$\frac{\partial \xi_{\mathbf{k}}}{\partial \boldsymbol{\theta}_{\mathbf{k}}} = \mathbf{J}_{\mathbf{k}}\left(f_{\mathbf{k}}^{(c)}(\mathbf{x}) - \mathbf{y}\right), \tag{12}$$

where

$$\mathbf{J}_{\mathbf{k}} = \frac{\partial f_{\mathbf{k}}^{(c)}(\mathbf{x})}{\partial \boldsymbol{\theta}_{\mathbf{k}}}. \tag{13}$$

The overall sensitivity measure $\Phi_{\mathbf{k}}$ of residual $\xi_{\mathbf{k}}$ w.r.t. parameter set $\boldsymbol{\theta}_{\mathbf{k}}$ is defined as

$$\Phi_{\mathbf{k}} = \left|\left|\frac{\partial \xi_{\mathbf{k}}}{\partial \boldsymbol{\theta}_{\mathbf{k}}}\right|\right|_2^2. \tag{14}$$

The Jacobian $\mathbf{J}_{\mathbf{k}}$ of $f_{\mathbf{k}}^{(c)}(\mathbf{x})$ in (13) can be calculated directly from the set of Jacobians $\mathbf{J}_g$, $g = 1..s$ of $f_{\mathbf{l}_g+\mathbf{l}^\star}(\mathbf{x})$ on full subgrid $\Omega_{\mathbf{l}_g+\mathbf{l}^\star}$:

$$\mathbf{J}_{\mathbf{k}} = \begin{pmatrix} (-1)^{q_1}\binom{d-1}{q_1}\mathbf{J}_1 \\ \vdots \\ (-1)^{q_s}\binom{d-1}{q_s}\mathbf{J}_s \end{pmatrix}, \tag{15}$$

$$\mathbf{J}_g = \frac{\partial f_{\mathbf{l}_g+\mathbf{l}^\star}(\mathbf{x})}{\partial \boldsymbol{\theta}_g}. \tag{16}$$

Here, q_g is corresponding to subgrid $\Omega_{\mathbf{l}_g+\mathbf{l}^\star}$ and $\boldsymbol{\theta}_g$ is the set of parameters of $f_{\mathbf{l}_g+\mathbf{l}^\star}(\mathbf{x})$.

Grid selection The grid which reveals highest sensitivity measure $\Phi_{\mathbf{k}}$ is selected as next step. Return to step (1) thereafter until a suitably predefined highest refinement step r is reached.

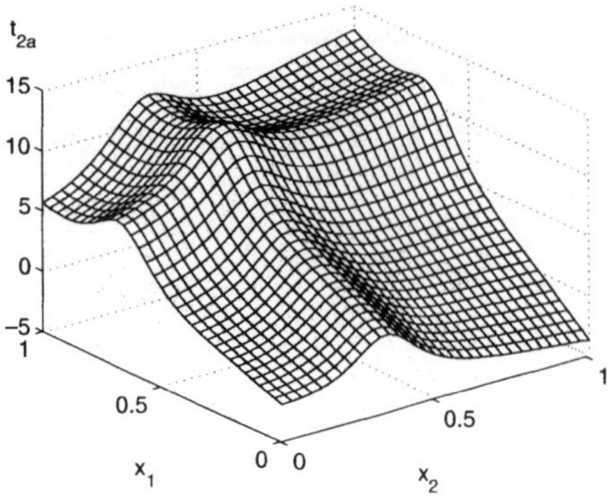

Fig. 1. Test function t_{2a}

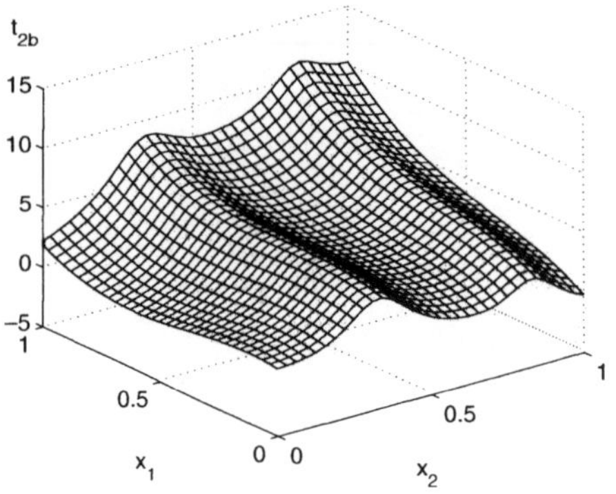

Fig. 2. Test function t_{2b}

4. REFINEMENT STEP SELECTION

Once the hierarchically refined grid sequence $\Gamma = \{\Omega_{\mathbf{k}_i}^{(c)}, i = 1, ..., r\}$ is identified, an appropriate refinement step needs to be selected. If it is chosen too high, this may lead to overfitting, where the error on the training set is very small, but a large error results if new data are presented. A too coarse resolution may result in poor representation of the training data. Here, two methods are presented aiming to balance the trade-off.

The *L-curve* (Hansen, 1998) is a criterion for selection of an appropriate level of regularization in the absence of knowledge on the data error level. The smoothing norm $||\mathbf{L}f_{\mathbf{k}}^{(c)}||$ is plotted versus the residual norm $||f_{\mathbf{k}}^{(c)} - \mathbf{y}||$ in a double-logarithmic scale, representing the tradeoff between data error and regularization error introduced by the discretization. Here, $\mathbf{L}$ is the second derivative operator. An optimal refinement step is located near the maximum curvature of the L-shaped plot.

Split-sample validation is another common technique where data are representatively split up into a training and a validation set. The generalization error of a certain discretization is evaluated on the validation set. The disadvantage of split-sample validation is that it reduces the amount of data available for both training and validation (Weiss and Kulikowski, 1991).

5. EXAMPLES

The approach of stepwise grid refinement and subsequent identification of an appropriate grid is examined on 3 multivariate test functions. To show the capability of the approach and still achieve clarity of illustration, dimensionality was restricted to bi- and trivariate functions, although higher dimensions can be treated accordingly.

5.1 Multivariate test functions

The first test function

$$t_{2a}(x_1, x_2) = \sin(x_1) + \cos(x_2) + x_1x_2 - 3x_2 + \frac{1}{0.2 + (x_1 - 4)^2} + \frac{1}{0.2 + (x_2 - 4)^2} \quad (17)$$

is bivariate with comparable smoothness properties in both dimensions. Here, an optimal approximation is expected to reveal equal resolution in each dimension. For the bivariate test function

$$t_{2b}(x_1, x_2) = 4\sin(x_1) + \cos(x_2) + x_1x_2 - 3x_2 + \frac{0.3}{0.1 + (x_2 - 5.5)^2} + \frac{1}{0.2 + (x_2 - 4)^2} \quad (18)$$

variable x_2 introduces a higher variability to data than x_1, an optimal discretization is consequently expected to show a higher discretization in the dimension of x_2. Plots of the test functions t_{2a} and t_{2b} are shown in Figure 1 and 2 respectively, with an arbitrarily chosen grid for visualization. In the 3-dimensional case,

$$t_{3b}(x_1, x_2, x_3) = x_3 + t_{2b}(x_1, x_2) \quad (19)$$

shows only linear dependency on variable x_3, a coarse discretization is likely to be adequate to account for changes in x_3.

The test functions are defined on the domain $(x_1, x_2, x_3) \in \Psi = [2, 5] \times [3, 6] \times [1, 2]$. Scaling to the domain $\Psi_s = [0, 1]^3$ is achieved by an affine transformation on homogeneous coordinates (see e.g. Rogers and Adams (1990)). For a parallely bounded original domain, the transformation of a single coordinate set $\mathbf{v}$ (in homogeneous coordinates) to $\mathbf{v}_s = \mathbf{Tv}$ is given by

$$\mathbf{T} = \begin{bmatrix} t_{11} & t_{12} & t_{13} & t_{14} \\ t_{21} & t_{22} & t_{23} & t_{24} \\ t_{31} & t_{32} & t_{33} & t_{34} \\ 0 & 0 & 0 & 1 \end{bmatrix}, \mathbf{v} = \begin{bmatrix} x_1 \\ x_2 \\ x_3 \\ 1 \end{bmatrix}, \quad (20)$$

with the transformation matrix parameters t_{ij} identified from the mapping of Ψ to Ψ_s.

For each of the test functions, a set of 20,000 data were generated, corrupted with normally distributed white noise with a standard deviation of $\sigma = 0.5$, resulting in sets S_{2a}, S_{2b} and S_{3b}.

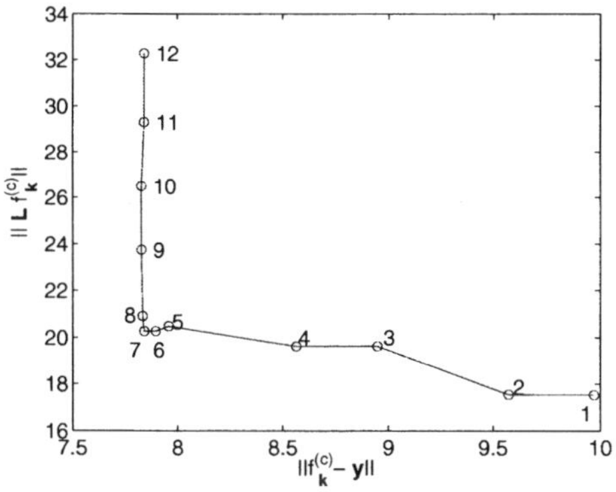

Fig. 3. L-curve for data set S_{2a}

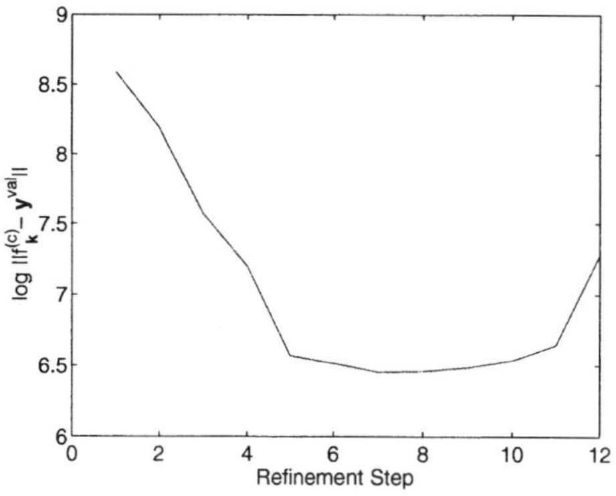

Fig. 4. Split-sample validation for data set S_{2a}

5.2 Grid sequences

Using the algorithm in Section 3.2, the hierarchical grid sequences Γ_{2a}, Γ_{2b} and Γ_{3b} were identified for each of the sets S_{2a}, S_{2b} and S_{3b} (Table 1).

5.3 Optimal approximations

Both L-curve criterion and split-sample validation were tested for identification of a suitable data approximation. In the first case, smoothness and residual to the data were calculated for the full data sets. Exemplarily, the L-curve resulting for the grid sequence of set S_{2a} is shown in Figure 3. The optimal refinement step is the one corresponding to the corner point of the L-curve.

In the latter case, data were split up into a randomly chosen 75% training set (corresponding to 15,000 data) and a 25% validation set (corresponding to 5,000 data). The grid sequences identified for the training sets of S_{2a}, S_{2b} and S_{3b} were identical to the ones determined for the full sets (when using the L-curve criterion). Exemplarily, the residual between approximation and validation set ($\log ||f_{\mathbf{k}}^{(c)}(\mathbf{x}) - \mathbf{y}^{\mathrm{val}}||$) is plotted in Figure 4, again for the grid sequence of test set S_{2a}. The optimal refinement is identified as the one where the residual has its smallest value.

The residual of the sparse grid function approximations to the uncorrupted data set can be plotted as $\log ||f_{\mathbf{k}}^{(c)}(\mathbf{x}) - \mathbf{y}^{\mathrm{true}}||$ as a function of the refinement step. This information is not available in real applications, but here it allows an assessment of the predictions made. Figure 5 shows the plot for the grid sequence Γ_{2a} of data set S_{2a}. Optimal function reconstruction is achieved when the residual has its smallest value.

Table 1. Grid sequence levels $\mathbf{k}$

Step	S_{2a}	S_{2b}	S_{3b}
1	(1,1)	(1,1)	(1,1,1)
2	(2,1)	(1,2)	(1,2,1)
3	(2,2)	(1,3)	(1,3,1)
4	(3,2)	(2,3)	(2,3,1)
5	(3,3)	(2,4)	(2,4,1)
6	(3,4)	(3,4)	(3,4,1)
7	(4,4)	(3,5)	(3,5,1)
8	(4,5)	(3,6)	(3,6,1)
9	(4,6)	(4,6)	(4,6,1)
10	(4,7)	(5,6)	(5,6,1)
11	(4,8)	(6,6)	(6,6,1)
12	(4,9)	(7,6)	(7,6,1)

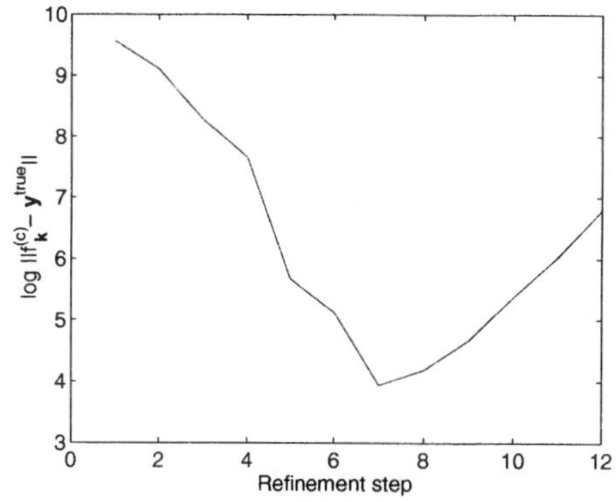

Fig. 5. Residual of approximation to true data (S_{2a})

The results for all three test sets are summarized in Table 2. The grids selected by L-curve and split-sample validation for the set S_{2a} correspond perfectly with the best possible grid $\Omega_{(4,4)}^{(c)}$ (step 7), taken from Figure 5. Generalization properties of the grids not covered in the grid sequence Γ_{2a} were checked additionally, confirming grid $\Omega_{(4,4)}^{(c)}$ to be the true optimum.

For the data set S_{2b}, the approximation with smallest residual to the true data is on grid $\Omega_{(3,5)}^{(c)}$, corresponding to step 7 within Γ_{2b}. The same grid is identified by the L-curve criterion. Using split-sample validation, grid $\Omega_{(3,4)}^{(c)}$ (step 6) is chosen as optimal. This might be accredited to the reduced training data set. However, the residuals of step 6 and 7 are approximately the same, with a 0.6% larger residual for step 6.

Table 2. Optimal approximations

	S_{2a}		S_{2b}		S_{3b}	
	Step	Level	Step	Level	Step	Level
Best	7	(4,4)	7	(3,5)	7	(3,5,1)
L-curve	7	(4,4)	7	(3,5)	7	(3,5,1)
Split	7	(4,4)	6	(3,4)	6	(3,4,1)

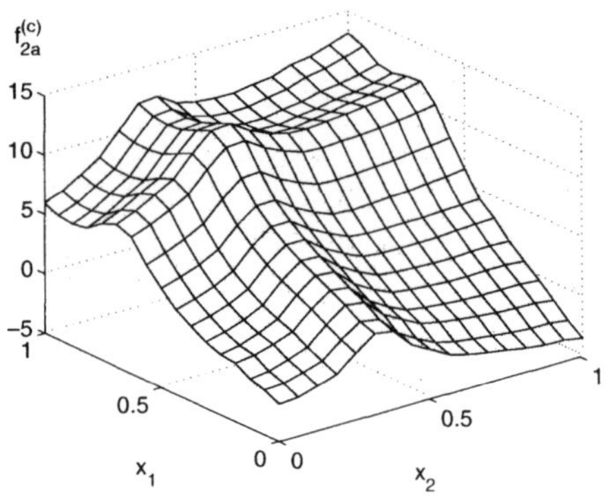

Fig. 6. Optimal approximation of t_{2a}

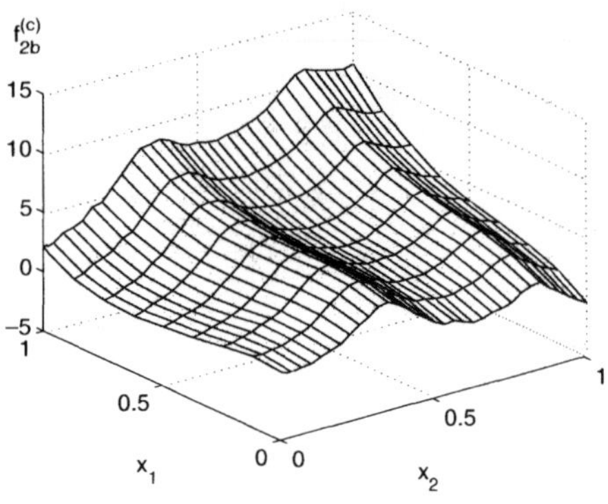

Fig. 7. Optimal approximation of t_{2b}

Plots of $f^{(c)}_{2a,(4,4)}$ and $f^{(c)}_{2b,(3,5)}$, selected as optimal for sets S_{2a} and S_{2b} are depicted in Figures 6 and 7 respectively.

For the trivariate set S_{3b}, the initially coarse resolution of the dependency on variable x_3 is not refined further. This is not surprising due to the linearity of both the x_3 dependency introduced in (19) and the hierarchical basis function set. The refinement steps in dimensions x_1 and x_2 correspond to the bivariate set S_{2b}.

6. CONCLUSIONS

An algorithm was proposed for functional approximation of a set of multivariate data using stepwise refinement of an initially coarse sparse grid and subsequent identification of a refinement step suited to resolve function details while suppressing measurement noise. Having restricted, in this work, to regularization inherent to projection of the sought function on a finite dimensional subspace, it was shown that the selected approximations describe the best possible function reconstructions within the mathematical framework. Contrary to competing methods such as feed-forward neural nets, the approach allows a simple adaption of both structure and parameters upon enlargement of the data set, while a reference feed-forward neural network with a Bayesian training algorithm revealed slightly superior performance in terms of generalization. Computational effort for both approaches lies in the same order of magnitude.

Future research will consequently focus on the improvement of the accuracy of prediction, which may be tackled by consideration of a Tikhonov regularization term as it is used by Garcke *et al.* (2001), enforcing smoothness of the approximation. As an alternative, the use of higher order basis functions will be subject to examination. Grid interpolation and sensitivity calculation are major contributions to computational expense. Further work will examine efficient techniques to gain sensitivity information, such as restriction to a subset of data.

ACKNOWLEDGEMENTS

This work was partially funded by the Deutsche Forschungsgemeinschaft (DFG) within the Collaborative Research Center (SFB 540) "Model-based experimental analysis of kinetic phenomena in fluid multiphase reactive systems". The authors thank A. Reusken for his reference to the work of M. Griebel.

REFERENCES

Binder, T., L. Blank, W. Dahmen and W. Marquardt (2002). On the regularization of dynamic data reconciliation problems. *J. Proc. Cont.* **12**(4), 557–567.

Bungartz, H.-J. and M. Griebel (1999). A note on the complexity of solving Poisson's equation for spaces of bounded mixed derivatives. *J. Complexity* **15**, 167–199.

Engl, H. W., M. Hanke and A. Neubauer (1996). *Regularization of Inverse Problems.* Kluwer Academic Publishers.

Garcke, J., M. Griebel and M. Thess (2001). Data mining with sparse grids. *Computing* **67**, 225–253.

Griebel, M., M. Schneider and C. Zenger (1992). A combination technique for the solution of sparse grid problems. In: *Iterative Methods in Linear Algebra* (P. de Groen and R. Beauwens, Eds.). Elsevier. Amsterdam. pp. 263–281.

Hansen, P. C. (1998). *Rank-deficient and Discrete Ill-posed Problems.* SIAM. Philadelphia.

Kirsch, A. (1996). *An Introduction to the Mathematical Theory of Inverse Problems.* Springer. New York.

Rogers, D. F. and J. A. Adams (1990). *Mathematical Elements for Computer Graphics.* McGraw-Hill. New York.

Weiss, S. M. and C. A. Kulikowski (1991). *Computer Systems That Learn.* Morgan Kaufmann.

Yserentant, H. (1986). On the multi-level splitting of finite-element spaces. *Numerische Mathematik* **49**, 379–412.

IFAC Publications
www.elsevier.com/locate/ifac

ITERATIVE IDENTIFICATION FOR CONTROL AND ROBUST PERFORMANCE OF BIOREACTOR

Kim Bøjstrup * Hans Henrik Niemann ** Niels Kjølstad Poulsen * Sten Bay Jørgensen ***

* *CAPEC, Departments of Chemical Engineering*
** *Ørsted•DTU*
*** *Informatics and Mathematical Modelling,*

Technical University of Denmark, DK-2800 Lyngby, Denmark

Abstract: A main result of control oriented identification is that, to minimize the closed-loop error that will ultimately result from a model-based controller, the identification of the model should use closed-loop data produced with the very same controller. However, the controller is not known prior to the actual experiment and therefore iterations between controller design and closed-loop identification are performed. This paper presents a modified procedure based upon the works of de Callafon (1998) and Tay et al. (1998). The performance of the above procedure is investigated on a continuous bioreactor which may exhibit unstable operation just beyond the critical dilution rate.

Keywords: Closed loop identification, robust control, iterative identification for control, model prediction control

INTRODUCTION

Chemical industry is under a constant pressure from the global competition and facing strict environmental, safety and health regulations. In order to ensure speed to market the companies must always seek towards the edge of development. Their new productions must be optimised fast not to delay the available production time during patent protection. The well established productions must undergo contuned improvements in order to compete on price and quality. In both perspectives the control of the production plant plays a crucial role. Today it is not sufficient only to rely on classical controllers such as a proportional, integral and differential (PID) controller, but it is necessary to implement multivariable controllers such as linear quadratic Gaussian (LQG) controller, an $\mathcal{H}_\infty$ controller or a model predictive controller (MPC). On top of this layer an optimising layer is required to render the plant operation optimal in face of varying disturbances and market conditions. Under such conditions the open loop plant may enter an unstable region Jørgensen and Jørgensen (1998) therefore the plant model representation must enable such behaviour. In addition any identified model represents an uncertain realisation of the plant, thus the uncertainty involved in the estimated behaviour also must be available in order to ensure a control performance which ensures robsutness.

It is therefore important to develop methods witch can perform closed loop model identification on a working plant with limited input perturbations, which do not jeopardize the quality of the product. Several different approaches have been discussed in the literature (e.g.Goodwin and Graham (2002); Ljung (1999a); Walter and Piet-Lahanier (1990)). Primarily the procedures proposed by de Callafon (1998) de Callafon (1998) and Tay & Moore (1998) Tay et al. (1998) will be discussed in this paper, since these approaches pocess the ability to handle both unstable plants and controllers.

The purpose of this paper is to describe results from investigation of iterative identification of a bioreactor which exhibits high sensitivity and also instability close to the optimal productivity.

1. A MODEL-BASED ITERATIVE PROCEDURE

The iterative identification procedure is called model-based since the performance level is evaluated on a model-basis instead of a data-basis. This point enables the model-based procedure to meet the requirement of finding a controller C for an unknown plant P_0 that can improve the performance of a controller currently implemented on the plant P_0. This problem can be reformulated into three basically different steps in the model-based iterative procedure as follows

(1) ***Initial identification:*** *Use experimental data from $T(P_0, C_i)$ and and prior information on data or plant, to estimate a set of models P_i such that γ_i is minimized while $P_0 \in \boldsymbol{P}_i$, where*

$$\|J(P, C_i)\|_\infty \leq \gamma_i \quad \forall P \in \boldsymbol{P}_i \tag{1}$$

(2) ***Control design:*** *Design C_{i+1} such that*

$$\|J(P, C_{i+1})\|_\infty \leq \gamma_{i+1} < \gamma_i \quad \forall P \in \boldsymbol{P}_i \tag{2}$$

(3) ***Re-identification:*** *Use new experimental data from $T(P_0, C_{i+1})$ and prior information to estimate a set of models $\boldsymbol{P}_{i+1}$ such that $P_0 \in \boldsymbol{P}_{i+1}$, subject to the condition*

$$\|J(P, C_{i+1})\|_\infty \leq \gamma_{i+1} \quad \forall P \in \boldsymbol{P}_{i+1} \tag{3}$$

In the model-based procedure step 1 can be regarded as an initialization step, while repeated execution of steps 2 and 3 will provide a design procedure in which the upper bound γ_i on a predetermined performance cost can be reduced progressively. This procedure is represented schematically in figure 1.

The methods combine an identification step and a controller design step. The identification procedure facilitates identification of a model set describing the unknown but closed-loop stable system based on a nominal model and an allowable model perturbation. The model set is based on a dual-Youla-Kučera parameterization and consists of the nominal model of limited complexity determined through a direct frequency domain identification and a model error for which an upper bound is estimated. The controller design step facilitates a design method for an enhanced robust controller based on the DK-iteration, which is an H_∞ optimization problem. The possiblities for enhancement of the closed loop performance are investigated through closed loop identification of model sets and subsequently of the control design.

2. BIOREACTOR CASE STUDY

The fermentation process is relevant, because it is still difficult to set up a model, which can describe the fermentation behaviour. This renders a black box model-based procedure interesting, because it is able to estimate a model set of an unknown plant, which can be used for robust control design, and progressively reduce the performance cost of the closed-loop system through an iterative identification and control design. In this case study a simple cell (biomass) producing continues fermentation process shown in figure 2 is used as the real plant. The aim of this first case study is to check if the model-based procedure is able to progressively reduce the performance cost of the closed-loop system.

The dynamics of the process are given by the state-space model

$$\frac{dX}{dt} = \mu(S)X - XD \tag{4}$$

$$\frac{dS}{dt} = -\frac{\mu(S)X}{Y} + (S_F - S)D \tag{5}$$

$$\mu(S) = \mu_{max}\frac{S}{K_2S^2 + S + K_1} \tag{6}$$

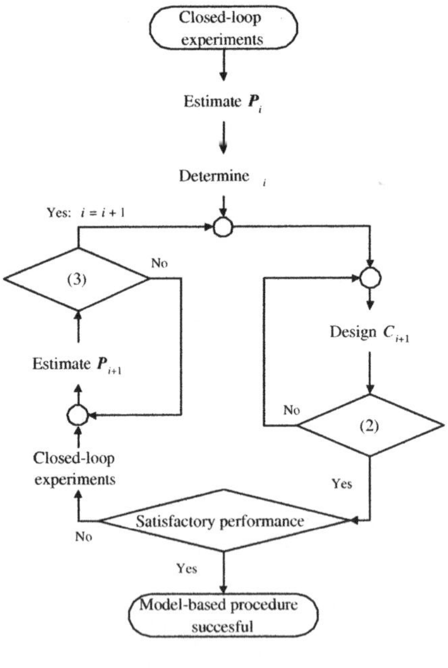

Fig. 1. Schematic flow chart for the steps in the model-based iterative procedure.

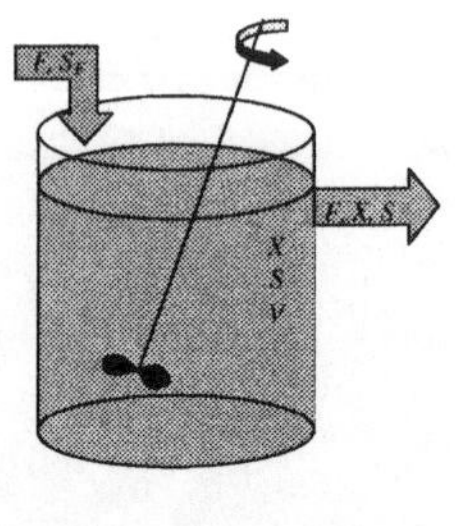

Fig. 2. Continuous fermentation process for biomass production

Where X is biomass, S substrate and $S_F 10$ g/L substrate feed concentration. $D = F/V$ is Dilution rate while the parameters are $Y = 0.5$ is yield coefficient for biomass while the kinetic parameters are $\mu_{max} = 1\ \text{h}^{-1}$, $K_1 = 0.03$ and $K_2 = 0.5$. The dilution rate is assumed to be time varying and is the control input. The biomass concentration X is assumed to be measurable and is the control output.

2.0.1. *Critical dilution rate* The production objective is to maximize biomass productivity. However this fermenter will exhibit multiple steady states and a fold bifurcation near the optimal productivity, even for relatively low inlet substrate concentrations.

In this case study, the model is used only to simulate a "true" system to be identified and controlled, and there are therefore no interest in operating the fermenter close to the critical dilution rate, which will cause unnecessarily complex behaviour. In this perspective, it is decided to operate the fermenter at 50% of the critical dilution rate. The critical dilution rate is located around 0.8 hr^{-1} and the operation point is therefore chosen to $D = 0.4\ \text{hr}^{-1}$.

2.1 *The initial controller*

The initial controller is chosen to be a simple PI-controller tuned with a simple Ziegler-Nichols method. The tuning of the controller ($K_c = 10$ and $T_i = 10$ hr).

The reference signals r_1 and r_2 used to excite the closed-loop system are two periodic signals that consists of a sum of sinusoids

$$r_k(t) := \sum_{j=1}^{n} \sin(\omega_j t + \phi_j), \quad k = 1, 2. \quad (7)$$

specified at a predefined frequency grid $\Omega = \{\omega | \omega = \omega_j,\ j = 1, 2, \ldots, n\}$. The phase shift ϕ_j of the sinusoids has to be chosen properly to avoid high signal amplitudes due to the cumulative effect of adding multiple sinusoids. In other words, it is important to have a low crest factor C_r Ljung (1999b). To avoid complications associated with identical reference signals, the phase shifts ϕ_j in the sequence $\{\phi_j\}$, $j = 1, 2, \ldots, n$ of (7) are chosen independently from a uniform distribution over the interval $[-\pi, \pi]$ to ensure uncorrelated reference signals.

A frequency grid Ω of 17 sinusoids is distributed logarithmically between 2.218 rad/hr and 18.85 rad/hr. There are computed 32 input-output signals with a sampling interval of 10 minutes. The output signal is added a 2% measurement error generated by a pseudo random binary sequence and the reference signals are multiplied with the steady state values of the feed rate and biomass concentration respectively, which makes the perturbations of the plant more that 50% of the steady state values.

Through closed-loop spectral analysis, a frequency response estimate $\hat{G}(\omega_j)$ of the unknown plant P_0 is obtained in 17 points within the frequency range between 2.218 rad/hr and 18.85 rad/hr. In closed-loop spectral analysis, a cross spectrum between a reference signal (either of the two) and the input signal and a cross spectrum between a reference signal (the same as before) and the output signal is estimated. An amplitude plot of the data $\hat{G}(\omega_j)$ is depicted in figure 3 by a dashed line.

2.2 *Identification of a nominal model*

The identification of the nominal model is done through the model based iterative scheme 1. The iterative procedure need an initial model estimate of the nominal model for the construction of the filter F_k prior to the estimation of $\hat{P}_{k+1}$. The initial model estimate is computed through a least-squares optimization. By means of a weighted two-norm minimization, a linear second order model is derived directly on the basis of the frequency response data.

2.3 *Identification through a min-max optimization*

For the control relevant estimation of a factorization $(\hat{N}, \hat{D})$ of the nominal model $\hat{P}$, the following minimization must be performed:

$$\min_{\theta \in \Theta} \left\| U_2 \left(\begin{bmatrix} N_{0,F} \\ D_{0,F} \end{bmatrix} - \begin{bmatrix} N(\theta) \\ D(\theta) \end{bmatrix} \right) F \begin{bmatrix} C & I \end{bmatrix} U_1 \right\|_\infty \quad (8)$$

Frequency domain data is helpful for the approximation of the $\mathcal{H}_\infty$ norm criterion by a point wise evaluation of (8). However the selected frequency grid Ω, must be sufficiently dense to represent the frequency response of the rcf $(N_{0,F}, D_{0,F})$.

The iterative scheme 1 is initialized with the model obtained from the least-squares optimization and a second order model is obtained. The frequency response is depicted in figure 3.

2.3.1. *Identification of model uncertainty* The set of models, used to reflect the limited knowledge of the plant P_0, is structured as follows:

$$\boldsymbol{P}_1(\hat{N}_1, \hat{D}_1, N_{c,1}, D_{c,1}, \hat{V}_{1,lk}) = \quad (9)$$

$$\{P | P = (\hat{N}_1 + D_{c,1}\Delta_{R_1})(\hat{D}_1 - N_{c,1}\Delta_{R_1})^{-1}\}$$

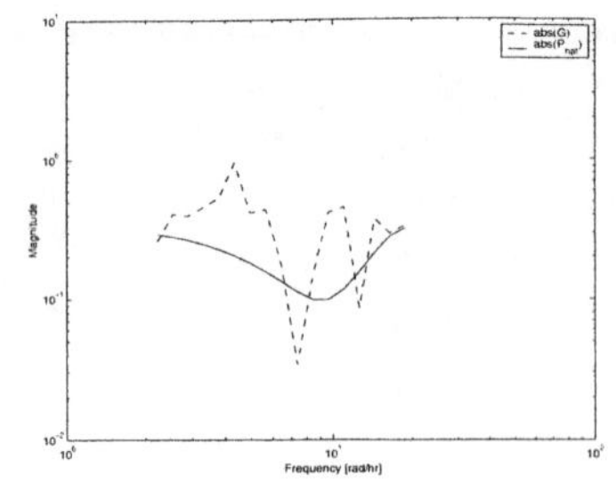

Fig. 3. Magnitude Bode plot of the experimentally obtained frequency response $\hat{G}_1(\omega_j)$ and the estimated 2nd order model $\hat{P}_1(\omega_j)$.

with

$$\Delta_{R_1} \in \mathbb{R}\mathcal{H}_\infty \text{ and } \Delta_1 := \hat{V}_{1,lk}\Delta_{R_1}, \quad \|\Delta_1\|_\infty < \gamma_1^{-1}$$

To estimate a set of models a nominal factorization $(\hat{N}_1, \hat{D}_1)$ and a weighting function $\hat{V}_1$ are needed (lk is omitted a because of the SISO case). The pair $(N_{c,1}, D_{c,1})$ is assumed to be known and is found by computing a $nrcf$ of the known controller used in the feedback connection during the closed-loop experiments. The estimation of a nominal model $\hat{P}_1 = \hat{N}_1\hat{D}_1^{-1}$ that satisfies $T(\hat{P}_1, C_1) \in \mathbb{R}\mathcal{H}_\infty$ has been presented above. Hence, to complete the characterization of this set of models $\boldsymbol{P}_1$, a frequency dependent weighting function $\hat{V}_1$ that upper bound the stable model perturbation Δ_{R_1} has to be estimated. In order to find this weighting function, first a frequency dependent (non-parametric) upper bound $\bar{\Delta}_{R_1}(\omega)$ for the unknown, but stable bounded model perturbation $\Delta_{R_1}(e^{i\omega})$ is estimated. Based on this upper bound estimate a low order frequency dependent weighting function $\hat{V}_1$ is determined. As it is seen in figure 4 the low order weighting filter does result in a narrow overbounding of the estimated probabilistic frequency dependent amplitude upper bound $\bar{\Delta}_{R_1}(\omega)$.

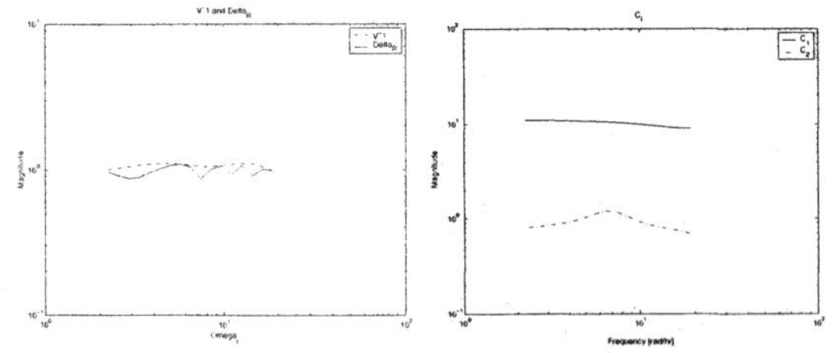

Fig. 4. Left: Magnitude Bode plot of stable and stably invertible weighting filter $\hat{V}_1(e^{i\omega})$(dashed) and estimated probabilistic frequency dependent amplitude upper bound $\bar{\Delta}_{R_1}(\omega)$(solid). Right: Magnitude Bode plot of C_1 (solid) and C_2 (dashed)

2.4 *Robust performance enhancement*

Given the estimated model set $\boldsymbol{P}_1$ performance robustness can be evaluated and an enhanced robust controller can be designed. As such, the estimated model set $\boldsymbol{P}_1$ is used to represent the knowledge of the plant P_0 that is currently available. This knowledge is used to actually evaluate the performance of the controller C_1 currently implemented on the plant P_0 and to redesign the feedback control.

2.4.1. *Performance evaluation* To compare the performance of a newly designed controller C_2 with C_1 the performance of $\|J(\boldsymbol{P}_1, C_1)\|$ is evaluated as the plant P_0 is unknown. Again it should be noted that the robust stability does not have to be evaluated for the controller C_1, as the set of models $\boldsymbol{P}_1$ has been constructed such that robust stability is ensured.
The performance of the controller C_1 is evaluated through the $\mathcal{H}_\infty$-norm

$$\|J(\boldsymbol{P}_1, C_1)\|_\infty = \|U_2 T(\boldsymbol{P}_1, C_1) U_1\|_\infty \qquad (10)$$

For this first evaluation we only calculate the performance level

$$\|J(\boldsymbol{P}_1, C_1)\| = \gamma_1 \qquad (11)$$

but in general robust performance is satisfied when

$$\|J(\boldsymbol{P}_i, C_i)\|_\infty \leq \gamma_{i-1} \qquad (12)$$

With the plant $P_0 \in \boldsymbol{P}_1$, the value of γ_1 gives an indication of the performance of C_1. This performance level can be compared with the performance γ_2 of a new controller C_2 that will be specifically designed on the basis of the set of models $\boldsymbol{P}_1$ to achieve

$$\|U_2 T(\boldsymbol{P}_1, C_2) U_1\|_\infty \leq \gamma_2 < \gamma_1$$

For the evaluation of γ_1, an LFT representation and the structured singular value $\mu\{M\}$ is used. The weighting filters U_1 and U_2 are identity matrices. Subsequently, the structured singular value is evaluated, and the performance level γ_1 for the controller C_1, applied to the set of models $\boldsymbol{P}_1$, is $\gamma_1 = 6.19$.

The weighting function $\hat{V}_1$ has been chosen to normalize the model uncertainty $\bar{\Delta}_{R_1}$, which means that the controller C_1 applied to the model set $\boldsymbol{P}_1$ does not satisfy robust performance. If that should be the case then $\mu\{M(e^{i\omega})\} < 1$, as the weighting function is chosen to normalize ($\gamma = 1$) the uncertainty. Lowering the value of γ to improve the performance is the task of the next controller design giving C_2.

2.4.2. *Controller design* There clearly is room for improving performance as C_1 is a simply tuned PI-controller. The model set $\boldsymbol{P}_1$ is used to design the new controller C_2. The controller C_2 is computed via μ-synthesis using a D-K iteration. The controller C_2 found by the μ-synthesis has

7 states. Compared to the initial controller C_1 figure 4 shows that C_2 has additional dynamics to account for the modelled uncertainties. In general, the complexity of a controller C_{i+1} generated by a μ-synthesis will be higher than the complexity of the coefficient matrix G being used. As G contains all the entries of the model set $\boldsymbol{P}_i$ and the weighting functions U_1 and U_2, the order of the controller C_{i+1} needs in general to be reduced significantly before implementation on the real system.

2.4.3. *Evaluation of performance* The robust stability and robust performance of the feedback connection $\boldsymbol{T}(P_0, C_2)$ needs to be evaluated before the actual implementation of the controller C_2, in order to verify, that an enhanced controller has been designed. Both robust stability and robust performance for the controller C_2 are checked. The new performance level γ_2 for the controller C_2, applied to the set of models $\boldsymbol{P}_1$, is found to be $\gamma_2 = 2.71$, which means that the controller C_2 has improved the performance and satisfies.

$$\|U_2 T(\boldsymbol{P}_1, C_2) U_1\|_\infty \leq \gamma_2 < \gamma_1 \qquad (13)$$

The controller C_2 does not satisfy robust performance, as $\gamma_2 > 1$, however, robust stability is guaranteed as $\mu(M_{11}) = 0.93 < 1$.

2.4.4. *Improving performance* In accordance with the iterative model-based procedure 1 the newly designed controller C_2 is implemented in the feedback connection $\boldsymbol{T}(P_0, C_2)$ and a new closed-loop experiment is conducted. This is labeled as "Step 3" in the model-based procedure. Basically, the same steps of nominal model identification, model uncertainty estimation and robust control design are performed, by a repeated execution of steps 2 and 3. To indicate the progressive performance improvement during the subsequent steps within the iterative procedure, the performance evaluation for four consecutive steps of subsequent model set estimation and robust controller design have been summarized in figure 5.

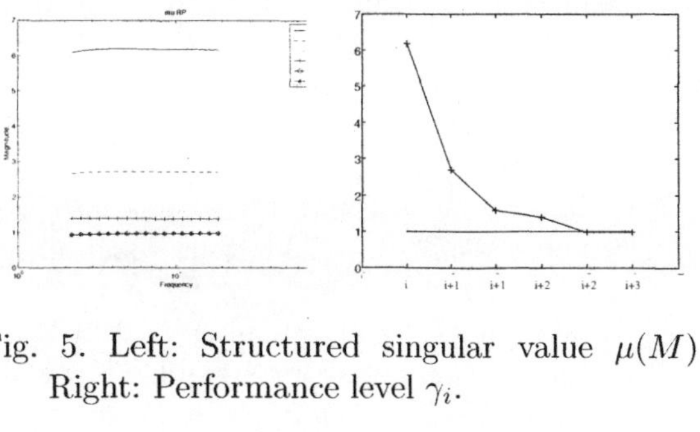

Fig. 5. Left: Structured singular value $\mu(M)$. Right: Performance level γ_i.

It can be observed in figure 5 that the performance improves progressively, and both controller C_3 ($\gamma_3 = 0.974$) and C_4 ($\gamma_4 = 0.969$) satisfies the robust performance criterion $\mu(M) < 1$, but the question will always be, if it is the best achievable. This question is even more interesting in this work, because of the lack of convergence in the estimation of the nominal model.

2.4.5. *Additional studies* In the previous sections it has been shown, that the model-based procedure is able to meet the problem description, and in order to test what limitations the procedure with the current programs are facing, additional studies have been performed. First a study of the continuous fermenter operating closer to the maximum production rate is performed ($D = 0.7$ hr^{-1}). The new PI-controller found above is implemented on the fermenter. The controller is able to stabilize the system with the settings as given for the fermentor above, but with perturbations ten times smaller (about 5% of the steady state values), which indicates the higher sensitivity in that operating region. A model set is identified and the performance level γ_1 calculated to be 0.426. A controller design based on the DK-iteration results in a 6th order controller. The evaluation of

$$\|U_2 T(\boldsymbol{P}_1, C_2) U_1\|_\infty$$

gives a performance level γ_2 equal to 0.428, which means that it is not possible to improve performance for the closed-loop system.

In this study it is shown, that the PI-controller found to be the best for $D = 0.4$ hr^{-1} is still the best achievable. The model-based procedure is not able to improve the performance, which means, either that C_1 is a very good guess or that the identified model set is too conservative, which makes it impossible to design an enhanced controller. The magnitude plot of the experimentally obtained frequency response $\hat{G}(\omega_j)$ and the estimated magnitude Bode plot of the model perturbation Δ_R is depicted in figures 6. In this figure it can be observed, that the nominal model and the uncertainty are at the same gain level, indicating, that the nominal model should be of higher order in order to make the model set less conservative.

It is studied next whether the model-based procedure is able to ensure robust performance for the closed-loop system operating at a dilution rate of $D = 0.8$ hr^{-1} which is 99.9% of the critical dilution rate. Again PI$_{new}$ is used in the feedback connection $\boldsymbol{T}(P_0, C)$. The controller is able to stabilize the system with the settings as given in section 2.1, but with perturbations which are 20 times smaller, i.e. about 3% of the steady state values. Again indicating the high sensitivity in

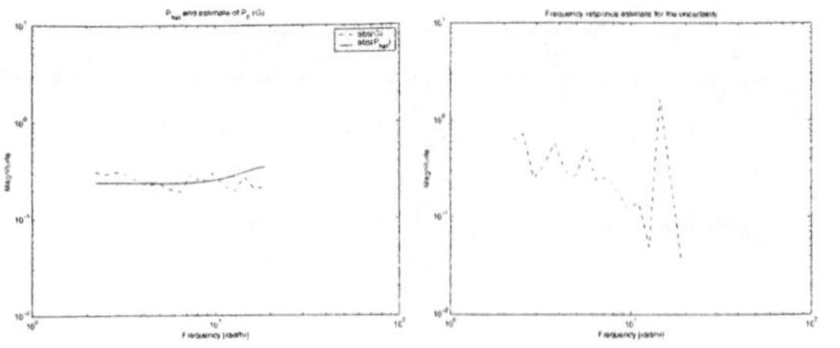

Fig. 6. Left: Magnitude Bode plot of the experimentally obtained frequency response $\hat{G}(\omega_j)$ and the estimated 2nd order model $\hat{P}(\omega_j)$. Right: Magnitude Bode plot of spectral estimate of model perturbation Δ_R.

that operating region.

The identification procedure was not able to estimate a nominal model with an accuracy needed in order to design a controller with lower performance cost that the implemented PI-controller. In figure 7 it is observed that the uncertainty has a larger magnitude than the nominal model.

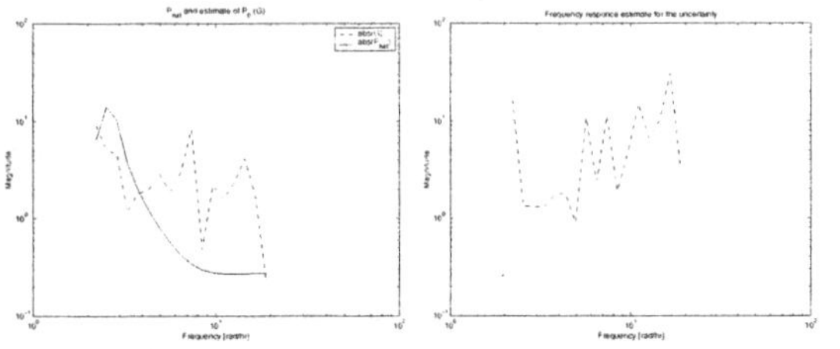

Fig. 7. Left: Magnitude Bode plot of the experimentally obtained frequency response $\hat{G}(\omega_j)$ and the estimated 2nd order model $\hat{P}(\omega_j)$. Right: Magnitude Bode plot of spectral estimate of model perturbation Δ_R.

Therefore the model-based procedure was stopped, as it was not able to satisfy (2).

$$\|J(P, C_{i+1})\|_\infty \leq \gamma_{i+1} < \gamma_i \quad \forall P \in \boldsymbol{P}_i$$

It is believed to be caused by the lack of convergence in the iterative scheme 1, as the routine was not able to satisfy (14).

$$D(\theta) + CN(\theta) = F^{-1} \tag{14}$$

On the other hand, the evaluation of the robust stability showed that the PI-controller will guarantee robust stability as $\bar{\mu}(M_{11_{C_1}}) = 0.31 < 1$ but not robust performance as seen in figure 8.(The two RP-plots are almost identical.)

3. CONCLUSIONS

This paper has presented the results of an iterative procedure for identification for control applied to a continuous fermentation process. The application has shown that a set of models $\boldsymbol{P}$ can be estimated on the basis of closed-loop data and used for robust control design. Furthermore, the

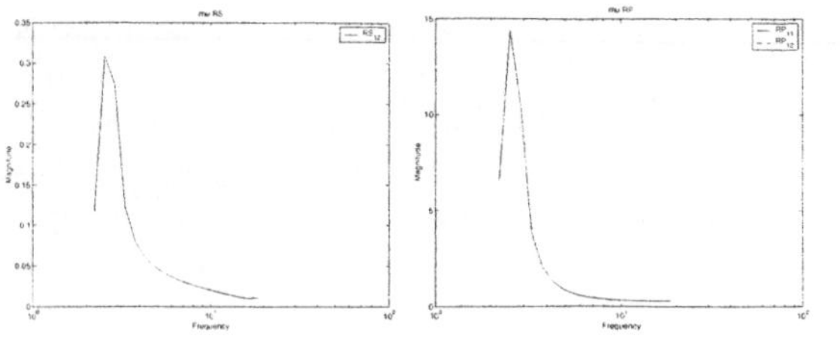

Fig. 8. Left: Structured singular value for $\mu(M_{11_{C_1}})$. Right: Structured singular value for $\mu(M_{C_1})$ (solid) and $\mu(M_{C_2})$ (dashed).

application showed how to improve the performance of the feedback control system by systematically monitoring the performance robustness. The resulting high order controller is easily reduced to a simple PI-controller with almost the same performance. This controller also ensured robust performance at higher dilution rates, but only ensured robust stability when operation very close to the critical dilution rate was attempted. Under these conditions it was not possible to improve the performance of the PI-controller, which is believed to be caused by the lack of convergence in the iterative scheme as the routine was not able to satisfy a key filter equation. The lack of improvement for the systems operating close to the critical dilution rate is believed to be caused by the lack of convergence in the identification of the nominal model. The identified model set is too conservative, which renders it impossible to design an enhanced controller.

References

R.A. de Callafon. *Feedback Oriented identification for Enhanced and Robust Control.* Phd thesis, Delft University of Technology, Delft, the Netherlands, 1998.

G.C. Goodwin and C. Graham. Non-stationary stochastic embedding for transfer function estimation. *Automatica*, 38, No. 1:47–62, 2002.

J.B. Jørgensen and S.B. Jørgensen. Operational implications of optimality. *AIChE Symposium Series*, ?, 1998.

L. Ljung. Comments on model validation as set membership identification. In *In: Robustness in Identification and Control*, volume 245, pages 7–16, 1999a.

L. Ljung. *System Identification.* Prentice Hall, New Jersey, USA, 1999b.

T.T. Tay, I. Mareels, and J.B. Moore. *High Performance Control.* Birkhäuser, Boston, USA, 1998.

E. Walter and H. Piet-Lahanier. Estimation of parameter bounds from bounded-error data. *Mathematics and Computers in Simulation*, 91: 449–68, 1990.

IFAC Publications
www.elsevier.com/locate/ifac

MODIFIED SUBSPACE IDENTIFICATION METHOD FOR BUILDING A LONG-RANGE PREDICTION MODEL FOR INFERENTIAL CONTROL

Yangdong Pan, Jay H. Lee [1]

School of Chemical Engineering, Georgia Institute of Technology, Atlanta, GA

Abstract: In a chemical plant involving a series of processing units, it is beneficial to have a model that can accurately forecast the behavior of downstream variables based on upstream measurements. Such a model can be useful in feedforward and inferential control of the downstream variables to compensate for various upstream disturbances. However, creating such a dynamic model can be very difficult. The conventional multivariable identification approach based on minimizing single-step-ahead prediction error, can result in models leading to poor prediction and control in the described context. To alleviate this difficulty, we propose a modification to the conventional subspace identification method geared towards accurate k-step-ahead prediction, where k is a number chosen according to the estimated dead time. It is shown that the modified subspace identification method can be used in conjunction with the k-step prediction error minimization (PEM). Using an illustrative examples involving six mixing units with a recycle loop, we demonstrate the improvement that is possible from adopting the suggested modification. *Copyright © 2003 IFAC*

1. INTRODUCTION

Most modern plants involve a large number of interconnected processing units, thus raising the need to consider the interactions and information flows among them. A typical plant setup involves measurements and manipulated variables located at the upstream and downstream property variables that need to be controlled. For disturbances occurring in the feed or upstream units, the upstream variables show more immediate responses. Their quick responses, if measured, can be used to manipulate upstream processing conditions in order to keep the downstream properties in control – as in feedforward control or inferential control. To realize this, the upstream measured process variables must be accurately related to the downstream property variables in a dynamic manner. The same situation appears in distributed parameter systems with a large residence time, such as a continuous pulp digester.

Developing a model that accurately captures the dynamic correlation between upstream and downstream variables presents a major challenge. Such models are likely to involve large time delays and dynamics of high order and possibly multiple time scales (due to recycle loops commonly found in industrial plants). Any one of the above features can pose difficulties for the existing system identification approaches. Furthermore, inferential control puts a higher demand on the model accuracy.

In the described problem's context, it is obvious that long-range prediction performance of the model is what ultimately matters. Since a large dead time is involved typically, the short-term predictions, however accurate they may be, are not useful. The importance of emphasizing the long-range prediction over the short-term prediction becomes more clear when one considers the significant model bias typical in most system identification carried out in practice. In the literature, the minimization of k-step-ahead prediction error in the prediction error minimization (PEM) method has been suggested and discussed [8][10] . In addition to the time-domain interpretation, Wahlberg and Ljung [6] formally showed that the

[1] To whom all correspondence should be addressed. phone (404)385-2148, fax (404)894-2866, e-mail:Jay.Lee@che.gatech.edu

use of k-step-ahead prediction methods amounts to emphasizing the accuracy of low-frequency dynamics more in distributing the bias, compared to the conventional one-step-ahead error minimization, which tends to put higher emphasis on the high frequency behavior.

In spite of these developments, understanding of where and how to use the more general k-step PEM in process control's context has been fairly limited. The few exceptions include papers by Shook *et al* [2], and Huang *et al* [3]. Still, a clear link between the method and situations or types of process applications, from which substantial benefits of the method are likely to be realized, is not there. Another reason for the lack of its use in practice is the numerical difficulty associated with using k-PEM for multivariable systems. In addition to the usual complexities (*e.g.*, local minima) associated with the standard PEM, the design of the prefilter necessary to turn the multi-step-ahead prediction error minimization into the one-step-ahead prediction error minimization requires the noise model, which is usually not known *a priori*. In many works, such as the long-range predictive identification (LRPI) approach advocated by Shook *et al* [2], the noise model is assumed to be fixed *a priori*. In this case, the quality of the identified model as well as the performance of the final predictive controller can be strongly influenced by the choice of the noise model.

For multivariable identification problems, the subspace identification method has many attractive features, including the numerical robustness and non-iterative nature of the algorithm [9]. However, the conventional subspace identification method is geared implicitly towards providing accurate one-step-ahead predictions. It is shown in this paper that, for those applications requiring accurate long-range predictions, the conventional method can perform poorly. Given the above-mentioned merits of the subspace method, however, it is useful to consider how the method can be extended to give higher emphasis on the long-range prediction performance.

The contribution of this paper can be two-fold. First, we bring to attention a situation ubiquitous in the process industries, for which the importance of fitting a model to optimize its long-range prediction performance is very high. Second, we present a modified version of subspace identification, in which the emphasis is given to the k-step-ahead prediction performance, where k is a general number chosen according to the process dead-time. We also show how a model obtained from the modified subspace method can be further improved through the k-step-ahead prediction error minimization (k-PEM). An example involving 6 mixing units with a recycle loop is chosen to show the importance of emphasizing the long-range performance through the proposed method.

2. PROPOSED MODIFICATIONS FOR EMPHASIZING THE K-STEP-AHEAD PREDICTION PERFORMANCE

Here we propose a modification to the conventional identification method with the aim of obtaining more accurate k-step-ahead predictions. We first show the modifications for the subspace identification method. After that, we discuss how the resulting model can be improved through the PEM method.

2.1 *Subspace Identification Based on Minimizing the k-Step-Ahead Prediction Error*

The conventional subspace ID approach, such as the N4SID method described in [9], implicitly assumes that the purpose of the model is to provide accurate one-step-ahead prediction. This is seen in the step where state space matrices A, B, C, D are estimated through least squares. In N4SID, data bank for one-step ahead Kalman state estimate $x_{t+1|t}$ is first created from the input/output data based on the following multi-step prediction equation:

$$\begin{bmatrix} y_{t+1} \\ y_{t+2} \\ \vdots \\ y_{t+\bar{n}} \end{bmatrix} = L_1 \begin{bmatrix} y_{t-\bar{n}+1} \\ y_{t-\bar{n}+2} \\ \vdots \\ y_t \end{bmatrix} + L_2 \begin{bmatrix} u_{t-\bar{n}+1} \\ u_{t-\bar{n}+2} \\ \vdots \\ u_t \end{bmatrix} + L_3 \begin{bmatrix} u_{t+1} \\ u_{t+2} \\ \vdots \\ u_{t+\bar{n}-1} \end{bmatrix} + \begin{bmatrix} \varepsilon_{t+1|t} \\ \varepsilon_{t+2|t} \\ \vdots \\ \varepsilon_{t+\bar{n}+1|t} \end{bmatrix} \quad (1)$$

Since we can write the optimal predictions in terms of the Kalman state estimate (i.e., the estimate by the nonstationary Kalman Filterinitialized at $t-\bar{n}+1$ as

$$\begin{bmatrix} y_{t+1|t} \\ y_{t+2|t} \\ \vdots \\ y_{t+\bar{n}|t} \end{bmatrix} = \begin{bmatrix} C \\ CA \\ \vdots \\ CA^{\bar{n}-1} \end{bmatrix} x_{t+1|t} + L_3 \begin{bmatrix} u_{t+1} \\ u_{t+2} \\ \vdots \\ u_{t+\bar{n}+1} \end{bmatrix}, \quad (2)$$

$$\begin{bmatrix} C \\ CA \\ \vdots \\ CA^{\bar{n}-1} \end{bmatrix} x_{t+1|t} = \begin{bmatrix} L_1 & L_2 \end{bmatrix} \begin{bmatrix} y_{t-\bar{n}+1} \\ \vdots \\ y_t \\ u_{t-\bar{n}+1} \\ \vdots \\ u_t \end{bmatrix} \quad (3)$$

Because state coordinates are not fixed *a priori*, one-step ahead state estimate $x_{t+1|t}$ can be created by estimating $[L_1 \quad L_2]$ through least squares and then finding a set of basis that spans its range space. In N4SID, this is done through a series of oblique matrix projections [9]. Once data for $x_{t+1|t}$ and $x_{t+2|t+1}$ are created, the state space matrices are obtained by solving the linear least squares problem

$$\begin{aligned} x_{t+2|t+1} &= Ax_{t+1|t} + Bu_{t+1} + w_{t+1|t} \\ y_{t+1} &= Cx_{t+1|t} + \varepsilon_{t+1|t} \end{aligned} \tag{4}$$

where the residuals w and ε are minimized. Hence, in this step of the subspace method, one-step-ahead prediction error is minimized. The covariance matrix for w and ε, $\begin{pmatrix} R_w & R_{w,\varepsilon} \\ R_{w,\varepsilon}^T & R_\varepsilon \end{pmatrix}$, is estimated from the residuals of the least squares and the Kalman filter is designed with the calculated system and covariance matrices to obtain the following innovation form of the model.

$$\begin{aligned} x_{t+2|t+1} &= Ax_{t+1|t} + Bu_{t+1} + K\varepsilon_{t+1|t} \\ y_{t+1} &= Cx_{t+1|t} + \varepsilon_{t+1|t} \end{aligned} \tag{5}$$

We may generalize N4SID to emphasize the k-step-ahead prediction in the following manner. To create k-step ahead state estimates, the optimal multi-step prediction equation of (1) can be modified to

$$\begin{aligned} \begin{bmatrix} y_{t+k} \\ y_{t+k+1} \\ \vdots \\ y_{t+k+\bar{n}} \end{bmatrix} &= L_1 \begin{bmatrix} y_{t-\bar{n}+1} \\ y_{t-\bar{n}+2} \\ \vdots \\ y_t \end{bmatrix} + L_2 \begin{bmatrix} u_{t-\bar{n}+1} \\ u_{t-\bar{n}+2} \\ \vdots \\ u_{t+k-1} \end{bmatrix} \\ &+ L_3 \begin{bmatrix} u_{t+k} \\ u_{t+k+2} \\ \vdots \\ u_{t+k+\bar{n}-1} \end{bmatrix} + \begin{bmatrix} \varepsilon_{t+k|t} \\ \varepsilon_{t+k+1|t} \\ \vdots \\ \varepsilon_{t+k+\bar{n}|t} \end{bmatrix} \end{aligned} \tag{6}$$

As before, it follows that

$$\begin{bmatrix} C \\ CA \\ \vdots \\ CA^{\bar{n}-1} \end{bmatrix} x_{t+k|t} = \begin{bmatrix} L_1 & L_2 \end{bmatrix} \begin{bmatrix} y_{t-\bar{n}+1} \\ \vdots \\ y_t \\ u_{t-\bar{n}+1} \\ \vdots \\ u_{t+k-1} \end{bmatrix} \tag{7}$$

Following the same procedure as before, data bank for k-step-ahead state estimates $x_{t+k|t}$ and $x_{t+k+1|t+1}$ can be obtained. Then, a state space model can be obtained by performing least squares on the following equations:

$$\begin{aligned} x_{t+k+1|t+1} &= Ax_{t+k|t} + Bu_{t+k} + w_{t+k|t} \\ y_{t+k} &= Cx_{t+k|t} + \varepsilon_{t+k|t} \end{aligned} \tag{8}$$

The residual $\varepsilon_{t+k|t}$ represents the k-step-ahead prediction error, which is minimized. Note that, if the data-based Kalman estimates were perfect, then

$$w_{t+k|t} = \underbrace{A^{k-1}K}_{\bar{K}} \varepsilon_{t+1|t} \tag{9}$$

Also,

$$\varepsilon_{t+k|t} = \underbrace{\sum_{i=0}^{k-1} q^{-i}\mathcal{H}_i}_{\bar{F}_k(q)} \varepsilon_{t+k|t+k-1} \tag{10}$$

where $\mathcal{H}_i$ is the ith Markov parameter of the noise model (A, K, C, I).

Based on these, the procedure for extracting $(A, B, C,)$ and K are as follows:

(1) Solve the least squares problem for the output equation to find C that minimizes $y_{t+k} - Cx_{t+k|t}$ in the 2-norm sense. The residuals represent the data for $\varepsilon_{t+k|t}$.
(2) Solve the least squares for the state equation to find A, B. The residual can be viewed as $w_{t+k|t}$.
(3) On the generated residual of $\varepsilon_{t+k|t}$, use a whitening filter to obtain one-step-ahead prediction error $\varepsilon_{t+k|t+k-1}$. A convenient way to do this is to apply subspace identification to the data. The ouput residual from this will be $\varepsilon_{t+1|t}$.
(4) Calculate the covariance matrix for $w(t+k|t)$ and using the whitened residual $\varepsilon(t+1|t)$. According to (9), the covariance matrix for the residual $w_{t+k|t}$ and $\varepsilon_{t+1|t)}$ has the form of

$$\begin{bmatrix} A^{k-1} & 0 \\ 0 & I \end{bmatrix} \begin{bmatrix} R_w & R_{w,\varepsilon} \\ R_{w,\varepsilon}^T & R_\varepsilon \end{bmatrix} \begin{bmatrix} A^{k-1} & 0 \\ 0 & I \end{bmatrix}^T \tag{11}$$

where $\begin{pmatrix} R_w & R_{w,\varepsilon} \\ R_{w,\varepsilon}^T & R_\varepsilon \end{pmatrix}$ represents the covariance for $w_{t+1|t}$ and $\varepsilon_{t+1|t}$. With the calculated system matrices and the extracted covariance matrix for w and ε, one can proceed to design the Kalman filter to put the model in the innovation form. The k-step ahead predictor can be easily derived from it.

It should be obvious to those familiar with the subspace identification method that the asymptotic properties of N4SID such as unbiasedness and consistency remain intact with the above modifications.

2.2 *k-Step Prediction Error Minimization*

Although the modified subspace ID method puts higher emphasis on the accuracy of the k-step ahead prediction in obtaining state space matrices, it does not directly minimize k-step-ahead prediction error for a finite data set. It has been suggested in Ljung [1] that the subspace method be used to initialize PEM, which generally requires a special parameterization and a good initial guess to be successful. Here we propose to use the model from the proposed k-step subspace ID method to start the k-step PEM.

A MIMO state space model,

$$\begin{aligned} x_{t+1} &= Ax_t + Bu_t + Ke_t \\ y_t &= Cx_t + Du_t + e_t \end{aligned} \tag{12}$$

can be represented in the following input/output form:

$$y_t = G(q)u_t + H(q)e_t \tag{13}$$

where

$$\begin{aligned} G &= C(qI - A)^{-1}B + D \\ H &= C(qI - A)^{-1}K + I \end{aligned} \tag{14}$$

The optimal one-step ahead predictor is given by Ljung [1]

$$\hat{y}_{t|t-1} = H^{-1}Gu_t + (1 - H^{-1})y_t \tag{15}$$

If parameterized models G_θ and H_θ are used, then the optimal one-step-ahead predictor can be written as

$$\hat{y}_{t|t-1} = H_\theta^{-1}G_\theta u_t + (1 - H_\theta^{-1})y_t \tag{16}$$

Optimal k-step-ahead predictor is

$$\hat{y}_{t|t-k} = W_k G_\theta u_t + (1 - W_k)y_t \tag{17}$$

where

$$W_k = F_k H_\theta^{-1} \tag{18}$$

and

$$F_k = \sum_{i=0}^{k-1} \mathcal{H}_i q^{-i} \tag{19}$$

Here, $\mathcal{H}_i$ is a $n_y \times n_y$ matrix representing the ith impulse response coefficient matrix of $H(q)$. The optimal k-step ahead predictor can also be viewed as the optimal one-step ahead predictor associated with the model

$$y_t = Gu_t + HF_k^{-1}\varepsilon_t \tag{20}$$

where ε_t is a white noise.

For a SISO system, F_k, if known, can be regarded as a prefilter and the k-step prediction error minimization is the same as the one-step prediction error minimization with the filtered I/O data. However, for a MIMO system, because matrices do not commute in multiplication, prefiltering the data before applying the one-step ahead PEM does not work. Therefore, F_k has to be embedded into the model structure when applying the PEM, resulting in a structured identification problem. Let us use the state-space representation of

$$\begin{aligned} F_k &= (\tilde{A}_F, \tilde{B}_G, \tilde{C}_G, \tilde{D}_G) \\ G &= (A, B, C, D) \\ H &= (A, K, C, I) \end{aligned} \tag{21}$$

First, the inverse system F_k^{-1} is,

$$\begin{aligned} F_k^{-1} = (&\tilde{A}_F - \tilde{B}_F\tilde{D}_F^{-1}\tilde{C}_F, \tilde{B}_F\tilde{D}_F^{-1}, \\ &-\tilde{D}_F^{-1}\tilde{C}_F, \tilde{D}_F^{-1}) \end{aligned} \tag{22}$$

Let us denote

$$F_k^{-1} = (\tilde{A}_{F^{-1}}, \tilde{B}_{F^{-1}}, \tilde{C}_{F^{-1}}, \tilde{D}_{F^{-1}}) \tag{23}$$

where $\tilde{D}_{F^{-1}} = I$. Then, the combined model structure HF_k^{-1} is,

$$HF_k^{-1} = \left(\begin{bmatrix} A & K\tilde{C}_{F^{-1}} \\ 0 & \tilde{A}_{F^{-1}} \end{bmatrix}, \begin{bmatrix} K\tilde{D}_{F^{-1}} \\ \tilde{B}_{F^{-1}} \end{bmatrix}, \begin{bmatrix} C & \tilde{C}_{F^{-1}} \end{bmatrix}, \tilde{D}_{F^{-1}} \right) \tag{24}$$

Now, the final combined model structure of both G and H is adopted as,

$$\begin{bmatrix} G & HF_k^{-1} \end{bmatrix} = \left(\begin{bmatrix} A & K\tilde{C}_{F^{-1}} \\ 0 & \tilde{A}_{F^{-1}} \end{bmatrix}, \begin{bmatrix} B & K \\ 0 & \tilde{B}_{F^{-1}} \end{bmatrix}, \begin{bmatrix} C & \tilde{C}_{F^{-1}} \end{bmatrix}, \begin{bmatrix} D & I \end{bmatrix} \right) \tag{25}$$

To solve this structured system identification problem, a grey box identification method, for example 'idgrey' in Matlab, can be used.

The overall iterative procedure can be described as follows.

(1) Use the proposed k-step-ahead subspace identification method to obtain the initial state space model (A, B, C, D, K).
(2) Obtain F_k from the noise model $H = C(qI - A)^{-1}K + I$.
(3) Apply the structured identification approach to minimize the prediction error for (25) in order to obtain new (A, B, C, D, K).
(4) Obtain a new prefilter F_k from the new noise model H.
(5) Go back to step 3. Continue until the model converges.

3. CASE STUDY

3.1 *CST Tanks in Series with A Recycle Loop*

The example chosen for illustrative purposes involves a 6 CST mixers and 1 plug flow pipe connected in series, as shown in Fig 1. In addition, there is a recycle flow, from mixer 6 back to mixer 1. The flowrate of the secondary inlet, represented by F_u, is assumed to be the manipulated input. The concentration of the main inlet flow C_{Ad}, is treated as an unknown disturbance variable. Outputs are C_{A1} and C_{A6}. The steady state condition is $F_u = 20$, $F_d = 100$, $F_r = 200$, $C_{Ad} = 2$, and $C_u = 20$. The volume of each mixer is 1000. The dynamics of the plug flow pipe between mixer 3 and mixer 4 are represented as a pure delay of 10 time units. We assume that C_{A1} is measured and we are interested in using this measurement to inferentially control the downstream concentration C_{A6}.

First identification data are generated by performing simulations with random input variables. Both the manipulated input and the unknown disturbance variable are drawn from uniform distributions with standard deviations of 15 and 0.5 respectively and switching probability of 0.2. 50 data sets are generated for identification, each with 4000 data points. First, to test the quality of the deterministic part of the identified models, two data sets are generated with manipulated input movement only, one with a step input change and the other with random input changes. Next, to test the model-based inferential prediction and control performance, additional 1000 data points are generated with the same type of input and disturbance variations as those used to generate the 50 modeling data sets.

3.2 *Simulation Results*

The conventional subspace ID method (N4SID) and the modified subspace ID method (k-N4SID) are applied to each of the 50 data sets, which resulted in 50 pairs of state-space models. For the both identification approaches, models with 8 states are identified, and k is chosen to be 50 in applying the k-N4SID algorithm.

The resulting 50 pairs of models are first tested on the two data sets with MV movement only. After that, the identified models are tested for their final purpose, inferential prediction and control. These are done with the validation data set involving the stochastic disturbances. The control objective is to regulate the concentration of the last mixer at the steady-state value. For this, model predictive controllers are designed based on the identified models. The controllers decide the adjustments in the MV based on the inferentially predicted values of the concentration of the last mixer. For every MPC controller, the prediction horizon is chosen to be 200 time units and the control horizon is chosen to be 10 time units. Also, the input and output weighting parameters are chosen to be 10^{-7} and 1, respectively.

Table 1. Comparison of inferential prediction performances of the k-N4SID and N4SID models obtained from the 50 modeling data sets

	1-step inference			k-step inference		
	mean	min	max	mean	min	max
N4SID	0.6259	0.1732	3.8915	0.7835	0.1593	6.6758
k-N4SID	0.4188	0.1812	0.9344	0.4183	0.1457	0.9691

The benefits of the proposed modification to the subspace identification method are clearly seen in the statistical comparison involving the 50 pairs models obtained with N4SID and k-N4SID. First, N4SID resulted in more unstable models, 28 compared to 23 by the k-N4SID. Unstable models for a stable system do not necessarily lead to bad prediction and control performance as long as a stable predictor is formed. However, depending on the location of the unstable eigenvalues, extremely poor prediction and control performance can result, even though the predictor may be stable. It was observed that none of the unstable models obtained by k-N4SID resulted in bad inferential prediction and control, whereas many unstable models obtained by N4SID led to very poor inferential prediction and control results, implying the unstable modes for the N4SID models were much faster growing than those found in the k-N4SID models. Table 1 shows the better performance by the k-N4SID models over the N4SID models, in terms of both one-step inferential prediction and k-step inferential prediction. The subsequent inferential control tests also confirmed the superior quality of the models by k-N4SID over those by N4SID.

To further scrutinize the differences, the identified models were grouped in four categories according to whether both or one of the N4SID and k-N4SID methods resulted in an unstable model. For all four categories, models obtained by k-N4SID method showed better overall inferential prediction and control performance than the corresponding models by the N4SID method. This includes the cases, where N4SID gave a stable model but k-N4SID gave an unstable model. Due to space limit, only the result from the first category, for which the data sets resulted in stable models with k-N4SID but unstable models with N4SID, is shown here. The unstable nature of the models from conventional N4SID can clearly be seen from Figure 2, which shows for one of the data sets the *open-loop* predictions of the two models for a step change in the MV. Figures 3 and 4 display the corresponding differences in the inferential prediction and control performances. We can see that significant improvements in inferential prediction and control performances could be achieved by using k-N4SID instead of N4SID.

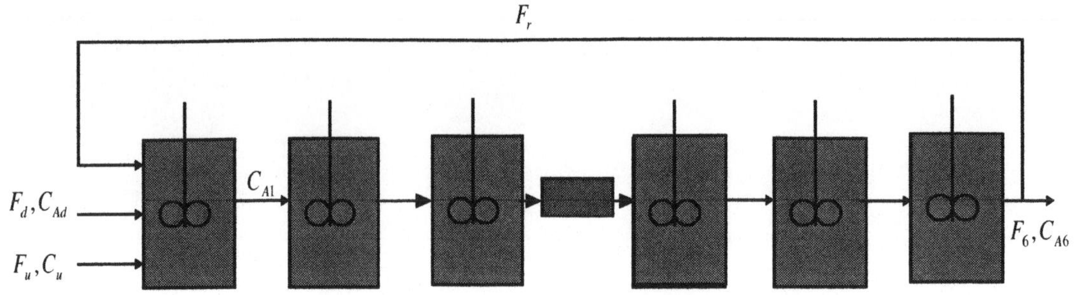

Fig. 1. The schematic for the example of 6 CST mixers in series with a recycle loop

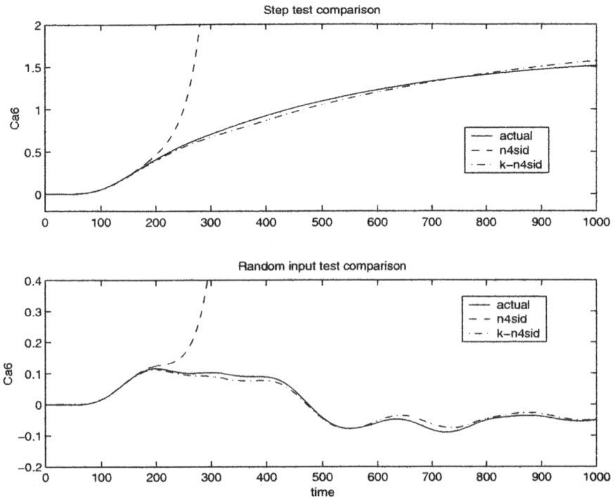

Fig. 2. Deterministic model comparison based on the MV movement data for case 1

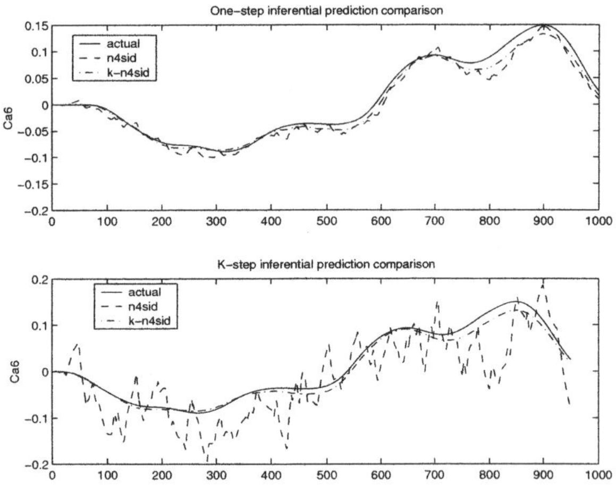

Fig. 3. Inferential prediction performance comparison for case 1

Acknowledgement: The Authors gratefully acknowledge the financial support from the National Science Foundation (CTS-#0096326).

4. REFERENCES

[1] Ljung, L., *System Identification: Theory For the User*, Prentice-Hall, 2nd edition, 1999.

[2] Shook, D., Mohtadi, S., C., and Shah, S. L., Identification for long-range predictive control, *IEE-Proceedings-*, 138, 75-84, 1991.

[3] Huang, B., Malhotra, A., and Tamayo, E.C., Model predictive control relevant identification and valida-

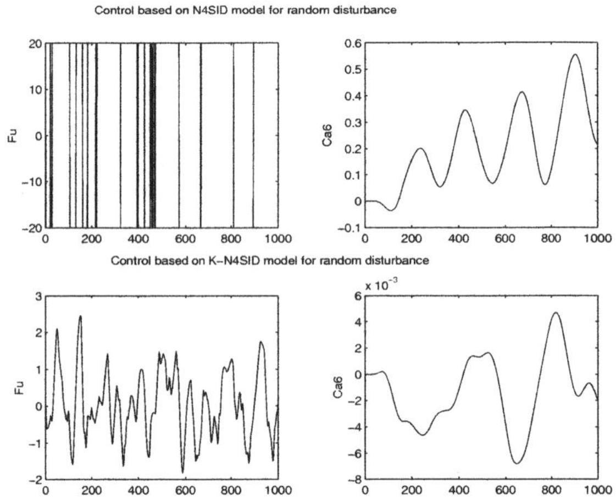

Fig. 4. Inferential control performance comparison for case 1

tion, submitted to *Chemical Engineering and Science*, 2001.

[4] Ljung, L., Estimation focus in system identification: prefiltering, noise models, and prediction, *Proceedings of the 38th Conference on Decision and Control*, Phoenix, Arizona, USA, 2810-2815, 1999.

[5] Chaplais, F., and Alaou, K., Two time scaled parameter identification by coordination of local identifiers, *Automatica*, Vol. 32, No. 9, 1303-1309, 1996.

[6] Wahlberg, B., and Ljung, L., Design variables for bias distribution in transfer function estimation, *IEEE Trans. Automatic Control*, AC-31, 134-144, 1986.

[7] Luse, D. W., and Khalil, H., Frequency domain results for systems with slow and fast dynamics,*IEEE Trans. Automatic Control*, AC-30, 1171-1179, 1985.

[8] Van Zee, G. A., and Bosgra, O. H., Validation of prediction error identification results by multivariable control implementation, in *Proc. 6th IFAC Symp. Ident. Syst. Parameter Estimation*, Washington, DC, 560-565, 1982.

[9] Van Overschee, P. and DeMoor, B., *Subspace Identification for Linear Systems: Theory, Implementation, Applications.*, Kluwer Academic, Dordrecht, The Netherlands, 1996.

[10] Åström, K. J., Maximum likelihood and prediction error methods, *Automatica*, Vol 16, 551-574, 1980.

IFAC Publications
www.elsevier.com/locate/ifac

IDENTIFICATION AND MODEL PREDICTIVE CONTROL OF AN INDUSTRIAL GLASS-FEEDER

L. Huisman ** **S.Weiland** *

* *Eindhoven University of Technology, Department of Electrical Engineering, P.O. Box 513, 5600 MB Eindhoven, s.weiland@tue.nl*
** *Eindhoven University of Technology, Department of Chemical Engineering, P.O. Box 513, 5600 MB Eindhoven, l.huisman@tue.nl*

Abstract: In this paper we discuss the use of reduced simulation models derived from first principles in the design of a model predictive controller of an industrial feeder. A linear reduced model is derived from a computational fluid dynamics model (CFD) using proper orthogonal decompositions (POD). This model is used in a model predictive controller to control the temperature as function of time and position in the feeder. It turns out that a relatively simple model captures the behaviour of the entire temperature profile quite well and that it is suitable for a well performing model predictive controller.

Keywords: Model reduction, POD, Subspace Identification, Glass melting

1. INTRODUCTION

A feeder is the end part of a glass melting furnace. After melting and conditioning in a melter and refiner the molten glass enters the feeder whose main purpose is to gradually cool down the glass to a desired output temperature. This temperature depends on the type of product that is being manufactured and must be maintained at a constant level during flow changes. Main disturbances occur when machines are changed to different product weights. In figure 1 a feeder layout is shown with four zones where different surface temperatures are applied. Relevant variables of this feeder are:

- *Manipulated variables (controlled inputs)* u which consist of 4 crown temperature offsets on the 4 different zones.
- *Measurements* y are derived from a 9 point temperature sensor where values are averaged into 3 measurements at 3 positions near the end of the feeder and 4 glass surface temperatures taken in the centers of the 4 feeder zones.
- *To be controlled variables* z which is u and y in this paper.
- *Disturbances* w which is the measured pull rate.

The purpose of this paper is to identify the relation between the inputs $u(t)$ and the temperature as function of position and time $T(\xi, t)$ and to design a controller which minimizes the effects of w on z in closed loop subject to constraints as discussed in section 4. The paper discusses the use of linear low order models, derived using a POD basis (Proper Orthogonal Decompositions, see e.g. (Glavaški *et al.*, 1998), (Volkwein, 1999), (Atwell and King, 2001), (Astrid *et al.*, 2002)) and a subspace identification algorithm ((Overschee, 1995), (Favoreel *et al.*, 2000)) in a model predictive controller ((Qin and Badgwell, 1997) gives an overview). In the next section the modelling part will be discussed, then the control problem is described and finally some conclusions will be given.

2. THE MODEL

2.1 *Computational fluid dynamics*

The behaviour of glass melt temperatures and velocities is often modeled using computational fluid dynamics models (CFD). The partial differential equations (Navier-Stokes) used to describe the behaviour

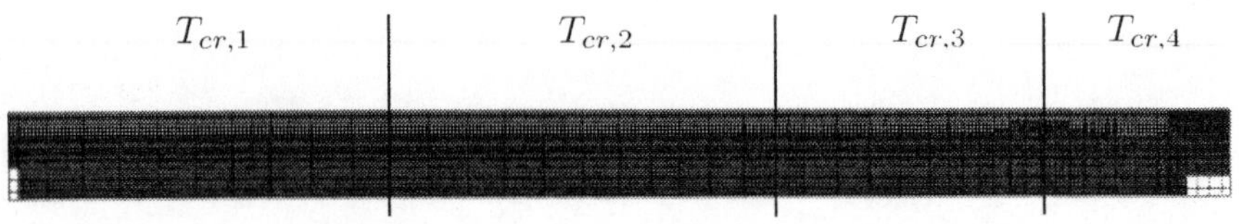

Fig. 1. Side view of an industrial feeder. The grid lines indicate the cells that are used in the discretisation of the simulation model. $T_{Cr,i}$ is the crown temperature applied in zone i.

of the glass melt (see e.g. (Loch and Krause, 2002)), are given by

$$\rho\left(\frac{\partial v}{\partial t}+v\cdot\nabla v\right)=\nabla\cdot\left[\mu\nabla v+\mu\left(\nabla v\right)^{T}\right]-\nabla p+\rho g \quad (1)$$

$$\rho\frac{\partial}{\partial t}\left(c_{p}T\right)+v\cdot\nabla\left(c_{p}T\right)=\nabla\cdot\left[\lambda\nabla T-q_{r}\right]+S_{e} \quad (2)$$

$$\nabla\cdot v=0 \quad (3)$$

and approximated through discretisation and then solved numerically. Here, ∇p is the pressure gradient (1st order tensor, see (Bird *et al.*, 1960)), ∇v is the velocity gradient (2nd order tensor), q_r is the radiative contribution to the heat flux (1st order tensor) and S_e is an additional source term (e.g. electrical boosting power in $\left[\mathrm{W\,m^{-3}}\right]$). The control inputs of this feeder, $u(t)$, are the changes in crown temperatures in the 4 feeder zones, where the entire crown temperature profile in a zone is changed by the same input (in K). That is, $u(t)=\mathrm{col}\left\{u_1(t),\ldots,u_4(t)\right\}$ where

$$u_i(t)=T\left(\xi_{u,i},t\right)$$

for all $\xi_{u,i}\in\Xi_i$ defining the crown grid point positions of the ith section of the feeder, $i=1,\ldots,4$. The measured outputs are

$$y(t)=\mathrm{col}\left\{T\left(\xi_{m,1|3},t\right),T\left(\xi_{m,4|6},t\right),T\left(\xi_{m,7|9},t\right),T\left(\xi_{m,10},t\right),\ldots,T\left(\xi_{m,13},t\right)\right\}$$

with $T\left(\xi_{m,p|q},t\right)=\frac{1}{q-p+1}\left(\sum_{i=p}^{q}T\left(\xi_{m,i},t\right)\right)$ is an average of measured temperatures at sensor locations $\xi_{m,p},\ldots,\xi_{m,q}$ and $\xi_{m,i}$ is the ith sensor position. Computationally demanding simulations with these models result in predictions of the time dependent behaviour of the glass melt temperature and velocity fields. Approximately 10^4 up 10^6 equations must be solved. In case of glass melting furnaces (which are very slow) the computation speed that can be achieved is in the order of 3 times faster than real time. A model predictive controller uses these models for both prediction and optimization of the input signals (e.g. air/fuel ratio, boosting power) such that the predicted temperature and velocity behaviour are satisfactory. Satisfactory performance could for instance mean that the redox state and the viscosity of the glass are within their specifications while the required energy is minimized. In order to use simulation models for this the models should at least be 100 times faster than real time. This section describes a procedure to find a very fast approximate simulation model starting from a CFD-model.

2.2 *Model reduction procedure*

Suppose that a numerical simulation model is available for the feeder that generates time dependent discretised temperature and velocity fields, $T\left(\xi_l,t_k\right)$ and $v\left(\xi_l,t_k\right)$ where $\xi_l\in\left\{\xi_1,\xi_2,\ldots,\xi_L\right\}\subseteq\mathbb{R}^3$ is the set of grid positions and $t_k=k\Delta t, k\in\mathbb{N}$,($\Delta t$ is a fixed sample interval) is discrete time. For these fields we define, with some abuse of notation, the temperature vector and the velocity vector:

$$T(t):=\mathrm{col}\left\{T\left(\xi_l,t\right)\right\}_{l=1}^{L}$$
$$v(t):=\mathrm{col}\left\{v\left(\xi_l,t\right)\right\}_{l=1}^{L}$$

which means that $T\in\mathbb{R}^L$ and $v\in\mathbb{R}^{3L}$ at any time instant t. In the sequel the meaning of T or v should be clear from the context and the number of arguments used. Here, only the temperature behaviour will be considered. Let $T_{\mathrm{snap},1}$ denote a matrix of snapshots of temperatures $\tilde{T}\left(\xi_i,t_j\right)$ taken from the simulation data at time instants t_k, $k=1,2,\ldots,K$ and stored as follows:

$$T_{\mathrm{snap},1}=\begin{pmatrix}\tilde{T}\left(\xi_1,t_1\right)&\ldots&\tilde{T}\left(\xi_1,t_K\right)\\ \vdots&\ddots&\vdots\\ \tilde{T}\left(\xi_L,t_1\right)&\ldots&\tilde{T}\left(\xi_L,t_K\right)\end{pmatrix} \quad (4)$$

and let its singular value decomposition be given by

$$T_{\mathrm{snap},1}=\Phi\Sigma\Psi^{T}. \quad (5)$$

Furthermore, express the temperature vector with any orthonormal basis $\left\{\phi_i\right\}_{i=1}^{L}$ of $\mathbb{R}^L$ by writing

$$T(t)=\sum_{i=1}^{L}a_i(t)\,\phi_i=\sum_{i=1}^{n}a_i(t)\,\phi_i+\varepsilon_T(t)$$

where $n\leq L$ and $\varepsilon_T:=\sum_{i=n+1}^{L}a_i(t)\,\phi_i$ denotes the error obtained by truncating the spectral decomposition. In the POD method an nth-order basis is determined such that $\left\|\varepsilon_T\right\|_2^2$ is minimized for a generated dataset $T_{\mathrm{snap},1}$. This minimum is obtained when $\phi_i=\varphi_i$, where φ_i is the ith column of Φ. By using Φ_n, the matrix consisting of the first n columns of Φ, the temperature vector is approximated by

$$T(t)=\sum_{i=1}^{n}a_i(t)\,\varphi_i+\varepsilon_n(t)=\Phi_n a(t)+\varepsilon_n(t) \quad (6)$$

where $a(t) := \begin{pmatrix} a_1(t) & a_2(t) & \cdots & a_n(t) \end{pmatrix}^T$ is the coefficient vector. Note that the truncation error $\varepsilon_n(t)$ was minimized for one realisation of the temperature vector $T(t)$ and is experiment dependent. It is assumed, however, that the POD-basis generated with a carefully generated snapshot matrix can be used to describe the temperature vector in other experiments as well. For a given temperature vector $T(t)$ the coefficient vector $a(t)$ can be calculated with

$$a(t) = \Phi_n^T T(t). \tag{7}$$

For notational convenience, $a(t) = a(k\Delta t)$ will be identified with $a(k)$. Given the model equations, e.g.

$$T(k+1) = f(T(k), u(k))$$

where $f(\cdot,\cdot)$ is a vectorfield obtained from the discretisations of 2, a reduced model can be derived by a similarity transformation and truncation using the POD-basis. That is,

$$a(k+1) = \Phi_n^T f(\Phi_n a(k), u(k)) \tag{8}$$

is taken as the reduced order model. For models used in the simulation of glass melting furnaces these manipulations turn out to be time consuming and tedious. Furthermore, the non-linearities, in the models are not that important in the case of the feeder. The next section gives an outline of the used strategy to find the reduced model through system identification.

3. SYSTEM IDENTIFICATION

Once the POD-basis Φ is determined, a dataset for the coefficient vector can be determined from a snapshots matrix (possibly the same as $T_{\text{snap},1}$) obtained from an experiment where the system is excited by an input sequence $\{u(k)\}_{k=1}^{K_2}$ and known pull rate $\{w(k)\}_{k=1}^{K_2}$:

$$A_{\text{snap},2} = \Phi_n^T T_{\text{snap},2} \tag{9}$$

In this paper, low pass filtered pseudo binary random signals (PRBNS) were used to excite the CFD-model of the feeder. The dataset

$$A_{\text{snap},2} = \begin{pmatrix} \tilde{a}_1(t_1) & \ldots & \tilde{a}_1(t_K) \\ \vdots & \ddots & \vdots \\ \tilde{a}_n(t_1) & \ldots & \tilde{a}_n(t_K) \end{pmatrix}$$

can be used together with the input data from the experiment $\{u(k)\}_{k=1}^{K_2}$ to determine a linear time invariant model $\hat{a}(k) = H(q)\,u(k)$, where q is the shift operator and $H(q)$ is a proper rational transfer function obtained, using identification techniques. Here, a subspace algorithm (the SMI toolbox (Haverkamp and Verhaegen, 1997)) has been used to determine the quadruple (A, B, C, D) in a state space realisation of $H(q)$:

$$x(k+1) = Ax(k) + B\hat{u}(k) \tag{10a}$$

$$\hat{a}(k) = Cx(k) + D\hat{u}(k) \tag{10b}$$

with $\hat{u}(k) = \text{col}\{u(k), w(k)\}$. This low order state space model can be used to predict dynamic temperature behaviour in the entire 3D-grid (see figure 2). Specifically, we define

$$x(k+1) = Ax(k) + B\hat{u}(k) \tag{11a}$$

$$\hat{T}(k) = \Phi_n Cx(k) + \Phi_n D\hat{u}(k) \tag{11b}$$

as a reduced model to predict temperatures at any position and at any time in the feeder. Two types of error occur in the prediction of $T(k)$: The truncation error $\varepsilon_n(k)$ and the model error in the state space model $e_M(k)$. The temperature vector can be written as:

$$T(k) = \Phi_n(\hat{a}(k) + e_M(k)) + \varepsilon_n(k) \tag{12}$$

where $\varepsilon_n(k) = \Phi_{\text{tail}} a_{\text{tail}}(k)$ is the tail in the sum (6), $\Phi = \begin{bmatrix} \Phi_n & \Phi_{\text{tail}} \end{bmatrix}$ and $a_{\text{tail}}(k)$ is the coefficient vector corresponding to the basis vectors $\{\varphi_i\}_{i=n+1}^{L}$ (columns of Φ_{tail}) that were thrown out in the truncation step. The total error introduced in the truncation of (6) and identification procedure is:

$$\begin{aligned} \varepsilon_T(k) &= T(k) - \hat{T}(k) \\ &= \varepsilon_M(k) + \varepsilon_n(k) \\ &= \Phi_n e_M(k) + \Phi_{\text{tail}} a_{\text{tail}}(k) \end{aligned} \tag{13}$$

Figure 3 and 4 show the estimated temperatures

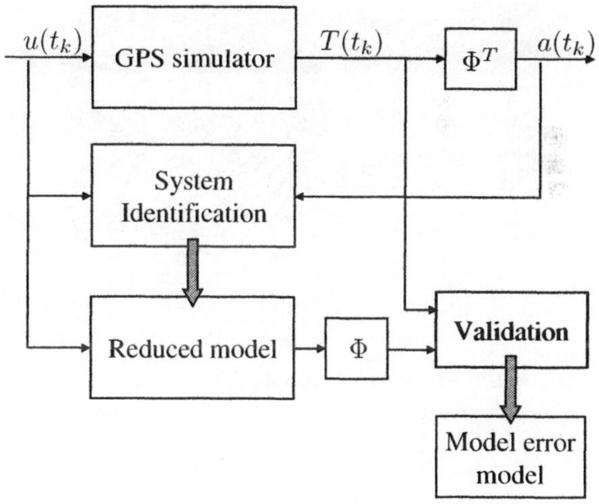

Fig. 2. Model identification strategy: 1. Find a low order basis $\{\varphi_i\}_{i=1}^{n}$. 2. Find a low order model for the coefficients. 3. Validate the model.

of a CFD model whose original order $L = 59976$ has been reduced to $n = 17$. The snapshot matrix was generated using filtered PRBNS signals on the 4 crown temperatures $u_1, \ldots, u_4$ (amplitude ~ 15 K) and the pull rate w (amplitude ~ 15 tons/day). The used sample time was 1 minute and the open loop bandwidth of the obtained model was approximately $2 \cdot 10^{-3}$ Hz. Although, significant errors can occur locally (near the walls), the reduced model turns out to be quite accurate. (Average absolute error is 0.15 K where the average has been taken over all grid points and over 500 time samples). The type of model that is

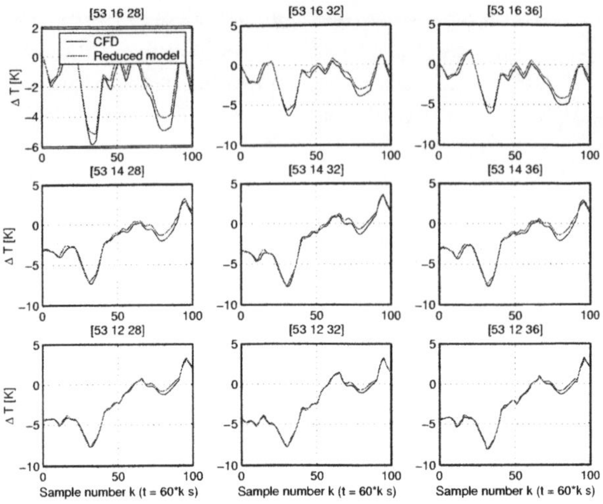

Fig. 3. Temperature reconstruction in a validation experiment in the 9 point grid, where 9 thermocouples are used to estimate temperature profiles in a plane.

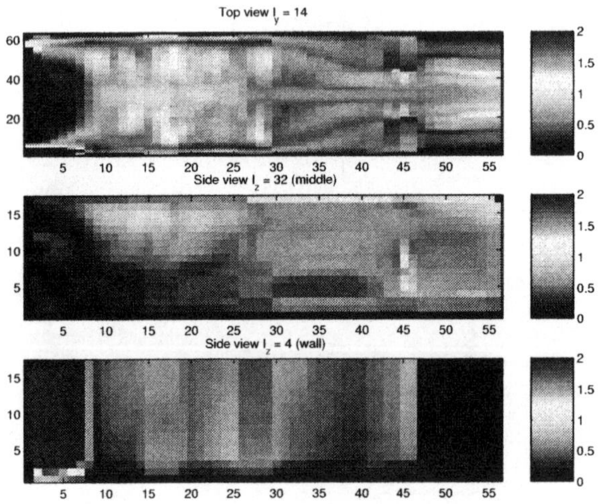

Fig. 4. Images of maxima in the time domain (500 time samples), that is, for each temperature in the grid the maximum of $\left|T_i(k) - \hat{T}_i(k)\right|$ was determined. In some regions the maximum error exceeds 2 K.

identified this way still has the possibility to simulate the entire 3D temperature profiles. This gives us a tool to perform very fast simulations (10^4 times faster than the CFD model) and tests, e.g. to determine smart sensor locations.

4. MODEL PREDICTIVE CONTROL

The model (11) can be used for control design so as to maintain temperatures at given setpoints in the presence of pull rate changes (the pull rate is the amount of glass withdrawn from the furnace). To account for the pull rate changes an extension of (11) is used:

$$x(k+1) = Ax(k) + B_u u(k) + B_w w(k) \quad (14a)$$
$$a(k) = Cx(k) + D_u u(k) + D_w w(k) \quad (14b)$$

where $w(k)$ is the measured pull rate. The pull rate is determined by the product weight and therefore it can not be freely manipulated. Once the type of product changes (shape and weight) the glass stream is not used by the machine for a short period of time. Certain parts of the machine are changed and then the new pull rate is introduced. When these changes in pull rate are large, significant temperature changes can occur (figure 5), which have to be annihilated. Since the process is operated under constraints, we use a model predictive controller (MPC, see (Qin and Badgwell, 1997) for an overview) which solves an optimisation problem online. In this MPC application the control problem amounts to minimizing a quadratic cost function:

$$J(u,k) := \sum_{i=1}^{N} \hat{x}^T(k+i|k)\, Q \hat{x}(k+i|k) + \sum_{i=0}^{N_c-1} u^T(k+i)\, Ru(k+i)$$

subject to constraints:

$$\begin{array}{ll} |\Delta u(k+i)| \le \Delta u_{\max} & i = 0,1,\ldots,N_c \\ u(k+i) \in \mathcal{U} & i = 0,1,\ldots,N_c-1 \\ u(k+i) = 0 & i = N_c,\ldots,N \end{array}$$

Here Q and R are weighting matrices, N is the prediction horizon, N_c is the control horizon, $\Delta u(k+i) = u(k+i) - u(k+i-1)$, $\Delta u_{\max}$ is the maximum increment, and $\mathcal{U}$, $\mathcal{X}$ are convex admissible sets containing the origin for the inputs and states, respectively. In this paper, three outputs are controlled, so here $Q = C_y^T Q_y C_y$, where $C_y = C_{\text{select}} \Phi C$ represents the sensor positions in the grid. The chosen outputs are the averages of the 3 bottom temperatures, the 3 middle temperatures and the 3 top temperatures in the 9-point grid (figure 3, $12 =$ bottom, $14 =$ middle, $16 =$ top). Figure 5 shows the response of the bottom temperature to a pull rate change. Measurement noise, estimated from real plant measurements, was added to the measurements taken from the CFD-model. A commercial model predictive controller has been used to control these 3 temperatures (high priority level) and 5 temperatures at the surface (low priority level). The used controller solves the optimisation problem sequentially starting at the highest priority level. If after finding a solution there still are degrees of freedom left it starts solving the lower priority level optimisation problem. As already mentioned, the crown temperature offsets in the 4 feeder zones are used as inputs and have to meet the following constraints:

$$\begin{array}{ll} -15\,\mathrm{K} \le u_i(k+j) \le 15\,\mathrm{K} & i = 1,\ldots,4 \\ & j = 1,\ldots,N_c \\ -5\,\mathrm{K} \le \Delta u_i(k+j) \le 5\,\mathrm{K} & i = 1,\ldots,4 \\ & j = 1,\ldots,N_c-1 \end{array}$$

Figure 6 shows a closed loop response of the averaged bottom temperature to a change in the pull rate. The

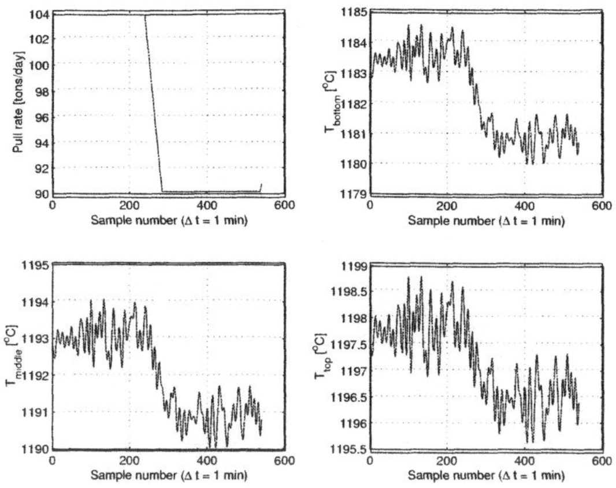

Fig. 5. Open loop response to a pull rate change (ramped)

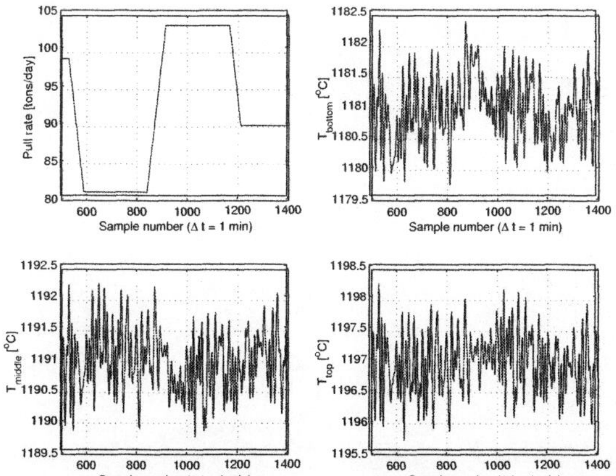

Fig. 6. Closed loop response to pull change (ramped). The ideal values for these temperatures were 1181 °C, 1191 °C and 1197 °C

effects of the pull changes on the outputs hardly exceed the measurement noise levels.

5. CONCLUSIONS

Starting from a CFD-type model for the temperature distribution in an industrial glass feeder, we derived a reduced order model by properly selecting an orthonormal basis of spatial functions from measured data. In turn, the parameters of the reduced order model have been identified using subspace methods. The resulting reduced order model shows excellent performance in validation tests and has been used to synthesize an MPC-type controller for the glass feeder. Simulation results can be obtained fast and it has been shown that the controller performs well in maintaining set-point temperatures in the face of pull rate changes of the feeder.

6. REFERENCES

Astrid, P., Huisman L., S. Weiland and A.C.P.M. Backx (2002). Reduction and predictive control design for a computational fluid dynamics model. *IEEE Conference on Decision and Control*.

Atwell, J. A. and B. B. King (2001). Proper orthogonal decomposition for reduced basis feedback controllers for parabolic equations. *Mathematical and Computer Modelling* **33**, 1–19.

Bird, R.B., W.E. Stewart and E.N. Lightfoot (1960). *Transport Phenomena*. Wiley.

Favoreel, Wouter, Bart de Moor and Peter van Overschee (2000). Subspace state space system identification for industrial processes. *Journal of Process Control* **10**, 149–155.

Glavaški, S., J.E. Marsden and R.M. Murray (1998). Model reduction, centering and the Karhunen Loève expansion. In: *Proceedings Conference on Decision and Control Vol. 37*. pp. 2071–2076.

Haverkamp, B. and M. Verhaegen (1997). SMI toolbox: State space model identification software for multivariable dynamical systems.

Loch, H. and Krause, D., Eds.) (2002). *Mathematical Simulation in Glass Technology*. Springer.

Overschee, Peter van (1995). Subspace Identification. PhD thesis. Katholieke Universiteit Leuven.

Qin, S.J. and T.A. Badgwell (1997). An overview of industrial model predictive control technology. *AIChE Symposium series* **93**(316), 232–257.

Volkwein, S. (1999). Proper orthogonal decomposition and singular value decomposition. Technical Report 153. Special Research Center F300 'Optimization and Control'.

IFAC Publications
www.elsevier.com/locate/ifac

COMPUTATIONALLY EFFICIENT ESTIMATION OF WAVE PROPAGATION FUNCTIONS OF VISCOELASTIC MATERIALS [1]

Kaushik Mahata[1] **Torsten Söderström**[1] and **Lars Hillström**[2]

[1] *Systems and Control, Department of Information Technology, Uppsala University. P O Box 337, SE-751 05 Uppsala, Sweden.*
Email: Kaushik.Mahata@it.uu.se, Torsten.Soderstrom@it.uu.se
[2] *Alfa Laval Tumba AB, SE-147 80 Tumba, Sweden.*
Email: lars.hillstrom@alfalaval.com

Abstract: Least squares based nonparametric estimation of the wave propagation functions of a viscoelastic material is considered in this paper. Widely used nonlinear least squares based algorithms are often computationally expensive and suffer from numerical problems. In this paper we propose a class of subspace estimators which assume equidistant sensor configuration. The proposed estimator is computationally economical and numerically robust. Analytical expressions for the estimation accuracy have been derived. It is also shown that the subspace estimator achieves optimal accuracy under optimal weighting. The algorithm is employed on real experimental data. *Copyright © 2003 IFAC*

Keywords: Subspace Identification, Viscoelasticity, Wave Propagation, Nonlinear Least Squares, Computationally Efficient

1. INTRODUCTION

Viscoelastic materials are used in wide range of applications (Lakes, 1999). In this work, we consider the non-parametric estimation of the wave propagation functions of a linearly viscoelastic material, which play a central role in the dynamics of stress and strain in such a material under dynamic loading. The estimates obtained here can further be used to estimate the state variables, *i.e.* stress, strain, shear force, moment, displacement etc, at any given cross-section of the solid (Hillström *et al.*, 2000), or determine the complex modulus of the material. A longitudinal wave experimental set-up is outlined in Figure 1. A bar specimen is impacted at one end giving rise

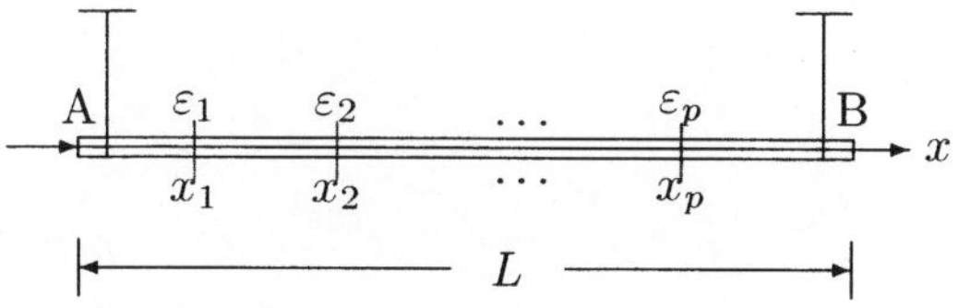

Fig. 1. Experimental set up.

to longitudinal waves traveling along the bar. The strains due to the wave propagation are registered using strain gauges at sections $\{x_i\}_{i=1}^{p}$. Similar types of experiments can be carried out for other types of waves like transversal waves (Hillström and Lundberg, 2001) or torsional waves. The analog strain data are passed through an anti-aliasing filter and discretized. The discretized data are transformed to the frequency domain using Discrete Fourier Transform (DFT). A non-parametric

[1] This work was supported by Swedish Research Council for Engineering Sciences under contract 2000-587.

identification of the wave propagation function is carried out for each frequency.

The paper is organized as follows. In the next section we review the nonlinear least squares algorithm and describe some subspace based methods. In Section 3 we present the experimental results. Section 4 gives a general discussion followed by conclusions in Section 5.

2. THEORY

2.1 *Background*

In frequency domain, the strain $\epsilon_0(x,\omega)$ at section x and frequency ω due to 1-D wave propagation in a viscoelastic solid is given by (Hillström and Lundberg, 2001; Hillström *et al.*, 2000),

$$\epsilon_0(x,\omega) = \sum_{k=1}^{n}\left[c_k^+(\omega)\mathrm{e}^{\gamma_k(\omega)x} + c_k^-(\omega)\mathrm{e}^{-\gamma_k(\omega)x}\right], \tag{1}$$

where for longitudinal (Mossberg *et al.*, 2001) and torsional waves, we have $n = 1$, while $n = 2$ for transversal waves (Hillström and Lundberg, 2001). The strain is measured at sections $x = x_i$ for $i = 1,\ldots,p$, where $p \geq 3n$. The measurements are assumed to be corrupted by additive complex-valued measurement noise $\tilde{\epsilon}(x_i,\omega)$, *i.e.* the measurement $\epsilon(x_i,\omega)$ at x_i is given by

$$\epsilon(x_i,\omega) = \epsilon_0(x_i,\omega) + \tilde{\epsilon}(x_i,\omega). \tag{2}$$

The problem we consider here is to estimate the complex valued wave propagation functions $\{\gamma_k(\omega)\}_{k=1}^n$ as a function of frequency ω from the measurements $\{\epsilon(x_i,\omega)\}_{i=1}^p$. Although, in the present context $n \leq 2$, the methods presented here can handle model structures where $n > 2$. Introduce

$$\boldsymbol{\gamma}_\omega = \begin{bmatrix}\gamma_1(\omega) & \ldots & \gamma_n(\omega)\end{bmatrix}^\top. \tag{3}$$

$$\mathbf{x} = \begin{bmatrix}x_1 & \ldots & x_p\end{bmatrix}^\top, \tag{4}$$

$$\boldsymbol{\epsilon}_{\omega 0}(\mathbf{x}) = \begin{bmatrix}\epsilon_0(x_1,\omega) & \ldots & \epsilon_0(x_p,\omega)\end{bmatrix}^\top, \tag{5}$$

$$\tilde{\boldsymbol{\epsilon}}_\omega = \begin{bmatrix}\tilde{\epsilon}(x_1,\omega) & \ldots & \tilde{\epsilon}(x_p,\omega)\end{bmatrix}^\top, \tag{6}$$

$$\boldsymbol{\epsilon}_\omega(\mathbf{x}) = \boldsymbol{\epsilon}_{\omega 0}(\mathbf{x}) + \tilde{\boldsymbol{\epsilon}}_\omega. \tag{7}$$

It can be shown under rather generalized assumptions (Mahata *et al.*, 2001) that the measurement noise $\tilde{\epsilon}(x_i,\omega)$ is zero-mean and circular (Brillinger, 1975) *i.e.*,

$$\mathbf{E}\tilde{\boldsymbol{\epsilon}}_\omega\tilde{\boldsymbol{\epsilon}}_\omega^* = \lambda\mathbf{I}_p, \quad \mathbf{E}\tilde{\boldsymbol{\epsilon}}_\omega\tilde{\boldsymbol{\epsilon}}_\omega^\top = \mathbf{0}, \tag{8}$$

where we have denoted the identity matrix of order p by $\mathbf{I}_p$. Then using (1) and (5) we can write the true strain vector $\boldsymbol{\epsilon}_{\omega 0}(\mathbf{x})$ as

$$\boldsymbol{\epsilon}_{\omega 0}(\mathbf{x}) = \boldsymbol{\Psi}(\mathbf{x},\boldsymbol{\gamma}_\omega)\mathbf{c}_\omega, \tag{9}$$

where $\boldsymbol{\Psi}(\mathbf{x},\boldsymbol{\gamma}_\omega)$ is a $p|2n$ matrix and $\mathbf{c}_\omega$ is a $2n$ dimensional vector given by

$$\boldsymbol{\Psi}(\mathbf{x},\boldsymbol{\gamma}_\omega) = \begin{bmatrix}\mathrm{e}^{\gamma_1(\omega)x_1} & \mathrm{e}^{-\gamma_1(\omega)x_1} & \ldots & \mathrm{e}^{\gamma_n(\omega)x_1} & \mathrm{e}^{-\gamma_n(\omega)x_1}\\ \vdots & \vdots & \ddots & \vdots & \vdots\\ \mathrm{e}^{\gamma_1(\omega)x_p} & \mathrm{e}^{-\gamma_1(\omega)x_p} & \ldots & \mathrm{e}^{\gamma_n(\omega)x_p} & \mathrm{e}^{-\gamma_n(\omega)x_p}\end{bmatrix} \tag{10}$$

$$\mathbf{c}_\omega = \begin{bmatrix}c_1^+(\omega) & c_1^-(\omega) & \ldots & c_n^+(\omega) & c_n^-(\omega)\end{bmatrix}^\top. \tag{11}$$

The expression (9) can now be combined with (7) to get

$$\boldsymbol{\epsilon}_\omega(\mathbf{x}) = \boldsymbol{\Psi}(\mathbf{x},\boldsymbol{\gamma}_\omega)\mathbf{c}_\omega + \tilde{\boldsymbol{\epsilon}}_\omega. \tag{12}$$

Given the measurement vector $\boldsymbol{\epsilon}_\omega(\mathbf{x})$, one way to obtain an estimate of $\boldsymbol{\gamma}_\omega$ is to solve the separable non-linear least squares problem (Hillström and Lundberg, 2001; Mossberg *et al.*, 2001; Golub and Pereyra, 1973), where the estimate $\hat{\boldsymbol{\gamma}}_\omega^l$ is given by

$$\hat{\boldsymbol{\gamma}}_\omega^l = \arg\min_{\boldsymbol{\gamma}} \boldsymbol{\epsilon}_\omega^*(\mathbf{x})\boldsymbol{\Pi}_{\boldsymbol{\Psi}}^\perp(\mathbf{x},\boldsymbol{\gamma})\boldsymbol{\epsilon}_\omega(\mathbf{x}). \tag{13}$$

$\boldsymbol{\Pi}_{\boldsymbol{\Psi}}^\perp(\mathbf{x},\boldsymbol{\gamma})$ is the projection operator (Golub and Loan, 1989) onto the null space of $\boldsymbol{\Psi}^*(\mathbf{x},\boldsymbol{\gamma})$, is given by

$$\boldsymbol{\Pi}_{\boldsymbol{\Psi}}^\perp(\mathbf{x},\boldsymbol{\gamma}) = \mathbf{I}_p - \boldsymbol{\Psi}(\mathbf{x},\boldsymbol{\gamma})\boldsymbol{\Psi}^\dagger(\mathbf{x},\boldsymbol{\gamma}). \tag{14}$$

$\boldsymbol{\Psi}^\dagger(\mathbf{x},\boldsymbol{\gamma})$ is the pseudoinverse of $\boldsymbol{\Psi}(\mathbf{x},\boldsymbol{\gamma})$. If $\tilde{\boldsymbol{\epsilon}}_\omega$ is Gaussian, then the estimate $\hat{\boldsymbol{\gamma}}_\omega^l$ in (13) is the maximum likelihood estimate of $\boldsymbol{\gamma}_\omega$ (Söderström and Stoica, 1989). Note that the maximum likelihood estimate $\hat{\boldsymbol{\gamma}}_\omega^l$ of $\boldsymbol{\gamma}_\omega$ is also an asymptotically efficient estimate as the signal to noise ratio is large. The principal demerit of (13) is the associated iterative numerical optimization procedure, which is computationally intensive. Moreover, the cost function is not well-behaved, *i.e*, there exists a possibility of ending up in one of the several local minima of the loss function (Mossberg *et al.*, 2001).

2.2 *Fast subspace based method*

In this section we introduce a subspace based method. The main idea is to parametrize the null-space of $\boldsymbol{\Psi}^*(\mathbf{x},\boldsymbol{\gamma}_\omega)$ by a new parameter vector $\boldsymbol{\theta}_\omega$, which is an invertible function of $\boldsymbol{\gamma}_\omega$ in such a way that the estimation problem in $\boldsymbol{\theta}_\omega$ can be solved in a relatively simple manner, *i.e*, in a computationally economical and non-iterative way. Subsequently $\boldsymbol{\gamma}_\omega$, the parameter vector of interest, can easily be recovered from the estimates of $\boldsymbol{\theta}_\omega$. As a consequence, the subspace based approach is computationally efficient and numerically more robust. The most crucial step here is to assume equidistant sensor configuration, *i.e.*

$$x_i = x_{i-1} + d, \qquad i = 2,\ldots,p. \tag{15}$$

This is not a restrictive assumption, since the sensor configuration is a part of the experiment design and is under control of the user. Note that under the assumption (15) the matrix $\mathbf{\Psi}(\mathbf{x}, \boldsymbol{\gamma}_\omega)$ in (10) is a Vandermonde matrix (Golub and Loan, 1989). The Vandermonde structure of $\mathbf{\Psi}(\mathbf{x}, \boldsymbol{\gamma}_\omega)$ will be exploited in what follows. Introduce the polynomial

$$A(z) \triangleq \sum_{k=0}^{2n} a_k z^k = \prod_{k=1}^{n} \left\{ \left(z - \mathrm{e}^{d\gamma_k}\right)\left(z - \mathrm{e}^{-d\gamma_k}\right) \right\} \tag{16}$$

It is readily verified by spectral factorization theorem (Söderström and Stoica, 1989) that that the coefficients $\{a_k\}_{k=0}^{2n}$ satisfy

$$a_i = a_{2n-i}, \qquad i = 0, \ldots, 2n, \tag{17}$$

$$a_0 = a_{2n} = 1, \tag{18}$$

where $1 \leq k \leq n$. Introduce the n dimensional vector $\boldsymbol{\theta}_\omega$ such that

$$\boldsymbol{\theta}_\omega = \begin{bmatrix} a_1 & \ldots & a_n \end{bmatrix}^\top \tag{19}$$

Note that by (16) and (17) the polynomial $A(z)$ is uniquely parameterized by the vector $\boldsymbol{\theta}_\omega$. Since the coefficients of the $2n$ order polynomial $A(z)$ in (16) is uniquely defined by its $2n$ roots, by (16) and (17) we can define a map $f : \mathbb{C}^n \to \mathbb{C}^n$ such that

$$\boldsymbol{\theta}_\omega = f(\boldsymbol{\gamma}_\omega). \tag{20}$$

Note that the map f is surjective but not injective (onto but not one to one) in the sense that

$$f(\mathbf{x}_1) = f(\mathbf{x}_2) \quad \Rightarrow \quad \mathbf{x}_1 = \mathbf{P}\mathbf{x}_2 \tag{21}$$

for an $n|n$ permutation matrix $\mathbf{P}$. But this is not restrictive for our problem, since we are not concerned with any particular permutation of the propagation functions in the parameter vector $\boldsymbol{\gamma}_\omega$. Note that from (16) it is evident that

$$A\{\mathrm{e}^{\pm d\gamma_i(\omega)}\} = \sum_{k=0}^{2n} a_k \mathrm{e}^{\pm kd\gamma_i(\omega)} = 0, \quad 1 \leq i \leq n. \tag{22}$$

Introduce the $p|(p-2n)$ Toeplitz matrix $\mathbf{M}$

$$\mathbf{M}(\boldsymbol{\theta}_\omega) = \begin{bmatrix} a_0 & \ldots & a_{2n} & 0 & \ldots & 0 \\ 0 & a_0 & \ldots & a_{2n} & \ddots & 0 \\ \vdots & \ddots & \ddots & \ddots & \ddots & \vdots \\ 0 & \ldots & 0 & a_0 & \ldots & a_{2n} \end{bmatrix}^* . \tag{23}$$

From (9), (10), (22) and (23) it is straightforward to see that under the assumption (15)

$$\mathbf{M}^*(\boldsymbol{\theta}_\omega)\mathbf{\Psi}(\mathbf{x}, \boldsymbol{\gamma}_\omega) = 0 \quad \Rightarrow \quad \mathbf{M}^*(\boldsymbol{\theta}_\omega)\boldsymbol{\epsilon}_{\omega 0} = 0. \tag{24}$$

Note that, since $a_0 = 1$, $\mathbf{M}(\boldsymbol{\theta})$ has full column rank for all $\boldsymbol{\theta}$ by construction. Since $\mathbf{\Psi}(\mathbf{x}, \boldsymbol{\gamma}_\omega)$ is assumed to be having full $2n$ column rank, the full $p-2n$ column-rank property of $\mathbf{M}(\boldsymbol{\theta}_\omega)$ together with (24) ensures that the $p-2n$ dimensional null-space of $\mathbf{\Psi}^*(\mathbf{x}, \boldsymbol{\gamma}_\omega)$ is uniquely parameterized by $\boldsymbol{\theta}_\omega$. We make a note by passing that a similar approach for parameterizing the orthogonal complement of the column space of a Vandermonde matrix has been used in the field of frequency estimation and array signal processing in the context of the iterative quadratic maximum likelihood (IQML) or MODE algorithm, see for example (Eriksson *et al.*, 1994; Kristensson *et al.*, 2001). Now for large signal-to-noise ratio, from (24) and (12) it follows that we can estimate $\boldsymbol{\theta}_\omega$ as

$$\begin{aligned} \hat{\boldsymbol{\theta}}_\omega^s &= \arg\min_{\boldsymbol{\theta}} \|\mathbf{\Delta}^{1/2}\mathbf{M}^*(\boldsymbol{\theta})\boldsymbol{\epsilon}_\omega(\mathbf{x})\|^2 \\ &= \arg\min_{\boldsymbol{\theta}} \boldsymbol{\epsilon}_\omega^*(\mathbf{x})\mathbf{M}(\boldsymbol{\theta})\mathbf{\Delta}\mathbf{M}^*(\boldsymbol{\theta})\boldsymbol{\epsilon}_\omega(\mathbf{x}), \end{aligned} \tag{25}$$

where $\mathbf{\Delta}$ is positive definite weighting matrix. Note that $\mathbf{\Delta}$ is user defined and can be data dependent also. Later in this section we will address the issue of choosing $\mathbf{\Delta}$. Now, by construction of $\mathbf{M}(\boldsymbol{\theta}_\omega)$ in (23) and (17) it follows that

$$\mathbf{M}^*(\boldsymbol{\theta}_\omega) = \sum_{k=0}^{n} a_k \mathbf{T}_k, \tag{26}$$

where the $(p-2n)|p$ Toeplitz matrices $\{\mathbf{T}_k\}_{k=0}^{n}$ are defined element-wise as

$$[\mathbf{T}_k]_{ij} = \begin{cases} 1, & j - i = k \\ 1, & j - i = 2n - k \\ 0 & \text{otherwise} \end{cases} . \tag{27}$$

At this point we introduce the following notations for simplicity of the expressions.

$$\begin{aligned} \hat{\mathbf{s}}_\omega(\mathbf{x}) &\triangleq \mathbf{T}_0\boldsymbol{\epsilon}_\omega(\mathbf{x}), \\ \hat{\mathbf{S}}_\omega(\mathbf{x}) &\triangleq \begin{bmatrix} \mathbf{T}_1\boldsymbol{\epsilon}_\omega(\mathbf{x}) & \ldots & \mathbf{T}_n\boldsymbol{\epsilon}_\omega(\mathbf{x}) \end{bmatrix}, \end{aligned} \tag{28}$$

Thus, according to the notations in (28) and by the structure of $\mathbf{M}(\boldsymbol{\theta})$ in (26) we have

$$\mathbf{M}^*(\boldsymbol{\theta})\boldsymbol{\epsilon}_\omega(\mathbf{x}) = \hat{\mathbf{s}}_\omega(\mathbf{x}) + \hat{\mathbf{S}}_\omega(\mathbf{x})\boldsymbol{\theta}. \tag{29}$$

Note that we have used (24) in the last equality. By the theory of weighted least squares, it follows from (29) that the optimization problem in (25) can be solved analytically and the corresponding minimizing argument is uniquely given by

$$\hat{\boldsymbol{\theta}}_\omega^s = -\left\{\hat{\mathbf{S}}_\omega^*(\mathbf{x})\,\mathbf{\Delta}\,\hat{\mathbf{S}}_\omega(\mathbf{x})\right\}^{-1}\hat{\mathbf{S}}_\omega^*(\mathbf{x})\,\mathbf{\Delta}\,\hat{\mathbf{s}}_\omega(\mathbf{x}). \tag{30}$$

The estimated $\hat{\boldsymbol{\theta}}_\omega^s$ can now be used to form an estimate of the polynomial $A(z)$ in (16), which in turn can be factorized and the natural logarithm of the roots yield the estimates of the wave propagation functions, *i.e.* in a more compact form, according to our notation, γ_ω can be estimated by finding a $\hat{\gamma}^s_\omega$ such that

$$\hat{\boldsymbol{\theta}}^s_\omega = f(\hat{\gamma}^s_\omega). \tag{31}$$

We have the following result concerning the accuracy of the estimates by subspace method introduced in this section.

Theorem 1: Define $\boldsymbol{\alpha} := \left[\operatorname{Re}(\hat{\gamma}^s_\omega)^\top \ \operatorname{Im}(\hat{\gamma}^s_\omega)^\top \right]^\top$. Then for large SNR the asymptotic second order statistics of $\boldsymbol{\alpha}$ is given by

$$\mathbf{C}_\theta \triangleq \operatorname{cov}(\boldsymbol{\alpha}) = \frac{\lambda}{2} \begin{bmatrix} \operatorname{Re}\{\mathbf{R}_\omega\} & -\operatorname{Im}\{\mathbf{R}_\omega\} \\ \operatorname{Im}\{\mathbf{R}_\omega\} & \operatorname{Re}\{\mathbf{R}_\omega\} \end{bmatrix}. \tag{32}$$

where $\mathbf{R}_\omega$ is a Hermitian matrix given by

$$\mathbf{R}_\omega = \mathbf{J}_f^{-1}(\gamma_\omega)\mathbf{Q}_\omega \mathbf{J}_f^{-1*}(\gamma_\omega), \ \gamma_\omega \neq \mathrm{i}\frac{k\pi}{d}, \ k \in \mathbb{Z}, \tag{33}$$

$$\mathbf{Q}_\omega = \{\mathbf{S}^*_\omega(\mathbf{x})\boldsymbol{\Delta}\mathbf{S}_\omega(\mathbf{x})\}^{-1}\mathbf{S}^*_\omega(\mathbf{x})\boldsymbol{\Delta}\mathbf{M}^*(\boldsymbol{\theta}_\omega) \\ \mathbf{M}(\boldsymbol{\theta}_\omega)\boldsymbol{\Delta}\mathbf{S}_\omega(\mathbf{x})\{\mathbf{S}^*_\omega(\mathbf{x})\boldsymbol{\Delta}\mathbf{S}_\omega(\mathbf{x})\}^{-1} \tag{34}$$

$\mathbf{J}_f(\gamma_\omega)$ is the Jacobian matrix of the function f evaluated at true parameters.

Proof: The theorem can be proved in the same way as deriving the covariance matrix of the instrumental variable estimates, which is well documented in the literature (see for example (Söderström and Stoica, 1989)). We omit the details here, due to page limitations. ■

Recall that the derivation of the optimal chioce of $\boldsymbol{\Delta}$ is well documented in the least squares literature (Söderström and Stoica, 1989). For this problem it can be shown that (we omit the details here)

$$\boldsymbol{\Delta}_o = \{\mathbf{M}^*(\boldsymbol{\theta}_\omega)\mathbf{M}(\boldsymbol{\theta}_\omega)\}^{-1} \tag{35}$$

is the optimal choice in the sense that, the variances of the parameter estimates are minimizad with respect to $\boldsymbol{\Delta}$. Note also that the optimal weight $\boldsymbol{\Delta}_o$ is a function of the true parameter vector $\boldsymbol{\theta}_\omega$. Since $\boldsymbol{\theta}_\omega$ is unknown, practically it is not possible to determine $\boldsymbol{\Delta}_o$. But a consistent estimate of $\boldsymbol{\theta}_\omega$, for example $\hat{\boldsymbol{\theta}}^s_\omega$ obtained from (30) for a known $\boldsymbol{\Delta}$, can be used to replace $\boldsymbol{\theta}_\omega$ in (35) to derive a consistent estimate $\widehat{\boldsymbol{\Delta}}$ of $\boldsymbol{\Delta}_o$. We can use $\boldsymbol{\Delta} = \widehat{\boldsymbol{\Delta}}$ back in (30) to obtain a refined estimate of γ_ω and $\boldsymbol{\theta}_\omega$. If the above iterative procedure is repeated twice and we denote the estimate by $\hat{\gamma}^r_\omega$, then it is possible to show that $\hat{\gamma}^r_\omega$ is asymptotically equivalent (in a statistical sense) to $\hat{\gamma}^l_\omega$.

3. EXPERIMENTS

The algorithm was tested with real experimental data. A longitudinal wave experiment was performed with equi-distant sensors at the locations

$$\{\, 0.2 \ 0.4 \ 0.6 \ 0.8 \,\} \ \ [\mathrm{m}].$$

The length of the Polypropelene bar was 2m. Detailed results using least squares estimation on the same experimental data have been reported in (Hillström *et al.*, 2000). Here we demonstrate the results using the proposed subspace algorithm. In

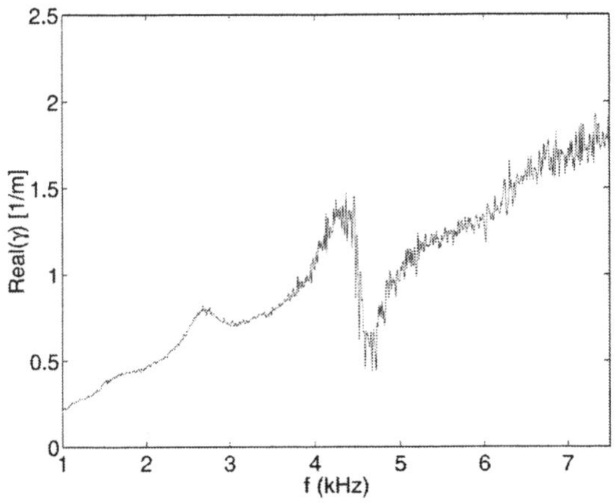

(a) $\operatorname{Re}(\gamma_1)$

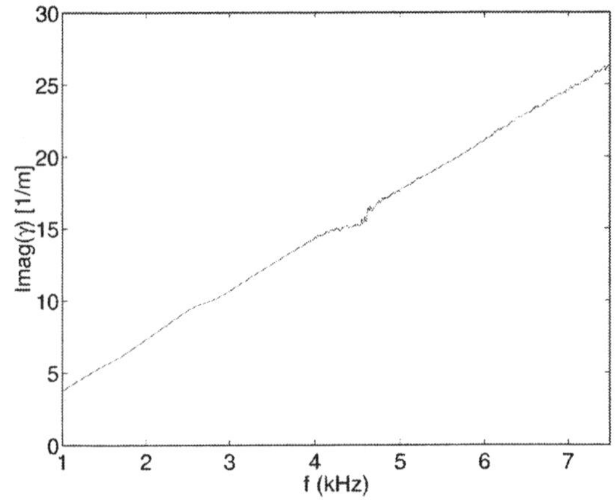

(b) $\operatorname{Im}(\gamma_1)$

Fig. 2. $\hat{\gamma}^r_\omega$ obtained from experiment. (a) The real part. (b) The imaginary part

Figure 2 we have shown the variation of the real and imaginary parts of the bi-iteration subspace estimate $\hat{\gamma}^r_\omega$ of the wave propagation function as a function of the frequency. We mention here that the least squares estimates $\hat{\gamma}^l_\omega$ is similar to $\hat{\gamma}^r_\omega$. One can notice here a regular bias effect, more prominent in the real part of $\hat{\gamma}^r_\omega$, around 4.5 kHz. This effect can be accounted for the finite signal to noise ratio as well as the Jacobian matrix $\mathbf{J}_f$ having relatively high condition number. A detailed analysis of such bias terms is beyond the scope of this paper. In Table 1 we have compared the

	$\hat{\gamma}^s_\omega$, $\mathbf{\Delta} = \mathbf{I}_p$	$\hat{\gamma}^r_\omega$	$\hat{\gamma}^l_\omega$
Matlab flops	466	2112	64660

Table 1. Comparison of computational load for different methods(Matlab version 5.3 is used).

Matlab flops required to compute the estimates for the subspace method with identity weighting, bi-iteration subspace approach and the nonlinear least squares method described in Section 2.1. It must be mentioned here that the nonlinear least squares cost function (13) was minimized using the function fminsearch in Matlab 5.3. As mentioned earlier, the computational advantage of the subspace method is significant compared to that of the least squares method. The bi-iteration subspace method is as accurate as the least squares method but nearly 50 times faster for the single wave case ($n = 1$). Even though the precise number of floating point operations will vary somewhat with the implementation, the striking difference in compuational load between the nonlinear least squares and the subspace algorithm will remain to be of the same order.

4. DISCUSSION

It is well known that the nonlinear least squares method in Section 2.1 suffers from numerical problems when equidistant sensor configuration is used. For a detailed discussion and experimental observations on this aspect we refer the reader to (Hillström *et al.*, 2000). The fundamental assumption of full rank $\mathbf{\Psi}(\mathbf{x}, \boldsymbol{\gamma})$ is not fulfilled if $d\gamma_k(\omega)$ is an integral multiple of $\mathrm{i}\pi$. In fact, for materials of substantially low damping, $\mathbf{\Psi}(\mathbf{x}, \boldsymbol{\gamma}_\omega)$ can become numerically ill-conditioned for certain frequencies. Note that, in order to compute $\hat{\gamma}^l_\omega$ we need to optimize the cost function in (13) numerically. For frequencies close to the critical frequencies it is difficult to evaluate the cost function accurately because of the ill-conditioning of $\mathbf{\Psi}(\mathbf{x}, \boldsymbol{\gamma}_\omega)$, for which it is not possible to compute $\mathbf{\Psi}^\dagger(\mathbf{x}, \boldsymbol{\gamma}_\omega)$ even with stable numerical techniques. We stress here that the alternative optimal estimate $\hat{\gamma}^r_\omega$ can be still be computed without any numerical problem at these critical frequencies, though the variance of the estimates is high. To illustrate this point let us consider the following simple but worst case scenario.

Example: Consider the case of longitudinal wave propagation in a bar of perfectly elastic material, *i.e* there is no damping and $n = 1$. Suppose the frequency ω is such that

$$\gamma(\omega) = \mathrm{i}\frac{k\pi}{d}, \qquad k \in \mathbb{Z}. \tag{36}$$

Thus from (10) we get

$$\mathbf{\Psi}(\mathbf{x}, \boldsymbol{\gamma}_\omega) = \left[\,\mathbf{e}_p \; \mathbf{e}_p\,\right], \tag{37}$$

where $\mathbf{e}_p$ denote the p dimensional column vector of all *ones*. Thus $\mathbf{\Psi}(\mathbf{x}, \boldsymbol{\gamma}_\omega)$ is clearly a rank-1 matrix, and it is not possible to compute the least squares cost function in (13) at the true parameter values.

To observe the consequences in the subspace approach, notice first that

$$\boldsymbol{\epsilon}_\omega = \left\{c^+(\omega) + c^-(\omega)\right\}\mathbf{e}_p + \tilde{\boldsymbol{\epsilon}}_\omega \tag{38}$$

By definition of $\mathbf{T}_0$ and $\mathbf{T}_1$ for $n = 1$ in (27) it readily follows from (28) that

$$\hat{\mathbf{s}}_\omega = 2\mathbf{e}_{p-2} + O(\|\tilde{\boldsymbol{\epsilon}}_\omega\|), \qquad \hat{\mathbf{S}}_\omega = \mathbf{e}_{p-2} + O(\|\tilde{\boldsymbol{\epsilon}}_\omega\|). \tag{39}$$

Thus from (30) it is easy to verify that

$$\hat{\boldsymbol{\theta}}^s_\omega = -2 - O(\|\tilde{\boldsymbol{\epsilon}}_\omega\|) \triangleq -(2+\delta) \tag{40}$$

where $\delta = O(\|\tilde{\boldsymbol{\epsilon}}_\omega\|) \geq 0$. Hence we get

$$A(z) = 1 - (2+\delta)z + z^2 \quad \Rightarrow$$
$$\mathrm{e}^{\hat{\gamma}(\omega)} = 1 + \frac{\delta}{2} + \sqrt{\delta\left\{1 + \frac{\delta}{4}\right\}} \approx 1 + \sqrt{\delta} \tag{41}$$

where we consider the root of $A(z)$ located outside the unit circle. From (41) we see that

$$\hat{\gamma}(\omega) = \mathrm{i}\frac{k\pi}{d} + O(\sqrt{\|\tilde{\boldsymbol{\epsilon}}_\omega\|}), \qquad k \in \mathbb{Z}, \tag{42}$$

which ensures consistent estimate as the signal-to-noise ratio approaches infinity. ∎

Next, we comment on identifiability issues. Here we consider the case when $n = 1$ to keep the discussion simple, but it is easy to extend the idea for $n > 1$. Consider that $\gamma = \alpha + \mathrm{i}\,\beta$ is a global minimum point of the loss function in (13). Then for equi-distant sensor configuration it is easy to verify that $\gamma = \pm\{\alpha + \mathrm{i}\,(\beta + \frac{2k\pi}{d})\}$, $k \in \mathbb{Z}$ is also gobal minimum point of the loss function. The subspace estimates also have the same property for the loss function (25). The ambiguity in sign can easily be resolved by imposing a restriction $|\mathrm{e}^{\gamma}| \geq 1$. The ambiguity in k can be resolved as follows. First the estimation of γ is carried out such that $-\pi \leq \arg\left[\mathrm{e}^{d\gamma}\right] \leq \pi$ and $|\mathrm{e}^{d\gamma}| \geq 1$ as a function of frequency. It is well known that that the wave propagation function of a linearly viscoelastic solid is a continuous function of frequency, since so is the complex modulus. Thus, the estimated phase of $\mathrm{e}^{d\gamma}$ is adjusted such that the resulting $\arg\left[\mathrm{e}^{d\gamma}\right]$ is a continuous function of frequency. This operation on the phase of a complex valued continuous function is very common in several other fields of signal processing, like the study of complex and real cepstrum (Oppenheim and Schafer, 1975) for example, is called *unwrapping* of the phase. In Matlab this operation can be acomplished by the function unwrap in Signal Processing Toolbox.

Finally, we comment on the effect of sensor separation d on the estimation methods as well as accuracy. For a given material, increasing d causes an increase in the real part of $d\gamma_\omega$. It can be shown that an increase in the real part of $d\gamma_\omega$ helps keeping the magnitudes of the elements $\mathbf{J}_f$ at a comparatively high value. Since at critical frequencies, the matrix $\mathbf{J}_f$ can be shown to be close to singular, an increase in d can lower the variance of the parameter estimates significantly at critical frequencies. Also the standard deviation of $\hat{\gamma}^s_\omega$ is inversely proportional to d^2 ($\mathbf{J}_f$ can be shown to be directly proportional to d). Hence we conclude that accuracy of the wave propagation function estimate improves significantly, particularly in the regions close to the critical frequencies with increasing d, as long as the signal-to-noise ratio remains constant. But we must recall that the signal to noise ratio of the data decrease with the sensor separation d if the magnitude of the excitation remains constant. Hence, increasing d to improve accuracy must be accompanied by a simultaneous increase in excitation.

5. CONCLUSIONS

In this paper, we have presented a computationally efficient algorithm for estimating the wave propagation function of a visco-elastic material from one dimensional wave experiments. The sensors are required to be equidistantly located. The special structure of the problem, due to the equidistant sensor locations, are exploited to develop a class of subspace based estimators. Further, it has been established that the subspace estimate with optimal weighting achieves optimal statistical accuracy. Apart from being computationally economical and statistically efficient, the class of subspace methods presented here are numerically robust and do not suffer from the problem of local minima like least squares. Thus, it is not required to have any prior knowledge about the parameters to apply the subspace based algorithms. The algorithm has been applied successfully to real experimental data. The results confirm the theoretical assertions.

6. ACKNOWLEDGEMENTS

The authors would like to thank one of the reviewers for his valuable comments.

7. REFERENCES

Brillinger, D. R. (1975). *Time Series: Data Analysis and Theory.* Holt, Rinehart and Winston, Inc. USA.

Eriksson, A, P Stoica and T Söderström (1994). Markov-based eigenanalysis method for frequency estimation. *IEEE Transactions on Signal Processing* **42:3**, 586–594.

Golub, G. H. and C. F. Van Loan (1989). *Matrix Computations, 2nd ed.* John Hopkins University Press. Baltimore, MD, USA.

Golub, G. H. and V. Pereyra (1973). The differentiation of pseudo-inverses and nonlinear least squares problems whose variables separate. *SIAM journal of Numerical Analysis* **10**(2), 413–432.

Hillström, L. and B. Lundberg (2001). Analysis of elastic flexural waves in non-uniform beams based on measurement of strains and accelarations. *Journal of Sound and Vibration* **246**, 227–242.

Hillström, L., M. Mossberg and B. Lundberg (2000). Identification of complex modulus from measured strains on an axially impacted bar using least squares. *Journal of Sound and Vibration* **230**, 689–707.

Kristensson, M., M. Jansson and B. Ottersten (2001). Further results and insights on subspace based sinusoidal frequency estimation. *IEEE Transactions on Signal Processing* **49**(12), 2962–2974.

Lakes, R. S. (1999). *Viscoelastic Solids.* CRC Press.

Mahata, K., T. Söderström, M. Mossberg, L. Hillström and S. Mousavi (2001). On the use of flexural wave propagation experiments for identification of complex modulus. Technical report. 2001-027, Department of Information Technology, Uppsala University. Uppsala, Sweden.

Mendel, J. M. (1995). *Lessons in Estimation Theory for Signal Processing, Communications and Control.* Prentice Hall. Englewood Cliffs, New Jersey.

Mossberg, M., L. Hillström and T. Söderström (2001). Non-parametric identification of viscoelastic materials from wave propagation experiments. *Automatica* **37**(4), 511–521.

Oppenheim, A. V. and R. W. Schafer (1975). *Discrete-Time Signal Processing.* Prentice Hall International. Englewood Cliffs, New Jersey.

Söderström, T. (2002). Using system identification for estimating material functions from wave propagation experiments. In: *3rd International conference on Identification in engineering systems.* Swansea, Wales.

Söderström, T. and P. Stoica (1989). *System Identification.* Prentice Hall International. Hemel Hempstead, UK.

IFAC Publications
www.elsevier.com/locate/ifac

IDENTIFICATION OF UNDERLYING INTENSITY PROCESSES OF INTERFERENCE PATTERNS

László Nádai, József Bokor and András Edelmayer

Computer and Automation Research Institute
Hungarian Academy of Sciences
H-1111 Budapest, Kende u. 13–17.
nadai@sztaki.hu

Abstract: Time-correlated single photon counting is a technique to record low level light signals with very high time resolution. In this paper we first describe the experimental apparatus and briefly outline the statistical interpretation of the measurement series. After, we propose a stochastic model for the formulation of interference patterns based on doubly stochastic autoregressive model class. We suggest the application of optimal filtering technique conceived in Elliott et. al. (1997) in order to estimate the underlying intensity process of the Poisson type measurement series. Finally, we present some results of the identification of realistic data that is although in coincidence with the general view of interference experiments, goes beyond the physical interpretation and may deeper our understanding of the quantum world. *Copyright © 2003 IFAC*

Keywords: Stochastic, experimentation, physical measurement systems.

1. INTRODUCTION

Time-correlated single photon counting (SPC) is based on the detection of single photons of a periodical light signal, the measurement of the detection times of the individual photons and the reconstruction of the waveform from the individual time measurements. The method was conceived in 1961 by Bollinger and Thomas, and a comprehensive overview of the field can be found in Connor (1984). Typical applications are:

- Ultra-fast recording of optical waveforms
- Fluorescence lifetime measurements
- Detection and identification of single molecules
- DNA sequencing
- Optical tomography
- Fluorescence lifetime imaging

The method makes use of the fact that for low level, high repetition rate signals the light intensity is usually so low that the probability of detecting one photon in one signal period is much less than one. Therefore, the detection of several photons can be neglected and the following principle can be used: the detector signal consists of a train of randomly distributed pulses due to the detection of the individual photons. There are many signal periods without photons, other signal periods contain one photon pulse. Periods with more than one photons are very rare. When a photon is detected, the time of the corresponding detector pulse is measured. The events are collected in memory by adding a '1' in a memory location with an address proportional to the detection time.

After many photons the histogram of the detection times, i.e. the waveform of the optical pulse, builds up in the memory. Although this principle looks complicated at the first glimpse, it is very efficient and accurate for the following reasons:

- The accuracy of the time measurement is not limited by the width of the detector pulse.
- Thus, the time resolution is much better then with the same detector used in front of an oscilloscope or another linear signal acquisition device.

- Furthermore, all detected photons contribute to the result of the measurement. There is no loss due to gating as in "Boxcar" devices or gated image intensified CCDs.

The SPC method differs from methods with analog signal processing in that the time resolution is not limited by the width of the detector impulse response. For the SPC method only the timing accuracy in the detection channel is essential. This accuracy is determined by the transit time spread of the single photon pulses in the detector and the trigger accuracy in the electronic system.

The standard quantum-optical approach involves only the calculation of (first- or higher-order) correlations, coincidences and momentums, but the temporal model for the photonic signal process – using the results of stochastic control theory – is ignored. In this paper we base our investigation theoretically on stochastic dependencies between absorption processes in the elements of the photo-detector array. As the interference picture (that is the average intensity distribution) can not be interpreted after the impact of a few photons, this paradox can be lift by applying *doubly stochastic* vector processes, which are well-known from modern probability theory. Moreover, we can give a statistical characterization for the dynamics of absorption time (events of photon detection) vector processes by assuming the existence of stochastic connection with "hidden" generating information processes.

The paper is organized as follows: first we describe the experimental technique called photon counting and briefly outline the statistical interpretation of the measurement series. After, we propose a stochastic model for interference patterns based on doubly stochastic autoregressive model class. We suggest the application of optimal filtering technique conceived in Elliott et. al. (1997) in order to estimate the underlying intensity process of the Poisson type measurement series. Finally, we present some results of the identification of realistic data that is although in coincidence with the general view of interference experiments, goes beyond the physical interpretation and may deeper our understanding of the quantum world.

2. FORMULATION OF THE PROBLEM

The experimental apparatus is shown schematically in Figure 1. Light from a highly coherent semiconductor laser (cf. Nádai et. al., 1995) can be considered as a source that is filtered to eliminate all but the 980nm emission line of the spectrum. The beam is split into two equal portions by a half-silvered mirror (BS). After, the two beams reflected on mirrors (M_1, M_2) are combined on the surface of a photodiode array (D_1, ..., D_M) and the fluctuations in the outputs of the detectors are under investigation. The description of the behavior of Mach–Zehnder type interferometers

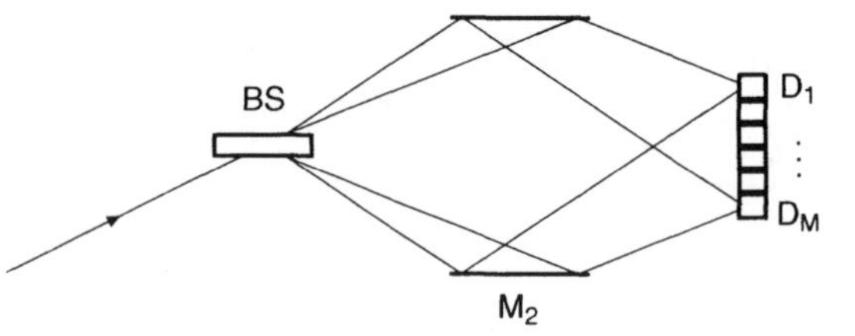

Fig. 1. Mach–Zehnder type interferometer (BS: Beam Splitter; M_1, M_2: Mirrors; D_1, ..., D_M: Detector Array)

(that is the one in the above experiment) using the so-called wave-packet model was presented elsewhere, see Papp et. al. (1998).

The output signal of the i-th pixel is denoted by $n_k^{(i)}$ ($i = 1,\dots,M$, $k = 1,2,\dots$) which – choosing an appropriate average of the signals of detectors be the unit – gives integer numbers during the reading time $\Delta\tau$, thus the signal of the detector array (neglecting the reading and detector noises) at k-th reading is

$$n_k = \left(n_k^{(1)}, n_k^{(2)}, \dots, n_k^{(M)}\right), \qquad k = 1,2\dots, \tag{1}$$

where M is the number of pixels. The counting time of the j-th photon on i-th pixel is denoted by $\tau_j^{(i)}$, $j = 1,2,\dots$. Counter vector process $z(t)$ is defined as

$$z(t) = \left(z^{(1)}(t), z^{(2)}(t), \dots, z^{(M)}(t)\right), \tag{2}$$

where

$$z^{(i)}(t) = \sum_{j=1}^{\infty} I\left(\tau_j^{(i)} \leq t\right)$$

(I is the indicator function.) $\{\tau_j^{(i)}, j = 1,2,\dots\}$ absorption time series can also be expressed using the measurement series $\{n_k^{(i)}, k = 1,2,\dots\}$, $1 \leq i \leq M$:

$$\begin{aligned} n_k^{(i)} &= \sum_{j=1}^{\infty} I\left((k-1)\Delta\tau < \tau_j^{(i)} \leq k\Delta\tau\right) \\ &= z^{(i)}(k\Delta\tau) - z^{(i)}((k-1)\Delta\tau). \end{aligned} \tag{3}$$

Remark 1. All of the above defined time series strongly depend on the setting of $\Delta\tau$ time intervals. Obviously, there are technical limitations in choosing $\Delta\tau$, on one hand the clock signal of the measurement card, and on the other hand the wake-up time and dark noise of detector pixels. Furthermore, the internal efficiency of the detector array and the reading efficiency constrain the performance of measurements.

The empirical statistical distribution of $z^{(i)}(t)$ contains information about the statistical properties of the light source. (It is assumed throughout that all the measured light beams are stationary.) We consider what this information is, and how it can be extracted from the results of a photon counting experiment.

3. STOCHASTIC MODELLING OF INTERFERENCE PATTERNS

According to Mandel (1958), the photon distribution $z^{(i)}(t)$ for the constant intensity case is a Poisson distribution. The counting statistics are the same as those found for the arrival of raindrops in a 'steady' downpour, or for the arrival of particles emitted during radioactive decay of a long-lived isotope. This fact motivates the model class selection.

Let us consider the scalar linear stochastic difference equation

$$x_{k+1} = A_{k+1}x_k + w_{k+1}, \tag{4}$$

where $x_k \in \mathbb{R}$ with x_0 is a Gaussian random variable with zero mean and nonzero variance Q_0, and $A_k \in \mathbb{R}$ is deterministic. Process $\{w_k\}$ is a sequence of independent, zero mean Gaussian random variables with nonzero variance Q_k. Further, process $\{w_k\}$ is assumed independent of x_0.

The observation process is a doubly stochastic, discrete time Poisson process $\{n_k\}$ $(k \geq 0)$ with rate $(C_k x_k)^2$, where $C_k \in \mathbb{R}$ is deterministic. Thus we have

$$E\{I(n_k = n) \mid x_k\} = \frac{(C_k x_k)^{2n}}{n!} e^{-(C_k x_k)^2}, \tag{5}$$

where E denotes expected value under P, and I is an indicator function.

Process x_k drives the scalar doubly stochastic autoregressive process

$$s_{k+1} = f_{k+1}(x_k)s_k + u_{k+1}, \qquad s_0 = 1, \tag{6}$$

where $s_k \in \mathbb{R}$, $f_k : \mathbb{R} \mapsto \mathbb{R}$, and $\{u_k\}$ is a sequence of zero mean random variables independent of x_0, $\{w_k\}$ and $\{n_k\}$.

Further, define the sigma field

$$\mathcal{Y}_k = \sigma\{n_0, \ldots, n_k\}$$

with the corresponding complete filtration $\{\mathcal{Y}_k\}$.

For convenience, define

$$\psi_k(x) = \frac{1}{\sqrt{2\pi Q_k}} e^{-x^2/(2Q_k)}, \quad x \in \mathbb{R}$$

$$\phi(y, n) = \frac{y^n}{n!} e^{-y}, \qquad y \in \mathbb{R},\ n = 0, 1, \ldots$$

Write

$$\lambda_0 = \frac{\phi((C_0 x_0)^2, n_0)}{\phi(1, n_0)}$$

$$\lambda_k = \frac{\phi((C_k x_k)^2, n_k)}{\phi(1, n_k)} \cdot \frac{\psi_k(x_k - A_k x_{k-1})}{\psi_k(x_k)}, \qquad k \geq 1,$$

and for $k \geq 0$ set

$$\Lambda_k = \prod_{l=0}^{k} \lambda_l. \tag{7}$$

4. EXACT FILTERING OF DOUBLY STOCHASTIC AR MODELS

In this section we derive recursive expressions for the conditional densities, then these will be applied to calculate the finite dimensional filters.

Let $\alpha_k(x)$ and $\rho_k(x)$ denote densities implicitly defined by

$$E\{\Lambda_k g(x_k) \mid \mathcal{Y}_k\} = \int_{\mathbb{R}} \alpha_k(x) g(x) dx \tag{8}$$

$$E\{\Lambda_k s_k g(x_k) \mid \mathcal{Y}_k\} = \int_{\mathbb{R}} \rho_k(x) g(x) dx \tag{9}$$

for any measurable function $g : \mathbb{R} \mapsto \mathbb{R}$.

The densities defined above obey the following recursions:

$$\alpha_k(x) = \frac{\phi((C_k x)^2, n_k)}{\phi(1, n_k)} \int_{\mathbb{R}} \psi_k(x - A_k z)\alpha_{k-1}(z) dz \tag{10}$$

$$\rho_k(x) = \frac{\phi((C_k x)^2, n_k)}{\phi(1, n_k)} \int_{\mathbb{R}} \psi_k(x - A_k z) f_k(z) \rho_{k-1}(z) dz, \tag{11}$$

for $k \geq 1$. The initial values for the recursions are given by

$$\alpha_0(x) = \frac{\phi((C_0 x)^2, n_0)}{\phi(1, n_0)} \psi_0(x)$$

$$\rho_0(x) = \alpha_0(x).$$

The proof of the recursions can be find e.g. in Evans et. al. (1999).

Theorem 2. (Solution for $\alpha_k(x)$). The filtered density of x_k is given by

$$\alpha_k(x) = x^{M_k} \left(\sum_{j=0}^{L_k} P_k(j) x^j \right) e^{-x^2/(2\Omega_k)}, \tag{12}$$

where the sufficient statistics

$$(M_k, L_k, P_k(0), \ldots, P_k(L_k), \Omega_k)$$

is recursively computed for $k \geq 1$ as follows

$$M_k = 2n_k, \qquad M_0 = 2n_0$$

$$L_k = L_{k-1} + M_{k-1}, \qquad L_0 = 0$$

$$P_k(j) = U_k \sum_{i=i^*}^{L_{k-1}} P_{k-1}(i) \xi(j + M_{k-1}, i)$$

$$P_0(0) = (2\pi Q_0)^{-1/2} C_0^{2n_0} e^{-1}$$

$$\Omega_k = (2C_k^2 + Q_k^{-1} - R_k A_k^2 Q_k^{-2})^{-1}$$

$$\Omega_0 = (2C_0^2 + Q_0^{-1})^{-1},$$

where $i^* = \max(0, j - M_{k-1})$ and

$$\xi(r, s) = \binom{r}{s} (R_k A_k Q_k - 1)^i \eta(r - s, R_K^{1/2})$$

$$R_k = (\Omega_{k-1}^{-1} + A_k^2 Q_k^{-1})^{-1}$$

$$U_k = R_K^{1/2} Q_K^{-1/2} C_k^{2n_k} e^{-1}.$$

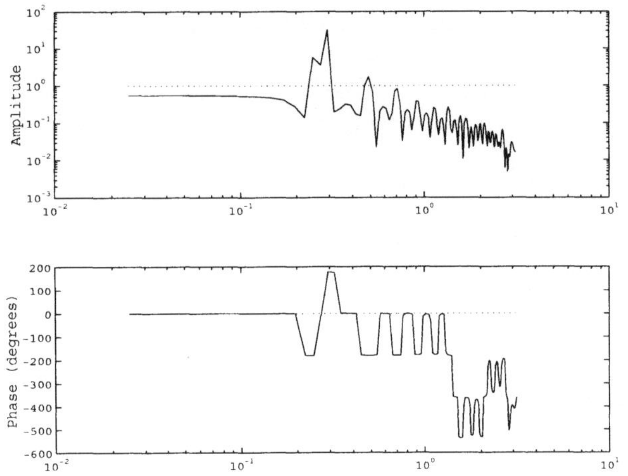

Fig. 2. Cross spectra data between two detector pixels ('$\cdots$' counting processes, '—' intensity processes)

Here $\eta(r,\sigma)$ denotes the r-th central moment of a normal random variable with variance σ (see Jazwinski, 1970)

$$\eta(r,\sigma) = \begin{cases} 1, & r=0 \\ 0, & r>0 \text{ and odd} \\ 1\cdot 3\cdots(r-1)\sigma^r, & r>0 \text{ and even.} \end{cases}$$

Finally, the filtered estimate of the Poisson rate is

$$E\{(C_k x_k)^2 \mid \mathscr{Y}_k\} = C_k^2 \frac{\sum_{j=0}^{L_k} P_k(j)\eta(j+M_k+2,\Omega_k^{1/2})}{\sum_{j=0}^{L_k} P_k(j)\eta(j+M_k,\Omega_k^{1/2})}.$$

PROOF. The filter was first proposed in Elliott et. al. (1997) where we refer for further details.

The above described filtering algorithm was implemented by the use of the System Identification and Control System Toolboxes of MATLAB. As a conclusion of this section we show the cross spectra data between two detector pixels (Fig. 2). It can be seen that the original counting processes $n_k^{(i)}$ and $n_k^{(j)}$ – although spatially separated – can not be distinguished in a statistical sense, while the intensity processes $(C_k x_k^{(i)})^2$ and $(C_k x_k^{(j)})^2$ have definitely different characteristics. An interpretation attempt for this result can be read in the next section.

5. INTERPRETATION OF EXPERIMENTAL RESULTS

We do not want to fell in the mistake of vanity and not take an attempt at explaining (or resolving) the Einstein–Podolsky–Rosen type paradoxes that are in focus of interest of the scientific community for more than 65 years. Instead, in this Section we point out some of the basic concepts behind our model and outline the parallelism with quantum-optical research.

As it is widely discussed in physical literature (see e.g. Mandel and Wolf, 1995), any *local realistic theory* which involves only readily measured quantities is expected to obey the Bell or the Clauser–Horne inequalities. However, these inequalities are violated in real experiments (Aspect et. al., 1982). It follows, says the standard conclusion, that the Einstein–Podolsky–Rosen type experiments can not be accommodated in a relativistic and deterministic universe.

However, there is a serious loophole in the real experiments: the original configuration contains "event-ready" detectors, which signal both arms that a pair of particles has been emitted. So, the statistics is assumed to be taken on the ensemble of particle pairs emitted by the source. It is obviously impossible to realize an "event-ready" detection (in practice all conceivable "event-ready" detectors depolarize or destroy the particles) and to perform the measurement on the unselected ensemble. In real experiments, instead of the "event-ready" detectors a four-coincidence circuit detects the "emitted particle-pairs". This method yields a *selected statistical ensemble*: only those pairs are taken into account, which coincidentally fire both arms of the interferometer. (It is to be emphasized that this claim is based on the logical schema of the experiment, independently of the detectors' inefficiency problem.)

Arthur Fine's "Prisms Model" (1982) resolves this contradiction. He argues that because of the random choice of the measured elements, the conditional probability of the outcome A_+, given that the measurement a has been performed, $p(A_+ \mid a)$ must be *equal* to the relative frequency of elements having property A_+ among those which are capable to produce an outcome of such a measurement, n_+^A/n^A. Let us realize that Fine's approach is closely related to *system identification*: given the noisy and incomplete measurement results, he formulates (or describe statistically) a model that can produce an outcome similar to that of the system. He knows that the system itself is "unrecognizable" and does not attempt to describe it, his only aim is to provide some verified statistical description.

6. CONCLUSIONS

In this paper we presented an approach based on the paradigm of mathematical system theory: in addition to the spatial information used in the quantum-optical formulation we investigated the temporal evolution of the interference pattern in order to identify the underlying intensity process. This is surely not a local hidden variable interpretation, instead it propose that there is some kind of global *hidden information pattern* that is only partially knowable.

We proposed a stochastic model for the formulation of interference patterns based on doubly stochastic autoregressive model class. We suggested the application of optimal filtering technique conceived in Elliott et.

al. (1997) in order to estimate the underlying intensity process of the Poisson type measurement series. Finally, we presented some results of the identification of realistic data that is although in coincidence with the general view of interference experiments, goes beyond the physical interpretation. It is obviously impossible to formalize this hidden information pattern, but it is possible to prove its existence. This is the main contribution of the paper.

REFERENCES

A. Aspect, J. Dalibard, G. Roger, 'Experimental test of Bell's inequalities using time-varying analyzers', *Phys. Rev. Lett.*, vol. 49, no. 25, p. 1804, 1982.

Bollinger, Thomas, *Rev. Sci. Instrum.* vol. 32, p. 1044, 1961.

D. V. O'Connor, D. Phillips, *Time-Correlated Single Photon Counting*, Academic Press, London, 1984.

A. Einstein, R. Podolsky, N. Rosen, 'Can quantum-mechanical description of physical reality be considered complete?', *Phys. Rev.*, vol. 47, p. 777, 1935.

R. J. Elliott, V. Krishnamurthy and J. H. Manton, 'Optimal estimation of Poisson rate from discrete time observations', in *Proc. Int. Conf. Communications*, Montreal, Canada, 1997, pp. 1392–1395.

J. Evans, V. Krishnamurthy, 'Exact filters for doubly stochastic AR models with conditionally Poisson observations', *IEEE Trans. Automat. Contr.*, vol. 44, no. 4, pp. 794–798, Apr. 1999.

A. Fine, 'Some local models for correlation experiments', *Synthese*, vol. 50, p. 279, 1982.

A. H. Jazwinski, *Stochastic Processes and Filtering Theory*, Academic Press, San Diego, 1970.

L. Mandel, *Proc. Phys. Soc.* vol. 72, p. 1037, 1958.

L. Mandel, E. Wolf, *Optical Coherence and Quantum Optics*, Cambridge University Press, Cambridge, MA, 1995.

L. Nádai, Zs. Papp, J. Bokor, P. Várlaki, 'On optimal frequency stabilization of semiconductor lasers', in: J. Shinar (Ed.), *Proc. 10th IFAC Workshop on Control Applications of Optimization*, IFAC, Technion, Haifa, Israel, 1995.

Zs. Papp, P. Várlaki, L. Nádai, 'Wave packet model and mach–zehnder type interferometers', in: G. Hunter, S. Jeffers, J.-P. Vigier (Eds.), *Causality and Locality in Modern Physics*, Fundamental Theories of Physics, Kluwer Academic Publishers, The Netherlands, 1998, pp. 373–382.

IFAC Publications
www.elsevier.com/locate/ifac

FRACTIONAL MULTIMODELS – APPLICATION TO HEAT TRANSFER MODELING

R. Malti[*], M. Aoun[+], J.-L. Battaglia[~], A. Oustaloup[+], K. Madani[*]

[*]*I2S – IUT de Sénart – Université Paris 12*
Avenue Pierre Point, 77127 Lieusaint, France
Tél: +33 (0)164 135 183 Fax: +33 (0)164 134 503
{malti, madani}@univ-paris12.fr

[+]*LAP – UMR 5131 CNRS – Université Bordeaux I - ENSEIRB*
351 cours de la Libération, 33405 Talence cedex, France
Tel : +33 (0)556 842 418 Fax : +33 (0)556 846 644
{aoun, oustaloup}@lap.u-bordeaux.fr

[~]*LEPT – UMR CNRS*
ENSAM Bordeaux
jlb@lept.u-bordeaux.fr

Abstract: This paper deals with identification of non linear systems using non linear fractional differentiation multimodels. All sub-models are described by fractional differentiation transfer functions. Performance of the newly proposed class of models is illustrated on a heat transfer process near a phase change temperature. *Copyright © 2003 IFAC*

Keywords: Identification, non-linear system, multimodel, fractional order differentiation, fractional dynamics

1. INTRODUCTION

Identification of non-linear systems is an important task for model based control, system design, simulation, prediction and fault diagnosis. It has widely been studied in the literature.

A large variety of **non-linear model** structures have already been proposed among which, Wiener and Hammerstein type models, Volterra series, neural networks, and Fuzzy logic based models. Applications of these models are numerous in dynamical system modeling.

On the other hand, Studies on real systems such as thermal or electrochemical (Battaglia, 2000), reveal inherent fractional differentiation behavior. The use of classical identification methods (based on integer order differentiation) is thus inappropriate in identifying these fractional systems. Thus a further class (called fractional) of mathematical models has been provided since 1983 by Oustaloup based on the concept of fractional differentiation. Different types of **linear fractional differentiation** (irrational) **models** have recently been studied. Models based on both equation and output errors were developed since 1998 (Le Lay et al. 1998), (Cois et al. 2000). Algorithms were implemented for optimizing differentiation orders, in addition to the usual transfer function's parameters. Applications of such linear fractional models have proven their utility in various domains such as thermal (Battaglia *et al.*, 2000, Cois *et al.*, 2000), viscoelastic (Bagley and Calico, 1991), acoustic (Matignon 1994) and electrochemical (Eckhard *et al.*, 2000). The concept of fractional differentiation will further be defined in the sense of Riemann-Liouville in section 3.

Our contribution is to use **fractional differentiation multimodels** for non-linear system modeling. Up to our knowledge, the only approach existing in the literature for identification of non-linear fractional systems is based on Hammerstein type models where the dynamical part is expressed using fractional differentiation transfer functions. It has recently been proposed by Aoun *et al.* (2002). Except this work, no other classes of models were ever developed for non-linear fractional systems.

In this paper, *fractional multimodels* are studied. The basis of this approach is to represent the system as an interpolation of simple fractional models. Each sub-model describes the behavior of the system in a limited part of the operating space. The local validity of a sub-model is specified by an associated weighting function. These models are also known as Takagi-Sugeno multimodels. They were impoved by Murray-Smith and Johansen (1997).

It will be shown that this class of models is well

suited for heat transfer process identification when temperature ranges in wide interval causing heat conductivity and diffusivity, no longer to be constants, but to depend on temperature. The introduced non linearity will entirely be modeled using two fractional sub-models.

2. MULTIMODEL APPROACH

The identification task using multimodels involves two essential steps: structure selection and parameter estimation. The objective of the former is to determine an appropriate characteristic variable set Z, to partition the operating space and to determine total number of local models and their respective orders. The latter aims to estimating an optimal parameter set w.r.t. a defined criterion, once the structure known. These two steps are linked and generally in black-box modeling iterative procedures are implemented in order to have the best compromise between error minimization and the total number of parameters in the global model.

2.1. Structure optimisation

Structure optimization has widely been addressed in the literature. Its main concern is to reduce the complexity of multimodels by eliminating the irrelevant parameters, thus finding an appropriate number of local models. Johansen and Foss, (1995) propose, for instance, to use k-d partitions where multimodel is gradually extended to allow new local models in regions most needed.

Another procedure for feature space decomposition consists of considering initially a grid partition. But as the feature space is rarely uniformly covered by training data and due to its combinatorial aspect, it is well known that the grid partition often produces useless models. This is generally due to neighboring local models that can provide the same description of the system but are arbitrarily separated by a lattice partition. Thus, removing the less important models and merging the redundant ones can reduce the number of local models. We refer the reader to (Boukhris et al. 2000) and (Gasso et al., 2002) for more details.

A third method consists of using Hingging planes (Ernst, 1998). It is based on partitioning according to hyperplanes oblique to characteristic variables. This type of modeling introduces a difficulty in choosing operating spaces.

In this paper, model structure (i.e. characteristic variables and the partition of operating space) will be assumed known. This assumption is not restrictive as techniques to achieve this goal were presented in this section. The choice of characteristic variables is usually done according to an *a priori* knowledge of the system.

2.2. Parameter tuning – Equation error versus Output error

Usually, autoregressive systems can be modeled with two different approaches: Equation Error models (also called series parallel model in (Narendra and Parthasmarathy, 1990) and (Nelles, 1997)) and Output Error (OE) (also known as parallel model) as presented in (Boukhriss *et al.*, 2000) and (Gasso *et al.* 2002).

As in the linear case, in EE modeling previous process outputs are used as regressors, while in OE models previous local model outputs are fed back. As a consequence, the former models can perform "only" few steps ahead predictions, whereas the latter are recurrent and therefore can predict an arbitrary number of steps into the future. Although most applications require a parallel model, since predictions of many steps ahead is needed, most papers use series-parallel modeling for identification. The reason is that EE modeling has easy-to-optimize parameters, all of which are obtained by simple least squares algorithm. It is worth emphasizing that running an EE model in parallel often leads to catastrophic results. A decreasing in EE quadratic criterion may lead to an increase of the OE quadratic criterion as shown in (Nelles, 1995) and (Gasso *et al.*, 2002).

A third alternative consists of optimizing sub-models using local criteria. As suggested by Nelles (1997) this approach leads to optimizing less parameter at a time. However, this technique leads to the increase of the global OE.

Since a multiple steps ahead predictor is needed in the case of heat transfer modeling, OE technique will be considered. It is based on non-linear optimization algorithms.

2.3. Model structure

In output error context, the global output model is directly expressed in terms of all local outputs y_i multiplied by their respective activation functions ρ_i as shown in figure(1):

$$y(t)=\sum_{i=1}^{M}\rho_i(z(t),\beta)\ y_i(t) \tag{1}$$

Local linear input-output models can be expressed as:

$$Y_i(s)=F_i(s)\,u(s)$$

where each transfer function $F_i(s)$ is fractional and hence irrational. Fractional differentiations will be presented in the next section.

This representation can directly be generalized to MISO systems by including all models between all inputs and local outputs.

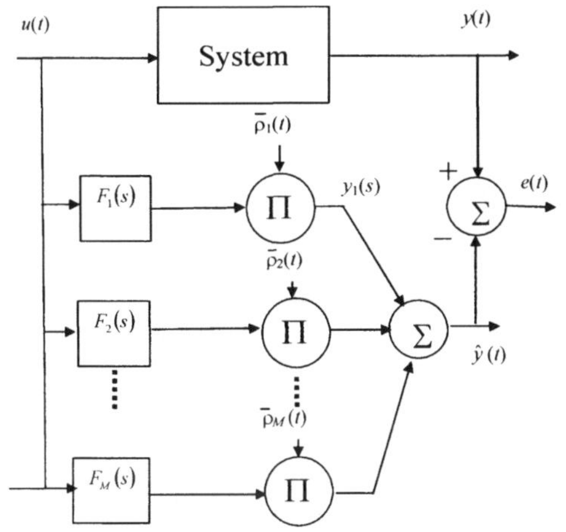

Figure 1 – Multimodel in output error context where each sub-model is a fractional transfer function

Activation functions depend directly on some characteristic variables $z(t)$ and parameters vector β, which are usually chosen according to an *a priori* knowledge of the system.

Once the partitioning done, a validity function is associated to each characteristic variable. In their initial paper, Takagi and Sugeno (1985) proposed to use trapezoidal validity functions. However, smoother validity functions (Gasso et al. 2002) are preferred such as Gaussian or hyperbolic tangents. Their general form is:

$$\mu_{l,j}\left(z_j(t)\right) = \exp\left(-\frac{\left(z_j(t) - c_{l,j}\right)}{2\beta_{l,j}^2}\right) \qquad l = 1, \dots p_j$$

$$\mu_{l,j}\left(z_j(t)\right) = \tanh\left(-\frac{\left(z_j(t) - c_{l,j}\right)}{2\beta_{l,j}^2}\right) \qquad (2)$$

As can be seen, validity functions are expressed w.r.t. the characteristic variable z_j. Each z_j can have a total number of p_j partitions.

As can be seen every validity function is characterized essentially by two parameters: a center $c_{l,j}$ and a dispersion coefficient β_j. These two parameters can be fixed either according to an *a priori* knowledge, if available, or some known algorithms as described in (Johansen and Foss, 1995), (Murray-Smith and Johansen, 1997).

To each sub-model will be associated an activation function determining the weight of sub-models in each region of the characteristic space. Activation functions are computed directly from validity functions by:

$$\rho_i = \prod_{j=1}^{n_z} \mu_{l_j^{(i)}} \qquad i = 1, \dots M$$

In $l_j^{(i)}$, the index j denotes the particular partition of z_j which intervenes in the construction of the model i.

Usually, it is prefered to use normalized activation functions $\bar{\rho}_i$:

$$\bar{\rho}_i = \frac{\rho_i}{\sum_{k=1}^{M} \rho_k}$$

which also correspond to the product of normalized validity functions:

$$\bar{\rho}_i = \prod_{j=1}^{n_z} \bar{\mu}_{l_j^{(i)},j} \qquad (3)$$

Let us now present local models, which in our paper are allowed to be fractional linear models.

3. FRACTIONAL DIFFERENTIATION SUB-MODELS – MATHEMATICAL BACKGROUND

The concept of differentiation to an arbitrary order (also called fractional differentiation) was defined in the 19th century by Riemann and Liouville. Their main concern was to extend differentiation by using not only integer but also non-integer (real or complex) orders.

Fractional differential-integration dates back to the work of Abel, Riemann and Liouville in the 19th century. It is an extension to the classical integer differentiation and integration. The generalized n^{th} order integral of a continuous real-time function $f(t)$ is defined by (Samko et al. 1993):

$$\left(I^n f\right)(t) \stackrel{\Delta}{=} \frac{1}{\Gamma(n)} \int_0^t \frac{f(\tau)}{(t-\tau)^{1-n}} d\tau \qquad (4)$$

The Riemann-Liouville n^{th} order fractional derivative of $f(t)$ is defined as an integer order derivative $\lfloor n+1 \rfloor$ (ceiling of n+1) of a fractional order integral of $f(t)$:

$$\mathbf{D}^n f(t) \stackrel{\Delta}{=} \left(\frac{d}{dt}\right)^{\lfloor n+1 \rfloor} \left(I^{1-n} f(t)\right) \qquad (5)$$

Numerical simulation of fractional systems is obtained using Grünwald's approximation of fractional derivatives (Oustaloup, 1995):

$$\mathbf{D}^n f(t) = \frac{1}{T_s} \sum_{k=0}^{K} (-1)^k \binom{n}{k} f\left((K-k)T_s\right) \qquad (6)$$

where (T_s) is the sampling period.

Moreover, it is proven (Oldham and Spanier, 1974) that the Laplace transform of $\mathbf{D}^n f(t)$ is:

$$L\left\{\mathbf{D}_0^n f(t)\right\} = s^n F(s) \qquad (7)$$

provided that $f(t)$ is relaxed at $t = 0$ (i.e. $f(t)$ and all derivatives of $f(t)$ are null for all $t < 0$).

Important results, obtained by Oustaloup and Matignon, permit the analytical expression of fractional system output. The impulse response of a fractional transfer function is an aggregation of two signals:

- an *exponential mode*, resulting from the computation of residue(s) on each pole of $H_l(s)$,

- an *aperiodic multimode*, the main characteristic of fractional systems, resulting from an integral along the negative real axis.

Moreover, stability of such systems was studied in, (Matignon, 1998) and other references. Stability condition is expressed as:

$$|\arg(\lambda_l)| > \frac{n\pi}{2}, \text{ for } l=1,..,\dim(\mathbf{x}), \tag{8}$$

where the λ_l are the eigenvalues of the system and n the commensurate order.

According to the presented definitions each local model can be written as:

$$F_i(s) = \frac{\sum_{m=0}^{m_A} a_{m,i} s^{\alpha_m}}{1 + \sum_{m=1}^{m_B} b_{m,i} s^{\beta_m}}$$

4. IDENTIFICATION ISSUES

Local models can have different structures, parameters and differentiation orders. However, for sake of simplicity, we will assume that all differentiation orders are multiples of a single value. This is not a constrained as long as, in thermal applications for instance, knowledge based models (Battaglia et al. 2000) show that the suited differentiation orders are multiples of 0.5.

When fractional differentiation orders cannot be fixed due to a lack of knowledge, techniques providing their estimation can be implemented as in the linear case (Cois, *et al.*, 2000).

4.1. Parameter optimization

Output error modeling is considered.

Centers ($c_{l,j}$ in (2)) of validity functions are fixed according to a uniform partitioning of activation space. Coefficients of each submodel together with dispersion parameter β in (2) are optimized.

Coefficients of ith sub-model are grouped in a vector form:

$$\theta_i = [a_{0,i},...,a_{m_A,i}; b_{1,i},...b_{m_B}; \beta_i]^T$$

θ_i of all models are as well grouped in:

$$\Theta = \left[\theta_1^T, \cdots \theta_M^T\right]$$

The optimal values of each transfer function $F_i(s)$ are obtained by minimizing the global quadratic criterion, based on the output error:

$$J(\Theta) = \int_0^K (e(t))^2 dt = \int_0^K (y(t) - \hat{y}(t))^2 dt$$

where:

$$\hat{y}(t) = \sum_{i=1}^{M} \mathcal{L}^{-1}\left(\frac{\sum_{m=0}^{m_A} a_{m,i} s^{\alpha_m}}{1 + \sum_{m=1}^{m_B} b_{m,i} s^{\beta_m}}\right) * u(t) \times \rho_i(z(t), \beta_i) \tag{9}$$

Optimal value of Θ is then obtained iteratively using Levenberg-Marquardt algorithm (Marquardt, 1963):

$$\Theta_{k+1} = \Theta_k - \left\{\left[\mathbf{J}''_{\Theta\Theta} + \xi \mathbf{I}\right]^{-1} \mathbf{J}'_{\Theta}\right\}_{\hat{\Theta}=\Theta_k}$$

where:

$$\begin{cases} \mathbf{J}'_{\Theta} = -2\int_0^K e(t)\mathbf{S}(t,\hat{\Theta})\,dt : \text{gradient} \\ \mathbf{J}''_{\theta\theta} \approx 2\int_0^K \mathbf{S}(t,\hat{\Theta})\mathbf{S}^{\mathrm{T}}(t,\hat{\Theta})\,dt : \text{hessian} \\ \mathbf{S}(t,\hat{\Theta}) = \dfrac{\partial \hat{y}(t,\hat{\Theta})}{\partial \Theta} : \text{output sensitivity function} \\ \xi : \text{Marquardt parameter} \end{cases}$$

All sensitivity functions are computed by a straightforward differentiation of expression (9) with respect to each parameter vector. Three types of sensitivity functions are distinguished:

- Sensitivity functions w.r.t. numerator coefficients $a_{m,i}$'s:

$$\frac{\partial \hat{y}}{\partial a_{m,i}} = \mathcal{L}^{-1}\left\{\frac{s^{\alpha_m}}{1 + \sum_{m=1}^{m_B} b_{m,i} s^{\beta_m}}\right\} * u(t) \times \rho_i$$

- Sensitivity functions w.r.t. denominator coefficients $b_{m,i}$'s

$$\frac{\partial \hat{y}}{\partial b_{m,i}} = \mathcal{L}^{-1}\left\{\frac{-s^{\beta_m}\sum_{m=0}^{m_A} a_{m,i} s^{\alpha_m}}{\left(1 + \sum_{m=1}^{m_B} b_{m,i} s^{\beta_m}\right)^2}\right\} * u(t) \times \rho_i$$

- Sensitivity of β depends on the kind of activation functions used and on the number of submodels. They are easy to evaluate by differentiating (3) knowing which validity function is used (2).

5. EXAMPLE

Heat transfer is studied in a plane wall of finite thickness of 5cm made of homogenous thermosetting medium. A heat flux is injected on one side of a perfectly isolated wall. A dynamical model is to be identified linking temperature at a given distance of the wall and the heat flux.

Knowledge based models show that heat transfer in governed by the following partial differential equation:

$$\frac{\partial T(x,t)}{\partial t} = \frac{1}{\alpha(T)} \frac{\partial}{\partial x} \left(k(T) \frac{\partial T(x,t)}{\partial x} \right) \quad 0 < x < e,\ t > 0 \tag{10}$$

having as limit conditions:

$$-\lambda(T) \frac{\partial T(x,t)}{\partial x} = \varphi(t) \ \text{ when } x = 0,\ t > 0 \tag{11}$$

$$\frac{\partial T(x,t)}{\partial x} = 0 \ \text{ when } \ x = e,\ t > 0 \tag{12}$$

$$T(x,t) = T_0 \quad \text{when } \ 0 \le x \ge e,\ t = 0 \tag{13}$$

(11) corresponds to the injected heat flux, (12) to the perfect isolation of the wall (there is no heat transfer outside the medium and (13) to the initial temperature all over the wall at the beginning of the experiment. Parameters α and λ are respectively known as heat conductivity and diffusivity.

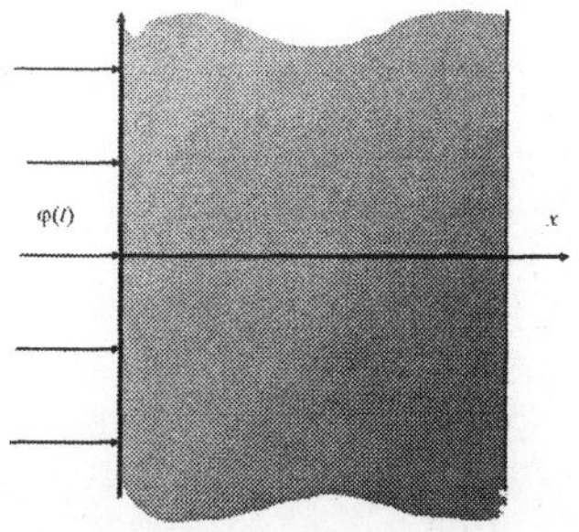

Figure 2 – plane wall

In the case of low temperatures (compared to the phase change) both parameters α and λ are considered as constants and (10) reduces to:

$$\frac{\partial T(x,t)}{\partial t} = \frac{k}{\alpha} \frac{\partial^2 T(x,t)}{\partial x^2} \tag{14}$$

Studies on the PDE (14) (Cois, *et al.* 2002) have shown that the exact solution of (14) is:

$$H(x,s) = \frac{\overline{T}(x,s)}{\overline{\varphi}(s)} = \frac{\cosh\left((e-x)\sqrt{\frac{s}{\alpha}} \right)}{\lambda\sqrt{\frac{s}{\alpha}} \sinh\left((e-x)\sqrt{\frac{s}{\alpha}} \right)}$$

Compact fractional differentiation models were used to approximate this solution using differentiation orders multiples of 0.5 (Cois *et al.* 2000).

However, when temperature range is high (up to fusion), conductivity and diffusivity are no longer constants. They depend on temperature. In the chosen thermosetting medium these parameters vary according to the following laws:

$$\lambda(T) = 0.523(T)^{-0.1316}$$

$$\alpha(T) = 300\log(T) - 857.3$$

The usual way of simulating (10) and (14) consists of discretizing the PDE using finite difference method. However this method is very time consuming and depends on the number of nodes in the discretization process. It is hence inappropriate for real-time simulations. However, in off-line simulations, it is a good mean for generating identification and validation data.

It is trivial that (10) can no longer be modeled by a single fractional differential equation, since the ruling partial differential equation is now non-linear.

Fractional multimodels are used to represent such a nonlinear system. The *a priori* knowledge has permitted to fix the characteristic variable as being the temperature T (since λ and α are known to vary in terms of T). Moreover, preliminary studies of heat transfer in small temperature ranges (the linear case) (Battaglia et al. 2000) show that inherent differentiation orders are multiples of 0.5. Hence, another *a priori* information was used to fix differentiation orders of all local models as multiples of 0.5.

Temperature is modeled at two distances: $x = 0$ and $x = e/6$ where e is the thickness of the wall. Local linear models were chosen on the basis of studies performed in (Battaglia et al. 2000) and (Cois *et al.* 2000) in the linear case i.e. when temperature variations are not significant:

at $x = 0$ ➔ $\widetilde{H}(0,s) = \dfrac{\overline{T}(0,s)}{\overline{\varphi}(s)} = \dfrac{a_1 s^{0.5} + a_0}{s}$

at $x = e/6$ ➔ $\widetilde{H}(e/6,s) = \dfrac{\overline{T}(e/6,s)}{\overline{\varphi}(s)} = \dfrac{a_1 s^{0.5} + a_0}{s^{1.5} + b_0 s}$

The input signal is shown in figure (3). System and model's outputs are shown in figures (3) and (4) on validation data.

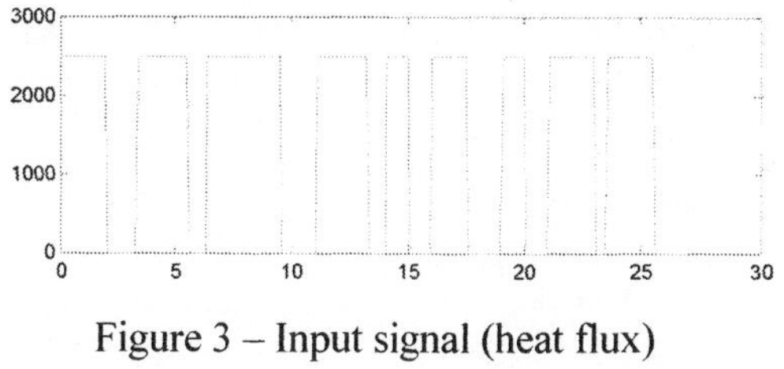

Figure 3 – Input signal (heat flux)

As it clearly appears, only two fractional models are required in order to fit system's output. Moreover, a small number of parameters is used in each sub-model. This proves the reliability of the model during the time concerned by the validation process.

6. CONCLUDING REMARKS

A novel class of models is presented, in this paper, for the identification of non-linear fractional systems. It is an extension of the well-known multimodels. This class is obtained by allowing sub-models to be of fractional type.

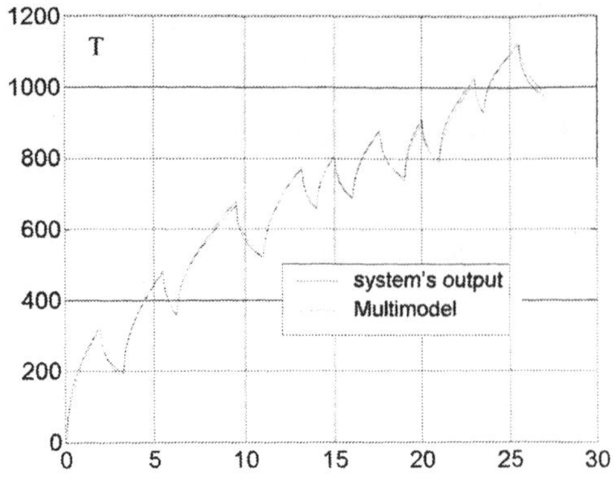

Figure 4 - Temperature versus time at $x = 0$

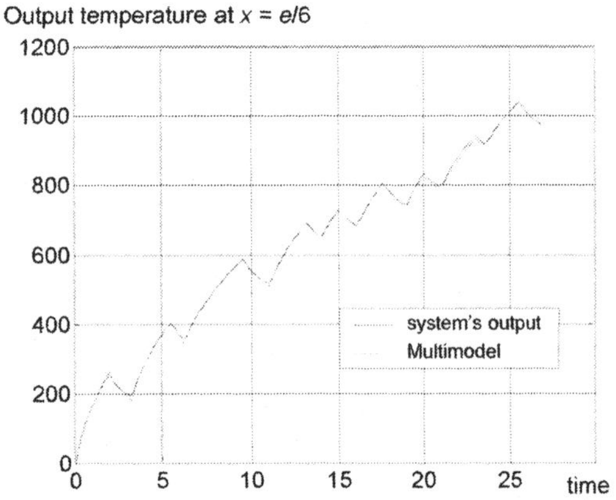

Figure 5 - Temperature versus time at x = e/6

Actually we are working on applications of this method to other systems and especially those representing stronger non linearity and a fractional behaviour.

7. REFERENCES

Aoun M. Malti R., Cois O. Oustaloup A. System identification using fractional Hammerstein models. 15th World IFAC Congress. Barcelona'02

Bagley, R.L. and Calico R.A. 1991. Fractional order state equations for the control of viscoelastically damped structures. J. of Guidance, Control and Dynamics. Vol. 14, pp. 304-311.

Battaglia J.-L., Le Lay L., Batsale J.-C., Oustaloup A., Cois O. (2000). Heat flux estimation through inverted non integer identification models. Int J. of thermal science. Vol 39, n° 3, pp. 374-389.

Boukhris, A, Mourot G. and Ragot J. (2000). Nonlinear dynamic system identification: a multiple-model approach. Int. J. of control, Vol. 72, N°7/8, pp. 591-604.

Cois O., Poinot T., Oustaloup A. (2002). Journées d'étude Automatique et Electronique, Angoulême, France, 2002

Cois, O., Oustaloup A., Battaglia E., Battablia J.-L. (2000) Non integer model from modal decomposition for time domain system identification. 5th SYSID congress. Santa Barbara.

Ernst S. (1998). Hinging hyperplane trees for appoximation and identification. 37th IEEE Conf. on Decision and Control, Tampa, Florida, USA.

Gasso K., Mourot G. and Ragot J. (2002). Structure identification of multiple models with output error local models. 15th world IFAC Congress. Barcelone b'02, Spain.

Johansen T. A. and Foss A.B. (1995). Identification of non-linear system structure and parameters using regime decomposition. Automatica, vol. 31, n°2, pp. 321-326.

Le Lay, (1998). Identification fréquentielle et temporelle par modèle non entire. Ph.D. thesis. University Bordeaux I, Talence, France.

Marquardt D.W. (1963). An algorithm for least-squares estimation of non-linear parameters. J. Soc. Industr. Appl. Math., Vol 11 (2), 431-441

Matignon D. (1994). Représentation en variables d'état de modèles de guides d'ondes avec dérivation fractionnaire. Ph.D. Thesis. University Paris-Sud Orsay.

Matignon D. (1998). Stability properties for generalized fractional differential systems. ESSAIM proceedings. Fractional differential systems: Models, Methods and applications. Vol 5, pp. 145-158.

Murray-Smith R. and Johansen T.A. (1997). Multiple model approaches to modelling and control. Edited by R. Murray-Smith and T.A. Johansen. Ed. Taylor and Francis.

Narendra, K.S. and Parthasarathy K. (1990). Identification and control of dynamical systems using neural networks, Vol. 1, No. 1.

Nelles O. (1995). On the identification with neural networks as series-parallel and parallel models, International Conference on Artificial Neural Networks (ICANN), Paris, France.

Nelles O. (1997) Orthonormal basis functions for nonlinear system identification with local linear model trees (LOLIMOT). Proc of SYSID'97, Fukuoka, Japan. Vol.2, pp. 667-672.

Oldham K. B. and J. Spanier (1974). *The fractional calculus*. Academic Press, New York and London.

Oustaloup, A. (1983) Systèmes asservis linéaires d'ordre fractionnaire. Masson

Samko S.G., Kilbas A.A. and Marichev O.I. (1993). Fractional integrals and derivatives: theory and applications. Gordon and Breach Science Publishers, Amsterdam.

Takagi T. and Sugeno M. (1985). Fuzzy identification of systems and its application to modeling and control. IEEE Trans. on Systems Man and Cyberneticc, Vol. 15, pp. 116-132.

IFAC Publications
www.elsevier.com/locate/ifac

A RECURSIVE ALGORITHM FOR ESTIMATING PARAMETERS IN A ONE DIMENSIONAL DIFFUSION SYSTEM

Bharath Bhikkaji, Torsten Söderström and **Kaushik Mahata**

Systems and Control,
Department of Information Technology, Uppsala University,
P O Box 337, SE 75105, Uppsala, Sweden.
Email: Bharath.Bhikkaji@It.uu.se,
Torsten.Soderstrom@It.uu.se, Kaushik.Mahata@It.uu.se

Abstract: A one-dimensional heat diffusion system with an associated partial differential equation (PDE) model is considered. The PDE model involves unknown parameters, which need to be estimated to model the physical diffusion system. A recursive algorithm in the frequency domain is deviced for an efficient estimation of the parameters involved. *Copyright © 2003 IFAC*

Keywords: Diffusion, System Identification, Fast Fourier Transform(FFT).

1. INTRODUCTION

In the recent past, the problem of estimating the parameters of a one dimensional heat diffusion system using the standard output error method, see Söderström and Stoica (1989), in the time domain was considered, see Söderström and Bhikkaji (2000) and Söderström and Remle (2000). Here, we device a recursive algorithm in the frequency domain to estimate the parameters involved.

2. PROBLEM DESCRIPTION

As in Söderström and Bhikkaji (2000) and Söderström and Remle (2000), we consider a homogeneous wall of thickness d units, see Figure 1. Let x be the co-ordinate across the wall, T_e be the exterior temperature, T_i be the interior temperature and q_i be the heat supplied from the interior.

The temperature dynamics across the wall is described by the PDE

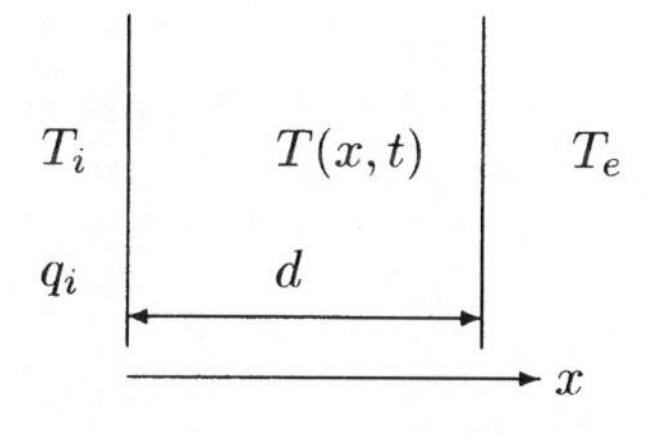

Fig. 1. Heat diffusion through a homogeneous wall.

$$\frac{\partial T(x,t)}{\partial t} = \alpha_0 \frac{\partial^2 T(x,t)}{\partial x^2}, \quad 0 < x < d \tag{1}$$

with boundary constraints

$$q_i(t) = -\kappa_0 \frac{\partial T(x,t)}{\partial x} |_{x=0}, \tag{2}$$

$$T_e(t) = T(d,t), \tag{3}$$

and initial condition

$$T(x,0) = 0, \quad 0 \le x \le d. \tag{4}$$

In (1) and (2), α_0 and κ_0, which are unknown, denote the diffusion coefficient and thermal conductivity, respectively.

The estimation problem considered here is the following:

Given: At any given time instant $t = Nh$, we are given the discrete time data $\{q_i(kh)\}_{k=0}^{N-1}$, $\{T_i(kh)\}_{k=0}^{N-1}$ and $\{T_e(kh)\}_{k=0}^{N-1}$ for a known sampling interval of h units.

Estimate: The parameters α_0 and κ_0, using a recursive algorithm and update the parameter estimates at each instant, as and when the data points are acquired.

3. SYSTEM DYNAMICS

In this section, we present a characterisation of the temperature dynamics of the PDE model in the frequency domain. For a more detailed discussion, see Söderström and Remle (2000).

The given diffusion system can be interpreted as an LTI system with inputs

$$u(t) = [q_i(t)\ , T_e(t)], \tag{5}$$

and the output

$$y(t) = T_i(t). \tag{6}$$

In Söderström and Remle (2000), the authors have shown that the given diffusion system can be rewritten in the Laplace domain as

$$Y(s) = G^{\infty}(s, \theta_0)U(s), \tag{7}$$

where $Y(s)$ and $U(s)$ are the Laplace transforms of the output $y(t)$ and the input $u(t)$ respectively, and the transfer-function

$$G^{\infty}(s,\theta_0) = \left[\frac{\tanh(d\sqrt{\frac{s}{\alpha_0}})}{\kappa_0\sqrt{\frac{s}{\alpha_0}}}\ , \ \frac{1}{\cosh(d\sqrt{\frac{s}{\alpha_0}})} \right]$$
$$\triangleq [G_1^{\infty}(s,\theta_0)\ ,\ G_2^{\infty}(s,\theta_0)] \tag{8}$$

and $\theta_0 = (\alpha_0, \kappa_0)$.

Note that both the system transfer-functions, $G_1^{\infty}(s,\theta_0)$ and $G_2^{\infty}(s,\theta_0)$ are transcendental functions having the same set of poles, which are real and negative.

4. RECURSIVE ALGORITHMS IN THE FREQUENCY DOMAIN

In this section, we present two different recursive algorithms, which are defined in the frequency domain, to estimate the parameters α_0 and κ_0.

Let

$$y^d(n) \triangleq y(nh) \tag{9}$$

and

$$u^d(n) \triangleq u(nh) \tag{10}$$

denote the inputs $u(t)$ and output $y(t)$ sampled at the instants $t = nh$, $n = 0, 1, 2, \ldots$.

Consider the following windowed data $\{y^d(n)\}_{n=N-L}^{N-1}$ and $\{u^d(n)\}_{n=N-L}^{N-1}$ of length L, where L is an integer such that $N \geq L$, at the time instant $t = Nh$.

Let

$$E_L^N(i\omega_k, \theta) \triangleq Y_L^N(e^{i\omega_k}) - \hat{Y}_L^N(e^{i\omega_k}, \theta), \tag{11}$$

where

$$Y_L^N(e^{i\omega_k}) \triangleq \sum_{m=0}^{L-1} y^d(m + N - L)e^{-i\omega_k m} \tag{12}$$

$$\hat{Y}_L^N(e^{i\omega_k}, \theta) \triangleq G^{\infty}(i\frac{\omega_k}{h}, \theta)U_L^N(e^{i\omega_k}), \tag{13}$$

$$U_L^N(e^{i\omega_k}) \triangleq \sum_{m=0}^{L-1} u^d(m + N - L)e^{-i\omega_k m} \tag{14}$$

$$\omega_k \triangleq \frac{2\pi k}{L} \quad k = 0, 1, 2, \ldots, \frac{L}{2}, \tag{15}$$

and

$$V_L^N(\theta) \triangleq \frac{2}{L}\sum_{k=0}^{\frac{L}{2}} \mid E_L^N(i\omega_k, \theta) \mid^2 . \tag{16}$$

Finally, note that $Y_L^N(e^{i\omega_k})$ is the Discrete Fourier transform (DFT) of the the windowed system output $\{y^d(n)\}_{n=N-L}^{N-1}$, $U_L^N(e^{i\omega_k}, \theta)$ is the DFT of windowed system input $\{u^d(n)\}_{n=N-L}^{N-1}$, $\hat{Y}_L^N(e^{i\omega_k}, \theta)$ is the DFT of windowed model output. $V_L^N(\theta)$ is an average over mean squared error in the DFT of windowed model output. Hence, $V_L^N(\theta)$ should be similar to the variance of the output error.

4.1 *Frequency Domain Recursive Algorithm (FD-R)*

Here, we present the first of the two recursive algorithms.

We define the cost function, for the first algorithm, at the instant $t = Nh$, as

$$W^N(\theta) \triangleq \frac{1}{N - L + 1}\sum_{m=L}^{N} V_L^m(\theta). \tag{17}$$

and the corresponding parameter estimate $\hat{\theta}(N)$ of θ_0 as

$$\hat{\theta}(N) = \arg\min_{\theta} W^N(\theta). \qquad (18)$$

Note that $W^N(\theta)$ is an average taken over all the averaged mean squared errors in windowed model output, $V_L^m(\theta)$'s, up to $t = (N-1)h$.

We now derive a recursive algorithm, which we call as the FD-R algorithm, to compute $\hat{\theta}(N)$, (18). Most of the derivation presented here is patterned after the derivation of the RPEM presented in page 26 of Ljung and Söderström (1983).

Let $\hat{\theta}(N-1)$ be the minimiser of $W^{N-1}(\theta)$ at the instant $t = (N-1)h$. A Taylor expansion of $W^N(\theta)$ around $\hat{\theta}(N-1)$ gives

$$\begin{aligned} W^N(\theta) = & W^N(\hat{\theta}(N-1)) \\ & + W^{N\,\prime}(\hat{\theta}(N-1))(\theta - \hat{\theta}(N-1)) \\ & + \frac{1}{2}(\theta - \hat{\theta}(N-1))^T W^{N\,\prime\prime}(\hat{\theta}(N-1)) \\ & \times (\theta - \hat{\theta}(N-1)) \\ & + o(|\, \theta - \hat{\theta}(N-1)\, |^2), \end{aligned} \qquad (19)$$

where the prime denotes the derivative with respect to θ and $o(x)$ denotes a function such that $\frac{o(x)}{x} \to 0$ as $|\, x \,| \to 0$. Minimisation of (19) with respect to θ gives

$$\begin{aligned} \hat{\theta}(N) = & \hat{\theta}(N-1) \\ & - [W^{N\,\prime\prime}(\hat{\theta}(N-1))]^{-1}[W^{N\,\prime}(\hat{\theta}(N-1))] \\ & + o(|\, \hat{\theta}(N) - \hat{\theta}(N-1)\, |). \end{aligned} \qquad (20)$$

Note that

$$\begin{aligned} W^N(\theta) = & \frac{N-L}{N-L+1} W^{N-1}(\theta) \\ & + \frac{1}{N-L+1} V_L^N(\theta). \end{aligned}$$

Hence

$$\begin{aligned} W^{N\,\prime}(\theta) = & \frac{N-L}{N-L+1} W^{N-1\,\prime}(\theta) \\ & + \frac{1}{N-L+1} V_L^{N\,\prime}(\theta). \end{aligned}$$

To evaluate (20), we introduce the following approximations

(1) First we assume that $\hat{\theta}(N-1)$ minimises $W^{N-1}(\theta)$, which implies that

$$W^{N-1\,\prime}(\hat{\theta}(N-1)) = 0. \qquad (21)$$

(2) Next we assume that $\hat{\theta}(N)$ lies in a small neighborhood of $\hat{\theta}(N-1)$. This assumption is reasonable for a large N, and implies that

$$o(|\, \hat{\theta}(N) - \hat{\theta}(N-1)\, |) \approx 0. \qquad (22)$$

(3) Finally, we assume L is large enough, such that

$$\begin{aligned} \sum_{m=L}^{N} V_L^{m\,\prime\prime}(\hat{\theta}(N-1)) \approx & (N-L+1) \\ & \times V_L^{N\,\prime\prime}(\hat{\theta}(N-1)). \end{aligned} \qquad (23)$$

Using (21)-(23) in (20), we have the following recursive algorithm

$$\begin{aligned} \hat{\theta}(N) = & \hat{\theta}(N-1) - \frac{1}{N-L+1}[V_L^{N\,\prime\prime}(\hat{\theta}(N-1))]^{-1} \\ & \times V_L^{N\,\prime}(\hat{\theta}(N-1)) \end{aligned} \qquad (24)$$

to estimate θ_0.

In order to implement the FD-R algorithm, (24), one has to compute the derivatives $V_L^{N\,\prime}(\theta)$ and $V_L^{N\,\prime\prime}(\theta)$ at each iteration of the algorithm. We discuss these implementation aspects of the FD-R algorithm in the sub-section 5.1.

4.2 *Windowed Frequency Domain Recursive Algorithm (WFD-R)*

In this subsection, we describe a recursive algorithm with $V_L^N(\theta)$, rather than $W^N(\theta)$, as the cost function. We call this algorithm as the "windowed frequency domain recursive algorithm (WFD-R)" to differentiate it from the FD-R algorithm defined in the previous subsection.

Here, we wish to obtain an estimate $\tilde{\theta}(N)$ of θ_0 at each instant $t = Nh$, such that

$$\tilde{\theta}(N) = \arg\min_{\theta} V_L^N(\theta). \qquad (25)$$

The derivation of the WFD-R algorithm is indeed similar to the derivation of the FD-R algorithm presented in the previous subsection.

Let $\tilde{\theta}(N-1)$ be the estimate of θ_0 at the time instant $t = (N-1)h$. Substituting $W^N(\theta)$ and its derivatives by $V_L^N(\theta)$ and its derivatives in (24) leads directly to

$$\begin{aligned} \tilde{\theta}(N) = & \tilde{\theta}(N-1) \\ & - [V_L^{N\,\prime\prime}(\tilde{\theta}(N-1))]^{-1}[V_L^{N\,\prime}(\tilde{\theta}(N-1))] \\ & + o(|\, \tilde{\theta}(N) - \tilde{\theta}(N-1)\, |). \end{aligned} \qquad (26)$$

As in the previous subsection, we assume that $\tilde{\theta}(N)$ lies in a small neighborhood of $\tilde{\theta}(N-1)$, and hence $o(|\, \tilde{\theta}(N) - \tilde{\theta}(N-1)\, |) \approx 0$. This implies

$$\begin{aligned} \tilde{\theta}(N) = & \tilde{\theta}(N-1) - [V_L^{N\,\prime\prime}(\tilde{\theta}(N-1))]^{-1} \\ & \times [V_L^{N\,\prime}(\tilde{\theta}(N-1))]. \end{aligned} \qquad (27)$$

Note that the FD-R algorithm, (24), and the WFD-R algorithm (27), differ only in their respective step sizes. This small but crucial difference makes the FD-R algorithm exhibit better convergence properties than the WFD-R algorithm, which will be demonstrated in the sub-section 5.2.

5. IMPLEMENTATION AND SIMULATIONS

In this section, we discuss the implementation of the FD-R and the WFD-R algorithms. Further, we also a consider a simulation example and test the properties of the proposed algorithms

5.1 *Implementation*

Since the FD-R and the WFD-R algorithms differ only in their step sizes, it is sufficient to discuss the implementation issues involved in only one of them. Therefore, here, we discuss the implementation of the FD-R algorithm alone.

Note that to use the FD-R algorithm, (24), one has to compute the first and second derivatives of $V_L^N(\theta)$, with respect to θ, at each instant $t = Nh$.

From, (16), we know that

$$V_L^N(\theta) = \frac{2}{L}\sum_{k=0}^{\frac{L}{2}} E_L^N(i\omega_k,\theta)E_L^{N\,*}(i\omega_k,\theta), \quad (28)$$

where * denotes the conjugate. Hence,

$$\begin{aligned} V_L^{N\,\prime}(\theta) = \frac{2}{L}\sum_{k=0}^{\frac{L}{2}} \{ & E_L^{N\,\prime}(i\omega_k,\theta)E_L^{N\,*}(i\omega_k,\theta) \\ & + E_L^N(i\omega_k,\theta)E_L^{N\,\prime *}(i\omega_k,\theta)\}. \end{aligned} \quad (29)$$

Moving on to the second derivative $V_L^{N\,\prime\prime}(\theta)$, from (29) we can see that

$$\begin{aligned} V_L^{N\,\prime\prime}(\theta) = & \frac{2}{L}\sum_{k=0}^{\frac{L}{2}} \{E_L^{N\,\prime\prime}(i\omega_k,\theta)E_L^{N\,*}(i\omega_k,\theta) \\ & + E_L^{N\,\prime}(i\omega_k,\theta)E_L^{N\,\prime *}(i\omega_k,\theta)\} \\ & + \frac{2}{L}\sum_{k=0}^{\frac{L}{2}} \{E_L^N(i\omega_k,\theta)E_L^{N\,\prime\prime *}(i\omega_k,\theta) \\ & + E_L^{N\,\prime}(i\omega_k,\theta)E_L^{N\,\prime *}(i\omega_k,\theta)\}. \end{aligned} \quad (30)$$

Note that, the matrix $V_L^{N\,\prime\prime}(\theta)$ has to be positive definite for (24) to converge. Hence, it would be better to force the positive definiteness on $V_L^{N\,\prime\prime}(\theta)$, than to assume it. Hence, we approximate

$$\begin{aligned} V_L^{N\,\prime\prime}(\theta) \approx & \frac{2}{L}\sum_{k=0}^{\frac{L}{2}} \{E_L^{N\,\prime}(i\omega_k,\theta)E_L^{N\,\prime *}(i\omega_k,\theta) \\ & + E_L^{N\,\prime}(i\omega_k,\theta)E_L^{N\,\prime *}(i\omega_k,\theta)\}. \\ = & \frac{2}{L}\sum_{k=0}^{\frac{L}{2}} E_L^{N\,\prime}(i\omega_k,\theta)E_L^{N\,\prime *}(i\omega_k,\theta) \end{aligned} \quad (31)$$

Note that, from (29) and (31), computing $V_L^{N\,\prime}(\theta)$ and $V_L^{N\,\prime\prime}(\theta)$ boils down to computing $E_L^N(i\omega_k,\theta)$ and $E_L^{N\,\prime}(i\omega_k,\theta)$, for each $k = 0, 1, \ldots, \frac{L}{2}$. While $E_L^N(i\omega_k,\theta)$ can be computed directly from (11)-(15), to compute $E_L^{N\,\prime}(i\omega_k,\theta)$ one has to analytically determine $\frac{dG_L^\infty}{d\theta}(i\frac{\omega}{h},\theta)$, and evaluate it at each ω_k. It can be checked that, by differentiating (8),

$$\begin{aligned} \frac{dG_1^\infty}{d\alpha}(s,\theta) = & \frac{1}{2\kappa}\left(\frac{\tanh(d\sqrt{\frac{s}{\alpha}})}{\sqrt{s\alpha}}\right. \\ & \left. - \frac{d}{\alpha}\frac{1}{\cosh^2(d\sqrt{\frac{s}{\alpha}})}\right) \end{aligned} \quad (32)$$

$$\frac{dG_1^\infty}{d\kappa}(s,\theta) = -\frac{G_1^\infty(s,\theta)}{\kappa} \quad (33)$$

$$\frac{dG_2^\infty}{d\alpha}(s,\theta) = \frac{d}{2}\frac{\sinh(d\sqrt{\frac{s}{\alpha}})}{\cosh^2(d\sqrt{\frac{s}{\alpha}})}\sqrt{\frac{s}{\alpha^3}} \quad (34)$$

$$\frac{dG_2^\infty}{d\kappa}(s,\theta) = 0. \quad (35)$$

Further, note that at any given instant one has to store the last L values of the data to compute (12) and (14), thereby incurring a storage cost. The computational complexity involved in computing the DFT's (12) and (14) can be considerably reduced if one notices the following recursions inherently present in them,

$$\begin{aligned} Y_L^{N+1}(e^{i\omega_k}) = & [Y_L^N(e^{i\omega_k}) - y^d(N-L)]e^{i\omega_k} \\ & + y^d(N)e^{-i(L-1)\omega_k} \end{aligned} \quad (36)$$

and

$$\begin{aligned} U_L^{N+1}(e^{i\omega_k}) = & [U_L^N(e^{i\omega_k}) - u^d(N-L)]e^{i\omega_k} \\ & + u^d(N)e^{-i(L-1)\omega_k}. \end{aligned} \quad (37)$$

5.2 *Simulation example*

Assume that the given heat diffusion system, (1)-(4), has parameters $\alpha_0 = 2$ m/s^2 and $\kappa_0 = 1$ J/msK. We first pass two independent white noise sequences $\{v(n)\}_{n=0}^{N-1}$ and $\{w(n)\}_{n=0}^{N-1}$, of variance one each, through the low-pass filter

$$H(q^{-1}) = \frac{1}{1 - 1.88q^{-1} + 0.9732q^{-2}}, \quad (38)$$

which is of finite bandwidth and has a resonance peak. The resulting filter outputs $H(q^{-1})v(n)$ and $H(q^{-1})w(n)$ are then further added with white noise sequences $\{v_2(n)\}_{n=0}^{N-1}$ and $\{w_2(n)\}_{n=0}^{N-1}$, each of variance 0.8, respectively, and the net signals $H(q^{-1})v(n)+v_1(n)$ and $H(q^{-1})w(n)+w_1(n)$ are then chosen as the inputs $\{q_i(nh)\}_{n=0}^{N-1}$ and $\{T_e(nh)\}_{n=0}^{N-1}$, respectively. The frequency contents of the above generated signals though have a large magnitude within the bandwidth of the filter (38), and they are also reasonably frequency rich in the high frequency regions.

The system output, $\{T_i(nh)\}_{n=0}^{N-1}$, is generated by multiplying the DFT of the generated inputs, with the transfer-function $G^{\infty}(i\frac{\omega}{h}, \theta_0)$, at the frequency points $\omega_k = \frac{2\pi k}{N}$, $k = 0, 1, 2, \ldots, N-1$, and then taking the inverse-DFT of the product.

In Figure 2, we have plotted the parameter estimates, averaged over 50 runs, obtained by using the the FD-R algorithm over the generated data with the window lengths $L = 100, 200$ and 300. As L increases the averaged estimates get closer to $\theta_0 = (2, 1)$ asymptotically.

In Figures 3 and 4, we have plotted the averaged parameter estimates obtained by using the WFD-R algorithm, for window lengths $L = 500$ and 700 respectively. Note that the averaged parameter estimates oscillate violently around θ_0, and an increase in L makes the oscillations less frequent.

A common strategy to remove the large variations in a given data set, is to filter the data using a low-pass filter. Here, in order to remove the large spikes in averaged WFD-R estimates, see Figures 3 and 4, we pass them through the low-pass filter

$$H(q^{-1}; \beta) = \frac{\beta}{1 - (1-\beta)q^{-1}}, \qquad (39)$$

where $0 < \beta < 1$. The resulting filtered WFD-R estimates are plotted, along with un-filtered WFD-R estimates, in Figures 5 and 6, for $L = 500$ and $\beta = 0.1, 0.01$ and 0.001. Note that the case $\beta = 1$ denotes the unfiltered WFD-R estimates.

From the Figures 5 and 6, it can be observed that the filtered and the un-filtered estimates do not differ much for a large β, and as $\beta \to 0$ the oscillations of the averaged WFD-R estimates are filtered out and one gets a smoother convergence.

6. CONCLUSIONS

In this paper, we have proposed two recursive algorithms in the frequency domain, called FD-R and WFD-R, to estimate the parameter $\theta_0 = (\alpha_0\ ,\ \kappa_0)^T$. Even though the algorithms have been

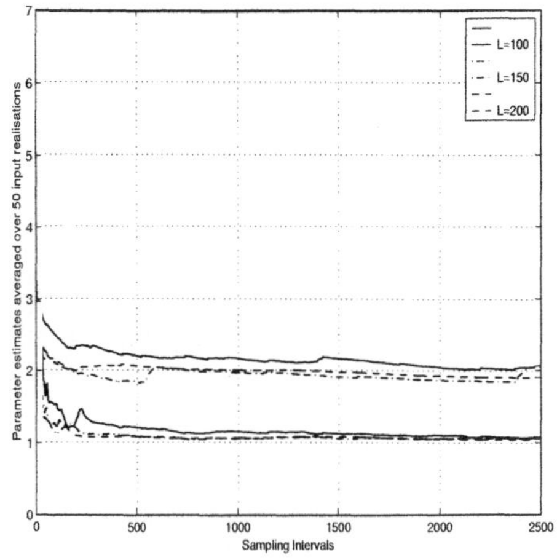

Fig. 2. Convergence of the the averaged parameter estimates to $\alpha_0 = 2$ m/s^2 and $\kappa_0 = 1$ J/msK, for different values of the window length L, $L = 100, 150, 200$.

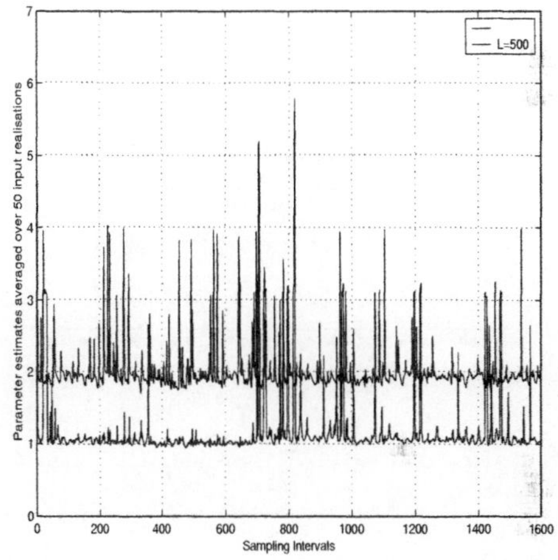

Fig. 3. Convergence of the averaged parameter estimates obtained by using WFD-R algorithm, to $\alpha_0 = 2$ m/s^2 and $\kappa_0 = 1$ J/msK respectively, for a window of length $L = 500$.

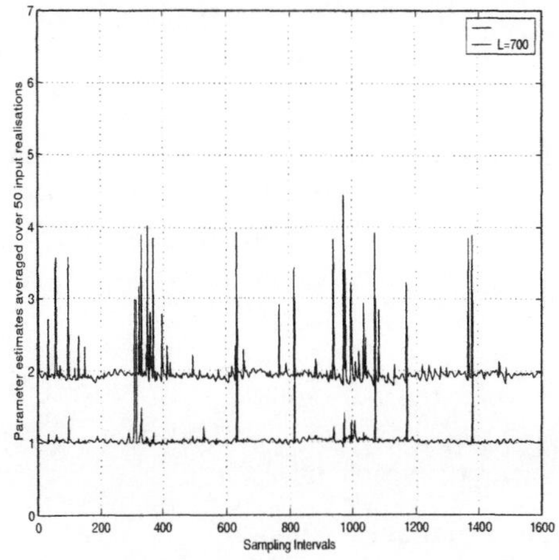

Fig. 4. Convergence of the parameter estimates obtained by using WFD-R algorithm, to $\alpha_0 = 2$ m/s^2 and $\kappa_0 = 1$ J/msK, for a window of length $L = 700$.

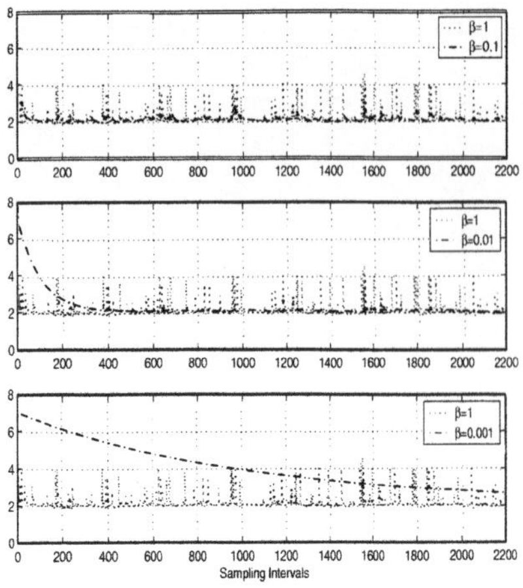

Fig. 5. Filtered WFD-R parameter estimates of $\alpha_0 = 2$ m/s^2 obtained by passing the WFD-R estimates, with and $L = 500$ thorough the filter $H(q^{-1};\beta)$, see (39), for different values of the filter parameter β.

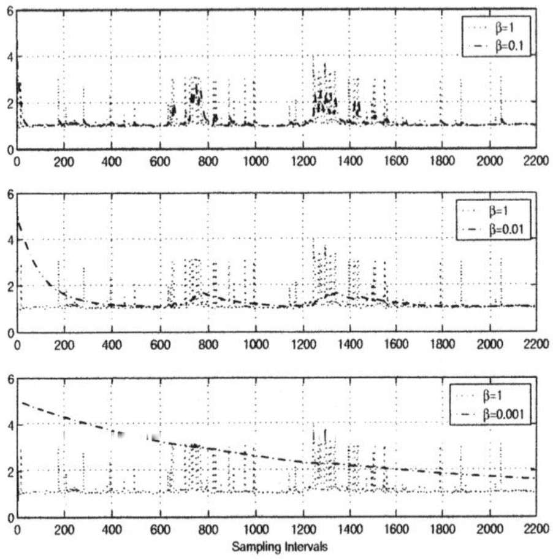

Fig. 6. Filtered WFD-R parameter estimates of $\kappa_0 = 1$ J/msk obtained by passing the WFD-R estimates, with and $L = 500$ thorough the filter $H(q^{-1};\beta)$, see (39), for different values of the filter parameter β.

devised in the frequency domain, the ordering of the data is in the time domain.

In the construction of the algorithms, we have exploited the facts that, the given system has a very simple and easily computable expression for the transfer function, (8), and over a sliding window one can compute the DFT's of the system inputs and output recursively. The recursive computation of the DFT's not only reduces the computational burden of the FD-R and the WFD-R algorithms, they also make the algorithms elegant and easy to implement.

From the simulation example presented in section 5, we can see that the FD-R algorithm converges quite rapidly to the true parameter θ_0 for all window lengths $L \geq 200$.

In the case of the WFD-R algorithm, the parameter estimates oscillate wildly around the true parameter θ_0, even for an L as large as $L = 700$. These oscillations were curtailed by passing the parameter estimates thorough a low-pass filter of the form (39). It was found that, if one chooses the filter parameter β, see (39), to be very small, say of the order of 10^{-2}, then the oscillations of the parameter estimates obtained by using the WFD-R are curtailed to a large extent, but at the expense of decreasing the rate of convergence of the parameter estimates.

Note that due to the finiteness of L, the estimates of the FD-R and the WFD-R algorithm will not converge to the true parameter θ_0 asymptotically. Convergence is possible only if $L \to \infty$. A future research problem would be to characterise the " asymptotic bias" due to a finite L.

In a related paper, see Bhikkaji and Söderström (2002), the parameters α and κ were estimated recursively in the time domain using the standard RPEM algorithm, from a Chebyshev Collocation approximation of the PDE (1)-(3). The simulations therein suggests that the time domain recursive algorithm (TD-R) has a slower but smoother convergence than the FD-R and the WFD-R algorithms. The TD-R estimates were found to have a small but noticeable bias even for a large order Chebyshev Collocation approximation of the PDE. Using a Chebyshev Collocation approximation of a very high order also leads to numerical instabilities.

References

Bhikkaji, B. and T. Söderström (2002). Recursive algorithm for estimating the parameters in one dimensional heat diffusion system. *Reglermöte.* Linköping, Sweden, 20-30 May.

Ljung, L. and T. Söderström (1983). *Theory and Practice of Recursive Identification.* The MIT Press. Cambridge, Massachusetts.

Söderström, T. and B. Bhikkaji (2000). Reduced Order Models for Diffusion Systems via Collocation Methods.. *Proc of 12th IFAC Symposium on System Identification.* Santa Barbara, CA, 21-23 June.

Söderström, T. and P. Stoica (1989). *System Identification.* Prentice Hall International. Hemel Hempstead, UK.

Söderström, T. and S. Remle (2000). Parameter Estimation for Diffusion Models. *Proc of 12th IFAC Symposium on System Identification.* Santa Barbara, CA, 21-23 June.

IFAC Publications
www.elsevier.com/locate/ifac

REGULARIZATION METHOD IN INFRARED IMAGE PROCESSING

S. Datcu*, L. Ibos*, Y. Candau*, S. Matteï, N. Ramdani***

(*) CERTES, Université Paris XII-Val de Marne, Créteil, France
(**) LTM, IUT du Creusot, Université de Bourgogne, Le Creusot, France

Abstract: Infrared images often present distortions induced by the measurement system. Thus, image processing is a vital part of infrared measurements. A distortion model based on a convolution product is presented. Image restoration is an ill-posed problem and its solution can be obtained using regularization methods. In this paper, image restoration is performed using a variation of Tikhonov regularization that makes use of the particular form of the convolution kernel matrix, which is built as a block-circulant matrix that admits a diagonal form in the two-dimensional Fourier space. The restoration procedure is used to restore a knife-edge infrared source image.

Keywords: regularization, deconvolution, image restoration, Fourier transforms, temperature measurement

1.INTRODUCTION

Thermography is a useful contactless way of quantifying the heat flux emitted by a surface. The first step consists of obtaining the surface temperature of the thermal scene. Generally, infrared images often present distortions that prevent from obtaining accurate values for the spatial temperature field of a target thermal scene. If the transfer function of the optical system (focal plan array infrared camera) that mathematically describes the distortions is known then the restoration of the infrared images is possible. In order to perform infrared image restoration, a deterministic approach based on Tikhonov regularization is used related to the use of Fourier transform to solving resulting big size block-circulant matrix (76000 x 76000 equations).

The present work is structured as follows: the physical background and the camera model are presented in the section 2. Then, the infrared image restoration procedure is developed in section 3. Section 4 is devoted to the description of the experimental set-up and, finally, the results obtained are presented in section 5.

2.IMAGE FORMATION THEORY

2.1.Physical background

The approach presented here is to study the link between all surface elements of an object M observed by an infrared camera and their images in the image plane (i.e. the detector matrix). The luminance of each surface element M_0 of co-ordinates (x_o, y_o) can be represented in the object plane by a two-dimensional Dirac distribution $\delta(x_0 - x, y_0 - y)$. The image of the element M_0 through an optical system is a blur image located around the corresponding elements $M'(x', y')$ in the image plane as shown in figure 1.

The irradiance distribution of this image can be represented by the impulse response h, also called the point-spread function (PSF). This function is often considered as position invariant in a limited area of the image plane (Frieden, 2001; Boreman, 2001; Gaussorgues, 1981; Papini, and Gallet, 1990).

Considering an object M with a luminance distribution $L_o(x, y)$, in Gaussian optics, the distribution of the irradiance $E(x',y')$ in the image plane (i.e. the detector matrix plane) is given by the convolution product between L_o and the impulse response h (Boreman, 2001; Gaussorgues, 1981; Papini, and Gallet, 1990):

$$E(x',y') = L_o(x,y) * h(x',y') \tag{1}$$

In real case, the optical system of the camera induces non-linear distortions of the measured signal, the luminance distribution on the thermal scene, which cannot be described by a convolution product. Nevertheless, the camera field of view (FOV) is split in several regions in order to make linear the optical distortion effects and to use a convolution model for the image formation.

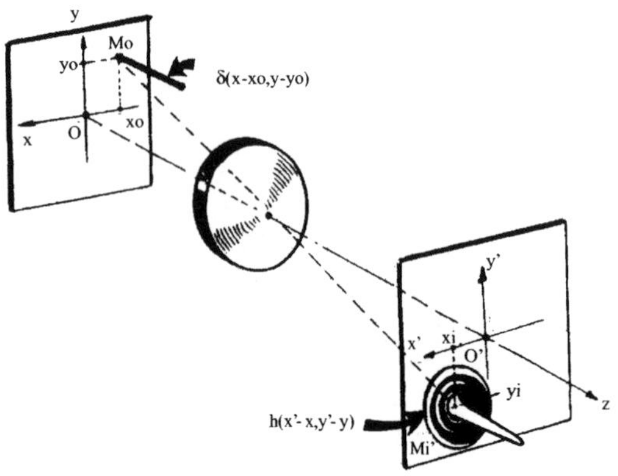

Fig. 1. Image Formation Physics

2.2.Infrared camera model

The spectrum of the impulse response $h(x',y')$, $OTF(\xi,\eta) = \Im\Im(h(x',y'))$ is the transfer function of the infrared viewing system; $\Im\Im$ is the two-dimensional Fourier transform operator and ξ,η are spatial frequencies in the Fourier's space. We assume that $OTF(\xi,\eta)$ has been normalized to have unit value at zero spatial frequency. This normalization yields a relative transmittance for the various frequencies and ignores attenuation factors that are independent of spatial frequency. With this normalization, $OTF(\xi,\eta)$ is referred the optical transfer function which is generally a complex function having both a magnitude (the modulation transfer function – MTF) and a phase (phase transfer function – PTF) portion:

$$\begin{aligned} OTF(\xi,\eta) &= |OTF(\xi,\eta)| \cdot e^{(-j\cdot\theta(\xi,\eta))} = \\ &= MTF(\xi,\eta)\cdot e^{(-j\cdot PTF(\xi,\eta))} \end{aligned} \tag{2}$$

In order to calculate the local camera transfer function, the response of the camera to a knife-edge infrared source is measured on each of the regions obtained by camera FOV split. A mean infrared image of the source is calculated to minimize the camera response noise. The experimental set-up used is presented in details in section 4.

The spatial first derivative of this edge spread function allows obtaining the line spread function which is the spatial integral of the impulsion response of the camera $h(x',y')$ on a second direction, orthogonal to the measurement direction. The magnitude of the Fourier transform of the line spread function allows accessing the one spatial-frequency component of the transfer function (Boreman, 2001):

$$MTF(\xi,0) = |\Im(LSF(x))| \tag{3}$$

Then the local camera transfer function $OTF(\xi,\eta)$ is obtained by parametric identification from the experimental MTF on the assumption of isotropy on the two orthogonal directions ξ,η of the Fourier space (Datcu, *et al.*, 2002). The local impulsion response is obtained by taking the two-dimensional spatial Fourier transform of the transfer function.

3.INFRARED IMAGE RESTORATION

Let us consider the degradation model given in equation 4:

$$g = Hf + n \tag{4}$$

where g is the measured signal, f is the real signal, H is the convolution kernel and n is a Gaussian additive noise. This discrete form of the degradation is obtained from the equation 1 by taking into account the camera noise.

The objective of image restoration is to estimate the original image f from a degraded image g and some knowledge of H and n. Central to the algebraic approach is the concept of seeking an estimate of f, denoted $\hat{f}$, that minimizes a predefined criterion of performance.

3.1.Tikhonov regularization

We consider the least square restoration problem as one of minimizing functions of the form $\|Q\hat{f}\|^2$, where Q is a linear operator on f, subject to the constraint $\|g - H\hat{f}\|^2 = \|n\|^2$. This approach introduces considerable flexibility in the restoration process because it yields different solutions for different choices of Q (Gonzales, *et al.*, 1993). The addition of an equality constraint in the minimization problem can be handled without difficulty by using the Lagrange multipliers method. Thus, we seek an estimation $\hat{f}$ that minimizes the objective function:

$$J(\hat{f}) = \left\|Q\hat{f}\right\|^2 + \alpha \cdot \left(\left\|g - H\hat{f}\right\|^2\right) \tag{5}$$

where α is the Lagrange multiplier. The estimate $\hat{f}$ is obtained by seeking a minimum of the objective function, which lead to:

$$\hat{f} = \left(H^T H + 1/\alpha \cdot Q^T Q\right)^{-1} H^T g \tag{6}$$

which is basically a Tikhonov regularization with $\lambda = 1/\alpha$ the regularization parameter.

The value of α must be adjusted so that the constraint is satisfied. This restoration procedure is optimal for each image and requires only knowledge of the noise mean and variance (Gonzales, *et al.*, 1993; Hansen, *et al.*, 2000). The solution also depends on the choice of the matrix Q. Owing to ill conditioning, direct deconvolution lead to solutions that are obscured by large oscillating values. Thus the matrix Q (equation 5) must minimize these adverse effects. One possibility is to formulate a criterion of optimality based on a measure of smoothness suitable for minimizing the second derivative of $\hat{f}$ (Gonzales, *et al.*, 1993). In the discrete case, the second derivative can be approximated using the discrete form of the Laplacian (which is identified to Q matrix):

$$l = \begin{bmatrix} 0 & -1 & 0 \\ -1 & 4 & -1 \\ 0 & -1 & 0 \end{bmatrix} \tag{7}$$

Then, the solution $\hat{f}$ is given by:

$$\begin{aligned} &\hat{f} = \left(H^T H + 1/\alpha \cdot L^T L\right)^{-1} H^T g \\ &\hat{f} = \left(\hat{f}_{ij}\right)_{1\times MN},\ g = \left(g_{ij}\right)_{1\times MN}, \\ &H = \left(H_{ij}\right)_{MN\times MN},\ L = \left(L_{ij}\right)_{MN\times MN} \end{aligned} \tag{8}$$

$\hat{f}$ and g are colon vectors obtained by colon concatenation of the matrix representing the restored and the degraded images. Basically, we have to deal with a circular convolution product and thus the entire operator's matrix was zero padded from its initial size to the $M \times N$ size in respect to the size of the convolution criteria.

3.2.Diagonal form of the deconvolution kernel

In the last equation, H is a block-circulant matrix derived from the discrete form of the convolution kernel h (equation 1) and L is a block-circulant matrix either, derived from the Laplacian kernel l (equation 7).

These considerations are important in order to derive a method for solving a large linear system of equations. A block-circulant matrix takes a diagonal form in the Fourier space (Gonzales, *et al.*, 1993). Consider the following orthonormal block-circulant matrix:

$$\begin{aligned} &W = w^x \otimes w^y,\quad W = \left(W_{ij}\right)_{MN\times MN} \\ &w^x(i,m) = e^{-j\cdot 2\cdot \pi/M \cdot i\cdot m},\ w^x = \left(w^x_{im}\right)_{M\times M} \\ &w^y(k,n) = e^{-j\cdot 2\cdot \pi/N \cdot i\cdot n},\ w^x = \left(w^x_{im}\right)_{M\times M} \end{aligned} \tag{9}$$

where $\otimes$ is a tensorial product. The matrix W has an inverse matrix W^{-1} that represents the discrete form of the two-dimensional space Fourier transform operator. It has been proved that if H is a block-circulant matrix, then it can be factorized as follows:

$$H = WDW^{-1} \tag{10}$$

where D is a diagonal matrix that contains the eigenvalues of H.

Using equation 10, system 8 can be written as:

$$\begin{aligned} &\hat{f} = \left(WD^* DW^{-1} + 1/\alpha \cdot WE^* EW^{-1}\right)^{-1} WD^* W^{-1} g, \\ &H = WDW^{-1},\ L = WEW^{-1} \end{aligned} \tag{11}$$

where D and E are the diagonal form of the matrix H and L. The operator * represents the complex conjugate.

Multiplying both sides by W^{-1} reduces system 11 to:

$$W^{-1}\hat{f} = \left(D^* D + 1/\alpha \cdot E^* E\right)^{-1} D^* W^{-1} g \tag{12}$$

which can be written as follows:

$$\begin{aligned} &\hat{F}(\xi,\eta) = \left(\frac{H^*(\xi,\eta)}{\left|H(\xi,\eta)\right|^2 + 1/\alpha \cdot \left|L(\xi,\eta)\right|^2}\right) \cdot G(\xi,\eta), \\ &\hat{F}(\xi,\eta) = \Im\Im\left(\hat{f}_{ij}\right),\ H(\xi,\eta) = \Im\Im\left(H_{ij}\right), \\ &L(\xi,\eta) = \Im\Im\left(L_{ij}\right),\ G(\xi,\eta) = \Im\Im\left(g_{ij}\right) \end{aligned} \tag{13}$$

and give us a feasible method to solve a large linear system of equations.

3.3.Regularization parameter calculation

The value of the Lagrange multiplier α is estimated by an iterative procedure tied to the noise norm. If we define a residual vector r as:

$$r = g - H\hat{f} \qquad (14)$$

From equation 8 we obtain:

$$r = g - H\left(H^T H + 1/\alpha \cdot L^T L\right)^{-1} H^T g \qquad (15)$$

and it has been proved that function $\varphi(1/\alpha) = \|r\|^2$ is a monotonically increasing function of $1/\alpha$ (Gonzales, *et al.*, 1993), which simplifies the search of an optimal value for the Lagrange multiplier in order to verify the minimization constraint:

$$\|r\|^2 = \|n\|^2 \pm a \qquad (16)$$

derived from the minimization constraint $\left\|g - H \cdot \hat{f}\right\|^2 = \|n\|^2$. a is an accuracy factor. The norm of the noise $\|n\|$ is derived from the definition of the noise variance:

$$\sigma_n^2 = E\left((n(x,y) - \mu_n)^2\right) = \|n\|^2 - \mu_n^2, \quad \|n\|^2 = \mu_n^2 + \sigma_n^2 \qquad (17)$$

where σ_n^2 is the noise variance, μ_n is the noise mean value and $E(\cdot)$ is the expectation operator.

4.EXPERIMENTAL SET-UP

Some of the major technical specifications of the camera used in this study (model AGEMA 570 Elite from FLIR Systems™) are listed in table 1. The experimental set-up is schematically presented in figure 2.

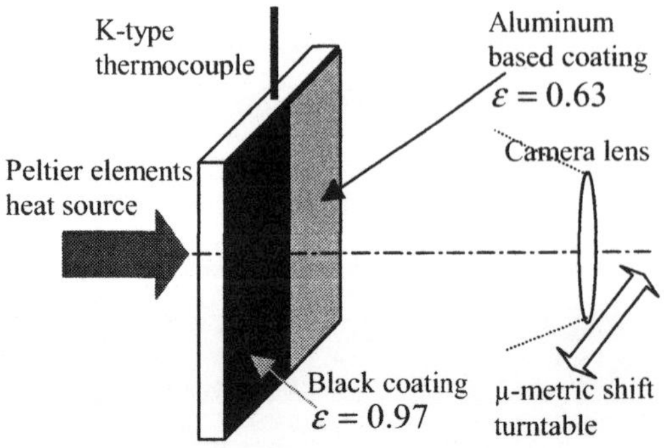

Fig. 2. Experimental set-up for transfer function camera measurement

The target scene is made of a 44 x 44 x 10 mm duralumin plate. In order to create an infrared knife-edge source, the viewed face of the plate was split into two equal vertical surfaces. One surface was coated with a black paint with 0.97 emissivity (Tang-Kwor, 1998), while the other one was coated with aluminum paint with 0.63 emissivity (Tang-Kwor, 1998). The high optical diffusivity of the used coatings prevents any specular reflections of infrared external source. The Duralumin plate is heated by thermoelectric elements, which allow keeping a stable plate temperature during measurements. The temperature of the plate is controlled by a K-type thermocouple. A fuzzy logic controller programmed under LabView™ regulates the heat flux supplied by the thermoelectric elements. The measurement of the transfer function of the camera was made for a fixed distance of 54.5 cm between the camera lens and the target scene. The acquisition of thermograms was made at a rate of about 7 images/second. The object parameters imposed to the camera during acquisition are listed in table 2. The camera was placed on a turntable, which allowed micrometric step translations with a total length of 2.5cm. Step accuracy is 10µm.

Table 1. Technical specifications of the Agema 570 camera from FLIR™.

Property	Value
Field of view	45° x 34°
IFOV	2.45 mrad
Spectral range	7.5 – 13 µm
Accuracy	±2% of range or ±2°C
Image size	320 x 240 pixels

Table 2. Object parameters used for infrared images acquisition.

Object parameter	Value
Relative humidity	50 %
Distance camera-object	54.5 cm
Ambient temperature	20 °C
Atmospheric temperature	20 °C
Emissivity ε	1
Atmospheric transmission factor τ	1

5.RESULTS

5.1.Calculation of the Infrared Camera Transfer Function

Seven ESF were measured on seven positions of the knife-edge source on the image plane. A thermogram obtained when the knife-edge source is placed in the center of the Field Of View (FOV) of the camera is presented on figure 3. For each position, the corresponding ESF was measured at about 20 points in order to obtain an accurate slope sampling of the ESF data. The edge spread and line spread functions obtained in the center of the field of view of the camera is presented on figure 4. The edge spread and line spread functions obtained when the knife-edge source is placed at 120 pixels from the center of the field of view of the camera is presented on figure 5. Two of the response functions (PSFs) are presented in figures 6 and 7.

Fig. 3. Thermogram of the knife-edge source obtained at the center of FOV.

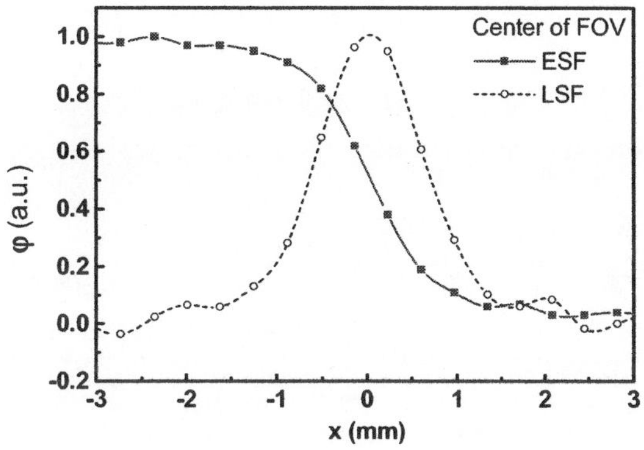

Fig. 4. Measured ESF and LSF in the center of the FOV.

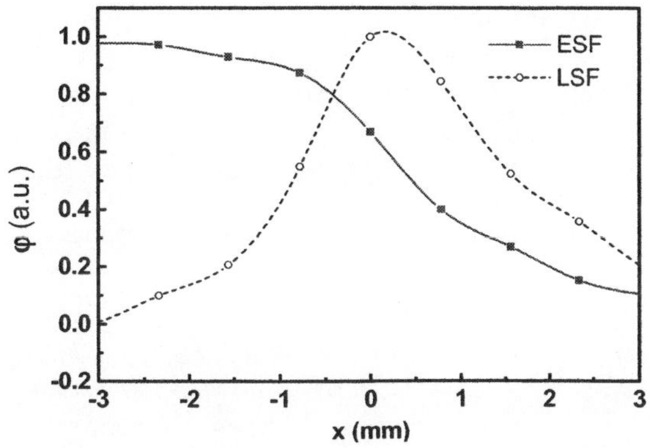

Fig. 5. Measured ESF and LSF at 120 pixels from the center of the FOV.

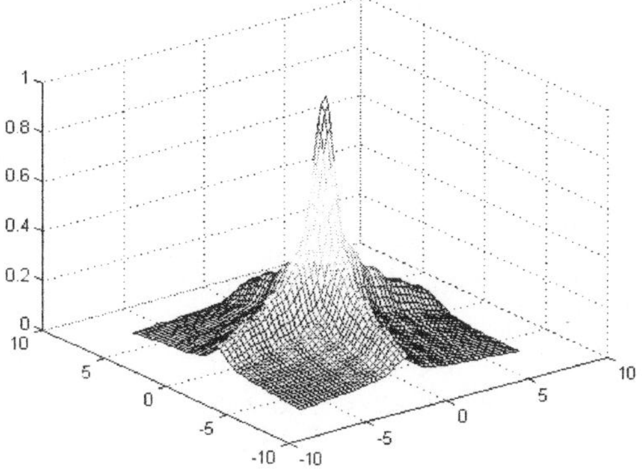

Fig. 6. PSF in the FOV center.

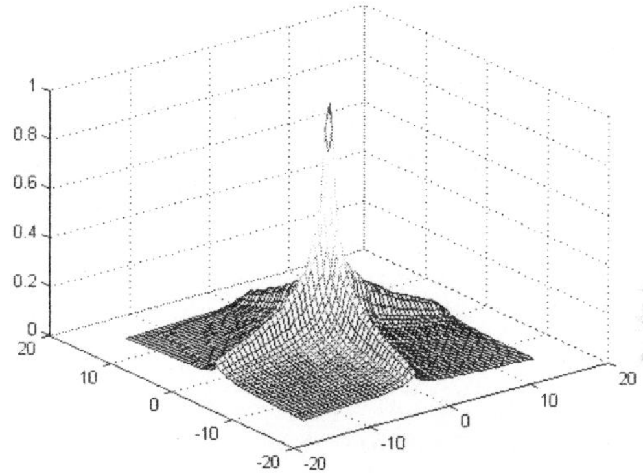

Fig. 7. PSF at 120 pixels from the FOV center.

5.2.Image restoration

Using the seven kernels obtained as described before, we have tried to locally restore the image of the knife-edge source. The system of equations 13 was used to reach the restored image. The deconvolution procedure was not performed on the entire image but on a local area that encloses the source image; for example, to restore the source image located at 120 pixels from the center of the camera FOV, we processed the area inside the dashed line rectangle on figure 8.

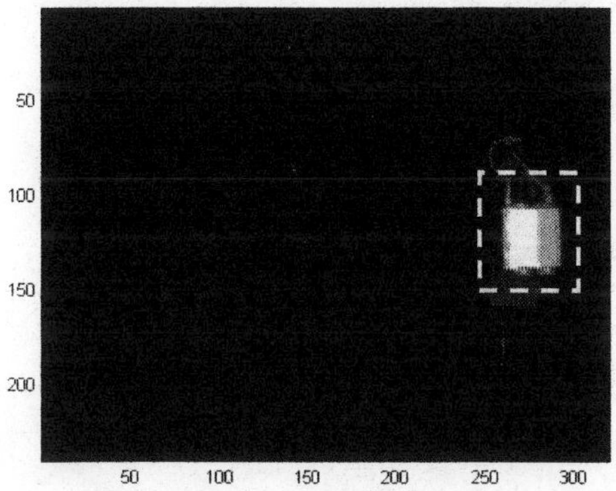

Fig. 8. Original thermogram of the infrared source located at 120 pixels from the FOV center.

Before processing the deconvolution, the work area was resampled in order to correlate the kernel sampling size (the local PSF) with the image sampling size. Obviously, the resampling factor is given by the ratio between the footprint size of the detector and the sampling interval size used to measure the corresponding ESF. The noise norm was calculated from about 100 thermograms. For each thermogram, a noise image was calculated by the difference between the thermogram and an average value (obtained by averaging the 100 thermograms). Mean noise was close to zero (about 10^{-3} aul, arbitrary units of luminance), and noise variance was about 1.7 aul^2. Thus, the noise norm was calculated using equation 17 and is given in table 3. The value of the regularization factor α was calculated in an iterative manner in order to come as close as possible to the noise norm with the residual norm (equation 16). The distorted and restored images are given in figure 9, represented using a gray palette with the following limits: [870; 1430] aul. These limit values were chosen in order to increase contrast in the images and to make the differences between the degraded and restored source images more visible. The luminance profiles obtained along the horizontal lines indicated on figure 9 before and after restoration are plotted on figure 10.

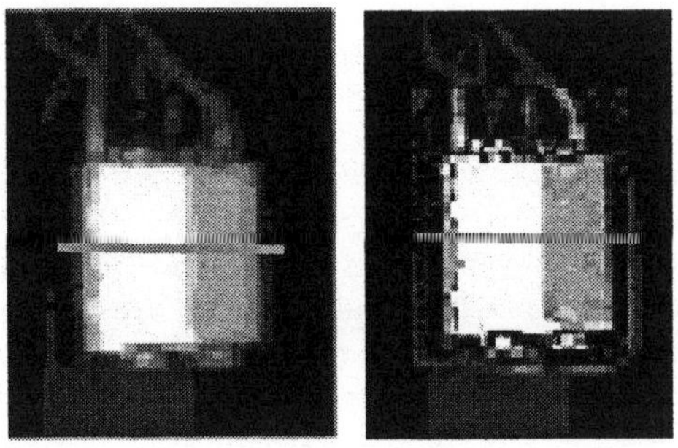

Fig. 9. Zoom view of the infrared source before restoration (left image) and of the restored image of the infrared source (right image).

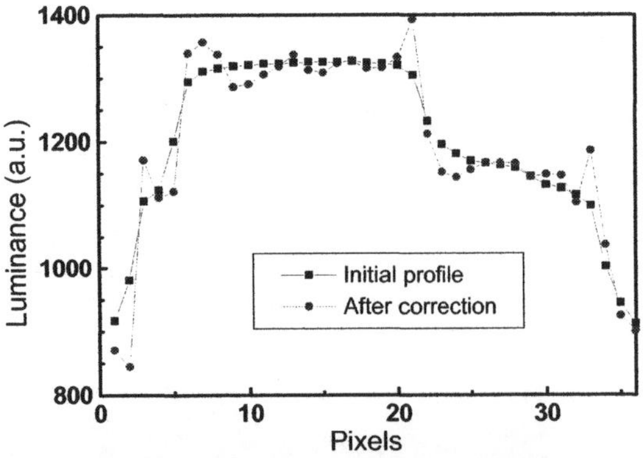

Fig. 10. Horizontal profile on the degraded initial image and on the restored image where the knife-edge source image is located at 120 pixels from the FOV center.

Table 3. Regularisation parameters for image restoration at 120 pixels from the FOV center.

$\|n\|^2$	$\|r\|^2$	α
1.7263	2.035	$5x10^{-3}$

CONCLUSION

The deconvolution method allowed to locally restoring a limited area of the image. It permitted to "de-blur" the original image, which improved the corresponding thermal field (on the restored image). The variations on the restored profile (figure 10) are due to the ill-posed nature of the deconvolution problem and of use of Fourier transform. The study of other methods can be used to perform deconvolution is one of the objectives for future works.

ACKNOWLEDGMENTS

This work was supported by the R&D Department of Electricité de France and we would like to thank Jean-Claude Frichet for his financial and technical support.

REFERENCES

Boreman, G.D. (2001). *Modulation Transfer Function in Optical and Electro-Optical Systems*, SPIE, Washington.

Datcu, S., L. Ibos, Y. Candau and S. Mattéï (2002). *On the FPA infrared camera transfer function calculation*. Proceedings of the 6th Quantitative InfraRed Thermography (QIRT'02) Conference, University of Zagreb, In press.

Frieden, B.R. (2001). *Probability, statistical optics, and data testing*. 3rd Edition, Springer, Berlin.

Gaussorgues, G. (1981). *La thermographie infrarouge*. Technique & documentation, Paris.

Gonzales, R.C. and R.E. Woods (1993). *Digital Image Processing*. Addison-Wesley, USA.

Hansen, P.C., B.H. Jacobsen and K. Mosegaard (2000). *Methods and Applications of Inversion*. Lecture Notes in Earth Sciences 92, Springer, Berlin.

Papini, F. and P. Gallet (1990). *Thermographie infrarouge*. Masson, Paris.

Tang-Kwor, E. (1998). *Contribution au développement de méthodes périodiques de mesure de propriétés thermophysiques des matériaux opaques*. Thesis, University Paris 12 Val de Marne.

IFAC Publications
www.elsevier.com/locate/ifac

FILTERING OF STOCHASTIC VOLATILITY MODEL

ShinIchi Aihara [1] Arunabha Bagchi **

* *Department of Mechanics and Systems Design, Tokyo University of Science, Suwa , 5000-1 Toyohira, Chino, Nagano, Japan*
E-mail:aihara@rs.suwa.tus.ac.jp
** *FELab and Department of Applied Mathematics, University of Twente, P.O.Box 217, 7500AE Enschede, The Netherlands*
E-mail:bagchi@math.utwente.nl

Abstract:
We study the filtering problem for the stochastic volatility model of Heston by using the nonlinear estimation theory. To solve the estimation problem for the stochastic volatility process, we use the random time change method. The derived basic equation for the filtering is the so-called Zakai equation and its numerically realized algorithm is proposed with the aid of the splitting-up method. Some numerical simulation studies are demonstrated to show the advantage of the proposed method. *Copyright © 2003 IFAC*

Keywords: Filtering, Heston's model, Stochastic Volatility, Zakai equation, Splitting up method

1. INTRODUCTION

The problem of nonlinear filtering plays an important role in stochastic control with many applications (see Bensoussan for details). As stated in (BENSOUSSAN, 1992, p.111) we have still an open problem that the observation mechanism is given by

$$dy_t = h(x_t)dt + \sigma(x_t)dw \qquad (1.1)$$

i.e., σ depends on the signal process x_t. For the simple scalar case, it is easy to show that

$$y_t^2 - 2\int_0^t y_s dy_s = \int_0^t \sigma^2(x_s)ds \ a.s. \qquad (1.2)$$

Hence if x_t is continuous in t a.s.,we have

$$\sigma^2(x_t) = \frac{d}{dt}(y_t^2 - 2\int_0^t y_s dy_s). \qquad (1.3)$$

[1] This work was partially supported by the MEXT, Grants-in-Aid for Scientific Research 14550456(c)

In practice we can realize (1.2) as

$$\sum |y_{t_{i+1}^{(n)}} - y_{t_i^{(n)}}|^2 \rightarrow \int_0^t \sigma^2(x_s)ds \text{ as } n \rightarrow \infty \text{ in probability} \qquad (1.4)$$

Hence the case when σ depends on x_t is out of the usual nonlinear filtering category.

Noting that the formula (1.4) is not numerically robust, we propose the alternative filtering equation by using the random-time change and the non linear filtering theory.

This paper is organized as follows. In Section II, we introduce the Heston model (HESTON, 1993) for a stock price and present a motivating numerical example to explain the idea of the introduction of random-time change. The Zakai equation for our setting is derived and the numerical implementation technique is proposed in Section III. Some numerical results are demonstrated in Section V.

2. MOTIVATING DISCUSSION

First we consider the following simple Heston model (HESTON, 1993) where a stock S_t is governed by

$$dS_t = \mu S_t dt + \sqrt{\sigma_t} S_t dW_t$$
$$d\sigma_t = \alpha(m - \sigma_t)dt + v\sqrt{\sigma_t}\{\rho dW_t' + \sqrt{1-\rho^2} dW_t\}$$

where W_t and W_t' are mutually independent Brownian motion processes. The stock price S_t is usually observable but the volatility σ_t is not. So we need to estimate σ_t from the observed process S_t. Now we define

$$Y_t = \log \frac{S_t}{S_0}.$$

Hence, by using Ito's lemma, we have

$$Y_t = \int_0^t (\mu - \frac{1}{2}\sigma_s)ds + \int_0^t \sqrt{\sigma_s} dW_s \qquad (2.1)$$

It is easy to show that

$$Y_t^2 - 2\int_0^t Y_s dY_s = \int_0^t \sigma_s ds \qquad (2.2)$$

Theoretically it follows from (2.2) that

$$\frac{d}{dt}\{Y_t^2 - 2\int_0^t Y_s dY_s\} = \sigma_t \text{ a.s,} \qquad (2.3)$$

because σ_t is continuous w.p.1. However as was explained in (AIHARA and BAGCHI, 2000), the algorithm (2.3) is only of theoretical interest and can not be applied in a practical situation. However (2.2) seems to be useful to obtain the $\int_0^{\cdot} \sigma_s^2 ds$-process. Here we present a simulation study to explain the point.

Example 2.1. We set

$$\alpha = 0.01, \beta = 1, \rho = -0.1, \mu = 0.3$$

and

$$dt \sim 10^{-4}$$

By using the usual finite difference method with respect to t, we simulate (2.2). From Figs.1 and 2, we find that the algorithm (2.2) seems to work well. However (2.3) does not work as is apparent from Fig.3.

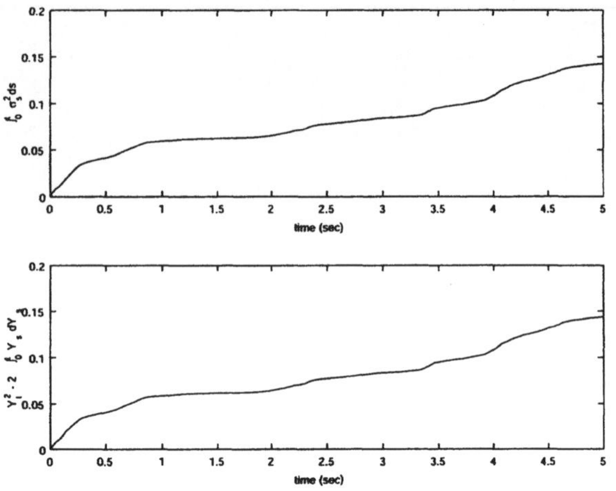

Fig. 1. True and estimated $\int_0 \sigma_s^2 ds$

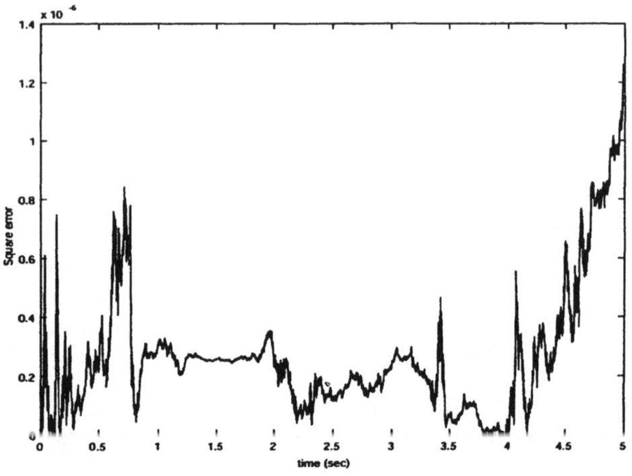

Fig. 2. Error of estimates

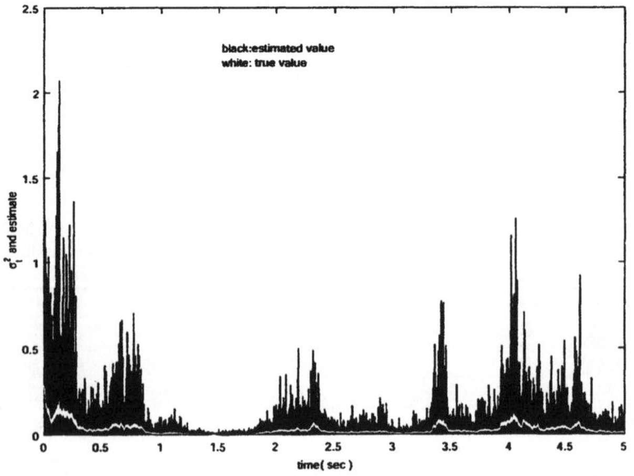

Fig. 3. Estimate from (2.3)

Now let ϕ_t be the inverse function of

$$t \rightarrow \int_0^t \sigma_s ds.$$

By using this random time change, (2.1) becomes

$$\tilde{Y}_t = Y_{\phi_t} = \int_0^{\phi_t} (a - \frac{1}{2}\sigma_s)ds + \int_0^{\phi_t} \sqrt{\sigma_s} dW_s$$

$$= \int_0^t (\frac{\mu}{\sigma_{\phi_s}} - \frac{1}{2})ds + W_t \quad (2.4)$$

From (2.4) we obtain $\tilde{Y}_t$-process shown in Fig.4.

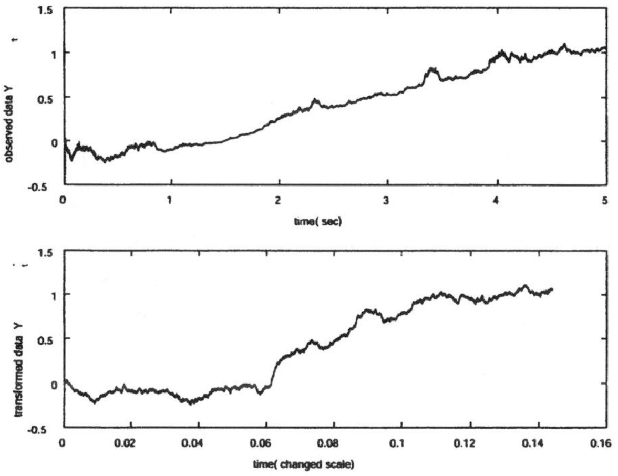

Fig. 4. Original and transformed data

Hence, denoting

$$\tilde{\sigma}_t = \sigma_{\phi_t}, \quad (2.5)$$

we have the following transformed system

$$d\tilde{\sigma}_t = \alpha(\frac{m}{\tilde{\sigma}_t} - 1)dt + v\{\rho dW_t' + \sqrt{1-\rho^2} dW_t\} \quad (2.6)$$

$$d\tilde{Y}_t = (\frac{\mu}{\tilde{\sigma}_t} - \frac{1}{2})dt + dW_t \quad (2.7)$$

3. THE FILTERING PROBLEM

3.1 *The Zakai equation*

Now we consider the usual filtering problem for (2.6) and (2.7). It is easy to show that

$$E\{f(\tilde{\sigma}_t)|\mathscr{F}_t^{\tilde{Y}}\} = \frac{p(t)(f)}{p(t)(1)} \quad (3.1)$$

where $p(t)$ is a solution of the Zakai equation

$$dp(t) + \mathscr{A}^* p(t)dt + \mathscr{B}^* p(t) d\tilde{Y}_t = 0 \quad (3.2)$$

and where

$$\mathscr{A}u = -\frac{1}{2}v^2\frac{\partial^2 u}{\partial x^2} + \alpha(\frac{m}{x} - 1)\frac{\partial u}{\partial x} - \frac{\alpha m}{x^2}u \quad (3.3)$$

$$\mathscr{B}u = -(\frac{\mu}{x} - \frac{1}{2})u + v\sqrt{1-\rho^2}\frac{\partial u}{\partial x}. \quad (3.4)$$

Hence using the inverse mapping $\phi^{-1}(t)$, we have

$$E\{f(\sigma_t)|\mathscr{F}_t^Y\} = \frac{p(\phi^{-1}(t))(f)}{p(\phi^{-1}(t))(1)} \quad (3.5)$$

3.2 *Approximation of the Zakai equation*

Theoretically we set the following discrete-time point

$$t_{i+1} - t_i = \frac{t_f}{n}.$$

By using the splitting-up method proposed by Benssousan et. al (BENSOUSSAN *et al.*, 1990) and (AIHARA and BAGCHI, 2001), the derived Zakai equation is approximated as follows: for $t \in [t_i, t_{i+1}[$

$$\frac{dp_{1i}^{(n)}(t)}{dt} + (A^* + \frac{1}{2}B^*B^*)p_{1,i}^{(n)}(t) = 0 \quad (3.6)$$

$$dp_{2,i}^{(n)}(t) = -B^* p_{2,i}^{(n)}(t) \circ d\tilde{Y}_t \quad (3.7)$$

where $\circ$ denotes the Stratonovich integral and

$$p_{1,i}^{(n)}(t_i) = p_i \quad (3.8)$$

$$p_{2,i}^{(n)}(t_i) = p_{i+1/2} = p_{1,i}^{(n)}(t_{i+1} - 0) \quad (3.9)$$

$$p_{i+1} = p_{2,i}^{(n)}(t_{i+1} - 0; n) \quad p_0 = p_o \quad (3.10)$$

The solution of (3.7) becomes

$$p_{2,i}^{(n)}(t) = \exp\{\int_{t_i}^t (\frac{\mu}{Z(s,t;x)} - \frac{1}{2}) \circ d\tilde{Y}_s\}$$
$$\times p_{2,i}^{(n)}(t_i - 0, Z(t_i, t; x)) \quad (3.11)$$

where

$$dZ(r,t;x) = -v\sqrt{1-\rho^2} \circ d\tilde{Y}_t \quad (3.12)$$

$$Z(r,r;x) = x \quad (3.13)$$

It is easy to show that

$$Z(r,t;x) = x - v\sqrt{1-\rho^2}(\tilde{Y}_t - \tilde{Y}_r) \quad (3.14)$$

and

$$\int_{t_i}^t (\frac{\mu}{Z(s,t;x)} - \frac{1}{2}) \circ d\tilde{Y}_s$$
$$= -\frac{\mu}{v\sqrt{1-\rho^2}} \log\{\frac{x - v\sqrt{1-\rho^2}(\tilde{Y}_t - \tilde{Y}_{t_i})}{x}\}$$
$$-\frac{1}{2}(\tilde{Y}_t - \tilde{Y}_{t_i}). \quad (3.15)$$

Hence the explicit solution of (3.7) becomes

$$p_{2,i}^{(n)}(t) = \{\frac{x - v\sqrt{1-\rho^2}(\tilde{Y}_t - \tilde{Y}_{t_i})}{x}\}^{-\frac{\mu}{v\sqrt{1-\rho^2}}}$$
$$\times \exp\{-\frac{1}{2}(\tilde{Y}_t - \tilde{Y}_{t_i})\}$$

$$\times p^{(n)}_{1,i}(t_{i+1}-0, x - \nu\sqrt{1-\rho^2}(\tilde{Y}_t - \tilde{Y}_{t_i})) \quad (3.16)$$

Equation (3.6) also becomes

$$\frac{\partial p^{(n)}_{1,i}(t)}{\partial t} - \frac{1}{2}\nu^2\rho^2\frac{\partial^2 p^{(n)}_{1,i}(t)}{\partial x^2}$$
$$+\{\alpha(\frac{m}{x}-1) - \nu\sqrt{1-\rho^2}(\frac{\mu}{x}-\frac{1}{2})\}\frac{\partial p^{(n)}_{1,i}(t)}{\partial x}$$
$$+\{\frac{1}{2}(\frac{\mu}{x}-\frac{1}{2})^2 + \frac{1}{2}\nu\sqrt{1-\rho^2}\frac{\mu}{x^2} - \alpha\frac{m}{x^2}\}p^{(n)}_{1,i}(t) = 0 \quad (3.17)$$

for $t \in [t_i, t_{i+1}[$.

3.3 *Convergence results*

It is possible to show the convergence property of the splitting up algorithm given by (3.16) and (3.17) from the basic results given by Benssousan et. al. (BENSOUSSAN *et al.*, 1990). Here we assume that for some $\varepsilon > 0$

$$\tilde{\sigma}(t) \in]\varepsilon, \infty[= R^1_\varepsilon \text{ a.s. for all } t \quad (3.18)$$

Now we set

$$V = \{\phi | \phi \in H^1(R^1_\varepsilon), \phi(\varepsilon) = 0\} \subset H = L^2(R^1_\varepsilon)$$
$$\subset V' = \text{dual of } V$$

In this subsection, let $(\Omega, \mathscr{F}, \mathscr{P})$ be a probability space on which $\tilde{Y}$ is a standard Brownian motion process and

$$\mathscr{F}^t = \mathscr{F}^{\tilde{Y}}_t .$$

Theorem 3.1. Assume that

$$p_o \in H \quad (3.19)$$

and

$$p_{1,i}(t;n,x)_{|x=\varepsilon} = 0 \quad (3.20)$$

Hence

$$p^{(n)}_{1i} \to p \text{ in } L^2_{\mathscr{F}}(T;V) \text{ strongly}$$
$$p^{(n)}_{2i} \to p \text{ in } L^2_{\mathscr{F}}(T;H) \text{ strongly}$$
$$p^{(n)}_{1i}(t), p^{(n)}_{2i}(t) \to p(t) \text{ in } L^2(\Omega;H)$$
$$\text{strongly } \forall t \in [0,T[$$

3.4 *Simulation studies*

In the real time world, we need to work with sampled data and set the sampling period Δ_s. Then at each time $i\Delta_s$ for $i = 1, 2, \cdots$, we get the observation data $Y_{i\Delta_s}$. Hence in the random transformed world, we can set

$$t_0 = 0 \quad (3.21)$$
$$t_{i+1} = t_i + \Delta_i(\omega) \quad (3.22)$$

where

$$\Delta_i(\omega) = Y^2_{(i+1)\Delta_s} - Y^2_{i\Delta_s} - 2\int_{i\Delta_s}^{(i+1)\Delta_s} Y_s dY_s \quad (3.23)$$

We will present some simulation results. We set

$$\alpha = 30, \mu = 0.1, m = 0.2, \nu = 0.2, \rho = -0.2$$

The spatial region is set as $x \in]0.01, 0.6[$ and Zakai equation (3.17) is discretized with respect to the spatial variable x such that

$$\Delta x = (0.6 - 0.01)/400$$

In Fig.5, the observation data Y_t is presented.

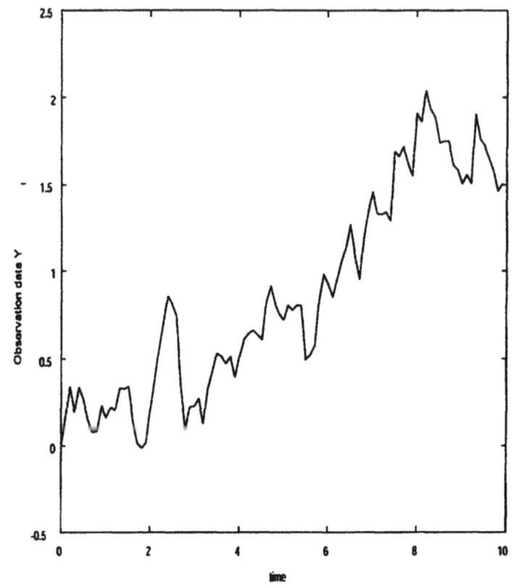

Fig. 5. Observation data Y_s

The original stock price S_t is shown in Fig.6.

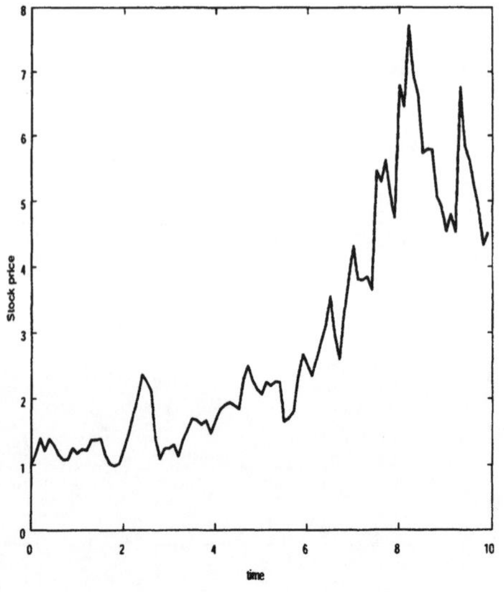

Fig. 6. Stock price S_t

The estimated process $E\{\sigma_t|\mathscr{F}_t^Y\}$ and true values are plotted in Fig.7. In Fig. 8 a sample process of $p(t,x)/\int_\varepsilon^\infty p(t,x)dx$ is also demonstrated.

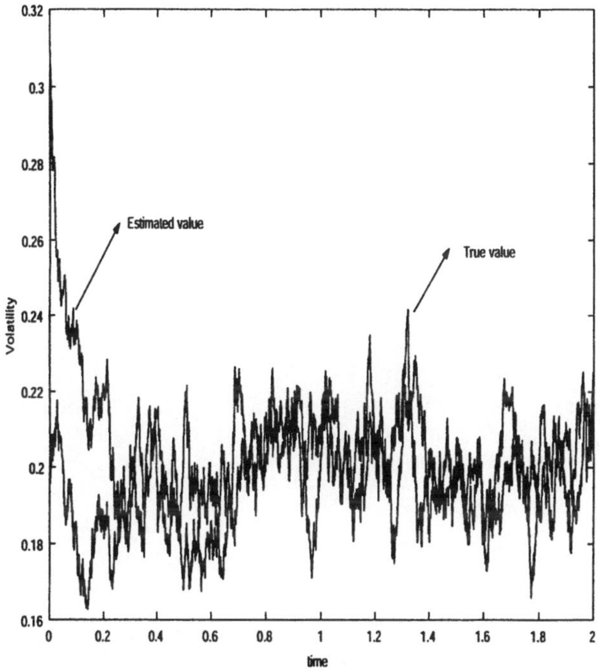

Fig. 7. True and estimated volatility processes

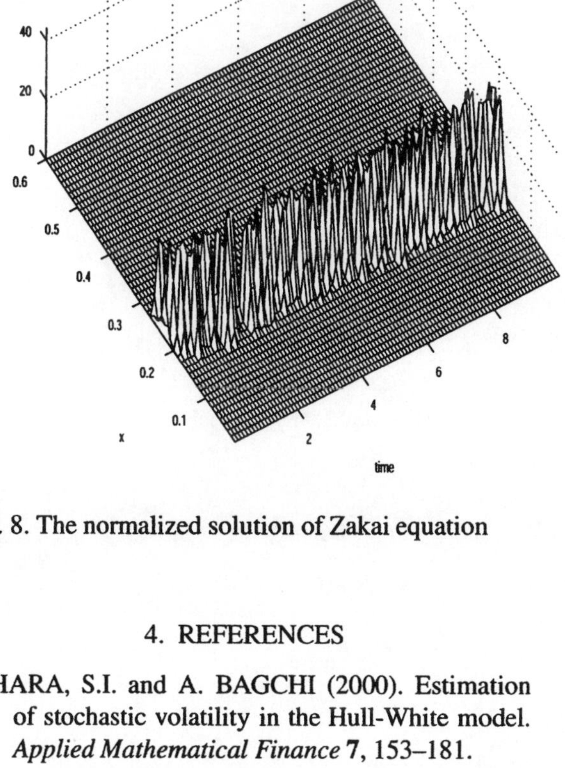

Fig. 8. The normalized solution of Zakai equation

4. REFERENCES

AIHARA, S.I. and A. BAGCHI (2000). Estimation of stochastic volatility in the Hull-White model. *Applied Mathematical Finance* **7**, 153–181.

AIHARA, S.I. and A. BAGCHI (2001). Robust nonlinear filtering of stochastic volatility in finance. *Proc. of ECC2001* pp. 1501–1506.

BENSOUSSAN, A. (1992). *Stochastic Control of Partially Observable Systems*. Cambridge University Press. Cambridge.

BENSOUSSAN, A., R. GLOWINSKI and A. RASCANU (1990). Approximation of the zakai equation by the splitting up method. *SIAM J. Control Optim.* **28**, 1420–1431.

HESTON, S.L. (1993). A closed-form solution for options with stochastic volatilities. *Review of Financial Studies* **6(2)**, 327–343.

Appendix A. APPENDIX A(PROPOSITIONS)

In order to prove Theorem 3.1, we need to the following propositions. As was used in (BENSOUSSAN, 1992), we introduce a convenient constant c in (3.6) and (3.7) such that

$$\frac{dp_{1i}^{(n)}(t)}{dt} + (A^* + \frac{1}{2}B^*B^* + \frac{c}{2})p_{1,i}^{(n)}(t) = 0 \quad (A.1)$$

$$dp_{2,i}^{(n)}(t) + \frac{c}{2}p_{2,i}^{(n)} = -B^* p_{2,i}^{(n)}(t) \circ d\tilde{Y}_t \quad (A.2)$$

The following propositions can be obtained form (AIHARA and BAGCHI, 2001).

Proposition A.1. The system (A.1) and (A.2) defines in a unique way $p_{1,i}^{(n)}$, $p_{2,i}^{(n)}$ in $L^2_{\mathscr{F}}(T;V)$,$L^2_{\mathscr{F}}(T;V)$ respectively.

Proposition A.2. The processes $p_{1i}^{(n)}, p_{2i}^{(n)}$ satisfy

$$E\int_0^T ||p_{1i}^{(n)}(t)||^2 dt \leq C, \quad (A.3)$$

$$E\int_0^T |p_{2i}^{(n)}(t)|^2 dt \leq C \quad (A.4)$$

$$E\{|p_{1i}^{(n)}(t)|^4\} \leq C, \quad (A.5)$$

$$E\{|p_{2i}^{(n)}(t)|^4\} \leq C, \forall t \in [0,t_f] \quad (A.6)$$

where C does not depend on t_f or i for a convenient choice of c.

Proposition A.3. $\xi = p$

Appendix B. PROOF OF THEOREM 3.1

From (A.2) we have

$$E\{|p_{2i}^{(n)}((i+1)\Delta)|^2\} - E\{|p_{2i}^{(n)}(i\Delta)|^2\}$$
$$+ E\{\int_{i\Delta}^{(i+1)\Delta} (c|p_{2i}^{(n)}(s)|^2 - (B_b p_{2i}^{(n)}(s), p_{2i}^{(n)}(s))ds\} = 0$$

where

$$B_b(\cdot) = \{2(\frac{\mu}{x} - \frac{1}{2})^2 - v\sqrt{1-\rho^2}\frac{\mu}{x^2}\}(\cdot) \in \mathscr{L}(H;H).$$

It follows from (A.6) that

$$E\{|p_{1i}^{(n)}(t)|^2\} - E\{|p_{1i}^{(n)}(i\Delta)|^2\}$$
$$+2E\{\int_{i\Delta}^{t} < (A^* + \frac{1}{2}B^*B^* + \frac{c}{2})p_{1i}^{(n)}(s), p_{1i}^{(n)}(s) > ds\} = 0$$

Summing up these equations, we obtain

$$E\{|p_{1i}^{(n)}(t)|^2\} - |p_o|^2$$
$$+2E\{\int_{i\Delta}^{t} < (A^* + \frac{1}{2}B^*B^* + \frac{c}{2})p_{1i}^{(n)}(s), p_{1i}^{(n)}(s) > ds\}$$
$$+E\{\int_0^{\Delta[t/\Delta]} (c|p_{2i}^{(n)}(s)|^2 - (B_b p_{2i}^{(n)}(s), p_{2i}^{(n)}(s))ds\} = 0 \quad \text{(B.1)}$$

and

$$E\{|p_{2i}^{(n)}(t)|^2\} - E\{|p_{1/2}^{(n)}|^2\}$$
$$+E\{\int_0^t (c|p_{2i}^{(n)}(s)|^2 - (B_b p_{2i}^{(n)}(s), p_{2i}^{(n)}(s))ds\}$$
$$+2E\{\int_{\Delta}^{\Delta[t/\Delta]+\Delta} < (A^* + \frac{1}{2}B^*B^*$$
$$+\frac{c}{2})p_{1i}^{(n)}(s), p_{1i}^{(n)}(s) > ds\} = 0$$

Here we shall show the strong convergence of $p_{1i}^{(n)}$-process. Introduce

$$\chi^{(n)}(t) = E\{|p(t) - p_{1i}^{(n)}(t)|^2\}$$
$$+2E\{\int_0^t < (A^* + \frac{1}{2}B^*B^*$$
$$+\frac{c}{2})(p(s) - p_{1i}^{(n)}(s)), p(s) - p_{1i}^{(n)}(s) > ds\}$$
$$+E\{\int_0^{\Delta[t/\Delta]} (c|p(s) - p_{2i}^{(n)}(s)|^2$$
$$-(B_b(p(s) - p_{2i}^{(n)}(s)), p(s) - p_{2i}^{(n)}(s)))ds\}$$
$$= \chi_1^{(n)}(t) + \chi_2^{(n)}(t) + \chi_3^{(n)}(t)$$

where

$$\chi_1^{(n)}(t) = E\{|p(t)|^2\}$$
$$+2E\{\int_0^t < (A^* + \frac{1}{2}B^*B^* + \frac{c}{2})p(s), p(s) > ds\}$$
$$+E\{\int_0^{\Delta[t/\Delta]} (c|p(s)|^2 - (B_b p(s), p(s)))ds\},$$

$$\chi_2^{(n)}(t) = -2E\{(p(t), p_{1i}^{(n)}(t))\}$$
$$-2E\{\int_0^t (< (A^* + \frac{1}{2}B^*B^* + \frac{c}{2})p(s), p_{1i}^{(n)}(s) >$$
$$+ < (A^* + \frac{1}{2}B^*B^* + \frac{c}{2})p_{1i}^{(n)}(s), p(s) >)ds\}$$
$$-2cE\{\int_0^{\Delta[t/\Delta]} (p(s), p_{2i}^{(n)}(s))ds\}$$
$$+2E\{\int_0^{\Delta[t/\Delta]} (B_b p(s), p_{2i}^{(n)}(s))ds\}$$

and

$$\chi_3^{(n)}(t) = |p_o|^2.$$

From

$$< (A^* + \frac{1}{2}B^*B^*)\phi, \phi > = \frac{1}{2}v^2\rho^2|\frac{\partial \phi}{\partial x}|^2 - \frac{\alpha m}{2}|\frac{\phi}{x}|^2$$
$$+\frac{1}{2}|(\frac{\mu}{x}^{\frac{1}{2}})\phi|^2 \geq \beta||\phi||^2 + \lambda|\phi|^2 \text{for } \phi \in V \text{(B.2)}$$

we get

$$< B^*B^*\phi, \phi > -(B_b\phi, \phi) = -\frac{1}{2}|B^*\phi|^2, \ \phi \in V \text{(B.3)}$$

Hence

$$\lim_{n\to\infty} \chi_1^{(n)}(t) = E\{|p(t)|^2\} + 2E\{\int_0^t (< A^*p(s), p(s) >$$
$$-\frac{1}{2}|B^*p(s)|^2 + c|p(s)|^2)ds\} = |p_o|^2$$

From the results of weak convergence of $p_{1i}^{(n)}$- process, we have

$$\lim_{n\to\infty} \chi_2^{(n)}(t) = -2E\{|p(t)|^2\} - 4E\{\int_0^t (< A^*p, p >$$
$$-\frac{1}{2}|B^*p|^2 + c|p|^2)ds\} = -2|p_o|^2.$$

From (B.2) and (B.3), choosing $c > \lambda$, we have $0 \leq \chi^{(n)}(t)$. Hence

$$p_{1i}^{(n)} \to p \text{ in } L^2_{\mathscr{F}}(T;V) \text{ strongly}$$
$$p_{1i}^{(n)}(t) \to p(t) \text{ in } L^2(\Omega;H) \ \forall t \geq 0 \text{ strongly.}$$

Similarly we can check the convergence of $p_{2i}^{(n)}$-process.

IFAC Publications
www.elsevier.com/locate/ifac

STATE ESTIMATION FOR NONLINEAR CONTINUOUS SYSTEMS IN A BOUNDED-ERROR CONTEXT

T. Raïssi, N. Ramdani and Y. Candau

Centre d'Etude et de Recherche en Thermique, Energétique et Systèmes, Université Paris XII- Val de Marne, ave G. de Gaulle, 94000 Créteil.
<raissi, ramdani, candau> @univ-paris12.fr

Abstract: The aim of this paper is to study state estimation for nonlinear continuous-time systems in a bounded-error context. A causal estimator based on prediction-correction approach is proposed. The prediction part consists on a validated integration of an Initial Value Problem for an Ordinary Differential Equation. The correction part uses set inversion. The main tools used are Taylor models and interval analysis. The derived estimator is illustrated on an example. *Copyright © 2003 IFAC*

Keywords: continuous time systems, state estimation, intervals, nonlinear systems, bounded noise, uncertain dynamic systems.

1. INTRODUCTION

In this paper, state estimation of nonlinear continuous-time systems in bounded error context is investigated.

In the literature, state estimation is usually resolved by stochastic or probabilistic methods where perturbations are assumed to be random white and Gaussian noise. In several cases, these hypotheses are not valid and it is more natural to assume that perturbations belong to a known set. In this case, bounded-error approaches allow the characterization of the set of all state vectors that are compatible with measured data, a model structure and some prior error bounds.

In the linear case, bounded-error approaches have been investigated and applied to state estimation and many authors have proposed outer- or inner-bounding techniques based on ellipsoïds, paralellotopes or boxes (see Maksarov and Norton, 2002 ; Milanese *et al.*, 1996 ; Durieu et *al.*, 2001 ; and the references therein). For nonlinear discrete systems, bounded-error state estimation is generally computed by an extension of the Kalman filtering to intervals (Becerra et *al.*, 2001; Magnus et *al.*, 2000), or with set inversion and constraint propagation (Jaulin *et al.*, 2001 ; Kieffer *et al.*, 2002).

For time-continuous nonlinear systems, Jaulin (2002) has proposed a causal and a non-causal state estimators by using interval analysis and the Picard theorem, which gives an enclosure of the solution of an ordinary differential equation (ODE). The derived algorithm makes use of interval constraint propagation. In order to reduce the wrapping effect of interval computations, Jaulin (2002) makes extensive use of bisections. Such a procedure is known to be time-consuming and is not practicable when the dimension of the state vector is large. The goal of this paper is to show that more accurate interval computations can make the bisections unnecessary.

2. PROBLEM STATEMENT

Consider a system represented by equations

$$\dot{\mathbf{x}} = \mathbf{f}(\mathbf{x}(t)),\ \mathbf{y} = \mathbf{g}(\mathbf{x}(t)),\ \mathbf{x}(t_0) \in [\mathbf{x}_0] \quad (1)$$

where $t \in [t_0, T]$, $\mathbf{f} \in C^{k-1}(D)$, and $D \subseteq \mathbb{IR}^n$ is an open set, $\mathbf{f} : D \rightarrow \mathbb{IR}^n$ and $[\boldsymbol{x_0}] \subseteq D$. $\boldsymbol{x} \in \mathbb{IR}^n$ is the state vector and $\boldsymbol{y} \in \mathbb{IR}^m$ is the output vector.

The goal is to estimate an interval vector $[x_j]$ that is guaranteed to contain the true solution of the system (1), i.e. the state $\boldsymbol{x}$ at some sampling times $\{t_1, t_2, \ldots, t_n\}$ in $[t_0, T]$ when measurements $\hat{y}_j$ at time t_j (j = 1,..., N) are available and an upper error bounds $\bar{e}_j$ are known (in such case the domain for y_j is the following box : $\left[y_j\right] = \left[\hat{y}_j - \underline{e}, \hat{y}_j + \bar{e}\right]$. When the functions **f** and **g** are linear, many efficient methods can be found (Shweppe, 1968). As noted by (Jaulin, 2002), if **f** and **g** are non-linear the problem is not well studied in the bounded-error time-continuous systems context.

The algorithm proposed in this paper is based on a classical predictor-corrector approach. The predictor part of the algorithm consists on a guaranteed integration of the continuous-time state equation which is in fact, equivalent to the resolution of an initial value problem for ordinary differential equation (IVP for ODE). The corrector part of the algorithm uses the inversion of measurement data in order to contract the predicted state vector.

3. VALIDATED INTEGRATION

Standard numerical methods for IVPs for ODEs attempt to compute an approximation for the solution that satisfies an acceptable tolerance specified by the user. Generally these methods give robust solutions, but it is easy to find many cases where they can give inaccurate solutions. To avoid inaccurate results one proposes to use interval methods for IVPs for ODEs (called validated methods).

These methods have two important advantages: if they return a solution to the problem, then the problem is guaranteed to have a unique one, and an enclosure of the solution is derived. However, validated methods for IVPs for ODEs have some disadvantages. They usually require considerably more computing time than does the computation of standard methods when all the variables are punctual. However, this is no longer the case when the state equations contain parameters which cannot be known exactly, but are known to lie in a given interval. Indeed, to compute solutions for such problems with standard methods, the same process is executed many times with different values of the parameters. In others problems where the initial state value is known in an interval domain, one has to study an infinite number of equations systems which is clearly numerically expensive. On the other hand, validated methods are known to introduce significant overestimation in the derived enclosures, mainly due to the wrapping effect of interval computations. To reduce this effect one usually uses mathematic techniques, some of which will be described in the following sections.

In this paper, it will be shown that Taylor models allow a better numerical integration and make state estimation for nonlinear continuous-time systems possible. The main tools used are interval analysis, Taylor models, and constraint propagation.

4. INTERVAL ANALYSIS

Interval analysis (IA) has been initially developed to account for the quantification errors induced by the rational representation of real number with computers (Moore, 1966). Then it has been extended to validated numerical and global optimisation (Hansen, 1992).

It is in fact set theory, which furnishes the foundations for interval computations. Functions are extended to sets as follows: Given two sets X and Y, the direct image of the set X is given by $Y = f(X) \equiv \{f(x) \mid x \in X\}$ and the reciprocal image of Y: $X = f^{-1}(Y) \equiv \{x \in X \mid f(x) \in Y\}$. Operations on numbers and Booleans are also extended to sets as follows: $X \Diamond Y \equiv \{x \Diamond y \mid x \in X, y \in Y\}$ *with* $\Diamond \in \{+, -, *, /\}$

A real interval x of $\mathbb{IR}$ is a non-empty, closed and bounded subset of the real numbers R, defined by

$$[\boldsymbol{x}] \equiv \left[x^-, x^+\right] \equiv \left\{x \in R \mid x^- \leq x \leq x^+\right\}.$$

Interval computations are generally pessimistic.

For an interval, the following entities are defined:
- the upper bound of x, $Sup(x) = x^+$;
- the lower bound of x, $Inf(x) = x^-$;
- the width of $\boldsymbol{x}$, $w(\boldsymbol{x}) = (x^+ - x^-)/2$;
- the centre of $\boldsymbol{x}$, $m(\boldsymbol{x}) = (x^+ + x^-)/2$.

The envelope of a set A is named the *interval hull*, and is denoted by $[\boldsymbol{A}]$. It is the smallest box of $\mathbb{IR}^n$ including $\boldsymbol{A}$.

Given a function on reals (-vectors) $\mathbf{f} : \mathbb{R}^n \rightarrow \mathbb{R}^n$, its natural extension to intervals can be obtained by replacing the real variables by their interval counterpart. The so-derived function $[\mathbf{f}] : \mathbb{IR}^n \rightarrow \mathbb{IR}^n$ is an inclusion function of **f**, if $\forall [\mathbf{x}] \in \mathbb{IR}^n, \mathbf{f}([\mathbf{x}]) \subseteq [\mathbf{f}]([\mathbf{x}])$.

Centred inclusion function - Mean value form :
$\mathbf{f} : \mathbb{IR}^n \rightarrow \mathbb{IR}$ is continuously differentiable on $D \subseteq \mathbb{IR}^n$, then:

$$\mathbf{f}(\mathbf{x}) \in \mathbf{f}_m([\mathbf{a}]) = \mathbf{f}(\mathbf{b}) + \mathbf{f}'([\mathbf{a}])([\mathbf{a}] - \mathbf{b}) \quad (2)$$

for any x, b $\in$ [a] et [a] $\subseteq D$. The expression $\mathbf{f}(\mathbf{b}) + \mathbf{f}'([\mathbf{a}])([\mathbf{a}] - \mathbf{b})$ is called the mean-value form of $\mathbf{f}$, $\mathbf{f}'$ is an evaluation of the gradient of **f**. When the interval $[\mathbf{a}]$ is not very large the wrapping effect of the inclusion function $\mathbf{f}_m$ is smaller than the one of the natural inclusion function.

5. TAYLOR MODELS

To solve a system of multidimensional ordinary differential equations

$$\dot{\mathbf{x}} = \mathbf{f}(\mathbf{x}(t)),\ \mathbf{x}(t_0) = \mathbf{x}_0 \qquad (3)$$

over long time with verification, the overestimation (called wrapping effect) adds a major difficulty. This effect is caused by the inflation of the size of the set at each time step containing the validated solution set. To reduce the wrapping effect Taylor model approaches, which combine high order polynomial techniques and interval analysis, are used. Any (n+1) times continuously differentiable function f in a domain D can be expressed by a Taylor polynomial $P_{n,f}$ at the expansion point $x_0 \in D$ and a remainder bounded by an interval $I_{n,f}$ via (Berz, 2001):

$$\forall \mathbf{x} \in D,\quad \mathbf{f}(\mathbf{x}) \in \mathbf{P}_{n,f}\ (\mathbf{x}-\mathbf{x}_0) + \mathbf{I}_{n,f} \qquad (4)$$

From Taylor's theorem the width of the remainder term is proportional to the size $(x\text{-}x_0)^{n+1}$. To reduce the wrapping effect we have to choose the size $|x\text{-}x_0|$ small, an order n sufficiently high and the polynomial $P_{n,f}$ with point coefficients (so there is no interval arithmetic inflation in the polynomial part). Thus it can optimally eliminate the overestimation, making possible not only the integration of the differential equation over long time, but also to deal with much larger domain for initial conditions.

5.1 Taylor Coefficients

Since one shall need to use point and interval Taylor coefficients, basic idea of the method is briefly described. Denote the ith Taylor coefficient of a (n+1) continuously differentiable function $\boldsymbol{u}(t)$ by:

$$(\mathbf{u}_j)_i = \frac{\mathbf{u}^i(t_j)}{i!} \qquad (5)$$

where $u^i(t)$ is the ith derivative of $u(t)$ at t_j. If one considers the differential system (1), the following sequence of functions is introduced:

$$\begin{aligned} \mathbf{f}^{[1]} &= \mathbf{f} \\ \mathbf{f}^{[i]} &= \frac{1}{i!}\frac{\partial \mathbf{f}^{[i-1]}}{\partial \mathbf{x}}\mathbf{f},\ i \geq 2 \end{aligned} \qquad (6)$$

Using equations (1) and (6), the Taylor coefficients of $\mathbf{x}(t)$ at t_j are:

$$\begin{aligned} (\mathbf{x}_j)_0 &= (\mathbf{x}_j) \\ (\mathbf{x}_j)_1 &= \mathbf{f}^{[1]}(\mathbf{x}_j) = \mathbf{f}(\mathbf{x}_j) \\ (\mathbf{x}_j)_i &= \mathbf{f}^{[i]}(\mathbf{x}_j) = \frac{1}{i}\left(\frac{\partial \mathbf{f}^{[i-1]}}{\partial \mathbf{x}}\mathbf{f}\right)(\mathbf{x}_j),\ \text{for } i \geq 2 \end{aligned} \qquad (7)$$

Let $\mathbf{x}(t_j) = \mathbf{x}_j \in [\mathbf{x}_j]$, to have interval Taylor coefficients, $\mathbf{x}_j$ is replaced by $[\mathbf{x}_j]$ in (7). Denote the i^{-th} interval Taylor coefficient of $\mathbf{x}(t)$ at t_j by $[\mathbf{x}_j]_i$, then:

$$\begin{aligned} [\mathbf{x}_j]_0 &= [\mathbf{x}_j] \\ [\mathbf{x}_j]_1 &= \mathbf{f}^{[1]}([\mathbf{x}_j]) = \mathbf{f}([\mathbf{x}_j]) \\ [\mathbf{x}_j]_i &= \mathbf{f}^{[i]}([\mathbf{x}_j]) = \frac{1}{i}\left(\frac{\partial \mathbf{f}^{[i-1]}}{\partial \mathbf{x}}\mathbf{f}\right)([\mathbf{x}_j]),\ \text{for } i \geq 2 \end{aligned} \qquad (8)$$

5.2. Integration

$$\dot{\mathbf{x}} = \mathbf{f}(\mathbf{x}(t)),\ \mathbf{x}(t_0) \in [\mathbf{x}_0] \qquad (9)$$

where $t \in [t_0, T]$, $f \in C^{k-1}(D)$, and $D \subseteq \mathbb{R}^n$ is an open set, $\mathbf{f}: D \to \mathbb{R}^n$ and $[\mathbf{x}_0] \subseteq D$, $\mathbf{x} \in \mathbb{R}^n$. (9) is a generalization of (3), it permits the initial value $x(t_0)$ to be interval. Since one assumes that $f \in C^{k-1}(D)$, we exclude functions that contains non differentiable parts like abs or sign. In most validated methods for initial value problem of ordinary differential equations (IVP for ODEs) each step of the algorithm contains two steps (Nedialkov, 1999 ; Nedialkov and Jackson, 2000):

First step: step size (when one chooses a non constant step size strategy) and an a priori enclosure $[\tilde{x}_j]$ of the solution are computed.

h_j is the step size. $h_j = t_{j+1} - t_j$. In this paper one considers constant step size strategy only, but adaptative or variable step size can also be used as well. To assure the stability of the algorithms one has to choose a step size small enough.

A priori enclosure must enclose the trajectory of the solution between t_j and t_{j+1}, the computation of $[\tilde{\mathbf{x}}_j]$ is based on the application of the Picard-Lindelöf operator and the Banach fixed-point theorem (for more details see Nedialkov and Jackson, 2000). It is equivalent to find a set $[\tilde{\mathbf{x}}_j]$ such as:

$$[\mathbf{x}_j] + [0,h]\mathbf{f}([\tilde{\mathbf{x}}_j]) \subseteq [\tilde{\mathbf{x}}_j] \qquad (10)$$

The computation of the solution of the previous equation is not easy to perform. Moore (1965) suggests that the interval $[\tilde{\mathbf{x}}_j]$ can be computed by the following formula:

$$[\tilde{\mathbf{x}}_j] = [\mathbf{x}_j] + \mathbf{f}([\mathbf{x}_j])[0,h] \qquad (11)$$

This expression will almost fail to enclose the trajectory of the solution. In this paper Moore formula is chosen but $[\tilde{\mathbf{x}}_j]$ is inflated so that it contains the solution.

Second step: one denotes the solution of (9) with an initial condition $\left[\mathbf{x}_j\right]$ at t_j by $\mathbf{x}\left(t; t_j, \left[\mathbf{x}_j\right]\right)$ for all $t \in [t_j, t_{j+1}]$, then $\mathbf{x}\left(t; t_j, \left[\mathbf{x}_j\right]\right)$. Using $\left[\tilde{\mathbf{x}}_j\right]$, in second step, a tighter enclosure $\left[\mathbf{x}_{j+1}\right] \subseteq \left[\tilde{\mathbf{x}}_j\right]$ is computed, for which $\mathbf{x}\left(t_{j+1}; t_j, \left[\mathbf{x}_j\right]\right) \in \left[\mathbf{x}_{j+1}\right]$. The main difficulty of the approach arises in this second step where the wrapping effect must be controlled. Moore (1965) proposes a method based on the Taylor models presented in the sections above. It computes $\left[\mathbf{x}_{j+1}\right]$ by:

$$\begin{aligned}\left[\mathbf{x}_{j+1}\right] &= \left[\mathbf{x}_j\right] + \left[\mathbf{x}_j\right]_i + \mathbf{f}^{[k]}\left(\left[\tilde{\mathbf{x}}_j\right]\right) \\ &= [\mathbf{x}]_j + \mathbf{f}^{[i]}\left(\left[\mathbf{x}_j\right]\right) + \mathbf{f}^{[k]}\left(\left[\tilde{\mathbf{x}}_j\right]\right)\end{aligned} \qquad (12)$$

A disadvantage of this approach is that the width of $[x_j]$ always increases with j. This is showed by the following expression:

$$w([x_{j+1}]) = w([x_j]) + \sum_{i=0}^{k-1} w([x_j]_i)\, h_j^i + w(f^{[k]}([\tilde{x}_j]))\, h_j^k$$

To avoid this problem, the mean value evaluation (2) should be used instead in order to reduce this effect. If such an evaluation is used for computing the enclosure of $\mathbf{f}^{[i]}\left(\left[\mathbf{x}_j\right]\right)$ instead of the direct evaluation, one can obtain enclosures with smaller widths than in (12). So, by applying the mean-value theorem to $f^{[i]}$ at some $\hat{x}_j \in [x_j]$, one obtains:

$$\mathbf{f}^{[i]}\left(\left[\mathbf{x}_j\right]\right) = \mathbf{f}^{[i]}(\hat{\mathbf{x}}_j) + \mathbf{J}\left(\mathbf{f}^{[i]}; \left[\mathbf{x}_j\right], \hat{\mathbf{x}}_j\right)\left(\left[\mathbf{x}_j\right] - \hat{\mathbf{x}}_j\right) \qquad (13)$$

Then from (12) and (13) :

$$\left[\mathbf{x}_{j+1}\right] = \mathbf{x}_j + \sum_{i=1}^{k-1} \mathbf{f}^{[i]}\left(\hat{\mathbf{x}}_j\right) h^i + \mathbf{f}^{[k]}\left(\left[\tilde{\mathbf{x}}_j\right]\right) h^k + \{\mathbf{I} + \mathbf{J}(\mathbf{f}^{[i]}; \left[\mathbf{x}_j\right])\}(\left[\mathbf{x}_j\right] - \hat{\mathbf{x}}_j) \qquad (14)$$

Let $\left[\mathbf{v}_{j+1}\right] = \hat{\mathbf{x}}_j + \sum_{i=1}^{k-1} \mathbf{f}^{[i]}\left(\hat{\mathbf{x}}_j\right) h^i + \mathbf{f}^{[k]}\left(\left[\tilde{\mathbf{x}}_j\right]\right) h^k$

and $\left[\mathbf{S}_j\right] = \mathbf{I} + \sum_{i=1}^{k-1} \mathbf{J}\left(\mathbf{f}^{[i]}; \left[\mathbf{x}_j\right]\right) h^i$

then $\left[\mathbf{x}_{j+1}\right] = \left[\mathbf{v}_{j+1}\right] + \left[\mathbf{S}_j\right]\left(\left[\mathbf{x}_j\right] - \hat{\mathbf{x}}_j\right)$.

There are many strategies to choose $\hat{\mathbf{x}}_j$, in this paper the medium of $\left[\mathbf{x}_j\right]$ is chosen.

The straightforward method to compute a tighter enclosure of the solution using the formula (14) is called the Direct method. The corresponding algorithm is as follow:

Algorithm Direct Method

Inputs :

$[x_0]$, h, $\hat{\mathbf{x}}_0 = m\left(\left[\mathbf{x}_0\right]\right)$

for j:= 0 to N-1 Compute :

$\left[\mathbf{x}_j\right]$, according to (11)

$\left[\mathbf{v}_{j+1}\right] = \hat{\mathbf{x}}_j + \sum_{i=1}^{k-1} \mathbf{f}^{[i]}\left(\hat{\mathbf{x}}_j\right) h^i + \mathbf{f}^{[k]}\left(\left[\tilde{\mathbf{x}}_j\right]\right) h^k$

$\left[\mathbf{S}_j\right] = \mathbf{I} + \sum_{i=1}^{k-1} \mathbf{J}\left(\mathbf{f}^{[i]}; \left[\mathbf{x}_j\right]\right) h^i$

$\left[\mathbf{x}_{j+1}\right] = \left[\mathbf{v}_{j+1}\right] + \left[\mathbf{S}_j\right]\left(\left[\mathbf{x}_j\right] - \hat{\mathbf{x}}_j\right)$

$\hat{\mathbf{x}}_{j+1} = m\left(\left[\mathbf{x}_{j+1}\right]\right)$

Outputs :

$[x_0], [x_1], \ldots, [x_N]$

Example: Consider the Lotka-Volterra predator-prey model represented by the following equations:

$$\begin{cases} \dot{x}_1 = (1 - 0.01 x_2) x_1 \\ \dot{x}_2 = (-1 + 0.02 x_1) x_2 \end{cases} \qquad (15)$$

with initial condition $x(0) = [49,51] \times [49,51]$, step size $h = 0.005$ and N = 400, the result of the direct method algorithm with a Taylor model order k = 4 is depicted in figure 1.

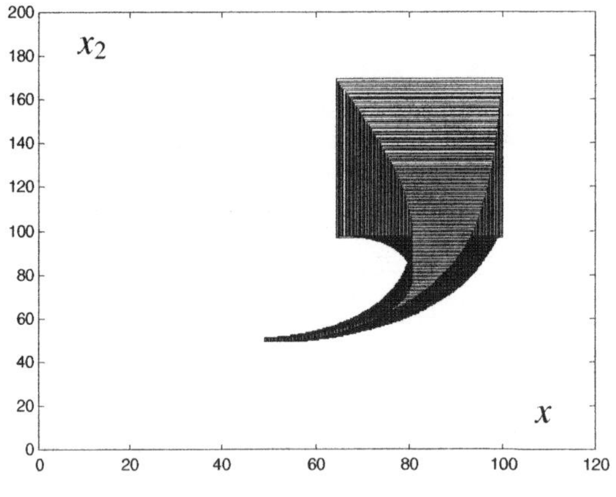

Fig.1. Superposition of boxes generated by the direct method algorithm.

As it was shown in figure 1, if the direct method is used to compute the enclosures of $\left[\mathbf{x}_{j+1}\right]$, one might obtain unacceptably solutions after few steps. The main disadvantage of this method is to compute $\left[\mathbf{S}_j\right]\left(\left[\mathbf{x}_j\right] - \hat{\mathbf{x}}_j\right)$ as an evaluation of the set $\left\{\mathbf{S}_j\left(\mathbf{x}_j - \hat{\mathbf{x}}_j\right) \mid \mathbf{S}_j \in \left[\mathbf{S}_j\right], \mathbf{x}_j \in \left[\mathbf{x}_j\right]\right\}$, which introduces a wrapping at each step.

To improve (14), Rihm (1994) makes a factorisation of $\left[\mathbf{S}_j\right]\left(\left[\mathbf{x}_j\right] - \hat{\mathbf{x}}_j\right)$. His method is called the Extended Mean-Value method (EMV), it is summarised in the following algorithm:

Algorithm: Extended Mean-Value method
Inputs :
$[x_0]$, $\hat{x}_0 = m([x_0])$, $[p_0] = 0$, $A_0 = I$, h

for $j = 0$ to N Compute :
$[\tilde{x}_j]$, according to (11)

$$[\mathbf{v}_{j+1}] = \hat{\mathbf{x}}_j + \sum_{i=1}^{k-1} \mathbf{f}^{[i]}(\hat{\mathbf{x}}_j) h^i + \mathbf{f}^{[k]}([\tilde{\mathbf{x}}_j]) h^k$$

$$[\mathbf{S}_j] = \mathbf{I} + \sum_{i=1}^{k-1} \mathbf{J}(\mathbf{f}^{[i]}; [\mathbf{x}_j]) h^i$$

$$[\mathbf{q}_{j+1}] = ([\mathbf{S}_j]\mathbf{A}_j)[\mathbf{p}_j] + [\mathbf{S}_j]([\mathbf{v}_j] - \hat{\mathbf{x}}_j)$$

$$[\mathbf{x}_{j+1}] = [\mathbf{v}_{j+1}] + [\mathbf{q}_{j+1}]$$

$$\mathbf{A}_{j+1} = m([\mathbf{S}_j]\mathbf{A}_j)$$

$$[\mathbf{p}_{j+1}] = (\mathbf{A}_{j+1}^{-1}([\mathbf{S}_j]\mathbf{A}_j))[\mathbf{p}_j] + (\mathbf{A}_{j+1}^{-1}[\mathbf{S}_j])([\mathbf{v}_j] - \hat{\mathbf{x}}_j)$$

$$\hat{\mathbf{x}}_{j+1} = m([\mathbf{v}_{j+1}])$$

Outputs :
$[\mathbf{x}_0]$, $[\mathbf{x}_1]$, ..., $[\mathbf{x}_N]$

Example: Consider the same example (15) as the direct method, with the same initial condition and step size. Extended Mean-Value method Algorithm with a Taylor model order k = 4, produces the sequence of boxes depicted on the following figures 2 and 3. Using this approach, the trajectory of the vector x have been considerably contracted. Figures 2 and 3 shows that the extended mean-value method gives good results even for long integration time, which is not possible with the direct method algorithm.

The good performance of the extended mean-value method is in agreement with Rihm (1994) statement; he claims that under some considerations (the choice of A_j, midpoints $\hat{x}_j$ and the *a priori* enclosures) this method gives good results.

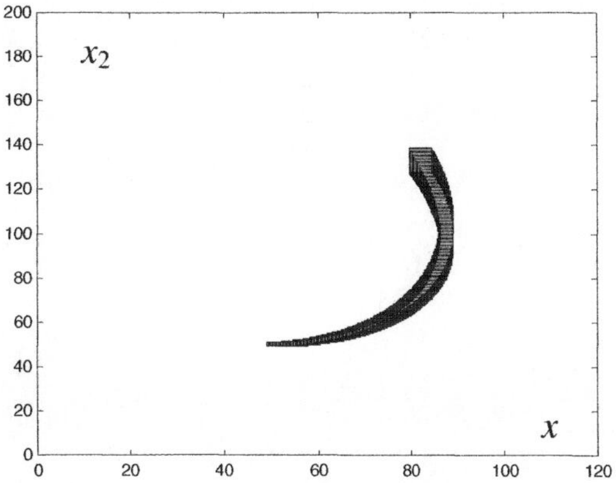

Fig.2. Superposition of boxes generated by the extended Mean-Value method algorithm. N=400.

In the following section, this method will be used to study state estimation for continuous systems.

6. STATE ESTIMATION

Consider a system represented by equations system (1). The aim of this section is estimating state $\boldsymbol{x}$ at some times $\{t_1, t_2, \ldots, t_n\}$. The algorithm proposed is based on predictor-corrector approach.

With forward propagation (integration of f), an enclosure domain for the state $\boldsymbol{x}$ at sample times $\{t_1, t_2, \ldots, t_n\}$ is computed; this step is called prediction phase. The extended mean value method is used to make this part. To simplify the state estimation algorithm, a constant step size strategy is chosen. At time (t_{k+1}), the measurement y_{k+1} becomes available which allows to contract the domain predicted for the state: correction phase. The state estimation approach proposed is summarised on the following algorithm:

Algorithm: State Estimation:

Inputs : $[x(0)]$, $[y_0]$, $[y_1]$,..., $[y_N]$

for j: = 1 to N
1. State prediction:
   compute $[x_{j+1}]$ with the Extended Mean-Value Method algorithm
2. State correction:
   $[x_{j+1}] = [x_{j+1}] \cap g^{-1}([y_{j+1}])$

Outputs :
$[x_1]$, $[x_2]$,..., $[x_N]$

7. EXAMPLE

Consider the Lotka-Volterra predatory-prey model given by equations (15).

For x_0 = [49;51]×[49;51], a constant step h = 0.005 and an order 4 for the Taylor model. We suppose that observation equation is $y = g(x_1, x_2) = x_1$. One takes $y = [y_m\text{-e}, y_m\text{+e}]$, where y_m denotes measurements obtained by model simulation and e the *prior* error bound. The pseudo-actual data was taken as a simulation run where the initial state is taken punctual x_0 = [50;50]×[50;50] and submitted to an additive uniform noise taken from the interval $[-0.05\mathbf{y}_m; 0.05\mathbf{y}_m]$. The prior error bound is taken as e = 0.05ym.

State estimation algorithm generates the set of boxes depicted in figure 4. As a basis for comparison, the prediction only steps are plotted in figure 3. Figure 4 indicates that state estimation is successful as one is able to efficiently bound the state vector. Note also that no bisections were necessary in the prediction part of the algorithm.

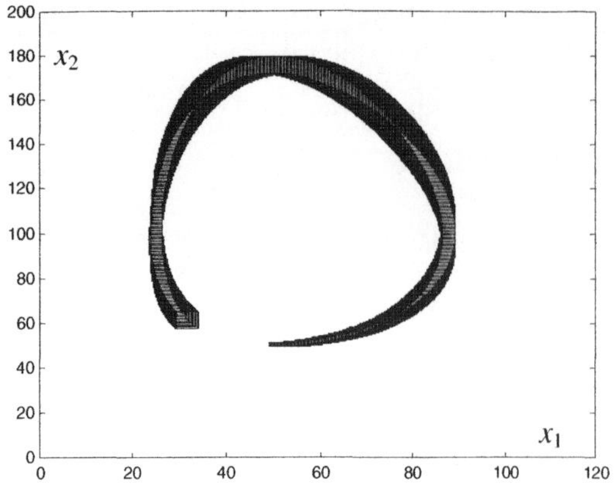

Fig.3. Superposition of boxes generated by the extended Mean-Value method algorithm. N=1000.

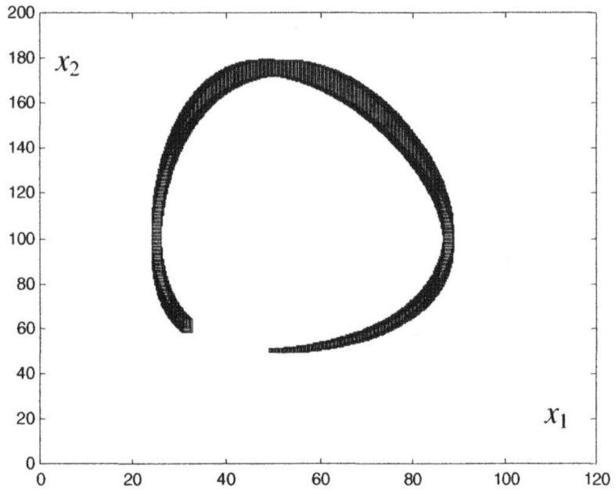

Fig.4. Set of boxes generated by state estimation algorithm when measurements are available. $x_0 = [49;51] \times [49;51]$.

8. CONCLUSION

In this paper, Taylor Models and interval analysis have been applied to study the guaranteed state estimation for nonlinear continuous-time systems in a bounded-error context. A predictor-corrector causal estimator, easy to implement, was proposed.

The prediction part of the state estimation algorithm consists on a validated integration of IVP for ODE with Taylor models. The latter need an *a priori* enclosure of state trajectory between sampling times. In addition, and in order to control the wrapping effect of interval computations, it was also necessary to use centred inclusion functions allied with pre-conditioning, the so-called extended mean-value method. In the examples used, the derived guaranteed estimator was successful even when the initial state is only approximately known.

In order to improve the quality of the estimator, one can opt for a variable step size strategy. A non-causal estimator can also be derived by backward propagation.

REFERENCES

Becerra, V.M., P.D. Roberts and G.W. Griffths (2001). Applying the extended Kalman filter to systems described by nonlinear differential-algebraic equations. *Control Engineering Practice* 9, pp 267-281.

Berz, M., K. Makino and J. Hoefkens (2001). Verified Integration of Dynamics in the Solar System. *Nonlinear Analysis 47,* pp 179-190.

Durieu, E. Walter and B. Polyak (2001). Multi-Input Multi-Output Ellipsoidal State Bounding. *Journal of Optimization Theory and Applications.* 111, No. 2, pp273-303.

Jaulin, L. (2002). Nonlinear bounded-error state estimation of continuous-time systems. *Automatica,* Vol 38, Issue 6, pp 1079-1082.

Jaulin, L. and E. Walter (1993). Set inversion via interval analysis for non linear bounded-error estimation. *Automatica,* 29(4) pp 1053-1064.

Jaulin, L., I Braems, M. Kieffer, and E. Walter (2000). Non linear state estimation using forward-backward propagation of intervals. In *SCAN 2000.*

Kieffer, M., L. Jaulin and E. Walter (2002), Guaranteed Recursive Non-Linear State Bounding Using Interval Analysis, *Int .J .Adapt. Control Signal Process* 16, pp. 193-218.

Magnus N., Niels K. Poulsen and Ole Ravn (2000). New developments in state estimation for nonlinear systems. *Automatica* 36, pp 1627-1638.

Maksarov, D.G. and J.P. Norton (2002). Computationally efficient algorithms for state estimation with ellipsoidal approximation, *Int. J. Adapt. Control Signal Process,* (in press).

Milanese, M., J.P. Norton, H. Piet-Lahanier and E. Walter, (1996). Bounding approaches to system identification. New York: Plenum.

Moore, R. E (1965). Automatic local coordinate transformations to reduce the growth of error bounds in interval computation of solutions of ordinary differential equations. *In Error in Digital Computation,* Vol. II, pp. 103-140, L. B. Rall, ed., Wiley, New York.

Moore, R. E. (1966). Interval analysis. Englewood Cliffs, NJ, Prentice-Hall.

Nedialkov, N. S. (1999). Computing Rigorous Bounds on the Solution of an Initial Value Problem for an Ordinary Differential Equation. *PhD,* University of Toronto.

Nedialkov, N. S. and K. R. Jackson (2000). A New Perspective on the Wrapping Effect in Interval Methods for Initial Value Problems for Ordinary Differential Equations. *SCAN2000*

Rihm, R. (1994). Interval methods for initial value problems in ODEs, in Topics in Validated computations: proceedings of the *IMACS-GAMM International Workshop on Validated Computations,* University of Oldenburg, J. Herzberger, ed. Elsevier Studies in Computational Mathematics, Elsevier, Amsterdam, New York.

Schweppe, F. C. (1968). Recursive State Estimation: unknown but bounded errors and system inputs. *IEEE Trans. On Automatic Control,* 13(1) pp 22-28.

IFAC
Publications
www.elsevier.com/locate/ifac

MULTIGRID DESIGN IN POINT-MASS APPROACH TO NONLINEAR STATE ESTIMATION

Miroslav Šimandl and Jakub Královec

Department of Cybernetics and
Center for Research of Cybernetic Systems
University of West Bohemia
Univerzitní 8, 306 14 Pilsen, Czech Republic
e-mail: simandl@kky.zcu.cz, kralove2@kky.zcu.cz

Abstract: Numerical solution of the Bayesian recursive relations in state estimation by the point-mass approach is treated. The stress is laid on the new grid design for multimodal probability density functions of state. A bank of grids is used for representation of the state space to cover different modes of the density. Splitting and merging techniques are designed for managing the bank of grids. *Copyright © 2003 IFAC*

Keywords: stochastic systems, state estimation, nonlinear filters, probability density function, estimation algorithms

1. INTRODUCTION

A general solution of the state estimation problem for discrete-time stochastic nonlinear non-Gaussian system is completely described by conditional probability density functions (pdf's) of state and it is given by the Bayesian recursive relations (BRR). A closed-form solution of the BRR is known only for linear Gaussian systems and a few special cases, e.g. (Sorenson, 1988; Söderström, 1994; Šimandl, 1996).

Existing global nonlinear filtering methods are based on the following approaches: analytical (Šimandl and Flídr, 1997), numerical (Bucy and Senne, 1971), and Monte Carlo (Liu and Chen, 1998) approaches. Each of these approaches approximates pdf's and/or state space differently.

This paper deals with the numerical approach, namely the point-mass (PM) method. This method was introduced by Bucy and Senne (1971) and it is based on covering the state space by a grid of isolated points. Values of conditional pdf's are computed only at these grid points. The method has been successfully applied e.g. in tracking (Bergman, 1997). The main advantages of the PM method are relative theoretical simplicity of filter design and possibility of reaching an arbitrary accuracy of pdf approximation. On the other hand, the original version of the method suffered from enormous numerical demands, a procedure for setting the number of the grid points was not specified, and the method was not appropriate for description of multimodal pdf's.

The PM method was further elaborated by Sorenson (1988) and Kramer and Sorenson (1988) where both filtering and prediction steps of estimation process were expressed and the formal presentation of the method was improved by introducing the p-vector approach and piece-wise representation of pdf's, but the main above mentioned disadvantages were not treated. Šimandl *et al.* (2002) designed an adaptive technique for setting the number of grid points, partially eliminating also numerical demands of the PM method.

This paper is focused on effective application of the PM method for filtering problems with multimodal pdf's. If the state space is covered by one grid in this case, as usual, then the grid may cover large areas with negligible probability as well. To avoid this, new algorithms for splitting and merging of grids should be designed.

The paper is organized as follows. The basic algorithm for PM estimator is introduced in Section 2. The goal of the paper is stated in Section 3. Section 4 deals with multigrid design and grid managing. The results of the paper are illustrated by a numerical example in Section 5.

2. POINT-MASS APPROACH TO NONLINEAR STATE ESTIMATION

Consider the nonlinear stochastic system

$$\mathbf{x}_{k+1} = \mathbf{f}_k(\mathbf{x}_k) + \mathbf{w}_k, \quad k = 0,1,2,\ldots \tag{1}$$

$$\mathbf{z}_k = \mathbf{h}_k(\mathbf{x}_k) + \mathbf{v}_k, \quad k = 0,1,2,\ldots \tag{2}$$

where the vectors $\mathbf{x}_k \in \mathscr{R}^n$, $\mathbf{z}_k \in \mathscr{R}^m$ represent the state of the system and the measurements at time k, respectively, $\mathbf{f}_k : \mathscr{R}^n \to \mathscr{R}^n$, $\mathbf{h}_k : \mathscr{R}^n \to \mathscr{R}^m$ are known vector functions, and $\mathbf{w}_k \in \mathscr{R}^n$, $\mathbf{v}_k \in \mathscr{R}^m$ are state and measurement zero-mean white noise sequences with positive definite covariance matrices $\mathbf{Q}_k$, $\mathbf{R}_k$, respectively, mutually independent and independent of $\mathbf{x}_0$. The pdf of the initial state $p(\mathbf{x}_0)$ is assumed to be known, as well as the pdf's of the noises $p(\mathbf{w}_k)$, $p(\mathbf{v}_k)$.

The filtering and predictive pdf's are given by the Bayesian recursive relations

$$p(\mathbf{x}_k|\mathbf{z}^k) = \frac{p(\mathbf{x}_k|\mathbf{z}^{k-1})\,p(\mathbf{z}_k|\mathbf{x}_k)}{\int p(\mathbf{x}_k|\mathbf{z}^{k-1})\,p(\mathbf{z}_k|\mathbf{x}_k)\,d\mathbf{x}_k} \tag{3}$$

$$p(\mathbf{x}_{k+1}|\mathbf{z}^k) = \int p(\mathbf{x}_k|\mathbf{z}^k)\,p(\mathbf{x}_{k+1}|\mathbf{x}_k)\,d\mathbf{x}_k \tag{4}$$

where $\mathbf{z}^k = \{\mathbf{z}_0, \ldots, \mathbf{z}_k\}$ and $p(\mathbf{x}_0|\mathbf{z}^{-1}) = p(\mathbf{x}_0)$. The transition pdf $p(\mathbf{x}_{k+1}|\mathbf{x}_k)$ can be expressed as $p(\mathbf{x}_{k+1}|\mathbf{x}_k) = p_{\mathbf{w}_k}(\mathbf{x}_{k+1} - \mathbf{y}_{k+1})$ where $\mathbf{y}_{k+1} = \mathbf{f}_k(\mathbf{x}_k)$. The measurement pdf can be written using (2) as $p(\mathbf{z}_k|\mathbf{x}_k) = p_{\mathbf{v}_k}(\mathbf{z}_k - \mathbf{h}_k(\mathbf{x}_k))$.

The key idea of the PM method (Bucy and Senne, 1971) for generating conditional pdf's of the state at k-th instant is to substitute a nonnegligible continuous support of the pdf by a grid of N_k isolated points. Values of the pdf are computed only at these grid points and thus the solution of (3), (4) is performed numerically over the grid instead of the continuous support. The nonnegligible support is a region in the state space where the actual state is probable to lie and hence values of the pdf are nonnegligible there.

The PM approach represents a pdf $p(\mathbf{x}_k)$ by a grid of points $\Xi_k(N_k) = \{\underline{\xi}_{ki}; \underline{\xi}_{ki} \in \mathscr{R}^n, i = 1, \ldots, N_k\}$, by the volume mass $\Delta\underline{\xi}_k$ of each grid point's neighbourhood and by a set of pdf values at the grid points, $\mathscr{P}_k = \{P_{k,i}; P_{k,i} = p_{\mathbf{x}_k}(\underline{\xi}_{ki}), \underline{\xi}_{ki} \in \Xi_k(N_k)\}$.

Algorithm of the PM method

Initialization: Define an initial grid $\Xi_0(N_0)$ in $\mathscr{R}^n$ for the prior pdf $p(\mathbf{x}_0|\mathbf{z}^{-1})$: $\Xi_0(N_0) = \{\underline{\xi}_{0i}; i = 1, \ldots, N_0\}$, volume $\Delta\underline{\xi}_0$ and set of pdf values $\mathscr{P}_{0|-1} = \{P_{0|-1,i}; i = 1, \ldots, N_0\}$.

The grid $\Xi_0(N_0)$ should be rectangular and its edges should be parallel with coordinate axes. Let an equally spaced axis grid for ℓ-th coordinate of $\mathbf{x}_0$ be denoted $\Xi_0^{(\ell)}(N_0^{(\ell)}) = \{\xi_0^{(\ell)}(i_\ell); i_\ell = 1, \ldots, N_0^{(\ell)}\}$ for $\ell = 1, \ldots, n$, where $N_0^{(\ell)}$ is the number of points for ℓ-th coordinate at time 0. Grid points $\underline{\xi}_{0,i} \in \Xi_0(N_0)$ are created by Cartesian product of the axis grids

$$\underline{\xi}_{0,j} = \left[\xi_0^{(1)}(i_1)\ \xi_0^{(2)}(i_2)\ \ldots\ \xi_0^{(n)}(i_n)\right]^{\mathrm{T}};$$

$$i = i_n + \sum_{\ell=1}^{n-1} (i_\ell - 1) \prod_{m=\ell+1}^{n} N_0^{(m)} \tag{5}$$

where $i_\ell = 1, \ldots, N_0^{(\ell)}$ for $\ell = 1, \ldots, n$.

The volume $\Delta\underline{\xi}_0$ can be enumerated by $\Delta\underline{\xi}_0 = \prod_{\ell=1}^{n}[\xi_0^{(\ell)}(2) - \xi_0^{(\ell)}(1)]$.

A grid with the described structure will be called *stiff grid*.

Then proceed for $k = 0, 1, \ldots$.

Step 1: At time k compute values of the approximate filtering pdf at points of the grid $\Xi_k(N_k)$ using (3), for $i = 1, \ldots, N_k$

$$P_{k|k,i} = c_k^{-1} P_{k|k-1,i}\, p_{\mathbf{v}_k}(\mathbf{z}_k - \mathbf{h}_k(\underline{\xi}_{ki})) \tag{6}$$

where $c_k = \Delta\underline{\xi}_k \sum_{i=1}^{N_k} P_{k|k-1,i}\, p_{\mathbf{v}_k}(\mathbf{z}_k - \mathbf{h}_k(\underline{\xi}_{ki}))$.

Step 2: Transform $\Xi_k(N_k)$ to a grid $H_{k+1}(N_k) = \{\underline{\eta}_{k+1,i}; i = 1, \ldots, N_k\}$ by the system dynamics

$$\underline{\eta}_{k+1,i} = \mathbf{f}_k(\underline{\xi}_{ki}) \tag{7}$$

Step 3: Redefine $H_{k+1}(N_k)$ to obtain a new grid $\Xi_{k+1}(N_{k+1})$ for $\mathbf{x}_{k+1}$ with the same structural properties as the original grid: $\Xi_{k+1}(N_{k+1}) = \{\underline{\xi}_{k+1,j}; j = 1, \ldots, N_{k+1}\}$.

Step 4: Compute values of the approximate predictive pdf for the new grid $\Xi_{k+1}(N_{k+1})$ using (4)

$$P_{k+1|k,j} = \Delta\underline{\xi}_k \sum_{i=1}^{N_k} P_{k|k,i}\, p_{\mathbf{w}_k}(\underline{\xi}_{k+1,j} - \underline{\eta}_{k+1,i}) \tag{8}$$

for $j = 1, \ldots, N_{k+1}$.

In the standard form of the PM method, the grid $\Xi_k(N_k)$ is usually designed by the *floating grid* technique (Bucy and Senne, 1971). The floating grid is a rectangular grid centered at the estimate of predictive mean of $\mathbf{x}_k$ and rotated according to the eigenvectors of predictive covariance matrix.

The discrete convolution (8) contains of $N_k N_{k+1}$ multiplications which makes it decisive for computational demands of the algorithm.

3. PROBLEM FORMULATION

The PM algorithm presented in Section 2 is not suitable for the case when the state is described by multimodal filtering or predictive pdf's. In such case there are regions in the state space which have negligible probability but they are covered by the grid. This phenomenon significantly increases numerical demands of the PM algorithm.

The aim of this paper is to generalize the standard PM algorithm by introducing several grids to cover the nonnegligible regions of the state space. This generalization requires to set up a handling strategy for the grids, namely to design procedures for splitting and merging of grids. The procedures should be simple enough not to burden the algorithm by too many additional operations.

4. MULTIGRID DESIGN AND HANDLING

This section consists of solutions of two key tasks in multigrid point-mass method. The first one is a generalization of the single grid representation to a multigrid version. The second one is to find procedures for splitting and merging of grids.

4.1 *Multigrid Representation of Multimodal Pdf*

The main reason for construction of several grids in the state space is consideration of a multimodal conditional pdf of state. The multimodality may arise in Bayesian inference as a result of multimodal pdf of state and/or measurement noise, prior pdf and nonlinear functions in state or measurement equations. It appears natural to cover each nonnegligible region of the state space by one grid the same way as in the general algorithm and to handle the grids independently.

An arbitrary multimodal pdf of state $p(\mathbf{x}_k)$ may be described by the following objects

$$\Xi_k[\mu](N_k[\mu]) = \{\underline{\xi}_{ki}[\mu]; \underline{\xi}_{ki}[\mu] \in \mathscr{R}^n, \quad i = 1,\ldots,N_k[\mu]\} \tag{9}$$

$$\mathscr{P}_k[\mu] = \{P_{k,i}[\mu]; P_{k,i}[\mu] = p_{\mathbf{x}_k}(\underline{\xi}_{ki}[\mu]), \quad \underline{\xi}_{ki}[\mu] \in \Xi_k[\mu](N_k[\mu])\} \tag{10}$$

$$\omega_k[\mu] = \Delta\underline{\xi}_k[\mu] \sum_{i=1}^{N_k[\mu]} P_{k,i}[\mu] \tag{11}$$

for $\mu = 1,\ldots,M_k$. The symbols in (9), (10) respect the notation in the basic PM algorithm in Section 2 and the bracketed index $[\mu]$ denotes pertinence of an object to μth grid. The new symbol $\omega_k[\mu]$ denotes the weight of μth grid $\Xi_k[\mu](N_k[\mu])$ and it represents probability of appearance of the state in the region covered by the μth grid and thus it holds that $\sum_{\mu=1}^{M_k} \omega_k[\mu] = 1$.

The generalized PM representation (9)–(11) can be used for approximation of filtering and predictive pdf's $p(\mathbf{x}_k|\mathbf{z}^k)$, $p(\mathbf{x}_{k+1}|\mathbf{z}^k)$.

For a μth grid with corresponding filtering pdf values it is possible to compute local predictive mean vector $\hat{\underline{\eta}}_{k+1}[\mu]$ and local predictive covariance matrix $\mathbf{C}_{k+1}[\mu]$ as follows

$$\hat{\underline{\eta}}_{k+1}[\mu] = \frac{\Delta\underline{\xi}_k[\mu] \sum_{i=1}^{N_k[\mu]} \underline{\eta}_{k+1,i}[\mu] P_{k|k,i}[\mu]}{\omega_{k|k}[\mu]} \tag{12}$$

$$\mathbf{C}_{k+1}[\mu] = \frac{\Delta\underline{\xi}_k[\mu] \sum_{i=1}^{N_k[\mu]} \underline{\eta}_{k+1,i}[\mu] \underline{\eta}^T_{k+1,i}[\mu] P_{k|k,i}[\mu]}{\omega_{k|k}[\mu]} - \hat{\underline{\eta}}_{k+1}[\mu] \hat{\underline{\eta}}^T_{k+1}[\mu] + \mathbf{Q}_k\,. \tag{13}$$

Global predictive mean $E(\mathbf{x}_{k+1}|\mathbf{z}^k)$ and covariance matrix $\mathrm{cov}(\mathbf{x}_{k+1}|\mathbf{z}^k)$ can be approximated using the local moments as follows:

$$\hat{\underline{\eta}}_{k+1} = \sum_{\mu=1}^{M} \omega_{k|k}[\mu] \hat{\underline{\eta}}_{k+1}[\mu] \tag{14}$$

$$\mathbf{C}_{k+1} = \sum_{\mu=1}^{M} \omega_{k|k}[\mu] \left(\mathbf{C}_{k+1}[\mu] + \hat{\underline{\eta}}_{k+1}[\mu] \hat{\underline{\eta}}^T_{k+1}[\mu]\right) - \hat{\underline{\eta}}_{k+1} \hat{\underline{\eta}}^T_{k+1}\,. \tag{15}$$

Moments of the filtering pdf can be enumerated analogously.

The general PM algorithm from Section 2 can be applied to each grid separately. Global probabilistic evaluation of the grid set is realized via the normalizing constant c_k at the filtering step which must be enumerated by the sum over all grids as

$$c_k = \sum_{\mu=1}^{M_k} \Delta\underline{\xi}_k[\mu] \sum_{i=1}^{N_k[\mu]} \left[P_{k|k-1,i}[\mu] \cdot p_{\mathbf{v}_k}\left(\mathbf{z}_k - \mathbf{h}_k(\underline{\xi}_{ki}[\mu])\right)\right]\,. \tag{16}$$

Parallel runs of the PM algorithm for each grid yields significant computational reduction because only nonnegligible regions of the state space are covered by grids and because the critical step 4 realized for a μth grid does not include points of other grids. However, it is necessary to treat situations when grids need to be split or merged.

4.2 *Splitting of Grid*

Probability of certain areas of the state space may drop significantly by including new information at the filtering step, causing that the pdf can be taken for multimodal. For that reason splitting of the grid should be considered at the filtering step. Moreover, it is possible to utilize the advantageous regular shape of the grid. Thus, the grids $\Xi_k[\mu](N_k[\mu])$, $\mu = 1,\ldots,M_k$ with the filtering pdf values $\mathscr{P}_{k|k}[\mu]$ will be examined for splitting.

It is necessary to find out if a grid $\Xi_k[\mu](N_k[\mu])$ covers separable areas with nonnegligible probabilities. To avoid searching separable areas in $\mathscr{R}^n$, which would be an extremely computationally expensive process, the separable areas will be examined individually for each state component by means of marginal pdf's and axis grids.

Points of axis grids and number of the points will be denoted without the grid index $[\mu]$ for notational simplicity. Thus, the axis grid of μth grid for ℓth state component $x_k^{(\ell)}$ will be written

$$\Xi_k^{(\ell)}[\mu](N_k^{(\ell)}) = \{\xi_k^{(\ell)}(i_\ell);\, i_\ell = 1,\ldots,N_k^{(\ell)}\}. \quad (17)$$

Before computation of marginal pdf's a decision rule must be set up to say what probability mass of a point is taken for negligible. The following rule is applied to all points of the grid $\Xi_k[\mu](N_k[\mu])$:

$$P_{k|k,i}[\mu] \leq \frac{\varepsilon}{N_k \Delta \underline{\xi}_k[\mu]}, \quad (18)$$

where N_k is the total number of points in all grids and ε is a threshold parameter, $0 \leq \varepsilon \ll N_k$. If the filtering pdf value $P_{k|k,i}[\mu]$ fulfills the condition (18), then it set to zero, i.e. $P_{k|k,i}[\mu] = 0$. With smaller value of ε fewer grid points are set to zero. Note that the grid points with zeroized pdf value are not deleted from the grid because they will be utilized during the computation of marginal pdf's.

The marginal conditional pdf of state component $x_k^{(\ell)}$, $\ell = 1,\ldots,n$, from the $p(\mathbf{x}_k|\mathbf{z}^k)$ covering an area $\Omega_k[\mu] = I_{k,\mu}^{(1)} \times \ldots \times I_{k,\mu}^{(n)}$ will be denoted as $p_\ell[\mu](x_k^{(\ell)}|\mathbf{z}^k)$. This pdf is defined by the $(n-1)$-fold integral:

$$\int\limits_{I_{k,\mu}^{(1)}} \cdots \int\limits_{I_{k,\mu}^{(\ell-1)}} \int\limits_{I_{k,\mu}^{(\ell+1)}} \cdots \int\limits_{I_{k,\mu}^{(n)}} p(x_k^{(1)},\ldots,x_k^{(n)}|\mathbf{z}^k) \cdot \mathrm{d}x_k^{(1)} \ldots \mathrm{d}x_k^{(\ell-1)} \mathrm{d}x_k^{(\ell+1)} \ldots \mathrm{d}x_k^{(n)} \quad (19)$$

where $x_k^{(\ell)} \in I_{k,\mu}^{(\ell)}$. Values of the marginal pdf $P_{k|k,i_\ell}^{(\ell)}[\mu] \approx p_\ell[\mu](\xi_k^{(\ell)}(i_\ell)|\mathbf{z}^k)$ at points of the ℓth axis grid $\xi_k^{(\ell)}(i_\ell) \in \Xi_k^{(\ell)}[\mu](N_k^{(\ell)})$ can be enumerated by a numerical approximation of (19) as

$$P_{k|k,i_\ell}^{(\ell)}[\mu] = \frac{\Delta\underline{\xi}_k[\mu]}{\Delta\xi_k^{(\ell)}[\mu]} \cdot \sum_{i_1=1}^{N_k^{(1)}} \cdots \sum_{i_{\ell-1}=1}^{N_k^{(\ell-1)}} \sum_{i_{\ell+1}=1}^{N_k^{(\ell+1)}} \cdots \sum_{i_n=1}^{N_k^{(n)}} P_{k|k,i}[\mu]$$

$$i = i_n + \sum_{j=1}^{n-1}(i_j - 1) \prod_{m=j+1}^{n} N_k^{(m)}, \quad (20)$$

where the relation $\Delta\underline{\xi}_k[\mu] = \prod_{\ell=1}^{n} \Delta\xi_k^{(\ell)}[\mu]$ has been used.

Marginal pdf's allow a simple procedure for grid splitting of the grid. If at least one point $\xi_k^{(\ell)}(i_\ell)$ of axis grid $\Xi_k^{(\ell)}[\mu](N_k^{(\ell)})$ has zero value $P_{k|k,i_\ell}^{(\ell)}[\mu]$ of marginal pdf and there exist two points $\xi_k^{(\ell)}(r), \xi_k^{(\ell)}(s) \in \Xi_k^{(\ell)}[\mu](N_k^{(\ell)})$ such that $\xi_k^{(\ell)}(r) < \xi_k^{(\ell)}(i_\ell)$, $\xi_k^{(\ell)}(s) > \xi_k^{(\ell)}(i_\ell)$ and $P_{k|k,r}^{(\ell)}[\mu] > 0$, $P_{k|k,s}^{(\ell)}[\mu] > 0$, then the grid can be split. However, a direct realization of this idea as an algorithm would be complicated and therefore the following procedure is preferred.

The points of the axis grid $\Xi_k^{(\ell)}[\mu](N_k^{(\ell)})$, which have non-zero marginal pdf value, are searched. The result of the searching is an index set $\mathscr{I}_\ell[\mu]$ defined as:

$$\mathscr{I}_\ell[\mu] = \{i_\ell : P_{k|k,i_\ell}^{(\ell)}[\mu] > 0\}. \quad (21)$$

The set $\mathscr{I}_\ell[\mu]$ contains all indexes i_ℓ of points $\xi_k^{(\ell)}(i_\ell)$ of the grid (17), whose values of the marginal filtering pdf are non-zero.

Let the index set $\mathscr{I}_\ell[\mu]$ be sorted in ascending order. Then divide $\mathscr{I}_\ell[\mu]$ into the smallest possible number m_ℓ of disjoint subsets $\mathscr{I}_\ell[\mu,r]$, $r = 1,\ldots,m_\ell$, $\mathscr{I}_\ell[\mu] = \bigcup_{r=1}^{m_\ell} \mathscr{I}_\ell[\mu,r]$, so that it holds:

$$\max_{i\in\mathscr{I}_\ell[\mu,r]} i = \min_{i\in\mathscr{I}_\ell[\mu,r]} i + \left|\mathscr{I}_\ell[\mu,r]\right| - 1, \quad (22)$$

where $\left|\mathscr{I}_\ell[\mu,r]\right|$ represents the number of elements of the subset $\mathscr{I}_\ell[\mu,r]$. Expression (22) means that the distance between two neighbouring indexes in the set $\mathscr{I}_\ell[\mu,r]$ is always 1. The index sets $\mathscr{I}_\ell[\mu,r]$, $r = 1,\ldots,m_\ell$, thus determine m_ℓ new axis grids which replace the original axis grid $\Xi_k^{(\ell)}[\mu]$:

$$\Xi_k^{(\ell)}[\mu,r] = \{\xi_k^{(\ell)}(i) : \xi_k^{(\ell)}(i) \in \Xi_k^{(\ell)}[\mu] \wedge i \in \mathscr{I}_\ell[\mu,r]\} \quad (23)$$

Note that simultaneously with the splitting of the axis grid, points with zero pdf value, representing borders of the pdf domain with negligible probability have been removed from the grid.

By the division of an axis grid $\Xi_k^{(\ell)}[\mu](N_k^{(\ell)})$, the grid $\Xi_k[\mu](N_k[\mu])$ is split into m_ℓ new grids as well. Then it is recommended to examine the new grids on the possibility of splitting again via all marginal pdf's except for the last one that was used for splitting.

4.3 *Merging of Grids*

The development of grids is determined by the system dynamics in Step 2 of the general PM method algorithm. This transformation may cause more grids to overlap. To keep the algorithm effective it is advantageous to use only one grid for covering each nonnegligible region of the state space. Contrary to the prediction step, overlapping of grids cannot happen at the filtering step because grid points are fixed in this step.

The following set of grids will be considered

$$\{H_{k+1}[\mu](N_k[\mu]); \mu = 1,\ldots,M_k\}, \quad (24)$$

where $H_{k+1}[\mu](N_k[\mu]) = \{\underline{\eta}_{k+1,i}[\mu]; i = 1,\ldots,N_k[\mu]\}$ and points of these grids were obtained by (7).

A decision whether two grids should be merged will be based on comparison of the distance of grids' local means with a certain limiting distance. The limiting distance will be determined by local covariance matrices for the two grids.

For each grid $H_{k+1}[\mu](N_k[\mu])$, $\mu = 1,\ldots,M_k$, local mean $\hat{\eta}_{k+1}[\mu]$ and local covariance matrix $\mathbf{C}_{k+1}[\mu]$ are computed by (12) and (13), respectively. The covariance matrix $\mathbf{C}_{k+1}[\mu]$ defines an n-axial ellipsoid with the center at $\hat{\eta}_{k+1}[\mu]$ as follows

$$(\mathbf{x} - \hat{\underline{\eta}}_{k+1}[\mu])^{\mathrm{T}} \mathbf{C}_{k+1}^{-1}[\mu](\mathbf{x} - \hat{\underline{\eta}}_{k+1}[\mu]) = 1\,. \quad (25)$$

Lengths of ellipsoid's half axes are equal to square roots of eigenvalues of the covariance matrix $\mathbf{C}_{k+1}[\mu]$. Let the limiting distance of the two local means depend on the sum of lengths of the shortest half axes of the two ellipsoids.

First, the smallest eigenvalue of $\mathbf{C}_{k+1}[\mu]$ should be found and it will be denoted as $\lambda_{k+1}[\mu]$, corresponding to the length of the shortest half axis of the ellipsoid (25).

Two grids $H_{k+1}[\mu](N_k[\mu])$, $H_{k+1}[\nu](N_k[\nu])$ will be merged if the following inequality is fulfilled

$$\left\|\hat{\underline{\eta}}_{k+1}[\mu], \hat{\underline{\eta}}_{k+1}[\nu]\right\| < K\left(\sqrt{\lambda_{k+1}[\mu]} + \sqrt{\lambda_{k+1}[\nu]}\right), \quad (26)$$

where

$$\left\|\hat{\underline{\eta}}_{k+1}[\mu], \hat{\underline{\eta}}_{k+1}[\nu]\right\| = [(\hat{\underline{\eta}}_{k+1}[\mu] - \hat{\underline{\eta}}_{k+1}[\nu])^{\mathrm{T}} \cdot (\hat{\underline{\eta}}_{k+1}[\mu] - \hat{\underline{\eta}}_{k+1}[\nu])]^{\frac{1}{2}} \quad (27)$$

and $K > 0$ is a design parameter.

Criterion (26), which decides about merging two grids, is quite simple but it does not cover all possible situations when two grids overlap. Simplicity of the rule is thus preferred to its generality because full-range verifying of grid overlapping would be computationally burdensome and algorithm efficiency would be decreased.

The merging of grids is realized by creating index sets, each of which contains indexes of grids that should be merged. A new grid is defined by the grids that are determined by one of the index sets. The index sets are constructed by the following algorithm (index sets have a vector structure).

Algorithm of index sets creation for grid merging

1. $\nu := 1, \mu := 1$
2. $\mathscr{M}_\nu := [\mu]$
3. $\mu := \mu + 1, i := 0$
4. $\quad i := i + 1, j := 0$
5. $\quad\quad j := j + 1$
6. $\quad\quad$ if $\left\|\hat{\underline{\eta}}_{k+1}[\mu], \hat{\underline{\eta}}_{k+1}[\mathscr{M}_i(j)]\right\| < K\left(\sqrt{\lambda_{k+1}[\mu]} + \sqrt{\lambda_{k+1}[\mathscr{M}_i(j)]}\right)$, $\mathscr{M}_i := \left[\mathscr{M}_i^{\mathrm{T}}, \mu\right]^{\mathrm{T}}$, go to 11
7. $\quad\quad$ if $j < \dim(\mathscr{M}_i)$, go to 5
8. $\quad$ if $i < \nu$, go to 4
9. $\quad \nu := \nu + 1$
10. $\quad \mathscr{M}_\nu := \mu$
11. $\quad$ if $\mu < M_k$, go to 3
12. end

The symbol := in the algorithm stands for the assignment statement. Denote the number of the index sets as M_{k+1}, i.e. $\nu = 1,\ldots,M_{k+1}$. The number of index sets is equal to the number of new grids $\Xi_{k+1}[\nu]$. Note that the algorithm ensures that $\mathscr{M}_{\nu_1} \cap \mathscr{M}_{\nu_2} = \emptyset$ for $\nu_1 \neq \nu_2$. The index μ of such a grid $H_{k+1}[\mu](N_k[\mu])$ which cannot merge with any other grid belongs to a one-element set $\mathscr{M}_\nu = [\mu]$.

New grids $\Xi_{k+1}[\nu](N_{k+1}[\nu])$ are designed separately by Step 3 of the PM method algorithm.

Finally, values of predictive pdf are enumerated for the points of each new grid $\Xi_{k+1}[\nu](N_{k+1}[\nu])$, $\nu = 1,\ldots,M_{k+1}$, using grids $\Xi_k[\mu](N_k[\mu])$ specified by the index set $\mathscr{M}_\nu$

$$P_{k+1|k,j}[\nu] = \sum_{\mu \in \mathscr{M}_\nu} \Delta\underline{\xi}_k[\mu] \sum_{i=1}^{N_k[\mu]} [P_{k|k,i}[\mu] \cdot p_{\mathbf{w}_k}(\underline{\xi}_{k+1,j}[\nu] - \underline{\eta}_{k+1,i}[\mu]) \quad (28)$$

for $j = 1,\ldots,N_{k+1}[\nu]$

5. NUMERICAL ILLUSTRATION

The above described algorithms of splitting and merging of grids were tested in a number of numerical examples. An illustration of grid merging will be shown in this section.

Consider the nonlinear stochastic system

$$x_{k+1}^{(1)} = x_k^{(1)} x_k^{(2)} + \frac{25 x_k^{(1)}}{1 + (x_k^{(1)})^2} + \cos(0.6k) + w_k^{(1)}$$

$$x_{k+1}^{(2)} = x_k^{(2)} + w_k^{(2)}$$

$$z_k = \frac{1}{20}\left(x_k^{(1)}\right)^2 + v_k\,,$$

where $p(\mathbf{x}_0) = \mathcal{N}(\mathbf{x}_0\colon [0\; 0.5]^{\mathrm{T}}, \mathrm{diag}\{9, 10^{-2}\})$, $p(\mathbf{w}_k) = \mathcal{N}(\mathbf{w}_k : [0\;\; 0]^{\mathrm{T}}, \mathrm{diag}\{2, 10^{-4}\})$, $p(v_k) = \mathcal{N}(v_k : 0, 1)$.

The state $\mathbf{x}_k$ is estimated by the Algorithm of the PM Method with grid splitting technique (17)–(23). The design parameters of the PM filter were set as follows $\varepsilon = 10^{-5}$, $K = 3$.

The development of filtering pdf generated by the filter for $k = 1, \ldots, 4$ is shown in Figure 1. The single grid is split at time $k = 4$ into two grids at the $x_4^{(1)}$ coordinate.

6. CONCLUSIONS

A significant part of the point-mass method is a design and management of grid covering the nonnegligible area of the state space because computational demands of the method are substantially affected by this step. Multimodal pdf's call for covering the state space by several isolated grids. New algorithms have been developed for managing multiple grids, namely for splitting and merging of grids, which bring a considerable reduction of overall number of grid points and consequently save computing time of the point mass algorithm.

ACKNOWLEDGEMENTS

This work was supported by the Grant Agency of the Czech Republic under project no. 102/01/0021.

7. REFERENCES

Bergman, N. (1997). A Bayesian approach to terrain-aided navigation. In: *Proceedings of the 11th IFAC Symp. on System Identification, SYSID'97*. Fukuoka, Japan. pp. 1531–1536.

Bucy, R. and K. Senne (1971). Digital synthesis of non-linear filters. *Automatica* **7**(3), 287–298.

Kramer, S. and H.W. Sorenson (1988). Recursive Bayesian estimation using piece-wise constant approximations. *Automatica* **24**(6), 789–801.

Liu, J.S. and R. Chen (1998). Sequential monte carlo methods for dynamic systems. *J. Amer. Statist. Assoc.* **93**(443), 1032–1044.

Söderström, T. (1994). *Discrete-Time Stochastic Systems*. Prentice Hall Series in Systems and Control Engineering.

Sorenson, H.W. (1988). Recursive estimation for non-linear dynamic systems. In: *Bayesian Analysis of Time Series and Dynamic Models* (J.C. Spall, Ed.). Marcel Dekker. New York.

Šimandl, M. (1996). State estimation for non-Gaussian models. In: *Proceedings of 13th IFAC World Congress*. Vol. H. San Francisco. pp. 463–468.

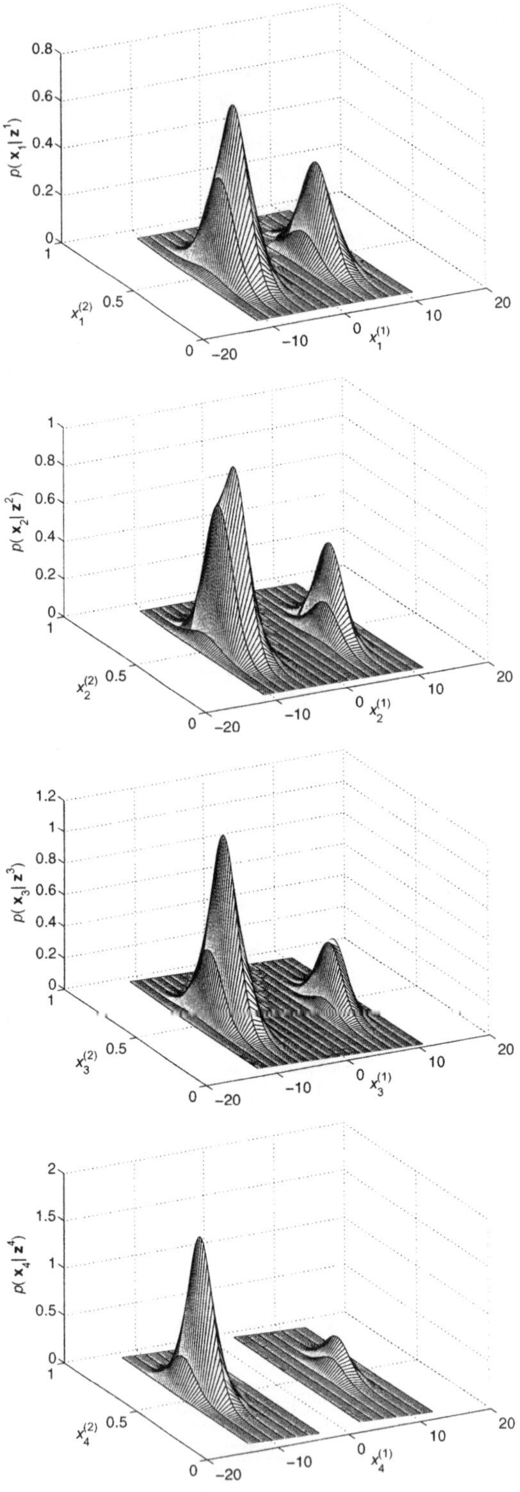

Fig. 1. Development of the filtering pdf for $k = 1, \ldots, 4$ with grid splitting.

Šimandl, M. and M. Flídr (1997). Nonlinear nonnormal dynamic models: State estimation and software. In: *Computer-Intensive Methods in Control and Signal Processing*. pp. 195–208. Birkhäuser. Boston.

Šimandl, M., J. Královec and T. Söderström (2002). Anticipative grid design in point-mass approach to nonlinear state estimation. *IEEE Transactions on Automatic Control* **47**(4), 699–702.

IFAC Publications
www.elsevier.com/locate/ifac

AN EFFICIENT NONLINEAR ADAPTIVE OBSERVER WITH GLOBAL CONVERGENCE

Qinghua Zhang[*], **Aiping Xu**[**] **and Gildas Besançon**[***]

[*] *IRISA-INRIA, Campus de Beaulieu, 35042 Rennes Cedex, France*
zhang@irisa.fr
[**] *Shandaong University, Jinan 250100, China*
aipingxu@math.sdu.edu.cn
[***] *LAG-ENSIEG, BP 46, 38402 Saint-martin d'Hères, France*
Gildas.Besancon@lag.ensieg.inpg.fr

Abstract: A nonlinear global exponential adaptive observer is proposed in this paper for joint state-parameter estimation. It is conceptually simple and numerically efficient. Its design is based on the combination of a nonlinear high gain observer and a linear adaptive observer. The considered class of systems is truly nonlinear in the sense that they cannot be linearized by coordinate change and output injection.

Keywords: nonlinear system, state and parameter estimation, adaptive observer, high gain observer, global convergence.

1. INTRODUCTION

Adaptive observers are recursive algorithms for joint state-parameter estimation in dynamic systems, or for state estimation only despite the presence of unknown parameters. Such algorithms have important applications in fault detection and isolation (FDI) and in adaptive control.

Some early works on adaptive observers for linear systems can be found in (Lüders and Narendra, 1973; Kreisselmeier, 1977). Adaptive observers for nonlinear systems then have drawn more attentions of researchers. Studies on global state and parameter estimation have been reported in (Bastin and Gevers, 1988; Marino and Tomei, 1995). However, these results are restricted to the nonlinear systems whose error dynamics can be linearized by coordinate change and output injection. More recently, some more general results on nonlinear systems have been published (Rajamani and Hedrick, 1995; Cho and Rajamani, 1997; Besançon, 2000). These methods do not require the considered nonlinear system to be linearizable, instead, they assume the existence of some Lyapunov function satisfying particular conditions.

A natural idea for joint estimation of states and parameters is to consider the extended system by appending the unknown parameters into the state vector. Such methods include the well known *extended Kalman filters* (EKF). See, for instance, (Reif and Unbehauen, 1999; Einicke and White, 1999). In general only local convergence of EKF is expected. Recently, a new method based on such extended systems has been proposed in (Poznyak and Martinez, 2001; Martinez and Poznyak, 2001). The algorithm design and convergence proof of this method are based on the analysis of the observability and inobservability manifolds in the extended state space. It requires a particular assumption on the form of the observability matrix of the extended system and an unusual persistent excitation condition.

Based on a recent result on adaptive observers for linear time varying (LTV) systems (Zhang, 2002) and on the techniques of *high gain observers* (Gauthier *et al.*, 1992; Gauthier and Kupka, 1994), a global implicit nonlinear adaptive observer (in the form of differential-algebraic equations) has been proposed in (Zhang and Xu, 2001), and an explicit nonlinear adaptive observer in (Xu and Zhang, 2002).

The main contribution of this paper is to simplify the algorithm of (Xu and Zhang, 2002) which was based on a multiple-delayed system and thus was numerically expensive. In this paper, it is shown that the multiple-delayed system is not necessary, and consequently the algorithm can be simplified to considerably reduce its numerical cost.

In this paper, the joint estimation of the state vector $x(t)$ and the parameter vector θ in nonlinear systems of the following form is considered

$$\dot{x}(t) = A_o x(t) + f(x(t)) + g(x(t))u(t) + \Psi(t)\theta \quad (1a)$$
$$y(t) = c_o x(t) \quad (1b)$$

where $x(t) \in \mathbb{R}^n$, $u(t) \in \mathbb{R}^l$ and $y(t) \in \mathbb{R}$ are system state, input and output,

$$A_o = \begin{bmatrix} 0 & 1 & \cdots & 0 \\ \vdots & \vdots & \ddots & \vdots \\ 0 & 0 & \cdots & 1 \\ 0 & 0 & \cdots & 0 \end{bmatrix} \qquad c_o = \begin{bmatrix} 1 & 0 & \cdots & 0 \end{bmatrix} \quad (2)$$

$f : \mathbb{R}^n \to \mathbb{R}^n$ and $g : \mathbb{R}^n \to \mathbb{R}^n \times \mathbb{R}^l$ are two nonlinear functions in the triangular form:

$$f(x) = \begin{bmatrix} f_1(x_1) \\ f_2(x_1, x_2) \\ \vdots \\ f_n(x_1 \ldots x_n) \end{bmatrix} \qquad g(x) = \begin{bmatrix} g_1(x_1) \\ g_2(x_1, x_2) \\ \vdots \\ g_n(x_1 \ldots x_n) \end{bmatrix} \quad (3)$$

$\Psi(t) \in \mathbb{R}^n \times \mathbb{R}^p$ is a matrix of known signals, possibly depending on u and y, $\theta \in \mathbb{R}^p$ is an unknown constant parameter vector. It is assumed that $u(t)$ and $\Psi(t)$ are both bounded.

Remark 1. System (1) without the term $\Psi(t)\theta$, that is,

$$\dot{x}(t) = A_o x(t) + f(x(t)) + g(x(t))u(t) \quad (4a)$$
$$y(t) = c_o x(t) \quad (4b)$$

is known as a standard triangular form for which a *high gain observer* can be designed for *global state estimation.* Typically, nonlinear systems observable for all inputs can be transformed into the form of (4) (Gauthier *et al.*, 1992). □

Remark 2. The main motivation for the study of systems in the form (1) is *fault detection and isolation* for which the term $\Psi(t)\theta$ models the faults to be detected and isolated. More precisely, if a component of θ is different from its nominal value, then it indicates that the fault modeled by the corresponding column of $\Psi(t)$ occurs. □

The paper is organized as follows. In Section 2 some background results are recalled. In Section 3 the proposed adaptive observer is described and its global exponential convergence is formally proved. Section 4 presents a numerical example for illustration. Finally, the paper is concluded by Section 5.

2. SOME BACKGROUND RESULTS

2.1 *A basic result on high gain observer*

Consider system (4) for state observer design. For any positive real number ρ, define

$$\Lambda = \text{diag}(1, \rho^{-1}, \rho^{-2}, \ldots, \rho^{-(n-1)}) \quad (5)$$

Let S be the solution of the matrix equation

$$A_o^T S + S A_o + S = c_o^T c_o \quad (6)$$

It is known that S is a positive definite matrix (Gauthier *et al.*, 1992).

Define

$$k_o = \frac{1}{2} S^{-1} c_o^T \quad (7)$$

Theorem 1. Consider system (4) with $A_o, c_o, f(x)$ and $g(x)$ as defined in (2) and (3). If the functions $f(x)$ and $g(x)$ are globally Lipschitz, and if the input $u(t)$ is bounded, then, for sufficiently large ρ, the ordinary differential equation (ODE)

$$\dot{\hat{x}}(t) = A_o \hat{x}(t) + f(\hat{x}(t)) + g(\hat{x}(t))u(t) + \rho \Lambda^{-1} k_o (y(t) - c_o \hat{x}(t)) \quad (8)$$

with Λ and k_o as defined in (5) and (7) is a global observer for the state of system (4) with exponential convergence, *i.e.*, for all initial conditions $x(t_0)$ and $\hat{x}(t_0)$, the error $\hat{x}(t) - x(t)$ tends to zero exponentially fast when $t \to \infty$. □

See (Gauthier *et al.*, 1992) for more details on the high gain observer.

2.2 *An adaptive observer for linear systems*

Consider linear time varying (LTV) systems of the form

$$\dot{x}(t) = A(t)x(t) + B(t)u(t) + \Psi(t)\theta \quad (9a)$$
$$y(t) = C(t)x(t) \quad (9b)$$

where $x(t) \in \mathbb{R}^n$, $u(t) \in \mathbb{R}^l$, $y(t) \in \mathbb{R}^m$ are respectively the state, input, output of the system, $A(t), B(t), C(t)$ are known time varying bounded matrices of appropriate sizes, $\theta \in \mathbb{R}^p$ is an unknown constant parameter vector, $\Psi(t) \in \mathbb{R}^n \times \mathbb{R}^p$ is a bounded matrix of known signals.

Assumption 1. Assume that there exists a bounded time-varying matrix $K(t) \in \mathbb{R}^n \times \mathbb{R}^m$ such that the system

$$\dot{\eta}(t) = [A(t) - K(t)C(t)]\eta(t) \quad (10)$$

is exponentially stable.

Assumption 2. Let $\Upsilon(t) \in \mathbb{R}^n \times \mathbb{R}^p$ be a matrix of signals defined as the solution of

$$\dot{\Upsilon}(t) = [A(t) - K(t)C(t)]\Upsilon(t) + \Psi(t) \quad (11)$$

Assume that $\Psi(t)$ is persistently exciting, so that the matrix of signals $\Upsilon(t)$ obtained by linearly filtering $\Psi(t)$ through (11) satisfies, for some positive constants α, T and for all $t \geq t_0$, the inequality

$$\int_t^{t+T} \Upsilon^T(\tau)C^T(\tau)C(\tau)\Upsilon(\tau)d\tau \geq \alpha I \quad (12)$$

Theorem 2. Let $\Gamma \in \mathbb{R}^p \times \mathbb{R}^p$ be any symmetric positive definite matrix. Under Assumptions 1 and 2, the ODE system

$$\begin{aligned} \dot{\Upsilon}(t) &= [A(t) - K(t)C(t)]\Upsilon(t) + \Psi(t) && (13a) \\ \dot{\hat{x}}(t) &= A(t)\hat{x}(t) + B(t)u(t) + \Psi(t)\hat{\theta}(t) \\ &\quad + K(t)[y(t) - C(t)\hat{x}(t)] \\ &\quad + \Upsilon(t)\Gamma\Upsilon^T(t)C^T[y(t) - C(t)\hat{x}(t)] && (13b) \\ \dot{\hat{\theta}}(t) &= \Gamma\Upsilon^T(t)C^T(t)\,[y(t) - C(t)\hat{x}(t)] && (13c) \end{aligned}$$

is a global exponential adaptive observer for system (9), *i.e.*, for any initial conditions $\Upsilon(t_0)$, $x(t_0)$, $\hat{x}(t_0)$, $\hat{\theta}(t_0)$ and $\forall\theta \in \mathbb{R}^p$, the errors $\hat{x}(t) - x(t)$ and $\hat{\theta}(t) - \theta$ tend to zero exponentially fast when $t \to \infty$. □

See (Zhang, 2002) for a proof of the theorem.

3. MAIN RESULT

In order to jointly estimate the state $x(t)$ and the parameter θ of system (1), the main idea is to combine the nonlinear high gain observer (8) and the linear adaptive observer (13). Essentially, the state estimation equation (13b) of the adaptive observer should incorporate the high gain observer (8). Obviously, $A(t)$ and $C(t)$ should be replaced by A_o and c_o of the nonlinear system. The gain $K(t)$ should be related to the gain k_o used in the high gain observer. Accordingly, some other modifications have to be made for the purpose of simplifying the error equation of the designed adaptive observer. This point will become clear later in the convergence analysis. Then the nonlinear adaptive observer for system (1) takes the form

$$\begin{aligned} \dot{\Upsilon}(t) &= \rho(A_o - k_oc_o)\Upsilon(t) + \rho\Lambda\Psi(t)\Omega && (14a) \\ \dot{\hat{x}}(t) &= A_o\hat{x}(t) + f(\hat{x}(t)) + g(\hat{x}(t))u(t) \\ &\quad + \Psi(t)\hat{\theta}(t) + \rho\Lambda^{-1}k_o[y(t) - c_o\hat{x}(t)] \\ &\quad + \Lambda^{-1}\Upsilon(t)\Gamma\Upsilon^T(t)c_o^T[y(t) - c_o\hat{x}(t)] && (14b) \\ \dot{\hat{\theta}}(t) &= \rho\Omega\Gamma\Upsilon^T(t)c_o^T[y(t) - c_o\hat{x}(t)] && (14c) \end{aligned}$$

where ρ is a positive parameter, Λ is as defined in (5), $\Gamma \in \mathbb{R}^p \times \mathbb{R}^p$ is a positive definite gain matrix, and $\Omega \in \mathbb{R}^p \times \mathbb{R}^p$ is a square invertible matrix depending on ρ in a way that is specified in the following remark.

Remark 3. The matrix Ω depending on ρ should be chosen such that, when $\rho \to \infty$, the matrix product $\Lambda\Psi(t)\Omega$ tends to a bounded limit, and that each column of $\Lambda\Psi(t)\Omega$ has at least one entry not tending to zero. The purpose of this requirement will become clear after the statement of the persistent excitation assumption. In order to choose such a matrix, it is sufficient to consider

$$\Omega = \text{diag}(\rho^{q_1}, \rho^{q_2}, \ldots, \rho^{q_p})$$

where $q_1, q_2, \ldots, q_p$ are integers to be chosen. Denote

$$\Psi = \begin{bmatrix} \psi_{1,1} & \psi_{1,2} & \cdots & \psi_{1,p} \\ \psi_{2,1} & \psi_{2,2} & \cdots & \psi_{2,p} \\ \vdots & \vdots & \vdots & \vdots \\ \psi_{n,1} & \psi_{n,2} & \cdots & \psi_{n,p} \end{bmatrix}$$

Then

$$\Lambda\Psi\Omega = \begin{bmatrix} \rho^{q_1}\psi_{1,1} & \cdots & \rho^{q_p}\psi_{1,p} \\ \rho^{q_1-1}\psi_{2,1} & \cdots & \rho^{q_p-1}\psi_{2,p} \\ \vdots & \vdots & \vdots \\ \rho^{q_1-(n-1)}\psi_{n,1} & \cdots & \rho^{q_p-(n-1)}\psi_{n,p} \end{bmatrix}$$

Consider each column of the matrix $\Lambda\Psi\Omega$. In the j-th column, some of the signals $\psi_{i,j}$ may be identically zero. Choose the value of q_j such that, for the entries with non identically zero $\psi_{i,j}$ in the j-th column, the highest exponent of ρ is zero. By choosing the values of q_j in this manner for all the columns, when $\rho \to \infty$, the entries of the product matrix $\Lambda\Psi\Omega$ with negative exponent of ρ tend to zero, the other entries are independent of ρ, and each column has one entry that does not tend to zero.

It is also clear that, when $\rho > 1$, there exists an upper bound of the matrix $\Lambda\Psi(t)\Omega$ independent of ρ. Now change the time scale with $s = \rho t$, then equation (14a) becomes

$$\frac{d\Upsilon}{ds} = (A_o - k_oc_o)\Upsilon + \Lambda\Psi\Omega$$

It is known that the gain vector k_o as defined in (7) ensures that all the eigenvalues of the matrix $(A_o - k_oc_o)$ have negative real part. It follows that there exists an upper bound of $\Upsilon(t)$ independent of ρ. □

Like in system identification, a persistent excitation condition is required for parameter estimation.

Assumption 3. The matrix of signals $\Psi(t)$ is assumed to be persistently exciting, in the sense that the matrix of signals $\Upsilon(t)$, obtained by linearly filtering $\Psi(t)$ through (14a) with some initial condition $\Upsilon(t_0)$, satisfies, for two positive constants α, T independent of ρ, for all $t \geq t_0$ and for sufficiently large ρ, the following inequality

$$\int_t^{t+T} \Upsilon^T(\tau)c_o^Tc_o\Upsilon(\tau)d\tau \geq \alpha I \quad (15)$$

Remark 4. It has been required that the constants α and T are independent of ρ when the value of ρ is sufficiently large. There are two ways to ensure this requirement. The first one is to assume that $\Psi(t)$ has sufficiently high frequency components (large band signals) so that the filtered $\Upsilon(t)$ remains persistently exciting even for a large value of ρ. The second one is based on the limiting behavior of $\Upsilon(t)$. Let us use the notation $\Upsilon_\rho(t)$ to make explicit the dependence of $\Upsilon(t)$ on ρ. Equation (14a) implies

$$\lim_{\rho\to\infty} \Upsilon_\rho(t) = -[A_o - k_o c_o]^{-1} \lim_{\rho\to\infty} \Lambda\Psi(t)\Omega$$

See Remark 3 for the limiting property of $\Lambda\Psi(t)\Omega$. By continuity, if there exist α and T for $\Upsilon_\infty(t)$ to satisfy (15), then, for sufficiently large ρ, the matrix $\Upsilon_\rho(t)$ will satisfy (15) with the same α and T. Note that, due to the particular form of $[A_o - k_o c_o]^{-1}$, for the limiting case to hold, the matrix $\Psi(t)$ must be filled with zero entries except the last row. The counterpart of this structural restriction is that $\Psi(t)$ is not necessarily large band in this case. □

Theorem 3. Let $\Gamma \in \mathbb{R}^p \times \mathbb{R}^p$ be any symmetric positive definite matrix, k_o as in (7). If $f(x)$, $g(x)$ are globally Lipschitz, and $u(t) \in \mathbb{R}^l$, $\Psi(t) \in \mathbb{R}^n \times \mathbb{R}^p$ are bounded, then, under Assumption 3, for sufficiently large $\rho > 0$, the ODE system (14) is a global exponential adaptive observer for system (1), *i.e.*, for any initial conditions $x(t_0), \hat{x}(t_0)$, $\hat{\theta}(t_0)$ and for all $\theta \in \mathbb{R}^p$, the errors $\hat{x}(t) - x(t)$ and $\hat{\theta}(t) - \theta$ tend to zero when $t \to \infty$, and the convergence is exponentially fast. □

Proof of Theorem 3.

Combine (14b) and (14c) to obtain

$$\dot{\hat{x}}(t) = A_o\hat{x}(t) + f(\hat{x}(t)) + g(\hat{x}(t))u + \Psi(t)\hat{\theta}(t) + \rho\Lambda^{-1}k_o[y(t) - c_o\hat{x}(t)] + \Lambda^{-1}\Upsilon(t)\rho^{-1}\Omega^{-1}\dot{\hat{\theta}}(t)$$

Let $\tilde{x}(t) = \hat{x}(t) - x(t)$, $\tilde{\theta}(t) = \hat{\theta}(t) - \theta$ and notice that $\dot{\theta} = 0$, then

$$\dot{\tilde{x}}(t) = (A_o - \rho\Lambda^{-1}k_o c_o)\tilde{x}(t) + f(\hat{x}(t)) - f(x(t)) + g(\hat{x}(t))u(t) - g(x(t))u(t) + \Psi(t)\tilde{\theta}(t) + \Lambda^{-1}\Upsilon(t)\rho^{-1}\Omega^{-1}\dot{\tilde{\theta}}(t)$$

Due to the special forms of Λ, A_o, c_o, it is easy to check that $\Lambda A_o = \rho A_o \Lambda$ and $c_o\Lambda = c_o$.

Define

$$\begin{aligned} z(t) &= \Lambda x(t) \\ \hat{z}(t) &= \Lambda \hat{x}(t) \\ \tilde{z}(t) &= \Lambda \tilde{x}(t) \\ \tilde{\vartheta}(t) &= \rho^{-1}\Omega^{-1}\tilde{\theta}(t) \end{aligned}$$

then

$$\dot{\tilde{z}}(t) = \rho(A_o - k_o c_o)\tilde{z}(t) + \xi(t) + \rho\Lambda\Psi(t)\Omega\tilde{\vartheta}(t) + \Upsilon(t)\dot{\tilde{\vartheta}}(t) \qquad (16)$$

where

$$\xi(t) = \Lambda[f(\Lambda^{-1}\hat{z}(t)) - f(\Lambda^{-1}z(t))] + \Lambda[g(\Lambda^{-1}\hat{z}(t)) - g(\Lambda^{-1}z(t))]u(t) \qquad (17)$$

Now define

$$\eta(t) = \tilde{z}(t) - \Upsilon(t)\tilde{\vartheta}(t)$$

then

$$\begin{aligned} \dot{\eta}(t) &= \rho(A_o - k_o c_o)[\eta(t) + \Upsilon(t)\tilde{\vartheta}(t)] + \xi(t) \\ &\quad + \rho\Lambda\Psi(t)\Omega\tilde{\vartheta}(t) - \dot{\Upsilon}(t)\tilde{\vartheta}(t) \\ &= \rho(A_o - k_o c_o)\eta(t) + \xi(t) \\ &\quad + [\rho(A_o - k_o c_o)\Upsilon(t) + \rho\Lambda\Psi(t)\Omega - \dot{\Upsilon}(t)]\tilde{\vartheta}(t) \end{aligned}$$

Because $\Upsilon(t)$ is generated by (14a), the last equation simply becomes

$$\dot{\eta}(t) = \rho(A_o - k_o c_o)\eta(t) + \xi(t) \qquad (18)$$

The definition of $\tilde{\vartheta}(t)$ implies $\dot{\tilde{\vartheta}}(t) = \rho^{-1}\Omega^{-1}\dot{\hat{\theta}}(t)$. Then, from (14c) and the fact that $\dot{\theta} = 0$,

$$\begin{aligned} \dot{\tilde{\vartheta}}(t) &= \Gamma\Upsilon^T(t)c_o^T[y(t) - c_o\hat{x}(t)] \\ &= -\Gamma\Upsilon^T(t)c_o^T c_o\tilde{z}(t) \\ &= -\Gamma\Upsilon^T(t)c_o^T c_o[\eta(t) + \Upsilon(t)\tilde{\vartheta}(t)] \end{aligned}$$

Putting together the equations of η and $\tilde{\vartheta}$ yields

$$\dot{\eta}(t) = \rho(A_o - k_o c_o)\eta(t) + \xi(t) \qquad (19a)$$

$$\dot{\tilde{\vartheta}}(t) = -\Gamma\Upsilon^T(t)c_o^T c_o[\eta(t) + \Upsilon(t)\tilde{\vartheta}(t)] \qquad (19b)$$

In (Xu and Zhang, 2002), a simple Lyapunov function has been used in the attempt to prove the convergence to zero of the error system (19). Because of the difficulties encountered in that attempt, the same algorithm was applied to a multiple-delayed system for which the algorithm convergence has been successfully proved in (Xu and Zhang, 2002). However, the use of the multiple-delayed system considerably increases the computational cost of the algorithm. In the following, an advanced Lyapunov function is designed in order to prove the convergence of the error system (19).

Consider the homogeneous part of the differential equation (19b), that is, the linear time varying system

$$\dot{e}(t) = -\Gamma\Upsilon^T(t)c_o^T c_o\Upsilon(t)e(t) \qquad (20)$$

Based on the persistent excitation condition (15) and a classical result on linear time varying system stability, system (20) is exponentially stable (Narendra and Annaswamy, 1989, page 72). This stability implies that, for any symmetric positive definite matrix $Q(t) \in \mathbb{R}^p \times \mathbb{R}^p$, there exists a symmetric positive definite matrix $P(t) \in \mathbb{R}^p \times \mathbb{R}^p$ satisfying the equation

$$\dot{P}(t) = [\Gamma\Upsilon^T(t)c_o^T c_o \Upsilon(t)]^T P(t) + P(t)[\Gamma\Upsilon^T(t)c_o^T c_o \Upsilon(t)] - Q(t)$$

In particular, let us choose $Q(t) = I$, then

$$\dot{P}(t) = [\Gamma\Upsilon^T(t)c_o^T c_o \Upsilon(t)]^T P(t) + P(t)[\Gamma\Upsilon^T(t)c_o^T c_o \Upsilon(t)] - I$$

Moreover, according to Lemma 1 stated in Appendix A, the matrix $P(t)$ has positive upper and lower bounds independent of ρ.

Consider the Lyapunov function candidate:

$$V(t, \eta, \tilde{\theta}) = \eta^T S \eta + \tilde{\vartheta}^T P(t) \tilde{\vartheta}$$

with S the solution of (6), a positive definite matrix. Then

$$\begin{aligned}\dot{V}(t) &= -\rho\eta^T(t)S\eta(t) + 2\eta^T(t)S\xi(t) + \dot{\tilde{\vartheta}}^T P(t)\tilde{\vartheta} \\ &\quad + \tilde{\vartheta}^T \dot{P}(t)\tilde{\vartheta} + \tilde{\vartheta}^T P(t)\dot{\tilde{\vartheta}} \\ &= -\rho\eta^T(t)S\eta(t) + 2\eta^T(t)S\xi(t) \\ &\quad - \eta^T[\Gamma\Upsilon^T(t)c_o^T c_o]^T P(t)\tilde{\vartheta} \\ &\quad - \tilde{\vartheta}^T P(t)[\Gamma\Upsilon^T(t)c_o^T c_o]\eta - \tilde{\vartheta}^T\tilde{\vartheta}\end{aligned}$$

where, for the first equality, the equations (6) and (7) have been used with (19a).

Because f and g are globally Lipschitz and are triangular, the nonlinear term $\xi(t)$ as defined in (17) satisfies the inequality

$$\begin{aligned}\|\xi(t)\| &\le \kappa(\rho^{-1})\|\tilde{z}(t)\| \\ &= \kappa(\rho^{-1})\|\eta(t) + \Upsilon(t)\tilde{\vartheta}(t)\|\end{aligned}$$

with $\kappa(\rho^{-1}) > 0$ a polynomial in ρ^{-1} depending on the Lipschitz constants of f and g and on the upper bound of $u(t)$. Therefore, there exist two polynomials $\mu_1(\rho^{-1})$ and $\mu_2(\rho^{-1})$ such that

$$\begin{aligned}2\eta^T(t)S\xi(t) &\le \mu_1(\rho^{-1})\|\eta(t)\|^2 \\ &\quad + \mu_2(\rho^{-1})\|\eta(t)\| \cdot \|\tilde{\vartheta}(t)\|\end{aligned}$$

Since the upper bounds of $P(t)$ and $\Upsilon(t)$ are independent of ρ, there exists a constant $c > 0$ independent of ρ such that

$$2\|P(t)\Gamma\Upsilon^T(t)c_o^T c_o\| \le c$$

The inequality $ab \le \frac{1}{2}(a^2 + b^2)$ with $a = [\mu_2(\rho^{-1}) + c]\|\eta\|$ and $b = \|\tilde{\vartheta}\|$ leads to

$$\begin{aligned}\dot{V}(t) &\le -\eta^T(t)[\rho S - \mu_1(\rho^{-1})I]\eta(t) - \|\tilde{\vartheta}\|^2 \\ &\quad + [\mu_2(\rho^{-1}) + c]\|\tilde{\vartheta}\| \cdot \|\eta\| \\ &\le -\eta^T(t)[\rho S - \mu_1(\rho^{-1})I]\eta(t) - \|\tilde{\vartheta}\|^2 \\ &\quad + \frac{(\mu_2(\rho^{-1}) + c)^2}{2}\|\eta(t)\|^2 + \frac{1}{2}\|\tilde{\vartheta}\|^2 \\ &= -\eta^T(t)\left[\rho S - \mu_1(\rho^{-1})I - \frac{(\mu_2(\rho^{-1}) + c)^2}{2}I\right] \\ &\quad \cdot \eta(t) - \frac{1}{2}\|\tilde{\vartheta}\|^2\end{aligned}$$

Let us choose a value of ρ sufficiently large, such that

$$\rho S - \mu_1(\rho^{-1})I - \frac{(\mu_2(\rho^{-1}) + c)^2}{2}I > \frac{1}{2}I$$

then

$$\begin{aligned}\dot{V}(t) &\le -\frac{1}{2}\|\eta\|^2 - \frac{1}{2}\|\tilde{\vartheta}\|^2 \\ &\le -\frac{1}{2}\left[\eta^T \frac{S}{\lambda_{\max}(S)}\eta + \tilde{\vartheta}^T \frac{P(t)}{\lambda_{\max}(P(t))}\tilde{\vartheta}\right] \\ &\le -\frac{1}{2}\min\left(\frac{1}{\lambda_{\max}(S)}, \frac{1}{\lambda_{\max}(P(t))}\right)V\end{aligned}$$

with $\lambda_{\max}(S)$ being the largest eigenvalue of S.

Then it is concluded that $\tilde{\vartheta}(t)$ and $\eta(t)$ tend exponentially to zero, so do $\tilde{\theta}(t)$, $\tilde{z}(t) = \eta(t) + \Upsilon(t)\tilde{\vartheta}(t)$ and $\tilde{x}(t) = \Lambda^{-1}\tilde{z}(t)$. □

4. NUMERICAL EXAMPLE

Let us illustrate the performance of the proposed adaptive observer with a simulation example. Consider the system

$$\begin{aligned}\dot{x}_1(t) &= -\arctan x_1(t) + x_2(t) + \psi_{1,1}(t)\theta_1 \\ \dot{x}_2(t) &= |x_1(t)| - 5\sin x_2(t) + \psi_{2,1}(t)\theta_1 + \psi_{2,2}(t)\theta_2 \\ y(t) &= x_1(t)\end{aligned}$$

with the parameters $\theta_1 = 1$, $\theta_2 = 1.5$. The excitation signals are $\psi_{1,1}(t) = \sin 2t + \cos 20t$, $\psi_{2,1}(t) = \sin 5t + \cos 16t$ and $\psi_{2,2}(t) = \sin 3t + \cos 27t$. The initial state $x(0) = [1, 1]^T$.

The parameters of the adaptive observer are: $\rho = 3$, $\Gamma = \mathrm{diag}([3, 0.5])$, $\Omega = \mathrm{diag}([1, 3])$. The initial values are $\hat{x}(0) = [0, 0]^T$, $\hat{\theta}(0) = [0, 0]^T$.

The state estimates and the corresponding simulated true states are plotted in Figure 1, and the parameter estimates in Figure 2. The convergence of the state and parameter estimation is practically achieved at the 5th second.

5. CONCLUSION

An adaptive observer has been proposed in this paper for joint estimation of state and parameters in nonlinear systems. The *global exponential convergence* of the adaptive observers for a class of nonlinear systems has been formally established. The proposed algorithm is *constructive* in the sense that its design is systematic and its global convergence is guaranteed under appropriate assumptions.

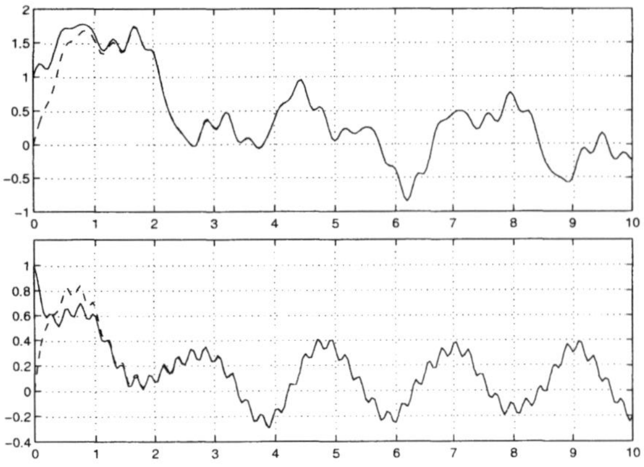

Figure 1. Upper figure: state estimate $\hat{x}_1(t)$ in dashed line and true state $x_1(t)$ in solid line. Lower figure: state estimate $\hat{x}_2(t)$ in dashed line and true state $x_2(t)$ in solid line.

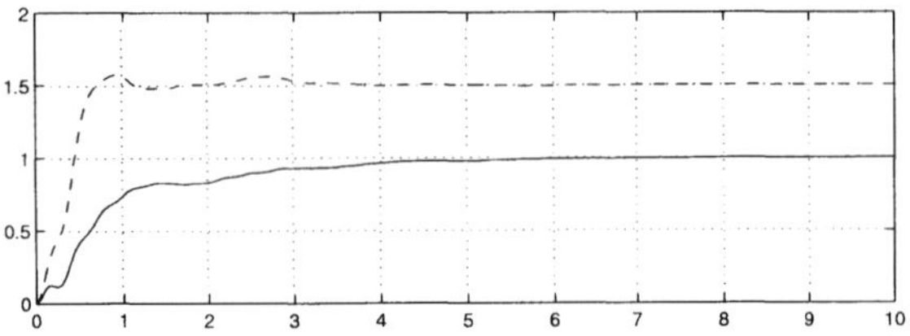

Figure 2. Parameter estimates $\hat{\theta}_1(t)$ (solid line) and $\hat{\theta}_2(t)$ (dashed line). The true parameter values are $\theta_1 = 1$ and $\theta_2 = 1.5$.

Appendix A. A LEMMA RELATED TO LINEAR TIME VARYING SYSTEM STABILITY

Lemma 1. Let $\phi(t) \in \mathbb{R}^m \times \mathbb{R}^p$ be a bounded and piecewise continuous matrix, and $\Gamma \in \mathbb{R}^p \times \mathbb{R}^p$ a symmetric positive definite matrix. If there exist three positive constants T, α, β such that for all $t \geq t_0$,

$$\alpha I \leq \int_t^{t+T} \phi^T(\tau)\phi(\tau)d\tau \leq \beta I$$

then for any symmetric positive definite $Q(t) \in \mathbb{R}^p \times \mathbb{R}^p$ satisfying $q_1 I \leq Q(t) \leq q_2 I$ with two positive constants q_1 and q_2, the solution $P(t)$ of the equation

$$\dot{P}(t) = [\Gamma\phi^T(t)\phi(t)]^T P(t) + P(t)[\Gamma\phi^T(t)\phi(t)] - Q(t).$$

satisfies, for some positive constants α_1 and β_1 depending only on T, α, β, q_1, q_2, Γ and on the upper bound of $\phi(t)$, the inequality

$$\alpha_1 I \leq P(t) \leq \beta_1 I$$

The proof of this lemma, as an adaptation of a classical result (Brockett, 1970), can be found in (Xu, 2002). □

Appendix B. REFERENCES

Bastin, Georges and Michel Gevers (1988). Stable adaptive observers for nonlinear time varying systems. *IEEE Trans. on Automatic Control* **33**(7), 650–658.

Besançon, Gildas (2000). Remarks on nonlinear adaptive observer design. *Systems and Control Letters* **41**(4), 271–280.

Brockett, Roger W. (1970). *Finite dimensional linear systems.* J. Wiley and sons. New York.

Cho, Young Man and Rajesh Rajamani (1997). A systematic approach to adaptive observer synthesis for nonlinear systems. *IEEE Trans. on Automatic Control* **42**(4), 534–537.

Einicke, Garry A. and Langford B. White (1999). Robust extended Kalman filtering. *IEEE Trans. on Signal Processing* **47**(9), 2596–2599.

Gauthier, J.-P. and I. A. K. Kupka (1994). Observability and observers for nonlinear systems. *SIAM Journal Control and Optimization* **32**(4), 975–994.

Gauthier, J-P., H. Hammouri and S. Othman (1992). A simple observer for nonlinear systems, application to bioreactors. *IEEE Trans. on Automatic Control* **37**(6), 875–880.

Kreisselmeier, Gerhard (1977). Adaptive observers with exponential rate of convergence. *IEEE Trans. on Automatic Control* **22**(1), 2–8.

Lüders, Gerd and Kumpati S. Narendra (1973). An adaptive observer and identifier for a linear system. *IEEE Trans. on Automatic Control* **18**, 496–499.

Marino, Riccardo and Patrizio Tomei (1995). *Nonlinear control design.* Information and system sciences. Prentice Hall. London, New York.

Martinez, Jorl Correa and Alex S. Poznyak (2001). Switching structure state and parameter estimator for mimo non-linear robust control. *International Journal of Control* **74**(2), 175–189.

Narendra, K. S. and A. M. Annaswamy (1989). *Stable adaptive systems.* Prentice Hall. Boston.

Poznyak, Alex S. and Jorl Correa Martinez (2001). Variable structure robust state and parameter estimator. *International Journal of Adaptive Control and Signal Processing* **15**(4), 179–208.

Rajamani, Rajesh and Karl Hedrick (1995). Adaptive observer for active automotive suspensions - theory and experiment. *IEEE Trans. on Control Systems Technology* **3**(1), 86–93.

Reif, Konrad and Rolf Unbehauen (1999). The extended Kalman filter as an exponential observer for nonlinear systems. *IEEE Trans. on Signal Processing* **47**(8), 2324–2328.

Xu, Aiping (2002). Observateurs adaptatifs non-linéaires et diagnostic de pannes. PhD thesis. Université de Rennes 1. Rennes, France. In French.

Xu, Aiping and Qinghua Zhang (2002). State and parameter estimation for nonlinear systems. In: *IFAC World Congress.* Barcelona.

Zhang, Qinghua (2002). Adaptive observer for multiple-input-multiple-output (MIMO) linear time varying systems. *IEEE Trans. on Automatic Control* **47**(3), 525–529.

Zhang, Qinghua and Aiping Xu (2001). Implicit adaptive observers for a class of nonlinear systems. In: *ACC'2001.* Arlington. pp. 1551–1556.

IFAC Publications
www.elsevier.com/locate/ifac

ADAPTIVE OBSERVER FOR DISCRETE TIME LINEAR TIME VARYING SYSTEMS

Arnaud Guyader * and Qinghua Zhang **

* *Université de Haute Bretagne, 35043 Rennes Cedex, France*
aguyader@irisa.fr
** IRISA-INRIA, *Campus de Beaulieu, 35042 Rennes Cedex, France*
zhang@irisa.fr

Abstract: For joint state-parameter estimation in discrete time stochastic multiple-input multiple-output linear time varying systems, an efficient adaptive observer is proposed in this paper. In the noise-free case, the global exponential convergence of the adaptive observer is first established. It is then proved that, in the noise-corrupted case, the state and parameter estimation errors remain bounded if the noises are bounded, and moreover, the estimation errors converge in the mean to zero if the noises have zero means. *Copyright © 2003 IFAC*

Keywords: state and parameter estimation, discrete time system, linear time varying system.

1. INTRODUCTION

The Luenberger observer and the Kalman filter are well known solutions for state estimation in linear dynamic systems, in continuous time as well as in discrete time. For joint estimation of state and unknown parameters, some results are also known under the name of *adaptive observer*, see, *e.g.*, (Kreisselmeier, 1977; Bastin and Gevers, 1988; Marino and Tomei, 1995; Besançon, 2000; Zhang, 2002). These adaptive observers have been known *in continuous time*, and there are relatively few results of similar nature for *discrete time* systems. For single-input single-output (SISO) time invariant discrete time systems, some results can be found in (Landau, 1979; Ioannou and Kokotovic, 1983).

An adaptive observer is proposed in this paper for joint estimation of state and parameters in discrete time stochastic multiple-input multiple-output (MIMO) linear *time varying* systems. It is a discrete time counterpart of the continuous time algorithm presented in (Zhang, 2002). As seen in the formulation of the discrete time persistent excitation condition and in the convergence analysis, this adaptation from continuous time to discrete time is not trivial.

Let us consider discrete time stochastic MIMO linear time varying systems of the form

$$\theta_{k+1} = \theta_k + e_k \tag{1a}$$
$$x_{k+1} = A_k x_k + B_k u_k + \Psi_k \theta_k + w_k \tag{1b}$$
$$y_k = C_k x_k + v_k \tag{1c}$$

where $x_k \in \mathbb{R}^n, u_k \in \mathbb{R}^l, y_k \in \mathbb{R}^m$ are respectively the state, input, output of the system, A_k, B_k, C_k are known time varying matrices of appropriate sizes, $\theta_k \in \mathbb{R}^p$ is an unknown parameter vector, $\Psi_k \in \mathbb{R}^n \times \mathbb{R}^p$ is a matrix of known signals, and e_k, w_k, v_k are noises of appropriate dimensions. The time varying matrices A_k, B_k, C_k, Ψ_k are all assumed bounded. The noises e_k, w_k, v_k are bounded and have zero mean. Note that *no* whiteness of the noises is required in this paper.

The purpose of this paper is to design a recursive algorithm for joint estimation of the state vector x_k and the parameter vector θ_k from the input u_k, the output y_k, the excitation Ψ_k and the system matrices A_k, B_k, C_k.

The study of such systems is mainly motivated by fault detection and isolation (FDI) for which the term $\Psi_k\theta_k$ models the faults to be detected and isolated. See (Xu and Zhang, 2002) for some related work in continuous time. Another motivation is adaptive control for which the term $\Psi_k\theta_k$ models some modeling uncertainties.

Remark 1. For the purpose of state and parameter estimation, it will not make any extra difficulty if, in (1b), the term $B_k u_k$ is replaced by any known nonlinear functions of u_k or of any other known variables. Such nonlinearities have been considered (in the continuous time case) in (Bastin and Gevers, 1988; Marino and Tomei, 1995). For presentation clearness, let us assume the linear inputs in this paper. □

Remark 2. In the proposed method, *no* particular form of the matrices A_k and C_k is required, whereas classical methods typically assume some (time invariant) canonical form of the two matrices. This feature is particularly important for time varying systems which would require some non trivial transformation to achieve a canonical form. □

A natural idea for joint state and parameter estimation is to apply the Kalman filter to the extended system obtained by appending the unknown parameters θ_k into the state vector. Note that, even in the case of constant matrices A, B, C, the extended system is typically time varying, since the matrix Ψ_k should sufficiently excite the system in order to estimate the unknown parameters. In general, it is not easy to guarantee the convergence of the Kalman filter for *time varying* systems. Application of classical results requires uniform complete observability (Jazwinski, 1970). In practice, it is difficult to check the uniform complete observability of the *extended* system. Therefore, the analysis of the Kalman filter applied to the extended system is not a trivial problem. In this paper, instead of assuming the observability of the extended system, the proposed method is essentially based on the observability of the matrix pair (A_k, C_k) and on some persistent excitation condition.

In section 2 we first establish the exponential convergence of the proposed adaptive observer in the noise-free case. The noise-corrupted case is considered in section 3 where the boundedness of state and parameter estimation errors and their convergence in the mean to zero are proved. Section 4 is devoted to a numerical example. Finally, some concluding remarks are drawn in section 5.

2. THE NOISE-FREE CASE

In this section, the proposed adaptive observer is described and the exponential convergence to zero of the estimation errors is established in the noise-free case. It will be the basis for the proofs in the noise-corrupted case presented in the next section.

Throughout the paper, the Euclidean norm is used for vectors and the spectral norm[1] is used for matrices.

Definition 1. The linear time varying system

$$\eta_{k+1} = F_k \eta_k$$

is said *exponentially stable* if there exist two constants $r > 0$ and $0 < q < 1$ such that

$$\|\eta_k\| \leq r q^{k-k_0} \|\eta_{k_0}\|$$

for all $k_0 \geq 0$, $k > k_0$ and for any value of η_{k_0}. □

Note that this definition implies

$$\left\| \prod_{i=k_0}^{k-1} F_i \right\| \leq r q^{k-k_0}$$

Assumption 1. The time varying matrices A_k and C_k are such that there exists a bounded time varying matrix $K_k \in \mathbb{R}^n \times \mathbb{R}^m$ so that the linear time varying system

$$\eta_{k+1} = (A_k - K_k C_k)\eta_k$$

is exponentially stable. □

Note that this assumption is equivalent to say that, when the term $\Psi_k\theta_k$ and the noises are absent in system (1), an exponential observer can be designed for the estimation of the state x_k. It is known that, if the time varying matrix pair (A_k, C_k) is completely uniformly observable, the Kalman gain will fulfill the requirement (Jazwinski, 1970).

As an adaptation of the continuous time adaptive observer presented in (Zhang, 2002), the proposed discrete time adaptive observer is as follows.

$$\Upsilon_{k+1} = (A_k - K_k C_k)\Upsilon_k + \Psi_k \quad (2a)$$

$$\hat{\theta}_{k+1} = \hat{\theta}_k + \mu_k \Upsilon_k^T C_k^T (y_k - C_k \hat{x}_k) \quad (2b)$$

$$\hat{x}_{k+1} = A_k \hat{x}_k + B_k u_k + \Psi_k \hat{\theta}_k + K_k (y_k - C_k \hat{x}_k) + \Upsilon_{k+1}(\hat{\theta}_{k+1} - \hat{\theta}_k) \quad (2c)$$

where $\Upsilon_k \in \mathbb{R}^n \times \mathbb{R}^p$ is a matrix sequence obtained by linearly filtering Ψ_k, the vector sequences $\hat{x}_k$ and $\hat{\theta}_k$ are respectively the state and parameter estimates, $\mu_k > 0$ is a bounded scalar gain sequence satisfying the following assumption.

Assumption 2. The scalar gain sequence $\mu_k > 0$ is small enough so that

$$\|\sqrt{\mu_k} C_k \Upsilon_k\| \leq 1 \quad (3)$$

[1] The spectral norm of a matrix is associated with the Euclidean norm and equal to the largest singular value of the matrix.

for all $k \geq 0$, where $\|\cdot\|$ denotes the matrix spectral norm.

Like in system identification, an assumption on persistent excitation is required for parameter estimation.

Assumption 3. The matrix of signals Ψ_k is persistently exciting so that the matrix sequence Υ_k (obtained by linearly filtering Ψ_k through (2a)) and the gain sequence μ_k satisfy, for some constant $\alpha > 0$, integer $L > 0$ and for all $k \geq 0$, the following inequality

$$\frac{1}{L}\sum_{i=k}^{k+L-1} \mu_i \Upsilon_i^T C_i^T C_i \Upsilon_i \geq \alpha I \tag{4}$$

Remark 3. Typically each term in the sum of (4) is rank deficient, since the number of outputs m (the number of rows of C_i) is typically smaller than the number of parameters p (the number of columns of Υ_i). Nevertheless, if Υ_i for different i vary in different "directions", the average stated in (4) can be positive definite. The positive definiteness of the sum of some matrices is typically required as a persistent excitation condition in system identification. See, *e.g.*, (Aström and Wittenmark, 1989). Note that Ψ_k is assumed deterministic in this paper, since its main motivation is FDI. Stochastic Ψ_k can also be considered in a similar way. □

The property of Algorithm (2) in the noise-free case is stated in the following theorem.

Theorem 1. If the noises are absent in system (1), that is, $e_k = 0, w_k = 0, v_k = 0$ for all $k \geq 0$, then, under Assumptions 1–3, Algorithm (2) is a global exponential adaptive observer, *i.e.*, the estimation errors $\hat{x}_k - x_k$ and $\hat{\theta}_k - \theta_k$ tend to zero exponentially fast when $k \to \infty$. □

The proof of this theorem requires the two following lemmas.

Lemma 1. If the linear time varying system $\eta_{k+1} = F_k \eta_k$, with $\eta_k \in \mathbb{R}^n, F_k \in \mathbb{R}^n \times \mathbb{R}^n$, is exponentially stable (see definition 1), then

(1) for any bounded sequence $g_k \in \mathbb{R}^n$, the sequence z_k defined by $z_{k+1} = F_k z_k + g_k$ is bounded;
(2) for any sequence g_k tending to zero exponentially fast, the sequence z_k defined as above tends also to zero exponentially fast. □

For the proof of the first part of this lemma, one can see (Freeman, 1965, page 168). The proof of the second part is a straightforward extension.

Lemma 2. Let $\phi_k \in \mathbb{R}^m \times \mathbb{R}^p$ be a matrix sequence such that its spectral norm $\|\phi_k\| \leq 1$ for all $k \geq 0$. If there exist a real constant $\alpha > 0$ and an integer $L > 0$ such that for all $k \geq 0$ the following inequality holds

$$\frac{1}{L}\sum_{i=k}^{k+L-1} \phi_i^T \phi_i \geq \alpha I \tag{5}$$

then the linear time varying system

$$z_{k+1} = (I - \phi_k^T \phi_k) z_k$$

is exponentially stable. □

A proof of this lemma can be found in Appendix A.

Now we are ready to prove Theorem 1.

Proof of Theorem 1. Define the error sequences

$$\tilde{x}_k = \hat{x}_k - x_k, \quad \tilde{\theta}_k = \hat{\theta}_k - \theta_k$$

In the absence of the noises in (1), following (1b) and (2c) it is easy to obtain the error dynamics

$$\tilde{x}_{k+1} = A_k \tilde{x}_k + \Psi_k \tilde{\theta}_k + K_k(y_k - C_k \hat{x}_k) + \Upsilon_{k+1}(\hat{\theta}_{k+1} - \hat{\theta}_k)$$

According to (1a) and (1c), $\theta_{k+1} = \theta_k$ and $y_k = C_k x_k$ (it is assumed $e_k = 0, v_k = 0$), then

$$\tilde{x}_{k+1} = (A_k - K_k C_k)\tilde{x}_k + \Psi_k \tilde{\theta}_k + \Upsilon_{k+1}(\tilde{\theta}_{k+1} - \tilde{\theta}_k) \tag{6}$$

The key step of the proof is to define the linearly combined error sequence

$$\eta_k = \tilde{x}_k - \Upsilon_k \tilde{\theta}_k \tag{7}$$

It is straightforward to compute the dynamic equation of η_k:

$$\eta_{k+1} = (A_k - K_k C_k)\eta_k + [(A_k - K_k C_k)\Upsilon_k + \Psi_k - \Upsilon_{k+1}]\tilde{\theta}_k$$

Because Υ_k is generated from (2a), the last equation simply becomes

$$\eta_{k+1} = (A_k - K_k C_k)\eta_k$$

According to Assumption 1, the sequence η_k tends to zero exponentially fast.

Now let us study the error $\tilde{\theta}_k = \hat{\theta}_k - \theta_k$. Following (2b) and (1a), (1c) with $e_k = 0$ and $v_k = 0$,

$$\tilde{\theta}_{k+1} = \tilde{\theta}_k - \mu_k \Upsilon_k^T C_k^T C_k \tilde{x}_k \tag{8}$$

Substitute $\tilde{x}_k$ with (7), then the error equation becomes

$$\begin{aligned}\tilde{\theta}_{k+1} &= \tilde{\theta}_k - \mu_k \Upsilon_k^T C_k^T C_k(\eta_k + \Upsilon_k \tilde{\theta}_k) \\ &= (I - \mu_k \Upsilon_k^T C_k^T C_k \Upsilon_k)\tilde{\theta}_k \\ &\quad - \mu_k \Upsilon_k^T C_k^T C_k \eta_k\end{aligned} \tag{9}$$

According to Assumptions 2, 3 and Lemma 2, the homogeneous part of (9), that is, the linear time varying system

$$z_{k+1} = (I - \mu_k \Upsilon_k^T C_k^T C_k \Upsilon_k) z_k \tag{10}$$

is exponentially stable.

The sequences μ_k, C_k have been assumed bounded, and the boundedness of Υ_k is a consequence of the boundedness of Ψ_k and of Assumption 1, following Lemma 1.

Then following Lemma 1, the sequence $\tilde{\theta}_k$ driven by the exponentially vanishing sequence

$$-\mu_k \Upsilon_k^T C_k^T C_k \eta_k$$

through (9) tends to zero exponentially fast.

Finally, $\tilde{x}_k = \eta_k + \Upsilon_k \tilde{\theta}_k$ tends also to zero exponentially fast. □

Remark 4. It is clear in the proof of Theorem 1 that Assumptions 2 and 3 are for the purpose of ensuring the exponential stability of the linear time varying system (10). There are other ways to ensure this stability. A direct condition is

$$\left\| \prod_{i=k}^{k+L-1} (I - \mu_k \Upsilon_k^T C_k^T C_k \Upsilon_k) \right\| \leq \gamma$$

for some $L > 0$, $0 < \gamma < 1$ and for all $k \geq 0$. The disadvantage of this condition is that it requires the computation of the matrix products. The condition stated with Assumptions 2 and 3 is probably not necessary, but it is sufficient and requires simple computations as stated in (3) and (4). □

3. THE NOISE-CORRUPTED CASE

Now let us study the properties of the same algorithm (2) applied to the noise-corrupted system (1).

Theorem 2. Under Assumptions 1–3, when algorithm (2) is applied to system (1) with bounded noises e_k, w_k, v_k, the estimation errors $\hat{x}_k - x_k$ and $\hat{\theta}_k - \theta_k$ remain bounded.

Moreover, if the noises e_k, w_k, v_k have zero mean, then the estimation errors tend to zero in the mean, that is, when $k \to \infty$, the mathematical expectations $\mathbf{E}(\hat{x}_k - x_k) \to 0$, $\mathbf{E}(\hat{\theta}_k - \theta_k) \to 0$, and the convergence is exponentially fast. □

Proof of Theorem 2. The proof of this theorem essentially relies on the result already established in the noise-free case. Like in the proof of Theorem 1, the equations of the errors $\tilde{x}_k = \hat{x}_k - x_k$ and $\tilde{\theta}_k = \hat{\theta}_k - \theta_k$ are first derived, but now the noises are involved:

$$\begin{aligned} \tilde{x}_{k+1} &= (A_k - K_k C_k)\tilde{x}_k + \Psi_k \tilde{\theta}_k \\ &\quad + \Upsilon_{k+1}(\tilde{\theta}_{k+1} - \tilde{\theta}_k) \\ &\quad - w_k + K_k v_k + \Upsilon_{k+1} e_k \qquad (11a) \\ \tilde{\theta}_{k+1} &= \tilde{\theta}_k - \mu_k \Upsilon_k^T C_k^T C_k \tilde{x}_k + \mu_k \Upsilon_k^T C_k^T v_k \qquad (11b) \end{aligned}$$

According to Theorem 1, the estimation errors tend to zero exponentially fast in the noise-free case. It means that the error system (11) is, in the absence of the terms involving the noises, exponentially stable (as a linear time varying system).

The sequences K_k, C_k are bounded by assumption, and Υ_k is bounded following the boundedness of Ψ_k. For bounded noises, the terms involving the noises in (11) are thus all bounded. Then according to Lemma 1, the errors $\tilde{x}_k$ and $\tilde{\theta}_k$ governed by (11) are also bounded.

Now let us take the mathematical expectation at both sides of (11). Notice that the sequences μ_k, C_k, K_k, Ψ_k are deterministic, and so is Υ_k. Then, following the zero mean assumption on the noises, equations (11) after the mathematical expectation becomes

$$\begin{aligned} \mathbf{E}\tilde{x}_{k+1} &= (A_k - K_k C_k)\mathbf{E}\tilde{x}_k + \Psi_k \mathbf{E}\tilde{\theta}_k \\ &\quad + \Upsilon_{k+1}(\mathbf{E}\tilde{\theta}_{k+1} - \mathbf{E}\tilde{\theta}_k) \\ \mathbf{E}\tilde{\theta}_{k+1} &= \mathbf{E}\tilde{\theta}_k - \mu_k \Upsilon_k^T C_k^T C_k \mathbf{E}\tilde{x}_k \end{aligned}$$

These equations are the same as the error equations (6) and (8) in the noise-free case, except that the errors $\tilde{x}_k$ and $\tilde{\theta}_k$ are respectively replaced by their mathematical expectation $\mathbf{E}\tilde{x}_k$ and $\mathbf{E}\tilde{\theta}_k$. Then following the same procedure as in the proof of Theorem 1, the mathematical expectations $\mathbf{E}\tilde{x}_k$ and $\mathbf{E}\tilde{\theta}_k$ tend to zero exponentially fast. □

4. NUMERICAL EXAMPLE

Let us illustrate the behavior of the proposed adaptive observer with the simulation of a controlled satellite. The classic linearized satellite model, in its continuous time version, can be found in (Brockett, 1970). The satellite nominal orbit is assumed to be circular with the radius normalized to 1. The nominal angular velocity of the satellite is 3.49×10^{-4}rad/s. The equations of motion of the satellite are linearized around the nominal orbit. In order to obtain a discrete time model, the linearized system is sampled with the period $T_s = 0.1$s and with zero-order holders applied at its inputs. The obtained discrete time model is

$$\begin{aligned} x_{k+1} &= \begin{bmatrix} 1 & 0.1 & 0 & 3.49\times 10^{-6} \\ 3.66\times 10^{-8} & 1 & 0 & 6.98\times 10^{-5} \\ -4.25\times 10^{-14} & -3.49\times 10^{-6} & 1 & 0.1 \\ -1.28\times 10^{-12} & -6.98\times 10^{-5} & 0 & 1 \end{bmatrix} x_k \\ &+ \begin{bmatrix} 0.1 & 0.005 & 0 & 1.16\times 10^{-7} \\ 1.83\times 10^{-9} & 0.1 & 0 & 3.49\times 10^{-6} \\ -1.06\times 10^{-15} & -1.16\times 10^{-7} & 0.1 & 0.005 \\ -4.25\times 10^{-14} & -3.49\times 10^{-6} & 0 & 0.1 \end{bmatrix} \\ &\cdot \begin{bmatrix} 0 & 0 \\ \theta^1 & 0 \\ 0 & 0 \\ 0 & \theta^2 \end{bmatrix} u_k \\ y_k &= \begin{bmatrix} 1 & 0 & 0 & 0 \\ 0 & 0 & 1 & 0 \end{bmatrix} x_k \end{aligned}$$

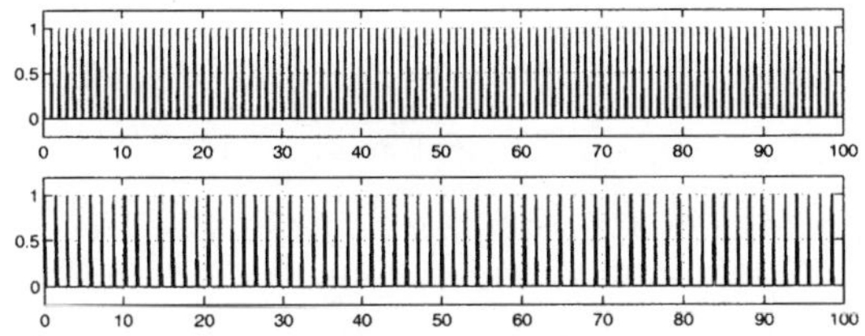

Figure 1. Input signals u_k^1 and u_k^2 used in the controlled satellite simulation. The time unit is second.

where the components of the state vector $x_k \in \mathbb{R}^4$ correspond to radial position, radial velocity, angular position and angular velocity, the components of the input vector $u_k \in \mathbb{R}^2$ are the radial and tangential thrusts, the output vector $y_k \in \mathbb{R}^2$ correspond to distance and angle observations, and the constant coefficients θ^1 and θ^2 represent the efficiencies of the radial and tangential thrusts.

Note that the 4×4 matrix in the term involving u_k is due to the discretization of the continuous time model, since the coefficients θ^1 and θ^2 were originally defined in the continuous time model. In order to put the model into the form of (1), this term is reformulated as

$$\Psi_k \theta = H \begin{bmatrix} 0 & 0 \\ u_k^1 & 0 \\ 0 & 0 \\ 0 & u_k^2 \end{bmatrix} \begin{bmatrix} \theta^1 \\ \theta^2 \end{bmatrix}$$

where H is the aforementioned 4×4 matrix.

In the simulation, the parameter values are set to $\theta^1 = 1$ and $\theta^2 = 1.5$. The square impulse signals shown in figure 1 are used as inputs. The two simulated outputs are both disturbed by a Gaussian white noise whose standard deviation is 0.01.

The initial values used in the simulation are $x_0 = [1, 0, 0, 3.49 \times 10^{-4}]^T$, $\hat{x}_0 = [0.9, 0, 0, 3.14 \times 10^{-4}]^T$, $\hat{\theta}_0 = [0.5, 0.5]$. The adaptive observer parameters are

$$\mu_k \equiv 4, \qquad K_k \equiv \begin{bmatrix} 0.1412 & 4.93 \times 10^{-6} \\ 0.0932 & 5.26 \times 10^{-5} \\ -4.93 \times 10^{-6} & 0.1412 \\ -5.26 \times 10^{-5} & 0.0932 \end{bmatrix}$$

In figures 1–3 are, respectively, plotted the input signals, the state estimation errors, and the parameter estimates. Notice that, due to the noises added to the outputs y_k, the estimation errors randomly oscillate around zero instead of tending to zero. According to Theorem 2, the means of the estimation errors tend to zero when $k \to \infty$.

5. CONCLUSION

A numerically efficient adaptive observer has been proposed in this paper for joint state-parameter estimation in discrete time stochastic multiple-input multiple-output linear time varying systems. Essentially, if an exponential state observer can be designed for a linear system, then an adaptive observer can be designed for the system obtained after adding additive terms with unknown coefficients. A persistent excitation condition is required in order to ensure the convergence of the adaptive observer. The boundedness and convergence in the mean of the estimation errors have been proved under the assumption of bounded and zero mean noises. The analysis of the covariances of the estimation errors would require some assumption on the decay of the noises correlations and will be reported elsewhere. Potential applications of the proposed algorithm are fault detection and isolation, and adaptive control.

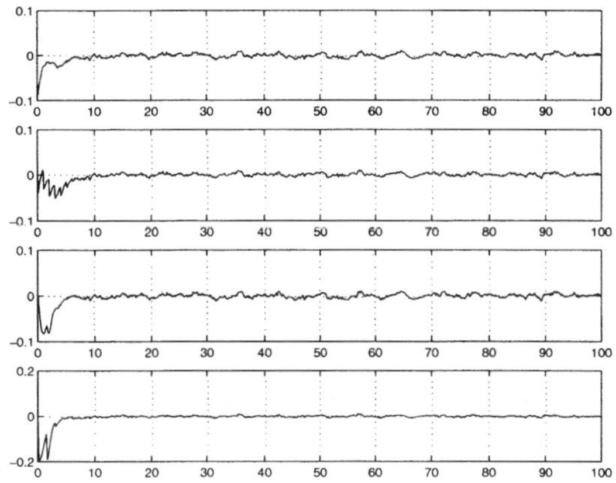

Figure 2. State estimation errors $\tilde{x}_k^1, \tilde{x}_k^2, \tilde{x}_k^3, \tilde{x}_k^4$. The time unit is second.

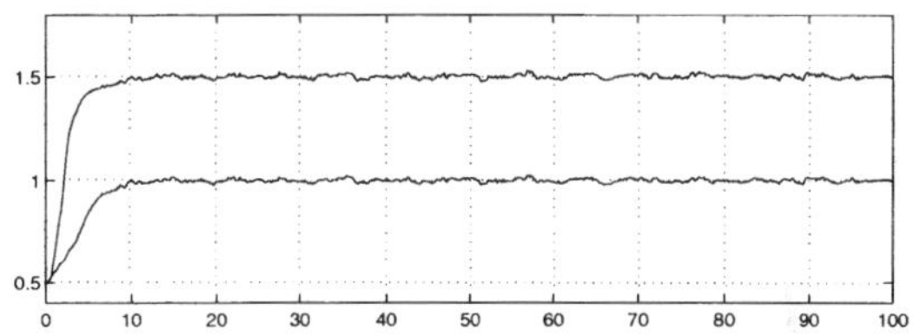

Figure 3. Parameter estimates $\hat{\theta}_k^1$ (lower) and $\hat{\theta}_k^2$ (upper). The true parameter values are $\theta^1 = 1$ and $\theta^2 = 1.5$. The time unit is second.

APPENDIX A. PROOF OF LEMMA 2

Though this result is considered classical, this appendix is provided for completeness. It is first observed that condition (5) is equivalent to

$$\frac{1}{L} \sum_{i=k}^{k+L-1} \|\phi_k w\| \geq \beta \qquad (12)$$

for some positive constant β and for all unitary vector $w \in \mathbb{R}^p$. This equivalence is based on the following simple inequalities (for any unitary w)

$$w^T \phi_k^T \phi_k w \leq \|\phi_k^T \phi_k w\| \leq \phi_{\max} \|\phi_k w\|$$

with

$$\phi_{\max} = \sup_k \|\phi_k\|$$

and on the Cauchy-Schwarz inequality

$$\sum_{i=k}^{k+L-1} w^T \phi_k^T \phi_k w \geq \frac{1}{L} \left(\sum_{i=k}^{k+L-1} \|\phi_i w\| \right)^2$$

Consider the Lyapunov function candidate

$$V_k = z_k^T z_k$$

then

$$\begin{aligned} V_k - V_{k+1} &= z_k^T (\phi_k^T \phi_k) z_k + z_k^T [\phi_k^T \phi_k - (\phi_k^T \phi_k)^2] z_k \\ &\geq z_k^T (\phi_k^T \phi_k) z_k \end{aligned}$$

where the inequality is due to the assumption that the matrix spectral norm $\|\phi_k\| \leq 1$. Then

$$\begin{aligned} V_k - V_{k+L} &\geq \sum_{i=k}^{k+L-1} z_i^T (\phi_i^T \phi_i) z_i \\ &\geq \frac{1}{L} \left(\sum_{i=k}^{k+L-1} \|\phi_i z_i\| \right)^2 \end{aligned}$$

where the last inequality follows Cauchy-Schwarz.

Now a lower bound of $\sum_{i=k}^{k+L-1} \|\phi_i z_i\|$ is needed. Let us proceed as follows.

$$\sum_{i=k}^{k+L-1} \|\phi_i z_i\| \geq \sum_{i=k}^{k+L-1} \|\phi_i z_k\| - \sum_{i=k}^{k+L-1} \|\phi_i (z_k - z_i)\|$$

The first term at the right hand side is bounded from below according to (12):

$$\sum_{i=k}^{k+L-1} \|\phi_i z_k\| \geq \beta L \|z_k\|$$

For the second term,

$$\sum_{i=k}^{k+L-1} \|\phi_i (z_k - z_i)\| \leq \phi_{\max} L \sup_{k \leq i \leq k+L-1} \|z_k - z_i\|$$

It turns out that

$$\begin{aligned} \sup_{k \leq i \leq k+L-1} \|z_k - z_i\| &\leq \sum_{i=k}^{k+L-1} \|z_i - z_{i+1}\| \\ &= \sum_{i=k}^{k+L-1} \|\phi_i^T \phi_i z_i\| \\ &\leq \phi_{\max} \sum_{i=k}^{k+L-1} \|\phi_i z_i\| \end{aligned}$$

Therefore,

$$\sum_{i=k}^{k+L-1} \|\phi_i z_i\| \geq \beta L \|z_k\| - \phi_{\max}^2 L \sum_{i=k}^{k+L-1} \|\phi_i z_i\|$$

or equivalently,

$$\sum_{i=k}^{k+L-1} \|\phi_i z_i\| \geq \frac{\beta L}{1 + \phi_{\max}^2 L} \|z_k\|$$

It then follows

$$V_k - V_{k+L} \geq \frac{\beta^2 L}{(1 + \phi_{\max}^2 L)^2} \|z_k\|^2$$

The inequality (12) implies $\beta \leq \phi_{\max}$, then

$$0 < \gamma = \frac{\beta^2 L}{(1 + \phi_{\max}^2 T)^2} < 1$$

It follows that

$$V_{k+L} \leq (1 - \gamma) V_k$$

The exponential convergence to zero of $V_k = \|z_k\|^2$ is thus established.

6. REFERENCES

Aström, Karl J and Bjorn Wittenmark (1989). *Adaptive control.* Addison-Wesley. Reading, MA.

Bastin, Georges and Michel Gevers (1988). Stable adaptive observers for nonlinear time varying systems. *IEEE Trans. on Automatic Control* **33**(7), 650–658.

Besançon, Gildas (2000). Remarks on nonlinear adaptive observer design. *Systems and Control Letters* **41**(4), 271–280.

Brockett, Roger W. (1970). *Finite dimensional linear systems.* J. Wiley and sons. New York.

Freeman, Herbert (1965). *Discrete-time systems – an introduction to the theory.* John Wiley and Sons. New York.

Ioannou, P. A. and P. V. Kokotovic (1983). *Adaptative systems with reduced models.* Vol. 47 of *Lecture Notes in Control and Information Sciences.* Springer. Berlin.

Jazwinski, A. H. (1970). *Stochastic Processes and Filtering Theory.* Vol. 64 of *Mathematics in Science and Engineering.* Academic Press. New York.

Kreisselmeier, Gerhard (1977). Adaptive observers with exponential rate of convergence. *IEEE Trans. on Automatic Control* **22**(1), 2–8.

Landau, Ioan Doré (1979). *Adaptive control: the model reference approach.* M. Dekker. New York.

Marino, Riccardo and Patrizio Tomei (1995). *Nonlinear control design.* Information and system sciences. Prentice Hall. London, New York.

Xu, Aiping and Qinghua Zhang (2002). Fault detection and isolation based on adaptive observers for linear time varying systems. In: *IFAC World Congress.* Barcelona.

Zhang, Qinghua (2002). Adaptive observer for multiple-input-multiple-output (MIMO) linear time varying systems. *IEEE Trans. on Automatic Control* **47**(3), 525–529.

IFAC
Publications
www.elsevier.com/locate/ifac

LINEAR DYNAMIC FILTERING WITH NOISY INPUT AND OUTPUT

Ivan Markovsky * and **Bart De Moor** *

* *ESAT, SCD-SISTA, K.U.Leuven,*
Kasteelpark Arenberg 10, B-3001 Leuven-Heverlee, Belgium
{Ivan.Markovsky,Bart.DeMoor}@esat.kuleuven.ac.be
http://www.esat.kuleuven.ac.be/sista-cosic-docarch
Tel: +32–16–32 17 09, *Fax: +32–16–32 19 70*

Abstract: We establish the equivalence between the optimal least-squares state estimator for a linear time-invariant dynamic system with noise corrupted input and output, and an appropriately modi£ed Kalman £lter. The approach used is algebraic and the result shows that the noisy input/output £ltering problem is not fundamentally different from the classical Kalman £ltering problem. The result is illustrated with a simulation example. *Copyright © 2003 IFAC*

Keywords: dynamic errors-in-variables model, Kalman £ltering, optimal smoothing, total least squares.

1. INTRODUCTION

Optimal least-squares state estimation for linear dynamical systems is a well developed topic with many practical applications. A central result is the Kalman £lter. In the discrete-time case, one considers the model

$$\begin{aligned} x(t+1) &= Ax(t) + Bu(t) + w(t), \\ y(t) &= Cx(t) + Du(t) + v(t), \end{aligned} \tag{1}$$

with $x(0) = x_0$ and $t = 0, 1, \ldots$ Here $u(t) \in \mathbb{R}^m$, $y(t) \in \mathbb{R}^l$, and $x(t) \in \mathbb{R}^n$ are the input, output, and state vectors at time instant t. When the model matrices A, B, C, and D are known, the state x depends linearly on the input u and the output y, so that it can be estimated from the input/output (I/O) data via the least-squares method. The unknown *process noise* w and the *measurement noise* v play the role of *equation errors* in (1). Even though v and w are unknown, we assume that they are zero mean, white, Gaussian noises, with covariance matrix

$$\mathbf{E}\begin{bmatrix} w(t) \\ v(t) \end{bmatrix} \begin{bmatrix} w(t+\tau) & v(t+\tau) \end{bmatrix} = \begin{bmatrix} Q(t) & S(t) \\ S^T(t) & R(t) \end{bmatrix} \delta(\tau).$$

The importance of the Kalman £lter is that it solves the least-squares problem recursively and in real time. Model (1), however, is in a certain sense asymmetric. The process noise w acts as an unobserved input while the measurement noise v represents measurement error on y. However, u is assumed to be noiseless! In the paper, we pose and answer the question:

> How should we modify the Kalman £lter when *both* the input and the output of the system are noisy?

The paradigm of treating the input and the output on an equal footing leads to the behavioral approach (Polderman and Willems, 1998). A derivation of the Kalman £lter in the behavioral context is given in (Fagnani and Willems, 1997).

In this paper, we consider the deterministic discrete-time LTI state-space system

$$\begin{aligned} x(t+1) &= Ax(t) + Bu(t), & x(0) &= x_0, \\ y(t) &= Cx(t) + Du(t), & t &= 0, 1, \ldots \end{aligned} \tag{2}$$

together with the *measurement errors model*

$$u_{\mathrm{d}}(t) = u(t) + \tilde{u}(t), \qquad y_{\mathrm{d}}(t) = y(t) + \tilde{y}(t). \tag{3}$$

and refer to the model (2) together with the measurement errors model (3) as the *noisy I/O model* (see Figure 1). We note that (2–3) is an *errors-in-variables*

(EIV) model in the sense of, *e.g.*, (Söderström *et al.*, 2002; Zheng, 2002), when the problem would be to estimate A, B, C, and D. But since the purpose of this paper is to estimate x, knowing A, B, C, and D, we call it a noisy I/O model.

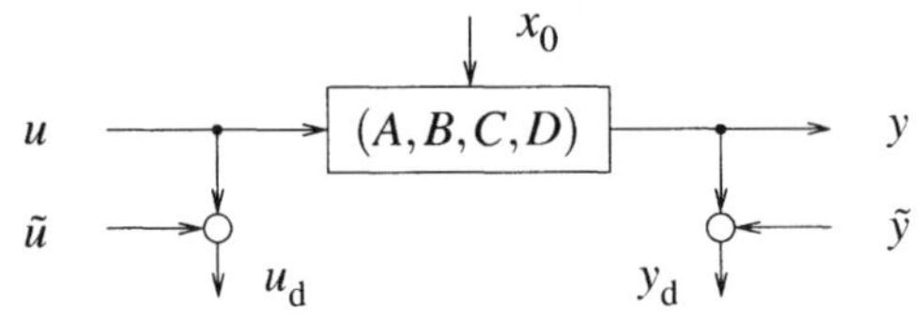

Fig. 1. Block scheme of the noisy I/O model.

A full analysis of a static EIV estimation problem, also known as the *total least squares* (TLS) problem, is given in the monograph (Van Huffel and Vandewalle, 1991). The dynamic equivalent of the TLS problem, see (Aoki and Yue, 1970; De Moor and Roorda, 1994), is a system identi£cation problem: given noisy I/O measurements, £nd the model (2). The dynamic TLS problem can be expressed as a static TLS problem with Toeplitz (or Hankel) structured data matrices. The *structured total least squares* problem is treated in (De Moor, 1993; Lemmerling, 1999).

Different type of estimation problem, in the EIV context, occurs when the model is exactly known and the state vector has to be estimated from the noisy I/O measurements. We refer to this latter problem as the *noisy I/O state estimation* problem. In (Guidorzi *et al.*, 2003; Diversi *et al.*, 2003), the noisy I/O state estimation problem has been considered, using the language of transfer functions. The authors there claim that it is fundamentally different problem from the Kalman £ltering problem and describe new recursive algorithms for its solution.

In this paper, we prove that the noisy I/O state estimation problem is *not* fundamentally different from the Kalman £ltering problem. Its solution boils down to the solution of a sequence of linear least-squares problems with a special structure coming from the state space equation (1). Our approach is linear algebraic. We represent the model over a £nite time horizon as a set of linear equations and apply standard linear algebra techniques for its analysis. The optimal solution is shown to be a Kalman £lter with correlated process and measurement noises. The continuous-time version of the noisy I/O state estimation problem is treated in (Markovsky *et al.*, 2002), where a completion of squares approach is used and the solution is also shown to be a Kalman £lter-type system.

In Section 2, we de£ne the smoothing and £ltering noisy I/O state estimation problems. Two explicit block solutions of the smoothing problem are derived in Section 3. They are weighted least-squares problems. In Section 4, we transform the noisy I/O model in the form (1) and de£ne the Kalman £lter for the resulting system as the modi£ed Kalman £lter. In Section 5, we prove the deterministic equivalence of the estimates of the noisy I/O £lter and the modi£ed Kalman £lter. Section 6 further establishes the optimal estimates of the true input/output signals. In Section 7, we con£rm the results on an example and in Section 8, give conclusions.

2. PROBLEM FORMULATION

An on-line, recursive estimation procedure can be realized by a causal dynamical system called a *£lter.* Such a £lter operates on previous and current measurements and produces an estimate of the to-be-estimated signal for the current moment of time. The problem of a £lter synthesis is referred to as a *£ltering problem.* A *smoothing problem* is an estimation problem, in which on the basis of the available measurements, an estimate for the to-be-estimated signal is produced for the whole (past) period of observation.

We introduce some notation used in the rest of the paper. A signal variable, without time index, denotes the vector obtained by stacking one over another the signal samples for the consecutive time instances. For example, over the time horizon $0, 1, \ldots, t_f - 1$, the vector of the consecutive input samples u is de£ned as

$$u := [u^T(0) \cdots u^T(t_f - 1)]^T$$

and the vector of the consecutive state samples is

$$x := [x^T(0) \cdots x^T(t_f - 1)\ x^T(t_f)]^T.$$

For a time indexed matrix sequence $\{V(t)\}_{t=0}^{t_f-1}$, we denote by V, without time argument, the block matrix

$$V := \operatorname{blk\,diag}\big(V(0), \ldots, V(t_f - 1)\big).$$

De£nition 1. (Optimal noisy I/O smoothing problem). Assume that the measurement errors $\tilde{u}$ and $\tilde{y}$ are random, centered, normal, uncorrelated, and white with known covariance matrices

$$\operatorname{cov}\big(\tilde{u}(t)\big) =: V_{\tilde{u}}(t), \qquad \operatorname{cov}\big(\tilde{y}(t)\big) =: V_{\tilde{y}}(t), \tag{4}$$

and that the initial condition x_0 is unknown. Then, given the matrices A, B, C, D, the optimal noisy I/O smoothing problem is de£ned as

$$\begin{aligned} &\min_{\hat{u},\hat{y},\hat{x}} \left\| \begin{bmatrix} V_{\tilde{u}} & \\ & V_{\tilde{y}} \end{bmatrix}^{-1/2} \begin{bmatrix} \hat{u} - u_{\mathrm{d}} \\ \hat{y} - y_{\mathrm{d}} \end{bmatrix} \right\|^2 \\ &\text{s.t.} \quad \begin{aligned} \hat{x}(t+1) &= A\hat{x}(t) + B\hat{u}(t) \\ \hat{y}(t) &= C\hat{x}(t) + D\hat{u}(t) \end{aligned} \end{aligned} \tag{5}$$

for $t = 0, 1, \ldots, t_f - 1$. The *optimal smoothed state estimate* $\hat{x}(\cdot, t_f)$ is the solution of (5).

Under the normality assumption for the noises, $\hat{x}(\cdot, t_f)$ is the minimum variance estimate, the maximum likelihood estimate, and the conditional expectation estimate of the state x. The equivalence is well known in the Kalman £lter case (Willems, 2002; Bryson and Ho, 1975; Anderson and Moore, 1979), and is shown in the noisy I/O case in (Guidorzi *et al.*, 2003).

De£nition 2. (Optimal noisy I/O £ltering problem). Given the model (2–3), satisfying the assumptions of De£nition 1, the optimal noisy I/O £ltering problem is to £nd a dynamical system,

$$\begin{aligned} z(t+1) &= A_{\mathrm{f}}(t)z(t) + B_{\mathrm{f}}(t)\begin{bmatrix} u_{\mathrm{d}}(t) \\ y_{\mathrm{d}}(t) \end{bmatrix}, \\ \hat{x}(t) &= C_{\mathrm{f}}(t)z(t) + D_{\mathrm{f}}(t)\begin{bmatrix} u_{\mathrm{d}}(t) \\ y_{\mathrm{d}}(t) \end{bmatrix}, \end{aligned} \tag{6}$$

such that $\hat{x}(t) = \hat{x}(t, t+1)$, where $\hat{x}(\cdot)$ is the solution of (6), *i.e.*, the *optimal £ltered state estimate*, and $\hat{x}(\cdot, t+1)$ is the optimal smoothed state estimate with a time horizon $t+1$.

3. SMOOTHING BY BLOCK PROCESSING

In this section, we write the optimal noisy I/O smoothing problem (5) as a weighted least-squares (WLS) problem. This representation is used as a conceptual tool for the analysis and not as a means to carry out the actual computations needed for the estimation.

We represent the I/O dynamics of the system (2), over the time horizon $0, \dots, t_f - 1$, explicitly as

$$y = \Gamma x_0 + Tu, \tag{7}$$

where

$$\Gamma := \begin{bmatrix} C \\ CA \\ \vdots \\ CA^{t_f-2} \end{bmatrix}, \quad \text{and} \quad T := \begin{bmatrix} H(0) & 0 & \cdots & 0 \\ H(1) & H(0) & \cdots & 0 \\ \vdots & \vdots & & \vdots \\ H(t_f-1) & H(t_f-2) & \cdots & H(0) \end{bmatrix}.$$

The matrix Γ is an extended observability matrix and T is a Toeplitz matrix formed from the Markov parameters

$$H(0) = D, \quad H(t) = CA^{t-1}B, \; t = 1, \dots, t_f - 1.$$

Using (7), we see that the optimal noisy I/O smoothing problem (5) is a weighted least-squares problem

$$\min_{\hat{x}_0, \hat{u}} \left\| \begin{bmatrix} V_{\tilde{u}} & \\ & V_{\tilde{y}} \end{bmatrix}^{-\frac{1}{2}} \left(\begin{bmatrix} u_{\mathrm{d}} \\ y_{\mathrm{d}} \end{bmatrix} - \begin{bmatrix} 0 & I \\ \Gamma & T \end{bmatrix} \begin{bmatrix} \hat{x}_0 \\ \hat{u} \end{bmatrix} \right) \right\|^2. \tag{8}$$

Alternatively, we represent the input/state/output dynamics of the system, over the time horizon $0, \dots, t_f - 1$, as

$$\begin{bmatrix} y(0) \\ 0 \\ y(1) \\ 0 \\ \vdots \\ y(t_f-1) \\ 0 \end{bmatrix} = \underbrace{\begin{bmatrix} C & 0 & & & & \\ A & -I & & & & \\ & C & 0 & & & \\ & A & -I & & & \\ & & \ddots & \ddots & & \\ & & & & C & 0 \\ & & & & A & -I \end{bmatrix}}_{\mathscr{A}} \begin{bmatrix} x(0) \\ x(1) \\ x(2) \\ \vdots \\ x(t_f) \end{bmatrix} + \underbrace{\begin{bmatrix} D & & & \\ B & & & \\ & D & & \\ & B & & \\ & & \ddots & \\ & & & D \\ & & & B \end{bmatrix}}_{-\mathscr{B}} \begin{bmatrix} u(0) \\ u(1) \\ \vdots \\ u(t_f-1) \end{bmatrix}.$$

Substituting $y_{\mathrm{d}} - \tilde{y}$ for y and $u_{\mathrm{d}} - \tilde{u}$ for u (see (3)), we have

$$\underbrace{\begin{bmatrix} y_{\mathrm{d}}(0) \\ 0 \\ y_{\mathrm{d}}(1) \\ 0 \\ \vdots \\ y_{\mathrm{d}}(t_f-1) \\ 0 \end{bmatrix}}_{\bar{y}_{\mathrm{d}}} + \mathscr{B} \begin{bmatrix} u_{\mathrm{d}}(0) \\ u_{\mathrm{d}}(1) \\ \vdots \\ u_{\mathrm{d}}(t_f-1) \end{bmatrix} =$$

$$= \mathscr{A}x + \mathscr{B} \begin{bmatrix} \tilde{u}(0) \\ \tilde{u}(1) \\ \vdots \\ \tilde{u}(t_f-1) \end{bmatrix} + \underbrace{\begin{bmatrix} I & & & \\ 0 & & & \\ & I & & \\ & 0 & & \\ & & \ddots & \\ & & & I \\ & & & 0 \end{bmatrix}}_{\mathscr{C}} \begin{bmatrix} \tilde{y}(0) \\ \tilde{y}(1) \\ \vdots \\ \tilde{y}(t_f-1) \end{bmatrix},$$

or with the de£nition of the new variables

$$\bar{y}_{\mathrm{d}} + \mathscr{B}u_{\mathrm{d}} = \mathscr{A}x + \mathscr{B}\tilde{u} + \mathscr{C}\tilde{y}. \tag{9}$$

Using (9), the optimal noisy I/O smoothing problem is equivalent to the following problem

$$\begin{aligned} &\min_{\hat{x}, \Delta u, \Delta y} \left\| \begin{bmatrix} V_{\tilde{u}} & \\ & V_{\tilde{y}} \end{bmatrix}^{-1/2} \begin{bmatrix} \Delta u \\ \Delta y \end{bmatrix} \right\|^2 \\ &\text{s.t.} \quad \bar{y}_{\mathrm{d}} + \mathscr{B}u_{\mathrm{d}} = \mathscr{A}\hat{x} + [\mathscr{B} \; \mathscr{C}] \begin{bmatrix} \Delta u \\ \Delta y \end{bmatrix}, \end{aligned} \tag{10}$$

which solution is given in Section 5.

The solutions (8) and (10) of the noisy I/O smoothing problem are not recursive and thus not practical for large data sets. See (Markovsky *et al.*, 2002) for recursive solution of an equivalent continuous-time problem. In the discrete-time case, the recursive solution is given by two time-varying £lters; one running backward in time and one running forward in time. The forward recursion is de£ned by a time-varying Riccati equation. The backward £lter produces the optimal smoothed state estimate.

4. THE MODIFIED KALMAN FILTER

We convert the noisy I/O model (2–3) in the form (1) by substituting $u_{\mathrm{d}}(t) - \tilde{u}$ for $u(t)$ and $y_{\mathrm{d}}(t) - \tilde{y}$ for $y(t)$ (see (3)) in (2)

$$\begin{aligned} x(t+1) &= Ax(t) + Bu_{\mathrm{d}}(t) - B\tilde{u}(t) \\ y_{\mathrm{d}}(t) &= Cx(t) + Du_{\mathrm{d}}(t) - D\tilde{u}(t) + \tilde{y}(t) \end{aligned}$$

and de£ne (fake) process noise w and measurement noise v by

$$w := -B\tilde{u} \quad \text{and} \quad v := -D\tilde{u} + \tilde{y}.$$

The resulting system

$$\begin{aligned} x(t+1) &= Ax(t) + Bu_{\mathrm{d}}(t) + w(t) \\ y_{\mathrm{d}}(t) &= Cx(t) + Du_{\mathrm{d}}(t) + v(t) \end{aligned} \tag{11}$$

is in the form (1), where

$$\begin{bmatrix} Q(t) & S(t) \\ S^T(t) & R(t) \end{bmatrix} = \begin{bmatrix} -B & 0 \\ -D & I \end{bmatrix} \begin{bmatrix} V_u(t) & \\ & V_y(t) \end{bmatrix} \begin{bmatrix} -B & 0 \\ -D & I \end{bmatrix}^T.$$

We call the Kalman £lter of the modi£ed system (11), *i.e.*, the system

$$z(t+1) = A_{\rm KF}(t)z(t) + B_{\rm KF}(t)\begin{bmatrix} u_{\rm d}(t) \\ y_{\rm d}(t) \end{bmatrix},$$
$$\hat{x}(t) = C_{\rm KF}(t)z(t) + D_{\rm KF}(t)\begin{bmatrix} u_{\rm d}(t) \\ y_{\rm d}(t) \end{bmatrix}, \qquad (12)$$

where

$$A_{\rm KF}(t) = \big(A - K(t)C\big), \quad B_{\rm KF}(t) = \big[B - K(t)D,\ K(t)\big],$$
$$C_{\rm KF}(t) = I - P(t)C^T\big(CP(t)C^T + R(t)\big)^{-1}C,$$
$$D_{\rm KF}(t) = P(t)C^T\big(CP(t)C^T + R(t)\big)^{-1}\big[-D\ \ I\big],$$
$$K(t) = \big(AP(t)C^T + S(t)\big)\big(CP(t)C^T + R(t)\big)^{-1},$$

and

$$P(t+1) = AP(t)A^T - \big(AP(t)C^T + S(t)\big)\times$$
$$\times\big(CP(t)C^T + R(t)\big)^{-1}\big(AP(t)C^T + S(t)\big)^T + Q(t),$$

the *modi£ed Kalman £lter*. It recursively solves (9) for the last block entry of the unknown x. The solution is in the sense of the weighted least-squares problem

$$\min_{\hat{x},\hat{e}} \|V_e^{-1/2}\hat{e}\|^2 \quad \text{s.t.} \quad \bar{y}_{\rm d} + \mathscr{B}u_{\rm d} = \mathscr{A}\hat{x} + \hat{e}. \qquad (13)$$

The variable $\hat{e}$ accounts for the cumulative noise

$$e := [\mathscr{B}\ \mathscr{C}]\begin{bmatrix} \tilde{u} \\ \tilde{y} \end{bmatrix}$$

added to the equation and the covariance matrix of e is $V_e = \mathscr{B}V_{\tilde{u}}\mathscr{B}^T + \mathscr{C}V_{\tilde{y}}\mathscr{C}^T$.

When the measurement noise covariances $V_{\tilde{u}}(t)$ and $V_{\tilde{u}}(t)$ does not depend on t, one can replace the time-varying Kalman £lter with the (suboptimal) time-invariant £lter, obtained by replacing $P(t)$ in (12) with the steady-state solution $\bar{P}$ of the algebraic Riccati equation

$$\bar{P} = APA^T - \big(A\bar{P}C^T + S\big)\big(C\bar{P}C^T + R\big)^{-1}\big(A\bar{P}C^T + S\big)^T + Q.$$

In the following section, we investigate the relation between the modi£ed Kalman £lter (12) state estimate and the noisy I/O £lter state estimate.

5. EQUIVALENCE OF THE MODIFIED KALMAN FILTER AND THE NOISY I/O FILTER

Consider the linear system of equations (9) and the two solution methods (10) and (13). Denote

$$z := \bar{y}_{\rm d} + \mathscr{B}u_{\rm d}, \quad \delta := \begin{bmatrix} \tilde{u} \\ \tilde{y} \end{bmatrix}, \quad V_\delta := \begin{bmatrix} V_{\tilde{u}} & \\ & V_{\tilde{y}} \end{bmatrix},$$
$$\hat{\delta} := \begin{bmatrix} \Delta u \\ \Delta y \end{bmatrix}, \quad \text{and} \quad \mathscr{D} := [\mathscr{B}\,\mathscr{C}].$$

We want to £nd a relation between the solutions of the following problems

$$\min_{\hat{x},\hat{e}} \|V_e^{-1/2}\hat{e}\|_2^2 \quad \text{s.t.} \quad z = \mathscr{A}\hat{x} + \hat{e}, \qquad (14)$$

where $V_e = \mathscr{D}V_\delta\mathscr{D}^T$ and

$$\min_{\hat{x},\hat{\delta}} \|V_\delta^{-1/2}\hat{\delta}\|_2^2 \quad \text{s.t.} \quad z = \mathscr{A}\hat{x} + \mathscr{D}\hat{\delta}. \qquad (15)$$

The £rst problem is a weighted least-squares problem and its solution is

$$\hat{x}_{\rm KF} = (\mathscr{A}^T V_e^{-1}\mathscr{A})^{-1}\mathscr{A}^T V_e^{-1} z. \qquad (16)$$

The noisy I/O estimation problem (15) is a minimum-norm type problem and its solution is (see Lemma 3 in the Appendix)

$$\hat{x} = [0\ \ I][V_e\mathscr{A}^\perp\ \ \mathscr{A}]^{-1}z, \qquad (17)$$

where $\mathscr{A}^\perp$ is a matrix which columns form a basis for the orthogonal complement of the range space of $\mathscr{A}$.

We transform (16) and (17) by the change of variables

$$\bar{\mathscr{A}} := V_e^{-1/2}\mathscr{A} \qquad \text{and} \qquad \bar{z} := V_e^{-1/2}z.$$

Then

$$\hat{x}_{\rm KF} = (\bar{\mathscr{A}}^T\bar{\mathscr{A}})^{-1}\bar{\mathscr{A}}^T\bar{z}, \qquad (18)$$

and

$$\hat{x} = [0\ \ I][\bar{\mathscr{A}}^\perp\ \ \bar{\mathscr{A}}]^{-1}\bar{z}. \qquad (19)$$

((19) follows from the identity $\bar{\mathscr{A}}^\perp = V_e^{1/2}\mathscr{A}^\perp$.) Now the question of the solutions' equivalence is answered by Theorem 4, see the Appendix, which states that $\hat{x}_{\rm KF} = \hat{x}$. Thus the two solutions are *deterministically* equal and the noisy I/O £ltering problem is solved by the modi£ed Kalman £lter (12), *i.e.* $A_{\rm f} = A_{\rm KF}$, $B_{\rm f} = B_{\rm KF}$, $C_{\rm f} = C_{\rm KF}$, and $D_{\rm f} = D_{\rm KF}$.

6. OPTIMAL ESTIMATION OF THE TRUE INPUT/OUTPUT SIGNALS

Up to now we were interested in the optimal £ltering in the sense of state estimation. In this section, we show how the optimal estimates of the input and the output can be derived from the modi£ed Kalman £lter.

The solutions $\hat{e}$ and $\hat{\delta}$ of (14) and (15), respectively, satisfy the following relation

$$\hat{e} = V_e\mathscr{A}^\perp[V_e\mathscr{A}^\perp\ \ \mathscr{A}]^{-1}z$$
$$= \mathscr{D}V_\delta\mathscr{D}^T\mathscr{A}^\perp[V_e\mathscr{A}^\perp\ \ \mathscr{A}]^{-1}z = \mathscr{D}\hat{\delta}.$$

This implies that the state estimate $\hat{x}$, the one-step-ahead prediction $z(t+1)$, and the optimal input estimate $\hat{u}$ satisfy the equation

$$z(t+1) = A\hat{x}(t) + B\hat{u}(t). \qquad (20)$$

Then we can £nd $\hat{u}$ *exactly* from $\hat{x}$ and $z(t+1)$, obtained from the modi£ed Kalman £lter (12). In fact, (20) and the Kalman £lter equations imply that

$$\hat{u}(t) = E(t)z(t) + F(t)\begin{bmatrix} u_{\rm d}(t) \\ y_{\rm d}(t) \end{bmatrix},$$

where

$$E(t) := -V_{\tilde{u}}D^T\big(CP(t)C^T + R(t)\big)^{-1}C \quad \text{and}$$
$$F(t) := \Big[I - V_{\tilde{u}}D^T\big(CP(t)C^T + R(t)\big)^{-1}D,\ V_{\tilde{u}}D^T\big(CP(t)C^T + R(t)\big)^{-1}\Big].$$

The optimal output estimate is $\hat{y}(t) = C\hat{x}(t) + D\hat{u}(t)$.

7. NUMERICAL EXAMPLE

In this section, we verify numerically the equivalence of the solutions established in Section 5. The particular system, we use, is

$$A = \begin{bmatrix} 0.6 & -0.45 \\ 1 & 0 \end{bmatrix}, \quad B = \begin{bmatrix} 1 \\ 0 \end{bmatrix},$$
$$C = [0.48429 \ \ -0.45739], \quad \text{and} \quad D = 0.5381.$$

The time horizon is $t_f = 100$, the initial state is $x_0 = 0$, and the input signal is a normal white noise sequence with unit variance. The input and the output of the system are corrupted by independent, centered, normal, white noises with variances $\text{var}(\tilde{u}(t)) = 0.4$ and $\text{var}(\tilde{y}(t)) = 0.4$ for all t.

The estimate of the noisy I/O £lter is computed directly from the de£nition, *i.e.*, we solve a sequence of optimal smoothing problems with increasing time-horizon. Every smoothing problem is a weighted least-squares problem that is solved explicitly according to (8). The last block entries of the obtained sequence of solutions form the noisy I/O £lter state estimate.

We compare the noisy I/O £lter estimate with the estimate of the modi£ed Kalman £lter (12). The experiment is carried out in MATLAB. The state estimate $\hat{x}_{\text{KF}}$ obtained by the modi£ed Kalman £lter is up to the numerical errors equal to the state estimate $\hat{x}_{\text{f}}$ obtained by the noisy I/O £lter, $||\hat{x}_{\text{KF}} - \hat{x}_{\text{f}}|| = 5.7723e-15$. This is the desired numerical veri£cation of the theoretical result of the paper. The absolute errors of estimation $||\hat{x}-x||^2$, $||\hat{u}-u||^2$, $||\hat{y}-y||^2$ for all estimation methods, discussed in the paper is given in Table 1.

Table 1. Comparison of the absolute errors of the state, input, and output estimates for all methods and the noisy data. (MKF — modi£ed Kalman £lter)

Method	$\|\|\hat{x}-x\|\|^2$	$\|\|\hat{u}-u\|\|^2$	$\|\|\hat{y}-y\|\|^2$
optimal smoothing	75.3981	29.2195	15.5409
optimal £ltering	75.7711	35.5604	16.4571
time-varying MKF	75.7711	35.5604	16.4571
time-invariant MKF	76.1835	35.7687	16.5675
noisy data	116.3374	42.4711	41.2419

8. CONCLUSIONS

We considered optimal noisy I/O estimation problems for discrete-time LTI systems. The £ltering problem is solved via a modi£ed Kalman £lter. The equivalence between the optimal noisy I/O £lter and the modi£ed Kalman £lter is proven algebraically using explicit state-space representation of the system.

ACKNOWLEDGEMENTS

Ivan Markovsky is a research assistant and Dr. Bart De Moor is a full professor at the Katholieke Universiteit Leuven, Belgium. Our research is supported by **Research Council KUL:** GOA-Me£sto 666, several PhD/postdoc & fellow grants; **Flemish Government: FWO:** PhD/postdoc grants, projects, G.0240.99 (multilinear algebra), G.0407.02 (support vector machines), G.0197.02 (power islands), G.0141.03 (Identi£cation and cryptography), G.0491.03 (control for intensive care glycemia), G.0120.03 (QIT), research communities (ICCoS, ANMMM); **AWI:** Bil. Int. Collaboration Hungary/Poland; **IWT:** PhD Grants, Soft4s (softsensors), **Belgian Federal Government:** DWTC (IUAP IV-02 (1996-2001) and IUAP V-22 (2002-2006), PODO-II (CP/40: TMS and Sustainibility); **EU:** CAGE; ERNSI; Eureka 2063-IMPACT; Eureka 2419-FliTE; **Contract Research/agreements:** Data4s, Electrabel, Elia, LMS, IPCOS, VIB.

9. REFERENCES

Anderson, B. D. O. and J. B. Moore (1979). *Optimal Filtering*. Prentice Hall.

Aoki, M. and P. C. Yue (1970). On certain convergence questions in system identi£cation. *SIAM J. Control* **8**(2), 239–256.

Bryson, A. and Y. Ho (1975). *Applied Optimal Control*. Hemisphere, Washington, D.C.

De Moor, B. (1993). Structured total least squares and L_2 approximation problems. *Lin. Alg. and Its Appl.* **188–189**, 163–207.

De Moor, B. and B. Roorda (1994). L_2-optimal linear system identi£cation structured total least squares for SISO systems. In: *In the proceedings of the CDC*. pp. 2874–2879.

Diversi, R., R. Guidorzi and U. Soverini (2003). Algorithms for optimal errors-in-variables £ltering. *Systems & Control Letters* **48**, 1–13.

Fagnani, F. and J. C. Willems (1997). Deterministic Kalman £ltering in a behavioral framework. *Systems & Control Letters* **32**, 301–312.

Guidorzi, R., R. Diversi and U. Soverini (2003). Optimal errors-in-variables £ltering. *Automatica* **39**, 281–289.

Lemmerling, P. (1999). Structured total least squares: analysis, algorithms and applications. PhD thesis. ESAT/SISTA, K.U. Leuven.

Markovsky, I., J. C. Willems and B. De Moor (2002). Continuous-time errors-in-variables £ltering. In: *Proc. of the Conference on Decision and Control*. pp. 2576–2581.

Meyer, C. D. (2000). *Matrix Analysis and Applied Linear Algebra*. SIAM.

Polderman, J. W. and J. C. Willems (1998). *Introduction to mathematical systems theory*. Springer-Verlag.

Söderström, T., U. Soverini and K. Mahata (2002). Perspectives on errors-in-variables estimation for dynamic systems. *Signal Processing* **82**, 1139–1154.

Van Huffel, S. and J. Vandewalle (1991). *The total least squares problem: Computational aspects and analysis*. SIAM, Philadelphia.

Willems, J. C. (2002). Deterministic Kalman £ltering. *J. of Econometrics*, to appear.

Zheng, W. X. (2002). A bias correction method for identi£cation of linear dynamic errors-in-variables models. *IEEE Trans. on Aut. Control* **47**(7), 1142–1147.

Appendix A. SOLUTION OF THE OPTIMIZATION PROBLEM (15) AND PROOF OF THE STATE ESTIMATES' EQUIVALENCE

Lemma 3. Assuming that $\mathscr{A}$ is full rank and V_δ is positive de£nite. The solution of the minimum-norm type problem (15) is

$$\begin{bmatrix}\hat{\delta}\\ \hat{x}\end{bmatrix} = \begin{bmatrix}V_\delta \mathscr{D}^T \mathscr{A}^\perp & 0\\ 0 & I\end{bmatrix} [V_e \mathscr{A}^\perp \;\; \mathscr{A}]^{-1} z.$$

PROOF. The Lagrangian of (15) is

$$L(\hat{x},\hat{\delta},\lambda) = \hat{\delta}^T V_\delta^{-1}\hat{\delta} + \lambda^T(\mathscr{A}\hat{x} + \mathscr{D}\delta - z).$$

The £rst order optimality condition

$$\frac{\partial L}{\partial \delta} = 2V_\delta^{-1}\hat{\delta} + \mathscr{D}^T\lambda = 0 \quad \Rightarrow \quad \hat{\delta} = -\frac{1}{2}V_\delta \mathscr{D}^T\lambda, \qquad (A.1)$$

$$\frac{\partial L}{\partial \hat{x}} = \mathscr{A}^T\lambda = 0 \quad \Rightarrow \quad \lambda = -2\mathscr{A}^\perp\bar{\lambda}, \qquad (A.2)$$

$$\frac{\partial L}{\partial \lambda} = \mathscr{A}\hat{x} + \mathscr{D}\hat{\delta} - z = 0 \quad \Rightarrow \quad z = \mathscr{A}\hat{x} + \mathscr{D}\hat{\delta} \qquad (A.3)$$

is a necessary and suf£cient condition for a global minimum. The matrix $\mathscr{A}^\perp$ is any matrix which columns form a basis for the range space of $\mathscr{A}^T$. Substituting (A.1) and (A.2) in (A.3), we have

$$[\mathscr{D}V_\delta \mathscr{D}^T \mathscr{A}^\perp \;\; \mathscr{A}] \begin{bmatrix}\bar{\lambda}\\ \hat{x}\end{bmatrix} = z.$$

Using $V_e = \mathscr{D}V_\delta\mathscr{D}^T$ and the assumption that $\mathscr{A}$ is full rank, the result follows. □

Theorem 4. For a full rank matrix $A \in \mathbb{R}^{m\times n}$, with m greater then n,

$$(A^TA)^{-1}A^T = [0_{n\times(m-n)} \;\; I_n]\begin{bmatrix}A^\perp & A\end{bmatrix}^{-1}, \qquad (A.4)$$

where $A^\perp$ is a matrix which columns form a basis for the orthogonal complement of the range space of A, *i.e.*, Range$(A^\perp)$ = Null(A^T), and rank$(A^\perp) = m - n$.

PROOF. In the proof, we use the SVD of the matrix A

$$A = U\Sigma V^T = [U_1 \; U_2]\begin{bmatrix}\Sigma_1\\ 0\end{bmatrix}V^T = U_1\Sigma_1 V^T.$$

Partition U as follows

$$U = [U_1 \; U_2] = \begin{array}{c} \begin{array}{cc} n & m-n \end{array} \\ \begin{bmatrix} U_{11} & U_{12} \\ U_{21} & U_{22}\end{bmatrix} \end{array} \begin{array}{c} m-n \\ n \end{array}.$$

The matrix U_2 satis£es Range(U_2) = Null(A^T) and rank$(U_2) = m - n$, so it serves as a particular $A^\perp$.

The left-hand side of the desired identity (A.4) is

$$(A^TA)^{-1}A^T = V\Sigma_1^{-1}U_1^T = V\Sigma_1^{-1}\begin{bmatrix}U_{11}^T & U_{21}^T\end{bmatrix},$$

and the right-hand side is

$$[0 \;\; I_n]\begin{bmatrix}A^\perp & A\end{bmatrix}^{-1} = [0 \;\; I_n]\begin{bmatrix}U_2 & U_1\Sigma_1V^T\end{bmatrix}^{-1} = [0 \;\; I_n]\begin{bmatrix}U_{12} & U_{11}\Sigma_1V^T\\ U_{22} & U_{21}\Sigma_1V^T\end{bmatrix}^{-1}. \qquad (A.5)$$

To £nd explicitly the inverse matrix in (A.5), we use the formula for inverse of a block matrix (Meyer, 2000, p.123)

$$\begin{bmatrix}B & C\\ D & E\end{bmatrix}^{-1} = \begin{bmatrix}B^{-1} + B^{-1}CS^{-1}DB^{-1} & -B^{-1}CS^{-1}\\ -S^{-1}DB^{-1} & S^{-1}\end{bmatrix},$$

where

$$S = E - DB^{-1}C$$

is the Schur complement of B in the matrix $\left[\begin{smallmatrix}B & C\\ D & E\end{smallmatrix}\right]$. For the block matrix in (A.5), the Schur complement of U_{12} is

$$S = U_{21}\Sigma_1V^T - U_{22}U_{12}^{-1}U_{11}\Sigma_1V^T = (U_{21} - U_{22}U_{12}^{-1}U_{11})\Sigma_1V^T.$$

Then

$$[0 \;\; I_n]\begin{bmatrix}A^\perp & A\end{bmatrix}^{-1} = V\Sigma_1^{-1}(U_{21} - U_{22}U_{12}^{-1}U_{11})^{-1}[U_{22}U_{12}^{-1} \;\; I]. \qquad (A.6)$$

Because of the orthogonality of U,

$$U_1^TU_2 = U_{11}^TU_{12} + U_{21}^TU_{22} = 0 \quad \Rightarrow \quad -U_{22}U_{12}^{-1} = U_{21}^{-T}U_{11}^T. \qquad (A.7)$$

Then

$$(U_{21} - U_{22}U_{12}^{-1}U_{11})^{-1} = (U_{21} + U_{21}^{-T}U_{11}^TU_{11})^{-1} = \Big(U_{21}^{-T}\underbrace{(U_{21}^TU_{21} + U_{11}^TU_{11})}_{U_1^TU_1 = I}\Big)^{-1} = U_{21}^T. \qquad (A.8)$$

Substitution of the expressions of (A.7) and (A.8) into (A.6) establishes the identity. □

IFAC
Publications
www.elsevier.com/locate/ifac

THE P-NORM GENERALIZATION OF THE LMS ALGORITHM FOR ADAPTIVE FILTERING

Jyrki Kivinen [*,1] **Manfred K. Warmuth** [**,2] **Babak Hassibi** [***]

[*] *Research School of Information Sciences and Engineering, Australian National University, Canberra, ACT 0200, Australia*
[**] *Computer Science Department, 237 Baskin Engineering, University of California, Santa Cruz, CA 95064, USA*
[***] *Department of Electrical Engineering, California Institute of Technology, Pasadena, CA 91125, USA*

Abstract: Recently much work has been done analyzing online machine learning algorithms in a worst case setting, where no probabilistic assumptions are made about the data. This is analogous to the H^∞ setting used in adaptive linear filtering. Bregman divergences have become a standard tool for analyzing online Machine Learning algorithms. Using these divergences, we motivate a generalization of the the Least Mean Squared (LMS) algorithm. The loss bounds for these so-called p-norm algorithms involve other norms than the standard 2-norm. The bounds can be significantly better if a large proportion of the input variables are irrelevant, *i.e.*, if the weight vector we are trying to learn is sparse. We also prove result for nonstationary targets. We only know how to apply kernel methods to the standard LMS algorithm (*i.e.*, $p = 2$). However even in the general p-norm case we can handle generalized linear models where the output of the system is a linear function combined with a nonlinear transfer function (*e.g.*, the logistic sigmoid). *Copyright © 2003 IFAC*

Keywords: adaptive filtering, online learning, H^∞ optimality, Least Mean Squares

1. INTRODUCTION

We focus on the following linear model of adaptive filtering:

$$y_t = \mathbf{u} \cdot \mathbf{x}_t + v_t. \tag{1}$$

Here $\mathbf{u}$ is the unknown target, $\mathbf{x}_t$ is a known input, v_t is unknown noise and y_t is the known output signal. We are interested in algorithms that maintain a weight vector $\mathbf{w}_t$ based on the past examples $(\mathbf{x}_j, y_j)$, $j = 1, \ldots, t$, and, over a sequence of T trials, get as close as possible to the target $\mathbf{u}$. As we shall see, closely related online problems have also been studied in machine learning.

More specifically, at trial t the algorithm receives $\mathbf{x}_t$ and y_t (in order) and has to commit to a weight vector at some point after seeing $\mathbf{x}_t$. We consider three problems depending on whether the algorithm needs to commit to its weight vector before or after seeing y_t and depending on how the loss of the algorithm is measured.

A priori filtering: Here we are interested in predicting the *uncorrupted* output $\mathbf{u} \cdot \mathbf{x}_t$ before the signal y_t is received. Therefore the algorithm needs to commit to its weight vector $\mathbf{w}_{t-1}$ right before seeing y_t and our loss is the energy of the a priori filtering error $\mathbf{u} \cdot \mathbf{x}_t - \mathbf{w}_{t-1} \cdot \mathbf{x}_t$, *i.e.*,

$$\sum_{t=1}^{T} (\mathbf{u} \cdot \mathbf{x}_t - \mathbf{w}_{t-1} \cdot \mathbf{x}_t)^2. \tag{2}$$

A posteriori filtering: Here we assume that for estimating the uncorrupted output $\mathbf{u} \cdot \mathbf{x}_t$, we also have

[1] Supported by the Australian Research Council
[2] Supported by grant NSF CCR 9821087

access to the measurement y_t. Thus, the algorithm needs to commit to its weight vector $\mathbf{w}_t$ only after seeing y_t and the loss is the square of the a posteriori error:

$$\sum_{t=1}^{T} (\mathbf{u} \cdot \mathbf{x}_t - \mathbf{w}_t \cdot \mathbf{x}_t)^2. \quad (3)$$

Prediction: Here we are interested in *predicting* the next observation y_t before receiving it. Thus the algorithm needs to commit to its weight vector $\mathbf{w}_{t-1}$ before seeing y_t. The prediction error is $y_t - \mathbf{w}_{t-1} \cdot \mathbf{x}_t$ and the loss

$$\sum_{t=1}^{T} (y_t - \mathbf{w}_{t-1} \cdot \mathbf{x}_t)^2. \quad (4)$$

The above prediction problem is also studied in machine learning. Note that in the filtering problems, the term v_t is regarded as a disturbance, so we are interested in estimating the *true* output $\mathbf{u} \cdot \mathbf{x}_t$ of the linear system for the input $\mathbf{x}_t$. In the prediction problem, however, we consider the y_t as the actual outcome of some event we are interested in predicting; in that case there is no particular value in matching the prediction $\mathbf{u} \cdot \mathbf{x}_t$ at those times where it is inaccurate.

In contrast to the loss function used by the prediction problem, the loss functions for the two filtering problems are not measurable by the online algorithm. This is because the true value of $\mathbf{u}$ is unknown. To partially alleviate this, one possibility is to consider the *worst-case* losses by maximizing (2) and (3) over $\mathbf{u}$. However, this is not really meaningful since the the losses can be made arbitrarily large by scaling $\mathbf{u}$. A much more reasonable route is to look at the *normalized* loss, normalized by the energy of the unknown sequence v_t and the unknown vector $\mathbf{u}$. Thus, we can consider

$$\frac{\sum_{t=1}^{T} (\mathbf{u} \cdot \mathbf{x}_t - \mathbf{w}_{t-1} \cdot \mathbf{x}_t)^2}{\sum_{t=1}^{T} (\mathbf{u} \cdot \mathbf{x}_t - y_t)^2 + \frac{1}{\eta} ||\mathbf{u}||_2^2},$$

where we have used $v_t = y_t - \mathbf{u} \cdot \mathbf{x}_t$ and where $\eta > 0$ is a normalization constant that weighs the relative contributions of $\sum_{t=1}^{T} (\mathbf{u} \cdot \mathbf{x}_t - y_t)^2$ and $||\mathbf{u}||_2^2$. In particular, we can look at the *worst-case* of this normalized loss by maximizing over $\mathbf{u}$:

$$\max_{\mathbf{u}} \frac{\sum_{t=1}^{T} (\mathbf{u} \cdot \mathbf{x}_t - \mathbf{w}_{t-1} \cdot \mathbf{x}_t)^2}{\sum_{t=1}^{T} (\mathbf{u} \cdot \mathbf{x}_t - y_t)^2 + \frac{1}{\eta} ||\mathbf{u}||_2^2}.$$

This choice of $\mathbf{u}$ is now well-behaved and can be determined when all examples are processed. In control-theoretic jargon, the above maximum energy gain is referred to as the H^∞ norm.

Our starting point is the Least Mean Squares (LMS) algorithm (also known as the Widrow-Hoff algorithm), defined by the update rule

$$\mathbf{w}_t = \mathbf{w}_{t-1} - \eta (\mathbf{w}_{t-1} \cdot \mathbf{x}_t - y_t) \mathbf{x}_t$$

where $\eta > 0$ is a learning rate parameter. The basic result for a priori filtering (Hassibi *et al.*, 1996) states that for the LMS algorithm,

$$\frac{\sum_{t=1}^{T} (\mathbf{u} \cdot \mathbf{x}_t - \mathbf{w}_{t-1} \cdot \mathbf{x}_t)^2}{\sum_{t=1}^{T} (\mathbf{u} \cdot \mathbf{x}_t - y_t)^2 + \frac{1}{\eta} ||\mathbf{u}||_2^2} \leq 1 \quad (5)$$

for all $\mathbf{u}$, provided $\eta \leq 1/\max ||\mathbf{x}_t||_2^2$. Further, no algorithm can in general guarantee a ratio less than one. Therefore we say that LMS is H^∞ *optimal.* (For the above, and as done throughout the paper, we assumed $\mathbf{w}_0 = \mathbf{0}$; if $\mathbf{w}_0 \neq \mathbf{0}$, then $||\mathbf{u}||_2^2$ must be replaced by $||\mathbf{u} - \mathbf{w}_0||_2$.)

To compare this with results from machine learning, assume there is a known upper bound X_2 such that $||\mathbf{x}_t||_2 \leq X_2$ for all t, and write $\eta = \alpha / X_2^2$. Then Cesa-Bianchi *et al.* (1996) have shown that for $0 < \alpha < 1$,

$$\sum_{t=1}^{T} (y_t - \mathbf{w}_{t-1} \cdot \mathbf{x}_t)^2 \leq \frac{1}{1-\alpha} \sum_{t=1}^{T} (\mathbf{u} \cdot \mathbf{x}_t - y_t)^2 + \frac{1}{\alpha} X_2^2 ||\mathbf{u}||_2^2. \quad (6)$$

To compare the prediction and filtering results, we write (5) as

$$\sum_{t=1}^{T} (\mathbf{u} \cdot \mathbf{x}_t - \mathbf{w}_{t-1} \cdot \mathbf{x}_t)^2 \leq \sum_{t=1}^{T} (\mathbf{u} \cdot \mathbf{x}_t - y_t)^2 + \frac{1}{\alpha} X_2^2 ||\mathbf{u}||_2^2 \quad (7)$$

where X_2 and η are as above and $0 < \alpha \leq 1$. The factor $1/(1-\alpha)$ in (6) is a source of many difficulties in machine learning, where the goal is to tune the learning rate so as to obtain the smallest possible bound. However, the filtering bound (7) is optimized at $\alpha = 1$. Thus we omit the α parameter from the filtering bounds when the norm of instances is bounded.

We are interested in taking machine learning techniques that have recently been used to generalize (6) and applying them in the filtering setting, leading to generalizations of (7) and new interpretations of the filtering algorithms. Techniques we are interested in include

(1) motivating algorithms in terms of minimization problems based on Bregman divergences (Kivinen and Warmuth, 1997; Kivinen and Warmuth, 2001),
(2) replacing the 2-norms in the bounds by other norms (Kivinen and Warmuth, 1997; Grove *et al.*, 2001; Gentile and Littlestone, 1999), and
(3) allowing for nonstationary targets (Herbster and Warmuth, 2001) and nonlinear predictors (Helmbold *et al.*, 1999).

In Section 2 we introduce Bregman divergences and show how a Bregman divergence can be used to derive two subtly different algorithms, the *implicit* and *explicit* algorithm. When the squared Euclidean distance is used as the Bregman divergence, these algorithms become the standard *LMS* and *normalized LMS* algorithm (Hassibi *et al.*, 1996), respectively. In Section 3 we give filtering loss bounds for the explicit and implicit algorithms in the case of Bregman divergences based on squared q-norms (Grove *et al.*, 2001). These bounds generalize the results of Hassibi *et al.* (1996) about the H^∞ optimality of LMS and normalized LMS for the a priori and a posteriori filtering problems. The generalization replaces the product $||\mathbf{x}||_2||\mathbf{u}||_2$ in the bound by another product of dual norms $||\mathbf{x}||_p||\mathbf{u}||_q$ where p and q are such that $1/p+1/q=1$ and $2 \le p < \infty$. The new bounds are significantly stronger when the target $\mathbf{u}$ is sparse, *i.e.*, has few nonzero components. In Section 4 we generalize the q-norm based algorithms to allow for nonstationary targets $\mathbf{u}_t$. The loss bounds in the nonstationary case include an extra term that depends on the total distance $\mathbf{u}_t$ travels during the whole sequence, as measured by the q-norm. Again there are no distribution assumptions about this movement. Finally, in Section 5 we give bounds for generalized linear regression where the linear predictor is fed through a nonlinear transfer function (such as the logistic sigmoid).

2. DERIVATION OF ALGORITHMS

In this section we give the basic definitions of Bregman divergences and explain their use in deriving online algorithms. See Azoury and Warmuth (2001) and references therein for more background on these divergences.

Assume that $F:\mathbf{R}^n \to \mathbf{R}^n$ is a strictly convex twice differentiable function. Denote its gradient by $\mathbf{f} = \nabla F$; notice that $\mathbf{f}$ is one-to-one. The *Bregman divergence* $d_F(\mathbf{u},\mathbf{w})$ (Bregman, 1967) is defined for $\mathbf{u},\mathbf{w} \in \mathbf{R}^n$ as the error in approximating $F(\mathbf{u})$ by its first order Taylor polynomial around $\mathbf{w}$. More formally,

$$d_F(\mathbf{u},\mathbf{w}) = F(\mathbf{u}) - F(\mathbf{w}) - \mathbf{f}(\mathbf{w})\cdot(\mathbf{u}-\mathbf{w}).$$

The Bregman divergence $d_F(\mathbf{u},\mathbf{w})$ is always nonnegative, and zero only for $\mathbf{u}=\mathbf{w}$. It is (strictly) convex in $\mathbf{u}$, but might not be convex in $\mathbf{w}$. Usually, d_F is not symmetric.

Example 1. Given $q>1$, define $F(\mathbf{w}) = \frac{1}{2}||\mathbf{w}||_q^2$ where $||\mathbf{u}||_q = (\sum_i |u_i|^q)^{1/q}$. We denote the corresponding Bregman divergence by d_q. Thus

$$d_q(\mathbf{u},\mathbf{w}) = \frac{1}{2}||\mathbf{u}||_q^2 - \frac{1}{2}||\mathbf{w}||_q^2 - \mathbf{f}(\mathbf{w})\cdot(\mathbf{u}-\mathbf{w})$$

where the gradient is given by

$$f_i(\mathbf{w}) = \frac{\operatorname{sign}(w_i)|w_i|^{q-1}}{||\mathbf{w}||_q^{q-2}}.$$

It is easy to verify that the inverse of the gradient is given by

$$f_i^{-1}(\theta) = \frac{\operatorname{sign}(\theta_i)|\theta_i|^{p-1}}{||\theta||_p^{p-2}}$$

where $1/p+1/q=1$. Grove *et al.* (2001) have shown that if $\mathbf{f}(\mathbf{w}') = \mathbf{f}(\mathbf{w}) + \mathbf{x}$, we have

$$d_q(\mathbf{w},\mathbf{w}') \le \frac{p-1}{2}||\mathbf{x}||_p^2. \tag{8}$$

The important special case $p=q=2$ gives $d_2(\mathbf{u},\mathbf{w}) = \frac{1}{2}||\mathbf{u}-\mathbf{w}||_2^2$, with $\mathbf{f}$ the identity function.

A second important family of Bregman divergences is the relative entropy and its variants.

Example 2. Define $F(\mathbf{w}) = \sum_i (w_i \ln w_i - w_i)$. Then

$$d_F(\mathbf{u},\mathbf{w}) = \sum_i (u_i \ln\frac{u_i}{w_i} - u_i + w_i)$$

is the unnormalized relative entropy. (When $\sum_i u_i = \sum_i w_i = 1$ this gives the standard relative entropy.) The gradient is given by $f_i(\mathbf{w}) = \ln w_i$, the inverse obviously being $f_i^{-1}(\theta) = \exp(\theta_i)$.

The following generalization of the Pythagorean Theorem follows directly from the definition of a Bregman divergence:

$$\begin{aligned} d_F(\mathbf{u},\mathbf{w}') = {} & d_F(\mathbf{u},\mathbf{w}) + d_F(\mathbf{w},\mathbf{w}') \\ & + (\mathbf{f}(\mathbf{w}') - \mathbf{f}(\mathbf{w}))\cdot(\mathbf{w}-\mathbf{u}). \end{aligned} \tag{9}$$

Since the dot product $(\mathbf{f}(\mathbf{w}') - \mathbf{f}(\mathbf{w}))\cdot(\mathbf{w}-\mathbf{u})$ can be positive, this shows in particular that d_F does not satisfy the triangle inequality. We recover the standard Pythagorean Theorem when the divergence is the squared Euclidean distance (*i.e.*, $\mathbf{f}$ is identity) and the dot product is zero (*i.e.*, $(\mathbf{w}'-\mathbf{w})$ and $\mathbf{w}-\mathbf{u}$ are orthogonal).

We now use a Bregman divergence d_F as a regularizer for deriving an update rule. This framework for motivating updates was introduced by Kivinen and Warmuth (1997) in the prediction setting. Bregman divergences based on the squared q-norm were first used by Grove *et al.* (2001) to analyze algorithms for learning linear threshold functions.

Suppose an example $(\mathbf{x}_t, y_t)$ has been observed, and we wish to update our hypothesis $\mathbf{w}_{t-1}$ based on this example. We wish to decrease the squared loss $(y_t - \mathbf{w}\cdot\mathbf{x}_t)^2$ (other convex loss functions can also be considered; see Section 5), but we should not make big changes based on just a single example. Thus, we define

$$C_t(\mathbf{w}) = d_F(\mathbf{w},\mathbf{w}_{t-1}) + \frac{1}{2}\eta(y_t - \mathbf{w}\cdot\mathbf{x}_t)^2$$

where $\eta > 0$ is a trade-off parameter, and tentatively set $\mathbf{w}_t = \arg\min_{\mathbf{w}} C_t(\mathbf{w})$. Since C_t is convex, we can minimize by setting $\nabla C_t(\mathbf{w}_t) = 0$. By substituting the definition of d_F, this becomes

$$\mathbf{f}(\mathbf{w}_t) = \mathbf{f}(\mathbf{w}_{t-1}) - \eta(\mathbf{w}_t \cdot \mathbf{x}_t - y_t)\mathbf{x}_t. \qquad (10)$$

Note that $\mathbf{w}_t$ appears on both sides of the equality. Hence, we call the algorithm defined by this equality the *implicit algorithm* for divergence d_F. Notice that (10) can be solved numerically by a line search since $\mathbf{w} = \mathbf{f}^{-1}(\mathbf{f}(\mathbf{w}_{t-1} + \alpha\mathbf{x}))$ for some scalar α, and the inverse $\mathbf{f}^{-1}$ is easy to compute in the cases we consider. Also in the special case of two-norm ($d_F = d_2$), with $\mathbf{f}$ the identity function, we can solve (10) in closed form, which turns out to give the the algorithm called *normalized LMS* by Hassibi *et al.* (1996).

Instead of solving (10) numerically, we often find it sufficient to notice that for reasonable values of η, the values $\mathbf{w}_t \cdot \mathbf{x}_t$ and $\mathbf{w}_{t-1} \cdot \mathbf{x}_t$ should be fairly close to each other. Thus, we may approximate the solution of (10) by

$$\mathbf{w}_t = \mathbf{f}^{-1}(\mathbf{f}(\mathbf{w}_{t-1}) - \eta(\mathbf{w}_{t-1} \cdot \mathbf{x}_t - y_t)\mathbf{x}_t). \quad (11)$$

We call this the *explicit algorithm* for divergence d_F. The special case $d_F = d_2$ gives the usual LMS algorithm.

3. BOUNDS IN TERMS OF DIFFERENT NORMS

Our interest in considering the generalization of LMS to the p-norm based algorithms comes from the fact that for these algorithms the term $||\mathbf{x}||_2||\mathbf{u}||_2$ in the LMS bound is replaced by another product of dual norms $||\mathbf{x}||_p||\mathbf{u}||_q$ (*i.e.*, $1/p + 1/q = 1$). We discuss the implications of this after giving the main result.

Theorem 3. Fix p and q such that $1/p + 1/q = 1$ and $2 \le p < \infty$. Assume that $||\mathbf{x}_t||_p \le X_p$ for all t. Then the explicit algorithm for d_q with learning rate $\eta = 1/((p-1)X_p^2)$ satisfies

$$\sum_{t=1}^{T}(\mathbf{u}\cdot\mathbf{x}_t - \mathbf{w}_{t-1}\cdot\mathbf{x}_t)^2 \le \sum_{t=1}^{T}(\mathbf{u}\cdot\mathbf{x}_t - y_t)^2 + (p-1)X_p^2||\mathbf{u}||_q^2$$

for any $\mathbf{u} \in \mathbf{R}^n$.

To see how the choice of p affects the bound, consider two cases: $p = 2$ (*i.e.*, LMS) and $p = 2\ln n$ (*i.e.*, p rather large). For the latter case, Gentile and Littlestone (1999) give the bound

$$(p-1)||\mathbf{x}||_p^2||\mathbf{u}||_q^2 \le (2e\ln n)||\mathbf{x}||_\infty^2||\mathbf{u}||_1^2$$

(where $||\mathbf{x}||_\infty = \max_i|x_i|$). Thus, we compare $||\mathbf{x}||_2^2||\mathbf{u}||_2^2$ with $(2e\ln n)||\mathbf{x}||_\infty^2||\mathbf{u}||_1^2$. The basic intuition is that the large p bound is better when the target $\mathbf{u}$ is sparse (few nonzero components) and the instances $\mathbf{x}_t$ are dense (many components of roughly equal size). As an extreme case, suppose $\mathbf{u} = (1,0,\ldots,0)$ and $\mathbf{x} = (1,\ldots,1)$. Then $||\mathbf{u}||_2^2||\mathbf{x}||_2^2 = n^2$ and $(2e\ln n)||\mathbf{x}||_\infty^2||\mathbf{u}||_1^2 = 2e\ln n$. So in this case, the large p case has a drastically better bound. On the other hand, when $\mathbf{u} = (1,\ldots,1)$ and the instances are permutations of $(1,0,\ldots,0)$, then $||\mathbf{x}||_2^2||\mathbf{u}||_2^2 = n^2$ and $(2e\ln n)||\mathbf{x}||_\infty^2||\mathbf{u}||_1^2 = 2en^2\ln n$. So now the $p = 2$ case is moderately better. See Kivinen and Warmuth (1997) for a more detailed discussion. They derive bounds of the $(\ln n)||\mathbf{x}||_\infty^2||\mathbf{u}||_1^2$ type for the prediction problem using an algorithm called the *exponentiated gradient* (EG) algorithm. EG uses (ignoring a normalization) the update (11) with $f_i(\mathbf{w}) = \ln w_i$. The analysis of EG can also be lifted to the filtering setting, giving again bounds of the $(\ln n)||\mathbf{x}||_\infty^2||\mathbf{u}||_1^2$ type; we omit the details.

For the a posteriori case we have the following theorem which generalizes the result about normalized LMS by Hassibi *et al.* (1996).

Theorem 4. Fix p and q such that $1/p + 1/q = 1$ and $2 \le p < \infty$. Assume that $||\mathbf{x}_t||_p \le X_p$ for all t. Then the implicit algorithm for d_q with learning rate $\eta = 1/((p-1)X_p^2)$ satisfies

$$\sum_{t=1}^{T}(\mathbf{u}\cdot\mathbf{x}_t - \mathbf{w}_t\cdot\mathbf{x}_t)^2 \le \sum_{t=1}^{T}(\mathbf{u}\cdot\mathbf{x}_t - y_t)^2 + (p-1)X_p^2||\mathbf{u}||_q^2.$$

4. NONSTATIONARY TARGETS

Following Herbster and Warmuth (2001), we now consider a variant of the algorithm that keeps the q-norm of the weight vector bounded by U_q, where $U_q > 0$ is a parameter to the algorithm. We call this two-step update the *bounded explicit update* for d_F:

Explicit update step: Let

$$\mathbf{w}_t' = \mathbf{f}^{-1}(\mathbf{f}(\mathbf{w}_{t-1}) - \eta(\mathbf{w}_{t-1}\cdot\mathbf{x}_t - y_t)\mathbf{x}_t).$$

Out of bound update step: If $||\mathbf{w}_t'||_q > U_q$, then $\mathbf{w}_t = U_q\mathbf{w}_t'/||\mathbf{w}_t'||_q$; otherwise $\mathbf{w}_t = \mathbf{w}_t'$.

Thus if the algorithm tries to increase the q-norm of its weight vector above U_q, then we scale it back.

We now let the target $\mathbf{u}_t$ vary with time (nonstationary model):

$$y_t = \mathbf{u}_t \cdot \mathbf{x}_t + v_t \qquad (12)$$

As previously, our bound will include a penalty for the (maximum) norm of $\mathbf{u}_t$. Additionally, there is now also a penalty for the total distance the target moves during the process.

Theorem 5. Fix p and q such that $1/p + 1/q = 1$ and $2 \le p < \infty$. Assume $||\mathbf{x}_t||_p \le X_p$ and $||\mathbf{u}_t||_q \le U_q$ for all t. Then the bounded explicit algorithm for d_q with learning rate $\eta = 1/((p-1)X_p^2)$ and parameter U_q satisfies

$$\sum_{t=1}^{T} (\mathbf{u}_t \cdot \mathbf{x}_t - \mathbf{w}_{t-1} \cdot \mathbf{x}_t)^2 \le \sum_{t=1}^{T} (\mathbf{u}_t \cdot \mathbf{x}_t - y_t)^2 + (p-1)X_p^2 U_q^2 + 2(p-1)X_p^2 U_q \sum_{t=1}^{T-1} ||\mathbf{u}_{t+1} - \mathbf{u}_t||_q.$$

In the special case $\mathbf{u}_t = \mathbf{u}_{t+1}$ for all t, the result becomes Theorem 3 with the exception that the norm bound U_q must now be defined before the algorithm can be run.

A similar result can be obtained for a bounded version of the implicit algorithm in a posteriori filtering; we omit the details.

5. NONLINEAR MODELS

The kernel method is a well known technique for generalizing linear learning algorithm to nonlinear problems. The idea is that if the relationship between the inputs $\mathbf{x}_t \in \mathbf{R}^n$ and outputs y_t is nonlinear, it is often possible to transform the inputs by a *feature map* Ψ into a higher dimensional space $\mathbf{R}^N$, $N > n$, such that the relationship between the *feature vectors* $\Psi(\mathbf{x}_t)$ and outputs y_t becomes linear. We can then apply linear learning methods on the data $(\Psi(\mathbf{x}_t), y_t)$. In many interesting cases, there is a *kernel* $k\colon \mathbf{R}^n \times \mathbf{R}^n \to \mathbf{R}$ such that $\Psi(\mathbf{x}) \cdot \Psi(\mathbf{x}') = k(\mathbf{x}, \mathbf{x}')$. The kernel can then be used to compute dot products of feature vectors implicitly without evaluating Ψ.

For example, if we run the LMS algorithm on the examples $(\Psi(\mathbf{x}_t), y_t)$ then its weight vector (now a vector in $\mathbf{R}^N$) can be written as

$$\mathbf{w}_T = \sum_{t=1}^{T-1} \alpha_t \Psi(\mathbf{x}_t), \tag{13}$$

where $\alpha_t = \eta(y_t - \mathbf{w}_{t-1} \cdot \Psi(\mathbf{x}_t))$. Thus $\mathbf{w}_T \cdot \Psi(\mathbf{x}) = \sum_{t=1}^{T-1} \alpha_t \Psi(\mathbf{x}_t) \cdot \Psi(\mathbf{x}) = \sum_{t=1}^{T-1} \alpha_t k(\mathbf{x}_t, \mathbf{x})$. Hence, instead of explicitly maintaining a high dimensional weight vector $\mathbf{w}_T \in \mathbf{R}^N$, it is sufficient to keep track of the coefficients α_t and compute the dot products using the kernel.

The key property for the kernel method is that the prediction of the algorithm for instance $\mathbf{x}_T$ can be written (using a feature map Ψ) in terms of the dot products $\Psi(\mathbf{x}_T) \cdot \Psi(\mathbf{x}_t)$ for the other instances $\mathbf{x}_t$. This in particular implies that the algorithm must be rotation invariant, *i.e.*, its predictions must remain unchanged if we rotate the feature vectors by replacing $\Psi(\mathbf{x}_t)$ by $A\Psi(\mathbf{x}_t)$ where A is a fixed orthonormal matrix. For example, the LMS algorithm and Support Vector Machines are rotation invariant. Not surprisingly, these algorithms are motivated using 2-norms, the only rotation invariant p-norm.

On the other hand, applying the feature map with the more general updates (10) and (11) (assuming $\mathbf{f}(\mathbf{w}_0) = \mathbf{0}$ as usual) gives us $\mathbf{w}_T = \mathbf{f}^{-1}(\sum_{t=1}^{T-1} \alpha_t \Psi(\mathbf{x}_t))$ where $\mathbf{f}$ is typically nonlinear. Thus, it is at least not immediately clear how kernels could be applied. Furthermore, in classification it is known that rotation invariant algorithms cannot really take advantage of sparse weight vectors (Kivinen *et al.*, 1997). There are also computational hardness results suggesting there is a trade-off between having easy evaluation via kernels and having fast convergence for sparse targets via a nonlinear $\mathbf{f}$ (Khardon *et al.*, 2002). (See also (Kivinen and Warmuth, 1997) for a related discussion). As obtaining improved bounds for sparse targets is the main motivation for considering the case $p > 2$, looking for a general kernel version of the p-norm algorithms does not seem promising. However, Takimoto and Warmuth (2002) have shown that some specific kernels can be efficiently combined with an EG type algorithm. (EG corresponds roughly to taking $f_i(\mathbf{w}) = \ln w_i$ and gives bounds similar to the p-norm algorithm with $p = O(\ln n)$.)

Another slightly extended framework is that based on *generalized linear regression.* Here we replace the model (1) by

$$y_t = \phi(\mathbf{u} \cdot \mathbf{x}_t + v_t) \tag{14}$$

where ϕ is a continuous, strictly increasing *transfer function.* The logistic sigmoid $\phi(r) = 1/(1+\exp(-r))$ is a typical example of such a transfer function. In the prediction setting (where the learner tries to match y_t), the prediction becomes $\hat{y}_t = \phi(\mathbf{w}_{t-1} \cdot \mathbf{x}_t)$. In the filtering setting, we would naturally also include the transfer function in the prediction, giving $\hat{y}_t = \phi(\mathbf{w}_{t-1} \cdot \mathbf{x}_t)$ for the a priori and $\hat{y}_t = \phi(\mathbf{w}_t \cdot \mathbf{x}_t)$ for the a posteriori case. The algorithm then tries to match $\hat{y}_t$ to $\phi(\mathbf{u} \cdot \mathbf{x}_t)$. One could in principle still use the squared error $(\phi(\mathbf{u} \cdot \mathbf{x}_t) - \phi(\mathbf{w} \cdot \mathbf{x}_t))^2$ as the performance measure, but this is nonconvex in $\mathbf{u}$ and $\mathbf{w}$ and actually leads to a very badly behaved optimization problem (Auer *et al.*, 1995). We obtain a better behaved problem by using the *matching loss* for ϕ (Auer *et al.*, 1995), defined for y and y' in the range of ϕ as

$$L(y, y') = \int_{\phi^{-1}(y)}^{\phi^{-1}(y')} (\phi(r) - y)dr.$$

(Notice that by our assumptions ϕ is one-to-one.) This definition of loss may seem arbitrary, but it is actually a one-dimensional Bregman divergence: if we let $\Phi(r) = \int \phi(r)dr$, then

$$L(\phi(a),\phi(a')) = d_\Phi(a',a). \qquad (15)$$

It is easy to see that for the identity transfer function $\phi(r) = r$ we get $L(y,y') = (y-y')^2/2$, and for the logistic sigmoid $\phi(r) = 1/(1+\exp(-r))$ we get the logarithmic loss

$$L(y,y') = y\ln\frac{y}{y'} + (1-y)\ln\frac{1-y}{1-y'}.$$

Directly from the definition of matching loss, we obtain a simple expression for its gradient:

$$\nabla_{\mathbf{w}} L(y,\phi(\mathbf{w}\cdot\mathbf{x})) = (\phi(\mathbf{w}\cdot\mathbf{x}) - y)\mathbf{x}.$$

Therefore, the explicit update (11) naturally generalizes to

$$\mathbf{w}_t = \mathbf{f}^{-1}(\mathbf{f}(\mathbf{w}_{t-1}) - \eta(\hat{y}_t - y_t)\mathbf{x}_t)$$

where $\hat{y}_t = \phi(\mathbf{w}_{t-1}\cdot\mathbf{x}_t)$. The implicit algorithm can be generalized similarly; for it we use $\hat{y}_t = \phi(\mathbf{w}_t \cdot \mathbf{x}_t)$. For these algorithms we can now prove bounds which have as an additional factor an upper bound on the slope of the transfer function. The techniques are essentially those introduced by Helmbold *et al.* (1999).

Theorem 6. Fix p and q such that $1/p + 1/q = 1$ and $2 \le p < \infty$. Let ϕ be strictly increasing and continuously differentiable with c such that $0 < \phi(r) \le c$ holds for all r, and let L be the matching loss for ϕ. Assume that $||\mathbf{x}_t||_p \le X_p$ for all t. Then both the explicit algorithm and implicit algorithm for d_q with learning rate $\eta = 1/((p-1)cX_p^2)$ satisfy

$$\sum_{t=1}^{T} L(\hat{y}_t,\phi(\mathbf{u}\cdot\mathbf{x}_t))$$
$$\le \sum_{t=1}^{T} L(y_t,\phi(\mathbf{u}\cdot\mathbf{x}_t)) + (p-1)cX_p^2||\mathbf{u}||_q^2$$

for any $\mathbf{u} \in \mathbf{R}^n$.

Because of how we defined $\hat{y}_t$, the theorem gives an a priori filtering bound for the explicit algorithm and a posteriori bound for the implicit algorithm.

When ϕ is the identity function, we get the results of Section 3 with $c = 1$. For the logistic sigmoid, $c = 1/4$. Thresholded transfer functions, such as $\phi(r) = \text{sign}(r)$, correspond to the limiting case $c \to \infty$, which makes the bound vacuous.

This result generalizes to the nonstationary case (Section 4) in the obvious manner; we omit the details.

6. REFERENCES

Auer, P., M. Herbster and M. K. Warmuth (1995). Exponentially many local minima for single neurons. In: *Proc. 1995 Neural Information Processing Conference*. MIT Press, Cambridge, MA. pp. 316–317.

Azoury, Katy S. and M. K. Warmuth (2001). Relative loss bounds for on-line density estimation with the exponential family of distributions. *Machine Learning* **43**(3), 211–246.

Bregman, L.M. (1967). The relaxation method of finding the common point of convex sets and its application to the solution of problems in convex programming. *USSR Computational Mathematics and Physics* **7**, 200–217.

Cesa-Bianchi, N., P. Long and M.K. Warmuth (1996). Worst-case quadratic loss bounds for on-line prediction of linear functions by gradient descent. *IEEE Transactions on Neural Networks* **7**(2), 604–619.

Gentile, Claudio and Nick Littlestone (1999). The robustness of the p-norm algorithms. In: *Proc. 12th Annu. Conf. on Comput. Learning Theory*. ACM Press, New York, NY. pp. 1–11.

Grove, Adam J., Nick Littlestone and Dale Schuurmans (2001). General convergence results for linear discriminant updates. *Machine Learning* **43**(3), 173–210.

Hassibi, Babak, Ali H. Sayed and Thomas Kailath (1996). H^∞ optimality of the LMS algorithm. *IEEE Transactions on Signal Processing* **44**(2), 267–280.

Helmbold, David P., Jyrki Kivinen and Manfred K. Warmuth (1999). Relative loss bounds for single neurons. *IEEE Transactions on Neural Networks* **10**(6), 1291–1304.

Herbster, Mark and Manfred K. Warmuth (2001). Tracking the best linear predictor. *Journal of Machine Learning Research* **1**, 281–309.

Khardon, R., D. Roth and R. A. Servedio (2002). Efficiency versus convergence of boolean kernels for on-line learning algorithms. In: *Advances in Neural Information Processing Systems 14* (T. G. Dietterich, S. Becker and Z. Ghahramani, Eds.). MIT Press. Cambridge, MA. pp. 423–430.

Kivinen, J. and M. K. Warmuth (1997). Additive versus exponentiated gradient updates for linear prediction. *Information and Computation* **132**(1), 1–64.

Kivinen, J. and M. K. Warmuth (2001). Relative loss bounds for multidimensional regression problems. *Machine Learning* **45**(3), 301–329.

Kivinen, J., M. K. Warmuth and P. Auer (1997). The Perceptron algorithm vs. Winnow: Linear vs. logarithmic mistake bounds when few input variables are relevant. *Artificial Intelligence* **97**, 325–343.

Takimoto, Eiji and Manfred K. Warmuth (2002). Path kernels and multiplicative updates. In: *Proceedings of the 15th Annual Conference on Computational Learning Theory* (Jyrki Kivinen and Bob Sloan, Eds.). Springer LNAI 2375. Berlin. pp. 74–89.

IFAC Publications
www.elsevier.com/locate/ifac

IDENTIFICATION OF LINEAR SYSTEMS WITH NONLINEAR DISTORTIONS

J. Schoukens (*), R. Pintelon (*), T. Dobrowiecki (**), Y. Rolain (*)

(*): Vrije Universiteit Brussel, dep. ELEC, Pleinlaan 2, B1050 Brussels, Belgium

(**): Budapest University of Technology & Economics, Dep. MIS, H-1521 Budapest

email: Johan.Schoukens@vub.ac.be

Abstract: In this paper the impact of nonlinear distortions on the linear system identification framework is studied. In the first part the nonlinear system is replaced by a linear model plus a nonlinear noise source. The properties of this representation are studied. Next a method to detect, qualify and quantify the nonlinear distortions is presented. In the second part, the (non)-parametric identification of the best linear approximation is studied. In the last part, the linear modelling approach is extended towards nonlinear modelling. A fast approximate nonlinear modelling framework is set up that is a natural extension of the linear framework, and bridges the gap between the linear and the nonlinear identification approaches. *Copyright © 2003 IFAC*

1. INTRODUCTION[1]

Identification of linear systems became a mature scientific discipline over the last decades (Ljung, 1999; Söderström and Stoica, 1989; Pintelon and Schoukens, 2001). It is a very successful method and is applied on a large variety of problems coming from a wide range of different fields. The basic reason for this success is the appealing simplicity of linear models. They give a lot of insight and are often used as the basis for many design techniques. The price for this 'simplicity' is the use of a strong assumption: the system is assumed to behave linearly. However, in practice many systems are not linear. If the linear modelling approach is maintained, the question arises if the whole framework is still valid, and if its results are still reliable. This leads to a first set of questions that are addressed in this paper:

- What is the validity of a linear model that is identified in the presence of nonlinear distortions?
- Can we detect, qualify and quantify the presence of nonlinear distortions?
- What is the best 'engineering practice' to obtain a linear model under these conditions?
- Can the convergence results of the linear identification framework be maintained under these conditions?

If the distortions are too large compared to the required accuracy of the model, the linear framework should be abandoned and a nonlinear model should be identified. Since all systems that are not linear are nonlinear systems, it is clear that there exist no general valid descriptions for the full class of nonlinear systems. For that reason, a very wide variety of models is proposed in the literature. Either a dedicated white box model is built for each specific situation, or a black box approach is made, extending the ideas that founded the success of linear modelling towards nonlinear modelling. While the first choice is very time consuming to construct the model, the second one requires very long experiments in order to collect a sufficient number of data to identify the black box model. So there exist a big gap between the effort needed to model linear and nonlinear systems. In practice there is often a need for a solution in between these two extremes: the quality of the linear model is too poor while the effort to built a nonlinear model is too large. In this paper we want to fill this gap by proposing a natural extension of the linear model class that allows to carry over many of the methods of the linear modelling approach to the nonlinear world, maintaining their simplicity, user friendliness, and 'short' experiment time to identify them. Although no perfect description of the nonlinearities will be obtained as would be possible with the advanced nonlinear modelling techniques, it will be possible in many cases to push down the remaining model errors with a factor 2 to 10. This gives the user the whole range of solutions: from very simple linear modelling techniques, over the approximate nonlinear methods proposed in this paper, to the complex but accurate full blown nonlinear models.

The paper consists of 4 parts. First the system setup

1. This work is supported by the Flemish government (FWO-ICCoS onderzoeksgemeenschap; Concerted Action ILiNoS; the Bilateral agreement program) and the Belgium government (IUAP-5/22).

will be defined, followed by a precise definition of the class of nonlinear systems that is considered. Next the impact of nonlinear distortions on the linear identification framework is analysed and eventually an identification method for approximative nonlinear modelling is proposed.

2. SETUP

Consider the time invariant, single-input, single-output (SISO), continuous or discrete time nonlinear dynamic system

$$y_0(t) = g_{NL}(u_0(t)), \quad (1)$$

and the discrete observations:

$$u_0(t) \text{ and } y(t) = y_0(t) + n_y(t),\ t = 0, 1, \ldots, N. \quad (2)$$

where $n_y(t)$ is zero mean noise. The sampling period is normalized to 1.

Assumption 1: *Noise model*
The input $u_0(t)$ is assumed to be known exactly. $n_y(t)$ is filtered white noise $n_y(t) = H_0(q)e_0(t)$ where $e_0(t)$ is a sequence of independent random variables, with zero mean values, variances λ_0, and bounded moments of order $4+\delta$, for some $\delta > 0$. $H_0(q)$ is an inversely stable, monic filter (Ljung, 1999).

3. A FORMAL FRAMEWORK TO DESCRIBE THE NONLINEAR SYSTEM

Describing nonlinear systems is a tedious job. Since there does not exist a single model structure that covers all possible nonlinear systems it is necessary to specify what subclass $\mathbb{S}$ of nonlinear systems will be considered. Because $\mathbb{S}$ depends also on the excitation signals that will be allowed, it is necessary to specify first the set of excitation signals $\mathbb{E}$ that will be used.

3.1 Class of excitation signals $\mathbb{E}$

In this paper normally distributed random excitations with a user defined power spectrum $S_u(f)$ will be used. Within this class we focus mostly on periodic signals, but the reader should be aware that all asymptotic results that are discussed in this paper are valid for the whole class of Gaussian excitations (Pintelon and Schoukens, 2002a). Hence the results can be applied without any restriction to time and frequency domain system identification methods. Periodic signals will allow to extract the information about the nonlinear distortions in the very same experiment that is used to measure/model the linear approximation to the nonlinear system without increasing the experiment time (Pintelon and Schoukens, 2001a, Schoukens *et al.*, 2002a). Two kinds of periodic signals are considered: those with a random amplitude and random phase spectrum (periodic noise), and those with a deterministic amplitude and a random phase spectrum (called random multisines). The overall structure of $\mathbb{E}$ is given in Fig. 1. As mentioned before, the models that

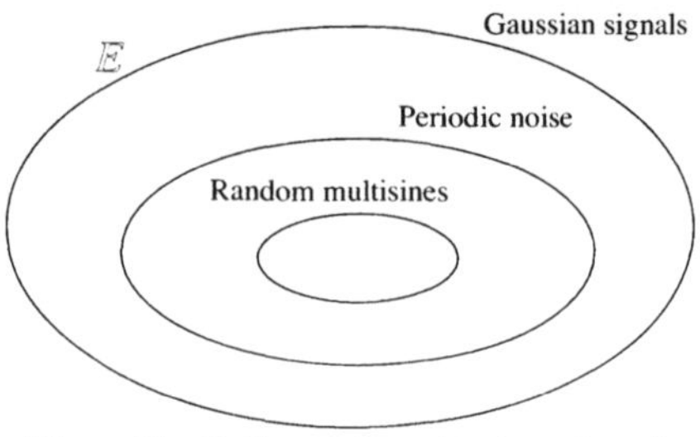

Fig. 1: Class $\mathbb{E}$ of allowed excitation signals.

are extracted using random multisines are valid for the whole class $\mathbb{E}$.

Below we give, for the interested reader, the precise definitions of the considered excitation signals followed by some remarks. All excitations are specified as discrete time signals. Although we do not elaborate on the behaviour of the excitation between the sample points, the reader should be aware that it affects the model that will be obtained from the sampled data.

Definition 2: Spectrum generating function: $S_u(f)$ is a uniformly bounded real positive function with a countable number of discontinuities.

Remark: $S_u(f)$ will be used as the power spectrum for noise excitations. For periodic excitations it will be used to set the amplitude of the discrete spectral components.

Definition 3: Gaussian noise excitation: A random sequence $u(t)$, $t = 0, 1, \ldots$ drawn from a zero mean normally distributed process with a user defined power spectrum $S_u(f)$.

Definition 4: Periodic noise: A signal $u(t)$ is a periodic noise excitation if

$$u(t) = N^{-1/2}\sum_{k=-N/2}^{N/2}\hat{U}\left(\frac{k}{N}\right)e^{j\left(2\pi k\frac{t}{N}+\varphi_k\right)} \quad (3)$$

with $\varphi_{-k} = -\varphi_k$, $\hat{U}(f) \geq 0$, and $\hat{U}(f=0) = 0$. $\hat{U}(k/N)$ and φ_k are the realisations of independent (jointly, and over k) random processes satisfying the following condition: $\hat{U}(f)$ has bounded moments of any order ($<\infty$), and $\mathcal{E}\{e^{j\varphi_k}\} = 0$, and $\mathcal{E}\{\hat{U}(f)^2\} = S_u(f)$.

Definition 5: Random (phase) multisine: A signal $u(t)$ is a random phase multisine (also called random multisine) if

$$u(t) = N^{-1/2}\sum_{k=-N/2}^{N/2}\hat{U}(\frac{k}{N})e^{j\left(2\pi k\frac{t}{N}+\varphi_k\right)} \quad (4)$$

with $\varphi_{-k} = -\varphi_k$, $\hat{U}(f) = \sqrt{S_u(f)} \geq 0$ and $\hat{U}(f=0) = 0$. The phases φ_k are the realisations of an independent (over k) uniformly distributed random process on $[0, 2\pi)$.

The set of excitation signals that is considered in this paper is the union of the signals defined before.

Definition 6: Set of excitation signals $\mathbb{E}$: A signal u with power spectrum $S_u(f)$ belongs to $\mathbb{E}$ if u is either a Gaussian noise excitation (Definition 3), a periodic noise (Definition 4), or a random phase multisine (Definition 5).

Remarks:

- The periodic signals (Definitions 4 and 5) are asymptotically ($N \to \infty$) normally distributed in the time domain.
- The phase condition in Definitions 4 and 5 can be relaxed. The phases can be restricted to a discrete set as long as $\mathcal{E}\{e^{j\varphi}\}$ is zero. This allows for example to include orthogonal frequency domain modulation (OFDM) where such random multisines are intensively used (Vandersteen *et al.*, 2000).
- The period length N will be sometimes indicated explicitly by using the subscript N, for example $\mathbb{E}_N$.

3.2 Class $\mathbb{S}$ of nonlinear systems

Very basic questions to be answered are: What systems can be approximated? What is the quality of the approximation? Below we give a simplified discussion, where it is shown that the properties of the approximation improve when the set of systems is more and more restricted.

The paper considers SISO nonlinear time invariant systems whose output can be approximated arbitrarily well in least squares sense by a Volterra series for the class of excitations $\mathbb{E}$. For these systems (called Wiener systems) (i) the influence of the initial conditions vanishes asymptotically, and (ii) the steady state response to a periodic input is a periodic signal with the same period as the input. Phenomena such as bifurcation, chaos, and sub harmonics are excluded, while strongly nonlinear phenomena such as saturation (e.g. amplifiers) and discontinuities (e.g. relays) are allowed. Only a point wise approximation of the output is obtained (see the Wiener theory in Schetzen, 1980; Doyle *et al.*, 2001). Hence, these models can be used to model for example the output spectrum, but they shouldn't be used to calculate the derivative of model characteristics. They also can not predict the output of the system at the discontinuities.

A stronger convergence result can be obtained if a subset of $\mathbb{S}$, restricted to continuous nonlinear systems is considered. Discontinuous nonlinear systems are excluded but hard saturating nonlinear systems can still be modelled. For these systems, called fading memory systems, uniform convergence of the model (output) to the system (output) is shown, while the model derivatives are still not guarantheed to converge. The approximation is valid for bounded inputs, where the bounds can be set by the user. The properties of fading memory systems (or approximately finite memory systems) are discussed in Boyd and Chua (1985), Borys (1985) and the work of Sandberg (1992, 1993, 2002).

Restricting the class of nonlinear systems even more, to those systems having a converging Volterra series around a given working point, similar to a convergent Taylor series for a static nonlinear system, gives the strongest convergence results. We call these systems Volterra systems. For these systems uniform convergence of the model (output) and the derivatives is guaranteed. Often these Volterra series exist only in a restricted input domain that can not be freely chosen by the user. Consider for example the Taylor series of $\arctan x$ that exists only for $|x| < 1$. A graphical overview of all the included systems with their convergence properties is given in Table 1.

Table 1 Convergence properties of the different nonlinear model classes

Model class	Properties
Wiener system	- output convergences in mean square sense, point wise convergence - discontinuities and saturation allowed - model valid for the set of Gaussian signals
Fading memory system	- output converges uniformly - saturation allowed - model valid for bounded inputs (bound set by the user)
Volterra system	- output converges uniformly - derivatives converge uniformly - saturation allowed - model valid for bounded inputs (bound can not be set by the user)

This leads eventually to the following formal definition:

Definition 7: Class $\mathbb{S}$ of nonlinear systems: $\mathbb{S}$ is the set of nonlinear systems for which there exists a Volterra series representation $y_Q(t) = \sum_{n=1}^{Q} y^n(t)$ that converges in least squares sense to $y(t)$ for all excitations $u \in \mathbb{E}$ with probability 1:

$$\lim_{Q\to\infty}\frac{1}{N}\sum_{t=1}^{N}\mathcal{E}\{|y(t)-y_Q(t)|^2\}=0, \quad (5)$$

with N the experiment length. The expected value $\mathcal{E}\{\ \}$ is the ensemble average over the considered class of random inputs u.

It should be emphasized that we are not interested in the identification of the Volterra kernels. The Volterra model is only used to allow for formal proofs of the claimed results for a given class of inputs $\mathbb{E}$ and systems $\mathbb{S}$. During the discussions the actual system is replaced at any moment by its Volterra approximation.

4. IMPACT OF NONLINEAR DISTORTIONS ON THE LINEAR FRAMEWORK

4.1 Why to use a linear model for a nonlinear system?

Linear models are very popular, even if it is well known that in practice many systems are not perfectly linear, because they offer important advantages: i) They result in useful models that give the user a lot of intuitive insight in the system behaviour; ii) Many design techniques are valid for linear models only; iii) Nonlinear model building is often difficult and time consuming and often the user can not afford or is not prepared to make this huge effort; iiii) No general framework is available for nonlinear systems. Dedicated models are needed, complicating the development/use of general software packages.
For all these reasons, there exists a strong need to use linear models even if it is known that they are erroneous. To allow extension of the linear framework to systems with a dominant linear behaviour, the nonlinear contributions are first considered to be a parasitic effect. In Section 5, the nonlinear distortions will be included in the model.

4.2 Non parametric linear framework

Although the discussion covers time domain and frequency domain identification methods, we will mostly use a frequency domain formulation using the discrete Fourier transform (DFT):

$$X(k) = N^{-1/2}\sum_{t=0}^{N-1} x(t)e^{-j2\pi kt/N}, \qquad (6)$$

where N is the considered record length. For simplicity we do not consider initial conditions and leakage effects because they have no fundamental impact on the interpretations that will be made. Leakage effects in the frequency domain are equivalent to initial- and end-condition effects in the time domain (Pintelon and Schoukens, 1997; Schoukens *et al.*, 1999).

A. Major result: intuitive presentation

A nonlinear system belonging to $\mathbb{S}$, excited with a random excitation $u \in \mathbb{E}$ can be represented by a linear system G_R plus a nonlinear noise source Y_S (see Fig. 2). This is an exact representation of the nonlinear system. All approximation errors are put into the second term Y_S. For N growing to infinity,

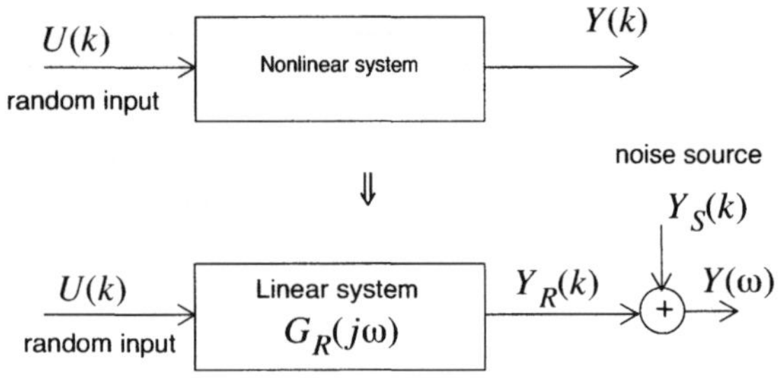

Fig. 2. Representation of a nonlinear system by a linear system for a random input.

Y_s behaves like noise and changes from one realization of the input to the other. G_R is the best linear approximation. It depends on the power spectrum S_u and the time domain distribution of the excitation (Schoukens et al., 1998; Pintelon and Schoukens, 2002; Evans, 1994b). The dependency on N disappears as $O(1/N)$.

Y_R, called the deterministic contributions, consist of all those output contributions that are 'coherent' with the input and can be written as:

$$Y_R(k) = G_R(k, S_u)U(k). \qquad (7)$$

It is important to note that: i) The complex proportionality factor $G_R(k, S_u)$ does not depend on the phases of U, only on its power spectrum; ii) Only the phase $\varphi_k = \text{phase}(U(k))$ comes into the output $Y_R(k)$, all the other input phases ($\varphi_l, l \neq k$) are eliminated in these contributions.
$G_R(k, S_u)$ will be called the best linear approximation and denoted from now on as $G_R(k)$. The reader should be aware that it depends on the power spectrum and the amplitude distribution (time domain) of the random excitation.

The stochastic contributions $Y_S(k)$ contain all the output contributions that depend on input phases $\varphi_l, l \neq k$ or on $m\varphi_k$ with $m \neq 1$. Hence it is impossible to write this output as $Y_S(k) = G_R(k, S_u)U(k)$, and hence it is also impossible to define an equivalent linear system for these terms. They should be considered as noise terms and will be characterized below in the formal description of the result.

B. Major result: formal presentation

In all the results presented below the expected value $\mathcal{E}\{\ \}$ is taken with respect to the random input phases (averaging over different realizations of the input).

Theorem 1: Linear representation of a nonlinear system:
The output of the nonlinear system belonging to $\mathbb{S}$ excited with $u \in \mathbb{E}$ can be written as:

$$Y(k) = G_R(s_k)U(k) + Y_S(k), \qquad (8)$$

with G_R the linear approximation and Y_S the nonlinear noise source.

Theorem 2: The best linear approximation G_R:
$G_R(s_k)$ is the best linear approximation in least square sense:

$$G_R(s_k) = \arg\min_G \mathcal{E}\{|Y(k) - GU(k)|^2\}, \text{ or}$$

$$G_R(s_k) = \frac{\mathcal{E}\{Y(k)\bar{U}(k)\}}{\mathcal{E}\{U(k)\bar{U}(k)\}} = S_{yu}(k)/S_u(k). \quad (9)$$

Note that this is the classical result for FRF measurements of linear systems (Bendat and Piersol, 1980). It is also connected to the early results of Bussgang (1952) and the work reported in Brillinger (1981).

Theorem 3: The nonlinear noise source:
The nonlinear noise source $Y_{S,N}$ has the following properties for $\mathbb{S}$ and $u \in \mathbb{E}$:

1) Zero mean: $\mathcal{E}\{Y_{S,N}(k)\} = 0$

2) Uncorrelated with the input $\mathcal{E}\{Y_{S,N}(k)\bar{U}(k)\} = 0$

3) $Y_{S,N}(k)$ is asymptotically independent from $U(l)$, $\forall k, l$.

4) $Y_{S,N}(k)$ is asymptotically circular complex normally distributed and mixing of arbitrarily order.

5) The even moments do not disappear:

$$\mathcal{E}\{N|Y_{S,N}(k)|^2\} = \sigma^2_{Y_{S,N}}(k) = O(N^0);$$

$$\mathcal{E}\{N^2(|Y_{S,N}(k)|^2 - \sigma^2_{Y_{S,N}}(k))(|Y_{S,N}(l)|^2 - \sigma^2_{Y_{S,N}}(l))\} = O(N^0)$$

6) The odd moments converge to zero ($k \neq l$)

$$\mathcal{E}\{NY_{S,N}(l)\overline{Y_{S,N}}(k)\} = O(N^{-1});$$

$$\mathcal{E}\{N^{3/2}Y_{S,N}(l)|Y_{S,N}(k)|^2\} = O(N^{-1})$$

$$\mathcal{E}\{N^2(|Y_{S,N}(k)|^2 - \sigma^2_{Y_{S,N}}(k))(|Y_{S,N}(l)|^2 - \sigma^2_{Y_{S,N}}(l))\} = O(N^{-1})$$

(Pintelon and Schoukens, 2001a; Pintelon and Schoukens, 2002a; Schoukens *et al.*, 1998).

Remarks:
- These observations are in agreement with the classical result, showing that the output of a nonlinear system can be split in two parts (Bendat, 1990 and 1998; Forsell and Ljung, 1999): a first part that is linearly related with the input (in our case leading to G_R), and a second part that is uncorrelated with the input (leading to Y_S). Theorem 3 states more about the properties of the uncorrelated part.
- The independency claims in the frequency domain ($Y_{S,N}(k)$, $U_N(l)$) are not in conflict with the obvious dependency of $y_{S,N}(t)$ on $u_N(t)$ ($y_{S,N}(t)$ is a periodic signal with the same period as $u_N(t)$). These asymptotic results are valid on a frequency by frequency basis. However, if arbitrary large numbers of such components are combined (as for the inverse Fourier transform to calculate the time domain signals), the independency is not necessarily maintained.
- All these properties can be transferred to $G_{S,N}$ by replacing $\sqrt{N}Y_{S,N}$ by $G_{S,N}$.

C. Experimental illustration

In Figure 3 these results are illustrated by experimental results on a nonlinear circuit (Pintelon and Schoukens, 2001a). The FRF is measured for different levels of the excitation signal (a random multisine). As the amplitude grows, it is seen that the FRF shifts due to the growing systematic contributions, and at the same time it looks more noisy due to the growing stochastic contributions. Since the small amplitude measurements are smooth, and the disturbing noise conditions do not change with the excitation level, it is clear that the dominant 'noise' effects are due to the nonlinearity.

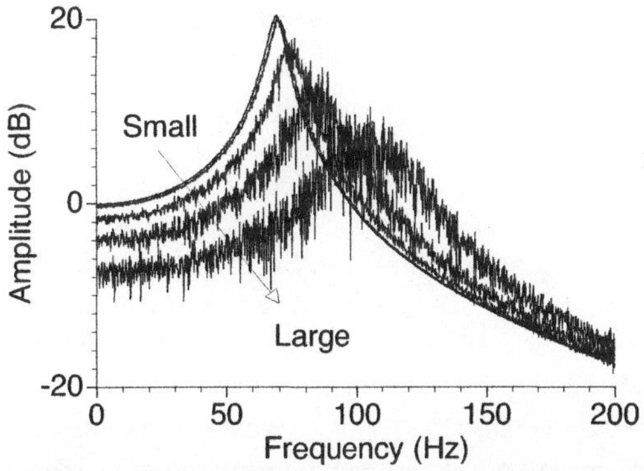

Fig. 3. Evolution of the FRF for growing excitation levels: 34, 54, 127, 253, and 507 mVRMS.

4.3 Best engineering practice for FRF measurements: reduction of the nonlinear noise source level.

In this section we show that a good choice of the excitation signal can significantly reduce the disturbances coming from the nonlinear noise source. In the beginning of the paper, we mentioned that all signals in the set $\mathbb{E}$ result in the same best linear approximation G_R (Pintelon and Schoukens, 2002a). This is formulated below more formally for the interested reader. Next we select within the class $\mathbb{E}$ the best signals for FRF measurements.

A. Formal result

Theorem 4: Equivalencies of the excitation signals: Consider random noise, periodic noise and the random multisine (Definitions 3 to 5) with power spectrum S_u. For these three classes of excitation signals and for a nonlinear system belonging to the class $\mathbb{S}$ (see Definition 7) we have that:

- $G_{R,N}(j\omega)$, converges ($N \to \infty$) at the rate $O(N^{-1})$ to the same limit value $G_R(j\omega)$.

- $G_R(j\omega)$ depends only on the odd nonlinear contributions, the best linear approximation of an even nonlinear distortion is zero.
- The variances $\text{var}(\sqrt{N}Y_{S,N}(k))$ of the stochastic nonlinear distortions converge ($N \to \infty$) at the rate $O(N^{-1})$ to the same limit value $\sigma_S^2(f)$.
- The moments specified in Theorem 3 converge ($N \to \infty$) at the rate $O(N^{-1})$ to the same limit value.
- $G_R(j\omega)$ is a continuous function of ω with a continuous higher order derivatives if the approximating Volterra system and its derivatives are continuous (Pintelon and Schoukens, 2002a).

B. Impact for the user

Once it is known that all these signals are equivalent, the question rises if one class of excitations out of all possible choices has a better behaviour than the others? Can a good choice reduce the effect of the nonlinear noise source on the measurement of G_R? The answer to this question is yes. By eliminating all even nonlinear contributions on the measurements using odd excitation signals, G_R will not change, while the nonlinear noise source Y_S will be reduced because no even nonlinear distortions are present at the odd lines in this case. So the uncertainty on G_R drops. These possibilities are extensively studied (Dobrowiecki and Schoukens, 2001; Pintelon and Schoukens, 2001a) where a number of possibilities are proposed to create such odd excitation signals (see also Godfrey, 1993). Replacing Gaussian noise with such excitations can reduce $\sigma_{G_{R,N}}$ up to a factor 10 or more.

C. Experimental illustration

The equivalency between the excitation signals and the possibility to reduce the uncertainty on $G_{R,N}$ due to the stochastic nonlinear distortions is illustrated in Figure 4 on a hair dryer device where the measured FRF and its uncertainty are shown for different kinds of excitation signals (Németh and Vargha, 1999). The

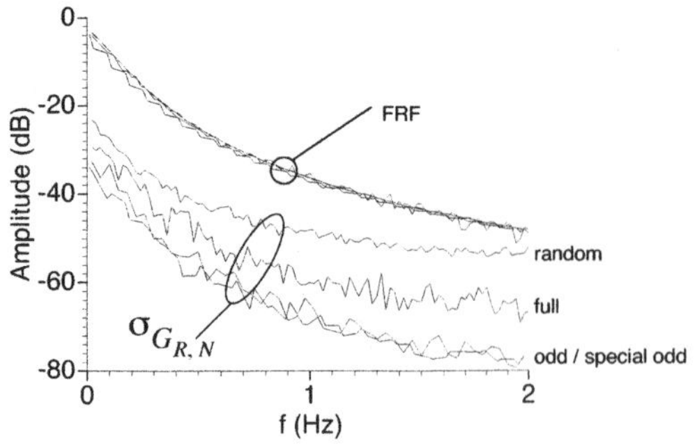

Fig. 4. Hair dryer experiment. Impact of the excitation signal on the uncertainty ($\sigma_{G_{R,N}}$) due to stochastic nonlinear distortions

expected value of the FRF's does not depend on the specific class of excitations (upper traces), while the uncertainties are significantly different.

4.4. Detection, qualification and quantification of nonlinear distortions

The level of the nonlinear distortions provides valuable information for the user, even in a linear modelling framework, because it will give natural bounds on the validity of the linear models. However, to be useful, not too much time should be lost to get this information, most of the time should be spent on the ultimate goal which is to obtain a (linear) model.

Two distortions are faced at the same time: the measurement/process noise $n_y(t)$ and the nonlinear noise source $y_S(t)$. It would be extremely useful to be able to separate the true signals, the disturbing noise, and the nonlinear distortions.

A large number of methods are developed to detect the presence of nonlinear distortions. An extensive overview is given in (Vanhoenacker *et al.*, 2002). Many of these are very time consuming and require dedicated experiments. Only few give detailed information about the distortion levels. Here we present two simple methods that allow to measure explicitly the nonlinear and the disturbing noise levels using periodic excitation signals (periodic random or random multisines) while most of the experiment time is still used to measure the FRF. The first method allows to detect the level of the nonlinear distortions, the second not only detects but also qualifies the nonlinearity (even or odd).

A. Detection of the level of the stochastic nonlinearities (Schoukens et al., 2002a).

The basic idea is to apply R realizations of a periodic signal and to measure the response to each input over $M \geq 2$ periods once the transients disappeared as shown in Fig. 5. Two variances, $\sigma_{G,p}^2(k)$ and

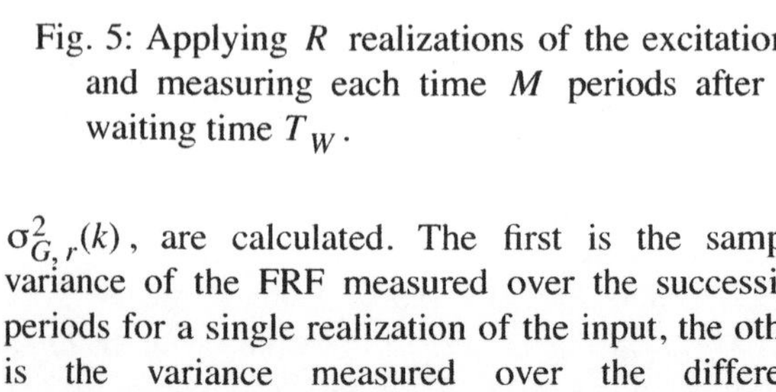

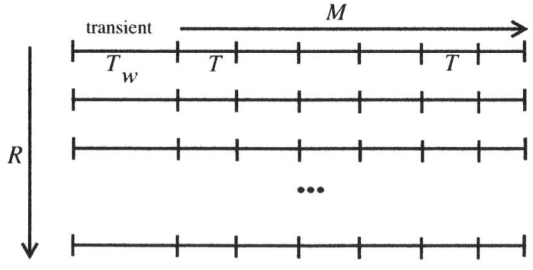

Fig. 5: Applying R realizations of the excitation, and measuring each time M periods after a waiting time T_W.

$\sigma_{G,r}^2(k)$, are calculated. The first is the sample variance of the FRF measured over the successive periods for a single realization of the input, the other is the variance measured over the different realizations. Since we consider systems for which a periodic input results in a periodic output with the same period, it is clear that $\sigma_{G,p}^2(k)$ depends only on the variations from one period to the other which are due to the disturbing noise n_y. $\sigma_{G,r}^2(k)$ is calculated

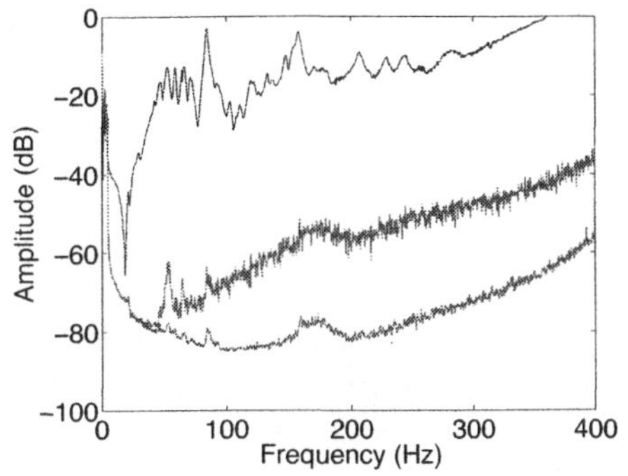

Fig. 6: Detection of nonlinear distortions on a body in white. Blue: the FRF, red: $\sigma^2_{G,\,p}$, green $\sigma^2_{G,\,r}$.

over the different input realizations and so it depends on both noise sources. Comparing $\sigma^2_{Y,\,p}(k)$ and $\sigma^2_{Y,\,r}(k)$ gives immediately an idea about the disturbing noise and the nonlinear noise levels (see Schoukens *et al.*, 2002a for the details).

Example: In Fig. 6, the method is applied on a car body in white (only the metal frame, without seats, etc.) with the following settings: $R = 8, M = 15$. The structure is excited with random multisines up to 400 Hz. The impact of the different noise sources is clearly visible. The reader should be aware that all nonlinearities are detected, also those that might be due to the measurement setup (Peeters *et al.*, 2003).

B. Separation of the even and the odd stochastic nonlinearities

A second possibility to detect nonlinear distortions makes explicitly use of the flexibility of random multisines: only a selected set of harmonics (called measurement lines) is excited and the nonlinearities at the output are detected by measuring the output levels at the non excited frequencies (called detection lines). This idea was already suggested by Evans *et al.* (1994a) and McCormack *et al.* (1994) and is further elaborated in (Vanhoenacker and Schoukens, 2001; Schoukens et al., 2002b) where the choice of the excitation and detection frequencies is studied. Consider for example a random multisine that excites the system at the frequencies k/N, $k = 1, 3, 9, 11, 17, 19, \ldots, 4p+1, 4P+3, \ldots$, $p \in \mathbb{N}$ (such a signal is called special odd). In that case the even nonlinearities are detected at the even lines, and the odd nonlinearities at the non excited odd lines.

Discussion: In practice some additional problems can occur during this test.

- The nonlinear interaction between generator and plant can also generate unwanted excitation lines at the detection frequencies which should remain zero in the ideal situation. Under these conditions it is no longer clear what part of the output should be assigned to the linear behaviour, and what part is due to the nonlinear distortions. A first order correction can be made to reduce the problem (Vanhoenacker and Schoukens, 2001), but this method is less robust compared to the previous method (Section A) where such interaction is not disturbing at all the results.
- It turns out that the level of the odd nonlinear distortions measured at the non excited odd frequencies underestimates the level of the odd nonlinear distortions at the measurement lines. A signal depended extrapolation factor is needed which is determined heuristically and should be used with care (Dobrowiecki and Schoukens 2001a; Vanhoenacker *et al.*, 2001). If the detection lines are randomly chosen, the extrapolation factor is 1.
- Another price to be paid is the loss in frequency resolution caused by the non-excited lines which increases the required measurement time if a given resolution should be respected.

Example: This method is illustrated on the hair dryer using a special odd multisine (Németh and Vargha, 1999). In Figure 7, the output amplitude spectrum is

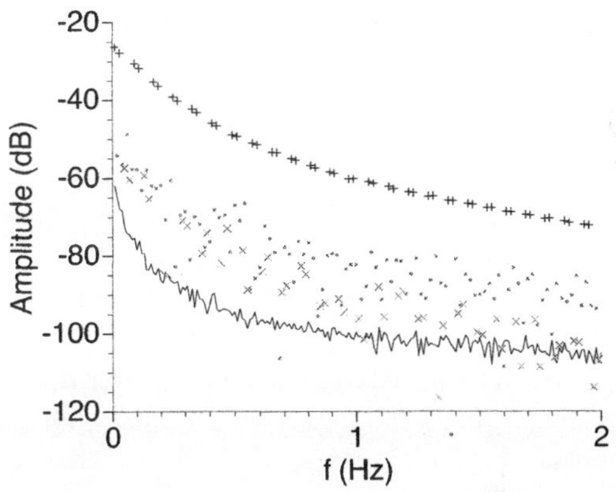

Fig. 7: Detection of nonlinear distortions: blue + output level at the excitation lines, red . even nonlinearities, green x odd nonlinearieties, thin line: noise level.

shown. Notice again that not only the level of the nonlinearities is detected, also the noise levels are available from the periodic repetitions. In this example it is clear that the even nonlinearities are the dominating distortion. For this reason odd excitations will reduce σ_{G_R} significantly, as was observed in Figure 4.

Conclusion: This test provides more information than the previous one, but an experienced user is needed to perform it well. Moreover, the test is less robust with respect to the experimental setup and heuristic extrapolation factors are needed during the interpretation of the results.

C. Estimating the level of the deterministic (bias) contributions

There is no direct access possible to the level of the bias contributions. The systematic nonlinear contributions can not be separated from the rest of the signal as was done for the stochastic contributions. It is tempting to use the levels of stochastic nonlinear contributions also to bound the bias contributions.

There is yet no formal theoretical framework available to support this idea, although some insights are available (Dobrowiecki and Schoukens, 2001a). Using the stochastic noise levels as an indication of the bias is an extrapolation. The bias is usually under estimated and the level depends upon the choice of the excitation and the system. Depending upon the situation extrapolation factors of ±3 dB up to ±20 dB are observed in the examples. The latter appear if only a small fraction of the total excitation power is in the pass band of the system.

4.5 Impact on the linear identification practice: parametric linear models

In this section we study the impact of nonlinear distortions on the parametric identification of linear models. Two approaches are considered. The first one is the classical prediction error approach (Söderström and Stoica, 1989; Ljung, 1999) where a parametric plant and noise model are simultaneously estimated. In the second approach a non parametric noise model is identified during the preprocessing of the raw data,. Next a parametric plant model is identified using the previous noise model as a weighting function (Schoukens *et al.*, 1997; Pintelon and Schoukens, 2001; Pintelon *et al.*, 2002).

A. Combined identification of plant and noise model

In the 'classical' identification approach, a linear model $G(q, \theta)$ is estimated together with a noise model $H(q, \theta)$

$$y(t) = G_0(q, \theta)u_0(t) + H_0(q, \theta)e(t) + T_G(\theta) + T_H(\theta). \tag{10}$$

$T_G(\theta)$ and $T_H(\theta)$ model the plant and noise filter transients. Often $T_G(\theta)$ is put equal to zero (no initial plant conditions are estimated), and in almost every method the noise model transient $T_H(\theta)$ is disregarded.

The model parameters θ are estimated by minimizing the prediction error, leading to the following definition for the estimates:

Definition 8: The prediction error estimates (estimated plant and noise model) are given by:

$$\hat{\theta}_{PE}(N) = \arg\min_{\theta} \frac{1}{N} \sum_{k=1}^{N} \left|H^{-1}(q, \theta)(y(t) - G(q, \theta)u(t) - T_G(\theta))\right|^2 \tag{11}$$

Because we deal here with identification in the presence of model errors, consistency should be replaced by convergence to the model (parameters) that would be obtained on the "exact" data G_R.

Definition 9 Best linear parametric approximation:

$$\hat{\theta}_*(N) = \arg\min_{\theta} \frac{2}{N} \sum_{k=1}^{N/2} \frac{\left|G_R(j2\pi\frac{k}{N}) - G(e^{j2\pi\frac{k}{N}}, \theta(N))\right|^2}{\sigma^2_{Y_S}(k) + \sigma^2_{Y_n}(k)} \tag{12}$$

Because Theorem 3 guarantees that G_R and $\sigma^2_{Y_S}(k)$ are smooth functions, these can be approximated arbitrary well by a rational form if the order of the models is large enough. For such a well selected model order and under the classical identifiability assumptions (Ljung, 1999), the following result is obtained:

Theorem 5: Consider a system belonging to the set $\mathbb{S}$, excited with an excitation $u_N \in \mathbb{E}$. If the noise Assumption 1 is met, $\hat{\theta}(N)$ converges in probability to $\hat{\theta}_*(N)$:

$$\operatorname*{plim}_{N \to \infty} (\hat{\theta}(N) - \hat{\theta}_*(N)) = 0. \tag{13}$$

From Theorem 5 it follows that the 'model errors' $G_R(j2\pi k/N) - G(e^{j2\pi k/N}, \hat{\theta}(N))|$ can be made arbitrarily small compared to the disturbing noise and the stochastic nonlinear contributions. Similar, the noise model $H(e^{j2\pi k/N}, \hat{\theta}(N))$ can follow arbitrarily well $\sigma^2_{Y_S}(k) + \sigma^2_{Y_n}(k)$. As a consequence this estimated plant/noise model will pass all 2nd order moment based validation tests. This is an unwanted and dangerous situation because the user gets no warning at all that a serious problem is hidden in the data. For example the uncertainty bounds that are calculated from this model are not valid. These bounds decrease to zero as an $O(N^{-1/2})$, while it is clear that this is not true for the nonlinear distortions and their induced errors (Pintelon and Schoukens 2001).

The basic reason for this failure is that the noise model H is shaped to whiten the sum of the disturbing noise and the stochastic nonlinearities, while the variance λ of the driving white noise source is scaled to match the observed levels. This situation changes completely if the noise model is obtained from a prior analysis of the data as is discussed in the next section.

B. Separated identification of the noise models

If periodic excitations are used, it is known from Section 4.4 that the disturbing noise variances $\sigma^2_Y(k)$ can be obtained separately from the variance of the stochastic nonlinear disturbances even before the identification process starts. Using this non parametric noise model $\sigma^2_{Y_S}(k) + \sigma^2_{Y_n}(k)$ as weighting function, the following frequency domain identification scheme can be defined:

$$V_{SML}(\theta, Z) = \frac{2}{N} \sum_{k=1}^{N/2} \frac{|Y(k) - G(j2\pi k/N, \theta)U(k)|^2}{\sigma^2_{Y_S}(k) + \sigma^2_{Y_n}(k)}. \tag{14}$$

There is a full equivalence with the time domain

identification framework (Schoukens *et al.*, 1999). Remark that in (14) the exact noise variance $\sigma_Y^2(k)$ is replaced by the estimated one $\hat{\sigma}_Y^2(k)$ obtained from measuring M successive periods of the input/output signals. The properties of this estimator, the sample maximum likelihood estimator (sample MLE) are known (Schoukens *et al.*, 1997). The estimator remains consistent (convergence to the noiseless solution in case of model errors) if $M \geq 4$. However, a small loss in efficiency appears: the covariance matrix on the parameters increases with $((M-2)/(M-3))$ ($M \geq 6$). The parameters are asymptotical normally distributed if $M \geq 7$ (Pintelon and Schoukens, 2002a).

Although both estimation procedures (prediction error, sample MLE) converge to the same limit model under very weak conditions, the behaviour of the model validation process is completely different. In the latter case, the cost function can be absolutely interpreted since the noise model is fixed. Consider the situation where the weighting in (14) is $\sigma_{Y_n}^2(k)$ (the nonparametric noise model is extracted from a number of over successive periods only, no averaging over different realizations of the excitation is made). In the absence of model errors, the expected value of the cost function is:

$$\mathcal{E}\{V_{\mathrm{SML}}(\theta, Z)\} = \frac{M-1}{M-2}\left(1 - \frac{n_\theta}{N}\right) , \qquad (15)$$

with n_θ the number of free parameters. A cost function that is too large indicates errors that are not explained by the observed noise levels. If these errors are white, the best linear approximation is found.

Table 2 Recommendations for the model selection process for V_{SML} with a weighting $\sigma_{Y_n}^2(k)$

	white residuals	coloured residuals
the cost function is too large	best linear approximation nonlinear distortions present it makes no sense to increase the model order	there are still unmodelled dynamics (model errors). Increase the model order to reduce them
the cost function is not significantly different from the expected value	this is the ideal situation best linear model no model errors detectable	good linear approximation check the noise analysis
the cost function is too small	good linear approximation check the noise analysis or reduce the model order	good linear approximation check the noise analysis

Correlated residuals point to unmodelled dynamics, hence it makes sense to increase the model order. This leads to the model selection/validation process given in Table 2. If the full weighting $\sigma_{Y_S}^2(k) + \sigma_{Y_n}^2(k)$ is available, the classical rules apply again (the nonlinearities are in that case detected from the procedure described in Section 4.4).

There exist a number of tools like the AIC and MDL criteria that are used to choose between different models (Akaike, 1974; Rissanen, 1978). These rules should be adapted to the situation of a fixed/estimated noise model. For a fixed noise model, the criterion should be reformulated to include the effect of model errors. For the standard identification methods this is done implicitly during the estimation of the noise variance λ starting from the residuals (Söderström and Stoica, 1989; Ljung, 1999). For the sample MLE, a modified criterion is needed. In the end the same criterion is found for both situations (Schoukens *et al.*, 2002c):

$$V_N(\theta, Z)\left(1 + 2\frac{n_\theta}{N}\right) , \qquad (16)$$

with $V_N(\theta, Z)$ equal to $V_{PE}(\theta, Z)$ or $V_{SML}(\theta, Z)$

5. APPROXIMATE MODELLING OF NONLINEAR SYSTEMS

5.1 Introduction

If a linear approximation proves insufficient, a nonlinear model is to be identified. The major problem of nonlinear system identification is the curse of dimensionality. The identification of a non parametric linear dynamic system requires $O(N)$ parameters for a system with 'a memory length' of N while for a parametric model $O(N^0)$ parameters are needed. The situation changes completely for nonlinear systems. The non parametric identification of a Volterra model of a pth order system (Schetzen, 1980) requires $O(N^p)$ parameters, resulting in an impractical large number of parameters, even if all Volterra kernel symmetries are exploited (Mathews and Sicuranza, 2000). This leads to very long experiments. Higher order spectra are a typical tool that is used to measure the non parametric Volterra representations (Mendel, 1991; Nikias and Mendel, 1993; Nikias and Petropulu, 1993). An alternative is to use multidimensional rational representations in the Laplace domain (the extension of the concept transfer function), but in this case the user faces an extremely complex model selection problem. Not only the order should selected, also the internal structure of the multidimensional polynomials needs to be retrieved. For all these reasons, the attention moved in an early stage to alternative approximative descriptions of the nonlinear system that are more convenient from the measurement and identification point of view (Billings, 1980; Doyle *et al.*, 2001;

Korenberg and Hunter, 1986; Worden and Tomlinson, 2001). It was realized (Schetzen, 1980; Mathews and Sicuranza, 2000; Rugh, 1981) that nonlinear systems can be approximated by parallel structures as shown in Fig. 8. The upper structure follows directly from

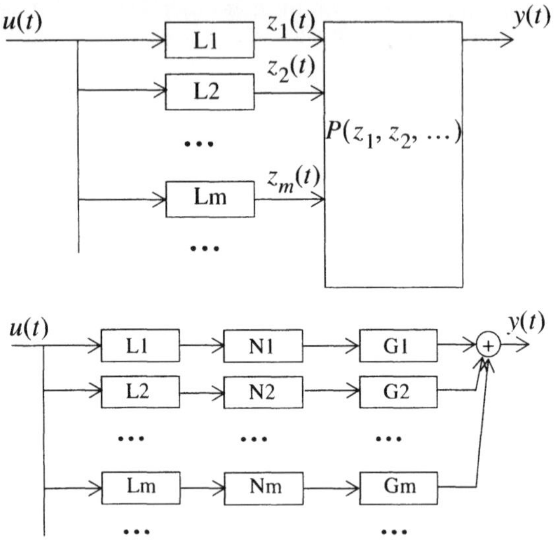

Fig. 8: Examples of some parallel cascade structures: Lk, Gk are linear systems; Nk and P represent static nonlinear systems.

the approximation theories, usually based on the Stone-Weierstrass theorem (Gallman and Narendra, 1976; Schetzen, 1980; Palm, 1979). The lower structure is obtained using matrix decomposition techniques like singular values decomposition or QR factorization, that are applied to the Volterra kernels (Matthews and Sicuranza, 2000; Korenberg and Hunter, 1986; Korenberg 1991). The second linear system Gk, $k = 1, 2, \ldots$ in the branches can be put equal to 1 to simplify the model structure even more. The number of required parallel branches to reach a given accuracy of the model can be very high, especially for the lower structure where it is shown that an exact representation of a pth order system with memory length N requires $O(N^{p-1})$ branches. Of course, in practice, the number of branches can be much smaller if an approximation error is tolerated. This allows also to reduce the required experiment length significantly. A large number of proposed identification methods is based on this idea (Baumgartner and Rugh, 1975; Bendat, 1990; Bendat, 1998; Billings and Fakhouri, 1980; Billings and Fakhouri, 1982; Chen, 1995; Korenberh and Hunter, 1986; Korenberg, 1991; Sjöberg *et al.*, 1995; Tan and Godfrey, 2002; Wysocki and Rugh, 1976; Wysocki and Wilson, 1979; Weiss *et al.*, 1998). Also neural network and wavelet modelling can be fitted in this framework (Sjöberg *et al.*, 1995; Juditsky *et al.*, 1995; Nelles, 2001).

5.2 A fast approximating nonlinear modelling/ identification technique

The identification of the parallel models still require large amounts of data. However, if the dominant behaviour of the system is linear, then it makes no sense to increase the experiment time dramatically to include the nonlinear distortions into the model since these represent only a fraction of the output power. For that reason it makes sense to look to even more restricted models if these can be identified with a similar effort (experiment time, computation time, model selection) as a linear system. Such structures can be considered as the natural extension of the linear model class, bridging the gap between 'simple' linear modelling techniques and 'very complex' nonlinear modelling approaches.

An attempt is made to propose such a structure (Schoukens *et al.*, 2003) is given in Fig. 9. It consists

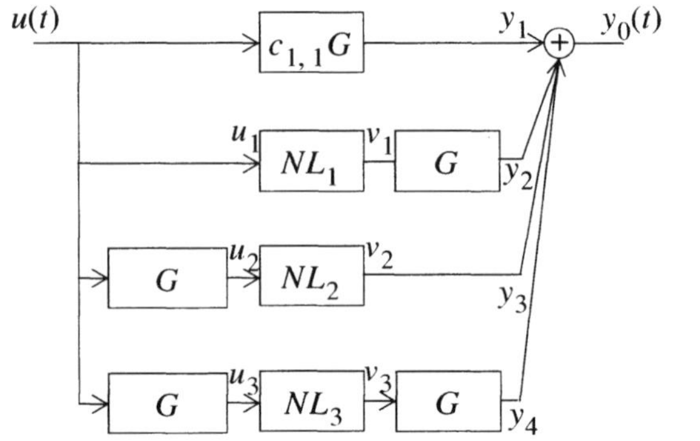

Fig. 9: Nonlinear structure consisting of a linear, a Hammerstein, a Wiener, and a 'feedback' branch.

of a number of parallel branches, each taking care of a typical nonlinear behaviour. Only one linear model block G is allowed. The major reason for this choice is the need for fast and simple identification methods. G is a rational form in s (continuous time systems), or in z^{-1} (discrete time systems), $c_{1,1}$ is a real number, NL_k is a static nonlinear system that is modelled linear-in-the-parameters. The last branch has a special structure. It also consists of a static nonlinearity, but this time it is pre- and post filtered by the linear system G. It is based on the p-th order inverse structure of a nonlinear system, and it turns out that it is very well suited to take care for nonlinear feedback dynamics. This branch is called the feedback branch.

A detailed discussion how to identify this structure using the classical linear identification methods in combination with a linear least squares approach is given in Schoukens *et al.* (2003).

This structure is experimentally tested on a number of case studies. For brevity, only one result is shown here. An electrical system with a cubic feedback is approximated using the simple structure of Fig. 9. Only the linear and the feedback branch are used. In Fig. 10 the validation results are shown. Clearly the nonlinear approximation gives much better results than the linear approximation while the experimental

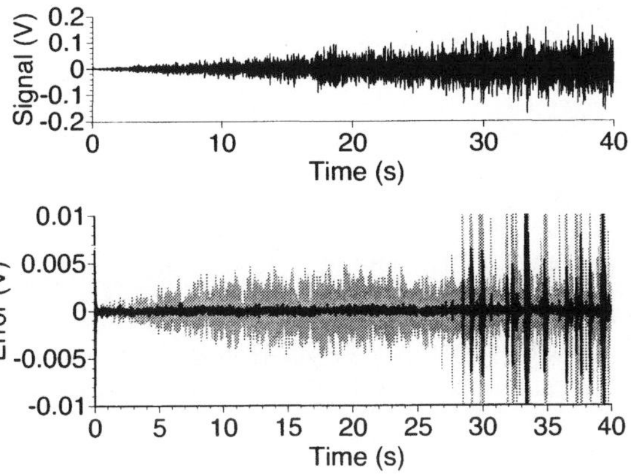

Fig. 10:Validation results. Top: the measured output. Bottom: gray: the linear simulated output error; black: the nonlinear simulated output error.

effort was exactly the same. The same data were also used to identify models that include explicitly the feedback structure using for example a NARMAX model (Chen and Billings, 1989) or another dedicated structure (Hjalmarsson, 2003). The NARMAX model ($y(t) = F(u(t-1), u(t-2), y(t-1), y(t-2))$, with F a polynomial function of order 3) was able to explain the full experiment, including the larger amplitude range, with a residual error on the output that is slightly larger than that of the approximate model shown in Fig. 10, without suffering from the large spikes that appear for times larger than about 30 s (Solomou, 2002).

Conclusion: The fast approximate nonlinear modelling/identification method bridges a gap between the linear and the full blown nonlinear identification machinery. It opens the possibility to make a fast check whether the linear model can be improved considerably with a minimum effort and no need for new experiments.

6. CONCLUSIONS

In this paper we proposed a framework to extend the linear system identification framework to systems with a dominant linear behaviour. We explicitly included nonlinear distortions into the framework. Three major results are presented.

A first result is that a nonlinear system can be replaced by a linear system plus a nonlinear noise source. The properties of this representation are studied for randomized excitations. It turns out that the linear system is the best linear approximation, while the moments of the nonlinear noise source has the same behaviour as normally distributed disturbing noise. These insights are used to line out a best engineering practice to measure the FRF of the best linear approximation.

The second major result is a better understanding of the behaviour of the classic identification schemes, for example the prediction error framework and the frequency domain identification schemes. It turns out that the properties of these identification schemes are maintained, but the validation process is strongly affected by the presence of the nonlinear distortions. If a parametric noise model is identified simultaneously with the plant model, the presence of the nonlinear distortions can be completely missed using the classical whiteness and cross-correlation tests. The user gets no indication at all that something is going seriously wrong. If a nonparametric noise model is extracted from a prior analysis, making use of periodic excitations, an alternative model selection and validation procedure is formulated that reveals the presence of nonlinear distortions.

The last major contribution of the paper is the presentation of a simplified nonlinear model structure that can be very easily identified using the linear identification methods plus a linear-least-squares step. This step can be considered as a natural extension of the linear framework, bridging the gap between the linear and nonlinear identification methods.

REFERENCES

Akaike, H. (1974). A new look at the statistical model identification. *IEEE Trans. Autom. Contr.*, 19, 716-723.

Baumgartner S.L. and W.J. Rugh (1975). Complete identification of a class of nonlinear systems fromsteady-state frequenc response. *IEEE Trans. Circuits Syst.*, 22, 753-759.

Bendat J. S., (1990), *Nonlinear System Analysis & Identification from Random Data*. New York: John Wiley & Sons.

Bendat J.S. and A.G. Piersol, (1980). *Engineering Applications of Correlation and Spectral Analysis*. New York: Wiley.

Bendat J.S., (1998). *Nonlinear systems. Techniques and applications*. John Wiley&Sons, Inc.

Billings S.A. , (1980). Identification of nonlinear systems: a survey, *IEE Proc.*, Vol. 127, pt. D, No. 6, 272-285.

Billings S.A. and S.Y. Fakhouri, (1980). Identification of non-linear systems using correlation analysis and pseudorandom inputs. *Int. J. Systems SCI.*, 11, 261-279.

Billings S.A. and S.Y. Fakhour, (1982). Identification of systems containing linear dynamic and static nonlinear elements, *Automatica*, 18, 15-26, 1982.

Billings S.A. and Q.H. Tao,(1991). Model validity tests for non-linear signal processing applications. *Int. J. Control*, 54, 157-194.

Borys A., (2000). *Nonlinear aspects of telecommunications. Discrete Volterra series and nonlinear echo cancellation*. CRC Press.

Boyd S. and L.O. Chua, (1985). Fading memory and the problem of approximating nonlinear operators with Volterra series. *IEEE Trans. Circuits Syst.*, 32, 1150-1161.

Brillinger D.R. (1981), *Time Series. Data Analysis and Theory.* New York: McGraw-Hill.

Bussgang J.J. (1952). *Crosscorrelation functions of amplitude-distorted Gaussian signals*. MIT Technical report. P.No 216.

Chen H.-W. (1995). Modeling and identification ofparallel nonlinear systems: structural classification and parameter estimation methods. *Proceedings of the IEEE*, 83, 39-66.

Chen S. and S.A. Billings (1989). Representations of nonlinear systems. - The Narmax model. *Int. J. Control*, 49, 1013-1032.

Dobrowiecki T. and J. Schoukens, (2001a). Bounds on Modelling Errors due to Weak Non-Linear Distortions. IMTC/01, *Proc. 18th IEEE Instrumentation and Measurement Technology Conference*, Budapest, Hungary, May 21-24.

Dobrowiecki T. and J. Schoukens (2001b). Practical choices in the

FRF measurement in presence of nonlinear distortions. *IEEE Trans. Instrum. Meas.*, 50, 2-7.
Doyle F.J., R.K. Pearson, and B.A. Ogunnaike, (2001). *Identification and control using Volterra models*. Springer.
Evans C., D. Rees, and D.L. Jones (1994a). Identifying linear models of systems suffering nonlinear distortions. *Proceedings IEE Control'94*, Coventry, pp. 288-296.
Evans C., D. Rees, and D.L. Jones (1994b). Nonlinear disturbance errors in system-identification using multisine test signals. *IEEE Trans. on Instr. and Meas.*, 43, 238-244.
Evans C, D. Rees (2000). Nonlinear distortions and multisine signals - Part I: Measuring the best linear approximation. *IEEE Trans. Instrum. Meas.*, 49, 602-609.
Forsell U. and Ljung (2000). A projection method for closed-loop identification. *IEEE Trans. Autom. Contr.*, 45, 2101-2105.
Gallman P.G. and K.S. Narendra, (1976). Representations of nonlienar systems via the Stone-Weierstrass theorem. *Automatica*, 12, 619-622.
Godfrey K. (1993). *Perturbation signals for system identification*. Prentice Hall.
Hjalmarsson H. (2003). *Identification of nonlinear feedback systems*. Technical report.
Juditsky A., H. Hjalmarsson, A. Beveniste, B. Deylon, L. Ljung, J. Sjöberg, and Q. Zhang (1995). Nonlinear black-box models in system identification: mathematical foundations. *AUTOMATICA*, 31, 1725-1750.
Korenberg M.J. and I.W. Hunter, (1986). The identification of nonlinear biological systems: LNL cascade models. *Biol. Cybern.*, 55, 125-134.
Korenberg M.J. (1991). Parallel cascade identification and kernel estimation for nonlinear systems. *Annals of Biomedical Engineering*, 19, 429-455.
Ljung, L. (1999). *System Identification. Theory for the User.* Englewood Cliffs: Prentice-Hall.
Mathews V.J. and G.L. Sicuranza, (2000). *Polynomial signal processing*. John Wiley.
McCormack A. S. , K. R. Godfrey, and J. O. Flower, (1994). The detection of and compensation for nonlinear effects using periodic input signals. *Proceedings IEE Control '94*, , Coventry, UK, pp. 297-302.
Mendel J.M., (1991). Tutorial on Higher-Order Statistics (Spectra) in Signal Processing and System Theory: Theoretical Results and Some Applications. *Proceedings of the IEEE*, 79, 278-305.
Németh J. and Vargha B., (1999). *Experiments with the hair dryer setup*. Internal report BIL99.
Nelles O. (2001). *Nonlinear system identification*. Springer.
Nikias C. L. and J.M. Mendel,(1993). Signal Processing with Higher-Order Spectra. *IEEE Signal Proc. Magazine*,. 10-36.
Nikias, C.L. and A.P. Petropulu, (1993). *Higher-Order Spectra Analysis*, New Jersey: Prentice Hall Signal Processing Series.
Palm G. (1979). On representation and approximation of nonlinear systems. Part II. Discrete time. *Biol. Cybernetics*, 34, 49-52.
Peeters B, H. Van der Auweraer, J. Schoukens, R. Pintelon. (2003). Using multisines to assess nonlinear distortions in vibrating mechanical structures. *20th Benelux-meeting on systems and control*, 2003.
Pintelon R., J. Schoukens, and G. Vandersteen (1997). Frequency domain system identification using arbitrary signals. *IEEE Trans. Autom. Contr.*, 42, 1717-1720.
Pintelon R. and J. Schoukens (2001). *System Identification. A frequency domain approach*. IEEE press, New Jersey.
Pintelon R., J. Schoukens, W. Van Moer, and Y. Rolain (2001). Identification of linear systems in the presence of nonlinear distortions. *IEEE Trans. Instrum. Meas.*, 50, 855-863.
Pintelon R. and J. Schoukens (2002a). Measurement and modelling of linear systems in the presence of non-linear distortions. *Mechanical Systems and Signal Processing*, 16, 785-801.
Pintelon R. and J. Schoukens (2002). Some peculiarities of identification in the presence of model errors. *AUTOMATICA* 38, 1683-1693.
Rissanen, J. (1978). Modeling by shortest data description. *AUTOMATICA*, 14, 465-471.
Rugh W.J. (1981). *Nonlinear system theory. The Volterra/Wiener approach*. Johns Hopkins University Press.
Sandberg I.W., (1992). Approximately-finite memory and input-output maps. *IEEE Trans. Circuits Syst.-I*, 39, 549-556.
Sandberg I.W., (1993). Uniform approximation and the circle criterion. *IEEE Trans. Autom. Contr.*, 38, 1450-1458.
Sandberg I.W. (2002). $I\!R_+$ Fading memory and extensions of input-output maps. *IEEE Trans. Circuits Syst.-I*, 49, 1586-1591.
Schetzen, M., (1980). *The Volterra and Wiener Theories of Nonlinear Systems*. New York: John Wiley & Sons.
Schoukens J., R. Pintelon, G. Vandersteen, and P. Guillaume (1997). Frequency-domain system identification using non-parametric noise models estimated from a small number of data sets. *AUTOMATICA* 33, 1073-1086.
Schoukens J. , T. Dobrowiecki, and R. Pintelon (1998). Identification of linear systems in the presence of nonlinear distortions. A frequency domain approach. *IEEE Trans. Autom. Contr.*, Vol. 43, No.2, pp. 176-190.
Schoukens J., R. Pintelon, and Y. Rolain (1999). Study of conditional ML estimators in time and frequency-domain system identification. *AUTOMATICA*, 35, 91-100.
Schoukens J., J. Swevers, R. Pintelon, and H. Van der Auweraer (2002a). Excitation design for FRF measurements in the presence of nonlinear distortions. *Proceedings of ISMA2002*, Vol. 2, 951-958.
Schoukens J., R. Pintelon, T. Dobrowiecki (2002b). Linear modeling in the presence of nonlinear distortions. *IEEE Trans. Instrum. Meas.*, 51, 786-792.
Schoukens J., Y. Rolain, and R.Pintelon (2002c). Modified AIC rule for model selection in combination with prior estimated noise models. *AUTOMATICA* 38, 903-906.
Schoukens J., J. Nemeth, P. Crama, Y. Rolain, and R. Pintelon (2003). Fast approximate identification of nonlinear systems. *SYSID 2003*, Rotterdam, ?-?.
Sjöberg J, Zhang QH, Ljung L, et al. (1995). Nonlinear black-box modeling in system identification: A unified overview. *AUTOMATICA* 31, 1691-1724.
Söderström, T. and P. Stoica. *System Identification*. Prentice-Hall, Englewood Cliffs, pp. 256., 1989.
Solomou M. (2002). *Narmax modelling for the mechanical nonlinear system*. School of Electronics, University of Glamorgan, technical report.
Tan AH, K. Godfrey (2002). Identification of Wiener-Hammerstein models using linear interpolation in the frequency domain (LIFRED). *IEEE Trans. Instrum. Meas.*, 51, 509-521.
Vandersteen G., J. Verbeeck, Y. Rolain, and J. Schoukens (2000). Accurate bit-error-rate estimation for OFDM based telecommunication systems schemes in the presence of nonlinear distortions. *Proc. 17th IEEE Instrum. and Meas. Conf. IMTC*, Baltimore, MD, 80-85.
Vanhoenacker K., T. Dobrowiecki, and J. Schoukens (2001). Design of multisine excitations to characterize the nonlinear distortions during FRF-measurements. *IEEE Trans. Instrum. Meas.*, 50, 1097-1102.
Vanhoenacker K., J. Schoukens, J. Swevers, and D. Vaes (2002). Summary and comparing overview of techniques fo the detection of non-linear distortions. *Proceedings of ISMA2002*, Vol. 2, 1241-1256.
Weiss M, C. Evans C, and D. Rees (1998). Identification of nonlinear cascade systems using paired multisine signals. *IEEE Trans. Instrum. Meas.*, 47, 332-336.
Worden K. and G.R. Tomlinson, (2001). *Nonlinearity in structural dynamics. Detection, Identification and Modelling*. IOP Publishing Ltd.
Wysocki E.M. andW.J. Rugh (1976). Further results on the identification problem for the class of nonlinear systems S_M. *IEEE Trans. Circuits Syst.*, 23, 664-670.
Wysocki E.M. and J.R. Wilson, (1979). An approximation approach to the identification of nonlinear systems based on frequency-response measurements. *Int. J. Control*, 29, 113-123.

IFAC Publications
www.elsevier.com/locate/ifac

SOME PROBLEMS IN STATISTICAL INFERENCE FOLLOWING MODEL SELECTION

Benedikt M. Pötscher *

* *Department of Statistics, University of Vienna*

Abstract: Statistical inference following a preliminary model selection step is common practice in most applied statistical analyses. In practice, statistical inference is frequently conducted in a classical manner thereby ignoring the model selection phase. Not surprisingly, this leads to invalid procedures. We review some recent attempts towards a coherent theory for statistical inference in the presence of model selection. *Copyright © 2003 IFAC*

Keywords: Model Selection, Statistical Inference

1. INTRODUCTION

In this talk we shall review some recent developments in the area of statistical model selection, with special emphasis on the difficulties one encounters when attempting to conduct valid statistical inference following the model selection phase. Although nowadays statistical model selection is the rule rather than the exception in statistical analyses, this problem has mainly be ignored in practical work and has received relatively little theoretical treatment except in recent years. Breiman (1992) has called this "a quiet scandal in the statistical community".

The traditional theory of parametric statistical inference is primarily concerned with statistical properties of estimators and inference procedures (like test or confidence procedures) under the central assumption of an a priori given model. That is, it is assumed that the model is known to the researcher prior to the statistical analysis, except for the value of the true parameter vector. In practice, however, the specification of the model (choice of functional form, choice of regressors, number of lags, etc.) is often also determined only after the data have been observed, violating the central assumption of an a priori given parametric model. As a consequence, the actual statistical properties of estimators or inference procedures following such a data-driven model selection step are not described by the traditional theory assuming an a priori given model; in fact, they may differ substantially from the properties predicted by the traditional theory, cf. (Pötscher, 1991, Section 3.3). Ignoring the additional uncertainty originating from the data-based model selection step and (inappropriately) applying traditional theory can hence result in very misleading conclusions.

In the following we shall discuss distributional properties of estimators that follow a model selection step (so-called post-model-selection estimators). For the sake of simplicity we shall confine the discussion to the framework of a normal linear regression model, although the issues discussed are relevant for any (perhaps more complex) model selection problem.

2. POST-MODEL-SELECTION ESTIMATORS

Consider as the overall model the linear regression model

$$Y = X\theta + \epsilon \qquad (1)$$

where the regressor matrix X is a non-stochastic $n \times P$ matrix with $rank(X) = P$. That is, there are P explanatory variables in the overall model. We assume that the errors are multivariate normal, i.e., $\epsilon \sim N(0, \sigma^2 I_n)$, $\sigma^2 > 0$. Here n denotes the sample size and we assume $n > P \geq 1$. Furthermore we assume that $Q = \lim_{n \to \infty} X'X/n$ exists and is non-singular.

If we have no doubt that all P explanatory variables affect the dependent variable Y, we would estimate the overall model (1) by least squares, i.e., we would estimated θ by $(X'X)^{-1} X'Y$. However, if we suspect that some of the regressors are superfluous, we may want to eliminate these regressors prior to estimation. More specifically, we may suspect that the true parameter value θ in fact belongs to one of the following submodels given by

$$M_p = \{(\theta_1, \ldots, \theta_P)' \in \mathbf{R}^P : \theta_{p+1} = \cdots = \theta_P = 0\}, \quad (2)$$

$0 \leq p \leq P$. Here model M_p corresponds to the situation where only the first p regressors have an effect on the dependent variable. Note that $M_0 = \{0\}$, and that $M_P = \mathbf{R}^P$ corresponds to the overall model. We call M_p the regression model of order p. Note that we here consider only the nested submodels $M_0 \subseteq M_1 \subseteq \cdots \subseteq M_P$. This amounts to the implicit assumption that the regressor variables are somehow ordered according to importance (as, e.g., in a polynomial regression). If we knew that the true parameter value belongs to M_p, say, we could of course estimate θ by the restricted least squares estimator which employs the restrictions defining M_p. I.e., we would use for $p \geq 1$

$$\hat{\theta}(p) = (X[p]'X[p])^{-1} X[p]'Y \quad (3)$$

to estimate the first p components of θ. Here $X[p]$ denotes the matrix consisting of the first p columns of X. In absence of this knowledge, we may want to use as a first step a model selection procedure and then fit the selected model by least squares. A model selection procedure in general is now nothing else than a *data-driven* rule $\hat{p} = \hat{p}(Y)$ that selects a value from $\{0, \ldots, P\}$ and thus selects a model from the list of candidate models $M_0, \ldots, M_P$. Consequently, the resulting estimator that is actually being used is given by $\hat{\theta}(\hat{p})$ (which estimates the first $\hat{p}$ components of θ) with the remaining $P - \hat{p}$ components being set equal to zero. For obvious reasons we shall refer to such an estimator as a *post-model-selection estimator.*

In practice one would typically obtain $\hat{p}$ from a model selection criterion like AIC or BIC. In the following we shall – for simplicity – use a general-to-specific hypothesis testing procedure as the model selection procedure defining $\hat{p}$. The main conclusions from this analysis will be seen to carry over to model selection by AIC or BIC; cf., Section 4. The model selection procedure we analyze in the sequel is now given as follows: The sequence of hypotheses $H_0^p : \theta \in M_{p-1}$ is tested against the alternative $H_1^p : \theta \in M_p \setminus M_{p-1}$ in decreasing order starting at $p = P$ until the first rejection occurs. If H_0^p is the first hypothesis in this process that is rejected, we set $\hat{p} = p$. If no hypothesis is rejected we set $\hat{p} = 0$. Each hypothesis in this sequence is tested by a kind of t-test where the error variance estimator is always taken from the overall model. More formally, we have

$$\hat{p} = \max\{p : |T_p| \geq c_p, 0 \leq p \leq P\} \quad (4)$$

where the test-statistics are given by

$$T_p = \frac{n^{1/2}\hat{\theta}_p(p)}{\hat{\sigma}\xi_{n,p}} \qquad (1 \leq p \leq P) \quad (5)$$

with

$$\xi_{n,p} = \left(\left[(X[p]'X[p]/n)^{-1}\right]_{p,p}\right)^{\frac{1}{2}}, \quad (6)$$

$1 \leq p \leq P$, being the square root of the p-th diagonal element of the matrix indicated and with

$$\hat{\sigma}^2 = \frac{(Y - X\hat{\theta}(P))'(Y - X\hat{\theta}(P))}{n - P}. \quad (7)$$

The critical values c_p are independent of sample size n and satisfy $0 < c_p < \infty$ for $1 \leq p \leq P$. We also set $T_0 = c_0 = 0$ in order to ensure a well-defined $\hat{p}$. Note that under the hypothesis H_0^p the statistic T_p is t-distributed with $n - P$ degrees of freedom for $1 \leq p \leq P$.

As mentioned above, the post-model-selection estimator $\tilde{\theta}$ is now given as follows: The first $\hat{p}$ coordinates of $\tilde{\theta}$ are given by the restricted least-squares formula based on the first $\hat{p}$ regressors and the remaining $P - \hat{p}$ coordinates are set equal to zero. More formally, $\tilde{\theta}$ is the $P \times 1$ vector given by

$$\tilde{\theta} = (0, \ldots, 0)' \mathbf{1}\{\hat{p} = 0\} + \sum_{p=1}^{P} (\hat{\theta}(p)' : 0)' \mathbf{1}\{\hat{p} = p\}. \quad (8)$$

We conclude this section by introducing some notation: Given a parameter value θ, its order is defined as

$$p_0(\theta) = \min\{p : 0 \leq p \leq P, \theta \in M_p\}. \quad (9)$$

Hence, if θ is the true parameter value, only models M_p with $p \geq p_0(\theta)$ are correct models. We stress that $p_0(\theta)$ is a property of a *single parameter value*, and hence needs to be distinguished from the notion of the order of the model M_p introduced earlier, which is a property of the *set* M_p.

Furthermore, for $p \geq 1$, the expected value of the restricted least-squares estimator $\hat{\theta}(p)$ given by (3) will be denoted by $\mu_n(p)$ and is given by

$$\theta[p] + (X[p]'X[p])^{-1}X[p]'X[\neg p]\theta[\neg p], \quad (10)$$

where $X[\neg p]$ denotes the matrix consisting of the last $P-p$ columns of X and $\theta[\neg p]$ denotes the vector containing the last $P-p$ elements of θ. (In case $p = P$, the second term on the r.h.s. of (10) and similar expressions are to be interpreted as zero; that is, $\mu_n(P) = \theta[P] = \theta$.) As usual, the j-th component of $\mu_n(p)$ will be denoted by $\mu_{n,j}(p)$. We note that $\mu_n(p)$ depends on θ, although this dependence is not shown explicitly in the notation.

3. THE DISTRIBUTION OF POST-MODEL-SELECTION ESTIMATORS

The finite-sample distribution of the restricted least squares estimator $\hat{\theta}(p)$ is clearly multivariate normal. More precisely, $n^{1/2}(\hat{\theta}(p) - \mu_n(p))$ is normal with mean zero and variance-covariance matrix $\sigma^2(X[p]'X[p]/n)^{-1}$. The cdf and pdf of this distribution, respectively, will be denoted by $\Phi_{n,p}$ and $\phi_{n,p}$. Also the asymptotic distribution of the restricted least-squares estimator is normal with mean zero and variance-covariance matrix $\sigma^2 \lim_{n\to\infty}(X[p]'X[p]/n)^{-1}$. The cdf and pdf of this distribution will be denoted by $\Phi_{\infty,p}$ and $\phi_{\infty,p}$, respectively.

However, when using the post-model-selection estimator $\tilde{\theta}$, it is the distribution of this estimator that is of interest and not the distribution of the restricted least squares estimator. (In practice often the analysis proceeds as if these two distributions were the same. As will be seen, this is not justified in any way and may mislead the entire analysis.) In this section we discuss the sampling distribution of the post-model-selection estimator. We first concentrate on the distribution of the post-model-selection estimator conditional on the event that model M_p has been chosen and comment on the unconditional distribution later.

Assume first that p satisfies $p \geq p_0(\theta)$, i.e., that the chosen model M_p is correct in the sense that it contains at least all the relevant regressors. The distribution function of the post-model-selection estimator $\tilde{\theta}$ (suitably scaled and centered) conditional on selecting the model of order p, i.e. conditional on the event $\hat{p} = p$, is then given by

$$P_{n,\theta,\sigma}\left(n^{1/2}(\tilde{\theta} - \theta) \leq (t_1, \ldots, t_P)' | \hat{p} = p\right) = P_{n,\theta,\sigma}\left(n^{1/2}(\hat{\theta}(p) - \theta[p]) \leq (t_1, \ldots, t_p)' \,|\, \hat{p} = p\right) \times \mathbf{1}_{\mathbf{R}_+^{P-p}}(t_{p+1}, \ldots, t_P) \tag{11}$$

in case $p \geq 1$, and is point mass at zero in case $p = 0$. Here $P_{n,\theta,\sigma}$ denotes the probability measure under the true parameters θ and σ. Since the effect of the indicator function in the above display is trivial, we shall concentrate on the first factor:

$$G_{n,\theta,\sigma}(t \,|\, p) = P_{n,\theta,\sigma}\left(n^{1/2}(\hat{\theta}(p) - \theta[p]) \leq t \,|\, \hat{p} = p\right) \tag{12}$$

with $t \in \mathbf{R}^p$, $p \geq 1$. For convenience we shall refer to (12) as the conditional distribution of $\tilde{\theta}$. It has been shown by Pötscher (1991) in a more general context including nonlinear models and dependent data (cf. also (Sen 1979)) that the asymptotic distribution corresponding to (12) exists and has a density that is up to a normalizing constant α_p given by

$$g_{\infty,\theta,\sigma}(t \,|\, p) = \phi_{\infty,p}(t)\, \frac{\mathbf{1}_{\mathbf{R}\setminus(-\sigma\xi_{\infty,p}c_p, \sigma\xi_{\infty,p}c_p)}(t_p)}{\alpha_p} \tag{13}$$

if $p > p_0(\theta)$, i.e., if $\theta \in M_{p-1}$, and by

$$g_{\infty,\theta,\sigma}(t \,|\, p) = \phi_{\infty,p}(t) \tag{14}$$

if $p = p_0(\theta)$, i.e., if $\theta \in M_p \setminus M_{p-1}$. Here $\xi_{\infty,p}$ is the limit of $\xi_{n,p}$ as $n \to \infty$ and the normal density $\phi_{\infty,p}(t)$ has been defined above.

Formula (13) shows that the asymptotic conditional distribution of the post-model-selection estimator is *not* normal, but rather is an 'excised' normal distribution. In particular, it does *not* coincide with the (asymptotic) distribution of the restricted least squares estimator based on model M_p (holding p fixed). In fact, both distributions can be quite different. For example, some of the marginal distributions of (13) can be bimodal, cf. (Pötscher 1991) and (Pötscher and Novak 1998). In contrast, (14) states that the asymptotic distribution conditional on having selected the 'minimal' true model coincides with the (asymptotic) distribution of the corresponding restricted least squares estimator. (As it turns out this latter asymptotic result has little finite sample relevance.) The asymptotic distribution of the post-model-selection estimator conditional on the event $\hat{p} = p$ with $p < p_0(\theta)$, i.e., if the selected model is missing some relevant regressors and hence is 'incorrect', can also be obtained, but requires a different centering of the estimator in order to stabilize the distribution; see (Leeb and Pötscher 2003a). Also the unconditional asymptotic distribution of the post-model-selection estimator can be derived (Pötscher 1991, Leeb and Pötscher 2003a), and – not surprisingly – turns out to be non-normal. The discussion so far certainly shows that ignoring the model selection step and proceeding with classical inference procedures that are based on normal approximations will typically be invalid and may be highly misleading as it does not take into account the additional uncertainty due to model selection.

One could now try to use the asymptotic (conditional or unconditional) distributions of the post-model-selection estimator, cf. (13), (14), as a basis for valid inference procedures. Unfortunately, it turns out that these asymptotic distributions can be very unreliable approximations to the corresponding finite-sample distributions of the post-model-selection estimator, and the same is true for the unconditional distributions (Pötscher and Novak 1998). The reason for this phenomenon is that in the asymptotic results sketched above the convergence of the finite-sample distributions to the asymptotic distributions is *not* uniform over the parameter space, i.e., the 'worst-case' error of the asymptotic distribution as an approximation to the finite-sample distribution does not converge to zero with increasing sample size; cf. (Pötscher 1991, Section 4, Remark (iii)), (Leeb and Pötscher 2003a, Corollary 4.6). This non-uniformity phenomenon clearly has a detrimental effect on any inference procedure based on the asymptotic distribution of the post-model-selection estimator (Kabaila 1995, Pötscher 1995).

Given the just mentioned drawbacks of the asymptotic distribution, one could try to fall back to the finite-sample distribution of the post-model-selection estimator. The closed form of the finite-sample distribution can be derived (Leeb and Pötscher 2003a). It turns out that the density of the finite-sample distribution function $G_{n,\theta,\sigma}(t\,|\,p)$ is given by

$$g_{n,\theta,\sigma}(t\,|\,p) = \phi_{n,p}(t)\, m_{n,p,\theta,\sigma}(t_p) \qquad (15)$$

where $\phi_{n,p}(t)$ is the multivariate normal density defined above and $m_{n,p,\theta,\sigma}$ is a (complicated) 'modification' factor that is given explicitly in (Leeb and Pötscher 2003a). (As written this formula holds for $p \geq p_0(\theta)$. It continues to hold for $p < p_0(\theta)$ with a different centering of the estimator.) Note that – as expressed in the notation – the modification factor depends on unknown parameters. Observe that both the finite-sample density as well as the density of the asymptotic distribution share a common structure: both are multivariate normal densities multiplied by modification factors that depend on the argument t only through the coordinate t_p. While the modification factor for the asymptotic distribution is a (constant times a) simple indicator function, this factor is much more complicated for the finite-sample density. The unconditional finite-sample distribution of the post-model-selection estimator $\tilde{\theta}$ can also be obtained and has been derived by Leeb and Pötscher (2003a).

As is apparent from (15), the finite-sample distribution of $\tilde{\theta}$ depends on unknown parameters. Hence, it cannot directly be used for the construction of inference procedures (e.g., confidence intervals). An obvious and standard idea is now to somehow *estimate* the finite-sample distribution (e.g., by replacing unknown parameters by suitable estimators) and to use the estimated finite-sample distribution as a basis for the construction of inference procedures. In order for this idea to work successfully, the estimated finite-sample distribution should be close to the true one, at least in large samples. Now, it is in fact possible to construct – for every $t \in \mathbf{R}^p$– estimators $\hat{G}_n(t\,|\,p)$ for $G_{n,\theta,\sigma}(t\,|\,p)$ that are consistent in the sense that

$$P_{n,\theta,\sigma}\left(\left|\hat{G}_n(t\,|\,p) - G_{n,\theta,\sigma}(t\,|\,p)\right| > \delta\right) \to 0 \qquad (16)$$

for $n \to \infty$ holds for every $\delta > 0$ and for every $\theta \in M_p$. (Note that for $\theta \notin M_p$ the probability of the event $\hat{p} = p$ converges to zero.) But is this good enough? Recall that the approximation error between the asymptotic distribution and the finite-sample distribution does not converge to zero uniformly in θ, a phenomenon that has led us away from basing inference procedures on the asymptotic distributions. We are hence confronted with the question whether or not we can find estimators $\hat{G}_n(t\,|\,p)$ such that the convergence in (16) is uniform with respect to θ. Surprisingly the answer is no. The following result is a special case of results given by Leeb and Pötscher (2003b).

Theorem 3.1. Let $0 < p \leq P$ and let $t \in \mathbf{R}^p$ be given. Suppose $\hat{G}_n(t\,|\,p)$ is a consistent estimator for $G_{n,\theta,\sigma}(t\,|\,p)$ in the above sense . Then

$$\sup_{\theta \in M_p} P_{n,\theta,\sigma}\left(\left|\hat{G}_n(t\,|\,p) - G_{n,\theta,\sigma}(t\,|\,p)\right| > \delta\right) \to 1 \qquad (17)$$

for $n \to \infty$ and for some $\delta > 0$.

The 'impossibility' result in this theorem says that, while an estimator for $G_{n,\theta,\sigma}(t\,|\,p)$ can be consistent, it can never be *uniformly* consistent. The result carries over to the unconditional distribution as well as to versions of the conditional distribution function using a different centering of the estimator. It also extends to linear transformations of the estimator. The theorem also continues to hold if the supremum over M_p is replaced by the supremum over strategically chosen shrinking balls in M_p, showing that the problem is a local rather than a global one. The practical upshot of this result is that inference procedures based on estimated versions of the distribution function of the post-model-selection estimator are lacking theoretical justification, and that the problem of inference following model selection is much more intricate as one might have expected.

4. EXTENSIONS

The results discussed above have been given in the framework of a linear regression model with normally distributed errors and when the model selection procedure is based on a sequence of general-to-specific hypothesis tests. Most of the results can be extended to more general model selection problems – nonlinear models, dependent data, and model selection based on AIC or BIC – although the development of the theory becomes much more laborious. Impossibility results like Theorem 3.1 have also been obtained for statistical procedures that are relatives of post-model-selection estimators, namely shrinkage-type estimators (e.g., Stein estimators, Lasso-type estimators, SCAD-estimators, etc.), see (Leeb and Pötscher 2002).

5. CONCLUSION

A systematic investigation into the structure and distributional properties of post-model-selection estimators has only begun in recent years. Progress has been made in understanding the structure of the asymptotic as well as of the finite-sample distribution of such estimators. Non-trivial difficulties in approximating the finite-sample distributions of post-model-selection estimators emerge, and impossibility results regarding the 'estimability' of the finite-sample distributions have been established. Much future work needs to be done to arrive at a full understanding of statistical inference following a model selection phase.

6. REFERENCES

Breiman, L. (1992). The little bootstrap and other methods for dimensionality selection in regression: X-fixed prediction error. *Journal of the American Statistical Association*, **87**, 738–754.

Kabaila, P. (1995). The effect of model selection on confidence regions and prediction regions. *Econometric Theory* **11**, 537–549.

Leeb, H. & B. M. Pötscher (2002). Performance limits for estimators of the risk or distribution of shrinkage-type estimators, and some general lower risk-bound results. Working Paper, Department of Statistics, University of Vienna.

Leeb, H. & B. M. Pötscher (2003a). The finite-sample distribution of post-model-selection estimators and uniform versus nonuniform approximations. *Econometric Theory* **19**, 100–142.

Leeb, H. & B. M. Pötscher (2003b). Can one estimate the distribution of post-model-selection estimators? Working Paper, Department of Statistics, University of Vienna.

Pötscher, B. M. (1991). Effects of model selection on inference. *Econometric Theory* **7**, 163–185.

Pötscher, B. M. (1995). Comment on 'The effect of model selection on confidence regions and prediction regions'. *Econometric Theory* **11**, 550–559.

Pötscher, B. M. & A. J. Novak (1998). The distribution of estimators after model selection: large and small sample results. *Journal of Statistical Computation and Simulation* **60**, 19–56.

Sen, P. K. (1979). Asymptotic properties of maximum likelihood estimators based on conditional specification. *Annals of Statistics* **7**, 1019–1033.

IFAC Publications
www.elsevier.com/locate/ifac

CHOOSING INTEGER PARAMETERS IN SUBSPACE METHODS: A SURVEY ON ASYMPTOTIC RESULTS

Dietmar Bauer *,1

* *Institute for Econometrics, Operations Research and System Theory,*
TU Wien, Argentinierstr. 8,
A-1040 Vienna, Austria

Abstract: When using subspace methods, the user has to specify a number of integer parameters. This paper surveys the literature on results relating to strategies for these choices and the consequences thereof. All results are asymptotic in nature and relate either to consistency questions or to the asymptotic covariance matrix of the estimated systems. *Copyright © 2003 IFAC*

Keywords: subspace algorithms, asymptotic theory, linear dynamical systems

1. INTRODUCTION

Subspace algorithms have been developed in the eighties and early nineties: The most popular algorithms are CCA (Larimore, 1983, sometimes referred to as CVA), N4SID (Van Overschee and DeMoor, 1994) and MOESP (Verhaegen, 1994). The asymptotic properties of these subspace algorithms have been investigated heavily in the past decade. This has led to a basic understanding of the asymptotic properties of the estimates of the system. For many different algorithms, conditions ensuring consistency of the obtained estimates have been identified (Deistler *et al.*, 1995; Peternell *et al.*, 1996; Jansson, 1997). Also asymptotic normality has been investigated (Bauer *et al.*, 1999; Bauer and Jansson, 2000). Recently also more explicit expressions for the asymptotic variance of the system matrix estimates have been obtained (Bauer and Ljung, 2002; Chiuso and Picci, 2002). (Bauer, 2000) shows, that for the case of no exogenous inputs, CCA for certain choices of the integer parameters f and p (to be defined below) is asymptotically equivalent to prediction error methods.

In the algorithm the user has to specify integer parameters including the order of the system. These choices turn out to be critical for the estimation accuracy. Nevertheless, the list of results relating to the effects of automated guesses for these integers on the asymptotic properties of the estimated systems is rather short.

[1] Support by the Austrian FWF under the project number P14438-INF is gratefully acknowledged.

2. MODEL SET AND ASSUMPTIONS

This paper deals with finite dimensional, discrete time, time invariant, linear, dynamical state space systems of the form

$$\begin{aligned} x_{t+1} &= Ax_t + Bu_t + K\varepsilon_t \\ y_t &= Cx_t + Du_t + \varepsilon_t \end{aligned} \qquad (1)$$

where y_t denotes the s-dimensional observed output, x_t the n-dimensional state, u_t the m dimensional observed exogenous input and ε_t the s-dimensional innovation sequence. $A \in \mathbb{R}^{n\times n}, B \in \mathbb{R}^{n\times m}, K \in \mathbb{R}^{n\times s}, C \in \mathbb{R}^{s\times n}, D \in \mathbb{R}^{s\times m}$ are real matrices. Throughout the paper it is assumed that the system is stable, i.e. all the eigenvalues of A are assumed to lie within the open unit disc, and strictly minimum-phase, i.e. all the eigenvalues of $A - KC$ are assumed to lie within the unit circle. The noise ε_t is for simplicity assumed to be i.i.d. with mean zero, variance $\Omega > 0$ and finite fourth moments. [2]

The input will be assumed to be an ergodic strictly stationary sequence, which has a representation as $u_t = \sum_{j=0}^{\infty} K_u(j)v_{t-j}$, where v_t is i.i.d. with finite fourth moments. Furthermore $\|K_u(j)\| \leq C\rho^j, 0 < \rho < 1$ is assumed and for $f_u(\lambda) = \sum_{j=0}^{\infty} K_u(j)e^{i\lambda j}$ we assume, that $0 < c < f_u(\lambda) < C < \infty$ for $\lambda \in [0, 2\pi]$ and suitable constants $c, C \in \mathbb{R}$. These assumptions on the input are overly strong for some of the results, however, they are required in some cases. For weaker requirements on the inputs we refer to the original articles.

3. SUBSPACE ALGORITHMS

In the following we will give a short description of the algorithms considered in the paper: Fix two integers f and p. Denoting $Z_{s|t} = [y_s', u_s', y_{s+1}', u_{s+1}', \cdots, y_t', u_t']'$, $Y_{s|t} = [y_s', \cdots, y_t']'$, $U_{s|t} = [u_s', \cdots, u_t']'$, we obtain the following equation ($\bar{A} = A - KC, \bar{B} = B - KD$):

$$Y_{t|t+f} = \Gamma_f \left(Q_p Z_{t-p|t-1} + \bar{A}^p x_{t-p}\right) + H_f^d U_{t|t+f} + N_{t|t+f} \qquad (2)$$

where $N_{t|t+f}$ summarizes the effects of the future of the noise, which is orthogonal to the two other terms due to the assumptions on ε_t and the assumptions on the input implying open loop operation. Further $\Gamma_f = [C', A'C', \cdots, (A^f)'C']'$, $Q_p = [\bar{A}^{p-1}K, \bar{A}^{p-1}\bar{B}, \bar{A}^{p-2}K, \bar{A}^{p-2}\bar{B}, \cdots, K, \bar{B}]$. H_f^d is block lower triangular and has a block Toeplitz structure. The exact definition is not relevant for this paper and hence not stated in detail.

In order to simplify notation, let

$$\langle a_t, b_t \rangle = T^{-1} \sum_{t=p+1}^{T-f} a_t b_t'.$$

Finally let an additional superscript Π denote residuals from a regression onto $U_{t|t+f}$. Therefore e.g.

$$Y_{t|t+f}^{\Pi} = Y_{t|t+f} - \langle Y_{t|t+f}, U_{t|t+f} \rangle \langle U_{t|t+f}, U_{t|t+f} \rangle^{-1} U_{t|t+f}$$

This basic equation (2) is central to all subspace algorithms, which operate in three basic steps:

- Estimate $\Gamma_f Q_p$ by LS regression of $Y_{t|t+f}$ onto $Z_{t-p|t-1}$ and $U_{t|t+f}$. This results in
$$L_{f,p} = \langle Y_{t|t+f}^{\Pi}, Z_{t-p|t-1}^{\Pi} \rangle \langle Z_{t-p|t-1}^{\Pi}, Z_{t-p|t-1}^{\Pi} \rangle^{-1}$$
estimating $\Gamma_f Q_p$.
- $L_{f,p}$ will be of full rank in general, whereas $\Gamma_f Q_p$ is of rank n, where n denotes the system order. Thus approximate
$$\hat{W}_+ L_{f,p} \hat{W}_- = \hat{U}\hat{\Sigma}\hat{V}' = \hat{U}_n \hat{\Sigma}_n \hat{V}_n' + \hat{R}_n$$
to obtain estimates $\hat{\Gamma}_f = (\hat{W}_+)^{-1}\hat{U}_n\hat{\Sigma}_n$ and $\hat{Q}_p = \hat{V}_n'(\hat{W}_-)^{-1}$. Here $\hat{U}\hat{\Sigma}\hat{V}'$ denotes the SVD of $\hat{W}_+ L_{f,p} \hat{W}_-$, where $\hat{W}_+$ and $\hat{W}_-$ are weighting matrices chosen by the user. Thus e.g. $\hat{\Sigma}$ is the diagonal matrix containing the singular values ordered decreasing in size as diagonal entries. $\hat{U}_n \in \mathbb{R}^{fs\times n}, \hat{V}_n \in \mathbb{R}^{ps\times n}$ and $\hat{\Sigma}_n \in \mathbb{R}^{n\times n}$ correspond to the submatrices obtained by neglecting the singular values numbered $n+1$ and higher. Therefore in this step the order is specified.
- Given the estimates $\hat{H}_f^d$ (from the regression in the first step), $\hat{\Gamma}_f$ and $\hat{Q}_p$ from the second step the system matrices are estimated. In this step the various subspace methods differ fundamentally. A distinction in MOESP type of estimates, which use the estimate $\hat{\Gamma}_f$ and neglect the estimate $\hat{Q}_p$, and the Larimore approach (also called the state approach), where $\hat{Q}_p$ is used and $\hat{\Gamma}_f$ is neglected, has been made in the literature.

Note, that the algorithm requires a number of details to be provided by the user: The integers f and p, the order of the system n and the weighting matrices $\hat{W}_+$ and $\hat{W}_-$ have to be chosen. Various subspace procedures differ in the actual choices of these matrices: The most popular version of CCA uses $\hat{W}_+ = \langle Y_{t|t+f}^{\Pi}, Y_{t|t+f}^{\Pi} \rangle^{-1/2}$, $\hat{W}_- = \langle Z_{t-p|t-1}^{\Pi}, Z_{t-p|t-1}^{\Pi} \rangle^{1/2}$, where $X^{1/2}$ denotes the uniquely defined symmetric square root of a matrix [3]. The original MOESP procedure (Verhaegen, 1994) uses the same choice for $\hat{W}_-$ but $\hat{W}_+ = I$. Both procedures could be used with a different set of weighting matrices subject basically only to conditions of nonsingularity of the weights. The effects of different choices for the weighting matrices have been studied, cf. e.g. the discussion in (Bauer, 1998, Chapter 5).

This paper focuses on surveying the results on the effects of choosing the integer parameters f and p as well as the order of the system n.

4. INTEGER PARAMETERS

The choices of the integer parameters f, p and n have an influence on the asymptotic accuracy of the procedure. In the following we will provide the results found in the literature for the various choices. The presentation will first deal with the choice of f and p and afterwards order estimation topics are discussed.

4.1 Choosing f and p

4.1.1. The MOESP *case* It has been recognized by (Jansson and Wahlberg, 1998) that choosing f and p such that $\min\{f+1, p\}s \geq n$ does in the MOESP case not ensure consistent estimation of the system matrices. The reason for this is the projection onto the orthogonal complement of the space spanned by the

[2] This assumption covers all results in the literature. Typically weaker assumptions in a martingale difference framework are sufficient.

[3] The choice of the square root of the matrix is not important. Any choice leads to numerically identical estimates.

future of the inputs, which in some cases results in a loss of information in the estimated state, i.e. the limit of $\mathbb{E}x_t^{\Pi}(Z_{t-p|t-1}^{\Pi})'$ is not of full rank n in these cases. (Bauer and Jansson, 2000) show, that this happens only in 'rare cases' (i.e. in a nongeneric set in the set of all systems of order n fulfilling the standard assumptions equipped with a proper topology, see (Bauer and Jansson, 2000)). Nevertheless, the phenomenon is an important one since problems for the estimation are expected also 'close' to these points.

The following result adapted from (Chui, 1997) provides sufficient conditions on the integers for consistency of the estimates, which in our setting can be stated as:

Theorem 1. Let y_t be generated by a system of the form (1). Further let r denote the degree of the A-annihilator, i.e. the polynomial $a(z)$ of minimal degree such that $a(A) = 0$. Let $p \geq 3r$ and $f \geq o$, where o denotes the observability index of the system. Then the estimates obtained using the MOESP procedure are consistent.

It always holds that $r \leq n, o \leq n$ and therefore $p \geq 3n$ and $f \geq n$ are sufficient conditions.

(Chui, 1997) also gives minimal requirements for the inputs u_t: Persistent excitation of order $f+p+r+1$ is required rather than the stronger assumption put forward in this paper. The first part of this result gives minimal requirements on the indices f and p, which depend on the system order n and the configuration of the eigenvalues of the matrix A and therefore on properties of the true system, which are only known after the estimation (if they can be determined at all). Hence, the practical content of this part may seem to be minor. The second part of the result, however, is valuable: It ascertains, that for choosing f and p large enough, no problems with consistency can occur. The meaning of large enough still depends on the unknown order of the true system. One problem in this respect is that the order of the system usually is specified only after the choice of f and p have been taken.

With respect to the effects on the asymptotic accuracy of the choice of these parameters only a number of calculations in simple test cases exist, which do not show any consistent behaviour. One problem in this respect could also be the interdependence between the effects of the choice of the integer parameters f and p and the choice of the weighting matrices $\hat{W}_+$ and $\hat{W}_-$. A number of such calculations can be found in (Bauer and Jansson, 2000) and (Jansson, 1997) respectively.

4.1.2. The CCA *case* In the CCA case (Deistler *et al.*, 1995) note that for the procedure as outlined above for consistency it is necessary, that p tends to infinity in general. Corresponding to f only $(f+1)s \geq n$ is required for consistency. (Bauer *et al.*, 1999) prove, that for asymptotic normality p has to tend to infinity at a minimum rate: Let $\rho_0 = |\lambda_{max}(A-KC)|$. From the minimum-phase assumption it follows that $0 < \rho_0 < 1$. Then for asymptotic normality it is sufficient that $p \geq \lceil -d\log T/2\log\rho_0 \rceil$ for $d > 1$ arbitrary. Again, this depends on quantities of the true system, which are not known at the time of the specification of the integer parameters. This led to the suggestion to use $p = 2\hat{p}_{AIC}$, where $\hat{p}_{AIC}$ denotes the estimate of the order in an autoregressive approximation of y_t using lagged values $y_{t-j}, j = 1, \cdots, p$ and lagged values of the input $u_{t-j}, j = 0, \cdots, p$. This suggestion is based on Theorem 6.6.3 of (Hannan and Deistler, 1988):

Theorem 2. Let y_t be generated by a system of the form (1), where $A - KC$ is not nilpotent (i.e. $\rho_0 > 0$). Then the estimate of the order of an autoregressive approximation of y_t, denoted as $\hat{p}_{BIC}$ has the following asymptotic behaviour:

$$\liminf_{T\to\infty} -\frac{\hat{p}_{BIC} 2\log\rho_0}{\log T} = 1 \quad \text{almost sure.}$$

Here $\hat{p}_{BIC} \leq \hat{p}_{AIC}$ denotes the BIC estimate of the autoregressive order. Therefore even if the true system is not known, the lower bound on the integer p can be estimated consistently prior to identification of the system. In many cases, an autoregressive approximation would be estimated prior to going to state space models. In these cases the estimation of the approximation order would not result in any additional costs in terms of computations.

With respect to the specification of f a minimal requirement is that $(f+1)s \geq n$ in order to ensure consistency. In this case, there are no additional requirements for asymptotic normality to hold. The following result clarifies the asymptotic properties for CCA estimation in the case of no exogenous inputs:

Theorem 3. Let y_t be generated by a system of the form (1). $m = 0$ is assumed. Let the integers f and p be chosen according to the restriction $\min\{f,p\} \geq -d\log T/(2\log\rho_0), \max\{f,p\} = o((\log T)^a)$ for some $d > 1, a < \infty$. Assume, that the true order of the system is known. Then the estimates of the system obtained using CCA under the assumption of correctly specified order with the weights according to the original algorithm are asymptotically equivalent to estimates obtained from minimizing the one step ahead prediction error. Therefore in the case of Gaussian innovations the CCA estimates are efficient.

The asymptotic variance of the estimated system (which has been transformed to the corresponding echelon canonical form) under the assumption of correctly specified order using a fixed f, $p \geq -d\log T/(2\log\rho_0)$ and the weighting matrices $\hat{W}_- = \langle Y_{t-p|t-1} Y_{t-p|t-1}\rangle^{1/2}$ and an arbitrary weighting $\hat{W}_+$, which converges to $W_+ > 0$, is of the form

$$M_1 M_1' + M_2\left[\Gamma_z \otimes \{O_f' \mathcal{N}_f O_f\}\right] M_2' \quad (3)$$

Here $\mathcal{N}_f = \mathbb{E}N_{t|t+f}N'_{t|t+f}, \Gamma_z = \mathbb{E}Y_{-\infty|t}Y'_{-\infty|t}$ and

$$O_f = W_2\Gamma_f(\Gamma'_f W_2 \Gamma_f)^{-1}$$

for $W_2 = (W_+)'W_+$. Here M_1 and M_2 do not depend on f and W_+. This expression is minimized for the CCA choice of the weighting for any fixed f. This minimal expression for fixed f decreases monotonically with $f \to \infty$.

For proofs of the result see (Bauer, 2000; Bauer and Ljung, 2002). The result also has been shown to be valid for white noise inputs. The first part of the result ensures, that especially the CCA subspace method having $f, p \to \infty$ at a moderate rate achieves optimal accuracy and cannot be outperformed using prediction error methods, which would be the traditional choice. The second part of the result implies, that it is always optimal with respect to asymptotic accuracy to choose the original CCA weighting. Secondly it also shows, that in the CCA case increasing f does not lead to a decrease in asymptotic accuracy (although in finite samples there has to be a tradeoff between sample size and size of f). This is not true for other weighting schemes: Choosing $\hat{W}_+ = I$ as in N4SID might lead to cases, where the choice $f = n - 1$ is optimal and the asymptotic accuracy decreases with increasing f. Finally the theorem also provides numerically feasible expressions for the asymptotic variance.
As a warning sign we stress, that all results obtained above relate to the case, that the true system is finite dimensional and the system order moreover is known. If this is not the case, the asymptotic bias might be of bigger concern. The influence of the weighting $\hat{W}_+$ on the asymptotic bias is discussed e.g. in (Bauer, 1998, Chapter 5). To the best of my knowledge, the dependence of the bias term in case of underestimation of the order on the integers f and p has not been investigated so far.
Finally in this section also a few words for systems including exogenous inputs are in order: In the case, that the inputs are not white, consistency and asymptotic normality have been proved. The effects of the user choices, however, have only been investigated in a limited set of examples. These investigations led to the conclusion, that none of the procedures proposed up to now is a candidate for being asymptotically equivalent to prediction error estimation in this case. Due to the small number of systems, on which the various algorithms have been tried, no results on the relative efficiency has been found. The theoretical evaluations contained in (Bauer and Ljung, 2002) cannot be generalized to the arbitrary colored noise case. No result concerning the choice of f and p seems to exist in the literature.

4.2 Order estimation

Despite the fact, that order estimation is a central part in the identification procedure, there is surprisingly few literature on the question of order estimation for subspace procedures. This is documented by comments in early papers on subspace algorithms, which suggest to 'look for a gap' in a plot of the estimated singular values. This idea also underlies the order selection procedure implemented in the N4SID version in MATLAB. From my experience on real data, such a gap in most cases is not visible.
The question of order estimation in the context of subspace methods has been addressed seldomly in the literature: (Peternell, 1995) was the first to investigate this question. His results heavily influenced the work of (Bauer, 2001). These two authors base the order estimation on the information contained in the estimated singular values. (Larimore, 1996) suggests to use the usual AIC criterion function, where the estimated system using CCA replaces the ML estimate. Another alternative approach is based on hypothesis testing, which is contained in (Sorelius, 1999) and (Camba-Mendez and Kapetanios, 2001). These three approaches will be discussed in the following.

4.2.1. Estimating the order based on estimated singular values The approach of (Peternell, 1995) is to imitate the information criteria. Let $\hat{R}_n$ denote the approximation error in the SVD of $\hat{W}_+ L_{f,p} \hat{W}_-$, if the order is specified to be equal to n. Then a criterion function is defined as

$$NIC(n) = \|\hat{R}_n\|^2 + \frac{d(n)C_T}{T}$$

where $d(n) = 2s(n+m)$ denotes the number of parameters corresponding to a model of order n (not counting parameters for D). $C_T > 0, C_T/T \to 0, C_T \to \infty$ denotes a penalty term. The order is estimated as the minimizing argument of this criterion, i.e. $\hat{n} = \arg\min_{0 \le n \le H_T} NIC(n)$, where H_T is an upper bound. Although this criterion has the same form as AIC or BIC, there is no connection yet established. The norm used in the definition is decisive for the properties of the corresponding estimates: (Peternell, 1995) proposed the Frobenius norm, (Bauer, 2001) suggests to use the two norm, which is shown in a number of simulations to be more robust with respect to the choice of f and p. The following result clarifies the properties of the order estimation using NIC(n), which is adapted from Theorem 3 in (Bauer, 2001):

Theorem 4. Let the process y_t be generated by a system of the form (1), where the true order of the system is denoted by n. Let $\min\{(f+1), p\} \ge n$ and $\max\{f, p\} = o((\log T)^a)$ for some $a < \infty$. Let the weightings $\hat{W}_+$ and $\hat{W}_-$ be chosen such that for fixed f, p both converge a.s. to nonsingular matrices. If f or p tend to infinity with growing sample size T, then only the CCA choice or $\hat{W}_+ = I$ are admitted.

In this case the conditions $C_T > 0, C_T/T \to 0$,

$$C_T/(fp \log\log T) \to \infty$$

are sufficient for consistency of NIC using either the Frobenius norm or the two norm.

The choice of $f = p = 2\hat{p}_{AIC}$ fulfills all the requirements a.s. For fixed f and p the choice $C_T = \log T$ leads to consistent order estimates. The theorem is not sharp in the sense that the bounds on the penalty term are only sufficient but not necessary. Therefore much smaller penalty terms might be feasible. The properties of different choices have not been analyzed. The interdependence between the choices of f and p and the order estimation has only been partly analyzed using simulation studies, where the two norm has been found to be more robust with respect to the specification of f and p than the Frobenius norm.
(Camba-Mendez and Kapetanios, 2001) motivate an estimation procedure for the order by pointing out that the SVD with the CCA choice for the weighting matrices can be interpreted as the estimation of the canonical correlations, where the distribution of $\sum_{j=n+1}^{m} -\log(1-\hat{\sigma}_j^2)$ is asymptotically χ^2 distributed under the assumption of i.i.d. sampling of the random vectors $Y_{t-p|t-1}$ and $Y_{t|t+f}$. This leads to the suggestion to use $\|\hat{R}_n\|^2 = \sum_{j=n+1}^{m} -\log(1-\hat{\sigma}_j^2)$ – although not being a norm – in NIC(n). Using the power series expansion of $\log(1+x)$ proves, that the asymptotic properties of this estimation method are identical to the ones obtained for $\|\hat{R}_n\|_{Fr}^2$. The finite sample properties of course could be very different.

4.2.2. Using estimates of the innovation variance (Larimore, 1996) proposes to use the AIC criterion directly for order estimation: If the usual pseudo maximum likelihood aproach is used, the criterion function can be written as

$$AIC(n) = \log\det\Sigma(\theta_n) + \frac{2d(n)}{T}$$

where θ_n denotes the pseudo ML estimate of the system using the order n. The innovation variance estimate corresponding to this parameter vector is denoted as $\Sigma(\theta)$. The order is estimated as the minimizing value of this criterion function. (Larimore, 1996) suggests to use the CCA estimate in the place of θ_n. Note, that while (Larimore, 1996) refers to this procedure as AIC, this claim has not been proven. Even in the cases, where the subspace estimates are asymptotically equivalent to pseudo ML estimates as stated above, this does not imply, that using them in the AIC criterion results in identical properties. Also, the computational load of such a procedure is higher than in the singular value based criteria. As CCA also directly produces an estimate of the innovation variance, it is tempting to use this estimate in the criterion in place of $\Sigma(\theta_n)$. (Bauer, 2001) shows, that such a procedure has some theoretical drawbacks.

4.2.3. Testing for the order The last method, which has been developed in this area is the sequential hypothesis testing approach of (Sorelius, 1999). The method has been developed for the case of no exogenous inputs. The main idea in this procedure is to use the fact, that in this case the Hankel matrix $\mathcal{H}_{f,f} = [\gamma(i+j-1)]_{i,j=1,\cdots,f}, \gamma(j) = \mathbb{E}y_t y_{t-j}$ of covariances has rank n for a system of order n and f large enough. In the statistical literature, there have been a number of tests proposed, which can be used for testing, whether a given matrix is of full rank with the alternative of a rank one deficiency. These tests are then used sequentially, as is most easily demonstrated for the case $s = 1$, i.e. the single output case:

(1) Choose $f = 1$ and test for nonsingularity of $\hat{\mathcal{H}}_{f,f}$.
(2) If the null of singularity is rejected, set $f = f+1$ and go to step (4).
(3) If the null of singularity is not rejected, the order is equal to $f-1$. Stop.
(4) Test for nonsingularity of the matrix $\hat{\mathcal{H}}_{f,f}$. Go to step (2).

An explicit expression for each test is given in (Sorelius, 1999). There also an extension to the multivariate case is presented. Finally also two competing testing procedures are discussed. Both asymptotic distributions of the test statistic are derived and the methods are tested on simulation examples. (Camba-Mendez and Kapetanios, 2001) contains a very similar sequential testing procedure.
Up to now, there has been no attempt to compare the three different methods for the specification of the order of the system. The knowledge on the properties of the procedures is limited mostly to the asymptotic results and a limited amount of simulation experiments. For the NIC case, these experiments show, that especially the finite sample behaviour is not satisfactory, whereas for large sample sizes, the performance seems to be good.

5. CONCLUSIONS

The discussion given above shows, that for many subspace procedures, the asymptotic properties in the open loop case have been clarified for given user choices. Also the effects of the user choices on the asymptotic accuracy are quite well understood for some procedures. In particular for the CCA method in the case of no exogenous inputs or for white noise exogenous inputs, the asymptotic theory is quite complete. Based on this theory in principle the development of fully automatic procedures for identification of linear dynamic models is straightforward.
On the contrary, for the case of exogenous inputs present, which are not white noise, there remain many questions unanswered also with respect to the asymptotic properties: In this case there still is no knowledge about the comparison of the accuracy of the various

algorithms. Only negative results have been found: Contrary to the no-inputs case no procedure attains the optimal accuracy in all cases, i.e. is a candidate for asymptotic equivalence to prediction error estimation. A classification of situations, where one of the algorithms works superior has not been found.
With respect to order estimation, it is striking, that there has not been much work done. The consistency results on the one hand and the development of the sequential testing approach only seem to be preliminary work. The remaining papers in this session will show, that with respect to finite sample still lots of work is required to obtain procedures, which are suitable in practice.

REFERENCES

Bauer, D. (1998). Some Asymptotic Theory for the Estimation of Linear Systems Using Maximum Likelihood Methods or Subspace Algorithms. PhD thesis. TU Wien.

Bauer, D. (2000). Asymptotic efficiency of the CCA subspace method in the case of no exogenous inputs. Technical report, Linköping University, Sweden.

Bauer, D. (2001). Order estimation for subspace methods. *Automatica* **37**, 1561–1573.

Bauer, D. and L. Ljung (2002). Some facts about the choice of the weighting matrices in larimore type of subspace algorithms. *Automatica* **38**, 763–773.

Bauer, D. and M. Jansson (2000). Analysis of the asymptotic properties of the MOESP type of subspace algorithms. *Automatica* **36**(4), 497–509.

Bauer, D., M. Deistler and W. Scherrer (1999). Consistency and asymptotic normality of some subspace algorithms for systems without observed inputs. *Automatica* **35**, 1243–1254.

Camba-Mendez, G. and G. Kapetanios (2001). Testing the rank of the Hankel covariance matrix: A statistical Approach. *IEEE Transactions on Automatic Control* **46**, 331–336.

Chiuso, A. and G. Picci (2002). Asymptotic variances of subspace identification by data orthogonalization and model decoupling. Technical report. University of Padua, Italy.

Chui, N. (1997). Subspace Methods and Informative Experiments for System Identification. PhD thesis. Cambridge University.

Deistler, M., K. Peternell and W. Scherrer (1995). Consistency and relative efficiency of subspace methods. *Automatica* **31**, 1865–1875.

Hannan, E. J. and M. Deistler (1988). *The Statistical Theory of Linear Systems*. John Wiley. New York.

Jansson, M. (1997). On Subspace Methods in System Identification and Sensor Array Signal Processing. PhD thesis. KTH, Stockholm.

Jansson, M. and B. Wahlberg (1998). On consistency of subspace methods for system identification. *Automatica* **34**(12), 1507–1519.

Larimore, W. (1996). Optimal order selection and efficiency of canonical variate analysis system identification. In: *Proceedings of the 1996 IFAC World Congress*. pp. 151–156.

Larimore, W. E. (1983). System identification, reduced order filters and modeling via canonical variate analysis. In: *Proc. 1983 Amer. Control Conference 2.* (H. S. Rao and P. Dorato, Eds.). Piscataway, NJ. pp. 445–451.

Peternell, K. (1995). Identification of Linear Dynamic Systems by Subspace and Realization-Based Algorithms. PhD thesis. TU Wien.

Peternell, K., W. Scherrer and M. Deistler (1996). Statistical analysis of novel subspace identification methods. *Signal Processing* **52**, 161–177.

Sorelius, J. (1999). Subspace-Based Parameter Estimation Problems in Signal Processing. PhD thesis. Uppsala University.

Van Overschee, P. and B. DeMoor (1994). N4sid: Subspace algorithms for the identification of combined deterministic-stochastic systems. *Automatica* **30**, 75–93.

Verhaegen, M. (1994). Identification of the deterministic part of mimo state space models given in innovations form from input-output data. *Automatica* **30**(1), 61–74.

IFAC Publications
www.elsevier.com/locate/ifac

ASYMPTOTIC VARIANCES OF SUBSPACE IDENTIFICATION BY DATA ORTHOGONALIZATION AND MODEL DECOUPLING

ALESSANDRO CHIUSO *,[1] GIORGIO PICCI **,[1]

* *Dipartimento di Ingegneria dell'Informazione, University of Padova, 35131 Padua, Italy; Email:* chiuso@dei.unipd.it
** *Dipartimento di Ingegneria dell'Informazione, University of Padova, 35131 Padua, Italy; Email:* picci@dei.unipd.it

Abstract: Subspace estimation methods based on a block-decoupled parametrization and orthogonal decomposition of the input-output data into a "deterministic" and a "stochastic" component, have been analyzed in a series of articles by the authors of this paper. In this paper we compare the asymptotic variances of the parameter estimates of the method based on the block-decoupled parametrization with those of traditional joint-model based methods. We show that, provided a suitable choice of weights is used, they perform better than the traditional methods. The improvement is more substantial in the circumstance of nearly parallel regressors. *Copyright © 2003 IFAC*

Keywords:
Subspace Identification; Exogenous inputs; Numerical conditioning; Stochastic realization; Stochastic state-space identification.

1. STATE SPACE MODELS FOR SUBSPACE IDENTIFICATION

Assume that the observed input-output data of the unknown system, which we want to identify

$$\{u_t\},\ \{y_t\},\quad u_t \in \mathbb{R}^p,\ y_t \in \mathbb{R}^m,\quad t \geq t_0 \tag{1.1}$$

are generated for $t \geq t_0$ by a linear finite-dimensional stochastic system called a *stationary stochastic realization* of $\mathbf{y}$ with input $\mathbf{u}$. This can always be taken in "innovation form":

$$\begin{cases} \mathbf{x}(t+1) = A\mathbf{x}(t) + B\mathbf{u}(t) + K\mathbf{e}(t) \\ \mathbf{y}(t) = C\mathbf{x}(t) + D\mathbf{u}(t) + \mathbf{e}(t) \end{cases} \tag{1.2}$$

where A, B, K, C, D are constant matrices, $\{\mathbf{x}(t)\}$ is the state process of dimensions n, and $\{\mathbf{e}(t)\}$ is the white stationary one step prediction error of $\{\mathbf{y}(t)\}$, given the infinite past history of $\{\mathbf{y}(t)\}$ $\{\mathbf{u}(t)\}$ up to time $t-1$ and t respectively. We shall assume that *there is no feedback from* $\mathbf{y}$ *to* $\mathbf{u}$ which implies that $\{\mathbf{u}(t)\}$ and $\{\mathbf{e}(t)\}$ are completely uncorrelated.

1.1 Notations

In this subsection we shall quickly set the notations used in the paper. For a thorough discussion of the background material we refer the reader to the articles (Chiuso and Picci, 2003; Chiuso and Picci, 2002*a*; Chiuso and Picci, 2002*b*).

Boldface symbols will denote random quantities. For $-\infty \leq t_0 \leq t \leq T \leq +\infty$ define the Hilbert spaces of scalar second order random variables, respectively the (finite) "past" [2] and "future" of

[1] This work has been supported in part through the national project *Identification and control of industrial systems* funded by the Italian ministry for higher education (MIUR). Part of this work has been supported by the TMR-European Research Network *System Identification* (ERB FMRX CT98 0206).

[2] By convention the past spaces do not include the present.

the process $\mathbf{y}$,

$$\mathcal{Y}_{[t_0,t)} := \overline{\text{span}}\,\{\mathbf{y}_k(s);\ k=1,\ldots,p,\ t_0 \le s < t\}$$
$$\mathcal{Y}_{[t,T)} := \overline{\text{span}}\,\{\mathbf{y}_k(s);\ k=1,\ldots,m,\ t \le s \le T\}$$

where the bar denotes closure in mean square, i.e. in the metric defined by the inner product $\langle \xi, \eta \rangle := E\{\xi, \eta\}$, the operator E denoting mathematical expectation. Similarly we define $\mathcal{U}_{[t_0,t)}$ and $\mathcal{U}_{[t,T]}$ and let $\mathcal{P}_{[t_0,t)} := \mathcal{U}_{[t_0,t)} \vee \mathcal{Y}_{[t_0,t)}$ (the $\vee$ denotes closed vector sum).

When $t_0 = -\infty$ we shall use the shorthand $\mathcal{Y}_t^-$ for $\mathcal{Y}_{[-\infty,t)}$, and similarly for $\mathcal{U}_{[t_0,t)}$ and $\mathcal{P}_{[t_0,t)}$.

We shall also use $\mathcal{Y}_t$ in place of $\mathcal{Y}_{[t,t]}$ while we shall use $\mathcal{Y}$ for the space generated by the whole time history of $\mathbf{y}$.

We shall assume that the input process is "sufficiently rich", i.e. $\mathcal{U}_{[t_0,T]}$ admits the direct sum decomposition

$$\mathcal{U}_{[t_0,T]} = \mathcal{U}_{[t_0,t)} + \mathcal{U}_{[t,T]}, \qquad t_0 \le t < T \quad (1.3)$$

the $+$ sign denoting direct sum of subspaces. The symbol $\oplus$ will be reserved for *orthogonal* direct sum.

We shall use indexed capitals, e.g. Y_t, to denote finite "tail" matrices, constructed at each time t from sample sequences of $\mathbf{y}$, by letting

$$Y_t := \begin{bmatrix} y_t & y_{t+1} & \ldots & y_{t+N-1} \end{bmatrix} \quad (1.4)$$

Symbols like $Y_{[t,T]}$ will denote a Hankel matrix, e.g.

$$Y_{[t,T]} := \begin{bmatrix} Y_t^\top & \ldots & Y_T^\top \end{bmatrix}^\top$$

and $\mathcal{Y}^N_{[t,T]}$ the corresponding (finite-dimensional) rowspace. The same will hold for $\mathbf{x}$ and $\mathbf{u}$.

We shall assume *second-order ergodicity* of all random processes involved. Introducing the notation $E_N X_t Y_t^\top := \frac{1}{N}\sum_{k=0}^{N-1} x_{t+k} y_{t+k}^\top$ second-order ergodicity means that, $E_N Y_{t+\tau} Y_t^\top \to E\mathbf{y}(t+\tau)\mathbf{y}(t)^\top$ for $N \to \infty$. In this sense we shall say that $Y_t \to \mathbf{y}(t)$ or $\mathcal{Y}^N_{[t,T]} \to \mathcal{Y}_{[t,T]}$. Finally we shall write the solution of the Least-Squares problem $\min_{A \in \mathbb{R}^{n\times m}} \|Y - AX\|$ as $E_N[X \mid Y] := E_N[XY^T]\, E_N[YY^T]^{-1} Y$. Let $\mathbf{x}$ and $\mathbf{y}$ be a jointly second-order ergodic, stationary processes, whose samples form the tails X and Y. We shall say that $E_N[X \mid Y]$ converges to $E[\mathbf{x} \mid \mathbf{y}] := E[\mathbf{x}\mathbf{y}^T]\, E[\mathbf{y}\mathbf{y}^T]^{-1}\mathbf{y}$ in the same sense as above.

1.2 A decoupled canonical form

Let $\mathcal{U}^\perp$ be the orthogonal complement of $\mathcal{U}$ in $\mathcal{U} \vee \mathcal{Y}$. The stochastic process $\mathbf{y}$ admits the orthogonal decomposition $\mathbf{y}(t) = \mathbf{y}_d(t) + \mathbf{y}_s(t)$, where $\mathbf{y}_d$ and $\mathbf{y}_s$, defined by

$$\begin{cases} \mathbf{y}_d(t) := E[\mathbf{y}(t) \mid \mathcal{U}] \\ \mathbf{y}_s(t) := E[\mathbf{y}(t) \mid \mathcal{U}^\perp] \end{cases} \quad (1.5)$$

called the *deterministic* and the *stochastic component* of $\mathbf{y}$, are obviously uncorrelated at all times.

Under absence of feedback, $\mathbf{y}_d$ is actually a *causal* linear functional of the input process (see (Picci and Katayama, 1996)), i.e. $\mathbf{y}_d(t) = E[\mathbf{y}(t) \mid \mathcal{U}_{t+1}^-]$, $t \in \mathbb{Z}$. The deterministic and stochastic components of a process represented by a state space realization of the type (1.2) are represented by (generally non minimal) individual state space realizations, obtained by setting $\mathbf{u} = 0$ and $\mathbf{e} = 0$ in (1.2).

Let us consider the (minimal w.l.o.g.) innovation model (1.2) and write the input output relation $\mathbf{y} = F(z)\mathbf{u} + G(z)\mathbf{e}$ with "stochastic" and "deterministic" transfer functions $F(z) = D + C(zI - A)^{-1}B$ and $G(z) = I + C(zI - A)^{-1}K$. Unless there is some knowledge that "physical" disturbances enter into the system through the same input channels as $\mathbf{u}$, it may be unnatural to parametrize $G(z)$ an $F(z)$ with the same dynamic parameters (A, C), as is done in the ARMAX philosophy. This may result in non-minimal representations, i.e. in the use of too many parameters and therefore in high variance of the estimates. Instead the innovation representation of $\mathbf{y}$ can be obtained by combining in parallel a "deterministic" state-space model for $\mathbf{y}_d$

$$\mathbf{x}_d(t+1) = A_d\mathbf{x}_d(t) + B_d\mathbf{u}(t) \quad (1.6a)$$
$$\mathbf{y}_d(t) = C_d\mathbf{x}_d(t) + D\mathbf{u}(t). \quad (1.6b)$$

and the innovation representation of $\mathbf{y}_s$

$$\mathbf{x}_s(t+1) = A_s\mathbf{x}_s(t) + K_s\mathbf{e}_s(t) \quad (1.7a)$$
$$\mathbf{y}_s(t) = C_s\mathbf{x}_s(t) + \mathbf{e}_s(t) \quad (1.7b)$$

where $\mathbf{e}_s(t)$ is the one-step prediction error of the stochastic component $\mathbf{y}_s$ based on its own past, i.e. the innovation process of $\mathbf{y}_s$. Hence, the innovation model of the process $\mathbf{y}$ with inputs, can be described by a "canonical" block-diagonal realization, i.e., there exist a suitable choice of basis[3] such that

$$\mathbf{x}(t) = \begin{bmatrix} \mathbf{x}_d(t) \\ \mathbf{x}_s(t) \end{bmatrix} \quad (1.8)$$

where $\mathbf{x}_d(t)$ and $\mathbf{x}_s(t)$, are the *deterministic* and *stochastic* components of the state, mutually uncorrelated at all times. In the basis of (1.8), the transfer functions can then be parametrized as $F(z) = D + C_d(zI - A_d)^{-1}B_d$ and $G(z) = I + C_s(zI - A_s)^{-1}K_s$. In this case the deterministic and stochastic models are parametrized independently and more parsimoniously than in the ARMAX model[4]. In fact, since $(p+m)(n_s + n_d) + (n_s + n_d)^2 > pn_d + mn_s + n_s^2 + n_d^2$, canonical forms for the decoupled model have always

---

[3] Just consider any choice of basis in the state space, coherent with the orthogonal direct sum decomposition $\mathbf{X} = \mathbf{X}_d \oplus \mathbf{X}_s$, where $\mathbf{X}_d$ and $\mathbf{X}_s$ are the reachable subspaces for $\mathbf{u}$ and $\mathbf{e}$ from $t = -\infty$.

[4] The so-called *Box-Jenkins* models in PEM identification (Ljung, 1999), seem to serve a similar purpose.

less free parameters than the jointly parametrized model. Henceforth, cases where there is common dynamics will be regarded as "non-generic" and excluded from our analysis.

Our main concern in this paper will be to compare the performance of subspace algorithms for the identification of decoupled models of the type (1.7),(1.6), with the standard subspace algorithms applied to the identification of the joint model (1.2) which is usually considered in the literature. We shall in fact only compare performances of the identification of the deterministic subsystem.

2. THE COMPLEMENTARY MODEL

If the data were available from the infinite past, we could in principle compute the projections (1.5) exactly and split the identification problem into two distinct subproblems with orthogonal data, as described by (1.6) and (1.7). But infinite data of course never occur in practice and the problem of recovering in a statistically consistent way, the stationary model (1.8) from *finite* input-output data is not completely trivial.

The finite-data algorithm we consider is based on a preliminary decomposition of the finite (random) data into finite-interval "deterministic" ($\hat{\mathbf{y}}_d(t) := E\left[\mathbf{y}(t) \mid \mathcal{U}_{[t_0,T]}\right]$) and a "stochastic" ($\hat{\mathbf{y}}_s(t) := \mathbf{y}(t) - \hat{\mathbf{y}}_d(t)$) components.

It is easy to see that the deterministic component $\hat{\mathbf{y}}_d(t)$ admits the finite-interval ($t \in [t_0, T]$) realization:

$$\begin{cases} \hat{\mathbf{x}}_d(t+1) = A_d\hat{\mathbf{x}}_d(t) + B_d\mathbf{u}(t) \\ \hat{\mathbf{y}}_d(t) = C_d\hat{\mathbf{x}}_d(t) + D\mathbf{u}(t) \end{cases} \quad (2.9)$$

with initial condition $\hat{\mathbf{x}}_d(t_0) := E\left[\mathbf{x}_d(t_0) \mid \mathcal{U}_{[t_0,T]}\right]$. Let $\hat{\mathcal{X}}_t^d := \text{span}\{\mathbf{x}_d(t)\}$ be the state space of the deterministic realization (2.9). A technical condition which will be needed in the following is the (deterministic) "consistency condition"

$$\hat{\mathcal{X}}_t^d \cap \mathcal{U}_{[t,T]} = \{0\} \quad (2.10)$$

which is similar (although a bit stronger) than the consistency condition of (Jansson and Wahlberg, 1998). Compare formula (3.2) in (Chiuso and Picci, 2002*a*).

Unfortunately (an approximation of) $\hat{\mathbf{x}}_d(t)$ is not directly constructible from finite input-output data (see a discussion for the joint case in (Chiuso and Picci, 2002*a*)). Therefore one may either resort to a stationary approximation, regarding $t - t_0$ is large, which leads to biased estimates of the parameters, or consider a more complicated recursion which involves the stationary parameters A_d, C_d. The PI-MOESP method of (Verhaegen and Dewilde, 1992) falls in this second category. The method is based on computing the complementary predictor $\hat{Y}^c_{[t,T-1]} = \hat{E}\left[Y_{[t,T-1]} \mid U^{\perp}_{[t,T]}\right]$, where $U^{\perp}_{[t,T]} := U_{[t_0,t-1]} - E\left[U_{[t_0,t-1]} \mid U_{[t,T]}\right]$.

Using the weighted SVD

$$W_d\hat{Y}^c_{[t,T-1]} = \underbrace{[\hat{U}\ \ \tilde{U}]}_{U} \underbrace{\begin{bmatrix} \hat{S} & 0 \\ 0 & \tilde{S} \end{bmatrix}}_{S} \underbrace{\begin{bmatrix} \hat{V}^\top \\ \tilde{V}^\top \end{bmatrix}}_{V^\top} \quad (2.11)$$

and retaining only the "most significant" n_1 singular values one computes a low-rank approximation

$$\hat{Y}^c_{[t,T-1]} \simeq \hat{\Gamma}^c_d \hat{X}^c_t. \quad (2.12)$$

where

$$\begin{aligned} \hat{\Gamma}^c_d &= W_d^{-1}\hat{U}\hat{S}^{1/2} \\ \hat{X}^c_t &= \hat{S}^{1/2}\hat{V}^\top = \hat{S}^{-1/2}\hat{U}^\top W_d\hat{Y}^c_{[t,T]} \end{aligned} \quad (2.13)$$

The following proposition holds.

Proposition 1. Assume that the rank determination step in the factorization (2.12) is statistically consistent (i.e. asymptotically $n_1 = n_d$). Let $\mathcal{U}^{\perp}_{[t,T]}$ denote the orthogonal complement of $\mathcal{U}_{[t,T]}$ in $\mathcal{U}_{[t_0,T]}$. Then as $N \to \infty$, the tail matrix $\hat{X}^c_t$ tends to the deterministic *complementary state* vector

$$\hat{\mathbf{x}}^c_d(t) := E\left[\hat{\mathbf{x}}_d(t) \mid \mathcal{U}^{\perp}_{[t,T]}\right] \quad (2.14)$$

which satisfies the state equation

$$\begin{cases} \hat{\mathbf{x}}^c_d(t+1) = A_d\hat{\mathbf{x}}^c_d(t) + B_d(t)\bar{\mathbf{v}}(t) \\ \hat{\mathbf{y}}^c_d(t) = C_d\hat{\mathbf{x}}^c_d(t) \end{cases} \quad (2.15)$$

where $\hat{\mathbf{y}}^c_d(t) := \mathbf{y}(t) - E\left[\mathbf{y}(t) \mid \mathcal{U}_{[t,T]}\right]$, $\bar{\mathbf{v}}(t) = \mathbf{u}(t) - E\left[\mathbf{u}(t) \mid \mathcal{U}_{[t,T]}\right]$ is the backward innovation process of $\mathbf{u}(t)$, $B_d(t) := A_dK_d(t) + B_d$, with $K_d(t) := E\left[\hat{\mathbf{x}}_d(t)\bar{\mathbf{v}}(t)^\top\right]\left(\left[\bar{\mathbf{v}}(t)\bar{\mathbf{v}}(t)^\top\right]\right)^{-1}$ and the initial condition is $\hat{\mathbf{x}}^c_d(t_0) = 0$.

Let $\Lambda_d := E\{\hat{\mathbf{y}}^c_{d\,[t,T-1]}(\hat{\mathbf{y}}^c_{d\,[t,T-1]})^\top\}$ and consider the reduced (full rank) factorization,

$$W_d\Lambda_dW_d^\top = US^2U^\top, \quad (2.16)$$

where $S = \text{diag}\{\sigma_1, \ldots, \sigma_{n_d}\}$, $U^\top U = I$, and σ_{n_d} is the smallest nonzero eigenvalue of $W_d\Lambda_dW_d^\top$. Then, for $N \to \infty$, $\hat{S} \to S$ and $\hat{X}^c_t$ "tends to "

$$\begin{aligned} \hat{\mathbf{x}}^c_d(t) &= S^{-1/2}U^\top W_d\hat{\mathbf{y}}^c_{d\,[t,T-1]} \\ &= (\Gamma^c_d)^{-L}\hat{\mathbf{y}}^c_{d\,[t,T-1]} \end{aligned} \quad (2.17)$$

where Γ^c_d is the observability matrix of the complementary realization (2.15)

Proof. See (Chiuso and Picci, 2002*b*).
□

Note that the covariance matrix of the complementary state $\hat{\mathbf{x}}^c_d(t)$ is the *conditional* covariance of $\hat{\mathbf{x}}_d(t)$ given the future inputs $\mathcal{U}_{[t,T]}$, namely $E\left[\hat{\mathbf{x}}_d^c(t)\,(\hat{\mathbf{x}}_d^c(t))^\top\right] = \Sigma_{\hat{\mathbf{x}}_d\,\hat{\mathbf{x}}_d|\mathbf{u}^+}$ etc. By (2.10) $\Sigma_{\hat{\mathbf{x}}_d\,\hat{\mathbf{x}}_d|\mathbf{u}^+}$ is nonsingular, so the matrices A_d and C_d are uniquely determined by the complementary state by the formulas

$$\left[\,A_d\ C_d\,\right] = \left[\,\Sigma_{\hat{\mathbf{x}}_{d,1}\,\hat{\mathbf{x}}_d|\mathbf{u}^+}\Sigma_{\hat{\mathbf{y}}_d\,\hat{\mathbf{x}}_d|\mathbf{u}^+}\,\right]\Sigma^{-1}_{\hat{\mathbf{x}}_d\,\hat{\mathbf{x}}_d|\mathbf{u}^+}. \quad (2.18)$$

An estimate of (A_d, C_d) can be computed using sample estimates of these covariances or shift invariance of $\hat{\Gamma}_d^c$ as is done in the MOESP class of algorithms. These two procedures are actually equivalent if $W_d = I$(see (?)) and therefore the conditioning of the computation of $(A_d,\, C_d)$ by the weighted PI-MOESP procedure described above, is actually governed by the condition number of the *conditional covariance matrix* $\Sigma_{\hat{\mathbf{x}}_d\,\hat{\mathbf{x}}_d|\mathbf{u}^+}$ (Chiuso and Picci, 2002*a*).

3. ASYMPTOTIC VARIANCE FORMULAS

Naturally the accuracy of identification of the estimated transfer function $\hat{F}(z)$, is the main criterion for comparing different algorithms. This can be easily obtained by standard linearization techniques (see (Jansson, 2000) and the appendix) once the asymptotic variance of the parameter (system matrices) estimates is available. We shall start with expressions for the variance of the estimates $\hat{A}_d$, $\hat{C}_d$. These will be similar, though not identical, to those holding for a jointly parametrized model derived in (Chiuso and Picci, 2003). A derivation and a discussion of the "weighted PI-MOESP type estimates" is given in (Chiuso and Picci, 2002*b*, Appendix A).
Later we shall provide formulas to compare the accuracy obtained using different kinds of parametrization.

Theorem 3.1. Let $\hat{A}_d$, $\hat{C}_d$ be the weighted PI-MOESP type estimates obtained from the sample version of (2.18) , expressed in the basis of the complementary state process $\hat{\mathbf{x}}_d^c$ defined by (2.17) and let $M_d := [(0_{n_d\times m}\ \ \Gamma_d^{-L}) - A_d(\Gamma_d^{-L}\ \ 0_{n_d\times m})]$, and $R_d := [(I_m\ \ 0) - C_d\Gamma_d^{-L}]$ Assume that $\{\mathbf{y}_s(t)\}$ (see (1.5)) is stationary, independent of $\mathbf{u}$ and has finite fourth order moments. Then the vectorized parameter estimates $[\mathrm{vec}(\hat{A}_d)^\top\ \mathrm{vec}(\hat{C}_d)^\top]^\top$ are asymptotically Gaussian with asymptotic covariance matrices

$$\mathrm{AsVar}\left(\sqrt{N}\mathrm{vec}(\hat{A}_d)\right) = \bar{M}_d\Sigma\bar{M}_d^\top \quad (3.19)$$

$$\mathrm{AsVar}\left(\sqrt{N}\mathrm{vec}(\hat{C}_d)\right) = \bar{R}_d\Sigma\bar{R}_d^\top \quad (3.20)$$

$$\mathrm{AsCov}\left(\sqrt{N}\mathrm{vec}(\hat{A}_d), \sqrt{N}\mathrm{vec}(\hat{C}_d)\right) = \bar{M}_d\Sigma\bar{R}_d^\top \quad (3.21)$$

where $\Sigma := \sum_{\tau=-\infty}^{+\infty}\Sigma_{\hat{\mathbf{x}}_d^c\hat{\mathbf{x}}_d^c}(\tau)\otimes\Sigma_{\bar{\mathbf{y}}_s^+\bar{\mathbf{y}}_s^+}(\tau)$, $\bar{M}_d := \Sigma^{-1}_{\hat{\mathbf{x}}_d^c\hat{\mathbf{x}}_d^c}\otimes M_d$, $\bar{R}_d := \Sigma^{-1}_{\hat{\mathbf{x}}_d^c\hat{\mathbf{x}}_d^c}\otimes R_d$ and

$$\begin{aligned}\Sigma_{\hat{\mathbf{x}}_d^c\hat{\mathbf{x}}_d^c}(\tau) &:= E\{\sigma^\tau\hat{\mathbf{x}}_d^c(t)\hat{\mathbf{x}}_d^c(t)^\top\}\\ \Sigma_{\bar{\mathbf{y}}_s^+\bar{\mathbf{y}}_s^+}(\tau) &= E\{\bar{\mathbf{y}}_s^+(t+\tau)\,[\,\bar{\mathbf{y}}_s^+(t)\,]^\top\}.\end{aligned} \quad (3.22)$$

The operator σ^τ is the τ-steps ahead stationary shift of the processes $\mathbf{y}, \mathbf{u}$, whereby

$$\sigma^\tau\hat{\mathbf{x}}_d^c(t) = E\left[\mathbf{x}(t+\tau)\mid\mathcal{U}^\perp_{[t+\tau,\,T+\tau]}\right] \quad (3.23)$$

and $\bar{\mathbf{y}}_s^+(t+\tau)^\top := [\,\mathbf{y}_s(t+\tau)^\top\ \ldots\mathbf{y}_s(T+\tau)^\top]^\top = \sigma^\tau[\,\bar{\mathbf{y}}_s^+(t)\,]^\top$. The orthogonal complement is defined by the identity $\mathcal{U}^\perp_{[t+\tau,\,T+\tau]}\oplus\mathcal{U}_{[t+\tau,\,T+\tau]} := \mathcal{U}_{[t_0+\tau,\,T+\tau]}$.

A few remarks are in order.

(1) There is no need to assume that $\{\mathbf{y}_s(t)\}$ is finite-dimensional.
(2) Since $\Sigma_{\hat{\mathbf{x}}_d^c\hat{\mathbf{x}}_d^c} = \Sigma_{\hat{\mathbf{x}}_d\,\hat{\mathbf{x}}_d|\mathbf{u}^+}$ the asymptotic covariances are roughly "proportional" to the inverse $\Sigma^{-1}_{\hat{\mathbf{x}}_d\,\hat{\mathbf{x}}_d|\mathbf{u}^+}$ and hence tend to be very large for ill-conditioned problems.
(3) The various terms in formulas (3.19), (3.20) (3.21), can be computed in terms of true system parameters and input spectrum.

Comparison with the joint approach In this section we shall outline a procedure which, for a given choice of the weighting matrix, W_d, in the SVD determining the complementary state $\hat{\mathbf{x}}_d^c$ (compare (2.17)), allows to compare the asymptotic variances of the parameter estimates with those obtained by the joint method. In order to perform this comparison we need to express the estimated matrices in a common basis. As we ar not interested in the "stochastic part" (usually referred to as the " noise model") we shall leave unspecified its basis.
Let (A_d, C_d) be the deterministic parameters expressed in the asymptotic basis given by (2.17) and let (A, C) correspond to a chosen weighted balanced canonical basis, $\hat{\mathbf{x}}^c(t)$, of the joint true model (see (Chiuso and Picci, 2002*a*) or (Chiuso and Picci, 2003) formula 4.21). There exists a change of basis T which brings (A, C) into the block diagonal form

$$\begin{bmatrix}A_d & 0\\ 0 & A_s\end{bmatrix}, \quad [\,C_d\ \ C_s\,].$$

If we apply T to the estimates $(\hat{A}, \hat{C})$ we obtain

$$T^{-1}\hat{A}T - \begin{bmatrix}A_d & 0\\ 0 & A_s\end{bmatrix} = \begin{bmatrix}\tilde{A}_{dJ} & \tilde{A}_{ds}\\ \tilde{A}_{sd} & \tilde{A}_{sJ}\end{bmatrix} \quad (3.24)$$

where the off diagonal terms $\tilde{A}_{ds}$, $\tilde{A}_{sd}$ are nonzero for finite N. Likewise, let $\hat{C}T := [C_{dJ}\ \ C_{sJ}]$. The off-diagonal term $\tilde{A}_{sd}$ induces an annoying "coupling" error in the estimate of the transfer function which explains the erratic behaviour discussed in (Chiuso and Picci, 2003). The asymptotic variance of this term is therefore also of interest.

For $t_0 \rightarrow -\infty$ the conditional state covariance $\Sigma_{\hat{\mathbf{x}}^c\hat{\mathbf{x}}^c}$ expressed in the decoupling basis (3.24), becomes block-diagonal, and equal to $\text{diag}\{\Sigma_{\hat{\mathbf{x}}_d^c\hat{\mathbf{x}}_d^c}, \Sigma_{\hat{\mathbf{x}}_s^c\hat{\mathbf{x}}_s^c}\}$. Hence, assuming $t - t_0$ is large, from the general variance expressions of Theorem 4.1 in (Chiuso and Picci, 2002*a*), the following approximate formulas:

$$\text{AsVar}\{\sqrt{N}\tilde{A}_{dJ}\} = \bar{M}_{dJ}\Sigma_d\bar{M}_{dJ}^\top \qquad (3.25)$$

$$\text{AsVar}\{\sqrt{N}\tilde{A}_{sd}\} = \bar{M}_{sd}\Sigma_d\bar{M}_{sd}^\top \qquad (3.26)$$

$$\text{AsVar}\{\sqrt{N}\tilde{C}_{dJ}\} = \bar{R}_{dJ}\Sigma_d\bar{R}_{dJ}^\top \qquad (3.27)$$

where $\Sigma_d = \sum_{\tau=-\nu}^{\nu} \Sigma_{\hat{\mathbf{x}}_d^c\hat{\mathbf{x}}_d^c}(\tau) \otimes \Sigma_{\bar{\mathbf{e}}^+\bar{\mathbf{e}}^+}(\tau)$, $\bar{M}_{dJ} := \Sigma_{\hat{\mathbf{x}}_d^c\hat{\mathbf{x}}_d^c}^{-1} \otimes [(0 \quad (\Gamma^{-L})_d H_s) - A_d((\Gamma^{-L})_d H_s \quad 0)]$, $\bar{M}_{sd} := \Sigma_{\hat{\mathbf{x}}_d^c\hat{\mathbf{x}}_d^c}^{-1} \otimes [(2K_s \quad (\Gamma^{-L})_s H_s) - A_s((\Gamma^{-L})_s H_s \quad 0)]$ and $\bar{R}_{dJ} := \Sigma_{\hat{\mathbf{x}}_d^c\hat{\mathbf{x}}_d^c}^{-1} \otimes \left[(I_m \quad 0) - C_d\left((\Gamma^{-L})_d H_s\right)\right]$ in which H_s is the block Toeplitz matrix of the stochastic subsystem and $(\Gamma^{-L})_d$ is defined by

$$[\Gamma_d \quad \Gamma_s]^{-L} = \begin{bmatrix} (\Gamma^{-L})_d \\ (\Gamma^{-L})_s \end{bmatrix}$$

and Γ_s and Γ_d are the extended obervability matrices attached to (A_s, C_s) and (A_d, C_d) respectively. Recall that $\Sigma_{\bar{\mathbf{e}}^+\bar{\mathbf{e}}^+}(\tau) = E\{\bar{\mathbf{e}}^+(t+\tau)\bar{\mathbf{e}}^+(t)^\top\}$ where $\bar{\mathbf{e}}^+(t+\tau) := [\mathbf{e}(t+\tau)^\top \ldots \mathbf{e}(T+\tau)^\top]^\top = \sigma^\tau[\bar{\mathbf{e}}^+(t)]^\top$, so that, the underlying finite dimensional model for $\mathbf{y}_s$, yields

$$\bar{\mathbf{y}}_s^+(t) = \bar{\Gamma}_s\mathbf{x}_s(t) + \bar{H}_s\bar{\mathbf{e}}(t)^+,$$

where $\bar{\Gamma}_s$ and $\bar{H}_s$ are defined as Γ_s and H_s with one block ro appended. From these expressions, recalling that[5] $(\Gamma^{-L})_d\Gamma_s = 0$, some manipulations yield

$$\begin{aligned}&[(0 \quad (\Gamma^{-L})_d) - A_d((\Gamma^{-L})_d \quad 0)]\,\bar{H}_s\bar{\mathbf{e}}^+(t) = \\ &= [(0 \quad (\Gamma^{-L})_d) - A_d((\Gamma^{-L})_d \quad 0)]\,\bar{\mathbf{y}}_s^+(t)\end{aligned} \qquad (3.28)$$

which implies that in the formulas for the asymptotic variance we may substitute $\bar{\mathbf{e}}^+$ with $\bar{\mathbf{y}}_s^+$, provided, say, $\bar{M}_{dJ}$ is redefined to be

$$\bar{M}_d := \Sigma_{\hat{\mathbf{x}}_d^c\hat{\mathbf{x}}_d^c}^{-1} \otimes [(0 \quad (\Gamma^{-L})_d) - A_d((\Gamma^{-L})_d \quad 0)]$$

Proposition 2. Assume that the same future horizon $\nu = T - t$, $(m\nu > n)$, is chosen in computing both the joint parametrization estimates $(\hat{A}, \hat{C})$, and the deterministic subsystem estimates $(\hat{A}_d, \hat{C}_d)$, then a weighting matrix W_d can be chosen in such a way that

$$\Gamma_d^{-L}\Gamma_s = 0 \qquad (3.29)$$

and with this choice the expressions (3.19), (3.20) coincide with (3.25), (3.27), respectively.

Proof. See (Chiuso and Picci, 2002*b*).
□

[5] No matter how the left-inverse Γ^{-L} is constructed

The proposition shows that the decoupled parametrization methods, with a suitable choice of the weighting matrix, perform no worse than the joint parametrization as far as estimation of the single parameters (A_d, C_d) is concerned. However, and this is really what matters, the latter do perform worse regarding the estimation of the transfer function, since, as it will be shown in the next Section the term $\tilde{A}_{sd}$ introduces an additional error in the transfer function estimate which is not introduced by the decoupled parametrization methods.

4. ON THE ESTIMATION OF (B, D)

In this section we shall study a version of the linear regression procedure to estimate (B, D) described in (Chiuso and Picci, 2002*a*) which estimates the parameters (B_d, D) of the "deterministic" subsystem. In order to facilitate the analysis, we shall assume that (A, C) have been estimated quite accurately in a previous step[6], so that these parameters can be considered as known quantities. Let us denote sample covariance matrices, e.g.

$$\hat{\Sigma}_{\mathbf{u}^+\mathbf{u}^+} := E_N\{U_{[t,T]}U_{[t,T]}^\top\}$$

where $\mathbf{u}^+$ is the vector of future inputs from time t to T. Other similar notations will be used without further comments.

It can be shown that (see (Chiuso and Picci, 2002*b*)), introducing $\hat{\Phi}_s = \hat{\Sigma}_{\mathbf{x}^s\mathbf{u}^+}\hat{\Sigma}_{\mathbf{u}^+\mathbf{u}^+}^{-1}$, $\hat{\Phi}_d = \hat{\Sigma}_{\mathbf{x}^d\mathbf{u}^+}\hat{\Sigma}_{\mathbf{u}^+\mathbf{u}^+}^{-1}$, $\hat{\Phi}_y = \hat{\Sigma}_{\mathbf{y}\mathbf{u}^+}\hat{\Sigma}_{\mathbf{u}^+\mathbf{u}^+}^{-1}$, there exist a suitable matrix L_d such that there holds

$$\begin{aligned}\text{vec}\left(\hat{\Phi}_y\right) &= [I_{km} \otimes (\bar{\Gamma}_d \quad \bar{\Gamma}_s)]\,\text{vec}\begin{pmatrix}\hat{\Phi}_d \\ \hat{\Phi}_s\end{pmatrix} + \\ &+ L_d\text{vec}\begin{pmatrix}B_d \\ D\end{pmatrix} + \left(\hat{\Sigma}_{\mathbf{u}^+\mathbf{u}^+}^{-1} \otimes \bar{H}_s\right)\text{vec}\left(\hat{\Sigma}_{\bar{\mathbf{e}}^+\mathbf{u}^+}\right)\end{aligned} \qquad (4.30)$$

The additive term $\text{vec}\left(\hat{\Sigma}_{\bar{\mathbf{e}}^+\mathbf{u}^+}\right)$ is regarded as a random perturbation vector with asymptotic variance W_0 (see (Hannan and Deistler, 1988, Chap. 3), and (Chiuso and Picci, 2003) for an explicit expression).

Proposition 3. Assume (A, C) are known. Introduce[7] $\Delta := W_0^{-\dagger/2}[I_{km} \otimes (\bar{\Gamma}_d \quad \bar{\Gamma}_s)]$, $\Delta^\perp = I - \Delta(\Delta^\top\Delta)^{-\dagger}\Delta^\top$ and define $\Theta_d := L_d^\top(W_0^{-\dagger/2})^\top\Delta^\perp$. Then for $N \rightarrow \infty$, the minimum variance unbiased (Markov) estimates of the parameters B_d, D (in vectorized form) from the regression equation (4.30) is

$$\text{vec}\begin{pmatrix}\hat{B}_d \\ \hat{D}_d\end{pmatrix} = \left(\Theta_d\Theta_d^\top\right)^{-\dagger}\Theta_d W_0^{-\dagger/2}\text{vec}\left(\hat{\Phi}_y\right) \qquad (4.31)$$

[6] We refer the reader to (Chiuso and Picci, 2003) for an analysis which accounts for uncertainties in (A, C).

[7] $-\dagger$ denotes Moore-Penrose pseudoinverse.

The asymptotic variance of the estimates is given by

$$\mathrm{AsVar}\left\{\begin{bmatrix}\sqrt{N}\mathrm{vec}(\hat{B}_d)\\ \sqrt{N}\mathrm{vec}(\hat{D}_d)\end{bmatrix}\right\} = \left(\Theta_d\Theta_d^\top\right)^{-\dagger}. \quad (4.32)$$

We now want to compare this estimate with that obtained by using the full parametrization of the joint model (1.2). In place of B_d the *full* $n \times p$ B matrix are estimated. The asymptotic variance of the joint estimate of $[B_d^\top \; B_s^\top]^\top, D$ is given by the same formula (4.32) with $[L_d \; L_s]$ in place of L_d (see (Chiuso and Picci, 2002*b*)). Then, by a standard block-inversion formula, it follows that

$$\mathrm{Var}\left\{\mathrm{vec}\begin{pmatrix}\hat{B}_d\\ \hat{D}\end{pmatrix}\right\} \leq \mathrm{Var}\left\{\mathrm{vec}\begin{pmatrix}\hat{B}_{dJ}\\ \hat{D}_J\end{pmatrix}\right\}$$

(a well-known fact in multiple linear regression) which shows that, assuming that in both settings the same future data horizon, ν, is used, *identification of* (B, D) *from orthogonalized data gives better estimates than the joint regression, based on a jointly parametrized model.*

Hence, not only the orthogonal decomposition approach guarantees that $\hat{B}_s = 0$ while in general this is not true using the full parametrization, but it also provides estimates of the matrices B_d, D which have lower error variance.

Transfer function analysis

Let us define the error matrices in the basis (3.24) conveniently partitined as

$$\tilde{A} = \begin{pmatrix}\tilde{A}_d & \tilde{A}_{ds}\\ \tilde{A}_{sd} & \tilde{A}_s\end{pmatrix} \; \tilde{B} := \begin{pmatrix}\tilde{B}_d\\ \tilde{B}_{sd}\end{pmatrix} \qquad (4.33)$$
$$\tilde{C} := \begin{pmatrix}\tilde{C}_d & \tilde{C}_s\end{pmatrix}$$

Let $\hat{F}(z) := \hat{C}\left(zI - \hat{A}\right)^{-1}\hat{B} + \hat{D}$. Introducing $F_{d1} := ((zI - A_d)^{-1}B_d)^\top \otimes I$, $F_{d2} := I \otimes C_d(zI - A_d)^{-1}$, $F_{s2} := I \otimes C_s(zI - A_s)^{-1}$, than an easy calculation (see (Jansson, 2000)) leads to the linearization

$$\begin{aligned}&\mathrm{vec}\left(\hat{F}(z) - F(z)\right) \simeq\\ &\simeq [F_{d1}F_{d2} \quad F_{d2} \quad F_{d1} \quad I \quad F_{d1}F_{s2} \quad F_{s2}] \cdot \quad (4.34)\\ &\cdot\mathrm{vec}\left([\tilde{A}_d \quad \tilde{B}_d \quad \tilde{C}_d \quad \tilde{D} \quad \tilde{A}_{sd} \quad \tilde{B}_{sd}]\right).\end{aligned}$$

(up to second order terms which are negligible in the computation of the asymptotic covariance, see (Ferguson, 1996, Thm. 7 p. 45)).

In particular as one may recognize some "stochastic" dynamics (from F_{s2} and $F_{d1}F_{s2}$) is attributed to the system due to the fact that $\tilde{A}_{sd}$, $\tilde{B}_{sd}$ affect the estimates (cfr. the results in (Chiuso and Picci, 1999). We stress instead that these term are zero by construction in the orthogonal decomposition approach.

5. CONCLUSIONS

In this paper we have presented a comparison of asymptotic variances between two classes of subspace identification methods. The first is based on a block-decoupled parametrization of the model combined with an orthogonal decomposition of the input-output data, while the other methods are based on the usual "joint model parametrization". The asymptotic variances of the (A, C) and (B, D) parameter estimates obtained by standard subspace algorithms have been derived. Since the asymptotic variance of the (A, C) estimates depends on the choice of a certain weighting matrix, no general comparison is possible; there is however enough evidence to conclude that subspace estimation based on a preliminary orthogonalization of the input-output data and block-diagonal model parametrization provides generically (i.e. for models with decoupled deterministic-stochastic dynamics), better estimates of the system parameters and transfer function than the standard "joint" input-output methods.

REFERENCES

Chiuso, A. and G. Picci (1999). Subspace identification by orthogonal decomposition. In: *Proc. 14th IFAC World Congress.* Vol. I. pp. 241–246.

Chiuso, A. and G. Picci (2002*a*). On the ill-conditioning of subspace identification with inputs. Technical report. *submitted to Automatica*, available from the authors.

Chiuso, A. and G. Picci (2002*b*). Subspace identification by data orthogonalization and model decoupling. Technical report. *submitted to Automatica*, available from the authors.

Chiuso, A. and G. Picci (2003). Asymptotic variances of subspace estimates. *To Appear, Journal of Econometrics.*

Ferguson, T.S. (1996). *A Course in Large Sample Theory.* Chapman and Hall.

Hannan, E.J. and M. Deistler (1988). *The Statistical Theory of Linear Systems.* Wiley.

Jansson, M. (2000). Asymptotic variance analysis of subspace identification methods. In: *Proceedings of SYSID2000.* S. Barbara Ca.

Jansson, M. and B. Wahlberg (1998). On consistency of subspace methods for system identification. *Autiomatica* **34**, 1507–1519.

Ljung, L. (1999). *System Identification : Theory for the User.* 2nd ed.. Prentice Hall.

Picci, G. and T. Katayama (1996). Stochastic realization with exogenous inputs and "subspace methods" identification. *Signal Processing* **52**, 145–160.

Verhaegen, M. and P. Dewilde (1992). Subspace model identification, part 1. the output-error state-space model identification class of algorithms; part 2. analysis of the elementary output-error state-space model identification algorithm. *Int. J. Control* **56**, 1187–1210 & 1211–1241.

IFAC
Publications
www.elsevier.com/locate/ifac

A FINITE SAMPLE COMPARISON OF AUTOMATIC MODEL SELECTION METHODS

Dietmar Bauer *,[1]

* *Institute for Econometrics, Operations Research and System Theory*
TU Wien
Argentinierstr. 8, A- 1040 Wien

Stijn de Waele **,[2]

** *Systems, Signals and Control Group*
Delft University of Technology
The Netherlands

Abstract: Three automatic model selection procedures are compared: The first one is the pem subroutine implemented in the MATLAB toolbox system identification. The second procedure is based on the CCA subspace procedure. The third procedure finally is the ARMAsel procedure. The three procedures have been studied using a simulation experiment. The accuracy of the subspace algorithms depends heavily on the choice of the integers f and p. Simple strategies for choosing f and p can yield very inaccurate models. For one simulation example, it is shown that a more complex strategy can yield a more accurate result. The ARMAsel algorithm provided an accurate model for all processes that have been considered. *Copyright © 2003 IFAC*

Keywords: subspace algorithms, automatic model selection, linear dynamical systems

1. INTRODUCTION

It has been one of the driving forces in research on the estimation of linear dynamical systems to obtain procedures, which supply satisfactory models given only the data in an automatic fashion. The typical process of finding a model for a data set at hand after some preliminary operations such as the detection of missing data and prefiltering have been performed, is to specify a model structure and estimate the model in the model structure, that provides the most accurate model of the process at hand. The specification of the model structure is in general a difficult task and one usually has to resort to the estimation of many models and to compare these models with respect to their performance. It is the model specification step, which is crucial in the automation of the estimation procedures.

One main concept in this respect are the information criteria proposed by (Akaike, 1975). In the state space framework, the model structure is fully specified by the order of the system. AIC or BIC could be used to specify the order of a state space system based on

[1] Support by the Austrian FWF under project number P-14438INF is gratefully acknowledged. The authors acknowledge the financial support in part by the European Commission through the program Training and Mobility of Researchers - Research Networks and through project System Identification (FMRX CT98 0206) and acknowledge contacts with the participants in the European Research Network System Identification (ERNSI).

[2] Support by the Dutch STW under project number DTN 4758 is gratefully acknowledged.

maximum likelihood estimation conditional on the order for a range of different orders. A detailed description of the asymptotic properties of fully automated procedures based on these paradigms can be found in (Hannan and Deistler, 1988). While BIC uses a too heavy penalty for the model order, AIC results in overly large order estimates. The order selection criterion CIC provides a close to optimal trade-off of underfit and overfit (Broersen, 2000).

In this paper three procedures for automatic estimation of a model providing only the data will be compared with respect to accuracy in finite sample as well as with respect to the number of computations involved. The three procedures are the ARMAsel algorithm (Broersen, 2002), a subspace procedure based algorithm, which uses the integer choices proposed in (Bauer *et al.*, 1999), and the procedure pem in the MATLAB toolbox system identification, which calculates the system using prediction error minimization starting from an initial guess obtained using the N4SID implementation in the MATLAB toolbox. The order estimation here is performed also in the subspace procedure using a heuristically motivated method.
The paper is organized as follows: In the next section we will describe the model set. Section 3 will describe in more detail the three algorithms used in the estimation. The results of the simulation study are presented in section 4, where the main emphasis is on the effects of the choice of the integer parameters in the subspace procedure. Finally section 5 concludes the paper.

2. MODEL SET AND PROCESS UNDER STUDY

In this paper we are considering the estimation of univariate, linear, time invariant ARMA processes, which are solutions to the difference equation

$$a(z)y_t = c(z)\varepsilon_t \qquad (1)$$

where $a(z) = 1 + a_1 z + \cdots + a_p z^p, c(z) = 1 + c_1 z + \cdots + c_q z^q$. Here z denotes both, the backward shift operator as well as a complex variable. It will be assumed throughout, that the zeros of $a(z)$ lie outside the unit circle (stability assumption) and that the zeros of $c(z)$ lie outside the unit circle (strict minimum-phase assumption). In this case there exists a stationary solution $y_t \in \mathbb{R}, t \in \mathbb{Z}$ for white noise inputs $\varepsilon_t, t \in \mathbb{Z}$, which are assumed to be i.i.d. with mean zero, $\mathbb{E}\varepsilon_t^2 > 0$ and finite fourth moments. We will always assume, that $a(z)$ and $c(z)$ are co-prime (i.e. they do not have common roots). The pair of polynomial matrices $(a(z), c(z))$ in this case will be called an ARMA system. Two special cases are AR systems ($c(z) = 1$) and MA systems ($a(z) = 1$).
The system equivalently can be written in state space form:

$$\begin{aligned} x_{t+1} &= Ax_t + B\varepsilon_t \\ y_t &= Cx_t + \varepsilon_t \end{aligned} \qquad (2)$$

where $A \in \mathbb{R}^{n\times n}, B \in \mathbb{R}^{n\times 1}, C \in \mathbb{R}^{1\times n}$. The state space system is called minimal, if there exists no other state space system $(\tilde{A}, \tilde{B}, \tilde{C})$ of order $\tilde{n} < n$, such that $CA^{j-1}B = \tilde{C}\tilde{A}^{j-1}\tilde{B}$ for all $j \in \mathbb{N}$. In this case n is called the order of the system. For a minimal system the stability assumption is equivalent to $|\lambda_{max}(A)| < 1$, where $\lambda_{max}(A)$ denotes an eigenvalue of maximum modulus of the matrix A. The minimum-phase assumption is equivalent to $|\lambda_{max}(A - BC)| < 1$. There exists a one-to-one relation between ARMA systems and state space systems, which moreover are in the so called echelon canonical form. This makes it possible to rewrite results achieved in one representation using the other representation. We will use this one-to-one relation, as the ARMAsel algorithm is formulated in the ARMA framework, whereas the subspace methods use extensively the state space representation.

3. ESTIMATION ALGORITHMS

In this section the three algorithms used in the comparison are described in more detail.

3.1 ARMAsel

The ARMAsel algorithm is an algorithm for model inference from stationary stochastic processes of an unknown structure (Broersen, 2002). It selects an accurate parametric model using the data only. This can be either an AR(p), MA(q) or ARMA(r,r-1) model, that is selected using statistical order selection and type selection. The algorithms that are used are selected based on their performance in finite samples. Since estimated models are used in order selection, a reasonable estimate must also be obtained for models, where the model order of the estimated model is greater than the true model order.

For AR models, the Burg algorithm is used, that is more accurate than the Yule-Walker estimator. The Yule-Walker estimate contains a bias that is a result of transients. The frequency domain representation of this bias is referred to as spectral leakage (Stoica and Moses, 1997). Spectral leakage is neglected in asymptotic theory. For MA and ARMA modelling, Durbin's estimators are used with a high order intermediate AR model.

So far, no example has been found where ARMAsel performs poorly. The automatically selected model is more accurate than the best windowed periodogram (Broersen, 2002). A freely available MATLAB implementation can be obtained from
http://www.tn.tudelft.nl/mmr.

3.2 The subspace based procedure

This algorithm uses the CCA procedure for estimating the state space model. This method has been pioneered by (Larimore, 1983). The algorithm hinges on the following basic fact: Fix two integers f and p. Denoting $Y_{s|t} = [y_s', y_{s+1}', \cdots, y_t']'$ we obtain the following equation:

$$Y_{t|t+f-1} = \Gamma_f Q_p Y_{t-p|t-1} + \Gamma_f \bar{A}^p x_{t-p} + N_f \quad (3)$$

where $\bar{A} = A - BC$ and N_f summarizes the effects of the future of the noise, which is orthogonal to the two other terms due to the assumptions on ε_t. Further $\Gamma_f = [C', A'C', \cdots, (A^{f-1})'C']'$, $Q_p = [(A-BC)^{p-1}B, \cdots, (A-BC)B, B]$. Finally let $\langle a_t, b_t \rangle = T^{-1}\sum_{t=p+1}^{T-f} a_t b_t'$. Let $L_p = \langle Y_{t-p|t-1}, Y_{t-p|t-1} \rangle$. Neglecting the second term in (3), since $(A-BC)^p$ tends to zero for $p \to \infty$, CCA obtains estimates of the system in the following three steps:

- Estimate $\Gamma_f Q_p$ by LS regression in (3) as $L_{f,p} = \langle Y_{t|t+f-1}, Y_{t-p|t-1} \rangle L_p^{-1}$.
- $L_{f,p}$ will be of full rank in general, whereas $\Gamma_f Q_p$ is of rank n, where n denotes the system order. Thus approximate

  $$\langle Y_{t|t+f-1}, Y_{t|t+f-1} \rangle^{-1/2} L_{f,p} L_p^{1/2} = \hat{U}\hat{\Sigma}\hat{V}' = \hat{U}_n \hat{\Sigma}_n \hat{V}_n' + \hat{R}_n$$

  to obtain estimates

  $$\hat{\Gamma}_f = \langle Y_{t|t+f-1}, Y_{t|t+f-1} \rangle^{1/2} \hat{U}_n \hat{\Sigma}_n$$

  and $\hat{Q}_p = \hat{V}_n' L_p^{-1/2}$. Here $\hat{U}\hat{\Sigma}\hat{V}'$ denotes the SVD of $\langle Y_{t|t+f-1}, Y_{t|t+f-1} \rangle^{-1/2} L_{f,p} L_p^{1/2}$. Thus e.g. $\hat{\Sigma}$ is the diagonal matrix containing the singular values ordered in decreasing size as diagonal entries. $\hat{U}_n \in \mathbb{R}^{fs \times n}, \hat{V}_n \in \mathbb{R}^{ps \times n}$ and $\hat{\Sigma}_n \in \mathbb{R}^{n \times n}$ correspond to the submatrices obtained by neglecting the singular values numbered $n+1$ and higher. The approximation error for system order n is denoted by $\hat{R}_n$. Therefore in this step the order of the estimated system is specified.
- Given the estimate $\hat{Q}_p$ from the second step the state is estimated as $\hat{x}_t = \hat{Q}_p Y_{t-p|t-1}$ and the system matrices are obtained using least squares regressions in the system equations (2), where the estimated state takes the place of the state.

The ARMA system corresponding to the estimated state space system can be calculated consecutively. This procedure requires the specification of three integer parameters: The indices f and p have to be chosen and the order of the system has to be specified. Estimation of the order can be performed using the information contained in the estimated singular values in a number of different ways (for a discussion see Bauer, 2001). Here we will deal with the criterion SVC. Let

$$SVC(n) = \hat{\sigma}_{n+1}^2 + \frac{C_T d(n)}{T}$$

where $d(n) = 2ns$ denotes the number of parameters and $C_T > 0, C_T/T \to 0$ denotes a penalty term. Here $\hat{\sigma}_i$ denotes the estimated singular values ordered decreasing in size. Under the assumptions put forward in this paper, $C_T/(fp \log\log T) \to \infty$ implies almost sure (a.s.) consistent estimates of the order $\hat{n} = \arg\min SVC(n), 0 \le n \le H_T, H_T = O((\log T)^a), a < \infty$ (provided $\min\{f,p\} > n$ a.s. asymptotically). The disadvantage of using a consistent estimator of the order is that the error as a result of underfit can become arbitrarily large with T(Broersen and Wensink, 1996).

In some simulation examples, the penalty term $C_T = \log T$ shows superior behaviour and hence is used also in this paper.
With respect to the specification of f and p asymptotic theory suggests to use $f = p = \lceil d\hat{p}_{AIC} \rceil, d > 1$, where $\hat{p}_{AIC}$ denotes the order estimate obtained using AIC in an autoregression. It has been suggested to use $d = 2$ in (Bauer *et al.*, 1999), which is however totally arbitrary.

With respect to the ARMAsel procedure the CCA algorithm has the advantage of being directly applicable to multivariable systems. Also generalizations to systems with observed inputs are available.

3.3 pem *as implemented in* MATLAB

The last procedure, which is seen to be an industry benchmark presently, is the procedure pem implemented in the system identification toolbox of MATLAB. The core of the procedure is a prediction error minimization done using a Gauss-Newton search procedure, which is started at an initial estimate obtained using the MATLAB implementation of the N4SID procedure of (Van Overschee and DeMoor, 1994) realized in the toolbox. In the fully automated pem for univariate data, the integer $f = 15$ is used in all cases. $p = \hat{p}_{AIC}$ is chosen. The order is estimated as the index of the smallest estimated singular value (contained in $\hat{\Sigma}$), which is larger than the geometric average between the largest and the smallest estimated singular value. There is no theoretical motivation for these choices. In this procedure, the subspace estimate is used in order to specify the order and as an initial estimate for the subsequent Gauss-Newton based prediction error optimization. The order of the system hence is estimated in the subspace step.

Since the basis of the algorithm is the N4SID subspace method, it also can handle multivariable processes as well as additional exogenous inputs.

4. SIMULATIONS

The automatic procedures are compared in a simulation experiment. The analysis of the behaviour of the resulting estimates for different strategies of choosing

Method	Mean error	max error	computation
ARMAsel	17	80	5.32
CCA	11	158	0.21
pem	28	779	2.52
CCAsel	13	230	14.10

Table 1. Kullback Leibler discrepancy (mean and maximal value, rounded to integers) of the estimated system and true system in 1000 trials of sample size $T = 200$. The right column gives the mean computation time for one simulation run.

f and p will be the main topic in the simulation exercise. The error measure used to evaluate the results is the exact Kullback-Leibler Discrepancy (Kullback, 1959). The Kullback-Leibler Discrepancy has an interpretation as the spectral distortion in the frequency domain, and as the normalized one-step ahead prediction error in the time domain (De Waele, 2003).

The first process, which is chosen for the comparison, is an ARMA(3,3) process given by $a(z) = 1 + 0.16z + 0.056z^2 + 0.8z^3, c(z) = 1 - 0.41z - 0.73z^2 + 0.5z^3$. The poles of the system are at $-1.0378, 0.4839 \pm 0.9850i$ and the zeros at $-1.0666, 1.2633 \pm 0.5283i$, hence the system has both poles and zeros close to the unit circle. 1000 realizations of this process of sample size $T = 200$ are generated and for each of the realizations an estimate using the three automatic procedures described above are obtained to result in one identified system for each procedure. The results of the simulations can be seen in Table 1: The table presents the Kullback Leibler discrepancy for the three procedures and also the average number of computations. In this example, the differences between the algorithms are minor: The subspace procedure shows the best performance, the ARMAsel procedure a slightly worse accuracy, while spending 20 times as much in terms of computations. The pem procedure shows the worst accuracy with approximately half the computation time needed with respect to ARMAsel.

Next the automatic procedures are applied to data simulated by an AR(100) process that is an accurate approximation of a Gaussian power spectrum. The Gaussian spectrum h with center frequency $f_C = 0.25$ and width $f_w = 0.04$ is given by:

$$h(f) = \exp(-\frac{1}{2(0.04)^2}(f - 0.25)^2). \quad (4)$$

This type of spectrum is used to describe radar reflections of sea waves due to weather disturbances (Haykin, 1979, p. 49). See (Broersen and de Waele, 2002) for the procedure that has been used for calculation of the AR(100) model. 100 realizations of this process are generated for sample size $T = 1000$. ARMAsel is compared to the subspace procedure using $f = p = 2\hat{p}_{AIC}$ and pem. The results can be seen in table 2. The table shows, that ARMAsel by far shows the best performance with an error of approximately 40 times better than the other two procedures. This

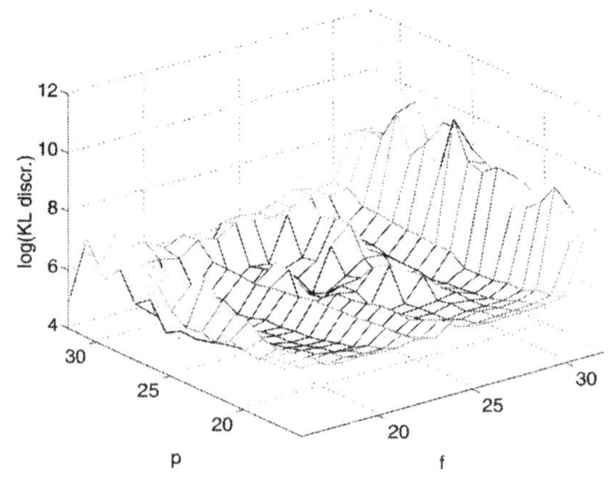

Fig. 1. Logarithm of the Kullback-Leibler discrepancy for the CCA estimates using $f = 16, \cdots, 32$ and $p = 16, \cdots, 32$.

comes at the cost of higher computational load. The pem method has a considerably bigger computational load as the subspace method. For the subspace method it is remarkable, that the maximal error is three times as huge as for the pem estimate, while the mean error is of the same order of magnitude. Looking at the individual values, one notices that for the subspace case 29 times the discrepancy is bigger than 1000 and in 9 cases even higher than 3000. For pem in 30 cases the error exceeds 1000, but never 3000. Hence the conclusion from this example is that the automatic procedure based on the subspace procedure is computationally much cheaper than ARMAsel, but the achieved accuracy is devastating.

The main reason for the bad accuracy of CCA in this case can be traced back to the sensitivity with respect to the choice of the integers f and p. This is documented by the plot in figure 1, which shows the logarithm of the Kullback-Leibler discrepancy for f and p in the range $\hat{p}_{AIC}$ to $2\hat{p}_{AIC}$ for one realization. Even in logs the huge differences are obvious. The values range from a minimum of 72.62 up to 23276. The value for the usual choice $f = p = 2\hat{p}_{AIC}$ in this case is equal to 2563.2.

This somewhat surprising sensitivity of the result with respect to the choice of f and p led to the implementation of a different strategy for the subspace based procedure using a search over different values of f and p, which can be described as follows:

(1) Obtain the estimate $\hat{p}_{AIC}$.
(2) For each $f = p$ in the range $2, 3, \cdots, 3\hat{p}_{AIC}$ calculate the CCA estimate, where the order is estimated using SVC with penalty $C_T = \log T$ for each value of $f = p$.
(3) Evaluate the in sample prediction error using the MATLAB function pe for each of the obtained systems and choose p according to the best fitting system.
(4) Compute the CCA estimate and select the order for this value of p and $f = 2, 3, \cdots, 3\hat{p}_{AIC}$

Method	Mean error	max error	computation
ARMAsel	25	56	88.45
CCA	984	6479	1.19
pem	989	2447	6.44
CCAsel	73	224	69.63

Table 2. Kullback Leibler discrepancy of the estimated system and true system in 100 trials of sample size $T = 1000$. The right columns gives the mean computation time for one simulation run.

(5) Evaluate the in sample prediction error using the MATLAB function pe for each of the obtained systems and choose f according to the best fitting system.
(6) Compute the CCA estimate and select the order for this value of f and $p = 2, 3, \cdots, 3\hat{p}_{AIC}$.
(7) For these systems find the best fitting one in terms of in sample prediction error using the MATLAB function pe: This system is the final estimate.

This procedure will by called CCAsel in the following. CCAsel leads to a potentially large number of computations due to the calculation of a big number of subspace estimates. The results can be found in table 1 and 2. Obviously the search strategy has a big impact: In the first example, CCAsel does not improve the accuracy, but the amount of computations is increased considerably, as on average in this case 64 times the CCA algorithm is used. The computational load is in this case considerably larger than for the ARMAsel procedure. Also there is a tendency to produce large errors in some cases. In the second example, the conclusions are adverse: On average approximately 128 times the CCA estimate for different values of f and p are estimated for one time series, resulting in a computational load, which is comparable to ARMAsel. With respect to the achieved accuracy, the CCAsel is still worse than the ARMAsel procedure, however the difference is small as compared to the difference for the original procedure.

Other processes can be found, where both the simple CCA and the more complex CCAsel selection procedures for p and f yield inaccurate results. A feature of the examples found so far is that the influence of spectral leakage is considerable. For the Gaussian spectrum discussed here, the Kullback-Leibler Discrepancy as a result of spectral leakage is 1500 (i.e. all procedures beat the expectation of the periodogram with respect to the accuracy in this example). Since finite sample effects have been taken into account in the ARMAsel algorithm, it is not affected by the triangular bias.

5. CONCLUSIONS

In this paper an automatic procedure for model estimation based on subspace procedures and the SVC order estimate is compared to two competitors, the ARMAsel method developed at Delft university and the pem procedure implemented in MATLAB. The main results are that the choice of f and p for the subspace procedure seems to be very critical for the resulting accuracy of the model. For the first example under study all three procedures produced similar accuracy, where therefore the smaller amount of computations can be seen as an advantage for CCA. In the second example, however, the CCA procedure based on recommendations from asymptotic theory leads to poor accuracy. A strategy to obtain acceptable model accuracy has been found. The drawback here is that the resulting procedure does not show advantages as compared to the ARMAsel procedure in terms of computational load, while still having worse precision. Hence in this case, ARMAsel is to be preferred undoubtedly. Also note, that the optimization strategy for the subspace case is not well studied at the moment: It is the opinion of the authors, that in this area more research seems to be valuable and also necessary. In any case, the potential of subspace procedures to provide a fully automatic identification procedure can be noticed.

REFERENCES

Akaike, H. (1975). Markovian representation of stochastic processes by canonical variables. *SIAM Journal of Control* **13**(1), 162–172.

Bauer, D. (2001). Order estimation for subspace methods. *Automatica* **37**, 1561–1573.

Bauer, D., M. Deistler and W. Scherrer (1999). Consistency and asymptotic normality of some subspace algorithms for systems without observed inputs. *Automatica* **35**, 1243–1254.

Broersen, P .M. T (2000). Finite sample criteria for autoregressive order selection. *IEEE Trans. Signal Processing* **48**(12), 3550–3558.

Broersen, P. M. T. (2002). Automatic spectral analysis with time series models. *IEEE Transactions on Instrumentation and Measurement* **51**(2), 211–216.

Broersen, P. M. T. and H. E. Wensink (1996). On the penalty factor for autoregressive order selection in finite samples. *IEEE Trans. Signal Processing* **44**(3), 748–752.

Broersen, P. M. T. and S. de Waele (2002). Generating data with a prescribed spectral density. In: *Proceedings IMTC Conference*. Anchorage, USA. pp. 781–786.

De Waele, S. (2003). Model Inference for Stationary Stochastic Processes Using Order Selection. PhD thesis. Delft University of Technology. Delft, The Netherlands.

Hannan, E. J. and M. Deistler (1988). *The Statistical Theory of Linear Systems*. John Wiley. New York.

Haykin, S. (1979). *Nonlinear Methods of Spectral Analysis*. Springer-Verlag. Berlin.

Kullback, S. (1959). *Information Theory and Statistics*. J. Wiley and Sons. London, England.

Larimore, W. E. (1983). System identification, reduced order filters and modeling via canonical variate analysis. In: *Proc. 1983 Amer. Control Conference 2.* (H. S. Rao and P. Dorato, Eds.). Piscataway, NJ. pp. 445–451. IEEE Service Center.

Stoica, P. and R. L. Moses (1997). *Introduction to Spectral Analysis*. Prentice Hall. Upper Saddle River.

Van Overschee, P. and B. DeMoor (1994). N4sid: Subspace algorithms for the identification of combined deterministic-stochastic systems. *Automatica* **30**, 75–93.

IFAC Publications
www.elsevier.com/locate/ifac

ON THE NUMBER OF ROWS AND COLUMNS IN SUBSPACE IDENTIFICATION METHODS

Bart L.R. De Moor *

* ESAT-SCD, K.U. Leuven, Kasteelpark Arenberg 10, B-3001 Leuven, Belgium
T: 32-(0)16321709, F: +32-(0)16321970, E:bart.demoor@esat.kuleuven.ac.be,
W:www.esat.kuleuven.ac.be/~demoor

Abstract: We discuss the role of the number of rows and columns in subspace identification algorithms. *Copyright © 2003 IFAC*

Keywords: Subspace identification, realization theory, Kalman filter, principal angles and directions between subspaces, canonical correlations, positive realness

1. INTRODUCTION

Subspace identification algorithms estimate state space models from inputs $u_k \in \mathbf{R}^m$ and outputs $y_k \in \mathbf{R}^l$, $k = 1, 2, \ldots, N$, 'generated' by linear, time-invariant MIMO systems of the form

$$x_{k+1} = Ax_k + Bu_k + w_k \,, \quad (1)$$

$$y_k = Cx_k + Du_k + v_k \,, \quad (2)$$

with $x_k \in \mathbf{R}^n$ the state vectors, A, B, C and D real matrices of appropriate dimensions and $v_k \in \mathbf{R}^l$ and $w_k \in \mathbf{R}^n$ the process, respectively measurement noise, assumed to be white, zero mean Gaussian, stationary with covariance matrix

$$\mathbf{E}\left[\begin{pmatrix} w_p \\ v_p \end{pmatrix}\begin{pmatrix} w_q^T & v_q^T \end{pmatrix}\right] = \begin{pmatrix} Q & S \\ S^T & R \end{pmatrix}\delta_{pq} \geq 0 \,. \quad (3)$$

We also assume that

$$\mathbf{E}\left[\begin{pmatrix} x_k \\ u_k \end{pmatrix}\begin{pmatrix} v_p^T & w_p^T \end{pmatrix}\right] = 0 \,, \forall p \geq k \,. \quad (4)$$

The analysis of subspace algorithms is not straightforward, even for this *ideal* case, let alone when only some or none of these assumptions are satisfied. Subspace methods start from $2li \times j$ block Hankel matrices constructed from the output vectors, which are partitioned as

$$Y_{0|2i-1} = \begin{pmatrix} Y_{0|i_p-1} \\ \hline Y_{i_p|2i-1} \end{pmatrix} = \begin{pmatrix} Y_p \\ \hline Y_f \end{pmatrix} \,. \quad (5)$$

Block Hankel matrices with the input vectors, denoted e.g. as $U_{0|2i-1}$ are similarly defined. In these block Hankel matrices, the number of block rows and columns are chosen by the user, in such a way that $N = 2i + j - 1$ and $j \gg 2i$. As can be seen in (5), the block Hankel matrices are partitioned in an upper part with i_p block rows, and a lower part with $2i - i_p$ block rows. The subscript p refers here to 'past', while f refers to future. The index i_p therefore represents a user defined splitting-up of the block Hankel matrices in two parts. While most often (see e.g.), i_p is taken to be $i_p = i$, in principle, all equations and results can be derived for any choice of i_p 'sufficiently large'.

The main steps in subspace algorithms are the following. **1.** Determine the model order n and estimates of the states $(\hat{x}_i \; \hat{x}_{i+1} \; \ldots \hat{x}_{i+j})$. They are typically found by first projecting (orthogonal or oblique) row spaces of data block Hankel matrices, and then applying a singular value decomposition (see e.g. for details); **2.** Solve a matrix least squares problem to obtain the state space matrices:

$$\begin{pmatrix} \hat{A} & \hat{B} \\ \hat{C} & \hat{D} \end{pmatrix} = \arg\min_{A,B,C,D} \left\| \begin{pmatrix} \hat{x}_{i+1} & \hat{x}_{i+2} & \cdots & \hat{x}_{i+j} \\ y_i & y_{i+1} & \cdots & y_{i+j-1} \end{pmatrix} - \begin{pmatrix} A & B \\ C & D \end{pmatrix}\begin{pmatrix} \hat{x}_i & \hat{x}_{i+1} & \cdots & \hat{x}_{i+j-1} \\ u_i & u_{i+1} & \cdots & u_{i+j-1} \end{pmatrix} \right\|_F^2 , \quad (6)$$

[1] Research supported by Research Council KUL; Flemish Government: FWO, AWI, IWT; Belgian Federal Government (IUAP); EU.

where $\|\cdot\|_F$ is the Frobenius-norm of a matrix. Define the least squares residuals $\rho_{w_k} = \hat{x}_{k+1} - \widehat{A}\hat{x}_k - \widehat{B}u_k$ and $\rho_{v_k} = y_k - \widehat{C}\hat{x}_k - \widehat{D}u_k$, $k = i, i+1, \ldots, i+j-1$, from which the noise covariance matrices can be estimated as

$$\begin{pmatrix} \widehat{Q} & \widehat{S} \\ \widehat{S}^T & \widehat{R} \end{pmatrix} = \frac{1}{j} \begin{pmatrix} \rho_{w_i} & \rho_{w_{i+1}} & \cdots & \rho_{w_{i+j-1}} \\ \rho_{v_i} & \rho_{v_{i+1}} & \cdots & \rho_{v_{i+j-1}} \end{pmatrix} (.)^T. \quad (7)$$

The main purpose of this short paper is to discuss some well and lesser known properties of subspace identification algorithms in terms of the indices i, i_p and j.

2. STOCHASTIC SYSTEMS

Problems and properties as a function of i and for $j \to \infty$ are best understood by analyzing stochastic systems where in (1)-(2) u_k is identically 0. We assume that the stochastic system that generated the data y_k is stationary (i.e. A stable and stationary process and measurement noise, i.e. as in (3) where Q, S and R are time-invariant). In that case $\mathbf{E}(x_k) = 0$ and from the orthogonality properties (4) we find that

$$\mathbf{E}\begin{pmatrix} x_{k+1} \\ y_k \end{pmatrix} (x_{k+1}^T \; y_k^T) = \begin{pmatrix} \Sigma & G \\ G^T & \Lambda_0 \end{pmatrix}$$
$$= \begin{pmatrix} A \\ C \end{pmatrix} \Sigma (A^T \; C^T) + \begin{pmatrix} Q & S \\ S^T & R \end{pmatrix}, \quad (8)$$

with obvious definitions for Σ, Λ_0 and G. Furthermore, one easily finds, using the orthogonality properties (4) and the stationarity assumption, that, $\forall t \geq 1$:

$$\Lambda_t = \mathbf{E}(y_{k+t}y_k^T) = CA^{t-1}G ,$$
$$\Lambda_{-t} = \mathbf{E}(y_{k-t}y_k^T) = G^T(A^{t-1})^T C^T . \quad (9)$$

2.1 *Rank deficiency of the cross-correlation of past and future*

We will use the averaging operator $\mathbf{E}_j(.)$ to indicate the sample average $\mathbf{E}_j(.) = \lim_{j\to\infty} \frac{1}{j}(.)$. Under appropriate conditions (e.g. ergodicity, finite fourth order moments,...), $\mathbf{E}_j(.) = \mathbf{E}(.)$. We can now easily derive that the cross-correlation matrix between future and past output block Hankel matrices [2] is block Toeplitz, and moreover rank deficient as can be seen from $\mathbf{E_j}(Y_{i|2i-1}Y_{0|i-1}^T)$

$$= \begin{pmatrix} \Lambda_i & \Lambda_{i-1} & \Lambda_{i-2} & \ldots & \Lambda_1 \\ \Lambda_{i+1} & \Lambda_i & \Lambda_{i-1} & \ldots & \Lambda_2 \\ \Lambda_{i+2} & \Lambda_{i+1} & \Lambda_i & \ldots & \Lambda_3 \\ \vdots & \vdots & \vdots & \vdots & \vdots \\ \Lambda_{2i-1} & \Lambda_{2i-2} & \ldots & \ldots & \Lambda_i \end{pmatrix} = \begin{pmatrix} C \\ CA \\ CA^2 \\ \vdots \\ CA^{i-1} \end{pmatrix}$$
$$\times (A^{i-1}G \; A^{i-2}G \; \ldots \; G) = \Gamma_i \, \Delta_i^G . \quad (10)$$

Hence it follows from the Theorem of Cayley-Hamilton that i needs to be larger than the controllability index of the matrix pair (A, G) and also needs to be larger than the largest observability index of the matrix pair (A, C). For a system with a scalar output for instance, this means that i needs to be larger than n. In principle, one could now obtain, via the rank of $\mathbf{E}_j(Y_{i|2i-1}Y_{0|i-1}^T)$, the order n of the system matrices A, C and G by invoking 'classical' realization theory and algorithms. The following formulas are also useful. First define the correlation matrices of past and future outputs as

$$L_i = \mathbf{E_j}(Y_{0|i-1}Y_{0|i-1}^T) = \mathbf{E_j}(Y_{i|2i-1}Y_{i|2i-1}^T) ,$$

where equality follows from stationarity. This matrix is block Toeplitz constructed with the output covariance matrices (9). The orthogonal projection of the row space of $Y_{i|2i-1}$ (the 'future') onto the row space of $Y_{0|i-1}$ (the 'past'), follows from the projection formula:

$$\mathbf{E_j}(Y_{i|2i-1}Y_{0|i-1}^T)(\mathbf{E_j}(Y_{0|i-1}Y_{0|i-1}^T))^{-1}Y_{0|i-1}$$
$$= \Gamma_i \Delta_i^G L_i^{-1} Y_{0|i-1} ,$$

where we have used equation (10). We can now take as a basis for the row space of this orthogonal projection the rows of the matrix $\widehat{X}_i$:

$$\widehat{X}_i = \Delta_i^G L_i^{-1} Y_{0|i-1} . \quad (11)$$

Similarly, the orthogonal projection of the row space of $Y_{0|i-1}$ (the 'past') onto the row space of $Y_{i|2i-1}$ (the 'future'), follows from the projection formula:

$$\mathbf{E_j}(Y_{0|i-1}Y_{i|2i-1}^T)(\mathbf{E_j}(Y_{i|2i-1}Y_{i|2i-1}^T))^{-1}Y_{i|2i-1}$$
$$= (\Delta_i^G)^T \Gamma_i^T L_i^{-1} Y_{i|2i-1} ,$$

where again we have used equation (10). We now take as a basis for the row space of this orthogonal projection, the rows of the matrix

$$\widehat{Z}_i = \Gamma_i^T L_i^{-1} Y_{i|2i-1} . \quad (12)$$

The fact that these projections are finite dimensional directly follows from the rank deficiency of (10). In principle, one can also determine the 'shifted' projections, which are obtained from the orthogonal projection of the row space of $Y_{i+1|2i-1}$ onto that of $Y_{0|i}$, with a basis given by the rows of $\widehat{X}_{i+1} = \Delta_{i+1}^G L_{i+1}^{-1} Y_{0|i}$ and the orthogonal projection of the row space of $Y_{0|i-2}$ onto that of $Y_{i-1|2i-1}$, with a basis given by the rows of $\widehat{Z}_{i+1} = \Gamma_{i+1}^T L_{i+1}^{-1} Y_{i-1|2i-1}$.

2.2 *Interpretations as banks of Kalman filters*

A least-squares state estimate $\hat{x}_k$ of the state x_k in (1)-(2) is generated by the non-steady state Kalman filter:

$$\hat{x}_k = A\hat{x}_{k-1} + K_{k-1}(y_{k-1} - C\hat{x}_{k-1}) \quad (13)$$
$$K_{k-1} = (G - AP_{k-1}C^T)(\Lambda_0 - CP_{k-1}C^T)^{-1} \quad (14)$$
$$P_k = AP_{k-1}A^T + (G - AP_{k-1}C^T)$$
$$\times(\Lambda_0 - CP_{k-1}C^T)^{-1}(G - AP_{k-1}C^T)^T . \quad (15)$$

In these recursions, K_k is the (non-steady state) Kalman gain and $P_k = \mathbf{E}(\hat{x}_k\hat{x}_k^T)$ is the estimated state covariance matrix [3]. When we apply these equations

[2] In what follows, we take $i_p = i$.

[3] Not to be confused with the state error covariance matrix which is more frequently described in the literature.

with an initial estimate $\hat{x}_0 = 0$ and initial state covariance $P_0 = 0$, we find, repeatedly using the matrix inversion lemma (see (, p.69)) for $\hat{x}_i$ the explicit formula

$$\hat{x}_i = \Delta_i^G . L_i^{-1} . (\, y_0^T \; y_1^T \; \dots \; y_{i-1}^T \,)^T \, , \tag{16}$$

which is easily seen to coincide with the first column of $\widehat{X}_i$. This state estimate will be denoted by $\hat{x}_i^{[0]}$. The second column of $\widehat{X}_i$ is generated by running a **second** Kalman filter, from time instant $k = 1, \dots, i+1$, also with initial state 0 and zero initial covariance matrix, using the data $(y_1, \dots, y_i)$, which generates an estimate of x_{i+1} that can be explicitly written as

$$\hat{x}_{i+1}^{[1]} = \Delta_i^G . L_i^{-1} . (\, y_1^T \; y_1^T \; \dots \; y_i^T \,)^T \, .$$

In this way, we can run j Kalman filters to generate the columns of $\widehat{X}_i$ in (11), which is visualized in Figure 1 [4]. From (11), it is also obvious that

$$P_i = \mathbf{E}_j(\widehat{X}_i \widehat{X}_i^T) = \Delta_i^G . L_i^{-1} . (\Delta_i^G)^T \, . \tag{17}$$

The columns in (12) can be interpreted as Kalman filter estimates from a backward (i.e. where time runs from future to past) Kalman filter: Let $\hat{z}_0 = 0$ be an initial state and $N_0 = \mathbf{E}[\hat{z}_0 \hat{z}_0^T] = 0$ an initial state covariance matrix. The backward Kalman filter is given by the following recursions:

$$\hat{z}_{-k} = A^T \hat{z}_{-k+1} + K_{-k+1}(y_{-k+1} - C\hat{z}_{-k+1}) \tag{18}$$

$$K_{-k+1} = (C^T - A^T N_{-k+1} G)(\Lambda_0 - G^T N_{-k+1} G)^{-1}$$

$$N_{-k} = A^T N_{-k+1} A + (C^T - A^T N_{-k+1} G) \times (\Lambda_0 - G^T N_{-k+1} G)^{-1} (C^T - A^T N_{-k+1} G)^T \tag{19}$$

from which the interpretation of the columns of $\widehat{Z}_i$ (12) follows as the result of j backward Kalman filters running backward in time over the future outputs. The explicit solution of the covariance matrix N_i now follows from (12) as

$$N_i = \Gamma_i^T . L_i^{-1} . \Gamma_i \, . \tag{20}$$

2.3 *How to estimate A, C, G and Q, S, R ?*

Having given the interpretations for the state matrices $\widehat{X}_i$ and $\widehat{X}_{i+1}$, it follows from the Kalman filter equations (13) that

$$\widehat{X}_{i+1} = A\widehat{X}_i + W_i \, , \tag{21}$$

$$Y_{i|i} = C\widehat{X}_i + V_i \, , \tag{22}$$

where W_i and V_i can be obtained as the least squares residuals of a matrix least squares problem as in (6) Similarly, G can be obtained from W_{i+1} and W_i. (For details we refer to). The matrices Q, S and R are estimated as

$$\begin{pmatrix} \hat{Q}_i & \hat{S}_i \\ \hat{S}_i^T & \hat{R}_i \end{pmatrix} = \mathbf{E}_j \left(\begin{pmatrix} W_i \\ V_i \end{pmatrix} (\, W_i^T \; V_i^T \,) \right) \, . \tag{23}$$

[4] It has been a little bit misleading that we have used in the past the terminology 'state sequence' to denote the columns of $\widehat{X}_i$, hereby misleadingly suggesting that they belong to some state sequence generated by only one Kalman filter, which is not the case (except when $i \rightarrow \infty$).

These estimates are biased for finite i, because of the fact that each of the Kalman filters in the bank is not in steady state. The bias decreases exponentially as a function of i, according to $\prod_{k=0}^{i}(A - K_k C)$.

2.4 *Principal angles between past and future for finite i*

One of the major observations in subspace identification is that the system matrices A, G, C, Q, R, S do not have to be known to determine the state matrices $\widehat{X}_i$ (11) and/or $\widehat{Z}_i$ (12) or their shifted versions $\widehat{X}_{i+1}$ and $\widehat{Z}_{i+1}$. These matrices can be obtained via orthogonal projections, but they can also be found as the so-called principal directions between the row spaces of Y_p and Y_f. In order to see this, define the 'past' and 'future' projectors:

$$\Pi_i = Y_{0|i-1}^T (\mathbf{E}_j (Y_{0|i-1} Y_{0|i-1}^T))^{-1} Y_{0|i-1} \, , \tag{24}$$

$$\Phi_i = Y_{i|2i-1}^T (\mathbf{E}_j (Y_{i|2i-1} Y_{i|2i-1}^T))^{-1} Y_{i|2i-1} \, . \tag{25}$$

Then, we easily find from (10), (11) and (12) that

$$\Phi_i \Pi_i = Y_{i|2i-1}^T L_i^{-1} \Gamma_i \Delta_i^G L_i^{-1} Y_{0|i-1} = \widehat{Z}_i^T \widehat{X}_i \, ,$$

leading to the conclusion that the row space of the product of the two projector matrices coincides with that of $\widehat{X}_i$, and its column space coincides with the row space of $\widehat{Z}_i$. Similarly, we have $\Phi_{i-1}\Pi_{i+1} = \widehat{Z}_{i+1}^T \widehat{X}_{i+1}$. Of course, the so-called principal angles and directions between the row spaces of $Y_{0|i-1}$ and $Y_{i|2i-1}$ can be obtained from the singular value decomposition of $\Phi_i \Pi_i$ (in principle, how to really do it numerically is explained in e.g.). The eigenvalues of this matrix will all be real between 0 and 1, corresponding to the cosines of the principal angles between the row space of Y_p and that of Y_f , and they derive from

$$\lambda(\Phi_i \Pi_i) = \lambda(L_i^{-1} \Gamma_i \Delta_i^G L_i^{-1} (\Delta_i^G)^T \Gamma_i^T)$$

$$= \lambda((\Gamma_i^T L_i^{-1} \Gamma_i)(\Delta_i^G L_i^{-1} (\Delta_i^G)^T)) = \lambda(P_i N_i) \, .$$

Hence, even for finite i, the cosines of the principal angles can be obtained, either from the output data, but also as the eigenvalues of the product of the solutions P_i and N_i of the Riccati difference equations (15) and (19). As $i \rightarrow \infty$, the matrices P_i and N_i converge to the steady state matrices P and N that are the solutions to the algebraic Riccati equations corresponding to the difference equations (15) and (19). The principal angles then correspond to the canonical correlations of the stochastic process, as defined in . A third order example is shown in Figures 2 and 3.

2.5 *State identification for combined-models*

When u_k is not identically zero, one can obtain a state matrix $\widehat{X}_i$ via an oblique projection of the row space of Y_f, along the row space of U_f, onto the row space of $(\, U_p^T \; Y_p^T)^T$. An additional requirement here for the formal derivations is the *quasi-stationarity* of the deterministic inputs. It can be shown that this projection is rank deficient, the rank being equal to the

$$\hat{X}_0 = \begin{bmatrix} 0 & \cdots & 0 & \cdots & 0 \end{bmatrix}$$
$$P_0 = 0$$

Bank of j
Kalman filters
in parallel

$$Y_p \begin{bmatrix} y_0 & & y_q & & y_{j-1} \\ \vdots & & \vdots & & \vdots \\ y_{i-1} & & y_{i+q-1} & & y_{i+j-2} \end{bmatrix}$$

$$\hat{X}_i \begin{bmatrix} \hat{x}_i^{[0]} & \cdots & \hat{x}_{i+q}^{[q]} & \cdots & \hat{x}_{i+j-1}^{[j-1]} \end{bmatrix}$$

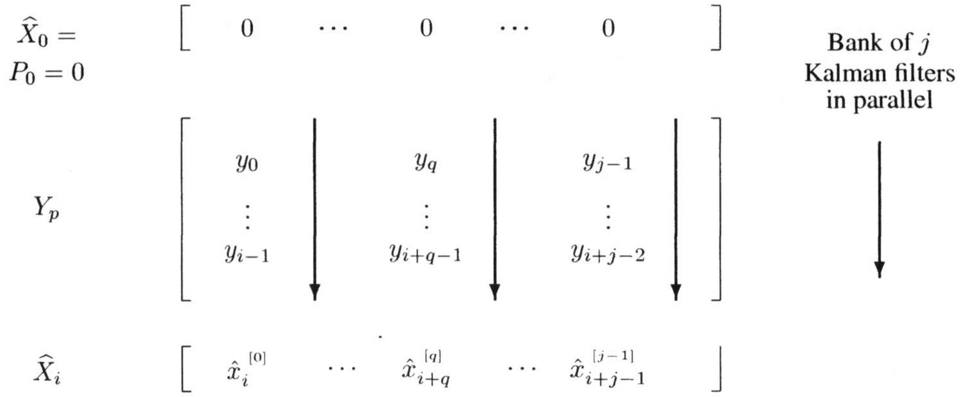

Fig. 1. Interpretation of the columns of the matrix $\hat{X}_i$ as the outputs of non-steady state Kalman filters, based upon i observations of y_k. If the system matrices A, G, C, Q, R, S were known, we denote by $\hat{x}_k^{[q]}$, the least squares estimate of the state x_k, by the $(q+1)$-th Kalman filter in the bank. For filter $q + 1$, start at time q, with an initial state estimate 0 and initial state covariance matrix 0. Now iterate the non-steady state Kalman filter over i time steps (the vertical arrow down). The Kalman filter will then return a state estimate $\hat{x}_{i+q}^{[q]}$. The j Kalman filters run in a *vertical* direction downwards (over the columns).

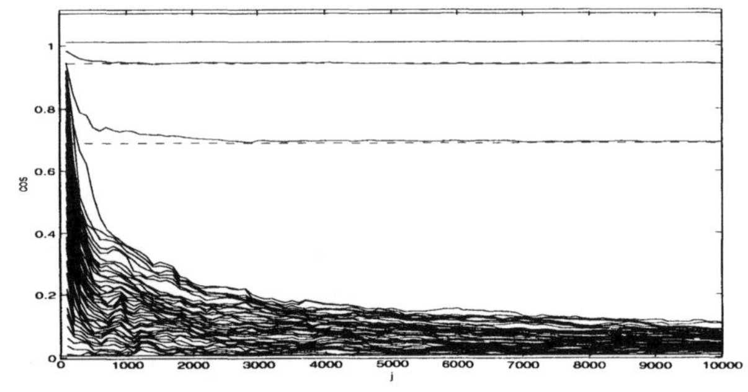

Fig. 2. As an example, we consider the 3-rd order stochastic system, with 2 inputs and 2 outputs, in innovation form $x_{k+1} = Ax_k + Ke_k, y_k = Cx_k + e_k$ where

$$A = \begin{pmatrix} -0.15 & -0.21 & -0.28 \\ 0.78 & 0.08 & -0.72 \\ 0.47 & 0.23 & -1.22 \end{pmatrix} \quad K = \begin{pmatrix} 1.65 & 0.09 \\ 1.50 & -0.54 \\ 0.05 & 1.07 \end{pmatrix}$$

$$C = \begin{pmatrix} -0.36 & 0.02 & 0.22 \\ 1.47 & 0.10 & -1.74 \end{pmatrix} \quad S_e = \begin{pmatrix} 0.38 & -0.15 \\ -0.15 & 1.33 \end{pmatrix}$$

where S_e is the input covariance matrix and K is the steady-state Kalman filter gain. The block Hankel matrices have $i = 20$ block rows and an increasing number of columns: $j = 100, 200, \ldots, 10\,000$. The cosines of the principal angles between the row spaces of Y_p and Y_f are shown in full line. The dashed lines represent the asymptotic values for $i = 20$ and $j = \infty$, which can be obtained from the eigenvalues of $(P_{i=20} N_{i=20})$. One observes the asymptotic convergence as $j \to \infty$: 3 of the cosines of the angles converge to the eigenvalues of $(P_{i=20} N_{i=20})$, while the 37 other ones (2 outputs $\times$ ($i = 20$ blockrows) - ($n = 3$)), converge to 0, corresponding to 'orthogonal' principal directions (no correlation between past and future).

2.6 Identification of the model: Role of i and j

Consistency and asymptotic normality: When using the averaging operator $\mathbf{E}_j(.)$, we 'do as if' the number of observations goes to infinity while in practice the number is always finite. Results on consistency, asymptotic normality of 'finite sample' errors and derived quantities, such as the system poles, transfer matrix estimates for a given frequency ω, etc..., can be found in   , also leading to statistically optimal weighting matrices. The plots, like the one in Figure 2, suggest that typically one needs 'enough' data (thousands rather than hundreds of time points). In particular, the estimates for A, B, C, D are consistent (asymptotically unbiased as $j \to \infty$); The 'size' of the error covariances on the estimates of A, B, C and D typically decrease as $1/j$. In some of these papers also suggestions are given on how to split up the available number of data N into i and j (e.g. in): The i required for consistency is a function of the available sample size, but also of the magnitude of the zero (eigenvalue of $A - KC$) of maximum modulus.

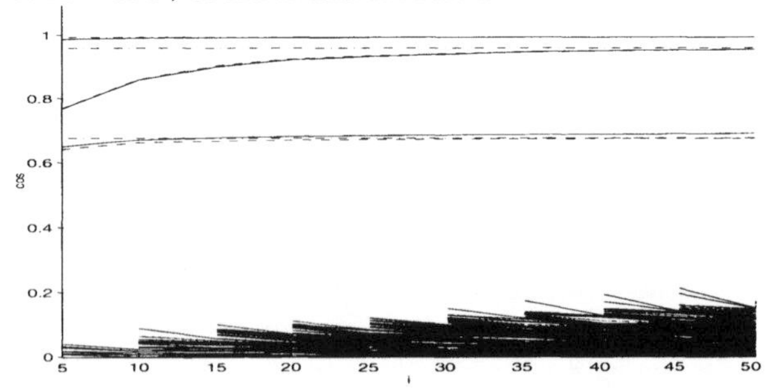

Fig. 3. The cosines of the principal angles between the row spaces of Y_p and Y_f. The full lines represent the cosines of the angles between the row spaces of the $2i \times 10\,000$ data block Hankel matrices, where i increases from 5 to 50 in steps of 5. The dashed lines show the cosines of the angles for the same values of i, but $j = \infty$. The cosines for $i = \infty$ and $j = \infty$ are given by the dash-dot lines. These values are the canonical correlations of the past and the future output process of the model.

order of the system. Again, there is an interpretation via a bank of j Kalman filters, the difference with the pure stochastic case being the fact that the initial state estimate and state covariance matrix are non-zero. Again, for complete details, see (, Chapter 4).

Statistics of angles and principal directions: It would be interesting to derive the 'finite' sample probability densities of principal angles and directions under the standing assumptions, for both finite i and j, as was done for the 'static' case in the classical statistical book .

Order estimation: The estimation of the order also becomes easier as $j \rightarrow \infty$ (see e.g. Figure 2); However, as is discussed in and , for stochastic systems, one has to be careful because the algebraic degree (the rank of the matrix (10)) can be smaller than the positive degree: The problem is that the realization A, C, G found from (10) is basically obtained from a deterministic realization algorithm (Ho-Kalman), which does not incorporate that the identified system represented by A, C, G and Λ_0 should be positive real. A thorough analysis is found in and , but further research is needed to come up with reliable algorithms that take this requirement of positive realness into account (see e.g.).

When $j < \infty$, several problems can occur:

Estimate of A unstable, even when the original system is known to be stable. Remedies are proposed in and . **Rank deficiency** of the estimate of the noise covariance matrix. In innovation form, the noise covariance matrix is of the form $\begin{pmatrix} K \\ I_l \end{pmatrix} S_e(\; K^T \; I_l \;)$, which is a rank l matrix. However, when estimated from the residuals as in (23), rank deficiency of the matrix is not guaranteed. A remedy is proposed in . **Noise model not positive real:** In order to ensure that the identified model is positive real (at least the stochastic part), one can start from the results obtained in .

Bias as a function of i: When $i < \infty$ and $j \rightarrow \infty$, estimates of A, B, C and D are consistent, with decreasing error covariance, but estimates of Q, S and R are biased (as indicated by the subscript i in formula (23)); The dynamics of this bias as a function of i, have been explained in terms of the behavior of the non-steady state Kalman filter. Given this insight, arbitrarily bad counter-examples can be constructed, in which one uses a pole-placement approach to place the poles of $A - KC$, given A and C, very close to the unit circle, in which case the 'convergence' of the Kalman filter equations, and hence also the bias on the estimates of Q, R, S and K will extremely slowly decrease as a function of i. This means that for those cases, one has to be careful with the asymptotic analysis as e.g. presented in , where a term of the form $(A - KC)^i$ appears in one of the starting equations.

Non-stationary white noise: When the noise sequences in (1)-(2) are white, zero-mean Gaussian, but *non-stationary*, subspace algorithms will still work in the sense that the matrices A and C can be estimated consistently as $j \rightarrow \infty$, but for obvious reasons, the identification of Q_k, S_k and R_k (these covariance matrices are a function of k due to non-stationarity) and the Kalman gain K_k is impossible. Background can be found in .

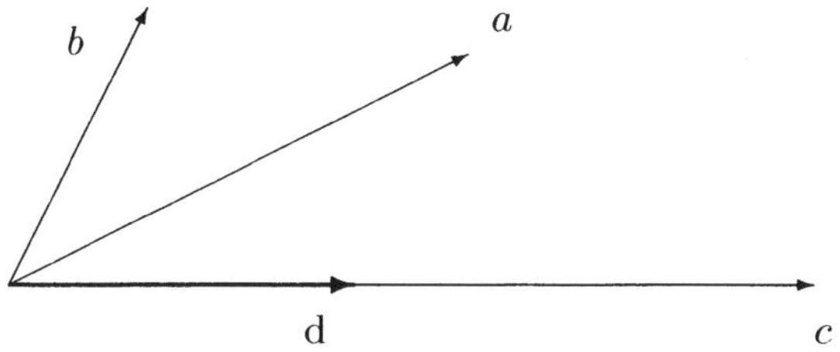

Fig. 4. *Visualization of an oblique projection in 2D: d is the oblique projection of a, along b onto c.*

Role of i_p: 'Asymmetric' choices (i.e. choices where $i_p \neq i$) for past and future, are studied in

Rank conditions: Subspace algorithms may also fail due to the fact that certain rank conditions (which often in papers on subspace research are only implicitly assumed to hold) are not satisfied. Typically, these conditions are reminiscent of more classical *persistancy-of-excitation* requirements. These conditions are necessary to estimate the order n correctly and also to guarantee the 'identifiability' of the state space model. An easy example of such a condition can be seen from the deterministic matrix input-output equation (put v_k and w_k identically to zero in (1)-(2): $Y_{0|i-1} = \Gamma_i X_0 + H_i U_{0|i-1}$, with an obvious definition for the convolution matrix H_i. Hence $\begin{pmatrix} Y_{0|i-1} \\ U_{0|i-1} \end{pmatrix} = \begin{pmatrix} \Gamma_i & H_i \\ 0 & I_{mi} \end{pmatrix} \begin{pmatrix} X_0 \\ U_{0|i-1} \end{pmatrix}$, from which it can be seen that $\text{rank} \begin{pmatrix} Y_{0|i-1} \\ U_{0|i-1} \end{pmatrix} = mi + n$, provided that i is larger than the largest observability index (so that $\text{rank}(\Gamma_i) = n$), the input is persistantly exciting of order mi (i.e. $U_{0|i-1}$ is of full row rank) and there is no intersection between the row space of X_0 and that of $U_{0|i-1}$ (which would be the case e.g. when there is (partial) state feedback). Of course, it is not completely obvious how to check conditions like this when one only has input-output data available. Examples of similar conditions for combined identification are discussed in and counterexamples are derived in . These examples and conditions certainly deserve further elaboration. Ideally, we should have tests that reveal potential bad conditioning of the data with respect to this type of (lack-of) persistancy-of-excitation or near ill-conditioning.

Geometric ill-conditioning: The (lack-of) persistancy of excitation conditions are but one extreme example of ill-conditioning that may occur in subspace identification. The geometry of the row spaces of U_f, Y_f and $(\; U_p^T \; Y_p^T \;)^T$ relative to each other, is also important. Consider the very simple example of a 2-D oblique projection as depicted in Figure 4. Here, d is the oblique projection of a along b into c. Hence, we can write $d = c\gamma = a - b\beta$ for certain real scalars β and γ. Defining the angles θ_{ab}, θ_{ac} and θ_{bc} between the vector pairs (a, b), (a, c) and (b, c) ($\theta_{bc} = \theta_{ab} + \theta_{ac}$), we easily find that $\|d\| = \|c\|\gamma = \|a\|(\cos\theta_{ac} - \cos(\theta_{bc} - \theta_{ac})\cos\theta_{bc})/(\sin^2\theta_{bc})$, and the ill-conditioning for small values of θ_{bc} and/or θ_{ab}

is apparent. The generalization to oblique projections of vector spaces, like the one we use in subspace identification, possibly in terms of principal angles between subspaces, remains to be done. The systematic and precise explicitation of the relation between geometry, conditioning, error covariances and numerical issues, is still a largely unexplored *terra incognita*. Helpful here may be the results on all kinds of principal angles that have been developed in .

References

Akaike H. *Markovian representation of stochastic processes by canonical variables.* SIAM Journal of Control, 13, 1, p.162-172, 1975.

Anderson T.W. *An introduction to multivariate statistical analysis.* Wiley, New York, 1984.

Baram Y. *Realization and reduction of Markovian models from nonstationary data.* IEEE Transactions on Automatic Control, AC-26, 6, pp.1225-1231, 1981.

Bauer D., Deistler M., Scherrer W. *Consistency and asymptotic normality of some subspace algorithms for systems without observed inputs.* Automatica 35, pp.1243-1254, 1999.

Bauer D., Jansson M. *Analysis of the asymptotic properties of the MOESP type of subspace algorithms.* Automatica, 36, 4, pp.497-509, 2000.

Bauer D., Ljung L. *Some facts about the choice of the weighting matrices in Larimore type of subspace algorithms.* Automatica, 38, pp.763-773, 2002.

Benveniste A., Fuchs J.J. *Single sample modal identification of a nonstationary stochastic process.* IEEE Transactions on Automatic Control, Vol.AC-30, 1, pp. 66-74, 1985.

Dahlen A., Lindquist A., Mari J. *Experimental evidence showing that stochastic identification methods may fail.* Systems and Control Letters, 34, pp.303-312, 1998.

De Cock K. *Principal angles in system theory, information theory and signal processing.* PhD Thesis, Department of Electrical Engineering, Katholieke Universiteit Leuven, Belgium, 293 pp., 2002. Downloadable from http://www.esat.kuleuven.ac.be/˜sistawww/cgi-bin/pub.pl

Deistler M., Peternell K., Scherrer W. *Consistency and relative efficiency of subspace methods.* Automatica, 31, pp.1865-1875, 1995.

Goethals I., Van Gestel T., Suykens J., Van Dooren P., De Moor B. *Identification of Positive Real Models in Subspace Identification by using Regularization*, Internal Report 02-104, ESAT-SISTA, K.U.Leuven (Leuven, Belgium), 2002.

Gustafsson T., Rao B.D. *Statistical analysis of subspace-based estimation of reduced-rank linear regressions.* IEEE Transactions on Signal Processing, 50, 1, pp.151-159, 2002.

Gustafsson T. *Subspace-based system identification: Weighting and pre-filtering of instruments.* Automatica 38, pp.433-443, 2002.

Jansson M., Wahlberg B. *A linear regression approach to state space subspace identification.* Signal Processing, 52, pp.103-129, 1996.

Jansson M., Wahlberg B. *On consistency of subspace methods for system identification.* Automatica, 34, 12, pp.1507-1519, 1998.

Knudsen T. *Consistency analysis of subspace identification methods based on a linear regression approach.* Automatica, 37, pp.81-89, 2001.

Lindquist A., Picci G. *Canonical correlation analysis, approximate covariance extension and identification of stationary time series.* Automatica, 32, 5, pp.709-733, 1996.

Ljung L., McKelvey T. *Interpretation of subspace methods: Consistency analysis.* Proc. of SYSID 97, pp.1125-1129, 1997.

Maciejowski J. *Guaranteed stability with subspace methods.* Systems and Control Letters, 26, pp.153-156, 1995.

Mari J., Stoica P., McKelvey T. *Vector ARMA estimation: A reliable subspace approach.* IEEE Transactions on Signal Processing, 48, 7, pp.2092-2104, 2000.

Peternell K., Scherrer W., Deistler M. *Statistical analysis of novel subspace identification methods.* Signal Processing, 52, pp.161-177, 1996.

Van Gestel T., Suykens J., Van Dooren P., De Moor B. *Identification of stable models in subspace identification by using regularization.* IEEE Transcations on Automatic Control, 46, 9, pp.1416-1420, 2001.

Van Gestel T., De Moor B., Van Overschee P. *On the Rank Deficiency of the Least Squares Residuals in Subspace Identification.* Proc. of the Symposium on System Identification (Sysid2000), Santa Barbara, California, pp. 247-252, 2000.

Van Overschee P., De Moor B. *Subspace algorithms for the stochastic identification problem.* Automatica 29, 3, pp.649-660, 1993.

Van Overschee P., De Moor B. *Subspace identification for linear systems: Theory, Implementation, Applications.* Kluwer Academic Publishers, 1996, 254 pp. Downloadable from http://www.esat.kuleuven.ac.be/˜sistawww/cgi-bin/pub.pl

Verhaegen M., Dewilde P. *Subspace model identification. Part I: The output-error state space model identification class of algorithms.* Int. J. Control, 56, pp.1187-1210, 1992.

Viberg M. *Subspace-based methods for the identification of linear time-invariant systems.* Automatica, 31, 12, pp.1835-1851, 1995.

Viberg M., Wahlberg B., Ottersten B. *Analysis of state space system identification methods based on instrumental variables and subspace fitting.* Automatica, Vol.33, no.9, pp.1603-1616, 1997.

IFAC Publications
www.elsevier.com/locate/ifac

ASPECTS AND EXPERIENCES OF USER CHOICES IN SUBSPACE IDENTIFICATION METHODS

Lennart Ljung *

* *Division of Automatic Control, Linköping University, SE-58183, Linköping, Sweden, email:* `ljung@isy.liu.se`

Abstract: Subspace identification methods, such as N4SID, MOESP, CVA etc have proven to be very successful for identification of multivariable, linear dynamic systems. These methods are associated with a number of design variables, or user choices. These include prediction horizons, weighting matrices and ways to perform the estimation. It is known that these choices may have a substantial influence on the model quality, at the same time as there is no comprehensive theory for rational decision making. In this contribution we study certain aspects of the choices, in particular their influence on both bias and variance. We also illustrate some aspects using a larger example, the Tennessee-Eastman identification challenge. *Copyright © 2003 IFAC*

1. INTRODUCTION

Subspace methods for identification of linear multivariable dynamical systems have been very successful. They can be traced back to the realization algorithms of Kalman and Ho, (Ho and Kalman 1965). One of the first descriptions in an identification context was given by (Akaike 1976). The methods were further developed by Larimore in a series of papers, e.g. (Larimore 1983), (Larimore 1990), by van Overschee and de Moor, see e.g. the book (Van Overschee and DeMoor 1996), by Verhaegen, e.g. (Verhaegen 1994) and others. The literature is quite extensive.

The methods estimate a state-space model of the type (in standard notation)

$$x(t+1) = Ax(t) + Bu(t) + Ke(t) \quad \text{(1a)}$$
$$y(t) = Cx(t) + Du(t) + e(t) \quad \text{(1b)}$$

Some variants estimate covariance matrices of process and measurement noise rather than the Kalman gain K.

The subspace algorithms are known under several names and acronyms, like CVA (Canonical Variate Analysis, Larimore), MOESP (Verhaegen), N4SID (van Overschee/de Moor) and have been incorporated in several commercial software packages, like SYSTEM IDENTIFICATION TOOLBOX for use with MATLAB, (Ljung 2000) and $ADAPT_X$, (Larimore 1997).

Despite the undeniable good performance of the algorithms, there is still no comprehensive theory for the (asymptotic) properties of the resulting estimates. This is a drawback, since the methods contain several important design issues and choices, so a theory for how these should be chosen is lacking. Important partial results are given in (among many papers), (Bauer *et al.* 1999), (Jansson and Wahlberg 2000), (Knudsen 2001), (Bauer and Ljung 2002), (Bauer 2000), (Chiuso and Picci 2003), and (Dahlen and Scherrer 2003). These results all deal with variance issues. The question of bias distribution when more complex systems are approximated by lower order ones has apparently not been dealt with in a comprehensive manner.

In this paper we shall review some of the basic choices in the algorithms and, by simulation, illustrate some aspects and experiences of how to select them. We shall then discuss both variance and bias aspects.

2. BASIC NOTION OF SUBSPACE METHODS

A conceptually fairly correct (but very simplified) first view of subspace methods is to say that they build a higher order ARX-model, which is then subjected to model reduction of weighted Hankel-norm type. The design variables are then associated with the orders (both number of past outputs, s_y and number of past inputs s_u) in this underlying ARX-model, and with the weighting matrices in the model reduction step. See the earlier mentioned references for details around this. A description with notation close to what we use here is given in pages 208–211 and 340–351 in (Ljung 1999).

2.1 *Choices and design variables*

The choices that have to be made can be listed as follows

(1) The number, r of forward predictions used in the model reduction step
(2) The number of past inputs, s_u and the number of past outputs, s_y used in the underlying ARX model. If $s_y = 0$ we may talk about an *output error model.*
(3) The matrices P_1 and P_2 that are used for pre- and post-multiplication of the predictor matrix in the model reduction step.
(4) The procedure for selecting model order. This amounts to analyzing the singular values of the above mentioned matrix.
(5) Once A and C are determined from the factorized matrix, the B and D matrices can be estimated in different ways.

2.2 *The subspace identification code in the* System Identification Toolbox

In this paper we shall perform the comparisons using the System Identification Toolbox, version 5, (Ljung 2000). The command is `N4SID`. This contains an implementation of a subspace algorithm with the following features:

- The horizons (property `N4Horizon`), $[r \; s_y \; s_u]$ can be chosen independently. Setting this property to `'auto'` (default) means that the Akaike criterion (AIC) is used to determine a suitable horizon, by estimating many ARX models of different orders. Then $s_u = s_y$, unless `DisturbanceModel` has been set to `'none'` (i.e. K set to zero in (1a)), in which case $s_y = 0$.
- Two different weightings (property `N4Weight`) (P_1, P_2), namely those that are known as CVA and MOESP are supported. (see page 351 in (Ljung 1999)). Earlier versions of the toolbox contained a third choice of weighting matrices, known as N4SID. These weightings, however seem to be inferior to CVA and MOESP. The routine has kept is name, `N4SID`, though.
- The A and C matrices are estimated from the shift properties of the estimated observability matrix (one of the factors obtained by the SVD step). The matrix K (in case `DisturbanceModel = Estimate`) is also directly estimated from the SVD-factors. For fixed A and C the estimation of B and D in (1a) is a linear regression problem. This is used in `N4SID`. The only question is whether K should be used or not:

$$y = (C(qI - A)^{-1}B + D)u \qquad (2a)$$

or

$$y = (C(qI - A + KC)^{-1} + D)u \qquad (2b)$$

Both these expressions are linear regressions in B and D (and the initial state $x(0)$ if estimated) for given A, C, K. The second case corresponds to a pre-whitening of the noise in the linear regression and is chosen if `focus = 'prediction'`. The former choice is made in case `focus = 'simulation'`.

2.3 *Format of comparisons*

In the following a number of comparisons of model qualities are made. The format for all these comparisons are as follows:

250 different, stable SISO systems or order 8 have been randomly generated. To the system was added a noise description, corresponding to a Kalman gain K. The systems were simulated with a random binary input signal (`u = idinput(N,'rbs',[0 0.2])`). $N = 1200$ points were used with added noise according to the noise model. The size of the noise source was always adjusted to that a signal-to-noise ratio of 10 (amplitude-wise) was obtained at the output.

For each of these 250 data sets a model was generated using various design options. The quality of each model was evaluated as the norm of the difference between the true frequency function and the model one. (
`f0 = freqfunc(m0)`,
`f1 = freqfunc(m)`
`fit = norm(squeeze(f0-f1))`. To illustrate the influence of the design variables, they were selected in pairs giving two models at a time to compare, one for each choice of design variable. The fit for the models for each data set were plotted against each other, thus creating a scatter plot with 250 points. The axes scale was chosen so that about 90% of these points were shown (to avoid strange scalings due to outliers.) A point below the bisecting line thus indicates that for that data set, the "y-axis-method" gave a better model than the "x-axis-method".

2.4 *The Prediction error method*

We shall occasionally compare models estimated using various subspace methods with those using prediction error methods. Such methods use a parameterization of the state-space matrices (canonical or full parameterizations) and minimize the prediction errors (the determinant of the sample covariance matrix of the prediction errors) with respect to these parameters. See (Ljung 1999) for a comprehensive treatment.

3. BIAS CONSIDERATIONS

When estimating models using data from complex systems, the resulting model suffers from errors of two different types:

- The *bias error*, which stems from the fact that the true system cannot be represented within the chosen model class. If the true system happens to be linear, one can think of the bias error as a difference between the true frequency function and that of the "best model" available in the model class.
- The *variance error*, which originates from the disturbances that affect the data. For noise free data, the variance error would be zero.

The total (mean square) model error is thus the sum of the variance error and (the square of) the bias error.

For many methods it is more difficult to analyze the bias error than the variance error. For prediction error methods, cf Section 2.4 the distribution of the bias error over frequencies can be characterized, see, e.g. (Ljung 1999), but for subspace methods this topic has apparently not been discussed. One comment can be made though: If the B and D matrices are identified separately (for fixed A, C, K) as a linear regression, this indeed is a prediction error method, and the bias distribution result can be applied. In particular it means that the choice (2a) corresponds to a weighting for the approximation that is given by the input spectrum, while (2b) puts more weight to frequencies where the predictor is important.

The simulations described in this section are performed as in Section 2.3. The true systems are of 8th order and the models are of second order.

3.1 *Influence of weighting matrices*

The first test is to see if there is a systematic difference between the CVA and MOESP choices of weighting matrices. The result is shown in Figure 1. This shows a clear advantage for CVA (by 68 %)

3.2 *Influence of Prediction Horizons*

There is no particular reason to believe that the prediction horizons (s_y and s_u) should affect the approximation qualities. In Figure 2 the use of long horizons ([30 30 30]) is compared to short horizons ([4 4 4]). The long horizon performs better. A possible explanation could be that the underlying ARX model with these orders have a better chance of picking up the dynamics of the complex system.

3.3 *Influence of* `'focus'`

Setting `'focus' = 'estimation'`, that is estimating B from (2a) rather than from (2b), should

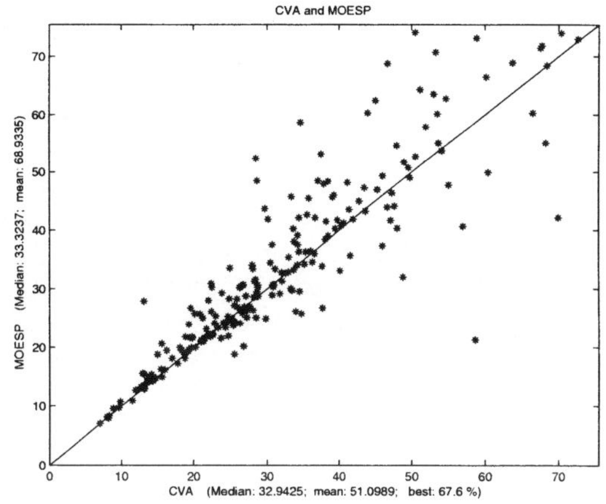

Fig. 1. CVA weighting (x-axis) versus MOESP weighting (y-axis). CVA is the better choice in 67 % of the cases.

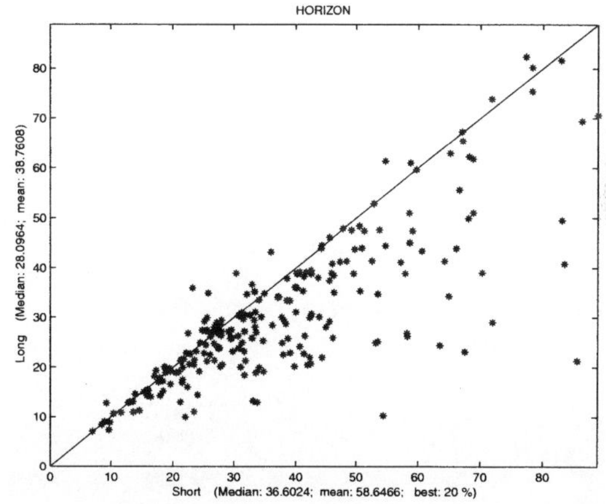

Fig. 2. A comparison between "long" and "short" prediction horizons. The long horizon is the better choice in 80% of the cases.

favor the fit of the low order model, according to our earlier discussion. This is confirmed in Figure 3. Note that the fit also depends on the input spectrum. If we had evaluated a fit between model and true system, that was weighted according to the input spectrum, the effect would have been even more pronounced.

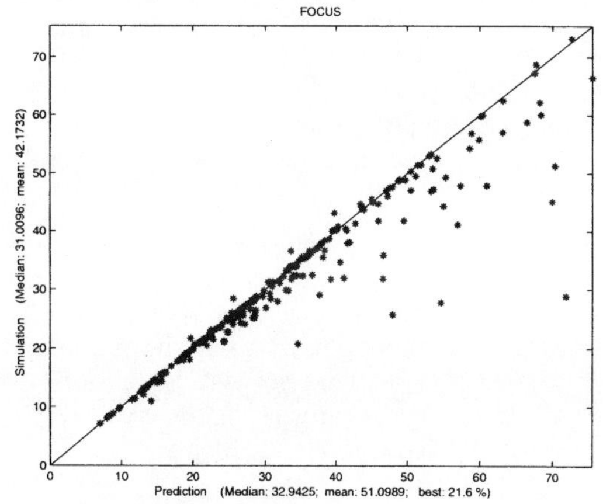

Fig. 3. A comparison between 'focus'='prediction' and 'focus'='simulation'. The simulation choice performs better in 79% of the cases.

3.4 *Influence of Subspace/PEM*

A further question is whether using further PEM iterations for the model obtained with the "simulate" focus will improve the approximation. It is expected that this should be the case, since the also the A and C matrices are adjusted to improve the fit as an optimization problem. As seen in Figure 4, the effect quite clear. Even more pronounced is a comparison between CVA/predict and PEM, where PEM performs better in 90% of the cases.

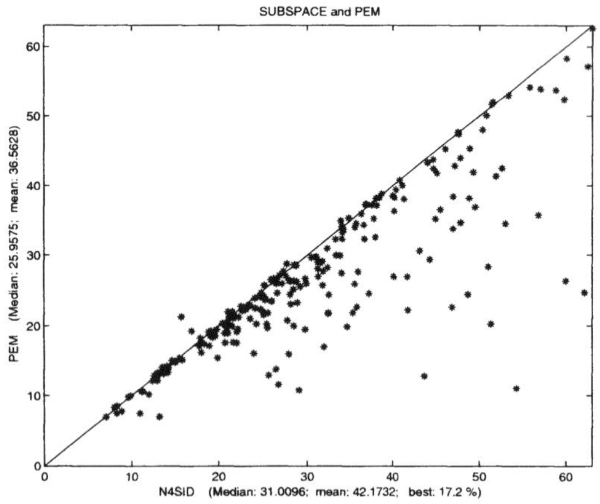

Fig. 4. A comparison between the CVA-subspace method (with "focus" = "simulation") and the model obtained by prediction error identification. PEM is the better choice in 82% of the cases.

4. VARIANCE CONSIDERATIONS

There have been many investigations of the statistical efficiency of subspace methods. The general experience is that the work well most of the time and may give models with variances that are not far from the Cramer-Rao bound. No general theory for how the variance of the estimate is affected by the various design variables has yet, been given though. A number of partial results are available, though. They indicate that the CVA-weightings have certain theoretical advantages, and will be optimal in certain cases, like the time-series case (no input) (Bauer 2000), and the case with a white noise input, (Bauer and Ljung 2002). Other relevant studies include (Gustafsson 2002) and (Deistler *et al.* 1995).

In this section we instigate the variance properties only by simulation. For that purpose, we ran Monte-Carlo simulations over 250 random models as described in Section 2.3. The simulated systems were of 8th order, and so were the models. In all cases, the direct term was set to zero ($D = 0$), i.e. a delay of 1 from input to output.

4.1 *Influence of weighting matrices*

The first question if there is a significant difference between the CVA and MOESP weighting. The simulation result is shown in Figure 5. For this case CVA has a clear edge over MOESP.

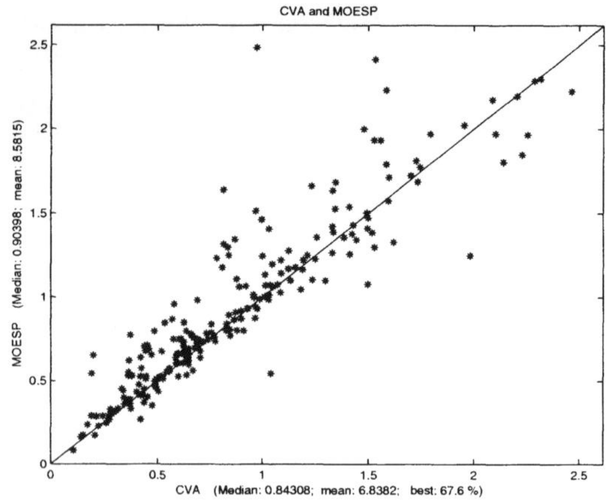

Fig. 5. Variance study: CVA (x-axis) compared to MOESP (y-axis). CVA performs better in 68% of the cases.

4.2 *Influence of Prediction Horizons*

The issue of the choice of the prediction horizons (s_y and s_u) is illustrated in Figure 6. The fixed choice [10 10 10] ("short" horizons) is compared to [30 30 30] ("long" horizons). The short horizons behave better than the long ones. It also turns out that both of them are (slightly) inferior to the AIC-based, automatic choice described in Section 2.2.

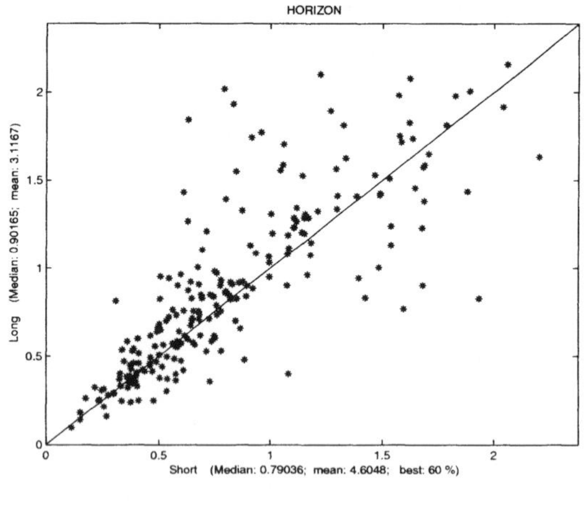

Fig. 6. A comparison between short (x-axis) and long (y-axis) prediction horizons. The short horizons are better in 60% of the cases.

4.3 *Influence of 'focus'*

Setting `focus = simulation` favors the fit of the frequency functions with weights that correspond to the input spectrum. This is at the possible loss of statistical efficiency, since the weighting/prewhitening in (2b) corresponds to the Markov (BLUE) estimate of a linear regression model. In Figure 7 the corresponding comparison is given. It is quite clear that the `prediction` choice gives estimates with better accuracy. It is interesting to compare Figure 7 with Figure 3. Depending on the degree of approximation involved in the model estimation, one might favor different foci for a fit of the model frequency function from u to y: If the bias error dominates the model error, it is better to choose `'focus'='simulation'`, otherwise `'focus'='prediction'` is the natural choice. Note also the comments on the input spectrum tha we gave in connection with the bias error.

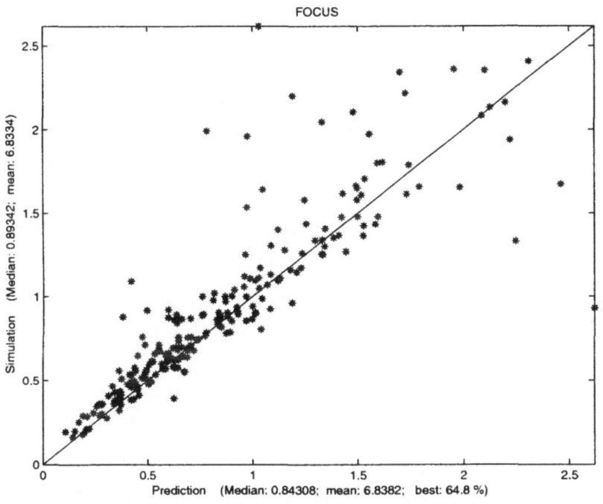

Fig. 7. The influence of the Focus: x-axis: Focus = 'prediction', y-axis: Focus='simulation'. 'Prediction' is the better choice in 65% of the cases.

4.4 *Influence of Subspace/PEM*

A key question for subspace methods is whether they really achieve the Cramer-Rao bound. Maximum-likelihood methods are know to do that asymptotically, so applying the prediction error method should give optimal accuracy (asymptotically). Now, that requires that the somewhat demanding minimization routine indeed gives the globally minimizing models, and numerical inaccuracies may hamper that. In Figure 8 the comparison is shown between the CVA-subspace estimate and the result from PEM (with `'lim'=0` and `'SSparameterization' = 'Canonical'`) (no robustification and canonical form parameterization). It is seen that PEM has a clear edge over CVA in these simulations. The result may be quite dependent on the input chosen. For a white noise Gaussian input, the two methods behave similarly. This is in line with the result of (Bauer and Ljung 2002). On the other hand, for band-limited (but still persistently exciting) inputs, PEM is considerably better than CVA.

5. THE TENNESSEE EASTMAN CHALLENGE PROCESS

The Tennessee Eastman identification and control challenge is described in (Downs and Vogel 1993). It consists of data from a chemical process, which

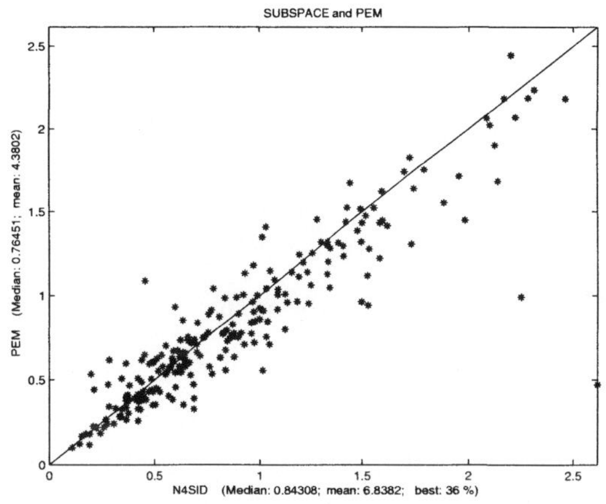

Fig. 8. A comparison between CVA-subspace method (x-axis) and the PEM-estimate (y-axis). PEM is the better choice in 64% of the cases.

is quite complex. The process has 7 inputs and 10 outputs and several non-linear features. The data themselves are simulated from a full scale non-linear model, and the challenge is to obtain good models from these data.

The data consists of 13141 samples. In in what follows, the first 6570 values were used for identification and the remaining 6571 values for validation purposes. A small portion of the data is shown in Figure 9.

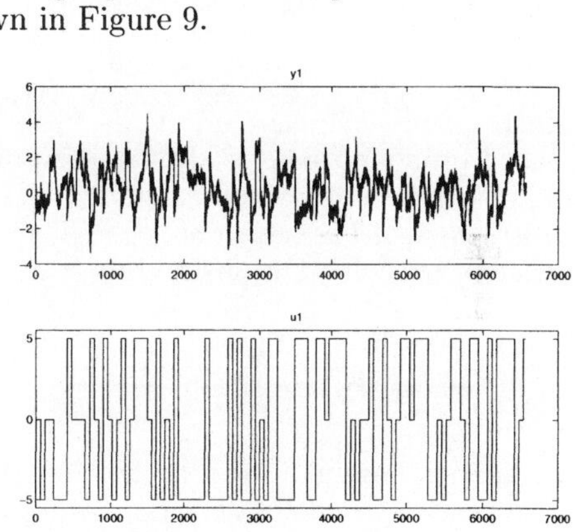

Fig. 9. A portion of input 1 and output 1 of the Tennessee Eastman data

The data has been analyzed, e.g. in (Juricek *et al.* 2001), They found that a CVA model of order 23 works well.

We have applied subspace methods with different options to the data, as well as the prediction error method. For direct comparisons with (Juricek *et al* 2001), we also chose the order $n = 23$, For the prediction horizons we used the defaults (`N4hor = auto`) in the code which gave the horizon [24 10 10]. For the different models we have tested their ability to reproduce the validation data in simulation. That is, we computed the fit in `[yh,fits] = compare(datv,mod)`.

The results are shown in Table 1. The different models correspond to subspace estimates with CVA and MOESP weightings, with focus='prediction' or focus='simulation' (CVA(s) and MOE(s) for simulation focus). The table also includes the prediction error estimated model (PEM) (with canonical parameterization and no robustification, 'lim'=0 and a simulation focus), as well as the model estimated in (Juricek *et al* 2001), ("Juri"). The measure shown is

$$fit = (1 - \frac{\text{norm}(y - \hat{y})}{\text{norm}(y - \text{mean}(y))}) * 100$$

It thus differs from the common R^2 measure in that it is the norm and not the squared norm that is used. This generally gives lower figures for the fit.

CVA	MOESP	CVA(s)	MOE(s)	PEM	Juri
72.98	70.94	79.62	79.67	81.30	80.38
73.42	69.09	79.78	79.44	81.22	80.72
72.57	50.57	89.24	89.39	88.92	82.87
84.97	81.82	91.44	89.52	91.93	91.87
79.51	67.82	90.85	89.68	91.32	88.54
72.74	50.80	89.23	89.37	88.92	82.89
72.74	51.03	89.50	89.64	89.13	83.15
73.38	20.36	65.88	51.29	88.84	88.13
76.04	78.48	82.40	83.54	86.76	85.40
81.66	76.74	87.37	88.55	89.72	89.15

Table 1. The simulation precision for the 10 outputs (the rows) for models corresponding to different design variables.

6. CONCLUSIONS

We have illustrated some aspects of user choices in subspace identification methods. The results are based on simulations and examples, and one should be careful to draw far-reaching conclusions. However, the features shown here could be of some guidance in choosing options and methods.

7. ACKNOWLEDGMENTS

This research was supported by the Swedish National Research Council (VR).

Ben Juricek kindly provided the Tennessee Eastman data, as well as the model ("Juri") that was estimated in the paper (Juricek *et al* 2001). I am also grateful to Dietmar Bauer who helped with several comments.

8. REFERENCES

Akaike, H. (1976). Canonical correlation. In: *Advances in System Identification* (R.K Mehra and D.G. Lainiotis, Eds.). Academic Press. New York.

Bauer, D. and L. Ljung (2002). Some facts about the choice of the weighting matrices in Larimore type of subspace algoithms. *Automatica* **38**(5), 763–774.

Bauer, D., M. Deistler and W. Scherrer (1999). Consistency and asymptotic normality of some subspace algorithms for systems without observed inputs.. *Automatica* **35**(10), 1243–1254.

Bauer, Dietmar (2000). Asymptotic efficiency of cca subspace methods in the case of no exogenous inputs. Technical Report LiTH-ISY-R-2262. Department of Electrical Engineering, Linköping University. SE-581 83 Linköping, Sweden.

Chiuso, A. and G. Picci (2003). Asymptotic variances of subspace identification by data orthogonalization and model decoupling. In: *Proc. IFAC Symposium on Identification, SYSID'03.* Rotterdam, The Netherlands. p. This Session.

Dahlen, A. and W. Scherrer (2003). The relation of the CCA subspace method to a balanced reduction of an autoregressive model. *J. of Econometrics* p. To appear.

Deistler, M., K. Peternell and W. Scherrer (1995). Consistency and relative efficiency of subspace methods. *Automatica* **31**, 1865–1875.

Downs, J. J. and E. Vogel (1993). A plant-wide industrial process control problem. *Comp. and Chem. Eng.* **17**, 245–255.

Gustafsson, T. (2002). Subspace-based system identification: weighting and pre-filtering of instruments. *Automatica* **38**, 433–443.

Ho, B. and R. E. Kalman (1965). Efficient construction of linear state variable models from input/output functions. *Regelungstechnik* **12**, 545–548.

Jansson, M. and B. Wahlberg (2000). Om consistency of subspace methods for system identification. *Automatica* **34**(12), 1507 – 1519.

Juricek, B. C., D. E. Seborg and W. E. Larimore (2001). Identification of the tennessee eastman challenge process with subspace methods. *Control Engineering Practice* **9**(12), 1337–1351.

Knudsen, T. (2001). Consistency analysis of subspace identification methods based on a linear regression approach. *Automatica* **37**, 81 – 89.

Larimore, W. E. (1983). System identification, reduced order filtering and modelling via canonical variate analysis. In: *Proc 1983 American Control Conference.* San Francisco.

Larimore, W. E. (1990). Canonical variate analysis in identification, filtering and adaptive control. In: *Proc. 29th IEEE Conference on Decision and Control.* Honolulu, Hawaii. pp. 596–604.

Larimore, W. E. (1997). *$Adapt_X$ Automated System Identification Software, Users Manual.* Adaptics, Inc.. 1717 Briar Ridge Road, McLean, VA.

Ljung, L. (1999). *System Identification - Theory for the User.* 2nd ed.. Prentice-Hall. Upper Saddle River, N.J.

Ljung, L. (2000). *System Identification Toolbox for use with* MATLAB. *Version 5.*. 5th ed.. The MathWorks, Inc. Natick, MA.

Van Overschee, P. and B. DeMoor (1996). *Subspace Identification of Linear Systems: Theory, Implementation, Applications.* Kluwer Academic Publishers.

Verhaegen, M. (1994). Identification of the deterministic part of MIMO state space models, given in innovations form from input-output data. *Automatica* **30**(1), 61–74.

IFAC Publications
www.elsevier.com/locate/ifac

INFERRING MULTIVARIABLE DELAY AND SEASONAL STRUCTURE FOR SUBSPACE MODELING

Wallace E. Larimore

Adaptics, Inc, 1717 Briar Ridge Road, McLean, VA 22101
Phone: 703 532-0062, Fax: 703 536-3319, Email: larimore@adaptics.com

Abstract. ARX models are used for modeling the multivariable delay structure of a system. A fast order-recursive algorithm is used for factorization of a generalized inverse of a covariance matrix that may be highly illconditioned or singular. The Akaike information criterion (AIC) and generalized likelihood ratio (GLR) tests are used to decide on the most likely multivariable delay structure. Additional synthetic inputs to the system can then be defined if necessary as delayed versions of the system inputs and outputs. Subspace methods such as canonical variate analysis (CVA) are used to identify a state space model. As a result, the state order and total number of estimated parameters of the identified system can be considerably decreased giving reduced parameter estimation and prediction errors. *Copyright © 2003 IFAC*

Key Words: Delay Estimation, Subspace identification, Generalized inverse, Canonical variate analysis, System identification.

1. OVERVIEW

Subspace methods such as canonical variate analysis (CVA) have been used to obtain optimal accuracy in the identification of multivariable systems (Larimore, 1990a, 1994). In a direct application of CVA, the input-output behavior of the system is automatically modeled implicitly including any delays in the system (Schaper, Larimore, Seborg and Mellichamp, 1994), potentially at the cost of adding additional states to the model. If the delay times are a large multiple of the sample time, then the state order of the system can be considerably increased. The cost of a higher state order is the associated increase in the number of estimated parameters and the proportionate increase in the prediction error for such a model (Larimore, 1996). Process delays occur in numerous applications including transport delays in pulp and paper as well as petrochemical processes, and wave propagation delays in communication and audio systems. In traditional time series analysis, periodic or seasonal processes exhibit a delayed feedback structure when expressed in autoregressive form. These include numerous economic data as well as demand for electric power and even the pattern of human births. The use of subspace models without consideration of the delay structure can often lead to high state order models with numerous parameters leading to reduced accuracy. The methods discussed in this paper make possible the parsimonious modeling of such processes using subspace methods once the delay structure has been determined.

The approach to modeling the delay structure of the system is to consider autoregressive with inputs (ARX) models of the process along with delay structure. The delay structure can involve arbitrary delays between the multivariable inputs and outputs as well as delayed internal feedback of the past outputs to the present outputs. Because computation of the autoregressive parameters is a linear least squares problem, computationally efficient and numerically stable methods are developed for determining the delay structure. This employs a new theory using generalized inverses, or g-inverse for brevity, for obtaining solutions to least squares estimation and prediction error problems (Larimore, 2002). These problems involve the computation of the inverse of a covariance matrix that may be illconditioned or even singular. The g-inverse approach is transparent to such illconditioning and singularity so that a fast order-recursive computation is completely stable and highly accurate.

Use of the Akaike information criteria (AIC) (Akaike, 1973) as well as more traditional generalized likelihood ratio tests makes possible the precise test of hypotheses for comparison of alternative model structures in selecting the delay structure. Once the delay structure is determined, delayed inputs and outputs are considered as additional synthetic inputs. A subspace system identification procedure such as CVA is then used to obtain a state space model with additional inputs that are explicitly delayed past system inputs and outputs. The approach can also be used to model seasonal time series using delayed variables. The procedure is applied to several examples including a process with transport delays and a seasonal time series. In many industrial processes, the determination of system delays in and of itself is of great value, particularly in the monitoring of system changes and the design of control systems.

2. AUTOREGRESSIVE MODELS WITH DELAY STRUCTURE

In this section, the way in which delay information could be used in subspace model fitting is first discussed, and then autoregressive models are used for determining the delay structure.

A disadvantabe of subspace modeling methods is that the presense of long delays in the process can require a high state order and consequencely a large number of parameters and reduced modeling accuracy. So first, the types of delays are discussed that could be considered in a state space structure within the notions of past and future used in developing subspace models. A state space model is identified in the form

$$x_{t+1} = \Phi x_t + G u_t + w_t \tag{1}$$

$$y_t = H x_t + A u_t + B w_t + v_t \tag{2}$$

where x_t is a k-order Markov state and w_t and v_t are white noise processes that are independent with covariance matrices Q and R respectively. The parameters to be estimated are the coefficient matrices Φ, G, H, A, and B along with the covariance matrices Q and R. The number of estimated parameters of a state space model are

$$M_k = k(2n + m) + nm + n(n - 1)/2 \tag{3}$$

where k is the state order, n is the number of outputs and m is the number of inputs. If the variables with long delays are included as additional inputs, then a significant reduction in the total number of parameters needed to model the system can result. In such cases, the increase in the number of inputs m is more than offset by the decrease in the state order k to produce a net reduction in the number of estimated parameters M_k.

Suppose that the true system is a 2-state system with 2 inputs and 2 outputs and with a pure delay of 10 sample times from input 1 to output 1. Direct subspace modeling will produce a model of 12 states. Then one possibility is to introduce an additional synthetic input, input 3, to the system that is input 1 delayed by 10 samples. Fitting such a subspace model will produce a 2-state system. The number of parameters will be 2*(2*2+3) + 3 = 17 rather than 12*(2*2+2) + 3 = 75 corresponding to the 12 state system with 2 inputs. Another type of delay is a internal feedback or recirculation delay where the output has a pure delay and then is input to the system. Such a feedback can be handled in the same way with an additional synthetic input consisting of the output delayed appropriately. These types of delays can thus be handled very simply and naturally by subspace modeling as additional synthetic inputs with no additional changes to the method. The main problem is then how to detect and decide the likely delay structure so that appropriate delayed inputs and outputs can be added as synthetic inputs.

A very powerful method for determining the delay structure is by fitting autoregressive models with inputs (ARX) to the data with various delays present in the inputs and outputs of the ARX model. Consider the fitting of ARX models for the observed inputs u_t and outputs y_t of the form

$$y(t) = \sum_{s=1}^{p} \alpha(s) y(t-s) + \sum_{s=0}^{q} \beta(s) u(t-s) + e(t) \tag{4}$$

where p and q are respectively the AR and X orders. The functions $\alpha(s)$ and $\beta(s)$ are respectively the pulse response functions from the past outputs to the present outputs and from the past and present inputs to the present outputs. The objective is to determine the delay times of these pulse response functions. For a matrix pulse response function $h(s)$, the delay from the $j-th$ input $u_j(t)$ to the $i-th$ output is $k\Delta t$ where Δt is the sample time if

$$h_{ij}(s) = 0 \quad \text{for} \quad 0 \leq s < k; \quad h_{ij}(k) \neq 0. \tag{5}$$

For a pulse response function $h(s)$, define the delay matrix $\Delta(h(s))$ as the matrix of nonnegative integers containing the above delay times for all pairs (i, j). In particular, define $\Delta(\beta(s))$ as the *input-output* delay. By replacing past inputs $u(t - s)$ with past outputs $y(t - s)$ in the above argument, the *internal feedback* delay matrix $\Delta(\alpha(s))$ is defined.

An internal feedback delay will often be manifest in a different way. Suppose that the system is a low order autoregression with no delay so that there is significant correlation between the last few outputs and the present output. Suppose that in addition to this structure, there is a pure delay of 12 of the output that is feed back internally in the system. In this case the determination of such a delay is more involved but as will be seen below can be done in a similar way.

Any $n \times m$ matrix pulse response function $h(s)$ containing delays can be written as a delay structure from n inputs to mn possibly delayed variables that cascade into a $n \times mn$ pulse response function with no delays. Depending of the particular delay structure of a system, the delay structure may involve far less than mn variables. The basic idea exploited in this paper is to determine the delay structure of the autoregressive model. Then only the delayed variables are used that contribute to a net reduction of the number of estimated parameters and a corresponding decrease in the prediction error of the corresponding model. There are a number of situations where this strategy will significantly improve the accuracy of the identified model. If the delay times are long relative to the state order of the system, then a considerable improvement will result. Also if only a few of the input-output pairs involve long delays, then a substantial improvement can result.

3. STABLE ORDER-RECURSIVE COMPUTATION

In this section, a stable order-recursive computation for fitting ARX models is discussed. This is key to efficient computational solution of the problem since the approach proposed above requires the fitting of all ARX models up to some maximum order for combinations of all input/output pairs. Below, the order-recursive theory based on g-inverses as developed in Larimore (1990b, 2002) is discussed in the context of fitting ARX models for delay determination.

The fitting of numerous ARX models as discussed above along with the computation of the AIC or equivalently likelihood ratio test of hypotheses can involve a considerable amount of computation frequently of highly illconditioned data. In particular if the true order is K, then the fitting of order $K+1$ can lead to illconditioning. The fitting of ARX models from order 1 through order K involve the inversion of matrices of order 1 through K with a total number of multiplications or order K^4. There do exist order recursive computational algorithm, but these are not computational stable for illconditioned matrices that occur not infrequently in fitting ARX models of unknown order. Fortunately, a stable order-recursive algorithm was developed in Larimore (1990b) that is accurate to machine precision even for singular data that involves the inversion of singular covariance matrices. This algorithm requires order K^3 multiplications.

This is in striking contrast with a number of reports in the literature that order-recursive fitting of ARX models becomes unstable if a high enough order is considered. The typical "fast" algorithms are based on the Levinson-Durbin recursion (Marple 1987) that has numerical accuracy and stability comparable to the Cholesky decomposition (Cybenko 1980) which is strongly dependent on the matrix condition number. Of course such algorithms are order K^2 rather than K^3 as in Larimore (1990b), so that there is a price to be paid for the accurate handling of illconditioned or colinear data. On the other hand, Larimore (1990b) does not require a Toeplitz covariance structure as in (Cybenko 1980).

The initial order-recursive theory and algorithm (Larimore 1990b) was developed using a pseudoinverse that has been shown to fail for a particular counter example (Zhou and Zhu, 2002). A revised theory using a g-inverse corrects the problem and provides a complete theory with no change to the order-recursive algorithm (Larimore, 2002). Examples of fitting ARX models in Larimore(2002) show that the order-recursive algorithm is accurate to machine percision in the case of singular covariances matrices and close to that for illconditioned matrices.

A major use of a g-inverse is in finding a consistent solution even if the matrix is singular. To this the following definitions and theorem are given (Rao, 1966):
Definition 1: Consistency. Consider a $p \times q$ matrix A of any rank. For a given p-dimensional vector y, the equation $y = Ax$ is *consistent* if there exists an q-dimensional vector x for which it is satisfied.
Definition 2: G-inverse. Consider a $p \times q$ matrix A of any rank. A *generalized inverse*, or *g-inverse* for brevity, of A is a $q \times p$ matrix A^- such that if y is any vector for which there exists at least one x satisfying $y = Ax$, then $\overline{x} = A^- y$ is a solution for x in $y = Ax$.
Theorem 1: G-inverse. A^- is a g-inverse of A if and only if $AA^-A = A$.

The g-inverse is the weakest definition of a matrix that will produce solutions x for all of the consistent y in the equation $y = Ax$, i.e. it is the weakest generalization of an inverse in this sense. Thus the equation $AA^-A = A$ must be satisfied by all such generalizations of an inverse. The pseudoinverse satisfies this equation along with three other equations that are not necessarily satisfied by the g-inverse.

A key idea in the order-recursive algorithm is the factorization of a g-inverse of a possibly singular covariance matrix (Larimore, 2002).
Theorem 2: G-inverse Factorization. If Σ_{xx} is a positive semidefinite and symmetric matrix of rank m and $M\Sigma_{xx}M^T = I_m$, then a g-inverse of Σ_{xx} is

$$\Sigma_{xx}^- = M^T M \tag{6}$$

so that M is a factor of a g-inverse of Σ_{xx}.

The significance of the factorization is that the usual computation involving the inverse of a full rank covariance matrix can be replaced by the above factorization in the case of a singular covariance matrix as stated in the following theorem (Larimore 2002).
Theorem 3: Estimation for Possibly Singular Case. If $\text{rank}(\Sigma_{xx}) = m$ and $M\Sigma_{xx}M^T = I_m$, then the minimum variance estimate of y given observation of x is

$$\hat{y}(x) = \mu_y + \Sigma_{yx}M^T M(x - \mu_x) \tag{7}$$

and the estimation error covariance matrix $\Sigma_{yy/x} = \Sigma_{(y-\hat{y})(y-\hat{y})}$ is

$$\Sigma_{yy/x} = \Sigma_{yy} - \Sigma_{yx}M^T M\Sigma_{xy} \tag{8}$$

To elaborate, the solution developed in Larimore (1990b) determines a factor M of a generalized inverse Σ^- of the covariance matrix that is block lower triangular of the form

$$M = \begin{pmatrix} M_{11} & 0 \\ M_{21} & M_{22} \end{pmatrix} \tag{9}$$

with $M_{11} = M_1$ the factor M_1 of $\Sigma_{11}^\dagger$ corresponding to the variables x_1. Then the minimum variance estimate (7) can be expressed as

$$\begin{aligned} \hat{y}\begin{pmatrix} x_1 \\ x_2 \end{pmatrix} &= \begin{pmatrix} \Sigma_{yx_1} & \Sigma_{yx_2} \end{pmatrix} \begin{pmatrix} M_{11}^T & M_{21}^T \\ 0 & M_{22}^T \end{pmatrix} \\ &\quad \begin{pmatrix} M_{11} & 0 \\ M_{21} & M_{22} \end{pmatrix} \begin{pmatrix} x_1 \\ x_2 \end{pmatrix} \\ &= \Sigma_{yx_1} M_{11}^T M_{11} x_1 + (\Sigma_{yx_1} M_{21}^T \\ &\quad + \Sigma_{yx_2} M_{22}^T)(M_{21}x_1 + M_{22}x_2) \end{aligned}$$

where the term $\Sigma_{yx_1} M_{11}^T M_{11} x_1$ is the estimate $\hat{y}(x_1)$ obtained from the variables x_1 alone. Note that the second term does not involve the blocks Σ_{11} and $M_{11} = M_1$ which are generally much larger in one dimension so that the first term involves most of the computation. Similarly the estimation error covariance matrix (8) can be expressed as

$$\begin{aligned} & E\{(y-\hat{y})(y-\hat{y})^T\} \\ = \; & \Sigma_{yy} \\ & - \begin{pmatrix} \Sigma_{yx_1} & \Sigma_{yx_2} \end{pmatrix} \begin{pmatrix} M_{11}^T & M_{21}^T \\ 0 & M_{22}^T \end{pmatrix} \\ & \begin{pmatrix} M_{11} & 0 \\ M_{21} & M_{22} \end{pmatrix} \begin{pmatrix} \Sigma_{yx_1} & \Sigma_{yx_2} \end{pmatrix}^T \cdot \\ = \; & \Sigma_{yy} - \Sigma_{yx_1} M_{11}^T M_{11} \Sigma_{yx_1}^T \\ & -(\Sigma_{yx_1} M_{21}^T + \Sigma_{yx_2} M_{22}^T) \\ & (\Sigma_{yx_1} M_{21}^T + \Sigma_{yx_2} M_{22}^T)^T \end{aligned}$$

where as before the second term corresponds to the solution obtained for x_1 alone and the last term requires much less computation when the dimension of x_2 is small compared with that of x_1.

Now the main theorem is given, the order recursive computation of the factor M (Larimore, 1990b). *Theorem 4: Order-Recursive Factorization.* Let the partitioned covariance matrix

$$\Sigma = \begin{pmatrix} \Sigma_{11} & \Sigma_{12} \\ \Sigma_{21} & \Sigma_{22} \end{pmatrix} \tag{10}$$

be given with M a factor of a g-inverse of the rank j matrix Σ_{11} satisfying

$$M \Sigma_{11} M^T = I_j \tag{11}$$

and with Σ_{22} of rank ℓ. Then a factor $\overline{M}$ of a g-inverse of Σ is given by

$$\overline{M} = \begin{pmatrix} M & 0 \\ -TS^T J & TL \end{pmatrix} \tag{12}$$

where the generalized singular value decomposition (GSVD) specifies J and L satisfying

$$\begin{aligned} J \Sigma_{11} J^T &= I_j, \quad L \Sigma_{22} L^T = I_\ell, \\ J \Sigma_{12} L^T &= S = Diag(s_1 \geq \cdots \geq s_t > 0, \ldots, 0) \end{aligned}$$

and where T satisfies

$$T(I - S^T S)T^T = I_t \tag{13}$$

with t the rank of $I - S^T S$.

The computation is done using a GSVD which is equivalent to a canonical variate analysis (CVA). The numerical computations involves two singular value decompositions with rank decisions to determine variables with zero variance or unity correlation.

The way the algorithm proceeds is to build up the rows of the factor M of the g-inverse of Σ as additional variables are added to the regression computation. If the additional variables are linearly dependent on the already chosen variables, then one or more singular values of S will be unity so $I - S^T S$ will be singular or very nearly so. Then the rank t is determined to be less than the the number of additional variables and T specifies a factor giving a basis for the linearly independent part.

It turns out that the GSVD that is equivalent to a CVA of the covariance structure is an optimal weighting since covariances are normalized to correlations. As discussed in Larimore (1997), this weighting produces generalized likelihood ratio (GLR) tests on rank that have optimal properties. So the rank decisions made by the procedure are statistically optimal. Also, the procedure is invariant to scaling of the multivariable data.

4. IDENTIFICATION OF DELAY STRUCTURE FOR SUBSPACE MODELING

In this section the procedure for determination of the delay structure and using it in the subspace modeling is discussed.

The first step in the computation is to determine the delay matrices $\Delta(\alpha_s)$ and $\Delta(\beta_s)$ from the observational data. This can be accomplished by the following steps:

- Choose Maximum ARX Order. Fit ARX models using the order recursive procedure described by Larimore (1990b) that is computationally stable and well conditioned even for collinear and illconditioned data. This procedure adds lags one at a time up to some chosen upper bound on ARX order. Compute the AIC for all order ARX models fitted and choose the ARX order corresponding to the minimum AIC. This defines the optimal order for ARX modeling and is used in the CVA computations for the length of the past and future.
- Choose Optimal Input-output Delays. Now for each output component $y_i(t)$, consider the multiinput single output ARX model. Consider the order recursive procedure where the terms associated with input j are removed from the ARX model starting with lags $s = 0$, then $s = 1$, $s = 2$, ... This is the equivalent of setting the coefficients $\beta_{ij}(s) = 0$ successively for the sets $\{s = 0\}$, $\{s = 0, 1\}$, $\{s = 0, 1, 2\}$.... For each such model, the AIC measure of model fit is computed and the model with the minimum AIC is the optimal delay for input j to output i.
- Choose Optimal Autoregressive Delays. Repeat the above step but consider lags of the output $y_j(s)$ instead of lags of the inputs $u_j(s)$ to determine the presence of pure autoregressive delays. In addition, where there is no pure delay, look for a delayed response after the initial impulse response becomes insignificant. This is done by successively removing autoregressive lags after the initial response dies out until a significant response is encountered at a longer lag. This is directly related to time series seasonal models as discussed below.
- Fit Subspace Models with Synthetic Delayed Inputs. The above procedure will flag the poten-

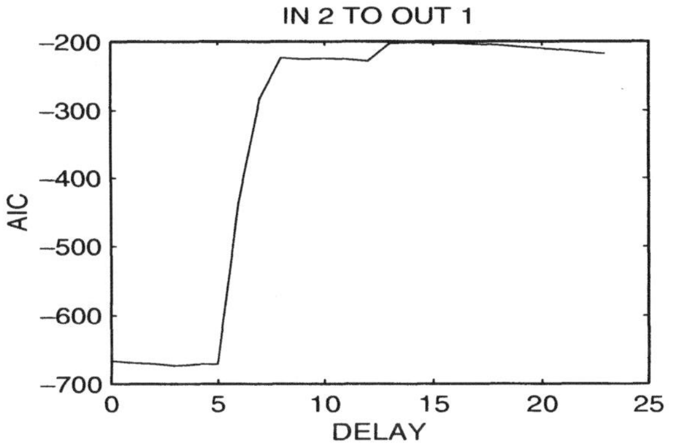

Figure 1: AIC verses delay.

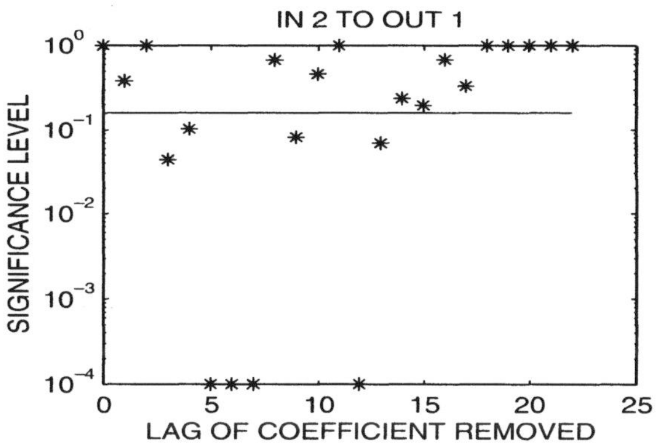

Figure 2: Significance level verses delay.

tially important delays in inputs and internal output feedback of the system. These can then be used to construct the synthetic delayed inputs to use in subspace modeling. There may be several combinations of such synthetic delayed inputs to try particularly if some of these are of low statistical significance. These alternative subspace models can be compared using the AIC to determine the more plausable models.

The above procedure has both computational and statistical advantages. The procedure is computationally efficient and numerically stable. Precise tests of hypotheses are performed to determine the optimal multivariable delay structure. Significant differential delays from one input to various outputs is handled by introducing delayed versions of that input as additional synthetic inputs to the system.

5. EXAMPLES

Several examples are given of systems with delays and delayed internal feedback. These include processes with transport delays common in the process industries as well as examples including some classical seasonal time series. It is shown that the delay and feedback structure is accurately identified. Incorporating the delay structure in the fitting of a state space model using CVA results in a marked improvement in the identified model accuracy over not using the delay structure.

5.1 Input/Output Delay Identification

The original system is a 2-input, 2-output system with input-output delay matrix $\Delta = [1\ 1; 1\ 1]$. The fitting of a CVA subspace model using the ADAPTx software gives a 6-state model estimating 43 parameters. This model was used as the basis for a simulation model with input-output delay matrix $\Delta_s = [1\ 5; 6\ 21]$. The outputs of the Delay computation are shown in Figures 1 and 2 for input 2 to output 1, but the numbers on the vertical axis should be negative in Fig. 1 and the exponents should be negative in Fig. 2.

Fig. 1 shows the AIC as a function of delay, while Fig. 2 shows the statistical significance of the test of the null hypotheses H_K that the coefficient at lag K is zero. For example, at lag 5, if the null hypothesis that the coefficient $\beta_{1,2}(5) = 0$ were true, then the probability of the observed test statistic is less than 10^{-4}, so that it is very unlikely. Thus it is concluded that the coefficient is nonzero and the pair has a delay of 5.

The plots of the other input-output pairs are similar, and imply that the best choice for the delay matrix is $\Delta_s = [1\ 5; 7\ 21]$. The input 1 to output 2 is slightly misestimated since the noise in this pair is high. Fitting of a CVA subspace model resulted in a 26 state model with minimum $AIC = -1079.45$ estimating 165 parameters.

The delay structure analysis suggests introducing an additional synthetic input of $u_3(t) = u_2(t-20)$. Identification of a CVA subspace for the 3-input, 2-output system results in a 6 state model with minimum $AIC = -1612.76$ estimating 53 parameters, only 10 more than the original system with no delays introduced. The AIC's indicate that the model with the synthetic input is much more accurate than that achieved by direct application with 26 states and 165 parameters.

5.2 Seasonal Time Series - Live Births

A delay in the internal feedback of the system is closely related to seasonal time series models or multiplicative models (Box and Jenkins, 1976). The attraction of these multiplicitive models is their parametric efficiency. The disadvantage is that the considerably flexibility is lost in modeling the actual dynamics that may not follow the imposed parametric form

Seasonal or periodic process have many important applications to rotating machinery and other quasi-periodic physical and engineering systems. One way to look at such processes is that the information of interest is clustered near lagged variables at lags slightly larger than a multiple of the seasonal period. Consider the data of the number of live human births in the U.S.

shown in thousands of births per month in Fig.3. This data has a distinct periodic behavior with a period of 12 months. Fig. 4 shows the result of fitting AR models of increasing order. While most of the information is contained in the lags 1 through 13, additional information is also contained in lags 25 and 37. Fig. 5 shows the fitting of a CVA subspace model using a past and future of 37 lags. The optimal state space order is 12 and corresponds to the minimum $AIC = 2787.77$ estimating 26 parameters.

Given the identified AR structure, a model is fitted with the three synthetic inputs $u_t = [y_{t-12}; y_{t-24}; y_{t-36}]$. The fitting of AR models of increasing order had a minimum AIC at AR order 1 so a past and future of 2 lags was used in the CVA subspace model fitting. The state order of 1 had minimum $AIC = 2671.72$ estimating 11 parameters, a considerably better fit than the original CVA model using 37 lags and no synthetic inputs.

REFERENCES

Akaike, H. (1973). Information theory and an extension of the maximum likelihood principle. *2nd International Symposium on Information Theory*, Eds. B.N. Petrov and F. Csaki, pp. 267-281. Budapest: Akademiai Kiado.

Box, G.E.P. and G.M. Jenkins (1976). *Time Series Analysis Forecasting and Control*, San Francisco: Holden-Day

Cybenko,George G. (1980). The numerical stability of the Levinson-Durbin algorithm for Toeplitz systems of equations. *SIAM J. Sci. Stat. Comput.*, **1**, pp. 303-319.

Larimore, W.E. (1990a). Canonical variate analysis for system identification, filtering, and adaptive control. *Proc. 29th IEEE Conference on Decision and Control*, Honolulu, Hawaii, December, **1**, pp. 635-9.

Larimore, W.E. (1990b). Order-recursive factorization of the pseudoinverse of a covariance matrix. *IEEE Trans. of Automatic Control*, **35**, pp. 1299-1303.

Larimore, W.E. (1994). The optimality of canonical variate identification by example. *10th IFAC Symposium on System Identification*, Copenhagen, 4-6 July.

Larimore, W.E. (1996). Statistical optimality and canonical variate analysis system identification. *Signal Processing*, **52**, pp. 131-144.

Larimore, W.E. (1999). Automated multivariable system identification and industrial applications. *Proc. American Contr. Conf.*, June 24, 1999, San Diego, CA, pp. 1148-1162.

Larimore, W.E. (2001). *ADAPTx Users Manual*, Adaptics, Inc, McLean, Virginia, USA.

Larimore, W.E. (2002). Reply to 'Comment on 'Order-recursive factorization of the pseudoinverse of a covariance matrix' '. *IEEE Trans. Automat. Contr.*, **47**, pp. 1953-7.

Marple, S.L. (1987). *Digital Signal Processing.* New Jersey: Prentice-Hall, Inc.

Rao, C.R. (1966). Generalized inverse for matrices and its application in mathematical statistics. In Festschrift Volume for J. Neyman, *Research Papers in Statistics* edited by F.N. David. New York: Wiley. pp. 263-299.

Schaper, C.D., W.E. Larimore, D.E. Seborg, D.A. Mellichamp (1994). Identification of chemical processes using canonical variate analysis. *Computers & Chemical Engineering*, **18**, pp. 55-69.

Zhou, J., and Zhu, Y. (2002). Note on 'Order-recursive factorization of the pseudoinverse of a covariance matrix'. *IEEE Trans. Automat. Contr.*, **47**, pp. 1952.

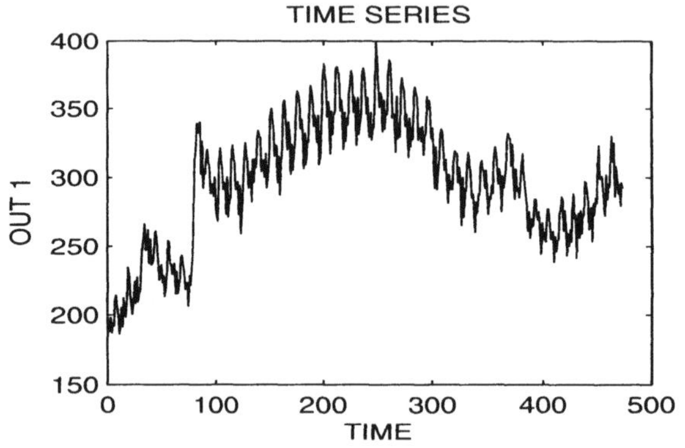

Figure 3: Live births time series.

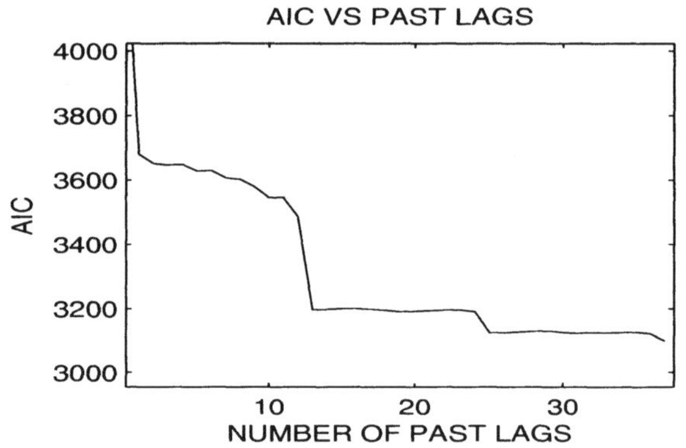

Figure 4: AIC verses order of AR model.

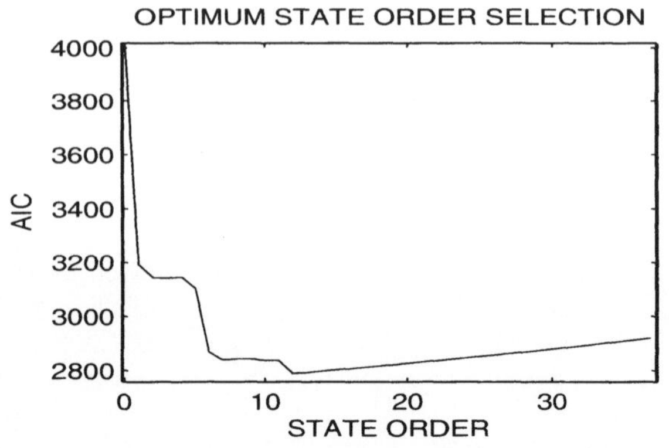

Figure 5: AIC verses order of state space model.

IFAC Publications
www.elsevier.com/locate/ifac

MATHEMATICAL RESULTS CONCERNING KERNEL TECHNIQUES

Robert Schaback *

* *Universität Göttingen*
Institut für Numerische und Angewandte Mathematik
Lotzestraße 16–18
D–37083 Göttingen
schaback@math.uni-goettingen.de

Abstract: Black–box models based on kernels K are written as mappings of the form

$$F_\alpha(x) = \sum_j \alpha_j K(x_j, x)$$

that are intended to reproduce observational input/output data pairs (x_j, y_j) in the sense $F(x_j) \approx y_j$. Such functions have been studied in a general mathematical context for quite some time, and this contribution reviews part of the known facts and provides links to a subset of the background literature. Special emphasis is given to questions of optimality and complexity within the context of black–box modelling. *Copyright © 2003 IFAC*

Keywords: Kernels. radial basis functions, reduction, optimality, black–box modelling.

1. MODELS

The goal of this contribution within SYSID 2003 is to make old and new mathematical results on kernels available to the system identification community, in particular within the context of black–box modelling (Sjoberg *et al.*, 1995). For easy alignment with the mathematical background, a black–box model is simplified here to be a parametrized nonlinear multivariate function

$$\begin{array}{c} I\!R^M \overset{F_p}{\rightarrow} I\!R^N \\ Input \mapsto Output \end{array} \quad (1)$$

where the model parameters p come from some parameter domain in $I\!R^P$. Usually there is a large number of input/output observations pairs $(x_j, y_j) \in I\!R^M \times I\!R^N$ such that $y_j = F(x_j)$ holds for the "true" transfer function or model F. The standard identification task is to find a parameter p such that

$$y_j \approx F_p(x_j)$$

holds for all observations $(x_j, y_j) \in I\!R^M \times I\!R^N$, where the parametrized class $\{F_p\}_p$ of models should be adequate for the application in question. Simulation then means the evaluation of $F_p(x)$ for new inputs x. In the context of learning theory the sample of the (x_j, y_j) is called the "training set", and finding p or F_p is called "learning" instead of "identification". From the mathematical point of view one has a nonlinear approximation problem (Braess, 1986), which turns into an interpolation problem, if exact reproduction of the input/output pairs is required. Without losing generality, one can assume the whole process to be scalar–valued, i.e. $N = 1$ in (1). Time series are a special case.

2. KERNELS

In many applications it is useful to map the input observations x_j first into an element $\Phi(x_j)$ of a larger "feature space" describing additional properties of x_j, in order to be able to use a specific distance

$dist(\Phi(x_j), \Phi(x_k))$ in feature space that allows to model the similarity of x_j with x_k much more closely than by the data x_j and x_k themselves. This key idea was termed the "kernel trick" by the community focusing on learning algorithms and "kernel machines" (**?**), but its mathematical roots date back to reproducing kernel Hilbert spaces (Meschkowski, 1962). The feature map Φ is supposed to map into some function space $\mathcal{F}$ that carries an inner product $(.,.)$ and a reproducing kernel K such that

$$\begin{aligned} \Phi(x) &= K(x, \cdot) \\ (\Phi(x), \Phi(y)) &= (K(x, \cdot), K(y, \cdot)) \\ &= K(x, y) \\ (\Phi(x), v) &= (K(x, \cdot), v) \\ &= v(x) \end{aligned} \quad (2)$$

for all $x, y \in I\!R^M$ and all $v \in \mathcal{F}$. Such kernels exist in many variations (see section 4), and the space $\mathcal{F}$ can be written as the Hilbert space closure of all images $\Phi(x) = K(x, \cdot)$ of the feature map under the above inner product. This "native" Hilbert space intrinsically belongs to the kernel in question, and each Hilbert space of functions with continuous point evaluation has a unique reproducing kernel. Thus there is a one–to–one correspondence between kernels and Hilbert spaces.

3. OPTIMAL MODELS

For modeling purposes there now is a surprising observation: one can find an optimal model without making specific assumptions about the model function F_p and its parametrization. This is what makes kernel techniques interesting for black–box modelling.

Theorem 1. Under all model functions F such that each component of F is an arbitrary function in $\mathcal{F}$ and which reconstructs the observational data exactly, the one with components of minimal norm, if it exists, is necessarily of the form

$$F_\alpha(x) = \sum_j \alpha_j K(x_j, x) \quad (3)$$

with coefficients $\alpha_j \in I\!R^N$.

Our proof sketch of this standard mathematical observation just minimizes the quadratic form (F, F) under the constraints of exact reproduction of the observational data. With Lagrange multipliers α_j one gets the variational equation

$$(F, v) - \sum_j \alpha_j v(x_j) = 0 \text{ for all } v \in \mathcal{F}$$

which leads to (3) via (2). □

In some sense this result eliminates the task of model selection, provided that a feature map and an inner product in feature space is chosen, and if one accepts the resulting minimal norm model.

A drawback of this optimality criterion is that it aims at some kind of "energy" of the model function itself, and not at the prediction quality of the model for new input/output pairs (x, y). But there is another optimality result that precisely aims at reproduction quality.

Theorem 2. Assume that the observational data come from an arbitrary "true" function F from a Hilbert space $\mathcal{F}$ with reproducing kernel K, and consider arbitrary models of the form

$$F_u(x) = \sum_j u_j(x) F(x_j) \quad (4)$$

that are linear in the observational data and use arbitrary weight functions u_j. Then for all x the specific weights $u_j^*(x) \in I\!R$ with

$$K(x, x_k) = \sum_j u_j^*(x) K(x_j, x_k) \quad (5)$$

lead to a minimal worst–case value of the relative error $|F(x) - F_u(x)|/\|F\|$.

The proof sketch for this fact uses Hilbert space duality. Starting from

$$\begin{aligned} |F(x) - F_u(x)|^2 &= |(\delta_x - \sum_j u_j(x)\delta_{x_j})F|^2 \\ &\le \|\delta_x - \sum_j u_j(x)\delta_{x_j}\|^2 \|F\|^2 \end{aligned}$$

and using (2) in the dual form $(\delta_x, \delta_y) = K(x, y)$ one can minimize the above quadratic form with respect to the real variables $u_j(x)$ to get (5) as normal equations

$$(\delta_x, \delta_{x_k}) = \sum_j u_j^*(x)(\delta_{x_j}, \delta_{x_k}). \quad \square$$

The connection of the model (4) with $u_j = u_j^*$ from (5) to the model (3) still needs explanation. If, as discussed in the next section, the matrix with entries $K(x_j, x_k)$ is nonsingular, the functions $u_j^*(x)$ are linear combinations of the $K(x, x_k)$ with the Lagrange interpolation property $u_j^*(x_k) = \delta_{jk}$. Thus the optimal version of (4) is nothing else than a rewriting of (3) in the Lagrange basis, proving that (3) is also minimizing the pointwise reproduction error at all other locations x in the sense of Theorem 2.

The "physical" meaning of the model function (3) can be illustrated by the trivial case

$$F_y(x) = \sum_j y_j \delta_{x_j}(x)$$

for a Dirac delta kernel, and (3) can be seen as a regularized version of the above, still maintaining exact reproduction, but with a different kernel like a Gaussian. Another interpretation views $K(x_j, x)$ as a similarity measure between the observations x_j and x, being large iff x is close to x_j, and then (3) lets the model return a result that lies close to y_j if x is close to x_j. This line of argument can be made more precise by stochastic assumptions, but one of the goals of this paper is to show that probabilistic arguments are irrelevant to the basics of kernel techniques. They are no more than "add-ons", and this also applies to the nondeterministic parts of learning theory.

The consequence of the arguments of this section is that after proper choice of Φ and K there is no way around working with the representation (3) in black–box modelling. But, after all, picking a very specific Φ and K that suits the application cannot be termed "black–box modelling" any more.

4. POSITIVE DEFINITE KERNELS

If (3) reconstructs the observational data exactly, one has to solve the system

$$F_\alpha(x_k) = y_k = \sum_j \alpha_j K(x_j, x_k) \qquad (6)$$

and this requires that the symmetric matrix with entries $K(x_j, x_k)$ should be nonsingular. Since it is a Gramian due to (2), it must always be positive semidefinite, but one needs additional information to prove its nondegeneracy. Fortunately, the notion of (strictly) positive definite kernels has a long history in mathematics (Stewart, 1976) and provides precisely the required positive definiteness of all such matrices. There also is the notion of conditional positive definiteness (Micchelli, 1986), but this extension is dropped here to simplify the presentation.

By the fundamental Micchelli paper of 1986, there was quite a number of useful (conditionally) positive definite functions on the market, including the radial kernels known as Gaussians, thin–plate splines, polyharmonic splines, and multiquadrics. The term "radial basis function" is used to describe kernels of the form $K(x,y) = \phi(\|x-y\|_2)$ with Euclidean invariance. Special cases are the Gaussian $\exp(-\|x-y\|_2^2)$, the multiquadrics $\sqrt{1+\|x-y\|_2^2}$, and the thin–plate splines $\|x-y\|_2^2 \log \|x-y\|_2$.

In 1995 the first smooth compactly supported positive definite radial kernels were constructed (Wu, 1995; Wendland, 1995) and Wendland's functions turned out to be polynomial and of least possible degree for fixed smoothness requirements. There is a survey (Schaback and Wendland, 2001) of the state–of–the–art of construction techniques for positive definite kernels, while other contributions (Schaback, 1999*b*; Schaback, 2000) provide specific properties and relations to integral equations (e.g. "Mercer" kernels in the language of learning theory).

It should be remarked on the side that kernels play a dominant part in integral and differential equations, since they occur in the context of fundamental solutions and Hilbert–Schmidt expansions, for instance. In approximation theory, kernels have a long history dating back to the Dirichlet kernel occurring in Fourier analysis. Finally, it should be kept in mind that one can construct conditionally positive definite kernels K without ever caring for Hilbert space arguments. Applications of kernel models need not care for Hilbert spaces at startup time. But the native Hilbert space for K will always come up later through the back door as a certain closure of the space of functions of the form $K(x,\cdot)$.

5. SUPPORT REDUCTION

If a modeling process involves an abundance of observational data, it is not reasonable to use the model (3) in its original form, because it involves a sum over the full data. Furthermore, in most applications there will be noise in the observations, and then it makes no sense to insist on exact reproduction. However, the two optimality properties of (3) suggest to stay within the overall form of (3), and to try to get away with some useful modifications.

Fortunately, there are strategies to combine both modifications into one. While allowing an error in the reproduction, the complexity of the sum in (3) can be reduced to a smaller set of observational data, called the "support vectors" in the context of learning algorithms. Between the two extreme cases

- using the full data with zero error or
- using no data ($F = 0$) with a huge error

there is a tradeoff between the complexity of a "thinned" sum in (3) and its reproduction quality. This tradeoff is still under investigation, and it remains a major challenge for the future.

The mathematical background for the mainstream of support reduction techniques is based on optimization theory and has nothing to do with modeling, learning theory, time series, probability, or statistics. Reduction simply results from imposing a Chebyshev–type (uniform) bound

$$\|y_j - F(x_j)\|_\infty \le \epsilon \qquad (7)$$

on the reproduction quality, while minimizing a suitable penalty function depending on F, for instance $\|F\|^2$ as in Theorem 1. The necessary Karush–Kuhn–Tucker conditions for the optimum will pick a set of indices j where (7) is attained with equality. This is called the "active set" in optimization theory, and it determines the (hopefully few) "support vectors" and nonzero contributions in (3) using only the indices j from the active set. The other observations are irrelevant for the optimal solution and could have been left out right from the start, if the calculation of the optimal model would ever be repeated.

So far, the modeling problem was stated in "regression" form, but the same argument applies for what is called 'classification" in learning theory. The additional ingredient is the "margin" that plays the part of (7) in a slightly different way that should be explained here for completeness of presentation. The observations x_j are grouped in two classes X^+ and X^-, and the easiest way to define a classification model is to ask for a real–valued function F that does something like

$$\begin{array}{ll} F(x_j) \geq +\gamma & \text{for all } x_j \in X^+ \\ F(x_j) \leq -\gamma & \text{for all } x_j \in X^- \end{array}$$

for some positive γ. Compared with (7) and using standard optimization theory arguments, this is perfectly fine to guarantee a reduction to a few active constraints, if some penalty function on F is minimized. Since the scaling and an additive shift do not matter, one could also ask for an F satisfying

$$\begin{array}{ll} F(x_j) \geq 2 & \text{for all } x_j \in X^+ \\ F(x_j) \leq 0 & \text{for all } x_j \in X^-. \end{array} \tag{8}$$

Unfortunately, the standard approach to classification in learning theory is somewhat more complicated, but essentially a particular case of the above straightforward attack, as is shown now. One looks for a hyperplane in feature space that optimally separates the sets $\Phi(X^+)$ and $\Phi(X^-)$. If a general hyperplane is written as $\{z \ : \ (u,z) = \beta\}$ with an element u of norm 1 in feature space (the normal vector on the hyperplane) and a real number β, then the signed distance of a point z from the hyperplane is $(z,u) - \beta$. Thus one wants a maximal positive "margin" μ with

$$\begin{array}{ll} (\Phi(x_j), u) - \beta \geq +\mu & \text{for all } x_j \in X^+ \\ (\Phi(x_j), u) - \beta \leq -\mu & \text{for all } x_j \in X^-. \end{array}$$

Introducing $z_j := \Phi(x_j)$ and $\sigma_j := \pm 1$ if $x_j \in X^\pm$ this amounts to use optimization to find a maximal positive number μ such that

$$((z_j, u) - \beta)\sigma_j \geq \mu \text{ for all } x_j.$$

The pair (u, β) allows a renormalization, and one can divide the inequalities by μ. Thus the above optimization is equivalent to a minimization of $\|u\|^2$ under the linear Chebyshev–type constraints

$$((z_j, u) - 1)\sigma_j \geq 1 \text{ for all } x_j$$

that allow a reduction argument to active sets based on the Karush–Kuhn–Tucker conditions. The connection to (8) is easily made when taking u as a scalar–valued F and applying (2) in the form

$$(z_j, u) = (z_j, F) = (K(x_j, \cdot), F) = F(x_j)$$

to get

$$(F(x_j) - 1)\sigma_j \geq 1 \text{ for all } x_j,$$

which turns out to be exactly the same as (8). Note that (8) works for any penalty on F, while the the classical geometric margin argument requires a penalty based on $\|F\|$.

6. SPECIAL REDUCTION TECHNIQUES

Besides using Chebyshev–type constraints as in (7) and (8) there are other techniques to reduce the complexity of (3) while allowing some tolerable reproduction error.

"Greedy" methods (DeVore and Temlyakov, 1996; Schaback and Wendland, 2000; Hon *et al.*, 2001) were used to solve (6) partially, working iteratively on the equations where the reproduction error $F_\alpha(x_k) - y_k$ is still too large. It turns out that exact reproduction of a small subset of the observational data often yields small errors on the rest, but research is still incomplete.

The other methods mentioned here are trying to localize the problem somehow. Using fast multipole expansions of kernels (Beatson and Newsam, 1992; Powell, 1993; Beatson and Greengard, 1997; Beatson and Light, 1997; Beatson and Newsam, 1998), one can lump "far" points x_j together and treat them computationally as one, leading to very good reductions in the complexity of solving the system (6) and evaluating the sum in (3). Another localization technique uses partitions of unity (Wendland, 2002) combined with rather arbitrary local models. This technique does not require expansions and allows fairly general applications in modelling. If the partitions of unity satisfy a certain stability property, the global reproduction error can be bounded by the local errors in the subproblems.

It is an interesting open research area to compare and to combine the various reduction techniques. In particular, on can possibly insert greedy, localization, and multipole techniques into advanced methods to solve the huge quadratic linearly constrained problems of section 5 by sophisticated optimization methods.

7. REPRODUCTION QUALITY AND STABILITY

There is a well–established mathematical literature (Duchon, 1978; Madych and Nelson, 1988; Madych and Nelson, 1990; Madych and Nelson, 1992; Wu and Schaback, 1993; Schaback, 1999*a*; Buhmann, 2000) on the error committed by kernel models of the form (3), long before learning machines and black–box modelling with kernels were fashionable. The results will be useful for modelling purposes, but for space limitations only a short summary is possible here, extending an earlier survey (Schaback, 1997).

When experimenting with kernel models (3) and corresponding systems (6) it turns out that results often depend crucially on the scaling of the kernel in relation to the density of the observational data. To quantify the latter, the fill distance

$$h_{X,\Omega} := \sup_{y \in \Omega} \min_{x_j \in X} \|y - x_j\|_2$$

of the set $X = \{x_j\}$ of observational data within an enclosing domain Ω is useful. Then one looks at sequences of observational data such that the fill distance $h_{X,\Omega}$ tends to zero, and the goal is to prove convergence in the sense that the reproduction error tends to zero as a function of $h_{X,\Omega}$. If the kernel K is fixed throughout this process, the situation is called "nonstationary" within approximation theory, while a "stationary" setting scales the kernel with $h_{X,\Omega}$ to keep the data distance and the kernel width proportional.

For the stationary setting, theory (Buhmann, 1989) says that integrable kernels like the Gaussian, the inverse multiquadric or the compactly supported Wendland kernels cannot yield convergence, though in practice one often observes that the errors are small enough to keep the user satisfied. In fact, the error usually decreases if the ratio of the kernel width and the fill density (i.e. the "bandwidth" in the matrix of (6)) is increased, and this is sufficient in many cases to keep the error below a tolerable level.

In the nonstationary setting the reproduction error always behaves like a power $h^k_{X,\Omega}$ where $k > 0$ increases with the smoothness of K. If the kernel is analytic (this occurs for the Gaussian and for multiquadrics, for instance), the error decreases even exponentially like $\exp(-c/h_{X,\Omega})$ with a positive constant c, at least in theory (Madych and Nelson, 1992).

However, this fantastic convergence behavior comes at a price. In practice, the linear systems (6) get more and more ill–conditioned (Narcowich and Ward, 1991; Narcowich and Ward, 1992) when $h_{X,\Omega}$ gets small, and this effect is dramatically increasing with the smoothness of K. It can be proven (Schaback, 1995) that good reproduction always comes with instability and vice versa, while additional smoothness of the kernel boosts both of them, unfortunately. If the user does not apply additional techniques like preconditioning or localization, the best choice of scale usually is the one that works close to the condition limits of the machine. Thus preconditioning is another important topic (Dyn *et al.*, 1983; Jetter and Stöckler, 1995; Beatson *et al.*, 1999; Fasshauer and Jerome, 1999; Hon and Kansa, 2000) in the context of making (3) work for large–scale application problems. Even for Gaussians, which show extremely good reconstruction quality and catastrophic instability, there was no well–established preconditioning technique so far, but the situation will improve (Schaback, 2002).

The above discussion was restricted to exact reconstruction of data from functions in the Hilbert space associated to the kernel. This "native" Hilbert space is very small when the kernel is very smooth, and thus in theory the user must make sure to use a kernel that is not too smooth. However, if the reconstruction is allowed to be not exact, or if there is contamination by noise, one can observe in practice and prove (Schaback, 1996) that the convergence rate in the nonstationary setting adapts in an optimal and local way to the smoothness of the unknown function supplying the observational data.

This is a partial result concerning the tradeoff between complexity of (3) and the reproduction quality, but the proof technique, when studied in detail, only treats the case where the actually used observational data still "covers" the whole data domain Ω. Compared to the reduction technique of section 5 this is a worst–case scenario that does not exploit specific features of the observational data.

8. ACKNOWLEDGEMENT

Special thanks go to Alexander Hornstein and Ulrich Parlitz for help with a first version of this manuscript.

REFERENCES

Beatson, R.K. and G.N. Newsam (1992). Fast evaluation of radial basis functions: I. Advances in the theory and applications of radial basis functions. *Comput. Math. Appl.* **24**(12), 7–19.

Beatson, R.K. and G.N. Newsam (1998). Fast evaluation of radial basis functions: moment based methods. *SIAM J. Comput.* **19**, 1428–1449.

Beatson, R.K. and L. Greengard (1997). A short course on fast multipole methods. In: *Wavelets, Multilevel Methods and Elliptic PDEs* (M. Ainsworth, J. Levesley, W. Light and M. Marletta, Eds.). pp. 1–37. Oxford University Press.

Beatson, R.K. and W.A. Light (1997). Fast evaluation of radial basis functions: Methods for two–dimensional polyharmonic splines. *IMA Journal of Numerical Analysis* **17**, 343–372.

Beatson, R.K., J.B. Cherrie and C.T. Mouat (1999). Fast fitting of radial basis functions: Methods based on preconditioned GMRES iteration. *Advances in Computational Mathematics* **11**, 253–270.

Braess, D. (1986). *Nonlinear Approximation Theory*. Springer, Berlin.

Buhmann, M.D. (1989). Multivariable interpolation using radial basis functions. PhD thesis. University of Cambridge.

Buhmann, M.D. (2000). Radial basis functions. *Acta Numerica* **10**, 1–38.

DeVore, R. A. and V. N. Temlyakov (1996). Some remarks on greedy algorithms. *Advances in Computational Mathematics* **5**, 173–187.

Duchon, J. (1978). Sur l'erreur d'interpolation des fonctions de plusieurs variables pas les D^m–splines. *Rev. Française Automat. Informat. Rech. Opér. Anal. Numer.* **12**(4), 325–334.

Dyn, N., D. Levin and S. Rippa (1983). Surface interpolation and smoothing by "thin plate" splines. *Journal of Approximation Theory* **38**, 445–449.

Fasshauer, G. and J. W. Jerome (1999). Multistep approximation algorithms: Improved convergence rates through postconditioning with smoothing kernels. *Advances in Computational Mathematics* **10**(1), 1–27.

Hon, Y.C. and E.J. Kansa (2000). Circumventing the ill-conditioning problem with multiquadric radial basis functions: applications to elliptic partial differential equations. *Comput. Math. Applic.* **39**, 123–127.

Hon, Y.C., R. Schaback and X. Zhou (2001). Adaptive greedy algorithms for solving large RBF collocation problems. manuscript.

Jetter, K. and J. Stöckler (1995). A generalization of de Boor's stability result and symmetric preconditioning. *Advances in Computational Mathematics* **3**, 353–367.

Madych, W.R. and S.A. Nelson (1988). Multivariate interpolation and conditionally positive definite functions. *Approximation Theory and its Applications* **4**, 77–89.

Madych, W.R. and S.A. Nelson (1990). Multivariate interpolation and conditionally positive definite functions II. *Mathematics of Computation* **54**, 211–230.

Madych, W.R. and S.A. Nelson (1992). Bounds on multivariate polynomials and exponential error estimates for multiquadric interpolation. *Journal of Approximation Theory* **70**, 94–114.

Meschkowski, H. (1962). *Hilbertsche Räume mit Kernfunktion*. Springer, Berlin.

Micchelli, C.A. (1986). Interpolation of scattered data: distance matrices and conditionally positive definite functions. *Constructive Approximation* **2**, 11–22.

Narcowich, F.J. and J.D. Ward (1991). Norm of inverses and condition numbers for matrices associated with scattered data. *Journal of Approximation Theory* **64**, 69–94.

Narcowich, F.J. and J.D. Ward (1992). Norm estimates for the inverses of a general class of scattered–data radial–function interpolation matrices. *Journal of Approximation Theory* **69**, 84–109.

Powell, M.J.D. (1993). Truncated Laurent expansions for the fast evaluation of thin plate splines. *Numer. Algorithms* **5**(1–4), 99–120.

Schaback, R. (1995). Error estimates and condition numbers for radial basis function interpolation. *Advances in Computational Mathematics* **3**, 251–264.

Schaback, R. (1996). Approximation by radial basis functions with finitely many centers. *Constructive Approximation* **12**, 331–340.

Schaback, R. (1997). On the efficiency of interpolation by radial basis functions. In: *Surface Fitting and Multiresolution Methods* (A. LeMéhauté, C. Rabut and L.L. Schumaker, Eds.). pp. 309–318. Vanderbilt University Press, Nashville, TN.

Schaback, R. (1999*a*). Improved error bounds for scattered data interpolation by radial basis functions. *Mathematics of Computation* **68**, 201–216.

Schaback, R. (1999*b*). Native Hilbert spaces for radial basis functions I. In: *New Developments in Approximation Theory* (M.D. Buhmann, D. H. Mache, M. Felten and M.W. Müller, Eds.). pp. 255–282. Number 132 In: *International Series of Numerical Mathematics*. Birkhäuser Verlag.

Schaback, R. (2000). A unified theory of radial basis functions (native Hilbert spaces for radial basis functions II). *J. Comp. Appl. Math.* **121**, 165–177.

Schaback, R. (2002). Multivariate interpolation by polynomials and radial basis functions. Manuscript.

Schaback, R. and H. Wendland (2000). Adaptive greedy techniques for approximate solution of large RBF systems. *Numer. Algorithms* **24**, 239–254.

Schaback, R. and H. Wendland (2001). Characterization and construction of radial basis functions. In: *Multivariate Approximation and Applications* (N. Dyn, D. Leviatan, D.and Levin and A. Pinkus, Eds.). pp. 1–24. Cambridge University Press.

Sjoberg, J., Q. Zhang, L. Ljung, A. Benveniste, B. Delyon, P. Glorennec, H. Hjalmarsson and A. Juditsky (1995). Nonlinear black-box modeling in system identification: a unified overview.

Stewart, J. (1976). Positive definite functions and generalizations, an historical survey. *Rocky Mountain J. Math.* **6**, 409–434.

Wendland, H. (1995). Piecewise polynomial, positive definite and compactly supported radial functions of minimal degree. *Advances in Computational Mathematics* **4**, 389–396.

Wendland, H. (2002). Fast evaluation of radial basis functions: Methods based on partition of unity. In: *Approximation Theory X: Wavelets, Splines, and Applications* (C. K. Chui, L. L. Schumaker and J. Stöckler, Eds.). pp. 473–483. Vanderbilt University Press.

Wu, Z. (1995). Multivariate compactly supported positive definite radial functions. *Advances in Computational Mathematics* **4**, 283–292.

Wu, Z. and R. Schaback (1993). Local error estimates for radial basis function interpolation of scattered data. *IMA Journal of Numerical Analysis* **13**, 13–27.

Manuscripts and preprints can be obtained from `http://www.num.math.uni-goettingen.de/schaback/research/group.html`

IFAC Publications
www.elsevier.com/locate/ifac

MULTI-OUTPUT SUPPPORT VECTOR REGRESSION

Emmanuel Vazquez * **and Eric Walter** *

* *Laboratoire des Signaux et Systèmes, 91192, Gif-sur-Yvette, France*
CNRS – Supélec – Université Paris-Sud

Abstract: Support vector regression builds a model of a process that depends on a set of factors. It traditionally considers one output at a time, which means that advantage cannot be taken of the correlations that may exist between outputs. The purpose of this paper is to show how the body of knowledge accumulated by geostatisticians on Kriging and its extensions over the last 40 years can help extend support vector regression to the multi-output case and provides guidance for the choice of a suitable kernel for a given application, a recurrent, fundamental and largely open question. *Copyright © 2003 IFAC*

Keywords: Black-box modeling, Kriging, MIMO systems, Nonlinear regression, Regularization, Reproducing kernels, Robust estimation, Support vector machines

1. INTRODUCTION

Support Vector Regression (SVR) belongs to the category of reproducing-kernel methods, just as *splines* (Wahba, 1990), *Kriging* (Chiles and Delfiner, 1999) (also known under the name of Gaussian processes (Williams, 1997)) and methods involving *radial basis functions* (Schaback, 1999). Based on the theory of Support Vector Machines, SVR is now a well-established method for designing *black-box* models in engineering. A frequently felt difficulty with SVR, however, is the lack of guiding rules to choose a kernel for a given application. Kriging is seldom mentioned in the literature in this context, although it turns out to be very useful for this task (the kernel is viewed as the covariance of a second-order Gaussian process).

The aim of SVR is to build a model f_{sv} of the output of a process or system that depends on a set of factors. The values taken by these factors will be stored in a vector $\mathbf{x} \in \mathbb{R}^d$. SVR is traditionally used with only one output, and the multi-output case is then dealt with by modeling each output independently of the others. The purpose of this paper is to extend SVR to multi-output systems by considering the so-called *Cokriging* method (Chiles and Delfiner, 1999), which is the multi-variable version of Kriging. This will make it possible to take advantage of the possible correlations between the outputs to improve the quality of the predictions provided by the model.

First, the SVR framework will be presented in order to make notation clear. The link between a given kernel and the covariance of the corresponding Gaussian process will then be recalled in the case of a single-output system. In order to extend SVR to the multi-output case we shall then present Cokriging and show how it can be adapted to the framework of SVR.

2. CLASSICAL SVR

Support vector regression is based on n elementary experiments during which the vector of factors $\mathbf{x}$ is assigned values specified by the user – within an admissible domain $\mathscr{X}$ of $\mathbb{R}^d$ – and the resulting system output is measured. The value taken by $\mathbf{x}$ during the i-th elementary experiment, $i \in \{1, \cdots, n\}$, will be denoted by $\mathbf{x}_i$, and the corresponding (scalar) system output by $f(\mathbf{x}_i)$.

SVR builds a function

$$f_{\mathrm{sv}} : \begin{array}{ccc} \mathbb{R}^d & \rightarrow & \mathbb{R} \\ \mathbf{x} & \mapsto & (w, \psi(\mathbf{x}))_{\mathscr{F}} + b \end{array} \quad (1)$$

minimizing the functional

$$\frac{1}{2}||w||^2_{\mathscr{F}} + C\sum_{i=1}^{n}[f_{\mathrm{SV}}(\mathbf{x}_i) - f(\mathbf{x}_i)]_{\varepsilon}\,. \tag{2}$$

In these equations, b is a *bias* term and $C \in \mathbb{R}_+$ is a parameter chosen by the user. $\psi(.) : \mathbb{R}^d \to \mathscr{F}$ is an application mapping the space of factors into a so-called *feature space* $\mathscr{F}$, typically a space of functions $\mathbb{R}^d \to \mathbb{R}$. Finally, $[\cdots]_\varepsilon$ is the *ε-insensitive* loss function. In practice, $\mathscr{F}$ and the functions $w \in \mathscr{F}$ and $\psi(\mathbf{x}) \in \mathscr{F}$ are deduced from a kernel $k(.,.) : \mathbb{R}^d \times \mathbb{R}^d \to \mathbb{R}$ involved in the computation of a scalar product:

$$w = \sum_i w_i k(\mathbf{x}_i, .)\,, \tag{3}$$

$$\psi(\mathbf{x}) = k(\mathbf{x}, .)\,, \tag{4}$$

$$(w, \psi(\mathbf{x}))_{\mathscr{F}} = \sum_i w_i k(\mathbf{x}_i, \mathbf{x})\,. \tag{5}$$

The functional (2) may be interpreted as a classical regularization approach to solving an ill-posed inverse problem, with its first term a *regularization term* (a functional of w). This kernel-based regularization has received a number of interpretations, within the theory of regularization itself (Smola *et al.*, 1998), and in the context of the theories of splines and Gaussian processes (Wahba, 1998; Jaakkola and Haussler, 1999). The second part of (2) is a *data fidelity term*, involving the ε-insensitive loss function. This loss function comes from the principle of best discrimination and the notion of optimal hyperplane (Vapnik, 1995) and a more direct interpretation is provided by the theory of robust regression (see, for instance, (Huber, 1981; Schölkopf and Smola, 2002)). The value given to $\varepsilon \in \mathbb{R}_+$ can be used to control the complexity of the model.

It is well known that the feature space $\mathscr{F}$ is a reproducing kernel Hilbert space (rkhs) where $k(.,.)$ plays the role of the reproducing kernel, with the following properties:

(1) $k(\mathbf{x}, .) \in \mathscr{F}$,
(2) $f(\mathbf{x}) = (f, k(\mathbf{x}, .))_{\mathscr{F}}, \forall f \in \mathscr{F}$,
(3) $k(\mathbf{x}, \mathbf{y}) = \langle \delta_{\mathbf{x}}, k(., \mathbf{y})\rangle_{\mathscr{F}^*, \mathscr{F}} = (\delta_{\mathbf{x}}, \delta_{\mathbf{y}})_{\mathscr{F}^*}$,

where $\mathscr{F}^*$ is the dual space of $\mathscr{F}$, and $\delta_{\mathbf{x}} \in \mathscr{F}^*$ the evaluation functional at $\mathbf{x}$.

The SVR feature space $\mathscr{F}$ will now be reinterpreted as a space of random variables in order to establish the link between SVR and Kriging.

3. THE LINK WITH KRIGING

3.1 *Kernels and covariances*

The connection between rkhs and random processes is now well documented. See (Wahba and Kimeldorf, 1970) in the context of splines, (Matheron, 1981) in the context of Kriging, and (Wahba, 1998) in the context of SVM.

Kriging (Chiles and Delfiner, 1999; Matheron, 1963) is a probabilistic method that can be used to approximate or interpolate data, just as SVR does but without the robust regression property provided by the ε-insensitive loss function. The observed data are considered as realizations of a second-order random process $F(\mathbf{x})$. Thus, for each experiment i, $f(\mathbf{x}_i)$ is a realization of the random variable $F(\mathbf{x}_i)$. The Kriging estimate is the *best linear unbiased prediction* (BLUP) of $F(\mathbf{x})$ in the space $\mathscr{H}_S$ generated by the random variables $F(\mathbf{x}_1), \cdots, F(\mathbf{x}_n)$. This estimate can be written as a linear combination

$$\hat{F}(\mathbf{x}) = \sum_{i=1}^{n} \lambda_{i,\mathbf{x}} F(\mathbf{x}_i)\,, \tag{6}$$

and is obtained by orthogonal projection of $F(\mathbf{x})$ onto $\mathscr{H}_S$ (in an appropriate Hilbert space).

Equation (6) can also be written as a stochastic integral

$$\hat{F}(\mathbf{x}) = \int_{\mathbb{R}^d} F(\mathbf{x}) d\lambda(\mathbf{x})\,, \tag{7}$$

where λ is a finite-support measure that can be formally written as $\lambda = \sum_{i=1}^{n} \lambda_i \delta_{\mathbf{x}_i}$, using the Dirac measure.

An rkhs $\mathscr{F}$ is a Hilbert space of functions in which the evaluation functional $\delta_{\mathbf{x}} : f \in \mathscr{F} \to f(\mathbf{x}) \in \mathbb{R}$ is continuous. A given element of $\mathscr{F}$ can be identified to an element λ of its dual, as well as to a random variable obtained as in (7), *i.e.*, an element of the Hilbert space $\mathscr{H}$ generated by linear combinations of $F(\mathbf{x})$.

The feature space $\mathscr{F}$ inherits the scalar product of $\mathscr{H}$. If the mean of $F(\mathbf{x})$ is zero, this scalar product is obtained from the covariance of $F(\mathbf{x})$ and $F(\mathbf{y})$, which is precisely the reproducing kernel of $\mathscr{F}$, *i.e.*,

$$k(\mathbf{x}, \mathbf{y}) = \mathrm{cov}(F(\mathbf{x}), F(\mathbf{y}))\,. \tag{8}$$

The feature space $\mathscr{F}$ can thus be viewed as a space of random variables endowed with a scalar product, *i.e.* a symmetric, strictly positive, bilinear functional equal to the covariance in this random space (remember that the mean of the random variables is assumed to be zero). The covariance structure (equivalently the kernel) $k(\mathbf{x}, \mathbf{y})$ indicates how two values of the process output to be modeled are correlated depending on the corresponding values $\mathbf{x}$ and $\mathbf{y}$ taken by the factor vectors.

3.2 *Kernel choice*

3.2.1. *The Matérn class of kernels* Choosing a suitable kernel for a given application is a recurrent, fundamental and largely open question. The asymptotic theory of Kriging (Stein, 1999) stresses the importance of taking care of the behaviour of the covariance near the origin. This behaviour is indeed linked with the quadratic mean regularity of the random process. For instance, if the covariance is continuous at the

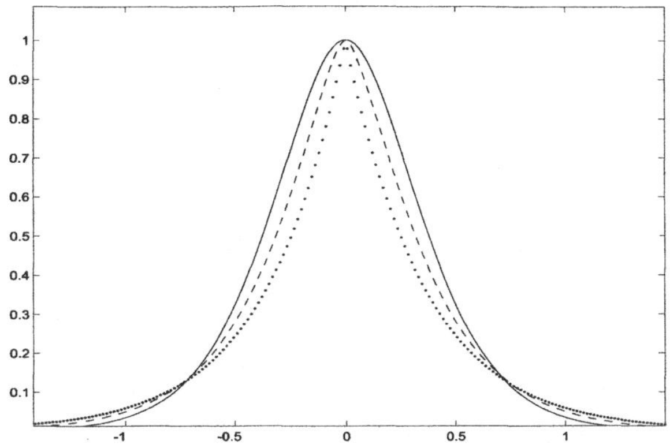

Figure 1. Matérn covariances for the parameterization (10) with $\rho = 0.5$, $\sigma^2 = 1$. Solid line corresponds to $\nu = 3$, dashed line to $\nu = 1$ and dotted line to $\nu = 0.5$.

origin, then the process will be continuous in quadratic mean. To address the issue of regularity, it is useful to consider a covariance model that is *isotropic*, *invariant by tranlation* and such that regularity can be adjusted with only one parameter. The *Matérn covariance* is a good candidate. Its Fourier transform is

$$\hat{k}(\omega) = \frac{\sigma^2}{(\omega_0^2 + \omega^2)^{\nu+1/2}}, \quad \omega \in \mathbb{R}, \tag{9}$$

where ν controls the decay of $\hat{k}(\omega)$ at infinity and therefore the regularity of the covariance. Stein (1999) advocates the use of the following parameterization of the Matérn model:

$$\begin{aligned} k(\mathbf{x},\mathbf{y}) &= R(||\mathbf{x}-\mathbf{y}||) = R(h) \\ &= \frac{\sigma^2}{2^{\nu-1}\Gamma(\nu)} \left(\frac{2\nu^{1/2}h}{\rho}\right)^{\nu} \mathscr{K}_\nu \left(\frac{2\nu^{1/2}h}{\rho}\right), \end{aligned} \tag{10}$$

where $\mathscr{K}_\nu$ is the modified Bessel function of the second kind. This parameterization allows an easy interpretation of the parameters, as ν controls the regularity, σ^2 is the variance ($R(0) = \sigma^2$), and ρ represents the *range* of the covariance, *i.e.*, the characteristic correlation distance. Figure 1 shows the influence of the ν parameter. Using this kernel in applications turns out to be an improvement over the blind choice of a kernel among those classically used in SVR, such as Gaussian kernels. The next section addresses the estimation of the parameters of a given kernel from the data.

3.2.2. *Parameter estimation* In a probabilistic framework such as Kriging, the maximum-likelihood approach is one of the standard routes that can be followed to estimate the parameters of the covariance model. This approach can also be applied in the context of SVR. If the outputs are observed without noise, SVR can be interpreted as Kriging where the observed outputs $\mathbf{f} = [f(\mathbf{x}_1), \cdots, f(\mathbf{x}_n)]^\mathsf{T}$ are assumed to be realizations of a Gaussian vector $[F(\mathbf{x}_1), \cdots, F(\mathbf{x}_n)]^\mathsf{T}$ (a *high* value of C and $\varepsilon = 0$ is consistent with this hypothesis).

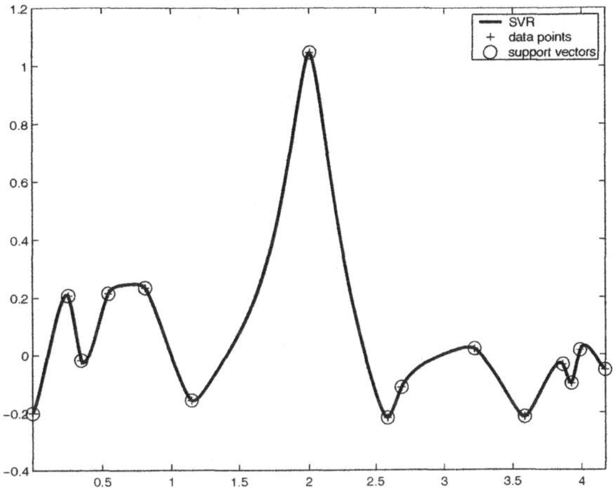

Figure 2. Plot of a SVR solution after maximum-likelihood estimation of the covariance parameters.

The maximum-likelihood estimate of the parameters of the covariance maximizes the probability density of the data. Using the joint probability density of the observed Gaussian vector, and assuming that the mean of $F(\mathbf{x})$ is zero, the maximum-likelihood estimate of the vector $\boldsymbol{\theta}$ of the covariance parameters is obtained (see for instance (Jones and Vecchia, 1993)) by minimizing the cost function

$$l(\boldsymbol{\theta}) = \frac{n}{2}\log 2\pi + \frac{1}{2}\log\det \mathbf{K}(\boldsymbol{\theta}) + \frac{1}{2}\mathbf{f}^\mathsf{T}\mathbf{K}(\boldsymbol{\theta})^{-1}\mathbf{f}, \tag{11}$$

where n is the number of observed output data and $\mathbf{K}(\boldsymbol{\theta})$ is the covariance matrix of the vector $[(F(\mathbf{x}_1), \cdots, F(\mathbf{x}_n)]^\mathsf{T}$.

Figure 2 illustrates an SVR with a Matérn kernel, the parameters of which have been estimated by maximum likelihood.

Assume now that the observations of the output are corrupted by noise. As a rough first approximation, assume moreover that the noise is white and Gaussian, although this hypothesis is not consistent with the use of the ε-insensitive function in SVR (see (Schölkopf and Smola, 2002)). For maximum-likelihood estimation, the observation noise must now be taken into account in the kernel structure as if it were part of the process. This simply means that a given kernel $k(\mathbf{h})$, where $\mathbf{h} = \mathbf{x} - \mathbf{y} \in \mathbb{R}^d$, is modified into:

$$k_\mathrm{n}(\mathbf{h}) = k(\mathbf{h}) + \sigma_\mathrm{n}^2 \delta_0(\mathbf{h}), \tag{12}$$

where $\delta_0(\mathbf{h})$ is the Dirac impulse at the origin and σ_n^2 is the noise variance. It is now possible to estimate the parameter σ_n^2 along with the other parameters of the covariance using the same maximum-likelihood formulation as above. Figures 3 and 4 illustrate the procedure. In this example, C is chosen by analogy with the splines–kriging equivalence where C is equal to $k(0)/\sigma_\mathrm{n}^2$ (Wahba, 1990). The choice of ε is related to the variance of the noise σ_n^2. We have set $\varepsilon = 0.3\sigma_\mathrm{n}$ ($0.3\sigma_\mathrm{n}$ corresponds to a confidence interval of approximatively 20% for a Gaussian variable).

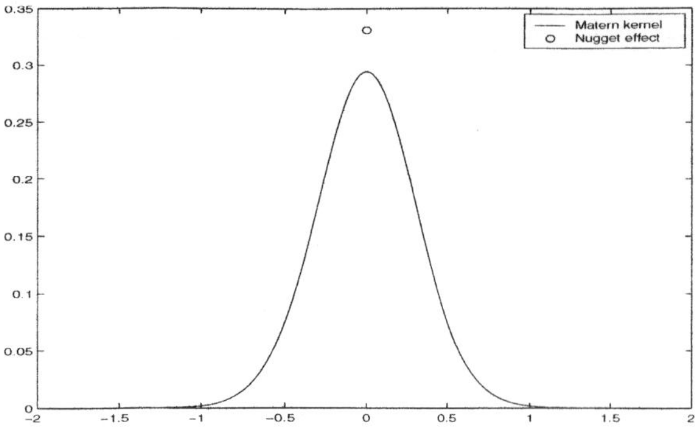

Figure 3. Matérn kernel with parameters estimated by maximum likelihood. The circle materializes the Dirac impulse at the origin corresponding to the white noise.

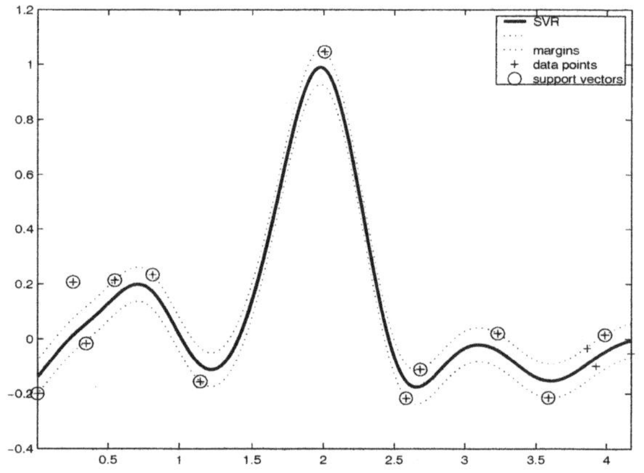

Figure 4. SVR solution after maximum likelihood estimation of the kernel parameters. The same data are used as in Figure 2.

To summarize, the properties of the reproducing kernel define the nature of the model. The Matérn covariance facilitates the tuning of the variance, regularity and range of correlation of the underlying random process describing the data. Moreover, these parameters can be estimated from the data in a unified framework by the maximum likelihood approach. Theses conclusions will be used again in the next sections when the multi-output case will be considered.

4. COKRIGING

Cokriging, the multi-output version of Kriging, exploits the correlations due to proximity in the space of factors (as Kriging does) but also the correlations between the outputs. Assume $F_1(\mathbf{x})$ is to be predicted from the observations of the random variables $F_\alpha(\mathbf{x}_i)$, $\alpha \in \Pi = \{1, \cdots, n\} \subset \mathbb{N}$ and $i \in S_\alpha \subset \mathbb{N}$ (we choose to index the random processes by Greek letters). With $n = 2$, prediction will based on the covariances and cross-covariances of $F_1(\mathbf{x})$ and $F_2(\mathbf{x})$, that is, on

$$k_{\alpha,\beta}(\mathbf{x},\mathbf{y}) = \mathrm{cov}\, F_\alpha(\mathbf{x}) F_\beta(\mathbf{y}), \quad \alpha, \beta \in \{1,2\}.$$

The estimator can again be written as a linear combination,

$$\hat{F}_1(\mathbf{x}) = \sum_{\alpha \in \Pi} \sum_{i \in S_\alpha} \lambda_{\alpha,i} F_\alpha(\mathbf{x}_i). \tag{13}$$

As for Kriging, $\hat{F}_1(\mathbf{x})$ is obtained as the best approximation of $F_1(\mathbf{x})$ on the space $\mathcal{H}_S$ generated by the variables $F_\alpha(\mathbf{x}_i)$. The random variables $F_\alpha(\mathbf{x})$ depend on $\alpha \in \Pi$, which can be considered as an additional factor. So with a slight change of notation we can write them as $F(\alpha, \mathbf{x})$ and write the covariance of $F(\alpha, \mathbf{x})$ and $F(\beta, \mathbf{y})$ as $k[(\alpha, \mathbf{x}), (\beta, \mathbf{y})]$. Thus, the multi-output case turns out to be formally equivalent to the single-output case. A single covariance function can then be used, since this function has been extended by means of the additional factor α to specify the nature of the random variable being considered.

5. MULTI-OUTPUT SVR

Using the above covariance $k[(\alpha, \mathbf{x}), (\beta, \mathbf{y})]$ as a kernel, it becomes straightforward to extend the feature space to deal with multi-output processes. If the bias term b is the same for each output of the process, then, a multi-output SVR builds a function

$$\begin{array}{rcl} f_{\mathrm{sv}} : \Pi \times \mathbb{R}^d & \rightarrow & \mathbb{R} \\ \alpha, \mathbf{x} & \mapsto & (w, \psi(\alpha, \mathbf{x}))_{\mathcal{F}} + b \end{array} \tag{14}$$

minimizing the functional

$$\frac{1}{2}\|w\|^2_{\mathcal{F}} + C \sum_{i,\alpha} [f_{\mathrm{sv}}(\alpha, \mathbf{x}_i) - f(\alpha, \mathbf{x}_i)]_\varepsilon \tag{15}$$

Note that the function $w \in \mathcal{F}$ will now be written as

$$w = \sum_{\beta,i} w_{\beta,i} k[(\beta, \mathbf{x}_i), (.,.)] \tag{16}$$

and the scalar product in (14) will be written as

$$(w, \psi(\alpha, \mathbf{x}))_{\mathcal{F}} = \sum_{\beta,i} w_{\beta,i} k[(\beta, \mathbf{x}_i), (\alpha, \mathbf{x})]. \tag{17}$$

SVR is thus easily extended to the multi-output case and the selection of the output to be predicted is performed with α in f_{sv}. This formulation has the additional advantage that existing optimization algorithms for SVR need not be modified, since the formalism remains unchanged. It suffices to design an appropriate kernel of the type of $k[(\alpha, \mathbf{x}), (\beta, \mathbf{y})]$, which amounts to choosing models of covariance and cross-covariance.

Linear combinations of the f_{sv} are also easily predicted. For instance,

$$f_{\mathrm{sv}}(1, \mathbf{x}) - f_{\mathrm{sv}}(2, \mathbf{x}) = (w, \psi(1, \mathbf{x}) - \psi(2, \mathbf{x}))_{\mathcal{F}}. \tag{18}$$

Until now, we have assumed that the bias term b was the same for each output of the process. To deal with unknown and possibly different means, a semi-parametric formulation of SVR is necessary (Smola *et al.*, 1999). Note that because of the presence of the term b in the classical formulation of SVR, this classical formulation can already be seen as a semi-parametric prediction. The easiest case is when the different means are assumed not to be correlated. Then, SVR is extended with parametric terms, such that,

$$\begin{aligned} f_{sv} : \Pi \times \mathbb{R}^d &\rightarrow \mathbb{R} \\ \alpha, \mathbf{x} &\mapsto (w, \psi(\alpha, \mathbf{x}))_{\mathscr{F}} + \sum_\beta b_\beta \delta_\beta(\alpha) \end{aligned} \quad (19)$$

where $\delta_\beta(\alpha)$ is equal to one if $\beta = \alpha$ and zero if $\beta \neq \alpha$.

If the means are correlated, the previous formulation can still be used if there are enough observations of each output to make it possible to estimate the means independently. Finally, in this multi-output context, it might be interesting to have different data fidelity terms in (15) because the nature of noise on the each output might be different. This means that the data fidelity functional could be written as

$$\sum_{i,\alpha} C_\alpha [f_{sv}(\alpha, \mathbf{x}_i) - f(\alpha, \mathbf{x}_i)]_{\varepsilon_\alpha} . \quad (20)$$

This formulation will be exploited in the next section, which will show some examples of multi-output SVR and briefly discuss methods to choose the covariance structure.

6. CROSS-COVARIANCE CHOICE

Choosing a cross-covariance is even more difficult than choosing a covariance. Consider first the proportional cross-covariance model, assuming again invariance by translation. Then,

$$\begin{aligned} k_{1,1}(\mathbf{h}) = k_{2,2}(\mathbf{h}) = k(\mathbf{h}) = k(-\mathbf{h}), \quad \mathbf{h} \in \mathbb{R}^d, \\ k_{1,2}(\mathbf{h}) = k_{2,1}(\mathbf{h}) = \gamma k(\mathbf{h}), \end{aligned} \quad (21)$$

where the scalar γ must be chosen in $]-1,1[$ since the Cauchy-Schwarz inequality implies

$$|k_{\alpha,\beta}(\mathbf{h})| \leq \sqrt{k_{\alpha,\alpha}(0) k_{\beta,\beta}(0)} . \quad (22)$$

Figure 5 illustrates the use of the proportional model. We have considered two (arbitrary) functions, representing hypothetic outputs of a process. The prediction was performed by adding the proportional kernel to an existing SV software (Gunn, 1997).

It is not necessary to have $k_{\alpha,\beta}(-\mathbf{h}) = k_{\alpha,\beta}(\mathbf{h})$ as in (21) (whereas $k_{\alpha,\beta}(\mathbf{h}) = k_{\beta,\alpha}(-\mathbf{h})$ always holds). Many possible covariance structures can actually be considered to take into account complex relations between outputs. A simple example of non-symmetric cross-covariance is obtained when considering a delay between the outputs, for instance, with a single factor x, $F_2(x) = F_1(x-\tau) + G(x)$, assuming $G(x)$ and $F_1(x)$ independent and $\text{cov}[F_1(x), F_1(y)] = k_{1,1}(|x-y|)$. In such a case, the cross-covariance is $k_{1,2}(x,y) = k_{1,1}(|x-y+\tau|) \neq k_{1,2}(y,x) = k_{1,1}(|x-y-\tau|)$.

Many other possibilities of covariance structures can be used (Chiles and Delfiner, 1999). Often, one has

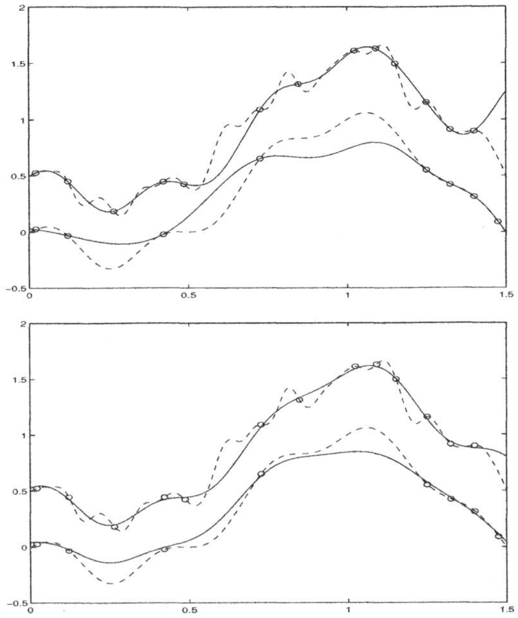

Figure 5. Multi-output SVR with a proportional model for the cross-covariance. The top plot shows independent predictions of the outputs (in solid lines). The circles materialize the data sampled from the dashed-line graphs. The bottom plot shows the improved results obtained with a cross-correlation coefficient $\gamma = 0.61$ estimated by maximum-likelihood.

some knowledge about the relations between the outputs. For instance the multi-output SVR framework can be used to take into account information about the derivatives of the process. This property is illustrated here in the simplest case, that is, when the space of factors is of dimension one, but generalization is straightforward. The inclusion of information about the derivatives is based on the consideration of two outputs, namely the one to be predicted, modeled by a random process $F(x)$, and its derivative. The derivative output is defined as

$$F'(x) = \lim_{h \to 0} \frac{F(x+h) - F(x)}{h} \quad (23)$$

From (23), and assuming that $F(x)$ is characterized by a twice differentiable radial covariance $k(h)$, $h \in \mathbb{R}_+$ the covariance structure of the multi-output model can be written as

$$\begin{aligned} \text{cov}[F(x_1), F'(x_2)] \\ &= \lim_{h \to 0} \frac{k(|x_1 - x_2 - h|) - k(|x_1 - x_2|)}{h} \\ &= -\text{sgn}(x_1 - x_2) k'(|x_1 - x_2|) \end{aligned} \quad (24)$$

$$\text{cov}[F'(x_1), F(x_2)] = \text{sgn}(x_1 - x_2) k'(|x_1 - x_2|) \quad (25)$$

$$\text{cov}[F'(x_1), F'(x_2)] = -k''(|x_1 - x_2|) \quad (26)$$

When Matérn covariance is used, the following property helps differentiate the modified Bessel function of the second kind :

$$\mathscr{K}'_\nu(x) = -\mathscr{K}_{\nu-1}(x) - \frac{\nu}{x} \mathscr{K}_\nu(x) . \quad (27)$$

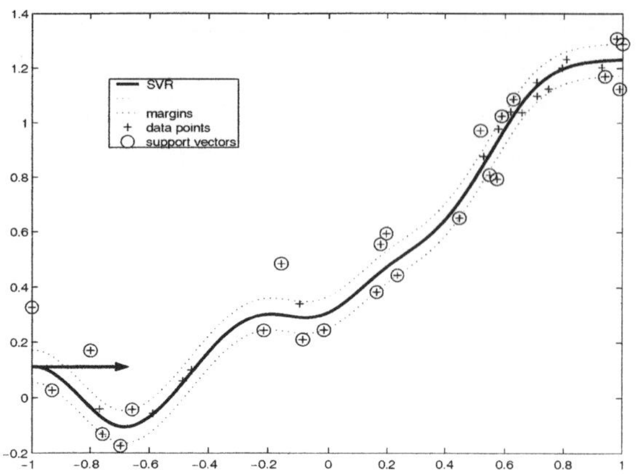

Figure 6. SVR with incorporation of prior information about the derivative of F. It is known that tangent at $x = -1$ should be horizontal.

This approach may use two types of data fidelity term, as in (20). Assume, for instance, that the available observations data on the output to be predicted are corrupted by noise, which leads to choosing $C \approx k(0)/\sigma_n^2$, but that it is known for physical reasons that the gradient must be strictly equal to zero at some boundary points. The prediction of the derivative of the output must then interpolate the constraints data, which implies $C \to \infty$, $\varepsilon = 0$. Figure 6 illustrates this approach.

7. CONCLUSIONS

The main contributions of this paper lie in the confrontation of the points of view of support vector machines and Kriging, and in the presentation of a multi-output support vector regression based on ideas that are now classical in the community of Kriging but perhaps less well known in that of SVMs.

We hope to have convinced the reader that the theory of Kriging provides useful guidelines for the solution of such fundamental problems as the choice of a kernel structure for support vector regression and the estimation of the parameters of this structure via a probabilistic interpretation of a given kernel as a correlation structure.

8. REFERENCES

Chiles, J.-P. and P. Delfiner (1999). *Geostatistics: Modeling Spatial Uncertainty*. Wiley. New York.

Gunn, S.R. (1997). Support vector machines for classification and regression. Technical report. Image Speech and Intelligent Systems Research Group. University of Southampton.

Huber, P. J. (1981). *Robust Statistics*. Wiley. New York.

Jaakkola, T. and D. Haussler (1999). Probabilistic kernel regression models. In: *Proceedings of the 1999 Conference on AI and Statistics*.

Jones, R. and A. Vecchia (1993). Fitting continuous ARMA models to unequally spaces spatial data. *J. of the Amer. Stat. Assoc.* **88**(423), 947–954.

Matheron, G. (1963). Principles of geostatistics. *Economic Geology* **58**, 1246–1266.

Matheron, G. (1981). Splines and Kriging: their formal equivalence. In: *Down-to-Earth Statistics: Solutions Looking for Geological Problems* (Merriam D.F., Ed.). Syracuse Univ. of geology contributions ed.. pp. 77–95.

Schaback, R. (1999). Native Hilbert spaces for radial basis functions I. *Inter. Series of Numerical Mathematics* **132**, 255–282.

Schölkopf, B. and A. Smola (2002). *Learning with Kernels*. MIT Press. Cambridge.

Smola, A. J., B. Schölkopf and K.-R. Müller (1998). The connection between regularization operators and support vector kernels. *Neural Networks* **11**(4), 637–649.

Smola, A., T. Friess and B. Schölkopf (1999). Semiparametric support vector and linear programming machines. In: *Advances in Neural Information Processing Systems* (M. Kearns, S. Solla and D. Cohn, Eds.). pp. 585 – 591. 11. MIT Press. Cambridge.

Stein, M.L. (1999). *Interpolation of Spatial Data: Some Theory for Kriging*. Springer. New York.

Vapnik, V.N. (1995). *The Nature of Statistical Learning Theory*. Springer. Heidelberg.

Wahba, G. (1990). *Spline Models for Observational Data*. Vol. 59 of *CBMS-NSF Regional Conference Series in Applied Mathematics*. SIAM. Philadelphia.

Wahba, G. (1998). SVM, RKHS, and randomized GACV. In: *Advances in Kernel Methods – Support Vector Learning* (B. Schölkopf, C.J.C. Burges and A.J. Smola, Eds.). Chap. 6, pp. 69–87. MIT Press. Boston.

Wahba, G. and G.S. Kimeldorf (1970). Spline functions and stochastic processes. *Sankhyä: the Indian Journal of Statistics: Series A* **32**(2), 173–180.

Williams, C.K.I (1997). Regression with Gaussian processes. In: *Mathematics of Neural Networks: Models, Algorithms and Applications* (S.W. Ellacott, J.C. Mason and I.J. Anderson, Eds.). Kluwer. Presented at the Mathematics of Neural Networks and Applications Conference, Oxford, 1995.

IFAC Publications
www.elsevier.com/locate/ifac

SET MEMBERSHIP IDENTIFICATION OF PIECEWISE AFFINE MODELS

Alberto Bemporad, Andrea Garulli, Simone Paoletti, Antonio Vicino

Dipartimento di Ingegneria dell'Informazione,
Università di Siena
Via Roma 56, 53100 Siena, Italy
{bemporad,garulli,paoletti,vicino}@dii.unisi.it

Abstract: This paper addresses the problem of identification of piecewise affine (PWA) models, which involves the joint estimation of both the parameters of the affine submodels and the partition of the PWA map from data. According to ideas from set-membership identification, the key approach is to characterize the model by its maximum allowed prediction error, which is used as a tuning knob for trading off between prediction accuracy and model complexity. At initialization, the proposed procedure for PWA identification exploits a technique for partitioning an infeasible system of linear inequalities into a (possibly minimum) number of feasible subsystems. This provides both an initial clustering of the datapoints and a guess of the number of required submodels, which therefore is not fixed a priori. A refinement procedure is then applied in order to improve both data classification and parameter estimation. The partition of the PWA map is finally estimated by considering multicategory classification techniques.

Keywords: Nonlinear identification, hybrid systems, bounded error, data classification, parameter estimation

1. INTRODUCTION

Black-box identification of nonlinear systems has been widely addressed in different contexts. A large number of model classes have been considered and their properties deeply investigated, see the survey papers (Sjöberg *et al.*, 1995; Juditsky *et al.*, 1995) and references therein. In this paper, we deal with the problem of identifying a piecewise affine (PWA) model of a discrete-time nonlinear system from input-output data. Recently, this problem has deserved more and more attention, given the equivalence of PWA systems with several classes of hybrid systems (Bemporad *et al.*, 2000; Heemels *et al.*, 2001). Identification of PWA models is a challenging problem, as it involves the simultaneous estimation of both the parameters of the affine submodels and the partition of the PWA map. It is crucial to point out that the first issue is closely related to the problem of classifying the data, *i.e.* the problem of correctly assigning each datapoint to one affine submodel. In (Ferrari-Trecate *et al.*, 2003) the classification problem is reduced to an optimal clustering problem, in which the number of clusters is fixed. Once the datapoints have been classified, linear regression is used to compute the final submodels. In (Münz and Krebs, 2002) the authors propose an identification algorithm consisting of analysis of the knowledge available a priori, data clustering, and optimization of the cluster shapes. The user can affect the accuracy of the identification by modifying the number of clusters. In (Bemporad *et al.*, 2001) the attention is focused on two subclasses of PWA models, for which the identification problem can be formulated as a suitable mixed-integer linear, or quadratic, programming problem that can be solved for the global optimum.

The procedure for PWA identification proposed in this paper does not fix the number of affine submodels a priori, rather this number is estimated from data, together with the parameters of the submodels and the partition of the PWA map. The key approach in this work is the selection of a bound on the prediction error, which induces a set of linear inequality constraints on the parameters of the PWA model to be estimated. Unless a single affine model fits all the data within the chosen bound, the whole set of constraints is, in general, infeasible. Hence, in Section 3 a suitable strategy is suggested for picking a number of submodels which is consistent with the available data and the bounded-error condition. In particular, a partition of the above system of linear inequalities into a minimum number of feasible subsystems (Amaldi and Mattavelli, 2002) is sought. Given any solution of this problem, the partition of the inequalities provides the initial classification of the datapoints, whereas a set of feasible parameter vectors for the corresponding affine submodel is associated to each feasible subsystem according to the bounded-error condition. In Section 4 a refinement procedure alternating between datapoint reassignment and parameter update is proposed in order to improve both data classification and parameter estimation, and to possibly reduce the number of submodels. Notice that the final number of submodels and the corresponding parameter vectors will depend on the selected bound on the prediction error, so that this allows one to trade off between the complexity and the accuracy of the model. The estimation of the partition of the PWA map is addressed in Section 5. The final clusters of regression vectors are separated via classification methods such as Linear (Vapnik, 1998) or Multicategory (Bredensteiner and Bennett, 1999) Support Vector Machines. The identified PWA model associates to each submodel a set of feasible parameters, thus allowing for evaluation of the parametric uncertainty associated with it (Milanese and Vicino, 1991).

2. PROBLEM STATEMENT

Consider the discrete-time nonlinear dynamic system

$$y_k = F(u^{k-1}, y^{k-1}) + e_k, \tag{1}$$

where F is a (possibly non-smooth) nonlinear function, u^{k-1} and y^{k-1} are, respectively, past system inputs and outputs up to time $k-1$, and e_k is additive noise. Assuming that a finite collection of input-output samples (u_k, y_k) is given, the aim is to estimate a PieceWise Affine (PWA) model $\hat{y}_k = f(x_k)$ of system (1), where f is the PWA map

$$f(x) = \begin{cases} \theta_1' \begin{bmatrix} x \\ 1 \end{bmatrix} & \text{if } x \in \mathscr{X}_1 \\ \vdots & \vdots \\ \theta_s' \begin{bmatrix} x \\ 1 \end{bmatrix} & \text{if } x \in \mathscr{X}_s, \end{cases} \tag{2}$$

$x_k \in \mathbb{R}^n$ is a suitable regression vector depending on u^{k-1} and y^{k-1}, $\theta_i \in \mathbb{R}^{n+1}$, $i = 1, \dots, s$, are parameter vectors, and $\{\mathscr{X}_i\}_{i=1}^s$ is a partition of the regressor set $\mathscr{X} \subseteq \mathbb{R}^n$, *i.e.* the region of validity of the PWA model[1]. Each *region* $\mathscr{X}_i$ is assumed to be a convex polyhedron, represented in the form

$$\mathscr{X}_i = \left\{ x \in \mathbb{R}^n : H_i \begin{bmatrix} x \\ 1 \end{bmatrix} \preceq 0 \right\},$$

where $H_i \in \mathbb{R}^{q_i \times (n+1)}$, $i = 1, \dots, s$. In this paper, the focus is on PWARX (PieceWise affine AutoRegressive eXogenous) models, for which x_k is the standard regression vector, *i.e.*

$$x_k = [y_{k-1} \dots y_{k-n_a} \; u_{k-1} \dots u_{k-n_b}]'.$$

In this case, $n = n_a + n_b$, and the parameter vectors θ_i contain the coefficients of the *ARX submodels*. For a more compact notation, hereafter the extended regression vector $\varphi_k = [x_k' \; 1]'$ will be considered. The key approach in this paper consists in selecting a bound on the *prediction error*,

$$|y_k - f(x_k)| \le \delta, \tag{3}$$

for a fixed $\delta > 0$. Accordingly, the considered identification problem can be stated as follows.

Problem 1. Given N datapoints (y_k, x_k), $k = 1, \dots, N$, and $\delta > 0$, estimate a minimum positive integer s, a partition $\{\mathscr{X}_i\}_{i=1}^s$, and parameter vectors $\{\theta_i\}_{i=1}^s$, such that the corresponding PWA model (2) of system (1) satisfies condition (3).

From the above formulation, it is clear that one seeks the "simplest" PWA model that is consistent with the data and condition (3), where, for a given δ, "simplicity" is measured in terms of the number of affine submodels. Notice that the bound δ is not necessarily given a priori, it is rather a tuning knob of the procedure. A reliable choice can often be made a posteriori by performing a series of trials for different values of δ, and then selecting a value that provides a suitable trade-off between model complexity (in terms of number of submodels) and quality of fit (in terms of mean square error).
The proposed procedure for solving Problem 1 consists of the following steps: *Initialization*, exploiting a greedy strategy for partitioning a (possibly infeasible) system of linear inequalities into a minimum number of feasible subsystems; *Refinement of the estimates*, alternating between datapoint reassignment and parameter update; *Estimation of the regions*, exploiting multicategory classification techniques.

3. INITIALIZATION

In the first part of the proposed identification procedure, the problem of estimating the hyperplanes defining the polyhedral partition of the regressor set is provisionally not addressed. The focus is only on classifying the datapoints according to the fact that

[1] $\bigcup_{i=1}^s \mathscr{X}_i = \mathscr{X}$ and $\mathscr{X}_i \cap \mathscr{X}_j = \emptyset$ if $i \neq j$

they are fitted by the same affine submodel. In this phase, it is reasonable to look for the minimum number of submodels (namely s) fitting all (or most of, due to possible outliers) the datapoints. By requiring condition (3), the initial classification problem can be stated as follows.

Problem 2. Given $\delta > 0$ and the system of N complementary inequalities

$$\left|y_k - \varphi_k'\theta\right| \leq \delta, \quad k = 1, \dots, N, \tag{4}$$

find a partition of this system into a minimum number s of feasible subsystems.

The above formulation enables to address simultaneously the two fundamental issues of data classification and parameter estimation. Given any solution of Problem 2, the partition of the complementary inequalities provides the classification of the datapoints, whereas each feasible subsystem defines the set of feasible parameter vectors for the corresponding affine submodel, according to the bounded-error condition. This naturally leads to a set-membership or bounded-error approach to the identification problem, see, *e.g.*, (Milanese and Vicino, 1991; Milanese *et al.*, 1996). Since each complementary inequality in system (4) corresponds to the pair of linear inequalities

$$\begin{cases} \varphi_k'\theta \leq y_k + \delta \\ \varphi_k'\theta \geq y_k - \delta\,, \end{cases} \tag{5}$$

Problem 2 turns out to be an extension of the combinatorial problem of finding a *Partition* of an infeasible system of linear inequalities into a *MINimum* number of *Feasible Subsystems* (MIN PFS problem), with the additional constraint that two paired linear inequalities (5) must be included in the same subsystem (*i.e.* they must be simultaneously satisfied by the same parameter vector θ). The MIN PFS problem is NP-hard. On the other hand, in order to initialize the identification procedure, one is interested in finding approximate solutions of Problem 2 rapidly. To this aim, the greedy approach proposed in (Amaldi and Mattavelli, 2002), which efficiently provides good approximate solutions, is used in this paper. This approach divides the overall partition problem into a sequence of subproblems, each subproblem consisting in finding one parameter vector θ that satisfies the maximum number of complementary inequalities. Starting from system (4), maximum feasible subsystems are iteratively extracted (and the corresponding inequalities removed), until the remaining subsystem is feasible. The problem of finding one θ that satisfies as many pairs of complementary inequalities (4) as possible, extends the combinatorial problem of finding a *MAXimum Feasible Subsystem* of an infeasible system of linear inequalities (MAX FS problem). Based on the consideration that also MAX FS is NP-hard, the approach proposed in (Amaldi and Mattavelli, 2002) tackles the above extension of MAX FS using a randomized and thermal variant of the classical Agmon-

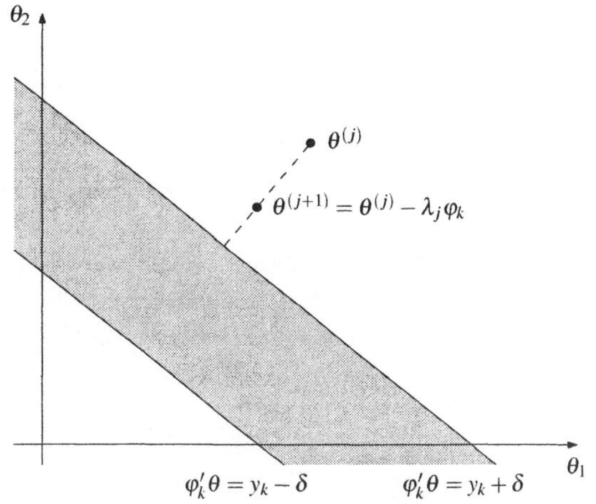

Fig. 1. Geometric interpretation in the parameter space of a single iteration of the relaxation method for the extended MAX FS (case $\theta \in \mathbb{R}^2$)

Motzkin-Schoenberg relaxation method for solving systems of linear inequalities. The proposed method consists in a simple iterative procedure that generates a sequence of estimates. Starting with an arbitrary initial guess $\theta^{(1)} \in \mathbb{R}^{n+1}$ (*e.g.*, randomly selected, or computed by least squares), at each iteration one complementary inequality is selected from the system at hand according to a prescribed rule (*e.g.*, cyclicly, or uniformly at random without replacement), while all the others are relaxed. Assume that at iteration j the complementary inequality $\left|y_k - \varphi_k'\theta\right| \leq \delta$ is considered. Then the current estimate $\theta^{(j)}$ is updated as follows:

if $\varphi_k'\theta^{(j)} > y_k + \delta$
  then $\theta^{(j+1)} = \theta^{(j)} - \lambda_j\varphi_k$
else if $\varphi_k'\theta^{(j)} < y_k - \delta$
  then $\theta^{(j+1)} = \theta^{(j)} + \lambda_j\varphi_k$
else $\theta^{(j+1)} = \theta^{(j)}$

with $\lambda_j > 0$. Geometrically, the complementary inequality $\left|y_k - \varphi_k'\theta\right| \leq \delta$ defines a hyperstrip in the parameter space (see Figure 1). If the current estimate $\theta^{(j)}$ belongs to the hyperstrip, then it is unchanged. Otherwise, $\theta^{(j+1)}$ is obtained by making a step along the line drawn from $\theta^{(j)}$ in the direction orthogonal to the hyperstrip. The size of the step λ_j decreases exponentially with the violation of the considered complementary inequality, which is computed as follows:

$$v_j^k = \begin{cases} \varphi_k'\theta^{(j)} - y_k - \delta & \text{if } \varphi_k'\theta^{(j)} > y_k + \delta \\ y_k - \varphi_k'\theta^{(j)} - \delta & \text{if } \varphi_k'\theta^{(j)} < y_k - \delta \\ 0 & \text{otherwise}. \end{cases}$$

The basic idea is indeed to favor updates of the current estimate $\theta^{(j)}$, which aim at correcting unsatisfied inequalities with a relatively small violation. Decreasing attention to unsatisfied inequalities with large violations (whose correction is likely to corrupt other inequalities that the current estimate satisfies) is obtained by introducing a decreasing temperature parameter T, which the violations are compared to.

The algorithm is stopped after a predefined maximum number of cycles C through all the inequalities at hand. The solution returned is the estimate that, during the process, has satisfied the largest number of complementary inequalities. Nevertheless, this is not guaranteed to be optimal, due to the randomness of the method. For the choice of the maximum number of cycles C and the initial temperature parameter T_0, as well as for more details concerning the implementation of the algorithm, the reader is defered to (Amaldi and Mattavelli, 2002).

3.1 *Comments about the initialization*

Denote by $\hat{s}$ the number of feasible subsystems of system (4) provided by the application of the greedy algorithm for the extended MIN PFS described in Section 3. Due to the suboptimality of the greedy strategy for MIN PFS, and the randomness of the method used to tackle the MAX FS, the estimate of the number of affine submodels needed to fit the data and the classification of the datapoints thus obtained, suffer two drawbacks. First, this strategy is not guaranteed to yield minimum partitions, *i.e.* the number of submodels $\hat{s}$ could be larger than the minimum number s. Indeed, due to the randomness of the method used to tackle the MAX FS, two subsets of complementary inequalities that could be satisfied by one and the same parameter vector, may be extracted at two different steps. Second, since some datapoints might be consistent with more than one submodel, the cardinality and the composition of the clusters could depend on the order in which the feasible subsystems are extracted. In order to cope with these drawbacks, a procedure for the refinement of the estimates will be proposed in the next section. As it will be shown, such a procedure improves both the data classification and the quality of fit by properly reassigning the datapoints, and selecting pointwise estimates of the parameter vectors that characterize each submodel. Notice that one could decide to stop the algorithm when the cardinalities of the extracted clusters become too small. This might be useful in order to penalize submodels that account for just a few datapoints (most likely outliers).

4. REFINEMENT OF THE ESTIMATES

The initialization of the identification procedure, described in Section 3, provides the clusters $\mathscr{D}_i^{(0)}$, which consist of all the datapoints (y_k, x_k) corresponding to the i-th extracted feasible subsystem of system (4), $i = 1, \ldots, \hat{s}$. Moreover, each feasible subsystem defines the set of feasible parameter vectors for the corresponding affine submodel. As discussed in Section 3.1, a procedure for the refinement of the estimates is necessary in order to improve both data classification and quality of fit, as well as to possibly reduce the number of submodels. The proposed basic procedure consists of two steps to be iterated. In the first step, using the current estimated parameter vectors, datapoints are grouped together in the same cluster only if they can be fitted by the same affine submodel. In the second step, new pointwise parameter estimates are computed for each submodel. The *projection estimate* is used, defined as:

$$\Phi_p(\mathscr{D}) = \arg\min_{\theta} \max_{(y_k, x_k) \in \mathscr{D}} \left| y_k - \varphi_k' \theta \right|, \qquad (6)$$

where $\mathscr{D}$ is a cluster of datapoints (y_k, x_k). Notice that the computation of the projection estimate can be formulated as a suitable *linear programming* (LP) problem. The basic refinement procedure can be formalized as follows.

(0) **Initialization**
Set $t = 1$ and select a termination threshold $\gamma \geq 0$
For $i = 1, \ldots, \hat{s}$, set $\hat{\theta}_i^{(1)} = \Phi_p(\mathscr{D}_i^{(0)})$

(1) **Datapoint reassignment**
For each datapoint (y_k, x_k), $k = 1, \ldots, N$:
- If $\left| y_k - \varphi_k' \hat{\theta}_i^{(t)} \right| \leq \delta$ for only one $i = 1, \ldots, \hat{s}$, then assign (y_k, x_k) to $\mathscr{D}_i^{(t)}$
- If $\left| y_k - \varphi_k' \hat{\theta}_i^{(t)} \right| > \delta$ for all $i = 1, \ldots, \hat{s}$, then mark (y_k, x_k) as *infeasible*
- Otherwise, mark (y_k, x_k) as *undecidable*

(2) **Parameter update**
For $i = 1, \ldots, \hat{s}$, compute $\hat{\theta}_i^{(t+1)} = \Phi_p(\mathscr{D}_i^{(t)})$

(3) **Termination**
If $\left\| \hat{\theta}_i^{(t+1)} - \hat{\theta}_i^{(t)} \right\| / \left\| \hat{\theta}_i^{(t)} \right\| \leq \gamma$ for all $i = 1, \ldots, \hat{s}$, then exit; Otherwise, set $t = t + 1$ and go to step 1

In order to avoid that the procedure does not terminate, a maximum number $t_{\max}$ of refinements can be predefined. The underlying idea is that, as the new parameter estimates $\hat{\theta}_i^{(t+1)}$ are computed based on the clusters $\mathscr{D}_i^{(t)}$, some infeasible, as well as undecidable, datapoint may become *feasible*, *i.e.* it may be assigned to one cluster $\mathscr{D}_i^{(t+1)}$, thus improving the quality of the classification. Notice that the use of the projection estimate in step 2 guarantees that no feasible datapoint at refinement t becomes infeasible at refinement $t + 1$, since

$$\max_{(y_k, x_k) \in \mathscr{D}_i^{(t)}} \left| y_k - \varphi_k' \hat{\theta}_i^{(t+1)} \right| \leq \max_{(y_k, x_k) \in \mathscr{D}_i^{(t)}} \left| y_k - \varphi_k' \hat{\theta}_i^{(t)} \right|,$$

and the right-hand side of the above inequality is less than or equal to δ. Motivations for the distinction among infeasible, undecidable, and feasible datapoints are twofold. Infeasible datapoints are not consistent with any submodel, and may be outliers (especially if the corresponding violation is large). Hence, it is reasonable to expect that neglecting them in the parameter update helps to improve the quality of fit. The undecidable datapoints are instead consistent with more than one submodel. Neglecting them helps to reduce the number of misclassifications. As it will be clarified in the next section, this will favorite a better estimation of the PWA partition.

When the greedy algorithm provides an overestimation of the number of submodels needed to fit the data (see the discussion in Section 3.1), further steps are required in order to possibly reduce the complexity of the model. To this aim, the similarity of the parameter vectors and the cardinality of the clusters can be exploited. Indeed, if two subsets of complementary inequalities can be satisfied by one and the same parameter vector, it is likely that the corresponding parameter estimates are very similar, so that they can be merged into one subset. Notice that, in this case, a large number of undecidable datapoints should show up. On the other hand, if during the refinement of the estimates the cardinality of one cluster becomes too small with respect to N, the corresponding submodel can be discarded, since it accounts only for few datapoints (most likely outliers). Additional steps to the basic refinement procedure are thus the following (α and β are fixed nonnegative thresholds).

- **Similarity of the parameter vectors**
  If $\alpha_{i^*,j^*} \triangleq \min_{1\le i<j\le \hat{s}} \mu(\hat{\theta}_i^{(t)}, \hat{\theta}_j^{(t)}) \le \alpha$, then merge the submodels i^* and j^*, and set $\hat{s} = \hat{s} - 1$
- **Cardinality of the clusters**
  If $\beta_{i^*} = \min_{1\le i\le \hat{s}} \mathrm{card}(\mathscr{D}_i^{(t)})/N \le \beta$, then discard the i^*-th submodel, set $\hat{s} = \hat{s} - 1$, and reassign only the undecidable datapoints as in step 1

The similarity of the parameter vectors is tested before step 1. Here $\mu(\hat{\theta}_i^{(t)}, \hat{\theta}_j^{(t)})$ is a suitable measure of the distance between $\hat{\theta}_i^{(t)}$ and $\hat{\theta}_j^{(t)}$, *e.g.*,

$$\mu(\hat{\theta}_i^{(t)}, \hat{\theta}_j^{(t)}) \triangleq \|\hat{\theta}_i^{(t)} - \hat{\theta}_j^{(t)}\| / \min\{\|\hat{\theta}_i^{(t)}\|, \|\hat{\theta}_j^{(t)}\|\}.$$

Two submodels i^* and j^* can be merged by computing the new parameter vector as $\Phi_p(\mathscr{D}_{i^*}^{(t-1)} \bigcup \mathscr{D}_{j^*}^{(t-1)})$. The cardinality of the clusters is instead tested after step 1. The thresholds α and β should be suitably chosen in order to possibly reduce the number of submodels still preserving a good fit of the data. Indeed, it is clear that, if such thresholds are too large, the number of submodels might decrease under s. In this case, the number of infeasible datapoints considerably increases, since some significant dynamics is no more in the model. One could use this information in order to adjust α and β, and then repeat the refinement. Current research is aimed at deriving rules for the automatic selection and update of α and β, in order to completely automatize the procedure.

5. ESTIMATION OF THE REGIONS

The final step of the identification procedure consists in estimating the partition of the regressor set. This step can be performed by considering pairwise the clusters $\mathscr{F}_i = \{x_k | (y_k, x_k) \in \mathscr{D}_i\}$ (where $\mathscr{D}_i$, $i = 1, \ldots, \hat{s}$, is the final classification of the feasible datapoints provided by the refinement procedure), and by computing a separating hyperplane for each of such pairs. If two clusters $\mathscr{F}_i$ and $\mathscr{F}_j$ are not linearly separable, it is reasonable to look for a *generalized* separating hyperplane of the two, *i.e.* a hyperplane

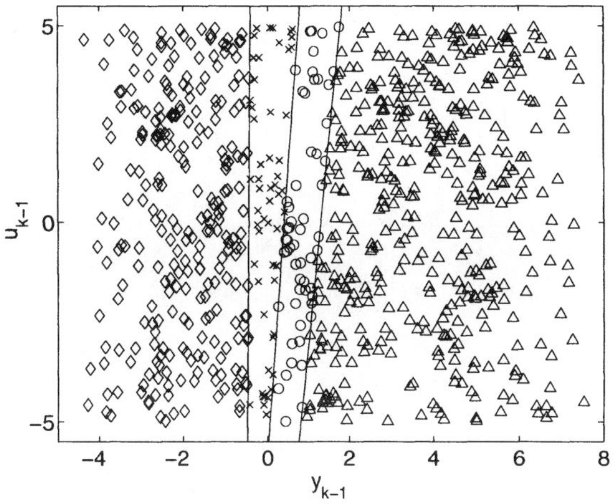

Fig. 2. Final classification of the regression vectors, and partition of the PWA map

that maximizes the number of well-separated points. This amounts to find a solution (a,b), with $a \in \mathbb{R}^n$ and $b \in \mathbb{R}$, of the MAX FS problem of system

$$\begin{cases} x_k' a + b \le -1 & \forall x_k \in \mathscr{F}_i \\ x_k' a + b \ge 1 & \forall x_k \in \mathscr{F}_j . \end{cases}$$

Support Vector Machines (Vapnik, 1998) with a linear kernel are a suitable tool to accomplish this task. Notice that, when the number of misclassified points is large, it likely means that at least one of the two clusters corresponds either to a nonconvex region (which then needs to be split into convex polyhedra), or to nonconnected regions where the submodel is the same (recall that the classification procedure groups together all the datapoints fitted by the same affine submodel). Efficient techniques for detecting and splitting the clusters corresponding to such situations, are currently under investigation. Each region $\hat{\mathscr{X}}_i$, $i = 1, \ldots, \hat{s}$, is finally defined by all the hyperplanes separating $\mathscr{F}_i$ from $\mathscr{F}_j$, with $j \ne i$. Although computationally appealing, this method has the major drawback of not guaranteeing that the estimated regions form a complete partition of the regressor set. In order to avoid "holes"in the partition, all clusters can be simultaneously involved in a computationally more demanding *multicategory* classification problem, for whose solution both linear and quadratic programming based methods have been proposed (Bredensteiner and Bennett, 1999). Once the partition of the regressor set has been estimated, the undecidable datapoints can be finally classified, and final parameter estimates for each submodel can be computed using (6).

6. NUMERICAL EXAMPLE

The PWA identification algorithm has been applied to fit the data generated by the nonlinear system

$$y_k = \sqrt{|y_{k-1}|} - u_{k-1} + e_k . \qquad (7)$$

The input signal u_k was drawn from a uniform distribution on $[-5,5]$, and the noise signal e_k was drawn

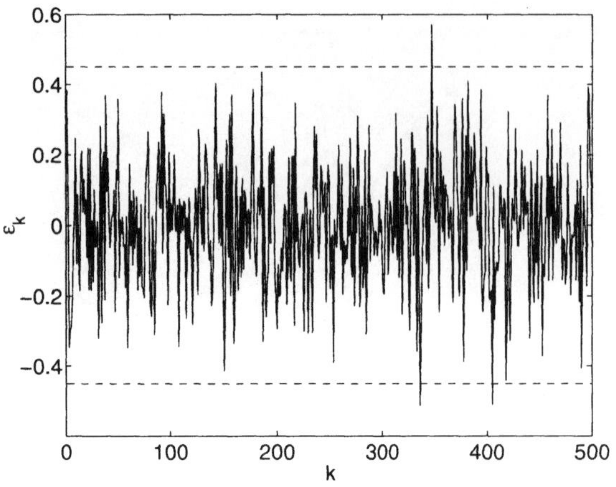

Fig. 3. Plot of the prediction error

from a zero-mean normal distribution with variance $\sigma^2 = 0.025$. $N = 1000$ and $N_V = 500$ datapoints were used for estimation and validation, respectively. The noise-to-signal ratio (computed as the ratio between the standard deviations of the noise and the output) was about 5%. The algorithm was run with δ equal to 0.2 (approximately, 1.26σ), and provided $s = 4$ submodels. The final classification of the regression vectors $x_k = [y_{k-1}\ u_{k-1}]'$ used for estimation, and the partition of the PWA map are shown in Figure 2. The model was validated by computing the prediction error ε_k, *i.e.* the difference between the measured and the predicted output, whose plot is depicted in Figure 3. It is noticeable that only 3 times out of 500 it falls outside the interval $[-3\sigma, 3\sigma]$. The Mean Square Error

$$MSE = \frac{1}{N_v}\sum_{k=1}^{N_v}\varepsilon_k^2$$

was equal to 0.030, which is very close to the variance of the noise. Recall that the prediction error is influenced by both the noise and the model error. The overall computation of the PWA model took about 9 seconds on a 1GHz Pentium III running Matlab 6.1.

7. CONCLUSIONS

This paper has addressed the problem of identifying a PWA model of a discrete-time nonlinear system from input-output data. The proposed two-stage procedure first classifies the data into clusters and estimates the parameters of the affine submodels, and then estimates the coefficients of the hyperplanes defining the partition of the PWA map. The key approach is the selection of a bound on the prediction error. This makes it possible to formulate the initial classification problem as an extension of the MIN PFS problem for infeasible systems of linear inequalities. The major capability of this formulation is that it also provides an estimate of the minimum number of submodels needed to fit the data. Other approaches could be used to initialize the identification procedure, *e.g.*, the *k*-plane clustering algorithm proposed in (Bradley and Mangasarian, 2000). A procedure for improving both data classification and parameter estimation was also proposed. It alternates between datapoint reassignment and parameter update. Moreover, the number of submodels is allowed to vary from iteration to iteration. The partition of the PWA map is finally estimated via multicategory classification techniques. Future studies will concern the convergence properties of the refinement procedure and the evaluation of the quality of the identified PWA models.

REFERENCES

Amaldi, E. and M. Mattavelli (2002). The MIN PFS problem and piecewise linear model estimation. *Discrete Applied Mathematics* **118**, 115–143.

Bemporad, A., G. Ferrari-Trecate and M. Morari (2000). Observability and controllability of piecewise affine and hybrid systems. *IEEE Trans. Automatic Control* **45**(10), 1864–1876.

Bemporad, A., J. Roll and L. Ljung (2001). Identification of hybrid systems via mixed-integer programming. In: *Proc. 40th IEEE Conf. on Decision and Control*. pp. 786–792.

Bradley, P. S. and O. L. Mangasarian (2000). *k*-plane clustering. *Journal of Global Optimization* **16**, 23–32.

Bredensteiner, E. J. and K. P. Bennett (1999). Multicategory classification by support vector machines. *Computational Optimization and Applications* **12**, 53–79.

Ferrari-Trecate, G., M. Muselli, D. Liberati and M. Morari (2003). A clustering technique for the identification of piecewise affine systems. *Automatica* **39**(2), 205–217.

Heemels, W.P.M.H., B. De Schutter and A. Bemporad (2001). Equivalence of hybrid dynamical models. *Automatica* **37**(7), 1085–1091.

Juditsky, A., H. Hjalmarsson, A. Benveniste, B. Delyon, L. Ljung, J. Sjöberg and Q. Zhang (1995). Nonlinear black-box models in system identification: mathematical foundations. *Automatica* **31**(12), 1725–1750.

Milanese, M. and A. Vicino (1991). Optimal estimation theory for dynamic systems with set membership uncertainty: an overview. *Automatica* **27**(6), 997–1009.

Milanese, M., J. P. Norton, H. Piet-Lahanier and E. Walter (eds.) (1996). *Bounding Approaches to System Identification*. Plenum Press. New York.

Münz, E. and V. Krebs (2002). Identification of hybrid systems using apriori knowledge. In: *Proc. 15th IFAC World Congress*.

Sjöberg, J., Q. Zhang, L. Ljung, A. Benveniste, B. Delyon, P. Glorennec, H. Hjalmarsson and A. Juditsky (1995). Nonlinear black-box modeling in system identification: a unified overview. *Automatica* **31**(12), 1691–1724.

Vapnik, V. (1998). *Statistical Learning Theory*. John Wiley.

IFAC Publications
www.elsevier.com/locate/ifac

PIECEWISE-LINEAR OUTPUT-ERROR MODELS

Fredrik Rosenqvist * **Anders Karlström** **

* *Department of Signals and Systems*
Chalmers University of Technology, SE-412 96 Göteborg, Sweden
** *Chalmers Industriteknik, SE-412 88 Göteborg, Sweden*

Abstract: Piecewise-linear systems in input/output form can have different switching schemes. In this paper two categories, instant and delayed switching, are analysed. Even though a general piecewise-linear state-space model cannot be converted into input/output form, it is shown that it is possible to find state-space models representing instant as well as delayed switching. In addition, a prediction-error minimisation (PEM) method for piecewise-linear output-error predictors is derived and it is concluded that the instant-switching model candidate is not necessarily the most suitable for the parameter estimation procedure.

Keywords: Parameter estimation, Piecewise-linear systems, Prediction-error method

1. INTRODUCTION

Piecewise-linear (PWL) models for parameter estimation have previously focused on systems in which all the parameters switch simultaneously as the conditions for switching are fulfilled (Bemporad *et al.*, 2001; Ferrari-Trecate *et al.*, 2001; Skeppstedt *et al.*, 1992). However, the way in which a piecewise-linear representation in input/output (*I/O*) form corresponds to the more commonly used piecewise-linear state-space form, can not be found in the literature.

PWL models are, on the one hand, commonly represented in state-space form (Bemporad *et al.*, 2000; Johansson, 1999; Liberzon and Morse, 1999). On the other hand, when it comes to parameter estimation, the utilisation of *I/O* forms is often seen (Bemporad *et al.*, 2001; Ferrari-Trecate *et al.*, 2001; Billings and Voon, 1987). Parameter estimation, where the models are in state-space form (Nakamura and Hamada, 1990), also occurs in the literature, but in this case an observer is usually needed to ensure satisfactory identification. Whereas the relationship between state-space and *I/O* is straightforward in the linear case, care must be taken in the PWL modelling, as explained in this paper.

2. FUNDAMENTALS

Let $k \in \{1,2,....N\}$ be the discrete time and consider the state+input space $\mathscr{Z} = \mathscr{X} \times \mathscr{U}$ [1], where $x(k) \in \mathscr{X} \subseteq \mathbb{R}^n$ is the state vector and $u(k) \in \mathscr{U} \subseteq \mathbb{R}^p$ is the input vector. As a result, $\mathscr{Z} \subseteq \mathbb{R}^{n+p}$ can be partitioned into s disjoint subsets $\mathscr{Z}_\sigma$ such that $\bigcup_{\sigma=1}^{s} \mathscr{Z}_\sigma = \mathscr{Z}$ and $\bigcap_{\sigma=1}^{s} \mathscr{Z}_\sigma = \{\emptyset\}$.

Following the notation above, the piecewise-linear system can be described in state-space form (Bemporad *et al.*, 2000)

$$\begin{aligned} x(k) &= F_\sigma x(k-1) + G_\sigma u(k-1) \\ y(k) &= H_\sigma x(k) \\ \text{for } &\begin{bmatrix} x \\ u \end{bmatrix} \in \mathscr{Z}_\sigma \end{aligned} \tag{1}$$

The *I/O* form of piecewise-linear systems (Bemporad *et al.*, 2001; Ferrari-Trecate *et al.*, 2001) is commonly represented as

$$y(k) = \frac{\sum_{j=1}^{n_b} b_{j,\sigma} q^{-j}}{1 + \sum_{i=1}^{n_a} a_{i,\sigma} q^{-i}} u(k) \tag{2}$$

[1] The product space is interpreted here as the space which is spanned by the factors $\mathscr{A} \times \mathscr{B} = \{[a^T \quad b^T]^T : a \in \mathscr{A}, b \in \mathscr{B}\}$.

where the inverse shift operator, q^{-1}, is defined such that $q^{-1}x(k) = x(k-1)$.

The parameter $\sigma : \mathscr{X} \rightarrow \{1,2,...,s\}$ depends on the system conditions at time k, i.e. $\sigma = \sigma(x(k),u(k))$. In other words, at a given point in time, σ corresponds to one of the linear modes, depending on the state and input at this time. Further, the notation $\sigma(k)$ will be used for $\sigma(x(k),u(k))$. This makes the interpretation of Eq. 1 clear, e.g. $F_\sigma x(k-1)$ for $[x \quad u]^T \in \mathscr{X}_\sigma$ becomes $F_{\sigma(k-1)}x(k-1)$. On the other hand, in Eq. 2, it is not obvious which k determines the active sub-model for a given parameter. In the literature, the interpretation is that $\sigma = \sigma(k-1)$ for all the parameters in Eq. 2 (Bemporad *et al.*, 2001; Ferrari-Trecate *et al.*, 2001; Billings and Voon, 1987).

For the PWL model in state-space form, a corresponding *I/O model*, in which each parameter corresponds to a specific sub-model, is not necessarily found. Example 1 illustrates how a parameter in an *I/O* model may represent more than one mode of the piecewise-linear state-space model.

Example 1. Consider the system described by Eq. 1, where

$$F_\sigma = \begin{bmatrix} f_{11,\sigma} & f_{12,\sigma} \\ 0 & f_{22,\sigma} \end{bmatrix} \quad G_\sigma = \begin{bmatrix} 0 \\ g_\sigma \end{bmatrix} \qquad H_\sigma = \begin{bmatrix} 1 & 0 \end{bmatrix} \tag{3}$$

Further, let $s = 2$ and the switching rule be according to

$$\sigma(u(k)) = \begin{cases} 1, & u(k) < 0 \\ 2, & u(k) \geq 0 \end{cases} \tag{4}$$

To derive the corresponding *I/O* model, $x_1(k)$ and $x_2(k)$ are first evaluated

$$x_1(k) = f_{11,\sigma(k-1)}x_1(k-1) + f_{12,\sigma(k-1)}x_2(k-1) \tag{5}$$

$$x_2(k) = f_{22,\sigma(k-1)}x_2(k-1) + g_{\sigma(k-1)}u(k-1) \tag{6}$$

It is now important to keep the notation $x_2(k-1)$, which is to be substituted

$$\begin{aligned} x_1(k) &= \frac{f_{12,\sigma(k-1)}x_2(k-1)}{1-f_{11,\sigma(k-1)}q^{-1}} \\ x_2(k-1) &= \frac{g_{\sigma(k-2)}u(k-2)}{1-f_{22,\sigma(k-2)}q^{-1}} \end{aligned} \tag{7}$$

The substitution of $x_2(k-1)$ and the fact that $x_1(k) = y(k)$ give, after simplification,

$$y(k) = \frac{b_{2,\sigma}q^{-2}}{1-a_{1,\sigma}q^{-1}+a_{2,\sigma}q^{-2}}u(k) \tag{8}$$

where

$$\begin{aligned} a_{1,\sigma} &= f_{11,\sigma(k-1)} + f_{22,\sigma(k-2)} \\ a_{2,\sigma} &= f_{11,\sigma(k-1)}f_{22,\sigma(k-2)} \\ b_{2,\sigma} &= f_{12,\sigma(k-1)}g_{\sigma(k-2)} \end{aligned} \tag{9}$$

Consequently, the parameters in the corresponding *I/O* model cannot be designated for only one mode each. □

A given piecewise-linear state-space model can therefore not necessarily be converted into an *I/O* model whose parameters belong to only one mode. There are, however, cases where this is possible.

Proposition 2. If a piecewise-linear state-space model (Eq. 1) is represented in the form

$$\begin{aligned} F &= \begin{bmatrix} -a_1 & \dots & -a_{n-1} & -a_n \\ 1 & \dots & 0 & 0 \\ \vdots & \ddots & \vdots & \vdots \\ 0 & \dots & 1 & 0 \end{bmatrix}_{\sigma(k-1)} \\ G &= \begin{bmatrix} 1 & \dots & 0 & 0 \end{bmatrix}^T \\ H &= \begin{bmatrix} b_1 & \dots & b_{n-1} & b_n \end{bmatrix}_{\sigma(k-1)} \end{aligned} \tag{10}$$

there is a corresponding *I/O* model, in which each parameter is associated with only one mode

$$y(k) = \frac{\sum_{j=1}^n b_{j,\sigma(k-1)}q^{-j}}{1+\sum_{i=1}^n a_{i,\sigma(k-1)}q^{-i}}u(k) \tag{11}$$

Moreover, if the state-space model is in the form

$$\begin{aligned} F &= \begin{bmatrix} -a_1 & 1 & \dots & 0 \\ \vdots & \vdots & \ddots & \vdots \\ -a_{n-1} & 0 & \dots & 1 \\ -a_n & 0 & \dots & 0 \end{bmatrix}_{\sigma(k-1)} \\ G &= \begin{bmatrix} b_1 & \dots & b_{n-1} & b_n \end{bmatrix}^T_{\sigma(k-1)} \\ H &= \begin{bmatrix} 1 & 0 & \dots & 0 \end{bmatrix} \end{aligned} \tag{12}$$

there is a corresponding *I/O* model, in which each parameter is associated with only one mode

$$y(k) = \frac{\sum_{j=1}^n b_{j,\sigma(k-j)}q^{-j}}{1+\sum_{i=1}^n a_{i,\sigma(k-i)}q^{-i}}u(k) \tag{13}$$

PROOF. The evaluation of Eq. 1 with the F and G matrices in Eq. 10, followed by time-shifting of the $x_2(k)$ to $x_n(k)$ equations, results in

$$\begin{aligned} x_2(k-1) &= x_1(k-2) \\ &\vdots \\ x_n(k-1) &= x_1(k-n) \end{aligned} \tag{14}$$

Insertion into the $x_1(k)$ equation

$$x_1(k) = -\left(\sum_{i=1}^{n} a_{i,\sigma(k-1)} q^{-i}\right) x_1(k) + u(k-1) \quad (15)$$

$$\Longrightarrow$$

$$x_1(k)\left(1 + \sum_{i=1}^{n} a_{i,\sigma(k-1)} q^{-i}\right) = u(k-1) \quad (16)$$

Further, the evaluation of Eq. 1 with the H matrix in Eq. 10 gives

$$y(k) = \left(\sum_{j=1}^{n} b_{j,\sigma(k-1)} q^{-j-1}\right) x_1(k) \quad (17)$$

The substitution of Eq. 16 into Eq. 17 results in Eq. 11.

The evaluation of Eq. 1 with the F and G matrices in Eq. 12 gives

$$\begin{aligned}
& x_1(k) = \\
& -a_{1,\sigma(k-1)} x_1(k-1) + x_2(k-1) + \\
& +b_{1,\sigma(k-1)} u(k-1) \\
& x_2(k-1) = \\
& -a_{2,\sigma(k-2)} q^{-1} x_1(k-1) + x_3(k-2) + \\
& +b_{2,\sigma(k-2)} q^{-1} u(k-1) \\
& \vdots \\
& x_{n-1}(k-n+2) = \\
& -a_{n-1,\sigma(k-n+1)} q^{-(n-2)} x_1(k-1) + \\
& +x_n(k-n+1) + \\
& +b_{n-1,\sigma(k-n+1)} q^{-(n-2)} u(k-1) \\
& x_n(k-n+1) = \\
& -a_{n,\sigma(k-n)} q^{-(n-1)} x_1(k-1) + \\
& +b_{n,\sigma(k-n)} q^{-(n-1)} u(k-1)
\end{aligned} \quad (18)$$

$$\begin{aligned}
\Longrightarrow \quad & x_1(k) = \\
& -\left(\sum_{i=1}^{n} a_{i,\sigma(k-i)} q^{-i}\right) x_1(k) + \\
& +\left(\sum_{j=1}^{n} b_{j,\sigma(k-1)} q^{-j}\right) u(k-1)
\end{aligned} \quad (19)$$

Since $y(k) = x_1(k)$, Eq. 13 follows. □

The *I/O* model in Eq. 11 will be called the *instant* switching model, as all the parameters switch simultaneously, and the *I/O* model in Eq. 13 will be called the *delayed* switching model, as the parameter switches are delayed for the high order parameters.

Example 3. The difference between instant and delayed switching is illustrated here by an example. Consider the piecewise-linear system

$$y(k) = -a_1 y(k-1) - a_{2,\tilde{\sigma}} y(k-2) + b_1 u(k-1) \quad (20)$$

whose switching boundary is at $u(k) = 0$:

$$\sigma(k) = \begin{cases} 1, & u(k) < 0 \\ 2, & u(k) \geq 0 \end{cases} \quad (21)$$

The notation $\tilde{\sigma}$ means that $\tilde{\sigma} = \sigma(k-1)$ for instant switching and that $\tilde{\sigma} = \sigma(k-2)$ for delayed switching respectively. Consider the following parameter values

$$\begin{aligned}
& a_1 = -0.5 \\
& a_{2,1} = 0.125 \; a_{2,2} = 1 \\
& b_1 = 1
\end{aligned} \quad (22)$$

The different behaviours for the two switching schemes are exemplified by the simulation in Fig. 1. □

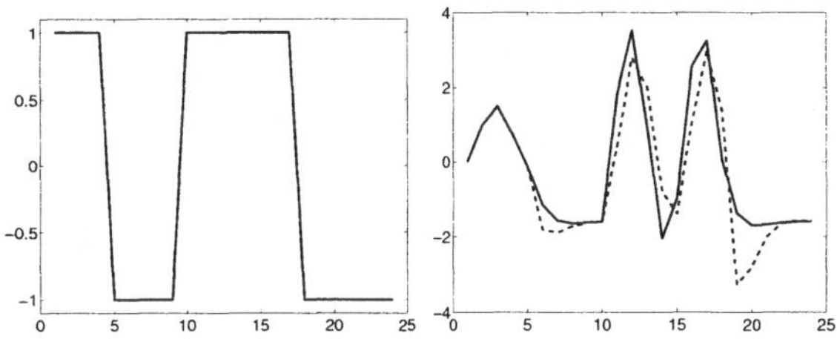

Fig. 1. Simulated responses for the system given in Example 3. The *left* plot shows the input signal, $u(k)$. The *right* plot shows the output signals, $y(k)$, for instant switching (solid) and delayed switching (dashed) respectively.

3. MODEL FORMULATION

To identify the parameters above, piecewise-affine *ARX* predictors for instant-switching systems have been derived by Bemporad *et al.* (2001) and Ferrari-Trecate *et al.* (2001). If the perturbations are assumed to be independent of the different sub-models, a predictor which adds the perturbation directly to the output would be more natural. A piecewise-linear *output-error* (*PWOE*) predictor of this kind, $\hat{y}^I(k)$, for the instant-switching model can be represented as

$$\begin{aligned}
\hat{y}^I(k) = & -a_{1,\sigma(k-1)} \hat{y}(k-1) - \ldots \\
& -a_{n_a,\sigma(k-1)} \hat{y}(k-n_a) + \\
& +b_{1,\sigma(k-1)} u(k-1) + \ldots \\
& +b_{n_b,\sigma(k-1)} u(k-n_b)
\end{aligned} \quad (23)$$

Similarly, the delayed-switching model has a *PWOE* predictor, $\hat{y}^D(k)$, which is described as

$$\begin{aligned}
\hat{y}^D(k) = & -a_{1,\sigma(k-1)} \hat{y}(k-1) - \ldots \\
& -a_{n_a,\sigma(k-n_a)} \hat{y}(k-n_a) + \\
& +b_{1,\sigma(k-1)} u(k-1) + \ldots \\
& +b_{n_b,\sigma(k-n_b)} u(k-n_b)
\end{aligned} \quad (24)$$

It is convenient to write the predictor in Eq. 23 and 24 in terms of a parameter vector, Θ, and a regression vector, $\Phi(k)$, when it comes to parameter estimation

$$\hat{y}(k|\Theta) = \Theta^T \Phi(k) \quad (25)$$

The parameter vector, Θ, for the system is an array consisting of the parameter vectors of all sub-models.

$$\Theta = \begin{bmatrix} \theta_1 \\ \theta_2 \\ \vdots \\ \theta_s \end{bmatrix} ; \; \theta_p = \begin{bmatrix} a_{1,p} \\ \vdots \\ a_{n_a,p} \\ b_{1,p} \\ \vdots \\ b_{n_b,p} \end{bmatrix} \tag{26}$$

The regression vector for the instant switching *PWOE* predictor, $\Phi^I(k)$, is based on the regression vector for the sub-models, which are all equal according to the description in Eq. 23.

$$\Phi^I(k) = \begin{bmatrix} \phi(k)\delta_{\sigma 1}(k) \\ \phi(k)\delta_{\sigma 2}(k) \\ \vdots \\ \phi(k)\delta_{\sigma s}(k) \end{bmatrix}$$

$$\phi(k) = \begin{bmatrix} -\hat{y}(k-1) \\ \vdots \\ -\hat{y}(k-n_a) \\ u(k-1) \\ \vdots \\ u(k-n_b) \end{bmatrix} \tag{27}$$

where $\delta_{\sigma p}$ is the Kronecker delta

$$\delta_{\sigma p}(k) = \begin{cases} 1 \text{ if } & \sigma(k) = p \\ 0 \text{ else} & \end{cases} \tag{28}$$

The regression vector in the delayed *PWOE* case, $\Phi^D(k)$, is formed by the regressors for each sub-model

$$\Phi^D(k) = \begin{bmatrix} \phi_1(k) \\ \phi_2(k) \\ \vdots \\ \phi_s(k) \end{bmatrix}$$

$$\phi_p(k) = \begin{bmatrix} -\hat{Y}_p(k-1) \\ \vdots \\ -\hat{Y}_p(k-n_a) \\ U_p(k-1) \\ \vdots \\ U_p(k-n_b) \end{bmatrix} \tag{29}$$

where $\hat{Y}_p(k-l)$ and $U_p(k-l)$ assume the value of $\hat{y}(k-l)$ and $u(k-l)$ respectively, only if the sub-model p is active at sample $k-l$. Otherwise they assume the value zero. In terms of Kronecker deltas, this can be written as:

$$X_p(k-l) = x(k-l)\delta_{\sigma p}(k-l) \tag{30}$$

This means that, if there are many sub-models, the regression vector is very sparse, but, in the case of frequent switching the terms correspond to different sub-models.

4. PARAMETER ESTIMATION

Assuming the predictor models from the previous section, an algorithm for estimating the parameters is to be derived (Ljung, 1999). Given the predictor $\hat{y}(k|\Theta)$, the prediction error is defined as

$$\varepsilon(k,\Theta) = y(k) - \hat{y}(k|\Theta) \tag{31}$$

where $y(k)$ is the measured output value at sample k. Given an estimation data set, Z^N, consisting of acquired input and output signals at N samples, a sufficiently good estimate of the parameters is to be found. According to the *prediction error method* (Ljung, 1999), the parameter estimate is the minimum of a cost function, $V_N(\Theta, Z^N)$

$$\hat{\Theta} = \arg\min V_N(\Theta, Z^N) \tag{32}$$

The estimation data set incorporates the measured values of the input and the output signal from the process to be identified. These values are needed when calculating the residuals, $\varepsilon(k,\Theta)$. The most common cost function that is used is the sum of the squares of the residuals

$$V_N(\Theta, Z^N) = \frac{1}{N}\sum_{k=1}^{N} \varepsilon^2(k,\Theta) \tag{33}$$

The algorithm for finding the parameter estimate must be iterative, as the cost function to be minimised is a function of the parameter vector itself. The search algorithm used here is a gradient method

$$\hat{\Theta}_{i+1} = \hat{\Theta}_i - \mu_i R_i^{-1}\frac{\partial}{\partial\Theta}V_N(\hat{\Theta}_i) \tag{34}$$

where i is the iteration step, μ_i is the step size and R_i, which corresponds to the second derivative of the cost function, determines the search direction (Sjöberg *et al.*, 1995). The Gauss-Newton algorithm utilises $\mu_i = 1$ and

$$R_i = \frac{1}{N}\sum_{k=1}^{N} \Psi(\Phi(k),\hat{\Theta}_i)\Psi^T(\Phi(k),\hat{\Theta}_i) \tag{35}$$

where

$$\Psi(\Phi(k),\Theta) = \frac{\partial}{\partial\Theta}\hat{y}(k) \tag{36}$$

Further,

$$\frac{\partial}{\partial\Theta}V_N(\hat{\Theta}_i) = -\frac{1}{N}\sum_{k=1}^{N} \Psi(\Phi(k),\hat{\Theta})\varepsilon(k,\hat{\Theta}) \tag{37}$$

Because the predictor is a function of previous predictions, the model becomes recurrent (Sjöberg *et al.*, 1995; Nerrand *et al.*, 1994). These models do not cause any problems for the algorithm itself, but some extra computational effort is required to approximate Eq. 35.

Proposition 4. Given the predictors in Eq. 23 and 24, their differentiations with respect to the parameter vector can be computed as

$$(1+\sum_{h=1}^{n_a} a_{h,\sigma(k-h^*)} q^{-h})\Psi = \Phi \tag{38}$$

where

$$k-h^* = \begin{cases} k & \text{if } (\Psi,\Phi) = (\Psi^I,\Phi^I) \\ k-h & \text{if } (\Psi,\Phi) = (\Psi^D,\Phi^D) \end{cases} \tag{39}$$

PROOF. $\Psi(k)$ consists of several time series ordered in an array in the same way as $\Phi(k)$, which is illustrated in Eq. 27 and 29. Let $p \in \{1,2,\ldots,s\}$ and $i \in \{1,2,\ldots,n_a,n_a+1,\ldots,n\}$, where $n = n_a + n_b$. The regression vector can be described as $\Phi_{(p-1)n+i}$, where p corresponds to the pth sub-model and i is the position in the parameter vector of the sub-model, θ_p.

In the case where $i \in \{1,2,\ldots,n_a\}$,

$$\Psi_{(p-1)n+i} = \frac{\partial \hat{y}(k)}{\partial a_{i,p}} \tag{40}$$

can be computed as

$$\begin{aligned}
&\frac{\partial}{\partial a_{i,p}}\hat{y}(k) = -\hat{y}(k-i)\delta_{\sigma p}(k-i^*) - \\
&\sum_{h=1}^{n_a} a_{h,\sigma(k-h^*)} \frac{\partial}{\partial a_{i,p}} \hat{y}(k-h) = \\
&= -\hat{y}(k-i)\delta_{\sigma p}(k-i^*) - \\
&\sum_{h=1}^{n_a} a_{h,\sigma(k-h^*)} \frac{\partial}{\partial a_{i,p}} q^{-h}\hat{y}(k) = \\
&= -\hat{y}(k-i)\delta_{\sigma p}(k-i^*) - \\
&\sum_{h=1}^{n_a} a_{h,\sigma(k-h^*)} q^{-h} \frac{\partial}{\partial a_{i,p}} \hat{y}(k) \\
&\Longrightarrow \\
&\frac{\partial}{\partial a_{i,p}}\hat{y}(k) + \\
&\sum_{h=1}^{n_a} a_{h,\sigma(k-h^*)} q^{-h} \frac{\partial}{\partial a_{i,p}} \hat{y}(k) = \\
&-\hat{y}(k-i)\delta_{\sigma p}(k-i^*) \\
&\Longrightarrow \\
&(1+\sum_{h=1}^{n_a} a_{h,\sigma(k-h^*)} q^{-h}) \frac{\partial}{\partial a_{i,p}} \hat{y}(k) = \\
&-\hat{y}(k-i)\delta_{\sigma p}(k-i^*)
\end{aligned} \tag{41}$$

where

$$-\hat{y}(k-i)\delta_{\sigma p}(k-i^*) = \\ = \begin{cases} \Phi^I_{(p-1)n+i} & \text{if the switching is instant} \\ \Phi^D_{(p-1)n+i} & \text{if the switching is delayed} \end{cases} \tag{42}$$

therefore

$$(1+\sum_{h=1}^{n_a} a_{h,\sigma(k-h^*)} q^{-h})\Psi_{(p-1)n+i} = \Phi_{(p-1)n+i} \tag{43}$$

Analogously, if $i \in \{n_a+1, n_a+2, \ldots, n\}$ and $j = i - n_a \in \{1,2,\ldots,n_b\}$

$$\Psi_{(p-1)n+i} = \Psi_{(p-1)n+n_a+j} \frac{\partial \hat{y}(k)}{\partial b_{j,p}} \tag{44}$$

can be computed as

$$\begin{aligned}
&\frac{\partial}{\partial b_{j,p}}\hat{y}(k) = \\
&-\sum_{h=1}^{n_a} a_{h,\sigma(k-h^*)} \frac{\partial}{\partial b_{j,p}} \hat{y}(k-h) + \\
&u(k-j)\delta_{\sigma p}(k-j^*) = \\
&= -\sum_{h=1}^{n_a} a_{h,\sigma(k-h^*)} q^{-h} \frac{\partial}{\partial b_{j,p}} \hat{y}(k) + \\
&u(k-j)\delta_{\sigma p}(k-j^*) \\
&\Longrightarrow \\
&(1+\sum_{h=1}^{n_a} a_{h,\sigma(k-h^*)} q^{-h}) \frac{\partial}{\partial b_{j,p}} \hat{y}(k) = \\
&u(k-j)\delta_{\sigma p}(k-j^*)
\end{aligned} \tag{45}$$

where

$$u(k-j)\delta_{\sigma p}(k-j^*) = \\ = \begin{cases} \Phi^I_{(p-1)n+i} & \text{if the switching is instant} \\ \Phi^D_{(p-1)n+i} & \text{if the switching is delayed} \end{cases} \tag{46}$$

Consequently, Eq. 43 is also true for $i \in \{n_a+1, n_a+2, \ldots, n\}$, which proves Proposition 4. □

Example 5. Parameter estimation using the predictors $\hat{y}^I$ in Eq. 23 and $\hat{y}^D$ in Eq. 24 respectively is illustrated in a model system. The true model is described by Eq. 1, with added white Gaussian noise and the following parameters

$$\begin{aligned}
&F_1 = \begin{bmatrix} 0.5 & 1 \\ 0 & 0.5 \end{bmatrix} \; F_2 = \begin{bmatrix} 0.7 & 1 \\ -0.5 & 0.5 \end{bmatrix} \\
&G_{1,2} = \begin{bmatrix} 0 & 1 \end{bmatrix}^T \\
&H_{1,2} = \begin{bmatrix} 1 & 0 \end{bmatrix}
\end{aligned} \tag{47}$$

where

$$\sigma(u(k)) = \begin{cases} 1, & u(k) < 0 \\ 2, & u(k) \geq 0 \end{cases} \tag{48}$$

The model orders of the predictors were chosen to be $n_a = n_b = 2$ and parameter estimation was performed using the data shown in Fig. 2. It can be observed that sub-model 2, i.e. when $u(k) \geq 0$, is oscillatory, whereas sub-model 1 is damped. It should also be

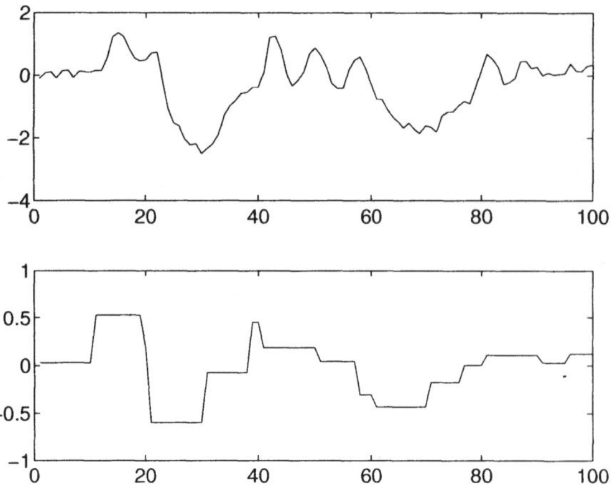

Fig. 2. Identification data: *Upper plot:* The output data, $y(k)$. *Lower plot:* The input data, $u(k)$.

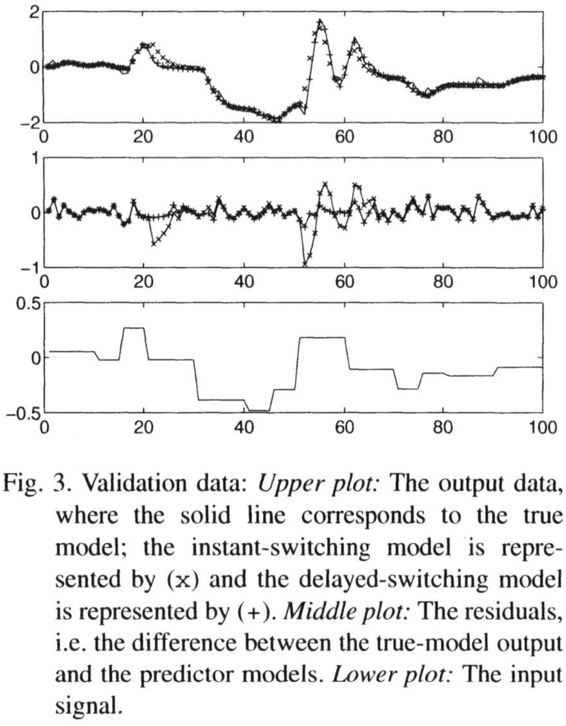

Fig. 3. Validation data: *Upper plot:* The output data, where the solid line corresponds to the true model; the instant-switching model is represented by (x) and the delayed-switching model is represented by (+). *Middle plot:* The residuals, i.e. the difference between the true-model output and the predictor models. *Lower plot:* The input signal.

noted that neither of the predictor models has the correct model structure, which means that a bias should occur.

The validation data and their residuals are shown in Fig. 3. Even though neither of the predictors has a model structure that perfectly matches the true model, it can be concluded that the predictor corresponding to delayed switching is a better model. □

5. CONCLUSIONS

An *I/O* representation of a piecewise-linear system can have different switching schemes according to Eq. 11 and 13. These representations are referred to as *instant* and *delayed* switching respectively. As illustrated in Example 1, a general PWL state-space system cannot be converted into an *I/O* form, where each parameter corresponds to only one mode. Nevertheless, it is possible to find state-space models representing instant and delayed switching.

The piecewise-linear output-error (PWOE) predictors represent model structures for the parameter estimation of PWL systems according to the prediction-error method. Finally, it is concluded that the instant-switching model, which is the one found in the literature, is not necessarily the most suitable for the parameter estimation of PWL models.

6. REFERENCES

Bemporad, A., G. Ferrari-Trecate and M. Morari (2000). Observability and controllability of piecewise affine and hybrid systems. *Transactions on Automatic Control* **45**(10), 1864 – 1876.

Bemporad, A., J. Roll and L. Ljung (2001). Identification of hybrid systems via mixed-integer programming. In: *Conference on Descision and Control*. pp. 786 – 792.

Billings, S.A. and W.S.F. Voon (1987). Piecewise linear identification of non-linear systems. *International Journal of Control* **46**, 215 – 235.

Ferrari-Trecate, G., M. Muselli, D. Liberati and M. Morari (2001). Identification of piecewise affine and hybrid systems. In: *American Control Conference*. pp. 3521 – 3526.

Johansson, M. (1999). Piecewise Linear Control systems. PhD thesis. Lund Institute of Technology.

Liberzon, D. and A. S. Morse (1999). Basic problems in stability and design of switched systems. *IEEE Control Systems Magazine* **19**(19), 59–70.

Ljung, L. (1999). *System Identification - Theory for the User*. Prentice Hall.

Nakamura, A. and N. Hamada (1990). Nonlinear dynamical system identification by piecewise-linear system. In: *IEEE International Symposium on Circuits and Systems*. pp. 1454 – 1457.

Nerrand, O., P. Roussel-Ragot, D. Urbani, L. Personnaz and G. Dreyfus (1994). Training recurrent neural networks: why and how? an illustration in dynamical process modeling. *Transactions on Neural Networks* **5**(2), 178 – 184.

Sjöberg, J., Q. Zhang, L. Ljung, A. Benveniste, B. Delyon, P-Y. Glorennec, H. Hjalmarsson and A. Juditsky (1995). Nonlinear black-box modeling in system identification: a unified overview. *Automatica* **31**(12), 1691 – 1724.

Skeppstedt, L., L. Ljung and M. Millnert (1992). Construction of composite models from observed data. *International Journal of Control* **55**(1), 141 – 152.

IFAC Publications
www.elsevier.com/locate/ifac

CMAC WITH LINEAR FUNCTIONAL WEIGHTS

Qiang Gan and Eric Rosales

Department of Computer Science
University of Essex
Colchester CO4 3SQ, UK
E-mail: jqgan@essex.ac.uk, emrosa@essex.ac.uk

Abstract: Cerebellar model articulation controller (CMAC) has been widely applied to modelling and control due to its attractive features such as fast training speed and parsimonious structure. This paper modifies the CMAC model by replacing its constant weights by linear functional weights, aiming not only to improve its efficiency in modelling smooth nonlinear processes but also to increase its interpretability and applicability. Following the reformulation of the CMAC model and its learning process, experimental results are given and analysed. The approximation ability and interpretability of the modified CMAC is especially investigated. *Copyright © 2003 IFAC*

Keywords: CMAC, data models, fuzzy modelling, interpretability, linearisation, local models, neural networks, nonlinearity.

1. INTRODUCTION

Local model approaches have found widespread applications in complex system/process modelling and control, due to the following advantages: high training speed, good approximation and generalisation capability, and good interpretability that is very useful for subsequent model applications. In general there are two types of local models, one is in the form of local basis function expansion, such as CMAC (Albus, 1975a, 1975b) and neurofuzzy systems (Brown and Harris, 1994), and the other is based on the Takagi-Sugeno (T-S) fuzzy model (Takagi and Sugeno, 1985), such as ANFIS (Adaptive fuzzy inference systems) (Jang, 1993). As far as approximation ability is concerned, the type of T-S fuzzy models can be viewed as a special case of local basis function expansion with excessive free parameters (Gan and Harris, 1999a). However, the type of T-S fuzzy models has extra advantages (Johansen, *et al.*, 2000): (i) Using local linear models brings together fuzzy and conventional control theories; (ii) The relatively complex consequence part allows the number of fuzzy rules to be quite small in many applications, leading to parsimonious structures; and (iii) The model structure and local model properties can, in some applications, be easily related to the physics of the underlying system, simplifying the model development and validation.

One critical issue in local model approaches is the curse of dimensionality problem, *i.e.*, the number of local models and the amount of training samples required increase exponentially with the input space dimension. Some heuristic construction algorithms and optimal structural search methods (Gan and Harris, 1999a, 2001; Nelles, 1999; Hoffman and Nelles, 2001) have been developed in recent years for optimal and automatic input space partitioning, alleviating the curse of dimensionality problem to some extent. CMAC has provided a good example to handle this problem by using a fixed number of layers of receptive fields and a hashing function, making itself attractive for real-time applications. Based on our previous work on the comparison of CMAC and ANFIS (Rosales and Gan, 2003), this paper proposes a modified CMAC model with linear functional (LF) weights and thus extends CMAC into a fuzzy local linear (FLL) model, aiming to combine the advantages of CMAC and ANFIS.

2. MODIFICATION OF CMAC

This section presents a brief introduction to CMAC and FLL models, and then proposes a modified CMAC with LF weights.

2.1 A Brief Description of CMAC

A CMAC model can be described by (Albus, 1975a, 1975b; Miller, *et al.*, 1990; Rosales and Gan, 2003):

$$y=\sum_{j=1}^{C}\frac{f(\delta_j(x_1,\cdots,x_n))}{\sum_{k=1}^{C}f(\delta_k(x_1,\cdots,x_n))}w_{A_j} \tag{1}$$

where $f(\delta_j(x_1,\cdots,x_n))$ represent receptive field (RF) functions, $\delta_j(x_1,\cdots,x_n)$ maps the input vector $\mathbf{x}$ into a RF address in the jth layer of receptive fields, w_{A_j} are adjustable weights, A_j represents the address of an adjustable weight associated with the RF in the jth RF layer and is determined by mapping δ_j using a hashing function, and C is the number of RF layers. In the CMAC model, only one receptive field in each RF layer is activated by the input, and only those weights associated with the activated RFs are updated using a least-mean-square (LMS) type algorithm. Therefore, there are only C weights to be updated in each training iteration. For many applications the required number of RF layers in CMAC is much smaller than the number of weights in most other commonly used neural networks. That is the main reason why the CMAC training process converges very fast.

CMAC provides a local modelling approach, in which the input space is split by receptive fields that are organised into multiple layers. The RFs in each layer do not overlap, but the RF layers are placed with offsets, as shown in Fig. 1 with 4 layers of triangle-shaped functions as RFs. Corresponding to an input, there will be only one RF activated in each layer. However, the activated RFs in all the layers have an overlapping effect in covering the local regions of the input space, similar as in fuzzy systems.

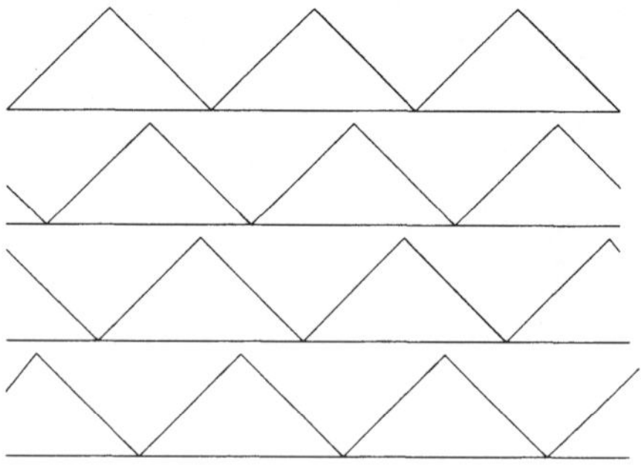

Figure 1 Input space partitioning in CMAC

2.2 Fuzzy Local Linear Modelling

The basic idea of FLL modelling was introduced by the T-S fuzzy model based on the following fuzzy rules:

R^j: If x_1 is A_{j1} and ... and x_n is A_{jn} then $y=g_j(x_1,\ldots,x_n)$, $j=1,2,\ldots,K$.

where A_{ji} represent fuzzy sets defined in the input space and K is the number of fuzzy rules. The output of the model with the input $(x_1,\ldots,x_n)$ is obtained as

$$y=\sum_{j=1}^{K}\frac{\mu_{A_j}(x_1,\cdots,x_n)}{\sum_{k=1}^{K}\mu_{A_k}(x_1,\cdots,x_n)}\cdot g_j(x_1,\cdots,x_n) \tag{2}$$

where $\mu_{A_j}(x_1,\cdots,x_n)=\mu_{A_{j1}}(x_1)\wedge\cdots\wedge\mu_{A_{jn}}(x_n)$, and $\mu_{A_{ji}}(x_i)$ represents the membership function of the fuzzy set A_{ji} and $\wedge$ the fuzzy AND operator. From (2) it is clear that the T-S fuzzy model is a weighted combination of local models, with normalised membership functions as the weights. The most popular function used in the consequence part, which can be regarded as a local model, is a linear function:

$$g_j(x_1,\cdots,x_n)=p_{j0}+p_{j1}x_1+\cdots+p_{jn}x_n,$$

which makes the T-S fuzzy model an FLL model.

Without sacrificing their approximation ability, FLL models provide good interpretability/transparency that allows the use of prior knowledge and well-developed linear systems theory and techniques in the subsequent applications (Fink, *et al.*, 2002; Gan and Harris, 1999b, 1999c; Harris and Gan, 2001; McGinnity and Irwin, 1996).

2.3 CMAC with Linear Functional Weights

Comparing the CMAC model with FLL models such as ANFIS (Rosales and Gan, 2003), it is clear that both CMAC and ANFIS models have good approximation ability, but CMAC has an absolute advantage in training speed and ANFIS provides better interpretability and applicability due to its linear local models. In order to combine the advantages of CMAC and ANFIS, this paper extends CMAC into an FLL model by replacing its constant weights with linear functional weights. The output of the CMAC with LF weights is defined by

$$y = \sum_{j=1}^{C} \frac{f(\delta_j(x_1,\cdots,x_n))}{\sum_{k=1}^{C} f(\delta_k(x_1,\cdots,x_n))} \cdot l(\mathbf{x}) \quad (3)$$

$$l(\mathbf{x}) = (w_{0,A_j} + w_{1,A_j}x_1 + \cdots + w_{n,A_j}x_n) \quad (4)$$

where w_{i,A_j} are adjustable weights and A_j represents the address of a set of adjustable weights associated with the RF in the *j*th RF layer. Now the number of weights to be updated in one training iteration becomes $(n+1)*C$, due to the addition of excessive free parameters. The learning rule for the modified CMAC can still be based on the LMS type algorithm, but modifications are introduced by deriving the error gradients with respect to the adjustable weights w_{i,A_j}. Model (2) and model (3) look similar, but there exist essential differences that are due to the two mappings adopted in the CMAC model: one maps the input into RF addresses by quantization and the other maps RF addresses into weight indexes by a hashing function.

The modification here aims to provide good interpretability and applicability without sacrificing the approximation ability and other advantages of the original CMAC model. In the following section, the properties of the CMAC with LF weights will be investigated by applying it to both non-smooth and smooth nonlinear function approximation.

3. EXPERIMENTAL RESULTS

The modified CMAC model has established a close link between CMAC and ANFIS models. Both models can be regarded as FLL models, but they use different approaches to input space partitioning. In this section, the performance of the modified CMAC in modelling nonlinear functions is investigated, in comparison with the original CMAC and ANFIS.

3.1 Approximation of a Non-smooth Function

In general, FLL models are good at modelling smooth functions, providing good interpretability. In order to investigate if the modification leads to any sacrifice of the approximation ability of the CMAC model, a non-smooth function is used in the first experiment. The training and testing data were generated using the following nonlinear function:

$$y = \frac{\sin(x_1)}{x_1} \cdot \frac{\sin(x_2)}{x_2} \quad (5)$$

A series of training data sets of 50x50 data points were generated with (x_1, x_2) randomly distributed and some noise added. Each set was used in a different training epoch, which means that in each epoch some new training data were firstly presented and some training data used in the previous epochs were repeated. The testing data were generated with (x_1, x_2) uniformly distributed. A typical training data set is shown in Fig. 2, and the testing data is shown in Fig. 3, where $-10<x_1, x_2<10$.

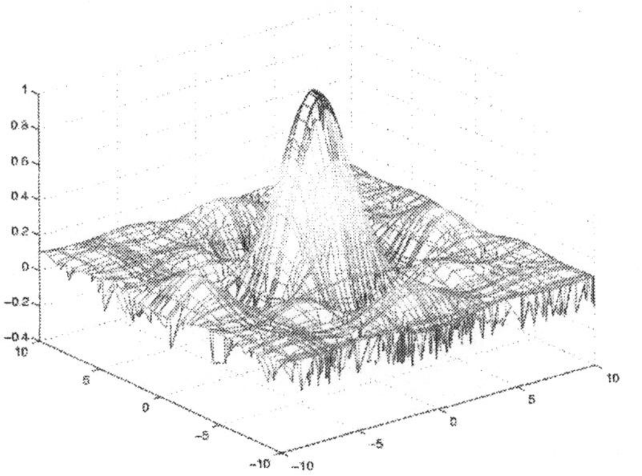

Figure 2 A typical training data set

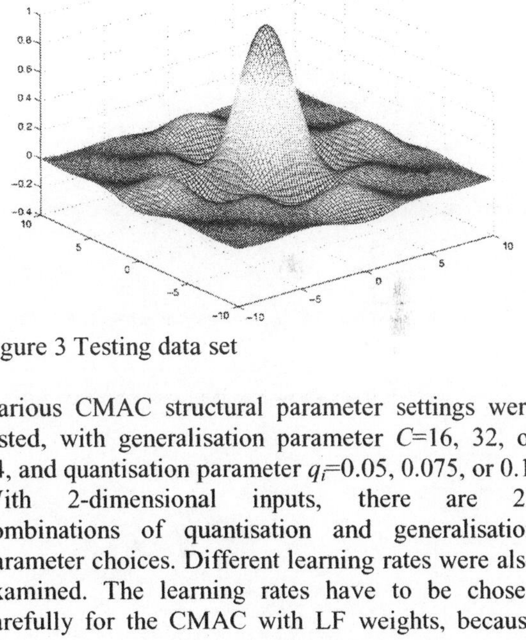

Figure 3 Testing data set

Various CMAC structural parameter settings were tested, with generalisation parameter C=16, 32, or 64, and quantisation parameter q_i=0.05, 0.075, or 0.1. With 2-dimensional inputs, there are 27 combinations of quantisation and generalisation parameter choices. Different learning rates were also examined. The learning rates have to be chosen carefully for the CMAC with LF weights, because the change in w_{0,A_j} has different influence on the model output from the change in w_{i,A_j}. In our experiment, different learning rates were used for updating w_{0,A_j} and w_{i,A_j}, respectively. To guarantee the convergence, small learning rates or normalised learning algorithms should be applied. In total, 108 runs were carried out for the original CMAC and modified CMAC, respectively. In the experiment with ANFIS, the number of univariate membership functions varies from 5, 8, to 10. The training processes for both CMAC and ANFIS models stopped when either the preset maximum number of epochs or root mean square error (RMSE) threshold of test-set validation was satisfied. While modelling the nonlinear function defined in (5), the maximum number of epochs was set to 80 and the RMSE threshold set to 0.003.

The modified CMAC has good approximation ability, in comparison to the original CMAC and ANFIS. Fig. 4 shows an approximation result of the modified CMAC on the testing data, with structural parameters set as C=16 and q_i=0.1. Its approximation error is shown in Fig. 5. In a similar way, an approximation result of the original CMAC is shown in Fig. 6 and Fig. 7, and an approximation result of the ANFIS in Fig. 8 and Fig. 9. Looking at the results given by Figs. 4-9, it is difficult to notice the difference of the three models from the network's outputs. The error figures also show similar approximation accuracy, CMAC models perform a bit better than ANFIS though. This can be noticed by looking carefully at the error range of each model.

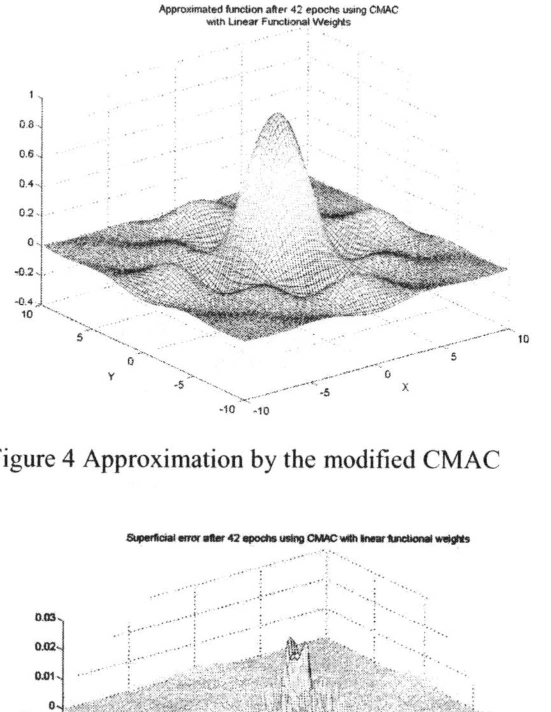

Figure 4 Approximation by the modified CMAC

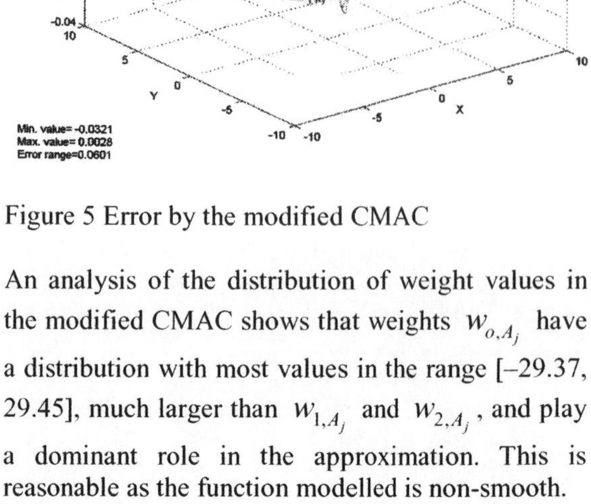

Figure 5 Error by the modified CMAC

An analysis of the distribution of weight values in the modified CMAC shows that weights w_{o,A_j} have a distribution with most values in the range [–29.37, 29.45], much larger than w_{1,A_j} and w_{2,A_j}, and play a dominant role in the approximation. This is reasonable as the function modelled is non-smooth.

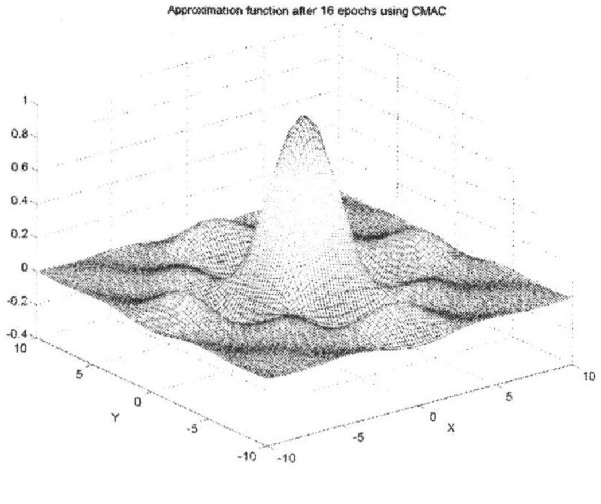

Figure 6 Approximation by the original CMAC

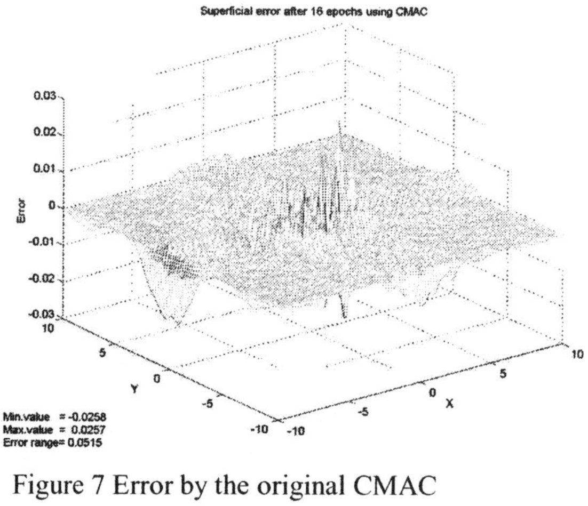

Figure 7 Error by the original CMAC

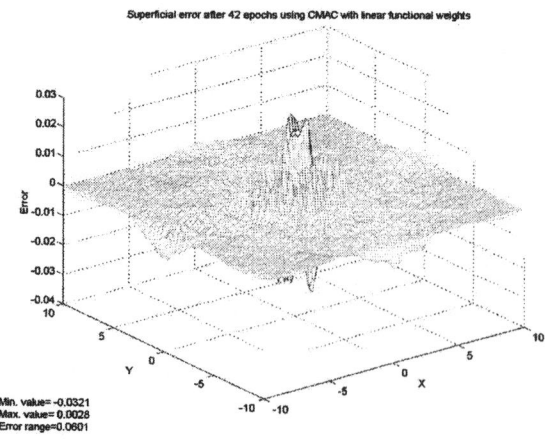

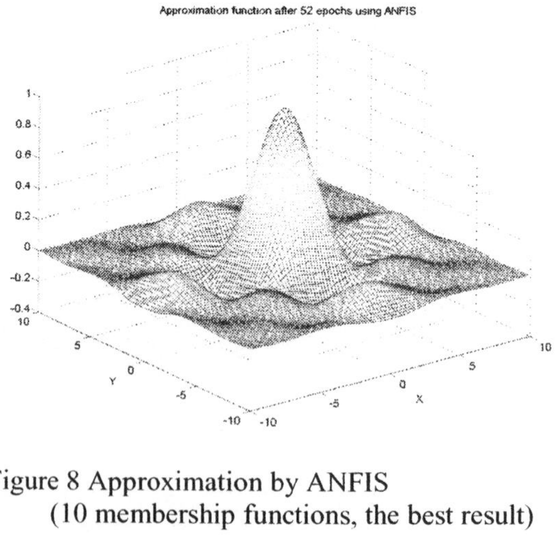

Figure 8 Approximation by ANFIS
(10 membership functions, the best result)

From this experiment it is clear that the modified CMAC with LF weights works very well even in modelling non-smooth functions. Although it shows no advantages over the original CMAC in terms of approximation accuracy and training speed, the CMAC with LF weights has an absolute advantage over ANFIS in terms of training speed. The main purpose of this modification is to make the modified CMAC have advantages in interpretation and subsequent applications. In the next section, a smooth function is to be modelled and the role of the linear function part is further examined.

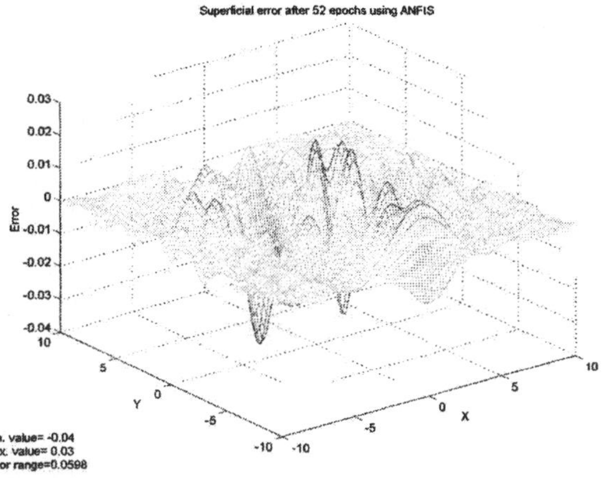

Figure 9 Error by ANFIS

3.2 *Approximation of a Smooth Function*

For many applications, it is very useful for a local linear model to be able to approximate the local linearisations of the nonlinear system to be modelled, *i.e.*, in a local region nearby a given input point x_0, the local linear model should fit within the linearisation of the nonlinear system:

$$y_{linearisation} = y(x_0) + y'(x_0)(x - x_0)\,.$$

The ability of the CMAC with LF weights to model local linearisations of smooth nonlinear systems is investigated in this section. For this purpose, another experiment was set up to approximate the following function:

$$Y = X^2 + 1, \qquad -1 < X < 3 \tag{6}$$

as shown in Fig. 10 with a solid line. One-dimensional function is considered here for the purpose of visualising experimental results. As in the first experiment, the training data were generated by choosing inputs randomly and adding some noise to the outputs, and the testing data were generated using uniformly distributed inputs.

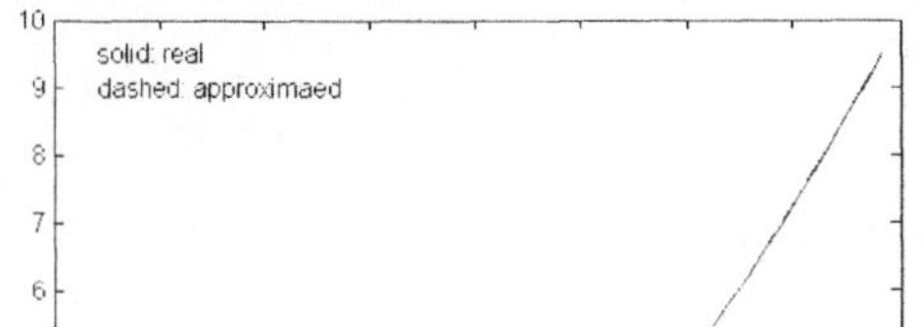

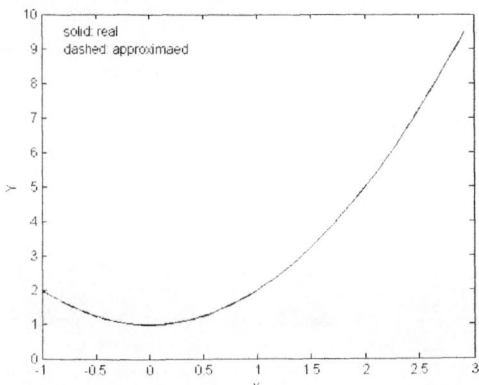

Figure 10 A one-dimensional smooth function

In the second experiment, the following structural parameters were tested: with quantisation parameter q=0.5, 0.75, 1.0, or 1.25, and generalisation parameter C=2, 4, 8, or 16. The maximum number of epochs was set to 80, and the RMSE threshold set to 0.01. The experimental results of the CMAC with constant weights and the CMAC with LF weights were compared using the same structural parameter settings. It was observed that the CMAC with LF weights outperforms the CMAC with constant weights in terms of their global approximation ability. This shows that the modified CMAC has an advantage over the original CMAC in terms of approximation accuracy in modelling smooth functions.

In the experiment of modelling non-smooth functions, it was observed that the values of w_{0,A_j} are much larger than those of w_{i,A_j}, that is, weights w_{0,A_j} play a dominant role in the approximation. In modelling the smooth function (6), an analysis of the distributions of the 2 classes of weights, w_{0,A_j} and w_{1,A_j}, shows that the value ranges of w_{0,A_j} and w_{1,A_j} are comparable. This indicates that in modelling smooth functions the linear function part in the local models does play an important role in the approximation, providing extra approximation power.

3.3 *On Local Linearisation Approximation*

At a given input x_0, the local linearisation of the function $Y=X^2+1$ is given by

$$Y_{linearisation} = (2x_0)X + (1 - x_0^2) \tag{7}$$

and the linear local model of the CMAC with LF weights is:

$$Y_{CMAC} = aX + b \tag{8}$$

where

$$a = \sum_{j=1}^{C} \frac{f(\delta_j(x_0))}{\sum_{k=1}^{C} f(\delta_k(x_0))} \cdot w_{1,A_j} \tag{9}$$

$$b = \sum_{j=1}^{C} \frac{f(\delta_j(x_0))}{\sum_{k=1}^{C} f(\delta_k(x_0))} \cdot w_{0,A_j} \tag{10}$$

If the CMAC model produces good local linearisation approximation, Y_{CMAC} should match $Y_{linearisation}$ very well. As an example, now consider the CMAC with LF weights using the parameter setting of $\{q$=1.25, C=2$\}$, whose global approximation result is shown as a dashed line in Fig. 10. Assuming x_0=2.5, from (7)-(10) it is easy to obtain that the real local linearisation at this point is $Y_{linearisation}$=5X-5.25, and the local linear model given by the CMAC with LF weights is Y_{CMAC}=2.373X+1.354 (a and b were calculated from the training program). Other input points have also been examined in a similar way. The real local linearisations (dashed lines) and local linear models (dash-dot lines) at three different points are shown in Fig. 11. It is clear that although the CMAC model with LF weights does produce very good global approximation ability, good local linearisation approximation is not guaranteed. With the excessive free parameters in the CMAC with LF weights, the model has powerful approximation ability, providing non-unique approximation solutions. Imposing constraints on the training objective would be a possible method to improve local linearisation approximation while keeping good global approximation.

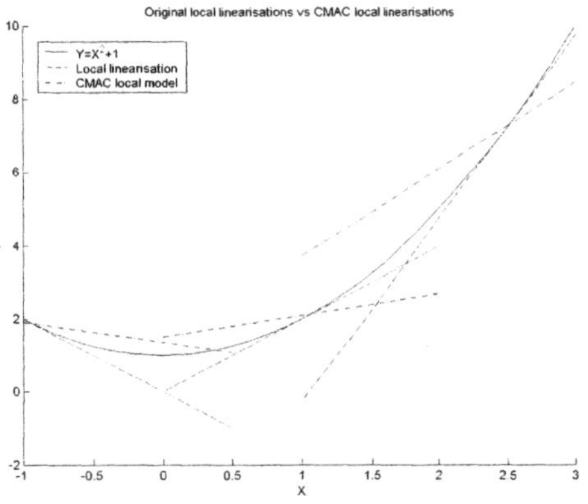

Figure 11 Real linearisation vs local model linearity

4. DISCUSSION AND CONCLUSION

This paper proposes a modified CMAC with LF weights and investigates its approximation ability and interpretability. The idea of the modification is natural, but it may have significant impact on FLL model applications, especially real-time applications, due to its obvious advantage in training speed over the existing T-S fuzzy models such as ANFIS. Furthermore, this is the first time to extend the CMAC model into an FLL model.

It has been illustrated that the modified CMAC works very well in modelling both non-smooth and smooth functions and outperforms the original CMAC in modelling smooth functions in terms of approximation ability. It is expected that the modified CMAC, with the advantage of fast training speed and the potential to approximate local linearisations of nonlinear functions, would find wide applications in nonlinear modelling, state estimation and control (*e.g.*, robot arm modelling and control). As far as its applications are concerned, it is highly desired that the local linearisation approximation ability of the CMAC with LF weights would be improved by using regularisation techniques (Gan and Rosales, 2003; Johansen, 1996; Johansen and Babuska, 2002). This will be further investigated in our future research.

REFERENCES

Albus, J.S. (1975a), "A New Approach to Manipulator Control: The Cerebellar Model Articulation Controller (CMAC)", *J. of Dynamic Sys., Measurement and Control, ASME*, pp. 220-227.

Albus, J.S. (1975b), "Data Storage in the Cerebellar Model Articulation Controller (CMAC)". *J of Dynamic Sys., Measurement and Control, ASME*, pp. 228-233.

Brown, M. and C.J. Harris (1994), *Neurofuzzy Adaptive Modelling and Control*. Prentice-Hall.

Fink, A., O. Nelles, and R. Isermann (2002), "Nonlinear internal model control for MISO systems based on local linear neuro-fuzzy models," *The 15th IFAC World Congress*, Barcelona, Spain.

Gan, Q. and E. Rosales (2003), "CMAC with linear functional weights and its approximation ability analysis," *Technical Report*, Dept. of Computer Science, University of Essex.

Gan, Q. and C.J. Harris (2001), "A hybrid learning scheme combining EM and MASMOD algorithms for fuzzy local linearization modeling," *IEEE Transactions on Neural Networks*, vol. 12, no. 1, pp. 43-53.

Gan, Q. and C.J. Harris (1999a), "Fuzzy local linearization and local basis function expansion in nonlinear system modeling," *IEEE Transactions on Systems, Man, and Cybernetics*, vol. 29: Part B, no. 4, pp. 559-565.

Gan, Q. and C.J. Harris (1999b), "Linearization and state estimation of unknown discrete-time nonlinear dynamic systems using recurrent neurofuzzy networks," *IEEE Transactions on Systems, Man, and Cybernetics*, vol. 29: Part B, no. 6, pp. 802-817.

Gan, Q. and C.J. Harris (1999c), "Neurofuzzy local linearisation for model identification and state estimation of unknown nonlinear processes," *International Journal of Intelligent Control and Systems*, vol. 3, no. 4, pp389- 407.

Harris, C.J. and Q. Gan (2001), "State estimation and multi-sensor data fusion using data-based neurofuzzy local linearisation process models," *Information Fusion*, vol. 2, no.1, pp. 17-29.

Hoffmann, F. and O. Nelles (2001), "Genetic programming for model selection of TSK-fuzzy systems," *Journal of Information Sciences*, vol. 136, pp. 7-28.

Jang, J.S.R. (1993), "ANFIS: Adaptive-network-based fuzzy inference systems," *IEEE Transactions on Systems, Man, and Cybernetics*, vol. 23, pp. 665-685.

Johansen, T.A. and R. Babuska (2002), "On multi-objective identification of Takagi-Sugeno fuzzy model parameters," *IFAC World Congress*, Barcelona.

Johansen, T.A., R. Shorten, and R. Murray-Smith (2000), "On the interpretation and identification of Takagi-Sugeno fuzzy models," *IEEE Transactions on Fuzzy Systems*, vol. 8, pp. 297-313.

Johansen, T.A. (1996), "Robust identification of Tagaki-Sugeno-Kang fuzzy models using regularisation," *Proc. IEEE Conf. On Fuzzy Systems*, New Orleans, pp.180-186.

McGinnity, S. and G.W. Irwin (1996), "Nonlinear state estimation using fuzzy local linear models," *International Journal of System Science*, vol. 28, pp.643-656.

Miller, W.T., F.H. Glanz, and L.G. Kraft (1990), "CMAC: An associative neural networks alternative to back propagation," *Proc. of the IEEE*, vol. 78, no. 10, pp. 1560-1567.

Nelles, O. (1999), *Nonlinear System Identification with Local Linear Neuro-Fuzzy Models*, Ph.D. Thesis, Darmstadt University.

Rosales, E. and Q. Gan (2003), "A comparative study on CMAC and ANFIS for nonlinear system modelling," *Int. Conf. on Computational Intelligence for Modelling, Control and Automation*, Vienna, Austria.

Takagi, T. and M. Sugeno (1985), "Fuzzy identification of systems and its applications to modeling and control," *IEEE Transactions on Systems, Man, and Cybernetics*, vol. 15, pp. 116-132.

IFAC
Publications
www.elsevier.com/locate/ifac

OPTIMAL EXPANSIONS OF DISCRETE-TIME VOLTERRA MODELS USING LAGUERRE FUNCTIONS

Ricardo J. G. B. Campello *, **Gérard Favier** *
Wagner C. Amaral **

* *I3S/CNRS/UNSA, B.P. 121, 06903 Sophia Antipolis Cédex, France*
** *DCA/FEEC/UNICAMP, 13083-970, Campinas-SP, Brazil*

Abstract: This paper is concerned with the optimization of Laguerre bases for the orthonormal series expansion of discrete-time Volterra models. Fu and Dumont (1993) approached this problem in the context of linear systems by minimizing an upper bound for the error resulting from the truncated Laguerre expansion of impulse response models, which are equivalent to first-order Volterra models. The present work generalizes the work mentioned above to Volterra models of any order. The main result is the derivation of analytic strict global solutions for the optimal expansion of the Volterra kernels either using an independent Laguerre basis for each kernel or using a common basis for all the kernels. *Copyright © 2003 IFAC*

Keywords: Volterra Models, Laguerre bases, Optimization.

1. INTRODUCTION

A discrete-time Volterra model is described as:

$$y(k)=\sum_{m=1}^{\infty}\sum_{k_1=0}^{\infty}\cdots\sum_{k_m=0}^{\infty}h_m(k_1,\cdots,k_m)\prod_{j=1}^{m}u(k-k_j) \quad (1)$$

where u, y and h_m are the input, the output and the mth order kernel, respectively. It is a non-linear generalization of the impulse response model as well as a specific realization of the following system representation:

$$y(k)=\mathscr{H}\left(\{u(\tau)\}_{\tau=-\infty}^{k}\right) \quad (2)$$

where $\mathscr{H}$ is a generic non-linear operator. Indeed, model (1) can be seen as a multidimensional Taylor development of $\mathscr{H}$ if this operator is analytic (Eykhoff, 1974). The Taylor series converges if the input signal u is bounded (and normalized so that $|u|<1$). An immediate consequence is that the higher-order terms can be neglected in such a way that only the first M Volterra kernels need to be taken into account, where M is the order of the resulting model. In addition, if the desired representation is stable, then the elements $h_m(k_1,\cdots,k_m)$ with $k_l>\varepsilon_m$ $(\forall l\in\{1,\cdots,m\})$ can be ignored and the Volterra model can be simplified as:

$$y(k)=\sum_{m=1}^{M}\sum_{k_1=0}^{\varepsilon_m}\cdots\sum_{k_m=0}^{\varepsilon_m}h_m(k_1,\cdots,k_m)\prod_{j=1}^{m}u(k-k_j) \quad (3)$$

Boyd and Chua (1985) showed that this model can approximate to desired accuracy any non-linear system in (2) that meets the following requirements: *i)* The operator $\mathscr{H}$ is causal, continuous, time-invariant and has fading memory; *ii)* The input u is (upper and lower) bounded. These requirements comprise a wide class of real-world systems. The main problem in practice is how to select the Volterra kernels which provide an adequate representation of the system to be modeled.

The estimation of Volterra kernels for modeling of non-linear systems has been investigated for decades (Eykhoff, 1974; Billings, 1980; Schetzen, 1980). In the 1990s, a set of mathematical findings pointed out new perspectives with respect to this theme. An important result was the establishment of equivalences between Volterra models and artificial neural networks (Wray and Green, 1994; Marmarelis and Zhao, 1997). This result implies that the Volterra kernels can be written as functions of neural network parameters and numerically computed after the training of these networks.

Once the Volterra kernels have been estimated, it is possible to significantly reduce the complexity of the resulting model by expanding the kernels using orthonormal bases of functions, such as the Laguerre basis (Dumont and Fu, 1993). Laguerre function based models have been widely used in the context of system identification and control (see (Dumont and Fu, 1993; Oliveira *et al.*, 2000) and references therein). The reasons include their tolerance to unmodeled dynamics, reduced number of free-design parameters and ability to deal with time-delays. An important issue concerns how to optimize the Laguerre basis in order to minimize the number of functions which provides the model with a given approximation accuracy, thus simplifying the corresponding identification and control problems. Fu and Dumont (1993) approached this problem in the context of linear systems by deriving an analytic optimal solution which forces a fast convergence of the function series by linearly increasing the cost assigned to each additional Laguerre coefficient. Later, Tanguy et al. (1995; 2000) showed that the analytic solution proposed by Fu and Dumont minimizes the upper bound of the squared norm of the error resulting from the truncation of the series into a finite number of functions. In the context of non-linear systems, Campello *et al.* (2001) generalized Fu and Dumont's approach to second-order Volterra models by proposing the use of two independent Laguerre bases for the expansion of the first and second order Volterra kernels. Such a strategy allows the application of Fu and Dumont's solution to accomplishing the optimal expansion of the first-order kernel, which is equivalent to the impulse response of a linear system. In addition, an analogous solution for the optimal expansion of the second-order kernel was derived. The present paper generalizes the above-mentioned works by deriving an analytic global strict solution for the optimal Laguerre series expansion of Volterra kernels of any order. This solution minimizes an upper bound for the series truncation error and relies on the use of an independent Laguerre basis for the expansion of each kernel. A solution to the problem of expanding all the kernels using a common basis is also provided.

2. LAGUERRE EXPANSION OF VOLTERRA KERNELS

Assuming that the kernels h_m in (1) are absolutely summable on $[0,\infty[$, which means that the corresponding Volterra model is stable, then they can be represented by means of orthonormal bases of functions. Considering the expansion of these kernels using Laguerre functions (Dumont and Fu, 1993), but using an independent basis for each kernel, yields [1] :

$$h_m(k_1,\cdots,k_m)=\sum_{i_1=1}^{\infty}\cdots\sum_{i_m=1}^{\infty}\alpha_{i_1,\ldots,i_m}\prod_{j=1}^{m}\phi_{m,i_j}(k_j) \quad (4)$$

where $\phi_{m,l}$ is the lth function of the mth Laguerre basis and $\alpha_{(\cdot)}$ are the corresponding expansion coefficients, given by:

$$\alpha_{i_1,\ldots,i_m}=\sum_{k_1=0}^{\infty}\cdots\sum_{k_m=0}^{\infty}h_m(k_1,\cdots,k_m)\prod_{j=1}^{m}\phi_{m,i_j}(k_j) \quad (5)$$

From equations (1) and (4), a Volterra model of order M can easily be rewritten as:

$$y(k)=\sum_{m=1}^{M}\sum_{i_1=1}^{\infty}\cdots\sum_{i_m=1}^{\infty}\alpha_{i_1,\ldots,i_m}\prod_{j=1}^{m}\bar{\phi}_{m,i_j}(k) \quad (6)$$

where $\bar{\phi}_{m,l}$ is the output of the lth Laguerre filter (with input u) of the mth basis, i.e.

$$\bar{\phi}_{m,l}(k)=\sum_{\tau=0}^{\infty}\phi_{m,l}(\tau)u(k-\tau) \quad (7)$$

The transfer functions of the corresponding Laguerre filters are given by:

$$\Phi_{m,l}(z)=Z\{\phi_{m,l}\}=\frac{\sqrt{1-p_m^2}}{z-p_m}\left(\frac{1-p_m z}{z-p_m}\right)^{l-1} \quad (8)$$

where $l=1,2,\cdots,\infty$, Z denotes the unilateral z-transform and $p_m\in(-1,1)$ is the real stable pole which parameterizes the mth Laguerre basis.

3. OPTIMIZATION OF THE LAGUERRE POLES

For computational reasons, the infinite-dimensional expansion of the mth-order kernel given by equation (4) is, in practice, truncated into a finite number of functions. The underlying problem considered here is how to select the Laguerre pole p_m in (8) so as to minimize the error resulting from this truncation.

3.1 Optimal expansion of the first-order kernel

The pole p_1 which parameterizes the Laguerre expansion of the first-order kernel can be obtained by solving the following optimization problem (Fu and Dumont, 1993):

$$\min_{-1<p_1<1} J_1\triangleq\sum_{i_1=1}^{\infty}i_1\,\alpha_{i_1}^2 \quad (9)$$

[1] This representation is exact if the mth-order kernel is completely *separable*, i.e., if it can be expressed as a product of m first-order kernels.

whose solution forces a fast convergence of the function series by linearly increasing the cost assigned to each additional Laguerre coefficient $\alpha_{(\cdot)}$. In the work cited above, the authors showed that problem (9) has an analytic strict global solution if the kernel satisfies the following conditions:

(1) $\sum_{k_1=0}^{\infty} |h_1(k_1)| < \infty$.

(2) $h_1(0) = 0$.

The first one is the BIBO stability condition for a linear system. The second condition refers to the absence of a direct connection between the input and the output of the system, i.e., it must be strictly proper.

It has been demonstrated that the solution of problem (9) globally minimizes the upper bound of the squared norm of the error resulting from the truncation of the Laguerre series into a finite number of functions (Tanguy *et al.*, 1995; Tanguy *et al.*, 2000). In the sequel, this idea is extended to the optimal expansion of Volterra models of any order.

3.2 Optimal expansion of the mth-order kernel

A generalization of the result outlined in section 3.1 with respect to the pole p_m associated with the Laguerre series expansion of the mth-order Volterra kernel is presented in the sequel.

Theorem 1. Let the mth-order Volterra kernel satisfy the stability and unit delay conditions, i.e.

(1) $\sum_{k_1=0}^{\infty} \cdots \sum_{k_m=0}^{\infty} |h_m(k_1,\cdots,k_m)| < \infty$.

(2) $h_m(k_1,\cdots,k_m) = 0$ if $\exists l \in \{1,\cdots,m\} : k_l = 0$.

Then, the following equality holds:

$$J_m \triangleq \sum_{i_1=1}^{\infty} \cdots \sum_{i_m=1}^{\infty} (i_1 + \cdots + i_m)\,\alpha^2_{i_1,\ldots,i_m} = \left(\frac{(Q_{1,m}-1)p_m^2+(Q_{2,m}-2Q_{1,m}+1)p_m+Q_{1,m}}{1-p_m^2}\right) m\,||h_m||^2 \quad (10)$$

where $Q_{1,m}$, $Q_{2,m}$ and $||h_m||$ are constant terms which depend exclusively on the mth-order Volterra kernel, as follows:

$$||h_m||^2 = \sum_{k_1=0}^{\infty} \cdots \sum_{k_m=0}^{\infty} h_m^2(k_1,\cdots,k_m) \quad (11)$$

$$Q_{j,m} = \frac{1}{m}\sum_{l=1}^{m} S_{j,l} \quad (12)$$

with $j = 1,2$ and

$$S_{1,l} = \frac{1}{||h_m||^2}\sum_{k_1=0}^{\infty} \cdots \sum_{k_m=0}^{\infty} k_l\, h_m^2(k_1,\cdots,k_m) \quad (13)$$

$$S_{2,l} = \frac{1}{||h_m||^2}\sum_{k_1=0}^{\infty} \cdots \sum_{k_m=0}^{\infty} k_l\,[\Delta_l h_m(k_1,\cdots,k_m)]^2 \quad (14)$$

where $l = 1,\cdots,m$ and the operator Δ_l is defined as

$$\Delta_l h_m(k_1,\cdots,k_m) = h_m(k_1,\cdots,k_l+1,\cdots,k_m) - h_m(k_1,\cdots,k_l,\cdots,k_m) \quad (15)$$

Proof: The proof is based on a strategy along the lines of (Fu and Dumont, 1993) generalized to the mth-dimensional kernel expansion given by equation (4). The detailed demonstration can be found in (Campello, 2002). ■

Remark 1. Although the first condition of theorem 1 implies $||h_m||^2 < \infty$ (Desoer and Vidyasagar, 1975), the same does not hold with respect to the terms $S_{1,l}$ and $S_{2,l}$ in (13) and (14), respectively. The convergence of these terms is necessary so that the cost function J_m in (10) be well-defined and can be optimized. The convergence of the terms associated with the first-order kernel is guaranteed in the modeling of sampled linear systems, but not necessarily in the present context. In the general case, the convergence depends on the rates of decay of the kernel $h_m(k_1,\cdots,k_m)$, i.e., the rates with which the kernel tends to zero as $k_1,\cdots,k_m$ go to infinity. A *sufficient* condition for convergence is to assume that the kernel is non-null ($||h_m||^2 \neq 0$) and has finite memory (Eykhoff, 1974), i.e., $h_m(k_1,\cdots,k_m) = 0$ if $k_l > \varepsilon_m$ ($\forall\, l \in \{1,\cdots,m\}$), where $\varepsilon_m < \infty$ (no matter how large it is). This non-restrictive assumption results in the usual approximate model representation given by equation (3).

Bearing in mind the above considerations and assuming that the terms $S_{1,l}$ and $S_{2,l}$ as defined in (13) and (14) are finite, it is possible to select the pole p_m which parameterizes the Laguerre expansion of the mth-order Volterra kernel h_m by solving the following optimization problem:

$$\min_{p_m \in \mathscr{P}} \bar{J}_m(p_m) \triangleq \frac{J_m}{m\,||h_m||^2} \quad (16)$$

where $\mathscr{P} \triangleq \{p_m \in \Re : |p_m| < 1\}$ and J_m is given by equation (10). Before proceeding to solve this problem, the following lemmas are necessary:

Lemma 1. If the kernel h_m satisfies the second condition stated in theorem 1, then $Q_{1,m} - 1 \geq 0$.

Proof: See appendix A. ■

Lemma 2. If the kernel h_m satisfies the conditions stated in theorem 1, then $4Q_{1,m}Q_{2,m} - Q_{2,m}^2 - 2Q_{2,m} > 0$.

Proof: See appendix A. ■

Theorem 2. Problem (16) has a strict global solution given by:

$$p_m^* = \frac{2Q_{1,m} - 1 - Q_{2,m}}{2Q_{1,m} - 1 + \sqrt{4Q_{1,m}Q_{2,m} - Q_{2,m}^2 - 2Q_{2,m}}} \quad (17)$$

if the kernel h_m satisfies the conditions stated in theorem 1.

Proof: Since the cost function $\bar{J}_m$ is a pseudo-convex function inside the domain $\mathscr{P}$ (see appendix B), then any solution to $d\bar{J}_m/dp_m = 0$ in $\mathscr{P}$ is a global solution to problem (16) (Bazaraa *et al.*, 1993). Computing the derivative of $\bar{J}_m$ with respect to p_m yields:

$$\frac{d\bar{J}_m}{dp_m} = \frac{\mu_m p_m^2 + (2Q_{1,m} - 1)2p_m + \mu_m}{(1 - p_m^2)^2} \quad (18)$$

where $\mu_m \triangleq (Q_{2,m} - 2Q_{1,m} + 1)$. Assuming that $\mu_m \neq 0$ (otherwise the solution will clearly be $p_m^* = 0$)[2], the derivative (18) will be equal to zero if and only if

$$p_m^2 + \frac{(2Q_{1,m} - 1)2p_m}{\mu_m} + 1 = 0 \quad (19)$$

The solutions of equation (19) are

$$p_{m,1} = -\frac{2Q_{1,m} - 1 - \sqrt{4Q_{1,m}Q_{2,m} - Q_{2,m}^2 - 2Q_{2,m}}}{\mu_m} \quad (20)$$

and

$$p_{m,2} = -\frac{2Q_{1,m} - 1 + \sqrt{4Q_{1,m}Q_{2,m} - Q_{2,m}^2 - 2Q_{2,m}}}{\mu_m} \quad (21)$$

Since $p_{m,1}\,p_{m,2} = 1$, then $p_{m,1}$ and $p_{m,2}$ have the same sign, which in turn depends on the sign of their common denominator μ_m. In addition, since $2Q_{1,m} - 1 > 0$ (from lemma 1) and $4Q_{1,m}Q_{2,m} - Q_{2,m}^2 - 2Q_{2,m} > 0$ (lemma 2), then

$$2Q_{1,m} - 1 + \sqrt{4Q_{1,m}Q_{2,m} - Q_{2,m}^2 - 2Q_{2,m}} > 0 \quad (22)$$

which implies

$$2Q_{1,m} - 1 - \sqrt{4Q_{1,m}Q_{2,m} - Q_{2,m}^2 - 2Q_{2,m}} > 0 \quad (23)$$

and, accordingly, $|p_{m,1}| < |p_{m,2}|$. Finally, since $p_{m,1}\,p_{m,2} = 1$, then $|p_{m,1}| < 1$ ($p_{m,1} \in \mathscr{P}$) and $|p_{m,2}| > 1$ ($p_{m,2} \notin \mathscr{P}$). Therefore, the strict global solution to problem (16) is $p_m^* = p_{m,1} = 1/p_{m,2}$, that is, equation (17). ■

Theorem 3. The optimal Laguerre pole given by equation (17) minimizes the upper bound of the squared norm of the error resulting from the truncation of the series expansion into a finite number N_m of functions.

Proof: Truncating the kernel expansion given by equation (4) into N_m Laguerre functions yields:

$$\bar{h}_m(k_1, \cdots, k_m) = \sum_{i_1=1}^{N_m} \cdots \sum_{i_m=1}^{N_m} \alpha_{i_1,\ldots,i_m} \prod_{j=1}^{m} \phi_{m,i_j}(k_j) \quad (24)$$

The function which describes the corresponding truncation error can then be determined from (4) and (24) as follows:

$$e_m(k_1, \cdots, k_m) \triangleq h_m(k_1, \cdots, k_m) - \bar{h}_m(k_1, \cdots, k_m) = \sum_{i_1=N_m+1}^{\infty} \cdots \sum_{i_m=N_m+1}^{\infty} \alpha_{i_1,\ldots,i_m} \prod_{j=1}^{m} \phi_{m,i_j}(k_j) \quad (25)$$

Using the orthonormality property of the Laguerre functions, i.e., $\sum_{k=0}^{\infty} \phi_{m,q}(k)\phi_{m,r}(k) = \delta_{q,r}$, where $\delta_{q,r}$ is the Kronecker delta, it can be shown that the squared norm of the error function in (25) is given by:

$$\begin{aligned} \|e_m\|^2 &\triangleq \sum_{k_1=0}^{\infty} \cdots \sum_{k_m=0}^{\infty} e_m^2(k_1, \cdots, k_m) \\ &= \sum_{i_1=N_m+1}^{\infty} \cdots \sum_{i_m=N_m+1}^{\infty} \alpha_{i_1,\ldots,i_m}^2 \end{aligned} \quad (26)$$

Then, using the above equation as well as the definition of J_m in (10) and the following inequality:

$$\begin{aligned} &\sum_{i_1=1}^{\infty} \cdots \sum_{i_m=1}^{\infty} (i_1 + \cdots + i_m)\, \alpha_{i_1,\ldots,i_m}^2 \geq \\ &\quad m\,(N_m + 1) \sum_{i_1=N_m+1}^{\infty} \cdots \sum_{i_m=N_m+1}^{\infty} \alpha_{i_1,\ldots,i_m}^2 \end{aligned} \quad (27)$$

one can easily conclude that:

$$\|e_m\|^2 \leq \frac{J_m}{m\,(N_m + 1)} \quad (28)$$

Therefore, since the optimal pole given by equation (17) globally minimizes the cost function J_m, then it also minimizes the upper bound of the sum of the squared errors resulting from the truncation of the series into N_m functions. Moreover, it is also evident from (28) that increasing the number N_m of functions causes this upper bound to decrease. ■

3.3 Optimal model expansion using a single basis

In previous sections the Laguerre expansion of an Mth-order Volterra model has been approached by adopting M independent bases of functions for the development of the kernels. For the sake of computational simplicity, but with the price of the corresponding series truncation error being larger, it is possible to expand all the kernels using a single Laguerre basis. This is equivalent to setting $p_1 = \cdots = p_M \triangleq p$ in (8).

---

[2] This result can also be derived from equation (17).

In this case, the optimization problem can be rewritten as[3]:

$$\min_{-1<p<1} \quad J \triangleq \sum_{m=1}^{M} \frac{J_m}{m} \tag{29}$$

where the cost function J is represented using the above definitions as well as equation (10) as:

$$J = \left(\frac{(\bar{Q}_1 - 1)p^2 + (\bar{Q}_2 - 2\bar{Q}_1 + 1)p + \bar{Q}_1}{1 - p^2} \right) \lambda \tag{30}$$

with $\lambda = \sum_{m=1}^{M} ||h_m||^2$ and

$$\bar{Q}_1 = \frac{1}{\lambda} \sum_{m=1}^{M} Q_{1,m} \, ||h_m||^2 \tag{31}$$

$$\bar{Q}_2 = \frac{1}{\lambda} \sum_{m=1}^{M} Q_{2,m} \, ||h_m||^2 \tag{32}$$

where $Q_{1,m}$ and $Q_{2,m}$ are given by equation (12).

Theorem 4. Problem (29) has a strict global solution given by:

$$p^* = \frac{2\bar{Q}_1 - 1 - \bar{Q}_2}{2\bar{Q}_1 - 1 + \sqrt{4\bar{Q}_1\bar{Q}_2 - \bar{Q}_2^2 - 2\bar{Q}_2}} \tag{33}$$

if the kernels h_m $(m = 1, \cdots, M)$ satisfy the conditions stated in theorem 1.

Proof: The proof is the same as that of theorem 2 (*mutatis mutandis*) since the cost functions in (30) and (10) are equivalent in form and the terms $\bar{Q}_1$ and $\bar{Q}_2$ exhibit the same relevant properties of $Q_{1,m}$ and $Q_{2,m}$ presented in lemmas 1 and 2. ■

It is important to notice from equations (31) and (32) that if the squared norm of a specific kernel tends to be much larger than those of the others, then solution (33) tends to the respective individual solution given by equation (17). This result is in conformity with the relative importance of the corresponding kernel to the overall model behavior.

Theorem 5. The optimal Laguerre pole given by equation (33) minimizes the upper bound of the squared norm of the error resulting from the truncation of the series expansion into a finite number N of functions.

Proof: Theorem 3 states that $||e_m||^2 \leq J_m \,/\, (N_m + 1)\, m$, where N_m is the number of functions used for expanding the mth-order kernel h_m. When using a single Laguerre basis for expanding all the M kernels of the model, with $N_1 = \cdots = N_M \triangleq N$, it is straightforward to conclude that $\sum_{m=1}^{M} ||e_m||^2 \leq \sum_{m=1}^{M} J_m \,/\, (N+1)\, m$ and from the definition of J in (29) that $\sum_{m=1}^{M} ||e_m||^2 \leq J/(N+1)$. It is clear from this relation that since the optimal pole given by equation (33) globally minimizes the cost function J, then it also minimizes the upper bound of the sum of the squared errors resulting from the truncation of the series into N functions. ■

[3] The selection of this optimization criterion will be justified in theorem 5.

4. CONCLUSION

The optimization of Laguerre bases for the orthonormal series expansion of discrete-time Volterra models of any order has been addressed. The main result is the derivation of analytic strict global solutions for the optimal expansion of the Volterra kernels either using an independent Laguerre basis for each kernel or using a common basis for all the kernels. These solutions minimize the upper bound of the squared norm of the error resulting from the truncation of the series into a finite number of functions, thus optimizing the performance of the resultant finite dimensional representation. Illustrative examples of the results presented herein can be found in (Campello, 2002). In future works the authors intend to extend these results with respect to generalized orthonormal bases (Van den Hof *et al.*, 1995). Multivariable orthonormal functions might also be considered in order to guarantee consistency of the problem formulation with respect to non-separable Volterra kernels.

ACKNOWLEDGMENTS

The first and third authors acknowledge CAPES (for fellowship BEX0467/02-2) and CNPq (for fellowship 301345184), respectively.

REFERENCES

Bazaraa, M. S., H. D. Sherali and C. M. Shetty (1993). *Nonlinear Programming: Theory and Algorithms*. 2nd ed.. John Wiley & Sons.

Billings, S. A. (1980). Identification of nonlinear systems - a survey. *IEE Proc. Pt D* **127**(6), 272–285.

Boyd, S. and L. O. Chua (1985). Fading memory and the problem of approximating nonlinear operators with Volterra series. *IEEE Trans. on Circuits and Systems* **32**(11), 1150–1161.

Campello, R. J. G. B. (2002). New Architectures and Methodologies for Modeling and Control of Complex Systems combining Classical and Modern Tools. PhD thesis. School of Electrical and Computer Engineering of the State University of Campinas (FEEC/UNICAMP). Campinas/SP, Brazil. In Portuguese.

Campello, R. J. G. B., W. C. Amaral and G. Favier (2001). Optimal Laguerre series expansion of discrete Volterra models. In: *Proc. European Control Conference*. Porto/Portugal. pp. 372–377.

Desoer, C. A. and M. Vidyasagar (1975). *Feedback Systems: Input-Output Properties*. Academic Press.

Dumont, G. A. and Y. Fu (1993). Non-linear adaptive control via Laguerre expansion of Volterra kernels. *Int. J. Adaptive Control and Signal Processing* **7**, 367–382.

Eykhoff, P. (1974). *System Identification: Parameter and State Estimation*. John Wiley & Sons.

Fu, Y. and G. A. Dumont (1993). An optimum time scale for discrete Laguerre network. *IEEE Transactions on Automatic Control* **38**(6), 934–938.

Marmarelis, V. Z. and X. Zhao (1997). Volterra models and three-layer perceptrons. *IEEE Trans. Neural Networks* **8**, 1421–1433.

Oliveira, G. H. C., W. C. Amaral, G. Favier and G. A. Dumont (2000). Constrained robust predictive controller for uncertain processes modeled by orthonormal series functions. *Automatica* **36**(4), 563–571.

Schetzen, M. (1980). *The Volterra and Wiener Theories of Nonlinear Systems*. John Wiley & Sons.

Tanguy, N., P. Vilbé and L. C. Calvez (1995). Optimum choice of free parameter in orthonormal approximations. *IEEE Transactions on Automatic Control* **40**, 1811–1813.

Tanguy, N., R. Morvan, P. Vilbé and L. C. Calvez (2000). Online optimization of the time scale in adaptive Laguerre-based filters. *IEEE Transactions on Signal Processing* **48**, 1184–1187.

Van den Hof, P. M. J., P. S. C. Heuberger and J. Bokor (1995). System identification with generalized orthonormal basis functions. *Automatica* **31**(12), 1821–1834.

Wray, J. and G. G. R. Green (1994). Calculation of the Volterra kernels of non-linear dynamic systems using an artificial neural network. *Biological Cybernetics* **71**, 187–195.

APPENDIX A

Proof of Lemma 1: From the definition of $Q_{1,m}$ in (12) it follows that $Q_{1,m} - 1 = \frac{1}{m}\sum_{l=1}^{m}(S_{1,l} - 1)$. The term $S_{1,l} - 1$ can be written using equation (13) and the second condition stated in theorem 1 as:

$$S_{1,l} - 1 = \frac{1}{||h_m||^2}\sum_{k_1=1}^{\infty}\cdots\sum_{k_m=1}^{\infty}(k_l - 1)\,h_m^2(k_1,\cdots,k_m)$$

Since this expression cannot take negative values, which means that $S_{1,l} - 1 \geq 0$, then $Q_{1,m} - 1 \geq 0$. ■

Proof of Lemma 2: Using elementary algebraic operations and the second condition stated in theorem 1 it is possible to show that:

$$\frac{1}{||h_m||^2}\sum_{k_1=0}^{\infty}\cdots\sum_{k_l=0}^{\infty}\cdots\sum_{k_m=0}^{\infty}k_l\Big(h_m(k_1,\cdots,k_l,\cdots,k_m)+ +h_m(k_1,\cdots,k_l+1,\cdots,k_m)\Big)^2 = 4S_{1,l} - S_{2,l} - 2 \tag{34}$$

Since the first condition of theorem 1 implies $\lim_{k_l\to\infty} h_m(k_1,\cdots,k_m) = 0,\ \forall l \in \{1,\cdots,m\}$, and provided that the kernel h_m is non-null by assumption, then equation (34) implies $4S_{1,l} - S_{2,l} - 2 > 0$. From this inequality as well as the definition of $Q_{1,m}$ and $Q_{2,m}$ in (12) it is straightforward to conclude that:

$$4Q_{1,m} - Q_{2,m} - 2 > 0 \tag{35}$$

In addition, $Q_{2,m}$ is non-negative by definition. Moreover, it is always positive since h_m is non-null and satisfies the first condition of theorem 1. Hence, inequality (35) results in $4Q_{1,m}Q_{2,m} - Q_{2,m}^2 - 2Q_{2,m} > 0$. ■

APPENDIX B

Theorem 6. Let a function $f : \mathscr{X} \to \Re$ be defined as

$$f(\mathbf{x}) \triangleq \frac{\omega(\mathbf{x})}{\gamma(\mathbf{x})} \tag{36}$$

where $\mathscr{X}$ is an open convex set such that $\mathscr{X} \subset \Re^n$ and

- ω is convex, differentiable and non-negative for all $\mathbf{x} \in \mathscr{X}$.
- γ is concave, differentiable and positive for all $\mathbf{x} \in \mathscr{X}$.

Then, f is a pseudo-convex function in $\mathscr{X}$, i.e., $\nabla f(\mathbf{x}_1)^T(\mathbf{x}_2 - \mathbf{x}_1) \geq 0$ implies $f(\mathbf{x}_2) \geq f(\mathbf{x}_1)$ for any pair of elements $\mathbf{x}_1, \mathbf{x}_2 \in \mathscr{X}$ (see *(Bazaraa* et al., *1993), ch. 3*).

Proposition 1. Function $\bar{J}_m(p_m) \triangleq J_m/m||h_m||^2$, with J_m given by equation (10), is pseudo-convex inside the domain $\mathscr{P} = \{p_m \in \Re : |p_m| < 1\}$.

Proof:

(1) $\mathscr{P}$ is an open convex set.

(2) $\gamma(p_m) \triangleq 1 - p_m^2$ is concave, differentiable and positive for all $p_m \in \mathscr{P}$.

(3) $\omega(p_m) \triangleq (Q_{1,m}-1)p_m^2 + (Q_{2,m} - 2Q_{1,m}+1)p_m + Q_{1,m}$ is convex since $Q_{1,m} - 1 \geq 0$ (lemma 1), differentiable and non-negative[4] for all $p_m \in \mathscr{P}$.

■

[4] Otherwise $\bar{J}_m$ (and J_m) would be negative, which is not possible by definition.

IFAC Publications
www.elsevier.com/locate/ifac

QUANTIFICATION OF THE VARIANCE OF ESTIMATED TRANSFER FUNCTIONS IN THE PRESENCE OF UNDERMODELING

Roland Hildebrand * **Michel Gevers** **

* *CORE, Université Catholique de Louvain, 34 voie du Roman Pays, 1348 Louvain-la-Neuve, Belgium, Phone +32(0)10-474337,* hildebrand@core.ucl.ac.be
** *CESAME, Université Catholique de Louvain, 4 av. Georges Lemaitre, 1348 Louvain-la-Neuve, Belgium, Phone +32(0)10-472590,* gevers@csam.ucl.ac.be

Abstract: We study the effect of undermodeling on the parameter variance for prediction error time-domain identification with linear model structures. We restrict our consideration to linear time-invariant discrete time single input single output systems. We examine the asymptotic expression for the variance as the number of data tends to infinity. This quantity is known to depend in general on the fourth order statistical properties of the applied input. However, we establish a sufficient condition under which the asymptotic variance is a function of the input power spectrum only. For this case we deliver exact expressions. We show that for a stochastic input the undermodeling contributes to the parameter variance due to the correlation between the prediction errors and its gradients. As an additional contribution we investigate the parameter variance under the assumptions of the stochastic embedding procedure. We show by means of a counterexample that in the framework of stochastic embedding the parameter variance is not necessarily monotonous with respect to the input power spectrum. *Copyright © 2003 IFAC*

Keywords: parametric identification, variance error, undermodeling

1. INTRODUCTION

Identification experiments should deliver along with an identified model also an uncertainty region, which specifies the quality of the model. Without this additional information the model is virtually useless for practical purposes. Within the framework of parametric model structures the uncertainty is usually expressed in terms of the covariance of the identified parameter vector. Often it is sufficient to consider the asymptotic variance as the number of data tends to infinity, since for common data record lengths these expressions are of satisfying accuracy.

In this contribution we consider asymptotic variance expressions for discrete time, linear and time-invariant systems. Under mild restrictions on input and noise and under the assumption that the true system dynamics can be exactly reproduced within the model structure, the asymptotic variance expressions are tractable functions of the input power spectrum (Ljung, 1999).

In the presence of undermodeling, however, the situation is considerably more complicated. Although closed-form expressions for the asymptotic variance are well-known (Ljung, 1999), they are in general intractable. Moreover, they depend on higher order sta-

[1] This contribution presents research results of the Belgian Programme on Interuniversity Poles of Attraction, Phase V, initiated by the Belgian State, Prime Minister's Office for Science, Technology and Culture. The scientific responsibility rests with its authors.

tistical properties of the noise and the input (Pintelon and Schoukens, 2001). Basically both the time domain and the frequency domain approach face the same difficulties when computing exact expressions. However, in the past decade much advance was made to overcome these problems.

Several results on the asymptotic variance for parametric frequency domain identification were obtained. In (Goodwin *et al.*, 1991) the contribution of the noise to the parameter variance was computed, while that of the undermodeling was neglected. In (de Vries and van den Hof, 1998) linear model structures and a deterministic input were assumed. In this case the expressions for the asymptotic parameter variance somewhat simplify. In (Pintelon *et al.*, 2002) the prediction errors were assumed to be uncorrelated with their gradients, which also facilitates computations.

For time-domain identification estimators of the asymptotic variance based on input-output data of the experiment were proposed. In (Hjalmarsson and Ljung, 1992),(Tjarnstrom and Ljung, 2002) different techniques were presented to obtain a sample estimate of the parameter variance from data gathered during the experiment. One method consisted in introducing an exponential forgetting factor in the expression for the parameter estimate. This led to a windowing effect, which in turn yielded certain ergodicity properties of the so-obtained sequence of parameter estimates. For long data records the sample covariance of this sequence was a good approximation of the true parameter covariance. Another method consisted in estimating the undermodeling with a high-order ARX model and using bootstrap techniques to obtain artificial noise realizations for a Monte-Carlo simulation.

These techniques thus assume the availability of input-output data. However, sometimes it is necessary to estimate the parameter variance prior to the experiment, e.g. for purposes of optimal input design. In this framework one is going to perform an identification experiment and wishes to choose the input sequence for this experiment in order to let the parameter variance have some desired properties. Therefore one has to know how the parameter variance depends on the input that is going to be applied. In most cases it is sufficient to describe the asymptotic parameter variance as a function of the input power spectrum.

We address the question of computing asymptotic variance expressions in the presence of undermodeling for prediction error time domain identification. We focus on the dependence of the variance on the input power spectrum. For ease of treatment we restrict our considerations to linear model structures with known noise properties and to the SISO case.

If a stochastic input is used, the asymptotic variance depends in general on the undermodeling as well as on higher order properties of the input. It is known that for parametric frequency domain identification these dependencies do not hold, if a deterministic input is applied (Pintelon and Schoukens, 2001). We establish a similar result for time domain identification.

For the case of a stochastic input we formulate a condition under which the asymptotic parameter variance does not depend on higher order properties of the input. This condition covers a wide class of input sequences and is satisfied e.g. for filtered Gaussian white noise. Under this condition we establish explicit expressions for the asymptotic parameter variance as a function of the input power spectrum. We show that the contribution of the undermodeling to the parameter covariance has its origin in the correlation between the prediction errors and its gradients. While this correlation vanishes at lag zero by the nature of the prediction error identification procedure, it is in general non-zero at the other lags if undermodeling is present.

In (Goodwin *et al.*, 1992) a method called *stochastic embedding* was introduced. Within this framework the undermodeling is treated as being stochastic with zero mean. Hence the undermodeling error can be treated as a variance error along with the error introduced by the noise. Frequency domain identification by means of frequency response function measurements within the stochastic embedding framework was investigated in (Schoukens and Pintelon, 1994).

In this contribution we also consider the parameter variance in the framework of stochastic embedding. We derive asymptotic expressions for the total variance and show by an example that the total parameter variance does not necessarily decrease when the input power is increased and can even increase. Thus the usual property of monotonicity of the variance with respect to the input power spectrum is not satisfied.

The remainder of the contribution is structured as follows. In Section 2 we give formal definitions and remind the expressions for the asymptotic variance as given in (Ljung, 1999). We prove that the undermodeling has no effect on the asymptotic parameter variance if a zero mean periodic input is used. In Section 3 we consider the case of a stochastic input. We establish a condition under which the parameter variance is independent of the higher order properties of the input. Assuming this condition, we deduce expressions for the variance as a function of the input power spectrum. Section 4 is devoted to the investigation of the variance expressions in the framework of stochastic embedding. In Section 5 we give an example, which shows that under adoption of the stochastic embedding paradigms the information matrix need not be monotonic with respect to the input power spectrum. Finally, we give some conclusions in Section 6.

2. GENERAL VARIANCE EXPRESSIONS

For simplicity we assume a linear model structure $G(\theta) = \theta^T \Lambda$. Here $\theta \in \mathbf{R}^n$ is the parameter vector, Λ

is an n-dimensional vector of stable transfer functions. Let the true system be given by

$$y = G_0 u + He.$$

Here u is the scalar input, y the scalar output, G_0 is the transfer function of the system, and e is white noise with variance λ_0, which is filtered through the monic stable and inversely stable noise filter H. We assume H to be known. The input u is assumed to be a quasistationary sequence with zero time average and with power spectrum Φ_u.

Identification of θ is performed by minimizing the squared deviation of the output y from the 1-step ahead predictor $\hat{y}(\theta) = (1 - H^{-1})y + \theta^T H^{-1}\Lambda u$. The prediction error is given by $\varepsilon(\theta) = y - \hat{y}(\theta) = H^{-1}y - \theta^T\psi$, where $\psi = -\frac{\partial\varepsilon}{\partial\theta} = H^{-1}\Lambda u$ is the predictor gradient. The identified parameter vector $\hat{\theta}_N$ minimizes the cost function $V_N(\theta) = \frac{1}{2N}\sum_{t=1}^{N}\varepsilon_t^2(\theta)$, where t indexes the time instants and N is the number of data samples: $\hat{\theta}_N = \arg\min_\theta V_N(\theta)$.

Under mild assumptions (see (Ljung, 1999) for details) the time average $\bar{V}(\theta) = \lim_{N\to\infty} V_N(\theta)$ is defined and $\hat{\theta}_N$ tends to the minimizer θ^* of $\bar{V}(\theta)$ as the number of data N tends to ∞:

$$\theta^* = \arg\min_\theta \bar{V}(\theta), \ \lim_{N\to\infty} \hat{\theta}_N = \theta^* \text{ with prob. 1.}$$

It is well-known that the vector θ^* is not completely determined by the properties of the system, but depends on the input u, specifically on its power spectrum Φ_u (Ljung, 1999),(Pintelon and Schoukens, 2001).

The vector θ^* minimizes the variance of $\varepsilon(\theta) - e = H^{-1}G_0 u - \theta^T\psi$. Thus we have

$$\theta^* = (\bar{E}(\psi\psi^T))^{-1}\bar{E}((H^{-1}G_0 u)\cdot\psi) \tag{1}$$
$$= \left(\frac{1}{2\pi}\int_{-\pi}^{\pi}\frac{\Phi_u}{|H|^2}\Lambda\Lambda^*\,d\omega\right)^{-1}\frac{1}{2\pi}\int_{-\pi}^{\pi}\frac{\Phi_u}{|H|^2}\Lambda G_0^*\,d\omega.$$

The vector θ^* admits the following frequency domain interpretation. Define a pseudoscalar product on the space $\mathcal{H}_\infty$ of stable transfer functions by

$$\langle A, B\rangle = \frac{1}{2\pi}\int_{-\pi}^{\pi}\frac{\Phi_u(\omega)}{|H(e^{j\omega})|^2}A(e^{j\omega})B^*(e^{j\omega})\,d\omega. \tag{2}$$

Then θ^* corresponds to the particular transfer function $(\theta^*)^T\Lambda$ within the model structure that realizes the minimal distance to the true transfer function G_0 with respect to the pseudoscalar product (2). In other words, the mismatch $G_0 - (\theta^*)^T\Lambda$ is orthogonal to the model structure with respect to (2).

Moreover, the quantity $\sqrt{N}(\hat{\theta}_N - \theta^*)$ is asymptotically normally distributed and its asymptotic covariance is given by (Ljung, 1999)

$$P_\theta = (\bar{V}'')^{-1}(\lim_{N\to\infty} NE(V_N'(\theta^*)V_N'^T(\theta^*))(\bar{V}'')^{-1} \tag{3}$$

with $\bar{V}'' = \bar{E}(\psi\psi^T)$. The central term on the right-hand side of (3) is given by

$$\lim_{N\to\infty} N^{-1}E\sum_{t=1}^{N}\sum_{s=1}^{N}\varepsilon_t(\theta^*)\varepsilon_s(\theta^*)\psi_t\psi_s^T$$
$$= \lambda_0\bar{E}(\psi\psi^T) + \lim_{N\to\infty} N^{-1}\sum_{t,s=1}^{N} E\left\{\tilde{\varepsilon}_t\tilde{\varepsilon}_s\psi_t\psi_s^T\right\},$$

where $\tilde{\varepsilon} = \varepsilon(\theta^*) - e$ and t, s index time instants. Let us denote the second term in the last expression by Ξ. Then (3) can be written as

$$P_\theta = \lambda_0(\bar{E}(\psi\psi^T))^{-1} + (\bar{E}(\psi\psi^T))^{-1}\Xi(\bar{E}(\psi\psi^T))^{-1}. \tag{4}$$

Thus the asymptotic covariance of the parameter estimate is the sum of two terms. While the first term in (4) is induced by the noise e, the second term is due to the undermodeling. A similar situation holds for parametric frequency domain identification. It is known (Goodwin *et al.*, 1991),(Pintelon *et al.*, 2002),(Pintelon and Schoukens, 2001, Section 7.11.4) that in this case the second contribution in fact is due to the variability of the input u. Hence for a stochastic input the variance of the parameter estimate in general does not vanish even in the absence of noise.

The term Ξ can be written as

$$\Xi = \lim_{N\to\infty} N^{-1}E\left\{\left(\sum_{t=1}^{N}\tilde{\varepsilon}_t\psi_t\right)^2\right\}. \tag{5}$$

Note that by (1) we have

$$\bar{E}(\tilde{\varepsilon}\psi) = \bar{E}(H^{-1}G_0 u\cdot\psi) - \bar{E}(\psi\psi^T)\theta^* = 0. \tag{6}$$

Proposition 1. If the input signal u is a multisine, then $\Xi = 0$ and $P_\theta = \lambda_0(\bar{E}(\psi\psi^T))^{-1}$.

We hereby assume that the period of the multisine remains constant when the number of data tends to infinity, i.e. the number of periods tends to infinity. Observe that this proposition also covers the case where u is a square wave signal.

Proof. Suppose u is a multisine. Then the signals $\tilde{\varepsilon}$ and ψ and their product $\tilde{\varepsilon}\cdot\psi$ are also multisines. Thus the signal $\tilde{\varepsilon}\cdot\psi$ is periodic and by (6) it has zero mean. But then its cumulative sum is also periodic, specifically bounded. The proposition now follows from (5). □

Proposition 1 states that in the case of a deterministic input the undermodeling has an impact only on the value of the asymptotic estimate θ^*, but not on the variance of $\hat{\theta}_N - \theta^*$. The latter is entirely due to the noise.

In the next section we quantify the impact of undermodeling on the parameter variance if a stochastic input signal is used.

3. THE PARAMETER VARIANCE IN THE CASE OF A STOCHASTIC INPUT

In this section we examine the asymptotic covariance matrix (4) for zero mean quasistationary stochastic inputs. We establish a condition under which the asymptotic covariance depends only on the second order properties of the involved signals. Assuming this condition, we derive an explicit frequency domain expression for the asymptotic parameter variance as a function of the input power spectrum Φ_u.

If u is filtered white noise, then the signal $\tilde{\varepsilon} \cdot \psi$ has by (6) zero mean, but the standard deviation of its cumulative sum grows proportionally to the square root of the number of summands. Therefore the term Ξ in (4) might be nonzero. By (5) the matrix Ξ is positive semidefinite and, as expected, undermodeling can only increase the asymptotic parameter variance.

It is known that in the presence of undermodeling the asymptotic covariance depends on the 4th order properties of the input and the noise (Pintelon and Schoukens, 2001, p.198). Indeed, the definition of Ξ involves 4th order products and powers of the input. Therefore the asymptotic variance cannot be described as a function of the second order properties of u alone. In general it will depend also on the 4th order cumulant spectrum (Rosenblatt, 1985). This poses serious difficulties e.g. for input design. However, if we restrict the 4th order cumulants of u to be zero, then the asymptotic variance is a function of the input power spectrum Φ_u only. Denote the autocorrelation function $\bar{E}(u_t u_{t-\tau})$ of u by $R_u(\tau)$. Then the vanishing of the 4th order cumulants of u can be equivalently rewritten as the condition

$$\bar{E}(u_{p+t}u_{q+t}u_{r+t}u_{s+t}) = R_u(p-r)R_u(q-s) \quad (7)$$
$$+R_u(p-s)R_u(q-r) + R_u(p-q)R_u(r-s)$$

for all p, q, r, s. Here the time average is taken with respect to t and the numbers p, q, r, s are assumed to be fixed. Condition (7) is in fact not very restrictive. It is satisfied for instance for filtered zero mean white noise, where the probability density function of the white noise has zero *kurtosis* "peakedness", see e.g. (Rosenblatt, 1985). This is equivalent to the condition that the 2nd and 4th moments m_2, m_4 of the probability density function satisfy the relation $m_4 = 3m_2^2$. This relation holds e.g. for a Gaussian distribution.

We now proceed to compute an expression of Ξ in terms of signal spectra for the case where condition (7) holds. Straightforward calculation yields

$$\Xi = \sum_{\tau=1}^{\infty}(R_{\psi\tilde{\varepsilon}}(\tau)+R_{\tilde{\varepsilon}\psi}(\tau))(R_{\psi\tilde{\varepsilon}}(\tau)+R_{\tilde{\varepsilon}\psi}(\tau))^T, \quad (8)$$

where $R_{gh}(\tau)$ denotes the cross-correlation of signals g, h at lag τ. Thus we have represented Ξ as a sum of squares for the case where the input satisfies condition (7). Representation (8) yields the following frequency domain expression for Ξ.

$$\Xi = \frac{1}{\pi}\int_{-\pi}^{\pi} Re\Phi_{\psi\tilde{\varepsilon}} \cdot Re\Phi_{\psi\tilde{\varepsilon}}^T \, d\omega \quad (9)$$

Here $\Phi_{\psi\tilde{\varepsilon}}$ is the cross-spectrum between the signals ψ and $\tilde{\varepsilon}$. We obtain the following result.

Proposition 2. Let the input u be a quasistationary stochastic zero mean process satisfying condition (7). Then undermodeling does not have an impact on the asymptotic parameter covariance if and only if the cross-spectrum between the signals $\psi = H^{-1}\Lambda u$ and $\tilde{\varepsilon} = H^{-1}G_0 u - (\theta^*)^T\psi = H^{-1}(G_0 - (\theta^*)^T\Lambda)u$ is a purely imaginary function of the frequency. Otherwise the undermodeling increases the asymptotic parameter covariance. □

Thus the condition $\Xi = 0$ admits the following interpretation. Suppose for some frequency ω we have $\Phi_u(\omega) \neq 0$. Then $Re\Phi_{\psi\tilde{\varepsilon}}(\omega) = 0$ implies $Re[(G_0(e^{j\omega}) - (\theta^*)^T\Lambda(e^{j\omega}))\Lambda^*(e^{j\omega})] = 0$, or equivalently, $\arg(G_0(e^{j\omega}) - (\theta^*)^T\Lambda(e^{j\omega})) - \arg\Lambda_k(e^{j\omega}) = \pm\frac{\pi}{2}$ for all $k = 1, \ldots, n$. This means that the bias $G_0 - (\theta^*)^T\Lambda$ is orthogonal to the model structure not only in the sense of the pseudoscalar product (2), but frequency-wise in the Nyquist plane at the excited frequency ranges.

In this section we considered the case of a stochastic input and derived expressions for the asymptotic parameter covariance as a function of the input power spectrum under assumption (7). As (8) shows, the contribution of the undermodeling to the parameter covariance is caused by the correlation between the prediction errors $\tilde{\varepsilon}_t + e_t$ and its negative gradients ψ_t at time lags other than zero.

4. COVARIANCE IN THE FRAMEWORK OF STOCHASTIC EMBEDDING

In this section we compute the asymptotic parameter variance in the framework of stochastic embedding. In this framework the parameter error is entirely described as a variance error. It is then interesting to compute that part of this variance error that is due to the undermodeling.

Let us assume $G_0 = \theta_0^T\Lambda + \eta^T Z$, where θ_0 is a fixed parameter vector, $\eta \in \mathbf{R}^L$ and $Z = (Z_1, \ldots, Z_L)^T$ is a vector of L stable transfer functions. Within the stochastic embedding framework, η is assumed to be a random variable with zero mean and covariance matrix C_η. For any given identification experiment the vector η assumes a fixed value, which is drawn according to its probability distribution. For details and a justification of the procedure see (Goodwin *et al.*, 1992). Hence under the assumptions of stochastic embedding the vector θ_0 reflects intrinsic properties of the system and can be considered as the "true" parameter vector.

Thus the asymptotic value θ^* of the parameter estimate as well as its asymptotic covariance P_θ become a function of η, $\theta^* = \theta^*(\eta)$, $P_\theta = P_\theta(\eta)$. We obtain from (1)

$$\theta^* - \theta_0 = \left(\bar{E}(\psi\psi^T)\right)^{-1} \left(\frac{1}{2\pi} \int_{-\pi}^{\pi} \frac{\Phi_u}{|H|^2} \Lambda Z^* \, d\omega \right) \eta,$$

$$\tilde{\varepsilon} = (\theta_0 - \theta^*)^T \psi + \eta^T H^{-1} Z u. \qquad (10)$$

Note that both $\tilde{\varepsilon}$ and $\theta^* - \theta_0$ are linear in η. Observe that by $E\eta = 0$ we have $E\theta^* = \theta_0$, where the expectation is taken over η. Similarly $E\tilde{\varepsilon} = 0$. Further, Ξ becomes a matrix-valued positive semidefinite quadratic form in η. Let us write this as $\tilde{\varepsilon} = \sum_i \eta_i \tilde{\varepsilon}^i$, $\theta^* - \theta_0 = \sum_i \eta_i \Delta\theta^i$ and $\Xi = \sum_{i,j} \eta_i \eta_j \Xi^{ij}$. Here

$$\tilde{\varepsilon}^i = \left(\left(\frac{1}{2\pi} \int_{-\pi}^{\pi} \frac{\Phi_u}{|H|^2} Z_i \Lambda^* \, d\omega \right) \left(\frac{1}{2\pi} \int_{-\pi}^{\pi} \frac{\Phi_u}{|H|^2} \Lambda\Lambda^* \, d\omega \right)^{-1} H^{-1}\Lambda + H^{-1} Z_i \right) u,$$

$$\Delta\theta^i = \left(\frac{1}{2\pi} \int_{-\pi}^{\pi} \frac{\Phi_u}{|H|^2} \Lambda\Lambda^* \, d\omega \right)^{-1} \frac{1}{2\pi} \int_{-\pi}^{\pi} \frac{\Phi_u}{|H|^2} \Lambda Z_i^* \, d\omega,$$

and Ξ^{ij} are matrices, which under condition (7) are given by

$$\Xi^{ij} = \frac{1}{\pi} \int_{-\pi}^{\pi} Re\Phi_{\psi\tilde{\varepsilon}^i} \cdot Re\Phi_{\psi\tilde{\varepsilon}^j}^T \, d\omega. \qquad (11)$$

Let us now compute the variance of the parameter estimate $\hat{\theta}_N$ after averaging over η. Since η has zero mean, the expectation of $\hat{\theta}_N$ is equal to θ_0. Assuming a normal distribution for η, we obtain after some transformations for the asymptotic covariance $\mathbf{P}_N$ of $\hat{\theta}_N$

$$E(\hat{\theta}_N - \theta_0)(\hat{\theta}_N - \theta_0)^T = \\ = \lambda_0 (N\bar{E}(\psi\psi^T))^{-1} + \sum_{i,j} (C_\eta)_{ij} \cdot \qquad (12) \\ \left[N^{-1} (\bar{E}(\psi\psi^T))^{-1} \Xi^{ij} (\bar{E}(\psi\psi^T))^{-1} + \Delta\theta^i (\Delta\theta^j)^T \right].$$

Here by $(C_\eta)_{ij}$ we denote the entries of the covariance matrix C_η. Besides the familiar variance term $\lambda_0(N\bar{E}(\psi\psi^T))^{-1}$ caused by the noise we have two different contributions from the undermodeling to the total variance. The term $\sum_{i,j}(C_\eta)_{ij}\Delta\theta^i(\Delta\theta^j)^T$ is due to the bias, i.e. the shift away from θ_0 of the asymptotic value θ^*. It is included into the variance only by the stochastic embedding procedure. The term $\sum_{i,j}(C_\eta)_{ij} N^{-1}(\bar{E}(\psi\psi^T))^{-1}\Xi^{ij}(\bar{E}(\psi\psi^T))^{-1}$ is due to the increase of the asymptotic parameter variance by the undermodeling for any fixed η.

Note that the covariance matrix $\mathbf{P}_N$ does not completely describe the distribution of $\hat{\theta}_N - \theta_0$, even if η is normally distributed. The distribution of $\hat{\theta}_N$ will not be Gaussian if Ξ is not identically zero, because its asymptotic covariance $P_\theta(\eta)$ is a function of the random vector η, as stated above. The same holds for the distribution of the transfer function in the frequency domain. The definition of uncertainty regions at certain probability levels will therefore face considerable difficulties. We stress that this property does in no way contradict the familiar theorems on asymptotic normality of the parameter estimate. The non-normality of $\hat{\theta}_N - \theta_0$ is an artifact introduced by randomizing the undermodeling, i.e. by averaging with respect to the probability distribution of the actually constant undermodeling parameter vector η. However, if Ξ is zero, i.e. when using multisines as input, then averaging over η yields a Gaussian probability distribution of $\hat{\theta}_N$, given η is normally distributed.

5. NON-MONOTONICITY OF THE INFORMATION MATRIX

In this section we investigate the monotonicity properties of the total parameter variance with respect to the input power spectrum under the assumptions of stochastic embedding. We shall work with the information matrix M, the inverse of the covariance matrix, $M = [E(\hat{\theta}_N - \theta_0)(\hat{\theta}_N - \theta_0)^T]^{-1}$. This information matrix depends on the input power spectrum Φ_u, and one would expect that the usual monotonicity property holds, namely $M(\Phi_u + \Phi_u') \succeq M(\Phi_u)$ for any power spectra Φ_u, Φ_u'. This is in general not true, as demonstrated by the following example.

Example. Consider prediction error identification in a stochastic embedding framework for the following system:

$$y = \theta_0 z^{-1} u + \eta z^{-2} u + e,$$

where the model has the structure $\theta\Lambda = \theta z^{-1}$. Thus $n = 1$, $L = 1$, $H \equiv 1$, $\Lambda = z^{-1}$, $Z = z^{-2}$. Let $x_k(\Phi_u)$, $k = 0, 1$, be the first trigonometric moments of the power spectrum Φ_u, i.e.

$$x_k(\Phi_u) = \frac{1}{2\pi} \int_{-\pi}^{\pi} \Phi_u(\omega) e^{-jk\omega} \, d\omega, \qquad k = 0, 1.$$

Suppose further that the inputs are multisines, so that $\Xi = 0$. Then direct computation yields

$$M(\Phi_u) = \frac{x_0^2(\Phi_u)}{\lambda_0 N^{-1} x_0(\Phi_u) + C_\eta x_1^2(\Phi_u)}.$$

Let us now consider the three multisine sequences $u_t = \frac{\sqrt{3}}{3}\cos(\pi t) + \frac{2\sqrt{3}}{3}\sin(\frac{\pi}{3}t)$, $u'_t = \frac{\sqrt{2}}{2}\cos(\pi t) + \sin(\frac{\pi}{2}t)$, $u''_t = \frac{\sqrt{30}}{6}\cos(\pi t) + \sin(\frac{\pi}{2}t) + \frac{2\sqrt{3}}{3}\sin(\frac{\pi}{3}t)$. Their respective power spectra Φ_u, Φ'_u and Φ''_u are related by the equality $\Phi''_u = \Phi_u + \Phi'_u$ and have moments $x_0(\Phi_u) = x_0(\Phi'_u) = 1$, $x_1(\Phi'_u) = x_1(\Phi''_u) = -\frac{1}{2}$, $x_1(\Phi_u) = 0$, $x_0(\Phi''_u) = 2$. Hence the information matrices of experiments performed with inputs u and u'' are given by

$$M(\Phi_u) = \frac{N}{\lambda_0},\ M(\Phi_u + \Phi'_u) = \frac{4}{2\lambda_0 N^{-1} + \frac{1}{4}C_\eta}.$$

Hence we have $M(\Phi_u) \preceq M(\Phi_u + \Phi'_u)$ if and only if $C_\eta N \leq 8\lambda_0$. Thus if the SNR is large enough (i.e. λ_0 is small), if the undermodeling effects begin to dominate the noise effects, or if the number of data becomes large, then the input u'' yields a smaller information matrix than the input u, although its power spectrum Φ''_u is larger or equal to Φ_u frequency-wise. □

This leads to the counterintuitive conclusion that an increase in input power at some frequencies without decrease at the other frequencies does not necessarily imply an increase of information, and may even lead to a decrease. This is an artifact caused by the stochastic embedding procedure, which randomizes the undermodeling by considering it as being of stochastic nature and lumping it together with the actual parameter variance. The increase of variance can therefore be attributed to an increase of the bias.

A weaker monotonicity property does hold, however. The following assertion is a consequence of (12).

Corollary 1. Let Φ_u be a power spectrum and let $\beta > 1$ be a constant. Let $M(\Phi_u)$ denote the information matrix when the input signal has power spectrum Φ_u and let $M(\beta\Phi_u)$ be the corresponding information matrix for inputs with power spectrum $\beta\Phi_u$. Assume further that the inputs satisfy condition (7). Then $M(\beta\Phi_u) \succeq M(\Phi_u)$. □

6. CONCLUSIONS

In this contribution we have investigated the asymptotic parameter variance under time-domain prediction error identification in the presence of undermodeling. It was shown that under identification with a zero mean periodic input the undermodeling does not influence the parameter variance. This result is summarized in Proposition 1.

For stochastic input, undermodeling leads to an increase of the variance. In general the amount of this increase is difficult to evaluate and depends on higher order properties of the input signal. Under assumption (7) on the input signal, however, it is proportional to the integral of the squared real part of the cross-spectrum between the prediction error and its gradient. This result is formalized in formulae (9), (4). Thus the undermodeling impacts the parameter variance through the correlation between the prediction error and its gradient.

Further we investigated the asymptotic parameter covariance within the stochastic embedding framework. An explicit expression is given by formula (12), but note that in general the distribution of the parameter vector is not normal. This can be attributed to the way the bias error is randomized under the assumptions of stochastic embedding, and does not contradict the standard asymptotic normality theorems of prediction error identification. Normality of the distribution is preserved under the conditions of Proposition 1. For the information matrix as a function of the input power spectrum a weak monotonicity property holds, which is formalized in Corollary 1. A counterexample to the usual monotonicity condition has also been given.

REFERENCES

de Vries, D. K. and P. M. J. van den Hof (1998). Frequency domain identification with generalized orthonormal basis functions. *IEEE Trans. Automat. Control* **AC-43**, 656–669.

Goodwin, G. C., M. Gevers and B. Ninness (1992). Quantifying the error in estimated transfer functions with application to model order selection. *IEEE Trans. Automat. Control* **AC-37**, 913–928.

Goodwin, G. C., M. Gevers and D. Q. Mayne (1991). Bias and variance distribution in transfer function estimation. In: *Proceedings of the 9-th IFAC Symposium on Identification*. Budapest.

Hjalmarsson, H. and L. Ljung (1992). Estimating model variance in the case of undermodeling. *IEEE Trans. Automat. Control* **AC-37**, 1004–1008.

Ljung, Lennart (1999). *System identification: Theory for the user*. Prentice-Hall Information and System Sciences Series. second ed.. Prentice Hall.

Pintelon, R. and J. Schoukens (2001). *System identification. A frequency domain approach*. IEEE Press.

Pintelon, R., J. Schoukens and Y. Rolain (2002). Uncertainty of transfer function modeling using prior estimated noise models. Technical Report RP042002. Vrije Universiteit Brussel, Dept. ELEC.

Rosenblatt, Murray (1985). *Stationary sequences and random fields*. Birkhaeuser.

Schoukens, J. and R. Pintelon (1994). Quantifying model errors of identified transfer functions. *IEEE Trans. Automat. Control* **AC-39**, 1733–1737.

Tjarnstrom, F. and L. Ljung (2002). Using the bootstrap to estimate the variance in the case of undermodeling. *IEEE Trans. Automat. Control* **AC-47**, 395–398.

IFAC Publications
www.elsevier.com/locate/ifac

RELIABLE PARAMETER ESTIMATION IN PRESENCE OF UNCERTAIN VARIABLES THAT ARE NOT ESTIMATED

Isabelle Braems * **Luc Jaulin** ** **Michel Kieffer** *
Nacim Ramdani *** **Eric Walter** *

* *L2S - CNRS - Supélec - Université Paris-Sud*
Plateau de Moulon, F-91192 Gif-sur-Yvette, France
** *LISA - Université d'Angers*
Avenue Notre-Dame du Lac, F-49000 Angers, France
*** *CERTES - Université Paris XII*
Avenue du Général de Gaulle, F-94000 Créteil, France

Abstract: In a bounded-error context, reliable set-inversion algorithms such as SIVIA provide guaranteed estimates of the set of all the parameters deemed compatible with the selected model and the collected data, assuming that all the uncertain variables of the model are those to be estimated. In this paper we propose a new approach to estimate the parameters of interest assuming that there are other parameters that will not be estimated. This leads to the idea of set projection. A new algorithm for set projection is proposed and applied to the estimation of thermal quantities via a new experimental device to be calibrated.

Keywords: bounded-error estimation, guaranteed estimation, interval analysis, nuisance parameters, parametric models, set-membership estimation

1. INTRODUCTION

Most methods for the estimation of physical parameters minimize a possibly weighted quadratic norm of the difference between the vector of collected data $\mathbf{y}$ and the corresponding model output. The success of such a minimization, usually performed by local iterative search (by, *e.g.*, the Newton, Gauss-Newton, Levenberg-Marquardt or conjugate gradients algorithm) is uncertain for nonlinear models, for two main reasons. First, the estimate obtained is very sensitive to the initial value given to the parameter vector. Second, the search method may be trapped near a local optimum or stop before reaching the actual global optimum.

Moreover, the estimation of physical parameters should be regarded in a same way as any technique for experimental measurement and an uncertainty region should always be provided for the estimate. For the maximum-likelihood estimator, under the very strong assumption that the structure used for the model is correct and that the measurement noise is additive white and Gaussian with zero mean and known variance, the *asymptotic* variance of the estimate is the inverse of the Fisher information matrix, which can be computed. Unfortunately, such hypotheses are seldom verified as the number of samples used might be small, the measurement error may include some deterministic systematic errors or be far from being normally distributed, and the model is in general a much simplified version of reality.

To face these problems, it has been proposed to describe the parameter vector estimate as a set containing all parameter vectors that are consistent with the experimental data and the model given some bounds on the acceptable errors. The size of the set quantifies the uncertainty of the estimate. The development of this approach, called *set-membership estimation* or *bounded-error estimation* started more than thirty years ago with the seminal work of (Schweppe, 1968) and (Witsenhausen, 1968).

This paper deals with the case where, in addition to the parameters of interest, *i.e.*, the parameters to be estimated, there are some non-essential parameters (or *nuisance* parameters) subject to bounded uncertainty. To keep the dimension of parameter space sufficiently low, these nuisance parameters are not going to be estimated. However, their uncertainty will affect both the estimates of the parameters of interest and the uncertainty of these estimates.

Bounded-error estimation is recalled in Section 2. Before presenting the new set projection algorithm PROJECT, Section 3 briefly describes interval analysis and constraint propagation techniques that will allow PROJECT to be reliable. As an application, Section 4 shows that these guaranteed techniques permit the simultaneous identification of thermal resistance and Fourier time of material by the periodic method developed at CERTES by (Tang-Kwor, 1998).

2. SET-MEMBERSHIP ESTIMATION

In the sequel two types of parameters will be distinguished. The parameters of interest, *i.e.*, those to be identified, are in the *parameter vector* $\mathbf{p}$. The other non-essential parameters are gathered in a vector $\mathbf{q}$ called the *nuisance parameter vector*. It is assumed that $\mathbf{p} \in \mathbb{P}$ and $\mathbf{q} \in \mathbb{Q}$, where $\mathbb{P}$ and $\mathbb{Q}$ are known prior domains.

Let $\mathbf{e}$ be the model output error $\mathbf{e} = \mathbf{y} - \mathbf{f}(\mathbf{p}, \mathbf{q})$, where $\mathbf{y}$ is the vector of the collected data and $\mathbf{f}(\cdot, \cdot)$ the corresponding model output. In bounded-error estimation (or *set-membership estimation*), one looks for the set of all parameter vectors such that the error stays within some known feasible domain $\mathbb{E}$, *i.e.*, $\mathbf{e} \in \mathbb{E}$ (see *e.g.* (Milanese *et al.*, 1996), (Norton, 1994), (Norton, 1995) and the references therein). The set estimate then contains all values of the parameter vector that are *acceptable*, *i.e.*, consistent with the model and the collected data $\mathbf{y}$, given what is deemed an acceptable error. The size of this set quantifies the uncertainty associated with the estimated parameters.

Assume first that the value $\mathbf{q}^*$ taken by the nuisance parameter vector $\mathbf{q}$ is known. The set $\mathbb{C}$ to be estimated is the set of all the acceptable parameter vectors $\mathbf{p}$

$$\mathbb{C} = \{\mathbf{p} \in \mathbb{P}, \quad \mathbf{f}(\mathbf{p}, \mathbf{q}^*) \in \mathbb{Y}\}, \tag{1}$$

where $\mathbb{Y} = \mathbf{y} + \mathbb{E}$. Characterizing $\mathbb{C}$ is a set-inversion problem, as (1) can be rewritten as

$$\mathbb{C} = \mathbf{g}^{-1}(\mathbb{Y}) \cap \mathbb{P}, \tag{2}$$

where $\mathbf{g}(\cdot) = \mathbf{f}(\cdot, \mathbf{q}^*)$. It can be solved in a guaranteed way using the algorithm SIVIA (Jaulin *et al.*, 2001), see Section 3.

Suppose now that $\mathbf{q}^*$ is unknown. One may of course choose to estimate the set

$$\mathbb{S} = \{(\mathbf{p}, \mathbf{q}) \in \mathbb{P} \times \mathbb{Q} \mid \mathbf{f}(\mathbf{p}, \mathbf{q}) \in \mathbb{Y}\}, \tag{3}$$

which can again be seen as a set-inversion problem. However, characterizing $\mathbb{S}$ will be much more difficult than estimating $\mathbb{C}$, since the dimension of $\mathbb{S}$ is larger than that of $\mathbb{C}$ and the volume of $\mathbb{S}$ may be very large, if the parameters in $(\mathbf{p}, \mathbf{q})$ are not identifiable.

Since the value of $\mathbf{q}$ is not considered essential, an alternative simpler approach is to characterize the set Π of all the acceptable parameter vectors $\mathbf{p}$ under the assumption that $\mathbf{q}$ belongs to its prior domain, *i.e.*,

$$\Pi = \{\mathbf{p} \in \mathbb{P} \mid \exists \mathbf{q} \in \mathbb{Q}, \mathbf{f}(\mathbf{p}, \mathbf{q}) \in \mathbb{Y}\}. \tag{4}$$

The estimation of the acceptable values of $\mathbf{q}$ is then given up to simplify computation.

While $\mathbb{C}$ is a cut of $\mathbb{S}$, Π is the projection of $\mathbb{S}$ onto the $\mathbf{p}$-space (see Figure 1)

$$\Pi = proj_{\mathbb{P}}\mathbb{S}. \tag{5}$$

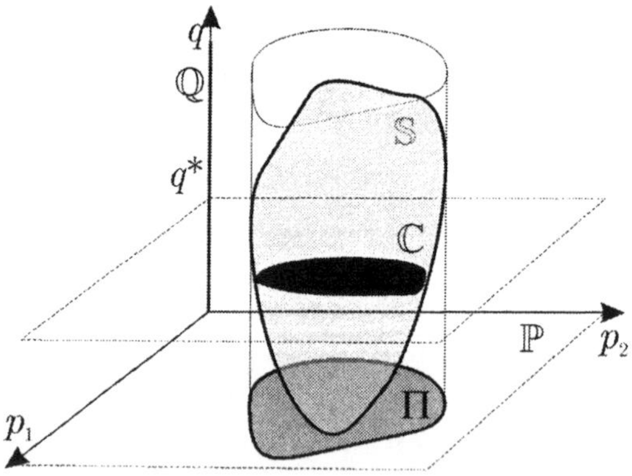

Fig. 1. The various set estimates with $dim\mathbf{p} = 2$ and $dim\mathbf{q} = 1$

Remark 1. The inclusion $\mathbb{C} \subset \Pi$ illustrates the fact that when $\mathbf{q}$ is uncertain, the uncertainty on $\mathbf{p}$ increases.

The basic tools for the characterization of Π will now be presented.

3. INTERVAL ANALYSIS

Several techniques have been developed to describe estimated sets, using various objects, such as zonotopes, ellipsoids, or vector intervals (Milanese *et al.*, 1996). When $\mathbf{g}$ is non-linear, the interval analysis techniques, using intervals and unions of intervals are particularly well suited. For any given function $\mathbf{g}(\mathbf{p})$, interval analysis permits the computation of an outer approximation $[\mathbf{g}]([\mathbf{p}])$ of $\mathbf{g}([\mathbf{p}])$ where $[\mathbf{p}]$ and $[\mathbf{g}]([\mathbf{p}])$ are boxes (or vector intervals), *i.e.*, Cartesian products of intervals, and where the image of a box $[\mathbf{p}]$ by $\mathbf{g}$ is

$$\mathbf{g}([\mathbf{p}]) \triangleq \{\mathbf{y} \mid \exists \mathbf{p} \in [\mathbf{p}],\ \mathbf{y} = \mathbf{g}(\mathbf{p})\}.$$

The function $[\mathbf{g}](\cdot)$ satisfying

$$\forall [\mathbf{p}],\ \mathbf{g}([\mathbf{p}]) \subset [\mathbf{g}]([\mathbf{p}]), \tag{6}$$

is called an *inclusion function* associated with $\mathbf{g}$. Inclusion functions provide reliable outer approximations of images of boxes, while other numerical approaches just evaluate the images of discrete sets of values.

3.1 *Set inverter*

Assume that the set to be estimated is $\mathbb{C}$ as defined by (2). Interval analysis allows us to obtain a reliable enclosure of $\mathbb{C}$ as defined by

$$\underline{\mathbb{C}} \subset \mathbb{C} \subset \overline{\mathbb{C}}. \tag{7}$$

The inner approximation $\underline{\mathbb{C}}$ of $\mathbb{C}$ consists of boxes $[\mathbf{p}]$ that have been proved acceptable, *i.e.* such that $\forall \mathbf{p} \in [\mathbf{p}]$, $\mathbf{p}$ is acceptable. To prove that $[\mathbf{p}]$ is acceptable, it suffices to check that $[\mathbf{g}]([\mathbf{p}]) \subset \mathbb{Y}$ and use (6). Else, if $[\mathbf{g}]([\mathbf{p}]) \cap \mathbb{Y} = \varnothing$, then the whole box $[\mathbf{p}]$ can be rejected. Otherwise, no conclusion is reached and the box $[\mathbf{p}]$ is said undetermined. The recursive algorithm SIVIA (*Set Inverter Via Interval Analysis*) (Jaulin *et al.*, 2001) partitions the prior space $\mathbb{P}$ into boxes $[\mathbf{p}]$ to be submitted to these tests. Any undetermined box is bisected and tested again, unless its size is less than a precision parameter ε to be tuned by the user, which ensures that the algorithm terminates after a finite number of iterations. The outer approximation is then computed as $\overline{\mathbb{C}} = \underline{\mathbb{C}} \cup \Delta\mathbb{C}$ where $\Delta\mathbb{C}$ is the union of all remaining undetermined boxes.

As SIVIA is a branch-and-bound algorithm, the computational time and memory space required increase exponentially with the dimension of $\mathbf{p}$. *Contractors* may also be used to reduce the size of the box to be tested without bisection (Jaulin *et al.*, 2001). For any given set $\mathbb{C}$, a contractor $\mathcal{C}_{\mathbb{C}}$ is an algorithm computing a box $[\mathbf{p}_0'] = \mathcal{C}_{\mathbb{C}}([\mathbf{p}_0])$ from any given box $[\mathbf{p}_0]$ in $\mathbb{P}$, such that the following properties hold

$$[\mathbf{p}_0'] \subset [\mathbf{p}_0], \tag{8}$$

$$[\mathbf{p}_0'] \cap \mathbb{C} = [\mathbf{p}_0] \cap \mathbb{C}. \tag{9}$$

(see Figure 2). When used in SIVIA before the tests, contractors may eliminate boxes such that $[\mathbf{p}] \cap \mathbb{C} = \varnothing$ without any bisection and may reduce the size of undetermined boxes for which $[\mathbf{p}] \cap \mathbb{C} \neq \varnothing$. Contractors are especially useful when the set to be estimated is small, or when the large dimension of $\mathbf{p}$ makes the use of the basic version of SIVIA intractable. The resulting algorithm to be used here is called SIVIAP and is a direct variation of SIVIA using contractors (see (Jaulin *et al.*, 2001) and the references therein).

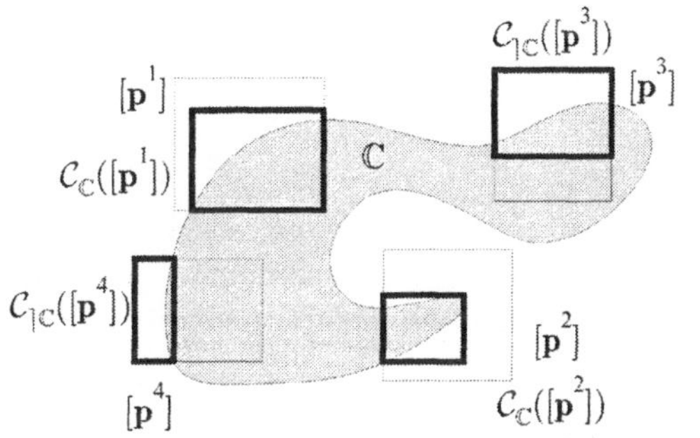

Fig. 2. Contractors $\mathcal{C}_{\mathbb{C}}$ for a given set $\mathbb{C}$ and $\mathcal{C}_{\rceil\mathbb{C}}$ for the complementary set $\rceil\mathbb{C}$; $\mathcal{C}_{\mathbb{C}}$ and $\mathcal{C}_{\rceil\mathbb{C}}$ are respectively pessimistic on $[\mathbf{p}^1]$ and $[\mathbf{p}^3]$, but optimal on $[\mathbf{p}^2]$ and $[\mathbf{p}^4]$

3.2 *Project*

When only Π is to be characterized, one can use another algorithm called PROJECT (Braems, 2002), (Jaulin *et al.*, 2002). This algorithm computes inner and outer approximations $\underline{\Pi}$ and $\overline{\Pi}$ of the set Π defined by (4). As only the $\mathbf{p}$-space is partitioned, the memory and computational time required are much smaller than for a full characterization of $\mathbb{S}$. Obviously, the main difference between PROJECT and SIVIA lies in the tests to be implemented. In SIVIA, the outer approximation $[\mathbf{g}]([\mathbf{p}])$ is directly used to test the acceptability of *all* elements of $[\mathbf{p}]$. Here, to characterize Π, $[\mathbf{p}]$ will be said acceptable if *there exists* $\mathbf{q} \in \mathbb{Q}$ such that $\mathbf{f}([\mathbf{p}], \mathbf{q}) \subset \mathbb{Y}$. Feasible point finders then require specific approaches. An algorithm based on partitioning can be found in (Jaulin and Walter, 1996). In order to allow consideration of higher dimensions, the procedure implemented in PROJECT uses contractors. Because of lack of space, PROJECT will not be presented in detail, and we shall only describe one of its constituants, INSIDE (Table 1), which for a given box $[\mathbf{p}_0]$ evaluates if there is a feasible $\mathbf{q}$ in $[\mathbf{q}_0]$.

In Table 1, $\mathcal{C}_{\mathbb{S}}$ is a contractor associated with $\mathbb{S}$ and $\mathcal{C}_{\rceil\mathbb{S}}$ a contractor associated with the complementary set $\rceil\mathbb{S}$ (see Figure 2). At Step 6, if $[\overline{\mathbf{p}}] \neq [\mathbf{p}_0]$, then the part $\mathcal{P}$ of $[\mathbf{p}_0]$ that has been eliminated

Table 1. INSIDE

Algorithm INSIDE (in:$[\mathbf{p}_0]$,$[\mathbf{q}_0]$; out: $[\overline{\mathbf{p}}]$)

1	$\mathcal{L} := \{[\mathbf{q}_0]\}$; $[\overline{\mathbf{p}}] := [\mathbf{p}_0]$;
2	do
3	take the first box $[\mathbf{q}]$ out of $\mathcal{L}$;
4	do
5	$\mathbf{q}_0 :=$ center $([\mathbf{q}])$;
6	$[\overline{\mathbf{p}}] := proj_{\mathbb{P}}\mathcal{C}_{\rceil\mathbb{S}}\left([\overline{\mathbf{p}}], \mathbf{q}_0\right)$;
7	$[\mathbf{p}_1] := [\overline{\mathbf{p}}]$;
8	$\left([\mathbf{p}_1'], [\mathbf{q}]\right) := \mathcal{C}_{\mathbb{S}}\left([\mathbf{p}_1], [\mathbf{q}]\right)$;
9	while contraction of $[\overline{\mathbf{p}}]$ is significant;
10	if $([\overline{\mathbf{p}}] = \emptyset)$, return;
11	if $w([\mathbf{q}]) > w([\mathbf{p}_0])$,
12	bisect $[\mathbf{q}]$ into $[\mathbf{q}_1]$ and $[\mathbf{q}_2]$;
13	store $[\mathbf{q}_1]$ and $[\mathbf{q}_2]$ at the end of $\mathcal{L}$;
14	while $\mathcal{L} \neq \varnothing$.

by the contractor for the complementary set satisfies $(\mathcal{P}, \mathbf{q}_0) \cap \rceil\mathbb{S} = \emptyset$, so $(\mathcal{P}, \mathbf{q}_0) \subset \mathbb{S}$, $\mathbf{q}_0$ is feasible and $\mathcal{P} \subset \Pi$. Otherwise, $\mathbf{q}_0$ has not been proved feasible, and other feasible points have to be found in $[\mathbf{q}]$. Step 8 then allows a contraction of the domain $[\mathbf{q}]$, and both contractors are applied as long as contraction of $[\overline{\mathbf{p}}]$ is significant. If the resulting box $[\overline{\mathbf{p}}]$ is empty, then the whole box $[\mathbf{p}_0]$ has been proved feasible. Otherwise, it is still undetermined, as some points in $[\overline{\mathbf{p}}]$ may either belong to Π or to $\rceil\Pi$. The box $[\mathbf{q}]$ is then bisected unless its size is less than some precision parameter, here chosen as the width of the initial box $[\mathbf{p}_0]$ to be tested. Finally, the resulting algorithm INSIDE returns a box $[\overline{\mathbf{p}}]$ that satisfies $[\mathbf{p}_0] \cap \rceil\Pi = [\overline{\mathbf{p}}] \cap \rceil\Pi$, so INSIDE is a contractor for $\rceil\Pi$. As it does not modify $[\mathbf{p}_0]$, another routine also based on contractors can be called before INSIDE to reduce the size of the boxes to be tested (see (Jaulin *et al.*, 2002)).

4. APPLICATION

Techniques such as the flash method (see (Navarette *et al.*, 2000), (Thermitus and Laurent, 1997) and the references therein) and the periodic methods (Mattei and Tang Kwor, 1999) have recently been developed for the simultaneous identification of several thermo-physical parameters. To test the set-up developed in (Tang-Kwor, 1998) and based on a periodic method, reliable set estimation will be used to identify the thermo-physical characteristics of a sample under study. For that purpose, the uncertainty intrinsic to this experimental set-up will have to be taken into account.

The experimental set-up is as shown on Figure 3. The sample under study is fixed within a metallic rack by a glue with very large conductivity. While the front side of the rack is fixed to a heating device, the rear side is in contact with air at ambient temperature. Radiative shields are used to reduce lateral heat losses. The heating sequence consists of five sinusoids of angular frequency ω_i, $i = 1, \ldots, 5$ and the temperatures of the rear and front sides are measured with thermocouples. The experimental temperature spectra are used

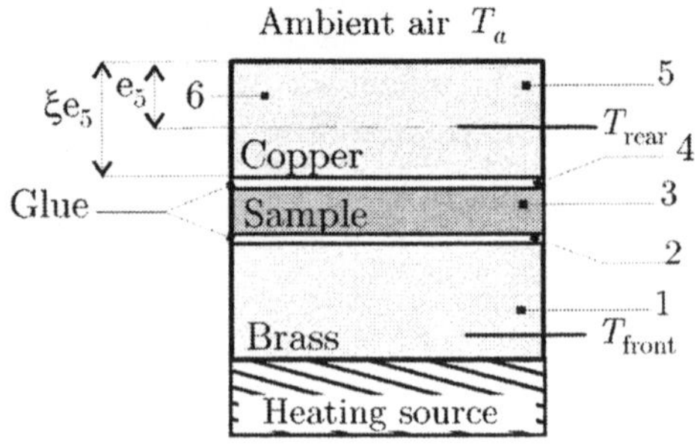

Fig. 3. The experimental set-up: 1: brass, 2: glue, 3: PVC, 4: glue, 5: copper

to estimate the experimental frequency response, taken as the ratio of the Fourier transforms of the output and the input

$$H^{\text{mes}}(j\omega_i) = H_i^{\text{mes}} = \frac{T_{\text{rear}}^{\text{mes}}(j\omega_i)}{T_{\text{front}}^{\text{mes}}(j\omega_i)}. \tag{10}$$

The data are corrupted by several measurement errors, including the device error, the error in the computation of the spectrum, and the reading error. As only 30 measurements have been performed, no statistical information can be reliably inferred, and a bounded-error context will be considered. For each input angular frequencies ω_i, $i = 1, \ldots, 5$ the output interval $[H_i^{\text{mes}}]$, is defined as

$$[H_i^{\text{mes}}] = \left[\inf_{j=1..30} H_i^{\text{mes}j}, \ \sup_{j=1..30} H_i^{\text{mes}j}\right].$$

to be gathered in the interval data vector $[\mathbf{H}^{\text{mes}}]$.

This system is modeled with a series of one-dimensional quadrupoles. For the k-th layer of homogeneous material, the relationship between the front pair temperature/flux $(T_{k-1}(s), \varphi_{k-1}(s))$ and the rear one $(T_k(s), \varphi_k(s))$ is given by

$$\begin{bmatrix} T_{k-1}(s) \\ \varphi_{k-1}(s) \end{bmatrix} = Z_k(s) \begin{bmatrix} T_k(s) \\ \varphi_k(s) \end{bmatrix},$$

where

$$Z_k(s) = \begin{bmatrix} \cosh\sqrt{\tau_k s} & \dfrac{R_k}{\sqrt{\tau_k s}} \sinh\sqrt{\tau_k s} \\ \dfrac{R_k}{\sqrt{\tau_k s}} \sinh\sqrt{\tau_k s} & \cosh\sqrt{\tau_k s} \end{bmatrix},$$

and R_k and τ_k are respectively the thermal resistance and the Fourier time of the k-th layer. For $k = 2, 4$, the glue is supposed with no inertia, so $Z_k(s)$ just depends on the resistance $R_2 = R_4 = R_g$. The i-th component of the output vector associated with $[\mathbf{H}^{\text{mes}}]$ is then given by

$$H_i(\mathbf{p}, \mathbf{q}) = \left.\frac{T_{\text{rear}}(s)}{T_{\text{front}}(s)}\right|_{s=j\omega_i}, \quad i = 1, \ldots, 5, \tag{11}$$

where the front temperature/flux pair is related to the ambient temperature/convective flux pair (T_a, φ_a) by

$$\begin{bmatrix} T_{\text{front}}(s) \\ \varphi_{\text{front}}(s) \end{bmatrix} = \bigotimes_{k=1,\dots,5} Z_k(s) \begin{bmatrix} T_a(s) \\ \varphi_a(s) \end{bmatrix},$$

and the rear pair temperature/flux is given by

$$\begin{bmatrix} T_{\text{rear}}(s) \\ \varphi_{\text{rear}}(s) \end{bmatrix} = Z_6(s) \begin{bmatrix} T_a(s) \\ \varphi_a(s) \end{bmatrix}.$$

Finally, heat transfers at ambient temperature are modeled by $\varphi_a = hT_a$, where the heat surface exchange coefficient h is generally assumed to be constant.

The thermal resistances R_k and Fourier time τ_k of each layer intervening in $H_i(\mathbf{p}, \mathbf{q})$ as defined by (11) depend on several uncertain physical quantities:
(i) the thermo-physical characteristics of the non-inertial materials (copper, brass and sample) cannot be identified without destroying the device because of the glue; their prior uncertainty then corresponds to the range of values found in literature;
(ii) the thermo-physical characteristic of the glue is provided with its uncertainty by the manufacturer;
(iii) the thickness e_k of each layer is measured with uncertainty;
Moreover, the exchange heat coefficient h is assumed to be constant but inside an interval range to take into account model error.

The parameter vector $\mathbf{p}$ to be identified is then

$$\mathbf{p} = \left(R_3 \ \sqrt{\tau_3} \right), \qquad (12)$$

while the nuisance parameter vector $\mathbf{q}$ contains all other uncertain quantities

$$\mathbf{q} = \left(R_1 \ \sqrt{\tau_1} \ R_g \ R_5 \ \sqrt{\tau_5} \ \xi \ h \right), \qquad (13)$$

where ξ is the position of the thermocouple in the copper layer (see Figure 3). Prior domains $\mathbb{Q}_i$ for the uncertain quantities q_i's are presented in Table 2, specifying their center q_{i0} and relative radius $\Delta_r q_i$.

Table 2. Uncertainty interval associated with each nuisance parameter

i	q_i	q_{i0}	$\Delta_r q_i$
1	$R_1(10^{-5}\text{SI})$	2.90	**38%**
2	$\sqrt{\tau_1}$	5.19×10^{-4}	**39%**
3	$R_g(10^{-5}\text{SI})$	3.05	**41%**
4	$R_5(10^{-5}\text{SI})$	2.5	4%
5	$\sqrt{\tau_5}$	0.92	2%
6	ξ	0.5005×10^{-3}	**40%**
7	h	7.5	**33%**

The sample under study is made of PVC for which only rough prior values for the thermo-physical values are available. The prior space for $\mathbf{p}$ is thus taken as $\mathbb{P} = [0.014, 0.047] \times [7.2, 23]$.

4.1 *Known nuisance parameters*

Assuming that the nuisance parameter q_i is known amounts to fixing its numerical value $\widehat{q}_i$. For the nuisance parameters q_i, $i = 1, 2, 4, 5, 6$ there exists an actual constant value q_i. As in (Tang-Kwor, 1998), it will be assumed for the time being that this value is equal to q_{i0}. The other nuisance parameters may actually vary during the experiment or characterize a structural uncertainty. Fixing their value is thus a strong hypothesis: in (Tang-Kwor, 1998), q_7 and q_3 have been arbitrarily chosen as $\widehat{q}_7 = \underline{\mathbb{Q}}_7$ and $\widehat{q}_3 = 0$, which corresponds to neglecting the glue layer. Contrary to the assumption $\widehat{q}_3 = 0$, the hypothesis $\widehat{q}_7 = \underline{\mathbb{Q}}_7$ has been invalidated as it could be proved that no acceptable solution exists in a bounded-error context (Braems, 2002). In the following, it will be assumed that

$$\widehat{q}_3 = 0 \text{ and } \widehat{q}_7 = \overline{\mathbb{Q}}_7.$$

Since $\widehat{\mathbf{q}}$ is fixed at a known numerical value,

$$\mathbb{C} = \{\mathbf{p} \in \mathbb{P} \mid \mathbf{H}(\mathbf{p}, \widehat{\mathbf{q}}) \in [\mathbf{H}^{\text{mes}}]\}.$$

can now be characterized by set inversion.

In 17.8 s on a Pentium IV at 1.7 GHz, for $\varepsilon = 0.01$, SIVIAP provides two non-empty sets $\underline{\mathbb{C}}$ and $\overline{\mathbb{C}}$ (see Figure 4). The projection of $\overline{\mathbb{C}}$ onto the p_1 and p_2 axes provides an outer approximation of the uncertainty interval associated with each parameter $p_1 \in [0.0290, 0.0308]$ and $p_2 \in [14.0468, 15.0469]$. So

$$p_{10} = 0.030, \ \Delta_r p_1 = 3.2\%,$$
$$p_{20} = 14.547, \ \Delta_r p_2 = 3.5\%.$$

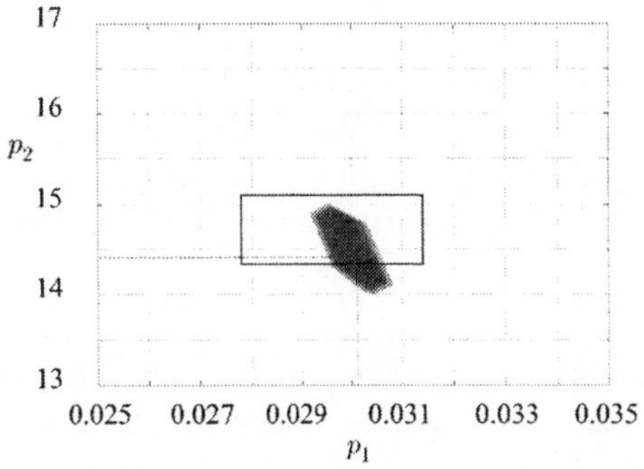

Fig. 4. Inner and outer approximations $\underline{\mathbb{C}}$ (black) et $\overline{\mathbb{C}}$ (grey) obtained by SIVIAP when the nuisance parameters are assumed to be known; the frame in black corresponds to the range found in the literature for the two parameters

4.2 *Unknown nuisance parameters*

Assume now that $\mathbf{q}$ is only known to belong to $\mathbb{Q}$ and take

$$\Pi = \{\mathbf{p} \in \mathbb{P} \mid \exists \mathbf{q} \in \mathbb{Q}, \mathbf{H}(\mathbf{p}, \mathbf{q}) \in [\mathbf{H}^{\text{mes}}]\}. \quad (14)$$

PROJECT provides inner and outer approximations of Π (see Figure 5) in 1h54mn on a Pentium IV at 1.7 GHz, for $\varepsilon = 0.05$. The smallest box containing the outer approximation of Π is

$$[\Pi] = [0.0265, 0.0336] \times [13.125, 15.5625]$$

and

$$p_{10} = 0.031, \quad \Delta_r p_1 = 11.9\%,$$
$$p_{20} = 14.354, \quad \Delta_r p_2 = 8.5\%.$$

Taking into account the uncertainty associated with $\mathbf{q}$ has thus significantly increased the uncertainty associated with the estimate of $\mathbf{p}$, from 3.2% to 11.9% for R_3 and from 3.5% to 8.5% for $\sqrt{\tau_3}$.

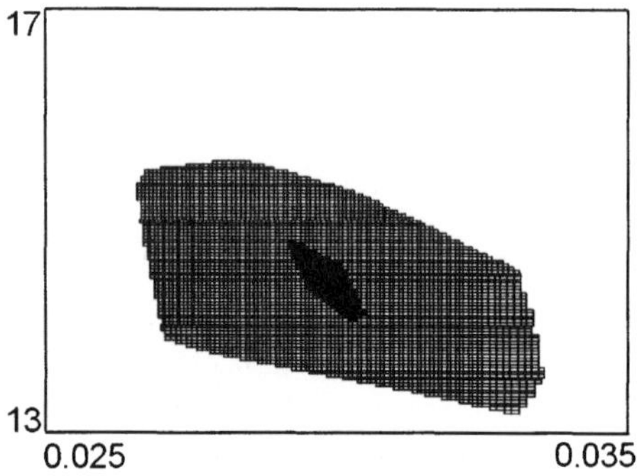

Fig. 5. Inner and outer approximations of the set Π obtained by PROJECT)

5. CONCLUSIONS

A new method for computing inner and outer approximations of the projection of a set over a subspace has been briefly presented. This method is based on interval analysis, which allows guaranteed results to be derived.

The bounded-error estimation of a vector $\mathbf{p}$ of parameters of interest when another vector $\mathbf{q}$ of nuisance parameters is only known to belong to a set is a direct application of this new method. The fact that the resulting set-estimation technique is guaranteed makes it possible to bypass any prior identifiability study and to characterize the set of all possible values for the parameters, whether these parameters are identifiable or not.

The feasibility of the approach has been demonstrated on a real-life problem of estimation of physical parameters, involving nine uncertain quantities. This would have been impossible without the use of contractors and of the new algorithm PROJECT that makes it possible to concentrate on the parameter of interest.

6. REFERENCES

Braems, I. (2002). Méthodes ensemblistes garanties pour l'estimation de grandeurs physiques. PhD thesis. University Paris-Sud. Orsay, France.

Jaulin, L. and E. Walter (1996). Guaranteed tuning, with application to robust control and motion planning. *Automatica* **32**(8), 1217–1221.

Jaulin, L., I. Braems and E. Walter (2002). Interval methods for nonlinear identification and robust control. In: *41st Conference on Decision and Control.* Las Vegas, Nevada, US. Accepted for publication.

Jaulin, L., M. Kieffer, O. Didrit and E. Walter (2001). *Applied Interval Analysis.* Springer-Verlag. London.

Mattei, S. and E. Tang Kwor (1999). A new periodic technique for thermal conductivity measurement. In: *15th European Conference on Thermophysical Properties.* Würzburg, Germany.

Milanese, M., Norton, J., Piet-Lahanier, H. and Walter, E., Eds.) (1996). *Bounding Approaches to System Identification.* Plenum Press. New York, NY.

Navarette, M., F. Serrania and M. Villagran (2000). Application of the flash method for the thermal characterisation of woven carbon fibre laminates, materials and designs. *Materials and Designs* **22**, 93–97.

Norton, J. P., Ed.) (1994). Special Issue on Bounded-Error Estimation: Issue 1. *International Journal of Adaptive Control and Signal Processing* **8**(1):1–118.

Norton, J. P., Ed.) (1995). Special Issue on Bounded-Error Estimation: Issue 2. *International Journal of Adaptive Control and Signal Processing* **9**(1):1–132.

Schweppe, F. C. (1968). Recursive state estimation: unknown but bounded errors and system inputs. *IEEE Transactions on Automatic Control* **13**(1), 22–28.

Tang-Kwor, E. (1998). Contribution au développement de méthodes périodiques de mesure de propriétés thermophysiques des matériaux opaques. PhD thesis. University Paris XII Val-de-Marne. Créteil, France.

Thermitus, M. A. and M. Laurent (1997). New logarithmic technique in the flash method. *Int. J. Heat Mass Transfer* **40**(17), 4183–4190.

Witsenhausen, H. S. (1968). Sets of possible states of linear systems given perturbed observations. *IEEE Transactions on Automatic Control* **13**(5), 556–558.

IFAC Publications
www.elsevier.com/locate/ifac

VALIDATION TEST BASED PARAMETER UNCERTAINTY VERSUS ANALYSIS-BASED CONFIDENCE BOUNDS

Sippe G. Douma * Xavier J.A. Bombois * Paul M.J. Van den Hof *

* *Delft Center for Systems and Control,*
Delft University of Technology, Mekelweg 2, 2628 CD Delft,
The Netherlands,
s.g.douma@dcsc.tudelft.nl x.j.a.bombois@dcsc.tudelft.nl
p.m.j.vandenhof@dcsc.tudelft.nl

Abstract: Standard Instrumental Variable system identification methods provide for a particular parameter confidence region under a Gaussian distribution. Alternatively, a parameter bounding technique based on a set of constraints on a cross-correlation function directly provides for a parameter uncertainty set. This paper relates the two uncertainty regions under the standard assumption of additive noise on the measured data, both in case undermodelling is considered and in case it is not. The ellipsoidal region associated with the IV estimation technique is strongly related to the polytopic region induced by the cross-correlation constraints as both techniques are based on the same set of regressors. However, they differ due to the fact that the former incorporates a covariance between errors, while the latter is limited to the variances of the errors only. The results are also discussed in terms of the standard model validation which is identical in nature to the cross-correlation parameter bounding technique. *Copyright © 2003 IFAC*

Keywords: parameter uncertainty, confidence region, set membership, model validation

1. INTRODUCTION

Given measured input and output data of a system any technique can be applied to yield a model describing the relation between the input and the output. However, a model is useful only when some indication is provided of its precision and accuracy. In turn, the measure of confidence in the model depends entirely on the assumption made on the data-generating system.

In standard prediction error identification the output data is assumed to be noise-corrupted by a stationary stochastic sequence with a particular distribution. Based on this assumption a confidence region of estimated parameters can be constructed, both when assuming that the data-generating system is in the model class (e.g. Ljung (1999)) and when undermodelling is taken into consideration (e.g. Hakvoort and Van den Hof (1997)).

Alternatively, parameter bounding methods or set-membership techniques pose a number of deterministic bounds on the noise (e.g. Milanese and Vicino (1991); Veres and Norton (1991)). These techniques result directly in a set of feasible parameters. While standard set membership techniques place a time domain bound on the additive noise, Hakvoort en Van den Hof (1995) have proposed a bound on the cross-correlation between the noise and some instrumental signal. The former methods are known to lead to conservative results, or even to no consistency, as the noise is allowed to take a worst-case realization within the noise bounds. Bounds on the cross-correlation of the noise with a instrumental signal give rise to a consistent parameter bounding method leading to less conservative results.

In this paper the standard prediction error confidence region is related to the set of parameters resulting from the cross-correlation parameter bounding technique. In particular, the IV estimation method without undermodelling (e.g. Ljung (1999); Söderström and

Stoica (1989)) and with undermodelling (Hakvoort and Van den Hof, 1997) is considered with respect to the cross-correlation method of Hakvoort and Van den Hof (1995). Since the cross-correlation method is in nature identical to a posteriori model validation tests used in standard system identification, the results will also be interpreted in light of model validation based on data.

After preliminary information is given in section 2, section 3 introduces the notation used in the paper and briefly discusses the different methods. Based on a particular set of assumptions, the methods are analyzed in section 4. The comparison itself is made in sections 5 and 6 for the most general case and for illustrative special cases, respectively. After a small section with further remarks, the conclusions are presented.

2. PRELIMINARIES

Here we consider discrete linear time-invariant finite dimensional models which, additionally, are linearly parametrized. These parametrized models $G(q,\theta)$ are represented by

$$G(q,\theta) = \sum_{k=0}^{n-1} g(k)F_k(q),$$

where θ denotes the parameter vector $\left[\, g(0)\ g(1)\ \cdots\ g(n-1)\,\right]^T \in \mathbb{R}^n$ and $F_k(q)$ the basis functions. A standard choice for $F_k(q)$ is to consider orthonormal basis functions (see e.g. Heuberger *et al.* (1995)). This class contains the well-known FIR, Laguerre and Kautz models. For example, for a FIR-model $F_k(q)$ is the standard pulse basis, i.e. $F_k(q) = q^{-k}$.

3. METHODS

3.1 *Method 1: Instrumental Variable estimate*

The procedure of the Instrumental Variable Estimation method is discussed here briefly, as it is documented in many references (e.g. Ljung (1999); Söderström and Stoica (1989)).

Procedure of the IV estimation

1. Choose an instrumental variable signal $r(t)$ (For properties of $r(t)$ see section 4.2).
2. Filter the measured input $u(t)$ and the instrumental variable $r(t)$ with the basis functions $F_k(q)$ to create signal sequences $x_k(t)$ and $z_k(t)$,

$$x_k(t) = F_k(q)u(t), \qquad z_k(t) = F_k(q)r(t),$$

and define the column-vectors
$\phi(t) := \left[\, x_0(t) \ldots x_{n-1}(t)\,\right]^T$
$\zeta(t) := \left[\, z_0(t) \ldots z_{n-1}(t)\,\right]^T.$

3. Obtain the instrumental variable estimate $\hat{\theta}_N^{IV}$ by

$$\hat{\theta}_N^{IV} := \underset{\theta\in\Theta}{sol} \left\{ \frac{1}{\tilde{N}} \sum_{t=t_s}^{N} \zeta(t)\varepsilon(t,\theta) = 0 \right\}, \quad (1)$$

with $\Theta \subset \mathbb{R}^n$ an appropriate parameter space, $t_s > 0$ representing the starting sample used in the estimation, and $\tilde{N} = N - t_s + 1$. The term $\varepsilon(t,\theta)$ denotes the residue-signal

$$\begin{aligned}\varepsilon(t,\theta) &= y(t) - G(q,\theta)u(t)\\ &= y(t) - \phi^T(t)\theta. \end{aligned} \quad (2)$$

Straightforward algebraic manipulation reveals the solution to (1) as

$$\hat{\theta}_N^{IV} = \left[\frac{1}{\tilde{N}} \sum_{t=t_s}^{N} \zeta(t)\phi^T(t) \right]^{-1} \frac{1}{\tilde{N}} \sum_{t=t_s}^{N} \zeta(t)y(t). \quad (3)$$

provided that the matrix to be inverted is nonsingular (which depends on the choice of $r(t)$). It is to be noted that when the input $u(t)$ itself is taken as instrumental variable $r(t)$, expression 3 gives the standard linear least squares estimate of the expansion coefficients.

Parameter confidence region of the IV estimate
A confidence region $\mathcal{D}^{IV}$ of the instrumental variable estimate is usually constructed based on the assumption that $\hat{\theta}_N^{IV}$ is Gaussian distributed with the true parameter vector θ_0 as mean (Ljung, 1999). In section 4 the underlying assumptions of this approach are discussed.

Theoretically, every estimate $\hat{\theta}_N^{IV} \in \mathcal{N}(\theta_0, P_\theta)$ is contained, with a probability of α, within the standard ellipsoidal region associated with the Gaussian distribution defined by all θ such that

$$(\theta-\theta_0)^T P_\theta^{-1} (\theta-\theta_0) \leq c_{\chi,\alpha}, \quad (4)$$

where $c_{\chi,\alpha}$ corresponds to a probability α in the chi-squared distribution with n degrees of freedom. This follows from the fact that (the spectral factor of) the inverse of the matrix P_θ transforms the parameters to a set of uncorrelated Gausian distributed parameter each having a standard deviation of 1. The norm of this transformed parameter vector then has a chi-squared distribution with n degrees of freedom. The shape of the ellipsoid is defined by the square root of the eigenvalues of P_θ and its orientation by the associated eigenvectors.

Using (4), a confidence region $\mathcal{D}^{IV}$ containing the unknown true parameter vector θ_0 at a certain probability level α can be constructed with a particular estimate $\hat{\theta}_N^{IV}$ and based on an estimation $\hat{P}_\theta$ of the covariance matrix as

$$\mathcal{D}^{IV} = \left\{ \theta \mid \left(\theta - \hat{\theta}_N^{IV}\right)^T \hat{P}_\theta^{-1} \left(\theta - \hat{\theta}_N^{IV}\right) \leq c_\alpha \right\}. \quad (5)$$

Note that the the probability level α corresponding to c_α now depends on the properties of the estimate $\hat{P}_\theta$ as well (Ljung, 1999).

3.2 *Method 2: Cross-correlation method*

Parameter bounding identification methods (e.g. Milanese and Vicino (1991); Veres and Norton (1991)) are based on a set of linear constraints, e.g. constraints on the noise signal or a cross-correlation function. Here we consider the procedure proposed in Hakvoort and Van den Hof (1995):

Procedure of the cross-correlation method

1-2 as in the IV method

3. Define the feasible *set* of parameters by

$$\mathcal{D}^{cc} = \left\{ \theta \mid b_l(k) \leq \frac{1}{\bar{N}} \sum_{t=t_s}^{N} \zeta_k(t)\varepsilon(t,\theta) \leq b_u(k) \right. ,$$
$$\left. k = 1, \ldots, n. \right\} \qquad (6)$$

This set of parameters defines a polytope in the parameter space $\mathbb{R}^n$ as $\varepsilon(t,\theta)$ is linear in the parameters. This polytope is defined by $2n$ half planes in $\mathbb{R}^n$ of which the n halfplanes induced by the lower bounds b_l are parallel to those induced by the upper bounds b_u. The corner points of this polytope are solutions to a convex optimization problem (e.g. Broman and Shensa (1990)). However, note that in (6) the number of constraints equals the number of parameters. In this case a simple analytical solution exists, which is formulated in the following proposition:

Proposition 1. The solution space $\mathcal{D}^{cc} \in \mathbb{R}^n$ of all parameter vectors θ satisfying the cross-correlation constraint of (6) defines a polytope with 2^n corner points induced by the solutions to

$$\frac{1}{\bar{N}} \sum_{t=t_s}^{N} \zeta(t)\varepsilon(t,\theta) = Be_j \quad , j = 1, \ldots, 2^n.$$

where the vector e_j denotes the j-th Euclidean vector taking the j-the column from the matrix $B \in \mathbb{R}^{n \times 2^n}$ containing all possible combination of $b_l(k)$ and $b_u(k)$. For example, when $n = 3$, $B =$

$$\begin{bmatrix} b_l(0) & b_l(0) & b_l(0) & b_l(0) & b_u(0) & b_u(0) & b_u(0) & b_u(0) \\ b_l(1) & b_l(1) & b_u(1) & b_u(1) & b_u(1) & b_l(1) & b_u(1) & b_l(1) \\ b_l(2) & b_u(2) & b_l(2) & b_u(2) & b_u(2) & b_u(2) & b_l(2) & b_l(2) \end{bmatrix}$$

In other words, B simply defines the corner points of a n-dimensional box.
In particular, with $R = \frac{1}{\bar{N}} \sum_{t=t_s}^{N} \zeta(t)\phi^T(t)$, the corner points of $\mathcal{D}^{cc}$ are given by the vectors

$$\theta_j^{cc} = R^{-1} \frac{1}{\bar{N}} \sum_{t=t_s}^{N} \zeta(t)y(t) - R^{-1}Be_j, \qquad (7)$$

for $j = 1, \ldots, 2^n$.
Proof is omitted as it is straightforward. □

Confidence region of cross-correlation method
The properties of the polytope defined by the solution set $\mathcal{D}^{cc}$ of (6) can be discerned from the operation on the 'box' B by the matrix R^{-1}. Consider the singular value decomposition $U_R \Sigma_R V_R^T$ of R^{-1}. The matrix R^{-1} is then seen to rotate the 'box' B by V_R^T, dilate the result along the Euclidean axes with the elements of Σ_R, after which the matrix U_R induces a second rotation (see figure 1 for $n = 2$).

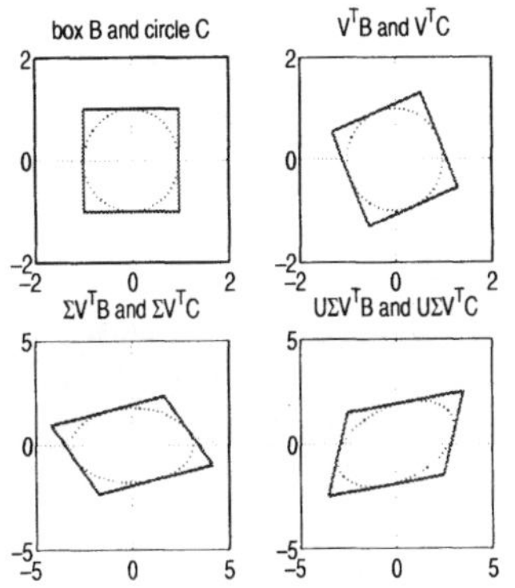

Fig. 1. *Transformation of a box B and a circle C induced by $R^{-1} = U_R \Sigma_R V_R^T$.*

4. ASSUMPTIONS AND ASSOCIATED ANALYSIS

Expressions (3) and (7) show that:

$$\theta_j^{cc} = \hat{\theta}_N^{IV} - R^{-1}Be_j. \qquad (8)$$

In other words, the nominal parameter vector of the cross–correlation based set $\mathcal{D}^{cc}$ equals the IV estimate $\hat{\theta}_N^{IV}$. The term B represents a confidence region: the possible deviations around the nominal estimate. In section 5 this confidence region $\mathcal{D}^{cc}$ will be compared with the confidence region $\mathcal{D}^{IV}$ of the IV-estimate $\hat{\theta}_N^{IV}$. But first the confidence regions $\mathcal{D}^{IV}$ and $\mathcal{D}^{cc}$ are analyzed in detail.

4.1 *Assumptions*

Measures of confidence in estimated models depend almost entirely on a priori assumptions. Here we consider the following assumptions on the data-generating system, based on Hakvoort and Van den Hof (1997):

$$y(t) = \tilde{G}_0(q)u(t) + \bar{G}_0(q)u(t) + G_0(q)u^-(t) + v(t),$$

where

i. $\tilde{G}_0(q)u(t) + \bar{G}_0(q)u(t)$, denotes the part of the output $y(t)$ directly related to the measured sequence u by means of a linear-time invariant operator $G_0(q)$, decomposed into a modelled ($\tilde{G}_0(q)$) and unmodelled part ($\bar{G}_0(q)$). That is,

$$G_0(q)u(t) = \tilde{G}_0(q)u(t) + \bar{G}_0(q)u(t)$$
$$= \phi^T(t)\theta_0 + \sum_{k=n+1}^{\infty} g_0(k)F_k(q)u(t),$$

($u(t) := 0$ outside the measurement interval $t = 1, \ldots, N$)

ii. $G_0(q)u^-(t)$ denotes the effect of initial conditions, i.e. the effect of a sequence u^- preceding the measured $u(t)$ ($u^-(t) := 0$ for $t > 0$) and

iii. $v(t)$ denotes a stochastic component of which only the ensemble properties are constant in time. Here we consider $v(t)$ to be uncorrelated with the sequences u and r.

4.2 *Analysis of the Instrumental Variable estimate*

With the assumptions given above the instrumental variable estimate (3) can be analyzed as

$$\hat{\theta}_N^{IV} = \theta_0 + R^{-1}\frac{1}{\tilde{N}}\sum_{t=t_s}^{N}\zeta(t)\gamma(t). \tag{9}$$

with $\gamma(t) = \bar{G}_0(q)u(t) + G_0(q)u^-(t) + v(t)$.
A confidence region around the IV estimate $\hat{\theta}_N^{IV}$ will be based on the error term $R^{-1}\frac{1}{\tilde{N}}\sum_{t=t_s}^{N}\zeta(t)\gamma(t)$.
In the following two possible assumptions on this error term are evaluated.

4.2.1. *System is in the model class* In case the true system $G_0(q)$ is considered to be in the model class, i.e. $\bar{G}_0(q)u(t) = 0$ and $u^-(t) = 0$ for all t, expression (9) becomes

$$\hat{\theta}_N^{IV} = \theta_0 + R^{-1}\frac{1}{\tilde{N}}\sum_{t=t_s}^{N}\zeta(t)v(t) \tag{10}$$

Using the assumptions on $v(t)$, it follows that

$$\hat{\theta}_N^{IV} \in \mathcal{N}(\theta_0, P_\theta) \tag{11}$$

since $R^{-1}\frac{1}{\tilde{N}}\sum_{t=t_s}^{N}\zeta(t)v(t)$ is obtained from linear combinations of $v(t)$. If $v(t)$ is assumed to have a distribution other than Gaussian, the term $R^{-1}\frac{1}{\tilde{N}}\sum_{t=t_s}^{N}\zeta(t)v(t)$ will be Gaussian distributed only asymptotically as a result of the Central Limit Theorem. The covariance matrix P_θ is given by

$$P_\theta = R^{-1}\Lambda^N\left(R^{-1}\right)^T \tag{12}$$

where

$$\Lambda^N = E\left[\left(\frac{1}{\tilde{N}}\sum_{t=t_s}^{N}\zeta(t)v(t)\right)\left(\frac{1}{\tilde{N}}\sum_{t=t_s}^{N}\zeta(t)v(t)\right)^T\right].$$

The covariance matrix P_θ is seen to consists of (co)variances of the term $\frac{1}{\tilde{N}}\sum_{t=t_s}^{N}\zeta(t)v(t)$, for which an exact expression is available (Hakvoort and Van den Hof, 1997):

$$\Lambda^N(i,j) = \frac{1}{\tilde{N}}\sum_{\tau=-N+t_s}^{N-t_s} R_{\zeta_i\zeta_j}^N(\tau)R_v(\tau), \tag{13}$$

where

$$R_{\zeta_i\zeta_j}^N(\tau) := \frac{1}{\tilde{N}}\sum_{t=t_s}^{N+\tau}\zeta_i(t)\zeta_j(t-\tau),\ \tau = [-N+t_s, 0]$$

$$R_{\zeta_i\zeta_j}^N(\tau) := \frac{1}{\tilde{N}}\sum_{t=t_s}^{N-\tau}\zeta_i(t+\tau)\zeta_j(t)\ ,\ \tau = [1, N-t_s]$$

The confidence region $\mathcal{D}^{IV}$ of expression (5) is obtained with the estimate of P_θ which follows from applying expressions (12) and (13) with an estimate $\hat{R}_v(\tau)$ of the noise properties.

4.2.2. *Undermodelling* The effects of undermodelling reflected in $\bar{G}_0(q)u(t)$ and of initial conditions reflected in $G_0(q)u^-(t)$ in expression (9) introduce a deterministic but unknown bias in the estimator $\hat{\theta}_N^{IV}$. This bias Δ can be incorporated by means of bounds on its absolute value. That is, it holds that

$$\hat{\theta}_N^{IV} \in \Delta + \mathcal{N}(\theta_0, P_\theta),\quad |\Delta| < \beta,$$

where $\beta = \beta_1 + \beta_2$ and

$$\left|R^{-1}\frac{1}{\tilde{N}}\sum_{t=t_s}^{N}\zeta(t)\bar{G}_0(q)u(t)\right| < \beta_1$$

$$\left|R^{-1}\frac{1}{\tilde{N}}\sum_{t=t_s}^{N}\zeta(t)G_0(q)u^-(t)\right| < \beta_2.$$

The calculation of the bounds is not elaborated here as they follow easily from adapting the (frequency domain) bounds proposed in Hakvoort and Van den Hof (1997). A confidence region in the parameter space is given by the union of ellipsoids of expression (5) and section (4.2.1) centred at all possible mean values $\hat{\theta}_N^{IV} + \Delta$, with $|\Delta| < \beta$. It can be said that this union of ellipsoids contains the true system with a probability *larger than* α.

Alternatively, the effect of initial conditions could be considered as a stochastic influence similar to $v(t)$ as it depends on the unknown $u^-(t)$. Though not elaborated here, in that case, the covariance matrix P_θ could be extended additively by a matrix containing the covariances of the term $R^{-1}\frac{1}{\tilde{N}}\sum_{t=t_s}^{N}\zeta(t)G_0(q)u^-(t)$. The incorporation of the undermodelling would proceed as above with $\beta = \beta_1$.

4.3 *Analysis of the cross-correlation method*

The IV-estimation method provides for a nominal estimate based on data and a confidence region based on assumptions. The cross-correlation method, on the other hand, directly provides for a confidence region $\mathcal{D}^{cc}$ based on a data set. However, the cross-correlation method is not without assumptions. The bounds used in the cross-correlation method have to be chosen based on assumptions on the underlying system. Or, alternatively, the bounds chosen in the cross-correlation method have to be interpreted with respect to assumptions if statements are to be made about the quality of the resulting parameter set.

Considering the assumptions of section 4.1, the cross-correlation term $\frac{1}{\bar{N}}\sum_{t=t_s}^{N}\zeta(t)\varepsilon(t,\theta)$ in (6) is evaluated as

$$\frac{1}{\bar{N}}\sum_{t=t_s}^{N}\zeta(t)\varepsilon(t,\theta) = R\left(\theta_0 - \theta\right) + \frac{1}{\bar{N}}\sum_{t=t_s}^{N}\zeta(t)v(t),$$

in case the system is assumed to be in the model class. Clearly, for $\mathcal{D}_{cc}$ to contain the true parameter vector θ_0, the bounds $b_l(k)$ and $b_u(k)$ should allow for the error term $\frac{1}{\bar{N}}\sum_{t=t_s}^{N}\zeta_k(t)v(t)$. They could be interpreted as (deterministic) upper bounds on the amplitude of $\frac{1}{\bar{N}}\sum_{t=t_s}^{N}\zeta_k(t)v(t)$, or, based on a stochastic noise assumption, as bounds on the standard deviation of $\frac{1}{\bar{N}}\sum_{t=t_s}^{N}\zeta_k(t)v(t)$ times a factor $\sqrt{c_\alpha}$ corresponding to a certain level of probability. That is, with $b_l(k) = -b_u(k) = b(k)$,

$$b(k) = \sqrt{c_\alpha \Lambda^N(k,k)}.$$

Similarly, in case of undermodelling, the bounds $b_l(k)$ and $b_u(k)$ should explain the effects of both the noise term $\frac{1}{\bar{N}}\sum_{t=t_s}^{N}\zeta_k(t)v(t)$ and the undermodelling, i.e.

$$b(k) = \sqrt{c_\alpha \Lambda^N(k,k)} + \beta,$$

where Λ^N and β are as in section 4.2.

5. COMPARISON OF THE CONFIDENCE REGIONS

From the discussion in the previous section and in particular from expressions (5), (12) and (7) it is seen that

$$\mathcal{D}^{IV} = \left\{\theta \mid \theta = \hat{\theta}_N^{IV} + \Delta_\theta\,, \Delta_\theta \in R^{-1}\mathcal{E}\left(c_\alpha \hat{\Lambda}^N\right)\right\}$$
$$\mathcal{D}^{cc} = \left\{\theta \mid \theta = \hat{\theta}_N^{IV} + \Delta_\theta\,, \Delta_\theta \in R^{-1}\mathcal{B}(B)\right\}, \quad (14)$$

where $\mathcal{E}(A)$ denotes the ellipsoid defined by all $x \in \mathbb{R}^{n\times 1}$ such that $x^T A^{-1} x < 1$ and $\mathcal{B}(B)$ denotes the interior of the n-dimensional box with corner points defined in the matrix $B \in \mathbb{R}^{n\times 2^n}$. That is, the polytope of $\mathcal{D}^{cc}$ is formed by the transformation induced by the matrix $R^{-1}$of the *hyperbox* B (cf. section 3.2). Similarly, the ellipsoid of $\mathcal{D}^{IV}$ according to the estimated matrix $\hat{P}_\theta$ can be interpreted as the transformation induced by R^{-1} of an *ellipsoid* defined by $\hat{\Lambda}^N$ based on expression (13) with an estimate $\hat{R}_v(\tau)$ of the noise properties. (see figure 2).

With expression (14) the strong connection between $\mathcal{D}^{cc}$ and $\mathcal{D}^{IV}$ is made clear. For example, the smallest hyperbox $\mathcal{D}^{cc}$ containing the ellipsoid $\mathcal{D}^{IV}$ can easily be found (see figure 2). The link between the two sets results from the fact that the same regressors are used. The essential difference between the two methods lies with the information used on the

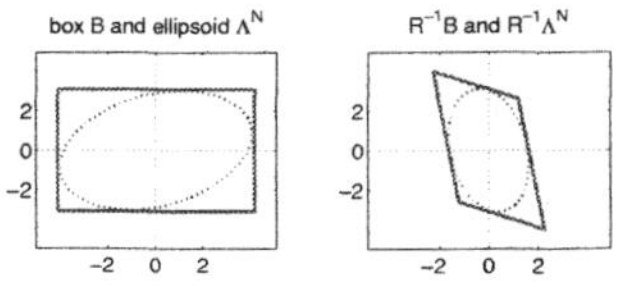

Fig. 2. *Confidence regions for $\mathcal{D}^{cc}$ and $\mathcal{D}^{IV}$: transformation by R^{-1} of a box B and of an ellipsoid associated with Λ^N.*

error term $\frac{1}{\bar{N}}\sum_{t=t_s}^{N}\zeta_k(t)v(t)$. While in $\mathcal{D}^{cc}$ only the size of the variance at each constraint k can be taken into account, in $\mathcal{D}^{IV}$ also the correlation of the errors $\frac{1}{\bar{N}}\sum_{t=t_s}^{N}\zeta_k(t)v(t)$ over all n constraints is incorporated by means of the covariance matrix Λ^N.

$\mathcal{D}^{IV}$ and $\mathcal{D}^{cc}$ are confidence regions containing θ_0 at a certain probability level when the true system is in the model class. However, in the presence of undermodelling or initial conditions similar confidence regions as (14) can be found. The size of the ellipsoid $\mathcal{E}\left(c_\alpha \hat{\Lambda}^N\right)$ and the hyperbox $\mathcal{B}(B)$ simply have to be increased as indicated in sections 4.2.2 and 4.3.

In the next section some special cases will be presented further illustrating the link between the two sets.

6. SPECIAL CASES

In some special cases the connection between the ellipsoidal $\mathcal{D}^{IV}$ and the boxed $\mathcal{D}^{cc}$ is particularly straightforward and elucidating.

6.1 Λ^N *is diagonal*

If the errors induced by $\frac{1}{\bar{N}}\sum_{t=t_s}^{N}\zeta_k(t)v(t)$ are uncorrelated over k, the two methods are both based on variances of the error term only. As such they would use the same information to construct the confidence region. For example, assume $i)$ $\bar{G}_0(q)u(t) = 0$ and $u^-(t) = 0$ for all t, $ii)$ $v(t)$ is white noise, i.e. $R_v(\tau) = 0$ for $\tau \neq 0$ and $iii)$ the signal $\zeta_k(t)$ is such that $R^N_{\zeta_i\zeta_j}(\tau) = 0$ for $i \neq j$. Expressions (13) then shows Λ^N to be diagonal. The third condition could be achieved exactly by a custom made sequence $\zeta_k(t)$, or asymptotically in N by a realization of a stochastic white noise sequence $\zeta_k(t)$ (e.g. $F_k(q) = q^{-k}$ (FIR) and $r(t)$ a white noise sequence).
Figure 3 illustrates how the smallest hyperbox $\mathcal{D}^{cc}$ embedding $\mathcal{D}^{IV}$ is obtained by taking $b_l(k) = -b_u(k) = \sqrt{c_\alpha \lambda_k}$, where the λ_k are the entries (variances) of the now diagonal matrix Λ^N.
Alternatively, the hyperbox $\mathcal{D}^{cc}$ could be chosen to have the same probability level of $\alpha\%$ as the ellipsoidal region $\mathcal{D}^{IV}$. To that end, take $b_l(k) = -b_u(k) = \left(c_{\mathcal{N},\sqrt[n]{\alpha}}\right)\sqrt{\lambda_k}$, where $c_{\mathcal{N},\sqrt[n]{\alpha}}$ corresponds to a probability in a one-dimensional standard Gaussian distribution $\mathcal{N}(0,1)$, such that for $x \in \mathcal{N}(0,1)$ prob$\left(|x| \leq c_{\mathcal{N},\sqrt[n]{\alpha}}\right) = \sqrt[n]{\alpha}$.

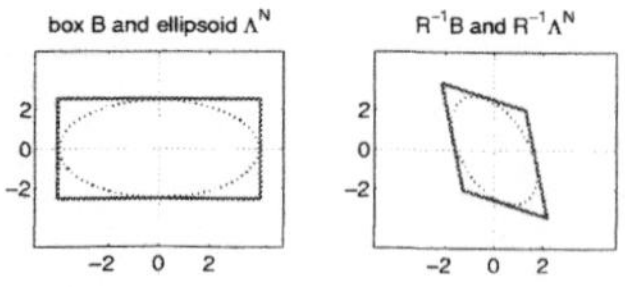

Fig. 3. *Confidence regions for $\mathcal{D}^{cc}$ and $\mathcal{D}^{IV}$ in the special case that Λ^N is diagonal.*

Two parameter case

Another special alignment of $\mathcal{D}^{cc}$ and $\mathcal{D}^{IV}$ occurs in case $i)$ $n = 2$, $ii)$ $\bar{G}_0(q)u(t) = 0$ and $u^-(t) = 0$ for all t, $iii)$ $v(t)$ is white noise, i.e. $R_v(\tau) = \sigma_v^2$ for $\tau = 0$ and zero otherwise, $iii)$ $r = u$ and $iv)$ $F_k(q)$ the FIR-basis. Then $\Lambda^N = \frac{\sigma_v^2}{N} R$ is a 2×2 Toeplitz matrix with $R_{12} = R_{21}$ and $R_{11} = R_{22}$. Due to this special structure, the eigenvectors of R are in the direction of $[1, 1]^T$ and $[1, -1]^T$. Consequently, the ellipsoid associated with Λ^N has its principal axis at 45 and -45 degrees. If B is chosen to define a square box by $b_l(1) = b_l(2) = -b_u(1) = -b_u(2) = b$, the diagonals of this box are aligned with the principal axis of the ellipsoid. Now, the subsequent operation on this square by R^{-1} in expression (14) consists of a rotation again over 45 degrees, a dilation and a subsequent rotation over 45 degrees. Figure 4 illustrates that this results into a rotated rhombus $\mathcal{D}^{cc}$ with its diagonals still aligned with the principal axis of the ellipsoid $\mathcal{D}^{IV}$. The ratio between the diagonals and the principal axis is $\lambda_R 2\sqrt{2}b : \frac{\sigma_v}{\sqrt{N}}\sqrt{\lambda_R}$.

7. SOME REMARKS AND FURTHER RESEARCH

Undermodelling can also be incorporated in the cross-correlation constraints similarly to section 4.2.2. If, on the other hand, the bounds $b(k)$ are chosen based on the properties of $v(t)$ only, the cross-correlation constraints reflect the standard validation test in system identification. From the arguments so far, it is clear that this validation test might suffer from the fact that correlation between the cross-correlation terms is not taken into account. The results also show that any undermodelling effect smaller than the effect of the noise $v(t)$ might not be detected for a particular noise realization. This in particular since the bounds are based on estimated properties of the noise.

8. CONCLUSIONS

Standard Instrumental Variable system identification and a cross-correlation parameter bounding method are strongly related. The ellipsoidal region associated with the IV estimation technique can be tightly linked to the polytopic region induced by the cross-correlation constraints as both techniques are based on the same set of regressors. However, they differ due to the fact that the former incorporates a covariance

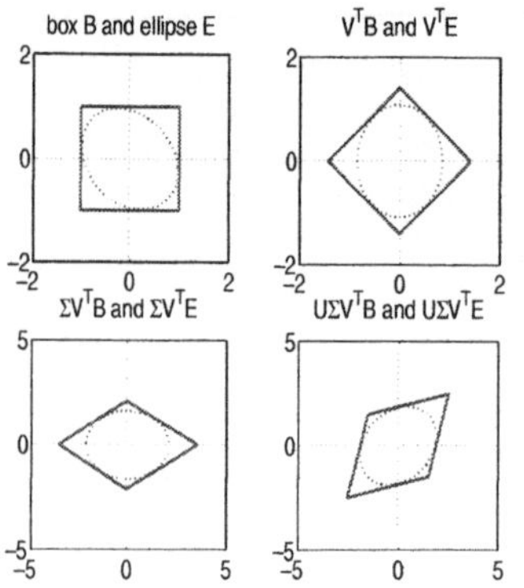

Fig. 4. *Confidence regions for $\mathcal{D}^{cc}$ and $\mathcal{D}^{IV}$ in the special case that $u(t) = r(t)$, $n = 2$ and Λ^N is diagonal.*

between errors, while the latter is limited to the variances of the errors only. Both methods can incorporate undermodelling effects by enlarging their confidence regions with upper bounds on the bias effects. The results hold asymptotically for ARX models too. The insights into the connection between the two methods also apply for the standard practice of model validation.

9. REFERENCES

Broman, V. and M.J. Shensa (1990). A compact algorithm for the intersection and approximation of n-dimensional polytopes. *Mathematics and Computers in Simulation* **32**, 469–480.

Hakvoort, R.G. and P.M.J. Van den Hof (1995). Consistent parameter bounding identification for linearly parametrized model sets. *Automatica* **31**(7), 957–969.

Hakvoort, R.G. and P.M.J. Van den Hof (1997). Identification of probabilistic uncertainty regions by explicit evaluation of bias and variance errors. *IEEE Trans. Autom. Control* **42**(11), 1516–1528.

Heuberger, P.S.C., P.M.J. Van den Hof and O.H. Bosgra (1995). A generalized orthonormal basis for linear dynamical systems. *IEEE Trans. Autom. Control* **40**, 451–465.

Ljung, L. (1999). *System Identification: Theory for the User (2nd ed.)*. Prentice-Hall, Upper Saddle River, New Jersey, USA.

Milanese, M. and A. Vicino (1991). Optimal estimation theory for dynamic systems with set membership uncertainty: an overview. *Automatica* **27**, 997–1009.

Söderström, T. and P. Stoica (1989). *System Identification*. Prentice-Hall, Hemel Hempstead, UK.

Veres, S.M. and J.P. Norton (1991). Structure selection for bounded parameter models: consistency conditions and selection criterium. *IEEE Trans. Autom. Control* **36**(7), 474–481.

IFAC Publications
www.elsevier.com/locate/ifac

EMPIRICAL ESTIMATION OF PARAMETER DISTRIBUTIONS IN SYSTEM IDENTIFICATION

Wayne J. Dunstan * **Robert R. Bitmead** *,[1]

* *Department of Mechanical & Aerospace Engineering, University of California San Diego, La Jolla CA 92093-0411, USA.*

Abstract: The distribution of parameter estimates from a finite data record is of concern for assessing the confidence in the resulting estimate. Our interest is in the development of nonlinear dynamical models from experimental data and the problem which arises in associating a degree of confidence with the estimated parameter values. If the distribution of the estimates were known then the variance might provide a sensible measure of confidence. Accordingly we consider here procedures for using a single data record to generate a distribution of parameter estimates.

Keywords: System Identification, Model Confidence, Parameter Distribution.

1. INTRODUCTION

We consider the empirical estimation of the distribution of a parameter estimate from a single data record. This is done by considering several data resampling schemes which are used to generate new data records or subsets of the experimental data from which sample parameter estimates are calculated. The resulting sample distribution of these estimates is used as an approximant of the underlying "true" distribution of the parameter estimate. This true distribution would be revealed in a sequence of independent Monte Carlo trials with repeated calculation of the parameter estimate. However, only the one data set is available and we seek to generate an approximate empirical distribution. The methods which we consider are Subsampling, Resampling and Bootstrapping techniques.

The contribution of this paper is in the reformulation of these techniques for empirical estimation of the distribution of the estimated parameters for **dynamical** system models. The consequence of this, is that our models are required to capture the correlation properties of the input-output data as well as possible. We shall see that this constrains the resampling schemes to preserve the underlying dynamical structure of the data. This results in using the prediction error residuals for the resampling scheme or, in subsampling, the data set. The significance of our study is that it treats these resampling techniques in the context of dynamical modelling.

Finite length data parameter estimates have a level of variability due to the stochastic nature of the data and due to the finite record lengths. Successive trials of the same size would yield different values of the parameter estimate and, if stationary, these values are distributed according to an underlying true distribution. Quantifying the properties of this distribution is an important step in assessing model confidence. A small variance of the parameter estimate indicates high confidence in the model parameters — or at least a high probability that a repeated trial with a similar but independent data set would yield a close parameter value.

This paper proceeds as follows. Section 2 defines the problem of interest. Section 3 describes different methods for empirically estimating the parameter estimate distributions. Section 4 is an example illustrating

[1] Research supported by USA National Science Foundation Grant ECS-0070146 and DARPA Grant N00014-00-1-0799

these methods. Section 5 discusses further extensions of these techniques in System Identification. Section 6 relates the parameter estimate distribution to the overall assessment of model confidence. Some concluding remarks set forth the authors' direction of future research in this area.

2. PROBLEM STATEMENT

A parametric model, $\hat{P}(\theta)$, may be fitted by minimizing a measure of fit, $V(\theta, \mathcal{Z})$, over a parameter set, $\theta \in \Theta$, and a data set, $\mathcal{Z}$. Formally,

$$\hat{\theta}(\mathcal{Z}) = \arg\min_{\theta \in \Theta} V(\theta, \mathcal{Z}), \tag{1}$$

where $\hat{\theta}(\mathcal{Z})$ is the minimizing parameter estimate and,

$$\mathcal{Z} = (z_1, z_2, ..., z_N), \tag{2}$$

and, $z_k = \begin{pmatrix} u_k \\ y_k \end{pmatrix}$ where u_k and y_k, $k = 1, 2, ..., n$ are the input and output vectors of the plant, $P(\theta)$, being identified.

Due to the stochastic nature of experimental data, independent data sets $(\mathcal{Z}^1, \mathcal{Z}^2, ...)$ may yield different minimizing parameter estimates $(\hat{\theta}^1, \hat{\theta}^2, ...)$. These estimates are assumed to originate from the same underlying distribution, $F(\hat{\theta})$. Our aim is to calculate estimates of the statistics of the distribution $F(\hat{\theta})$, such as the variance. For some of the schemes discussed, this will involve computing explicitly an empirical estimate of the distribution function $\hat{F}(\hat{\theta})$.

The approach to producing an $\hat{F}(\hat{\theta})$ is to create a sequence of new data sets $\mathcal{X}^i, i = 1, 2, ..., M$, from which to compute extra parameter estimates which, under some assumptions, should be distributed as the underlying $\hat{\theta}$. That is, we seek to produce from the experimental data set, $\mathcal{Z}$, new input-output sets which we combine into each $\mathcal{X}^i$. Each of these sets should possess similar underlying correlation structure to $\mathcal{Z}$ but should explore a different stochastic variation — thereby generating an estimate of the distribution of identified dynamic model parameters.

3. METHODS FOR PARAMETER DISTRIBUTION ESTIMATION

The approach to produce new sets $\mathcal{X}^i$ will be based on the single data record $\mathcal{Z}$ of length N. To achieve this we shall consider four techniques,

1. Monte Carlo,
2. Subsampling,
3. Resampling, and
4. Bootstrapping.

Because we shall be using a sole $\mathcal{Z}$, we need to make the following assumptions.

Assumption 1. The data is derived from a stationary process.

Assumption 2. The data are representative of the process variability.

That is, we need to make an ergodic type assumption that the experimental data set is distributed according to the underlying process and that it is sufficiently long to display both the correlation required for system identification and the stochastic variability in the data.

3.1 Monte-Carlo

The Monte-Carlo procedure involves repeating the the experiment M times with independent data sets $(\mathcal{Z}^1, ..., \mathcal{Z}^M)$, which yield different parameter estimates $(\hat{\theta}^1, ..., \hat{\theta}^M)$. These estimates can then be used to create the empirical distribution $F(\hat{\theta})$ or an estimate of its variance.

We introduce Monte Carlo methods here to provide a point of reference for the subsequent schemes. The parameter estimates, $\hat{\theta}^i$, are from the underlying distribution. However, the conduct of such a number of independent trials would require more data than our single set. Accordingly, we consider three different approaches to this problem of generating new data sets based on the single data set $\mathcal{Z}$.

3.2 Subsampling

From the experimental data set $\mathcal{Z} = (z_1, ..., z_N)$ we construct M subsampled data sets $\mathcal{X}_S^1, ..., \mathcal{X}_S^M$ of length W by selecting W-long contiguous subsets of $\mathcal{Z}$. Thus, if $(j_1, j_2, ..., j_M)$ are members of the set $(1, 2, ..., N-W+1)$ then

$$\mathcal{X}_S^i = (z_{j_i}, z_{j_i+1}, ..., z_{j_i+W-1}). \tag{3}$$

There are $N-W+1$ possible such sequences extractable from $\mathcal{Z}$.

Since the subsampled sets $\mathcal{X}_S^i$ are taken from the actual data set $\mathcal{Z}$, they are distributed identically to the original data and possess identical correlation properties. Their length is reduced, however, since $W < N$. Normally one would select $W \sim \mathcal{O}(\sqrt{N})$.

For each $\mathcal{X}_S^i$ a corresponding $\hat{\theta}_S^i$ is identified. The resulting set of parameters $(\hat{\theta}_S^1, ..., \hat{\theta}_S^M)$ are distributed differently from $\hat{\theta}$ because the sample sizes are different. However, subject to assumptions to be clarified shortly, the distributions of $\hat{\theta}_S^i$ and $\hat{\theta}$ are related. Indeed, their means should be the same and their standard deviations scaled by $\frac{1}{\sqrt{W}}$ and $\frac{1}{\sqrt{N}}$ respectively. Provided N and W are large enough, a Central Limit Theorem holds which characterizes the large sample parameter estimate distribution (Ljung, 1999; Soderstrom and Stoica, 1989; Hjalmarsson and Ljung, 1992).

The competing phenomena of the subsampling approach are the limited number $(N-W+1)$ of subsampled trials possible and the requirement for W to be sufficiently long for $\hat{\theta}_S^i$ to exhibit the correct mean and variance behavior. In terms of the data properties, this requires that the correlation and cross-correlation matrices derived from the W-long subsampled sets be close to their stationary values with an error described by the Central Limit Theorem. Thus a sufficiently large W would be determined by the convergence rates for the Law of Large Numbers and the Central Limit Theorem associated with the data. This in turn is linked to strong mixing properties, see (Politis and Romano, 1994).

With subsampling, it is apparent that many subsampled sequences overlap strongly and will yield highly correlated parameter estimates. However, the mixing over the total sample length N together with the order-invariant calculation of the empirical distribution, should see the $\hat{\theta}_S^i$ display the scaled variance properties of the N-sample Monte Carlo distribution. The balance struck between the competing phenomena of sample size, sample number and asymptotic independence is usually to take $W \sim \mathcal{O}(\sqrt{N})$. This is further discussed by Politis (1998).

3.3 Resampling

In resampling we create M new *pseudo-data* sets by randomly sampling from the original N data to form records of length N. Thus,

$$\mathcal{X}_R^i = (z_{i_1}, z_{i_2}, ..., z_{i_N}), \tag{4}$$

where the indices i_j are selected independently from a uniform distribution of the integers 1 through N. This resampling may be done with replacement or without replacement, in which the latter case corresponds to a permutation set of samples from the original data, referred to as the *Jackknife* (Turkey, 1958; Efron, 1982).

A precept of the resampling method and of the subsequent bootstrapping method is that they require an additional assumption.

Assumption 3. The data set is composed of independent and identically distributed (i.i.d.) samples.

This requirement precludes the possibility of directly applying such a scheme to the sampled input-output data because they are (hopefully) correlated over time. Indeed, our interest is in identifying a dynamical model for the system generating the data. Inclusion of dependent data structures in the resampling and upcoming bootstrap methods normally involves deriving an i.i.d. sequence from the data set (Freedman, 1944; Ljung, 1999; Tjänström, 2000*a*). For this we use the prediction error residual sequence derived from a high-order model.

Suppose that the data, u_k and y_k, are generated by a linear system with zero mean i.i.d. excitation process e_k. That is

$$y_k = G(z)u_k + H(z)e_k,$$

where $G(z)$ and $H(z)$ are causal, stable linear systems with $H(z)$ stably invertible. Then we may define the associated one-step-ahead prediction of y_k,

$$\hat{y}_{k|k-1} = \hat{H}(z)^{-1}\hat{G}(z)u_k + (1 - \hat{H}(z)^{-1})y_k$$

and its associated prediction error sequence of residuals,

$$\epsilon_k = \hat{H}(z)^{-1}[y_k - \hat{G}(z)u_k].$$

This residual sequence should be asymptotically white and should yield $\epsilon_k \to e_k$ as $k \to \infty$. The sequence ϵ_k will form the basis of the resampling process. We note that the prediction error generation process is simply invertible;

$$y_k = \hat{G}(z)u_k + \hat{H}(z)\epsilon_k.$$

Denote the computed residual set $\mathcal{E} = (\epsilon_1, ..., \epsilon_N)$, and consider a resampling of this nominally i.i.d. process to yield a new set $\mathcal{E}_R = (\epsilon_{i_1}, ..., \epsilon_{i_N})$. Here the indices i_j are chosen randomly from a uniform distribution over the integers 1 through N. This then is our resampled residual sequence, whose elements we denote ϵ_k^R.

From $\mathcal{E}_R$ we compute the resampled output sequence,

$$y_k^R = \hat{G}(z)u_k + \hat{H}(z)\epsilon_k^R. \tag{5}$$

Define the pseudo-data element, $z_k^R = \begin{pmatrix} u_k \\ y_k^R \end{pmatrix}$, and construct the pseudo-data set $\mathcal{X}_R = (z_1^R, ..., z_N^R)$. This data record is then used to produce a parameter estimate $\hat{\theta}_R$.

Repeating this residual resampling process M times yields a sequence of M parameter estimates, which then may be used to compute the mean and variance of $\hat{\theta}_R^i$. Because the residual sequence tends to the true driving noise and this noise is the source of variability of the parameter estimate, this procedure yields another way to generate an approximate distribution of $\hat{\theta}$.

An interesting feature of this approach is that it explicitly relies on knowing a high fidelity model of the underlying process. This, in turn would seem to obviate the need for system identification. However, we shall see that (following the work of Tjänström (2000*b*)) one may sensibly combine low and high order models to gain an appreciation of model confidence.

3.4 Bootstrapping

Bootstrapping, first proposed by Efron (1979), is another procedure for generating pseudo-data sets from a single data set and is akin to resampling. It also requires an i.i.d. candidate data set. However, in bootstrapping one computes explicitly the distribution of the residuals ϵ_i, $Q(\varepsilon) = \text{Prob}(\epsilon_i \leq \varepsilon)$, where as before this vector inequality is taken componentwise.

Rather than resampling the residuals from the original population, bootstrapping uses the empirical distribution function, $Q(\varepsilon)$, to generate completely new pseudo-residuals ϵ_k^B with the same distribution. Theoretically, since we are generating completely new i.i.d. pseudo-residual sets, there is no limit to the number of possible new sequences ϵ^B. This should be compared with resampling where there is an absolute (but usually large) limit of N^N possible sequences with replacement and $N!$ sequences without replacement. This is further discussed by Politis (1998) and Zoubir and Boashash (1998).

With bootstrapped sequence ϵ^B we now generate bootstrapped output data,

$$y_k^B = \hat{G}(z)u_k + \hat{H}(z)\epsilon_k^B.$$

Define the pseudo-data element $z_k^B = \begin{pmatrix} u_k \\ y_k^B \end{pmatrix}$, and construct the pseudo-data set $\mathcal{X}_B = (z_1^B, ..., z_N^B)$. This data record is then used to produce a parameter estimate $\hat{\theta}_B$, which in turn is used to compute the empirical distribution of $\hat{\theta}$.

Similarly to resampling, bootstrapping requires a high-fidelity model of the underlying process. It also necessitates the computation of an explicit estimate of the residual distribution function $Q(\varepsilon)$ as an intermediate step.

4. EXAMPLE

We use a System Identification example to compare the following methods,
1. Monte-Carlo,
2. Resampling, and,
3. Subsampling.

Consider a **plant** which is a second order Auto-Regressive (AR) process with measurement noise,

$$\begin{bmatrix} x_{1,k+1} \\ x_{2,k+1} \end{bmatrix} = \alpha \begin{bmatrix} cos(\omega) & -sin(\omega) \\ sin(\omega) & cos(\omega) \end{bmatrix} \begin{bmatrix} x_{1,k} \\ x_{2,k} \end{bmatrix} + (1-\alpha^2)^{\frac{1}{2}} \begin{bmatrix} sin(\omega) \\ -cos(\omega) \end{bmatrix} e_{1,k} \quad (6)$$

$$y_k = \begin{bmatrix} 1 & 0 \end{bmatrix} \begin{bmatrix} x_{1,k} \\ x_{2,k} \end{bmatrix} + r.\, e_{2,k}$$

where $e_{1,k}$ and $e_{2,k}$ are $\mathcal{N}(0,1)$ sequences of length N. The plant parameters are, $\alpha = 0.8$, $\omega = 0.2$, $r = 1.0$, and $N = 250,000$. A single output data record was generated, $(y_1, ..., y_N)$. The first 1000 samples and Power Spectral Density estimate (PSD) of the entire record are shown in Figure (1). The **model** is of a second order AR process,

$$y_k = \frac{1}{1 + a_1 z^{-1} + a_2 z^{-2}} e_k, \quad (7)$$

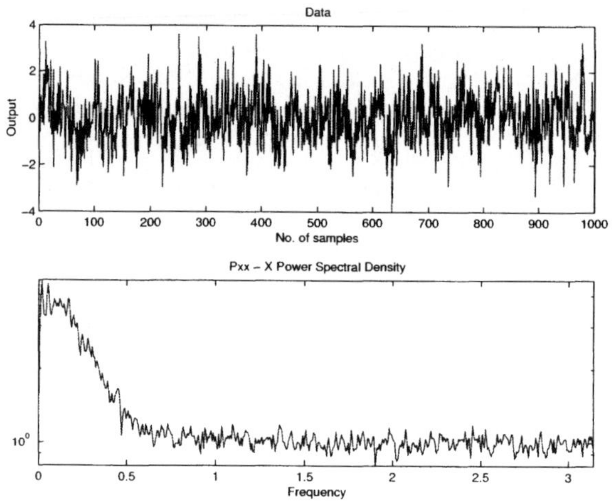

Fig. 1. First 1000 output samples (upper) and PSD of data record (lower).

with $\theta = \begin{bmatrix} a_1 \\ a_2 \end{bmatrix}$. Our objective is to compute empirical estimates of the variance of $\hat{\theta}$.

4.1 *Monte-Carlo*

The plant was used to generate $M = 10,000$ independent data sets of length N. A least squares measure was used to identify each $\hat{\theta}^i$. The empirical distribution estimates for $\hat{\theta}_i$ are shown the upper plots in Figure (2) and the variances in Table (1).

4.2 *Resampling*

The model was fitted to the single plant data set using a least squares estimator. The residual set was resampled $10,000 (= M)$ times. Each resampled residual set of length N was simulated using the parametrized model to produce a resampled output set. New parametrizations were fitted to each resampled output set to identify $\hat{\theta}_R^i$. The empirical distribution estimates for $\hat{\theta}_R^i$ are shown the middle plots in Figure (2) and the variances in Table (1).

4.3 *Subsampling*

The segment size, W, was set at $500 (= \sqrt{N})$, which created $249,501 (= N - W + 1)$ contiguous segments from the single plant data record. Parametrizations were fitted to each data segment using a least squares estimator. The empirical distribution estimates for $\hat{\theta}_S^i$ are shown the lower plots in Figure (2) with the x-axis scaled for visual comparison with the other distributions. The scaled variances are in Table (1).

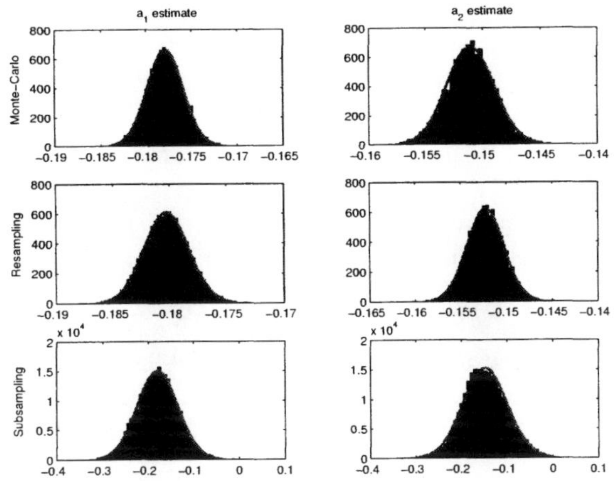

Fig. 2. Empirical distribution estimates for $\hat{\theta}^i$, $\hat{\theta}^i_R$ and $\hat{\theta}^i_S$.

Table 1. Variances of $\hat{\theta}^i$, $\hat{\theta}^i_R$ and $\hat{\theta}^i_S$.

Method	a_1 var	a_2 var
Monte-Carlo	0.0000039	0.0000043
Resampling	0.0000039	0.0000039
Subsampling	0.0000040	0.0000046

Notice that subsampling, requiring no assumptions of i.i.d. data, produced a distribution as close to the Monte Carlo distribution as when resampling was used. This will be paramount to establishing model confidence in systems with no access to white residuals.

5. EXTENSIONS TO SYSTEM IDENTIFICATION

System Identification typically involves producing a low order (L.O.) model that captures the system dynamics present in the data. This objective appears inconsistent with resampling or bootstrapping techniques, which use a high order (H.O.) model to generate i.i.d. pseudo-data. However, we will now outline how L.O. and H.O. models can be used in a *two-stage* fashion to satisfy both objectives.

5.1 L.O./H.O. Method

A two-stage method by Tjänström (2000*b*) makes connections to Model Error Modelling (Ljung, 1999) by calculating $\hat{F}(\hat{\theta})$ as follows.

The data, (u_k, y_k), is fitted to a candidate L.O. model which results in non-white residuals,

$$\zeta_k = \hat{H}_1(z)^{-1}(y_k - \hat{G}_1(z)u_k). \tag{8}$$

The set (u_k, ζ_k) is then fitted a H.O. model which results in white (i.i.d.) residuals,

$$\epsilon_k = \hat{H}_2(z)^{-1}(\zeta_k - \hat{G}_2(z)u_k), \tag{9}$$

The residual set $\mathcal{E} = (\epsilon_1, ..., \epsilon_N)$ is bootstrapped to produce the set $\mathcal{E}_B = (\epsilon_1^B, ..., \epsilon_N^B)$.

From $\mathcal{E}_B$ the bootstrapped output sequence is computed in two steps,

$$\zeta_k^B = \hat{G}_2(z)u_k + \hat{H}_2(z)\epsilon_k^B \tag{10}$$

$$y_k^B = \hat{G}_1(z)u_k + \hat{H}_1(z)\zeta_k^B. \tag{11}$$

The resulting pseudo-data set $\mathcal{X}_B = (z_1^B, ..., z_N^B)$ where $z_k^B = \begin{pmatrix} u_k \\ y_k^B \end{pmatrix}$, is used to produce a parameter estimate $\hat{\theta}_B$.

Repeating this residual bootstrapping process M times yields a sequence of M parameter estimates, which then may be used to compute the mean and variance of $\hat{\theta}^i_B$.

Because this method fits the L.O. model first, the mean of $\hat{\theta}^i_B$ will be constrained to the $\hat{\theta}$ of the first L.O. model fit. This however may be of little concern if the goal is estimation of the variance.

5.2 H.O./L.O. Method

This method is similar to the previous one, but does not constrain the mean of $\hat{F}(\hat{\theta})$. It also uses two models, but in the reverse order.

The data, (u_k, y_k), is fitted to a H.O. model which results in white (i.i.d.) residuals,

$$\epsilon_k = \hat{H}_1(z)^{-1}(y_k - \hat{G}_1(z)u_k), \tag{12}$$

The residual set $\mathcal{E} = (\epsilon_1, ..., \epsilon_N)$ is bootstrapped (or resampled) to produce the set $\mathcal{E}_B = (\epsilon_1^B, ..., \epsilon_N^B)$.

From $\mathcal{E}_B$ the bootstrapped output sequence is computed,

$$y_k^B = \hat{G}_1(z)u_k + \hat{H}_1(z)\epsilon_k^B. \tag{13}$$

The resulting pseudo-data set $\mathcal{X}_B = (z_1^B, ..., z_N^B)$ where $z_k^B = \begin{pmatrix} u_k \\ y_k^B \end{pmatrix}$, is fitted to a L.O. model,

$$y_k^B = \hat{G}_2(z)u_k + \hat{H}_2(z)\varepsilon_k, \tag{14}$$

where ε_k is an independent white noise sequence. This produces a parameter estimate $\hat{\theta}_B$.

Repeating this "residual bootstrapping and L.O. model fitting" process M times yields a sequence of M parameter estimates, which then may be used to compute the mean and variance of $\hat{\theta}^i_B$.

6. MODEL CONFIDENCE

Our ultimate aim in using these techniques so far discussed is in the assessment of model confidence. Model confidence is related to how uniquely identifiable the best-fit parameter values are from the data and this is captured in part by the distribution of the parameter estimate.

It can be shown under some quite general conditions (P. Stoica and Kay, 1987; Hjalmarsson, 1993) that $\hat{\theta}(V)$ converges to a point $\theta^*(V)$, defined,

$$\theta^*(V) = \lim_{N \to \infty} \hat{\theta}(V, \mathcal{Z}). \qquad (15)$$

Hjalmarsson (1993) shows that

$$\sqrt{N}(\hat{\theta}(V, \mathcal{Z}) - \theta^*(V)) \sim As\,\mathcal{N}(0, P) \qquad (16)$$

where,

$$P = [\bar{V}''(\theta^*)]^{-1} \sum_{\tau=-\infty}^{\infty} R(\tau)\ [\bar{V}''(\theta^*)]^{-1} \qquad (17)$$

where $R(\cdot)$ is the covariance function of the loss function, $\{l'(t, \theta^*)\}$.

The $[\bar{V}''(\theta^*)]^{-1}$ terms in Equation(17) are the inverse sensitivity of the minimizer of V_N at the parameter value $\hat{\theta}$.

It would seem reasonable to want high measure sensitivity to variations in the parameter values, i.e. $\partial^2 V / \partial \theta^2$ large, as this would mean small perturbations in the minimizer estimate adversely affect the measure. This property is considered in Dunstan and Bitmead (2002).

The covariance function, $R(\tau)$, in Equation(17) is the sensitivity of $\hat{\theta}$ to variabilities in the data set, $\mathcal{Z}$. This variability, due to finite length data and the underlying stochastic processes, we have explored in the distributions of the parameter estimate.

Hence these two properties seem a reasonable beginning in assessment of model confidence in System Identification.

7. CONCLUSION

This paper has formulated some methods for empirical estimation of parameter distributions from finite data records for dynamic models. These methods have been shown applicable in the overall assessment of model confidence.

Future work by the authors' will attempt to apply these results to the practical problem of combustion instability modelling. The model considered for this application is both nonlinear and stochastically driven, and so empirical quantification of the confidence measures seems most to hold the most promise.

REFERENCES

Dunstan, W.J. and R.R. Bitmead (2002). Model confidence fo nonlinear systems. *IFAC World Congress on Automatic Control, Barcelona, Spain.*

Efron, B. (1979). Bootstrap methods: Another look at the jackknife. *The Annals of Statistics* **7**, 1–26.

Efron, B. (1982). The jackknife, the bootstrap, and other resampling plans. *SIAM NSF-CBMS, Monograph 38.*

Freedman, D. (1944). On bootstrapping two-stage least squares estimates in stationary linear models. *The Annals of Statistics* **12**(3), 827–842.

Hjalmarsson, H. (1993). Aspects on Incomplete Modeling in System Identification. Phd thesis. Linköping University. 409 Department of Electrical Engineering, Linköping, Sweden.

Hjalmarsson, H. and L. Ljung (1992). Estimating model variance in the case of undermodeling. *IEEE Transactions on Automatic Control* **37**(7), 1004–1008.

Ljung, L. (1999). *System Identification: theory for the user*. 2nd ed.. Prentice-Hall, Englewood Cliffs.

P. Stoica, A. Nehorai and S.M. Kay (1987). Statistical analysis of the least squares autoregressive estimator in the presence of noise. *IEEE Transactions on Acoustics, Speech and Signal Processing* **35**(9), 1273–1281.

Politis, D.N. (1998). Computer-intensive methods in statistical analysis. *IEEE Signal Processing Magazine* **15**, 39–55.

Politis, D.N. and J.P. Romano (1994). Large sample confidence regions based on subsamples under minimal assumptions. *The Annals of Statistics* **22**(4), 2031–2050.

Soderstrom, T. and P. Stoica (1989). *System Identification*. Prentice-Hall International (UK).

Tjänström, F. (2000*a*). Computing uncertainty regions with simulataneous confidence degree using bootstrap. *IFAC System Identification, Santa Barbara, California, USA.*

Tjänström, F. (2000*b*). Quality Estimation of Approximate Models. Phd thesis. Linköping University. Department of Electrical Engineering, Linköping, Sweden.

Turkey, J.W. (1958). Bias and confidence in not quite large samples. *Annals of Mathematical Statistics* **29**, 614.

Zoubir, A.M. and B. Boashash (1998). The bootstrap and its application in signal processing. *IEEE Signal Processing Magazine* **15**, 56–76.

IFAC Publications
www.elsevier.com/locate/ifac

UNCERTAINTY OF TRANSFER FUNCTION MODELING USING PRIOR ESTIMATED NOISE MODELS

R. Pintelon, J. Schoukens, and Y. Rolain

Vrije Universiteit Brussel, Department ELEC, Pleinlaan 2, 1050 Brussel, Belgium

Abstract: Assuming small model errors (unmodelled dynamics and/or nonlinear distortions) and large signal-to-noise ratios we derive in this paper explicit expressions for the covariance matrix $\mathrm{Cov}(\hat{\theta}(Z))$ of a frequency domain estimator using prior estimated noise models. These analytic expressions (i) give a clear insight in the behaviour of $\mathrm{Cov}(\hat{\theta}(Z))$ as a function of the signal-to-noise ratio, the plant model errors (unmodelled dynamics and the nonlinear distortions), and (ii) allow to predict accurately the order of magnitude of the actual uncertainty of the estimates. The link with the classical prediction error approach is also established. *Copyright © 2003 IFAC*

Keywords: frequency domain, non-parametric noise model, model errors, uncertainty.

1. INTRODUCTION

Since real life systems are mostly distributed and/or nonlinear, the model errors (unmodelled dynamics and/or the nonlinear distortions) are often the limiting factor in transfer function modeling problems. The influence of model errors on the asymptotic properties (amount of data going to infinity) of the estimated plant model parameters $\hat{\theta}(Z)$ have been studied in Ljung (1999) and Pintelon and Schoukens (2001), and general closed form expressions for the asymptotic covariance matrix $\mathrm{Cov}(\hat{\theta}(Z))$ are available. The difficulty is that these expressions are not tractable in the presence of model errors. Numerical methods for calculating $\mathrm{Cov}(\hat{\theta}(Z))$ are described in Hjalmarsson and Ljung (1992) and Tj rnstr m and Ljung (1999), and a qualitative study of the influence of model errors on $\mathrm{Cov}(\hat{\theta}(Z))$ is available in Pintelon and Schoukens (2002).

This work is supported by the Belgium National Fund for Scientific Research; the Belgian Government as a part of the Belgian Programme on Interuniversity Poles of Attraction (IUAP5) initiated by the Belgian State, Prime Minister's Office, Science Policy Programming; and the Flemish Community (GOA-IMMI).

The main contribution of this paper is to derive, under some suitable assumptions concerning the disturbing noise and the plant model errors, approximate, easy to use analytic expressions for the asymptotic covariance matrix $\mathrm{Cov}(\hat{\theta}(Z))$ of a frequency domain estimator using prior estimated noise models. These analytic expressions (i) give a clear insight in the behaviour of $\mathrm{Cov}(\hat{\theta}(Z))$ as a function of the disturbing noise level, and the plant model errors, (ii) allow accurate prediction of the order of magnitude of the actual uncertainty of the estimates; and (iii) establish a link between transfer function modeling using prior estimated noise models and the classical prediction error framework.

Due to space limitations the proof of the theorem is not included in this paper. The reader is referred to Pintelon *et al.* (2002) for the complete version. It includes a precise definition of the nonlinear systems for which the theory applies.

2. THE IDENTIFICATION FRAMEWORK

The asymptotic covariance matrix $\mathrm{Cov}(\hat{\theta}(Z))$ is analysed for two frequency domain estimators: (i) the maximum likelihood (ML) method which assumes

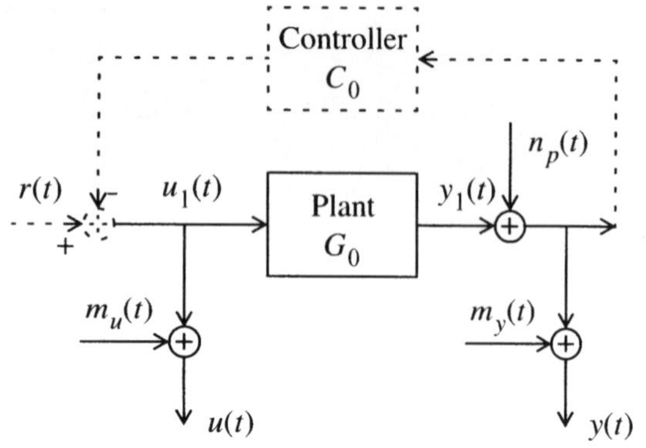

Fig. 1.: Identification of a plant in open loop (solid line) or in closed loop (solid and dashed lines). $r(t)$ is the reference signal; $u_0(t)$, $y_0(t)$ the input-output signals; $m_u(t)$, $m_y(t)$ the input-output measurement errors; and $n_p(t)$ the process noise.

that the noise model is known, and (ii) the sample maximum likelihood (SML) method which estimates the noise model from the data in a pre-processing step (see Pintelon and Schoukens, 2001 for the details). Both methods estimate the model parameters θ from input-output observations Z of the steady state response of the system to a periodic excitation $r(t)$ (see Fig. 1.). For the open loop experiment (see Fig. 1., solid line) the plant may be nonlinear, so that the plant model errors are due to unmodelled dynamics and/or nonlinear distortions. For the closed loop experiment (see Fig. 1., solid and dashed lines) we explicitly assume that the plant is linear so that in this case the plant model errors are due to unmodelled dynamics only. The following two classes of excitations are considered.

Definition 1 (excitation signals): (i) Deterministic excitations: periodic signals with a deterministic amplitude and phase spectrum (also called multisines); and (ii) random excitations: periodic signals with a random amplitude and phase spectrum (also called periodic noise).

The stochastic framework is a frequency domain errors-in-variables model

$$Y(k) = Y_0(k) + N_Y(k), \; U(k) = U_0(k) + N_U(k) \quad (1)$$

with $k = 1, 2, \ldots, F$; F the number of frequencies; $X(k)$ the DFT spectrum of $x(t)$

$$X(k) = N^{-1/2}\sum_{t=0}^{N-1} x(t)\exp(-j2\pi tk/N) \quad (2)$$

with $X = U, Y, \ldots, N_Y$, $x = u, y, \ldots, n_y$, and N the number of time domain samples. Since any deviation from the periodic behaviour is considered as noise, $U_0(k)$ and $Y_0(k)$ in (1) are the DFT spectra of the PERIODIC PART of the input-output signals $u_1(t)$ and $y_1(t)$. Hence,

$$\text{open loop: } \begin{cases} U_0 = U_1 \\ Y_0 = Y_1 \end{cases} \text{ and } \begin{cases} N_U = M_U \\ N_Y = N_P + M_Y \end{cases} \quad (3)$$

with M_U, M_Y the input/output measurement errors and N_P the process noise, and

$$\text{closed loop: } \begin{cases} U_0 = R/(1+G_0C_0) \\ Y_0 = RG_0/(1+G_0C_0) \end{cases} \text{ and}$$

$$\begin{cases} N_U = -N_PC_0/(1+G_0C_0) + M_U \\ N_Y = N_P/(1+G_0C_0) + M_Y \end{cases} \quad (4)$$

with R the reference signal, and G_0, C_0 the true plant and controller respectively (see Fig. 1.). Note that the input-output errors $N_U(k)$, $N_Y(k)$ in (1) are ALWAYS independent of $U_0(k)$ and $Y_0(k)$, even for a feedback experiment.

The maximum likelihood estimate $\hat{\theta}_{ML}(Z)$ minimizes

$$V_{ML}(\theta, Z) = \sum_{k=1}^{F} |\varepsilon(\Omega_k, \theta, Z(k))|^2 = \sum_{k=1}^{F} \frac{|Y(k) - G(\Omega_k, \theta)U(k)|^2}{\sigma_\varepsilon^2(\Omega_k, \theta)} \quad (5)$$

$$\sigma_\varepsilon^2(\Omega_k, \theta) = \sigma_Y^2(k) + \sigma_U^2(k)|G(\Omega_k, \theta)|^2 - 2\mathrm{Re}(\sigma_{YU}^2(k)\bar{G}(\Omega_k, \theta))$$

w.r.t. θ (overbar denotes the complex conjugate). Z contains the input/output DFT spectra at the excited frequencies

$$Z^T = [Z^T(1)\; Z^T(2)\; \ldots\; Z^T(F)],\; Z^T(k) = [Y(k)\; U(k)]$$

$G(\Omega_k, \theta)$ is the plant model (rational form in Ω); Ω the generalised frequency variable: $\Omega = z^{-1}$ for discrete-time systems, and $\Omega = s$ for continuous-time systems; and $\sigma_U^2(k)$, $\sigma_Y^2(k)$ and $\sigma_{YU}^2(k)$ are the known noise (co-)variances

$$\sigma_Y^2(k) = \mathcal{E}\{|N_Y(k)|^2\},\; \sigma_U^2(k) = \mathcal{E}\{|N_U(k)|^2\} \text{ and}$$
$$\sigma_{YU}^2(k) = \mathcal{E}\{N_Y(k)\bar{N}_U(k)\} \quad (6)$$

(given or obtained in a previous experiment).

Assuming that $M \geq 4$ independent repeated experiments are available (in practice M consecutive periods of the steady state response to a periodic excitation), the sample maximum likelihood estimate $\hat{\theta}_{SML}(Z)$ minimizes

$$V_{\mathrm{SML}}(\theta, Z) = \sum_{k=1}^{F} |\hat{\varepsilon}(\Omega_k, \theta, \hat{Z}(k))|^2$$
$$= \sum_{k=1}^{F} \frac{|\hat{Y}(k) - G(\Omega_k, \theta)\hat{U}(k)|^2}{\hat{\sigma}_{\hat{Y}}^2(\Omega_k, \theta)} \quad (7)$$
$$\hat{\sigma}_{\hat{Y}}^2(\Omega_k, \theta) = \hat{\sigma}_{\hat{Y}}^2(k) + \hat{\sigma}_{\hat{U}}^2(k)|G(\Omega_k, \theta)|^2 - 2\mathrm{Re}(\hat{\sigma}_{\hat{Y}\hat{U}}^2(k)\bar{G}(\Omega_k, \theta))$$

w.r.t. θ. Z contains the sample means of the M input/output DFT spectra at the excited frequencies

$$Z^T = [\hat{Z}^T(1)\ \hat{Z}^T(2)\ \dots\ \hat{Z}^T(F)],\ \hat{Z}^T(k) = [\hat{Y}(k)\ \hat{U}(k)]$$

where

$$\hat{Y}(k) = \frac{1}{M}\sum_{m=1}^{M} Y^{[m]}(k),\ \hat{U}(k) = \frac{1}{M}\sum_{m=1}^{M} U^{[m]}(k) \quad (8)$$

with $U^{[m]}(k)$, $Y^{[m]}(k)$ the input/output DFT spectra of the mth independent experiment (mth signal period), and $U_0^{[m]}(k) = U_0(k)$. $\hat{\sigma}_{\hat{U}}^2(k)$, $\hat{\sigma}_{\hat{Y}}^2(k)$ and $\hat{\sigma}_{\hat{Y}\hat{U}}^2(k)$ in (7) are the sample noise covariances (= non-parametric noise model)

$$M\hat{\sigma}_{\hat{Y}}^2(k) = \frac{1}{M-1}\sum_{m=1}^{M} |Y^{[m]}(k) - \hat{Y}(k)|^2$$
$$M\hat{\sigma}_{\hat{U}}^2(k) = \frac{1}{M-1}\sum_{m=1}^{M} |U^{[m]}(k) - \hat{U}(k)|^2 \quad (9)$$
$$M\hat{\sigma}_{\hat{Y}\hat{U}}^2(k) = \frac{1}{M-1}\sum_{m=1}^{M} [(Y^{[m]}(k) - \hat{Y}(k))\overline{(U^{[m]}(k) - \hat{U}(k))}]$$

Note that the prior estimated noise model (9) is independent of the estimated plant model.

3. ASYMPTOTIC UNCERTAINTY FOR INPUT/ OUTPUT MEASUREMENTS

3.1. Introduction

The maximum likelihood (ML) and the sample maximum likelihood (SML) estimators minimize a quadratic like cost function

$$V_F(\theta, Z) = \varepsilon^H(\theta, Z)\varepsilon(\theta, Z)/F \quad (10)$$

with superscript H the Hermitian transpose, $\varepsilon_{[k]}(\theta, Z) = \varepsilon(\Omega_k, \theta, Z(k))$ for the ML estimator (see eq. (5)), and $\varepsilon_{[k]}(\theta, Z) = \hat{\varepsilon}(\Omega_k, \theta, \hat{Z}(k))$ for the SML estimator (see eq. (7)). The noisy observations Z are related to the true values Z_0 as $Z = Z_0 + N_Z$ with N_Z a zero mean, circular complex ($\mathcal{E}\{N_Z N_Z^T\} = 0$) normally distributed disturbance with covariance matrix $\mathrm{Cov}(N_Z) = \mathcal{E}\{N_Z N_Z^H\}$

$$N_Z = [N_Y(1)\ N_U(1)\ \dots\ N_Y(F)\ N_U(F)]^T \text{ for ML}$$
$$N_Z = [\hat{N}_Y(1)\ \hat{N}_U(1)\ \dots\ \hat{N}_Y(F)\ \hat{N}_U(F)]^T \text{ for SML} \quad (11)$$

where N_U, N_Y are defined in (3) and (4). Define $\tilde{\theta}(Z_0)$ as the minimizer of the expected value of the cost function (10)

$$\tilde{\theta}(Z_0) = \arg\min_{\theta} V_F(\theta) \quad (12)$$

with $V_F(\theta) = \mathcal{E}\{V_F(\theta, Z)\}$. Under some suitable assumptions it can be shown that $\hat{\theta}(Z)$, the minimizer of (10), converges ($F \to \infty$) with probability 1 to $\tilde{\theta}(Z_0)$, and that $\sqrt{F}(\hat{\theta}(Z) - \tilde{\theta}(Z_0))$ is asymptotically ($F \to \infty$) normally distributed with zero mean and covariance matrix P_F

$$\sqrt{F}P_F^{-1/2}(\hat{\theta}(Z) - \tilde{\theta}(Z_0)) \in AsN(0, I_{n_\theta})$$
$$P_F = V_F''^{-1}(\tilde{\theta}(Z_0))Q_F(\tilde{\theta}(Z_0))V_F''^{-1}(\tilde{\theta}(Z_0)) \quad (13)$$
$$Q_F(\tilde{\theta}(Z_0)) = F\mathcal{E}\{V_F'^T(\tilde{\theta}(Z_0), Z)V_F'(\tilde{\theta}(Z_0), Z)\}$$

with ' the derivative w.r.t. θ, I_{n_θ} an $n_\theta \times n_\theta$ identity matrix, and $n_\theta = \dim(\theta)$ (Pintelon and Schoukens, 2001). As will be shown in the sequel of the paper, the fact that the expectations in (12) and (13) are taken w.r.t. the disturbing noise AND the excitation signal has an essential impact on the behaviour of P_F.

3.2. Main result

To approximate P_F in (13), suitable assumptions concerning the plant model errors $\Delta\tilde{G}(k) = G_0(\Omega_k) - G(\Omega_k, \tilde{\theta}(Z_0))$ and the disturbing noise N_Z must be made.

Assumption 1 (disturbing noise): The approximation of P_F is calculated for small noise levels: N_Z and $\mathrm{Cov}(N_Z)$ in (13) are replaced by υN_Z and $\upsilon^2\mathrm{Cov}(N_Z)$ with $\upsilon \to 0$. The noise $N_Z(k)$ is independent (over the frequency k) circular complex normally distributed. At least $M \geq 6$ independent repeated experiments are available where $Z^{[m]}(k) = Z_0(k) + N_Z^{[m]}(k)$, with $N_Z^{[m]}(k)$ independent of $Z_0(k)$, and where $Z_0(k)$ and $\mathrm{Cov}(N_Z^{[m]}(k))$ are independent of m.

Assumption 2 (plant model errors): The approximation of P_F is calculated for small plant model errors: $\Delta\tilde{G}(k)$ in (13) is replaced by $\mu\Delta\tilde{G}(k)$ with $0 < \mu \ll 1$. For random excitation signals we assume in addition that (the expectations are taken w.r.t. the noise AND the excitation):
(i) the model error $\varepsilon(\tilde{\theta}(Z_0), Z_0)$ is independent of the noise $\varepsilon(\tilde{\theta}(Z_0), N_Z)$,
(ii) $\varepsilon(\tilde{\theta}(Z_0), Z_0)$ is circular complex distributed: $\mathcal{E}\{\varepsilon(\tilde{\theta}(Z_0), Z_0)\varepsilon^T(\tilde{\theta}(Z_0), Z_0)\} = 0$,

(iii) the model error is white: $\mathcal{E}\{\varepsilon(\tilde{\theta}(Z_0), Z_0)\varepsilon^H(\tilde{\theta}(Z_0), Z_0)/F\} = m^2 I_F$ with I_F the $F \times F$ identity matrix and

$$m^2 = \frac{1}{F}\sum_{k=1}^{F} \mathcal{E}\{|\varepsilon(\Omega_k, \tilde{\theta}(Z_0), Z_0(k))|^2\}$$
$$= \frac{1}{F}\sum_{k=1}^{F} \frac{\mathcal{E}\{\Delta\tilde{G}(k)^2\}}{\sigma_Y^2(\Omega_k, \tilde{\theta}(Z_0))}\mathcal{E}\{|U_0(k)|^2\} \quad (14)$$

(iv) $\varepsilon(\tilde{\theta}(Z_0), Z_0)$ is uncorrelated with $\varepsilon'(\tilde{\theta}(Z_0), Z_0)$ ($' = \partial / \partial\theta$):

$$\mathcal{E}\{\varepsilon'^H \varepsilon\varepsilon^T \bar{\varepsilon}'\} = 0$$
$$\mathcal{E}\{\varepsilon'^H \varepsilon\varepsilon^H \varepsilon'\} = Fm^2\mathcal{E}\{\varepsilon'^H \varepsilon'\}. \quad (15)$$

The assumptions concerning the noise are standard in frequency domain identification and are asymptotically (for the number of time domain samples going to infinity) valid for a broad class of time domain disturbances, for example, filtered white noise (Brillinger, 1981 and Pintelon and Schoukens, 2001). The assumptions concerning the plant model errors are quite hard (uncorrelated, white and circular complex), but necessary for obtaining tractable approximations of $Q_F(\tilde{\theta}(Z_0))$ (13) in case of random excitations. Note, however, that Assumption 2 is reasonable for models that pass the whiteness test of the residuals. Note also that Assumption 2 should not be satisfied for $\Delta\tilde{G}(k)$ but for

$$\varepsilon(\Omega_k, \tilde{\theta}(Z_0), Z_0(k)) = \frac{\Delta\tilde{G}(k)U_0(k)}{\sigma_Y(\Omega_k, \tilde{\theta}(Z_0))} \quad (16)$$

For random $U_0(k)$ this is less severe than initially thought.

Theorem 1 (approximation uncertainty for input/ output measurements): Under Assumptions 1 and 2, the covariance matrix $\mathrm{Cov}(\hat{\theta}(Z)) \approx P_F/F$ can be approximated as,

(i) for deterministic excitations

$$\mathrm{Cov}(\hat{\theta}_{\mathrm{ML}}(Z)) \approx \hat{C}$$
$$\mathrm{Cov}(\hat{\theta}_{\mathrm{SML}}(Z)) \approx \left(1 + \frac{\hat{m}^2}{M-2}\right)\frac{M-2}{M-3}\hat{C} \quad (17)$$

(ii) for random excitations

$$\mathrm{Cov}(\hat{\theta}_{\mathrm{ML}}(Z)) \approx (1 + \hat{m}^2)\hat{C}$$
$$\mathrm{Cov}(\hat{\theta}_{\mathrm{SML}}(Z)) \approx (1 + \hat{m}^2)\frac{M-2}{M-3}\hat{C} \quad (18)$$

where $\hat{C}$ is an estimate of the Cram r-Rao lower bound

$$\hat{C} = [2\mathrm{Re}(\varepsilon'^H(\hat{\theta}_{\mathrm{ML}}, Z)\varepsilon'(\hat{\theta}_{\mathrm{ML}}, Z))]^{-1}$$
$$\approx \left(\frac{M-1}{M-2}\right)[2\mathrm{Re}(\hat{\varepsilon}'^H(\hat{\theta}_{\mathrm{SML}}, Z)\hat{\varepsilon}'(\hat{\theta}_{\mathrm{SML}}, Z))]^{-1}, \quad (19)$$

$\hat{m}^2$ is an estimate of m^2 (14), the contribution of the plant model errors to the expected value of the normalised cost function,

$$\hat{m}^2 = \frac{V_{\mathrm{ML}}(\hat{\theta}_{\mathrm{ML}}, Z)}{F} - 1$$
$$\approx \left(\frac{M-2}{M-1}\right)\frac{V_{\mathrm{SML}}(\hat{\theta}_{\mathrm{SML}}, Z)}{F} - 1 \quad (20)$$

Proof: see Pintelon *et al.* (2002). □

To understand the results of Theorem 1 for the ML estimates, the noiseless cost function ((5) with $N_Z = 0$) is written as

$$V_F(\theta, Z_0) = \sum_{k=1}^{F} \frac{|G_0(\Omega_k) - G(\Omega_k, \theta)|^2}{F\sigma_Y^2(\Omega_k, \theta)}|U_0(k)|^2 \quad (21)$$

In case of plant model errors ($\Delta\tilde{G} \neq 0$) it follows that the minimizer of (21) depends on the particular realisation of the random amplitude spectrum $|U_0(k)|^2$, which is not the case for a deterministic excitation. This observation explains why the uncertainty of the estimates in the presence of plant model errors ($\Delta\tilde{G} \neq 0 \to m^2 \neq 0$, see eq. (14)) strongly depends on the type of excitation signal: for random inputs it can be much larger than for deterministic inputs ($m^2 \gg 1$). Indeed, each realisation of a random excitation hits in a different way the plant model errors, which results in a different estimate $\hat{\theta}(Z)$, even in the absence of disturbing noise (see eq. (21)). This is not the case for a deterministic excitation. The additional term in the uncertainty for the SML estimate (17) is due to the variability of the non-parametric noise model (9) over different experiments.

Note that expressions (17) to (18) can easily be calculated from data.

3.3. Connection with prediction error framework

From eq. (18) it follows that the covariance matrix for random excitations equals the Cram r-Rao lower bound multiplied by the value of the normalised cost function in its minimum

$$1 + \hat{m}^2 = \frac{1}{F}V_{\mathrm{ML}}(\hat{\theta}_{\mathrm{ML}}, Z)$$
$$= \frac{1}{F}\sum_{k=1}^{F}|\varepsilon(\Omega_k, \hat{\theta}_{\mathrm{ML}}, Z(k))|^2 \quad (22)$$

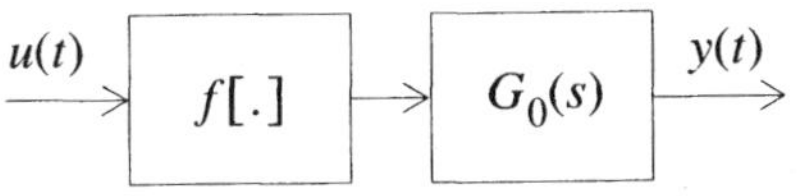

Fig. 2.: Hammerstein system consisting of the cascade of a static nonlinearity $f[.]$ and a linear dynamic bloc $G_0(s)$.

This result is completely similar to that of the prediction error (PE) framework where a parametric noise model $H(z^{-1}, \theta)$ is estimated simultaneously with the plant model $G(z^{-1}, \theta)$. Indeed the uncertainty of the prediction error estimate $\hat{\theta}_{PE}$ is calculated as the inverse of an estimate of the covariance of the predictor gradients multiplied by an estimate $\hat{\sigma}^2$ of the true innovations variance σ^2

$$\begin{aligned}\hat{\sigma}^2 &= \frac{1}{N}\sum_{t=0}^{N-1} \varepsilon^2(t, \hat{\theta}_{PE}) \\ &= \frac{1}{N}\sum_{t=0}^{N-1} [H^{-1}(q, \hat{\theta}_{PE})(y(t) - G(q, \hat{\theta}_{PE})u(t))]^2\end{aligned} \quad (23)$$

where q stands for the backward shift operator (see Ljung, 1999). Clearly, beside the true innovation variance σ^2, $\hat{\sigma}^2$ also contains the plant model errors. Hence, $\hat{\sigma}^2/\sigma^2$ plays exactly the same role as $1 + \hat{m}^2$ in (18).

The results of Theorem 1 for the ML estimate are also valid for the PE estimate with a fixed (given) noise model (the proof follows exactly the same lines).

4. SIMULATION EXAMPLE

The true plant is a Hammerstein system (see Fig. 2.) consisting of a saturating static nonlinearity $f[x] = 3\tanh(x/3)$, and an 8th order continuous-time bandpass Butterworth filter $G_0(s)$ with -3 dB cutoff frequencies of $[0.18, 0.25]$Hz. The Hammerstein system is approximated by a 6th order continuous time model

$$G(s, \theta) = \left(\sum_{r=0}^{4} b_r s^r\right) / \left(\sum_{r=0}^{6} a_r s^r\right) \quad (24)$$

For random (phase) multisines it has been shown that the best linear approximation of the Hammerstein system is asymptotically (for the number of frequencies going to infinity) given by $KG_0(s)$ where K depends on the static nonlinearity and the power spectrum of the excitation (see Pintelon and Schoukens, 2001). Hence, model (24) is identified in the presence of unmodelled dynamics and stochastic nonlinear distortions. Fig. 3. compares the linear part $G_0(s)$ of the Hammerstein system with the low order approximation (24). Observe in the passband the gain difference (due to the constant K) and the good phase match.

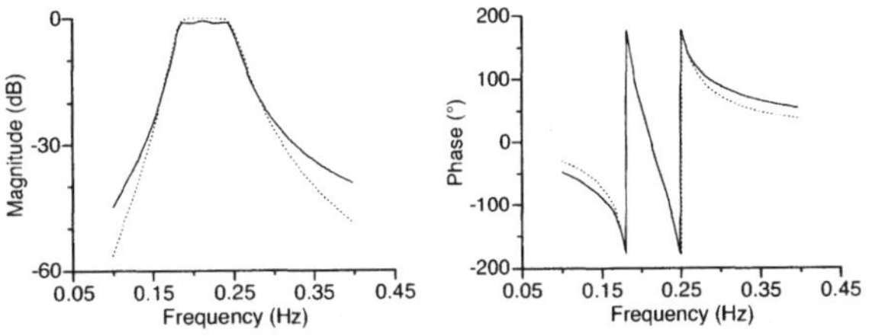

Fig. 3.: Amplitude and phase characteristics of the linear dynamic bloc $G_0(s)$ in Fig. 2. (dashed line), and of the low order linear approximation $G(s, \theta)$ of the Hammerstein system (solid line).

The plant is excited with deterministic and random (phase) multisines (see Definition 1) consisting of 511 frequencies uniformly distributed in the band $(0, 0.5]$Hz. They have a flat power spectrum with an rms value of 1. The random phase multisine has uniformly distributed phase in $[0, 2\pi)$, and the random multisine is periodic Gaussian noise. Independent, white normally distributed noise with standard deviation $\sigma_u = 0.03$ and $\sigma_y = 0.02$ is added to $M = 6$ periods of the sampled input and output signals (sampling rate $f_s = 32$ Hz). Frequencies number 102 (0.1 Hz) to 408 (0.4 Hz) are used for the identification of (24) ($F = 307$ in total).

A Monte-Carlo simulation consisting of 400 runs is performed for a deterministic, a random phase, and a random excitation (see Definition 1). For each run and each excitation signal the ML (5) and SML (7) estimators are calculated. Fig. 4. and Table 1 show the results. It follows that Theorem 1 predicts accurately the order of magnitude of the actual uncertainty: a maximal deviation of 6 dB is observed in the passband (see Fig. 4., second row, first column). Although the error on the predicted uncertainty strongly depends on how well Assumption 2 is satisfied, the improvement of Theorem 1 w.r.t. the classical result

$$\mathrm{Cov}(\hat{\theta}_{ML}(Z)) \approx \hat{C}\,,\ \mathrm{Cov}(\hat{\theta}_{SML}(Z)) \approx \frac{M-2}{M-3}\hat{C} \quad (25)$$

is significant. Indeed, using eq. (25) one would predict that the uncertainty is the same for all excitations and equal to that in Fig. 4., first row, first column (errors of about 25 dB). From the third column of Fig. 4. it follows that Theorem 1 predicts quite accurately the increase in uncertainty of the SML w.r.t. the ML estimates. Table 1 shows that the ML and SML estimates of the model errors coincide.

Table 1: Estimate of the plant model errors.

	ML determ.	ML random	SML determ.	SML random
$\hat{m}^2$	289	278	286	274

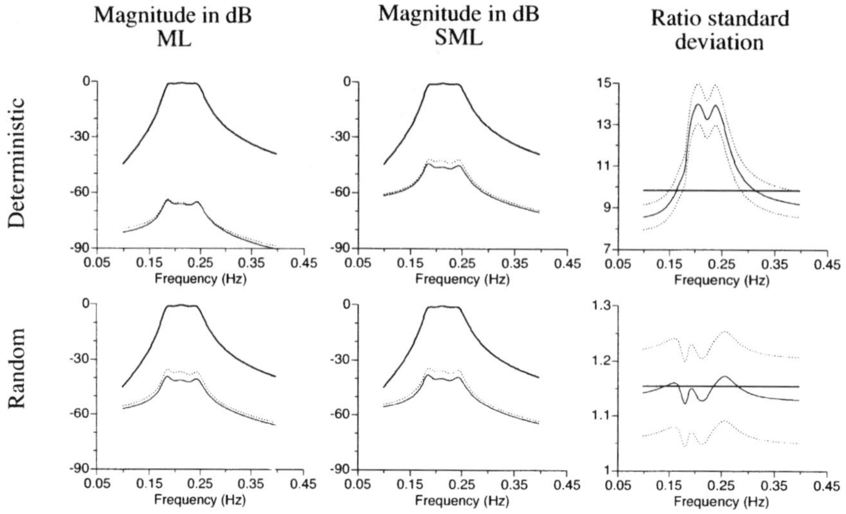

Fig. 4.: Results Monte-Carlo simulation estimated transfer function model. Left column: ML estimates (bold line) with the actual (dashed line) and the predicted (solid line) standard deviation. Middle column: SML estimates (bold solid line) with the actual (dashed line) and the predicted (solid line) standard deviation. Right column: ratio standard deviation SML estimate over standard deviation ML estimate, actual ratio (solid line) with its 95% uncertainty bounds (dashed lines) and predicted ratio (bold horizontal line). First row: deterministic input, and second row random input.

5. CONCLUSION

The approximate expressions for the covariance matrix of the plant model parameters (see Theorem 1) predict quite accurately (i) the order of magnitude of the uncertainty and (ii) the increase in uncertainty of the SML estimates w.r.t. the ML estimates. They can easily be calculated from data and establish a clear link between the parameter uncertainty, the type of model errors (unmodelled dynamics and nonlinear distortions), the type of excitation signal (deterministic, random phase, and random), and the noise level. Although the predicted uncertainty can be several dB off the actual uncertainty, the improvement w.r.t. the classical formula is significant and good enough for most applications. If more precise estimates of the uncertainty are required then one must rely one numerical techniques as, for example, in Tj rnstr m and Ljung (2002).

Theorem 1 also establishes a nice connection between the classical prediction error framework and the frequency domain approach using prior estimated noise models.

REFERENCES

Brillinger, D. R. (1981). *Time Series: Data Analysis and Theory.* McGraw-Hill, New York.

Giri, N. (1965). On the complex analogues of T^2 and R^2 tests. *Annals of Mathematical Statistics,* **vol. 36**, pp. 664-670.

Hjalmarsson, H. and L. Ljung (1992). Estimating model variance in the case of undermodeling. *IEEE Trans. Autom. Contr.,* **vol. 37**, no. 7, pp. 1004-1008.

Ljung, L. (1999). *System Identification: Theory for the User.* Prentice-Hall, Upper Saddle River, NJ.

Pintelon, R. and J. Schoukens (2001). *System Identification: a Frequency Domain Approach.* IEEE Press, Piscataway, NJ.

Pintelon, R. and J. Schoukens (2002). Some peculiarities of identification in the presence of model errors, *Proceedings of the 15th IFAC World Congress,* Barcelona, Spain, July 21-26.

Pintelon, R., J. Schoukens, and Y. Rolain (2002). Uncertainty of transfer function modeling using prior estimated noise models. *Internal report no RP042002,* ReportRP042002.pdf @ ftp://elecftp.vub.ac.be/Papers/RikPintelon/.

Schoukens J., G. Vandersteen, R. Pintelon and P. Guillaume (1997). Frequency Domain System Identification Using Non-parametric Noise Models Estimated from a Small Number of Data Sets, *Automatica,* **vol. 33**, no. 6, pp. 1073-1086.

Schoukens J., T. Dobrowiecki and R. Pintelon (1998). Parametric Identification of Linear Systems in the Presence of Nonlinear Distortions. A Frequency Domain Approach, *IEEE Trans. Autom. Contr.,* **vol. AC-43**, no. 2, pp. 176-190.

Schoukens J., G. Vandersteen, R. Pintelon and P. Guillaume (1999). Frequency-domain identification of linear systems using arbitrary excitations and a nonparametric noise model, *IEEE Trans. Autom. Contr.,* **vol. AC-44**, no. 2, pp. 343-347.

Tj rnstr m, F. and L. Ljung (2002). Using the bootstrap to estimate the variance in the case of undermodeling, *IEEE Trans. Autom. Contr.,* **vol. 47**, no. 2, pp. 395-398.

IFAC Publications
www.elsevier.com/locate/ifac

THE SIZE OF THE MEMBERSHIP-SET IN A PROBABILISTIC FRAMEWORK

Hüseyin Akçay *

* *Anadolu University, Department of Electrical and Electronics Engineering, 26470 Eskişehir, Turkey. E-mail: huakcay@anadolu.edu.tr*

Abstract: In this paper, we study the size of the membership-set for system identification in a probabilistic framework. Assuming that the regressors are persistently exciting and the measurement noise is a sequence of independent, identically distributed bounded random variables, tight lower and upper non-asymptotic probability bounds on the membership-set diameter are obtained. These bounds are used in the computation of the confidence intervals for interpolatory estimators including the Chebyshev center, the analytic center, the constrained least-squares estimator (projection estimator), and the minmax estimator. *Copyright © 2003 IFAC*

Keywords: Parameter estimation, Identification, Bounded noise, Membership-set, Diameter, Error analysis, Confidence interval.

1. INTRODUCTION

Consider a discrete-time scalar system represented by the model

$$y(t) = \phi^T(t)\theta + \eta(t), \qquad t = 1, \cdots, N \quad (1)$$

where $y(t) \in \mathbf{R}$ is the system output, $\phi(t) \in \mathbf{R}^n$ is the measurable regression vector, $\theta \in \mathbf{R}^n$ is the unknown parameter vector, and $\eta(t) \in \mathbf{R}$ is the additive measurement noise. Here, we assume that $\phi(t)$ is a deterministic signal.

The model sets captured by (1) are rather large. For example, by taking $\phi(t)$ as the regression of the delayed inputs, $[u(t-1) \cdots u(t-n)]^T$, one obtains impulse-response models; and by taking $\phi(t)$ as the convolution of the fixed-pole rational basis functions with the input, $[g_1(t) * u(t) \cdots g_n(t) * u(t)]^T$, the model sets considered in (Akçay and Ninness, 1998), which generalize the Laguerre and the Kautz models.

The goal of system identification is to find an algorithm that maps available input-output data: $y(t), \phi(t)$, $t = 1, \cdots, N$ into the estimate $\widehat{\theta}_N$ of θ. Clearly, based on different assumptions on $\eta(t)$ and the intended use of the model, different algorithms can be used to determine $\widehat{\theta}_N$.

Most methods in system identification assume the disturbances to be random. This leads to instrumental variable and prediction error methods, see for example (Ljung, 2000). The least-squares method is the archetype for such methods. A drawback of this approach is that in order to obtain reliable estimates from relatively small data sets much *a priori* statistical information on the noise is to be given.

On the other hand, in bounded–error–parameter estimation the disturbances are only assumed to be unknown–but–bounded:

$$\max_{1 \le t \le N} |\eta(t)| \triangleq \|\eta\|_\infty \le \varepsilon. \quad (2)$$

The motivation for this noise description is that in many applications, the noise is amplitude bounded. A further motivation stems from recent interest in deterministic worst-case identification. There is also a large body of research devoted to the bounded-error-parameter estimation area.

Given input-output data $y(t), \phi(t)$, $t = 1, \cdots, N$, the set of all possible parameter vectors which are consistent with (1) and (2) is

$$S_N \triangleq \bigcap_{t=1}^{N} \left\{ \widehat{\theta} \in \mathbf{R}^n : |y(t) - \phi^T(t)\widehat{\theta}| \leq \varepsilon \right\}. \quad (3)$$

This set is referred to as the *membership-set* and it represents all parameter vectors $\widehat{\theta}$ which are unfalsified with the data $y(t), \phi(t)$, $t = 1, \cdots, N$ and the bounded noise assumption $\|\eta\|_\infty \leq \varepsilon$.

In bounded error parameter estimation, there are two directions. The first one aims obtaining the exact membership-set. Due to the complexity of this set, an approximate outer bounding of the membership set by a simply shaped set such as an ellipsoid, an orthotope, or a parellotope is widely used. It is not our purpose finding some new algorithms to describe S_N.

The other direction in the bounded error parameter estimation is to compute a specific estimate in the membership-set that enjoys some optimality properties. A well-known example is the *Chebyshev center* of S_N

$$\theta_N^{\mathrm{CH}} \triangleq \arg \min_{\widehat{\theta}_1 \in S_N} \max_{\widehat{\theta}_2 \in S_N} \|\widehat{\theta}_1 - \widehat{\theta}_2\|_2. \quad (4)$$

Here, $\|x\|_2 \triangleq (\sum_{k=1}^{n} x_k^2)^{\frac{1}{2}}$ denotes the Euclidian norm on $\mathbf{R}^n$. The Chebyshev center is optimal in the sense that it minimizes parameter estimation error. A more recent estimator proposed in (Bai *et al.*, 1999) is the *analytic center*

$$\theta_N^{\mathrm{A}} \triangleq \arg \max_{\widehat{\theta} \in S_N} \sum_{t=1}^{N} \ln(\varepsilon^2 - (y(t) - \phi^T(t)\widehat{\theta})^2). \quad (5)$$

If $\eta(t)$ is a sequence of zero mean independently and identically distributed truncated Gaussian random variables, then θ_N^{A} happens to be the maximum likelihood estimate of θ (Bai *et al.*, 1999). The third example is the *constrained least-squares* estimator (the *projection* estimator)

$$\theta_N^{\mathrm{P}} \triangleq \arg \min_{\widehat{\theta} \in S_N} \sum_{t=1}^{N} (y(t) - \phi^T(t)\widehat{\theta})^2. \quad (6)$$

This is the best estimate in the membership set that minimizes the arithmetic average output error. This projection estimator minimizes the average error at a price that the maximum error can be large. On the other hand, the *minmax* estimate

$$\theta_N^{\mathrm{MM}} \triangleq \arg \min_{\widehat{\theta} \in \mathbf{R}^n} \max_{1 \leq t \leq N} (y(t) - \phi^T(t)\widehat{\theta})^2 \quad (7)$$

minimizes the maximum error at a price that the average error can be large.

The estimators (4)–(7) all lie in the membership-set. Such estimators are called *interpolatory*. Thus, studying the size of the membership-set in both worst-case and stochastic settings is utmost interest. A measure of the size for S_N is its diameter defined by

$$\mathcal{D}(S_N) \triangleq \sup_{\widehat{\theta}_1, \widehat{\theta}_2 \in S_N} \|\widehat{\theta}_1 - \widehat{\theta}_2\|_2. \quad (8)$$

The diameter of S_N depends on the assumed noise amplitude as well as the realized values of the noise. In information based complexity theory, the quantity

$$\sup_{\|\eta\|_\infty \leq \varepsilon} \mathcal{D}(S_N)$$

is known as the worst-case diameter of S_N, and it provides a lower bound on the best achievable error of optimal estimation algorithms. We refer the interested reader to (Tse *et al.*, 1991; Milanese and Tempo, 1985) for applications of this concept to worst-case identification/estimation. A well-known inequality is the following

$$\mathcal{D}(S_N) \leq 2 \max_{\widehat{\theta} \in S_N} \|\widehat{\theta} - \theta_N^{\mathrm{CH}}\|_2.$$

Thus, any interpolatory algorithm is suboptimal within a factor of two with respect to an optimal algorithm.

On the other hand, when $\eta(t)$, $t = 1, \cdots, N$ is a sequence of independent identically distributed random variables, $\mathcal{D}(S_N)$ is a random variable, and less is known about it. Recently, in (Bai *et al.*, 1995; Bai *et al.*, 1998*a*) convergence properties of S_N were investigated in a probabilistic set-up. In this paper, we also study $\mathcal{D}(S_N)$ in a probabilistic framework.

The main result of this paper is to derive in Section 2 upper and lower probability bounds on $\mathcal{D}(S_N)$ assuming that the regressors are persistently exciting and $\eta(t)$ is a sequence of independent, identically distributed bounded random variables. These conditions are weak. In particular, the regressors are not assumed to be periodic. Thus, the main result is the generalization of a result in (Bai *et al.*, 1995) for periodic regressors to persistently exciting regressors. These bounds are tight and non-asymptotic. Then, they are used in the derivation of confidence intervals for the estimators (4)–(7). The utility of these inequalities is illustrated in establishing formal conditions for the least-squares parameter estimate in its raw to lie in the membership-set. The proofs are omitted due to the space constraints.

2. THE DIAMETER OF THE MEMBERSHIP-SET IN A PROBABILISTIC FRAMEWORK

In this section, we present two inequalities bounding the probability $\text{Prob}(\mathcal{D}(S_N) > \delta)$, $\delta > 0$ from above and below. These inequalities will be used in section 3 to obtain confidence intervals for the interpolatory parameter estimators and a common upper bound on the covariances of such estimators. These inequalities also provide formal conditions when the least-squares parameter estimate in its raw form lies in the membership-set. First, we need the definition of *persistent excitation* of the regressors in a worst-case setting.

Definition 1. (Persistent excitation). The regressor sequence $\{\phi(t)\}$ is persistently exciting of order m if for some $\alpha, \beta > 0$ and all t_0,

$$\alpha^2 I_n \leq \frac{1}{m} \sum_{t=t_0}^{t_0+m-1} \phi(t)\phi^T(t) \leq \beta^2 I_n. \quad (9)$$

Here, I_n is the n by n identity matrix and the matrix inequality $A \geq B$ means that $x^T A x \geq x^T B x$ for all x.

This definition of persistence of excitation has already appeared in the literature (Dasgupta and Huang, 1987; Bai *et al.*, 1995; Bai *et al.*, 1998*b*; Bai *et al.*, 1998*a*). Note that periodic signals are persistently exciting. In worst-case identification, much stronger conditions are imposed on regressors (see (Makila, 1991; Harrison *et al.*, 1998; Akçay and Ninness, 1998) and the references therein).

Let $\mathcal{N}(A)$ denote the number of elements of a given set A and let $\aleph$ denote the set of positive integers. The following lemma will be instrumental in deriving an upper bound on $\text{Prob}(\mathcal{D}(S_N) > \delta)$.

Lemma 2. (Separation lemma). Suppose $N \in 2m\aleph$. Let $\{\phi(t)\}$ be as in Definition 1. Then there is a set $\mathcal{K}$ of subsequences $\{t_k\}$ with the properties

$$\mathcal{N}(\{t_k\}) \geq \frac{N}{2m}, \qquad \forall \{t_k\} \in \mathcal{K}, \quad (10)$$

$$\mathcal{N}(\mathcal{K}) \leq 2 + 2q(q-1)^{n-2} \triangleq \mathcal{P} \quad (11)$$

where

$$q \triangleq \min_k \{k \in \aleph : k\alpha \geq \pi\beta\sqrt{m(n-1)}\}. \quad (12)$$

For each $x \in \mathbf{R}^n$, there exists a $\{t_k\} \in \mathcal{K}$ such that either one of the following holds

$$\begin{aligned} (i) \quad & x^T\phi(t_k) \geq \frac{\alpha}{2}\|x\|_2, \ \forall k \\ (ii) \quad & x^T\phi(t_k) \leq -\frac{\alpha}{2}\|x\|_2, \ \forall k. \end{aligned} \quad (13)$$

In the proof of Lemma 2, the idea of extracting subsequences from regressors, which satisy (13), is taken from (Bai *et al.*, 1995). The essence of Lemma 2 is the inequalities (10) and (11), *that is*, as $N \to \infty$, $\mathcal{N}(\mathcal{K})$ remains bounded, and for each $\{t_k\} \in \mathcal{K}$, $\mathcal{N}(\{t_k\})$ is proportional to N. These two properties will be crucial in deriving the main result of this paper in the following.

Theorem 3. Consider the system represented by (1). Suppose $\{\eta(t), t = 1, \cdots, N\}$ is a collection of independent identically distributed random variables bounded by ε and $N \geq 2mM$, $M \in \aleph$. Let $\{\phi(t)\}$ be persistently exciting of order m. Let $\mathcal{D}(S_N)$ and $\mathcal{P}$ be as in (8) and (11), respectively. Let

$$\varepsilon'_M \triangleq \min_{1\leq t\leq M} \eta(t), \quad \varepsilon''_M \triangleq \max_{1\leq t\leq M} \eta(t), \quad (14)$$

and

$$\begin{aligned} Q_M(\delta) \triangleq & \, \text{Prob}\,(\varepsilon + \varepsilon'_M > \delta) \\ & + \text{Prob}\,(\varepsilon - \varepsilon''_M > \delta), \end{aligned} \quad (15)$$

$$T_N(\delta) \triangleq \text{Prob}\,(\varepsilon + \varepsilon'_N > \delta, \varepsilon - \varepsilon''_N > \delta). \quad (16)$$

Then

$$\begin{aligned} \text{Prob}\,(\mathcal{D}(S_N) > \delta) & \geq T_N\left(\beta m^{\frac{1}{2}}\delta\right) \\ \text{Prob}\,(\mathcal{D}(S_N) > \delta) & \leq \mathcal{P} Q_M(\alpha\delta/4). \end{aligned} \quad (17)$$

The conditions imposed on the regressors and the noise are minimal. In particular, the regressors are not assumed to be periodic. Therefore, it generalizes a result of (Bai *et al.*, 1995) for periodic regressors to persistently exciting regressors.

Let M be the integer part of $N/2m$. Then (15) and (16) tell us that $Q_M(\delta)$ and $T_N(\delta)$ have the same (asymptotic) convergence rates since m is a fixed integer. It follows that $\mathcal{P}Q_M$ and T_N have the same asymptotic order. This shows that the upper and the lower bounds on $\text{Prob}(\mathcal{D}(S_N) > \delta)$ derived in Theorem 3 are tight.

Suppose now that

$$\varepsilon_1 \leq \eta(t) \leq \varepsilon_2, \qquad t = 1, \cdots, N \qquad (18)$$

where $\varepsilon_2 \neq -\varepsilon_1$. In this case, the membership-set is given by

$$S_N = \bigcap_{t=1}^{N} \left\{ \widehat{\theta} \in \mathbf{R}^n : \varepsilon_1 \leq y(t) - \phi^T(t)\widehat{\theta} \leq \varepsilon_2 \right\}.$$

However, this set can be put into the form (3) by defining

$$\varepsilon \triangleq \frac{\varepsilon_2 - \varepsilon_1}{2}, \qquad \overline{\varepsilon} \triangleq \frac{\varepsilon_1 + \varepsilon_1}{2}$$

and

$$\widetilde{y}(t) \triangleq y(t) - \overline{\varepsilon}, \qquad t = 1, \cdots, N.$$

Then

$$S_N = \bigcap_{t=1}^{N} \left\{ \widehat{\theta} \in \mathbf{R}^n : |\widetilde{y}(t) - \phi^T(t)\widehat{\theta}| \leq \varepsilon \right\}.$$

Thus, without loss of generality we may assume that the noise bounds in (18) satisfy $\varepsilon_2 = -\varepsilon_1$ and then the preceding analysis applies to this case as well.

Theorem 3 greatly simplifies stochastic analysis of the membership-set. For example, the problem of bounding the membership-set diameter in probability is reduced to the problem of the *minmax* estimation of the noise amplitude. The detailed analysis will be carried out in the next section.

Recall that the result expressed in Theorem 3 was derived in two stages. In the first stage, we used the persistence of excitation condition on regressors to extract a collection $\mathcal{K}$ that separates points in $\mathbf{R}^n$. In the second stage, we exploited the fact that $\eta(t)$ additively affects $y(t)$.

3. CONFIDENCE INTERVALS AND CONVERGENCE IN L^P

Given a random variable x, we define the distribution function of x by

$$F_x(\delta) \triangleq \text{Prob}\left(|x| \geq \delta\right), \qquad \delta > 0. \qquad (19)$$

To simplify the discussion, we assume that $\eta(t)$ is symmetrically distributed on the interval $[-\varepsilon, \varepsilon]$. Then, $Q_M(\delta)$ in (15) can be written as

$$Q_M(\delta) = [2 - F_\eta(\varepsilon - \delta)]^M. \qquad (20)$$

Likewise, $T_N(\delta)$ in (16) can be written as

$$T_N(\delta) = [1 - F_\eta(\varepsilon - \delta)]^N. \qquad (21)$$

Hence from (20), (21), and (17)

$$\text{Prob}\left(\mathcal{D}(S_N) > \delta\right) \leq \mathcal{P}[2 - F_\eta(\varepsilon - \frac{\alpha\delta}{4})]^{\frac{N}{2m}} \quad (22)$$

$$\text{Prob}\left(\mathcal{D}(S_N) > \delta\right) \geq [1 - F_\eta(\varepsilon - \beta m^{\frac{1}{2}}\delta)]^N \quad (23)$$

Note that the upper and lower bounds in (22) and (23) are the same type since m is a fixed number. The lower bound shows that the diameter of S_N will not shrink to zero in probability as $N \to \infty$ unless ε is a *tight bound* on $\eta(t)$.

A bound ε on $\eta(t)$ is said to be tight, if for each $\nu > 0$ satisfying $\varepsilon > \nu$,

$$\text{Prob}(\eta(t) \geq \nu) > 0 \quad \text{and} \quad \text{Prob}(\eta(t) \leq -\nu) > 0.$$

Thus, if ε is a *tight bound* on $\eta(t)$, from (22) we have

$$\lim_{N\to\infty} \text{Prob}\left(\mathcal{D}(S_N) > \delta\right) = 0, \quad \delta > 0 \quad (24)$$

and therefore, $\mathcal{D}(S_N) \to 0$ in probability as $N \to \infty$. Since convergence in probability implies almost everywhere convergence along a subsequence $\mathcal{D}(S_{N_k})$ and $\mathcal{D}(S_N)$ is non-increasing (with N), it follows that

$$\lim_{N\to\infty} \mathcal{D}(S_N) = 0 \qquad \text{a. e.} \qquad (25)$$

The convergence results expressed in (24) and (25) were obtained in (Bai *et al.*, 1995; Bai *et al.*, 1998*a*) by different techniques.

Since the volume of S_N, denoted by $\mathcal{V}(S_N)$, is bounded above by $[\mathcal{D}(S^N)]^n$, it follows from (25)

$$\lim_{N\to\infty} \mathcal{V}(S_N) = 0 \qquad \text{a. e.} \qquad (26)$$

However, (26) does not necessarily imply (25). In the quantification of the error, $\mathcal{D}(S_N)$ is a more useful measure.

3.1 *Confidence Intervals*

Since the true parameter θ lies in S_N for all N, we have from (22) for every interpolatory estimator $\widehat{\theta}_N$,

$$\begin{aligned} \text{Prob}(|\widehat{\theta}_N^{(k)} - \theta^{(k)}| > \delta) &\leq \text{Prob}(\|\widehat{\theta}_N - \theta\|_2 > \delta) \\ &\leq \text{Prob}\left(\mathcal{D}(S_N) > \delta\right) \qquad (27) \\ &\leq \mathcal{P}[2 - F_\eta(\varepsilon - \frac{\alpha\delta}{4})]^{\frac{N}{2m}}. \end{aligned}$$

In this way, we may form the *confidence intervals* for the true parameters $\theta^{(k)}$, $k = 1, \cdots, n$.

Thus, (27) furnishes the confidence intervals for the Chebyshev center, the analytic center, the constrained least-squares, and the minmax estimators.

We will compute the right-hand side of (27) for some well-known distributions.

(1) *Uniform distribution.*

$$F_\eta^{\mathrm{u}}(\delta) \triangleq \begin{cases} 1-\dfrac{\delta}{\varepsilon}, & 0 \le \delta \le \varepsilon \\ 0, & \delta > \varepsilon. \end{cases} \quad (28)$$

Then, from (27) for $k = 1, 2, \cdots, n$,

$$\mathrm{Prob}(|\widehat{\theta}_N^{(k)} - \theta^{(k)}| > \delta) \le 2\mathcal{P}(1-\frac{\alpha\delta}{8\varepsilon})^{\frac{N}{2m}}.$$

Thus, the lengths of the confidence intervals decrease to zero geometrically as $N \to \infty$.

(2) *Binomial distribution.*

$$F_\eta^{\mathrm{b}}(\delta) \triangleq \begin{cases} 1, \; 0 \le \delta \le \varepsilon \\ 0, \; \delta > \varepsilon. \end{cases} \quad (29)$$

Then, for $k = 1, 2, \cdots, n$,

$$\mathrm{Prob}\left(|\widehat{\theta}_N^{(k)} - \theta^{(k)}| > \delta\right) \le 2\mathcal{P}\, 2^{-\frac{N}{2m}}. \quad (30)$$

Hence, the lengths of the confidence intervals decrease to zero geometrically but faster than those of the uniform distribution as $N \to \infty$.

The distributions F_η^{u} and F_η^{b} are charecterized by heavy tails at the extremes $\pm\varepsilon$. If, on the other hand, the distribution has very thin tails at $\pm\varepsilon$, then the convergence towards zero is extremely slow (see (Akçay *et al.*, 1996) for sample distributions).

3.2 *Convergence in L^p*

Given a measurable function g on a measure space (X, μ), we define the distribution function of g by

$$G(\delta) \triangleq \mu(\{x \in X : |g(x)| > \delta\}), \qquad \delta > 0.$$

For all $0 < p < \infty$, the following identity holds:

$$\int |g|^p \mathrm{d}\mu = \int_0^\infty p\delta^{p-1} G(\delta) \mathrm{d}\delta.$$

Let X be the sample space and μ the probability measure on X. Put $g = \mathcal{D}(S_N)$. Let $\mathbf{E}$ denote the expectation operator. Then

$$\mathbf{E}[\mathcal{D}(S_N)]^p \doteq \int_0^\infty p\delta^{p-1} \mathrm{Prob}(\mathcal{D}(S_N) > \delta)\, \mathrm{d}\delta \quad (31)$$

Note that

$$\lim_{N\to\infty} \mathbf{E}[\mathcal{D}(S_N)]^p = 0, \qquad p > 0. \quad (32)$$

This, is due to (24) and the fact that the random variables $\mathcal{D}(S_N)$ are uniformly bounded. Thus, we have shown that every interpolatory estimator converges to the true parameter vector in probability, almost everywhere, and in L^p. This recovers the convergence result in (Bai *et al.*, 2000) for the analytic center estimator (5).

In the rest of this section, we determine how fast $\mathcal{D}(S_N)$ tends to zero in L^2 for the uniform and binomial distributions. Let ν be a positive number less than the minimum of ε and $\alpha\varepsilon/4$. From (31) and (22), we have

$$\begin{aligned} \mathbf{E}[\mathcal{D}(S_N)]^2 \le\; & 4\mathcal{P} \int_0^{\frac{4\nu}{\alpha}} \delta \left[1 - \frac{1}{2}F_\eta\left(\varepsilon - \frac{\alpha\delta}{4}\right)\right]^{\frac{N}{2m}} \mathrm{d}\delta \\ & + 4\mathcal{P} \int_{\frac{4\nu}{\alpha}}^{\varepsilon} \delta \left[1 - \frac{1}{2}F_\eta\left(\varepsilon - \frac{\alpha\delta}{4}\right)\right]^{\frac{N}{2m}} \mathrm{d}\delta \\ & \triangleq \Sigma_1 + \Sigma_2. \end{aligned}$$

The term Σ_2 is bounded as

$$\Sigma_2 < 2\varepsilon^2 \mathcal{P} \left[1 - \frac{1}{2}F_\eta(\varepsilon - \nu)\right]^{\frac{N}{2m}}.$$

As $N \to \infty$, for each fixed ν this bound converges to zero exponentially and independently of F_η provided that ε is tight bound on $\eta(t)$. Hence, we may concentrate only on the term Σ_1.

For the uniform distribution (28), we have

$$\begin{aligned} \Sigma_1 &\le 4\mathcal{P} \int_0^{\frac{4\nu}{\alpha}} \delta \left(1 - \frac{\alpha\delta}{8\varepsilon}\right)^{\frac{N}{2m}} \mathrm{d}\delta \\ &< \frac{1024\varepsilon^2 m^2 \mathcal{P}}{\alpha^2 (N+2m)(N+4m)}. \end{aligned}$$

Thus, for some constant $C_1 > 0$ that does not depend on N

$$\mathbf{E}[\mathcal{D}(S_N)]^2 \le C_1 N^{-2}.$$

Likewise, for some constant C_2 that does not depend on N

$$\mathbf{E}[\mathcal{D}(S_N)]^2 \ge C_2 N^{-2}.$$

It follows that

$$\mathbf{E}[\mathcal{D}(S_N)]^2 = O\left(N^{-2}\right). \quad (33)$$

For the binomial distribution (29) we have from

$$\Sigma_1 \le 2\mathcal{P}\varepsilon^2 2^{-\frac{N}{2m}} \le C_3 \gamma_1^N$$

for some constants $C_3 > 0$ and $0 < \gamma_1 < 1$ which do not depend on N. Therefore, for some constant $0 < \gamma < 1$

$$\mathbf{E}[\mathcal{D}(S_N)]^2 = O\left(\gamma^N\right). \qquad (34)$$

It is clear that noise distributions which agree with F_η^u and F_η^b only in some neighborhoods of $\pm\varepsilon$ yield the same equations (33) and (34). Thus, we may conclude that noise distributions with heavy tails make the mean-squared membership-set diameter shrink to zero at least as fast as $O(N^{-2})$. On the other hand, distributions with thin tails result in the mean-squared membership-set diameter shrinking to zero rather slowly.

Since $\theta \in S_N$ for all N, we have for every interpolatory estimator $\widehat{\theta}_N$,

$$\mathbf{E}[\|\widehat{\theta}_N - \theta\|_2^2] \leq \mathbf{E}\left[\mathcal{D}(S_N)\right]^2. \qquad (35)$$

It is well known (Ljung, 2000) that the least-squares parameter estimate

$$\theta_N^{\text{LS}} \triangleq \arg \min_{\widehat{\theta} \in \mathbf{R}^n} \sum_{t=1}^{N} \left(y(t) - \phi^T(t)\widehat{\theta}\right)^2 \qquad (36)$$

has a covariance matrix decreasing to zero as $O(N^{-1})$. Therefore, from (35), (33), and (34) we note that if the distribution of $\eta(t)$ has heavy tails, θ_N^{LS} can not lie in the membership-set, *that is*, it is not interpolatory. Instead, we may use the constrained least-squares estimator (6). If on the other hand, the distribution of $\eta(t)$ has very thin tails, the mean-squared membership-set diameter will shrink to zero slower than $O(N^{-1})$, thus θ_N^{LS} will be likely to lie in the membership-set for all large N. (See (Bai *et al.*, 1998*b*), for a detailed discussion of the convergence properties of the weighted least-squares estimates in membership-set identification).

It should be noted that although we derived a tight lower bound on $\mathrm{E}[\mathcal{D}(S_N)]^2$, it is not necessarily a lower bound on the covariance of the interpolatory estimator under consideration. This subject requires a more detailed analysis tailored to that specific estimator.

4. CONCLUSIONS

In this paper, we study the diameter of the membership-set assuming that the regressors are persistently exciting and the measurement noise is a sequence of independent, identically distributed bounded random variables. Then we obtained tight lower and upper non-asymptotic probability bounds on the membership-set diameter. These bounds were then used in the computation of the confidence intervals for the interpolatory estimators and provided a criterion whether a given estimator is likely to lie in the membership-set or not. The results derived in this paper should provide insights in further statistical analysis of some well-known membership-set based estimators.

5. REFERENCES

Akçay, H. and B. Ninness (1998). Rational basis functions for robust identification from frequency and time-domain measurements. *Automatica* **34**, 1101–1117.

Akçay, H., H. Hjalmarsson and L. Ljung (1996). On the choice of norms in system identification. *IEEE TAC* **41**, 1367–1372.

Bai, E. W., H. Cho and R. Tempo (1998*a*). Convergence properties of the membership set. *Automatica* **34**, 1245–1249.

Bai, E. W., L. Qui and R. Tempo (1998*b*). Unfalsified weighted least squares estimates in set-membership identification. *IEEE Tr. CAS–Part I* **45**, 41–49.

Bai, E. W., M. Fu, R. Tempo and Y. Ye (2000). Convergence results of the analytic center estimator. *IEEE TAC* **45**, 569–572.

Bai, E. W., R. Tempo and H. Cho (1995). Membership set estimators: size, optimal inputs, complexity and relations with least squares. *IEEE Tr. CAS–Part I* **42**, 266–277.

Bai, E. W., Y. Ye and R. Tempo (1999). Bounded error parameter estimation: a sequential analytic center approach. *IEEE TAC* **44**, 1107–1117.

Dasgupta, S. and Y. F. Huang (1987). Asymptotically convergent modified recursive least-squares with data dependent updating and forgetting factor for systems with bounded noise. *IEEE Tr. IT* **33**, 383–392.

Harrison, K. J., J. R. Partington and J. A. Ward (1998). Complexity of identification of linear systems with rational transfer functions. *MCSS* **11**, 265–288.

Ljung, L. (2000). System identification: Theory for the user.

Makila, P. M. (1991). Robust identification and Galois sequences. *IJC* **54**, 1189–1200.

Milanese, M. and R. Tempo (1985). Optimal algorithms theory for robust estimation and prediction. *IEEE TAC* **AC-30**, 730–738.

Tse, D., M. A. Dahleh and J. Tsitsiklis (1991). Optimal asymptotic identification under bounded disturbance. *IEEE TAC* **38**, 1176–1190.

IFAC Publications
www.elsevier.com/locate/ifac

CONNECTIONS BETWEEN L_2-MODEL REDUCTION AND BALANCED TRUNCATION ⋆

Wolfgang Scherrer *,1 **Fredrik Tjärnström** **,2

* *Institut für Ökonometrie, Operations Research und Systemtheorie, Technische Universität Wien, Argentinierstr. 8/119, A-1040, Vienna, Austria, e-mail:* `W.Scherrer@tuwien.ac.at`
** *Automatic Control, Department of Electrical Engineering, Linköpings universitet, SE-58183 Linköping, Sweden, e-mail:* `fredrikt@isy.liu.se`

Abstract: In this paper we investigate the connection between model reduction by balanced truncation and by L_2 reduction. We show that locally, i.e., close to the set of lower order systems, balanced truncation and (unweighted) L_2 model reduction produce models that are almost identical. This implies that high order estimated models can be reduced by either L_2 reduction or balanced truncation, both methods giving a low order model with the same asymptotic varaiance, if the true data generating model is in the class of low order models.
Copyright © 2003 IFAC

Keywords: Model reduction, Balanced truncation, L_2 reduction, System identification

1. INTRODUCTION

Model reduction has been a widely studied problem since the 70's. The purpose of model reduction is to find a good low order approximation of a high order model. The reason for this is typically that the simpler low order model is more attractive for simulation and controller design. The L_2 reduction problem has been studied by several authors, e.g., (Wilson, 1970; Spanos *et al.*, 1992; Ferrante *et al.*, 1999). All L_2 reduction algorithms generally suffer from the need of numerical optimization routines which cannot guarantee convergence. On the other hand model reduction by balanced truncation (Moore, 1981) has achieved a widespread use due to its ease of implementation with good numerical properties. See also (Pernebo and Silverman, 1982; Glover, 1984)

In this paper we present some results connecting L_2 model reduction and balanced truncation. We show that for a sequence of n-th order models converging to an m-th order model, the L_2 reduced model and the balanced truncated model are "close" to each other. This paper takes a similar approach to that in (Scherrer, 2002), where the performance of minimum phase balanced truncation is analyzed with respect to the Kullback-Leibler distance rate.

The closeness of the two reduction methods can exemplified from a system identification perspective where both reduced models will have asymptotically the same covariance. This is in accordance with simulation results in (Tjärnström, 2002).

The outline of the paper is as follows. In Section 2 the problem formulation and the necessary notation is given. Section 3 presents the L_2 model reduction problem and a linear approximation of the L_2 problem for models "close" to the set of lower order models. By comparing the balanced truncated model in Section 4 with the linear approximation of the L_2 reduced

⋆ The authors acknowledge support by the program 'Training and Mobility of Researchers - Research Networks' through project System Identification (FMRX CT98 0206) and acknowledge contacts with the participants in the European Research Network System Identification (ERNSI).
[1] The author acknowledges support by the Austrian 'Fond zur Förderung der wissenschaftlichen Forschung' through project P-14438INF.
[2] The author acknowledges ECSEL for financial support.

model, it follows that balanced truncated models are "close" to optimal with respect to the L_2 norm, if the model is "close" to the set of lower order models. In Section 5 we apply these results to the system identification framework and in Section 6 we give some conclusions.

2. PROBLEM FORMULATION AND NOTATION

A stable state space model with n states, q inputs and p outputs is a quadruple $\theta = \{A, B, C, D\}$, where $A \in \mathbb{R}^{n\times n}$, $B \in \mathbb{R}^{n\times q}$, $C \in \mathbb{R}^{p\times n}$, $D \in \mathbb{R}^{p\times q}$ and A is asymptotically stable. The set of all such stable state space models will be denoted by $\mathbb{S}_{n,p,q} \subseteq \mathbb{R}^{(n+p)(n+q)}$. For $T \in \mathbb{R}^{n\times n}$ nonsingular and $\theta = \{A,B,C,D\}$ we will use $T\cdot\theta = \{TAT^{-1}, TB, CT^{-1}, D\}$ to denote the state space model obtained by the state space transformation T.

A model $\theta = \{A,B,C,D\} \in \mathbb{S}_{n,p,q}$ is a realization of the stable $p\times q$ transfer function $k(z) = C(z^{-1}I - A)^{-1}B + D$, where z denotes the backward shift. Let $\mathbb{M}_{n,p,q}$ ($\bar{\mathbb{M}}_{n,p,q}$) denote the set of all stable $p \times q$ transfer functions of McMillan degree n ($\leq n$). Furthermore $\Pi : \mathbb{S}_{n,p,q} \rightarrow \bar{\mathbb{M}}_{n,p,q}$ denotes the mapping attaching a transfer function $k = \Pi(\theta)$ to a state space model θ. Of course $\Pi(T\cdot\theta) = \Pi(\theta)$.

We will consider state space models $\theta = \{A,B,C,D\}$ of order n that are close to an m-th order model ($m < n$). These models are of the form

$$
\begin{aligned}
A =& A_0 + \varepsilon A_1 = \begin{pmatrix} A_{0,11} + \varepsilon A_{1,11} & \varepsilon A_{1,12} \\ A_{0,21} + \varepsilon A_{1,21} & A_{0,22} + \varepsilon A_{1,22} \end{pmatrix} \\
B =& B_0 + \varepsilon B_1 = \begin{pmatrix} B_{0,1} + \varepsilon B_{1,1} \\ B_{0,2} + \varepsilon B_{1,2} \end{pmatrix} \\
C =& C_0 + \varepsilon C_1 = \begin{pmatrix} C_{0,1} + \varepsilon C_{1,1} & \varepsilon C_{1,2} \end{pmatrix} \\
D =& D_0 + \varepsilon D_1
\end{aligned} \tag{1}
$$

where all the matrices are partitioned conformingly, e.g., $A_{0,11} \in \mathbb{R}^{m\times m}$, $A_{0,21} \in \mathbb{R}^{(n-m)\times m}$. Note that by construction, the limiting model $\theta_0 = \{A_0, B_0, C_0, D_0\}$ (as $\varepsilon \rightarrow 0$) is not minimal, and the truncated model $\{A_{0,11} + \varepsilon A_{1,11}, B_{0,1} + \varepsilon B_{1,1}, C_{0,1} + \varepsilon C_{1,1}, D_0 + \varepsilon D_1\}$ is a "good" approximation of the full order model for small ε. We impose the assumption that $A_{0,11}$ and $A_{0,22}$ are asymptotically stable matrices, and that the limiting transfer function $k_0 = \Pi(\theta_0)$ has McMillan degree m, i.e., $k_0 \in \mathbb{M}_{m,p,q}$.

The L_2 norm of the transfer function $k(z)$ is defined as

$$\|k\|_2 = \sqrt{\mathbb{E}\,\mathrm{tr}\, y(t)y(t)'} \tag{2}$$

where $y(t) = k(z)u(t)$ and $(u(t)|t \in \mathbb{Z})$ is a q-dimensional white noise process with mean zero and a unit variance.

In the sections to come we will compare the L_2 reduced model $\bar{\theta}_{L_2} = \{\bar{A}, \bar{B}, \bar{C}, \bar{D}\}$ and the balanced truncated model $\tilde{\theta}_{b.t.} = \{\tilde{A}, \tilde{B}, \tilde{C}, \tilde{D}\}$ and show that there exist a state transformation $T(\varepsilon)$ such that

$$\bar{\theta}_{L_2} = T(\varepsilon)\cdot\tilde{\theta}_{b.t.} + O(\varepsilon^2). \tag{3}$$

3. L_2 MODEL REDUCTION

The L_2-model reduction problem may be stated as follows: Given $k \in \mathbb{M}_{n,p,q}$, find a transfer function $\bar{k} \in \mathbb{M}_{m,p,q}$, $m < n$ such that

$$J(\bar{k}, k) = \frac{1}{2}\|\bar{k} - k\|_2^2 \tag{4}$$

is minimal.

Let $\bar{\theta} = \{\bar{A}, \bar{B}, \bar{C}, \bar{D}\} \in \mathbb{S}_{m,p,q}$ and $\theta = \{A,B,C,D\} \in \mathbb{S}_{n,p,q}$ be two minimal realizations of $\bar{k} = \Pi(\bar{\theta})$ and of $k = \Pi(\theta)$, respectively. Furthermore let $e(t) = (\bar{k}(z) - k(z))u(t)$. With a slight abuse of notation, the L_2 criterion as a function of the system matrices is then given by

$$J(\bar{\theta}, \theta) := \frac{1}{2}\mathbb{E}\,\mathrm{tr}\, e(t)e(t)' \tag{5}$$

We denote an optimal L_2 reduced model by $\bar{\theta}_{L_2}$, i.e.,

$$J(\bar{\theta}_{L_2}, \theta) = \min_{\bar{\theta}} J(\bar{\theta}, \theta) \tag{6}$$

Of course one has to keep in mind that optimal models are only unique up to state transformations, i.e., $J(\bar{\theta}_{L_2}, \theta) = J(T\cdot\bar{\theta}_{L_2}, \theta)$.

By defining $\bar{x}(t) = (z^{-1}I - \bar{A})^{-1}\bar{B}u(t)$, the derivative of the "residual" $e(t)$ with respect to the matrices $\{\bar{A}, \bar{B}, \bar{C}, \bar{D}\}$ is given by:

$$
\begin{aligned}
de(t) =& d\bar{C}\bar{x}(t) - \bar{C}(z^{-1}I - \bar{A})^{-1}d\bar{A}\bar{x}(t) \\
&+ \bar{C}(z^{-1}I - \bar{A})^{-1}d\bar{B}u(t) + d\bar{D}u(t)
\end{aligned} \tag{7}
$$

The derivative of the criterion J with respect to the matrices $\{\bar{A}, \bar{B}, \bar{C}, \bar{D}\}$ is

$$
\begin{aligned}
dJ =& \mathbb{E}\,\mathrm{tr}\, de(t)e(t)' \\
=& -\mathrm{tr}\sum_{j>0}\bar{C}\bar{A}^{j-1}d\bar{A}\mathbb{E}\bar{x}(t-j)e(t)' \\
&+ \mathrm{tr}\sum_{j>0}\bar{C}\bar{A}^{j-1}d\bar{B}\mathbb{E}u(t-j)e(t)' + \\
&+ \mathrm{tr}\, d\bar{C}\mathbb{E}\bar{x}(t)e(t)' + \mathrm{tr}\, d\bar{D}\mathbb{E}u(t)e(t)'
\end{aligned} \tag{8}
$$

The quantities u, $\bar{x}$, e may be computed by the following state space model

$$
\left[\begin{array}{c|c} \hat{A} & \hat{B} \\ \hline \hat{C} & \hat{D} \end{array}\right] = \left[\begin{array}{cc|c} A & 0 & B \\ 0 & \bar{A} & \bar{B} \\ \hline 0 & 0 & I \\ 0 & I & 0 \\ -C & \bar{C} & \bar{D} - D \end{array}\right]. \tag{9}
$$

As easily can be seen, $\hat{y}(t) := (\hat{C}(z^{-1} - \hat{A})^{-1}\hat{B} + \hat{D})u(t) = (u(t)', \bar{x}(t)', e(t)')'$. Furthermore let us define

$$\hat{P} = \hat{A}\hat{P}\hat{A}' + \hat{B}\hat{B}' \tag{10}$$

$$\hat{M} = \hat{A}\hat{P}\hat{C}' + \hat{B}\hat{D}' \tag{11}$$

Then $\hat{\Omega}_0 := \mathbb{E}\hat{y}(t)\hat{y}(t)' = \hat{C}\hat{P}\hat{C}' + \hat{D}\hat{D}'$ and $\hat{\Omega}_j := \mathbb{E}\hat{y}(t-j)\hat{y}(t)' = \hat{M}'(\hat{A}')^{j-1}\hat{C}'$ for $j > 0$. Using these definitions the derivative dJ may be further simplified

$$\begin{aligned} dJ = & -\operatorname{tr} d\bar{A} S_x \hat{M}' \sum_{j>0} (\hat{A}')^{j-1} \hat{C}' S_e' \bar{C} \bar{A}^{j-1} \\ & + \operatorname{tr} d\bar{B} S_u \hat{M}' \sum_{j>0} (\hat{A}')^{j-1} \hat{C}' S_e' \bar{C} \bar{A}^{j-1} \\ & + \operatorname{tr} d\bar{C} S_x \hat{\Omega} S_e' + \operatorname{tr} \bar{D} S_u \hat{\Omega} S_e' \end{aligned} \quad (12)$$

Here $S_u = [I, 0, 0]$, $S_x = [0, I, 0]$ and $S_e = [0, 0, I]$ are suitably defined selection matrices. The sum $\sum_{j>0} (\hat{A}')^{j-1} \hat{C}' S_e' \bar{C} \bar{A}^{j-1}$ is the solution of the Lyapunov equation:

$$\hat{X} = \hat{A}' \hat{X} \bar{A} + \hat{C}' S_e' \bar{C} \quad (13)$$

and thus we finally obtain

$$\begin{aligned} dJ = & -\operatorname{tr} d\bar{A} S_x \hat{M}' \hat{X} + \operatorname{tr} d\bar{B} S_u \hat{M}' \hat{X} \\ & + \operatorname{tr} d\bar{C} S_x \hat{\Omega}_0 S_u' + \operatorname{tr} d\bar{D} S_u \hat{\Omega}_0 S_e'. \end{aligned} \quad (14)$$

In particular note: If $\bar{\theta} = \bar{\theta}_{L_2}$ is an optimal reduced order system, then we must have

$$\begin{aligned} & D_{\bar{A}} := S_x \hat{M}' \hat{X} = 0,\ D_{\bar{B}} := S_u \hat{M}' \hat{X} = 0, \\ & D_{\bar{C}} := S_x \hat{\Omega}_0 S_u' = 0,\ D_{\bar{D}} := S_u \hat{\Omega}_0 S_e' = 0 \end{aligned} \quad (15)$$

3.1 *Linear approximation of the optimal reduced order model*

In this section we compute a first order approximation of the optimal reduced order model. We make the assumption that the reduced order model $\bar{\theta}$ takes the following form

$$\begin{aligned} \bar{A} = & A_{0,11} + \varepsilon \bar{A}_1 + O(\varepsilon^2) \\ \bar{B} = & B_{0,1} + \varepsilon \bar{B}_1 + O(\varepsilon^2) \\ \bar{C} = & C_{0,1} + \varepsilon \bar{C}_1 + O(\varepsilon^2) \\ \bar{D} = & D_0 + \varepsilon \bar{D}_1 + O(\varepsilon^2). \end{aligned} \quad (16)$$

As mentioned above, (minimal) realizations of the optimal reduced order model are only unique up state transformations. By the above assumption w.l.o.g. we restrict ourselves to realizations which are "close" to the truncated models and thus in the limit converge to the truncation of the limiting model $\theta_0 = \{A_0, B_0, C_0, D_0\}$.

The strategy now is as follows. First a Taylor series expansion of the "derivatives" (15)

$$\begin{aligned} D_{\bar{A}} = & 0 + \varepsilon D_{1,\bar{A}} + O(\varepsilon^2) \\ D_{\bar{B}} = & 0 + \varepsilon D_{1,\bar{B}} + O(\varepsilon^2) \\ D_{\bar{C}} = & 0 + \varepsilon D_{1,\bar{C}} + O(\varepsilon^2) \\ D_{\bar{D}} = & 0 + \varepsilon D_{1,\bar{D}} + O(\varepsilon^2) \end{aligned} \quad (17)$$

is computed. (Note that the derivatives are zero for $\varepsilon = 0$.) Setting $D_{1,\bar{A}} = 0$, $D_{1,\bar{B}} = 0$, $D_{1,\bar{C}} = 0$, and $D_{1,\bar{D}} = 0$ gives four equations, which have to be solved for the unknowns $\bar{A}_1$, $\bar{B}_1$, $\bar{C}_1$, and $\bar{D}_1$.

In the course of these evaluations Taylor series expansions for the solutions of Lyapunov equations have to be computed. Consider e.g., the controllability Grammian of the model $\theta = \{A, B, C, D\}$, which is defined as the solution of the Lyapunov equation $P = APA' + BB'$. By the asymptotic stability assumption, we can compute a Taylor series expansion of P by the following recursions:

$$\begin{aligned} P = & P_0 + \varepsilon P_1 + \varepsilon^2 P_2 + \cdots \\ P_0 = & A_0 P_0 A_0' + B_0 B_0' \\ P_1 = & A_0 P_1 A_0' + B_1 B_0' + B_0 B_1' + A_1 P_0 A_0' + A_0 P_0 A_1' \\ P_2 = & A_0 P_2 A_0' + B_1 B_1' + A_1 P_1 A_0' + A_0 P_1 A_1' + A_1 P_0 A_1' \\ \cdots = & \cdots \end{aligned} \quad (18)$$

Furthermore due to the lower triangular block structure of A_0, it is easy to see that P has the form:

$$P = \begin{pmatrix} P_{0,11} + O(\varepsilon) & P_{0,12} + O(\varepsilon) \\ P_{0,21} + O(\varepsilon) & O(1) \end{pmatrix}, \quad (19)$$

where e.g., $P_{0,11} = A_{0,11} P_{0,11} A_{0,11}' + B_{0,1} B_{0,1}'$. In a similar manner, we obtain for the observability Grammian $Q = A'QA + C'C$:

$$Q = \begin{pmatrix} Q_{0,11} + O(\varepsilon) & \varepsilon Q_{1,12} + O(\varepsilon^2) \\ \varepsilon Q_{1,21} + O(\varepsilon^2) & O(\varepsilon^2) \end{pmatrix} \quad (20)$$

Note that by the assumption $k_0 \in \mathbb{M}_{m,p,q}$, it follows that $P_{0,11} > 0$ and $Q_{0,11} > 0$ holds.

Using the same strategy, one obtains

$$\hat{P} \doteq \begin{pmatrix} P_{0,11} + O(\varepsilon) & P_{0,12} + O(\varepsilon) & P_{0,11} + \varepsilon \hat{P}_{1,13} \\ P_{0,21} + O(\varepsilon) & O(1) & P_{0,21} + O(\varepsilon) \\ P_{0,11} + \varepsilon \hat{P}_{1,31} & P_{0,12} + O(\varepsilon) & P_{0,11} + \varepsilon \hat{P}_{1,33} \end{pmatrix} \quad (21)$$

and

$$\hat{X} \doteq \begin{pmatrix} Q_{0,11} + \varepsilon \hat{X}_{1,1} \\ \varepsilon Q_{1,21} \\ -Q_{0,11} + \varepsilon \hat{X}_{1,3} \end{pmatrix}. \quad (22)$$

Here and in the following $\doteq$ means that only terms up to order ε are considered.

Next these expressions are plugged into the "derivatives" (15), and then the first order terms are set equal to zero. This leads to the following four equations:

$$\begin{aligned} & Q_{1,12} A_{0,21} P_{0,11} + (\hat{X}_{1,1}' + \hat{X}_{1,3}) A_{0,11} P_{0,11} + \\ & Q_{0,11} A_{0,11} (\hat{P}_{1,13} - \hat{P}_{1,33}) + Q_{0,11} A_{1,12} P_{0,21} + \\ & Q_{1,12} A_{0,22} P_{0,21} + Q_{0,11} (A_{1,11} - \bar{A}_1) P_{0,11} = 0 \end{aligned} \quad (23)$$

$$Q_{1,12} B_{0,2} + (\hat{X}_{1,1}' + \hat{X}_{1,3}) B_{0,1} + Q_{0,11} (B_{1,1} - \bar{B}_1) = 0 \quad (24)$$

$$C_{0,1} (\hat{P}_{1,13} - \hat{P}_{1,33}) + C_{1,2} P_{0,21} + (C_{1,1} - \bar{C}_1) P_{0,11} = 0 \quad (25)$$

$$(D_1 - \bar{D}_1) = 0 \quad (26)$$

Finally, by some simple but tedious algebraic manipulations, we may eliminate $\hat{P}_{1,13}$, $\hat{P}_{1,33}$, $\hat{X}_{1,1}$ and $\hat{X}_{1,3}$, and we obtain the solutions as:

$$\begin{aligned}
\bar{A}_1 &= A_{1,11} + X_1 A_{0,11} - A_{0,11} X_1 \\
&\quad + Q_{0,11}^{-1} Q_{1,12} A_{0,21} + A_{1,12} P_{0,21} P_{0,11}^{-1} \\
&\quad + Q_{0,11}^{-1} Q_{1,12} A_{0,22} P_{0,21} P_{0,11}^{-1} \\
&\quad - A_{0,11} Q_{0,11}^{-1} Q_{1,12} P_{0,21} P_{0,11}^{-1} \\
\bar{B}_1 &= B_{1,1} + X_1 B_{0,1} + Q_{0,11}^{-1} Q_{1,12} B_{0,2} \\
\bar{C}_1 &= C_{1,1} - C_{0,1} X_1 + C_{1,2} P_{0,21} P_{0,11}^{-1} \\
&\quad - C_{0,1} Q_{0,11}^{-1} Q_{1,12} P_{0,21} P_{0,11}^{-1} \\
\bar{D}_1 &= D_1
\end{aligned} \tag{27}$$

where $X_1 \in \mathbb{R}^{m \times m}$ is arbitrary. All terms on the right hand side of the above equations are linear functions of $\{A_1, B_1, C_1, D_1\}$ except for the terms involving X_1. These terms correspond to a state space transformation $(I + \varepsilon X_1 + O(\varepsilon^2))$ of the reduced order model and reflect the non uniqueness of the realizations. (See also the discussion above and the next section.) Apart from this, the above relations correspond to the linearization of the mapping, attaching the optimal reduced order model to θ, in the point θ_0.

4. BALANCED TRUNCATION

Balanced truncation is a simple model reduction algorithm, which works as follows. Let T be a state space transformation, which renders a state space model $\theta = \{A, B, C, D\}$ into balanced form, i.e., such that the transformed Grammians are equal and diagonal: $TPT' = T^{-T} Q T^{-1} = \text{diag}(\sigma_1, \ldots, \sigma_n)$. Then simply take the truncation of the transformed system matrices as the reduced order model. To be more precise let $S = (I, 0) \in \mathbb{R}^{m \times n}$ be a selection matrix which picks the first m rows of a matrix. Then the reduced order model is obtained as $\tilde{\theta} = \{STAT^{-1}S', STB, CT^{-1}S', D\}$.

For our purpose, it suffices to consider "block" balanced truncations, where the transformation T is chosen such that the transformed Grammians are "block" diagonal and not necessarily equal. It is easy to see, that the corresponding truncation is equivalent to the "full" balanced truncation, in the sense that both represent the same transfer function. In other words they are related via a state transformation to each other.

Let

$$\begin{aligned}
T &= T_0 + \varepsilon T_1 + O(\varepsilon^2) \\
&\doteq \begin{pmatrix} I + \varepsilon T_{1,11} & \varepsilon T_{1,12} \\ -T_{0,21} + \varepsilon T_{1,21} & I + \varepsilon T_{1,22} \end{pmatrix}
\end{aligned} \tag{28}$$

be a state space transformation, which renders $\{A, B, C, D\}$ into a block balanced form. By easy calculations one may see that the transformed system matrices are of the form:

$$\begin{aligned}
TAT^{-1} &= \begin{pmatrix} \tilde{A} & O(\varepsilon) \\ O(1) & O(1) \end{pmatrix}, \quad TB = \begin{pmatrix} \tilde{B} \\ O(1) \end{pmatrix} \\
CT^{-1} &= (\tilde{C} \; O(\varepsilon)), \quad D.
\end{aligned} \tag{29}$$

The (block) balanced truncation is therefore given by:

$$\begin{aligned}
\tilde{A} &\doteq A_{0,11} + \varepsilon(A_{1,11} + T_{1,11} A_{0,11}) \\
&\quad - \varepsilon(A_{0,11}(T_{1,12} T_{0,21} + T_{1,11})) \\
&\quad + \varepsilon(T_{1,12} A_{0,21} + A_{1,12} T_{0,21}) \\
&\quad + \varepsilon(T_{1,12} A_{0,22} T_{0,21}) \\
\tilde{B} &\doteq B_{0,1} + \varepsilon(B_{1,1} + T_{1,11} B_{0,1} + T_{1,12} B_{0,2}) \\
\tilde{C} &\doteq C_{0,1} + \varepsilon(C_{1,1} - C_{0,1} T_{1,11} + C_{1,2} T_{0,21}) \\
&\quad - \varepsilon C_{0,1} T_{1,12} T_{0,21}
\end{aligned} \tag{30}$$

The transformed Grammians $\tilde{P} = TPT'$, $\tilde{Q} = T^{-T} Q T^{-1}$ are of the form

$$\tilde{P} = \begin{pmatrix} P_{0,11} + O(\varepsilon) & \tilde{P}_{12} \\ \tilde{P}_{21} & O(1) \end{pmatrix} \tag{31}$$

$$\tilde{Q} = \begin{pmatrix} Q_{0,11} + O(\varepsilon) & \tilde{Q}_{12} \\ \tilde{Q}_{21} & O(\varepsilon^2) \end{pmatrix} \tag{32}$$

where

$$\tilde{P}_{21} \doteq (P_{0,21} - T_{0,21} P_{0,11}) + O(\varepsilon) \tag{33}$$

$$\tilde{Q}_{21} \doteq \varepsilon(Q_{0,21} - T'_{1,12} Q_{0,11}) \tag{34}$$

If T is a block balancing transformation, then the off diagonal blocks of $\tilde{P}$ and $\tilde{Q}$ must be zero and thus $T_{1,12} = Q_{0,11}^{-1} Q_{1,12}$ and $T_{0,21} = P_{0,21} P_{0,11}^{-1}$ must hold. (Note that both P and Q are symmetric by construction.) Comparing (30) with (27) reveals that the (block) balanced truncation of $\{A, B, C, D\}$ is up to terms of $O(\varepsilon^2)$ equal to the optimal reduced order model given in (27), i.e.,

$$\begin{aligned}
\bar{A} &= \tilde{A} + O(\varepsilon^2), \quad \bar{B} = \tilde{B} + O(\varepsilon^2) \\
\bar{C} &= \tilde{C} + O(\varepsilon^2), \quad \bar{D} = \tilde{D} + O(\varepsilon^2),
\end{aligned} \tag{35}$$

if $T_{1,11} = X_1$ is chosen.

In order to make the above results independent of the choice of a particular realization, we need a (local) parametrization of the set $\mathbb{M}_{m,p,q}$ as follows: Let two smooth maps

$$\begin{aligned}
\Phi : \mathbb{S}_{m,p,q} &\to \mathbb{R}^{n(p+q)} \\
\theta &\mapsto \mu
\end{aligned} \tag{36}$$

and

$$\begin{aligned}
\Phi^{-1} : \mathbb{R}^{n(p+q)} &\to \mathbb{S}_{m,p,q} \\
\mu &\mapsto \theta
\end{aligned} \tag{37}$$

be given such that $\Phi(\theta_1) = \Phi(\theta_2)$ iff $\Pi(\theta_1) = \Pi(\theta_2)$ and $\Phi(\Phi^{-1}(\mu)) = \mu$. Thus μ may be interpreted as a vector of (free) parameters describing the transfer function $k = \Pi(\theta)$. Note that Φ and Φ^{-1} only need to be defined in suitably chosen neighborhoods of $\bar{\theta}_0 = \{A_{0,11}, B_{0,1}, C_{0,1}, D_0\}$ and of $\mu_0 = \Phi(\bar{\theta}_0)$.

An example of such a parametrization for the SISO case $p=q=1$ is given in the next section. For the general case canonical forms may be used to define Φ and Φ^{-1}.

Next let $\bar{\mu}_{L_2}(\theta)=\Phi(\bar{\theta}_{L_2})$ and $\tilde{\mu}_{b.t.}(\theta)=\Phi(\tilde{\theta}_{b.t.})$.

Using these definitions the main result now may be stated as follows:

Theorem 1. Let $\theta=\theta_0+\varepsilon\theta_1\in\mathbb{S}_{n,p,q}$ be a state space model of the form (1). Then

$$\bar{\mu}_{L_2}(\theta)=\tilde{\mu}_{b.t.}(\theta)+O(\varepsilon^2) \tag{38}$$

and

$$\left.\frac{\partial\bar{\mu}_{L_2}}{\partial\theta}\right|_{\theta_0}=\left.\frac{\partial\tilde{\mu}_{b.t.}}{\partial\theta}\right|_{\theta_0}. \tag{39}$$

Sloppy speaking this means that balanced truncation is "close" to the optimal reduced order model if the high order model is "close" to the class of low order models.

5. APPLICATIONS TO SYSTEM IDENTIFICATION

The result presented above can be applied to a system identification problem. Assume that a high order model has been estimated and we want to reduce this model to a good low order approximant. When doing this, it is of great interest to minimize the mean square error of the low order model. What is presented below is basically that the mean square error for L_2 model reduction and balanced truncation are asymptotically the same, when the high order model is reduced to a "correct" lower order. This result confirms the simulation results and ideas in (Tjärnström, 2002). See also (Wahlberg, 1987). For simplicity we describe the results in the SISO case, i.e., $p=q=1$.

Assume that data are generated from

$$y(t)=\bar{k}_0(z)u(t)+e(t), \tag{40}$$

where u is the input signal and e is white noise with variance λ. We assume that

$$\bar{k}_0(z)=\frac{b_1^0z^{n_k}+\cdots+b_{n_b}^0z^{n_k+n_b-1}}{1+f_1^0z+\cdots+f_{n_f}^0z^{n_f}}, \tag{41}$$

and for simplicity we also assume that $n_k=1, n_b=n_f=m$. (More general cases can be treated in a similar fashion.) Let $\mu_0=(b_1^0,\ldots,b_m^0,f_1^0,\ldots,f_m^0)'$ be the stacked vector of the coefficients of the denominator and nominator polynomials.

The considered high order models ($n>m$) are of the form

$$k(z,\eta)=\frac{b_1z^1+\cdots+b_nz^n}{1+f_1z^1+\cdots+f_nz^n} \tag{42}$$

where $\eta=(b_1,\ldots,b_n,f_1,\ldots,f_n)'$ again is the stacked coefficient vector.

Moreover, for the model $y(t)=k(z,\eta)u(t)+e(t)$ we define the prediction error as

$$\xi(t,\eta)=y(t)-\hat{y}(t|\eta)=y(t)-k(z,\eta)u(t). \tag{43}$$

Due to the overmodeling there exist a set $\mathscr{D}=\{\eta : k(e^{i\omega},\eta)=\bar{k}_0(e^{i\omega})\}$ and in particular

$$\eta_0=(b_1^0\cdots b_m^0\,0\cdots0\,f_1^0\cdots f_m^0\,0\cdots0)'\in\mathscr{D}. \tag{44}$$

Since the model uses too many parameters we usually lack from identifiability. To avoid this we estimate η from

$$\hat{\eta}=\arg\min_{\eta}V_N(\eta)+\frac{\delta}{2}\|\eta-\eta_0\|_2^2, \tag{45}$$

where δ is a positive scalar (usually called a regularization parameter), N is the number of data, and

$$V_N(\eta)=\frac{1}{N}\sum_{t=1}^{N}\xi^2(t,\eta). \tag{46}$$

We know that (Ljung, 1999) the asymptotic covariance of $\hat{\eta}$ equals

$$\operatorname{Cov}\hat{\eta}\approx\frac{\lambda}{N}(\mathbb{E}\Psi(t,\eta_0)\Psi'(t,\eta_0)+\delta I)^{-1}, \tag{47}$$

where

$$\Psi(t,\eta_0)=\left.\frac{\partial}{\partial\eta}\xi(t,\eta)\right|_{\eta_0}. \tag{48}$$

Writing $k(z,\eta_0)$ in controller form gives

$$\begin{aligned}x(t+1)&=A_0x(t)+B_0u(t)\\ y(t)&=C_0x(t)+D_0u(t),\end{aligned} \tag{49}$$

where

$$\begin{aligned}A_0&=\left(\begin{array}{cccc|ccc}-f_1^0&-f_2^0&\cdots&-f_m^0&0&\cdots&0\;0\\1&0&\cdots&0&0&\cdots&0\;0\\ \vdots&&&&&&\vdots\\0&0&\cdots&0&0&\cdots&1\;0\end{array}\right)\\ B_0'&=(1\;0\;\cdots\;0|0\;\cdots\;0)\\ C_0'&=(b_1^0\;b_2^0\;\cdots\;b_m^0|0\;\cdots\;0)\\ D_0&=0\end{aligned} \tag{50}$$

It is now clear that writing $k(z,\hat{\eta})$ in controller form gives a state space model of the form (1) with the following matrices equal to zero

$$A_{1,21},\ A_{1,22},\ B_{1,1},\ B_{0,2},\ B_{1,2},\ D_1. \tag{51}$$

Now we can construct two estimates of the low order model:

(1) Compute the balanced truncation $\tilde{\theta}_{b.t.}$ of the above mentioned controller form state space model $\hat{\theta}$ of $k(z,\hat{\eta})$, and let $\tilde{\mu}_{b.t.}$ denote the stacked vector of coefficients of the polynomial representation of the corresponding transfer function.

(2) Compute the optimal L_2 reduced order model $\bar{\theta}_{L_2}$ of the state space model $\hat{\theta}$, and let $\bar{\mu}_{L_2}$ denote the stacked vector of coefficients of the

polynomial representation of the corresponding transfer function. (Of course the L_2 reduction can also be directly performed for the corresponding polynomial representations, without the need to transform forth and back from state space to polynomial representations.)

Note that here the mapping Φ, used in section 4, maps a state space model $\bar{\theta} \in \mathbb{S}_{m,1,1}$ to the coefficient vector of the corresponding polynomial representation, and Φ^{-1} gives the controller form state space model of a polynomial representation.

Both $\tilde{\mu}_{b.t.}$ and $\bar{\mu}_{L_2}$ are functions of the estimates $\hat{\eta}$ of the high order model and by Theorem 1, and in particular from (39), it follows for the Jacobians that

$$H_0 := \left.\frac{\partial \bar{\mu}_{L_2}}{\partial \hat{\eta}}\right|_{\eta_0} = \left.\frac{\partial \tilde{\mu}_{b.t.}}{\partial \hat{\eta}}\right|_{\eta_0} \tag{52}$$

This in turn implies that

$$\mathrm{Cov}\,\bar{\mu}_{L_2} = H_0 \mathrm{Cov}\,\hat{\eta} H_0' = \mathrm{Cov}\,\tilde{\mu}_{b.t.} \tag{53}$$

i.e., both methods have asymptotically the same covariance.

6. CONCLUSIONS

In this paper we have discussed model reduction of an n-th order system that is "close" to an m-th order approximant. We have shown that reducing such a model by minimizing an unweighted L_2 model reduction criteria essentially gives the same result as reducing the model by balanced truncation. This result has also been applied in a system identification setting. Here we showed that estimating a high order model and then reducing it to correct order by subjecting it to L_2 reduction or balanced truncation gives low order models with the same covariance.

From the results presented here it is clear that balanced truncation can be used instead of or as an initial guess for L_2 reduction in certain situations. This will avoid problems with local minima that are present in the numerical optimization problems for computing the L_2 reduced model.

7. REFERENCES

Ferrante, A., W. Krajewski, A. Lepschy and U. Viaro (1999). Convergent algorithm for l_2 model reduction. *Automatica* **35**, 75–79.

Glover, K. (1984). All optimal hankel-norm approximations of linear multivariable systems and their L^∞-error bounds. *International Journal of Control* **39**(6), 1115–1193.

Ljung, L. (1999). *System Identification: Theory for the User*. 2nd ed.. Prentice-Hall. Upper Saddle River, NJ.

Moore, B. (1981). Principal component analysis in linear systems: Controllability, observability and model reduction. *IEEE Transactions on Automatic Control* **26**, 17–31.

Pernebo, L. and L. M. Silverman (1982). Model reduction via balanced state space representations. *IEEE Transactions on Automatic Control* **27**(2), 382–387.

Scherrer, W. (2002). Local optimality of minimum phase balanced truncation. In: *Proceedings ofthe 15th IFAC World Congress*. Barcelona, Spain.

Spanos, J. T., M. H. Milman and D. L. Mingori (1992). A new algorithm for L_2 optimal model reduction. *Automatica* **28**(5), 897–909.

Tjärnström, F. (2002). Variance Expressions and Model Reduction in System Identification. PhD thesis 730. Department of Electrical Engineering, Linköpings Universitet. Linköing, Sweden.

Wahlberg, B. (1987). On the Identification and Approximation of Linear Systems. PhD thesis 163. Department of Electrical Engineering, Linköping University.

Wilson, D. A. (1970). Optimum solution of model reduction problem. *Proceedings IEE* **117**(6), 1161–1165.

IFAC Publications
www.elsevier.com/locate/ifac

RECURSIVE EXACT H∞ IDENTIFICATION FROM IMPULSE RESPONSE MEASUREMENTS

Osamu Kaneko * **Paolo Rapisarda** **

* *Graduate school of Engineering Science, Osaka University, Machikaneyama, Toyonaka, Osaka, 560-8531, Japan*
** *Department of Mathemetics, University of Maastricht, P.O. Box 616, 6200 MD Maastricht, The Netherlands*

Abstract: We study the H_∞-partial realization problem from a behavioral point of view; we give necessary and sufficient conditions for solvability, and a characterization of all solutions. Instrumental in such analysis is the notion of time- and space-symmetrization of the data, which allows to transform the realization problem with metric- and stability constraints into an unconstrained behavioral modeling one.

Keywords: Behavioral approach; most powerful unfalsified model; quadratic difference forms; H_∞-partial realization; data symmetrization.

1. INTRODUCTION

In this paper we deal with the following problem: *Let w_0, w_1, ..., w_N be real numbers; find polynomials a and b such that*

(1) $\frac{a(\xi)}{b(\xi)} = \sum_{j=0}^{N} w_j \xi^{-j} + \varphi(\xi)\xi^{-(N+1)}$ *for some proper rational function* φ;
(2) *b is Hurwitz*;
(3) $\| \frac{a}{b} \|_\infty < 1$.

This is the Schur interpolation problem of analytic interpolation theory; we refer to (Schur, 1917),(Schur, 1918) for a treatment along the classical lines.

In this paper we call $(1)-(3)$ the H_∞*−partial realization problem.* In order to solve it, we adopt the point of view of partial realization as exact modeling of time series in the behavioral framework (see (Antoulas and Anderson, 1989),(Antoulas and Willems, 1993),(Antoulas, 1994), (Kuijper, 1997), (Kuijper and J.C.Willems, 1997),(Willems, 1986)). We model a finite number of time series derived from the impulse-response measurement, with the *Most Powerful Unfalsified Model* (MPUM in the following), which can be constructed iteratively; from a suitable representation of the MPUM, a solution to the H_∞−partial realization problem is easily computed. The novel aspect of our approach lies in how the metric and stability constraints are accommodated in such framework: we transform the H_∞−partial realization problem into an *unconstrained* behavioral modeling problem as follows. Besides modeling the "primal" data derived from $\{w_i\}_{i=0,1,\dots,N}$, we implicitly model also the *dualized data*, on which a special structure, symmetric in time and space with respect to the primal ones, has been imposed. The result of such modeling procedure is a kernel representation of the MPUM $\mathfrak{B}^*$ for the primal data, from which a kernel representation of the MPUM $\mathfrak{B}'^*$ for the dualized data is easily obtained. In this paper we also re-derive the well-known necessary and sufficient condition for the existence of a solution to the H_∞-realization problem, based on the positivity of a certain Stein matrix derived from the data (see (Bruinsma, 1991)). The last result presented in this paper is a characterization of all solutions to the H_∞-partial realization problem by means of a special representation of the MPUM $\mathfrak{B}^*$.

In (Antoulas and Anderson, 1989), the stable partial realization problem with metric constraints is solved. The main difference between this approach and ours lies in the fact that in our case, identification is performed in the time-domain. Moreover, the proof of the correctness of our procedure for performing H_∞-identification (see Remark 4.1 in section 4) is entirely self-contained, and need not refer to the algorithm of Schur.

The paper is organized as follows: in section 2 we illustrate the basics of exact behavioral modeling. In section 3 we introduce the concept of dualization of the data and the notion of Stein matrix associated with the data. Necessary and sufficient conditions for the existence of a solution are stated in section 4, while section 5 provides a characterization of all solutions, based on a special representation of the MPUM. We conclude the paper with section 6.

Due to the limitation of the space of the manuscript, we omit the detailed discussions and proofs. As for more details, see our forthcoming paper (Kaneko and Rapisarda, 2003).

Notation. We denote the set of nonnegative (nonpositive) integers with $\mathbb{Z}_+$ ($\mathbb{Z}_-$, respectively). We denote with $\mathbb{D}_e = \{z \in \mathbb{C} \mid |z| \geq 1\}$ the exterior of the open unit disk. The space of n dimensional real vectors is denoted by $\mathbb{R}^n$, and the space of $m \times n$ real matrices, by $\mathbb{R}^{m\times n}$. If $A \in \mathbb{R}^{m\times n}$, then $A^T \in \mathbb{R}^{n\times m}$ denotes its transpose. The set consisting of all sequences from $\mathbb{Z}_+$ to $\mathbb{R}^q$ (from $\mathbb{Z}_-$ to $\mathbb{R}^q$) is denoted with $(\mathbb{R}^q)^{\mathbb{Z}_+}$ ($(\mathbb{R}^q)^{\mathbb{Z}_-}$, respectively). On such space we define the *left*, i.e. *backward, shift* $(\sigma w)(t) := w(t+1)$ for all $t \in \mathbb{N}$. On $(\mathbb{R}^q)^{\mathbb{Z}_-}$ we define the *right*, i.e. *forward, shift* σ^*, defined as $(\sigma^* w)(t) = w(t-1)$. The ring of polynomials with real coefficients in the indeterminate ξ is denoted by $\mathbb{R}[\xi]$; the ring of two-variable polynomials with real coefficients in the indeterminates ζ and η is denoted by $\mathbb{R}[\zeta,\eta]$. The space of all $n \times m$ polynomial matrices in the indeterminate ξ is denoted by $\mathbb{R}^{n\times m}[\xi]$. Given a polynomial matrix $R(\xi) := R_0 + \ldots + R_L\xi^L \in \mathbb{R}^{n\times m}[\xi]$ with $R_L \neq 0$, we define its *reciprocal matrix* $R^r(\xi)$ as $R^r(\xi) := R_0\xi^L + \ldots + R_L \in \mathbb{R}^{n\times m}[\xi]$.

2. MODELING WITH BEHAVIORS

In this paper we consider linear, shift-invariant behaviors with time axis $\mathbb{Z}_+$ or $\mathbb{Z}_-$, in other words, subspaces of $(\mathbb{R}^q)^{\mathbb{Z}_+}$ (respectively, $(\mathbb{R}^q)^{\mathbb{Z}_-}$) consisting of trajectories $w : \mathbb{Z}_+ \to \mathbb{R}^q$ (respectively, $w : \mathbb{Z}_- \to \mathbb{R}^q$) such that if w belongs to the behavior, then also $\sigma^k w$ ($\sigma^{*k} w$) belong to the behavior for all $k \in \mathbb{Z}_+$. We denote the set of such behaviors with $\mathcal{L}^q(\mathbb{Z}_+)$ ($\mathcal{L}^q(\mathbb{Z}_-)$, respectively).

In (Willems, 1986) a framework has been developed for modeling in a behavioral framework; in such approach, the notion of most powerful unfalsified model is of fundamental importance, and we briefly review it now. In the following we limit ourselves to the treatment of sequences in $(\mathbb{R}^q)^{\mathbb{Z}_+}$, the case of $(\mathbb{R}^q)^{\mathbb{Z}_-}$ being completely analogous.

In this paper we consider sets $\mathcal{D} = \{d_i\}_{i=0,\ldots,N}$ of *data*, where $d_i \in (\mathbb{R}^q)^{\mathbb{Z}_+}$ for $0 \leq i \leq N$, and we pick our models from the *model class* $\mathcal{M} = \mathcal{L}^q(\mathbb{Z}_+)$. A model $\mathfrak{B}$ is called an *unfalsified model* for $\mathcal{D}$ if $\mathcal{D} \subseteq \mathfrak{B}$. We call a model $\mathfrak{B}^*$ the *most powerful unfalsified model* (abbreviated *MPUM*) for $\mathcal{D}$ if $\mathfrak{B}^* \supseteq \mathcal{D}$ and any other unfalsified model $\mathfrak{B}$ for $\mathcal{D}$ satisfies $\mathfrak{B}^* \subseteq \mathfrak{B}$. Observe that the MPUM restricts the possible outcomes of the phenomenon under study to the smallest possible set not refuted by the data.

The MPUM does not need to exist in general; it can be shown, however, that for the model classes $\mathcal{L}^q(\mathbb{Z}_+)$ and $\mathcal{L}^q(\mathbb{Z}_-)$ considered in this paper, the MPUM always exists and is uniquely determined; we denote such behavior with $\mathfrak{B}^*$. It can also be proved that $\mathfrak{B}^*$ can be represented as the kernel of a matrix polynomial operator $R(\sigma)$ in the shift σ, with the property that R is square and nonsingular as a polynomial matrix.

3. DATA DUALIZATION, AND THE STEIN MATRIX

In this paper, in addition to the notion of MPUM, we require the stability- and metric constraints. For this purpose, we introduce the concept of dualization of the data in this section. From the impulse response samples w_i, $i = 0,\ldots,N$ we define the following time series with time axis $\mathbb{Z}_+$:

$$d = \{\begin{pmatrix} w_N \\ 0 \end{pmatrix}, \ldots, \begin{pmatrix} w_1 \\ 0 \end{pmatrix}, \begin{pmatrix} w_0 \\ 1 \end{pmatrix}, \begin{pmatrix} 0 \\ 0 \end{pmatrix}, \ldots\} \tag{1}$$

We call (1) the *primal data*, and we refer to the problem of finding a representation of the MPUM for such time series and its left (backwards) shifts $\sigma^k d$, $k = 0, 1, \ldots$, as the *primal modeling problem.* We associate to the primal time series (1) its *dual*, defined as the time series with time axis $\mathbb{Z}_-$

$$d' = \{\ldots, \begin{pmatrix} 0 \\ 0 \end{pmatrix}, \begin{pmatrix} 1 \\ w_0 \end{pmatrix}, \begin{pmatrix} 0 \\ w_1 \end{pmatrix}, \ldots, \begin{pmatrix} 0 \\ w_N \end{pmatrix}\} \tag{2}$$

Observe that d' is obtained from d by reversing the direction of time and multiplying the result by $\Pi = \begin{pmatrix} 0 & 1 \\ 1 & 0 \end{pmatrix}$. We refer to the problem of finding (a representation of) the MPUM for (2) and its right (forward) shifts $\sigma^{*k} d'$, $k = 0, 1, \ldots$, as the *dual modeling problem.* Observe that the primal (dual) problem is a behavioral modeling problem with model class $\mathcal{L}^2(\mathbb{Z}_+)$ ($\mathcal{L}^2(\mathbb{Z}_-)$,respectively).

Finally, we introduce the notion of Stein matrix associated with the impulse-response data w_i, $i = 0, \ldots, N$.. Denote the successive left (backwards) shifts of the primal time series d as

$$d_k := \sigma^{N-k}d =$$

$$\{\begin{pmatrix} w_k \\ 0 \end{pmatrix}, \ldots, \begin{pmatrix} w_1 \\ 0 \end{pmatrix}, \begin{pmatrix} w_0 \\ 1 \end{pmatrix}, \begin{pmatrix} 0 \\ 0 \end{pmatrix}, \begin{pmatrix} 0 \\ 0 \end{pmatrix}, \ldots\} \tag{3}$$

$k = 0, 1, \ldots$, and consider the MPUM $\mathfrak{B}^*$ for $\{d_k\}_{k=0,1,\ldots,N}$; observe that $\mathfrak{B}^*$ consists of all linear combination of the d_k's. Let $v, w \in \mathfrak{B}^*$, and let J denote the 2×2 matrix

$$J = \begin{pmatrix} -1 & 0 \\ 0 & 1 \end{pmatrix}. \tag{4}$$

Such matrix induces the indefinite inner product on $\mathfrak{B}^*$ defined by $< v, w >_J := \sum_{k=0}^{\infty} v(k)^T Jw(k)$; such inner product is well-defined, since all trajectories in $\mathfrak{B}^*$ have compact support. The *Stein matrix* associated with $\{w_i\}_{i=0,\ldots,N}$ is the symmetric $(N+1) \times (N+1)$ matrix defined as

$$S_{\{d_i\}_{i=0,\ldots,N}} = (< d_i, d_j >_J) \tag{5}$$

Whenever the dimensions of the Stein matrix will be clear from the context, we will simply denote it with $S_{\{d_i\}}$.

4. NECESSARY AND SUFFICIENT CONDITIONS FOR SOLVABILITY

In this section, we provide our main result of this paper. That is, necessary and sufficient conditions for the existence of a solution to the H_∞-partial realization problem and the iterative algorithm to obtain the solution to this problem.

Theorem 4.1. The following three statements are equivalent:

(1) The Stein matrix $S_{\{d_i\}_{i=0,\ldots,N}}$ is positive definite;
(2) The MPUM $\mathfrak{B}^* \in \mathcal{L}^2(\mathbb{Z}_+)$ for the primal data $\{\sigma^k d\}_{k=0,1,\ldots} \subseteq (\mathbb{R}^2)^{\mathbb{Z}_+}$ and the MPUM $\mathfrak{B}'^* \in \mathcal{L}^2(\mathbb{Z}_-)$ for the dual data $\{\sigma^{*k} d'\}_{k=0,1,\ldots} \subseteq (\mathbb{R}^2)^{\mathbb{Z}_-}$ have kernel representations $\mathfrak{B}^* = \ker R(\sigma)$ and $\mathfrak{B}'^* = \ker R^r(\sigma^*)$ respectively, induced by a matrix $R := (r_{ij})_{i,j=1,2}$ satisfying the following properties:
   (a) $\begin{pmatrix} r_{11} & r_{12} \end{pmatrix} = \begin{pmatrix} r_{21} & r_{22} \end{pmatrix}^r \Pi$;
   (b) r_{22} is a Hurwitz polynomial;
   (c) $R(\xi)JR(\xi^{-1})^T = R(\xi^{-1})^T JR(\xi) = J = R(\xi^{-1})JR(\xi)^T = R(\xi)^T JR(\xi^{-1})$;
   (d) $\| \frac{r_{21}}{r_{22}} \|_\infty < 1$;
   (e) $\| \frac{r_{12}}{r_{22}} \|_\infty < 1$;
(3) There exists a solution to the H_∞−partial realization problem.

(*The outline of the Proof:*) Due to the limitation of the space, we focus on the outline of the proof of (1) $\Longrightarrow$ (2), which is used to derive an iterative algorithm for our recursive H_∞ idetification from impulse data. As for the proofs of (2) $\Longrightarrow$ (3) and (3) $\Longrightarrow$ (1), please see our forthcoming paper (Kaneko and Rapisarda, 2003).

Now we prove the implication (1) $\Longrightarrow$ (2), using induction on the number K of time series to be modeled. For $K = 0$, consider the model $\mathfrak{B}_0 \in \mathcal{L}^q(\mathbb{Z}_+)$ represented in kernel form by

$$R_0(\xi) = -\frac{1}{1-w_0^2}\begin{pmatrix} -1 & w_0 \\ -w_0 & w_0^2 \end{pmatrix} + \frac{1}{1-w_0^2}\begin{pmatrix} -w_0^2 & w_0 \\ -w_0 & 1 \end{pmatrix}\xi \tag{6}$$

Observe that $R_0(\sigma)d_0 = 0$ and consequently $\ker R_0(\sigma)$ is an unfalsified model for $\{\sigma^k d_0\}_{k=0,1,\ldots}$. Since $\det(R_0(\xi)) = \xi$, the dimension of $\ker R_0(\sigma)$ is one, and consequently $\{\sigma^k d_0\}_{k=0,1,\ldots} \supseteq \ker R_0(\sigma)$. This shows that R_0 is a representation of the MPUM $\mathfrak{B}_0$ for d_0. It is easy to verify in an analogous way, that the model $\mathfrak{B}_0' = \ker R_0^r(\sigma^*) \in \mathcal{L}^q(\mathbb{Z}_-)$ represented in kernel form by $R_0^r(\xi)$ is the MPUM for $\{\sigma^{*k} d_0'\}_{k=0,1,\ldots}$. Moreover, by using the Stein matrix $1 - w_0^2 > 0$ and so on, we can prove that the matrix R_0 defined in (6) satisfies (2*a*) - (2*d*).

Before proceeding with the inductive step, we prove a property of the error time-series $\varepsilon_1 = R_0(\sigma)d_1$ which is essential in order to apply the one-step model (6) iteratively. We claim that $S_{\{d_i\}} > 0$ implies that the second component of $\varepsilon_1(0)$ is nonzero; in such case ε_1 can be multiplied by a suitable constant so that the one-step model (6) can be applied iteratively. We call such property of the error time-series the *nondegeneracy property*. In order to prove such claim, observe that the second component of $\varepsilon_1(0) = (R_0(\sigma)d_1)(0)$ equals $\frac{1}{1-w_0^2}w_0 w_1 + 1$. Assume by contradiction that it is zero, equivalently, $-w_0 w_1 = 1 - w_0^2$. Observe that $-w_0 w_1$ equals the $(1,2)-$ and $(2,1)-$entry of $S_{\{d_i\}}$. Now substitute $1 - w_0^2$ for such entries, and transform $S_{\{d_i\}}$ as $T^T S_{\{d_i\}} T$ with the nonsingular matrix T defined as

$$T = \begin{pmatrix} 1 & -(1-w_0^2) & 0_{1\times(N-1)} \\ 0 & 1-w_0^2 & 0_{1\times(N-1)} \\ 0_{(N-1)\times 1} & 0_{(N-1)\times 1} & I_{N-1} \end{pmatrix}$$

It can be verified that the upper-left 2×2 submatrix of $T^T S_{\{d_i\}} T$ is $\mathrm{diag}(1 - w_0^2, -w_1^2(w_0^2 - 1)^2)$; since the $(2,2)-$entry of such submatrix is ≤ 0, this contradicts the positivity of $S_{\{d_i\}}$. We conclude that the nondegeneracy property holds.

We now go to the inductive step. Define the new set of data $\hat{d}_{k-1} := \sigma^{N-k} R_0(\sigma)d$, $k = 1, 2, \ldots K - 1$ and assume that if necessary, the normalization of the second component of $\hat{d}_0(0) = (\sigma^{N-1}R_0(\sigma)d)(0) = (R_0(\sigma)d_1)(0)$ has been carried out. In order to apply the inductive assumption, we now show that the Stein matrix $S_{\{\hat{d}_k\}_{k=0,1,\ldots K-2}}$ of the $\hat{d}_k$'s is positive definite.

In order to do this, denote the $K \times K$ Stein matrix $S_{\{d_k\}_{k=0,1,\ldots K-1}}$ of the original data with

$$S_{\{d_k\}_{k=0,1,\ldots K-1}} = \begin{pmatrix} <d_0, d_0>_J & b^T \\ b & S' \end{pmatrix}$$

where $b^T = (<d_0, d_1>_J \cdots <d_0, d_{K-1}>_J)$ and $S'_{ij} = <d_i, d_j>_J$, $i, j = 1, \ldots, K-1$. Let

$$T = \begin{pmatrix} 1 & -<d_0, d_0>_J^{-1} b^T \\ 0_{(K-1)\times 1} & I_{(K-1)\times(K-1)} \end{pmatrix}$$

and observe that T is well-defined and nonsingular because $<d_0, d_0>_J > 0$. Under the congruence transformation induced by T, the matrix $S_{\{d_k\}_{k=0,1,\ldots K-1}}$ becomes

$$T^T S_{\{d_k\}_{k=0,1,\ldots K-1}} T = \begin{pmatrix} <d_0, d_0>_J & 0_{1\times(K-1)} \\ 0_{(K-1)\times 1} & -<d_0, d_0>_J^{-1} bb^T + S' \end{pmatrix}. \quad (7)$$

Consider that $\hat{d}_i = \sigma^{N-i-1} R_0(\sigma) d = R_0(\sigma) d_{i+1}$, $i = 0, \ldots, K-2$ and consequently $<\hat{d}_i, \hat{d}_j>_J = \sum_{k=0}^{\infty} L_\Phi(d_{i+1}, d_{j+1})(k)$, where L_Φ is the bilinear difference form induced by $\Phi(\zeta, \eta) := R_0(\zeta)^T J R_0(\eta) \in \mathbb{R}^{2\times 2}[\zeta, \eta]$. It follows from (2c) that $\Phi(\xi^{-1}, \xi) = J$; applying the argument used in the necessity part of Lemma 3.1 of (Kaneko and Fujii, 2000) we conclude that $\Phi(\zeta, \eta) - J$ is divisible by $\zeta\eta - 1$, and consequently there exists $\Psi(\zeta, \eta) \in \mathbb{R}^{2\times 2}[\zeta, \eta]$ such that $\Phi(\zeta, \eta) - J = (\zeta\eta - 1)\Psi(\zeta, \eta)$, or equivalently $L_\Psi(v, w)(t+1) - L_\Psi(v, w)(t) = L_\Phi(v, w)(t) - L_J(v, w)(t)$ for all v and $w \in (\mathbb{R}^2)^Z$. Indeed, such equality holds with $\Psi(\zeta, \eta)$ defined as

$$\Psi(\zeta, \eta) = \frac{1}{1 - w_0^2} \begin{pmatrix} w_0^2 & -w_0 \\ -w_0 & 1 \end{pmatrix}.$$

Appplying $\hat{d}_i$ and $\hat{d}_j$ to this bilinear form and summing it up from 0 to ∞. After some algebraic computations, we obtain $<\hat{d}_i, \hat{d}_j>_J = -<d_0, d_{i+1}>_J <d_0, d_0>^{-1} <d_0, d_{j+1}>_J + <d_{i+1}, d_{j+1}>_J$ $(i, j = 0, \ldots, K-2)$, which is equal to the $(i+1, j+1)$th entry of $-<d_0, d_0>_J^{-1} bb^T + S'$. This yields immediately the positivity of the matrix $S_{\{\hat{d}_k\}_{k=0,1,\ldots K-2}}$.

Having shown this, we apply the inductive assumption, and conclude that there exists a representation R' of the MPUM for the $\hat{d}_k$'s, $k = 0, \ldots, K-2$, satisfying properties $(2a) - (2e)$ above, and such that the nondegeneracy property holds, i.e. that the second component of $(R'(\sigma)\hat{d}_{K-1})(0)$ is nonzero; observe that by inductive assumption, such representation induces also a representation of the MPUM for the dual problem. A representation of the MPUM for d_i, $i = 0, \ldots, K-1$, is obtained as

$$\underbrace{\begin{pmatrix} r_{11} & r_{12} \\ r_{21} & r_{22} \end{pmatrix}}_{R} := \underbrace{\begin{pmatrix} r'_{11} & r'_{12} \\ r'_{21} & r'_{22} \end{pmatrix}}_{R'} \underbrace{\begin{pmatrix} r^0_{11} & r^0_{12} \\ r^0_{21} & r^0_{22} \end{pmatrix}}_{R_0}$$

We now prove that such representation satisfies properties $(2a) - (2d)$ of the Theorem and the nondegeneracy property. In order to prove that $(2a)$ is satisfied, verify first that $\Pi R_0 = R_0^r \Pi$. Now use the inductive assumption $(2a)$ on R' and note the fact that the reciprocal of the product of two matrix polynomials is the product of the reciprocals. As a result, we can see that $\begin{pmatrix} r_{11} & r_{12} \end{pmatrix} = \begin{pmatrix} r_{21} & r_{22} \end{pmatrix}^r \Pi$. In order to prove $(2b)$, assume by contradiction that there exists $\mu \in \mathbb{D}_e$ such that $0 = r_{22}(\mu) = r'_{21}(\mu) r^0_{12}(\mu) + r'_{22}(\mu) r^0_{22}(\mu)$. Since by inductive assumption r'_{22} and r^0_{22} are Hurwitz, $r'_{22}(\mu) r^0_{22}(\mu) \neq 0$ and consequently we can write $\left|\frac{r'_{21}(\mu)}{r'_{22}(\mu)}\right| \left|\frac{r^0_{12}(\mu)}{r^0_{22}(\mu)}\right| = 1$, which yields a contradiction with $\| \frac{r^0_{12}}{r^0_{22}} \|_\infty < 1$, $\| \frac{r'_{21}}{r'_{22}} \|_\infty < 1$. Consequently, r_{22} is Hurwitz. In order to prove $(2c)$, observe that $R(\xi) J R(\xi^{-1})^T = R'(\xi) R_0(\xi) J R_0(\xi^{-1})^T R'(\xi^{-1})^T = R'(\xi) J R'(\xi^{-1})^T = J$ Analogous computations yield the other equalities of $(2c)$. In order to prove $(2d)$, and $(2e)$, proceed with the same argument used in the proof of the $K = 0$ step. Finally, we prove the nondegeneracy property. This follows from $R(\sigma) d_K = R'(\sigma) \hat{d}_{K-1}$ and from the inductive assumption. This concludes the proof of $(1) \Longrightarrow (2)$.□

This theorem connects the Stein matrix, the solvability of the $H_\infty-$partial realization problem, and the notion of MPUM. The first connection is well-known, see for example (Bruinsma, 1991), while the second is reminiscent of the results of (Rapisarda and Willems, 1997), where a "symmetrization" procedure on the data is used in place of the dualization considered above.

Remark 4.1. From the proof of the implication $(1) \Longrightarrow (2)$ in Theorem 4.1, an iterative algorithm arises, that takes as inputs the impulse response samples and provides as output a representation R of the MPUM satisfying $(a) - (e)$, from which a solution to the $H_\infty-$partial realization problem is easily computed. We state such algorithm explicitly:

Input: $N + 1$ impulse response samples

$$w_0, w_1, \ldots, w_N$$

Output: an MPUM representation satisfying $(a)-(e)$, if such representation exists
Compute the Stein matrix S of the data, and check whether $S > 0$; if no, then exit: no representation satisfying $(a) - (e)$ exists.
Let $R_{-1}(\xi) = I_{2\times 2}$;
For $i = 0, \ldots, N$ do
  Compute $i-$th error series $\varepsilon_i := R_{i-1}(\sigma) d_i$, with d_i defined as in (3);
  Normalize second component of $\varepsilon_i(0)$ to 1;
  Compute one-step model $V_i(\xi)$ for ε_i as in (6);
  Define $R_i(\xi) := V_i(\xi)\ R_{i-1}(\xi)$;
end for;
Return $R_N(\xi)$;

Example 4.1. Consider the numbers $\{w_0 = 0, w_1 = \frac{1}{6}, w_2 = \frac{1}{36}\}$, corresponding to the data

$$d = \{ \begin{pmatrix} \frac{7}{216} \\ 0 \end{pmatrix}, \begin{pmatrix} \frac{1}{36} \\ 0 \end{pmatrix}, \begin{pmatrix} \frac{1}{6} \\ 0 \end{pmatrix}, \begin{pmatrix} 0 \\ 1 \end{pmatrix}, \begin{pmatrix} 0 \\ 0 \end{pmatrix}, \ldots \}$$

The Stein matrix corresponding d is easily seen to be positive definite. From (1) of Theorem 4.1 we conclude that there exists a solution to the H_∞–partial realization problem. We now use the algorithm above in order to compute one such solution.

Since $d_0 = \varepsilon_0 = \{ \begin{pmatrix} 0 \\ 1 \end{pmatrix}, \begin{pmatrix} 0 \\ 0 \end{pmatrix}, \ldots \}$, the model for such time-series is given by (6):

$$R_0(\xi) = \begin{pmatrix} 1 & 0 \\ 0 & \xi \end{pmatrix}$$

Such model satisfies the properties $(2a) - (2e)$ of Theorem 4.1. Next, from equation (6) we devise a model for

$$\varepsilon_1 = R_0(\sigma) d_1 = \{ \begin{pmatrix} 1/6 \\ 1 \end{pmatrix}, \begin{pmatrix} 0 \\ 0 \end{pmatrix}, \ldots \}$$

Observe that there is no need to normalize the second component of $\varepsilon_1(0)$. We obtain

$$V_1(\xi) = \frac{1}{35} \begin{pmatrix} 36 - \xi & -6 + 6\xi \\ 6 - 6\xi & -1 + 36\xi \end{pmatrix}$$

This implies

$$R_1(\xi) := V_1(\xi) R_0(\xi) = \frac{1}{35} \begin{pmatrix} 36 - \xi & -6\xi + 6\xi^2 \\ 6 - 6\xi & -\xi + 36\xi^2 \end{pmatrix}$$

By computations, one casn see that such model satisfies properties $(2a) - (2e)$ of Theorem 4.1. Moreover one can see that the Markov parameters of $-\frac{r_{21}}{r_{22}}$ are given by

$$-\frac{r_{21}(\xi)}{r_{22}(\xi)} = 0\ \xi^0 + \frac{1}{6}\xi^{-1} - \frac{35}{216}\xi^{-2} + \ldots$$

We now consider the modeling of

$$d_2 = \{ \begin{pmatrix} 1/36 \\ 0 \end{pmatrix}, \begin{pmatrix} 1/6 \\ 0 \end{pmatrix}, \begin{pmatrix} 0 \\ 1 \end{pmatrix}, \begin{pmatrix} 0 \\ 0 \end{pmatrix}, \ldots \}. \quad (8)$$

The error time series $\varepsilon_2 = R_1(\sigma) d_2$ equals

$$\varepsilon_2 = \{ \begin{pmatrix} 41/216 \\ 211/216 \end{pmatrix}, \begin{pmatrix} 0 \\ 0 \end{pmatrix}, \ldots \}$$

Normalizing the second component of $\varepsilon_2(0)$ yields the to-be-modeled time series

$$\{ \begin{pmatrix} 41/211 \\ 1 \end{pmatrix}, \begin{pmatrix} 0 \\ 0 \end{pmatrix}, \ldots \}$$

After computing from (6) a model represented by V_2 for such time series, a representation for the model for d_2 is obtained as

$$R_2(\xi) = \frac{1}{1499400} \begin{pmatrix} -50225\xi^2 - 1225\xi + 1550850 & 50225\xi^3 - 42875\xi^2 - 258415\xi \\ 50225 - 42875\xi - 258415\xi^2 & -50225\xi - 1225\xi^2 + 1550850\xi^3 \end{pmatrix} \quad (9)$$

It is a matter of rather tedious computations to check that the model satisfies property $(2a)$. As for property $(2b)$, observe that the roots of the $(2,2)$–element of such model are 0, -0.179565, and 0.180355. Property $(2c)$ is a matter of straightforward verification. As for properties $(2d) - (2e)$, observe that using `Matlab` it can be computed that $\| \frac{r_{21}}{r_{22}} \|_\infty = 0.35068 = \| \frac{r_{12}}{r_{22}} \|_\infty$. Finally, observe that

$$-\frac{r_{21}(\xi)}{r_{22}(\xi)} = 0\ \xi^0 + \frac{1}{6}\xi^{-1} + \frac{1}{36}\xi^{-2} + \frac{1229}{45576}\xi^{-3} + \ldots$$

as it is expected from the proof of the implication $(2) \Longrightarrow (3)$ of Theorem 4.1.

5. CHARACTERIZING ALL SOLUTIONS

It is easy to see that if a representation R of the MPUM for d is available, then all unfalsified models M for the data can be represented as $M = R'R$ for some polynomial matrix R'. We use this fact in order to derive a characterization of all solutions to the H_∞-partial realization problem, given in terms of an MPUM representation satisfying $(2a) - (2e)$ of Theorem 4.1.

Proposition 5.1. Let a kernel representation R of the MPUM for the data be given as in Theorem 4.1, and let $p, q \in \mathbb{R}[\xi]$. Then $\frac{p}{q}$ is a solution to the H_∞-partial realization problem if and only if there exists π, φ, with φ Hurwitz and $\| \frac{\pi}{\varphi} \|_\infty < 1$, such that

$$\begin{pmatrix} p & -q \end{pmatrix} = \begin{pmatrix} \pi & -\varphi \end{pmatrix} R \quad (10)$$

(*The outline of the proof*): *(Necessity)* Without loss of generality, assume that p and q are coprime; observe that this implies that $\begin{pmatrix} -q(0) & p(0) \end{pmatrix} \neq 0$. We can see that if $\frac{p}{q}$ has the expansion $w_0 + w_1\xi^{-1} + w_2\xi^{-2} + \ldots + w_N\xi^{-N} + \ldots$ then $\begin{pmatrix} -q & p \end{pmatrix}^r$ represents an unfalsified model for the data d of (1). Since R is an MPUM for d, $\begin{pmatrix} -q & p \end{pmatrix}^r$ represents an unfalsified model for the data d of (1) if and only if there exist polynomials π' and φ' such that $\begin{pmatrix} -q & p \end{pmatrix}^r = \begin{pmatrix} -\varphi' & \pi' \end{pmatrix} R$. Now take the reciprocal of both sides of such equality, and observe that since $\begin{pmatrix} -q(0) & p(0) \end{pmatrix} \neq 0$, it holds $(\begin{pmatrix} -q & p \end{pmatrix}^r)^r = \begin{pmatrix} -q & p \end{pmatrix}$. Conclude that

$$\begin{pmatrix} -q & p \end{pmatrix} = \begin{pmatrix} -\varphi' & \pi' \end{pmatrix}^r R^r \quad (11)$$

Now multiply both sides of equation (11) on the right by Π and note that $R^r\Pi = \Pi R$ holds if R satisfies property $(2a)$ of Theorem 4.1. Then we conclude that equation (10) holds with polynomials π and φ defined by $\begin{pmatrix} \pi & -\varphi \end{pmatrix} := \begin{pmatrix} -\varphi' & \pi' \end{pmatrix}^r \Pi$.

We proceed to prove that (10) and $\| \frac{p}{q} \|_\infty < 1$ imply that $\| \frac{\pi}{\varphi} \|_\infty < 1$. Use (10) and property $(2c)$ of Theorem 4.1 to write $q(\xi)q(\xi^{-1}) - p(\xi)p(\xi^{-1})$ as

$$\begin{pmatrix} \pi(\xi) & -\varphi(\xi) \end{pmatrix} \underbrace{R(\xi) J R(\xi^{-1})^T}_{J} \begin{pmatrix} \pi(\xi^{-1}) \\ -\varphi(\xi^{-1}) \end{pmatrix}$$
$$= \varphi(\xi)\varphi(\xi^{-1}) - \pi(\xi)\pi(\xi^{-1}).$$

Now let $\xi = e^{i\omega}$ and conclude that $q(e^{i\omega})q(e^{-i\omega}) - p(e^{i\omega})p(e^{-i\omega}) = \varphi(e^{i\omega})\varphi(e^{-i\omega}) - \pi(e^{i\omega})\pi(e^{-i\omega})$. Consequently, $\| \frac{\pi}{\varphi} \|_\infty < 1$ iff $\| \frac{p}{q} \|_\infty < 1$.

We now prove that φ is Hurwitz. Observe first that since $\| \frac{\pi}{\varphi} \|_\infty < 1$ and $\| \frac{r_{12}}{r_{22}} \|_\infty < 1$, the function $\frac{1}{\frac{\pi}{\varphi}\frac{r_{12}}{r_{22}}-1}$ is well-defined. It follows from $-q = \pi r_{12} - \varphi r_{22}$ that $-\frac{1}{q} = \frac{1}{\varphi}\frac{1}{\frac{\pi}{\varphi}\frac{r_{12}}{r_{22}}-1}\frac{1}{r_{22}}$. And then, by computing its winding number *wno* of this rational function, we can see that φ is Hurwitz(for the limitationof the space, we omit the detailed discussion here), which completes the proof of necessity.

(Sufficiency) Take the reciprocals of both sides of the equality (10) and multiply on the right by Π in order to show that

$$\begin{pmatrix} p & -q \end{pmatrix}^r \Pi = \begin{pmatrix} -q & p \end{pmatrix}^r = \begin{pmatrix} \pi & -\varphi \end{pmatrix}^r R^r \Pi$$

Again, by noting that $R^r \Pi = \Pi R$, we can see that $\begin{pmatrix} -q & p \end{pmatrix}^r = \begin{pmatrix} \pi' & -\varphi' \end{pmatrix} R$ for suitable π', $\varphi' \in \mathbb{R}[\xi]$. It follows that $\ker \begin{pmatrix} -q & p \end{pmatrix}^r (\sigma)$ is an unfalsified model for $\{\sigma^k d\}_{k=0,\ldots}$, which implies $\frac{p}{q}$ is a solution to the partial realization problem.

In order to prove that $\| \frac{\pi}{\varphi} \|_\infty < 1$ implies $\| \frac{p}{q} \|_\infty < 1$, use the same argument applied to the converse implication in the "necessity" part above.

We conclude the proof showing that φ Hurwitz implies q is Hurwitz. First, use (10) to conclude that $q = \varphi r_{22} - \pi r_{12}$. Assume by contradiction that there exists $\mu \in \mathbb{D}_e$ such that $q(\mu) = 0$; then $\varphi(\mu) r_{22}(\mu) - \pi(\mu) r_{12}(\mu) = 0$. Since φ and r_{22} are Hurwitz, $\varphi(\mu) \neq 0$ and $r_{22}(\mu) \neq 0$; then the last equality implies $\frac{\pi(\mu)}{\varphi(\mu)}\frac{r_{12}(\mu)}{r_{22}(\mu)} = -1$ from which $\left| \frac{\pi(\mu)}{\varphi(\mu)}\frac{r_{12}(\mu)}{r_{22}(\mu)} \right| = \left| \frac{\pi(\mu)}{\varphi(\mu)} \right| \left| \frac{r_{12}(\mu)}{r_{22}(\mu)} \right| = 1$ follows. The last equality, however, is a contradiction with $\| \frac{\pi}{\varphi} \|_\infty < 1$, with statement (2*e*) of Theorem 4.1, and with the maximum modulus theorem. This concludes the proof of the Proposition.

Example 5.1. Again consider the same data described by Eq.(8) in Example 4.1 and the model R_2 descried by Eq.(9). It has been observed already that another solution of the H_∞–partial realization problem is $\frac{\xi}{6\xi^2-\xi-1}$. This solution can be obtained by premultiplying $R_2(\xi)$ by $[\pi \; \varphi]$ where

$$\pi(\xi) = 1.2848\xi^2 - 0.16127\xi - 0.18802$$
$$\varphi(\xi) = 5.6765\xi^2 - 0.96685\xi + 0.9671$$

Observe that φ is Hurwitz. and $\| \frac{\pi}{\varphi} \|_\infty = \frac{1}{4} < 1$ Hence the above π and φ satisfy the conditions stated in Proposition 5.1.

6. CONCLUSION

We have considered the H_∞-partial realization problem in a behavioral framwork. The main result of this paper is Theorem 4.1, which connects the solutions to the H_∞-partial realization problem with the MPUM and the Stein matrix associated with the data. As a ramification of such result, we obtained the iterative algorithm presented in section 4, which computes a special representation for the MPUM from which a solution to the H_∞-partial realization problem is easily obtained. By means of such representation of the MPUM, one can characterize all solutions to the H_∞-partial realization problem (Proposition 5.1).

In this manuscript, we have omitted the detailed proofs and discussions. They are given in our forthcoming paper(Kaneko and Rapisarda, 2003).

7. REFERENCES

Antoulas, A.C. (1994). Recursive modeling of discrete-time time series. *Linear Algebra for Control Theory, P. Van Dooren and B. Wyman (eds.), Spinger-Verlag, IMA* **62**, 1–20.

Antoulas, A.C. and B.D.O Anderson (1989). On the problem of stable rational interpolation. *Linear Algebra and its Applications* **122/123/124**, 301–329.

Antoulas, A.C. and J.C. Willems (1993). A behavioral approach to linear exact modeling. *IEEE Trans. Aut. Contr.* **38**, 1776–1802.

Bruinsma, P. (1991). *Interpolation problems for Schur and Nevanlinna pairs.* Ph.D. Thesis, University of Groningen.

Kaneko, O. and P. Rapisarda (2003). Recursive exact $h\infty$ idetification from impulse response measurement. *Systems and control letters* **-to appear**, –.

Kaneko, O. and T. Fujii (2000). Discrete time average positivity and spectral factorization in a behavioral framework. *Syst. and Contr. Lett.* **39**, 31–44.

Kuijper, M. (1997). An algorithm for constructing a minimal partial realization in the multivariable case. *Systems and Control Letters* **31**, 225–233.

Kuijper, M. and J.C.Willems (1997). On constructing a shortest linear recurrence relation. *IEEE Trans. Aut. Contr.* **42**, 1554–1558.

Rapisarda, P. and J.C. Willems (1997). The subspace nevenlinna interpolation problem and the most powerful unfalsified model. *Syst. and Contr. Lett.* **32**, 291–300.

Schur, J. (1917). Uber potenzreihen, die im innern des einheitskreises beschrankt sind-i. *J. Reine Angew. Math.* **147**, 205–232.

Schur, J. (1918). Uber potenzreihen, die im innern des einheitskreises beschrankt sind-ii. *J. Reine Angew. Math.* **148**, 122–145.

Willems, J.C. (1986). From time series to linear system, part ii: Exact modeling. *Automatica* **22**, 675–694.

IFAC Publications
www.elsevier.com/locate/ifac

PROPERTIES OF OPTIMAL SOLUTIONS IN L_1 IDENTIFICATION PROBLEM

Mehrzad Namvar * , **Alina Besançon-Voda** **

* *Canadian Space Agency, 6767 route de l'Aéroport, St-Hubert, Québec, J3Y 8Y9, Canada, mehrzad.namvar@space.gc.ca*
** *Laboratoire d'Automatique de Grenoble, ENSIEG, Bp.46, 38402, St.Martin d'Heres, France, Alina.Besancon@inpg.fr*

Abstract: This paper presents primal-dual formulation of a convex optimization problem related to l_1 interpolatory-projection algorithm. The use of duality theory permits to obtain information on the structure of optimal solutions for l_1 identi£cation problem and gives extra analytical tools for the analysis of identi£cation error and the estimation of a lowerbound for noise amplitude. The proposed analysis applies also to the case of l_1 model validation problem and characterizes properties of uncertainty block and disturbance signal. A simulation example illustrates the results.

Keywords: Identi£cation, Model validation, Duality theory, Optimization

1. INTRODUCTION

Identi£cation of nominal models and quanti£cation of model uncertainty in different norms have been widely studied in the literature during the past several years either through non-parametric approach (Gu and Khargonekar 1992), (Mäkilä 1991), (Jacobson *et al.* 1992), (Zhou and Kimura 1995) or parametric approach (Giarré *et al.* 1997), (Milanese 1996), (Kacewicz 1999), (Wang *et al.* 1999), (Namvar and Besançon-Voda 2001). Depending on *a priori* assumptions on noise, model sets and based on different types of algorithms, the identi£cation problems are usually formulated and solved by convex or non-convex optimizations. While in most approaches the main focus is on the analysis and synthesis of identi£cation algorithms and their informational complexity, the analysis of the resulting optimization problems has received less attention. From optimization point of view, however it is known that primal-dual formulation of an optimization problem can give more information on the structure of optimal solutions that are not necessarily unique. For example, in l_1 control problem, the analysis of primal-dual optimization problems has resulted in the characterization of all optimal controllers (Elia and Dahleh 1997), (Luenberger 1969).

The proposed analysis of this paper has many applications among which is £nding the structure of uncertainty block and disturbance signal in l_1 model validation problem (Smith 1992). Another application is computation of an upperbound for modelling error in l_1 identi£cation problem. The upperbound normally depends on *a priori* knowledge of plant relative degree and noise amplitude, however we will show how this upperbound can be approximated by a quantity that its computation does not need any *a priori* information.

As another application of the proposed analysis we will demonstrate how a lowerbound for noise amplitude can be computed. In fact, in many identi£cation formulations the difference between plant and model output, is attributed to both unmodeled dynamics and noise. Unmodeled dynamics is caused by the mismatch between the plant-model structures and noise is caused by different sources such as sensors, quantization, non-zero initial conditions, etc. Many identi£cation algorithms such as interpolatory or optimal algorithms require *a priori* knowledge of noise level (Kacewicz 1999).

Let u be plant input and y be plant output which is contaminated additively by a bounded noise $v(t)$ with amplitude σ_v

$$y(t) = Gu(t) + v(t) \tag{1.1}$$

where G represents the plant belonging to the space of stable LTI systems, $\mathcal{B}$. We assume that $\|u\|_\infty \le 1$. The identi£cation problem in this paper is stated as follows:

Problem 1.1. Given the model structure

$$y(t) = (\hat{G} + \Delta)u(t) + w(t) \qquad (1.2)$$

identify the model $\hat{G}$ and compute a sequence like w with $\|w\|_\infty \leq \sigma$ together with coef£cients of impulse response of an operator like Δ with minimum l_1 norm, such that the model structure is consistent with data. Moreover, compute an upperbound for the modelling error $\|G - \hat{G}\|_1$.

The model $\hat{G}$ is assumed to belong to the set of linearly parameterized models with £xed order (Heuberger *et al.* 1995)

$$\mathcal{G}^n = \{\hat{G} \mid \hat{G} = \sum_{k=0}^{n} F_k(z)\theta_k, \; F_k(z) \in \mathcal{B}\} \qquad (1.3)$$

where $F_k(z)$s are given basis functions. θ_ks are model parameters. Moreover, Δ is a bounded linear operator expanded by

$$\Delta = \sum_{k=0}^{\infty} \vartheta_k z^{-1} \qquad (1.4)$$

where ϑ_ks are coef£cients of impulse response of Δ. The sequence w in (1.2) represents the effect of output noise v. The solutions of the identi£cation problem (1.1) are denoted by model parameters θ^o, the coef£cients of impulse response of Δ^o (ϑ_k^os) and the sequence w^o. The minimum achieved value of $\|\Delta^o\|_1 = \sum_{k=0}^{\infty} |\vartheta_k^o|$, is denoted by μ^o. In problem 1.1, σ is a parameter whose value should be £xed each time the problem is going to be solved. Therefore, all optimal solutions such as μ^o are functions of σ. We will analyze the dependence of optimal solutions on σ. Note the difference between σ and σ_v, the true noise amplitude. It will be shown that Problem 1.1 can be solved via an in£nite dimensional linear programming.

It is seen from (1.2) that $\Delta u + w$ is the difference between the plant output y and model output $\hat{G}u$. Therefore, both Δ and w represent the mismatch between the plant and model outputs. Intuitively, we expect that by choosing a large value for σ, the mismatch is more explained by w and consequently the resulting μ^o becomes small. In fact, it will be shown that μ^o is a monotonically decreasing function of σ or more precisely $\frac{d\mu^o}{d\sigma} \leq -1$. Moreover, we will show that under some assumptions on u and y, Δ^o has *£nite* impulse response despite the fact that data sequences are in£nite dimensional.

The above mentioned properties of optimal solutions help to analyze the modelling error $\|G - \hat{G}\|_1$. Assuming that the distance of plant from $\mathcal{G}^n$ is not larger than a given value μ^*, the upperbound for modelling error can be expressed in terms of μ^* and σ_v. However, it will be shown that this upperbound can be approximated by a quantity that only depends on μ^o and σ and consequently its computation does not need knowledge of μ^* and σ_v. A measure of the approximation error will also be given.

Notations: The following notations are used throughout the paper.

$v^{[N]}$ is de£ned by $\{v(t) \mid v(t) \in \mathcal{R}, t = 0, ..., N\}$. A lower triangular Toeplitz matrix constructed from v is denoted by T_v:

$$T_v \triangleq \begin{bmatrix} v(0) & 0 & \cdots & 0 \\ v(1) & v(0) & \cdots & 0 \\ v(2) & v(1) & \cdots & 0 \\ \vdots & \vdots & \ddots & \vdots \end{bmatrix}$$

T_v^i consists of the £rst i columns of T_v and $T_v(k)$ denotes k'th column of T_v.

l_∞ denotes the space of all one-sided real sequences v with bounded norm de£ned by $\|v\|_\infty = \sup_{t\geq 0} |v(t)| < \infty$ and l_1 denotes the space of all one-sided real sequences g with norm de£ned by $\|g\|_1 = \sum_{t=0}^{\infty} |g(t)| < \infty$. If $q, p \in l_\infty$ and $p(t) = \Delta q(t)$ then $\|\Delta\|_1 \triangleq \sup_{q\neq 0} \frac{\|p\|_\infty}{\|q\|_\infty}$. $\mathcal{B}$ denotes the space of all stable LTI systems. $c_0 \triangleq \{v \in l_\infty \mid \lim_{t\to\infty} v(t) = 0\}$. The dual space of c_0 is l_1.

The remainder of this paper is as follows. In section 3 we present primal-dual formulation of identi£cation problem and analyze the structure of optimal solution. Some applications of this analysis are also given. In section 3, a simulation example is presented and section 4 is the conclusion of the paper.

2. STRUCTURAL ANALYSIS OF OPTIMAL SOLUTIONS

In this section the linear programming related to Problem 1.1 and its dual are formulated and some applications of this primal-dual formulation are presented. We consider the case where u and y are in£nite dimensional sequences. Nonetheless, the interpretation of results for £nite dimensional case will be straightforward. The results of this section are based on the application of duality theory in linear vector spaces, summarized brie¤y in Appendix A.

The following linear programming solves the identi£cation problem 1.1 (We assume that $\hat{G}$ and Δ are expanded by (1.3) and (1.4), respectively).

$$\min_{\vartheta,\theta,w} \; \|\vartheta\|_1 \qquad (2.5)$$

$$\text{subject to}: \; y = T_u \bar{F}\theta + T_u\vartheta + w, \qquad (2.6)$$

$$\|w\|_\infty \leq \sigma \qquad (2.7)$$

where $\bar{F}$ is a matrix whose k'th column consists of samples of impulse response of basis function $F_k(z)$ and y, ϑ, w are in£nite dimensional vectors with $\vartheta = [\vartheta_0, \vartheta_1, \cdots]^T$. It can be shown that the dual of the previous linear programming problem with no duality gap is given by

$$\max_{\tau\geq 0, \delta\in l_1} \; \langle y, \delta\rangle - \tau\sigma \qquad (2.8)$$

$$\text{subject to}: \|T_u^*\delta\|_\infty \le 1, \quad (2.9)$$

$$\|\delta\|_1 \le \tau, \quad (2.10)$$

$$(T_u\bar{F})^*\delta = 0 \quad (2.11)$$

where $\delta \in l_1$ is the functional of y and $\langle y, \delta\rangle$ represents operation of δ on y. T_u^* is adjoint of the Toeplitz operator T_u. In £nite dimensional case, $\langle y, \delta\rangle = \delta^T y^{[N]}$ and $T_u^* = T_u^T$. The absence of duality gap means that the results of the minimization problem (2.5) and maximization problem (2.8) are equal.

Note that here we assume that u and y are bounded sequences converging to zero (or elements of c_0) because in this case their functional will belong to l_1. According to (1.1), for any input signal $u \in c_0$ and any plant in $\mathcal{B}$, the output y will belong to c_0 if the noise signal v is an element of c_0.

The next result demonstrates the relationship between the optimal solutions of primal-dual problems. Note that these solutions are not necessarily unique.

Theorem 2.1. Let θ^o, ϑ^o, w^o and δ^o, τ^o be the optimal solutions to the primal-dual problems (2.5) and (2.8). Then

$$|\langle T_u(k), \delta^o\rangle| < 1 \Rightarrow \vartheta_k^o = 0 \quad (2.12)$$

$$\langle T_u(k), \delta^o\rangle\vartheta_k^o \ge 0, \quad \forall k \ge 0 \quad (2.13)$$

$$\delta^o(k) \ne 0 \Rightarrow |w^o(k)| = \sigma \quad (2.14)$$

$$\delta^o(k)w^o(k) \ge 0 \quad \forall k \ge 0 \quad (2.15)$$

$$\tau^o(\sigma) \ge 1 \quad \forall\sigma \quad (2.16)$$

where $T_u(k)$ is the k'th column of T_u. The above conditions, usually called as alignment conditions, characterize all properties of optimal solutions.

The previous result reveals an interesting property of l_1 identi£cation problem when the model $\hat{G}$ belongs to the space of FIR systems i.e. $\hat{G} = \sum_{k=0}^{n} \theta_k z^{-k}$. In that case, $T_u\bar{F}$ in dual problem reduces to $T_u\bar{F} = T_u^n$ (where T_u^n consists of the £rst n columns of T_u) and consequently, (2.11) becomes $T_u^{n*}\delta = 0$ that is equivalent to

$$\langle T_u(k), \delta\rangle = 0, \text{ for } 0 \le k \le n$$

According to alignment condition (2.12),

$$\vartheta_k^o = 0, \text{ for } 0 \le k \le n$$

which means that the *£rst n samples of the impulse response of Δ^o are always zero*, for any σ, u and y. This implies that the mismatch between the £rst n samples of the plant-model outputs are only represented by w.

Another interesting property of the in£nite dimensional identi£cation problem (2.5) is that for any σ, the optimal Δ^o has *£nite* impulse response.

Theorem 2.2. Assume that u, y belong to c_0, then in the in£nite dimensional optimization problem (2.5), Δ^o has £nite impulse response.

This means that despite in£nite dimensionality of problem 1.1, only £nite number of parameters are needed to be considered in Δ.

It is worthwhile to indicate that if the identi£cation problem 1.1 is stated in $\mathcal{H}_\infty$ topology, under the same assumptions, the optimal Δ^o does not necessarily have £nite impulse response (Zhou and Kimura 1995), (Namvar and Besançon-Voda 2001). Theorem 2.2 states that the order of Δ^0 is £nite, however the estimation of this order before solving the optimization problem still remains an open subject.

In the sequel, based on primal-dual formulation of the identi£cation problem, we will analyze the dependence of optimal solutions on σ.

Theorem 2.3. For any σ

$$\frac{d\tau^o(\sigma)}{d\sigma} \le 0 \quad (2.17)$$

$$\frac{d\mu^o(\sigma)}{d\sigma} = -\tau^o(\sigma) \quad (2.18)$$

From (2.18) and (2.16) it is inferred that μ^o is a decreasing function of σ or $\frac{d\mu^o}{d\sigma} \le -1$, however since μ^o is non-negative there exists a £nite σ_0 for which $\mu^o(\sigma_0) = 0$. Similarly, it is seen that τ^o is a non-increasing function of σ and is lower bounded by 1, which implies that $\lim_{\sigma\to\sigma_0} \tau^o(\sigma) = 1$. This means that *$\mu^o(\sigma)$ reaches zero at σ_0 with a slope of -1.*

Finally, differentiating (2.18) with respect to σ yields

$$\frac{d^2\mu^o}{d\sigma^2} \ge 0 \quad (2.19)$$

that indicates $\mu^o(\sigma)$ is a convex function of σ.

2.1 Identi£cation error

In this section we compute an asymptotic upperbound for modelling error $\|G - \hat{G}^o\|_1$ and show how the upperbound can be approximated by an easily computable quantity.

Consider the model set $\mathcal{G}^n$ de£ned by (1.3) and let $\mathcal{G}(\mu^*)$ be a subset of $\mathcal{B}$ in which every element has a minimum distance of no larger than a *given* value μ^* from $\mathcal{G}^n$

$$\mathcal{G}(\mu^*) \stackrel{\triangle}{=} \{\bar{G} \mid \inf_{\hat{G}\in\mathcal{G}^n} \|\bar{G} - \hat{G}\|_1 \le \mu^*\} \quad (2.20)$$

This set can also be described by

$$\mathcal{G}(\mu^*) = \{\bar{G} \mid \bar{G} = \hat{G} + \Delta, \hat{G} \in \mathcal{G}^n, \Delta \in \mathcal{B}, \|\Delta\|_1 \le \mu^*\}$$

Clearly $\mathcal{G}(\mu^*)$ is a subset of *convex* and *balanced* set $\mathcal{B}$. We assume that the plant G belongs to $\mathcal{G}(\mu^*)$.

Given u and y satisfying (1.1), the set of all unfalsi£ed systems in $\mathcal{G}(\mu^*)$ is de£ned by (Tse *et al.* 1993)

$$S_\infty(\mathcal{G}(\mu^*), u, y, \sigma_v) \stackrel{\triangle}{=} \{\bar{G} \in \mathcal{G}(\mu^*) \mid \|y - \bar{G}u\|_\infty \le \sigma_v\}$$

Assuming that input signal u contains all possible sequences of -1 and 1, it can be shown that the worst-case distance between any element in $S_\infty(\mathcal{G}(\mu^*), u, y, \sigma_v)$ and the plant G is asymptotically bounded by $2\sigma_v$ (Tse *et al.* 1993), (Milanese 1996). Nonetheless, the identi£cation problem (1.1) is equivalent to £nding an element of $S_\infty(\mathcal{G}(\mu), u, y, \sigma)$ like $\bar{G}^o = \hat{G}^o + \Delta^o$ with the minimum value of μ or equivalently an element of $S_\infty(\mathcal{G}(\mu^o), u, y, \sigma)$.

An asymptotic upperbound for $\|G - \hat{G}^o\|_1$ is given in the following result.

Theorem 2.4. Let σ^* be a value of σ for which $\mu^o(\sigma^*) = \mu^*$, then

$$\|G - \hat{G}^o\|_1 \le O_B(\sigma) \triangleq \begin{cases} 2\sigma_v + \mu^o(\sigma) & \text{if } \sigma \le \sigma^*, \\ 2\sigma_v + \mu^* & \text{if } \sigma^* \le \sigma \le \sigma_v, \\ 2\sigma + \mu^* & \text{if } \sigma \ge \sigma_v, \end{cases}$$

The derivative of O_B with respect to σ is given by

$$\frac{dO_B}{d\sigma} = \begin{cases} -\tau^o & \text{if } \sigma \le \sigma^*, \\ 0 & \text{if } \sigma^* \le \sigma \le \sigma_v \\ 2 & \text{if } \sigma \ge \sigma_v \end{cases}$$

Moreover, $\frac{d^2 O_B}{d\sigma^2} \ge 0$ which shows that O_B is a convex function of σ and its minimum is achieved for any $\sigma \in (\sigma^*, \sigma_v)$.

It can also be seen that

$$O(\sigma) \triangleq 2\sigma + \mu^o(\sigma) \tag{2.21}$$

is a lowerbound for O_B and that $O(\sigma)$ is also a convex function of σ and has a *unique* minimum. In fact, from (2.21), (2.18) and (2.19) it is inferred that

$$\frac{dO}{d\sigma} = 2 - \tau^o \le 1 \tag{2.22}$$

$$\frac{d^2 O}{d\sigma^2} \ge 0 \tag{2.23}$$

Unfortunately the upperbound O_B depends on both σ_v and μ^* and even if μ^* is given, since σ_v is unknown, O_B is not computable. However, computation of $O(\sigma)$ does not obviously need any *a priori* knowledge of σ_v and μ^*. Therefore, in this case, it is reasonable to choose $O(\sigma_{min})$ as an upperbound for the modelling error where σ_{min} is a value of σ for which $O(\sigma)$ is minimized. It can be shown that the maximum error between $O(\sigma_{min})$ and $O_B(\sigma_{min})$ is bounded by

$$O_B(\sigma_{min}) - O(\sigma_{min}) \le \max(2\sigma_v, \mu^*, 2\sigma_v - 2\sigma^* + \mu^*) \tag{2.24}$$

where σ^* is a value of σ for which $\mu^o(\sigma)$ equals μ^*. The above inequality shows that when model set $\mathcal{G}(\mu^*)$ is enlarged (e.g. by choosing a large n) such that μ^* becomes small, and when noise amplitude σ_v is small, $O(\sigma_{min})$ becomes an acceptable upperbound for modelling error.

It is also possible to obtain a lowerbound for σ_v provided that μ^* is known.

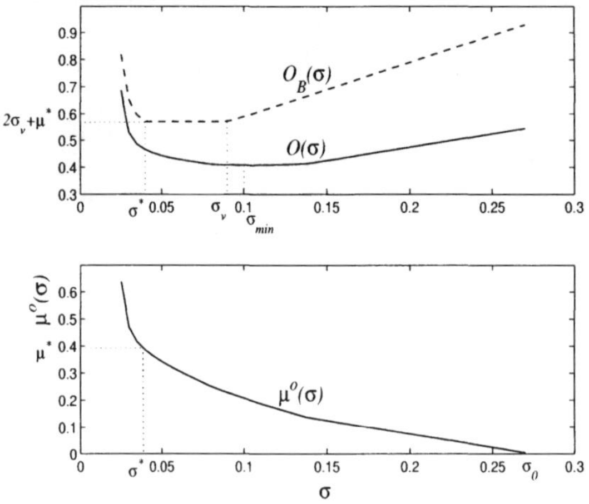

Fig. 1. $O_B(\sigma)$: Upperbound for modelling error and its approximation $O(\sigma)$ together with $\mu^o(\sigma)$

Theorem 2.5.

$$\sigma_v \ge \max(\sigma^*, \mu^o(\sigma_{min}) - \mu^* + \sigma_{min})$$

3. EXAMPLE

A PRBS of amplitude one was applied to input of the plant

$$G(z) = 0.1 \frac{(z+0.2)(z+0.3)(z+0.4)}{(z-0.4)(z-0.5)(z^2 - z + 0.74)}$$

and an additive noise with amplitude $\sigma_v = 0.091$ was added to plant output such that $\frac{\|v\|_\infty}{\|y\|_\infty} = 8\%$ as in (1.1). $N = 100$ samples of data were used for identi£cation.

The model $\hat{G}$ belongs to the set of £nite impulse response systems with degree $n = 9$. The l_1 norm of the tail of the impulse response of plant for $9 < k < N$ is $\mu^* = 0.39$. Therefore, the plant is assumed to belong to $\mathcal{G}(\mu^*)$ de£ned by (2.20). It is possible to calculate μ^* by knowing a bound on the relative stability and gain of plant (Milanese 1996). For 50 values of σ in the interval $[0.02, 0.27]$, the dual problem (2.8) was solved, and $O(\sigma)$ and $\mu^o(\sigma)$ were evaluated and are shown in Fig. 1. $O_B(\sigma)$ was evaluated by knowing σ_v and μ^*, however the computation of $\mu^o(\sigma)$ and $O(\sigma)$ obviously did not require knowledge of σ_v and μ^*.

To verify Theorem 2.2, it was observed that despite the fact that data sequence length was $N = 100$, the order of Δ^o (for $\sigma = 0.1$) was 35 i.e. $\vartheta_k^o = 0$ for $k > 35$.

As predicted by Theorem 2.3, the slope of $\mu^o(\sigma)$ is smaller than -1 and μ^o reaches zero for a £nite σ_0 and with a slope of -1.

Using the knowledge of μ^*, the lowerbound for σ_v, as stated in Theorem 2.5, is given by 0.0397.

It was veri£ed that by choosing $O(\sigma_{min})$ as an upperbound for modelling error, one makes an error of $O_B(\sigma_{min}) - O(\sigma_{min}) = 0.18$. Noting that $O_B(\sigma_{min})$ equals 0.59, this error is equivalent to 30%.

4. CONCLUSION

Formulation of primal-dual of optimization problems arising in l_1 identi£cation and model validation problems, gives a novel analytical tool for analyzing structural properties of optimal solutions. All properties of optimal solutions were characterized by alignment conditions. Moreover, we demonstrated two applications of this analysis in £nding a lowerbound for noise amplitude and computation of upperbounds for modelling error. However, the applicability of the analysis is not limited to these cases and can be explored further. Moreover, the results can be extended to other norms.

REFERENCES

Elia, N. and M. Dahleh (1997). Controller design with multiple objectives. *IEEE Transactions on Automatic control* **42**(5), 596–611.

Giarré, L., M. Milanese and M. Taragna (1997). H_∞ identi£cation and model quality evaluation. *IEEE Transactions on Automatic control* **42**(2), 188–199.

Gu, G. and P. Khargonekar (1992). Linear and nonlinear algorithms for identi£cation in H_∞ with error bounds. *IEEE Transactions on Automatic control* **37**(7), 953–963.

Heuberger, P., P. Van den Hof and O. Bosgra (1995). A generalized orthonormal basis for linear dynamical systems. *IEEE Transactions on Automatic Control* **40**(3), 451–465.

Jacobson, C., C. Nett and J. Partington (1992). Worst-case identi£cation in l_1: Optimal algorithms and error bounds. *Systems and Control Letters* **19**, 419–424.

Kacewicz, B. (1999). Worst-case conditional system identi£cation in a general class of norms. *Automatica* (35), 1049–1058.

Luenberger, D. (1969). *Optimization by vector space methods*. John Wiley and sons, Inc.

Mäkilä, P. (1991). Robust identi£cation and Galois sequences. *Int.J. Control* **54**(5), 1189–1200.

Milanese, M. (1996). *Worst-case l_1 identi£cation in Bounding approaches to system identi£cation, edited by M. Milanese and J. Norton and H. Piet-Lahanier and E. Walter*. Pleunum Press. New York and London.

Namvar, M. and A. Besançon-Voda (2001). A near-optimal algorithm for H_∞ identi£cation of £xed order rational models. *International Journal of Control* **74**(13), 1370–1381.

Smith, R. (1992). Model validation and parameter identi£cation in H_∞ and l_1. *Proc. of American Control Conference* pp. 2852–2856.

Tse, D., M. Dahleh and J. Tsitsiklis (1993). Optimal asymptotic identi£cation under bounded disturbances. *IEEE Transactions on Automatic control* **38**(8), 1176–1190.

Wang, S., J. Dai and M. Tanaka (1999). A parametric approach for l_1 robust identi£cation. *IEEE Transactions on Automatic control* **44**(6), 1282–1286.

Zhou, T. and H. Kimura (1995). Structure of model uncertainty for a weakly corrupted plant. *IEEE Transactions on Automatic control* **40**(4), 639–655.

Appendix A: Preliminary results in duality theory

Some de£nitions and preliminary results in duality theory in linear vector spaces are recalled. For a complete discussions see (Luenberger 1969), (Elia and Dahleh 1997).

If X is a Banach vector space, X^* is the space of all bounded linear functionals on X. The operation of the functional x^* on an element $x \in X$ is shown as $\langle x, x^* \rangle$. If $\mathcal{A}$ is a bounded linear operator mapping the normed space X into Y then the adjoint operator $\mathcal{A}^*$ maps Y^* to X^* as

$$\langle x, \mathcal{A}^* y^* \rangle = \langle \mathcal{A}x, y^* \rangle \quad \text{(A.1)}$$

For two bounded linear operators $\mathcal{A}$ and $\mathcal{B}$, $(\mathcal{AB})^* = \mathcal{B}^*\mathcal{A}^*$. When $\mathcal{A}$ is a £nite dimensional operator then $\mathcal{A}^* = \mathcal{A}^T$. If $x^* \in l_\infty$ and $x \in l_1$ then the action of x^* on x is uniquely represented by $\langle x, x^* \rangle = \sum_{t=0}^{\infty} x(t)x^*(t)$. The element x is said to be *aligned* with x^* when

$$|x^*(t)| < \|x^*\|_\infty \Rightarrow x(t) = 0 \quad \text{(A.2)}$$

$$\text{and } x(t)x^*(t) \geq 0 \quad \text{(A.3)}$$

Appendix B

Proof of Theorem 2.1: Assuming that the initial conditions are zero, from the model structure (1.2) we have

$$y = T_u\vartheta + T_u\bar{F}\theta + w \quad \text{(B.1)}$$

Since there is no duality gap between the primal and dual problems (2.5) and (2.8), then μ^o will be the optimum value in dual problem

$$\mu^o = \langle y, \delta^o \rangle - \tau^o\sigma$$

Replacing for y yields

$$\mu^o = \langle T_u\vartheta^o, \delta^o \rangle + \langle (T_u\bar{F})\theta^o, \delta^o \rangle + \langle w^o, \delta^o \rangle - \tau^o\sigma$$

Using properties of adjoint operators described by (A.1), we obtain

$$\mu^o = \langle T_u^*\delta, \vartheta^o \rangle + \langle (T_u\bar{F})^*\delta^o, \theta^o \rangle + \langle \delta^o, w^o \rangle - \tau^o\sigma \text{(B.2)}$$

and observing (2.11), we obtain

$$\mu^o = \langle T_u^*\delta^o, \vartheta^o \rangle + [\langle \delta^o, w^o \rangle - \tau^o\sigma] \quad \text{(B.3)}$$

From the constraints $\|\delta^o\|_1 \leq \tau^o$ and $\|w^o\|_\infty \leq \sigma$ we conclude that δ^o (as an element of l_1) must be aligned with $w^o \in c_0$. Writing the alignment condition yields (2.14).

On the other hand since $\|T_u^*\delta^o\|_\infty \leq 1$, when the optimal solution is achieved, $T_u^*\delta$ (as an element of c_0) must be aligned with $\vartheta^o \in l_1$. Writing the alignment condition yields (2.12).

Next, we show that $\tau^o \geq 1$. Using (A.1), the £rst term in RHS of (B.3) becomes

$$\langle T_u^*\delta^o, \vartheta^o \rangle = \langle T_u\vartheta^o, \delta^o \rangle$$

On the other hand by considering the Toeplitz structure of T_u and by recalling that $\|u\|_\infty \leq 1$, it is inferred that $\|T_u \vartheta^o\|_\infty \leq \mu^o$. Observing dual constraint (2.10) then yields

$$\langle \delta^o, T_u \vartheta^o \rangle \leq \|\delta^o\|_1 \|T_u \vartheta^o\|_\infty \leq \tau^o \mu^o$$

Therefore, μ^o from (B.3) can be bounded by

$$\begin{aligned} \mu^o &\leq \tau^o \mu^o + \|\delta\|_1 \|w\|_\infty - \tau^o \sigma && \text{(B.4)} \\ &\leq \tau^o \mu^o + \tau^o \sigma - \tau^o \sigma && \text{(B.5)} \\ &\leq \tau^o \mu^o && \text{(B.6)} \end{aligned}$$

which implies $\tau^o \geq 1$.

Proof of Theorem 2.2: Recall that ϑ is the impulse response of Δ. Since δ is an element of l_1 there exists a finite L such that

$$\sum_{k=L}^{\infty} |\delta(k)| < 1$$

On the other hand T_u^* is an infinite dimensional LT matrix constructed from $u(k)$'s (In finite dimensional case $T_u^* = T_u^T$). Recalling that $\|u\|_\infty \leq 1$ one has

$$|(T_u^* \delta)(k)| \leq \sum_{k=L}^{\infty} |\delta(k)| < 1, \text{ for } k \geq L$$

where $(T_u^* \delta)(k)$ is the k'th column of $T_u^* \delta$. From the alignment condition (2.12), this implies $\vartheta_k = 0$, for $k \geq L$

Proof of Theorem 2.3: Assume that $\mu_1^o, \tau_1^o, \delta_1^o, \theta_1^o, \vartheta_1^o, w_1^o$ and $\mu_2^o, \tau_2^o, \delta_2^o, \theta_2^o, \vartheta_2^o, w_2^o$ are two sets of solutions for $\sigma = \sigma_1$ and $\sigma = \sigma_2$, respectively. Clearly we have

$$\begin{aligned} y &= T_u \vartheta_1^o + T_u \bar{F} \theta_1^o + w_1^o && \text{(B.7)} \\ y &= T_u \vartheta_2^o + T_u \bar{F} \theta_2^o + w_2^o && \text{(B.8)} \end{aligned}$$

For $\sigma = \sigma_1$, the dual problem (2.8) yields

$$\mu_1^o = \langle y, \delta_1^o \rangle - \tau_1^o \sigma_1$$

Substitution for y from (B.8) yields (see proof of Theorem 2.1)

$$\mu_1^o = \langle T_u^* \delta_1^o, \vartheta_2^o \rangle + [\langle \delta_1^o, w_2^o \rangle - \tau_1^o \sigma_1]$$

which implies that

$$\begin{aligned} \mu_1^o &\leq \|T_u^* \delta_1^o\|_\infty \mu_2^o + \|\delta_1^o\|_1 \sigma_2 - \tau_1^o \sigma_1 && \text{(B.9)} \\ &\leq \mu_2^o + \tau_1^o(\sigma_2 - \sigma_1) && \text{(B.10)} \end{aligned}$$

Similarly one can obtain

$$\mu_2^o \leq \mu_1^o + \tau_2^o(\sigma_1 - \sigma_2) \qquad \text{(B.11)}$$

From (B.10) and (B.11) it is inferred that

$$(\tau_2^o - \tau_1^o)(\sigma_2 - \sigma_1) \leq 0$$

Taking $\sigma_2 = \sigma_1 + d\sigma$ and $\tau_2^o = \tau_1^o + d\tau^o$ implies (2.17).

Let $\mu_2^o = \mu_1^o + d\mu^o$, then from (B.11) and (B.10) it results that

$$-\tau_1^o d\sigma \leq d\mu^o \leq -(\tau_1^o + d\tau^o) d\sigma$$

which implies (2.18).

Proof of Theorem 2.4: First, note that since the plant G is an element of $\mathcal{G}(\mu^*)$, there exist $G_n \in \mathcal{G}^n$ and $\Delta_r \in \mathcal{B}$ such that $G = G_n + \Delta_r$ and $\|\Delta_r\| \leq \mu^*$. Therefore, from (1.1) it is inferred that

$$y = T_u \bar{F} \theta_r + T_u \vartheta_r + v$$

where θ_r is parameter vector for G_n and ϑ_r is impulse response of Δ_r. Replace for y in (2.8) to obtain

$$\mu^o = \langle T_u \vartheta_r, \delta^o \rangle + \langle (T_u \bar{F}) \theta_r, \delta^o \rangle + \langle v, \delta^o \rangle - \tau^o \sigma$$

Using (A.1) together with (2.11), a similar argument described in proof of Theorem 2.1 shows that

$$\mu^o(\sigma) \leq \mu^* + \tau^o(\sigma_v - \sigma) \qquad \text{(B.12)}$$

Replacing $\sigma = \sigma^*$ in the inequality and noting that $\mu^o(\sigma^*) = \mu^*$ and $\tau^o \geq 1$, implies that

$$\sigma^* \leq \sigma_v \qquad \text{(B.13)}$$

Next, recall that by solving Problem 1.1, we find an element like $\bar{G}^o$ in $S_\infty(\mathcal{G}(\mu^o), u, y, \sigma)$. Moreover, the plant G is always an element of $S_\infty(\mathcal{G}(\mu^*), u, y, \sigma_v)$. Now consider three cases:

1) $\sigma \leq \sigma^*$: From Theorem 2.3 it is clear that $\mu^o(\sigma) \geq \mu^*$ which implies $\mathcal{G}(\mu^*) \subset \mathcal{G}(\mu^o)$. Hence both G and $\bar{G}^o$ belong to $\mathcal{G}(\mu^o)$. From (B.13) it is inferred that $\sigma \leq \sigma_v$. This implies that $S_\infty(\mathcal{G}(\mu^o), u, y, \sigma) \subset S_\infty(\mathcal{G}(\mu^o), u, y, \sigma_v)$. Hence, both G and $\bar{G}^o$ belong to $S_\infty(\mathcal{G}(\mu^o), u, y, \sigma_v)$. When u contains all possible sequences of -1 and 1, the worst-case distance between any two elements like $\bar{G}^o = \hat{G}^o + \Delta^o$ and G in $S_\infty(\mathcal{G}(\mu^o), u, y, \sigma_v)$ is then bounded by $2\sigma_v$ and we have

$$\begin{aligned} \|G - \hat{G}^o\|_1 &\leq \|G - \bar{G}^o\|_1 + \|\Delta^o\|_1 && \text{(B.14)} \\ &\leq 2\sigma_v + \mu^o && \text{(B.15)} \end{aligned}$$

2) $\sigma^* \leq \sigma \leq \sigma_v$: From Theorem 2.3 it is inferred that $\mu^o(\sigma) \leq \mu^*$. In this case both G and $\bar{G}^o$ belong to $\mathcal{G}(\mu^*)$. Since $\sigma \leq \sigma_v$, both G and $\bar{G}^o$ belong to $S_\infty(\mathcal{G}(\mu^*), u, y, \sigma_v)$ and consequently,

$$\begin{aligned} \|G - \hat{G}^o\|_1 &\leq \|G - \bar{G}^o\|_1 + \|\Delta^o\|_1 && \text{(B.16)} \\ &\leq 2\sigma_v + \mu^* && \text{(B.17)} \end{aligned}$$

3) $\sigma \geq \sigma_v$: From (B.13) it is inferred that $\sigma \geq \sigma^*$ and consequently from Theorem 2.3 it is inferred that $\mu^o(\sigma) \leq \mu^*$. Since $\sigma \geq \sigma_v$, both G and $\bar{G}^o$ belong to $S_\infty(\mathcal{G}(\mu^*), u, y, \sigma)$ and consequently,

$$\begin{aligned} \|G - \hat{G}^o\|_1 &\leq \|G - \bar{G}^o\|_1 + \|\Delta^o\|_1 && \text{(B.18)} \\ &\leq 2\sigma + \mu^* && \text{(B.19)} \end{aligned}$$

Proof of Theorem 2.5: Since the plant G is an element of $\mathcal{G}(\mu^*)$, there exist $G_n \in \mathcal{G}^n$ and $\Delta_r \in \mathcal{B}$ such that $G = G_n + \Delta_r$ and $\|\Delta_r\| \leq \mu^*$. In order that Problem (1.1) has a solution, it is necessary that $\mu^o(\sigma_{min}) + \sigma_{min} \leq \sigma_v + \mu^*$. Another lowerbound for σ_v is given by (B.13).

IFAC Publications
www.elsevier.com/locate/ifac

OPTIMAL APPROXIMATION AND MODEL QUALITY ESTIMATION FOR NONLINEAR SYSTEMS

P.M. Mäkilä

Tampere University of Technology, Automation and Control Institute, P.O. Box 692, FIN-33101 Tampere, FINLAND

Abstract:
An optimal squared error based approximation problem for static polynomial models is solved. This problem is similar to an optimal approximation problem for linear time-invariant (LTI) models. Corresponding absolute error based approximation problems are also studied. Model quality estimation is typically based on sample variance analysis of the squared error criterion. Error squaring, however, results in increased sample variability especially for error and noise distributions with heavy tails. Error analysis based on the use of the sum of the absolute values of the errors has advantages in such situations. Two model quality estimation methods for static polynomial models are suggested based on similar techniques for LTI models. These techniques extend to many other linear-in-parameters regression problems. *Copyright © 2003 IFAC*

Keywords: model quality estimation, optimal approximation, static nonlinearity

1. INTRODUCTION

There are many realistic ways to introduce quantitative measures of model quality for linear and nonlinear systems. It is the purpose of this paper to analyse squared and absolute error based optimal approximation problems and related model quality estimation methods for nonlinear systems. Specifically, we consider static models and linear time-invariant (LTI) models, respectively.

Least squares methods dominate system identification (Söderström and Stoica 1996, Ljung 2001). There are of course alternatives to least squares methods, such as least absolute deviations (LAD) methods (Denoël and Solvay 1985, Gustafsson and Mäkilä 1996). An advantage with LAD methods over least squares (LS) methods is that LAD methods are more robust for problems with heavy-tailed noise, see for example (An and Chen 1982). Estimation problems with heavy-tailed noise have received growing interest in recent years due to their many potential applications (Ilow and Hatzinakos 1998, Masry 2000).

In the present work we are interested in the situation that the sample average of the absolute value of the error (noise) is significantly smaller than the square root of the sample average of the squared values of the error (noise). In fact, if $\{v(t)\}_{t=0}^{n-1}$ is any sequence of n real numbers, it holds that

$$\frac{1}{n}\sum_{t=0}^{n-1}|v(t)| \leq \left(\frac{1}{n}\sum_{t=0}^{n-1}v(t)^2\right)^{1/2}$$

It is possible that the two quantities in this inequality differ from each other by many orders of magnitude (for large n). It is the sample average of the absolute value of the error (noise) that exhibits a smaller sample variability. This is an important fact.

We shall use the absolute error to determine intrinsic model quality estimates. This step is based on set membership and worst-case estimation concepts (Milanese and Vicino 1993). The sample average of the absolute value of the error has been used in this manner in (Mäkilä *et al.* 1995, Gustafsson and Mäkilä 1995, 2001) to obtain strong model quality estimates in the LTI case.

Recently there has been considerable interest to study optimal LTI modelling for nonlinear systems, see (Sastry 1999, Pintelon and Schoukens 2001, Ljung 2001, Enqvist and Ljung 2002, Mäkilä and Partington 2003). Quasistationary signals (Ljung 2001) form an interesting (nonlinear) signal space in this context.

2. PRELIMINARIES

Let $\mathbb{N}$, $\mathbb{Z}$, and $\mathbb{R}$ denote the nonnegative integers, the integers, and the reals, respectively. The space $s(\mathbb{N})$ is the linear space of all real sequences $\{x = \{x(k) \in \mathbb{R}\}_{k\in\mathbb{N}}$ over $\mathbb{N}$. The linear normed space $\ell_\infty(\mathbb{N})$ is the space of all real sequences $x = \{x(k) \in \mathbb{R}\}_{k\in\mathbb{N}}$ such that

$$\|x\|_\infty \equiv \sup_{k\in\mathbb{N}} |x(k)| < \infty.$$

Furthermore, the linear normed space $\ell_1(Q)$ is the space of all real sequences $x = \{x(k) \in \mathbb{R}\}_{k\in Q}$ such that

$$\|x\|_1 \equiv \sum_{k\in Q} |x(k)| < \infty,$$

where $Q = \mathbb{N}, \mathbb{Z}$.

Let $x \in s(\mathbb{N})$, i.e. let x be a signal, and introduce the autocovariance of the signal x as

$$R_{xx}(k) \equiv \lim_{n\to\infty} \frac{1}{n} \sum_{t=0}^{n-1} x(t)x(t+k), k \in \mathbb{Z}.$$

(Here we put $x(t) = 0$ for $t < 0$.) The autocovariance sequence $R_{xx} = \{R_{xx}(k) \in \mathbb{R}\}_{k\in\mathbb{Z}}$ need not exist for a general signal x.

We use the terminology in (Ljung 2001) and we say that a signal x is *quasistationary* if $x \in \ell_\infty(\mathbb{N})$ and x possesses an autocovariance sequence R_{xx}. Furthermore, we say that the two real sequences v and x possess a crosscovariance sequence if

$$R_{vx}(k) \equiv \lim_{n\to\infty} \frac{1}{n} \sum_{t=0}^{n-1} v(t)x(t+k)$$

exists for any $k \in \mathbb{Z}$. We write $R_{vx} = \{R_{vx}(k) \in \mathbb{R}\}_{k\in\mathbb{Z}}$.

3. AN OPTIMAL APPROXIMATION PROBLEM

In this section we study optimal approximation of nonlinear systems using both squared errors and absolute errors for models of the static nonlinear type.

Let $G : D(G; s(\mathbb{N})) \to s(\mathbb{N})$ denote a nonlinear system, i.e. a nonlinear input-output operator

$$y = Gu,$$

where y is the output and $u \in D(G; s(\mathbb{N}))$ is the input. Here $D(G; s(\mathbb{N}))$ denotes the domain of definition of G in $s(\mathbb{N})$, i.e. the set of all $u \in s(\mathbb{N})$ such that $y = Gu \in s(\mathbb{N})$.

We may have a fuzzy idea that the nonlinear system G is nearly static, so that the output $y(t) = (Gu)(t)$ of G could nearly be given by some static relationship $y(t) = g(u(t))$, where $g : D(g; \mathbb{R}) \to \mathbb{R}$ is defined in a subset $D(g; \mathbb{R})$ of $\mathbb{R}$. Let us study the optimal approximation of G within a given class of static nonlinear models *without* assuming that G is static.

We consider here static models of the form

$$(Pu)(t) = \sum_{i\geq 0} b_i u(t)^i, \; t \in \mathbb{N},$$

where $b = \{b_i\}_{i\geq 0}$ is a sequence of real numbers. We shall now make the convention that the input u has been so normalized (e.g. by the choice of a suitable physical unit for the input) that $|u(t)| \leq 1$ for all $t \geq 0$.

Let p denote the function

$$p(x) = \sum_{i\geq 0} b_i x^i, \tag{1}$$

with domain of definition given by the set of those real numbers x for which the above sum converges to a (real, proper) limit.

Now if $b \in \ell_1(\mathbb{N})$, it follows that

$$|(Pu)(t)| \leq \|b\|_1 < \infty, \; t \in \mathbb{N},$$

and thus $Pu \in \ell_\infty(\mathbb{N})$. Then $p(x)$ is a continuous function in $[-1, 1]$. In fact, $p(x)$ is continuously differentiable, an arbitrary number of times, in the open interval $(-1, 1)$. (The polynomial expression in (1) is the Taylor series of p around $x = 0$.) So we are interested in approximating the nonlinear system G as a smooth static nonlinearity.

3.1 The Squared Error Problem

An optimal squared error approximation problem for G is now introduced as follows. Let $u \in D(G; s(\mathbb{N}))$ and $|u(t)| \leq 1$ for $t \in \mathbb{N}$. Consider

$$\inf_P \lim_{n\to\infty} \frac{1}{n}\sum_{t=0}^{n-1}[(Gu)(t)-(Pu)(t)]^2, \quad (2)$$

where the infimum is taken over all smooth static nonlinearities P such that $b \in \ell_1(\mathbb{N})$.

It is clear that additional conditions are required for the limit in (2) to exist. Introduce the notation

$$r_{yu}(i) = \lim_{n\to\infty} \frac{1}{n}\sum_{t=0}^{n-1} y(t)u(t)^i,\ i \in \mathbb{N},$$

when the indicated limit exists. We put $r_{yu} = [r_{yu}(0), r_{yu}(1), \ldots]^T$, where the symbol ^{T} denotes vector transpose. Let the limit in

$$m_u(i) = \lim_{n\to\infty} \frac{1}{n}\sum_{t=0}^{n-1} u(t)^i,\ i \in \mathbb{N},$$

exist. We put $m_u = [m_u(0), m_u(1), \ldots]^T$. We say that u possesses moments of arbitrary order when $m_u(i)$ exists for any $i \in \mathbb{N}$. Furthermore, introduce the matrix

$$[M_u]_{ij} = m_u(i+j),\ i,j \in \mathbb{N}.$$

We shall also interpret M_u as a linear operator from $\ell_1(\mathbb{N})$ into $\ell_1(\mathbb{N})$ with domain of definition $D(M_u;\ell_1(\mathbb{N}))$. Let $R(M_u;\ell_1(\mathbb{N}))$ denote the range of M_u.

Theorem 1. Let the nonlinear system G, with a nonempty domain of definition, be given. Let $u \in D(G; s(\mathbb{N}))$ possess moments of arbitrary order, and let $\|u\|_\infty \le 1$. Let $D(M_u;\ell_1(\mathbb{N})) = \ell_1(\mathbb{N})$. Let M_u be a strictly positive definite matrix in the sense that the associated quadratic forms satisfy

$$x^T M_u x > 0 \quad \text{for any} \quad x \ne 0, x \in \ell_1(\mathbb{N}). \quad (3)$$

Let $R_{yy}(0)$ and let $r_{yu} \in R(M_u;\ell_1(\mathbb{N}))$. Then the optimal approximation problem (2) has a unique solution P^* with coefficients $b^* \in \ell_1(\mathbb{N})$ given by

$$b^* = M_u^{-1} r_{yu}.$$

Standard conventions for matrix and vector operations are used here. Note that the assumption $D(M_u;\ell_1(\mathbb{N})) = \ell_1(\mathbb{N})$ holds if and only if (iff) $m_u \in \ell_1(\mathbb{N})$, i.e. iff m_u is an absolutely summable real sequence. Clearly by (3) the null space $N(M_u;\ell_1(\mathbb{N}))$ of M_u consists only of the zero element, i.e.

$$N(M_u;\ell_1(\mathbb{N})) = \{x \in \ell_1(\mathbb{N}) \mid M_u x = 0\} = \{0\}.$$

Hence M_u possesses an inverse $M_u^{-1} : R(M_u;\ell_1(\mathbb{N})) \to \ell_1(\mathbb{Z})$ satisfying $M_u^{-1}M_u x = x$ for any $x \in D(M_u;\ell_1(\mathbb{N})) = \ell_1(\mathbb{N})$.

Note that the assumption $R_{yy}(0) \in \mathbb{R}$ does not imply that the output $y \in \ell_\infty(\mathbb{N})$, see also (Mäkilä *et al.* 1998).

Let now the nonlinear system G in Theorem 1 be a smooth static nonlinearity such that

$$(Gu)(t) = \sum_{i\ge 0} a_i u(t)^i,$$

where $a = \{a_i\}_{i\ge 0}$ is such that $a \in \ell_1(\mathbb{N})$. Thus G is well-defined for any input u satisfying $|u(t)| \le 1$ for $t \ge 0$. Furthermore, let m_u satisfy the assumptions of Theorem 1. Then

$$R_{yy}(0) = \sum_{i,j\ge 0} a_i a_j m_u(i+j) = a^T M_u a \in \mathbb{R},$$

and furthermore,

$$r_{yu} = M_u a \in R(M_u;\ell_1(\mathbb{N})),$$

and so all the assumptions of Theorem 1 are satisfied. Hence by this theorem

$$M_u b^* = M_u a,$$

and so as M_u is invertible (in $\ell_1(\mathbb{N})$), it follows that the best smooth approximation $P^* = G$. In this case the infimal value in (2) is equal to zero, and in fact also

$$\|Gu - P^*u\|_\infty = 0.$$

So in this case a zero average squared error means that the model is perfect also pointwise and uniformly in time.

3.2 The Absolute Error Problem

An optimal absolute error approximation problem for G is now introduced as follows. Let $u \in D(G; s(\mathbb{N}))$ and $\|u\|_\infty \le 1$. Consider

$$\inf_P \lim_{n\to\infty} \frac{1}{n}\sum_{t=0}^{n-1}|(Gu)(t)-(Pu)(t)|, \quad (4)$$

where the infimum is taken over all smooth static nonlinearities P such that $b \in \ell_1(\mathbb{N})$. That is, the only difference to the problem (2) is that here the absolute value of the error between the system output and the model output is used.

Theorem 2. Let the nonlinear system G be a smooth static nonlinearity so that $y = Gu$ is given by

$$y(t) = \sum_{i\ge 0} a_i u(t)^i,$$

where $a = \{a_i\}_{i\geq 0} \in \ell_1(\mathbb{N})$. Let the input u possess moments of arbitrary order, and let $\|u\|_\infty \leq 1$. Let $D(M_u; \ell_1(\mathbb{N})) = \ell_1(\mathbb{N})$ and let

$$x^T M_u x > 0 \quad \text{for any} \quad x \neq 0, x \in \ell_1(\mathbb{N}).$$

Then the optimal approximation problem (4) has a unique solution $\bar{P}$ with coefficients $\bar{b} \in \ell_1(\mathbb{N})$ given by

$$\bar{b} = a,$$

and so $\bar{P} = G$ in the sense that, it holds for any u satisfying $\|u\|_\infty \leq 1$ that

$$\|Gu - \bar{P}u\|_\infty = 0.$$

4. AN OPTIMAL LTI MODEL APPROXIMATION PROBLEM

Let $G : D(G; s(\mathbb{N})) \rightarrow s(\mathbb{N})$ denote a nonlinear system, i.e. a nonlinear input-output operator

$$y = Gu,$$

where y is the output and $u \in D(G; s(\mathbb{N}))$ is the input. Here $D(G; s(\mathbb{N}))$ denotes the domain of definition of G in $s(\mathbb{N})$, i.e. the set of all $u \in s(\mathbb{N})$ such that $y = Gu \in s(\mathbb{N})$.

An optimal LTI approximation problem for G is now introduced as follows. Let $u \in D(G; s(\mathbb{N}))$. Consider

$$\inf_F \lim_{n\rightarrow\infty} \frac{1}{n} \sum_{t=0}^{n-1} [(Gu)(t) - (Fu)(t)]^2, \qquad (5)$$

where the infimum is taken over all linear time-invariant (LTI) systems F having an absolutely summable kernel $f = \{f(k)\}_{k\in\mathbb{Z}} \in \ell_1(\mathbb{Z})$. Here

$$(Fu)(t) = \sum_{k\in\mathbb{Z}} f(k)u(t-k),\ t \in \mathbb{N}.$$

(We put $u(t) = 0$ for $t < 0$.) Note that we allow for noncausal G and F.

Let us introduce the matrix

$$[S_u]_{kl} = R_{uu}(k-l),\ k, l \in \mathbb{Z},$$

when u is quasistationary. We shall also interpret S_u as a linear operator from $\ell_1(\mathbb{Z})$ into $\ell_1(\mathbb{Z})$ with domain of definition $D(S_u; \ell_1(\mathbb{Z}))$.

Theorem 3. (Modified from (Mäkilä 2002).) Let the nonlinear system $G : D(G; s(\mathbb{N})) \rightarrow s(\mathbb{N})$, with a nonempty domain of definition, be given. Let $u \in D(G; s(\mathbb{N}))$ be quasistationary. Let $D(S_u; \ell_1(\mathbb{Z})) = \ell_1(\mathbb{Z})$. Let the matrix S_u be strictly positive definite in the sense that

$$x^T S_u x > 0 \quad \text{for any} \quad x \neq 0, x \in \ell_1(\mathbb{Z}).$$

Let $y = Gu \in \ell_\infty(\mathbb{N})$ and let $R_{yy}(0)$ exist. Furthermore, let $R_{uy} \in R(S_u; \ell_1(\mathbb{Z}))$. Then the optimal approximation problem (5) has a unique solution F^*, called the best LTI approximation of G, with kernel $f^* \in \ell_1(\mathbb{Z})$ given by

$$f^* = S_u^{-1} R_{uy}.$$

Standard conventions for matrix and vector operations are used here. Note that the assumption $D(S_u; \ell_1(\mathbb{Z})) = \ell_1(\mathbb{Z})$ holds if and only if (iff) $R_{uu} \in \ell_1(\mathbb{Z})$, i.e. iff R_{uu} is an absolutely summable real sequence. Note that by the strict positive definiteness assumption of S_u, it holds that S_u has an inverse $S_u^{-1} : R(S_u; \ell_1(\mathbb{Z})) \rightarrow \ell_1(\mathbb{Z})$ satisfying $S_u^{-1} S_u x = x$ for any $x \in \ell_1(\mathbb{Z})$.

The assumptions $y = Gu \in \ell_\infty(\mathbb{N})$ and $R_{yy}(0) \in \mathbb{R}$ are, in general, milder than the assumption that y is quasistationary, see (Mäkilä *et al.* 1998, pp. 107). Note that u being quasistationary does not imply that y is quasistationary nor that $R_{yy}(0)$ exists, see (Mäkilä *et al.* 1998, pp. 104-105).

Let G be a bounded-input, bounded-output (BIBO) stable LTI system, so that

$$y = (Gu)(t) = \sum_{k\in\mathbb{Z}} g(k)u(t-k),$$

where $g = \{g(k)\}_{k\in\mathbb{Z}} \in \ell_1(\mathbb{Z})$. Then clearly under the input conditions of Theorem 3, it holds that

$$R_{yy}(0) = g^T S_u g,\ R_{uy} = S_u g \in R(S_u; \ell_1(\mathbb{Z})).$$

Hence the optimal BIBO stable LTI approximation F^*, with kernel $f^* \in \ell_1(\mathbb{Z})$, of G satisfies $f^* = g$, i.e. $F^* = G$.

We can establish a corresponding result when an absolute error criterion is used. An optimal LTI approximation problem for G based on absolute errors is then as follows. Let G be a BIBO stable LTI system with kernel $g \in \ell_1(\mathbb{Z})$. Let $u \in D(G; s(\mathbb{N}))$. Consider

$$\inf_F \lim_{n\rightarrow\infty} \frac{1}{n} \sum_{t=0}^{n-1} |(Gu)(t) - (Fu)(t)|, \qquad (6)$$

where the infimum is computed over all BIBO stable LTI systems. Then the unique solution $\bar{F}$ to Problem (6), under the input conditions of Theorem 3, is given by $\bar{f} = g$, where $\bar{f}$ is the kernel of $\bar{F}$.

5. MODEL QUALITY ESTIMATION

How can we estimate the model accuracy when the number of data points is finite and there is additive noise corrupting the output measurements?

Let G be a nonlinear system, which is nearly static as in section 3. Let P_m denote a static polynomial model of finite degree $m \in \mathbb{N}$

$$(P_m u)(t) = \sum_{i=0}^{m} b_i u(t)^i.$$

Introduce the degree of polynomial approximation of G as

$$e_m(G) \equiv \inf_{P_m} \sup_{\|u\|_\infty \leq 1} \|Gu - P_m u\|_\infty,$$

where the infimum is computed over all polynomial models of degree m. Let P^* be a polynomial of degree m such that

$$\sup_{\|u\|_\infty \leq 1} \|Gu - P^* u\|_\infty = e_m(G).$$

Let G and m be such that $e_m(G)$ is a small nonnegative number.

Let the input-output data be generated as

$$y(t) = (Gu)(t) + v(t),$$

where u is the input, y is the output, and $v = \{v(t) \in \mathbb{R}\}_{t \geq 0}$ is an additive noise term.

Clearly for any input u such that $\|u\|_\infty \leq 1$ and any $n \geq 1$, it now holds that

$$\frac{1}{n}\sum_{t=0}^{n-1} |y(t) - (P^* u)(t)| \leq e_m(G) + \frac{1}{n}\sum_{t=0}^{n-1} |v(t)|.$$

Let $\delta_m \geq 0$ and $\{\rho_n \geq 0\}_{n \geq 1}$ be given. Interpret δ_m as an estimate of $e_m(G)$ and ρ_n as an estimate of $(1/n)\sum_{t=0}^{n-1} |v(t)|$. If

$$e_m(G) + \frac{1}{n}\sum_{t=0}^{n-1} |v(t)| \leq \delta_m + \rho_n, \tag{7}$$

it then holds that

$$\frac{1}{n}\sum_{t=0}^{n-1} |y(t) - (P^* u)(t)| \leq \delta_m + \rho_n.$$

Introduce the problem

$$r_1 = \inf_{d \in \mathbb{R}^{m+1}} \sup_{b \in \mathbb{R}^{m+1}} \|b - d\|_1 \tag{8}$$

$$\text{s.t.}$$

$$\frac{1}{n}\sum_{t=0}^{n-1} |y(t) - (Pu)(t)| \leq \delta_m + \rho_n,$$

where $(Pu)(t) = \sum_{i=0}^{m} b_i u(t)^i$. Let $b_{cc} \in \mathbb{R}^{m+1}$ satisfy

$$r_1 = \sup_{b \in \mathbb{R}^{m+1}} \|b - b_{cc}\|_1$$

$$\text{s.t.}$$

$$\frac{1}{n}\sum_{t=0}^{n-1} |y(t) - (Pu)(t)| \leq \delta_m + \rho_n.$$

The coefficient vector b_{cc} is a Chebyshev center of the set of all such $(m+1)$ dimensional vectors, which are coefficient vectors of unfalsified polynomial models of degree m of G.

If $e_m(G) \leq \delta_m$ and $(1/n)\sum_{t=0}^{n-1} |v(t)| \leq \rho_n$, it is clear that

$$\sup_{\|u\|_\infty \leq 1} \|Gu - P_{cc} u\|_\infty \leq \delta_m + r_1.$$

The polynomial model $P_{cc}u$ minimizes, among polynomial models of degree m, the uncertainty about G for the given input-output data.

An analogous model quality estimate has been studied in (Mäkilä *et al.* 1995, Gustafsson and Mäkilä 1995, 2001) for LTI systems. So our discussion below applies almost unchanged to the noise corrupted version of the setup in section 4.

How does one compute r_1 and b_{cc} when they are defined? This is essentially the same computational problem as the problem of computing the uncertainty estimate for LTI systems studied in (Mäkilä *et al.* 1995). Hence it is clear that we can compute r_1 by solving a set of standard linear programming problems.

What about replacing the absolute error with the squared error? This gives the model quality estimate

$$r_2 = \min_{d \in \mathbb{R}^{m+1}} \max_{b \in \mathbb{R}^{m+1}} \sum_{i=0}^{m} |b_i - d_i|$$

$$\text{s.t.}$$

$$\frac{1}{n}\sum_{t=0}^{n-1} (y(t) - (Pu)(t))^2 \leq \beta(n, m), \tag{9}$$

where $\beta(n, m) \geq 0$ is some suitable bound for the squared error. Now the inequality (9) is quadratic in b, but computing r_2 would seem to be at least as hard as computing r_1.

Let us now assume that $e_m(G) = 0$. In this case we can interpret $\sqrt{\beta(n, m)}$ as an estimate of

$$s_2(v, n) \equiv \left(\frac{1}{n}\sum_{t=0}^{n-1} v(t)^2\right)^{1/2}$$

It is easy to see that

$$1 \leq \frac{s_2(v,n)}{s_1(v,n)} \leq \sqrt{n}, \quad \text{for any } v \text{ and } n \geq 1, (10)$$

where

$$s_1(v,n) \equiv \frac{1}{n} \sum_{t=0}^{n-1} |v(t)|.$$

The lower bound 1 is a greatest lower bound and the upper bound $\sqrt{n}$ is a least upper bound (i.e. they are both attained for some v which may depend on n). It is quite exotic that the signal size measure $s_2(v,n)$ can differ so much from the signal size measure $s_1(v,n)$ for large n. This is the situation for heavy-tailed noise!

An interesting consequence of the above inequalities is as follows. Let v be a realization of a sequence of independent and identically distributed random variables $\{v(t)\}_{t \geq 0}$. Let $f : \mathbb{R} \to \mathbb{R}_+$ denote the probability density function of this distribution. Here $\mathbb{R}_+ = \{x \in \mathbb{R} \mid x \geq 0\}$. Let us assume for simplicity that f is a continuous function. By the Cauchy-Schwarz inequality

$$\int_{-\infty}^{\infty} |x| f(x) dx \leq \left(\int_{-\infty}^{\infty} x^2 f(x) dx \right)^{1/2}.$$

It is easy to find $f(x)$ such that the expected value of $|x|$ is finite, while the square root of the expected value of x^2 is not finite. (Take e.g. $f(x) = c/(1 + a|x|^3)$, where $a > 0$ and c is a normalization coefficient (so that the integral of f over the real line equals one).) Is there any reasonable way to choose $\beta(m,n)$ in (9) under such conditions? The problem is that $s_2(v,n)$ will then not tend to any real number when $n \to \infty$ for a typical realization v.

6. CONCLUSIONS

We have studied several optimal approximation problems based on squared and absolute errors for nonlinear systems. We have also discussed two model quality estimation methods for nonlinear systems modelled approximately by static polynomial models. An analogous model quality estimation method, based on absolute errors, has been discussed for stable LTI systems in (Mäkilä *et al.* 1995, Gustafsson and Mäkilä 1995).

ACKNOWLEDGEMENTS

Financial support to the author from the Academy of Finland under grant number 50991 is gratefully acknowledged.

REFERENCES

An, H.-Z. and Z.-G. Chen (1982). On convergence of LAD estimates in autoregression with infinite variance. *J. Multivariate Analysis*, **12**, 335–345.

Denoël, E. and J.-P. Solvay (1985). Linear prediction of speech with a least absolute error criterion. *IEEE Trans. Acoustics, Speech, and Signal Processing*, **ASSP-33**, 1397–1403.

Enqvist, M. and L. Ljung (2002). Estimating nonlinear systems in a neighborhood of LTI-approximants. *Proc. 41st IEEE Conf. Decision and Control*, Las Vegas, 1005–1010.

Gustafsson, T.K. and P.M. Mäkilä (1995). Black-box methods for identification of set-valued models for control. *Proc. 3rd European Control Conf.*, Rome, Italy, vol. 2, 943–948.

Gustafsson, T.K. and P.M. Mäkilä (1996). Modelling of uncertain systems via linear programming. *Automatica*, **32**, 319–335.

Gustafsson, T.K. and P.M. Mäkilä (2001). Modelling of uncertain systems with application to robust process control. *J. Process Control*, **11**, 251–264.

Ilow, J. and D. Hatzinakos (1998). Analytic alpha-stable noise modeling in a Poisson field of interferers or scatterers. *IEEE Trans. Signal Processing*, **46**, 1601–1611.

Ljung, L. (2001). Estimating linear time-invariant models of nonlinear time-varying systems. *European J. Control*, **7**, 203–219.

Mäkilä, P.M. (2002) On optimal LTI approximation of nonlinear systems. Submitted. Oct. 2002.

Mäkilä, P.M. and J.R. Partington (2003). On linear models for nonlinear systems. *Automatica*, **39**, 1–13.

Mäkilä, P.M., J.R. Partington and T.K. Gustafsson (1995). Worst-case control-relevant identification. *Automatica*, **31**, 1799–1819.

Mäkilä, P.M., J.R. Partington and T. Norlander (1998). Bounded power signal spaces for robust control and modelling. *SIAM J. Control & Optim.*, **37**, 92–117.

Masry, E. (2000). Alpha-stable signals and adaptive filtering. *IEEE Trans. Signal Processing*, **48**, pp. 3011-3016, 2000.

Milanese, M. and A. Vicino (1993). Information based complexity and nonparametric worst-case system identification. *J. Complexity*, **9**, 427–446.

Pintelon, R. and J. Schoukens (2001). *System Identification: A Frequency Domain Approach*, Piscataway, NJ: IEEE Press.

Sastry, S. (1999). *Nonlinear Systems: Analysis, Stability, and Control*, Springer.

Söderström, T. and P. Stoica (1989). *System Identification*, London: Prentice Hall.

IFAC Publications
www.elsevier.com/locate/ifac

LINEAR MODELS OF NONLINEAR FIR SYSTEMS WITH GAUSSIAN INPUTS

Martin Enqvist * **Lennart Ljung** *

* *Division of Automatic Control,*
Department of Electrical Engineering,
Linköpings universitet, SE-58183 Linköping, Sweden
e-mail: maren,ljung@isy.liu.se

Abstract: We present a result that can be viewed as a generalization of Bussgang's classical theorem about static nonlinearities with Gaussian inputs. This result is used to characterize the best linear approximation of a nonlinear finite impulse response (NFIR) system with a Gaussian input. The best linear approximation is here defined as the causal and stable LTI system that minimizes the mean-square error. Furthermore, we discuss how this characterization can be used for structure identification and for identification of generalized Hammerstein and Wiener systems. *Copyright © 2003 IFAC*

Keywords: System identification, Nonlinear systems, Linearization, Gaussian processes

1. INTRODUCTION

System identification deals with the problem of how to estimate a model of a dynamical system from measurements of the input and output signals. In practice, linear system models are very common and they are often used also when the true system is nonlinear. It is therefore interesting to understand how an estimated linear model depends on the properties of the true nonlinear system and of the input signal.

This question is hard to answer in general but it is possible to prove results for certain special cases. If the classes of systems and input signals are restricted to nonlinear finite impulse response (NFIR) systems with Gaussian inputs it is actually possible to characterize the "best" linear model completely. An NFIR system can be written as

$$y_t = f(u_t, u_{t-1}, \dots, u_{t-M_0}) + e_t$$

where u_t and y_t are the input and output signals, respectively, and where e_t is measurement noise.

A general linear time-invariant (LTI) model can be written as

$$y_t = G(q)u_t + H(q)e_t \qquad (1)$$

where H describes how the output depends on the noise e_t and where q is the shift operator, $qu_t = u_{t+1}$. In general, both G and H contain parameters that must be estimated. However, we will only consider models where H is fixed to 1, i.e. *output error* models. If G is causal this implies that only past inputs u_t, u_{t-1}, ... are used to predict the output y_t (Ljung, 1999).

All signals are here assumed to be stationary stochastic processes with zero means. Let $R_{yu}(\tau) = \mathrm{E}\{y_t u_{t-\tau}\}$ and $R_u(\tau) = \mathrm{E}\{u_t u_{t-\tau}\}$ and assume that these functions have z-transforms $\Phi_{yu}(z)$ and $\Phi_u(z)$ that are well-defined on the unit circle.

We will here define the best linear approximation of an NFIR system to be the causal and stable LTI system G_0 that minimizes the mean-square error, $\mathrm{E}\{(y_t - G(q)u_t)^2\}$. We will call G_0 the LTI second order equivalent (LTI-SOE) of the nonlinear system with respect to the input u. Note that LTI-SOE:s with the model structure (1) also can be defined for $H \neq 1$ (Ljung, 2001).

It is shown e.g. in (Ljung, 2001) that the output error LTI-SOE G_0 of a nonlinear system will be equal to the following expression

$$G_0(z) = \frac{1}{L(z)}\left[\frac{\Phi_{yu}(z)}{L(z^{-1})}\right]_{\text{causal}} \tag{2}$$

where $[\dots]_{\text{causal}}$ denotes taking the causal part and where $L(z)$ is a scaled version of the canonical spectral factor of Φ_u. (This means that $L(z)$ is a causal minimum phase LTI system with $L(\infty) = c > 0$ such that $\Phi_u(z) = L(z)L(z^{-1})$, see (Kailath *et al.*, 2000)). Note that $G_0(z) = \Phi_{yu}(z)/\Phi_u(z)$ if this ratio defines a stable and causal LTI system.

Assume that the prediction error method is used to estimate an output error model from input output data that come from an NFIR system. It can then be shown that this model will converge to the LTI-SOE of the system when the number of measurements tends to infinity, provided that the model order is sufficiently high (Ljung, 2001).

An example of the LTI-SOE of a simple nonlinear system can be found in Example 1.1. Note that the LTI-SOE is dynamical in this example even though the nonlinear system is static.

Example 1.1. Consider the static nonlinear system

$$y_t = u_t^3 \tag{3}$$

with the input

$$u_t = e_t + 0.5e_{t-1}$$

where e_t is a sequence of independent random variables with uniform distribution over the interval $[-1, 1]$. Straightforward calculations give

$$\Phi_{yu}(z) = \frac{1}{240}(34z + 91 + 46z^{-1})$$
$$\Phi_u(z) = \frac{1}{12}(2z + 5 + 2z^{-1})$$

Hence, the LTI-SOE of the system (3) for this input is

$$G_0(z) = \frac{\Phi_{yu}(z)}{\Phi_u(z)} = \frac{0.85 + 0.575z^{-1}}{1 + 0.5z^{-1}}$$

Note that although the nonlinear system is static the LTI-SOE is not.

LTI-SOE:s are also discussed in (Ljung, 2001) and in (Enqvist and Ljung, 2002). Other aspects of the use of linear models of nonlinear systems can be found, e.g., in (Mäkilä and Partington, 2002) and (Pintelon and Schoukens, 2001).

2. BACKGROUND

The following theorem is a classical result about Gaussian processes (see (Bussgang, 1952) for the original report and, e.g., (Papoulis, 1984) for a more recent reference).

Theorem 2.1. (Bussgang). Let y_t be the output from a static nonlinearity f with a Gaussian input u_t, i.e. $y_t = f(u_t)$. Assume that the expectations $\mathrm{E}\{y_t\} = \mathrm{E}\{u_t\} = 0$. Then

$$R_{yu}(\tau) = \mathrm{E}\{f'(u_t)\}R_u(\tau)$$

where

$$R_{yu}(\tau) = \mathrm{E}\{y_t u_{t-\tau}\}$$
$$R_u(\tau) = \mathrm{E}\{u_t u_{t-\tau}\}$$

Bussgang's theorem has turned out to be very useful for the theory of Hammerstein and Wiener system identification. (A Hammerstein systems consists of a static nonlinearity followed by an LTI system while a Wiener system is an LTI system followed by a static nonlinearity). The reason for this is that Bussgang's theorem explains why it is possible to estimate the linear and nonlinear parts of a Wiener or Hammerstein system separately when the input is Gaussian (Billings and Fakhouri, 1982; Korenberg, 1985; Bendat, 1998).

3. A GENERALIZATION OF BUSSGANG'S THEOREM

We will now present a result that can be viewed as a generalization of Bussgang's theorem. Related results can be found in (Scarano *et al.*, 1993) and have also previously been used in the research area of stochastic mechanical vibrations (see, e.g., Chapter 9 in (Lutes and Sarkani, 1997)). We will however present a complete, independent derivation here and we begin with the following technical assumption.

Assumption A1: The real-valued functions $f(x)$ and $\varphi(\tilde{x})$, where $x \in \mathbb{R}^N$ and $\tilde{x} = (x^T, v)^T \in \mathbb{R}^{N+1}$, are such that $f \cdot \varphi$, $f'_{x_i} \cdot \varphi$ and $f \cdot \tilde{x}_i \cdot \varphi$, $i = 1, \dots, (N+1)$ all belong to $\mathscr{L}^1(\mathbb{R}^{N+1})$ and that $f(x)\varphi(\tilde{x}) \to 0$ when $|\tilde{x}| \to +\infty$. (Here, f'_{x_i} is the partial derivative of f with respect to x_i).

Assumption A1 is used in the following theorem.

Theorem 3.1. Let

$$\tilde{x} = (x^T, v)^T = (x_1, x_2, \dots, x_N, v)^T \tag{4}$$

be a jointly Gaussian distributed random vector with zero mean and covariance matrix C with $\det C \neq 0$. Let $f : \mathbb{R}^N \to \mathbb{R}$ be a differentiable function of x with $\mathrm{E}\{f(x)\} = 0$ and let φ denote the probability density function of $\tilde{x}$. Furthermore, assume that f and φ satisfy Assumption A1. Then

$$\mathrm{E}\{f(x)\tilde{x}\} = Cw \tag{5}$$

where

$$w = \begin{pmatrix} \mathrm{E}\{f'_{x_1}(x)\} \\ \mathrm{E}\{f'_{x_2}(x)\} \\ \vdots \\ \mathrm{E}\{f'_{x_N}(x)\} \\ 0 \end{pmatrix}$$

Proof: Factorize C as $C = \tilde{M}\tilde{M}^T$ and define a new stochastic vector z as $z = \tilde{M}^{-1}\tilde{x}$. Then z is jointly normally distributed with zero mean and a covariance matrix that is equal to the identity matrix. Let M denote the matrix that is obtained from $\tilde{M}$ by removing the last row. Then $x = Mz$ and we get

$$\begin{aligned} E\{f(x)\tilde{x}\} &= \tilde{M}E\{f(x)\tilde{M}^{-1}\tilde{x}\} = \\ &= \tilde{M}E\{f(Mz)z\} = \\ &= \tilde{M}\begin{pmatrix} E\{\frac{\partial f(Mz)}{\partial z_1}\} \\ E\{\frac{\partial f(Mz)}{\partial z_2}\} \\ \vdots \\ E\{\frac{\partial f(Mz)}{\partial z_{N+1}}\} \end{pmatrix} = \\ &= \tilde{M}\tilde{M}^T \begin{pmatrix} E\{f'_{x_1}(x)\} \\ E\{f'_{x_2}(x)\} \\ \vdots \\ E\{f'_{x_N}(x)\} \\ 0 \end{pmatrix} = \\ &= Cw \end{aligned}$$

(The third equality follows from the fact that

$$E\{h(z)z_i\} = E\{h'_{z_i}(z)\}$$

when z has an $N(0,I)$ distribution).

□

Theorem 3.1 gives the following corollary.

Corollary 3.1. Let $y_t = f(u_t, u_{t-1}, \dots, u_{t-M_0})$ be an NFIR system with a stationary Gaussian process $(u_t)_{t=-\infty}^{\infty}$ with zero mean as input.

Form random vectors

$$\omega_\sigma = (u_t, u_{t-1}, \dots, u_{t-M_0}, u_{t-\sigma})^T \qquad (6)$$

with $\sigma < 0$ or $\sigma > M_0$. Let C_σ and φ_σ denote the covariance matrices and joint probability density functions of these vectors, respectively. Assume that $\det C_\sigma \neq 0$ for all $\sigma < 0$ or $\sigma > M_0$.

Furthermore, assume that $E\{y_t\} = 0$ and that f and φ_σ satisfy Assumption A1 for all $\sigma < 0$ or $\sigma > M_0$. Then

$$R_{yu}(\tau) = \sum_{k=0}^{M_0} b_k R_u(\tau - k) \quad \forall \tau \in \mathbb{Z} \qquad (7)$$

where

$$\begin{aligned} b_k &= E\{f'_{u_{t-k}}(u_t, u_{t-1}, \dots, u_{t-M_0})\} \\ R_{yu}(\tau) &= E\{y_t u_{t-\tau}\} \\ R_u(\tau) &= E\{u_t u_{t-\tau}\} \end{aligned}$$

Proof: Choose an arbitrary $\sigma < 0$ or $\sigma > M_0$ and let $x = (u_t, u_{t-1}, \dots, u_{t-M_0})^T$ and $v = u_{t-\sigma}$ in Theorem 3.1. Then Equation (5) gives

$$E\{y_t \begin{pmatrix} u_t \\ u_{t-1} \\ \vdots \\ u_{t-M_0} \\ u_{t-\sigma} \end{pmatrix}\} =$$

$$\begin{pmatrix} R_u(0) & R_u(1) & \dots & R_u(M_0) & R_u(\sigma) \\ R_u(1) & R_u(0) & \dots & R_u(M_0-1) & R_u(\sigma-1) \\ \vdots & \vdots & \ddots & \vdots & \vdots \\ R_u(M_0) & R_u(M_0-1) & \dots & R_u(0) & R_u(\sigma-M_0) \\ R_u(\sigma) & R_u(\sigma-1) & \dots & R_u(\sigma-M_0) & R_u(0) \end{pmatrix} w \qquad (8)$$

where $w_{i+1} = E\{f'_{u_{t-i}}\}$ for $i = 0, \dots, M_0$ and $w_{M_0+2} = 0$. Equation (8) can be written more compactly as

$$R_{yu}(\tau) = \sum_{k=0}^{M_0} b_k R_u(\tau - k)$$
$$\tau = 0, 1, \dots, M_0 \ \vee \tau = \sigma$$

where $b_k = w_{k+1} = E\{f'_{u_{t-k}}\}$. As σ was chosen arbitrarily, this relation must hold $\forall \tau \in \mathbb{Z}$.

□

Let Φ_{yu} and Φ_u denote the z-transforms of R_{yu} and R_u, respectively. Provided that these transforms are well-defined, (7) can also be written as

$$\Phi_{yu}(z) = B(z)\Phi_u(z) \qquad (9)$$

where $B(z) = \sum_{k=0}^{M_0} b_k z^{-k}$.

4. LTI-SOE:S FOR GAUSSIAN INPUTS

The generalization of Bussgang's theorem in the previous section can be used to characterize the LTI-SOE of an NFIR system with Gaussian inputs. As previously mentioned the LTI-SOE is in general obtained by the Wiener filter construction in Equation (2). However, from (9) we see that the quotient $\Phi_{yu}(z)/\Phi_u(z)$ is causal if the nonlinear system is an NFIR system with a Gaussian input. Hence, we can state the following theorem.

Theorem 4.1. Consider an NFIR system

$$y_t = f(u_t, u_{t-1}, \dots, u_{t-M_0})$$

with a Gaussian input u_t such that the requirements in Corollary 3.1 are fulfilled. Then the LTI-SOE is the linear FIR system

$$G_0(z) = \frac{\Phi_{yu}(z)}{\Phi_u(z)} = \sum_{k=0}^{M_0} b_k z^{-k} \qquad (10)$$

where $b_k = E\{f'_{u_{t-k}}\}$.

In general, it is quite possible that the LTI-SOE of an NFIR system with a non-Gaussian input will have an infinite impulse response (cf. Example 1.1) and it is usually hard to give a detailed characterization of it. However, as we have shown here, when the input is

Gaussian the LTI-SOE is always an FIR system and the coefficients of this system can be characterized exactly by (10).

The property that makes Gaussian processes special is that for a Gaussian process the conditional expectation $E\{u_{t-\tau}|u_t,\ldots,u_{t-M_0}\}$ is always a linear combination of $u_t,\ldots,u_{t-M_0}$. Note that there also exist non-Gaussian processes that have this property (Nuttall, 1958; Enqvist, 2003).

5. GEOMETRIC INTERPRETATION

In many cases, it is possible to shed some light on a theoretical result by interpreting it in a geometrical framework. This can as a matter of fact be done also in our case. For a fixed t, we can view the output y_t and the components of the input signal u_τ, $\tau \in \mathbb{Z}$ as vectors in an infinite dimensional inner-product space with the inner product $< u,v >= E\{uv\}$ (cf. (Brockwell and Davis, 1987)).

The LTI-SOE of the NFIR system will in this framework be the orthogonal projection of y_t into the linear subspace that is spanned by u_t, u_{t-1}, $\ldots u_{t-\infty}$. From (10) we can draw the conclusion that this projection actually lies in the finite dimensional linear subspace that is spanned by $u_t, u_{t-1},\ldots,u_{t-M_0}$.

6. APPLICATIONS

The characterization (10) of the LTI-SOE of an NFIR system with a Gaussian input is not only theoretically interesting but can also be useful in some applications of system identification. We will here briefly discuss three such applied identification problems.

6.1 *Structure Identification of NFIR Systems*

The most obvious application of the result (10) is perhaps to use it for guidance when an NFIR system is going to be identified. However, linear models are not useful for all types of NFIR systems. Any function $g : \mathbb{R}^N \to \mathbb{R}$ can be divided in an even part g_e and an odd part g_o

$$g(x) = \underbrace{\frac{g(x)+g(-x)}{2}}_{g_e(x)} + \underbrace{\frac{g(x)-g(-x)}{2}}_{g_o(x)}$$

such that $g_e(-x) = g_e(x)$ and $g_o(-x) = -g_o(x)$. This partitioning can be done also for the function f in an NFIR system. As all Gaussian probability density functions with zero mean are even functions and as the FIR coefficients in (10) are expectations of the partial derivatives of f it is easy to verify that the LTI-SOE of an NFIR system only is influenced by the odd part of the system.

Hence we will here only consider odd NFIR systems, i.e. NFIR systems $y_t = f(u_{t-n_k}, u_{t-n_k-1},\ldots,u_{t-n_k-M_0})$ where

$$\begin{aligned} &f(u_{t-n_k}, u_{t-n_k-1},\ldots,u_{t-n_k-M_0}) = \\ &\quad -f(-u_{t-n_k}, -u_{t-n_k-1},\ldots,-u_{t-n_k-M_0}) \end{aligned}$$

When such an odd NFIR system is going to be identified it is in general not obvious how the time delay n_k and order M_0 should be estimated in an efficient way. However, if the input is Gaussian and sufficiently many measurements can be collected, n_k and M_0 can both be obtained from an impulse response estimate. Such an estimate can be computed very efficiently by means of the least squares method.

Furthermore, if only a few of the input terms u_{t-n_k}, $u_{t-n_k-1},\ldots,u_{t-n_k-M_0}$ enter the system in a nonlinear way it might be interesting to know which these terms are. If a nonlinear model of the system is desired this knowledge can be used to reduce the complexity of the proposed model. A coefficient b_j in (10) will be invariant of the input properties if the corresponding input term u_{t-j} only affects the system linearly, while an input term that affects the system in a nonlinear way will have an input dependent b-coefficient in (10).

This fact makes it possible to extract information about which nonlinear terms that are present in the system simply by looking at the differences between FIR models that have been estimated with different Gaussian input signals. The coefficients that correspond to an input term that enters the system in a nonlinear way will be different in these estimates, provided that the covariance functions of the inputs are different.

6.2 *Identification of Generalized Hammerstein Systems*

In Section 2 we mentioned that Bussgang's theorem has been used to show important results concerning the identification of Hammerstein and Wiener systems. In principle, these results give that an estimated LTI model will converge to a scaled version of the linear part of a Hammerstein or Wiener system when the number of measurements tends to infinity, provided that the input is Gaussian. These results simplify the identification of Wiener and Hammerstein systems significantly.

Hence, it is interesting to investigate if the result (10) about the LTI-SOE:s of NFIR systems can be used to prove similar results for extended classes of systems. In this section we will study a type of systems that we will call generalized Hammerstein systems, while we will consider generalized Wiener systems in the next section.

More specifically, we will call a nonlinear system a generalized Hammerstein system if it consists of an

NFIR system $v_t = f(u_t, u_{t-1}, \dots, u_{t-M_0})$ followed by an LTI system $y_t = G(q)v_t$. If G is causal it can be written as

$$y_t = \sum_{k=0}^{\infty} g_k v_{t-k} \quad (11)$$

If we multiply both sides of Equation (11) with $u_{t-\tau}$ and take the expectation we get

$$R_{yu}(\tau) = \sum_{k=0}^{\infty} g_k R_{vu}(\tau - k), \quad (12)$$

provided that a term by term calculation of the expectation is allowed. If all z-transforms are well-defined, Equation (12) can also be written as

$$\Phi_{yu}(z) = G(z)\Phi_{vu}(z)$$

If u is Gaussian and f is such that Corollary 3.1 can be applied we thus get

$$\Phi_{yu}(z) = G(z)B(z)\Phi_u(z)$$

where $B(z) = \sum_{k=0}^{M_0} b_k z^{-k}$ and $b_k = \mathrm{E}\{f'_{u_{t-k}}\}$.

Hence, the LTI-SOE of a generalized Hammerstein system with a Gaussian input will be $G(z)B(z)$ and an estimated output error model will approach this system as the number of measurements tends to infinity. In particular, as $B(z)$ is an FIR system, this shows that the denominator of the estimated model will approach the denominator of G if the degree of the model denominator polynomial is correct.

We will thus get consistent estimates of the poles of G despite the presence of the NFIR system. This result is verified experimentally in Example 6.1 for a particular system.

Example 6.1. Consider a generalized Hammerstein system

$$y_t = G(q)f(u_t, u_{t-1}) + e_t$$

where

$$G(q) = \frac{1}{1 + 0.6q^{-1} + 0.1q^{-2}}$$

$$f(u_t, u_{t-1}) = \arctan(u_t) \cdot u_{t-1}^2$$

and where e_t is white Gaussian noise with $\mathrm{E}\{e_t\} = 0$ and $\mathrm{E}\{e_t^2\} = 1$.

Let the input u_t be generated by linear filtering of a white Gaussian process x_t with $\mathrm{E}\{x_t\} = 0$ and $\mathrm{E}\{x_t^2\} = 1$ such that

$$u_t = \frac{1 + q^{-1} + q^{-2}}{1 - 0.2q^{-1}} x_t$$

and assume that x_t and e_s are independent $\forall t, s \in \mathbb{Z}$.

This input signal has been used in an identification experiment where a data set consisting of 10000 measurements of u_t and y_t was collected. A linear output error model $\hat{G}_{oe}$ with $n_b = n_f = 2$ and $n_k = 0$ has been estimated from this data set and the result was

$$\hat{G}_{oe} = \frac{1.13 + 2.61q^{-1}}{1 + 0.573q^{-1} + 0.0954q^{-2}} \quad (13)$$

As can easily be seen from (13), the denominator of $\hat{G}_{oe}$ is indeed close to the denominator of G. This is exactly what one would expect as the previous theoretical discussion give that the LTI-SOE of the generalized Hammerstein system is the product between $G(q)$ and an FIR system $B(q)$.

6.3 *Identification of Generalized Wiener Systems*

We will call a nonlinear system a generalized Wiener system if it consists of an LTI system $n_t = G(q)u_t$ followed by an NFIR system $y_t = f(n_t, \dots, n_{t-M_0})$.

Assume that u_t is Gaussian and that the linear and nonlinear parts of the system are such that Theorem 3.1 can be applied with $x = (n_t, n_{t-1}, \dots, n_{t-M_0})^T$ and $v = u_{t-\tau}$ for any $\tau \in \mathbb{Z}$. (Note that n_t will be Gaussian as it is a linearly filtered Gaussian signal). The last row of Equation (5) then gives

$$R_{yu}(\tau) = \sum_{k=0}^{M_0} b_k R_{nu}(\tau - k) \quad (14)$$

where $b_k = \mathrm{E}\{f'_{n_{t-k}}\}$. Equation (14) can also be written as

$$\Phi_{yu}(z) = B(z)\Phi_{nu}(z) = B(z)G(z)\Phi_u(z)$$

and hence the LTI-SOE of a generalized Wiener system with a Gaussian input will be $B(z)G(z)$.

This implies, just as in the case with generalized Hammerstein systems, that consistent estimates of the poles of G can be obtained by estimating an output error model. The following example verifies this result for a particular generalized Wiener system.

Example 6.2. Consider a generalized Wiener system consisting of the same linear and nonlinear blocks as the generalized Hammerstein system in Example 6.1 but with the linear block before the nonlinear, i.e.

$$y_t = f(n_t, n_{t-1}) + e_t$$
$$n_t = G(q)u_t$$

where

$$G(q) = \frac{1}{1 + 0.6q^{-1} + 0.1q^{-2}}$$

$$f(n_t, n_{t-1}) = \arctan(n_t) \cdot n_{t-1}^2$$

and where e_t is white Gaussian noise with $\mathrm{E}\{e_t\} = 0$ and $\mathrm{E}\{e_t^2\} = 1$.

Let the input u_t be generated in the same way as in Example 6.1 i.e.

$$u_t = \frac{1+q^{-1}+q^{-2}}{1-0.2q^{-1}}x_t$$

where x_t is a white Gaussian process with $\mathrm{E}\{x_t\} = 0$ and $\mathrm{E}\{x_t^2\} = 1$ such that x_t and e_s are independent $\forall t, s \in \mathbb{Z}$.

An identification experiment has been performed on this generalized Wiener system with a realization of this u_t as input and 10000 measurements of u_t and y_t have been collected. A linear output error model $\hat{G}_{oe}$ with $n_b = n_f = 2$ and $n_k = 0$ has been estimated from the measurements and the result was

$$\hat{G}_{oe} = \frac{1.01+0.874q^{-1}}{1+0.565q^{-1}+0.0975q^{-2}} \tag{15}$$

From (15) we can see that the denominator of $\hat{G}_{oe}$ is close to the denominator of G also when the data has been generated by a generalized Wiener system.

---

7. CONCLUSIONS

In the previous sections we have given a characterization of the LTI second order equivalent (LTI-SOE) of a nonlinear FIR system with a Gaussian input. We have shown that this LTI-SOE will be an FIR system and described how the coefficients of this FIR system depends on the properties of the NFIR system and the input signal. Furthermore, we have discussed some applications of these results in structure identification and identification of generalized Hammerstein and Wiener models.

The LTI-SOE will only depend on the odd part of the NFIR system. This is due to the fact that the FIR coefficients are expectations of the partial derivatives of the NFIR system. However, a model that is estimated from a relatively small data set can be heavily influenced by the even part of the nonlinear system. Hence, there is a need to investigate the influence of even nonlinearities further.

8. ACKNOWLEDGMENTS

This work has been supported by the Swedish Research Council, which is hereby gratefully acknowledged.

9. REFERENCES

Bendat, J. S. (1998). *Nonlinear Systems Techniques and Applications*. John Wiley & Sons. New York.

Billings, S. A. and S. Y. Fakhouri (1982). Identification of systems containing linear dynamic and static nonlinear elements. *Automatica* **18**(1), 15–26.

Brockwell, P. J. and R. A. Davis (1987). *Time Series: Theory and Methods*. Springer. New York.

Bussgang, J. J. (1952). Crosscorrelation functions of amplitude-distorted Gaussian signals. Technical Report 216. MIT Laboratory of Electronics.

Enqvist, M. (2003). Nonlinear FIR systems with separable input processes. Technical Report (To appear). Department of Electrical Engineering, Linköping University. SE-581 83 Linköping, Sweden.

Enqvist, M. and L. Ljung (2002). Estimating nonlinear systems in a neighborhood of LTI-approximants. In: *Proc. of the 41st IEEE Conference on Decision and Control*. Las Vegas, NV. pp. 1005–1010.

Kailath, T., A. H. Sayed and B. Hassibi (2000). *Linear Estimation*. Prentice Hall. Upper Saddle River, NJ.

Korenberg, M. J. (1985). Identifying noisy cascades of linear and static nonlinear systems. In: *Proc. 7th IFAC Symp. on Identification and System Parameter Identification*. York, U.K.. pp. 421–426.

Ljung, L. (1999). *System Identification: Theory for the User*. 2nd ed.. Prentice Hall. Upper Saddle River, NJ.

Ljung, L. (2001). Estimating linear time-invariant models of nonlinear time-varying systems. *European Journal of Control* **7**(2-3), 203–219.

Lutes, L. D. and S. Sarkani (1997). *Stochastic Analysis of Structural and Mechanical Vibrations*. Prentice Hall. Upper Saddle River, NJ.

Mäkilä, P. M. and J. R. Partington (2002). On linear models for nonlinear systems. *Automatica* **39**(1), 1–13.

Nuttall, A. H. (1958). Theory and Application of the Separable Class of Random Processes. PhD thesis. MIT.

Papoulis, A. (1984). *Probability, Random Variables and Stochastic Processes*. 2nd ed.. McGraw Hill.

Pintelon, R. and J. Schoukens (2001). *System Identification: A Frequency Domain Approach*. IEEE Press. New Jersey.

Scarano, G., D. Caggiati and G. Jacovitti (1993). Cumulant series expansion of hybrid nonlinear moments of n variates. *IEEE Transactions on Signal Processing* **41**(1), 486–489.

IFAC Publications
www.elsevier.com/locate/ifac

AN ALGEBRAIC METHOD FOR SYSTEM REDUCTION OF STATIONARY GAUSSIAN SYSTEMS

Dorina Jibetean and Jan H. van Schuppen

CWI, P.O. Box 94079, 1090 GB Amsterdam, The Netherlands

Abstract: System identification for a particular approach reduces to system reduction, determining for a system with a high state-space dimension a system of low state-space dimension. For Gaussian systems the problem of system reduction is considered with the divergence rate criterion. The divergence or Kullback-Leibler pseudo-distance corresponds to the expected value of the negative natural logarithm of the likelihood function. System reduction for Gaussian systems is thus a certainty equivalent way of maximum likelihood identification. An algebraic method is proposed for system reduction. The results are a theorem that this problem reduces to an infimization problem for a rational function for which programs are available and a procedure for computing the best approximant w.r.t. the divergence rate criterion. As illustration two examples of system reduction are presented.

Keywords: System identification, system reduction, Gaussian system, divergence, maximum likelihood method, algebraic method, global optimization, local minima.

1. INTRODUCTION

The aim of this paper is to introduce an algebraic approach to system reduction for Gaussian systems by the divergence rate criterion.

The motivation for this paper is system identification of Gaussian systems. A finite-dimensional Gaussian system is a linear system driven by a Gaussian white noise process. In this paper attention is limited to discrete-time systems. As is well known, a stationary Gaussian system is a mathematical model for an observed stationary Gaussian process. The system identification problem is to construct from observed data and from assumptions a mathematical model, here a Gaussian system, such that the observed processes of the system approximate the observed data as well as possible according to an approximation criterion.

Methods of system identification for Gaussian systems often used include the maximization of the likelihood function, the subspace identification algorithm, and the least-squares method. The divergence between two probability measures is a well known pseudo-distance. It equals the expectation of the negative of the natural logarithm of the likelihood function. The divergence rate is derived from divergence and is needed because of the consideration of a stationary process.

The approximation problem of system identification is one of the major problems of this area. The main questions of parameter estimation include: How to find the global infimum? How to derive the first-order conditions? How to compute the local minima? How many local minima are there? Is the global minimum unique?

The aim of the paper is: (1) To present an algebraic approach and an algorithm for the infimization of the divergence rate criterion of Gaussian systems. (2) To show for several low order Gaussian systems that system reduction leads for the divergence rate criterion to two or more local minima. These examples and the method have serious implications for system identification of Gaussian systems by the maximum likelihood method. Though it is known from theoretical investigations and from numerical experiments with examples of system identification problems that two or more local minima exist, the consequences of this for system identification practice seem not to be widely known.

The results of the paper include a procedure to determine the best approximant w.r.t. the divergence rate criterion by an algebraic method. Determining the best approximant is proven to be equivalent to infimization of a rational function for which recently an algorithm was determined by the first-named author, see (Jibetean, 2001; Jibetean, 2003). The approach is illustrated in Example 2 with the reduction from a third order Gaussian system to a second order one. The set of local minima is not completely determined in this case although an upper bound on its cardinality is provided. Example 1 treats model reduction for a Gaussian system of state-space dimension 2 to one of state-space dimension 1. In this case there are two potential minima, one is the global minimum and the other a local one. The criterion values are quite close.

The novelty of the paper is in: (1) The combination of the divergence rate criterion for system reduction with algebraic methods; (2) The equivalence of the system reduction problem w.r.t. the divergence rate criterion to optimization of a rational function of the system parameters; (3) The application of algorithms for infimization of rational functions (based on LMI relaxations) to system reduction w.r.t. the divergence rate criterion; (4) Two examples which illustrate that two or more local minima of the criterion exist.

In comparison with the literature, algebraic approaches for computing *globally* best approximants have been applied to system reduction w.r.t. the H_2 criterion, see e.g. (Hanzon and Maciejowski, 1996), (Jibetean and Hanzon, 2002). The main interest to system identification of the proposed method is that it provides an algebraic way to determine the global minimum of the criterion and to circumvent the case of multiple local minima.

This paper is a sequel to those of (Stoorvogel and van Schuppen, 1996; Stoorvogel and van Schuppen, 1998).

2. PROBLEM FORMULATION

The motivating engineering problem is to determine a simple mathematical model for a time series. One speaks of the system identification problem or of the approximate realization problem. Examples of such a problem are the modeling of a signal in a noisy communication channel, of messages in a digital communication network, and of the traffic flow on a motorway.

Mathematical notation for the problem is summarized below. See the appendices for further details. Let (Ω, F) be a measurable space and $T = \mathbb{Z}$ denote the time index set and let $\mathbb{N} = \{0, 1, \ldots\}$ denote the set of natural numbers. Let P_1 be a probability measure on (Ω, F) induced by a stationary Gaussian process $y : \Omega \times T \to \mathbb{R}^p$ with zero mean value function and covariance function $W : T \to \mathbb{R}^{p \times p}$.

A time-invariant finite-dimensional Gaussian system on a probability space (Ω, F, P) is a stochastic system with representation

$$\begin{aligned} x(t+1) &= Ax(t) + Bv(t), \\ y(t) &= Cx(t) + Dv(t), \\ &\quad (A, B, C, D) \in LSP(p, n, p), \\ &\quad x : \Omega \times T \to \mathbb{R}^n, \quad y : \Omega \times T \to \mathbb{R}^p, \end{aligned}$$

see Appendix A for the full specification of the system. If the parameters of the system are in the set $SGSP_{min}(p, n, p)$ then the output process is a stationary Gaussian process. The probability measure induced by this system on the output process y is denoted by $P(q)$ where $q \in QD$ represents the parameter of a selected parametrization.

In this paper attention is restricted to the approximation problem of the system identification procedure.

Procedure 1. (1) Determine from a finite time series a high-order Gaussian system.
(2) System reduction: Determine from a high-order Gaussian system a low-order Gaussian system.

In this paper attention for the approximation problem is restricted to the divergence rate criterion. The concept of divergence of two probability measures is used in information theory. In probability theory divergence corresponds to the Kullback-Leibler measure, see (Cover and Thomas, 1991). For a stationary stochastic process the concept of divergence rate of two probability measures has been defined. In Section 3 an expression is provided for the divergence rate of two measures induced by stationary Gaussian processes which are outputs of two time-invariant finite-dimensional Gaussian systems. Denote this divergence rate by $D_r(P_1 \| P_2)$.

Let $n_2 \in \mathbb{N}$ denote an upper bound on the dimension of the Gaussian system to be determined.

Problem 1. Solve

$$\inf_{n \le n_2,\ q \in SGSP_{min}(p,n,p)} D_r(P_1 \| P(q)). \tag{1}$$

The problem involves establishing whether or not a minimum exists, if a minimum exists to characterize the set of minima, and to construct a procedure to compute a minimum or to approximate an infimum.

3. PROCEDURE FOR INFIMIZATION OF DIVERGENCE RATE

Recall the formula for the divergence rate of two measures induced by stationary Gaussian processes which are outputs of two time-invariant finite-dimensional Gaussian systems

System 1 $n_1 \in \mathbb{N}$,

$(A_1, B_1, C_1, D_1) \in SGSP_{min}(p, n_1, p)$,

System 2 $n_2 \in \mathbb{N}$,

$(A_2, B_2, C_2, D_2) \in SGSP_{min}(p, n_2, p)$.

The expression is available from the literature (Stoorvogel and van Schuppen, 1996; Stoorvogel and van Schuppen, 1998) in terms of a realization, (A_4, B_4, C_4, D_4), of the series interconnection of System 3 and System 1, where System 3 is the inverse of System 2. The relation between System 2 and System 3 is expressed by $n_3 = n_2$ and

$$\begin{aligned}&(A_3, B_3, C_3, D_3)\\&= (A_2 - B_2 D_2^{-1} C_2, B_2 D_2^{-1}, -D_2^{-1} C_2, D_2^{-1})\\&\in SGSP_{min}(p, n_3, p).\end{aligned}$$

Procedure 2. The divergence rate associated with the Systems 1 and 2 is computed by the following steps:

(1) Construct (A_4, B_4, C_4, D_4) according to the formulas $n_4 = n_1 + n_3$,

$$\begin{aligned}&(A_4, B_4, C_4, D_4)\\&= \left(\begin{pmatrix} A_1 & 0 \\ B_3 C_1 & A_3 \end{pmatrix}, \begin{pmatrix} B_1 \\ B_3 D_1 \end{pmatrix}, \right.\\&\left. \begin{pmatrix} D_3 C_1 & C_3 \end{pmatrix}, D_3 D_1 \right) \in SLSP(p, n_4, p).\end{aligned}$$

(2) Solve the discrete-time Lyapunov equation for the matrix $Q_4 \in \mathbb{R}^{n_4 \times n_4}$,

$$Q_4 = A_4 Q_4 A_4^T + B_4 B_4^T. \quad (2)$$

(3) Calculate

$$\begin{aligned}f_c &= D_r(P_1 \| P_2) \quad (3)\\&= \frac{1}{2} \mathrm{tr}(C_4 Q_4 C_4^T + D_4 D_4^T - I)\\&- \frac{1}{2} \ln \det(D_4 D_4^T).\end{aligned}$$

Algorithm 1. Infimization of the divergence rate of stationary Gaussian processes.
Input: System 1 representing the first probability measure and $n_2 \in \mathbb{N}$, the desired order of the approximant.
Output: System 2 representing the probability measure associated to the approximant

(1) Parametrize System 3 by a canonical parametrization map $f_p : QD \to SGSP_{min}(p, n_3, p)$,

$$q \overset{f_p}{\mapsto} (A_3(q), B_3(q), C_3(q), D_3(q)).$$

Note that $QD \subseteq \mathbb{R}^r$, where r is the dimension of the manifold $SGSP_{min}(p, n, p)$. Here f_q is restricted to functions which are rational in the entries of the parameter vector q.

(2) Determine, if it exists, a parameter value $\hat{q}_3 \in QD$ such that

$$\begin{aligned}\hat{q}_3 &= \mathrm{argmin}_{q \in QD} f_c(q), \quad (4)\\f_c(q) &:= D_r(P_1 \| P_2(q)), \quad (5)\end{aligned}$$

where $f_c(q)$ is determined according to Procedure 2.

(3) Set $(\hat{A}_3, \hat{B}_3, \hat{C}_3, \hat{D}_3) = f_p(\hat{q}_3)$ according to the parameterization map f_p.

(4) Compute the approximant System 2 according to

$$\begin{aligned}(\hat{A}_2, \hat{B}_2, \hat{C}_2, \hat{D}_2) &= \left(\hat{A}_3 - \hat{B}_3 \hat{D}_3^{-1} \hat{C}_3, \right. \quad (6)\\&\left. \hat{B}_3 \hat{D}_3^{-1}, -\hat{D}_3^{-1} \hat{C}_3, \hat{D}_3^{-1} \right).\end{aligned}$$

4. ALGEBRAIC METHOD

For the divergence infimization an algebraic method will be used. The *algebraic method* refers to the use of abstract algebra, computer algebra, and the use of the computer programs like MAPLE and MATHEMATICA. The difficulties to be overcome in the algebraic methods are to organize the calculations and to find an approach that is of low complexity.

Theorem 1. Consider the infimization problem

$$\inf_{q \in QD} f_c(q),$$

where $f_c(q)$ is computed by Procedure 2 and the matrices (A_3, B_3, C_3, D_3) depend on the parameter vector $q \in QD$.

(a) The infimization of the criterion with respect to the matrix C_3 is reached at the matrix

$$C_3 = -D_3 C_1 Q_2 Q_3^{-1}, \text{ where,} \quad (7)$$

$$Q_4 = \begin{pmatrix} Q_1 & Q_2 \\ Q_2^T & Q_3 \end{pmatrix} \in \mathbb{R}^{n_4 \times n_4}$$

is the solution of (2),

$$Q_2 \in \mathbb{R}^{n_1 \times n_2}, \; Q_3 \in \mathbb{R}^{n_2 \times n_2}.$$

Hence the criterion depends on the matrices A_3, B_3 and D_3.

(b) The infimization with respect to $D_3 \in \mathbb{R}^{p \times p}$ is reached for D_3 satisfying

$$D_3^T D_3 = M^{-1}, \quad (8)$$

and the criterion simplifies to,

$$\begin{aligned}f_c(q) &= -\frac{1}{2} \ln \det(D_1^T M^{-1} D_1)\\&= \frac{1}{2} \ln \det M - \ln \det D_1 \text{ , where}\end{aligned}$$

$$M = C_1 \left(Q_1 - Q_2 Q_3^{-1} Q_2^T\right) C_1^T + D_1 D_1^T.$$

The simplified criterion is a natural logarithm of a function which is a rational function with respect to entries of the (A_3, B_3). Thus the infimization problem is reduced to an infimization problem for a rational function.

PROOF. See (Jibetean, 2003). □

Procedure 3. (1) Select a parameterization for the matrices of System 3, A_3, B_3, C_3, and D_3, see

Algorithm 1. In view of Theorem 1, choose a parameterization with $C_3,\ D_3$ fully parametrized, independent from $A_3,\ B_3$. For example, the control canonical form gives such a parameterization.

(2) Solve by computer algebra the discrete-time Lyapunov equation (2) for the symbolic matrix $Q_4(q)$.
(3) Calculate the value of the criterion according to formula (3). The criterion f_c is the sum of a rational function and of a natural logarithm of the parameters of the system matrices $A_3,\ B_3,\ C_3$, and D_3.
(4) Apply the reduction technique formulated in Theorem 1 to solve analytically for the matrices C_3 and D_3 and to derive the simplified formula for the criterion. There remains then an infimization problem for a rational function, $\det M$.
(5) Determine the value of the infimum. If, moreover, the infimum is attained, i.e. the global minimum exists, then determine its location as well. For this use the approach of (Jibetean, 2001; Jibetean, 2003). Once M is determined, compute D_3 by (8) and C_3 by (7).
(6) If this is of interest then the infimization of the rational function also can provide information on the local minima. Derive the first order conditions of the simplified criterion with respect to the elements of the parameter vector $q \in QD$. Computer algebra provides programs for this.
(7) Determine all solutions in the set of the real numbers of the equation obtained by setting to zero the first derivative of the criterion with respect to the parameter vector. This is the most difficult and demanding part of the procedure.
(8) Calculate for each solution the second derivative of the criterion. Discard all points for which the second derivative is not positive semi-definite.
(9) For each of the remaining points calculate the value of the criterion $f_c(q)$. By comparing the different values numerically determine the global minimum or the set of global minima if there exist two or more parameter vectors which attain exactly the same value.

Note that steps (6)-(9) are optional. They should be executed only if there is interest in local minima.
The problem of finding the global optimum of a rational function at step (5), is solved in (Jibetean, 2001; Jibetean, 2003) by constructing an LMI relaxation of the original problem. The LMI relaxation returns in general a lower bound on the sought infimum but it is sharp and the sharpness of the lower bound can be checked under certain conditions. In case it is not sharp one constructs a sequence of LMI relaxations of the original problem which returns a sequence of increasing lower bounds, converging to the infimum of the rational function. Then the computations are more expensive.

5. EXAMPLES

Example 1. Consider a Gaussian system of order 2 with representation in the control canonical form as

$$A_1 = \begin{pmatrix} -0.4 & -0.32 \\ 1 & 0 \end{pmatrix}, \quad B_1 = \begin{pmatrix} 1 \\ 0 \end{pmatrix},$$
$$C_1 = \begin{pmatrix} 0 & -0.28 \end{pmatrix}, \quad D_1 = \begin{pmatrix} 1 \end{pmatrix}.$$

An approximant will be determined in the form of a Gaussian system of order 1, according to the divergence rate criterion. The class of Gaussian systems in which an approximant is to be sought is taken to be $SGSP_{min}(1,1,1)$. This class is parameterized by the control canonical form, hence

$$(A_2, B_2, C_2, D_2) = (a_2, 1, c_2, d_2).$$

If $d_2 > 0,\ |a_2| < 1,\ |a_2 - c_2 d_2^{-1}| < 1,\ c_2 \neq 0$, then $(A_2, B_2, C_2, D_2) \in SGSP_{min}(1,1,1)$.

Construct the quadruple, in control canonical form

$$(A_3, B_3, C_3, D_3) = (a_2 - c_2 d_2^{-1}, 1, -c_2 d_2^{-2}, d_2^{-1})$$

and compute the criterion to be minimized. As remarked, the optimum with respect to c_3 and d_3 can be computed analytically. The criterion becomes

$$f_c(q) = -\frac{1}{2} \ln \left(\frac{-34\,(25 + 10\,a_3 + 8\,a_3{}^2)\ldots}{(731\,a_3{}^2 + 1801\,a_3 + 19500)\ldots} \right)$$

The critical points equation with respect to a_3 is a univariate polynomial in a_3 whose roots are computed by numerical approximation. It turns out that in the stability region there exist two points of minimum of the criterion $f_c((\hat{a}_3, \hat{b}_3, \hat{c}_3, \hat{d}_3))$ such that

$$f_c((0.6353, 1, 0.1059, 0.9631)) = 0.0376,$$
$$f_c((-0.7835, 1, -0.1269, 0.9693)) = 0.0312.$$

In consequence, the second point is a global minimum, while the first returns a local minimum, although their values are close. As in Step (4) of Algorithm 1, compute the approximants

$$(\hat{a}_2, \hat{b}_2, \hat{c}_2, \hat{d}_2) = (0.5253, 1.0383, -0.1142, 1.0383),$$
$$\text{respectively } (-0.6525, 1.0317, 0.1351, 1.0317).$$

However the two approximant systems have a very different behavior. Below, the impulse response of the global approximant (dashed line) together with the impulse response of the original system (solid line) are plotted.

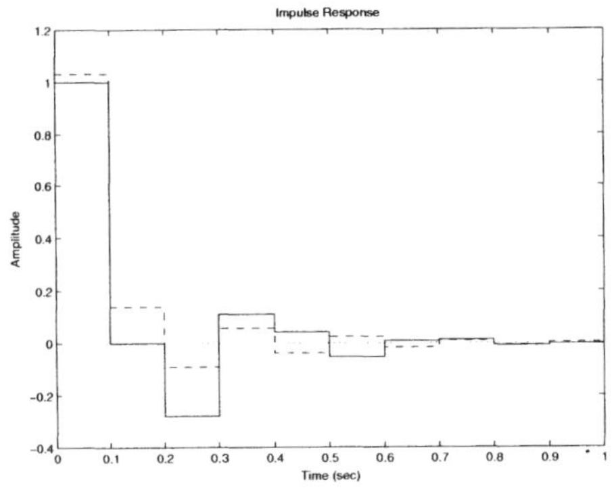

Example 2. Consider also a model reduction from order 3 to order 2, for the system

$$A_1 = \begin{pmatrix} -1/4 & 1/2 & 1/3 \\ 1 & 0 & 0 \\ 0 & 1 & 0 \end{pmatrix}, \quad B_1 = \begin{pmatrix} 1 \\ 0 \\ 0 \end{pmatrix},$$
$$C_1 = \begin{pmatrix} 1 & 2 & 1 \end{pmatrix}, \quad D_1 = \begin{pmatrix} 2 \end{pmatrix}.$$

The approximant is taken in the control canonical form, parameterized by $\alpha_1, \alpha_2, \gamma_1, \gamma_2, \delta$. After optimizing analytically with respect to $\gamma_1, \gamma_2, \delta$, an optimization of a logarithm of a rational function remains

$$-\frac{1}{2} \ln \left(\frac{(5640\alpha_1{}^3 + 85896\alpha_1{}^2 + 201240\alpha_1 + \ldots)}{(376\alpha_2{}^3 - 618\alpha_2{}^2 + \ldots)} \right),$$

which reduces, due to the monotonicity of the logarithm function, to optimization of a rational function. Using Gröbner bases methods for the first order conditions, one can notice that the function above has at most 100 complex critical points, including multiplicities. The computer failed to compute all of them. However, the methods of (Jibetean, 2001; Jibetean, 2003) can be employed for computing global optima of rational functions.

6. CONCLUSIONS

The main result of the paper is Procedure 3 with an algebraic method for infimization of the divergence rate between a Gaussian system and a class of such systems of lower state-space dimension. Theorem 1 establishes that the infimization problem reduces to an infimization problem for a rational function. Two examples illustrate the approach. In general a system reduction problem with this criterion and, by analogy, the parameter estimation with the likelihood function, will have many local minima. Further research is required to make the algebraic method more efficient and to streamline the computer algebra.

The authors advise the use of the proposed approach but caution the reader that at the current state of computer algebra, the method can handle effectively only low order systems. The authors also recommend for system identification of multi-output Gaussian systems the subspace identification algorithm which is based on stochastic realization theory.

ACKNOWLEDGMENTS

The authors gratefully acknowledge comments of Dr. B. Hanzon on the problem and approach of the paper.

Appendix A. LINEAR SYSTEMS

In the body of the paper concepts and results for time-invariant finite-dimensional linear systems are needed.

A *discrete-time time-invariant finite-dimensional linear system* is a dynamical system with the representation

$$x(t+1) = Ax(t) + Bu(t), \;\; x(t_0) = x_0,$$
$$y(t) = Cx(t) + Du(t),$$

where $T = \{t_0, t_0 + 1, \ldots\}$ is called the time axis, $n, m, p \in \mathbb{N}$, $x_0 \in \mathbb{R}^n$ is called the initial state, $u : T \to \mathbb{R}^m$ is called the input function, $x : T \to \mathbb{R}^n$ is called the state function, $y : T \to \mathbb{R}^p$ is called the output function, and $A \in \mathbb{R}^{n \times n}$, $B \in \mathbb{R}^{n \times m}$, $C \in \mathbb{R}^{p \times n}$, $D \in \mathbb{R}^{p \times m}$. The parameters of this system will be denoted by

$$(A, B, C, D) \in LSP(p, n, m).$$

Denote the reachability matrix and the observability matrix of this system respectively by

$$\mathcal{R}(A, B) = \begin{pmatrix} B & AB & \ldots & A^{n-1}B \end{pmatrix} \in \mathbb{R}^{n \times mn},$$
$$\mathcal{O}(A, C) = \begin{pmatrix} C \\ CA \\ \vdots \\ CA^{n-1} \end{pmatrix} \in \mathbb{R}^{np \times n}.$$

It is said that (A, B) is a *reachable pair* if $\operatorname{rank}(\mathcal{R}(A, B)) = n$ and that (A, C) is an *observable pair* if $\operatorname{rank}(\mathcal{O}(A, C)) = n$. Denote the spectrum of the matrix $A \in \mathbb{R}^{n \times n}$ by $\operatorname{spec}(A)$ and let $\mathbb{C}^- = \{\lambda \in \mathbb{C} | \; |\lambda| < 1\}$ denote the interior of the unit disc in the complex plane. Define the subclasses of linear systems

$$LSP_{min}(p, n, m) = \left\{ \begin{array}{l} (A, B, C, D) \in LSP(p, n, m) | \\ (A, B) \text{ reachable pair}, \\ (A, C) \text{ observable pair} \end{array} \right\},$$

$$SLSP(p, n, p) = \left\{ \begin{array}{l} (A, B, C, D) \in LSP(p, n, p) | \\ \operatorname{rank}(D) = p, \ \operatorname{spec}(A) \subset \mathbb{C}^-, \\ \operatorname{spec}(A - BD^{-1}C) \subset \mathbb{C}^- \end{array} \right\},$$

$$SLSP_{min} = SLSP(p, n, p) \cap LSP_{min}(p, n, p).$$

Appendix B. GAUSSIAN SYSTEMS

A time-invariant finite-dimensional *Gaussian system* (without inputs) is a stochastic system with representation

$$x(t+1) = Ax(t) + Bv(t), \tag{B.1}$$
$$y(t) = Cx(t) + Dv(t), \tag{B.2}$$

where $r, n, p \in \mathbb{N}$, $p \geq 1$, $v : \Omega \times T \rightarrow \mathbb{R}^r$ is a Gaussian white noise process, thus an independent sequence of random variables with for each $t \in T$, $v(t) \in G(0, V)$ ($v(t)$ has a Gaussian probability distribution function with parameters 0 and V), $V \in \mathbb{R}^{r \times r}$, $V = V^T > 0$; $A \in \mathbb{R}^{n \times n}$, $B \in \mathbb{R}^{n \times r}$, $C \in \mathbb{R}^{p \times n}$, $D \in \mathbb{R}^{p \times r}$; $x : \Omega \times T \rightarrow \mathbb{R}^n$, $y : \Omega \times T \rightarrow \mathbb{R}^p$ are stochastic processes satisfying the recursions (B.1,B.2).

Below a canonical form is used for Gaussian systems with respect to the covariance function of the output of the Gaussian system. For this purpose the reader is reminded of the theorem that a Gaussian system is a minimal stochastic realization of its output process iff it is stochastically observable and stochastically reconstructible, see (Lindquist and Picci, 1996). Consider a Gaussian system that is stable, with $\text{spec}(A) \subset \mathbb{C}^-$. Let $Q \in \mathbb{R}^{n \times n}$ be the solution of the discrete Lyapunov equation $Q = AQA^T + BB^T$, and let $G = AQC^T + BVD^T \in \mathbb{R}^{n \times p}$. Then the Gaussian system is a minimal stochastic realization of its output process iff (A, C) is an observable pair and (A, G) is an observable pair. A time-invariant finite-dimensional Gaussian system is said to be a *Kalman realization* if in addition to being of minimal state-space dimension it satisfies $r = p$, $\text{rank}(D) = p$, $\text{spec}(A) \subset \mathbb{C}^-$, and $\text{spec}(A - BD^{-1}C) \subset \mathbb{C}^-$.

Define the set of parameters of Gaussian systems with $p, n, r, \in \mathbb{N}$ by

$$SGSP_{min}(p, n, p) = \left\{ \begin{array}{l} (A, B, C, D) \in SLSP(p, n, p) \mid \\ V = I,\ (A, B) \text{ reachable pair,} \\ (A, C),\ (A, G) \text{ observable pairs,} \end{array} \right\}.$$

Appendix C. DIVERGENCE RATE

The *divergence* or the *Kullback-Leibler pseudo-distance* on the set of probability measures of a measurable space (Ω, F) is defined by the formula

$$D(P_1 \| P_2) = E_Q[r_1 \ln(\frac{r_1}{r_2}) I_{(r_2>0)}] = \int_\Omega r_1(\omega) \ln\left(\frac{r_1(\omega)}{r_2(\omega)}\right) I_{(r_2(\omega)>0)} Q(d\omega),$$

where Q is a σ-finite measure on (Ω, F) such that

$$P_1 \ll Q,\ \frac{dP_1}{dQ} = r_1,\ P_2 \ll Q,\ \frac{dP_2}{dQ} = r_2,$$

see (Cover and Thomas, 1991, Ch. 16) and (Stoorvogel and van Schuppen, 1996, Def. C.7).

Let $y_1 : \Omega \times T \rightarrow \mathbb{R}^p$ be a stationary stochastic process on $T = \mathbb{Z}$. Denote by P_1, P_2 two measures for process y_1 on $(\mathbb{R}^p)^T$. The *divergence rate* between P_1, P_2 is defined by the formula

$$D_r(P_1 \| P_2) = \lim_{n \rightarrow \infty} \frac{1}{2n+1} D(P_1|_{[-n,n]} \| P_2|_{[-n,n]}), \tag{C.1}$$

if the limit exists, where $P_1|_{[-n,n]}, P_2|_{[-n,n]}$ denote the restrictions of P_1, P_2 respectively to probability measures of processes defined on the time index set $\{-n, \ldots, -1, 0, 1, \ldots, n\}$, see (Stoorvogel and van Schuppen, 1996, Def. E.4). It is shown in (Stoorvogel and van Schuppen, 1998), based on a theorem of (Stoorvogel and van Schuppen, 1996), that $D_r(P_1 \| P_2)$, the divergence rate of two probability measures induced by the output processes of two time-invariant finite-dimensional Gaussian systems, can be expressed in terms of a realization of the series interconnection between the first system and the inverse of the second system, as described in Procedure 2.

Appendix D. REFERENCES

Cover, T.M. and J.A. Thomas (1991). *Elements of information theory*. John Wiley & Sons. New York.

Hanzon, B. and J.M. Maciejowski (1996). Constructive algebra methods for the L_2−problem for stable linear systems. *Automatica* **32**(12), 1645–1657.

Jibetean, D. (2001). Global optimization of rational multivariate functions. Report PNA-R0120. CWI. Amsterdam.

Jibetean, D. (2003). Algebraic optimization with applications in system theory. PhD thesis. Vrije Universiteit Amsterdam.

Jibetean, D. and B. Hanzon (2002). Linear matrix inequalities for global optimization of rational functions and H_2 optimal model reduction. In: *Proc. of the 15th International Symposium on MTNS* (D.S. Gilliam and J. Rosenthal, Eds.).

Lindquist, A. and G. Picci (1996). Geometric methods for state space identification. In: *Identification, adaption, learning* (S. Bittanti and G. Picci, Eds.). pp. 1–69. Springer. London.

Stoorvogel, A.A. and J.H. van Schuppen (1996). System identification with information theoretic criteria. In: *Identification, adaptation, learning* (S. Bittanti and G. Picci, Eds.). pp. 289–338. Springer. Berlin.

Stoorvogel, A.A. and J.H. van Schuppen (1998). Divergence rate approximation of a stationary gaussian process by the output of a gaussian system. In: *Mathematical Theory of Networks and Systems* (A. Beghi, L. Finesso and G. Picci, Eds.). Il Poligrafo. Padova. pp. 879–882.

IFAC Publications
www.elsevier.com/locate/ifac

SEPARABLE LEAST SQUARES DATA DRIVEN LOCAL COORDINATES

Thomas Ribarits *,1,2 **Manfred Deistler** *,2 **Bernard Hanzon** *,2

* *Institute for Econometrics, OR and System Theory*
Vienna University of Technology
Argentinierstrasse 8, 1040 Vienna, Austria
{Thomas.Ribarits@,Manfred Deistler@,bhanzon@eos.}tuwien.ac.at

Abstract: In this paper, the parametrization of state-space systems by *data driven local coordinates* as introduced by (McKelvey *et al.*, 2003) is modified. This modification leads to an alternative analogous parametrization which can be used for a suitable concentrated likelihood-type criterion function, where the concentration step can be done by a generalized least squares step. An obvious consequence is the reduced number of parameters resulting in less computational burden, but, of course, the criterion function itself is changed by the concentration step. The resulting new parametrization is called `slsDDLC`, and its topological and geometrical properties are investigated in detail. *Copyright © 2003 IFAC*

Keywords: Parametrization, Linear multivariable systems, State-space models, Identifiability, System Identification.

1. INTRODUCTION

In this paper a new parametrization for classes of linear systems is introduced and analyzed: *Data driven local coordinates combined with separable least squares methods* (`slsDDLC`). Data driven local coordinates (DDLC) have been introduced by (McKelvey *et al.*, 2003), where similar ideas can already be found in (Wolodkin *et al.*, 1997) in an LFT-type parametrization setting. We will provide a modification of DDLC by applying the main idea to a suitably concentrated likelihood-type criterion function. A theorem stating the main topological and geometrical results for this new parametrization will also be given.

We will be concerned with linear, time invariant, discrete-time stochastic state-space systems of the following form

$$\begin{aligned} x_{t+1} &= Ax_t + Bu_t + K\varepsilon_t \\ y_t &= Cx_t + Du_t + \varepsilon_t \end{aligned} \qquad (1)$$

Here, x_t is the n-dimensional state vector, $A \in \mathbb{R}^{n\times n}$, $B \in \mathbb{R}^{n\times m}$, $C \in \mathbb{R}^{s\times n}$, $D \in \mathbb{R}^{s\times m}$ and $K \in \mathbb{R}^{n\times s}$ are parameter matrices; y_t and u_t are the observed s-dimensional outputs and the m-dimensional exogenous inputs, respectively. In addition, (ε_t) is an s-dimensional white noise process, i.e. $\mathbb{E}\varepsilon_t = 0$ and $\mathbb{E}\varepsilon_s\varepsilon_t' = \delta_{s,t}\Sigma$ for all $s,t \in \mathbb{Z}$, where $\Sigma > 0$ is assumed throughout.

The *impulse reponse* of the linear system (1) is given by

$$(L_j,K_j)_{j\in\mathbb{N}} = ((D,I),C(B,K),CA(B,K),CA^2(B,K),\dots)$$

where $\mathbb{N} = \{0,1,\dots\}$. The corresponding *transfer function* from (u_t,ε_t) to y_t is given by

$$(l(z),k(z)) = C(z^{-1}I - A)^{-1}(B,K) + (D,I) \qquad (2)$$

[1] Support by the Austrian 'Fonds zur Förderung der wissenschaftlichen Forschung', Project P-14438 is acknowledged.
[2] The authors acknowledge support by the program 'Training and Mobility of Researchers' through project System Identification (ERB FMRX CT98 0206) and contacts with the participants in the European Research Network System Identification (ERNSI).

where z denotes a complex variable. For $|z|$ sufficiently small, $(l(z),k(z))$ coincides with its power series expansion $\sum_{j=1}^{\infty} CA^{j-1}(B,K)z^j + (D,I)$.

Let $\mathbb{M}(n)$ be the set of all rational and causal $s \times (m+s)$ transfer functions $(l(z),k(z))$ of fixed McMillan degree n, where $k(0) = I$. $\mathbb{M}(n)$ is endowed with the pointwise topology T_{pt} corresponding to the relative topology in the product space $(\mathbb{R}^{s\times(m+s)})^{\mathbb{N}}$ for the coefficients $((L_j,K_j)|j \in \mathbb{N})$.

It is well known that $\mathbb{M}(n)$ is a real analytic manifold of dimension $2ns + m(n+s)$; see e.g. (Hannan and Deistler, 1988). $\bar{\mathbb{M}}(n)$ denotes the closure of $\mathbb{M}(n)$ in the set of all rational and causal transfer functions and satisfies: $\bar{\mathbb{M}}(n) = \cup_{i \leq n}\mathbb{M}(i)$. Note that $\mathbb{M}(n)$ consists of $n+1$ pathwise connected components in the SISO case ($s = 1$, $m = 0$) and is pathwise connected otherwise; see (Brockett, 1976) and (Glover, 1975). Finally, note that the same holds true if we restrict ourselves to stable and strictly minimum phase systems; see (Chou and Hanzon, 1995).

For given m, s and n, one can embed the matrices (A,B,C,D,K) in $S(n) := \mathbb{R}^{n^2+2ns+m(n+s)}$. We always identify (A,B,C,D,K) with $(vec(A)', vec(\tilde{B})', vec(C)', vec(D)')'$ where $\tilde{B} = (B,K)$ and $vec(X)$ stacks the last, second last, etc. column of the matrix X on top of each other. Note that for $m = 0$, i.e. in the case where no exogenous inputs are present and the state-space system is given by (A,K,C,I), the embedding described above simplifies to $(vec(A)', vec(K)', vec(C)')'$. $S(n)$ is endowed with the Euclidean norm. Finally, the set $S_m(n) = \{(A,\tilde{B},C,D) \in S(n)|(A,\tilde{B},C,D) \text{ is minimal}\}$ is introduced.

The mapping π, which is evidently continuous, attaches transfer functions to the system matrices:

$$\begin{array}{llll} \pi: & S(n) & \rightarrow & \bar{\mathbb{M}}(n) \\ & (A,B,C,D,K) & \mapsto & C(z^{-1}I-A)^{-1}(B,K)+(D,I) \end{array}$$

It is well known that for every $(l,k) \in \mathbb{M}(n)$, the (l,k)-equivalence class $\mathcal{E}(A,\tilde{B},C,D)$ of minimal systems, i.e. the inverse image $\pi^{-1}(l,k)$ in $S_m(n)$, is of the form

$$\mathcal{E}(A,\tilde{B},C,D) = \left\{(TAT^{-1}, T\tilde{B}, CT^{-1}, D), T \in GL(n)\right\} \tag{3}$$

where $GL(n)$ denotes the set of real non singular $n \times n$ matrices and (A,B,C,D,K) is any minimal realization of (l,k), i.e. $\pi(A,B,C,D,K) = (l,k)$. This set constitutes an n^2 dimensional real analytic manifold consisting of two pathwise connected components; see (Ribarits, 2002).

The paper is organized as follows: In section 2, maximum likelihood estimation and the separable least squares method will be briefly dealt with, and in section 3 `slsDDLC` is introduced. A list of results for the special case $s = n = 1$ and $m = 0$ is presented in section 4. These results are then generalized to yield the main theorem in section 5. Finally, section 6 contains the conclusions.

2. MAXIMUM LIKELIHOOD ESTIMATION AND SEPARABLE LEAST SQUARES

In the sequel, we consider the case $\Sigma = \Sigma(\sigma)$ and $(A,B,C,D,K) = (A(\tau),B(\tau),C(\tau),D(\tau),K(\tau))$, where $\sigma \in \mathbb{R}^{s(s+1)/2}$ is the vector of on- and above-diagonal elements of Σ and τ is a vector of parameters appearing in the state-space matrices (A,B,C,D,K).

In this section we will also assume that (1) is in ('strict') innovations representation, i.e. $|\lambda_{max}(A)| < 1$ and $|\lambda_{max}(A-KC)| < 1$, and that the white noise process (ε_t) in (1) is Gaussian. Here, $\lambda_{max}(X)$ denotes an eigenvalue of the square matrix X of maximum modulus. Then $-2T^{-1}$ times the logarithm of the (Gaussian) likelihood function is given by:

$$L_T(Y_1^T;U_1^T,(\tau,\sigma)) = \frac{1}{T}\sum_{t=1}^{T}\log\det\Sigma_{t|t-1}(\tau,\sigma) + \frac{1}{T}\sum_{t=1}^{T} tr\left\{e_t(\tau,\sigma)e_t(\tau,\sigma)'\Sigma_{t|t-1}(\tau,\sigma)^{-1}\right\} \tag{4}$$

Here, $Y_1^T = (y_1',\ldots,y_T')'$ and $U_1^T = (u_1',\ldots,u_T')'$ denote the stacked vectors of observed outputs and inputs, respectively. Note that the quantities $e_t(\tau,\sigma)$ and $\Sigma_{t|t-1}(\tau,\sigma)$ have to be computed, e.g. using some *Kalman filtering procedure*. If the Kalman filter is initialized appropriately, $\Sigma_{t|t-1}$ depends only on the covariance parameters σ and e_t depends only on system parameters τ. The e_t's are then given by

$$\begin{aligned} \hat{x}_{t+1|t}(\tau) &= \bar{A}\hat{x}_{t|t-1}(\tau) + \bar{B}u_t + \bar{K}y_t \\ e_t(\tau) &= \bar{C}\hat{x}_{t|t-1}(\tau) + \bar{D}u_t + y_t \end{aligned} \tag{5}$$

corresponding to the inverse system where $\hat{x}_{1|0}(\tau) = 0$,

$$(\bar{A},\bar{B},\bar{C},\bar{D},\bar{K}) = (A-KC, B-KD, -C, -D, K) \tag{6}$$

and $\hat{x}_{t|t-1}(\tau)$ are given by the filter recursions (5) for $t \geq 2$. The likelihood then becomes

$$L_T(Y_1^T;U_1^T,(\tau,\sigma)) = \log\det\Sigma(\sigma) + \frac{1}{T}\sum_{t=1}^{T} tr\left\{e_t(\tau)e_t(\tau)'\Sigma^{-1}(\sigma)\right\} \tag{7}$$

and if there are no cross restrictions between the unknown parameters σ and the system parameters τ and no restrictions on σ, one can 'concentrate out' the parameters in Σ to obtain

$$L_T^c(Y_1^T;U_1^T,\tau) = \log\det\sum_{t=1}^{T} e_t(\tau)e_t(\tau)' \tag{8}$$

For any given τ, the corresponding Σ minimizing (7) is given by $\Sigma = \frac{1}{T}\sum_{t=1}^{T} e_t(\tau)e_t(\tau)'$.

Let us now consider a criterion function which is slightly different from (8):

$$\tilde{L}_T^c(Y_1^T;U_1^T,\tau) = tr\sum_{t=1}^{T} e_t(\tau)e_t(\tau)'\hat{\Sigma}^{-1} \tag{9}$$

where $\hat{\Sigma}$ is assumed to be a consistent estimate for the true innovations covariance.

Remark 1. We conjecture that using a consistent estimate $\hat{\Sigma}$ in (9) suffices to guarantee that the estimates obtained by minimizing (9) are asymptotically equivalent to the estimates obtained from (7) or (8) in the sense that they are also consistent and have the same asymptotic distribution. Note that if the identity matrix is used instead of $\hat{\Sigma}$ in (9)– see, e.g.(Bruls *et al.*, 1997) – the estimates are known to lack the property of asymptotic efficiency if the true innovations covariance is not a scalar multiple of the identity matrix.

Using (5), one gets after tedious calculations

$$\overbrace{\hat{\Sigma}_T^{-\frac{1}{2}}\begin{pmatrix} e_1 \\ e_2 \\ e_3 \\ \vdots \\ e_T \end{pmatrix}}^{\hat{E}_1^T} = \overbrace{\hat{\Sigma}_T^{-\frac{1}{2}}\begin{pmatrix} y_1 \\ y_2 \\ y_3 \\ \vdots \\ y_T \end{pmatrix}}^{\hat{Y}_1^T} - \tag{10}$$

$$\underbrace{\hat{\Sigma}_T^{-\frac{1}{2}}\begin{pmatrix} -u_1'\otimes I_s & \\ -u_2'\otimes I_s & -(u_1',y_1')\otimes\bar{C} \\ -u_3'\otimes I_s & -(u_1',y_1')\otimes\bar{C}\bar{A}-(u_2',y_2')\otimes\bar{C} \\ \vdots & \vdots \\ -u_T'\otimes I_s & -(u_1',y_1')\otimes\bar{C}\bar{A}^{T-2}-\cdots-(u_{T-1}',y_{T-1}')\otimes\bar{C} \end{pmatrix}}_{\tilde{X}=\tilde{X}\left(Y_1^{T-1},U_1^T,\bar{A},\bar{C},\hat{\Sigma}\right)}\begin{pmatrix} vec(\bar{D}) \\ vec(\bar{B}) \\ vec(\bar{K}) \end{pmatrix}$$

where $\hat{\Sigma}_T^{-1/2} = I_T \otimes \hat{\Sigma}^{-1/2}$ and $\hat{\Sigma}^{-1/2}$ denotes the inverse of any square root of $\hat{\Sigma}$ and $\tilde{X} \in \mathbb{R}^{sT\times ms+n(m+s)}$.

Let us now assume that we *parametrize the inverse system* $(\bar{A},\bar{B},\bar{C},\bar{D},\bar{K})$ using a parameter vector of the form $\tau = (\tau^o,\tau^u)$ where the parameters in $\tau^o \in \mathbb{R}^d$ only appear in $(\bar{A},\bar{C}) = (\bar{A}(\tau^o),\bar{C}(\tau^o))$ and the parameters in $\tau^u \in \mathbb{R}^{ms+n(m+s)}$ are simply taken to be the matrix entries of $(\bar{D},\bar{B},\bar{K})$. Then we have the following

Lemma 2. Let $\hat{\tau}^o \in \mathbb{R}^d$, $\hat{\Sigma}$ and the observations Y_1^T, U_1^T be given, where $T > (m+n)+mn/s$. Assume that $(\bar{A}(\tau^o),\bar{C}(\tau^o))$ are analytic functions of τ^o and that the matrix $\tilde{X}(Y_1^{T-1};U_1^T,\bar{A}(\hat{\tau}^o),\bar{C}(\hat{\tau}^o),\hat{\Sigma})$ in (10) has full column rank. Then $\tilde{X}$ has full column rank for all $\tau^o \in T^o$ where T^o is an open and dense subset of $\mathbb{R}^d$ and the unique $\hat{\tau}^u = ((vec\bar{D})',(vec\bar{B})',(vec\bar{K})')'$ that solves the optimization problem

$$\min_{\tau^u\in\mathbb{R}^{ms+n(m+s)}} \tilde{L}_T^c(Y_1^T;U_1^T,(\hat{\tau}^o,\tau^u)) \tag{11}$$

where $\tilde{L}_T^c$ is given in (9), is given by

$$\hat{\tau}^u = ((vec\bar{D})',(vec\bar{B})',(vec\bar{K})')' = (\tilde{X}'\tilde{X})^{-1}\tilde{X}'\tilde{Y}_1^T \tag{12}$$

PROOF. Clearly, $\tilde{X}(Y_1^{T-1},U_1^T,\bar{A}(\tau^o),\bar{C}(\tau^o),\hat{\Sigma})$ is an analytic function of τ^o. Hence, the function $\Delta_{\tilde{X}} : \mathbb{R}^d \to \mathbb{R}$ attaching $\det\tilde{X}'\tilde{X}$ to τ^o is also analytic (and therefore trivially continuous). The inverse image $\Delta_{\tilde{X}}^{-1}(\mathbb{R}\setminus\{0\})$ is an open subset of $\mathbb{R}^d$. Moreover, as $\Delta_{\tilde{X}}(\hat{\tau}^o) \neq 0$, $\Delta_{\tilde{X}}(\tau^o) = 0$ can only hold true on a thin subset of $\mathbb{R}^d$, showing that T^o is open and dense in $\mathbb{R}^d$. The fact that $\hat{\tau}^u$ in (12) is the unique solution to the optimization problem above is clear as (11) can be rewritten as

$$\min_{\tau^u\in\mathbb{R}^{ms+n(m+s)}} \|\tilde{Y}_1^T - \tilde{X}\cdot\tau^u\|_2^2 \tag{13}$$

and the unique solution $\hat{\tau}^u$ of (13) is given by (12).

We can immediately use the lemma above to obtain a new concentrated criterion function from (9):

$$L_T^{cc}(Y_1^T;U_1^T,\tau^o) = tr\sum_{t=1}^{T} e_t(\bar{A}(\tau^o),\bar{C}(\tau^o))e_t(\bar{A}(\tau^o),\bar{C}(\tau^o))'\hat{\Sigma}^{-1} \tag{14}$$

Here, $e_t(\bar{A}(\tau^o),\bar{C}(\tau^o))$ is given by (10) (without premultiplication with $\hat{\Sigma}_T^{-1/2}$) where the matrices $(vec(\bar{D})'$, $vec(\bar{B})', vec(\bar{K})')'$ are replaced by (12). The following result is obtained in (Golub and Pereyra, 1973); note that a critical point is a point where the gradient is the zero vector:

Lemma 3. Let $\tilde{L}_T^c(Y_1^T;U_1^T,(\tau^o,\tau^u)) = \tilde{L}_T^c(\tau^o,\tau^u)$ and $L_T^{cc}(Y_1^T;U_1^T,\tau^o) = L_T^{cc}(\tau^o)$ be given, let the assumptions of lemma 2 be valid and let $T^o \subseteq \mathbb{R}^d$ be as described in lemma 2. Then we have:

(i) If $\hat{\tau}^o \in T^o$ is a critical point (or a global minimizer in T^o) of $L_T^{cc}(\tau^o)$ and $\hat{\tau}^u$ is given by (12), then $(\hat{\tau}^o,\hat{\tau}^u)$ is a critical point of $\tilde{L}_T^c(\tau^o,\tau^u)$ (or a global minimizer for $\tau^o \in T^o$) and $\tilde{L}_T^c(\hat{\tau}^o,\hat{\tau}^u) = L_T^{cc}(\hat{\tau}^o)$.

(ii) If $(\hat{\tau}^o,\hat{\tau}^u)$ is a global minimizer of $\tilde{L}_T^c(\tau^o,\tau^u)$ for $\tau^o \in T^o$, then $\hat{\tau}^o$ is a global minimizer of $L_T^{cc}(\tau^o)$ in T^o and $L_T^{cc}(\hat{\tau}^o) = \tilde{L}_T^c(\hat{\tau}^o,\hat{\tau}^u)$. Furthermore, if there is an unique $\hat{\tau}^u$ among the minimizing pairs of $\tilde{L}_T^c(\tau^o,\tau^u)$, then $\hat{\tau}^u$ must satisfy (12).

PROOF. The proof is very simple and indeed a special case of theorem 2.1 in (Golub and Pereyra, 1973). The last statement in (ii) refers to the fact that in (Golub and Pereyra, 1973) the set $T^o \subseteq \mathbb{R}^d$ is replaced by some open subset of $\mathbb{R}^d$ where $\tilde{X}$ in (10) has *constant*, but not necessarily full, column or row rank.

Note that the original nonlinear optimization problem in $\tau = (\tau^o,\tau^u)$ in (9) has been reduced to a nonlinear optimization problem in the parameters τ^o in (14). In the literature, approaches of this type are termed *separable least squares methods*; see (Golub and Pereyra, 1973) or (Bruls *et al.*, 1997) for an application in a system identification context, where the separable least squares method is used to obtain a canonical form with *ns* parameters.

3. PARAMETRIZATION BY SLSDDLC

Let us start by considering the function $L_T^{cc}(Y_1^T;U_1^T,\tau^o)$ in (14). We take $\tau^o = ((vec\bar{A})',(vec\bar{C})')' \in \mathbb{R}^{n^2+ns}$ to be a vector consisting of all matrix entries of the pair $(\bar{A},\bar{C})$. Thus, we may write $L_T^{cc}(Y_1^T;U_1^T,\bar{A},\bar{C})$. As a special case of lemma 2 we then immediately get

Lemma 4. Let the pair $(\bar{A}_0,\bar{C}_0)$ be given. Moreover, let $\hat{\Sigma}$ and (Y_1^T,U_1^T) be given where $T > (m+n)+mn/s$ and let the matrix $\tilde{X}(Y_1^{T-1},U_1^T,\bar{A}_0,\bar{C}_0,\hat{\Sigma})$ in (10) have full column rank. Then there exists an open and dense subset S of $\mathbb{R}^{n^2+ns}$ such that

$$\begin{aligned} \Delta_Y^U &: S \to S(n) \qquad (15)\\ (\bar{A},\bar{C}) &\mapsto (A,B,C,D,K) = (\bar{A}-\bar{K}\bar{C},\bar{B}-\bar{K}\bar{D},-\bar{C},-\bar{D},\bar{K}) \end{aligned}$$

defines an analytic function, where $\bar{D}$, $\bar{B}$ and $\bar{K}$ are given by (12).

The following assumption will be needed frequently:

Assumption 5. Let the pair $(\bar{A},\bar{C})$ be observable. Moreover, let $\hat{\Sigma}$ and (Y_1^T,U_1^T) be given where $T > (m+n)+mn/s$. Assume that the matrix $\tilde{X}(Y_1^{T-1},U_1^T,\bar{A},\bar{C},\hat{\Sigma})$ in (10) has full column rank and that $\Delta_Y^U(\bar{A},\bar{C})$ as given in (15) is minimal.

We are now able to continue with the following

Definition 6. (The L_T^{cc}-equivalence class $\mathcal{E}^{cc}(\bar{A},\bar{C})$). Let ASSUMPTION 5 be satisfied. The set $\mathcal{E}^{cc}(\bar{A},\bar{C}) = \{(\hat{\bar{A}},\hat{\bar{C}}) \in S : \Delta_Y^U(\hat{\bar{A}},\hat{\bar{C}}) \in \mathcal{E}(\Delta_Y^U(\bar{A},\bar{C}))\}$ is called the the *L_T^{cc}-equivalence class*.

The following lemma clarifies the structure of $\mathcal{E}^{cc}(\bar{A},\bar{C})$:

Lemma 7. (Structure of $\mathcal{E}^{cc}(\bar{A},\bar{C})$). Let ASSUMPTION 5 be satisfied. Then the L_T^{cc}-equivalence class $\mathcal{E}^{cc}(\bar{A},\bar{C})$ constitutes an n^2 dimensional real analytic submanifold of $\mathbb{R}^{n^2+ns}$ consisting of two pathwise connected components and is given by

$$\mathcal{E}^{cc}(\bar{A},\bar{C}) = \{(T\bar{A}T^{-1},\bar{C}T^{-1}), \quad T \in GL(n)\} \quad (16)$$

PROOF. If $(\bar{A},\bar{C})$ and $(\hat{\bar{A}},\hat{\bar{C}})$ are in the same L_T^{cc}-equivalence class, it is straightforward to see that there exists a unique $T \in GL(n)$ relating the L_T^{cc}-equivalent systems to each other according to (16). For the converse direction and the remaining statements, see lemma 3.6.2 in (Ribarits, 2002).

Remark 8. For $n = s = 1$ and arbitrary m, the L_T^{cc}-equivalence classes $\mathcal{E}^{cc}(\bar{a},\bar{c})$ in $\mathbb{R}^2$ are vertical straight lines given by a fixed $\bar{a}$, but excluding the points where $\bar{c} = 0$; see figure 1.

We will now consider the tangent space to $\mathcal{E}^{cc}(\bar{A},\bar{C})$ at $(\bar{A},\bar{C})$ which will be denoted by $Q^{cc}_{(\bar{A},\bar{C})}$. Note that we view the tangent space as an affine space (instead of a linear vector space) in the sequel. Analogously, by the ortho-complement to the tangent space $Q^{cc}_{(\bar{A},\bar{C})}$, being a bit sloppy, we mean the ortho-complement to the linear space corresponding to $Q^{cc}_{(\bar{A},\bar{C})}$ shifted back by the initial $(\bar{A},\bar{C})$. Note that orthogonality is to be understood with respect to the 'usual' inner product in the Euclidean space $\mathbb{R}^{n^2+ns}$:

Lemma 9. (Tangent space $Q^{cc}_{(\bar{A},\bar{C})}$ to $\mathcal{E}^{cc}(\bar{A},\bar{C})$). Let ASSUMPTION 5 be satisfied. Then $(\bar{A},\bar{C}) + (\bar{A}_s,\bar{C}_s) \in Q^{cc}_{(\bar{A},\bar{C})}$ if and only if $(\bar{A}_s,\bar{C}_s) = (\dot{T}\bar{A}-\bar{A}\dot{T},-\bar{C}\dot{T})$ for some $\dot{T} \in \mathbb{R}^{n\times n}$.

Remark 10. For $n = s = 1$ and arbitrary m, $Q^{cc}_{(\bar{a},\bar{c})}$ is the vertical straight line described in remark 8, now including the point where $\bar{c} = 0$; see figure 1.

PROOF. Consider a differentiable path $T = T(\theta)$ in $GL(n)$ where $\theta \in [0,1]$, $\dot{T} = \dot{T}(\theta)$ is the derivative with respect to θ, $T(1/2) = I$ and $\dot{T}(1/2) = \dot{T}$ (with slight abuse of notation). Differentiation of (16) yields

$$\{(\dot{T}\bar{A}T^{-1} - T\bar{A}T^{-1}\dot{T}T^{-1},-\bar{C}T^{-1}\dot{T}T^{-1}), \quad \dot{T} \in \mathbb{R}^{n\times n}\} \quad (17)$$

and at $\theta = 1/2$, i.e. at $(\bar{A},\bar{C})$, (17) reduces to

$$\{(\dot{T}\bar{A}-\bar{A}\dot{T},-\bar{C}\dot{T}), \quad \dot{T} \in \mathbb{R}^{n\times n}\} \quad (18)$$

showing the result.

Remark 11. Clearly, (18) can be vectorized to yield

$$\{Q_{cc}\cdot vec(\dot{T}),\dot{T} \in \mathbb{R}^{n\times n}\},Q_{cc} = \begin{pmatrix} \bar{A}'\otimes I_n - I_n\otimes\bar{A} \\ -I_n\otimes\bar{C} \end{pmatrix} \quad (19)$$

The matrix $Q_{cc} \in \mathbb{R}^{(n^2+ns)\times n^2}$ can be shown to have full column rank for any observable pair $(\bar{A},\bar{C})$.

A direct consequence of lemma 7 is the fact that for any given observable pair $(\bar{A},\bar{C})$ there are n^2 essentially unnecessary coordinates when using all entries in $(\bar{A},\bar{C})$ as parameters. The idea now is to avoid this drawback by only considering the ns dimensional ortho-complement, $Q^{cc,\perp}_{(\bar{A},\bar{C})}$ say, to the tangent space $Q^{cc}_{(\bar{A},\bar{C})}$ at a given $(\bar{A},\bar{C})$ as a parameter space. Clearly, the parameter space will then be of dimension ns and thus has no unnecessary coordinates.

By $Q_{cc}^{\perp}$ we denote a matrix with columns which span the orthogonal complement to the tangent space spanned by the columns of Q_{cc} in (19); $Q_{cc}^{\perp}$ can be obtained e.g. from a singular value decomposition or a QR-factorization of Q_{cc}. `SlsDDLC` is now defined as follows:

Definition 12. (SlsDDLC). Let $(A,B,C,D,K) \in S_m(n)$ be given where $(\bar{A},\bar{B},\bar{C},\bar{D},\bar{K})$ is the corresponding minimal inverse system in (6) and let ASSUMPTION 5 be satisfied. The `slsDDLC` are given by the mapping

$$\varphi_D^o : T_D^o \to S_m(n) \tag{20}$$
$$\tau_D^o \mapsto \Delta_Y^U \begin{pmatrix} vec(\bar{A}(\tau_D^o)) \\ vec(\bar{C}(\tau_D^o)) \end{pmatrix} = \Delta_Y^U \left(\begin{pmatrix} vec(\bar{A}) \\ vec(\bar{C}) \end{pmatrix} + Q_{cc}^{\perp} \tau_D^o \right)$$

Here, $T_D^o \subseteq \mathbb{R}^{ns}$ denotes the parameter space for `slsDDLC`, i.e. the set of all $\tau_D^o \in \mathbb{R}^{ns}$ such that $\Delta_Y^U(\bar{A}(\tau_D^o),\bar{C}(\tau_D^o))$ is well defined and minimal; Δ_Y^U is defined in (15) in lemma 4. Let $V_D^o = \pi(\varphi_D^o(T_D^o))$.

Remark 13. With slight abuse of notation we will use the symbol $\pi(T_D^o)$ for $\pi(\varphi_D^o(T_D^o))$.

Note that for any fixed minimal (A,B,C,D,K), the mapping φ_D^o from the parameter vectors τ_D^o to the state-space matrices of the original system is highly non linear due to the transformation Δ_Y^U.

4. THE SISO CASE WITH $N = 1$

We consider the trivial case where $n = s = 1$ and $m = 0$ in order to provide some insights into geometrical and topological properties of the `slsDDLC` parametrization. Commencing from a minimal system $(a,k,c) \in S_m(1)$ where $(\bar{a},\bar{k},\bar{c}) = (a-kc,k,-c)$ is the corresponding minimal inverse system, we assume that ASSUMPTION 5 is satisfied. Then we have $Q_{cc} = (0,-\bar{c})'$ and we can choose $Q_{cc}^{\perp} = (1,0)'$. Using this $Q_{cc}^{\perp}$, the `slsDDLC` parametrization is given by

$$\Delta_Y^U \begin{pmatrix} \bar{a}(\tau_D^o) \\ \bar{c}(\tau_D^o) \end{pmatrix} = \Delta_Y^U \begin{pmatrix} \bar{a}+\tau_D^o \\ \bar{c} \end{pmatrix} \tag{21}$$

Note that $\bar{c} \neq 0$ because of minimality. Consequently, $\tilde{X} = \tilde{X}(Y_1^{T-1},\bar{a}(\tau_D^o),\bar{c}(\tau_D^o),\hat{\sigma}^2)$ given by

$$\tilde{X} = -\frac{1}{\hat{\sigma}} \begin{pmatrix} 0 \\ \bar{c}y_1 \\ \bar{c}\bar{a}(\tau_D^o)y_1 + \bar{c}y_2 \\ \vdots \\ \bar{c}\bar{a}(\tau_D^o)^{T-2}y_1 + \cdots + \bar{c}y_{T-1} \end{pmatrix}$$

has full column rank for all $\tau_D^o \in \mathbb{R}$ unless $y_1 = y_2 = \cdots = y_{T-1} = 0$. The latter case is ruled out by ASSUMPTION 5, and therefore, using the notation of lemma 2, we have $T^o = \mathbb{R}$. Making use of figure 1, we will now discuss *geometrical and topological properties* of `slsDDLC`, where the statements in italics refer to the general case discussed in theorem 14 below:

(i) Although $(\bar{a}(\tau_D^o),\bar{c}(\tau_D^o))$ is observable for all $\tau_D^o \in \mathbb{R}$, $\Delta_Y^U(\bar{a}(\tau_D^o),\bar{c}(\tau_D^o))$ may contain non minimal systems. Consider, for instance, the case $(\bar{a},\bar{c}) = (0,1)$ depicted in figure 1, where additionally $Y_1^3 = (1,1,1)'$ and $\hat{\sigma}^2 = 1$, implying that $\tilde{X} = -(0,1,\tau_D^o+1)'$ and $\hat{\tau}^u = \bar{k}(\tau_D^o) = -\frac{\tau_D^o+2}{1+(\tau_D^o+1)^2}$ (see (12)) such that

$$(\bar{a}(\tau_D^o),\bar{k}(\tau_D^o),\bar{c}(\tau_D^o)) = (\tau_D^o, -\frac{\tau_D^o+2}{1+(\tau_D^o+1)^2}, 1) \tag{22}$$

For $\tau_D^o = -2$, we get $(\bar{a}(-2),\bar{k}(-2),\bar{c}(-2)) = (-2,0,1)$ and therefore $\Delta_Y^U(\bar{a}(-2),\bar{c}(-2)) = (-2,0,-1)$ which is clearly non minimal. Hence, $T_D^o = \mathbb{R} \setminus \{-2\}$, which is obviously an *open and dense subset of* $\mathbb{R}$.

(ii) It is straightforward to see from figure 1 that there exists an open interval $T_D^{o,loc}$ containing $0 \in T_D^o$ such that each L_T^{cc}-equivalence class given by the vertical line described in remark 8 intersects the straight horizontal line $(\bar{a}+\tau_D^o,\bar{c})$ only once and this intersection corresponds to a minimal original system. The interval $T_D^{o,loc}$ can be chosen to be $T_D^{o,loc} = (-2,\infty)$. In other words: *$T_D^o$ is locally identifiable at $\tau_D^o = 0$.*

(iii) Clearly, the boundary points of T_D^o (which do not belong to T_D^o by statement (i)) correspond to systems $(\bar{a}(\tau_D^o),\bar{c}(\tau_D^o))$ where $\Delta_Y^U(\bar{a}(\tau_D^o),\bar{c}(\tau_D^o))$ is either not defined or non minimal. Here, the point $\tau_D^o = -2$ is the only boundary point and $\Delta_Y^U(\bar{a}(-2),\bar{c}(-2))$ is non minimal. Note that $T_D^o = \mathbb{R}$ is also possible, implying that $\pi(\bar{T}_D^o) = \pi(\mathbb{R})$ *does* not *contain any transfer functions of lower McMillan degree.* To see this, consider the same example, but this time with $Y_1^4 = (1,1,1,1)'$, i.e. with one more observation. The same calculations as above reveal that $(a(\tau_D^o),k(\tau_D^o),c(\tau_D^o))$ is well defined and minimal for all $\tau_D^o \in \mathbb{R}$.

(iv) The intersection of the vertical thin lines with the horizontal line $(\bar{a}+\tau_D^o,\bar{c}))$ always yields a single point, i.e. T_D^o is globally identifiable and this trivially implies the weaker statement that *every transfer function in V_D^o has just a finite number of representatives within T_D^o.* It is not difficult to see that in general `slsDDLC` need not be locally identifiable at every point in T_D^o and hence it need not be globally identifiable; see (Ribarits, 2002).

(v) *V_D^o is clearly open in $\pi(\bar{T}_D^o)$.* Note that the dimension of T_D^o is smaller than the dimension of the manifold $\mathbb{M}(n)$. Hence, V_D^o cannot be open in $\mathbb{M}(n)$.

(vi) We show that $\pi(\bar{T}_D^o) = \bar{V}_D^o$, i.e. the same transfer functions are described in the closure of the parameter space T_D^o and in the closure of the corresponding transfer function space V_D^o. First, note that $\pi \circ \varphi_D^o$ is well defined and continuous on $\mathbb{R}$, such that $\pi(\bar{T}_D^o) \subseteq \bar{V}_D^o$ follows. Clearly, if there existed a transfer function $k_0 \in \bar{V}_D^o \setminus \pi(\bar{T}_D^o)$, then for every sequence of transfer functions $k_t \to k_0$, $k_t \in V_D^o$ the corresponding sequence $\tau_{D,t}^o$ in T_D^o would have to satisfy $\|\tau_{D,t}^o\| \to \infty$. However, inverting (22) yields $(a(\tau_D^o),k(\tau_D^o),c(\tau_D^o))$ of the form

$$(\tau_D^o + \frac{\tau_D^o+2}{1+(\tau_D^o+1)^2}, -\frac{\tau_D^o+2}{1+(\tau_D^o+1)^2}, -1)$$

corresponding to

$$\pi(a(\tau_D^o),k(\tau_D^o),c(\tau_D^o)) = \frac{1-\tau_D^o z}{1-\left(\tau_D^o+\frac{\tau_D^o+2}{1+(\tau_D^o+1)^2}\right)z} \quad (23)$$

As is easily seen, $(a(\tau_D^o),k(\tau_D^o),c(\tau_D^o)) \to (\pm\infty,0,-1)$ for $\tau_D^o \to \pm\infty$ and the corresponding sequence of transfer functions also *diverges* in T_{pt}:

$$(K_j)_{j\in\mathbb{N}} = \left(1,\frac{\tau_D^o+2}{1+(\tau_D^o+1)^2},K_1a(\tau_D^o),K_1a(\tau_D^o)^2,\dots\right)$$
$$\to (1,0,1,\pm\infty,\dots)$$

because the degree of the numerator polynomial of K_3 is seven, whereas the denominator polynomial is of degree six only. This implies that no additional transfer function can be obtained in $\bar{V}_D^o$, i.e. $\pi(\bar{T}_D^o) = \bar{V}_D^o$, containing lower degree transfer functions. Note that we have divergence in T_{pt} despite the fact that there is a pole-zero cancellation *at* $z=0$ in the limit; see (23). However, it is not difficult to find (MIMO) examples where $\pi(\bar{T}_D^o) \subset \bar{V}_D^o$ and the inclusion is strict; see section 4.10.2 in (Ribarits, 2002).

5. TOPOLOGICAL AND GEOMETRICAL RESULTS

The following theorem generalizes the discussion in section 4, stating the main topological and geometrical properties of `slsDDLC`:

Theorem 14. Let a minimal (A,B,C,D,K) be given where $(\bar{A},\bar{B},\bar{C},\bar{D},\bar{K})$ is the corresponding minimal inverse system in (6) and let ASSUMPTION 5 be satisfied. The parametrization by `slsDDLC` as given in (20) has the following properties:

(i) T_D^o is an open and dense subset of $\mathbb{R}^{ns}$.
(ii) There exists an open neighborhood $T_D^{o,loc}$ of the parameter vector $0 \in T_D^o$ such that $T_D^{o,loc}$ is identifiable and the mapping $\psi_D^{o,loc} : V_D^{o,loc} \to T_D^{o,loc}$ defined by $\psi_D^{o,loc}(\pi(\tau_D^o)) = \tau_D^o$ is a homeomorphism where $V_D^{o,loc} = \pi(T_D^{o,loc})$.
(iii) $\pi(\bar{T}_D^o)$ may (but need not necessarily) contain transfer functions of lower McMillan degree.
(iv) For 'almost every' $(l,k) \in V_D^o$, the corresponding (l,k)-equivalence class in T_D^o consists of a finite number of isolated points.
(v) V_D^o is open (and trivially dense) in $\pi(\bar{T}_D^o)$.
(vi) $\pi(\bar{\bar{T}}_D^o) \subseteq \bar{V}_D^o$ where equality may hold, but the inclusion may also be strict, and where $\bar{\bar{T}}_D^o$ is the set of all points in $\bar{T}_D^o$ where φ_D^o is well-defined.

PROOF. See the proof of theorem 4.10.1 in (Ribarits, 2002).

6. CONCLUSIONS

In this paper, a new parametrization for state-space systems is introduced: The main idea of parametrizing the orthogonal complement to the tangent space to the equivalence class of a given initial system is combined with the method of separable least squares for a likelihood-type criterion function. The resulting parametrization is called `slsDDLC`, and its topological and geometrical properties are investigated in detail.

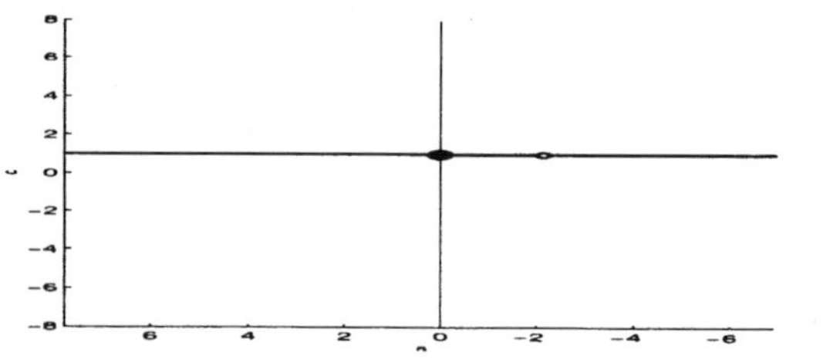

Fig. 1. The slsDDLC construction in the $(\bar{a},\bar{c})$-space for $(\bar{a},\bar{c}) = (0,1)$ (dark point). The thin vertical line corresponds to the L_T^{cc}-equivalence class $\mathcal{E}^{cc}(\bar{a},\bar{c})$, excluding the point where $\bar{c}=0$. The tangent space $Q^{cc}_{(\bar{a},\bar{c})}$ is given by the same line (including the point where $\bar{c}=0$) and its orthogonal complement is given by the thick horizontal line, representing the slsDDLC parameter space T_D^o. Note the exclusion of the point $(\bar{a}(-2),\bar{c}(-2)) = (-2,1)$ corresponding to $\tau_D^o = -2$.

7. REFERENCES

Brockett, R. W. (1976). Some geometric questions in the theory of linear systems. *IEEE Transaction on Automatic Control* **21**(4), 449–455.

Bruls, J., C.T. Chou and M. Verhaegen (1997). Linear and non-linear system identification using separable least squares. In: *Proceedings of the IFAC Conference 'SYSID' 1997.* Fukuoka, Japan. pp. 715–720.

Chou, C. T. and B. Hanzon (1995). Diffeomorphisms between classes of linear systems. *Systems and Control Letters* **26**, 289–300.

Glover, K. (1975). Some geometric properties of linear systems with implications in identification. In: *Proceedings of the 6th IFAC World Congress.* Boston, Massachusetts.

Golub, G. H. and V. Pereyra (1973). The differentiation of pseudo-inverses and nonlinear least squares problems whose variables separate. *SIAM J. Numerical Analysis* **10**(2), 413–432.

Hannan, E. J. and M. Deistler (1988). *The Statistical Theory of Linear Systems.* John Wiley & Sons. New York.

McKelvey, T., A. Helmersson and T. Ribarits (2003). Data driven local coordinates for multivariable linear systems and their application to system identification. *submitted to Automatica.*

Ribarits, T. (2002). The Role of Parametrizations in Identification of Linear Dynamic Systems. PhD thesis. TU Wien.

Wolodkin, G., S. Rangan and K. Poolla (1997). An lft approach to parameter estimation. In: *Preprints to the 11th IFAC Symposium on System Identification.* Vol. 1. Kitakyushu, Japan. pp. 87–92.

IFAC Publications
www.elsevier.com/locate/ifac

OPTIMAL YULE WALKER METHOD FOR POLE ESTIMATION OF ARMA SIGNALS

Magnus Jansson [*,1] **Petre Stoica** [**]

[*] *Royal Inst. of Technology (KTH), Stockholm, Sweden*
[**] *Uppsala University, Uppsala, Sweden*

Abstract: In this paper we reconsider the analysis and implementation of weighted Yule Walker or instrumental variable methods for estimating the AR parameters of ARMA signals. We present a simplified analysis and propose a new estimate of the optimal weighting matrix leading to more accurate parameter estimates compared to previous approaches. *Copyright © 2003 IFAC*

Keywords: time series models, parameter estimation

1. INTRODUCTION

The problem of parametric spectral estimation arises in numerous applications. In this paper we consider the estimation of the AR-part of an ARMA signal (Gersch, 1970). Once the AR parameters have been found, the MA-parameters or the corresponding spectrum could be estimated in a second stage (see, e.g., (Stoica *et al.*, 2000; Kay, 1980; Cadzow, 1982)). In many applications it is also in fact the AR-part that is of most interest. This is for example the case when we want to do frequency estimation (or AR modeling) in white (or even finitely autocorrelated) noise (Chan and Langford, 1982; Kay, 1980; Cadzow, 1982).

The estimation of AR parameters of ARMA signals by instrumental variable or Yule-Walker (YW) methods has been studied in for example (Friedlander, 1983; Stoica *et al.*, 1985; Stoica *et al.*, 1987). It is clear from these references that the performance of the standard YW method is poor in most cases. Moreover, there is no clear relation between the user choice of the instrumental variable vector length, m, and the obtained accuracy; increasing m may well degrade the accuracy. Therefore, the choice of m is hard. Optimally weighted YW methods may have, at least theoretically, much better accuracy. Additionally, the accuracy now improves as m increases and asymptotically in the number of data samples and in m it achieves the Cramér Rao bound. See (Stoica *et al.*, 1985; Stoica *et al.*, 1987) for a detailed discussion.

One of the goals of this contribution is to revisit the analysis in (Stoica *et al.*, 1985) and derive the optimal weight in a very simple manner. The main objective is however to reconsider the implementation of the optimally weighted YW method in (Stoica *et al.*, 1987). In (Stoica *et al.*, 1987) the optimal weighting matrix (which depends on unknown quantities) has been estimated either (i) parametrically or (ii) nonparametrically. Approach (i) requires initial estimates of the MA parameters and will not be considered here. Instead we focus on (ii) to present a new nonparametric estimate of the optimal weighting matrix that is consistent and guaranteed to be non-negative definite. We rely on the recent result in (Stoica and Jansson, 2001) for this. We compare our results with those obtained by the method in (Stoica *et al.*, 1987) in a simulation example. This clearly shows the improved estimation and robustness obtained by using the new weighting matrix estimate.

[1] Corresponding author: Magnus Jansson, Dept. of Signals, Sensors and Systems, Signal Processing, Royal Inst. of Technology (KTH), SE-100 44 Stockholm, Sweden. Email: magnusj@s3.kth.se, fax +46 8 7907260.

2. PROBLEM STATEMENT

As mentioned in the introduction we consider an ARMA signal $y(t)$ satisfying

$$A(q)y(t) = C(q)e(t) \triangleq \varepsilon(t) \quad (1)$$

where

$$A(q) = 1 + a_1 q^{-1} + \cdots + a_{n_a} q^{-n_a}$$
$$C(q) = 1 + c_1 q^{-1} + \cdots + c_{n_c} q^{-n_c}$$

q^{-1} is the unit delay operator and $e(t)$ is a zero mean white noise. We will assume that the orders n_a and n_c are known and that $A(q)$ and $C(q)$ are stable. The model can equivalently be written in a linear regression form

$$y(t) + \phi^T(t)\theta_0 = \varepsilon(t)$$

where $\varepsilon(t)$ was defined in (1) and

$$\phi(t) = \left[y(t-1)\ y(t-2)\ \ldots\ y(t-n_a)\right]^T$$
$$\theta_0 = \left[a_1\ a_2\ \ldots\ a_{n_a}\right]^T$$

The problem of interest is to estimate θ_0 from a batch of observations $\{y(t)\}_{t=1}^N$.

3. YULE WALKER APPROACH

Let

$$\varepsilon(t,\theta) \triangleq y(t) + \phi^T(t)\theta$$

where θ is a generic parameter vector. (Note that $\varepsilon(t,\theta_0) = \varepsilon(t)$.) Let

$$\mathbf{z}(t) = \left[y(t-n_c-1)\ y(t-n_c-2)\ \ldots\ y(t-n_c-m)\right]^T$$

be the vector of instrumental variables (IV) (Ljung, 1987) for some $m \geq n_a$, and note that

$$\mathrm{E}[\mathbf{z}(t)\varepsilon(t)] = 0 \quad (2)$$

where E is the statistical expectation operator. This property can be used to estimate the unknown parameters by minimizing the following function with respect to θ

$$\mu^T(\theta)\mathbf{W}^{-1}\mu(\theta) \quad (3)$$

where $\mathbf{W} \in \mathbb{R}^{m\times m}$ is a positive definite weighting matrix, and

$$\mu(\theta) = \frac{1}{\sqrt{N}} \sum_{t=1}^{N} \mathbf{z}(t)\varepsilon(t,\theta).$$

As seen from (2) the main property of the IV vector $\mathbf{z}(t)$ is that it is uncorrelated with $\varepsilon(t,\theta_0)$. To keep the paper concise we omit any detailed discussion on the instrumental variable or Yule Walker approach to parameter estimation (for such discussions, see (White, 1984; Söderström and Stoica, 1989)). Let $\hat{\theta}$ denote the parameter estimate obtained by minimizing the *quadratic* criterion function in (3). The solution can be written explicitly as

$$\hat{\theta} = \left(\hat{\mathbf{R}}_{z\phi}^T \mathbf{W}^{-1} \hat{\mathbf{R}}_{z\phi}\right)^{-1} \hat{\mathbf{R}}_{z\phi}^T \mathbf{W}^{-1} \hat{\mathbf{R}}_{zy} \quad (4)$$

where

$$\hat{\mathbf{R}}_{z\phi} = \frac{1}{N} \sum_{t=1}^{N} \mathbf{z}(t)\phi^T(t)$$
$$\hat{\mathbf{R}}_{zy} = \frac{1}{N} \sum_{t=1}^{N} \mathbf{z}(t)y(t).$$

The accuracy of $\hat{\theta}$ depends heavily on the weighting matrix $\mathbf{W}$ used in (3). Note that the "standard" YW is obtained for $\mathbf{W} = \mathbf{I}$ and that it typically has a poor performance. The optimal weight that leads to the $\hat{\theta}$ with the smallest covariance matrix in the class of unbiased estimates is given by the Aitken-Markov theory of best linear unbiased estimation (BLUE) (Ljung, 1987; Söderström and Stoica, 1989):

$$\mathbf{W} = \mathrm{E}[\mu(\theta_0)\mu^T(\theta_0)]. \quad (5)$$

or any (root-N) consistent estimate thereof. Furthermore, the covariance matrix of the asymptotic distribution of the parameter estimates obtained from (3) and (5) is well known to decrease monotonically with increasing m (see, e.g., (Stoica *et al.*, 1985)). From now on we will use the symbol $\mathbf{W}$ to denote the optimal weighting matrix in (5).

4. OPTIMAL WEIGHT ESTIMATION

As usually $\mathbf{W}$ in (5) is unknown, the problem of estimating it is the main step of the parameter estimator. Before discussing how to estimate the weighting matrix, we will derive an explicit expression for $\mathbf{W}$ in (5). In the following analysis we assume for simplicity that $e(t)$ is Gaussian. We have

$$\mathbf{W} = \mathrm{E}[\mu(\theta_0)\mu^T(\theta_0)]$$
$$= \frac{1}{N}\mathrm{E}\left[\sum_{t=1}^{N}\sum_{s=1}^{N} \mathbf{z}(t)\mathbf{z}^T(s)\varepsilon(t)\varepsilon(s)\right]$$

Let us study the generic term in the sum using the fact that data are Gaussian:

$$\mathrm{E}[\mathbf{z}(t)\mathbf{z}^T(s)\varepsilon(t)\varepsilon(s)] = \mathrm{E}[\mathbf{z}(t)\mathbf{z}^T(s)]\,\mathrm{E}[\varepsilon(t)\varepsilon(s)]$$
$$+ \mathrm{E}[\mathbf{z}(t)\varepsilon(t)]\,\mathrm{E}[\mathbf{z}^T(s)\varepsilon(s)]$$
$$+ \mathrm{E}[\mathbf{z}(t)\varepsilon(s)]\,\mathrm{E}[\mathbf{z}^T(s)\varepsilon(t)]$$

The last two terms are zero due to the construction of $\mathbf{z}(t)$ (see (2)). Using this observation leads to the following expression for the optimal weight

$$\mathbf{W} = \frac{1}{N}\sum_{t=1}^{N}\sum_{s=1}^{N} \mathbf{R}_z(t-s) r_\varepsilon(t-s)$$
$$= \frac{1}{N}\sum_{k=-N}^{N} (N-|k|)\mathbf{R}_z(k) r_\varepsilon(k) \quad (6)$$

where

$$\mathbf{R}_z(k) = \mathrm{E}[\mathbf{z}(t+k)\mathbf{z}^T(t)]$$
$$r_\varepsilon(k) = \mathrm{E}[\varepsilon(t+k)\varepsilon(t)]$$

Next note that $r_\varepsilon(k) = 0$ for all $k > n_c$ and, hence, we can truncate the sum in the expression for $\mathbf{W}$. If we also assume that $N \gg n_c$, we get the following expression for the asymptotically optimal weighting matrix

$$\mathbf{W} = \sum_{k=-n_c}^{n_c} \mathbf{R}_z(k) r_\varepsilon(k) \tag{7}$$

Since the sequence $\{\mathbf{z}(t)\}$ is available (for $t = 1, \ldots, N$) we can estimate $\mathbf{R}_z(k)$ as

$$\begin{cases} \hat{\mathbf{R}}_z(k) = \dfrac{1}{N} \sum_{t=k+1}^{N} \mathbf{z}(t)\mathbf{z}^T(t-k) \\ \hat{\mathbf{R}}_z(-k) = \hat{\mathbf{R}}_z^T(k) \end{cases}$$

for $k = 0, 1, \ldots, N-1$. The sequence $\{\varepsilon(t)\}$ is not directly available. However, given an initial consistent estimate of θ_0, $\tilde{\theta}$, we can estimate $\varepsilon(t)$ as the residual of the linear regression

$$\tilde{\varepsilon}(t) \triangleq y(t) + \phi^T(t)\tilde{\theta} \tag{8}$$

where $\tilde{\theta}$, for example, can be obtained by using the identity weighting $\mathbf{W} = \mathbf{I}$ in (3). Now we can estimate $r_\varepsilon(k)$ as $(k = 0, 1, \ldots, N-1)$

$$\begin{cases} \hat{r}_\varepsilon(k) = \dfrac{1}{N} \sum_{t=k+1}^{N} \tilde{\varepsilon}(t)\tilde{\varepsilon}(t-k) \\ \hat{r}_\varepsilon(-k) = \hat{r}_\varepsilon(k). \end{cases}$$

A natural nonparametric estimate of the weighting matrix corresponding to (7) is given by

$$\hat{\mathbf{W}} = \sum_{k=-n_c}^{n_c} \hat{\mathbf{R}}_z(k)\hat{r}_\varepsilon(k). \tag{9}$$

This is in effect the nonparametric weighting matrix estimate proposed in (Stoica *et al.*, 1987). It is a consistent estimate of **W** but it may be ill-conditioned or even indefinite. Therefore, instead of using (9), we propose to use the following estimate, which corresponds to the untruncated expression for **W** in (6)

$$\hat{\mathbf{W}} = \frac{1}{N} \sum_{k=-N}^{N} (N - |k|)\hat{\mathbf{R}}_z(k)\hat{r}_\varepsilon(k). \tag{10}$$

The idea of estimating the optimal weight in this way was recently proposed in (Stoica and Jansson, 2001), where it was shown that (10) is a root-N consistent estimate of **W** and it is guaranteed to be positive semidefinite. Thus we expect (10) to yield a more reliable and accurate estimate of the weighting matrix than (9) and hence more accurate AR parameter estimates. In the next section we confirm this by comparing the accuracies of the resulting AR parameter estimates obtained by using the two alternative weighting matrix estimates described above.

5. NUMERICAL EXAMPLES

In the first example, data were generated according to

$$\begin{aligned} y(t) &= \frac{C(q)}{A(q)} e(t) \\ A(q) &= 1 - 1.5q^{-1} + 0.7q^{-2} \\ C(q) &= 1 - 1.8q^{-1} + 0.95q^{-2} \end{aligned}$$

where $e(t)$ was a zero mean white Gaussian process with unit variance. 500 independent Monte Carlo runs were performed each based on 1000 samples of $y(t)$. The root mean square (RMS) errors for the estimated AR-parameters are displayed in Figure 1. The instrumental variable methods (the OIV-1 method from (Stoica *et al.*, 1987) using (9), and the method of this paper based on (10)) were applied with different values of m. Both methods were initialized by the unweighted Yule-Walker method (i.e., by using $\mathbf{W} = \mathbf{I}$ in (4)) and then iterated three times using the most recent estimates of the AR parameters in an attempt to get refined estimates of the residuals in (8). For comparison, Figure 1 also displays the RMS errors for the unweighted YW method and for a prediction error method (PEM) initialized at the true parameter values. Note that the PEM curves essentially can be seen as the Cramér Rao bounds (CRBs). We see that the performance of the proposed method improves as m increases and approaches the CRB as the theory suggests. Apparently, for the current example, the OIV-1 method is less robust. It had difficulties to estimate the weighting matrix accurately for large m in some of the 500 Monte Carlo runs, which led to erratic estimates.

In the second example we simulated a second order narrowband AR process in white noise:

$$y(t) = x(t) + e(t) \tag{11}$$

where

$$x(t) = 1.4x(t-1) - 0.95x(t-2) + w(t)$$

and where $e(t)$ and $w(t)$ are independent zero mean white Gaussian noise processes. The variances of these processes were chosen such that $\mathrm{E}[x^2(t)] = \mathrm{E}[e^2(t)]$. Note that $y(t)$ in (11) has a second order equivalent ARMA(2,2) representation. Similar as before, 500 independent Monte Carlo runs were performed, each based on 500 samples of $y(t)$ in (11). The root mean square (RMS) errors for the estimated AR-parameters are displayed in Figure 2. Again, we see that OIV-1 had severe difficulties to produce reliable weighting matrix estimates leading to a poor AR estimation performance.

6. CONCLUSIONS

In this paper we reconsidered the analysis and implementation of weighted Yule Walker or instrumental variable methods for estimating AR parameters of ARMA signals. We presented a simplified analysis and proposed a new estimate of the optimal weighting matrix leading to more accurate parameter estimates compared to previous approaches. The availability of this new accurate nonparametric estimate of the optimal weighting matrix is expected to promote the use of optimal instrumental variable methods for parameter estimation.

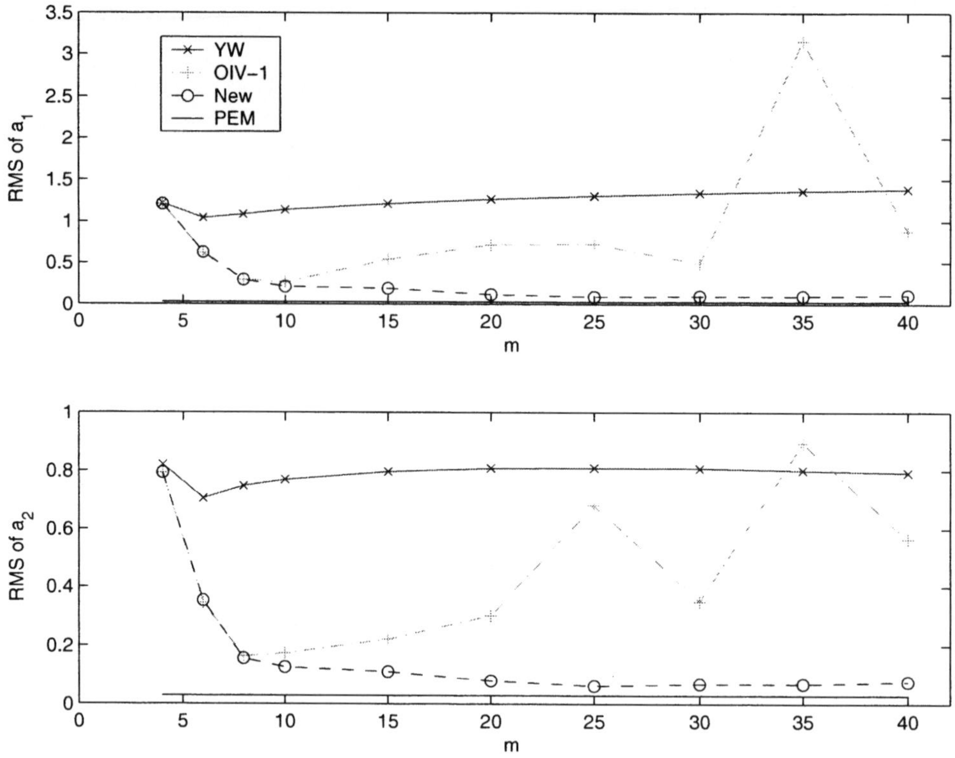

Fig. 1. Results for the first example. The graphs show the RMS errors of the estimated AR-parameters obtained by applying the YW, OIV-1, and the proposed method with different m values. The horizontal line corresponds to the prediction error method.

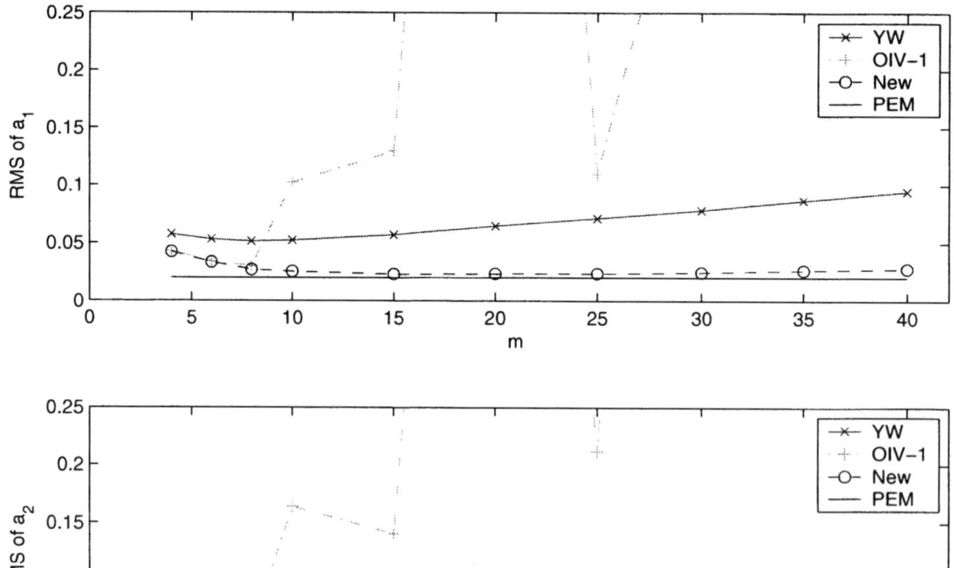

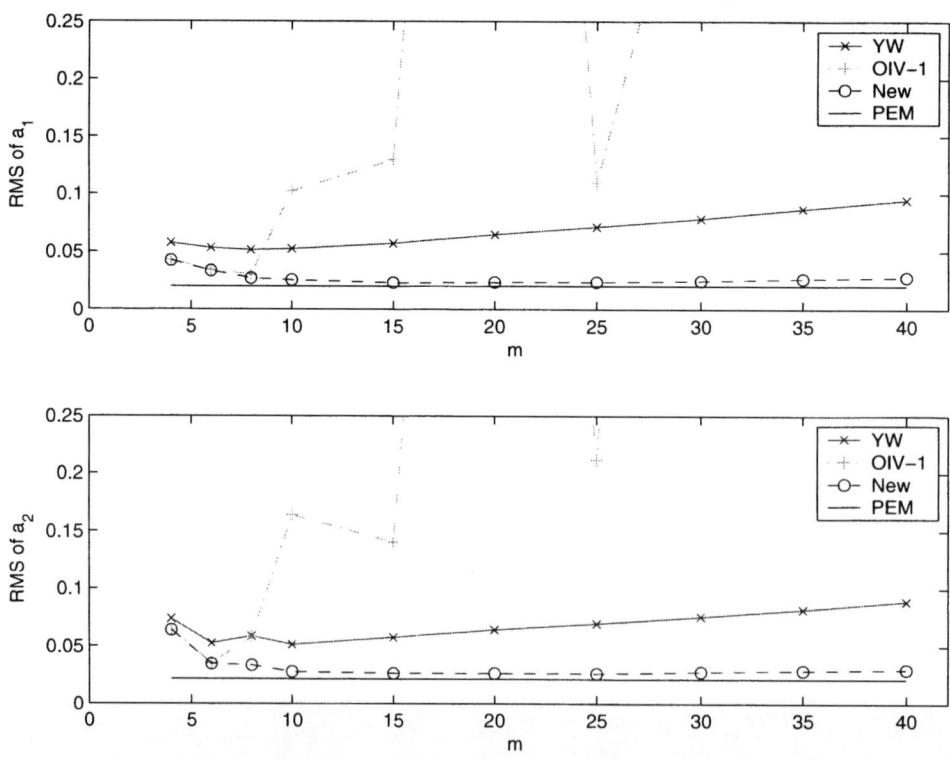

Fig. 2. Results for the second example. The graphs show the RMS errors of the estimated AR-parameters obtained by applying the YW, OIV-1, and the proposed method with different m values. The horizontal line corresponds to the prediction error method. Note that the results for OIV-1 are out of scale for some values of m.

7. REFERENCES

Cadzow, J. A. (1982). Spectral estimation: An overdetermined rational model equation appproach. *Proc. IEEE* **70**(9), 907–939.

Chan, Y. T. and R. P. Langford (1982). Spectral estimation via the high-order yule-walker equations. *IEEE Trans. Acoustics, Speech, and Signal Processing* **30**(5), 689–698.

Friedlander, B. (1983). Instrumental variable methods for ARMA spectral estimation. *IEEE Trans. Acoustics, Speech, and Signal Processing* **31**(2), 404–415.

Gersch, W. (1970). Estimation of the autoregressive parameters of a mixed autoregressive moving-average time series. *IEEE Trans. Automatic Control* **15**, 583–588.

Kay, S. M. (1980). A new ARMA spectral estimator. *IEEE Trans. Acoustics, Speech, and Signal Processing* **28**(5), 585– 588.

Ljung, L. (1987). *System Identification: Theory for the User*. Prentice-Hall. Englewood Cliffs, NJ. 2nd ed. 1999.

Söderström, T. and P. Stoica (1989). *System Identification*. Prentice-Hall International. Hemel Hempstead, UK.

Stoica, P. and M. Jansson (2001). Estimating optimal weights for instrumental variable methods. *Digital Signal Processing* **11**(3), 252–268.

Stoica, P., B. Friedlander and T. Söderström (1987). Optimal instrumental variable multistep algorithms for estimation of the AR parameters of an ARMA process. *Int. J. Control* **47**(6), 2083–2107.

Stoica, P., T. McKelvey and J. Mari (2000). MA estimation in polynomial time. *IEEE Trans. on Signal Processing* **48**(7), 1999–2012.

Stoica, P., T. Söderström and B. Friedlander (1985). Optimal instrumental variable estimates of the AR parameters of an ARMA process. *IEEE Trans. on Automatic Control* **30**(11), 1066–1074.

White, H. (1984). *Asymptotic Theory for Econometricians*. Academic Press. Orlando, FL. (esp. Chap. VI).

IFAC Publications
www.elsevier.com/locate/ifac

INITIALIZING PARAMETER ESTIMATION ALGORITHMS UNDER SCARCE MEASUREMENTS

Albertos P. * **Sanchis R.** ** **Peñarrocha I.** *

* *Departamento de Ingeniería de Sistemas y Automática (DISA)*
Universidad Politécnica de Valencia
Apdo. 22012, E-46071, Valencia, Spain.
e-mail: pedro@isa.upv.es, igpeaal@doctor.upv.es

** *Departament de Tecnologia, Universitat Jaume I*
Campus de Riu Sec, 12071 Castellón, Spain.
e-mail: rsanchis@tec.uji.es

Abstract: In this paper, the problem of initializing identification algorithms with non-regular sampling is addressed. When a recursive identification algorithm is used to estimate the parameters, the convergence of the parameters is affected by the existence of wrong attractors. The initialization of the algorithm is studied in different situations. First, the algorithm starts without past information about the model parameters. An interpolation method is used to estimate the missing data. If a change of the control action updating rate is planned, the new model parameters are initialized by estimations obtained either by interpolation (if the periods are multiple) or by approximate δ modelling using the measurements taken under current operating conditions. Some examples illustrate the attractors avoidance and some conclusions are drafted. *Copyright © 2003 IFAC*

Keywords: Missing-Data, Unconventional Sampling, Pseudo-Linear Recursive Identification, Algorithm Initialization, Convergence Analysis, Delta Operator

1. INTRODUCTION

In many industrial applications the control signal is updated at a fixed rate, but the output is not available at every sampling time due to communication errors, shared or slow sensors, or the use of destructive measuring methods. Different authors have dealt with the identification of such systems when the measurement pattern is regular (periodic). This allows the implementation of a standard RLS algorithm where the regression vector is constructed with only measured variables.

If the pattern of data availability is not regular, the multirate approach can not be used. In that case, the regression vector can not be filled with only measured values. It is then necessary to include estimated values of the non measured outputs on the regression vector. This results in a pseudo-linear recursive algorithm (PLR). This algorithm has a convergence problem due to the existence of wrong attractors.

In (Isaksson, 1993) the problem of identification with missing-data is addressed by means of Expectation-Maximization offline algorithms. A recursive version of that algorithm is also studied on (Isaksson, 1994), but no initialization neither convergence analysis is carried out.

The study of the pseudo-linear identification algorithms for estimating the parameters of the discrete

[1] This project has been partially granted by the CICYT project number DPI2002-04432, and through FPI grant number CTBPRB/2002/245 from Technology and Science Office from Generalitat Valenciana.

transfer function of the process from scarcely sampled output measurements has been addressed by the authors in previous works (Sanchis *et al.*, 1997), assuming that the availability of data may be irregular. A basic PLR algorithm was introduced, being later generalized by the introduction of different possible predictors to obtain the estimates of the unmeasured outputs, in (Albertos *et al.*, 1999). The existence of wrong attractors is demonstrated in (Sanchis and Albertos, 2002).

This paper can be considered as an extension of those works, presenting as a new contribution the analysis of different approaches for the initialization of pseudo-linear identification algorithms, with the objective of avoiding the wrong attractors.

The layout of the paper is as follows: the problem statement, defining the PLR algorithm, its convergence problem, and different solutions for its initialization, is presented in Section 2. The method of interpolation for the initialization of the algorithm is described in Section 3. The initialization method if a change in the control action rate is planned is presented in Section 4. Some illustrative examples of the application of the initialization method to systems with convergence problems are worked out in Section 5. Some draft conclusions are summarized in the last section.

2. PROBLEM STATEMENT

2.1 *Pseudo-linear recursive identification algorithm*

Consider a SISO continuous time linear system of order n whose input is updated at period T by a computer with a zero-order hold, the output being measured synchronously with the input update. Also assume that there is a disturbance such that the discrete difference equation at period T can be written as $y_k = \psi_k^\top \theta + v_k$, where $\theta = [a_1 \dots a_n \, b_1 \dots b_n]^\top$ is the parameter vector, $\psi_k = [-y_{k-1} \dots -y_{k-n} \, u_{k-1} \dots u_{k-n}]^\top$ (with $y_i = y(iT)$, $u_i = u(iT)$) is the regression vector and v_k is the disturbance.

Now, let us assume that only a subset of output measurements are available, the measurement data pattern being scarce (the probability of getting n consecutive measurements is null, see (Sanchis and Albertos, 2002) for a precise definition), and the availability is not regular. In this case, the regression vector of the standard RLS algorithm can not be constructed with only measured outputs. Therefore an output predictor based on the available parameters must be used in order to obtain the estimates of the missing outputs to complete the regression vector. This leads to a pseudo-linear algorithm where the regression vector depends on the available model parameters. In (Sanchis and Albertos, 2002), the following general form of the PLR algorithm has been proposed:

$$\hat{\psi}_k = [-\hat{y}_{k-1} \dots -\hat{y}_{k-n} \, u_{k-1} \dots u_{k-n}]^\top \tag{1a}$$

$$\gamma_k = \frac{P_{k-1}\hat{\psi}_k}{\lambda + \hat{\psi}_k^\top P_{k-1}\hat{\psi}_k} \tag{1b}$$

$$\hat{\theta}_k = \hat{\theta}_{k-1} + \gamma_k \left(y_k - \hat{\psi}_k^\top \hat{\theta}_{k-1} \right) r_k \tag{1c}$$

$$\hat{y}_k = f(\hat{\theta}_k, y_j, r_j, u_j, \hat{y}_j) \tag{1d}$$

$$P_k = \frac{1}{\lambda}\left(I - \gamma_k \hat{\psi}_k^\top \right) P_{k-1} r_k + P_{k-1}(1 - r_k) \tag{1e}$$

where r_k defines the availability of measurements ($r_k = 1$ if y_k is available and $r_k = 0$ if it is not), and λ is the forgetting factor.

Two changes are introduced w.r.t. the RLS algorithm: the components of the regression vector (1a), and the use of a predictor, (1d), to obtain the estimates of the missing outputs. This predictor will use in general the available parameters of the process ($\hat{\theta}_k$) and all previous inputs, measurements and output estimates. In (Albertos *et al.*, 1999) different alternatives have been described for this predictor. A very simple linear predictor may be considered. The idea is to update the whole regression vector when a measurement is obtained. The predictors used in (Sanchis *et al.*, 1997) and (Adams *et al.*, 1994) are particular cases of this one. The equation (1d) will split now into n equations:

$$\begin{aligned} \hat{y}_k &= \hat{\psi}_k^\top \hat{\theta}_k + l_1 (y_k - \hat{\psi}_k^\top \hat{\theta}_k) r_k \\ \hat{y}_{k-i} &= \hat{y}_{k-i} + l_{i+1} (y_k - \hat{\psi}_k^\top \hat{\theta}_k) r_k \end{aligned} \tag{2}$$

where $i = 1, \dots, n-1$, and the gains l_i can be selected to achieve a desired predictor dynamics (see (Sanchis, 1999)).

The PLR algorithm (with the predictor (2)) represents a very complex non-linear stochastic difference equation. The convergence around the actual process parameters is not guaranteed for all the initial vector of parameters $\hat{\theta}_0$ and matrix P_0. On one hand the equilibrium point near the process parameters may be unstable. The stability of the predictor is a necessary condition for the local convergence of the identification algorithm, but simulations have shown that this is not a sufficient one. There can be several local attraction points where the identification algorithm can be caught, far from the process parameters, as is demonstrated in (Sanchis and Albertos, 2002).

The objective of the paper is to analyse the initialization of PLR algorithms that use the measurements of scarce data patterns to estimate the parameters online.

2.2 *Approaches*

The problem of convergence to local wrong attractors can be solved if the identification algorithm starts from a vector of parameters $\hat{\theta}_0$ inside the attraction region of the correct parameters. The P_0 matrix is also needed to convey sufficient information about past signals. If this is not the case, $\hat{\theta}_k$ could move away from the right attraction zone due to initial transient. Different approaches can be followed to get a correct set of initial values.

In this paper, different scenarios are considered, always assuming the scarce measurement condition. First of all, it is assumed to start from scratch, without any information about the model. As previously mentioned, the initialization of the PLR algorithm with $\hat{\theta}_0 = 0$ and $P_0 = MI$, M being a large scalar value and I the identity matrix, will cause convergence problems. Thus, some initial information about the process behaviour should be collected.

It will be considered off-line and online initialization, and the typical scenario of the variation of the sampling period.

3. INTERPOLATION

This is the technique to be used when no information is available before the start-up of the PLR algorithm.

3.1 *Off-line initialization*

The off-line initialization involves the completion of the regression vectors in order to apply the standard LS algorithm. Let us define the following vectors and matrices:

$$Y_k = \left[y_1 \dots y_k\right]^\top; \, X_k = \left[\psi_1^\top \dots \psi_k^\top\right]^\top \quad (3)$$

where ψ_i stands for the regression vector, and k refers to the number of input updates during data compilation. If a scarce measurement data pattern is taking place, we must define the above variables in terms of available and estimated data. First let us define the output as

$$\bar{y}_i = \begin{cases} y_i, & r_i = 1 \\ y_i^s, & r_i = 0 \end{cases} \quad (4)$$

where y_i^s is the estimated output by an interpolation function. The proposed method of interpolation is a cubic spline one. This is a compromise between computational cost and fitting error.

The estimated regression vector is modified as $\bar{\psi}_i = \left[-\bar{y}_{i-1} \dots -\bar{y}_{i-n}\, u_{i-1} \dots u_{i-n}\right]^\top$, and so do $\bar{X}_k$ and $\bar{Y}_k$. The estimated initial vector of parameters is given by

$$\begin{aligned}\hat{\theta}_0 &= \left(\sum_{i=0}^{k} \bar{\psi}_i \bar{\psi}_i^\top\right)^{-1} \left(\sum_{i=0}^{k} \bar{\psi}_i \bar{y}_i\right) \\ &= \left(\bar{X}_k^\top \bar{X}_k\right)^{-1} \bar{X}_k^\top \bar{Y}_k = \bar{P}_0 \bar{X}_k^\top \bar{Y}_k \quad (5)\end{aligned}$$

A value of P_0 matrix is also obtained during this computation (approximated with $\bar{P}_0$), as it is shown. Algorithm described in Section 2.1 is ready to work with these initial values.

Equation (5) is not the off-line version of algorithm (1) with (1d) as the interpolation function. The difference is that in PLR algorithm, the parameters are updated only when there is an output available. This would result in an off-line equation of the form $\bar{\theta}_0 = \left(\sum_{i=0}^{k} \bar{\psi}_i \bar{\psi}_i^\top r_i\right)^{-1} \left(\sum_{i=0}^{k} \bar{\psi}_i \bar{y}_i r_i\right)$. This equation may converge to a trivial model independent of the process and related to the interpolation function. See (Sanchis, 1999) for details about this question.

3.2 *On-line initialization*

The online initialization may lead to a trivial model related to the interpolation function. In order to avoid this, the full set of outputs (some measured and other estimated by interpolation) are fed into a standard RLS algorithm (thus the parameters and P matrix are updated every input period). If a periodic sampling pattern is assumed and an output is measured every N input updates, the interpolator can be expressed by:

$$\hat{y}_{k-i} = f(y_k, y_{k-N}, y_{k-2N}, \dots, y_0) \quad (6)$$

with $i = 1, \dots, N-1$. This can be easily modified to take into account irregular availability patterns. The interpolation function may be the same as the off-line case (cubic spline). The whole function is updated with every measurement. Nevertheless, only the last piece of the smooth function between the last two measurements (from y_{k-N} to y_k) is used to compute the last $N-1$ missing outputs.

This method leads to a biased convergence near the process parameters. The simulations show that the bias is usually lower than the attraction region of the right parameters. When the initial phase of interpolation has converged (after k samples), the reached value of $\hat{\theta}_k$ is used to obtain the predictor gains l_i in (2) to make the predictor stable. Now, the predictor based on interpolation is changed by the linear model based predictor (2), and algorithm (1) is applied using as initial values for $\hat{\theta}_0$ and P_0 those reached during the interpolation start-up.

4. CHANGE IN THE INPUT RATE

When the system is being operated under a certain control action rate (at period T), and is going to be operated under a different one (T'), the parameters of the PLR algorithm for the DT model at the new period need to be obtained. This circumstance can appear when the processor that carries out the control task changes its computing load. At this time, the algorithm must be tuned for the new period of input updates (T') with the available data, whilst working under T-period input updates. The output measurement pattern is assumed to be scarce at both periods.

4.1 *Multiple period case*

Two different situations (with different solutions) can be distinguished when the periods T and T' are integer multiples, depending on whether the period of the input update is increased or decreased.

The case when the input updating rate will be increased ($T' < T = mT'$), is solved applying a variation of the technique shown in Section 3.2. Let us assume that the algorithm (1) has converged while operating at period T, and the output data (measured or estimated) is given by predictor (2). Then, interpolation is used to estimate $m-1$ outputs between input updating periods, leading to a sequence of outputs at period T'. The sequence of inputs is obtained taking into account that the input is constant during period T. The strategy explained in Section 3.2 is used.

When an increase in the control period is planned ($T' = mT$), another variation of the interpolation strategy should be used. In order to apply the technique detailed in Section 3.2, the inputs and outputs at period T' are needed. The outputs are directly given by (2). However, the inputs are not available at period T', because they are updated at period T. Therefore, an "equivalent" input $\bar{u}(iT')$ must be obtained from the applied ones $u(jT)$. This situation is shown in figure 1.

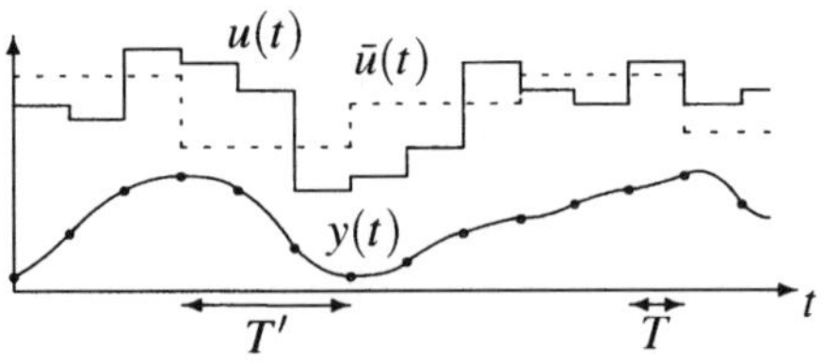

Fig. 1. Equivalent input (- -) in multiple case $T' = 3T$.

An approximate relationship between inputs at different rates can be obtained from the impulse response and convolution expressions. The output at period T can be expressed as

$$y(k) = \sum_{i=1}^{k} g_i u(k-i) = \sum_{i=1}^{m} g_i u(k-i) + \sum_{i=m+1}^{k} g_i u(k-i) \tag{7}$$

where g_i stands for the impulse response of the discrete system at period T. This can also be expressed in terms of virtual inputs updated at period T' as

$$y(k) = \sum_{i=1}^{k/m} \bar{g}_i \bar{u}(k-im) = \bar{g}_1 \bar{u}(k-m) + \sum_{i=2}^{k/m} \bar{g}_i \bar{u}(k-im) \tag{8}$$

where $\bar{g}_i$ is the impulse response at period T'. It is well known that $\bar{g}_1 = \sum_{i=1}^{m} g_i$. The idea is to obtain the value of a constant input $\bar{u}(k-m)$ that has the same contribution to $y(k)$ as the m inputs $\{u(k-1), \ldots, u(k-m)\}$. Taking into account that the contribution of the older inputs ($u(k-j)$, $j > m$) is lower than the newer ones, the value given by

$$\bar{u}(k-m) = \frac{\sum_{i=1}^{m} g_i u(k-i)}{\sum_{i=1}^{m} g_i} \tag{9}$$

is a good approximation of the "equivalent" constant input (that strictly speaking, does not exist). The first m values of the impulse response are easily obtained from the available model at period T.

During the previous instants to the period change an RLS algorithm is computed with the sequences $\bar{u}(k)$ and $y(k)$ in order to obtain the initial values for θ and P for the new period. Since that time, the PLR algorithm is computed with the usual input $u(k)$ updated at period T'.

4.2 *Rational period case*

The above methods can not be applied if there is no integer value that relates both periods. One possibility is to calculate the CT plant from the DT one at period T using the ZOH method and then to obtain the ZOH equivalent DT model at period T'. This approach has the drawback of a high computational complexity due to the matrix exponentials involved. This is a critic problem when dealing with embedded control systems.

The strategy presented here is more general and also applicable in the multiple case. The idea is to obtain two δ-models ($\delta = (q-1)/T$) when the change of input updating rate is incoming. These δ-models are then used to get the parameters of the DT model for the new rate. It is based on the delta modelling treated in (Albertos, 1993).

Let us assume that the PLR algorithm has converged to the correct parameters $\theta(T)$ while working at period T. Therefore, the missing outputs ($\hat{y}(kT)$) are predicted by (2). The DT model at period T is available. The period $0.5T$ is chosen to obtain the second DT model needed. The values of the output at this period $0.5T$ ($\hat{y}(kT + 0.5T)$) are estimated by interpolation (cubic spline), as it is shown in figure 2.

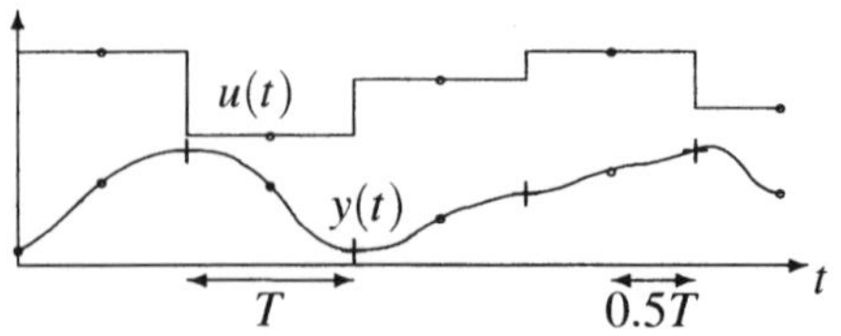

Fig. 2. Obtaining two DT models. $\hat{y}(kT)$ (+) and $\hat{y}(kT + 0.5T)$ (o)

With the interpolated data and taking into account that the input remains unchanged between sampling periods, a standard RLS algorithm is run to estimate the parameters of the DT model at $0.5T$ ($\theta(0.5T)$). Discrete δ-transfer function polynomial coefficients, $\tilde{\theta}(T)$, are linearly related to those of the q-transfer function $\theta(T)$ by

$$\tilde{a}_j = \frac{1}{T^{n-j}} \sum_{i=j}^{n} \binom{i}{j} a_{n-i}, \; j = 0, \ldots, n-1 \quad (10)$$

where $a_0 = 1$. A similar expression is obtained for $\tilde{b}_i$, with $b_0 = 0$. Applying this formula to the DT models, two δ-model parameters are obtained ($\tilde{\theta}(T)$ and $\tilde{\theta}(0.5T)$). The parameters of these δ-models vary with the period in an approximately linear way (see (Albertos, 1993)) and therefore the parameters of the δ-model at period T' can be approximated by

$$\hat{\tilde{\theta}}_i(T') = \left(\frac{2T'}{T} - 1\right) \tilde{\theta}_i(T) + \left(2 - \frac{2T'}{T}\right) \tilde{\theta}_i(0.5T)$$
$$i = 1, \ldots, 2n \quad (11)$$

The DT model parameters at period T' are obtained from the δ-model by inverting the expression (10)

$$a_j = \sum_{i=0}^{j} \binom{n-i}{n-j} \tilde{a}_{n-i} T'^i, \; j = 1, \ldots, n \quad (12)$$

where $\tilde{a}_n = 1$. A similar expression is obtained for b_i, with $\tilde{b}_n = 0$. The result is the initial vector of parameters θ_0 for the PLR algorithm at period T'. However, the initialization of the PLR algorithm also requires the P_0 matrix. Two approaches can be followed at this point. The first one consists of initializing the P matrix with a very high value, and then to execute the algorithm (1) without upgrading the vector of parameters (eq. (1c)) until P converges. The parameters will then be updated applying the full algorithm (1). This will prevent the parameters to leave the attraction region due to a wrong transient.

The second approach consists of obtaining an estimation of the P matrix at the new sampling rate. If the input is a random independent discrete signal of zero mean, then P^{-1} contains elements such as $\sum_{i=0}^{k} y_i^2 \lambda^{k-i}, \sum_{i=0}^{k} y_i y_{i-d} \lambda^{k-i}, \sum_{i=0}^{k} y_i u_{i-d} \lambda^{k-i}, \sum_{i=0}^{k} u_i^2 \lambda^{k-i}$ and so on. These values could be approximated by the mathematical expectations, that can be easily obtained from the available model and input properties (see (Ljung, 1999)). As an example

$$\sum_{i=0}^{k} y_i^2 \lambda^{k-i} \approx \frac{1}{1-\lambda} E(y^2) = \frac{1}{1-\lambda} E(u^2) \sum_{i=0}^{\infty} g_i^2 \quad (13)$$

Nevertheless, the assumption of a white noise control input is very restrictive and therefore this approach is not very useful in practice.

5. EXAMPLES

To illustrate the validity of the proposed algorithms, the procedures described in previous sections will be applied to a second order system with different sampling conditions. The different algorithms will be tested to solve the different situations.

5.1 *Interpolation start-up*

Consider a system that can be approximated by the second order model $G(s) = 100/(s^2 + 2s + 2)$. The input is assumed to be updated at constant period $T = 0.1$ seconds. The DT transfer function with a ZOH is: $G(q) = (0.4675q + 0.4373)/(q^2 - 1.801q + 0.8187)$. The discrete parameter vector is $\theta = [-1.801 \; 0.8187 \; 0.4675 \; 0.4373]^\top$. Assume that there is no disturbance, and the input is an independent discrete signal of zero mean and variance σ_u^2. Assume that the output is measured every $N = 3$ input periods. An interpolator is applied to estimate de missing outputs during initialization and a standard RLS algorithm is applied with the initial values $\theta_0 = [0.1 \; 01. \; 0 \; 0]$ and $P_0 = 1000I$, and a forgetting factor $\lambda = 0.98$. When matrix P has converged, the denominator of the available model is $q^2 - 1.87q + 0.9$. With this value, the gain of the predictor (2) is designed for stable eigenvalues (0.3,0.3) leading to $l_1 = 0.8765$ and $l_2 = 0.8018$. Since then, the PLR algorithm is applied. The results of this experiment are shown in figure 3. The previous procedure can be compared

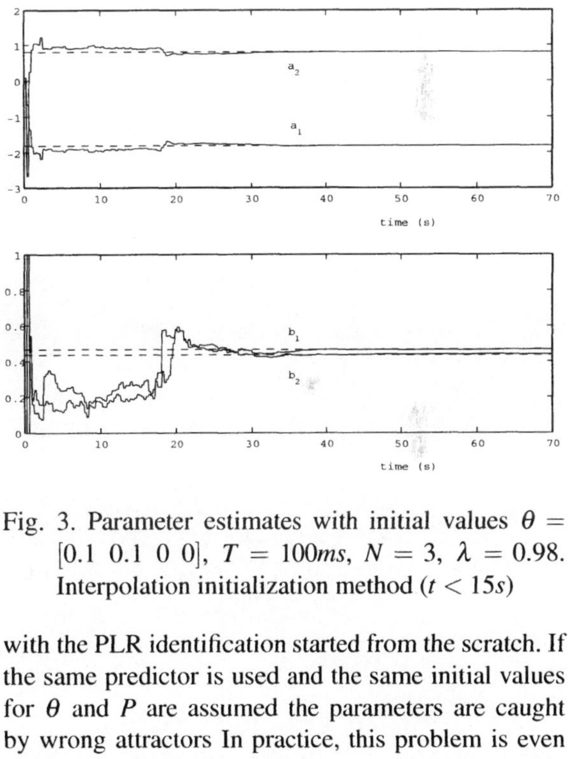

Fig. 3. Parameter estimates with initial values $\theta = [0.1 \; 0.1 \; 0 \; 0]$, $T = 100ms$, $N = 3$, $\lambda = 0.98$. Interpolation initialization method ($t < 15s$)

with the PLR identification started from the scratch. If the same predictor is used and the same initial values for θ and P are assumed the parameters are caught by wrong attractors In practice, this problem is even more serious as far as there is no model available before starting the identification algorithm in order to design the predictor (2). Therefore, it is not possible to guarantee the predictor stability.

5.2 *Change on input rate*

Consider the same system being operated at an input updating period of $T = 0.05$ *s*. Assume that one measurement is taken every $N = 3$ input updates. Along the first $t_1 = 15$ *s*, the available data is used for online initialization with $\lambda = 0.98$, giving as a result the initial values of $\theta_0(T)$ and $P_0(T)$. The value of an stable predictor (eigenvalues 0.3,0.3) is calculated for this available model at period T ($l_1 = 0.9035, l_2 =$

0.7564). Since then, the PLR algorithm is run in normal operation. At time $t_2 = 90\ s$ a change of input period is planned, so that the new period from instant $t_3 = 180\ s$ will be $T' = 0.15\ s$. At instant t_2, the technique described in Section 4.1 is run in order to get the initial values of the PLR algorithm at period T'. When the instant t_3 has arrived, a new predictor is calculated for the new available model at period T' ($l'_1 = 0.9685, l'_2 = 0.9965$). Now the right initial values of the parameters ($\theta_0(T')$ and $P_0(T')$) for identification at period T' are available. From instant t_3 the measurement pattern is changed, so that an output is available for every $N' = 2$ input updates. This scenario is shown in figure 4. The value $\hat{\theta}_k(T)$ is shown from 0 to t_2 , while $\hat{\theta}_k(T')$ is shown from t_2 on. The initialization of parameters at period T' for the incoming change of rate at instant t_3 is shown between t_2 and t_3.

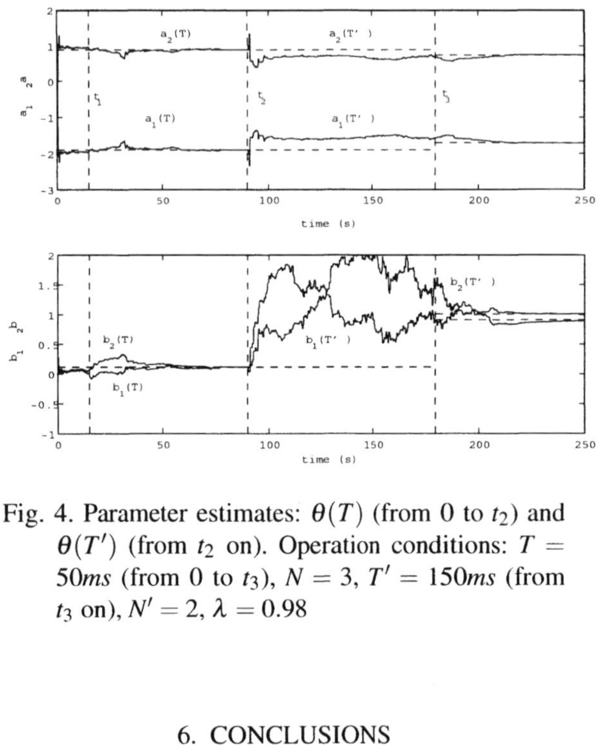

Fig. 4. Parameter estimates: $\theta(T)$ (from 0 to t_2) and $\theta(T')$ (from t_2 on). Operation conditions: $T = 50ms$ (from 0 to t_3), $N = 3$, $T' = 150ms$ (from t_3 on), $N' = 2$, $\lambda = 0.98$

6. CONCLUSIONS

In this paper the problem of convergence of a pseudo-linear online identification algorithm to wrong attractors in systems with scarce measurements has been addressed.

In order to avoid the wrong attractors, different initialization strategies of this recursive identification algorithm have been analyzed. The proposed algorithms can be applied to regular or irregular data availability, and also in the presence of an incoming change of control rate.

The first considered case is the start-up of the algorithm from the scratch. In the initial phase of initialization the missing data are obtained by interpolation to fill in the regression vector of the PLR algorithm (1). This leads to a matrix of covariances and a biased vector of parameters that are near enough the correct one to avoid wrong attractors. This solution can also be applied off line.

Another situation studied is the case of a change on the control action rate when the algorithm has already converged to the right parameters. In this situation several solutions have been proposed for the different scenarios. When the periods are multiple, interpolators have been applied to obtain the parameters for the new rate.

This can not be applied in the rational periods case. In that case there is the need to go through a CT model in order to extract the DT model for the new period (by means of the ZOH conversion). To reduce the computational complexity of the ZOH method, δ-operator modelling has been applied in order to directly obtain the DT model at the new period from the δ-model calculated by extrapolation from other two δ models (corresponding to the original rate and a multiple obtained by interpolation).

Finally, the validity of the proposed methods of initialization to prevent the PLR algorithm to be caught by wrong attractors has been checked by simulations.

7. REFERENCES

Adams, G.J., P. Albertos, G.C. Goodwin and A.J. Isaksson (1994). Parameter estimation for systems with missing data in the presence of white measurement noise. *SYSID IFAC Symposium on System Identification. Copenhagen.*

Albertos, P. (1993). Continuous-time modelling by multirate processing of sampled data. *12th IFAC World Congress, Area R-11, Sydney, Australia,.*

Albertos, P., R. Sanchis and A. Sala (1999). Output prediction under scarce data operation. control applications. *Automatica* **35**, 1671–1681.

Isaksson, A.J. (1993). Identification of arx-models subject to missing-data. *IEEE Transactions of Automatic Control* **AC-38**, 813–819.

Isaksson, A.J. (1994). Recursive em algorithm for identification subject to missing-data. *SYSID IFAC Symposium on System Identification. Copenhagen* **2**, 679–684.

Ljung, L. (1999). *System Identification. Theory for the user. Second Edition.* Prentice-Hall.

Sanchis, R. (1999). *Control of Industrial Processes with Scarce Measurements.* Doctoral Thesis (in spanish). Universidad Politécnica de Valencia, Spain. Main results submitted to a journal paper.

Sanchis, R., A. Sala and P. Albertos (1997). Scarce data operating conditions: Process model identification. *SYSID IFAC Symposium on System Identification.*

Sanchis, R. and P. Albertos (2002). Recursive identification under scarce measurements. convergence analysis. *Automatica* **38**, 535–544.

IFAC Publications
www.elsevier.com/locate/ifac

ROBUST PARAMETER ESTIMATION FOR UNCERTAIN GROSS-ERROR MODELS [1]

Katsuji Uosaki * Koichi Saito *,[2] Toshiharu Hatanaka *

* *Department of Information and Physical Sciences*
Graduate School of Information Science and Technology
Osaka University
Suita, Osaka 565–0871, Japan

Abstract: Many assumptions commonly made in science and engineering problems are at most approximations to reality and they do not always hold unfortunately. Recognizing this fact, the concept of robustness has been attracted and robust procedures have been developed to cover this fact. One of the most important ideas on this direction is Huber's M-estimator (maximum likelihood type estimator). Identifying the neighbourhoods of stochastic models in terms of the class of ε-contaminated probability distribution, he derived an estimator that minimizes the maximum degradation of performance possible for an ε-deviation from the assumption. This idea, however, is not applicable in practice since the exact value of the gross error ε is not known. In this paper, an M-estimator applicable in such situations is derived. A numerical example is presented for illustrating the proposed idea. *Copyright © 2003 IFAC*

Keywords: Stochastic systems; parameter estimation; robust procedure; M-estimator; gross-error.

1. INTRODUCTION

Many assumptions commonly made in science and engineering problems are at most approximations to reality and they do not always hold unfortunately. Recognizing this fact, many people have given their attentions to the concept of robustness and developed robust procedures in design of control systems, detection systems, filters, etc. (see, for examples, (Staudte and Sheather, 1990), (Schick and Mitter, 1994), (Zhou et al., 1995), (Mangoubi and Grimble, 1998), (Chen and Gu, 1998), and (Kim et al., 2001))
Among these, a robust parameter estimation problem is addressed here. Consider, for example, a parameter estimation problem of a finite autoregressive (AR) model of p-th order

$$\begin{aligned} y_n &= a_1 y_{n-1} + a_2 y_{n-2} + \cdots + a_p y_{n-p} + e_n \\ &= \theta^T \phi_n + e_n \quad n = 1, 2, \cdots, N \end{aligned} \tag{1}$$

where $\phi_n = (y_{n-1}, y_{n-2} \cdots, y_{n-p})^T$ and $\{e_n\}$ is an i.i.d noise sequence with probability distribution F. Commonly used estimation approaches such as the maximum likelihood (ML) and the maximum a posterior probability (MAP) approaches require the knowledge of the exact form of F, which would never been known in practice. Though, another common approach, the least squares (LS) approach, is applicable without the knowledge of the distribution, it loses the efficiency in the presence of more heavy-tailed distribution than the normal. For example, when noise sequence obeys the Cauchy law, whose probability density function looks similar as the normal distribution, consistency property of the estimate is not assured and the behavior of

[1] Partially supported by the Grant-in-Aid for Scientific Research from the Japan Society for the Promotion of Science (C)(2)14550447.
[2] Now with Access Service Systems Laboratories, NTT, Chiba 261–0023, Japan

the estimate becomes quite unstable. This observation has led to the necessity of the development of robust procedure which prevents disasters due to somewhat larger deviations from the model and avoidable efficiency loss due to small deviations as well.

One of the most important works on this direction is Huber's M-estimator (maximum likelihood type estimator)(Huber, 1981). He, first, identified the neighborhoods of stochastic models which are supposed to contain the true distribution generating the data as

$$\begin{aligned}\mathscr{F}_\varepsilon = \{F|F = &(1-\varepsilon)F_0 + \varepsilon F_1,\\ &F_0 \text{ is a known symmetric distribution,}\\ &F_1 \text{ is an unknown arbitrary symmetric}\\ &\text{distribution,}\\ &0 < \varepsilon < 1 \text{ is a known fixed constant}\}, \quad (2)\end{aligned}$$

which he called an ε-contaminated probability distribution model (gross-error model). Then, he derived the estimator that minimizes the maximum degradation of performance possible for an ε-deviation from the assumption. Therefore, the estimator behaves optimal over the whole neighborhood in this minimax sense. The estimator, however, depends on the value of the gross-error ε, which is never known, and hence, this approach cannot be applied in exact manner. This is one of the main objections against this approach. This paper will discuss and provide a solution; a robust estimation procedure is developed, which is applicable to the uncertain gross-error situation where the exact value of the gross-error ε is not available but its upper bound $\alpha < 1$ is known.

2. M-ESTIMATOR WITH KNOWN GROSS-ERROR

In parameter estimation of the AR process (1), the M-estimator $\hat{\theta}_N$ is a solution of the following equation,

$$\prod_{n=1}^{N} \psi^*(\varepsilon_n) = 0 \tag{3}$$

where

$$\varepsilon_n = y_n - \hat{\theta}^T \phi_n^T \tag{4}$$

A function ψ^* is chosen to minimize the maximal asymptotic variance of the estimator $\hat{\theta}_N$ over $\mathscr{F}_\varepsilon$, that is, to satisfy

$$\sup_{F\in\mathscr{F}_\varepsilon} V(\psi^*, F) = \min \sup_{F\in\mathscr{F}_\varepsilon} V(\psi, F) \tag{5}$$

where the asymptotic variance $V(\psi, F)$ of an M-estimator defined by some function ψ at a distribution F is given by

$$V(\psi, F) = \frac{\int \psi^2 dF}{(\int \psi' dF)^2} \tag{6}$$

Let F_ε^* be the least favorable distribution, i.e., the distribution minimizing the Fisher information

$$J(F) = \int (\frac{f'}{f})^2 dF \tag{7}$$

over all $F \in \mathscr{F}_\varepsilon$, where f is the density of F. Then the M-estimator is given by

$$\psi^* = -\frac{f_\varepsilon'^*}{f_\varepsilon^*} \tag{8}$$

with f_ε^* being the density of F_ε^*. Corresponding robust recursive parameter estimation method is given by (Polyak and Tsypkin, 1980),(Nakamizo, 1984)

$$\begin{aligned}\hat{\theta}_n &= \hat{\theta}_{n-1} + P_{n-1}\phi_n \psi^*(\varepsilon_n)\\ P_n &= P_{n-1} - \frac{\psi^*(\varepsilon_n) P_{n-1}\phi_n \phi_n^T P_{n-1}}{1 + \psi(\varepsilon_n)\phi_n^T P_{n-1}\phi_n}\end{aligned} \tag{9}$$

For the class of ε-contaminated normal distributions, where F_0 is normal with expectation 0 and variance σ^2, respectively, the optimal ψ^* is given by

$$\psi^*(v) = \begin{cases} \dfrac{v}{\sigma^2} & |v| \le v_0 \\ \dfrac{v_0}{\sigma^2}\mathrm{Sgn}(v) & |v| > v_0 \end{cases} \tag{10}$$

where v_0 is a solution of the following equation

$$2F_0(v_0) - 1 + \frac{2\sigma^2}{v_0} f_0(v_0) = \frac{1}{1-\varepsilon} \tag{11}$$

which varies with gross-error ε. Thus, this approach is dependent on its exact value, and is not applicable without its exact knowledge. The problem considered here is to develop a robust estimation procedure applicable under more relaxed assumptions of the gross-error, i.e., the case only with the knowledge of the upper bound of the uncertain gross-error.

3. M-ESTIMATOR FOR UNCERTAIN GROSS-ERROR

Assume here that the exact value of the gross-error ε is not available but its upper bound α is known. In this case, the class of ε-contaminated probability distributions is defined by

$$\begin{aligned}\mathscr{F}_\varepsilon^* = \{F|F = &(1-\varepsilon)F_0 + \varepsilon F_1,\\ &F_0 \text{ is a known symmetric distribution,}\\ &F_1 \text{ is an unknown arbitrary symmetric}\\ &\text{distribution,}\\ &0 \le \varepsilon \le \alpha \text{ is an unknown fixed constant,}\\ &0 < \alpha < 1 \text{ is a known fixed constant}\}, \quad (12)\end{aligned}$$

The M-estimator for the class of $\mathscr{F}_\varepsilon^*$ of ε-contaminated normal distribution with an uncertain gross-error ε is given in the following theorem.

Theorem 1. If F_0 in $\mathscr{F}_\varepsilon^*$ is normal with expectation 0 and variance σ^2, then the M-estimator for the class $\mathscr{F}_\varepsilon^*$ is given by

$$\psi^*(v) = \begin{cases} \dfrac{v}{\sigma^2} & |v| \le v_0^* \\ \dfrac{v_0^*}{\sigma^2}\mathrm{Sgn}(v) & |v| > v_0^* \end{cases} \tag{13}$$

where v_0^* is a solution of the following equation

$$2F_0(v_0^*) - 1 + \frac{2\sigma^2}{v_0^*} f_0(v_0^*) = \frac{1}{1-\alpha} \tag{14}$$

Proof. Let F^* be the least favorable distribution for the class of $\mathscr{F}_\varepsilon^*$. Then

$$\begin{aligned} J(F^*) &= \min_{F \in \mathscr{F}_\varepsilon^*} J(F) = \min_{\varepsilon \in [0,\alpha]} \min_{F \in \mathscr{F}_\varepsilon} J(F) \\ &= \min_{\varepsilon \in [0,\alpha]} J(F_\varepsilon^*) \end{aligned} \tag{15}$$

where F_ε^* is the least favorable distribution for the class $\mathscr{F}_\varepsilon$.
If F_0 is normal with expectation 0 and variance σ^2, then the density of F_ε^* is given by

$$f_\varepsilon^*(v) = \begin{cases} \dfrac{1-\varepsilon}{\sqrt{2\pi}\sigma}\exp(-\dfrac{v^2}{2\sigma^2}) & |v| \le v_0 \\ \dfrac{1-\varepsilon}{\sqrt{2\pi}\sigma}\exp(-\dfrac{2v_0|v| - v_0^2}{2\sigma^2}) & |v| > v_0 \end{cases} \tag{16}$$

with v_0 satisfying (11), or

$$\frac{2(1-\varepsilon)}{\sqrt{2\pi}\sigma}\left(\int_0^{v_0} \exp(-\frac{v^2}{2\sigma^2})dv + \frac{\sigma^2}{v_0}\exp(-\frac{v_0^2}{2\sigma^2})\right) = 1, \quad v_0 > 0 \tag{17}$$

Corresponding Fisher information is given by

$$J(F_\varepsilon^*) = \frac{2(1-\varepsilon)}{\sqrt{2\pi}\sigma^5}\left(\int_0^{v_0} v^2 \exp(-\frac{v^2}{2\sigma^2})dv + \int_{v_0}^{\infty} v_0^2 \exp(-\frac{2v_0 v - v_0^2}{2\sigma^2})dv\right) \tag{18}$$

Since

$$\begin{aligned} \int_0^{v_0} v^2 \exp(-\frac{v^2}{2\sigma^2})dv &= -\sigma^2 v_0 \exp(-\frac{v_0^2}{2\sigma^2}) \\ &+\sigma^2 \int_0^{v_0} \exp(-\frac{v^2}{2\sigma^2})dv \end{aligned}$$

$$\int_{v_0}^{\infty} v_0^2 \exp(-\frac{2v_0 v - v_0^2}{2\sigma^2})dv = \sigma^2 v_0 \exp(-\frac{v_0^2}{2\sigma^2}) \tag{19}$$

the Fisher information $J(F_\varepsilon^*)$ can be rewritten by

$$J(F_\varepsilon^*) = \frac{2(1-\varepsilon)}{\sqrt{2\pi}\sigma^3}\int_0^{v_0} \exp(-\frac{v^2}{2\sigma^2})dv \tag{20}$$

Using the relation (17),

$$J(F_\varepsilon^*) = \frac{1}{\sigma^2 g(v_0)}\int_0^{v_0} \exp(-\frac{v^2}{2\sigma^2})dv \tag{21}$$

with

$$\begin{aligned} g(x) &= \int_0^x \exp(-\frac{v^2}{2\sigma^2})dv + \frac{\sigma^2}{x}\exp(-\frac{x^2}{2\sigma^2}) \\ &= \sqrt{2\pi}\sigma\left(F_0(x) - \frac{1}{2} + \frac{\sigma^2}{x}\exp(-\frac{x^2}{2\sigma^2})\right) \end{aligned} \tag{22}$$

is derived. Then

$$\frac{\sqrt{2\pi}\sigma}{2} \le g(v_0) = \frac{\sqrt{2\pi}\sigma}{2(1-\varepsilon)} \le \frac{\sqrt{2\pi}\sigma}{2(1-\alpha)} \tag{23}$$

since $0 \le \varepsilon \le \alpha \le 1$, and hence it is easy to show that

$$g(x) > 0, \quad \lim_{x \to +0} g(x) = +\infty, \quad \lim_{x \to \infty} g(x) = \sqrt{\frac{\pi}{2}}\sigma \tag{24}$$

Furthermore,

$$g'(x) = -\frac{\sigma^2}{x^2}\exp(-\frac{x^2}{2\sigma^2}) < 0 \quad x > 0 \tag{25}$$

i.e., $g(x)$ is monotone decreasing for $x > 0$. Let v_0 and v_0^* be the unique solution of the following equations

$$g(v_0) = \frac{\sqrt{2\pi}\sigma}{2(1-\varepsilon)}, \quad g(v_0^*) = \frac{\sqrt{2\pi}\sigma}{2(1-\alpha)} \tag{26}$$

respectively, then

$$v_0^* \le v_0 < \infty \tag{27}$$

and hence

$$g(v_0^*) = \min_{\varepsilon \in [0,\alpha]} g(v_0) \tag{28}$$

(see, Fig,1).
The Fisher information $J(F_\varepsilon^*)$ is a monotone increasing function of v_0 since its numerator and the denominator are non-negative monotone increasing and decreasing functions of v_0, respectively. Thus, the value of v_0 minimizing $J(F_\varepsilon^*)$ should be v_0^*, which correspond to the upper bound α of the gross-error. This implies

$$J(F^*) = \min_{\varepsilon \in [0,\alpha]} J(F_\varepsilon^*) = J(F_\alpha^*) \tag{29}$$

Theorem 1 indicates that the M-estimator for the class $\mathscr{F}_\varepsilon^*$ of ε-contaminated normal distributions with uncertain gross-error is obtained by simply substituting

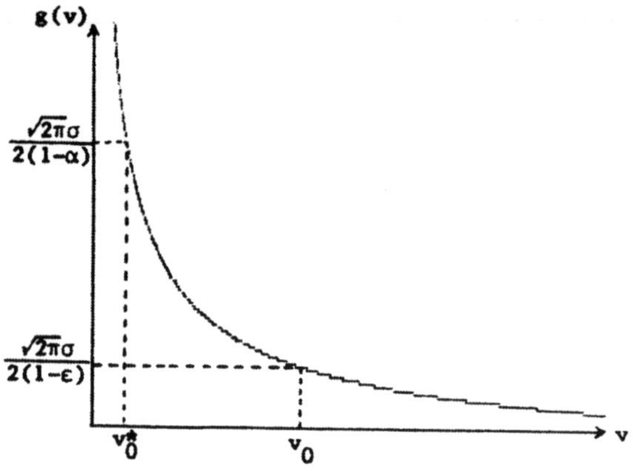

Fig. 1. Relation between gross-error ε and point v_0

the upper bound α into ε of the M-estimator for the class $\mathscr{F}_\varepsilon$ of ε-contaminated normal distributions with known gross-error. It is shown in the following theorems that this result holds for the class of more general ε-contaminated distributions.

Theorem 2. Assume that the probability distribution F_0 of the class $\mathscr{F}_\varepsilon$ of the ε-contaminated distributions with known gross-error, satisfies

C1: F_0 is three times continuously differentiable
C2: The density f_0 of the distribution F_0 is monotone decreasing for $v > 0$
C3: $-\log f_0(v)$ is concave
C4: $\lim_{v\to+0} \frac{f_0^2(v)}{f_0'(v)} \leq -\frac{1}{2(1-\varepsilon)}$

Then, the M-estimator for the class $\mathscr{F}_\varepsilon$ is given by

$$\psi^*(v) = \begin{cases} -\dfrac{f_0'(v)}{f_0(v)} & |v| \leq v_0 \\ -\dfrac{f_0'(v_0)}{f_0(v_0)}\mathrm{Sgn}(v) & |v| > v_0 \end{cases} \tag{30}$$

where v_0 is the unique solution of

$$2F_0(v_0) - 1 - \frac{2f_0^2(v_0)}{f_0'(v_0)} = \frac{1}{1-\varepsilon} \tag{31}$$

Proof. By applying the method of variations to the Fisher information $J(F)$, it is shown that the least favorable density $f^*(v)$ is a solution of the following equation

$$\frac{2f^*(v)f^{*\prime\prime 2}(v) - f^{*\prime 2}(v)}{f^{*2}(v)} = \lambda^2 + \eta(v) \tag{32}$$

where λ is some positive constant and

$$\begin{aligned} \eta(v) &= 0 \quad \text{if} \quad f^*(v) > (1-\varepsilon)f_0(v) \\ &\leq 0 \quad \text{if} \quad f^*(v) = (1-\varepsilon)f_0(v) \end{aligned} \tag{33}$$

Let $f^*(v) = \exp(-h^*(v))$ and $f_0(v) = \exp(-h_0(v))$. Then, the relations (32) and (33) and that $h_0(v)$ is monotone increasing for $v > 0$ by C2 lead the following relations:

$$\begin{aligned} &-2h^{*\prime\prime}(v) + h^{*\prime 2}(v) = \lambda^2 \\ &\qquad \text{if} \quad f^*(v) > (1-\varepsilon)f_0(v) \\ &-2h_0''(v) + h_0'^2(v) \leq \lambda^2 \\ &\qquad \text{if} \quad f^*(v) = (1-\varepsilon)f_0(v) \end{aligned} \tag{34}$$

If $f^*(v) > (1-\varepsilon)f_0(v)$, then $h^*(v) = \lambda|v| + K_1$ with some constant K_1 by C2. Thus

$$f^*(v) = C\exp(-\lambda|v|) \tag{35}$$

is a solution of (32) if $f^*(v) > (1-\varepsilon)f_0(v)$. In order that $f^*(v)$ is differentiable, two curves $f^*(v)$ given by (35) and $f^*(v) = (1-\varepsilon)f_0(v)$ should be smoothly connected at some point, say v_0. This implies that $\lambda|v| - \log C$ and $h_0(v) - \log(1-\varepsilon)$ should be connected smoothly at v_0. Hence

$$\begin{aligned} \lambda &= h_0'(v_0) = -\frac{f_0'(v_0)}{f_0(v_0)} \\ C &= (1-\varepsilon)f_0(v_0)\exp(\lambda v_0) \end{aligned} \tag{36}$$

Substitution of this relation into (34) leads

$$-2h_0''(v) + h_0'^2(v) \leq \lambda^2 = h_0'^2(v_0) \tag{37}$$

for $|v| \leq v_0$ or $|v| > v_0$. Proposition 2 (see Appendix) indicates that the relation (37) holds for $\forall|v| \leq v_0$, while it does not hold for some point $|v_1| > v_0$. This implies that the least favorable distribution $f_\varepsilon^*(v)$ is given by

$$f^*(v) = \begin{cases} (1-\varepsilon)f_0(v) & |v| \leq v_0 \\ C\exp(-\lambda|v|) & |v| > v_0 \end{cases} \tag{38}$$

and then the M-estimator is given by (30). The relation (31) comes from (36) and

$$\int_{-\infty}^{\infty} f^*(v)dv = 1 \tag{39}$$

Theorem 3. Assume C1 through C3 and

C4': $\lim_{v\to+0} \frac{f_0^2(v)}{f_0'(v)} \leq -\frac{1}{2(1-\alpha)}$

instead of C4 of Theorem 2. Then the M-estimator for the class $\mathscr{F}_\varepsilon^*$ is given by

$$\psi^*(v) = \begin{cases} -\dfrac{f_0'(v)}{f_0(v)} & |v| \leq v_0 \\ -\dfrac{f_0'(v_0)}{f_0(v_0)}\mathrm{Sgn}(v) & |v| > v_0 \end{cases} \tag{40}$$

with v_0^* being the unique solution of

$$2F_0(v_0^*) - 1 - \frac{2f_0^2(v_0^*)}{f_0'(v_0^*)} = \frac{1}{1-\alpha} \tag{41}$$

The Fisher information for the least favorable distribution F^* in this class is given by

$$J(F_\varepsilon^*) = \min_{\varepsilon\in[0,\alpha]} J(F_\varepsilon^*) = J(F_\alpha^*) \tag{42}$$

where $J(F_\varepsilon^*)$ is the Fisher information for the least favorable distribution F_ε^* in the class of $\mathscr{F}_\varepsilon$.

Proof. For any $\varepsilon \in [0,\alpha]$, the least favorable distribution $f_\varepsilon^*(v)$ has same form as in (38) and corresponding Fisher information for the class $\mathscr{F}_\varepsilon$ is given by

$$J(F_\varepsilon^*) = \frac{\int_0^{v_0} \frac{f_0'^2(v)}{f_0(v)} dv - f_0'(v_0)}{\int_0^x f_0(v)dv - \frac{f_0^2(x)}{f_0'(x)}} = \frac{h(v_0)}{g(v_0)} \quad (43)$$

where

$$h^*(x) = \int_0^x \frac{f_0'^2(v)}{f_0(v)} dv - f_0'(x)$$

$$g^*(x) = \int_0^x f_0(v)dv - \frac{f_0^2(x)}{f_0'(x)}$$

$$= F_0(x) - \frac{1}{2} - \frac{f_0^2(x)}{f_0'(x)} \quad (44)$$

Since

$$h^*(0) = -f_0'(0) > 0, \quad g^*(0) = -\frac{f_0^2(0)}{f_0'(0)} > 0,$$

$$h^{*\prime}(x) = \frac{f_0'(x)^2 - f_0(x)f_0''(x)}{f_0(x)}$$

$$= (-\log f_0(x))'' f_0(x) > 0$$

$$g^{*\prime}(x) = \frac{f_0^2(x)f_0''(x) - f_0(x)f_0'^2(x)}{f_0'^2(x)}$$

$$= -\frac{(-\log f_0(x))''}{(-\log f_0(x))'^2} f_0(x) > 0 \quad (45)$$

It is shown that $h^*(x)$ and $g^*(x)$ are nonnegative monotone increasing and positive monotone decreasing functions of x, respectively, and hence $J(F_\varepsilon^*)$ is positive and monotone increasing for $x \geq 0$.
Let v_0 and v_0^* be the solution of the following equations for $0 \leq \varepsilon \leq \alpha < 1$,

$$g^*(v_0) = \frac{1}{2(1-\varepsilon)}, \quad g(v_0^*) = \frac{1}{2(1-\alpha)} \quad (46)$$

respectively. By $0 \leq \varepsilon \leq \alpha < 1$,

$$0 < v_0^* \leq v_0 < \infty \quad (47)$$

holds. Since $J(F_\varepsilon^*)$ is monotone increasing, it is found

$$v_0^* = \arg \min_{0<v_0^* \leq v_0 < \infty} J(F_\varepsilon^*) \quad (48)$$

which corresponds to the upper bound α of the gross-error. This leads to the conclusion of the theorem.

Remark. In addition to the normal distribution, an example of the distribution satisfying conditions C1 through C4 (C4') is the normal distribution the logistic distribution

$$f_0(v) = \frac{\beta \exp(-\lambda|v|)}{(1+\exp(-\beta|v|))^2} \quad (49)$$

with positive constants β and λ, and corresponding M-estimator for the class $\mathscr{F}_\varepsilon{}^*$ is

$$\psi^*(v) = \begin{cases} \dfrac{\beta(1-\exp(-\beta|v|))}{1+\exp(-\beta|v|)} & \text{if } |v| \leq \dfrac{1}{2}\log\dfrac{2-\alpha}{\alpha} \\ \dfrac{\beta(1-\sqrt{\alpha(2-\alpha)})}{1-\alpha}\mathrm{Sgn}(v) & \text{if } |v| > \dfrac{1}{2}\log\dfrac{2-\alpha}{\alpha} \end{cases}$$

4. NUMERICAL EXAMPLE

Consider the following AR model of order 1,

$$y_n = \theta y_{n-1} + e_n \quad (50)$$

with true value of θ is 0.5, and i.i.d noise sequence $\{e_n\}$. The distribution of $\{e_n\}$ is ε-contaminated normal distribution

$$F = (1-\varepsilon)\Phi(v) + \varepsilon F_1 \quad (51)$$

where ε=0.05 and $F_1(v) = \Phi(v/10)$ (unknown to the experimenter), and Φ is the standard normal cumulative distribution function. Following three recursive estimation procedures are applied to estimate θ.

(1) ordinary least squares method (9) with $\psi^*(v) = v$
(2) robust estimation method (9) with (10) and (11) using the exact value of the gross-error $\varepsilon = 0.5$
(3) robust estimation method (9) with (13) and (14) using the value of the upper bound $\alpha = 0.5$ of the uncertain gross-error ε

Figure 2 shows the mean squared error of 200 simulation runs for each estimation methods. It indicates that the proposed robust estimation method for uncertain gross-error case works well compared to other two estimation methods.

5. CONCLUSIONS

Robust estimation procedure for ε-contaminated probability model has been discussed for the case where the exact value of the gross-error ε is not available but only its upper bound is known. It is shown by simulation studies that the proposed robust procedure obtained by simply substituting the upper bound of the gross-error into the M-estimator for the exactly known gross-error case works well.

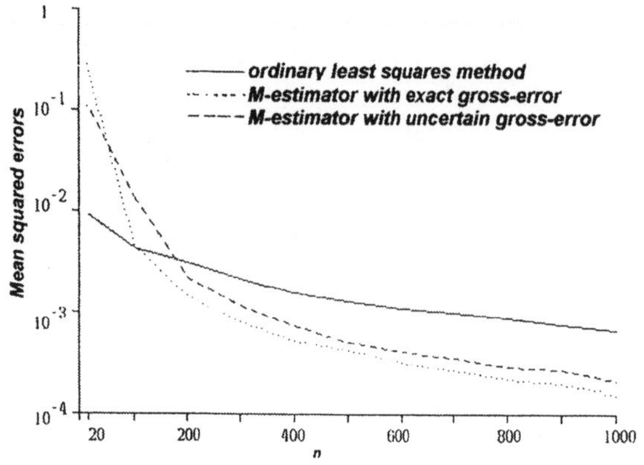

Fig. 2. Results of parameter estimation

6. REFERENCES

Chen, J. and G. Gu. (1998). *Control Oriented System Identification: An H^∞ Approach*, J. Wiley, New York.

Huber, P. J. (1981). *Robust Statistics*, J. Wiley, New York.

Kim, H-S., J-S. Lim, S-J. Baek and K-M. Sung. (2001). Robust Kalman filtering with variable forgetting factor against impulsive noise. *IEICE Trans. Fundamentals*. **E84-A**, No.1, 363–36.

Mangoubi,R. S. and M. J. Grimble (Eds.). (1998). *Robust Estimation and Failure Detection: A Concise Treatment*, Springer, Berlin.

Nakamizo, T. (1984). Robust estimation. *J. Soc. Instr. Contr. Eng.*, **23** 541–549.

Polyak, B. T. and Ya. Z. Tsypkin. (1980). Robust identification. *Automatica*. **16**, 53–63.

Saito, K. (1989). *Robust Estimation for ε-contaminated Probability Distribution Models with Uncertain Gross-error*, M. S. Thesis, Department of Applied Physics, Osaka University.

Schick, I. C. and S. K. Mitter. (1994). Robust recursive estimation in the presence of heavy-tailed observation noise. *Ann. Statist.*. **22**, No.2, 1045–1080.

Staudte, R. G. and S. J. Sheather. (1990). *Robust Estimation and Testing*, Wiley.

Zhou, K., K. Glover and John Doyle (1995). *Robust and Optimal Control*, Prentice-Hall.

APPENDIX

Proposition 1. If a function $f(x)$ satisfies

$$f(x) \geq 0, \; f'(x) > 0, \; f^2(x) \leq 2f'(x) \quad \text{for } x \geq \exists x_0 > 0$$

then

$$\lim_{x \to +\infty} f(x) = \infty$$

Proof. By the assumptions,

$$0 < f^2(x) < f^2(y) \leq 2f'(y) \quad \text{for } x_0 < x < y \tag{A.1}$$

Suppose

$$f(x) < \infty \quad \text{for } \forall x > 0 \tag{A.2}$$

Then if there exists a set of (ε, x_1), $\varepsilon > 0$, $x_1 > x_0$ such that

$$f'(\eta) > \varepsilon \quad \text{for } \forall \eta > x_1 \tag{A.3}$$

and integration of $f'(x)$ leads

$$\lim_{x \to \infty} f(x) = \infty$$

This contradicts (A.1). Hence for $\forall(\varepsilon, x_1)$, $\varepsilon > 0$, $x_1 > x_0$

$$f'(\eta) \leq \varepsilon \quad \text{for } \exists \eta > x_1 \tag{A.4}$$

This implies that there exists a real number sequence $\{\eta_n\}$ such that

$$\lim_{n \to \infty} \eta_n = \infty, \quad \lim_{n \to \infty} f'(\eta_n) = 0 \tag{A.5}$$

Then

$$0 < f^2(x) < f^2(y) \leq 2f'(\eta_n) \quad \text{for } x_0 < \forall x < \eta_n \tag{A.6}$$

and it is obtained by taking the limit to ∞ that $f(x) \leq 0$ for $x_0 < \forall x < \infty$. This means $f(x) \equiv 0$. It contradicts the assumption $f'(x) > 0$ and this indicates the unboundness of $f(x)$.

Proposition 2. If a function $f(x)$ satisfies

$$f'(x) \geq 0, \; f''(x) > 0, \quad \text{for } x \geq 0 \tag{B.1}$$

then there exists a constant $x_0 > 0$ such that

$$f'^2(x) - f'^2(x_0) > 2f''(x) \quad \text{for } \exists x > x_0 \tag{B.2}$$

Proof. It is assumed

$$f'^2(x) - f'^2(x_0) \leq 2f''(x) \quad \text{for } \forall x \geq x_0 \tag{B.3}$$

Since $f'(x)$ is monotone increasing,

$$f'^2(x) - f'^2(x_0) \geq (f'(x) - f'(x_0))^2 \quad \text{for } \forall x \geq x_0 \tag{B.4}$$

Let $g(x) = f'(x) - f'(x_0)$, then by (B.3) and (B.4)

$$0 < g^2(x) \leq 2g'(x) \quad \text{for } x > x_0 \tag{B.5}$$

This indicates

$$\lim_{x \to \infty} g(x) = \infty \tag{B.6}$$

by Proposition 1. Equation (B.4) implies that there exists a suitable function $q(x) \geq 0$ for $x > x_0$ such that

$$\frac{2g'(x)}{g^2(x)} = 1 + q(x) \quad \text{for } x > x_0 \tag{B.7}$$

Integrating the both hands of (B.6) from $x_1 (> x_0)$ to x

$$\int_{x_1}^{x} \frac{2g'(v)}{g^2(v)} dv = \int_{x_1}^{x} (1 + q(v)) dv \tag{B.8}$$

By taking the limit of x to ∞, the LHS tends to $1/h(x_1) < \infty$ while RHS tends to ∞. This contradiction comes from the assumption (B.3) and then it can be concluded that there exists x_0 such that (B.2) holds.

IFAC
Publications
www.elsevier.com/locate/ifac

LIMIT COVARIANCE OF ESTIMATION ERROR FOR QUASISTATIONARY FUNCTIONS

Andrey E. Barabanov

Saint Petersburg State University
Universitetskij pr., 28, 198504 St.-Petersburg, Russia
Andrey.Barabanov@pobox.spbu.ru

Abstract: The linear parameter estimation problem is studied without assumption on the existence of probability measure. Instead, averaged second order statistics are assumed to converge. The limit matrix of the estimation error covariances is computed and an upper bound for the rate of the parameter estimates convergence is obtained. *Copyright © 2003 IFAC*

Keywords: discrete time, parameter estimation, convergence analysis

1. INTRODUCTION

A useful mathematical model contains not only mathematical equations and relations but also a measure of their accuracy. Probabilistic models cannot give a guaranteed result because the underlying assumption of independence or of existence of a density of probability cannot be verified in detail. For this reason some new approach have been developed for model structure and parameter identification algorithms. A notion of quasistationary signals was introduced by Ljung (2001). In this paper, a measure of the accuracy of parameter estimates in the quasistationary signals approach is proposed. Attention is restricted to the second-order theory, and it is based on mean values and a correlation function. Similar results in the framework of the first–order approach and convex technique were proposed by Barabanov (1996).

2. PROBLEM STATEMENT

Let $(x(t))_{t=1}^{\infty}$, $(y(t))_{t=1}^{\infty}$ be such sequences that the joint sequence $z(t) = \text{col}(x(t), y(t))$ satisfies the following condition:

Condition A1: there exists a limit

$$\lim_{N\to\infty} \frac{1}{N} \sum_{t=1}^{N} z(t) z^T(t) = R_z,$$

and $\|z(t)\| \leq C$. Notice that no probability measure is assumed.

The main special case of such functions is given by a quasistationary function $(y(t))_{t=1}^{\infty}$ with the regressors $x(t) = (y(t-i))_{i=0}^{n-1}$.

For the sake of simplicity assume that y is scalar and x is a vector of the dimension n. Split the matrix R_z as

$$R_z = \begin{pmatrix} R_x & R_{xy} \\ R_{yx} & R_y \end{pmatrix}.$$

The matrix R_x is assumed to be nonsingular.

Then a vector of parameters $\theta^0 \in \mathrm{R}^n$ can be determined for which the residual $\varepsilon^0(t) = y(t) - x^T(t)\theta^0$ is orthogonal to $x(t)$, that is,

$$\lim_{N\to\infty} \frac{1}{N} \sum_{t=1}^{N} x(t)\varepsilon^0(t) = 0.$$

Obviously, $\theta^0 = R_x^{-1} R_{xy}$.

Fix a number $\lambda \in (0,1)$ and consider the functional

$$J_t(\theta) = \sum_{k=0}^{t-1} \lambda^k (\varepsilon(t-k,\theta))^2 + \theta^T P^0 \theta$$

where $\varepsilon(s,\theta) = y(s) - x^T(s)\theta$ and P^0 is a regularizing positive definite matrix. Minimum of $J_t(\theta)$ is achieved for

$$\theta_\lambda(t) = P_\lambda(t)^{-1} Y_\lambda(t),$$

where

$$P_\lambda(t) = \sum_{k=1}^{t} \lambda^{t-k} x(k) x^T(k) + P^0,$$

$$Y_\lambda(t) = \sum_{k=1}^{t} \lambda^{t-k} x(k) y(k).$$

It is required to find conditions under which it holds

$$\lim_{N\to\infty,\ \lambda\to 1} \frac{1}{N} \sum_{t=1}^{N} \theta_\lambda(t) = \theta^0,$$

and to estimate the rate of convergence in terms of the limit covariance matrix

$$S_N(\lambda) = \frac{1}{N} \sum_{t=1}^{N} (\theta_\lambda(t) - \theta^0)(\theta_\lambda(t) - \theta^0)^T.$$

In particular, it is required to find conditions of existence and to evaluate the limit

$$\lim_{N\to\infty,\ \lambda\to 1} (1-\lambda)^{-1} S_N(\lambda).$$

3. REMARKS

Remark 1. Even if the process is described by the equation

$$y(t) = \theta^0 x(t) + \varepsilon^0(t)$$

where the pair of random processes $(x(\cdot), \varepsilon^0(\cdot))$ is independent, the random values $\theta_\lambda(t)$ with fixed $\lambda \in (0,1)$ are not consistent estimates of θ^0. They are not unbiased for the general case.

The values $\mathsf{E}\theta_\lambda(t)$ differ from θ^0 and their limit, if it exists, may differ from θ^0 depending on the probability distribution function. For this reason for any $\lambda \in (0,1)$ the condition

$$\lim_{N\to\infty} \frac{1}{N} \sum_{t=1}^{N} \theta_\lambda(t) = \theta^0$$

does not hold for the general case. Therefore an asymptotic behavior of the estimation error as $\lambda \to 1$ is to be studied.

Remark 2. Let $0 < \lambda < 1$. Then the terms in the sums $P_\lambda(t)$ and $Y_\lambda(t)$ that have indices k near the value of t have a greater weight than those with smaller indices. Therefore the condition of quasistationarity seems to be too weak for analysis of convergence of functions defined by $P_\lambda(t)$ and $Y_\lambda(t)$.

In the sequel it will be assumed that the limit in the quasistationarity condition for the regressors $x(t)$ is uniform, namely,

Condition A2: there exists a limit

$$\lim_{N\to\infty} \frac{1}{N} \sum_{t=m+1}^{m+N} x(t) x^T(t) = R_x$$

uniformly for $m \geq 0$.

The uniformity condition for the limit means that there exists a decreasing sequence $(\bar{\alpha}_x(N))_{N=1}^{\infty}$ such that it holds

$$\left\| \frac{1}{N} \sum_{t=m+1}^{m+N} x(t) x^T(t) - R_x \right\| \leq \bar{\alpha}_x(N)$$

for all N, and $\bar{\alpha}_x(N) \to 0$ as $N \to \infty$. Such a function $\bar{\alpha}_x$ is called an approximation bound of the uniform convergence of the sequence $x(t)x^T(t)$ in mean.

The condition A2 of the uniform convergence is a hard assumption. It does not hold for almost all trajectories of a random process. This assumption can be weakened for the main assertion formulated in this note, but this is a subject for future research.

Remark 3. The value of $\theta_\lambda(t)$ is equal to the ratio of two quadratic forms of the variables (x,y). Hence, the covariance of $\theta_\lambda(t)$ includes forms of the forth degree, which did not appear in the previous assumptions. A convergence of the forms of the forth degree can not be derived from a convergence of quadratic forms. The following assumption is necessary for analysis of forms in the covariance matrix.

Condition A3: For any integer τ there exists a limit

$$\lim_{N\to\infty} \frac{1}{N} \sum_{t=1}^{N} Z(t) Z^T(t+\tau) = Q(\tau)$$

where $Z(t) = x(t)\varepsilon^0(t)$.

This condition can be derived from existence of limits of averaged forth moments of all entries of the vector $z(t)$. Nevertheless, the condition A3 seems to be more compact.

If (x,y) is a random process and it is stationary in the narrow sense and regular, then it holds $Q(\tau) = 0$ for $\tau \neq 0$ and $Q(0) = R_x \sigma_0^2$ where σ_0^2 is the variance of the process $\varepsilon^0(t)$.

4. MAIN RESULT

Assume condition A2 holds. Denote the approximation bound of the uniform convergence of the sequence $x(t)x^T(t)$ in mean by $\bar{\alpha}_x$. For any $\lambda \in (0,1)$ define the value

$$r_x(\lambda) = \min_{1 \leq M \leq 2/(1-\lambda)} \left\{ \bar{\alpha}_x(M) + C\eta_\lambda^2(M) + \eta_\lambda(M) \frac{C + \|R_x\| + \bar{\alpha}_x(M)}{2 - \eta_\lambda(M)} \right\}$$

where $\eta_\lambda(M) = (M-1)(1-\lambda)$ and C is defined in Condition A1. Assume the minimum is achieved for $M = M_\lambda$. Define

$$t(\lambda) = 2M_\lambda - 3 + \left| \frac{\log \eta_\lambda(M_\lambda)}{\log \lambda} \right|.$$

The difference between the sampling average and its limit will be denoted by

$$\gamma_{m_1}(M,\tau) = \frac{1}{M}\sum_{m=m_0}^{m_0+M-1} Z(m)Z(m-\tau) - Q(\tau)$$

with $m_1 = m_0$ for $\tau \le 0$ and $m_1 = m_0 - \tau$ for $\tau > 0$.

The next assertion gives an explicit bound of the rate of convergence of the sampling covariance matrices to their limit. The limit matrix is given explicitely in the corollary after theorem.

Theorem. *Assume the conditions A1, A2, A3 hold. Define the function*

$$q_m(\lambda) = (1-\lambda)\sum_{k,l=0}^{m-1} \lambda^{k+l} Q(k-l).$$

Then for any $\lambda \in (0,1)$, $t_0 \ge t(\lambda)$ and $N>0$ it holds

$$\|(1-\lambda)^{-1}\frac{1}{N}\sum_{t=t_0}^{t_0+N-1}(\theta_\lambda(t)-\theta^0)(\theta_\lambda(t)-\theta^0)^T -$$
$$R_x^{-1}q_{t_0+N-1}(\lambda)R_x^{-1}\| \le C_1(\lambda,N,t_0)(1-\lambda)+$$
$$(2K(\lambda)r_x(\lambda)+K^2(\lambda)r_x^2(\lambda))C_2(\lambda,N,t_0)$$

where the matrix norm is a norm of the corresponding linear operator,

$$K(\lambda) = \sup_{\|d\|\le r_x(\lambda)} \|(R_x+d)^{-1}\|,$$

$$C_1(\lambda,N,t_0) = \|R_x^{-1}\|^2 \sum_{k,l=0}^{t_0+N-2} \lambda^{k+l}\|G(k,l,t_0,N)\|,$$

$$C_2(\lambda,N,t_0) = \mathrm{tr}\left(R_x^{-1}\left[q_{t_0+N-1}(\lambda)+(1-\lambda)\right.\right.$$
$$\left.\left.\sum_{k,l=0}^{t_0+N-2}\lambda^{k+l}G(k,l,t_0,N)\right]R_x^{-1}\right),$$

$$G(k,l,t_0,N) = \left[\gamma_{h_+(k,l,t_0)}(N+h_-(k,l,t_0),l-k)\right.$$
$$\left(1+\frac{h_-(k,l,t_0)}{N}\right)+$$
$$\left.\frac{h_-(k,l,t_0)}{N}Q(l-k)\right],$$

$$h_+(k,l,t_0) = \max\{t_0-\max\{k,l\},1\},$$
$$h_-(k,l,t_0) = \min\{t_0-\max\{k,l\}-1,0\}.$$

Proof is given in Section 6.

Corollary. *Assume the conditions A1, A2, A3 hold. Define the function*

$$\bar{q}(\lambda) = \frac{1}{1+\lambda}\sum_{j=-\infty}^{\infty}\lambda^{|j|}Q(j).$$

Then for any $\lambda \in (0,1)$ it holds

$$\limsup_{N\to\infty}\left\|(1-\lambda)^{-1}\frac{1}{N}\sum_{t=1}^{N}(\theta_\lambda(t)-\theta^0)(\theta_\lambda(t)-\theta^0)^T\right.$$
$$\left.-R_x^{-1}\bar{q}(\lambda)R_x^{-1}\right\| \le$$
$$(2K(\lambda)r_x(\lambda)+K^2(\lambda)r_x^2(\lambda))\,\mathrm{tr}\left(R_x^{-1}\bar{q}(\lambda)R_x^{-1}\right).$$

It is easy to see that $r_x(\lambda) \to 0$ as $\lambda \to 1$ because $\eta_\lambda(M) \to 0$ with fixed M and $\bar{\alpha}_x(M) \to 0$ as $M \to \infty$. Hence, the limit matrix of the estimation error covariances is equal to $R_x^{-1}\bar{q}(\lambda)R_x^{-1}$.

The expression for the limit value $\bar{q}$ of the limit matrix was derived by Xie and Ljung (2000).

Proof. It follows from condition A3 that $\gamma_m(N,\tau) \to 0$ as $N \to \infty$ for any fixed m, τ. Hence, $G(k,l,t_0,N) \to 0$ as $N \to \infty$ for any fixed k, l, t_0. For any fixed $\lambda \in (0,1)$ all series in the conclusion of theorem converge uniformly as $N \to \infty$ because they are bounded by the geometric progression. For this reason the summation and limit can be interchanged. It remains to notice that

$$\sum_{k,l=0}^{\infty}\lambda^{k+l}Q(k-l) = \sum_{l=0}^{\infty}\sum_{j=-l}^{\infty}\lambda^{j+2l}Q(j) =$$
$$\sum_{j=-\infty}^{\infty}Q(j)\sum_{l=(-j)_+}^{\infty}\lambda^{j+2l} =$$
$$\sum_{j=-\infty}^{\infty}Q(j)\frac{\lambda^{j+2(-j)_+}}{1-\lambda^2} = \frac{1}{1-\lambda^2}\sum_{j=-\infty}^{\infty}Q(j)\lambda^{|j|}.$$

5. THE BASIC LEMMA

Lemma. *Assume a sequence of matrices $(\xi_t)_{t=1}^{\infty}$ converges in mean uniformly and $(\bar{\alpha}_\xi(N))_{N=1}^{\infty}$ is its approximation bound of uniform convergence in mean, that is,*

$$\frac{1}{N}\sum_{t=m}^{m+N-1}\xi_t = A+\alpha_\xi(m,N),$$

$\|\alpha_\xi(m,N)\| \le \bar{\alpha}_\xi(N)$ for all m, N, and it holds $\bar{\alpha}_\xi(N) \to 0$ as $N \to \infty$. Assume also that $\|\xi_t\| \le C$ for all $t \ge 1$.

For $\lambda \in (0,1)$ define a function r_ξ as

$$r_\xi(\lambda) = \min_{1\le M\le 2/(1-\lambda)}\left\{\bar{\alpha}_\xi(M)+C\eta_\lambda^2(M)\right.$$
$$\left.+\eta_\lambda(M)\frac{C+\|A\|+\bar{\alpha}_\xi(M)}{2-\eta_\lambda(M)}\right\}$$

where $\eta_\lambda(M) = (M-1)(1-\lambda)$.

Assume the minimum in the definition of $r_\xi(\lambda)$ is achieved for $M = M_\lambda$. Define

$$N(\lambda) = 2M_\lambda - 3 + \left|\frac{\log\eta_\lambda(M_\lambda)}{\log\lambda}\right|.$$

Then for any $N \geq N(\lambda)$ it holds

$$\left\| (1-\lambda)\sum_{i=0}^{N-1}\lambda^i\xi_{N-i} - A \right\| \leq r_\xi(\lambda).$$

By condition A2 it holds $\bar{\alpha}_\xi(M) \to 0$ as $M \to \infty$. Assume $\lambda \to 1$. Then it follows from the definition of $r_\xi(\lambda)$ that $\eta_\lambda(M_\lambda) \to 0$ and $M_\lambda \to \infty$. Hence, $r_\xi(\lambda) \to 0$, and this is an explicit estimate of the rate of convergence.

Proof. Fix $\lambda \in (0,1)$ and denote $\varepsilon = 1-\lambda$, $M \doteq M_\lambda$ and $\eta = \eta_\lambda(M_\lambda)$. The next auxiliary inequality will be used for several times below:

$$\frac{\varepsilon M}{1-\lambda^M} - 1 = \frac{\varepsilon M}{1-(1-\varepsilon)^M} - 1 =$$
$$\frac{1}{1-\frac{M-1}{2}\varepsilon + \frac{(M-1)(M-2)}{6}\varepsilon^2(1-\vartheta\varepsilon)^{M-3}} - 1 \leq$$
$$\frac{1}{1-\frac{M-1}{2}\varepsilon} - 1 = \frac{1}{1-\frac{1}{2}\eta} - 1 = \frac{\eta}{2-\eta}.$$

Here the number $\vartheta \in (0,1)$ appeared from the residual term in the Taylor expansion that is written in the Lagrange form.

Let $N = LM + K$, $0 \leq K < M$. Make the algebraic transformations

$$\varepsilon\sum_{i=0}^{N-1}\lambda^i\xi_{N-i} = \varepsilon\sum_{l=0}^{L-1}\lambda^{lM}\sum_{m=0}^{M-1}\lambda^m\xi_{N-lM-m} +$$
$$\varepsilon\sum_{i=LM}^{N-1}\lambda^i\xi_{N-i} = \varepsilon\sum_{l=0}^{L-1}\lambda^{lM}\left(\sum_{m=0}^{M-1}(\lambda^m-1)\xi_{N-lM-m}\right.$$
$$\left. + M(A+\alpha_{N-lM}(M))\right) + \varepsilon\sum_{i=LM}^{N-1}\lambda^i\xi_{N-i}.$$

Each term in the last expression is estimated separately.

1°. Derive an upper bound for the first sum.

$$\varepsilon\left\|\sum_{l=0}^{L-1}\lambda^{lM}\sum_{m=0}^{M-1}(\lambda^m-1)\xi_{N-lM-m}\right\|$$
$$\leq C\varepsilon\sum_{l=0}^{\infty}\lambda^{lM}\sum_{m=0}^{M-1}(1-\lambda^m)$$
$$= C\frac{1-\lambda}{1-\lambda^M}\left(M - \frac{1-\lambda^M}{1-\lambda}\right)$$
$$= C\left(\frac{\varepsilon M}{1-(1-\varepsilon)^M} - 1\right) \leq C\frac{\eta}{2-\eta}.$$

2°. The second term approximates the limit value:

$$\left\|\varepsilon\sum_{l=0}^{L-1}\lambda^{lM}MA - A\right\|$$
$$= \|A\|\left|\frac{M\varepsilon}{1-\lambda^M}(1-\lambda^{LM}) - 1\right|.$$

Let us show that for $N > N(\lambda)$ the last expression is bounded as

$$\left|\frac{M\varepsilon}{1-\lambda^M}(1-\lambda^{LM}) - 1\right| \leq \left|\frac{M\varepsilon}{1-\lambda^M} - 1\right|.$$

Then it follows directly that

$$\left\|\varepsilon\sum_{l=0}^{L-1}\lambda^{lM}MA - A\right\| \leq \|A\|\,\frac{\eta}{2-\eta}.$$

It is required to prove that

$$\frac{M\varepsilon}{1-\lambda^M} - 1 \geq \frac{1}{2}\frac{M\varepsilon}{1-\lambda^M}\lambda^{LM}.$$

Algebraic transformations give the equivalent inequality

$$\lambda^{LM} \leq 2\frac{\varepsilon M - 1 + \lambda^M}{\varepsilon M}.$$

The Taylor expansion of the first order is

$$(1-\varepsilon)^M = 1 - \varepsilon M + \frac{M(M-1)}{2}\varepsilon^2(1-\vartheta\varepsilon)^{M-2}$$

where $0 < \vartheta < 1$. The inequality

$$\lambda^{LM} \leq (M-1)\varepsilon(1-\vartheta\varepsilon)^{M-2}$$

holds if it holds after a substitution of 1 in it instead of ϑ. Thus, a sufficient condition is

$$\lambda^{LM-M+2} \leq (M-1)\varepsilon = \eta,$$

that is equivalent to

$$LM \geq M - 2 + \left|\frac{\log\eta}{\log\lambda}\right|.$$

Since $LM \geq N - M + 1$, a sufficient condition for the last inequality can be written as

$$N \geq 2M - 3 + \left|\frac{\log\eta}{\log\lambda}\right| = N(\lambda),$$

which was assumed in lemma.

3°. The third term is bounded similarly.

$$\left\|\varepsilon\sum_{l=0}^{L-1}\lambda^{lM}M\alpha_{N-lM}(M))\right\| \leq \bar{\alpha}(M)\frac{M\varepsilon}{1-\lambda^M}$$
$$\leq \bar{\alpha}(M)\left(1+\frac{\eta}{2-\eta}\right).$$

4°. Notice that it was proved that $\lambda^{LM} \leq \eta$ under the condition $N \geq N(\lambda)$. It helps to bound the forth term.

$$\|\varepsilon\sum_{i=LM}^{N-1}\lambda^i\xi_{N-i}\| \leq \varepsilon C\frac{\lambda^{LM}-\lambda^N}{1-\lambda}$$
$$\leq C\lambda^{LM}(1-\lambda^{N-LM}) \leq C\eta(1-\lambda^{M-1})$$
$$\leq C\eta(1-\lambda)(M-1) = C\eta^2.$$

The last inequality follows from $x^k - 1 \geq k(x-1)$ for all $k \geq 0$ and $x \geq 0$.

Thus, it has been proved that if $N \geq N(\lambda)$ then

$$\left\|(1-\lambda)\sum_{i=0}^{N-1}\lambda^i\xi_{N-i}-A\right\| \le \bar{\alpha}_{\xi}(M)\left(1+\frac{\eta}{2-\eta}\right)$$
$$+\eta\frac{C+\|A\|}{2-\eta}+C\eta^2 = r_{\xi}(\lambda),$$

that completes the proof of lemma.

6. PROOF OF THEOREM

Define

$$\tilde{P}_{\lambda}(t) = (1-\lambda)(\sum_{k=0}^{t-1}\lambda^k x(t-k)x^T(t-k)+P^0),$$
$$\tilde{Y}_{\lambda}(t) = (1-\lambda)\sum_{k=0}^{t-1}\lambda^k x(t-k)\varepsilon^0(t-k).$$

Then $\Delta\theta_{\lambda}(t) = \theta_{\lambda}(t)-\theta^0 = \tilde{P}_{\lambda}^{-1}(t)\tilde{Y}_{\lambda}(t)$.

Denote $\alpha_{x,t}(\lambda) = \tilde{P}_{\lambda}(t)-R_x$. It follows from lemma that if $t \ge t(\lambda)$ then $\|\alpha_{x,t}(\lambda)\| \le r_x(\lambda)$. It is easy to check that

$$\tilde{P}_{\lambda}^{-1}(t)-R_x^{-1} = -\tilde{P}_{\lambda}^{-1}(t)\alpha_{x,t}(\lambda)R_x^{-1}.$$

Introduce the notation

$$s_{\lambda}(t) = R_x^{-1}\tilde{Y}_{\lambda}(t)\tilde{Y}_{\lambda}^T(t)R_x^{-1},$$
$$\beta_{x,t}(\lambda) = \tilde{P}_{\lambda}^{-1}(t)\alpha_{x,t}(\lambda).$$

Then

$$(\theta_{\lambda}(t)-\theta^0)(\theta_{\lambda}(t)-\theta^0)^T = s_{\lambda}(t)-\beta_{x,t}(\lambda)s_{\lambda}(t)$$
$$-s_{\lambda}(t)\beta_{x,t}^T(\lambda)+\beta_{x,t}(\lambda)s_{\lambda}(t)\beta_{x,t}^T(\lambda).$$

The last three terms represent an error of approximation. They are obviously bounded by the value

$$(2\|\beta_{x,t}(\lambda)\|+\|\beta_{x,t}(\lambda)\|^2)\|s_{\lambda}(t)\|.$$

It follows from the definition of the function $\beta_{x,t}(\lambda)$ and from the inequality $\|\tilde{P}_{\lambda}-R_x\| \le r_x(\lambda)$ that

$$\|\beta_{x,t}(\lambda)\| \le \sup_{\|d\|\le r_x(\lambda)} \|(R_x+d)^{-1}\|\, r_x(\lambda)$$
$$= K(\lambda)r_x(\lambda).$$

for $t \ge t(\lambda)$. In addition,

$$\|s_{\lambda}(t)\| = \|R_x^{-1}\tilde{Y}_{\lambda}(t)\|^2 = \operatorname{tr} s_{\lambda}(t).$$

Finally,

$$\|\Delta\theta_{\lambda}(t)\Delta\theta_{\lambda}^T(t)-s_{\lambda}(t)\|$$
$$\le (2K(\lambda)r_x(\lambda)+K^2(\lambda)r_x^2(\lambda))\operatorname{tr} s_{\lambda}(t).$$

The last inequality can be averaged by t. For any $t_0 \ge t(\lambda)$ it holds

$$\left\|\frac{1}{N}\sum_{t=t_0}^{t_0+N-1}\Delta\theta_{\lambda}(t)\Delta\theta_{\lambda}^T(t)-\frac{1}{N}\sum_{t=t_0}^{t_0+N-1}s_{\lambda}(t)\right\|$$
$$\le (2K(\lambda)r_x(\lambda)+K^2(\lambda)r_x^2(\lambda))\frac{1}{N}\operatorname{tr}\sum_{t=t_0}^{t_0+N-1}s_{\lambda}(t).$$

It follows from the definition of $\tilde{Y}_{\lambda}(t)$ that

$$(1-\lambda)^{-2}\frac{1}{N}\sum_{t=t_0}^{t_0+N-1}\tilde{Y}_{\lambda}(t)\tilde{Y}_{\lambda}^T(t)$$
$$=\frac{1}{N}\sum_{m=0}^{N-1}\sum_{k,l=0}^{t_0+m-1}\lambda^{k+l}Z_{t_0+m-k}Z_{t_0+m-l}^T$$
$$=\sum_{k,l=0}^{t_0+N-2}\lambda^{k+l}\frac{1}{N}\sum_{m=m_{k,l}}^{N-1}Z_{t_0+m-k}Z_{t_0+m-l}^T.$$

where $m_{k,l} = (\max\{k,l\}-t_0+1)_+$. In this expression and further the symbol $(x)_+$ denotes the positive part of a number x, that is, x if $x \ge 0$ else 0. Respectively, $(x)_- = x-(x)_+$. For the sake of simplicity introduce the notation $h_+(k,l,t_0) = 1+(t_0-\max\{k,l\}-1)_+$, $h_-(k,l,t_0) = (t_0-\max\{k,l\}-1)_-$.

The difference between the averaged quadratic form of (Z_t) and its limit was denoted by γ in the Section 4: for any $M>0$, $m_0>0$ and $\tau<m_0$ it holds

$$\frac{1}{M}\sum_{m=m_0}^{m_0+M-1}Z_mZ_{m-\tau}^T = Q(\tau)+\gamma_{m_0-(\tau)_+}(M,\tau).$$

Extract terms that include the limit value $Q(\cdot)$.

$$\sum_{k,l=0}^{t_0+N-2}\lambda^{k+l}\frac{1}{N}\sum_{m=m_{k,l}}^{N-1}Z_{t_0+m-k}Z_{t_0+m-l}^T =$$
$$\sum_{k,l=0}^{t_0+N-2}\lambda^{k+l}[Q(l-k)+\frac{h_-(k,l,t_0)}{N}Q(l-k)$$
$$+\gamma_{h_+(k,l,t_0)}(N+h_-(k,l,t_0),l-k)(1+$$
$$N^{-1}(h_-(k,l,t_0))] = q_{t_0+N-1}(\lambda)(1-\lambda)^{-1}$$
$$+\sum_{k,l=0}^{t_0+N-2}\lambda^{k+l}G(k,l,t_0,N).$$

Notice that the first term in the last expression tends to $\bar{q}(\lambda)$ and the second term with the sum tends to zero as $N \to \infty$ as it was shown in the proof of Corollary.

Conclusion of theorem follows from a direct substitution of the last expression in the expression of the averaged value of $s_{\lambda}(t)$:

$$\|(1-\lambda)^{-1}\frac{1}{N}\sum_{t=t_0}^{t_0+N-1}\Delta\theta_\lambda(t)\Delta\theta_\lambda^T(t)-$$

$$R_x^{-1}q_{t_0+N-1}(\lambda)R_x^{-1}\|$$

$$\leq \|R_x^{-1}\|^2\left((1-\lambda)\sum_{k,l=0}^{t_0+N-2}\lambda^{k+l}\|G(k,l,t_0,N)\|\right)$$

$$+(2K(\lambda)r_x(\lambda)+K^2(\lambda)r_x^2(\lambda))\,\mathrm{tr}\left[R_x^{-1}(q_{t_0+N-1}(\lambda)\right.$$

$$\left.+(1-\lambda)\sum_{k,l=0}^{t_0+N-2}\lambda^{k+l}G(k,l,t_0,N))R_x^{-1}\right]$$

$$=C_1(\lambda,N,t_0)(1-\lambda)+(2K(\lambda)r_x(\lambda)$$

$$+K^2(\lambda)r_x^2(\lambda))C_2(\lambda,N,t_0)$$

that coincides with the assertion of theorem.

The work was partially supported by Russian Foundation for Basic Researches, grant 01-01-00306.

REFERENCES

Barabanov, A.E. (1996). *Design of Minimax Regulators.* St.Petersburg University Publ.

Ljung, L. (2001). Estimating Linear Time-Invariant Models of Nonlinear Time–Varying Systems. *European Journal of Control*, **7**, 203–219.

Xie, L.L. and Ljung L. (2000). *Aspects on the Interpretation of Disturbances in System Identification.* Report no.: LiTH-ISY-R-2290, Linkoping University, Sweden.

IFAC Publications
www.elsevier.com/locate/ifac

STRUCTURAL IDENTIFICATION OF MULTIVARIATE NEURAL NETWORKS FOR RAINFALL RUNOFF MODELLING

Giorgio Corani * **Giorgio Guariso** * **Simone Castelli**

* *Dipartimento di Elettronica ed Informazione*
Politecnico di Milano
Via Ponzio 34/5, 20133 Milano
email: corani@elet.polimi.it

Abstract: Designing neural networks predictors by pruning instead of trial and errors significantly reduces the amount of guesswork required to select the optimal architecture. Furthermore, the obtained model is partially connected and hence very parsimonious in the number of parameters, leading to relevant operational advantages in the hydrological forecast practice: in fact, removing from the model redundant measurement stations results in an improved forecast availability and in the reduction of the costs of the data acquisition system. We exploited pruning algorithms to design the network, providing also a better basin state representation in comparison to existing schemes. Thanks to this modelling improvement, the obtained pruned networks overperform some fully connected ones published in previous works on the same basins, while requiring a significantly smaller set of measurement stations.

Keywords: neural network models, identification algorithms, prediction methods, forecasts.

1. INTRODUCTION

An efficient flood alarm system may significantly improve public safety, and mitigate economical damages caused by inundations. Since the rainfall-runoff relationship has been recognized to be non linear, artificial neural networks (ANN) have been increasingly used in the hydrological forecasting practice (see, for instances (Maier and Dandy, 2000), where tens of paper on the topic are quoted). Typically, such applications use only only rainfall and past flows measurements as input variables, without providing any representation of the catchment state, as will be explained in greater detail later in the paper. Neural networks are however criticized from different standpoints. First, finding the optimal network architecture for a given problem is not a trivial task and modellers usually work by trial and error, which is often very time consuming and requires a great amount of experience and guesswork. Second, working by trial and error allows to take into account only fully connected architectures, since testing also partially connected models would lead to a combinatorial explosion of the number of trials needed, although it is quite likely that some of the many weights contained in the network architecture are irrelevant. As a matter of fact, fully connected networks may easily contain few hundreds parameters, while often parsimonious modelling is required, as in the case of recently gauged basins.

We address the criticisms mentioned above by means of the Optimal Brain Surgeon (Hassibi and Stork, 1993) pruning algorithm . The basic idea of pruning algorithms is to start from a fully connected network, considered large enough to capture the desired input-output relationship. Then, they compute some measure of the contribution of each parameter to the problem solution, and consequently prune the less influential one from the network, to generate a new partially connected model, containing one parameter less. In this way, weights and neurons considered redundant are eliminated, significantly *reducing the amount of guesswork needed for the model selection.* Pruned network architectures may contain one order of magnitude less parameters than the fully connected ones, and are hence *very parsimonious*. In the hydrological forecasting practice, this allows to remove from the model those measuring stations whose informative content is redundant (while modellers are used to include in the model all the gauges available on the basin), without worsening the representation of the phenomenon: polling a smaller set of gauges makes the predictor less subject to downtimes due to missing data in real time operations, thus improving the forecast availability. Moreover, basin Authorities may decide to remove measurement tools and manpower from the insensitive stations, i.e. those pruned from the model, lowering the maintenance costs of the system.

A further issue investigated in this paper regards the description of the state of the catchment, which is given by the saturation degree of the soil, which determines system outputs (i.e., observed runoffs) given the input patterns (i.e., observed rainfalls). Depending on the basin saturation state, a different part of rainfall is routed to the river, thus increasing the observed runoff, or is infiltrated by the soil without join the river. It should be remarked that the basin saturation is not directly observable: given its great spatial and temporal variability, its measure would require a huge number of real-time "in situ" samplings. Hence, we adopt a proxy constituted by the rainfall measured on the basin in the days antecedent the prediction time. Since neural networks do not allow for an explicit representation of the state of the dynamic system to be modeled, we treat such antecend precipitations data as further input variables, enriching the input dataset with respect to many papers devoted to hydrological application of neural networks.

The paper is organized as follows: first we describe the neural network model of the rainfall runoff relationship, and the pruning algorithm exploited in order to select its optimal architecture. Then we show the results obtained in two quite different Italian catchments.

2. NEURAL NETWORK MODELLING OF THE RAINFALL RUNOFF RELATIONSHIP

The proposed forecast system is based, according to a typical hydrological modelling approach, on the availability of hydrometers and rain gauges distributed within or near the watershed. The forecast are issued after the arrival of the rainfall events; since rainfall-runoff response times are in order of few hours for small and medium sized basins (i.e. under or about 1000 km^2), this is a natural bound on the forecast lead times.

We model the rainfall-runoff process through a *feed forward* neural network with one hidden layer. We assume the water level as variable y to be predicted. The model input set comprises an autoregressive part of order p (i.e., p past water level measurement taken at the hydrometer), and several rainfall variables, associated with m rain gauges $r_1, r_2, \ldots, r_m$. Further exogenous terms in the input set are constituted by the observed rainfall on the basin over the few days before forecast time instant, which behave as proxies for the saturation state of the basin. Let us define u the vector containing all the input terms.

At forecast time, the j-th node of the hidden layer computes the following sum:

$$z_j = \sum_k w'_{kj} u_k - \sigma_j \qquad (1)$$

where w'_{kj} is the weight of the $k-th$ term in the input vector u, and σ_j is the neuron bias, which from a computational point of view can be treated as a weight of an additional constant input with unit value. Then, the neuron processes the signal z_j through its activation function (namely, hyperbolic tangent):

$$a_j = 1 - \frac{2}{(\exp(2z_j) + 1)} \qquad (2)$$

Values returned by the hidden neurons are sent to the output layer, which contains several nodes, corresponding to predictions at different time horizons. We thus use a multivariate architecture as shown in Fig.1. Each different node of the output layer returns a forecast, weighting the values a_j received from the hidden layer as in formula (1).

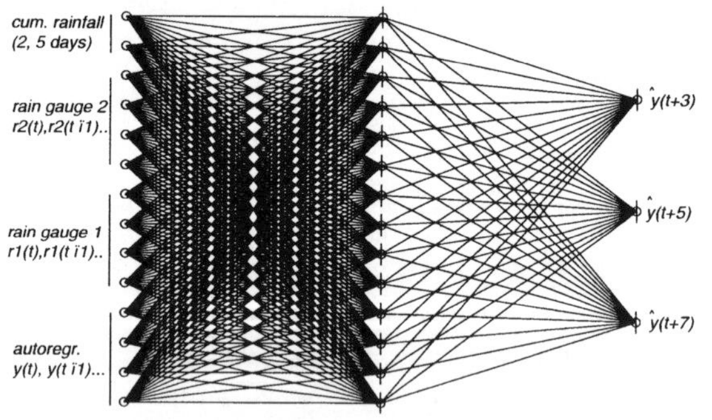

Figure 1. Sample of multivariate neural network architecture.

Multivariate direct predictors avoid the intermediate forecast steps required by recursive models, which may result in errors propagation; in fact, under mild assumptions, direct predictors can be demonstrated to be more reliable than recursive ones (Atiya *et al.*, 1999) . A further advantage of such kind of architecture is that the information available up to prediction time reaches each output node (i.e., it is exploited on each forecast horizon). In this way, the training algorithm will automatically adjust the relevance of each input information with respect to the different forecast horizons, avoiding the need of estimating rainfall delays, which is not a trivial task.

2.1 *Generalization criterion*

We calibrate the network through the Levenberg-Marquardt algorithm, adopting a regularized training objective function, i.e. adding to the mean squared error criterion a term (*weight decay*) proportional to the norm of the weights, in order to improve the generalization of the model (Girosi *et al.*, 1995).

The available data have been divided in three different subsets, according to the split sample approach (Bishop, 1995): a *training set*, used to estimate the parameters of the neural networks architectures; a *validation set*, used to compare the architecture performances, and hence to select the optimal one (the union of training and validation sets corresponds to the whole calibration data, in that the two sets are jointly exploited in the architecture identification); a *testing set*, used to assess the model performances on previously unused data. We standardize the three data sets using means and variances of training time series.

Some authors suggest to choose, between the considered architectures, that which minimizes the error on the validation set. In our experiments, however, such an approach resulted in model underfitting to the training set. Hence, we decided to select the architecture which minimizes the sum of squared error J over training and validation set:

$$J = \frac{1}{2N}\sum_{k}\sum_{i\in train.}(y_{ik}-\hat{y}_{ik})^2 + \\ + \frac{1}{2M}\sum_{k}\sum_{i\in val.}(y_{ik}-\hat{y}_{ik})^2 \tag{3}$$

where *k* refers to the different output nodes, i.e. to the different prediction horizons.

2.2 *Optimal Brain Surgeon* (OBS)

In this paper, we exploit the Optimal Brain Surgeon (*OBS*) algorithm to identify the optimal model structure (Hassibi and Stork, 1993). In the following we describe the algorithm assuming, for the sake of simplicity, to adopt a non-regularized training function; the extension for the regularized case can be found in (Norgaard *et al.*, 2000).

The training error function is approximated by means of a second order Taylor series expansion around the current parameter estimate $\overline{\theta}$:

$$E_{tr}(\theta) = E_{tr}(\overline{\theta}) + \nabla E_{tr}(\overline{\theta})\delta\theta + \\ + \frac{1}{2}[\theta-\overline{\theta}]^T H(\overline{\theta})[\theta-\overline{\theta}] \tag{4}$$

where $\nabla E_{tr}(\overline{\theta})$ is the gradient of the objective function evaluated in $\overline{\theta}$, and H is its Hessian, i.e. $H = \frac{\partial^2 E_{tr}}{\partial\theta^2}$. Assuming the network is already trained, the gradient term may be neglected since $\overline{\theta}$ represents a minimum for E_{tr}. Therefore, we can estimate the error surface around $\overline{\theta}$ looking just at the quadratic term; the change in E_{tr} is given by:

$$\partial E_{tr} = E_{tr}(\theta) - E_{tr}(\overline{\theta}) = \delta\theta^T \frac{H}{2}\delta\theta \tag{5}$$

The weight whose elimination leads to the minimum increase of the training error function can be identified finding the $\delta\theta$ which minimizes ∂E_{tr}, subject to constraint $e_i^T\delta\theta + \theta_i = 0$, where e_i is a unit vector in the weight space parallel to the w_i axis. According to such constrain, the only allowed adjustments in θ is the deletion of a unique weight θ_i. The problem can be solved by means of Lagrange multipliers and returns the following estimate of the increase error, due to the elimination of the j-th weight:

$$\partial E_{tr}(j) = \frac{1}{2}\frac{\hat{\theta}_j^2}{[H^{-1}]_{jj}} \tag{6}$$

where $[H^{-1}]_{jj}$ is the (j,j) element of the inverse of the Hessian matrix. Formula (6) is actually the *saliency* estimate for the $j-th$ parameter. The consequent change to be applied to the weights vector is given by:

$$\partial\theta = -\frac{\theta_j}{[H^{-1}]_{jj}}H^{-1}e_j \tag{7}$$

The overall architecture selection task can be automated as follows:

(1) training of the initial fully connected architecture;
(2) ranking of parameters on the base of their saliences;
(3) elimination of the weight with the lowest saliency and generation of a new architecture;
(4) re-training of the obtained network, and evaluation of its performances on training and validation set;
(5) back to step 2, until there are parameters left.

In order to reduce the probability that the algorithm falls into a local minimum, it is required to train several times the initial network and to run hence several pruning session. At the end of the procedure, we chose the architecture which minimizes the generalization criterion J and then we retrain it without weight decay, since the low number of parameters contained in the pruned model no longer requires regularization.

Neural networks have been implemented, calibrated and pruned through the Neural Network Based System Identification Toolbox for Matlab (Norgaard *et al.*, 2000), available as free software from the site `www.iau.dtu.dk/research/control/nnsysid.html`.

Computation time required by a pruning session is about half an hour on a 600 Mhz PC.

3. RESULTS

Two different Italian catchments (Olona and Tagliamento) have been considered as case studies. Specialized time series, containing data sampled at hourly steps, have been built on the two basins, containing those episodes showing a water level increase higher than 100% within a 24-*hours* time window.

Besides the classical measures of the root mean square error (*RMSE*) and correlation ρ between predicted and real flows, we compute the model efficiency R^2, defined as:

$$R^2 = \frac{F_0 - F}{F_0} \tag{8}$$

where F_0 is the variance of water levels $\sum_i (y_i - \mu(y))^2$ and F is the mean square error $\sum_i (y_i - \hat{y}_i)^2$. The "prediction" of the average value, which can be considered as the prediction available even in the worst case, has then an efficiency of 0. An efficiency value of 90% indicates a very satisfactory model performance while a value in the range 80 – 90% indicates a fairly good model (Shamseldin *et al.*, 2001). In the following, we compare the performances of fully connected networks, identified according to a classical procedure, (i.e. selecting the architecture through trial and error, using all the gauges, without any representation of the saturation state of the basin) with those of the pruned predictors, with and without the information regarding the antecedent precipitations.

	Training	Testing
Classical network		
3 gauges		
no antecedent rainfall		
43 parameters		
RMSE	.012	.046
R^2	.92	.84
ρ	.96	.92
Pruned network (a)		
2 gauges		
no antecedent rainfall		
10 parameters		
RMSE	.015	.047
R^2	.87	.84
ρ	.94	.92
Pruned network (b)		
2 gauges		
with antecedent rainfall (2 and 5 days)		
24 parameters		
RMSE	.013	.029
R^2	.91	.86
ρ	.96	.93

Table 1. Results for the 3 hours ahead prediction on the Olona basin.

Olona River The first case study refers to the basin of river Olona , located in Lombardia, Northern Italy. The average flow is about $2.5\,m^3/sec$, while the maximum expected flow over a time period of 10 years is about $108\,m^3/sec$. The basin size is about $200\,km^2$, and is gauged through one hydrometer, and three rain gauges. Data refer to 13 events (with an overall length of about 1100 hourly steps) occurred within the period 1999-2001; training, validation and testing sets contain respectively about 500, 200, 400 patterns. We evaluate the model performances (Table 1) on a three hours forecast horizon, judged as suitable by Civil Protection technicians.

Models identified through pruning recognize one of the three rain gauges as insensitive, removing from the network architectures all the related connections.

This is very convenient in real time operations, since the less the number of gauges to poll, the higher the probability that the model will have all the required measures, thus increasing the forecast availability. The complete predictor performs better on the training set, because of some overfitting due to the larger number of parameters employed; however, looking at the testing set, which is the most suitable for performance assessments, the pruned predictor with cumulated rainfall is the best performing one. Furthermore, the comparison between pruned predictors (a) and (b) allows to perceive the improvement given by the additional input (which is obviously available at no cost, since it is simply the integral of the other inputs).

Thus, pruning allows to substantially decrease the number of parameters and also of measurement stations in the model, without performance worsening. Furthermore, the introduction of the cumulated rainfall information results in a performance improvement.

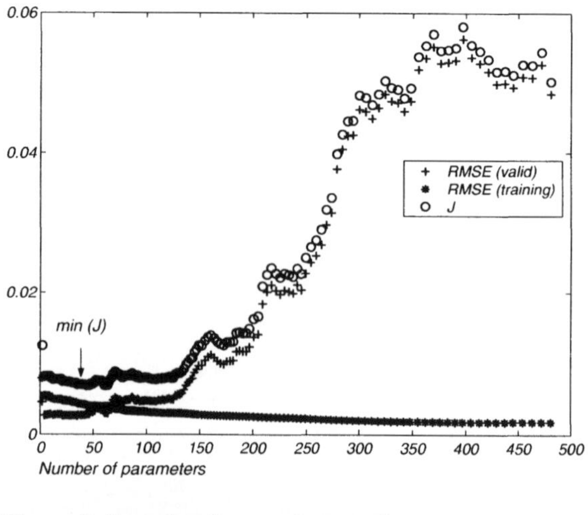

Figure 2. Sample of a pruning session

Tagliamento River The basin sizes about $1950\,km^2$, and is gauged through one hydrometer, at the end of the mountain district, and five rain gauges. The dataset comprises 20 flood events occurred over the years 1978-1996, for an overall length of 2000 hourly time steps. Training, validation and testing sets contain

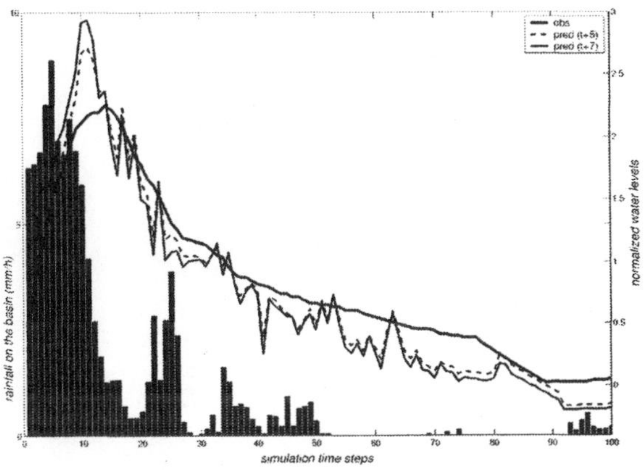

Figure 3. Flood sample taken form the testing set, simulated by means of the network with cumulated rainfall

respectively about 1000, 600, 400 time steps.A feed forward neural network for flood forecasting in this basin has been already proposed in (Campolo *et al.*, 1999). Such network is completely connected, and its input variables set comprises all the gauges on the basin. The model architecture comprises 10 nodes in the hidden layer and contains 5 autoregressive terms and 15 terms for each rain gauge in the input layer, thus totalizing about 800 parameters. The paper also states that a forecast horizon of 5 hours in advance can be considered satisfactory on this basin.

Table 2 shows the performances of classical and pruned networks on the five hours ahead prediction.

Classical network (Campolo)		
fully connected, 5 gauges		
no cumulated rainfall		
800 parameters		
	Training	Testing
RMSE	-	-
R^2	.89	.85
ρ	-	-
Pruned network (a)		
partially connected, 3 gauges		
fully connected, 5 gauges		
93 parameters		
	Training	Testing
	.007	.013
	.94	.85
	.97	.92
Pruned network (b)		
partially connected, 3 gauges		
cumulated rainfall (2, 5 and 10 days)		
188 parameters		
RMSE	Training	Testing
R^2	.005	.013
ρ	.96	.86
	.98	.93

Table 2. Results for the 5 hours ahead prediction on the Olona basin.

Models identified through pruning recognize two of the five rain gauges as redundant, and its terms are no more connected into the pruned models; Fig.2 shows a sample of behavior of training and validation error during a pruning session. Once more, pruned models behave better or as than the fully connected one, although the significant reduction in the input data.

It should be remarked that the advantages of the models with the cumulated rainfall increase on longer time horizons. The 7-hours ahead prediction shows an efficiency of 76.4 for the fully connected model and for the pruned one without cumulated rainfall, and 79.6 for the pruned model with cumulated rainfall.

4. CONCLUSIONS

We investigated the Optimal Brain Surgeon pruning algorithm in order to design the model architectures for real time floods forecasting. Results showed that pruned networks constitute a convenient alternative with respect to those designed by trial and errors, strongly reducing the number of model parameters, while providing at the same time an equal generalization capacity (i.e., performance observed on the testing set). Furthermore, using a smaller set of gauges lowers the downtimes of the forecast system due to data acquisition failures, which may be very welcomed by the Basin Authorities running the forecast system.

In a further series of experiments we use the rainfall observed on the basin over the few days before forecast time as additional input for the network, in order to provide an indirect estimate of the basin state. Results confirm in two ways the correctness of such an approach. First, the indirect measure of the state of the catchment is retained in the optimal model architectures, selected through pruning, while many other input terms are removed. Second, the information is clearly valuable in that it results in a forecast improvement over all the performed experiments, which may reach up to 4% of efficiency. Such an information does not require any further equipment to be installed, since data are directly available from the rain gauges records. Hence, it seems advisable to always consider its application in almost any neural networks flood forecasting system.

Acknowledgments The authors thank M. Molari and N. Quaranta, Civil Protection Service of Regione Lombardia, for supplying the data of Olona river; M. Campolo and A. Soldati, University of Udine, for the data of Tagliamento river.

5. REFERENCES

Atiya, A.F., S.M. El-Shoura, I.S. Shaheen and M.S. El-Sherif (1999). A comparison between neural-network forecasting techniques case study: River flow forecasting,. *IEEE Trans. on Neural Networks,* **10**(2), 402–409.

Bishop, C.M. (1995). *Neural networks for pattern recognition*. Oxford University Press.

Campolo, M., P. Andreussi and A. Soldati (1999). River flood forecasting with a neural network model. *Water Resour. Res.* **35**(4), 1191–1197.

Girosi, F., M. Jones and T. Poggio (1995). Regularization theory and neural networks architectures. *Neural Comput.* **7**, 219–269.

Hassibi, B. and D.G. Stork (1993). Second order derivatives for network pruning: Optimal brain surgeon. In: *Proceedings of Advances in Neural Information Processing System* (C.L. Giles S.J. Hanson, J.D. Cowan, Ed.). pp. 164–171.

Maier, H.R. and C.G. Dandy (2000). Neural networks for the prediction and forecasting of water resources variables: a review of modelling issues and applications. *Environmental Mod. & Soft.* (15), 101–124.

Norgaard, M., O. Ravn, N.K. Poulsen and L.K. Hansen (2000). *Neural networks for modelling and control of dynamic systems*. Springer-Verlag. London.

Shamseldin, A.Y., K.M. O'Connor and K.M. Liang (2001). Methods for combining the outputs of different rainfall-runoff models. *J. Hydrol.* **245**, 196–217.

IFAC Publications
www.elsevier.com/locate/ifac

PARAMETER AND STATE REGULARIZATION FOR PREDICTION OF DISTRIBUTED HYDROLOGIC SYSTEMS

E.E. van Loon * **K.J. Keesman** **,[1]

* *Hydrology and Quantitative Water Management Group, Wageningen University*
** *Systems and Control Group, Wageningen University*

Abstract: State and parameter regularization schemes are applied to a distributed hydrologic system for discharge prediction at different locations in the catchment. Experimental field data from a 200 ha catchment in Costa Rica have been used to evaluate the prediction method. It appears that, in general, state regularization leads to better predictions at finer resolutions than parameter regularization, and parameter regularization to better performance at coarse resolutions.

Keywords: State/parameter estimation, regularization, hydrologic systems

1. INTRODUCTION

In general the identification problem for distributed hydrologic systems focuses on a single catchment and a limited observation period. For a hydrological model to be a useful tool in planning, it should be possible to apply that same model to other locations or time periods without redoing the full identification step or having to collect large amounts of observations. Another reason for not changing the model structure at different situations is to keep model results compatible with earlier results and therewith reasonably straightforward to interpret. On the other hand, it has become clear from numerous studies that a certain amount of re-calibration is always required in catchment modelling. The question remains however, which parameters in the model are best suited for this purpose, how to determine the parameter values from data and at which resolution to define the model. In view of previous work on model (re-)calibration, especially the last question is interesting since the choice for the resolution at which a hydrological problem is defined is often not discussed nor explained. This is a remarkable situation, since a simple count of variables in any realistic hydrologic problem shows that the available observations alone cannot contain sufficient information to determine a distributed-parameter model to a reasonable degree, even not if the problem is limited to the re-calibration of a few parameters. Hence, the problem is *ill-posed.* When seeking a way to identify a model in face of ill-posedness, a common approach in many engineering disciplines is the development of an alternative model structure, with fewer degrees of freedom and a more suitable parameterization. However, as pointed out above, in hydrological problems the change of model structure is often undesirable since it would render the model incompatible with other applications or make the model results difficult to interpret. Therefore the problem is usually handled by using additional information, in the form of assumptions about parameter values and relations between parameters. Here the aim is now to provide such a follow-up by considering a *regularization* approach to combine a set of calibrated catchment scale models of over-

[1] Corresponding author.
E-mail address: karel.keesman@wur.nl

land flow with additional observations. Section 2 presents the regularization approach to solve the estimation problem for a general class of ill-posed time-varying linear models of overland flow. In Section 3 regularization is applied to predict overland flow in the study catchment. This is followed by an inter-comparison of various ways to re-calibrate models in Section 4. The implications of the results are discussed and conclusions are drawn in Section 5.

2. CALIBRATING A HYDROLOGICAL SYSTEM THROUGH REGULARIZATION

Conversion into standard linear time varying form

It has been shown by van Loon and Keesman [2000] that a large class of hydrological systems, which vary in time and space and are often non-linear, can be represented by a linear time-varying state-space model of the following form

$$\mathbf{x}_k = \mathbf{A}_k\mathbf{B}_k\mathbf{C}_k\mathbf{x}_{k-1} + \mathbf{A}_k\mathbf{u}_k \qquad (1)$$

$$\mathbf{y}_k = \mathbf{H}_k\mathbf{x}_k + \mathbf{e}_k \qquad (2)$$

where $\mathbf{x}_k$ contains M state variables for each of the L spatial units and the input vector $\mathbf{u}_k$ contains at least L elements at time instant k (but $\mathbf{u}_k$ may be larger, depending on the structure of $\mathbf{A}_k$ and the amount of input variables). The structured matrices $\mathbf{A}_k$, $\mathbf{B}_k$ and $\mathbf{C}_k$ (transition matrices) contain time-varying coefficients θ_k which are stochastic functions of $\mathbf{x}_{k-1}$ or $\mathbf{u}_k$ and can only take values between zero and one $(0 \leq \theta_k = f(\mathbf{x}_{k-1}, \mathbf{u}_k) \leq 1)$. In what follows the states are the stored soil water $(s_{k,l})$, the overland flow due to infiltration excess $(r_{k,l})$, the overland flow due to saturation excess $(t_{k,l})$ and the cumulative soil moisture content $(w_{k,l})$ all in mm. The effective rainfall $(p_{k,l})$, that is the difference between rain and evapotranspiration, is the input to the hydrologic system. The vector $\mathbf{y}_k$ contains P observations and the matrix $\mathbf{H}_k$ is an observation matrix which relates the state variables to the observations, and the output error vector $\mathbf{e}_k$ contains modelling as well as observation errors. In the following, the focus is on either on-line estimating the unknown coefficients in the transition matrices or estimating the state variables at time instant k from observations available at time instant k (parameter and state estimation respectively).

To apply the estimation techniques the above equations are converted to the following standard regression form

$$\mathbf{y}_k = \mathbf{D}_k\mathbf{m}_k + \mathbf{e}_k \qquad (3)$$

where $\mathbf{y}_k$ is a vector with measured values, $\mathbf{D}_k$ the design or data matrix with known values, $\mathbf{m}_k$ a vector with values to be estimated, and $\mathbf{e}_k$ a noise vector. The way in which (1) and (2) have to be transformed to obtain (3) depends on whether we consider *parameter* or *state estimation*.

For *parameter estimation* substitute (1) into (2), so that

$$\mathbf{y}_k = [\mathbf{H}_k\mathbf{A}_k\mathbf{B}_k\mathbf{C}_k]\,\mathbf{x}_{k-1} + [\mathbf{H}_k\mathbf{A}_k]\,\mathbf{u}_k + \mathbf{e}_k \qquad (4)$$

where $\mathbf{x}_{k-1}$ will be substituted by its estimate $\widehat{\mathbf{x}}_{k-1}$, which is common practice in nonlinear parameter estimation problems with output error structure. In what follows $\mathbf{x}_{k-1}$ is estimated using (1)

$$\widehat{\mathbf{x}}_{k-1} = \mathbf{A}_{k-1}\mathbf{B}_{k-1}\mathbf{C}_{k-1}\widehat{\mathbf{x}}_{k-2} + \mathbf{A}_{k-1}\mathbf{u}_{k-1} \qquad (5)$$

and $\mathbf{x}_0$ is assumed to be known.

Hence, the known vectors $\widehat{\mathbf{x}}_{k-1}$ and $\mathbf{u}_k$ are put into the data matrix $\mathbf{D}_k$, and all the unknown parameters in $\mathbf{A}_k$, $\mathbf{B}_k$, $\mathbf{C}_k$ and $\mathbf{H}_k$ are put in the parameter vector $\mathbf{m}_k$. A detailed derivation is given in Appendix C of van Loon [2001]. Consequently, (4) can be rewritten as (3) with observation vector $\mathbf{y}_k \doteq \left[\, y_k(1)\; y_k(2)\; \ldots\; y_k(P)\,\right]^T$, $\mathbf{D}$ the design matrix which embodies the geometry as well as inputs to the system, and $\mathbf{m}_k \doteq \left[\, \theta_k(1)\; \theta_k(2)\; \ldots\; \theta_k(Q)\,\right]^T$ with $Q = MPL + PL$ the size of the model parameter vector. Besides the P observations there are some additional constraints. These arise from the fact that the columns of the transition matrices in the original model (1) sum to unity. These constraints form a set of additional linear equations

$$\mathbf{y}_{cons,k} = \mathbf{D}_{cons,k}\mathbf{m}_k \qquad (6)$$

Even with these additional hard constraints there are, for any realistic distributed hydrological problem, relatively few observations available compared to the number of model parameters, so that information is lacking to determine uniquely all the model parameters,i.e. the problem has an almost singular regression matrix since $\mathbf{D}_k$ has many more columns than rows.

When considering *state estimation* we propose to rewrite (1) and (2) as follows,

$$\mathbf{y}_k - [\mathbf{H}_k\mathbf{A}_k]\,\mathbf{u}_k = [\mathbf{H}_k\mathbf{A}_k\mathbf{B}_k\mathbf{C}_k]\,\mathbf{x}_{k-1} + \mathbf{e}_k \qquad (7)$$

which again can be written into the regression form of (3), where now $\mathbf{m}_k$ is defined as $\mathbf{m}_k = [x_{k-1}(1) \ldots x_{k-1}(4L)]$. Similar to (6) equality constraints on the states can be formulated, e.g. derived from the global mass balance

$$\sum_{l=1}^{L} (s_{k,l} + r_{k,l} + t_{k.l} - p_{k,l}) = 0 \qquad (8)$$

$\forall\ k = 1, \ldots, K$ or after some manipulation (see Appendix A in van Loon and Keesman [2000])

$$\mathbf{v}^T \left[\mathbf{x}_{k-1} - \widehat{\mathbf{x}}_{k-1}\right] = 0 \tag{9}$$

with $\mathbf{v} \doteq \left[\ 1(1)\ \ldots\ 1(3L)\ 0(1)\ \ldots\ 0(L)\ \right]^T$ and $\widehat{\mathbf{x}}_{k-1} \doteq \mathbf{A}_{k-1}\mathbf{B}_{k-1}\mathbf{C}_{k-1}\widehat{\mathbf{x}}_{k-2} - \mathbf{A}_{k-1}\mathbf{u}_{k-1}$ Note that the vector $\mathbf{v}$, comprising a $(1, 3L)$ sub-vector with ones and a $(1, L)$ sub-vector with zeros, simply allows a summation of the first three variables (each defined at L spatial units) of $\mathbf{x}$, viz. $\mathbf{s}$, $\mathbf{r}$, and $\mathbf{t}$. By defining $\mathbf{D}_{cons,k} = \mathbf{v}^T$ and $\mathbf{y}_{cons,k} = \mathbf{v}^T\widehat{\mathbf{x}}_{k-1}$, (9) can be written in the form of (6). Notice that in this approach the dynamic hydrologic system equations are added as equality constraints, see (9), to the regression-type model (3). A more conventional to to the aforementioned problems would be based on observer-theory where design and updating of the gain matrix plays a central role. The reasons for not using conventional techniques for our problem is to avoid the tuning of the gain matrix, which is a laborious task for the set of models under study.

Discrete inverse theory

The concept of the generalized or pseudo-inverse ($\mathbf{G}$, see [Rao and Mitra, 1971]) is used to find a solution to (3). The exact form of the generalized inverse depends on the problem at hand. Some frequently used generalized inverses are $\mathbf{G} = \left(\mathbf{D}^T\mathbf{D}\right)^{-1}\mathbf{D}^T$ (least squares solution) or $\mathbf{G} = \mathbf{D}^T\left(\mathbf{D}\mathbf{D}^T\right)^{-1}$ (minimum length solution). For a given generalized inverse, the unknowns at each time instant k can be estimated from

$$\widehat{\mathbf{m}} = \mathbf{G}\mathbf{y} \tag{10}$$

where the subscript k is omitted for ease of notation.

The relation between the estimated and the true model parameters ($\mathbf{m}_{true}$) follows from inserting (3) in (10), so that

$$\widehat{\mathbf{m}} = \mathbf{m}_{true} + (\mathbf{GD} - \mathbf{I})\,\mathbf{m}_{true} + \mathbf{Ge} \tag{11}$$

where the $Q \times Q$ matrix $\mathbf{GD}\,(\equiv \mathbf{R})$ is an orthogonal projection matrix, which is often referred to as *model resolution matrix*. Since in general $E(\widehat{\mathbf{m}}) \neq \mathbf{m}_{true}$, except when $\widehat{\mathbf{m}}$ is the least squares solution so that $\mathbf{GD} = \mathbf{I}$ with $E(\mathbf{Ge}) = 0$, it follows that it is likely to be biased. Notice that the estimated parameter vector is a function of the true parameter vector, the deviation of the model resolution matrix from the identity matrix, and some mapping of the output error.

Similarly, the estimated model parameters $\widehat{\mathbf{m}}$ may be used to evaluate how well predictions by the model correspond to the observed data through

$$\widehat{\mathbf{y}} = \mathbf{D}\widehat{\mathbf{m}} \tag{12}$$

where $\widehat{\mathbf{y}}$ is the predicted output $\mathbf{y}$. By substituting (10) into (12) and applying the same ordering as in (11) the following expression is obtained

$$\widehat{\mathbf{y}} = \mathbf{y} + (\mathbf{DG} - \mathbf{I})\,\mathbf{y} + \mathbf{e} \tag{13}$$

Here the $P \times P$ matrix $\mathbf{DG}(\equiv \mathbf{N})$ is called the *data information matrix*. This matrix describes how well the predictions match the original data, apart from the observation errors in $\mathbf{e}$. The diagonal elements in the information matrix indicate how much weight a datum has in its own prediction.

Another measure for model quality is the covariance matrix of the estimated model parameters ($\mathbf{C}_{\widehat{m}}$). The resolution matrix, information matrix and model covariance matrix are useful to define the criteria of a good inverse and thus implicitly good measures of the model quality, e.g. [Backus and Gilbert, 1968]. On the basis of some norm of $(\mathbf{R} - \mathbf{I})$, $(\mathbf{N} - \mathbf{I})$, or $\mathbf{C}_{\widehat{m}}$ trade-off curves can be constructed (see e.g. [Hansen, 1992, Menke, 1989]). A rather popular technique, due to its robustness and the avoidance of using the covariance matrix, is *generalized cross-validation* [Golub et al., 1979, Hansen, 1998, Wahba, 1977], which amounts to the minimization of the generalized cross validation function

$$f_{gcv} = \frac{\|\mathbf{y} - \mathbf{D}\widehat{\mathbf{m}}\|_2^2}{(trace(\mathbf{DG} - \mathbf{I}))^2} \tag{14}$$

where $\mathbf{G}$ is the matrix to be chosen. It is based on the philosophy that if an arbitrary element y_i is left out, the corresponding solution should predict this observation well; and that the choice of the solution should be independent of an orthogonal transformation of $\mathbf{y}$ [Hansen, 1998]. This function f_{gcv} will be used in what follows to find a proper generalized inverse $\mathbf{G}$.

Solution method

The generalized inverse $\mathbf{G}$ for our ill-posed problem can only be formed by including prior information to reconstruct unknown system properties and/or regularization (dampening) factors to reduce instabilities. Therefore a set of soft constraints is added to (3), that is

$$\begin{bmatrix} \mathbf{y} \\ \rho\mathbf{m}_{pri} \end{bmatrix} = \begin{bmatrix} \mathbf{D} \\ \rho\mathbf{I} \end{bmatrix} \mathbf{m} + \begin{bmatrix} \mathbf{e} \\ \mathbf{e}_{pri} \end{bmatrix} \tag{15}$$

where $\mathbf{I}$ is a $Q \times Q$ identity matrix and ρ is a so-called regularization parameter that will be explained later. Since

$$\mathbf{rank}(\begin{bmatrix} \mathbf{D} \\ \rho\mathbf{I} \end{bmatrix}) = \mathbf{rank}(\begin{bmatrix} \mathbf{D} & \mathbf{y} \\ \rho\mathbf{I} & \rho\mathbf{m}_{pri} \end{bmatrix}) \quad (16)$$

the problem can now be solved in a least-squares sense.

At this point differences arise between the parameter and state estimation solutions. In the case of parameter estimation we use the averages of the stochastic parameters obtained from previous experiments as prior information ($\mathbf{m}_{pri}$), whereas in the case of state estimation we use the estimates at the previous time instant ($\mathbf{m}_{k-1}$).

Depending on the regularization parameter ρ, the solution to (15) will vary between the minimum length solution (i.e. the solution based on prior information) or the least squares solution in the observation space. As noted above, generalized cross-validation is used here to determine the desired value of ρ. The generalized inverse therefore gets of the following form

$$\mathbf{G} = \left(\mathbf{D}^T\mathbf{D} + \rho\mathbf{I}\right)^{-1}\mathbf{D}^T \quad (17)$$

The solution is found by applying a number of steps. The mathematical operations involved in each step at each time instant k are fully described in van Loon [2001], Appendix D. Since (4) varies over time, at each time instant (k) a different solution is obtained. In this study ρ varied between $3.4\ 10^{-4}$ and $4.7\ 10^{-2}$.

3. DESCRIPTION AND USE OF MODELS AND DATA SETS

The regularization algorithm described in Section 2 requires a prior model in combination with a data set. In this study a data set from the 200 ha Horizontes catchment in Costa Rica is used. The data set contains observations of rain, ground water levels, overland flow, soil properties, vegetation properties and topography (not discharge - since this is to be predicted). The data was collected from April 1996-August 1998 [van Loon, 2001]. In van Loon and Keesman [2000] a set of models (with different structures and parameter values) has been introduced to represent the hydrology of the catchment. This set with prior models is used here. The prediction is done for 15 individual rainfall events. Predictions are made in three different ways: 1) in open loop form (i.e. only using the required model inputs), 2) by applying parameter regularization, and 3) by applying state regularization. The prediction algorithm proceeds by first rewriting the prior models in the form of (3) and subsequently applying the algorithms for open loop calcualtions, parameter and state regularization for discharge prediction per rainfall event. Not all the all the available data is used for regularization but rather subsets of various sizes. The largest subset only contains 75% of the data, while the remaining 25% is used for validation purposes. The nine subsets are established by a latin-hypercube sampling scheme, where 25, 50 and 75% of the observation time instants are combined with 25, 50 and 75% of the observation locations (thus yielding 9 combinations). In all cases 25% of the observation time instants and locations are used for validation. The naming of the subsets is shown in Table 1. The relative root mean squared error (RRMSE) of the discharge prediction is considered for models at various resolutions and for the various sub-sets.

Table 1. Naming of nine sub-sets.

coverage in time	coverage in space 25%	50%	75%
25%	1.1	1.2	1.3
50%	2.1	2.2	2.3
75%	3.1	3.2	3.3

4. RESULTS

Overall model performance

The results for the open-loop prediction (without using observations) are shown in Figure 1. The figure shows the time-averaged RRMSE of the discharge prediction for Horizontes as function of upstream area, for different model resolutions. In the same figure, also the RRMSE of the discharge prediction for the calibration period is shown. In both the calibration and the prediction periods the RRMSE appears to decline for increasing catchment size. In addition, the RRMSE for the calibration period is considerably smaller than that in the prediction period. The fact that a larger RRMSE is found when considering a smaller area, implies that the heterogeneity of the small area is not represented well by the model; the model units are too coarse to capture the system heterogeneity at this scale.

Performance of state and parameter regularization

In Figure 2 the time-averaged RRMSE of the discharge prediction is shown for both parameter and state regularization, when using sub-set 1.1 of Horizontes (see Table 1). It shows that the RRMSE for these situations is considerably lower than that in the case of an open-loop prediction,

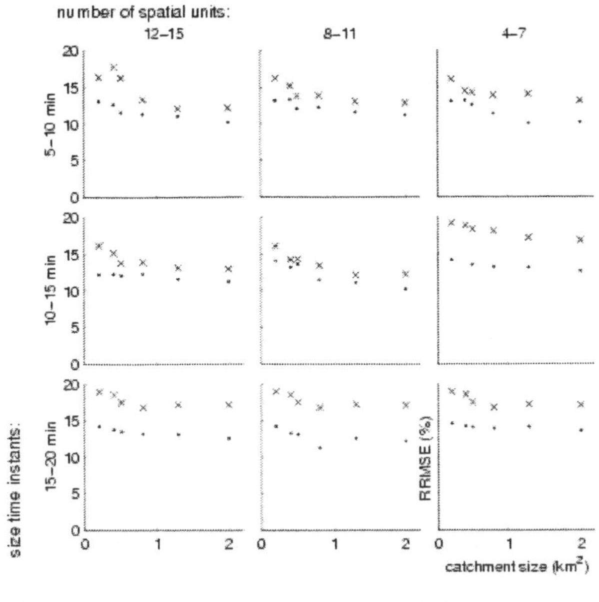

Fig. 1. Time-averaged RRMSE of discharge prediction in Horizontes at six different measurement locations for open-loop models: calibration (•) and prediction (×).

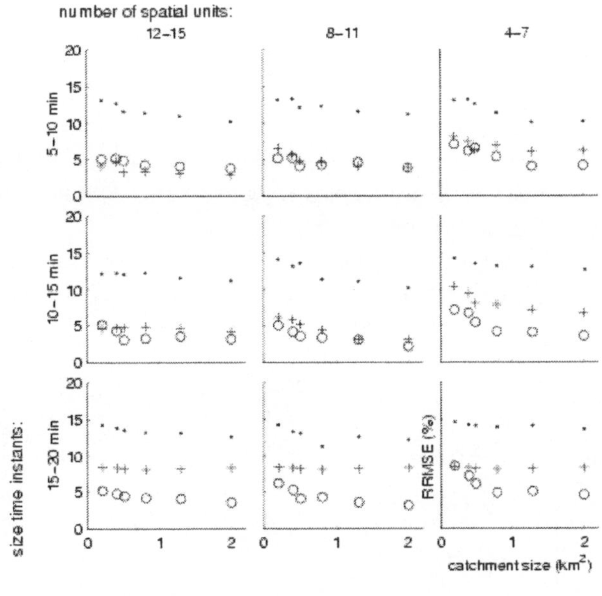

Fig. 2. Time-averaged RRMSE of discharge prediction in Horizontes at different measurement locations for models with parameter (+) and state (o) regularization and compared with calibration results (•).

at all resolutions but especially at the finer resolutions. State regularization generally leads to better predictions than parameter regularization (the o lay below the + symbols). For the subcatchment Horicajo similar results appear (not shown here).

In what follows the RRMSE of the predictions will be averaged over the six measurement locations. This allows to study the relation between the amount of data used for conditioning, the model resolution where the minimum RRMSE is found,

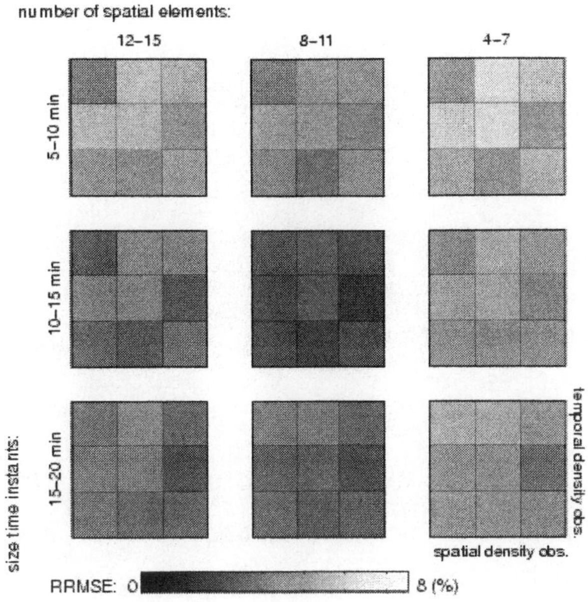

Fig. 3. RRMSE of discharge predictions by applying *parameter* regularization, for different resolutions and observation densities.

and the value of the RRMSE itself. In Figure 3 the RRMSE is shown for different observation densities (using the Horizontes data) when applying parameter regularization. The average RRMSE values of each plot in Figure 2 correspond to the values in the upper left corners of the nine plots in Figure 3 (+ in Figure 2) and Figure 4 (o in Figure 2). The cells in each of the nine sub-plots of the figures are defined by Table 1, e.g. lowest observation densities are in the upper left corner. It appears that for increasing observation densities in general the RRMSE decreases, and in addition the optimum RRMSE is found at the medium space-time resolutions. Figure 4 shows the same information when applying state estimation. By comparing Figures 3 and 4 it appears that the minimum RRMSE value for state regularization is slightly smaller than that for parameter regularization, but only at high observation densities and at finer resolutions. The plots of RRMSE when using the Horicajo observations show a very similar pattern. In general, state regularization leads to better predictions at finer resolutions (that is with a large number of spatial elements and small sampling time) than parameter regularization, and parameter regularization to better performance at coarse resolutions.

5. DISCUSSION AND CONCLUSIONS

Recall that in this study a so-called Tikhonov regularization is applied to the problem of combining model results with observations, in this way leading to predictions that are far better

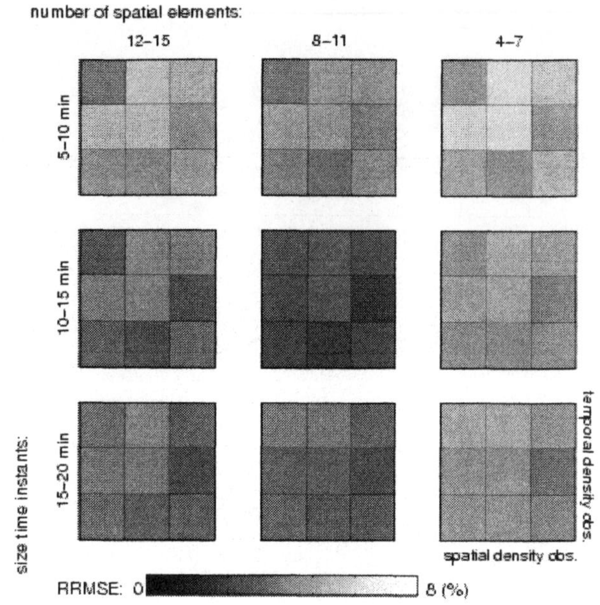

Fig. 4. RRMSE of discharge predictions by applying *state* regularization, for different resolutions and observation densities.

than the open-loop predictions with the same models. An objective and structured weighting of both components has been achieved through generalized cross-validation. Two regularization strategies were compared, i.e. parameter and state regularization. It was shown that both techniques lead to similar results, which differ in some details. Parameter regularization leads to better results at low data availability, whereas state regularization leads to better results at high data availability. This result is consistent with the fact that there are fewer parameters than states to be estimated. State regularization realizes its best predictions at finer resolutions than parameter regularization. This can be explained from the fact that parameter regularization maintains the *relative* structure of the kernel functions over different spatial units, and is in that way less flexible than state regularization. The comparative advantage of this flexibility appears at higher data availability and finer resolutions.

It is important to note that the methodology employed in this paper can relatively easily be adapted to different models or observations. If a hydrologic model can be written in state space form (according to (1) and (2)), the solution algorithm of Section 2 can be applied without any adjustments. This is an important asset since it means that different parameter distributed dynamic models can be evaluated with the same or different data for regularization relatively easily, without the requirement to include exactly the same state variables in the model dynamics. This point has been demonstrated by using a set of models instead of a single model to generate predictions. The technique used in this study only requires the solution of a series of independent constrained linear regression problems, for which there are numerous efficient solution algorithms available.

REFERENCES

G. Backus and F. Gilbert. The resolving power of gross earth data. *Geophys. J. R. Astr. Soc.*, 16: 169–205, 1968.

G. H. Golub, M. T. Heath, and G. Wahba. Generalized cross-validation as a method for choosing a good ridge parameter. *Technometrics*, 21:215–223, 1979.

P. C. Hansen. Analysis of discrete ill-posed problems by means of the l-curve. *SIAM Review*, 34: 561–580, 1992.

P. C. Hansen. *Rank-Deficient and Discrete Ill-Posed Problems: Numerical Aspects of Linear Inversion.* SIAM, Philadelphia, 1998.

W. Menke. *Geophysical Data Analysis: Discrete Inverse Theory*, volume 45 of *International Geophysiscs Series.* Academic Press, rev. edition, 1989.

C. R. Rao and S. K. Mitra. *Generalized Inverse of Matrices and its Applications.* John Wiley & Sons, 1971.

E. E. van Loon. *Overland Flow: Interfacing Models with Measurements.* PhD thesis, Wageningen University, 2001.

E. E. van Loon and K. J. Keesman. Investigating scale-dependent models: the case of overland flow at the hillslope scale. *Wat. Resour. Res.*, 36(1):243–255, 2000.

G. Wahba. Practical approximate solutions to linear operator equations when the data are noisy. *SIAM J. Numer. Anal.*, 14:651–667, 1977.

IFAC Publications
www.elsevier.com/locate/ifac

TIME–DELAY ESTIMATION OF A MANAGED RIVER REACH FROM SUPERVISORY DATA

Magalie Thomassin, * **Thierry Bastogne,** * **Alain Richard** *
and Antoine Libaux **

** Centre de Recherche en Automatique de Nancy (CRAN)*
CNRS UMR 7039
Université Henri Poincaré, Nancy 1, BP 239
54506 Vandœuvre–lés–Nancy Cedex, France
Phone: +33 (0) 3 83 68 44 61 – Fax: +33 (0) 3 83 68 44 62
`forename.name@cran.uhp-nancy.fr`
`http://www.cran.uhp-nancy.fr`
*** EDF CIH FCC, 73 373 Le Bourget–du–Lac Cedex, France*
Phone: +33 (0) 4 79 60 63 89
`antoine.libaux@edf.fr`

Abstract: The problem addressed in this article is the estimation of the time–delay between the inflow rate and the downstream water level of a managed river reach from data collected in imposed experimental conditions. The modelling of the managed river reach shows that the feedforward control performed by the operator "hides" the reach time–delay in the transfer function of the closed–loop system. Therefore, most of the classical time–delay estimation methods are inappropriate. A time–delay estimation approach devoted to managed rivers is proposed, in which a time–day description, composed of estimated impulse responses, allows to clearly observe the evolution of the time–delays over one year. In practice, this approach is particulary interesting since it is based on production data and hence does not disturb the daily management of the process by any experimental protocol. Moreover, the use of the suggested approach does not require any synthesis parameter. *Copyright © 2003 IFAC*

Keywords: Time–delay estimation, impulse responses, hydroelectric systems, process models.

1. INTRODUCTION

The problem addressed in this paper is the estimation of the time–delay between the inflow rate and the downstream water level of a managed river reach in a discrete–time context. The time–delay estimation errors can have serious consequences on the general performances of the control system and in particular on its stabilization. Several works dealing with modelling and identification of such processes have been already published (Cuno and Theobald, 1998; Litrico, 1999). However, few applications concern the time–delay estimations in this type of process (Bastogne *et al.*, 2002).

A theoretical approach consists in using *Barré de Saint–Venant*'s equations (1871) to define a theoretical hydrologic model of surface waves. But the variety of waves (kinematic, dynamic, diffusive, translation, etc.), added to that of the discretization techniques (Muskingum–Cunge, Crank–Nicholson, explicit methods, etc.) yield large variations of the results. Moreover, it is not only necessary to know *a priori* all the physical parameters of the river reaches (wetted area, wetted perimeter, width, length, etc.), but these parameters also must be constant in time and space. For these reasons, in practice, the time–delays are estimated empirically (operator knowledge) or experimentally (measurement of intumescences propagation time). But in both cases, the estimates are still characterized by a wide uncertainty. In this respect, identification methods aims to reduce this uncertainty on the time–delay estimates.

A frequently used identification method for discrete–time linear systems with unknown time–delay is based on a step–by–step search (Hsia, 1969). This two–step procedure (Elnaggar *et al.*, 1989) first assumes a known delay and estimates the other transfer function parameters, then minimizes a cost function with respect to the delay value. Another commonly used approach for estimating time–delays consists in using a general parametric model structure with an expanded numerator polynomial without time–delay (Kurz and Goedecke, 1981; De Keyser, 1986). This method is called the augmented B–parameters method. The time–delay is obtained by finding the first non–zero coefficient of the numerator. Correlation analysis provides an alternative method for the delay estimation. The idea of using the correlation analysis was proposed in (Isermann and Bauer, 1974) with a white noise input, and was taken up in (Zheng and Feng, 1990) where the delay is directly deduced from a generalized cross–correlation function. Other discrete–time delay estimation techniques exist, e.g. the ones based on specific input signals. Nevertheless, generally for economic reasons and safety precautions (flood risk), they cannot be used in this application. Herein, the estimation data set is daily collected from the nominal operation of the river reach. Accordingly, the problem amounts to give an estimate of delays and its evolution over one year of daily measurements.

For almost all existing time–delay estimation methods, the system is described by an open–loop discrete–time model with an input delay. As a consequence, the time–delay estimation amounts to find the first non null parameter of the augmented numerator polynomial or of the impulse response. In this application, available data are measured in a closed–loop context, the system being control by a human operator. The modelling of the managed river reach shows that the feedforward control performed by the operator "hides" the reach time–delay in the transfer function of the closed–loop system. The first coefficients of the impulse response are not equal to zero. So, most of the classical time–delay estimation methods are inappropriate. Nevertheless, as shown in this article, a two–step procedure based on a cost function of the time–delay can be used, but is not conclusive. Consequently, a nonparametric time–delay estimation approach devoted to managed rivers is proposed.

The paper is organized as follows. Section 2 is devoted to the managed river reach modelling. In section 3, two time–delay estimation methods are presented. The first method is a two–step procedure based on an output error approach. The second one is based on the correlation analysis to estimate the process impulse response from which the time–delay is estimated. A time–day description of these estimated impulse responses is then presented. This original description allows to clearly observe the time–delay evolution over one year. Finally, the application results are analyzed and compared with empirical estimates in section 4.

2. THE RIVER REACH PROCESS

2.1 Description

The system is a managed river reach producing hydroelectric power. As depicted in figure 1, it is composed of a river portion with upstream and downstream barrages. The downstream level of the reach $z(k)$ [1] is the controlled variable. The control variable is the outflow rate reference $F_o^*(k)$. The inflow rate reference $F_i^*(k)$ is considered as a known disturbance. Note that the inflow and outflow rates are not measured. The outflow rate reference $F_o^*(k)$ is given by a human operator (manual control mode) via a *supervisory control station*. The process parameters are supposed to be constant over one day. A one–day data set is presented in figure 2. Tables 1 and 2 give the main process variables and notations used in this article.

In this application, the system is composed of a series of four managed river reaches which exploit the hydraulic resources of the 'Basse–Isère' river (France). In this study, the application of two time–delay estimation methods to one river reach is developed. The application results for the four river reaches are given in section 4.

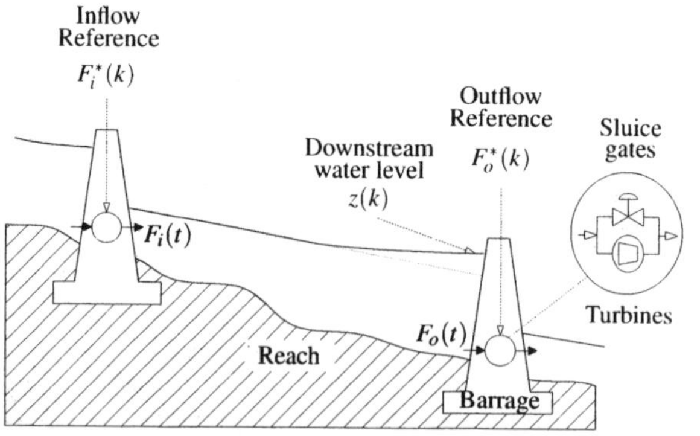

Fig. 1. A managed river reach

2.2 Managed river reach modelling

Barrage model. A barrage is composed of a dam (sluices) and a hydroelectric power plant (turbines).

---

[1] The discrete–time variables are noted $x(k)$ and correspond to the temporal sampling with a constant sampling period $T_s = 133s$ of the continuous–time variable $x(t)$: $x(k) = x(kT_s)$.

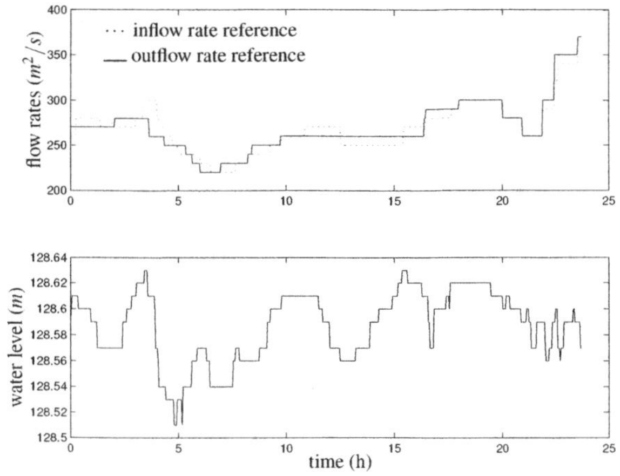

Fig. 2. One–day estimation data set

Table 1. Main process variables.

Reference	Description
$F^*(k)$	Flow rate reference of a barrage (discrete–time)
$F(t)$	Real flow rate of a barrage (continuous–time)
Subscript i	Variable of the upstream barrage (inflow)
Subscript o	Variable of the downstream barrage (outflow)
$z(t)$ and $z(k)$	Downstream water level
$z^*(k)$	Water level reference (discrete–time)

Table 2. Main notations.

Reference	Description
q^{-1}	Shift operator
$x(s)$	Laplace transform of the signal x
δx	Modelling errors and measurement uncertainties of x
$\hat{x}$	Estimate of x

The response time of its flow control loop is negligible as compared to the sampling period T_s (133s). Consequently, its loop may be modelled by a constant gain and a zero–order–hold (ZOH):

$$F(s) = K\frac{(1-e^{-T_s s})}{s}F^*(s) + \delta F(s), \quad (1)$$

where K corresponds to the static gain of the control flow loop.

Reach model. The *Barré de Saint–Venant*'s equations have always given a correct estimate of the water level as a function of time and distance (Hervouet and Péchon, 1991). Nevertheless, these equations are too complex and computationally expensive only to estimate the time–delays. However, to describe the dynamic behaviour of the water level at a given distance of the reach, a solution consists in reducing and linearizing the previous equations around an operating point (Cuno and Theobald, 1998; Litrico, 1999). As the measurement of the water level is filtered, the propagation and reflection effects of waves can be neglected. Consequently, in the absence of affluent, the water level is mainly determined by the volume conditions as shown in equation (2):

$$z(s) = \frac{1}{As}\left(e^{-\tau_i s}F_i(s) - e^{-\tau_o s}F_o(s)\right) + \delta z(s), \quad (2)$$

where A is the water surface of the reach (m^2). τ_i denotes the unknown time–delay between the inflow rate and the water level, while the delay between the outflow rate and the water level is represented by τ_o. Note that both are time–varying parameters.

2.3 Level control modelling

In the supervisory control station, the operator adjusts the outflow rate reference in order to control the water level of the reach and reject the disturbances due to the inflow rate variations. Consequently, the operator action on the outflow rate reference can be approximatively described by two controllers:

$$F_o^*(s) = C_f(s)F_i^*(s) + C_z(s)\left[z^*(s) - z(s)\right] + \delta F_o^*(s) \quad (3)$$

where $C_f(s)$ and $C_z(s)$ represent respectively the feedforward control between inflow and outflow rate references and the feedback control between the water level and the outflow rate reference; they are defined by:

$$C_f(s) = K_f e^{-\tau_f s} \quad \text{and} \quad C_z(s) = K_z, \quad (4)$$

where K_f and K_z are the proportional gains and τ_f denotes the time–delay of the controller C_f. Theoretically, τ_f is not equal to zero and has to be chosen in function of the time–delay τ_i in specifical tables devoted to the river reach control.

2.4 Discrete–time modelling of the process

The previous models have been put together to form the block diagram depicted in figure 3. Assuming that time–delays are integer multiples of the sampling period:

$$\tau_j = n_j T_s \text{ with } j \in \{f, i, o\}, \quad (5)$$

where n_f, n_i and n_o denote delay indices ($\in \mathbb{N}$), the continuous–time model of the water level can be approximated by the following discrete–time model:

$$\begin{aligned} z(k) = {} & L_{z,z^*}(q^{-1})z^*(k) + L_{z,\delta z}(q^{-1})\delta z(k) \\ & + L_{z,\delta F_o}(q^{-1})\delta F_o(k) + L_{z,\delta F_i}(q^{-1})\delta F_i(k) \\ & + L_{z,\delta F_o^*}(q^{-1})\delta F_o^*(k) + L_{z,F_i^*}(q^{-1})F_i^*(k). \end{aligned} \quad (6)$$

The level reference $z^*(k)$ is equal to zero due to the disturbance rejection control of the system. Note that, in this equation, only the signals $F_i^*(k)$ and $z(k)$ are known. The parameter to estimate is n_i, representing the time–delay between the inflow rate reference ($F_i^*(k)$) and the downstream water level ($z(k)$). It appears in the polynomial rational transfer $L_{z,F_i^*}(q^{-1})$ describing the dynamical behaviour between $z(k)$ and $F_i^*(k)$ of the closed–loop system:

$$L_{z,F_i^*}(q^{-1}) = \frac{\Gamma_i(q^{-1}) + C_f(q^{-1})\Gamma_o(q^{-1})}{1 + C_z(q^{-1})\Gamma_o(q^{-1})}, \quad (7)$$

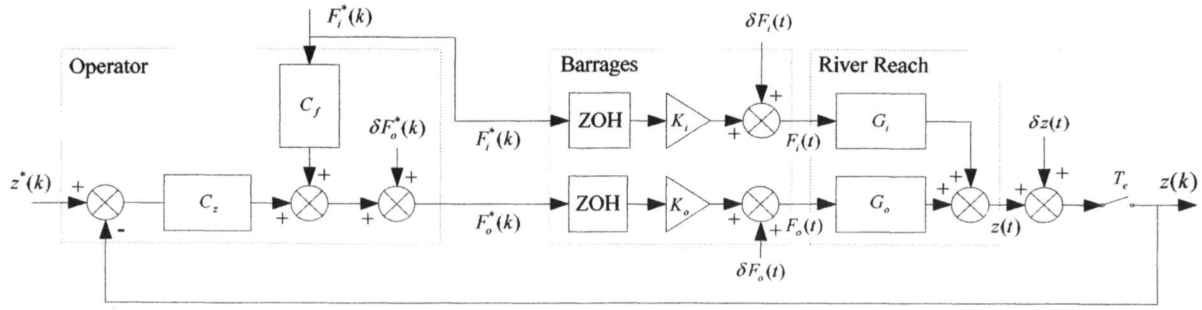

Fig. 3. Block diagram of the managed river reach control

with:

$$C_f(q^{-1}) = K_f q^{-n_f}, \tag{8}$$

$$C_z(q^{-1}) = K_z, \tag{9}$$

$$\Gamma_i(q^{-1}) = \frac{K_i T_s}{A} \frac{q^{-n_i-1}}{1-q^{-1}}, \tag{10}$$

$$\Gamma_o(q^{-1}) = \frac{-K_o T_s}{A} \frac{q^{-n_o-1}}{1-q^{-1}}. \tag{11}$$

In this application, since the water level measurement station is close to the downstream barrage, n_o is fixed to zero. Furthermore, the operator usually tends to reject the disturbance effects by acting on the control variable without waiting for its effect on the water level, so it is assumed that $n_f = 0$. Thus, the rational transfer (Eq. 7) becomes:

$$L_{z,F_i^*}(q^{-1}) = \frac{\alpha_i q^{-n_i-1} - \alpha_o K_f q^{-1}}{1-(1+\alpha_o K_z)q^{-1}}, \tag{12}$$

with $\alpha_i = K_i T_s / A$ and $\alpha_o = K_o T_s / A$, the coefficients α_i, α_o, K_f and K_z being not null. From this transfer, it is easy to note that the time–delay between $F_i^*(k)$ and $z(k)$ of the closed–loop system is equal to 1 and is different of n_i. The time–delay n_i does not correspond to the one between $F_i^*(k)$ and $z(k)$ in a closed–loop context because of the feedforward controller. So, this prevents us from successfully estimating n_i in a closed–loop context with classical time–delay estimation methods.

3. PRESENTATION OF TWO TIME–DELAY ESTIMATION METHODS

3.1 An output error approach

In the output error approach, an estimate of the time–delay is the solution of the following minimization problem:

$$\hat{n}_i = \arg \min_{\{n_i \in \mathbb{N};\ 0 \le n_i \le n_{i\max}\}} \left[\sum_k (z(k) - \hat{z}(k|n_i, \hat{\theta}))^2 \right],$$

where $\hat{\theta} = [\widehat{\alpha_i} \quad \widehat{\alpha_o K_f} \quad \widehat{\alpha_o K_z}]^T$ is the estimated parameter vector and

$$\hat{z}(k|n_i, \hat{\theta}) = \frac{\widehat{\alpha_i} q^{-n_i-1} - \widehat{\alpha_o K_f} q^{-1}}{1-(1+\widehat{\alpha_o K_z})q^{-1}} F_i^*(k). \tag{13}$$

This method is applied to one–day data sets supplied over one year and for $n_{i\max} = 30$ (*a priori* knowledge). Results are illustrated by a histogram in figure 4. They show that in most cases $\hat{n}_i$ is greater than 30; what is in contraction with the empirical knowledge. Therefore, this method is not conclusive. A solution consists in using a method which takes the modelling errors into account. However, parameterizing the modelling errors increases the parameters number and, as a consequence, the method becomes computationally expensive. So, a simpler method devoted to the time–delay estimation of a river reach is proposed in the next section.

3.2 The correlation analysis

Because of the uncertainty about the model structure of the process, it is proposed to use a nonparametric estimation method: the *correlation analysis* (Ljung, 1999; Söderström and Stoica, 1989) in order to estimate a truncated discrete–time impulse response. This estimation is directly obtained from the normalized cross–correlation function between the input and the output of the system. In a first step, a pre–whitening filter is applied to equation (6) to give:

$$\begin{aligned} z_f(k) = {} & L_{z,z^*}(q^{-1}) z_f^*(k) + L_{z,\delta_z}(q^{-1}) \delta z_f(k) \\ & + L_{z,\delta F_o}(q^{-1}) \delta F_{o,f}(k) + L_{z,\delta F_i}(q^{-1}) \delta F_{i,f}(k) \\ & + L_{z,\delta F_o^*}(q^{-1}) \delta F_{o,f}^*(k) + L_{z,F_i^*}(q^{-1}) F_{i,f}^*(k), \end{aligned} \tag{14}$$

with:

$$[z_f(k), z_f^*(k), \delta z_f(k), \delta F_{o,f}(k), \delta F_{i,f}(k), F_{i,f}^*(k)] = \widehat{A}_i(q^{-1})[z(k), z^*(k), \delta z(k), \delta F_o(k), \delta F_i(k), F_i^*(k)],$$

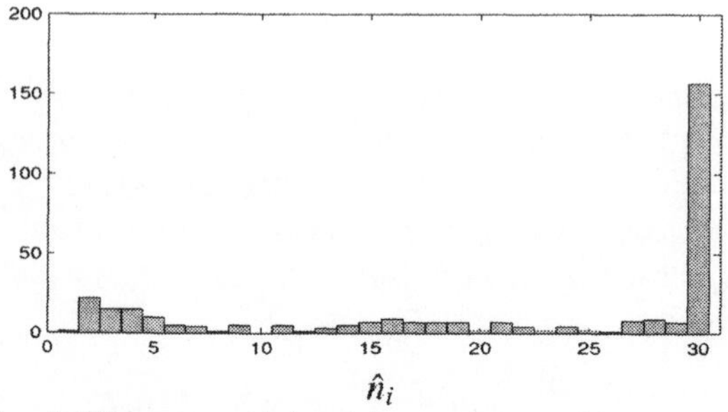

Fig. 4. Histogram of the delay estimates $\hat{n}_i$.

where $\widehat{A}_i(q^{-1})$ represents an estimate of the whitening filter corresponding to an autoregressive model of $F_i^*(k)$:

$$A_i(q^{-1})F_i^*(k) = e(k), \tag{15}$$

where $e(k)$ is a stationary zero–mean white noise sequence with variance σ^2. By multiplying equation (14) by $F_{i,f}^*(k-l)$ where l is the discrete–time lag, then by taking the expectation of each term, and under the hypothesis that the inputs are mutually uncorrelated, equation (14) becomes:

$$\begin{aligned} R_{z_f,F_{i,f}^*}(l) &= L_{z,F_i^*}(q^{-1})R_{F_{i,f}^*,F_{i,f}^*}(l) \\ &= \sum_{j=0}^{\infty} h_{z,F_i^*}(j)R_{F_{i,f}^*,F_{i,f}^*}(l-j), \end{aligned} \tag{16}$$

where $R_{x,y}$ denotes the cross–correlation function between the signals x and y. $h_{z,F_i^*}(l)$ is the impulse response between z and F_i^* at time l. Since $F_{i,f}^*$ is a white noise signal, then:

$$R_{F_{i,f}^*,F_{i,f}^*}(l) = \begin{cases} \sigma^2 & l = 0, \\ 0 & l \neq 0. \end{cases} \tag{17}$$

Therefore, from equations (16) and (17) the lth coefficient of the impulse response is given by:

$$h_{z,F_i^*}(l) = R_{z_f,F_{i,f}^*}(l)/\sigma^2. \tag{18}$$

So, an estimate of the impulse response is obtained from an estimate of $\widehat{R}_{z_f,F_{i,f}^*}(l)$.

An example of a theoretical impulse response (IR) for the model given by equation (12) is presented in figure 5 with $n_i = 7$, $\alpha_i = 1,2.10^{-4}$, $\alpha_o \cdot K_f = 5,32.10^{-5}$ and $\alpha_o \cdot K_z = -0,16$. This figure also shows another way to represent the IR, the *daily impulse response bar*. The value of each IR coefficient is represented by a grey level. This representation will be interesting for what comes next. This figure permits us to see that the presence of the time–delay n_i introduces a discontinuity in the IR. So, a delay estimation can be obtained from the detection of the first point of this discontinuity. Figure 6 presents the impulse response estimated with the correlation analysis from the daily measured signals z and F_i^*. To examine the evolution and the variation of the delay index estimates over one year, a new form of representation of the impulse responses is proposed: the *time–day representation* (TDR). It consists in piling up the M daily impulse response bars, as shown in figure 7, where M denotes the number of one–day data sets supplied over one year. One can easily observe that the transition between 7 and 9, so the time–delay estimates varies between 6 and 8.

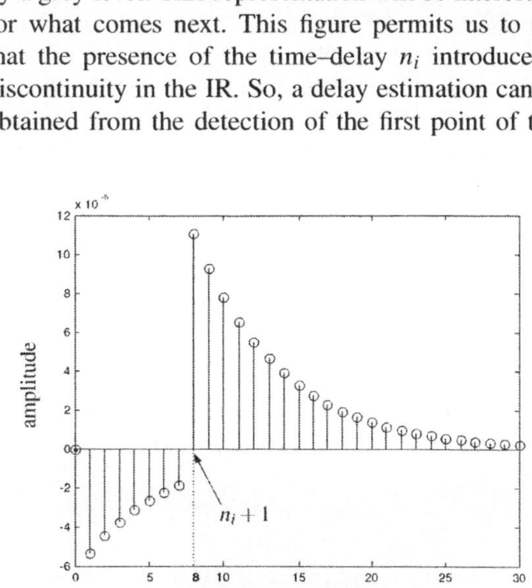

Fig. 5. Theoretical impulse response between z and F_i^* and its *daily impulse response bar*.

4. ANALYSIS OF THE RESULTS

Now, the relevance of the application results should be assessed by comparing the time–delay estimates to

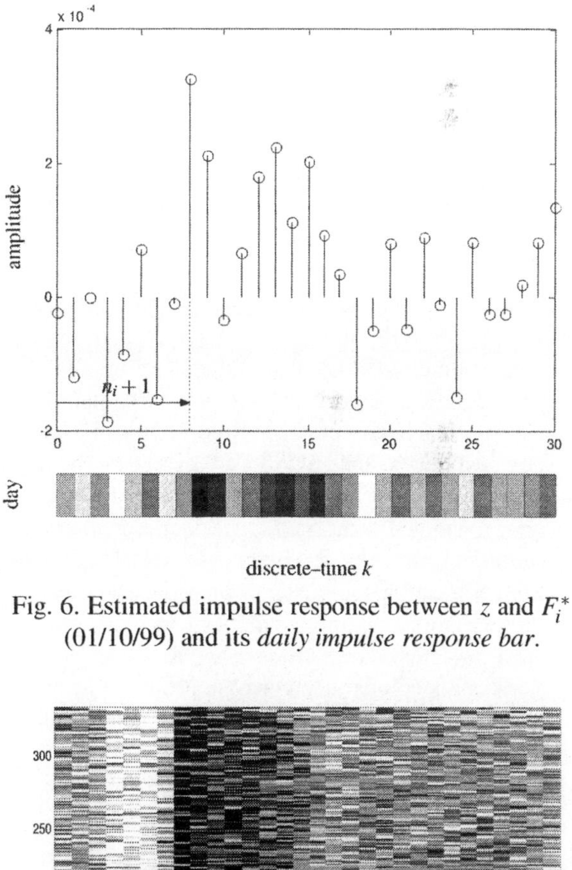

Fig. 6. Estimated impulse response between z and F_i^* (01/10/99) and its *daily impulse response bar*.

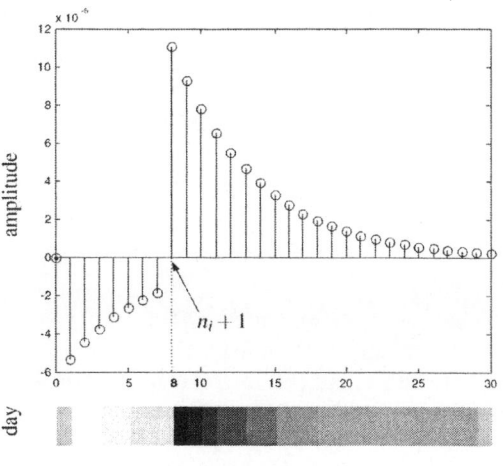

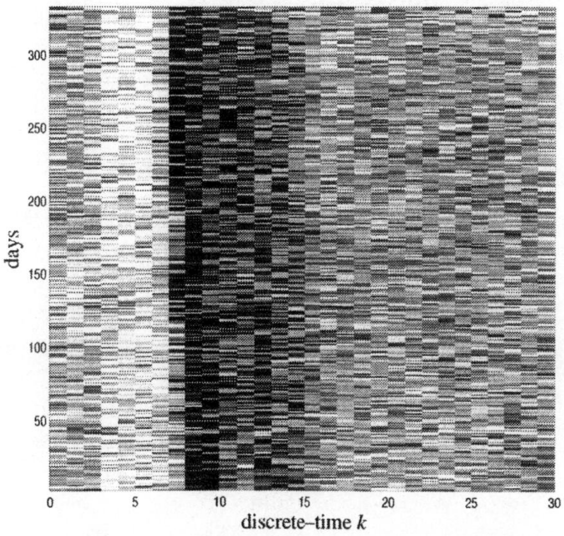

Fig. 7. Time–day representation of the impulse responses

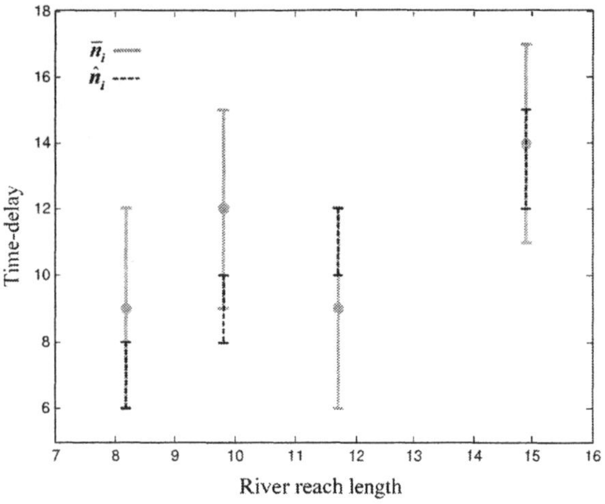

Fig. 8. Empirical time–delays ($\bar{n}_i$) and estimated time–delays ($\hat{n}_i$) variation intervals as a function of the river reach length

physical criteria and to previous empirical estimates of time–delays.

With a constant flow, the delay is proportional to the river reach length. Of course, in our case, the flow is not constant but its limited influence makes the river reach length a rather reliable indicator allowing to evaluate the relevance of the time–delay estimates. Figure 8 represents time–delay variation intervals (estimated from TDR) as a function of river reaches lengths for four different river reaches. It reveals a proportionality between the interval variations of the time–delays estimated by our proposed method and the river reach length; what reinforces the relevance of these estimates. This figure also represents the uncertainty intervals of the delays estimated from empirical knowledge and used by the operators for the manual control of the river reach process, which are distant from a linear characteristic besides. Finally, the figure shows significant improvement of the new estimates variation intervals which were at least divided by two. The results comparison corroborates the relevance of the estimates provided by the experimental approach developed in this article.

5. CONCLUSION

The problem addressed in this paper is the estimation of time–delays between the inflow rate and the downstream water level of a managed river reach from data sets collected in experimental conditions imposed by production modalities. In this application, it is shown that the feedforward control performed by the operator "hides" the reach time–delay in the transfer function of the closed–loop system. Consequently, classical time–delay estimation methods are inappropriate. A three–step time–delay estimation approach devoted to managed rivers is proposed. Firstly, a theoretical impulse response is analyzed to point out an estimate of the time–delay. Secondly, an impulse response is estimated by a correlation analysis for each one–day data set. Finally, a time–day description, formed by the previous estimated impulse responses, allows to clearly observe the time–delay evolution over one year. In practice, this approach is particulary interesting since it is based on production data and hence does not disturb the daily management of the process by any experimental protocol. Moreover, the proposed approach is easily to use since it is based on a nonparametric estimation method.

REFERENCES

Bastogne, T., M. Thomassin, H. Garnier, A. Richard and A. Libaux (2002). Modélisation expérimentale de biefs de rivière : une estimation appropriée des retards. In: *Proceedings of the 2ème Conférence Internationale Francophone d'Automatique*. Nantes, France.

Cuno, B. and S. Theobald (1998). The relationship between control requirements, process complexity and modelling effort in the design process of river control systems. *Mathematics and Computers in Simulation* **46**, 611–619.

De Keyser, R. M. C. (1986). Adaptive dead–time estimation. In: *Proceedings of the 2nd IFAC Workshop on Adaptative Systems in Control and Signal Processing*. Lund, Sweden. pp. 385–389.

Elnaggar, A., G. A. Dumont and A. L. Elshafei (1989). Recursive estimation for systems of unknown delay. In: *Proceedings of the 28th IEEE Conference on Decision and Control*. Tampa, Florida, USA. pp. 1809–1810.

Hervouet, J. M. and P. Péchon (1991). Modélisation numérique des écoulements à surface libre. L'état de l'art. *La Houille Blanche* **2**, 96–106.

Hsia, T. C. (1969). A discrete method for parameter identification in linear systems with transport lags. *IEEE Transactions Aerospace and Electronic Systems* **AES–5**, 236–239.

Isermann, R. and U. Bauer (1974). Two–step process identification with correlation analysis and least–squares parameter estimation. *Transaction of the ASME, Series G* **96**, 426–432.

Kurz, H. and W. Goedecke (1981). Digital parameter adaptative control of processes with unknown dead time. *Automatica* **17**, 245–252.

Litrico, X. (1999). Modélisation, identification et commande robuste de systèmes hydrauliques à surface libre. Phd thesis. ENGREF. Montpellier, France.

Ljung, L. (1999). *System identification: theory for the user*. 2nd ed.. Prentice Hall.

Söderström, T. and P. Stoica (1989). *System identification*. University Press, Cambridge, Prentice Hall.

Zheng, W.-X. and C.-B. Feng (1990). Identification of stochastic time lag systems in the presence of colored noise. *Automatica* **26**(4), 769–779.

IFAC Publications
www.elsevier.com/locate/ifac

GEOHYDROLOGICAL APPLICATION OF A NONLINEAR PHYSICALLY BASED TIME SERIES MODEL

W.L. Berendrecht * **A.W. Heemink** * **F.C. van Geer** **
J.C. Gehrels ***

* *Department of Applied Mathematical Analysis, Faculty of Electrical Engineering, Mathematics and Computer Science, Delft University of Technology, P.O. Box 5031, 2600 GA Delft, the Netherlands*
** *Netherlands Institute of Applied Geoscience TNO, National Geological Survey, P.O. Box 80015, 3508 TA Utrecht, the Netherlands*
*** *Department of Watermanagement, Faculty of Civil Engineering and Geosciences, Delft University of Technology, P.O. Box 5048, 2600 GA Delft, the Netherlands*

Abstract: This paper presents a physically based time series model that relates groundwater level fluctuations to precipitation and evapotranspiration. The model is based on the nonlinear relation between the degree of water saturation of the subsoil and the groundwater recharge. The model is written in state-space form, while the extended Kalman filter is used to estimate the state equation. An example application shows that the model performs very well and that the addition of physical knowledge is a valuable extension to standard linear transfer function-noise models. *Copyright © 2003 IFAC*

Keywords: Extended Kalman filters, Maximum likelihood estimators, Nonlinear systems, Physical models, State-space models, Time series analysis, Transfer functions

1. INTRODUCTION

During the past decades, time series models and especially linear transfer function-noise (TFN) models (Box and Jenkins, 1970) have frequently been applied in the field of hydrology (Gehrels *et al.*, 1994; Hipel and McLeod, 1994). In geohydrology, TFN modeling is mainly used to separate natural influences (often represented by precipitation and evapotranspiration) from human influences (e.g. groundwater withdrawal) in time series of groundwater levels.

A disadvantage of these models is that the relation between input and output is assumed to be linear. This assumption is often not satisfied in geohydrological applications. In fact, the temporal variation of groundwater recharge not only depends on the input variables, but also on the degree of water saturation of the subsoil. The degree of saturation determines both the downward flux and the evapotranspiration.

This paper presents a model that incorporates the degree of saturation to calculate the fluctuations of the groundwater level. The nonlinear model is based on well-known physical concepts. In order to calibrate the model to a time series, it is written in state-space form. The state equation is then combined with an extended Kalman filter. Subsequently, model parameters are estimated using a maximum likelihood criterion.

The presented model is tested on a time series of groundwater level data observed in the Netherlands.

The results show that the physically based model clearly performs better than a linear TFN model. An additional advantage of the physically based model is that some of the parameters can be obtained directly from field data, whereas the estimates of the other parameters can be compared with the expected range of these parameters.

The paper is organized as follows. In section 2 the physical model is presented. Section 3 rewrites this model in state-space form and describes how the parameters are estimated. Section 4 uses a real-world example to test the model. Here, the results are compared with a linear TFN model. Finally, section 5 gives some concluding remarks.

2. CONCEPTUAL GROUNDWATER MODEL

A widely applied schematization of a groundwater system is the reservoir model as illustrated in Figure 1. The first (upper) reservoir is the root zone. Here, the input variables enter the system. The second reservoir is the percolation zone, describing the downward flux from the root zone to the third reservoir, the saturated zone. This last reservoir is the lower boundary of the model and represents the observed groundwater level.

2.1 *Root zone*

The main parameter of the root zone is the effective degree of water saturation S_e [-], $0 < S_e \leq 1$, described by the following differential equation:

$$\frac{\mathrm{d}S_e}{\mathrm{d}t} = \frac{1}{D_e}(P_e - E_a - R_p). \tag{1}$$

Here, D_e is the effective thickness of the root zone [L]. P_e is the net precipitation [LT^{-1}], which is assumed

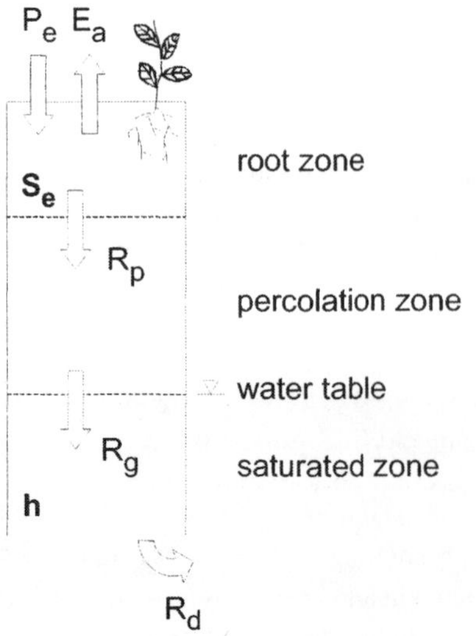

Fig. 1. Schematization of groundwater system

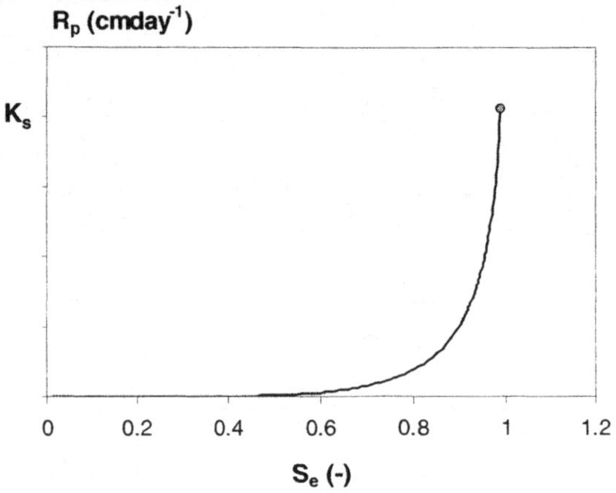

Fig. 2. A typical R_p-S_e relation, with $\lambda = -1$ and $m = 0.3$

to be related to the observed gross precipitation P_G [LT^{-1}] as

$$P_e = f_i P_G, \tag{2}$$

where f_i is an interception factor [-]. E_a is the actual evapotranspiration [LT^{-1}], depending on the saturation and the observed reference evapotranspiration E_r [LT^{-1}] following the relation

$$E_a = S_e E_p = S_e f_c E_r, \tag{3}$$

where E_p is the potential evapotranspiration [LT^{-1}], and f_c is an empirical crop factor [-] relating the rate of evapotranspiration to the type of vegetation. R_p is a downward flux [LT^{-1}] connecting the root zone and the percolation zone. Assuming that the pressure head remains constant with depth, R_p is related to the degree of saturation via the following nonlinear equation (Mualem, 1976):

$$R_p = K_s S_e^{\lambda} \left[1 - \left(1 - S_e^{1/m}\right)^m\right]^2, \tag{4}$$

where the constant K_s denotes the saturated hydraulic conductivity [LT^{-1}] and λ and m are empirical shape factors [-]. A typical curve of this R_p-S_e relation is given in Figure 2.

2.2 *Percolation zone*

Basically, the percolation zone redistributes the incoming flux R_p into an outgoing flux R_g [LT^{-1}]. This redistribution is calculated using a transfer function that is based on the general form of a convolution integral (Gehrels, 1999):

$$R_g(t) = \int_0^{\infty} R_p(t-\tau)F(\tau)\mathrm{d}\tau, \tag{5}$$

where F is a transfer function representing the percolation zone. In many practical applications, F has the form of a convection-dispersion equation.

2.3 Saturated zone

The saturated zone is fed from the percolation zone with a flux R_g. The lower boundary consists of a drainage flux R_d [LT^{-1}], which is assumed to have a linear relation with the groundwater level h [L]:

$$R_d = \frac{h - z_r}{\gamma}, \tag{6}$$

where z_r is the drainage base level [L] and γ is the drainage resistance [T]. The groundwater fluctuation is then described by the following reservoir function (Knotters and Bierkens, 2000):

$$\varphi \frac{\mathrm{d}h}{\mathrm{d}t} = R_g - R_d, \tag{7}$$

where φ is a storage coefficient [-].

In many situations, large-scale withdrawal of groundwater has a significant influence on the groundwater level. The drawdown at the observation point, s_{obs} [L], is written as

$$s_{obs} = \beta s_c, \tag{8}$$

where β is a correction factor [-], and s_c is the cumulative effect of all groundwater abstractions [L]: $s_c = \sum_i s_i$. The equation for s is given in the appendix. It is important to notice that this equation is only an approximation of the real drawdown.

Assuming that the processes in the saturated zone are linear, the cumulative drawdown can be superimposed on the groundwater level:

$$z = h - s_{obs}, \tag{9}$$

where z is the observed groundwater level [L].

3. METHODOLOGY

3.1 State-space model

In order to calibrate the described model to a groundwater time series, it is rewritten in the following nonlinear stochastic difference equation (Jazwinsky, 1970):

$$\mathbf{x}_t = \mathbf{f}\left[\mathbf{x}_{t-1}, \mathbf{u}_t\right] + \mathbf{G}\mathbf{w}_t. \tag{10}$$

The terms of this equation will be discussed separately.

The vector $\mathbf{u}_t$ represents the system input and is written as

$$\mathbf{u}_t^{\mathrm{T}} = \left[\, P_{G,t} \;\; E_{r,t} \;\; s_{d,t} \,\right], \tag{11}$$

with $s_{d,t} = s_{c,t} - s_{c,t-1}$.

The nonlinear function $\mathbf{f}$ relates the state at the previous time step $t-1$ to the state at the current time step t. The first state of equation 10 represents the root zone: $x_1 = S_e$. Before rewriting equation 1 into an explicit difference equation, one needs to bound x_1 between 0 and 1. This is because equation 4 only has a solution for $0 < S_e \leq 1$. Also it prevents that the actual evapotranspiration as given in equation 3 becomes larger than the potential evapotranspiration. Therefore, the following function is introduced, which is continuous and differentiable in $\mathbb{R}$:

$$\tilde{x}_1 = \begin{cases} 0.05\exp\left(20x_1 - 1\right), & x_1 < 0.05 \\ 1 - 0.05\exp\left(19 - 20x_1\right), & x_1 > 0.95 \\ x_1, & \text{otherwise.} \end{cases} \tag{12}$$

The first element of the state equation is then written as

$$x_{1,t} = x_{1,t-1} + \\ + \frac{1}{D_e}\left(f_i P_{G,t} - f_c E_{r,t}\tilde{x}_{1,t-1} - R_p^{t-1}\right), \tag{13}$$

with

$$R_p^{t-1} = K_s \tilde{x}_{1,t-1}^{\lambda}\left[1 - \left(1 - \tilde{x}_{1,t-1}^{1/m}\right)^m\right]^2. \tag{14}$$

The percolation-zone equation is approximated by a coupled set of ARMA-type equations, with $x_4 = R_g$:

$$\begin{bmatrix} x_{2,t} \\ x_{3,t} \\ x_{4,t} \end{bmatrix} = \begin{bmatrix} \delta_1 x_{2,t-1} + (1-\delta_1) R_p^{t-1} \\ \delta_1 x_{3,t-1} + (1-\delta_1) x_{2,t-1} \\ \delta_1 x_{4,t-1} + (1-\delta_1) x_{3,t-1} \end{bmatrix}, \tag{15}$$

where δ_1 is an autoregressive parameter [-]: $0 \leq \delta_1 < 1$. At every time step an impulse of the flux from the root zone R_p enters at x_2 and is damped while passing through the states. Notice that the gain of each equation is equal to 1. Obviously, the number of states can be reduced or extended depending on the length of the percolation zone, but in most practical applications satisfactory results have been obtained with the use of three states.

Equations 7 and 8, describing the processes in the saturated zone, are rewritten in the following vector notation:

$$\begin{bmatrix} x_{5,t} \\ x_{6,t} \end{bmatrix} = \begin{bmatrix} \delta_2 x_{5,t-1} + \omega x_{4,t-1} \\ x_{6,t-1} + \beta s_d \end{bmatrix}, \tag{16}$$

with $x_5 = h$, $x_6 = s_{obs}$, $\delta_2 = \exp(-1/\varphi\gamma)$, and $\omega = \gamma(1-\delta_2)$.

The system noise is represented by the vector $\mathbf{w}_t \simeq N(0, \mathbf{Q})$, with

$$\mathbf{Q} = \begin{bmatrix} q_{11} & 0 \\ 0 & q_{22} \end{bmatrix}. \tag{17}$$

The matrix $\mathbf{G}$ relates q_{11} and q_{22} to S_e and h, respectively:

$$\mathbf{G}^{\mathrm{T}} = \begin{bmatrix} 1 & 0 & 0 & 0 & 0 & 0 \\ 0 & 0 & 0 & 0 & 1 & 0 \end{bmatrix}. \tag{18}$$

The state $\mathbf{x}_t$ is not observed directly. Instead, the measurement z_t is observed, which is a function of $\mathbf{x}_t$ (see equation 9) and the measurement noise v_t:

$$z_t = \mathbf{c}\mathbf{x}_t + z_r + v_t, \tag{19}$$

where $\mathbf{c} = [0,0,0,0,1,-1]$, and v_t is a zero-mean white Gaussian noise process with variance r.

3.2 State and parameter estimation

The system is estimated with the well-known extended Kalman filter (EKF) (Jazwinsky, 1970). The EKF linearizes around the previous state estimate at each time instant. This approximation is only useful for weakly nonlinear models. Verlaan and Heemink (2001) propose a measure for nonlinearity, expressed by the nondimensional number

$$V \equiv \sqrt{N^{-1}\mathbf{b}^{\mathrm{T}}\mathbf{P}^{-1}\mathbf{b}}, \tag{20}$$

where N is the number of observations, $\mathbf{b}$ is the bias for EKF, and $\mathbf{P}$ is the error covariance matrix. It is likely that the bias is insignificant if $V \ll 1$.

Estimation of the parameter set α is based on the innovations n_t and innovation variance f_t obtained with the EKF, using the following maximum likelihood criterion (Schweppe, 1973).

$$\log L(N;\alpha) = -\frac{N}{2}\log 2\pi - \frac{1}{2}\sum_{t=1}^{N}\log f_t(\alpha) - \frac{1}{2}\sum_{t=1}^{N}\frac{n_t^2(\alpha)}{f_t(\alpha)}. \tag{21}$$

where N is the number of observations. The covariance matrix of parameter estimation errors is estimated using the Cramer-Rao lower bound (Schweppe, 1973).

4. AN EXAMPLE APPLICATION

4.1 Description of data set

The model is tested on a groundwater time series (period 1960-1999, $N = 819$) obtained from an observation well (code 27cp0002) in the center of the Netherlands. The observation frequency is two observations per month. Since an accurate modeling of the root zone requires a small modeling interval, daily observations of the input variables are used. Gross precipitation is obtained by spatial interpolation of observations from surrounding meteorological stations, whereas the reference evapotranspiration is obtained from the main meteorological station of the Royal Netherlands Meteorological Institute at De Bilt.

4.2 Model calibration

First, the performance of the EKF is tested by evaluating the nonlinearity of the model (based on V). Several runs of the model show that $V \ll 1$ only if $q_{11} \ll 1$. The reason for this is twofold. First, the noise works on x_1 and is passed forward through the nonlinear system and is thus linearized. Second, the modeling interval is 1 day, while the average interval between the observations is 14 days. Consequently, there are not enough observations available to control the linearization, which means that the time update diverges. Since for computational reasons the EKF is preferable to more accurate filters, the noise parameter q_{11} is not estimated but fixed at a sufficiently small value ($q_{11} = 1 \times 10^{-9}$).

The interception factor f_i and crop factor f_c are determined by the land use. For location 27cp0002 the dominant land use is mixed forest. Corresponding factors for this type of land use are:

$$f_i = 0.80,$$
$$f_c = 0.72.$$

All other 11 parameters are estimated and given in table 1.

Table 1. Calibrated parameters of physically based model with associated first-order estimates of standard deviation

	Estimated value	Standard deviation
D_e, cm	50.3	6.69
K_s, cmday^{-1}	0.809	0.209
λ	-4.63	0.462
m	0.333	0.0411
δ_1	0.663	0.0333
ω, day	4.57	0.162
δ_2	0.999	0.0000738
β	0.803	0.0931
z_r, cm	506	26.8
q_{22}, cm^2	0.283	0.0172
r, cm^2	1.21	0.146

From a physical point of view, the order of magnitude of the calibrated parameters and of the derived parameter values (drainage resistance $\gamma = 3515$ days and storage coefficient $\varphi = 0.22$) is realistic.

4.3 Results and comparison with linear model

Figure 3 gives the prediction of the groundwater level for the calibration period, using the calibrated parameters. Besides, the figure shows the prediction based on a linear transfer function-noise (TFN) model. This type of linear model has been used for many years in the analysis of groundwater time series and predicts the groundwater level through a linear relation between precipitation excess ($P_e - E_p$) and the groundwater level. The difference between the TFN model and the physically based model is that the first does

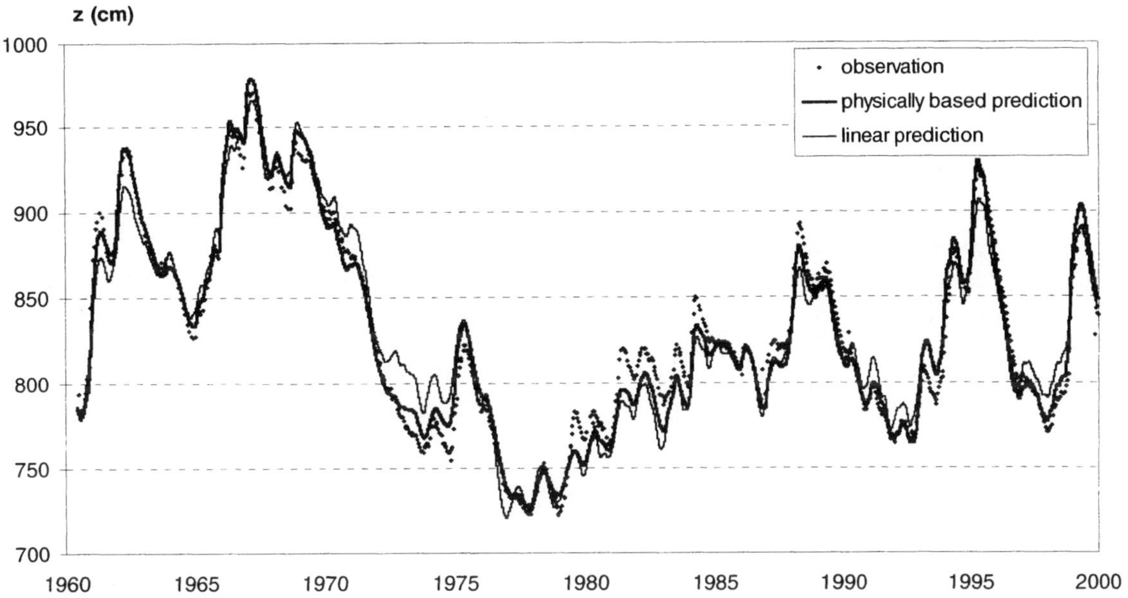

Fig. 3. physically based (nonlinear) and linear model predictions and observations at location 27cp0002. Model is calibrated on the period 1960-1999 with a modeling interval of 1 day.

not model the water saturation in the root zone: precipitation and potential evapotranspiration are directly added to the percolation zone. As a result, the actual evapotranspiration is equal to the potential evapotranspiration and the flux to the percolation zone does not depend on the saturation of the root zone. Since the root zone is not incorporated in the TFN model, only 8 parameters have to be estimated (see Table 2).

Table 2. Calibrated parameters of linear model with associated first-order estimates of standard deviation

	Estimated value	Standard deviation
ω_P, day	1.29	0.172
ω_E, day	0.610	0.0914
δ_1	0.949	0.00233
δ_2	0.999	0.000115
β	0.811	0.198
z_r, cm	463	46.1
q_{22}, cm^2	0.637	0.0300
r, cm^2	0.139	0.161

The difference between the predictive performance of both models is clear: the physically based model fits the observations much better (see Figure 3). Especially in predicting the extremes, the physically based model is superior. This can be easily understood when considering equation 4. The flux from the root zone to the percolation zone depends on the saturation of the root zone. Consequently, after a dry period the root zone buffers precipitation, resulting in a delayed response of the groundwater level. This can be clearly seen in the period between 1970 and 1975. On the other hand, after a wet period, the flux from the root zone is high, resulting in higher peaks in the groundwater level (e.g. 1962 and 1995). A long-term bias in both the TFN model and the physically based model as observed between 1980 and 1985 is very likely to be caused by the fact that the calculated drawdown is only an approximation of the real drawdown.

The estimated system noise q_{22} in the physically based model is more than 50% less than q_{22} in the linear model. This corresponds with the better predictions of the physically based model in Figure 3.

Another important aspect is the accuracy of the estimated correction factor β. The physically based model gives an estimated standard deviation of 0.09, whereas for the linear model, the estimated standard deviation is more than twice as high: 0.20. Although the standard deviation only is a first-order estimate, the difference between both estimates indicates that the physically based model determines the drawdown more accurately.

Since the available amount of data was too small for creating a validation set, the prediction performances were evaluated on the same data used for model calibration. Table 3 shows some important criteria and statistics for evaluating both models. The well-known Akaike Information Criterion (AIC) and Bayes Information Criterion (BIC) evaluate the maximum likelihood criterion in relation to the number of parameters. The mean error

$$\mathrm{ME} = \frac{1}{N}\sum_{i=1}^{N}(z_i - \tilde{z}_i), \tag{22}$$

and the root mean square error

$$\mathrm{RMSE} = \sqrt{\frac{1}{N}\sum_{i=1}^{N}(z_i - \tilde{z}_i)^2}, \tag{23}$$

measure the difference between the observations z and the predictions $\tilde{z}$.

Table 3. Comparison of physically based model and linear model

	physically based model	TFN model
AIC	3915.1	4279.7
BIC	3966.9	4317.4
ME, cm	-0.211	0.421
RMSE, cm	9.67	15.57

The statistics presented in Table 3 confirm that the physically based model gives better results. For example, the RMSE shows a reduction of 38%. Also the absolute value of the ME, which was already small, reduces with 50%.

5. CONCLUSIONS

This paper presented a physically based model for analyzing time series of groundwater level data. The model incorporates the degree of saturation of the root zone to model the groundwater recharge. The nonlinear model was written in state-space form and combined with an extended Kalman filter. The model was tested on a groundwater time series observed in the Netherlands. Comparison with a linear transfer function-noise model showed that the physically based model predicts the groundwater level much better. The results also showed that it can estimate the drawdown more accurately.

Besides better model performance, the presented model has the advantage that the physical basis of the model parameters enables comparison of the calibrated parameters with physical knowledge of the system. Further research may evaluate the influence of a priori parameter estimates and covariances, based on field data.

Finally, since the extended Kalman filter is a forward scheme, the model can be easily applied for online processing of groundwater level data.

REFERENCES

Box, G.E.P. and G.M. Jenkins (1970). *Time series analysis, forecasting and control.* Holden-Day. San Fransisco.

Gehrels, J.C. (1999). Groundwater level fluctuations; seperation of natural from anthropogenic influences and determination of groundwater recharge in the Veluwe area, the Netherlands. PhD thesis. Vrije Universiteit. Amsterdam.

Gehrels, J.C., F.C. van Geer and J.J. de Vries (1994). Decomposition of groundwater level fluctuations using transfer modelling in an area with shallow to deep unsaturated zones. *J. Hydrol.* **157**, 105–138.

Hipel, K.W. and A.I. McLeod (1994). *Time series modelling of water resources and environmental systems, Dev. Water Sci. Ser., vol. 45.* Elsevier. New York.

Huisman, L. (1972). *Groundwater Recovery.* MacMillan. London.

Jazwinsky, A.H. (1970). *Stochastic Processes and Filtering Theory.* Acedemic Press. New York.

Knotters, M. and M.F.P. Bierkens (2000). Physical basis of time series models for water table depths. *Water Resour. Res.* **36**(1), 181–188.

Mualem, Y. (1976). A new model for predicting the hydraulic conductivity of unsaturated porous media. *Water Resour. Res.* **12**(3), 513–522.

Schweppe, F.C. (1973). *Uncertain dynamic systems.* Prentice-Hall. New Jersey.

Verlaan, M. and A.W. Heemink (2001). Nonlinearity in data assimilation applications: a practical method for analysis. *Monthly Weather Review* **129**, 1578–1589.

APPENDIX

The drawdown s [L] after a time lag t [T] of a groundwater withdrawal Q [LT^{-3}] at distance r [L] is calculated as (Huisman, 1972)

$$s = \frac{Q}{4\pi T} W\left(u^2, \frac{r}{\sqrt{Tc}}\right) \qquad (24)$$

with

$$u^2 = \frac{\mu}{4T}\frac{r^2}{t}. \qquad (25)$$

This equation is applied for unsteady unidirectional flow in an unconfined aquifer with transmissivity T [L^2T^{-1}], above a semi-pervious layer with resistance c [T]. The function W is a logarithmic integral:

$$W\left(u^2, \frac{r}{\sqrt{Tc}}\right) = \int_{u^2}^{\infty} \frac{1}{v} \exp\left(-v - \frac{r^2}{4Tcv}\right) dv. \qquad (26)$$

IFAC
Publications
www.elsevier.com/locate/ifac

ON PHYSICAL AND DATA DRIVEN MODELLING OF IRRIGATION CHANNELS

Su Ki Ooi[1] M.P.M. Krutzen[2] E. Weyer[1]

[1] *CSSIP, Department of Electrical and Electronic Engineering, The University of Melbourne, Parkville, VIC 3010, Australia.*
Email: {skoo, e.weyer}@ee.mu.oz.au
[2] *DMP - Control Engineering Group, TNO Institute of Applied Physics, P.O. Box 155, 2600 AD DELFT, The Netherlands.*
Email: krutzen@tpd.tno.nl

Abstract: In this paper we compare the St. Venant equations against real data in order to examine their accuracy and capability to describe the relevant dynamics of an irrigation channel. The St. Venant equations are simulated using the Preissmann scheme, and the simulated and real measured water levels are compared. In addition, a comparison with system identification models is also performed in order to examine which model is more suitable for control design and prediction purposes. The results show that the St. Venant equations can adequately capture the dynamics of the real channels. However, system identification methods are as accurate as the St. Venant equations and are preferred over the St. Venant models for control and prediction purposes since they are much simpler to use.

Keywords: Modelling, system identification, physical modelling, parameter estimation, environmental systems, irrigation channel.

1. INTRODUCTION

Due to the sharp rise in demand for water in many parts of the world, water is becoming an increasingly scarce resource. It is therefore important to manage the water resources well and minimize the losses. This applies particularly to networks of irrigation channels, where large amounts of water are wasted due to poor management and control. These losses can be reduced by improving the decision support and control systems in the channels.

In order to design a good controller or a decision support system, most design methods require a good model that closely describes the relevant dynamics of the irrigation channel. Traditionally, the dynamics are modelled by the St. Venant equations, see e.g. (Chaudhry, 1993). The St. Venant equations are commonly used for prediction and control design for irrigation channels, see (Malaterre and Baume, 1998) for an overview. Despite widespread use, their accuracy are largely unknown which is surprising, taking into account the large amount of practically oriented research which uses the St. Venant equations as a starting point. A natural question is therefore whether the St. Venant equations are capable of describing the relevant dynamics of an irrigation channel accurately? From laboratory experiments (Brutsaert, 1971), it is known that the St. Venant equations are an accurate representation of a small scale laboratory channel. To the authors' knowledge, very few, if any of the models based on the St. Venant equations have been tested against data from real channels, which can be several kilometers long and the channel geometry is often nonuniform. In this paper, comparison of the St. Venant equations against real data is considered in order to examine their accuracy.

Previous works (see e.g. (Weyer, 2001) and (Ooi and Weyer, 2001)) showed that simple models that describe the dynamics of the channel adequately can be obtained using system identification method based on operational data from the channel. The St. Venant equations are hyperbolic partial differential equations and much more complex to use than the system identification models, and an open question is whether the St. Venant equations are significantly more accurate than the system identification models. We therefore also compare the performance of the models based on the St. Venant equations and the system identification models. In particular we examine their suitability for control design and prediction purposes.

The best known and most used finite difference method for solving the St. Venant equations is the Preissmann scheme (see e.g. (Chaudhry, 1993)). In order to examine the accuracy of the St. Venant equations, the water level in the channel is simulated using the Preissmann scheme based on physical data from the channel, such as length, width, etc, and the simulated water level is compared against the measured water level. There are parameters in the St. Venant equations which are not exactly known. In order to examine the effect of those parameters on the accuracy of the St. Venant equations, these parameters are estimated from the observed data, and the accuracy of the St. Venant equations with estimated and physical parameters is examined. This is followed by a comparison with system identification models, comparing the simulated water levels of the respective models with the measured water level.

In Section 2 a description of the irrigation channel is given. Then, the St. Venant equations are presented and the Preissmann scheme is briefly explained. The accuracy of the St. Venant equations is investigated in Section 4, followed by a comparison with system identification models. Finally, conclusions are given in Section 6.

2. CHANNEL DESCRIPTION

The channel considered in this paper is the Haughton Main Channel (HMC) in Queensland, Australia. Figure 1 shows a schematic side view of the channel. The channel is automated with overshot gates as shown in Figure 1. We refer to the stretch of the channel between two gates as a pool. We name the pool according to the number of the upstream gate, e.g. the pool in Figure 1 is pool i. y_i and y_j are the upstream water level of gate i and j respectively, and p_i and p_j are the position of gates. The amount of water above the gate is called the head over the gate, and denoted by h_i and h_j.

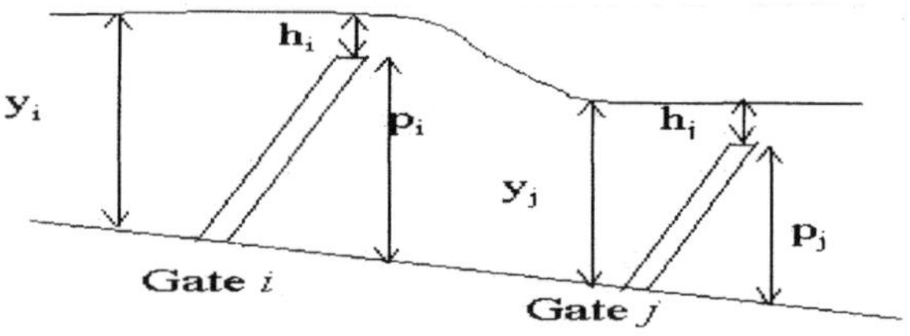

Fig. 1. Schematic of channel with overshot gates

The water levels, given in mAHD (meter Australia Height Datum), and the gate positions are the measured variables. The head over gate is computed from these variables. A fully shut gate has position of 0 meter and a positive value when the gate is open. The measured gate position, $\bar{p} = p_{max} - p$, where p_{max} is the position when the gate is fully shut. The head over the gate i and j is calculated as $h_i = y_i + \bar{p}_i - a_i$ and $h_j = y_j + \bar{p}_j - a_j$, where a_i and a_j are the gate adjustment constants necessary to convert from mAHD to meter.

3. ST. VENANT EQUATIONS

The St. Venant equations are derived from a mass and momentum balance, see e.g. (Chaudhry, 1993) and given by

$$\frac{\partial A}{\partial t} + \frac{\partial Q}{\partial x} = 0 \tag{1}$$
$$\frac{\partial Q}{\partial t} + \left(\frac{gA}{B} - \frac{Q^2}{A^2}\right)\frac{\partial A}{\partial x} + \frac{2Q}{A}\frac{\partial Q}{\partial x} + gA(S_f - \bar{S}) = 0$$

where A is the cross sectional area of the channel, B is the width of the water surface, $g = 9.81m/s^2$ is the gravity, $\bar{S}$ is the bottom slope, Q is the flow (discharge), and S_f is the friction slope. A commonly used relationship between the flow and the head over gate is $Q = ch^{3/2}$ (see e.g. (Chaudhry, 1993)), where c is the gate constant. The gate constant of the upstream and downstream gate are labelled as c_{in} and c_{out}. From (Fenton, 2001), for a sharp-edged rectangular weir, $c \approx 0.6\sqrt{g}b$ where b is the gate width.

According to the Manning equation, $S_f = \frac{n^2Q^2}{A^2R^{\frac{4}{3}}}$ where n is the Manning coefficient, which mainly depends on the surface roughness. Table values of n for different flow surfaces are available (see e.g. http://www.lmnoeng.com/manningn.htm). $R = \frac{A}{P}$ is the hydraulic radius, where the wetted perimeter, P, is defined as the length of line of intersection of the channel's wetted surface with a cross-sectional plane normal to the flow (see (Chaudhry, 1993)).

The pools we study are pool 9 and 10 of the HMC. The physical data are given in Table 1. The Manning coefficients are taken from *www.lmnoeng.com/manningn.htm*

for clean excavated earth channels. Obviously, the physical data listed in Table 1 are approximate values since the real condition of the channel will change with time and along the channel, and some parameters like the gate constant cannot be accurately measured or computed.

	Pool 9	Pool 10
Length, L	853m	3129m
Bottom width	6m	6m
Side slope	2	2
Bottom slope	1.993×10^{-4}	9.907×10^{-5}
Gravity, g	9.81 m/s^2	9.81 m/s^2
Gate width, b	4.4m	4.4m
Manning coefficient, n	0.02	0.02
Upstream gate constant, c_{in}	8.3	8.3
Downstream gate constant, c_{out}	8.3	8.3

Table 1. Physical data of pool 9 and 10

3.1 *Preissmann Scheme*

Finite difference methods have been extensively used for simulation of complex dynamical systems, see e.g. (Chaudhry, 1993). In these methods the time, t and spatial variable, x are discretised into a grid on which the dynamical model is solved and partial derivatives are approximated in an explicit or implicit way. The approximations are based on Taylor series expansions and are called explicit if the expansion is expressed in variables available at the current time instant, it is called implicit if it involves future time instants. In the Preissmann scheme the function, $f(x,t)$ and its partial derivatives are approximated as follows ($f = A$ or Q):

$$f = \frac{1}{2}\alpha(f_{i+1}^{k+1} + f_i^{k+1}) + \frac{1}{2}(1-\alpha)(f_{i+1}^k + f_i^k)$$

$$\frac{\partial f}{\partial t} = \frac{(f_i^{k+1} + f_{i+1}^{k+1}) - (f_i^k + f_{i+1}^k)}{2\Delta t}$$

$$\frac{\partial f}{\partial x} = \frac{\alpha(f_{i+1}^{k+1} - f_i^{k+1})}{\Delta x} + \frac{(1-\alpha)(f_{i+1}^k - f_i^k)}{\Delta x}$$

Subscript i is the ith spatial grid point and superscript k is the kth time grid point, and Δx and Δt are the grid intervals along the x-axis and t-axis. An advantage of the Preissmann scheme is that we can have variable spatial grid. The parameter α is a weighting coefficient; the scheme is totally explicit if $\alpha = 0$. The commonly used values are $0.6 \leq \alpha \leq 0.7$ (see e.g. (Chaudhry, 1993)), and in this paper we have used $\alpha = 0.6$. The boundary condition, i.e. the equations describing the end conditions of the pool, is included directly in the system of equations that needs to be solved. The boundary condition is expressed in the following equations

$$Q_{i=1}^{k+1} - Q_a^{k+1} = 0$$

$$h_{out}^{k+1} = y_{i=np}^{k+1} - p_{max} + p_{out}^{k+1}$$

$$Q_{i=np}^{k+1} = 0 \quad \text{if} \quad h_{out}^{k+1} < 0$$

$$Q_{i=np}^{k+1} = c(h_{out}^{k+1})^{3/2} \quad \text{otherwise}$$

where $Q_a^{k+1} = c_{in}(h_{in}^{k+1})^{3/2}$ is the discretised inflow function, where h_{in}^{k+1} and h_{out}^{k+1} are the discretised head over upstream and downstream gate, and p_{out}^{k+1} is the discretisation of the downstream gate opening. The subscript np refers to the last spatial grid point, i.e. the downstream end of the pool. Applying the approximations to the St. Venant equations, together with the boundary equations, a set of nonlinear algebraic equations is obtained. These algebraic equations can be solved using iterative search techniques, and the Newton-Raphson method is used in this paper.

4. ACCURACY OF THE ST. VENANT EQUATIONS

In this section, the St. Venant equations are compared against real data to see if they can adequately capture the dynamics of the irrigation channel. For pool 9 two data sets are available. Data set 1 is collected under low flow condition, where the channel is operated at around 25% of its capacity, while data set 2 is collected at around 75% of channel capacity.

4.1 *Pool 9*

Data set 1 is plotted in Figure 2 (data set 2 is not shown). The sampling interval is one minute and the gate adjustment constant is $a_9 = 23.97$. The gate constants and

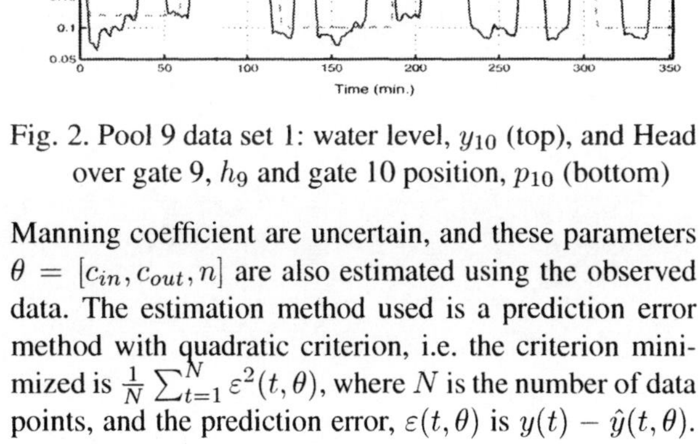

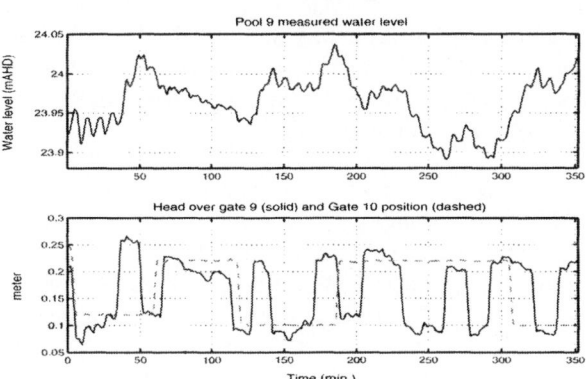

Fig. 2. Pool 9 data set 1: water level, y_{10} (top), and Head over gate 9, h_9 and gate 10 position, p_{10} (bottom)

Manning coefficient are uncertain, and these parameters $\theta = [c_{in}, c_{out}, n]$ are also estimated using the observed data. The estimation method used is a prediction error method with quadratic criterion, i.e. the criterion minimized is $\frac{1}{N}\sum_{t=1}^{N}\varepsilon^2(t,\theta)$, where N is the number of data points, and the prediction error, $\varepsilon(t,\theta)$ is $y(t) - \hat{y}(t,\theta)$.

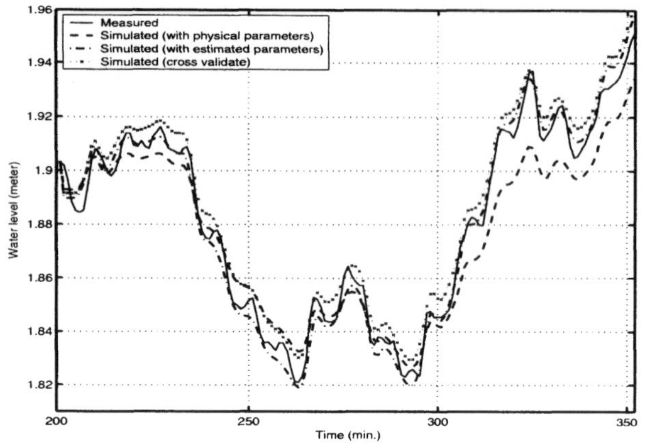

Fig. 3. Simulated and measured water levels of Pool 9 (Data set 1)

$y(t)$ and $\hat{y}(t,\theta)$ are the measured and simulated water level obtained by numerically solving the St. Venant equations. The first 200 data points in each data set were used for estimation and the rest were used for validation. The estimated gate constants and Manning coefficients are tabulated in Table 2. We next simulated the water

Estimated parameters	Measured Data 1	Measured Data 2
Manning coefficient, n	5.876×10^{-6}	2.022×10^{-2}
c_{in}	12.396	11.647
c_{out}	11.881	10.403

Table 2. Estimated parameters of pool 9

level using the St. Venant equations and compared it to the measured one. As the first 200 data points are used for estimation purposes, the comparison is based on the validation set only. The initial values of the intermediate water levels within the pool are the solution of the St. Venant equations in steady state with the last spatial grid point equalling the measured water level. The simulation is run from the beginning of the validation set where the first point of the measured water level and all the measured heads over upstream gate and measured downstream gate positions are used.

In addition to the standard validation, we also cross-validated by simulating data set 1 using the parameters estimated from data set 2 and vice-versa. Figures 3 and 4 show the simulated and measured water levels for data set 1 and 2. The Mean Squared Error (MSE) between the measured and simulated water levels are tabulated in Table 3.

Estimation set	Validation set	MSE (physical)	MSE (estimated)
Data set 1	Data set 1	1.077×10^{-4}	0.202×10^{-4}
Data set 2	Data set 2	11.11×10^{-4}	3.375×10^{-4}
Data set 1	Data set 2	-	4.690×10^{-4}
Data set 2	Data set 1	-	0.481×10^{-4}

Table 3. Mean squared errors for Pool 9

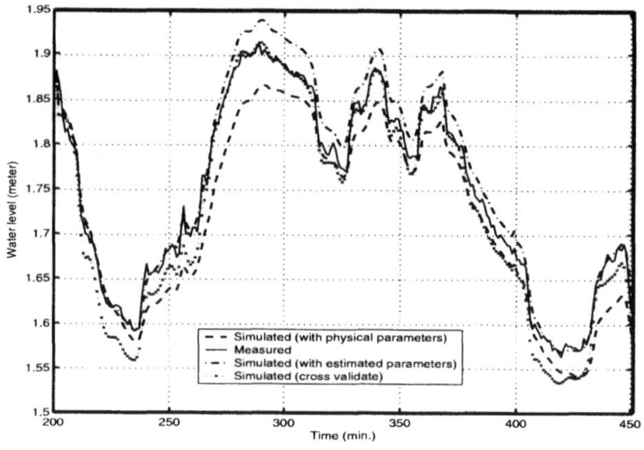

Fig. 4. Simulated and measured water levels of Pool 9 (Data set 2)

4.2 *Pool 10*

For pool 10, we only considered data collected under high flow condition. The sampling interval is one minute. The estimated Manning coefficient is 1.848×10^{-2}, and the estimated upstream and downstream gate constants are 11.078 and 12.382. The simulated and measured water levels are plotted in Figure 5. The MSE are 16.62×10^{-4} and 3.996×10^{-4} with physical and estimated parameters respectively.

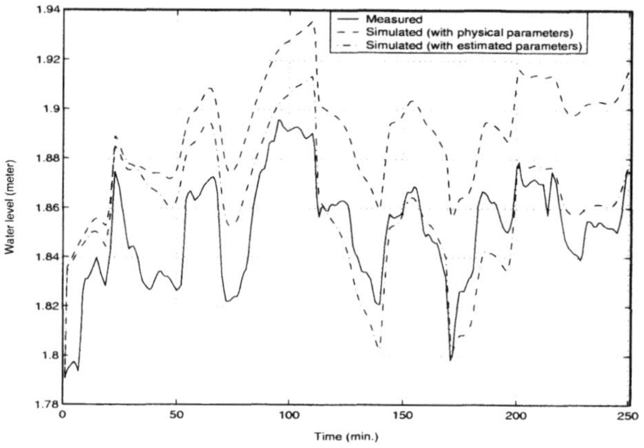

Fig. 5. Simulated and measured water levels of Pool 10

4.3 *Discussion*

There is good agreement between the Manning coefficient obtained from physical knowledge and the estimated ones, except the estimation using data set 1 for pool 9, where the estimated Manning coefficient is very small. Further investigations have shown that the simulated water level is insensitive to the value of the Manning coefficient. The gate constants estimated using measured data are similar to each other, but they are quite different from those computed based on physical data.

As expected, simulations using the St. Venant equations with estimated parameters give smaller MSE and the simulated water levels tracks the measured ones closer than those with physical parameters. As expected, simulations using data under low flow condition (data set 1) give smaller MSE than those under high flow condition. This is due to that larger variations in water levels are expected in the high flow condition than in the low flow condition.

In pool 9, the St. Venant equations with physical parameters are able to capture the main trends and the waves in the water level but with an offset error. The largest offset is around $3cm$ and $6cm$ for data set 1 and 2. As expected the performance is better with estimated parameters, where the models are able to accurately predict the water level with virtually no offset for data set 1, and with a small error of less than $4cm$ over small time periods for data set 2. Pool 10 is much longer than pool 9 and as expected the models are not as accurate as for pool 9 (larger MSE), and the largest offset is around $6cm$ with physical parameters. The offset is smaller with estimated parameters, around $4cm$. Note that the models only have access to the initial water level, hence they are able to predict the water level ahead of time for at least $2\frac{1}{2}$ hours. The offsets can be easily corrected by integral action in a feedback controller. Therefore, apart from the offset, the St. Venant equations based purely on the physical data of the channel seem to be able to capture the relevant dynamics, at least for control purposes.

5. COMPARISON WITH SYSTEM IDENTIFICATION MODELS

In this section the system identification models and the St. Venant equations are compared. From previous works (Weyer, 2001) and (Ooi and Weyer, 2001), it is known that a first order nonlinear model is able to capture the main trends in the water level well and a third order nonlinear model is able to give very accurate predictions of the water level. The predictors associated with the first and third order models for pool i are

$$\hat{y}_{i+1}(t+1,\theta) = \hat{y}_{i+1}(t,\theta) + c_{i,1}h_i^{3/2}(t-\tau) - c_{i+1,1}(\hat{y}_{i+1}(t,\theta) + \bar{p}_{i+1}(t) - a_{i+1})^{3/2} \quad (2)$$

where $\theta = [c_{i,1}, c_{i+1,1}]$, and

$$\begin{aligned}\hat{y}_{i+1}(t+1,\theta) &= c_{i,1}h_i^{3/2}(t-\tau) + c_{i,2}h_i^{3/2}(t-\tau-1) \\ &+ c_{i,3}h_i^{3/2}(t-\tau-2) + \hat{y}_{i+1}(t,\theta) \\ &+ c_{i+1,1}(\hat{y}_{i+1}(t,\theta) + \bar{p}_{i+1}(t) - a_{i+1})^{3/2} \\ &+ c_{i+1,2}(\hat{y}_{i+1}(t-1,\theta) + \bar{p}_{i+1}(t-1) - a_{i+1})^{3/2} \\ &+ c_{i+1,3}(\hat{y}_{i+1}(t-2,\theta) + \bar{p}_{i+1}(t-2) - a_{i+1})^{3/2} \\ &+ \alpha_1(\hat{y}_{i+1}(t,\theta) - 2\hat{y}_{i+1}(t-1,\theta) + \hat{y}_{i+1}(t-2,\theta)) \\ &+ \alpha_2(\hat{y}_{i+1}(t,\theta) - \hat{y}_{i+1}(t-1,\theta)) \quad (3)\end{aligned}$$

where $\theta = [c_{i,1}, c_{i,2}, c_{i,3}, c_{i+1,1}, c_{i+1,2}, c_{i+1,3}, \alpha_1, \alpha_2]$ and τ is the time delay, and the sampling interval is one minute.

5.1 *Pool 9*

The unknown parameters were obtained using system identification techniques (see (Weyer, 2000) and (Weyer, 2001)) using data set 1 (see Figure 2) and 2. The first 200 and 240 data points of data set 1 and 2 respectively were used for estimation, and the rest were used for validation. The water level of the validation set is simulated using the St. Venant equations and predicted using the discrete time first and third order nonlinear models. The results are shown in Figure 6 for data set 1, and similar result are obtained for data set 2 (plot not shown). The predictor is a simulation model since it uses the predicted water level at time t to predict the level at time $t+1$, hence the comparison is fair. The mean squared errors between the measured water level and the St. Venant equations (MSE St.V), and the identification model (MSE SI) are given in Table 4. The MSE St.Vs are different from Table 3 since the simulations started at a different time instant.

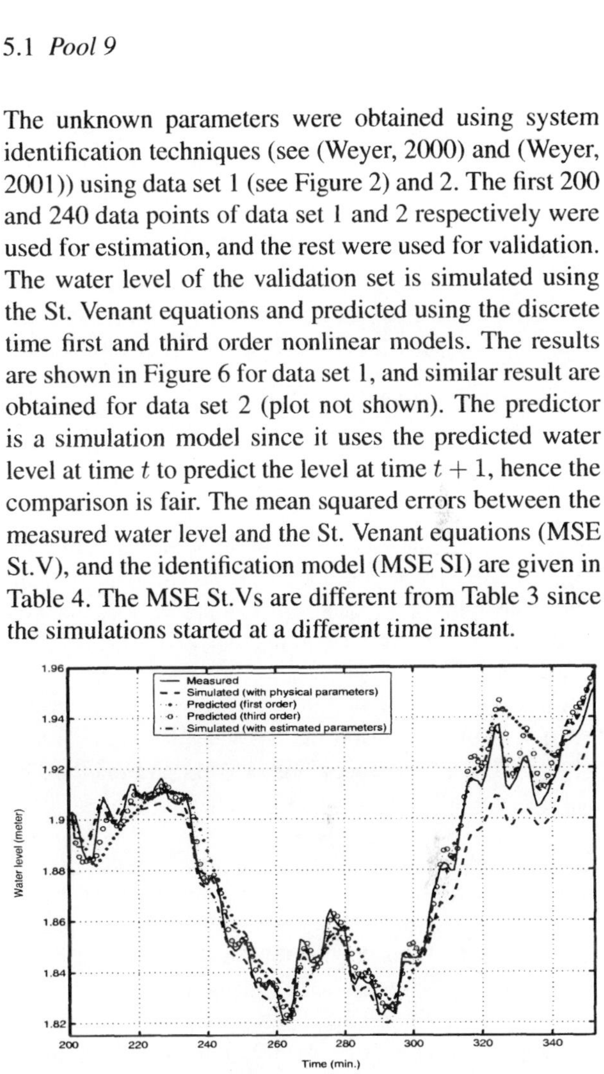

Fig. 6. Data set 1: Simulated, predicted and measured water level of pool 9

MSE St.V (physical)	MSE St.V (estimated)	MSE SI (first order)	MSE SI (third order)
1.077×10^{-4}	0.202×10^{-4}	1.107×10^{-4}	0.362×10^{-4}
10.15×10^{-4}	3.946×10^{-4}	11.652×10^{-4}	7.781×10^{-4}

Table 4. Pool 9 MSE of data set 1 (second row) and 2 (third row)

5.2 *Pool 10*

The same procedure is repeated for pool 10. The results are shown in Figure 7. The MSEs are given in Table 5.

MSE St.V (physical)	MSE St.V (estimated)	MSE SI (first order)	MSE SI (third order)
11.00×10^{-4}	1.96×10^{-4}	4.44×10^{-4}	1.43×10^{-4}

Table 5. MSE of pool 10

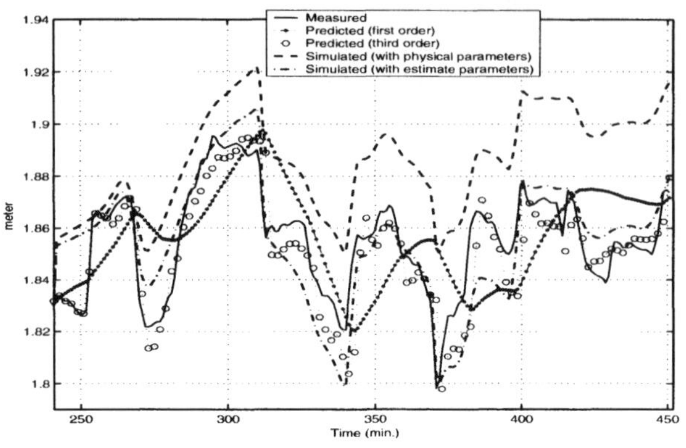

Fig. 7. Simulated, predicted and measured water level of pool 10

5.3 *Discussion*

In pool 9, the St. Venant equations with estimated parameters give the smallest MSE. However, the St Venant equations with physical parameters are only as accurate as the first order identification model, and the third order nonlinear identification model gives smaller MSE than the St Venant equations with physical parameters. In pool 10, the model based on the St. Venant equations with physical parameters gives largest MSE, and the third order identification model gives the smallest MSE.

Overall, there is not much difference between the St. Venant equations with estimated parameters and the third order nonlinear identification model, and both require observed data. However, the third order nonlinear model is much simpler to use for control and prediction purposes. Even a simple first order model is good enough for control design (Weyer, 2002). In the event that a new automated control scheme is to be implemented in a channel where no operational data is available, then models based on the St. Venant equations must be used. The St. Venant equations also give the water levels at the intermediate grid points, while the identification models only give the downstream water level. However, for control design purposes only the downstream water level is needed, and if there are operational data available, the system identification models are as accurate as the St. Venant equations with estimated parameters and much easier to use for control design and prediction purposes.

6. CONCLUSION

In this paper the accuracy of the St. Venant equations is examined. Water levels simulated using the St. Venant equations with both physical and estimated parameters are compared against the measured water level. The results show that the St. Venant equations can adequately capture the dynamics of the real channels. However, the third order nonlinear identification models are as accurate as the St. Venant equations with estimated parameters but much simpler to use. Therefore, due to the complexity of the St. Venant equations, simple first and third order nonlinear models obtained using identification methods are preferred for control design and prediction purposes.

Patent: A patent has been applied for to cover the developments that are described in this paper.

Acknowledgement: This research is part of a collaborative research project between the Department of Electrical and Electronic Engineering and Rubicon Systems on modelling and control of irrigation channels. The authors would like to thank Matthew Ryan at Rubicon System in Queensland for helping carrying out the identification experiments. They would also like to thank John Fenton and Iven Mareels for many fruitful discussions on modelling and simulation of irrigation channels. This work was supported by Rubicon System Pty Ltd under the auspices of an AusIndustry Grant.

7. REFERENCES

Brutsaert, W. (1971). De Saint-Venant equations experimentally verified. *Journal of Hyd. Div. ASCE* **97**, 1387–1401.

Chaudhry, M. Hanif (1993). *Open-Channel Flow*. Prentice Hall.

Fenton, J. D. (2001). *421-423: River Hydraulics (lecture note)*. The University of Melbourne.

Malaterre, P.-O. and J.-P Baume (1998). Modeling and regulation of irrigation canals: existing applications and ongoing researches. *IEEE International Conference on Systems, Man and Cybernetics, San Diego, California* pp. 3850–3855.

Ooi, Su Ki and E. Weyer (2001). Closed loop identification of an irrigation channel. *Procedings of the 40th IEEE CDC, Orlando, USA* pp. 4338–4343.

URL (2000). *http://www.lmnoeng.com/manningn.htm*. LMNO Engineering, Research, and Software, Ltd., Athens, Ohio, USA.

Weyer, E. (2000). Analysis of September data from the Haughton Main Channel. Internal report. Department of Electrical and Electronic Engineering, University of Melbourne.

Weyer, E. (2001). System identification of an open water channel. *Control Engineering Practise* **Vol. 9**, pp. 1289–1299.

Weyer, E. (2002). Decentralised PI controller of an open water channel. *Procedings of the 15th IFAC World Congress, Barcelona, Spain*.

IFAC Publications
www.elsevier.com/locate/ifac

IDENTIFICATION AND ON-LINE ESTIMATION OF THE UNSATURATED HYDRAULIC CONDUCTIVITY IN PRESENCE OF FORCED AIR CONVECTION BASED ON A DISTRIBUTED-PARAMETER MODEL

[1]Schoefs O., [1]Dochain D., [2]Chapuis R. [3]Samson R. and [3]Perrier M.

[1]CESAME, Université Catholique de Louvain, Ave Georges Lemaître 4-6, B-1348 Louvain-la-Neuve, Belgium

[2]Dept. of Civil, Geological and Mining Engineering, École Polytechnique de Montréal, P.O. Box 6079, Station Centre-ville, Montreal (Qc) H3C 3A7, Canada

[3]Dept. of Chemical Engineering, École Polytechnique de Montréal, P.O. Box 6079, Station Centre-ville, Montreal (Qc) H3C 3A7, Canada

Abstract: The goal of this study was to calibrate a numerical model aimed at describing simultaneous air and water flow in soil and, to build a model-based estimator of the unsaturated hydraulic conductivity. The experimental approach consisted of infiltration tests in a 1.5m high column of loamy sand. The numerical model correctly described water percolation without air convection, provided parameter adjustment of the predictive model of the unsaturated hydraulic conductivity was performed. An observer-based estimator was able to estimate on-line the hydraulic conductivity in steady-state but faced oscillation problems in unsteady-state. The difficulty in implementing the software-sensor can be explained by the strong nonlinearity of the dynamical model. *Copyright © 2003 IFAC*

Keywords: Distributed-parameter systems, nonlinear systems, modelling, parameter identification, parameter estimation.

1. INTRODUCTION

Biological processes are widely used for the treatment of petroleum-contaminated soils. They enhance contaminant biodegradation by providing air and water to the indigenous soil microorganisms. If on one hand air is used to provide oxygen to microorganisms, on the other hand, water is necessary to microbial growth and to the bioavailability of the contaminant. Treatment bioprocesses are then equipped with aeration and irrigation systems capable of providing air and water to the contaminated zone. Figure 1 illustrates two widely used treatment bioprocesses in unsaturated soils, the biopile and the *in situ* bioventilation. From the geometrical configuration of such processes, it clearly appears that an undesired water content gradient can be established within the soil, due to hydraulic gradients. In some cases, air convection prevents water from infiltrating soils and leads to a local drying or, on the contrary, to local saturation of the soil. Optimal operating conditions are therefore a compromise between the water and air levels within the soil pores, and control of air and water flows allows to optimize soil treatment bioprocesses. While the main states governing water and air flows through soils (air pressure and flow rate, water suction and water content) can be measured on-line, control of the two fluids is confronted to weak knowledge of the dynamics.

Transient flow of water into a non-swelling unsaturated soil is well understood and has been largely studied in the literature (Bear, 1972; Hillel, 1980; Kovács, 1981; Mualem, 1986; Fredlund, and Rahardjo, 1993). Nevertheless, prediction using mathematical models is still confronted to two important features: boundary conditions, which are usually not constant, and prediction of hydraulic conductivity in unsaturated soil. In particular, the unsaturated hydraulic conductivity, which is the main water-content dependent function that governs the fluid dynamics, is hardly identifiable a priori despite the existence of semi-empirical relations that allow to describe satisfactorily water infiltration through soils (Mualem, 1976; van Genuchten, 1980; Brutsaert, 2000; Schaap, and Leij, 2000; Aubertin, *et al.*, 1998; Arya, *et al.*, 1999; Poulsen, *et al.*, 1999). Several studies have successfully calibrated these semi-empirical relations but simulation results were very sensitive to the parameters identified. Such models revealed then to be more a descriptive tool than a predictive one. Moreover, to our knowledge, no study has been conducted to predict and validate the influence of forced air convection on transient water flow in unsaturated soils.

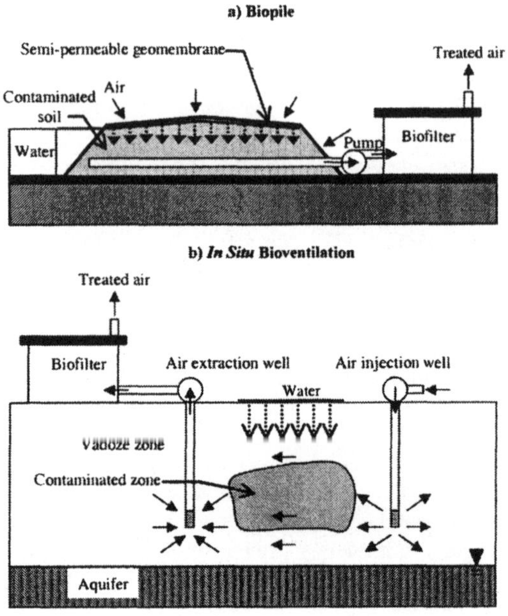

Fig. 1. Two typical soil treatment bioprocesses.

In this paper, the influence of forced air flow on water percolation is characterized and a mathematical model based on Darcy's law is developed and compared to experimental data. The goal of the study described in this paper was therefore to develop a numerical model capable of describing simultaneous water and air flow, to calibrate it using laboratory experiments and, to build a model-based estimator of the unsaturated hydraulic conductivity. Such estimator aims at converting the descriptive numerical model to an adaptive one that can be used to optimize soil treatment bioprocesses by controlling air and water flows.

This paper is organized as follows. The numerical model, based on the Darcy's law, will be first presented in section 2. The laboratory experiments, consisting of infiltration tests in column, will be described in section 3. The mathematical development involved in the on-line estimation of the unsaturated hydraulic conductivity will be presented in section 4. Finally, results on the model identification and on the parameter estimation will be described and discussed in section 5.

2. THE MATHEMATICAL MODEL

Transient flow of a fluid through a porous media is commonly described by Darcy's law :

$$\upsilon = -k\frac{\partial h}{\partial z} \qquad (1)$$

where υ is the flow rate of the fluid, k is the permeability coefficient and $\partial h/\partial z$ is the hydraulic head gradient in the z-direction. Darcy's law, when applied to air and water and combined with the continuity equations, leads to the two following partial differential equations:

$$C(\psi)\frac{\partial \psi}{\partial t} = \frac{\partial}{\partial z}\left[K_w(\theta_w)\left(\frac{\partial \psi}{\partial z} - 1\right)\right] \qquad (2)$$

$$\frac{\partial P}{\partial t} = \frac{1}{2\cdot\theta_a}\frac{\partial}{\partial z}\left[K_a(\theta_w)\frac{\partial P^2}{\partial z}\right] \qquad (3)$$

where $$\psi = \frac{P_w + P}{\gamma_w} \qquad (4)$$

where Ψ is the soil suction, K_w and K_a are respectively the water and air conductivities, P_w and P are respectively the water and the air pressures, γ_w is the volumetric weight of water, θ_a and θ_w are the volumetric air and water contents, and C is the specific water capacity defined as:

$$C(\psi) = \frac{d\theta_w}{d\psi} \qquad (5)$$

Functions $C(\psi)$ and $K_w(\psi)$ can be determined from the soil-water retention curve obtained experimentally. Van Genuchten (1980) proposed the following equations for this determination:

$$\theta_e = \left[\frac{1}{1+|\alpha\psi|^n}\right]^m \qquad (6)$$

where θ_e is the degree of saturation, and α, m and n are constants.

$$K_w = K_{sat}\sqrt{\theta_e}\left[1-\left(1-\theta_e^{1/m}\right)^m\right]^2 \qquad (7)$$

where K_{sat} is the saturated hydraulic conductivity, determined experimentally.

3. MATERIALS AND METHODS

Soil used for the study was a natural loamy sand (6 % clay, 7 % silt and 87 % sand). A 1.5 m high column was equipped with an irrigation and aeration system capable of providing co-current and counter-current water and air flow (Figure 2). Time Domain reflectometry (TDR) probes for water content measurement and tensiometers for water and air pressure measurement were placed at six different depths. A data acquisition system was installed to collect measurement at every minute. A basic control system (C) was implemented to ensure constant water content at the top of the column. A mass flow controller (FC) was used to maintain a constant air flow through the soil.

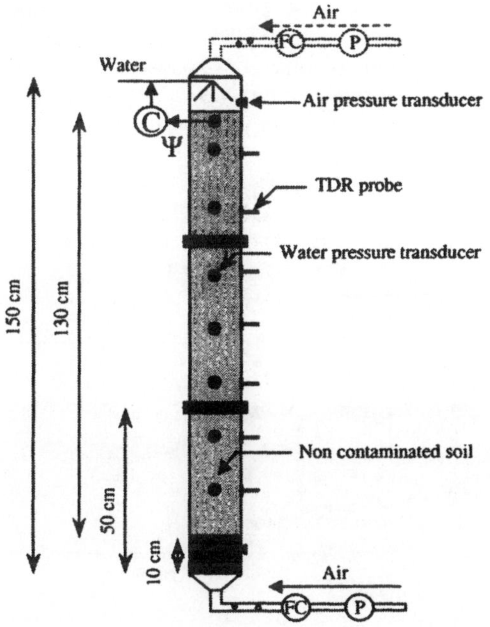

Fig. 2. Schematic representation of the experimental column.

The soil-water retention curve was built using tests in a 20 cm high column, equipped with a tensiometer and filled with soil whose humidity was known. The soil-water retention curve was then built point by point.

Details of the experimental procedure are available in Schoefs (2002).

4. MATHEMATICAL DEVELOPMENT

This section focuses on the on-line estimation of the unsaturated hydraulic conductivity without forced air convection. Provided that air pressure is measured, the mathematical development can be easily extrapolated to the case in presence of forced air convection.

Without forced air convection, equation (2) becomes:

$$C(\psi_w)\frac{\partial \psi_w}{\partial t} = \frac{\partial}{\partial z}\left[K_w(\theta_w)\left(\frac{\partial \psi_w}{\partial z} - 1\right)\right] \quad (8)$$

where $$\psi_w = \frac{P_w}{\gamma_w} \quad (9)$$

Using the finite difference method, equation (8) can be rewritten as follows:

$$\frac{d\psi_w(t,z)}{dt} = D(t,z)\cdot \ldots [K_w(t,z)\cdot H(t,z+1) - K_w(t,z-1)\cdot H(t,z)] \quad (10)$$

with $$D(t,z) = \frac{1}{dz}\cdot\frac{\psi_w(t,z)-\psi_w(t,z-1)}{\theta_w(t,z)-\theta_w(t,z-1)} \quad (11)$$

and $$H(t,z) = \frac{\psi_w(t,z)-\psi_w(t,z-1)}{dz} - 1 \quad (12)$$

From the on-line measurement of the suction, ψ_w, it is possible to build an observer-based estimator, as follows (Bastin, and Dochain, 1990):

$$\frac{d\hat{\psi}_w(t,z)}{dt} = D(t,z)\cdot \ldots \left[\hat{K}_w(t,z)\cdot H(t,z+1) - \hat{K}_w(t,z-1)\cdot H(t,z)\right] + \ldots \omega\cdot(\psi_w(t,z) - \hat{\psi}_w(t,z)) \quad (13)$$

$$\frac{d\hat{K}_w(t,z)}{dt} = \gamma\cdot D(t,z)\cdot H(t,z+1)\cdot \ldots (\psi_w(t,z) - \hat{\psi}_w(t,z)) \quad (14)$$

where ψ_w is the suction measured, $\hat{\psi}_w$ and $\hat{K}_w$ denote the on-line observation of ψ_w and the on-line estimate of K_w respectively and, ω and γ are two gains to be tuned.

In discrete time, by considering a first-order Euler approximation for the time derivatives, equations (13) and (14) are rewritten as follows:

$$\begin{bmatrix}\hat{\psi}_w(i+1,z)\\ \hat{K}_w(i+1,z)\end{bmatrix} = A_\psi(i,z)\cdot\begin{bmatrix}\hat{\psi}_w(i,z)\\ \hat{K}_w(i,z)\end{bmatrix} + B_\psi(i,z) \quad (15)$$

with

$$A_\psi(i,z) = \begin{bmatrix} 1 - dt(i)\cdot\omega(i,z) & dt(i)\cdot D(i,z)\cdot H(i,z+1) \\ -\dfrac{dt(i)\cdot\gamma(i,z)}{D(i,z)\cdot H(i,z+1)} & 1 \end{bmatrix} \quad (16)$$

and

$$B_\psi(i,z) = \begin{bmatrix} -dt(i)\cdot D(i,z)\cdot K_w(i,z-1)\cdot H(i,z) + \omega(i,z)\cdot\psi_w(i,z) \\ \dfrac{dt(i)\cdot\gamma(i,z)}{D(i,z)\cdot H(i,z+1)}\cdot\psi_w(i,z) \end{bmatrix} \quad (17)$$

where dt(i) is the sampling period at time step i.

The calibration of the design parameters ω and γ is performed on the basis of the tuning rules developed in Pomerleau (1990) and Perrier *et al.* (2000). These consist in assigning the dynamics of the estimator to a double pole λ. From the characteristic polynomial of the observer-based estimator, the following tuning relations are obtained:

$$\omega(i,z) = \frac{2 \cdot (1-\lambda)}{dt(i)} \quad . (18)$$

and

$$\gamma(i,z) = \frac{(\lambda - 1)^2}{(dt(i))^2} \quad (19)$$

with $\lambda \geq 0$.

5. RESULTS

5.1 Experimental results.

Three tests were performed on the column corresponding to three operating conditions. The first test consisted of water infiltration without forced air convection. The second and third tests consisted of water percolation with respectively co-current and counter-current forced air convection. At the beginning of the tests, the soil water content was equal to 0,10 g/g dry soil, which corresponded to a soil suction of –250 cm of water. The air flow rate was fixed at 5 L/mn and the water content was maintained at 0.26 g/g dry soil at the top of the column. This water content corresponded to a soil suction of –15 cm of water. Results, presented in figure 3, show that air convection has a significant impact on water percolation. The co-current air flow increased water infiltration, and the counter-current air flow decreased it dramatically.

5.2 Simulation results.

The soil-water retention curve obtained experimentally was used to identify parameters of the van Genuchten equations. A least-square method based on the Marquard-Levenberg approach was used and led to the following result: α = 0.0853 1/cm, m = 1.8057 and n = 1-1/m = 0.4462. This parameter identification was validated by comparing the simulated curves to the experimental data corresponding to the water infiltration test without forced air convection. However, by using this set of values and taking K_{sat} equal to 9.36 cm/h (determined experimentally), the model did not describe correctly the experimental result and underestimated the infiltration rate, as illustrated by figure 4a. The parameters were then re-identified by fitting the simulated values to the experimental data corresponding to the suction captor at the bottom of the column, i.e. at 118 cm from the top surface. This led to the following set of parameter values: K_{sat} = 10 cm/h, α = 0.07 1/cm, m = 2.5 and n = 1-1/m = 0.6. Results are illustrated in figure 4a-bis. It results that the model correctly described water percolation without air convection, provided parameter adjustment of the predictive model of the unsaturated hydraulic conductivity was performed The corresponding soil-water retention and hydraulic conductivity curves are presented in figure 5a and 5b respectively. On the one hand, the simulation curves obtained from the parameterization of the soil-water retention curve (default simulation curves) underestimated the infiltration rate (see figure 4a). On the other hand, the hydraulic conductivity corresponding to the adjusted simulation curves is higher than the one corresponding to the default simulation curves only for high suctions (above –35 cm of water), i.e. high water contents (see figure 5b). This observation demonstrates that the portion of the hydraulic conductivity curve corresponding to high water contents mainly governs the water infiltration rate. Note that the soil-water retention curves corresponding to the default and adjusted simulation curves are very close (see figure 5a). This reflects the high sensitivity of the water infiltration rate with respect to the parameterization of the soil-water retention curve.

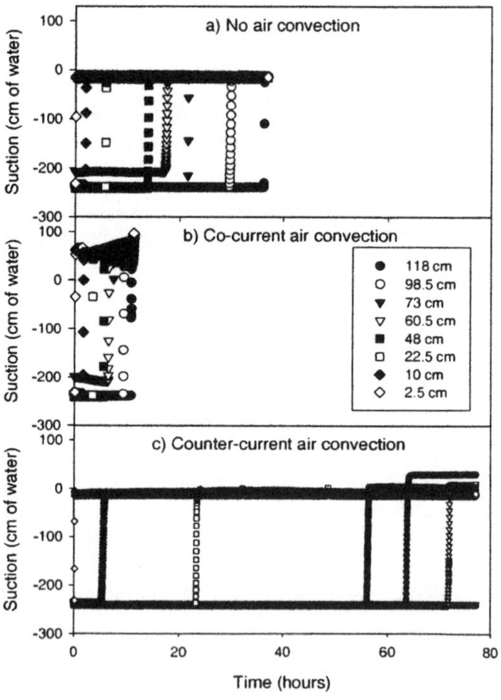

Fig. 3. Experimental results for the three infiltration tests.

Figure 4b and 4c represent the simulation curves in presence of forced air convection. The adjusted values of the parameters K_{sat}, α, m and n were taken. The model did not fit accurately the data even if it predicted qualitatively the influence of air convection on water infiltration. In particular, the model does not increase sufficiently the infiltration rate in co-current and decrease it too much in counter-current. This discrepancy between simulation and experimental data is partially attributed to the model's sensitivity with respect to the parameter identification of the van Genuchten relationships and the omission of the friction forces exerted by the air to the water.

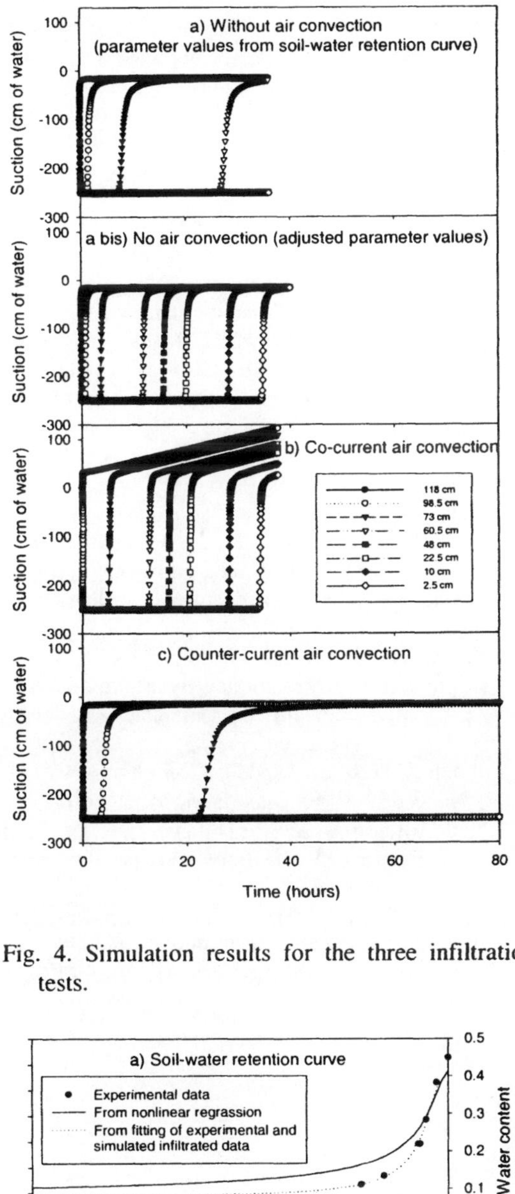

Fig. 4. Simulation results for the three infiltration tests.

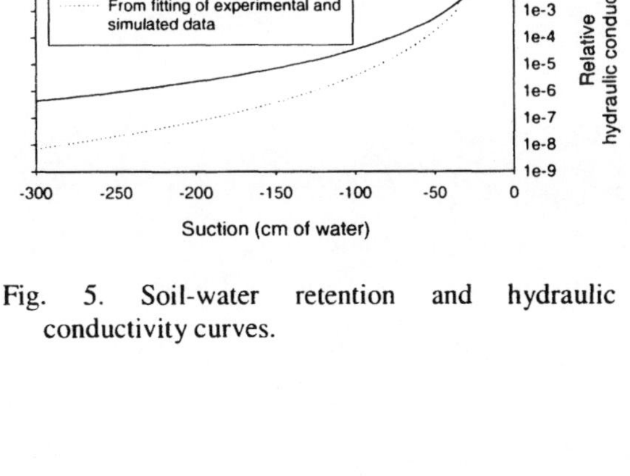

Fig. 5. Soil-water retention and hydraulic conductivity curves.

In order to cope with the identification difficulties mentioned above, development of estimators of the unsaturated hydraulic conductivity was investigated and is the purpose of the next sub-section.

5.3 On-line estimation of the unsaturated hydraulic conductivity.

The numerical model, described above, was chosen as a reference model and the adjusted values of the parameters were taken.

The observer-based estimator was first tested in steady-state (figure 6). The water content was maintained constant in the column, corresponding to a suction of –235 cm of water. The corresponded value of the hydraulic conductivity given by the van Genuchten relationships was equal to $3.6526 \cdot 10^{-7}$ cm/h. The initial value of the estimated hydraulic conductivity was taken equal to $3.6526 \cdot 10^{-5}$ cm/h and the double pole, λ, was fixed at 0.95. The observer-based estimators converge to the predicted values within ten hours. Note that the estimators do not converge simultaneously but from the top to the bottom of the column and that decreasing the value of the double pole, λ, can increase the convergence rate.

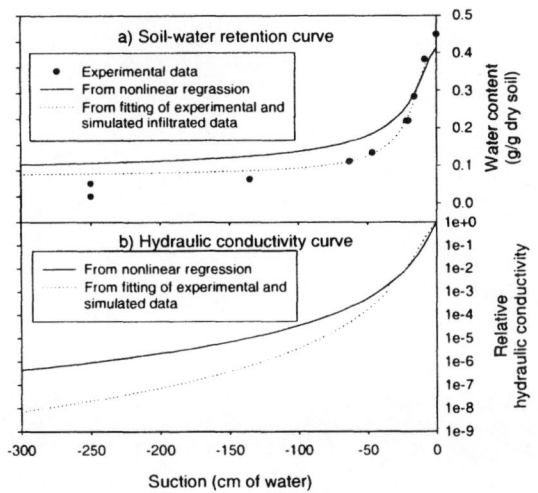

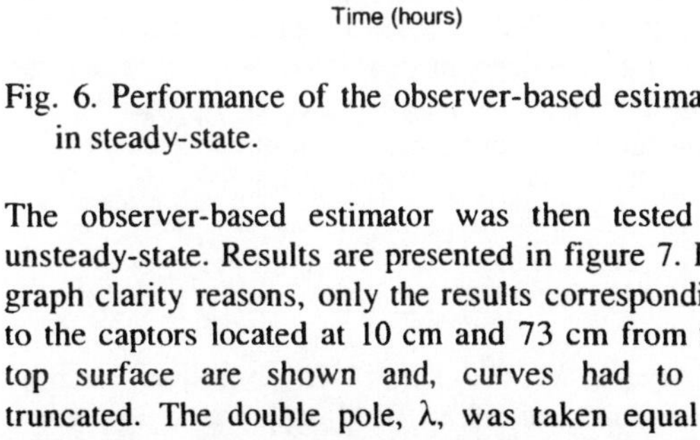

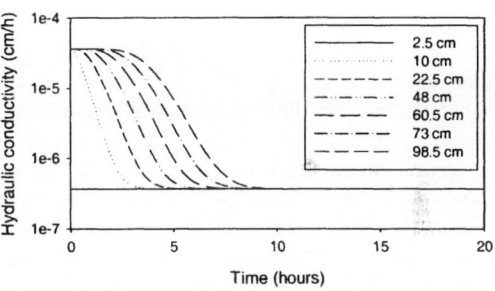

Fig. 6. Performance of the observer-based estimator in steady-state.

The observer-based estimator was then tested in unsteady-state. Results are presented in figure 7. For graph clarity reasons, only the results corresponding to the captors located at 10 cm and 73 cm from the top surface are shown and, curves had to be truncated. The double pole, λ, was taken equal to 0.95. The observer-based estimator faces large oscillations in unsteady-state, i.e. before the passage of the waterfront, and converges afterwards. Implementing a first-order filter can decrease the oscillation range. However this leads to an unacceptable decrease of the convergence rate. The weak performance of the estimator in unsteady-state can be attributed to the strong nonlinearity of the model and to the intrinsic characteristics of distributed parameter systems. Improvement of the estimator of the unsaturated hydraulic conductivity is presently under investigation.

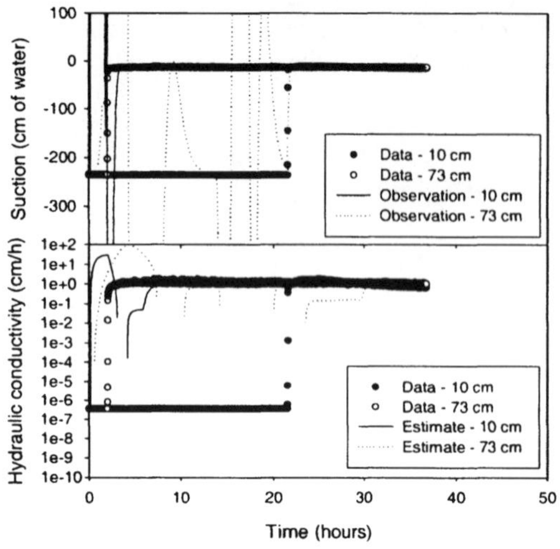

Fig. 7. Performance of the observer-based estimator in unsteady-state.

6. CONCLUSION

The simulation results revealed that the numerical model, based on the Darcy's law and predictive relationships for the unsaturated hydraulic conductivity, was able to correctly describe the water infiltration without air convection, provided parameter adjustment is performed. It also appeared that the model is very sensitive with respect to the parameter identification from the soil-water retention curve. Implementation of an on-line estimator of the hydraulic conductivity has then been investigated to cope with this difficulty. Its performance was satisfactory in steady-state but faced large oscillations in unsteady-state. Further studies are required to improve the developed estimator but the results obtained in this study are very encouraging with respect to the potential for controlling air and water flows and then, for optimising soil treatment bioprocesses.

REFERENCES

Arya, L. M., F. J. Leij, M.T. van Genuchten and P. J. Shousse (1999). Scaling parameter to predict the soil water characteristic from particle-size distribution data. *Soil Sci. Soc. Am. J.*, **63**, 510-519.

Aubertin, M., J.-F. Ricard and R. Chapuis (1998). A predictive model for the water retention curve: application to tailings from hard-rock mines. *Can. Geotech. J.*, **35**, 55-69.

Bastin G. and D. Dochain (1990). *On-line estimation and adaptive control of bioreactors*. Elsevier, Amsterdam.

Bear, J. (1972). *Dynamics of fluids in porous media*, Elsiever, New York.

Brutsaert, W. (2000). A concise parametrization of the hydraulic conductivity of unsaturated soils. *Advances in Water Resources*, **23**, 811-815.

Fredlund, D. G. and H. Rahardjo (1993). *Soil mechanics for unsaturated soils*, John Wiley & Sons, New York.

Hillel, D. (1980). *Fundamentals of soil physics*, Academic Press, New York.

Kovács, G. (1981). *Seepage Hydraulics*, Elsiever, New York.

Mualem, Y. (1976). A new model for predicting the hydraulic conductivity of unsaturated porous media. *Water Resources Research*, **12**, 513-522.

Mualem, Y. (1986). *Methods of soil analysis. Part 1: physical and mineralogical methods* (K.A. Madison), 799-823, American Society of Agronomy, New York.

Perrier, M., S. Feyo de Azevedo, E. C. Ferreira and D. Dochain (2000). Tuning of observer-based estimators: theory and application to the on-line estimation of kinetic parameters. *Control Engineering Practice 8*, 377-388.

Pomerleau, Y. (1990). Modélisation et contrôle d'un procédé fed-batch de culture des levures à pain Saccharomyces cerevisiae. *Ph.D Thesis.* Dpt. of chemical engineering, École Polytechnique de Montréal, Montréal (Qc).

Poulsen, T. G., P. Moldrup, T. Yamaguchi and O. H. Jacobsen (1999). Predicting saturated and unsaturated hydraulic conductivity in undisturbed soils from soil water characteristics. *Soil Science*, **164**, 877-887.

Schaap, M. G. and F. J. Leij (2000). Improved prediction of unsaturated hydraulic conductivity with the Mualem-van Genuchten model. *Soil Sci. Soc. Am. J.*, **64**, 843-851.

Schoefs O. (2002). Modélisation et observation des procédés de biodegradation d'un polluant dans un sol non-saturé. *Ph.D Thesis*, Dpt. of chemical engineering, École Polytechnique de Montréal, Montréal (Qc).

van Genuchten (1980). A closed-form equation for predicting the hydraulic conductivity of unsaturated soils. *Soil Sci. Soc. Am. J.*, **44**, 892-898.

IFAC Publications
www.elsevier.com/locate/ifac

CONFIDENCE REGIONS FOR NON-PARAMETRIC ERRORS-IN-VARIABLES ESTIMATES

W. P. Heath *

* *Centre for Integrated Dynamics and Control,*
School of Electrical Engineering and Computer Science,
University of Newcastle, NSW 2308, Australia.
Email: wheath@ee.newcastle.edu.au;
Tel: +61 2 4921 5997; Fax: +61 2 4960 1712.

Abstract: We construct a confidence region in the complex plane for the pointwise frequency response measurement of a plant subject to periodic excitation with noise on both the input and the output. The region is constructed via the Minkowski division of circular confidence regions for the output and input spectra. While correct, the resulting confidence region is conservative. *Copyright © 2003 IFAC*

Keywords: Frequency response functions, linear systems, system identification.

1. INTRODUCTION

Recently the bias and variance (Pintelon and Schoukens, 2001; Heath, 2002) and probability density function (Pintelon *et al.*, 2002) of FRF (frequency response function) measurements using the non-parametric errors-in-variables estimate have been quantified. The measured system is assumed to have input and output noise that may be correlated. In (Pintelon *et al.*, 2002) it is suggested that the probability density function might be used for constructing confidence regions. Strictly speaking this is only legitimate when the specific frequency values of the input and output noise spectra (and cross-spectra) are known (nevertheless the simulations in (Pintelon *et al.*, 2002) indicate the confidence regions are often very good approximations even when the noise spectra and cross-spectra are unknown).

Confidence regions for Welch's spectral estimate, which is appropriate when the input is known but non-periodic, are given in (Brillinger, 1981). These regions are asymptotically correct, and exact in the special case where the input is periodic, the output noise is white and there is no input noise (Bayard, 1993). The effects of initial transients on such periodic data have been considered (De Vries and Van den Hof, 1995). Elliptical confidence regions may be constructed for the case where the input is periodic with no noise and the output noise is coloured Gaussian (Heath, 2001*b*; Heath, 2001*a*). If a plant is operating in closed-loop, confidence regions for indirect plant estimates can be formed via Möbius transformation (Wellstead, 1981).

In this paper we consider the more general case where there is input noise, but the input and output noise spectra and cross-spectra are unknown. We construct a confidence region from the Minkowski division of confidence regions for the output and input spectra. The construction is straightforward following (Farouki *et al.*, 2001; Farouki and Pottmann, 2002), but the resulting confidence region is rather conservative, in the sense that we only provide a loose bound on the confidence percentage value.

The paper is structured as follows. In Section 2 we summarise our assumptions about the construction of the FRF measurement and the corresponding experimental conditions. In Section 3.1 we review the construction of confidence regions for input and output spectra. In Section 3.2 we show how these regions can be used to construct confidence regions for the FRF measurement. We show the corresponding region is bound by a Cartesian oval and review the geometric construction of such a boundary. In Section 4 we show some results from Monte-Carlo simulations which confirm the validity of such confidence regions. Finally in Section 5 we discuss the results and draw some conclusions.

2. SUMMARY OF ASSUMPTIONS

The following is based on (Pintelon and Schoukens, 2001). We are concerned with the EIV (errors-in-variables) frequency response measurement for the case where the excitation signal is periodic

$$\hat{G}(j\omega_k) = \frac{\hat{Y}_k}{\hat{U}_k} = \frac{\frac{1}{M}\sum_{m=1}^{M} Y^{[m]}(k)}{\frac{1}{M}\sum_{m=1}^{M} U^{[m]}(k)} \tag{1}$$

Here $U^{[m]}(k)$ and $Y^{[m]}(k)$, $m = 1, 2, \ldots, M$ are the DFT (discrete Fourier transform) spectra of the M synchronized input/output records $u(t_m + nT_s)$ and $y(t_m + nT_s)$, $n = 0, 1, \ldots, N-1$, with $m = 1, 2, \ldots, M$, $t_{m+1} = t_m + k_m NT_s$, k_m a natural number and NT_s an integer multiple of the known excitation period T_0. See (Pintelon and Schoukens, 2001) and references therein for a discussion of experimental conditions under which the EIV method has advantages over alternative measurements.

The measured input/output DFT spectra $U^{[m]}(k)$, $Y^{[m]}(k)$ are related to the true values $U_0(k)$, $Y_0(k)$ by

$$\begin{aligned} Y^{[m]}(k) &= Y_0(k) + N_Y^{[m]}(k) \\ U^{[m]}(k) &= U_0(k) + N_U^{[m]}(k) \end{aligned} \tag{2}$$

The following assumptions are made:

(1) the device under test is time-invariant;
(2) the excitation is a periodic signal with known time period T_0;
(3) an integer number periods T_0 of the steady-state response of the system are measured;
(4) the repeated measurements $Y^{[m]}(k)$, $U^{[m]}(k)$, $m = 1, 2, \ldots, M$ have the same true complex values $Y_0(k)$, $U_0(k)$;
(5) $N_U^{[m]}(k)$, $N_Y^{[m]}(k)$, $m = 1, 2, \ldots, M$ are independent and identically distributed (over the repeated measurements m), jointly correlated, zero mean, circular complex random variables with finite second-order moments

$$\begin{aligned} \text{var}\left(N_U^{[m]}(k)\right) &= \mathcal{E}\left\{\left|N_U^{[m]}(k)\right|^2\right\} = \sigma_U^2(k) \\ \text{var}\left(N_Y^{[m]}(k)\right) &= \mathcal{E}\left\{\left|N_Y^{[m]}(k)\right|^2\right\} = \sigma_Y^2(k) \\ \text{covar}&\left(N_Y^{[m]}(k), N_U^{[m]}(k)\right) \\ &= \mathcal{E}\left\{N_Y^{[m]}(k)\bar{N}_U^{[m]}(k)\right\} = \sigma_{YU}^2(k) \end{aligned} \tag{3}$$

and $\mathcal{E}\left\{Z_1^{[m]}(k)Z_2^{[m]}(k)\right\} = 0$ for Z_1, $Z_2 = N_U$ and/or N_Y;
(6) the input/output errors $N_U^{[m]}(k)$, $N_Y^{[m]}(k)$ are linearly correlated;
(7) $N_U^{[m]}(k)$, $N_Y^{[m]}(k)$, $m = 1, 2, \ldots, M$, are complex normally distributed.

3. CONSTRUCTION OF CONFIDENCE REGIONS

3.1 *Confidence regions for input and output spectra*

We begin by constructing confidence regions for $Y_0(k)$ and $U_0(k)$. The construction is classical (Brillinger, 1981). Let

$$s_Y^2(k) = \frac{1}{M}\sum_{m=1}^{M}\left|Y^{[m]}(k) - \hat{Y}_k\right|^2 \tag{4}$$

Then

$$(M-1)\frac{1}{s_Y^2(k)}\left|\hat{Y}_k - Y_0(k)\right|^2 \sim F(2, 2M-2) \tag{5}$$

where $F(2, 2M-2)$ denotes the Fisher distribution with 2 and $2M-2$ degrees of freedom. Hence

$$\begin{aligned} P&\left(\frac{1}{s_Y^2(k)}\left|\hat{Y}_k - Y_0(k)\right|^2 < x^2\right) \\ &= 1 - \left(1 + x^2\right)^{(1-M)} \end{aligned} \tag{6}$$

Thus the disc $\mathbb{D}(c_Y, r_Y)$ with centre c_Y and radius r_Y forms a $100 \times p_Y\%$ confidence region for $Y_0(k)$ when

$$\begin{aligned} c_Y &= \hat{Y}_k \\ r_Y^2 &= s_Y^2(k)\left[(1-p_Y)^{\frac{1}{1-M}} - 1\right] \end{aligned} \tag{7}$$

Similarly the disc $\mathbb{D}(c_U, r_U)$ with centre c_U and radius r_U forms a $100 \times p_U\%$ confidence region for $U_0(k)$ when

$$\begin{aligned} c_U &= \hat{U}_k \\ r_U^2 &= s_U^2(k)\left[(1-p_U)^{\frac{1}{1-M}} - 1\right] \end{aligned} \tag{8}$$

3.2 *Confidence region for transfer function estimate*

Let $\otimes$ and $\oslash$ denote Minkowski multiplication and division (Farouki *et al.*, 2001). Specifically

$$\begin{aligned} \mathbb{A} \otimes \mathbb{B} &= \{a \times b : a \in \mathbb{A} \text{ and } b \in \mathbb{B}\} \\ \mathbb{A} \oslash \mathbb{B} &= \{a/b : a \in \mathbb{A} \text{ and } b \in \mathbb{B}\} \end{aligned} \tag{9}$$

Then the region $\mathbb{G}_k = \mathbb{G}_k(c_Y, r_Y, c_U, r_U)$ defined as

$$\mathbb{G}_k = \mathbb{D}(c_Y, r_Y) \oslash \mathbb{D}(c_U, r_U) \tag{10}$$

is a $100 \times p_Y \times p_U\%$ confidence region for $G_0(k) = Y_0(k)/U_0(k)$. The result follows immediately from the definition of $\mathbb{G}_k(c_Y, r_Y, c_U, r_U)$, since $Y_0(k) \in \mathbb{D}(c_Y, r_Y)$ and $U_0(k) \in \mathbb{D}(c_U, r_U)$ is sufficient for $G_0(k) \in \mathbb{G}_k(c_Y, r_Y, c_U, r_U)$.

The geometric construction of $\mathbb{G}_k(c_Y, r_Y, c_U, r_U)$ follows from (Farouki *et al.*, 2001; Farouki and Pottmann, 2002). Let $\partial\mathbb{D}$ denote the boundary of $\mathbb{D}$. We have for $r_U \neq |c_U|$

$$\begin{aligned} &\mathbb{F}(c_Y, r_Y, c_U, r_U) = \partial\mathbb{D}(c_Y, r_Y) \oslash \partial\mathbb{D}(c_U, r_U) \\ &= \partial\mathbb{D}(c_Y/c_U, r_Y/|c_U|) \oslash \partial\mathbb{D}(1, r_U/|c_U|) \\ &= \left\{ \frac{|c_U|^2}{|c_U|^2 - r_U^2} \right\} \otimes \partial\mathbb{D}(c_Y/c_U, r_Y/|c_U|) \\ &\quad \otimes \partial\mathbb{D}(1, r_U/|c_U|) \\ &= \left\{ \frac{1}{|c_U|^2 - r_U^2} \right\} \otimes \partial\mathbb{D}(c_Y, r_Y) \otimes \partial\mathbb{D}(c_U^\dagger, r_U) \end{aligned} \tag{11}$$

where $\dagger$ denotes complex conjugate.

Furthermore, when $c_Y \neq 0$ and $c_U \neq 0$, the region $\mathbb{F}(c_Y, r_Y, c_U, r_U)$ is the region between the two loops of the Cartesian oval defined as follows (see Fig 1). We can write

$$\mathbb{F} = \mathbb{S} \otimes \partial\mathbb{D}(1, r_Y/|c_Y|) \otimes \partial\mathbb{D}(1, r_U/|c_U|) \tag{12}$$

with

$$\mathbb{S} = \{\sigma\}, \sigma = \frac{c_Y c_U^\dagger}{|c_U|^2 - r_U^2} \tag{13}$$

Note that since the set $\mathbb{S}$ has a single member σ, Minkowski multiplication by $\mathbb{S}$ corresponds to a scaling and rotation operation. Set

$$\begin{aligned} r_1 &= \min\left\{ \frac{r_Y}{|c_Y|}, \frac{r_U}{|c_U|} \right\} \\ r_2 &= \max\left\{ \frac{r_Y}{|c_Y|}, \frac{r_U}{|c_U|} \right\} \end{aligned} \tag{14}$$

and

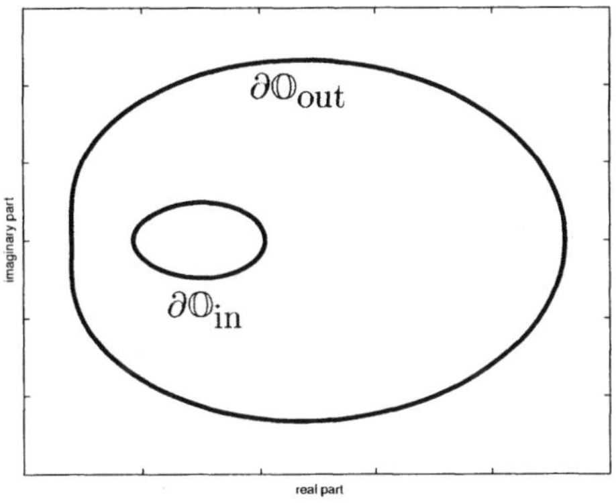

Fig. 1. *Cartesian oval. The area between the inner and outer loop is the Minkowski product of two circles. If the set formed as the Minkowski division of two discs has a boundary (it may cover the infinite plane) then the boundary is either the outer or inner loop of a Cartesian oval.*

$$\begin{aligned} h_1 &= |z-1|^2 + r_1^2 - (r_1^2+1)r_2^2 \\ h_2 &= 2r_1|z - 1 + r_2^2|^2 \end{aligned}$$

with z in the complex plane $z \in \mathbb{C}$. Then $\mathbb{F}$ is the Cartesian oval bound by

$$\partial\mathbb{F} = \mathbb{S} \otimes \partial\mathbb{O} \tag{15}$$

with

$$\partial\mathbb{O} = \{z \in \mathbb{C} : h_1^2 = h_2^2\} \tag{16}$$

Such a Cartesian oval has two loops, the outer and inner given respectively by

$$\begin{aligned} \partial\mathbb{F}_{\text{out}} &= \mathbb{S} \otimes \partial\mathbb{O}_{\text{out}} \\ \partial\mathbb{F}_{\text{in}} &= \mathbb{S} \otimes \partial\mathbb{O}_{\text{in}} \end{aligned} \tag{17}$$

with

$$\begin{aligned} \partial\mathbb{O}_{\text{out}} &= \{z \in \mathbb{C} : h_1 - h_2 = 0\} \\ \partial\mathbb{O}_{\text{in}} &= \{z \in \mathbb{C} : h_1 + h_2 = 0\} \end{aligned} \tag{18}$$

Observing the relation

$$\mathbb{G}(c_Y, r_Y, c_U, r_U) = \bigcup_{\rho_Y \le r_Y} \bigcup_{\rho_U \le r_U} \mathbb{F}(c_Y, \rho_Y, c_U, \rho_U) \tag{19}$$

we may distinguish four cases:

1/ When $r_Y < |c_Y|$ and $r_U < |c_U|$ then $\mathbb{G}$ is the area enclosed by the outer loop $\partial\mathbb{F}_{\text{out}}$ of the Cartesian oval $\mathbb{F}$. We may express $\mathbb{G}$ as

$$\mathbb{G} = \mathbb{S} \otimes \mathbb{O}_1 \tag{20}$$

where

$$\mathbb{O}_1 = \{z \in \mathbb{C} : h_1 - h_2 \le 0\} \tag{21}$$

2/ When $r_Y > |c_Y|$ and $r_U < |c_U|$ then $\mathbb{G}$ may be expressed exactly as for case 1/ (although its construction differs—see below).

3/ When $r_Y < |c_Y|$ and $r_U > |c_U|$ then $\mathbb{G}(c_Y, r_Y, c_U, r_U)$ is the whole complex plane less the region enclosed by the inner loop $\partial\mathbb{F}_{\text{in}}$ of the Cartesian oval $\mathbb{F}$. We may express $\mathbb{G}$ as

$$\mathbb{G} = \mathbb{S} \otimes \mathbb{O}_2 \tag{22}$$

where

$$\mathbb{O}_2 = \{z \in \mathbb{C} : h_1 + h_2 \geq 0\} \tag{23}$$

4/ When $r_Y > |c_Y|$ and $r_U > |c_U|$ then $\mathbb{G}$ is the whole complex plane.

We will find it useful to define

$$\begin{aligned} f_{11}(\phi) &= r_1^2 \cos(\phi) + r_1\sqrt{1 - r_1^2 \sin^2(\phi)} \\ f_{12}(\phi) &= r_1^2 \cos(\phi) - r_1\sqrt{1 - r_1^2 \sin^2(\phi)} \\ f_{21}(\phi) &= r_2^2 \cos(\phi) + r_2\sqrt{1 - r_2^2 \sin^2(\phi)} \\ f_{22}(\phi) &= r_2^2 \cos(\phi) - r_2\sqrt{1 - r_2^2 \sin^2(\phi)} \end{aligned} \tag{24}$$

and

$$g_{mn}(\phi) = 1 - r_m^2 + f_{mn}(\phi)e^{i\phi} \tag{25}$$

for $m = 1, 2$ and $n = 1, 2$. The boundary of $\mathbb{G}(c_Y, r_Y, c_U, r_U)$ may be constructed as follows:

1/ When $r_Y < |c_Y|$ and $r_U < |c_U|$ then

$$\partial\mathbb{F}_{\text{out}} = \{z : z = \sigma g_{11}(\phi) g_{22}(\phi)\} \tag{26}$$

with $-\pi \leq \phi \leq \pi$.

2/ When $r_Y > |c_Y|$ and $r_U < |c_U|$ then

$$\begin{aligned} \partial\mathbb{F}_{\text{out}} &= \{z : z = \sigma g_{11}(\phi) g_{21}(\phi)\} \\ &\cup \{z : z = \sigma g_{11}(\phi) g_{22}(\phi)\} \end{aligned} \tag{27}$$

with $-\arcsin(1/r_2) \leq \phi \leq \arcsin(1/r_2)$.

3/ When $r_Y < |c_Y|$ and $r_U > |c_U|$ then

$$\begin{aligned} \partial\mathbb{F}_{\text{in}} &= \{z : z = \sigma g_{12}(\phi) g_{21}(\phi)\} \\ &\cup \{z : z = \sigma g_{12}(\phi) g_{22}(\phi)\} \end{aligned} \tag{28}$$

with $-\arcsin(1/r_2) \leq \phi \leq \arcsin(1/r_2)$.

Note that in cases 3/ and 4/, which correspond to $r_U > |c_U|$, the confidence region has infinite area. In particular the boundary in case 3/ is an *inner* boundary, while there is no boundary for case 4/. A similar phenomenon has been noted for the special case of indirect closed-loop estimates where confidence regions can be found via Möbius transformation (Wellstead, 1981).

4. SIMULATION EXAMPLE

As a simulation example a random frequency response was generated (a single complex number), and 5000 Monte-Carlo tests were performed where 10 input values were passed through, with noise on both the input and output.

For the results shown in Figs 2 to 4 the input and output noise were uncorrelated. Figs 2 to 4 indicate the proportional number of tests p_N where the true response lies within the confidence region, with confidence values $p_U^2 = p_Y^2 = p = 0.75$, 0.85, and 0.95 respectively. The values are plotted against the number of tests N. In each case the top plot shows results for the input and output spectra. According to the theory p_N will tend to the corresponding confidence value as the number of experiments increases in this case. The bottom plot shows the results for the frequency response estimate. In this case the theory guarantees the proportion of tests p_N will be greater than the corresponding confidence value ($p_U \times p_Y$) as the number of experiments tends to infinity. In fact in all the tests the proportion seems to be above or near $p_U = p_Y$.

Figs 5 to 7 show results from similar experiments, but where there is correlation between the input and output noise. Figs 8 to 10 show further results from similar experiments. This time there is no correlation between the input and output noise, but p_U and p_Y are chosen to be unequal. In all tests the proportion of tests that the frequency response lies within the confidence region seems to be above or near $\min(p_U, p_Y)$. Note that while in the experiments shown in Figs 8 to 10 this proportion lies near $\max(p_U, p_Y)$, in other experiments (not shown) the proportion lay below $\max(p_U, p_Y)$.

5. DISCUSSION AND CONCLUSION

We have constructed confidence regions for FRF measurements using the non-parametric errors-in-variables estimate. The regions are valid even when the input and output noise spectra (and the correlation between them) are unknown.

The confidence region is conservative on at least two counts:

(1) The condition $Y_0(k) \in \mathbb{D}(c_Y, r_Y)$ and $U_0(k) \in \mathbb{D}(c_U, r_U)$ is not *necessary* for $G_0(k) \in \mathbb{G}_k(c_Y, r_Y, c_U, r_U)$.

(2) No account is taken of any possible correlation between the output spectra $Y^{[m]}(k)$ and the input spectra $U^{[m]}(k)$.

Simulations confirm that the confidence regions are rather loose. From the simulation results presented in the paper, and also similar results, we make the following two conjectures (without any further supporting technical results):

Conjecture 1. When $p_U = p_Y$ then $\mathbb{G}_k$ is a $100 \times p_Y\%$ confidence region (we have shown that it is a $100 \times p_Y \times p_U\%$ confidence region).

Conjecture 2. For any choice of p_U, p_Y we have $\mathbb{G}_k(c_Y, r_Y, c_U, r_U)$ is a $100 \times \min(p_Y, p_U)\%$ confidence region.

It may also be possible to construct exact confidence regions for $G_0(k)$. If these were circular and centred on $\hat{G}(j\omega_k)$, we would expect them to approximate the regions defined in (Pintelon *et al.*, 2002) for large M.

6. REFERENCES

Bayard, D. S. (1993). Statistical plant set estimation using Schroeder-phased multisinusoidal input design. *Applied Mathematics and Computation* **58**, 169–198.

Brillinger, D. R. (1981). *Time Series: Data Analysis and Theory (expanded edition).* Holden-Day, San Francisco.

De Vries, D. K. and P. M. J. Van den Hof (1995). Quantification of uncertainty in transfer function estimation. *Automatica* **31**, 543–557.

Farouki, R. T. and H. Pottmann (2002). Exact Minkowski products of n complex disks. *Reliable Computing* **8**, 43–66.

Farouki, R. T., H. P. Moon and B. Ravani (2001). Minkowski geometric algebra of complex sets. *Geometrica Dedicata* **85**, 283–315.

Heath, W. P. (2001*a*). Confidence regions and experiment design for non-parametric estimation. CDC, Orlando.

Heath, W. P. (2001*b*). Confidence regions for non-parametric estimators with periodic excitation and coloured noise. European Control Conference, Porto.

Heath, W. P. (2002). The variance of non-parametric errors-in-variables estimates. *EE Report 02027, University of Newcastle, Australia.* Submitted for Publication.

Pintelon, R. and J. Schoukens (2001). Measurement of frequency response functions using periodic excitations, corrupted by correlated input/output errors. *IEEE Trans. on Instrumentation and Measurement* **50**, 1753–1760.

Pintelon, R., Y. Rolain and W. Van Moer (2002). Probability density function for frequency response function measurements using periodic signals. IEEE Instrumentation and Measurement Technology Conference, Anchorage, Ak, USA, 21-23 May.

Wellstead, P. E. (1981). Non-parametric methods of system identification. *Automatica* **17**, 55–69.

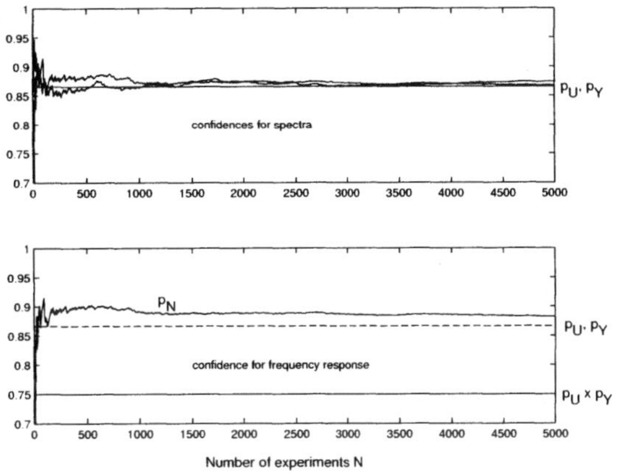

Fig. 2. *Monte Carlo tests with $p_U^2 = p_Y^2 = 0.75$. There is no correlation between input and output noise. In this and subsequent figures, the top plot shows the proportion of tests for which the input and output spectra lie within their respective confidence regions. Similarly the bottom plot shows the proportion of tests for which the frequency response lies within its confidence region.*

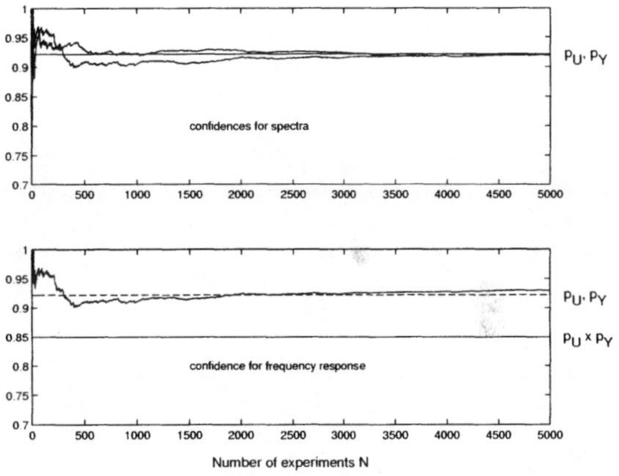

Fig. 3. *Monte Carlo tests with $p_U^2 = p_Y^2 = 0.85$. There is no correlation between input and output noise.*

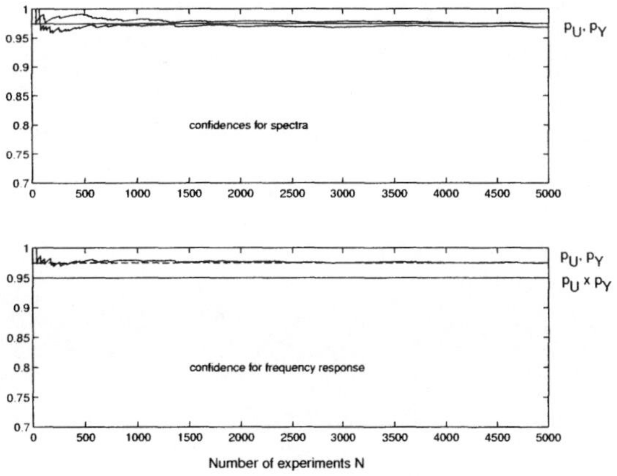

Fig. 4. *Monte Carlo tests with $p_U^2 = p_Y^2 = 0.95$. There is no correlation between input and output noise.*

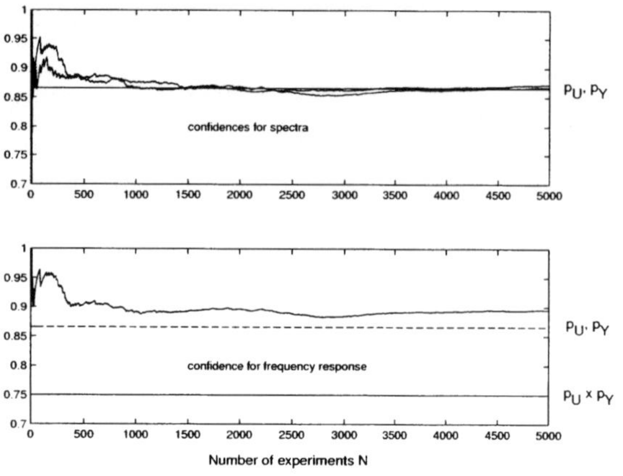

Fig. 5. *Monte Carlo tests with* $p_U^2 = p_Y^2 = 0.75$. *The input and output noises are correlated.*

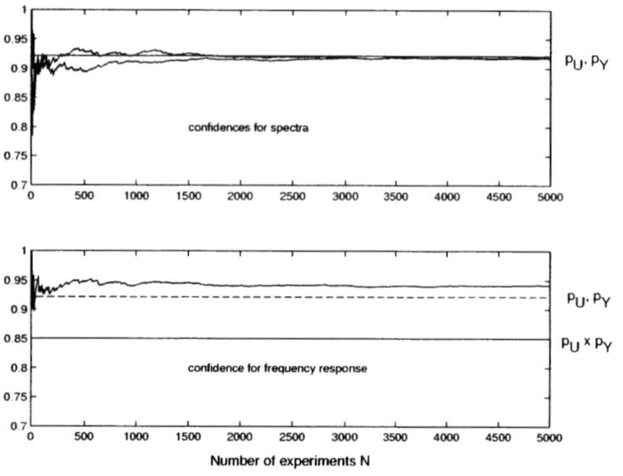

Fig. 6. *Monte Carlo tests with* $p_U^2 = p_Y^2 = 0.85$. *The input and output noises are correlated.*

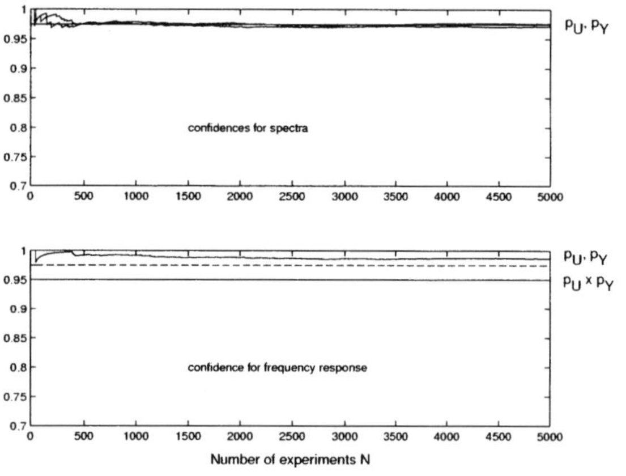

Fig. 7. *Monte Carlo tests with* $p_U^2 = p_Y^2 = 0.95$. *The input and output noises are correlated.*

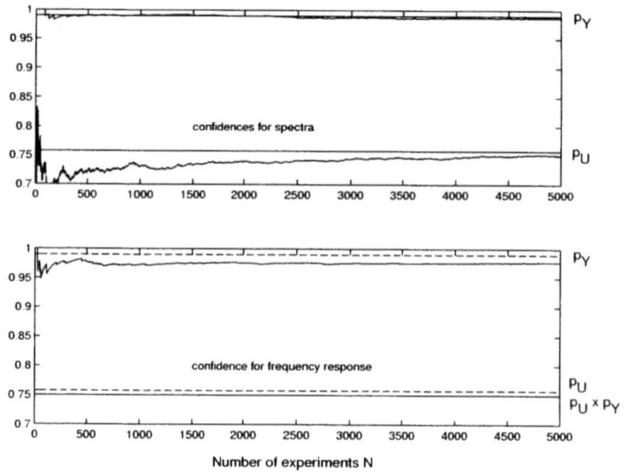

Fig. 8. *Monte Carlo tests with* $p = 0.75$, $p_U = 1.01 \times p$ *and* $p_Y = p/p_U$. *The input and output noises are uncorrelated.*

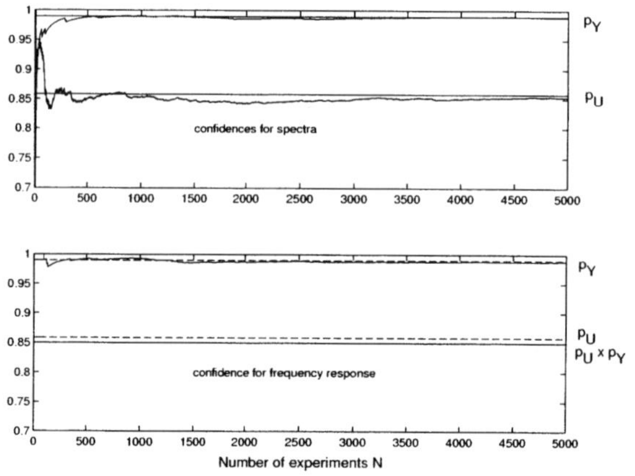

Fig. 9. *Monte Carlo tests with* $p = 0.85$, $p_U = 1.01 \times p$ *and* $p_Y = p/p_U$. *The input and output noises are uncorrelated.*

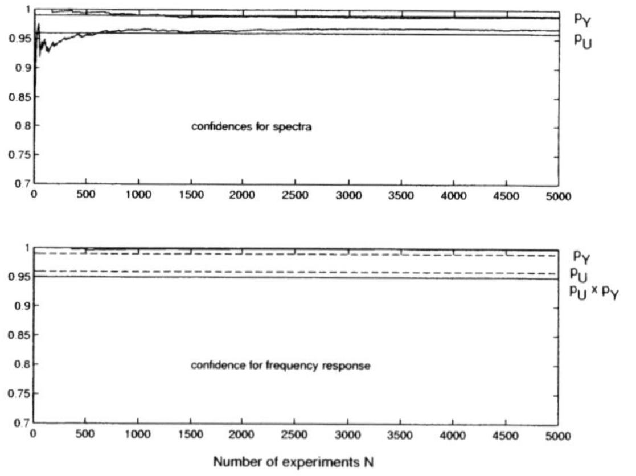

Fig. 10. *Monte Carlo tests with* $p = 0.95$, $p_U = 1.01 \times p$ *and* $p_Y = p/p_U$. *The input and output noises are uncorrelated.*

IFAC Publications
www.elsevier.com/locate/ifac

A NEW CRITERION IN EIV IDENTIFICATION AND FILTERING APPLICATIONS

Roberto Diversi * Roberto Guidorzi * Umberto Soverini *

* *Dipartimento di Elettronica, Informatica e Sistemistica*
Università di Bologna
Viale del Risorgimento 2, 40136, Bologna, Italy
e–mail: {rdiversi, rguidorzi, usoverini}@deis.unibo.it

Abstract: One of the advantages of errors–in–variables (EIV) models consists in the symmetrical description of all variables. These models, on the other hand, are characterized by more complex identification schemes that require, when applied to real data, the definition of suitable criteria. This paper introduces a new efficient and robust criterion based on covariance–matching properties and tests its performance by means of a Monte Carlo simulation concerning also the application of the identified models in EIV filtering. *Copyright © 2003 IFAC*

Keywords: System identification, linear models, errors–in–variables models, covariance–matching techniques, errors–in–variables filtering.

1. INTRODUCTION

The errors–in–variables (EIV) environment enjoys some peculiarities not shared by other classes of models; among them symmetry, that frees from model orienting, and the, equally symmetrical, presence of additive noise on all variables (inputs and outputs if a model orientation is, on the basis of external considerations, anyway introduced). Moreover, EIV models are more suitable when the interest is not focused on prediction but on the description of the true system behind noise–affected observations.

After some pioneer works (Levin, 1964; Aoki and Yue, 1970), only a limited number of EIV identification procedures has been proposed (Söderström, 1981; Anderson and Deistler, 1984; Fernando and Nicholson, 1985; Beghelli *et al.*, 1990; Zheng and Feng, 1989; Zheng, 1999; Söderström *et al.*, 2002). One of the main difficulties concerns the application of the algorithms to real data that do not satisfy exactly the assumptions behind identification schemes, possibly because of the limited number of available samples. In this case it is necessary to introduce suitably consistent and robust cost functions. In fact, the application of the theoretical results described in (Beghelli *et al.*, 1990) to real processes has been possible only after the introduction of cost functions whose robustness under marginal identifiability conditions can exhibit remarkable differences.

This paper proposes a new criterion whose robustness is equal or superior to that of previous ones (Beghelli *et al.*, 1994) and is endowed with an higher level of efficiency. This new approach has been developed on the basis of some results on EIV identification introduced in (Beghelli *et al.*, 1990) and some others concerning optimal EIV filtering (Diversi *et al.*, 2003). The effectiveness of the criterion has been then tested by means of a Monte Carlo simulation and by verifying the excellent performance of the identified models in EIV filtering. This is the first application of identified models in the context of the recently introduced EIV filtering (Guidorzi *et al.*, 2002; Guidorzi *et al.*, 2003).

The paper is organized as follows. Section 2 briefly recalls the EIV environment and defines the problem of EIV identification. Section 3 introduces a new covariance–matching criterion whose testing is performed in Section 4 by means of a Monte Carlo simulation. Section 5 compares the filtering performances of the models obtained in this simulation with the theoretical performance associated with the true model. Some concluding remarks can be found in Section 6.

2. EIV ENVIRONMENT AND PROBLEM STATEMENT

Consider the linear, discrete–time, time–invariant model described by the difference equation

$$q(z)\,\hat{y}(t) = p(z)\,\hat{u}(t) \quad (1)$$

where z denotes the unitary advance operator and

$$q(z) = z^n - \alpha_n\, z^{n-1} - \ldots - \alpha_2\, z - \alpha_1 \quad (2)$$

$$p(z) = \beta_{n+1}\, z^n + \beta_n\, z^{n-1} + \ldots + \beta_2\, z + \beta_1. \quad (3)$$

In an errors–in–variables environment the input and output observations are affected by additive noise, i.e.

$$y(t) = \hat{y}(t) + \tilde{y}(t) \quad (4)$$

$$u(t) = \hat{u}(t) + \tilde{u}(t). \quad (5)$$

Assumptions. It will be assumed that $\tilde{y}(t), \tilde{u}(t)$ are zero–mean ergodic white processes, mutually uncorrelated and uncorrelated with $\hat{u}(t)$, with unknown variances $\tilde{\sigma}_y^{2*}$, $\tilde{\sigma}_u^{2*}$. It will also be assumed that $\hat{u}(t)$ is a zero-mean ergodic process with rational spectral density.

Model (1) can also be written in the form

$$\hat{y}(t+n) = \sum_{k=1}^{n} \alpha_k\, \hat{y}(t+k-1) + \sum_{k=1}^{n+1} \beta_k\, \hat{u}(t+k-1). \quad (6)$$

Define now the Hankel matrix

$$H_k(\hat{y}) = \begin{bmatrix} \hat{y}(1) & \ldots & \hat{y}(k) \\ \hat{y}(2) & \ldots & \hat{y}(k+1) \\ \vdots & & \vdots \\ \hat{y}(N-k+1) & \ldots & \hat{y}(N) \end{bmatrix}, \quad (7)$$

where N is the number of available samples, and the analogous matrices $H_k(y)$, $H_k(\tilde{y})$, $H_k(\hat{u})$, $H_k(u)$ and $H_k(\tilde{u})$. Define also the matrices

$$\hat{H}_k = \left[\, H_k(\hat{y})\; H_k(\hat{u}) \,\right] \quad (8)$$

$$H_k = \left[\, H_k(y)\; H_k(u) \,\right] \quad (9)$$

$$\tilde{H}_k = \left[\, H_k(\tilde{y})\; H_k(\tilde{u}) \,\right]. \quad (10)$$

Relation (6) leads to the following set of $N-n$ linear equations

$$\hat{H}_{n+1}\,\theta^* = 0, \quad (11)$$

where

$$\theta^* = \left[\alpha_1 \cdots \alpha_n \;\; -1 \;\; \beta_1 \cdots \beta_{n+1}\right]^T. \quad (12)$$

By defining the $(2n+2)\times(2n+2)$ covariance matrix $\hat{\Sigma}_n$ as

$$\hat{\Sigma}_n = \lim_{N\to\infty} \frac{1}{N-n}\, \hat{H}_{n+1}^T \hat{H}_{n+1}, \quad (13)$$

relation (11) implies that

$$\hat{\Sigma}_n\,\theta^* = 0. \quad (14)$$

The assumptions on $\tilde{y}(t)$ and $\tilde{u}(t)$ lead to the relations

$$\Sigma_n = \hat{\Sigma}_n + \tilde{\Sigma}_n^* \quad (15)$$

$$\tilde{\Sigma}_n^* = \begin{bmatrix} \tilde{\sigma}_y^{2*}\, I_{n+1} & 0 \\ 0 & \tilde{\sigma}_u^{2*}\, I_{n+1} \end{bmatrix}, \quad (16)$$

where Σ_n and $\tilde{\Sigma}_n^*$ have been obtained as $\hat{\Sigma}_n$ in (13) starting from H_{n+1} and $\tilde{H}_{n+1}$ respectively.

The problem considered in this paper is the following.

Problem 1. Given a set of N samples of the observations $y(t), u(t)$, estimate the coefficients α_k $(k = 1, \ldots, n)$, β_k $(k = 1, \ldots, n+1)$ of $q(z)$, $p(z)$ and the noise variances $\tilde{\sigma}_y^{2*}$, $\tilde{\sigma}_u^{2*}$.

In the sequel, the conditions of system identifiability are assumed to be satisfied (Anderson and Deistler, 1984; Stoica and Nehorai, 1987).

3. A COVARIANCE–MATCHING CRITERION IN EIV IDENTIFICATION

Let us consider the family of non–negative definite diagonal matrices $\tilde{\Sigma}_n = \operatorname{diag}[\tilde{\sigma}_y^2\, I_{n+1}, \tilde{\sigma}_u^2\, I_{n+1}]$ such that

$$\Sigma_n - \tilde{\Sigma}_n \geq 0, \quad \det(\Sigma_n - \tilde{\Sigma}_n) = 0. \quad (17)$$

This set is described by the following theorem (Beghelli *et al.*, 1990).

Theorem 1. The set of all matrices $\tilde{\Sigma}_n$ satisfying relation (17) is defined by the points of a convex curve $\mathcal{S}(\Sigma_n)$, belonging to the first quadrant of the $(\tilde{\sigma}_y^2, \tilde{\sigma}_u^2)$ noise plane $\mathcal{R}^2$. Every point $P = (\tilde{\sigma}_y^2, \tilde{\sigma}_u^2)$ of this curve defines a model $\theta(P)$ satisfying the relation:

$$\hat{\Sigma}_n(P)\,\theta(P) = 0, \quad (18)$$

where

$$\hat{\Sigma}_n(P) = \Sigma_n - \operatorname{diag}\left[\tilde{\sigma}_y^2\, I_{n+1}, \tilde{\sigma}_u^2\, I_{n+1}\right] \geq 0. \quad (19)$$

The points of $\mathcal{S}(\Sigma_n)$ can be conveniently characterized as follows (Guidorzi and Pierantoni, 1995).

Theorem 2. Let $\xi = (\xi_1, \xi_2)$ be a generic point of the first quadrant of $\mathcal{R}^2$ and r the straight line from the origin through ξ. Its intersection with $\mathcal{S}(\Sigma_n)$ is given by the point $P = (\tilde{\sigma}_y^2, \tilde{\sigma}_u^2)$ defined by the relations

$$\tilde{\sigma}_y^2 = \frac{\xi_1}{\lambda_M} \qquad \tilde{\sigma}_u^2 = \frac{\xi_2}{\lambda_M}, \quad (20)$$

where

$$\lambda_M = \max \operatorname{eig}\left(\Sigma_n^{-1} \operatorname{diag}\left[\xi_1 I_{n+1}, \xi_2 I_{n+1}\right]\right). \quad (21)$$

Every point P of the curve can thus be associated with a direction r.

The selection of the correct solution on $\mathcal{S}(\Sigma_n)$ is possible because only the point $P^* = (\tilde{\sigma}_y^{2*}, \tilde{\sigma}_u^{2*})$ is associated with the true model θ^*.

Define, for this purpose, the process $\gamma(t)$ as

$$\begin{aligned}\gamma(t) &= p(z)\,u(t) - q(z)\,y(t) = p(z)\,\tilde{u}(t) - q(z)\,\tilde{y}(t)\\ &= \alpha_1\,\tilde{y}(t-n) + \cdots + \alpha_n\,\tilde{y}(t-1) - \tilde{y}(t)\\ &\quad + \beta_1\,\tilde{u}(t-n) + \cdots + \beta_{n+1}\,\tilde{u}(t); \qquad (22)\end{aligned}$$

$\gamma(t)$ is the sum of two MA processes driven by the white noise $\tilde{y}(t)$ and $\tilde{u}(t)$. Because of the hypotheses on $\tilde{y}(t)$ and $\tilde{u}(t)$ the autocorrelations of $\gamma(t)$, $r_\gamma(k) = \mathrm{E}[\gamma(t)\,\gamma(t-k)]$, are given by

$$r_\gamma(0) = \tilde{\sigma}_y^{2*} \sum_{i=1}^{n+1} \alpha_i^2 + \tilde{\sigma}_u^{2*} \sum_{i=1}^{n+1} \beta_i^2 \quad (23)$$

$$r_\gamma(k) = \tilde{\sigma}_y^{2*} \sum_{i=1}^{n-k+1} \alpha_i\,\alpha_{i+k} + \tilde{\sigma}_u^{2*} \sum_{i=1}^{n-k+1} \beta_i\,\beta_{i+k} \quad (24)$$
$$\text{for } k = 1, \ldots, n$$

$$r_\gamma(k) = 0 \quad \text{for } k > n, \quad (25)$$

where $\alpha_{n+1} = -1$. Consider now the Hankel matrix

$$H_k(\gamma) = \begin{bmatrix} \gamma(n+1) & \ldots & \gamma(n+k) \\ \gamma(n+2) & \ldots & \gamma(n+k+1) \\ \vdots & & \vdots \\ \gamma(N-k+1) & \ldots & \gamma(N) \end{bmatrix}, \quad (26)$$

with $k > n$, and define the covariance matrix

$$\Pi_{k-1}^* = \lim_{N\to\infty} \frac{1}{N-n-k+1} H_k^T(\gamma) H_k(\gamma) \quad (27)$$

$$= \begin{bmatrix} r_\gamma(0) & r_\gamma(1) & \cdots & r_\gamma(n) & 0 & \cdots & \cdots & \cdots & 0 \\ r_\gamma(1) & r_\gamma(0) & \cdots & r_\gamma(n-1) & r_\gamma(n) & \cdots & \cdots & \cdots & 0 \\ \vdots & & \ddots & \vdots & \vdots & \vdots & \vdots & & \vdots \\ r_\gamma(n) & r_\gamma(n-1) & \cdots & r_\gamma(0) & r_\gamma(1) & \cdots & r_\gamma(n) & 0 \cdots & 0 \\ 0 & \ddots & & \ddots & \ddots & \cdots & \vdots & & \vdots \\ \vdots & & & & \vdots & \cdots & r_\gamma(n) & \cdots \cdots & r_\gamma(0) \end{bmatrix},$$

that exhibits a band–toeplitz structure.

It is similarly possible to define, for every point $P = (\tilde{\sigma}_y^2, \tilde{\sigma}_u^2)$ of $\mathcal{S}(\Sigma_n)$, a band–toeplitz matrix $\Pi_{k-1}(P)$ with entries computed by means of (23)–(25) using the variances $(\tilde{\sigma}_y^2, \tilde{\sigma}_u^2)$ and the coefficients $\theta(P) = [\alpha_1(P) \cdots \alpha_n(P)\ -1\ \beta_1(P) \cdots \beta_{n+1}(P)]^T$. It can be immediatly noted that

$$\Pi_{k-1}(P^*) = \Pi_{k-1}^*. \quad (28)$$

It is now possible to introduce a covariance–matching criterion for the estimation of model coefficients and noise variances when only a finite number N of input–output observations is available. In this case Σ_n must be replaced with the sample covariance matrix $\bar{\Sigma}_n$. Then, for every point P of the curve $\mathcal{S}(\bar{\Sigma}_n)$ consider the model $\theta(P)$, the matrix $\Pi_{k-1}(P)$, the process $\gamma_P(t)$ $(t = n+1, \ldots, N)$ and the sample covariance matrix

$$\Pi_{k-1}^s(P) = \frac{1}{L} H_k^T(\gamma_P) H_k(\gamma_P), \quad (29)$$

where $L = N - n - k + 1$. The covariance–matching cost function is defined as

$$J(P) = \left\| \Pi_{k-1}^s(P) - \Pi_{k-1}(P) \right\|_F, \quad (30)$$

where $\|\cdot\|_F$ denotes the Frobenius norm; Problem 1 can be solved by minimizing $J(P)$ along $\mathcal{S}(\bar{\Sigma}_n)$. Condition (28) assures the consistency of the proposed method, since

$$\lim_{N\to\infty} \min_{P\in\mathcal{S}(\bar{\Sigma}_n)} J(P) = 0, \quad (31)$$

and this minimum is associated to P^*.

The following algorithm assumes that the order n of model (1) is known.

Algorithm 1.

(1) Estimate Σ_n as

$$\bar{\Sigma}_n = \frac{1}{N-n} H_{n+1}^T H_{n+1}. \quad (32)$$

(2) Start from a generic point ξ (a generic direction r) in the first quadrant of $\mathcal{R}^2$ and compute, by means of (20)–(21), the corresponding point $P = (\tilde{\sigma}_y^2, \tilde{\sigma}_u^2)$ on the curve $\mathcal{S}(\bar{\Sigma}_n)$.

(3) Compute $\bar{\bar{\Sigma}}_n(P)$ and $\theta(P)$ by means of the relations

$$\bar{\bar{\Sigma}}_n(P) = \bar{\Sigma}_n - \begin{bmatrix} \tilde{\sigma}_y^2 I_{n+1} & 0 \\ 0 & \tilde{\sigma}_u^2 I_{n+1} \end{bmatrix} \quad (33)$$

$$\bar{\bar{\Sigma}}_n(P)\theta(P) = 0. \quad (34)$$

(4) Compute $\gamma_P(t)(t = n+1, \ldots, N)$:

$$\begin{bmatrix} \gamma_P(n+1) \\ \gamma_P(n+2) \\ \vdots \\ \gamma_P(N) \end{bmatrix} = H_{n+1}\,\theta(P). \quad (35)$$

(5) Select $k > n$ (e.g. $k = n+1$), compute the matrix

$$\Pi^s_{k-1}(P) = \frac{1}{L} H^T_k(\gamma_P) H_k(\gamma_P) \qquad (36)$$

and construct the band–toeplitz matrix $\Pi_{k-1}(P)$ by means of the variances $\tilde{\sigma}^2_y, \tilde{\sigma}^2_u$ and of the entries of $\theta(P)$.

(6) Compute the value of the cost function $J(P)$ (30).

(7) Perform a search on the curve $\mathcal{S}(\bar{\Sigma}_n)$ in order to obtain the point P° associated with the minimum of $J(P)$.

4. EXPERIMENTAL RESULTS

The robustness of the proposed criterion has been tested on sequences with length $N = 500$ generated by the model

$$q(z) = z^2 - 0.5\,z + 0.3125$$

$$p(z) = 1.2808\,(z^2 - 0.6\,z - 0.27),$$

where the gain assures an output with the same variance as the input in order to simplify the description of the results. The input sequence $\hat{u}(\cdot)$ used for the simulation has been obtained by stretching a Pseudo Random Binary Sequence of length 100 to reduce its suitability for identification.

A Monte Carlo simulation of 100 independent runs has been performed by adding to the noise–free sequences $\hat{u}(\cdot)$, $\hat{y}(\cdot)$ different gaussian white noise realizations with variances

$$\tilde{\sigma}^{2*}_u = 0.0625, \qquad \tilde{\sigma}^{2*}_y = 0.1225,$$

corresponding to amounts of noise of 25% and 35% in standard deviation i.e. to signal–to–noise ratios of approximately 12 and 9 dB respectively.

EIV models have then been identified by means of Algorithm 1 assuming the index k, in step 5), equal to $n + 1$. The results are summarized in Tables 1–3 that report the true values of the parameters and noise variances, the means of their estimates and the corresponding standard deviations.

Table 1. True and estimated parameters of $q(z)$.

	α_1	α_2
true	0.3125	−0.5000
ident.	0.3157 ± 0.047	-0.5043 ± 0.081

Table 2. True and estimated parameters of $p(z)$.

	β_1	β_2	β_3
true	−0.3458	−0.7685	1.2808
ident.	-0.3298 ± 0.127	-0.7974 ± 0.181	1.2952 ± 0.088

Table 3. True and estimated noise variances.

	$\tilde{\sigma}^{2*}_u$	$\tilde{\sigma}^{2*}_y$
true	0.0625	0.1225
ident.	0.0625 ± 0.010	0.1196 ± 0.022

These results can be considered as excellent with respect to the noise context that has been adopted; in particular, the low dispersion of the obtained values indicates the absence of outliers.

The good selectivity of the cost function (30) can be observed in Figure 1 where the value of $J(P)$ is plotted, against the output noise variance $\tilde{\sigma}^2_y$, along the the curve $\mathcal{S}(\bar{\Sigma}_n)$ in a typical run of the Monte Carlo simulation.

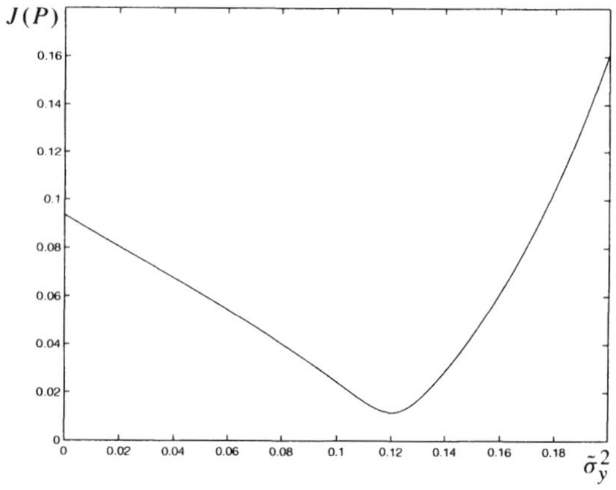

Fig. 1. Values of $J(P)$ along $\mathcal{S}(\bar{\Sigma}_n)$ in a typical run.

5. FILTERING APPLICATIONS

The purpose of this section is to test the performance of the models identified by means of Algorithm 1 in the EIV filtering problem:

Problem 2. Given the process model (1), the noise variances $\tilde{\sigma}^{2*}_u$, $\tilde{\sigma}^{2*}_y$ and a sequence of observations $\{u(1), y(1), \ldots, u(t), y(t)\}$ increasing with t, determine the optimal (minimal variance) estimate of $\hat{u}(t)$, $\hat{y}(t)$.

For the solution of this problem, it is worth recalling the efficient implementation of EIV filtering developed in (Diversi *et al.*, 2003).

Process $\gamma(t)$ can be represented by the following time–varying innnovation model of order n

$$\gamma(t) = \ell(t,t)\,\varepsilon(t) + \ell(t,t-1)\,\varepsilon(t-1) + \cdots + \ell(t,t-n)\,\varepsilon(t-n), \qquad (37)$$

with coefficients obtainable by means of the Bauer algorithm

$$\ell(t,t) = \sqrt{r_\gamma(0) - \sum_{k=1}^{n} \ell(t,t-k)^2} \qquad (38)$$

$$\ell(t,t-k) = \frac{1}{\ell(t-k,t-k)} \Bigg[r_\gamma(k) \qquad (39)$$

$$- \sum_{i=1}^{n-k} \ell(t,t-n+i-1)\,\ell(t-k,t-n+i-1) \Bigg],$$

where $k = 1, \ldots, n$ and $\ell(1,1) = \sqrt{r_\gamma(0)}$.

Starting from the innovation representation (37) it is possible to obtain optimal (minimal variance) estimates $\hat{y}(t|t)$, $\hat{u}(t|t)$ of $\hat{y}(t)$ and $\hat{u}(t)$ as follows.

Algorithm 2.

(1) Start at time $t = n + 1$.

(2) Compute the coefficients $\ell(t, t-n), \ldots, \ell(t, t-1)$, $\ell(t,t)$ by means of the Bauer algorithm (38)–(39).

(3) Compute $\gamma(t)$:

$$\gamma(t) = \alpha_1\, y(t-n) + \cdots + \alpha_n\, y(t-1) - y(t)$$
$$+ \beta_1\, u(t-n) + \cdots + \beta_{n+1}\, u(t)\,. \qquad (40)$$

(4) Compute the innovation $\varepsilon(t)$:

$$\varepsilon(t) = \frac{1}{\ell(t,t)} \left[\gamma(t) - \ell(t,t-1)\,\varepsilon(t-1) - \cdots \right.$$
$$\left. - \,\ell(t,t-n)\,\varepsilon(t-n) \right] \cdot \qquad (41)$$

(5) Compute the filtered input and output samples:

$$\hat{y}(t|t) = y(t) + \frac{\tilde{\sigma}_y^2}{\ell(t,t)}\, \varepsilon(t) \qquad (42)$$

$$\hat{u}(t|t) = u(t) - \frac{\beta_{n+1}\, \tilde{\sigma}_u^2}{\ell(t,t)}\, \varepsilon(t)\,. \qquad (43)$$

(6) Set $t \leftarrow t + 1$ and return to step 2.

The expected performance of the filtering algorithm is given by the expressions

$$\tilde{\sigma}_{ey}^2(t) = \mathrm{E}\left[\left(\hat{y}(t) - \hat{y}(t|t) \right)^2 \right]$$
$$= \tilde{\sigma}_y^2 - \left(\frac{\tilde{\sigma}_y^2}{\ell(t,t)} \right)^2, \qquad (44)$$

$$\tilde{\sigma}_{eu}^2(t) = \mathrm{E}\left[\left(\hat{u}(t) - \hat{u}(t|t) \right)^2 \right]$$
$$= \tilde{\sigma}_u^2 - \left(\frac{\beta_{n+1}\, \tilde{\sigma}_u^2}{\ell(t,t)} \right)^2. \qquad (45)$$

Remark 1. Under very mild conditions, the Cholesky coefficients (38)–(39) converge for $t \to \infty$ (Rissanen and Barbosa, 1969), so that $\tilde{\sigma}_{ey}^2(t)$ and $\tilde{\sigma}_{eu}^2(t)$ converge to constant scalars $\tilde{\sigma}_{ey}^2$ and $\tilde{\sigma}_{eu}^2$ (Diversi *et al.*, 2003).

The EIV models identified in the Monte Carlo simulation have been used to compute, by means of Algorithm 2, the filtered sequences $\hat{y}(\cdot|\cdot)$, $\hat{u}(\cdot|\cdot)$. Table 4 reports the theoretical values of $\tilde{\sigma}_{eu}^2$, $\tilde{\sigma}_{ey}^2$, the means of the values obtained from the sequences filtered with the identified models and the corresponding standard deviations.

Table 4. Theoretical and actual values of $\tilde{\sigma}_{eu}^2$, $\tilde{\sigma}_{ey}^2$.

	$\tilde{\sigma}_{eu}^2$	$\tilde{\sigma}_{ey}^2$
true	0.0373	0.636
ident.	0.0386 ± 0.002	0.0671 ± 0.005

These results show the excellent performance of the identified EIV models; they can thus be profitably used in filtering applications when neither $q(z)$, $p(z)$ nor $\tilde{\sigma}_y^{2*}$, $\tilde{\sigma}_u^{2*}$ are *a priori* known.

6. CONCLUSIONS

This paper has introduced a new robust and efficient criterion for the identification of errors–in–variables models and has tested its effectiveness by means of a Monte Carlo simulation that has evidentiated the precision of the estimated parameters and their reduced dispersion in presence of unbalanced and high levels of noise.

The paper has also proposed the first application of identified models to the recently introduced EIV filtering problem, showing that the performance obtained with these models is almost undistinguishable from the theoretical performance obtainable with the true model.

7. REFERENCES

Anderson B.D.O. and M. Deistler, (1984). Identifiability of dynamic errors–in–variables models. *Journal of Time Series Analysis*, 5, 1–13.

Aoki M. and P.C. Yue, (1970). On a priori estimates of some identification methods. *IEEE Transactions on Automatic Control*, 15, 541–548.

Beghelli S., R. Guidorzi and U. Soverini, (1990). The Frisch scheme in dynamic system identification. *Automatica*, 26, 171–176.

Beghelli S., P. Castaldi, R. Guidorzi and U. Soverini, (1994). A comparison between different model selection criteria in Frisch scheme identification. *Systems Science*, 20, 77–84.

Deistler M., (1986). Linear errors-in-variables models. In S. Bittanti (Ed.), *Time Series and Linear Systems* (pp. 37–68). Berlin: Springer–Verlag.

Diversi R., R. Guidorzi and U. Soverini, (2003). Algorithms for optimal errors–in–variables filtering. *Systems & Control Letters*, 48, 1–13.

Fernando K. V. and H. Nicholson, (1985). Identification of linear systems with input and output noise: the Koopmans–Levin method. *IEE Proceedings*, 132, 30–36.

Guidorzi R., R. Diversi and U. Soverini, (2002). Errors–in–variables filtering in behavioural and state–space contexts. In S. Van Huffel and P. Lemmerling (Eds.), *Total Least Squares and Errors–in–Variables Modelling: Analysis, Algorithms and Applications*, (pp. 281–291). Dordrecht: Kluwer Academic Publishers.

Guidorzi R., R. Diversi and U. Soverini, (2003). Optimal errors–in–variables filtering. *Automatica*, 39, 281–289.

Guidorzi R. and M. Pierantoni, (1995). A new parametrization of Frisch scheme solutions. *Proceedings of 12th International Conference on Systems Science*, Wroclaw, Poland (pp. 114–120).

Levin M.J., (1964). Estimation of a system pulse transfer function in the presence of noise. *IEEE Transactions on Automatic Control*, 9, 229–235.

Rissanen J. and L. Barbosa, (1969). Properties of infinite covariance matrices and stability of optimum predictors. *Information Sciences*, 1, 221–236.

Söderström T., (1981). Identification of stochastic linear systems in presence of input noise. *Automatica*, 17, 713–725.

Söderström T., U. Soverini and K. Mahata, (2002). Perspectives on errors–in–variables estimation for dynamic systems. *Signal Processing*, 82, 1139–1154.

Stoica P. and A. Nehorai, (1987). On the uniqueness of prediction error models for systems with noisy input–output data. *Automatica*, 23, 541–543.

Zheng W.X., (1999). On least–squares identification of stochastic linear systems with noisy input–output data. *International Journal of Adaptive Control and Signal Processing*, 13, 131–143.

Zheng W.X. and C.B. Feng, (1989). Unbiased parameter estimation of linear systems in the presence of input and output noise. *International Journal of Adaptive Control and Signal Processing*, 3, 231–251.

IFAC Publications
www.elsevier.com/locate/ifac

STRONGLY CONSISTENT PARAMETER ESTIMATE FOR ERROR-IN-VARIABLES MODEL

Han-Fu Chen [*,1]

[*] Institute of Systems Science
Academy of Mathematics and Systems Science
Chinese Academy of Sciences
Beijing 100080, P. R. China

Abstract: The paper considers the SISO error-in-variables systems. It is assumed that the input of the system is a stable ARMA process and that the driven noise of the ARMA process and the observation noise are jointly Gaussian. Under a factorization assumption the two-dimensional observation $z_k = [y_k, u_k]^T$ is presented by an ARMA process whose parameters are recursively estimated by an overparameterization technique. It is proved that the estimate for parameters contained in the ARMA system including those in $A(z)$ and $B(z)$ is strongly consistent, and the convergence rate is derived as well.

Keywords: Error-in-variables, coefficient estimate, strong consistency

1. INTRODUCTION

We consider the problem of identifying a linear single-input-single-output system described by the difference equation

$$A(z)y_k^0 = B(z)u_k^0, \qquad (1)$$

where $A(z)$ and $B(z)$ are unknown polynomials and z denotes the backward shift operator $zy_k = y_{k-1}$.

The measurements u_k and y_k of the system input u_k^0 and output y_k^0 are corrupted by noises ξ_k and η_k, respectively:

$$y_k = y_k^0 + \xi_k, \quad u_k = u_k^0 + \eta_k. \qquad (2)$$

Estimation of parameters of $A(z)$ and $B(z)$ from observed data $\{y_k\}$ and $\{u_k\}$ is referred as an "error-in-variables" problem.

[1] This work is supported by the National Natural Science Foundation of China and the Ministry of Science and Technology of China.

It is well-known (Anderson, B. D. O. and M. Deistler, (1984); Anderson, B. D. O. (1985); Deistler, M. (1986); Scherrer, W. and M. Deistler (1998)) that there does not exist a unique solution in general, if only second-order statistics are exploited. However, if some additional assumptions are imposed, for example, if high order cumulant statistics can be used, then it is, in principle, possible to achieve consistent estimates (Scherrer, W. and M. Deistler (1998); Nikias, C. L. and R. Pan (1988); Tugnait, J. R. (1992)).

By assuming the input is non-Gaussian and the noises are Gaussian in (Tugnait, J. R. (1992)), the square root of the magnitude of the fourth cumulant of a generalized error signal is taken as a performance criterion for parameter estimation, and the global minimizer of the proposed criterion $\sqrt{J_N(\theta)}$ is proved to be strongly consistent as $N \to \infty$. For a fixed N, a numerical algorithm is also proposed in (Tugnait, J. R. (1992)) to search the minimizer of $\sqrt{J_N(\theta)}$. But, it is not clear how to guarantee the algorithm to converge to the desired global minimizer. Besides, it would be of interest to recursively estimate unknown parameters with increasing data size N.

There have been developed many interesting numerical identification algorithms by using various methods, e.g., (Stoica, P., M. Cedervall and A. Eriksson (1995); Mahata, K. and T. Söderström (2002); Söderström,T., R. Mahata and U. Soverini (2002)) among others. By using the innovation representation of the observed data, both parametric and non-parametric identification methods are proposed in (Söderström,T., R. Mahata and U. Soverini (2002)). A survey of different approaches is given in (Söderström,T., U. Soverini and R. Mahata (2001)).

In this paper we propose a recursive identification algorithm based on the ARMA representation of the observation process, and give conditions to guarantee the estimate given by the algorithm to be strongly consistent as the number of iterations (or sample size) tends to infinity. The convergence rate is also obtained in the paper.

The rest of the paper is organized as follows. The basic assumptions and the ARMA representation of the observation process are given in Section 2. The main result is presented in Section 3, and a few remarks are given at the end of the paper.

2. REPRESENTATION OF OBSERVATION PROCESSES

The objective of the paper is to design a recursive algorithm based on the noise-corrupted observations $\{u_k\}$ and $\{y_k\}$ to consistently estimate coefficients of $A(z)$ and $B(z)$.

We first list conditions to be imposed on the system, input, and observation noises.

A1. Polynomials $A(z) = 1 + a_1 z + \cdots + a_s z^s$ and $B(z) = b_0 + b_1 z + \cdots + b_s z^s$ are coprime, and $A(z)$ is stable, i.e., all zeros of $A(z)$ are outside the closed unit disk.

A2. The input $\{u_k^0\}$ is an ARMA process

$$P(z)u_k^0 = Q(z)\epsilon_k, \tag{3}$$

$$P(z) = 1 + p_1 z + \cdots + p_s z^s,$$

$$Q(z) = 1 + q_1 z + \cdots + q_s z^s,$$

where $P(z)$ is stable, and $Q(z)$ has no common root with $P(z)$ and $A(z)$.

A3. $\Delta_k \triangleq [\xi_k, \eta_k, \epsilon_k]^T$ is a sequence of iid Gaussian random vectors $\Delta_k \in \mathcal{N}(0, R)$, where ξ_k, η_k and ϵ_k may be correlated..

Here, all polynomials and the covariance matrix R are unknown. The only available information is the observed data $\{u_k\}$ and $\{y_k\}$.

By (1) and (3) it is clear that

$$A(z)P(z)y_k^0 = B(z)Q(z)\epsilon_k$$

and y_k^0 is Gaussian stationary by A1 and A3.

Denote the two-dimensional observation vector by z_k:

$$z_k = \begin{bmatrix} y_k \\ u_k \end{bmatrix} = \begin{bmatrix} y_k^0 \\ u_k^0 \end{bmatrix} + \begin{bmatrix} \xi_k \\ \eta_k \end{bmatrix}. \tag{4}$$

By A1–A3, z_k is a Gaussian stationary process.

Let

$$G(z) \triangleq \begin{bmatrix} A(z) & -B(z) \\ 0 & P(z) \end{bmatrix} = I + G_1 z + \cdots + G_s z^s, \tag{5}$$

where I is the 2×2 identity matrix.

Then by (1) and (3)

$$G(z)z_k = \zeta_k, \tag{6}$$

where

$$\zeta_k \triangleq \begin{bmatrix} A(z)\xi_k - B(z)\eta_k \\ Q(z)\epsilon_k + P(z)\eta_k \end{bmatrix} \tag{7}$$

which is a Gaussian stationary process by A2 and A3.

Let

$$S(z) \triangleq \begin{bmatrix} A(z) & -B(z) & 0 \\ & P(z) & Q(z) \end{bmatrix}$$

$$= S_0 + S_1 z + \cdots + S_s z^s, \tag{8}$$

where

$$S_0 = \begin{bmatrix} 1 & b_0 & 0 \\ 0 & 1 & 1 \end{bmatrix}. \tag{9}$$

From (7) it follows that

$$\zeta_k = S(z)\Delta_k, \tag{10}$$

and the spectral density of $\{\zeta_k\}$ is

$$f_\zeta(\lambda) = f(e^{-i\lambda}),$$

where

$$f(z) = \frac{1}{2\pi} S(z) R S^*(z). \tag{11}$$

We need a technical condition to guarantee regularity of the presentation of $\{\zeta_k\}$.

A4. $f(z)$ can be factorized such that

$$f(z) = \phi(z)\phi^*(z), \tag{12}$$

where $\phi(z)$ is a stable 2×2-matrix polynomial with real coefficients. In addition, $[G_s \vdots\, C_s]$ is of row-full-rank, where C_s is the last matrix coefficient in

$$C(z) \triangleq \phi(z)\phi^{-1}(0) = I + C_1 z + \cdots + C_s z^s.$$

Remark. For A4 to be satisfied it suffices to require that there are 2×2 matrices C_i such that $S_i = C_i S_0$, $i = 1, \ldots, s$ and the polynomial $C(z) \triangleq 1 + C_1 z + \cdots + C_s z^s$ is stable. In this case $f(z) = (C(z)S_0 R^{1/2})(R^{1/2} S_0^T C^*(z))$, and $\phi(z) = C(z)S_0 R^{1/2}$.

Let $\psi(z)$ be the 2×2 matrix of rational functions such that

$$\psi(e^{-i\lambda})\phi(e^{-i\lambda}) = I. \tag{13}$$

Let the spectral representation (Rozanov, Yu. A. (1963)) of $\{\zeta_k\}$ be

$$\zeta_k = \int_{-\pi}^{\pi} e^{ik\lambda} \Phi_\zeta(d\lambda), \tag{14}$$

where $\Phi_\zeta(d\lambda)$ is the corresponding spectral measure such that

$$E\Phi_\zeta(d\lambda)\Phi_\zeta^*(d\lambda) = f_\zeta(\lambda)d\lambda, \tag{15}$$

where $f_\zeta(\lambda) = f(e^{-i\lambda})$ is the spectral density of $\{\zeta_k\}$, and $f(e^{-i\lambda})$ is given by (11).

Defining

$$\Lambda(\Delta) = \int_\Delta \psi(e^{-i\lambda})\Phi_\zeta(d\lambda), \tag{16}$$

we have

$$\begin{aligned} &E\Lambda(d\lambda)\Lambda^*(d\lambda) \\ =&\psi(e^{-i\lambda})\phi(e^{-i\lambda})\phi^*(e^{-i\lambda})\psi^*(e^{-i\lambda})d\lambda = d\lambda. \end{aligned}$$

Then defining

$$\bar{w}_k = \int_{-\pi}^{\pi} e^{ik\lambda}\Lambda(d\lambda), \tag{17}$$

we find that

$$E\bar{w}_k \bar{w}_j^* = 2\pi I \delta_{kj}, \tag{18}$$

where $\delta_{kj} = \begin{cases} 1, \text{ if } k = j \\ 0, \text{ if } k \neq j \end{cases}$.

Further, noticing

$$\begin{aligned} &(\phi(e^{-i\lambda})\psi(e^{-i\lambda}) - I)f_\zeta(e^{-i\lambda}) \\ &\cdot(\psi^*(e^{-i\lambda})\phi^*(e^{-i\lambda}) - I) = 0, \end{aligned}$$

we have

$$\begin{aligned} \zeta_k &= \int_{-\pi}^{\pi} e^{ik\lambda}\phi(e^{-i\lambda})\psi(e^{-i\lambda})\Phi_\zeta(d\lambda) \\ &= \int_{-\pi}^{\pi} e^{ik\lambda}\phi(e^{-i\lambda})\Lambda(d\lambda) = \phi(z)\bar{w}_k. \end{aligned} \tag{19}$$

Combining (6) and (19) leads to

$$G(z)z_k = \phi(z)\bar{w}_k,$$

It is clear that $\bar{w}_k$ must be real since $G(z)$ and $\phi(z)$ are with real coefficients, and both u_k and y_k are real.

Setting $w_k = \phi(0)\bar{w}_k$ and $C(z) = \phi(z)\phi^{-1}(0)$, we then have

$$G(z)z_k = C(z)w_k, \tag{20}$$

From (18) it follows that

$$Ew_k = 0,$$

$$Ew_k w_j = \begin{cases} 2\pi\phi(0)\phi^T(0) \triangleq R_w, \text{ if } k = j \\ 0, \text{ if } k \neq j \end{cases}$$

Since $\{y_k, u_k\}^T$ is Gaussian and C(z) is invertible, $\{w_k\}$ is Gaussian and iid.

3. CONSISTENT PARAMETER ESTIMATE

We now consistently estimate the coefficients of the two-dimensional ARMA process (20). By stability of $C(z)$, we have

$$C^{-1}(z) = \sum_{j=0}^{\infty} \Gamma_j z^j, \quad \forall |z| \leq 1,$$

where $\|\Gamma_j\| \leq M\lambda^j$, for some $M > 0$ and $\lambda \in (0, 1)$.

Let m be a sufficiently large integer such that

$$m > [\log[\|C(z)\|_\infty^{-1} M^{-1}(1 - \lambda)]/\log\lambda] - 1,$$

where $\|C(z)\|_\infty = \max_{|z|=1} |C(z)C(z^{-1})|$. Denote

$$\Gamma(z) \triangleq \sum_{i=0}^{m} \Gamma_j z^j, \quad \Gamma_0 = I.$$

It is known (Chen, H. F. and L. Guo (1991)) that

$$[\Gamma(z)C(z)]^{-1} - \frac{1}{2}I$$

is strictly positive real (SPR) and $\Gamma(z)$ is stable.

From (20) it follows that

$$\Gamma(z)G(z)z_k = \Gamma(z)C(z)w_k.$$

Let

$$M(z) \triangleq \Gamma(z)G(z) = I + M_1 z + \cdots + M_p z^p, \quad p = ms,$$

$$F(z) \triangleq \Gamma(z)C(z) = I + F_1 z + \cdots + F_p z^p.$$

According to the overparameterization technique proposed in (Guo, L. and D. Huang (1989)) (see also (Chen, H. F. and L. Guo (1991))), we first estimate w_k. By setting

$$\theta^T = [-M_1, \ldots, -M_p, F_1, \ldots, F_p],$$

the estimate $\hat{w}_k$ for w_k is given by the following algorithm:

$$\hat{w}_{k+1} = z_{k+1} - \bar{\theta}_{k+1}^T \bar{\phi}_k, \tag{21}$$

$$\bar{\theta}_{k+1} = \bar{\theta}_k + \bar{a}_k \bar{P}_k \bar{\phi}_k (z_{k+1}^T - \bar{\phi}_k^T \bar{\theta}_k), \tag{22}$$

$$\bar{P}_{k+1} = \bar{P}_k - \bar{a}_k \bar{P}_k \bar{\phi}_k \bar{\phi}_k^T \bar{P}_k, \ \bar{a}_k, = (1 + \bar{\phi}_k^T \bar{P}_k \bar{\phi}_k)^{-1}, \tag{23}$$

$$\bar{P}_0 = \alpha I,$$

$$\bar{\phi}_k^T = [z_k^T, \cdots, z_{k-p+1}^T, \hat{w}_k^T, \cdots, \hat{w}_{k-p+1}^T], \tag{24}$$

where $\alpha > 0$ and $\bar{\theta}_0$ is arbitrary.

Using $\hat{w}_k$, we now estimate the unknown coefficients

$$\theta^T = [-G_1, \cdots, -G_s, C_1, \ldots, C_s] \quad (25)$$

by the following algorithm:

$$\begin{aligned} \phi_k^T &= [z_k^T, \cdots, z_{k-s+1}^T, \hat{w}_k^T, \ldots, \hat{w}_{k-s+1}^T], \\ a_k &= (1 + \phi_k^T P_k \phi_k)^{-1} \end{aligned} \quad (26)$$

$$\begin{aligned} \theta_{k+1} &= \theta_k + a_k P_k \phi_k (z_{k+1}^T - \phi_k^T \theta_k), \\ P_{k+1} &= P_k - a_k P_k \phi_k \phi_k^T P_k, \end{aligned} \quad (27)$$

where $\{\hat{w}_k\}$ is given by (21)-(24).

Set

$$\phi_k^{0T} = [z_k^T, \ldots, z_{k-s+1}^T, w_k^T, \ldots, w_{k-s+1}^T]^T$$

and denote by $\lambda_{\max}^0(n)$ and $\lambda_{\min}^0(n)$ the maximum and minimum eigenvalue of $\sum_{i=0}^{n} \phi_i^0 \phi_i^{0T}$, respectively, where $z_j = w_j = 0$ for $j < 0$ by setting.

We now reformulate Theorem 4.8 of (Chen, H. F. and L. Guo (1991)) for the present case as Theorem 1. For its proof we refer to (Chen, H. F. and L. Guo (1991)).

Theorem 1. If i) $\{w_n, \mathcal{F}_n\}$ is a martingale difference sequence with $\sup_n E(\|w_{n+1}\|^{2+\delta}|\mathcal{F}_n) < \infty$ for some $\delta > 0$, and ii)

$$\log \lambda_{\max}^0(n) = o(\lambda_{\min}^0(n)) \text{ a.s. as } n \to \infty,$$

then for θ_n given by (21)-(27)

$$\|\theta_{n+1} - \theta\|^2 = O\Big(\frac{\log \lambda_{\max}^0(n)}{\lambda_{\min}^0(n)}\Big) \text{ a.s.}$$

We now formulate and prove our main result.

Theorem 2. Assume A1–A4. Let $\hat{w}_k$ be generated by (21)-(24). Then θ_n given by (25)-(27) is strongly consistent with the following rate of convergence:

$$\|\theta_{n+1} - \theta\|^2 = O(\frac{\log n}{n^\alpha}) \quad \text{a.s., } \forall \alpha \in (\frac{1}{2}, 1). \quad (28)$$

Proof. Let $\mathcal{F}_k$ be the smallest σ-algebra generated by $\{\zeta_j, j \le k\}$, i.e., $\mathcal{F}_k = \sigma\{\zeta_j, j \le k\}$.

By (6), (7) and the stability of $G(z)$ it follows that $z_k \in \mathcal{F}_k$. Further, by stability of $C(z)$ from (20) it is seen that $w_k \in \mathcal{F}_k$.

We now show that

$$E(w_{k+1}|\mathcal{F}_k) = 0. \quad (29)$$

From (14)-(17) it follows that for $j \le k$

$$\begin{aligned} &E\zeta_j \bar{w}_{k+1}^* \\ &= E\int_{-\pi}^{\pi} e^{ij\lambda}\Phi_\zeta(d\lambda)\Phi_\zeta^*(d\lambda)\psi^*(e^{-i\lambda})e^{-i(k+1)\lambda} \\ &= \int_{-\pi}^{\pi} e^{-i(k+1-j)\lambda} f_\zeta(d\lambda)\psi^*(e^{-i\lambda})d\lambda \\ &= \int_{-\pi}^{\pi} e^{-i(k+1-j)\lambda}\phi(e^{-i\lambda})\phi^*(e^{-i\lambda})\psi^*(e^{-i\lambda})d\lambda \\ &= \int_{-\pi}^{\pi} e^{-i(k+1-j)\lambda}\phi(e^{-i\lambda})d\lambda = 0, \end{aligned}$$

where the last but one equality follows from (13), while the last equality is because of analyticity of $\phi(z)$ in $|z| \le 1$. Thus, we have for $j \le k$

$$E\zeta_j w_{k+1}^T = E\zeta_j \bar{w}_{k+1}^* \phi^T(0) = 0.$$

By A3 from (7) and (20) it is seen that (ζ_j, w_{k+1}) is Gaussian. Therefore, w_{k+1} is independent of $\mathcal{F}_j$, $\forall j \le k$, and (29) is verified.

Therefore, if we can show $\lambda_{\max}^0(n) = O(n)$ and $\liminf_{n\to\infty} \lambda_{\min}^0(n) \ge rn^\alpha$ for $\alpha \in (\frac{1}{2}, 1)$ with $r > 0$ but possibly depending on ω, then the desired result (28) follows from Theorem 1.

Since $\lambda_{\max}^0(n) \le \sum_{i=0}^{n} \|\phi_i^0\|^2$ and $\{w_k\}$ is iid and Gaussian, by stability of $A(z)$ and $P(z)$ it follows that $\lambda_{\max}^0(n) = O(n)$.

We now show that

$$\liminf_{n\to\infty} \lambda_{\min}^0(n) \ge rn^\alpha. \quad (30)$$

We proceed in a way similar to that used in the proof of Theorem 6.2 of (Chen, H. F. and L. Guo (1991)).

Set

$$f_k = \det G(z)\phi_k^0.$$

Noticing

$$\begin{aligned} \lambda_{\min}(\sum_{j=1}^{n} f_j f_j^T) &\triangleq \inf_{\|x\|=1} \sum_{j=1}^{n} (x^T f_j)^2 \\ &\le (s+1)\sum_{j=0}^{s} a_j^2 \lambda_{\min}^0(n), \quad a_0 \triangleq 1, \end{aligned}$$

we find that to show

$$\liminf_{n\to\infty} \lambda_{\min}^0(n) \ge rn^\alpha \quad (31)$$

it suffices to prove

$$\liminf_{n\to\infty} n^{-\alpha}\lambda_{\min}(\sum_{j=1}^{n} f_j f_j^T) \ne 0 \text{ a.s.,} \quad (32)$$

where by $\lambda_{min}(X)$ we denote the minimum eigenvalue of X.

If (32) were not true, then there would exist a sequence of unit vectors $\{\chi_{n_k}\}$

$$\chi_{n_k} \triangleq [\alpha_{n_k}^{(1)T}, \ldots, \alpha_{n_k}^{(s)T}, \beta_{n_k}^{(1)T}, \ldots, \beta_{n_k}^{(s)T}]^T$$

such that

$$\lim_{k\to\infty} n_k^{-\alpha}(\sum_{j=1}^{n_k}(\chi_{n_k}^T f_j)^2) = 0 \tag{33}$$

Let

$$D_{n_k}(z) = \sum_{j=1}^{s} \alpha_{n_k}^{(j)T} z^{j-1} Ad_j G(z)[C(z)] + \sum_{j=1}^{s} \beta_{n_k}^{(j)} z^{j-1} \det G(z) I \triangleq \sum_{j=0}^{3s-1} [d_{n_k}^{(j)T}] z^j. \tag{34}$$

Noticing that

$$\chi_{n_k}^T f_j = \{\alpha_{n_k}^{(1)T} z^0 [Ad_j G(z) C(z)] + \cdots + \alpha_{n_k}^{(s)} z^{s-1} [Ad_j G(z) C(z)] + \beta_{n_k}^{(1)T} z^0 [\det G(z) I] + \cdots + \beta_{n_k}^{(s)} z^{s-1} [\det G(z) I]\} w_j$$
$$= D_{n_k}(z) w_j = \sum_{t=0}^{3s-1} d_{n_k}^{(t)T} w_{j-t},$$

from (33) we have

$$\lim_{n\to\infty} n_k^{-\alpha} \sum_{j=1}^{n_k} (\chi_{n_k}^T f_j)^2 = \lim_{k\to\infty} n_k^{-\alpha} \sum_{j=1}^{n_k} (d_{n_k}^{(0)T} w_j + \cdots + d_{n_k}^{(3s-1)T} w_{j-3s+1})^2 = 0 \tag{35}$$

Since $\{w_k\}$ is an iid Gaussian sequence with nondegerate covariance matrix, we find that

$$d_{n_k}^{(j)} \xrightarrow[k\to\infty]{} 0 \qquad \forall j = 0, \ldots, 3s-1.$$

Therefore, $D_{n_k}(z) \xrightarrow[k\to\infty]{} 0$. Since $\{\chi_{n_k}\}$ is a bounded sequence, from $\{\chi_{n_k}\}$ we may extract a convergent subsequence tending to a unit vector

$$\chi \triangleq [\alpha_1^{(1)T}, \ldots, \alpha^{(s)T}, \beta^{(1)T}, \ldots, \beta^{(s)T}]^T.$$

By $D_{n_k}(z) \to 0$, from (34) it follows that

$$\sum_{j=1}^{s} \alpha^{(j)T} z^{j-1} (Ad_j G(z)) C(z) + \sum_{j=1}^{s} \beta^{(j)T} z^{j-1} \det G(z) I) = 0. \tag{36}$$

We now show that $G(z)$ and $C(z)$ are left coprime. Since they both are with real coefficients, it suffices to show $\det(G(z)G^*(z) + C(z)C^*(z)) > 0$ for any complex z. For this it is sufficient to prove that $\det G(z)$ and $\det C(z)$ have no common root.

Assume the converse. Then $\det G(z)$ and $\det C(z)$ have at least one common root. Since $\det G(z) = A(z)P(z)$, a root of $\det G(z)$ is a root of either $A(z)$ or $P(z)$. Let z^0 be a root of $A(z)$. By A1 and A2, $B(z^0) \neq 0$, $Q(z^0) \neq 0$. Then $\det C(z^0)C^*(z^0) = [A(z^0)A^*(z^0) + B(z)B^*(z^0)] \cdot [P(z^0)P^*(z^0) + Q(z^0)Q^*(z^0)] - P(z^0)B(z^0)B^*(z^0) \cdot P^*(z^0) = B(z^0)B^*(z^0)Q(z^0)Q^*(z^0) > 0$. Similarly, for a root z^0 of $P(z)$ either $A(z^0) \neq 0$ or $B(z^0) \neq 0$ by A1, and $Q(z^0) \neq 0$ by A2, so we also have $\det C(z^0)C^*(z^0) > 0$. Thus, $\det G(z)$ and $\det C(z)$ have no common root, and $G(z)$ and $C(z)$ are left-coprime.

Therefore, there are polynomial matrices $U(z)$ and $V(z)$ such that

$$G(z)U(z) + C(z)V(z) = I$$

By (36) it follows that

$$\sum_{j=1}^{s} \alpha^{(j)T} z^{j-1} AdjG(z)$$
$$= \sum_{j=1}^{s} \alpha^{(j)T} z^{j-1} AdjG(z)(G(z)U(z) + C(z)V(z))$$
$$= \sum_{j=1}^{s} (\alpha^{(j)T} z^{j-1} U(z) - \sum_{j=1}^{s} \beta^{(j)T} z^{j-1} V(z)) \det G(z)$$
$$= \det G(z) \sum_{j=1}^{s} (\alpha^{(j)T} U(z) - \beta^{(j)T} V(z)) z^{j-1}$$
$$\triangleq \det G(z) \sum_{j=0}^{\rho} \nu_j^T z^j. \tag{37}$$

Multiplying (37) from right by $G(z)$ and $C(z)$, respectively, we derive

$$\sum_{j=1}^{s} \alpha^{(j)T} z^{j-1} = \sum_{j=0}^{\rho} \nu_j^T G(z) z^j, \tag{38}$$

and

$$\sum_{j=1}^{s} \beta^{(j)T} z^{j-1} = -\sum_{j=0}^{\rho} \nu_j^T C(z) z^j, \tag{39}$$

which is obtained by invoking (36).

Since $[G_s : C_s]$ is of row-full-rank for any non-zero ν_j we must have $\nu_j^T [G_s : C_s] \neq 0$. If $\nu_j^T G_s \neq 0$, then by comparing orders of both sides of (38), we see that the right-hand side of (37) is of order greater than or equal to s, while its left-hand side is of order $s-1$. This is impossible. Therefore, $\nu_j^T G_s = 0$, $\forall j = 0, 1, \ldots, \rho$. If $\nu_j^T C_s \neq 0$ for some j, then by (39) we arrive at a similar conclusion: $\nu_j^T C_s = 0$, $\forall j = 0, \ldots, \rho$. Consequently, $\nu_j = 0$, $j = 0, 1, \ldots, \rho$, and from (38) and (39), $\alpha^{(j)} = 0$, $\beta^{(j)} = 0$, $j = 1, \ldots, s$. But, this impossible, because $\|\chi\| = 1$. The contraction shows (31). Therefore, we have

$$\frac{\log \lambda_{\max}^0(n)}{\lambda_{\min}^0(n)} = O(\frac{\log n}{n^\alpha}) \xrightarrow[n\to\infty]{} 0,$$

and the desired result follows from Theorem 1. □

Remark If it is known that $C(z)$ is SPR, then the estimation algorithm can be simplified. To be precise, in this case (21)-(24) are not needed, $\hat{w}_k$

in (26) is simply set by $\hat{w}_{k+1} = y_{k+1} - \theta_{k+1}^T \phi_k$, and the theorem remains valid.

4. CONCLUDING REMARKS

The consistent parameter estimate is derived for SISO error-in-variables model under the Gaussian assumption. For further work it is desirable to extend results to the MIMO systems to weaken the factorization assumption, and to remove the Gaussian assumption. Furthermore, it is of interest to consider the input to be a general feedback control rather than an ARMA process.

REFERENCES

B. D. O. Anderson and M. Deistler(1984). Identifiability in dynamic errors-in-variables models, J. Time Series Analysis, Vol. 5, 1984, 1–13.

Anderson, B. D. O.(1985). Identification of scalar errors-in-variables models with dynamics. *Automatica,* **Vol. 21,** 709–716.

Chen, H. F. and L. Guo(1991). Identification and stochastic adaptive control. Birkhäuser, Boston.

Deistler, M. (1986). Linear errors-in-variables, Time Series and Linear Systems. *Lecture Notes in Control and Information Sciences, (S. Bittanti, Ed.),* **Vol. 86,** 37–86.

Guo, L. and D. Huang(1989). Least squares identification for ARMAX models without the positive real condition. *IEEE Trans. Autom. Control,* **Vol. 34,** 1094-1098.

Mahata, K. and T. Söderström(2002). Identification of dynamic errors-in-variables model using prefiltered data. *Preprints of the 15th Triennial World Congress of IFAC.*

Nikias, C. L. and R. Pan(1988). Time delay estimation in unknown Gaussian spatially correlated noise. *IEEE Trans. Acoustics, Speech, Signal Processing,* **Vol. ASSP-36,** 1706–1714.

Rozanov, Yu. A.(1963). Stationary Random Processes, State Publisher of Physics Mathematical Literature, (in Russian).

Scherrer, W. and M. Deistler(1998). A structure theory for linear dynamic errors-in-variables models, *SIAM J. Control and Optimization,* **Vol. 36,** No. 6, 2148–2175.

Stoica, P., M. Cedervall and A. Eriksson(1995). Combined instrumental variable and subspace fitting approach to parameter estimation of noisy input-output systems. *IEEE Trans. Signal Processing,* **43,** 2386–2397.

Söderström,T., R. Mahata and U. Soverini(2002). Identification of dynamic errors-in-variables model using a frequency domain Frisch scheme. *Preprints of the 15th Triennial World Congress of IFAC.*

Söderström.T., U. Soverini and R. Mahata(2001). Perspectives on errors-in-variables estimation. 3rd International Workshop on TLS and Errors-in-Variables Modelling, Leuven, , Belgium.

Tugnait, J. R.(1992). Stochastic system identification with noisy input using cumulant statistics, IEEE Trans. Automatic Control, **Vol. 37,** No. 4, 476–485.

IFAC Publications
www.elsevier.com/locate/ifac

ELLIPSOID SET REFINEMENT BY SIMULTANEOUS USE OF MULTIPLE HYPERPLANE CUTS

D. Joachim * J.R. Deller, Jr. **,[1]

* *Tulane University, New Orleans USA*
** *Michigan State University, East Lansing USA*

Abstract: Let $(\mathcal{H}_n^+, \mathcal{H}_n^-)$, $n = 1, 2, \ldots$ denote a sequence of pairs of parallel hyperplanes in $\mathbb{R}^m$, and, for each n, let $\mathcal{H}_n$ denote the set of points on the open "hyperstrip" between the planes $\mathcal{H}_n^+$ and $\mathcal{H}_n^-$. It is assumed that the polytopic set $\mathcal{P}_k = \cap_{n=1}^k \mathcal{H}_n$ is not empty for any k. Further let $\hat{\mathcal{E}}_n$, $n = 1, 2, \ldots$ denote a corresponding sequence of hyperellipsoids, and let $\mathcal{E}_n \in \mathbb{R}^m$ denote the open hyperellipsoidal set bounded by $\hat{\mathcal{E}}_n$. This paper derives a mathematical formulation for fitting a new, smaller hypervolume, hyperellipsoid, $\hat{\mathcal{E}}_k$, to the intersection $\mathcal{E}_1 \cap \mathcal{H}_1 \cap \mathcal{H}_2 \cap \cdots \cap \mathcal{H}_k = \mathcal{E}_1 \cap \mathcal{P}_k$. The prevailing method for solving this problem involves sequential refinement of ellipsoids as hyperstrips are considered one at a time. The simultaneous use of multiple hyperstrips results in a better fit to the polytope $\mathcal{P}_k$ than that achieved by sequential refinement. The problem arises in numerous control and signal processing problems – in particular the broad class of ellipsoid bounding algorithms used for identification and classification.

Keywords: Bounding ellipsoid, polytopes, bounded disturbances, system identification, guaranteed parameter estimation, worst-case criterion

1. INTRODUCTION

This work is motivated by recent developments in the area of set-based signal and system identification, but the results are relevant to many other problems in engineering and mathematics.

The bounded-error identification [2] algorithms are concerned with the estimation of linear-in-parameters models of the form [3]

$$y_n = \boldsymbol{\theta}^T \boldsymbol{x}_n + \varepsilon_n \tag{1}$$

with $y_n, \varepsilon_n \in \mathbb{R}^1$ and $\boldsymbol{\theta}, \boldsymbol{x}_n \in \mathbb{R}^m$, in which the model disturbances are assumed bounded: $|\varepsilon_n| < \gamma_n$, for each n. If the model has "true" but unknown parameters $\boldsymbol{\theta}_*$, then the bound γ_n, in conjunction with the observations $(y_n, \boldsymbol{x}_n)$, imply two parallel hyperplanes, say $\mathcal{H}_n^+$ and $\mathcal{H}_n^-$, which in turn bound an open "hyperstrip," say $\mathcal{H}_n$, guaranteed to contain $\boldsymbol{\theta}_*$. Since $\boldsymbol{\theta}_* \in \mathcal{H}_n$ for every n, the *feasibility set*, $\mathcal{P}_k = \cap_{n=1}^k \mathcal{H}_n$, contains all estimates that are consistent with the observations $\{(y_n, \boldsymbol{x}_n)\}_{n=1}^k$ and error bounds $\{\gamma_n\}_{n=1}^k$. However, the polytope, say $\hat{\mathcal{P}}_k$, that bounds $\mathcal{P}_k$ is very difficult and computationally expensive to track as the process evolves (Milanese *et al.* 1996). In contrast, an hyperellipsoidal outer bound to $\hat{\mathcal{P}}_k$ is relatively straightforward to compute and manipulate.

Accordingly, the *optimal bounding ellipsoid* (OBE) algorithms sequentially estimate each feasibility set by optimally outer-bounding $\hat{\mathcal{P}}_k$ by an hyperellipsoid, say $\hat{\mathcal{E}}_k$, where optimality is determined according to

[1] This work was supported by the National Science Foundation of the United States under Cooperative Agreement No. IIS-9817485. Any opinions, findings, and conclusions or recommendations expressed in this material are those of the authors and do not necessarily reflect the views of the NSF.

[2] Numerous references to this emerging field are available [e.g., see past SYSID proceedings, or tutorials in (Deller, Jr. *et al.* 1993) or (Deller, Jr. and Huang 2002) which include many seminal references]. Further, similar methods are used for bounded-error filter design under a error tolerance constraint [e.g., (Deller, Jr. and Huang 2002)].

[3] We consider a scalar signal model here for simplicity. The vector-signal, matrix-parameter case is a straightforward generalization (Deller, Jr. *et al.* 1993).

some measure of the ellipsoid's size. Let $\mathcal{E}_k$ denote open hyperellipsoidal set internal to $\hat{\mathcal{E}}_k$. $\mathcal{E}_k$ is often called the *membership set* since it contains all elements of $\mathcal{P}_k$, including $\boldsymbol{\theta}_*$. At each k OBE algorithms sequentially refine the membership set by computing a new hyperellipsoid which optimally bounds the intersection of the current ellipsoid, $\mathcal{E}_{k-1}$, with the new hyperstrip, $\mathcal{H}_k$. A "better" solution at time k could generally be obtained by optimally outer-bounding aggregate polytope $\mathcal{P}_k$, that is, the intersection of all hyperstrip sets known to time k (see Figure 1). However, refining an ellipsoid into another using multiple hyperstrips is a challenging undertaking. The novelty presented in this paper is a method for simultaneously incorporating multiple sets of hyperplane cuts in the ellipsoid update, thus improving the feasibility set refinements with respect to the conventional one hyperstrip at-a-time intersection methods.

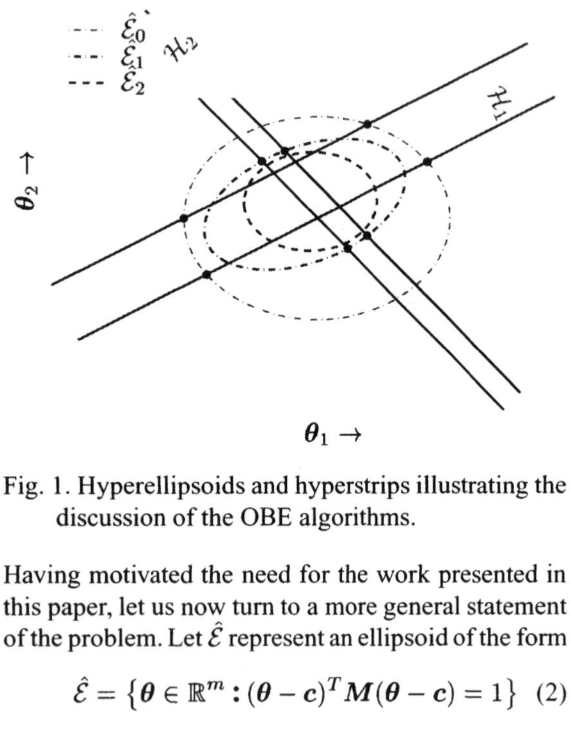

Fig. 1. Hyperellipsoids and hyperstrips illustrating the discussion of the OBE algorithms.

Having motivated the need for the work presented in this paper, let us now turn to a more general statement of the problem. Let $\hat{\mathcal{E}}$ represent an ellipsoid of the form

$$\hat{\mathcal{E}} = \{\boldsymbol{\theta} \in \mathbb{R}^m : (\boldsymbol{\theta} - \boldsymbol{c})^T \boldsymbol{M} (\boldsymbol{\theta} - \boldsymbol{c}) = 1\} \quad (2)$$

where $\boldsymbol{M} \in \mathbb{R}^{m \times m}$ is a positive definite matrix. The open set bounded by $\hat{\mathcal{E}}$ is denoted $\mathcal{E}$. In addition, we represent the intersection of k dissecting hyperstrips by

$$\mathcal{P} = \left\{\boldsymbol{\theta} \in \mathbb{R}^m : \left|\boldsymbol{A}^T\boldsymbol{\theta} - \boldsymbol{b}\right| < \boldsymbol{u}\right\} \quad (3)$$

where $\boldsymbol{A} = \left[\boldsymbol{a}_1 \ \boldsymbol{a}_2 \ \cdots \ \boldsymbol{a}_k\right] \in \mathbb{R}^{m \times k}; \boldsymbol{a}_1, \boldsymbol{a}_2, \cdots, \boldsymbol{a}_k \in \mathbb{R}^m$, and $\boldsymbol{b}, \boldsymbol{u} \in \mathbb{R}^k$ with $\boldsymbol{u} > \boldsymbol{0}$. The vector operators "| |," ">," and "<" are understood to be applied element-wise. Equation (3) describes either hyperplanes, hyperstrips or convex polytopes. The first step in the method to be presented is the geometric mapping of the ellipsoid and hyperstrips to transform the original ellipsoid into the unit sphere, and the trimming of hyperstrips to the "bare essentials." The transformation is followed by the generation of a bounding ellipsoid around the intersected unit sphere. The resulting ellipsoid is then transformed back to the original coordinate system.

2. PREPROCESSING: UNIT SPHEROID AND HYPERSTRIP SELECTION AND TRIMMING

For ease of manipulation, $\overline{\mathcal{E}}$ is transformed to the unit hypersphere centered at the origin, with appropriately mapped hyperstrips, by the change in variable

$$\boldsymbol{\varphi} = \boldsymbol{Q}^{-1}(\boldsymbol{\theta} - \boldsymbol{c}) \quad (4)$$

in which $\boldsymbol{Q}$ is a square-root matrix, $\boldsymbol{M} = \boldsymbol{Q}\boldsymbol{Q}^T$. The corresponding aggregate of hyperstrips $\mathcal{P}$ becomes

$$\overline{\mathcal{P}} = \{\boldsymbol{\varphi} \in \mathbb{R}^m : |\bar{\boldsymbol{A}}\boldsymbol{\varphi} - \bar{\boldsymbol{b}}| < \bar{\boldsymbol{u}}\} \quad (5)$$

where $\bar{\boldsymbol{A}} = \left[\bar{\boldsymbol{a}}_1 \ \bar{\boldsymbol{a}}_2 \ \cdots \ \bar{\boldsymbol{a}}_k\right] = \boldsymbol{Q}^T\boldsymbol{A}\boldsymbol{L}^{-1}$, $\bar{\boldsymbol{b}} = \boldsymbol{L}^{-1}(\boldsymbol{b} - \boldsymbol{A}^T\boldsymbol{c})$ and $\bar{\boldsymbol{u}} = \boldsymbol{L}^{-1}\boldsymbol{u}$, and $\boldsymbol{L}$ is a diagonal scaling matrix with i^{th} diagonal element $\sqrt{\boldsymbol{a}_i^T\boldsymbol{M}\boldsymbol{a}_i}$, such that $\|\bar{\boldsymbol{a}}_i\|_2 = 1$. The transformation $\boldsymbol{L}$ does not change the hyperstrips but it will simplify further developments.

Now let us denote by $\boldsymbol{v}(i)$ the ith element of a vector $\boldsymbol{v}$. A hyperplane $\left\{\boldsymbol{\varphi} \in \mathbb{R}^k : \bar{\boldsymbol{a}}_i^T\boldsymbol{\varphi} < \bar{\boldsymbol{b}}(i)\right\}$ cuts the unit sphere when its minimum distance to the origin $|\bar{\boldsymbol{b}}(i)|$ (corresponding to point $\bar{\boldsymbol{b}}(i)\bar{\boldsymbol{a}}_i$) is less than one. Similarly, the intersection of a hyperstrip

$$\overline{\mathcal{H}}_i = \left\{\boldsymbol{\varphi} \in \mathbb{R}^k : |\bar{\boldsymbol{a}}_i^T\boldsymbol{\varphi} - \bar{\boldsymbol{b}}(i)| < \bar{\boldsymbol{u}}(i)\right\} \quad (6)$$

and the unit sphere is non-empty when the minimum distance from the hyperstrip to the origin, $\left||\bar{\boldsymbol{b}}(i)| - \bar{\boldsymbol{u}}(i)\right|$ is less than one, meaning that at least one (hyperplane) boundary crosses the unit sphere. At least one hyperstrip boundary must cut through the unit sphere if the set $\overline{\mathcal{P}} \cap \overline{\mathcal{E}}$ is non-trivial.

Hyperstrips with a hyperplane boundary outside the unit sphere can be trimmed by moving these outer boundaries tangential to the unit sphere (Belforte *et al.* 1990). These trimming modifications discard all irrelevant portions of the hyperstrips without changing the set $\overline{\mathcal{P}} \cap \overline{\mathcal{E}}$. For all non-redundant hyperstrips, if $\left||\bar{\boldsymbol{b}}(i)| + \bar{\boldsymbol{u}}(i)\right| > 1$ (hyperplane boundary is outside the unit spheroid), define

$$\tilde{\boldsymbol{b}}(i) = \frac{1 - \bar{\boldsymbol{u}}(i) + |\bar{\boldsymbol{b}}(i)|}{2} \quad (7)$$

$$\tilde{\boldsymbol{u}}(i) = \frac{1 + \bar{\boldsymbol{u}}(i) - |\bar{\boldsymbol{b}}(i)|}{2}. \quad (8)$$

The preprocessing steps are illustrated in Figs. 2 and 3. Figure 2 shows an ellipse to be refined by strips $\overline{\mathcal{H}}_1$ through $\overline{\mathcal{H}}_3$. Strip $\overline{\mathcal{H}}_3$ has a redundant intersection with strip $\overline{\mathcal{H}}_3$ and therefore is discarded in Fig. 3. Strip

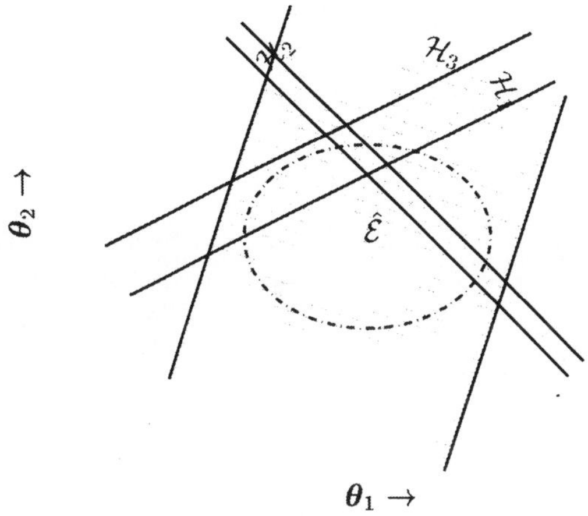

Fig. 2. Original ellipsoid and hyperstrips.

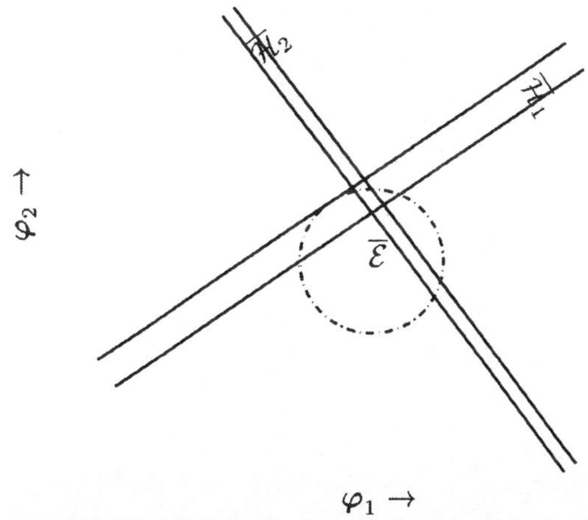

Fig. 3. Ellipsoid of Fig. 2 is mapped to the unit sphere with equally mapped and fitted hyperstrips. The redundant strip $\overline{\mathcal{H}}_3$ is removed.

$\overline{\mathcal{H}}_2$ has a boundary outside the unit sphere which is moved to the unit sphere boundary in Fig. 3.

The preprocessing steps are summarized in Table 2 and 2. In light of the preprocessing, we may now proceed under the following assumptions: **(i)** The membership set $\overline{\mathcal{E}}$ is bounded by the unit sphere. **(ii)** Each hyperstrip cuts through $\overline{\mathcal{E}}$. **(iii)** Matrix $\overline{A}$ is such that $\bar{a}_i^T \bar{a}_i = 1$. **(iv)** $|\overline{b}(i)| < 1$. **(v)** $0 < \overline{u}(i) < 1 - \overline{b}(i)$.

3. FORMULATION

Let $D = \text{diag}\{d_i\} \in \mathbb{R}^{m \times m}$ be a positive definite diagonal matrix. Then the set $\overline{\mathcal{P}}$ of (5) is equivalent to

$$\underline{\overline{\mathcal{P}}} = \left\{\varphi \in \mathbb{R}^k : \left|D\left(\overline{A}\varphi - \overline{b}\right)\right| < D\overline{u}\right\} \quad (9)$$

and is therefore a subset of

$$\left\{\varphi \in \mathbb{R}^k : \|D\left(\overline{A}^T\varphi - \overline{b}\right)\|_2^2 < \|D\overline{u}\|_2^2\right\}. \quad (10)$$

The set defined by inequality (10), re-written as

$$\varphi^T \overline{A} D^2 \overline{A}^T \varphi - 2\overline{b}^T D^2 \overline{A}^T \varphi + \overline{b}^T D^2 \overline{b} < \overline{u}^T D^2 u, \quad (11)$$

is equivalent to the ellipsoidal set

$$\underline{\overline{\mathcal{E}}} = \left\{\varphi \in \mathbb{R}^k : (\varphi - \underline{\overline{c}})^T Q (\varphi - \underline{\overline{c}}) < f\right\} \quad (12)$$

centered around a point $\underline{\overline{c}} \in \underline{\overline{\mathcal{P}}}$ where the scalar

$$f = \underline{\overline{c}}^T Q \underline{\overline{c}} + 2(\overline{b} - \overline{A}^T \underline{\overline{c}})^T D^2 \overline{A}^T \varphi + (\overline{u} + \overline{b})^T D^2 (\overline{u} - \overline{b}) \quad (13)$$

and the matrix $Q = I + AD^2A^T$, where I represents the "ellipsoid matrix" for the unit sphere. By choosing the ellipsoid center $\underline{\overline{c}}$ such that $\overline{b} = \overline{A}^T \underline{\overline{c}}$, we obtain

$$f(D) = 1 + \overline{u}^T D \overline{u} - \overline{b}^T [D^{-1} + G]^{-1} \overline{b} \quad (14)$$

where $G = \overline{A}^T \overline{A}$, $G(i,i) = 1$ and $|G(i,j)| < 1$ (see Section 2)].

Original variables	Normalized domain
A	$\overline{A} = Q^T A L^{-1}$
b	$0 < \overline{b} = L^{-1}(b - A^T c) < 1$
$u > 0$	$0 < \overline{u} = L^{-1} u < 1$
M	$\overline{M} = I$
c	$\overline{c} = 0$

Table 1. Transformations from the original to the normalized domain. These transformations precede the computation of a new bounding set.

Updated variables (Normalized domain)	Original domain
$\underline{\overline{A}}$	$\underline{A} = Q^{-1} \underline{\overline{A}}$
$\underline{\overline{b}}$	$\underline{b} = \overline{b} + L^{-1} A^T c$
$\underline{\overline{u}}$	$\underline{u} = \underline{\overline{u}} > 0$
$\underline{\overline{M}}$	$\underline{M} = Q \underline{\overline{M}} Q^T$
$\underline{\overline{c}}$	$\underline{c} = Q \underline{\overline{c}} + c$

Table 2. Transformations re-mapping the modified bounding ellipsoid back to the original domain.

4. ELLIPSOID MINIMIZATION

Given the bounding ellipsoid formulation above, an "optimal" volume is obtained by minimizing the trace of $\{fQ^{-1}\}$. The derivation of this optimal volume is presented in the following Lemma.

Lemma 1. The minimum ellipsoidal set $\bar{\mathcal{E}}$ resulting from m hyperplane refinements of the unit circle is obtained by solving the equation

$$g(\alpha) = (\boldsymbol{b} - \boldsymbol{S}\bar{\boldsymbol{u}})^T \boldsymbol{G}^{-1} (\boldsymbol{b} - \boldsymbol{S}\bar{\boldsymbol{u}}) + m\alpha - 1 \quad (15)$$

for the scalar α in $\bar{\boldsymbol{u}} = \sqrt{\boldsymbol{u}^2 - \alpha \boldsymbol{1}}$.

Proof: The volume of $\bar{\mathcal{E}}$ is inversely proportional to $\mathrm{tr}\left\{f^{-1}\boldsymbol{Q}\right\}$ in which tr indicates the trace. Therefore minimizing the volume of $\bar{\mathcal{E}}$ becomes equivalent to minimizing the ratio

$$v(\boldsymbol{D}) = \frac{f(\boldsymbol{D})}{\mathrm{tr}\{\boldsymbol{Q}(\boldsymbol{D})\}}. \quad (16)$$

Clearly

$$\mathrm{tr}\{\boldsymbol{Q}(\boldsymbol{D})\} = m + \sum_i d_i. \quad (17)$$

and therefore the gradient of $v(\boldsymbol{D})$ with respect to $\boldsymbol{D}$

$$\frac{1}{\mathrm{tr}^2\{\boldsymbol{Q}(\boldsymbol{D})\}}\left[(m + \sum_i d_i)\frac{\partial f(\boldsymbol{D})}{\partial \boldsymbol{D}} - f(\boldsymbol{D})\right]$$

is zero when

$$\frac{\partial f(\boldsymbol{D})}{\partial \boldsymbol{D}} = \frac{f(\boldsymbol{D})}{m + \sum_i d_i}\boldsymbol{I} = \alpha \boldsymbol{I} \quad (18)$$

where the scalar $\alpha > 0$ is such that

$$f(\boldsymbol{D}) = \alpha\left(m + \sum_i d_i\right). \quad (19)$$

The diagonal elements of (18) can be expressed in vector form as (Joachim *et al.* 1998)

$$\frac{\partial f(\boldsymbol{d})}{\partial \boldsymbol{d}} = \boldsymbol{u}^2 - \left[(\boldsymbol{I} + \boldsymbol{G}\boldsymbol{D})^{-1}\boldsymbol{b}\right]^2 \quad (20)$$

where the squaring operation (e.g., $\boldsymbol{u}^2$) is applied component-wise to each vector. Combining (20) and (18) yields (Joachim *et al.* 1998)

$$(\boldsymbol{I} + \boldsymbol{G}\boldsymbol{D})^{-1}\boldsymbol{b} = \boldsymbol{S}\bar{\boldsymbol{u}} \quad (21)$$

where $\boldsymbol{S}$ is a diagonal matrix of "sign" elements ± 1 and $\bar{\boldsymbol{u}} = \sqrt{\boldsymbol{u}^2 - \alpha\boldsymbol{1}}$. Therefore,

$$\boldsymbol{d} = \boldsymbol{U}^{-1}\boldsymbol{S}\boldsymbol{G}^{-1}(\boldsymbol{b} - \boldsymbol{S}\bar{\boldsymbol{u}}). \quad (22)$$

The function $f(\boldsymbol{D})$ [see (14)] may therefore be expressed as a function of the variable α (through $\bar{\boldsymbol{u}}$ and $\boldsymbol{S}$) by

$$f(\alpha) = 1 + \bar{\boldsymbol{u}}^T\boldsymbol{D}\bar{\boldsymbol{u}} - \boldsymbol{b}^T\boldsymbol{G}^{-1}(\boldsymbol{b} - \boldsymbol{S}\bar{\boldsymbol{u}}) \quad (23)$$

and the right hand side of (19) as

$$\alpha m + \boldsymbol{u}^T\boldsymbol{D}\boldsymbol{u} - \bar{\boldsymbol{u}}^T\boldsymbol{S}\boldsymbol{G}^{-1}(\boldsymbol{b} - \boldsymbol{S}\bar{\boldsymbol{u}})$$

to generate the volume-minimizing α at the zero crossing points of the function (15). ■

Once the solution α of $g(\alpha) = 0$ is computed, the "weights" $\boldsymbol{D}$ are easily generated by inserting $\bar{\boldsymbol{u}} = \sqrt{\boldsymbol{u}^2 - \alpha\boldsymbol{1}}$ into (22) to obtain $\boldsymbol{d}$. The new ellipsoid matrix and center in the normalized domain are then found by

$$\bar{\boldsymbol{c}}_{\text{updated}} = \bar{\bar{\boldsymbol{M}}}\boldsymbol{A}\boldsymbol{D}\boldsymbol{b}$$
$$\bar{\boldsymbol{M}}_{\text{updated}} = \beta\bar{\bar{\boldsymbol{M}}}$$

where $\bar{\bar{\boldsymbol{M}}} = \boldsymbol{I} + \boldsymbol{A}\left(\boldsymbol{D}^{-1} + \boldsymbol{G}\right)^{-1}\boldsymbol{A}^T$ and $\beta = 1 + \boldsymbol{u}^T\boldsymbol{D}\boldsymbol{u} - \boldsymbol{b}^T\left(\boldsymbol{D}^{-1} + \boldsymbol{G}\right)\boldsymbol{b}$.

Finally, mapping back to the original coordinate system is accomplished by use of Table 2 with $\bar{\boldsymbol{M}} \equiv \bar{\boldsymbol{M}}_{\text{updated}}$ and $\bar{\boldsymbol{c}} \equiv \bar{\boldsymbol{c}}_{\text{updated}}$.

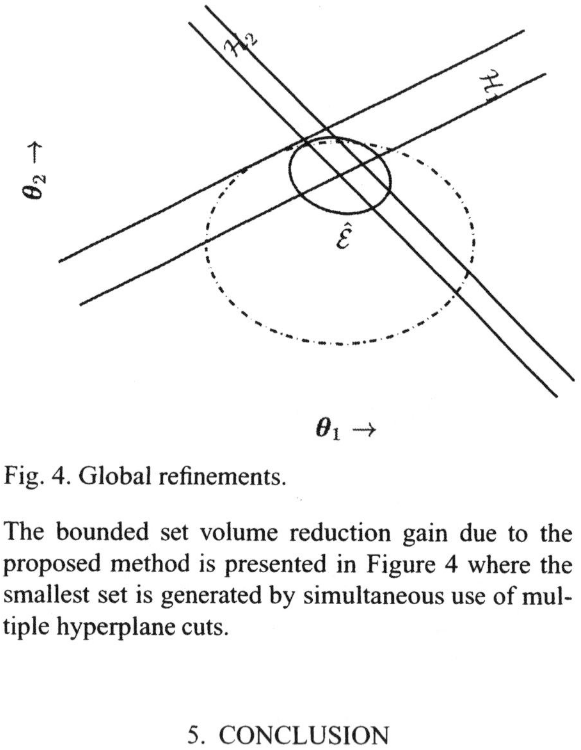

Fig. 4. Global refinements.

The bounded set volume reduction gain due to the proposed method is presented in Figure 4 where the smallest set is generated by simultaneous use of multiple hyperplane cuts.

5. CONCLUSION

Theoretical and experimental results have demonstrated the feasibility and benefits of employing simultaneous hyperplane cuts in ellipsoid set refinement. Future work will include seeking conditions for the existence of a solution that optimizes (15).

6. APPENDIX

Lemma 2. Let $\boldsymbol{A} \in \mathbb{R}^{n \times n}$ be positive definite, and let $\boldsymbol{b}$ denote an n-vector. Define a "sign vector" $\boldsymbol{s} \in \mathbb{R}^n$ to be such that each of its elements is ± 1. There exists at most one sign vector with corresponding diagonal matrix $\boldsymbol{S} = \mathrm{diag}\{\boldsymbol{s}\}$ such that

$$S(b - As) > 0. \tag{24}$$

Proof: Let $s_1 \neq s_2$ be sign vectors with corresponding diagonal matrices S_1 and S_2 both satisfying inequality (24). Without loss of generality, we assume that the mismatched elements of S_1 and S_2 are consecutively arranged in the top left quadrants as S_{1a} and S_{2a}, since we may re-order the basis and preserve the positive definitiveness of A. The partitioned matrices (including A appropriately partitioned) are

$$S_1 = \left[\begin{array}{c|c} S_{1a} & 0 \\ \hline 0 & S_{1b} \end{array}\right], \quad S_2 = \left[\begin{array}{c|c} -S_{1a} & 0 \\ \hline 0 & S_{1b} \end{array}\right],$$
$$A = \left[\begin{array}{c|c} A_a & A_c \\ \hline A_c^T & A_b \end{array}\right] \tag{25}$$

We now add the inequalities corresponding to S_1 and S_2,

$$0 < (S_1 + S_2)b - (S_1 A s_1 + S_2 A s_2)$$

and incorporate (25), to obtain

$$0 < \left[\begin{array}{c} -2S_{1a}A_a s_{1a} \\ \hline 2(S_{1b}b_b - S_{1b}A_b s_{1b}) \end{array}\right]. \tag{26}$$

The cross-product of the top partition inequality with the vector $S_{1a}s_{1a} > 0$ maintains the inequality, therefore

$$0 < -2s_{1a}^T A_a s_{1a}, \tag{27}$$

a contradiction since A_a is positive definite. Hence a sign vector s satisfying (24) is unique. ■

Lemma 3. Let A and b be as in Lemma 2 with a sign vector s_1 and corresponding diagonal matrix S_1 such that

$$S_1(b - As_1) > 0. \tag{28}$$

Then, for any other sign vector $s_2 \neq s_1$,

$$\|A^{-\frac{1}{2}}(b - As_1)\|_2^2 < \|A^{-\frac{1}{2}}(b - As_2)\|_2^2 \tag{29}$$

Proof: Let $s_1 \neq s_2$ be sign vectors with corresponding diagonal matrices S_1 and S_2. Without loss of generality, we assume that the mismatched elements of S_1 and S_2 are consecutively arranged in the top left quadrants as S_{1a} and S_{2a} as in Lemma 2 with appropriately partitioned $b^T = \left[\, b_a^T \; b_b^T \,\right]$. Then, the norm difference $\|A^{-\frac{1}{2}}(b - As_2)\|_2^2 - \|A^{-\frac{1}{2}}(b - As_1)\|_2^2$ develops into

$$\begin{aligned} &2b^T(s_1 - s_2) - (s_1 + s_2)^T A(s_1 - s_2) \\ &= 2b_a^T(2s_{1a}) - (2s_{1b})^T A_c^T(2s_{1a}) \\ &= 4(s_{1a}^T b_a^T - s_{1a}^T A_a s_{1a} - s_{1a}^T A_c s_{1b}) \\ &\quad + 4s_{1a}^T A_a s_{1a} \end{aligned} \tag{30}$$

The term $(s_{1a}^T b_a^T - s_{1a}^T A_a s_{1a} - s_{1a}^T A_c s_{1b})$ is positive since $S_1(b - As_1) > 0$. By the positive definitiveness of A, $s_{1a}^T A_a s_{1a}$ is also positive. Therefore (29) is verified. ■

7. REFERENCES

Belforte, G., B. Bona and V. Cerone (1990). Parameter estimation algorithms for a set-membership description of uncertainty. *Automatica* **26**, 887–898.

Deller, Jr., J.R. and Y.F. Huang (2002). Set-membership identification and filtering in signal processing. *Circuits, Systems, and Signal Processing.*

Deller, Jr., J.R., Majid Nayeri and S. F. Odeh (1993). Least-square identification with error bounds for real-time signal processing and control. *Proc. IEEE* **81**(6), 813–849.

Joachim, D., J.R. Deller, Jr. and M. Nayeri (1998). Multiweight optimization in OBE algorithms for improved tracking and adaptive identification. In: *Proc. Int. Conf. on Acoustics, Speech, and Signal Processing*. Vol. 4. Seattle. pp. 2201–2204.

Milanese, M., J. Norton, H. Piet-Lahanier and E. Walter (editors) (1996). *Bounding Approaches to System Identification*. Plenum. London.

IFAC Publications
www.elsevier.com/locate/ifac

IDENTIFICATION METHODS IN A UNIFIED FRAMEWORK

István Vajk

Department of Automation and Applied Informatics
Budapest University of Technology and Economics
and
HAS-BUTE Control Research Group
H-1521 Budapest, Hungary
fax : (361) 463-2871 and e-mail : vajk@aut.bme.hu

Abstract: The paper derives a framework suitable to discuss the errors-in-variables (EIV) and the maximum likelihood (ML) estimation algorithms to estimate linear system parameters in a unified way. Using the capability of the unified approach a new parameter estimation algorithm is presented offering flexibility to ensure acceptable variance in the estimated parameters. The developed algorithm is based on the application of Hankel matrices of variable size and can be considered as an extended version of the EIV method.

Keywords: Errors-in-variables, linear systems, maximum likelihood estimation, system identification, subspace methods.

1. INTRODUCTION

The classical errors-in-variables (EIV) method is one of the most commonly used way to estimate model parameters from noisy input-output records. In this paper a generalization of the EIV method for linear systems will be derived. The EIV method was exposed by (Castaldi and Soverini, 1996). Much attention has been recently attracted to the Subspace State Space Identification (4SID) and the Total Least Squares methods, as well, see (Chou and Verhaegen, 1997) and (Heij and Scherrer, 1999), respectively. Relation of the new method, called extended errors-in-variables (EEIV) to the 4SID will be discussed in the paper. In some sense EEIV can be considered as a scaleable algorithm between the EIV and the maximum likelihood (ML) estimation.

The paper is organized as follows. Sections 2 and 3 will give a short summary of the EIV and the ML estimation methods pointing out the difference in the loss functions whose minimization exhibit the essence of the estimation strategies these methods are based on. Section 4 contains the main contribution of the paper, namely a common framework to discuss the EIV and ML estimations will be developed here. Once the common framework has been set up, it will be shown that the estimation procedure can well be kept under control to achieve relatively small variance in the estimated parameters with limited computing power requested. The key factor to achieve this flexibility is to freely choose the size of the Hankel matrices involved in the estimation algorithm. Section 5 will show a simple way to derive the 4SID estimation algorithm using the EIV terminology. Section 6 presents simulation results, while Section 7 concludes the paper.

2. ERRORS-IN-VARIABLES METHOD (EIV)

Consider linear, discrete-time systems with noiseless input-output observations u_k^o and y_k^o, respectively:

$$y_k^o = -a_1 y_{k-1}^o - a_2 y_{k-2}^o - ... - a_n y_{k-m}^o$$
$$+ b_1 u_{k-1}^o + b_2 u_{k-2}^o + ... + b_n u_{k-m}^o$$

Introducing the $A(q^{-1})$ and $B(q^{-1})$ polynomials of the backward shift operator q^{-1} as

$$A(q^{-1}) = 1 + a_1 q^{-1} + a_2 q^{-2} + ... + a_m q^{-m}$$

and

$$B(q^{-1}) = b_1 q^{-1} + b_2 q^{-2} + ... + b_m q^{-m}$$

we have

$$A(q^{-1}) y_k^o - B(q^{-1}) u_k^o = 0$$

or equivalently

$$y_k^o = \frac{B(q^{-1})}{A(q^{-1})} u_k^o$$

see Fig. 1. Also, the parameter vector by

$$\mathbf{g} = [b_m, b_{m-1}, ..., b_1, b_0, -a_m, -a_{m-1}, ..., -a_1, -1]^T$$

and the observation vector by

$$\mathbf{x}_k^o = [u_{k-m}^o, u_{k-m+1}^o, ..., u_k^o, y_{k-m}^o, y_{k-m+1}^o, ..., y_k^o]^T$$

allow to write the system equation into an implicit form as $\mathbf{g}^T \mathbf{x}_k^o = 0$, setting $b_0 = 0$. Observe the extra zero element added to the parameter vector. This extension is actually unnecessary, however, it will support the unified problem treatment later on.

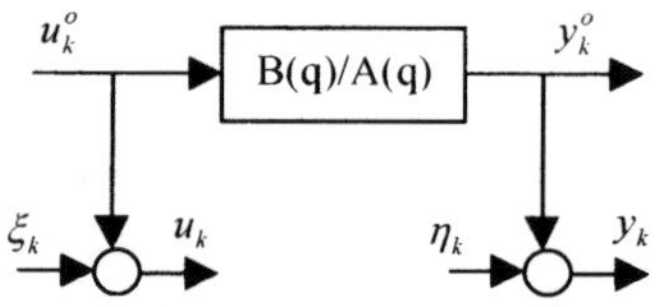

Fig.1 The structure of the investigated system

For the sake of simplicity assume independent white noise components acting on the input-output observations the system equation turns to

$$\mathbf{g}^T \mathbf{x}_k = 0,$$

where

$$\mathbf{x}_k = [u_{k-m}, u_{k-m+1}, ..., u_k, y_{k-m}, y_{k-m+1}, ..., y_k]^T$$

and

$$u_k = u_k^o + \xi_k \ , \ \text{var}(\xi_k) = \mu\sigma_u^2$$
$$y_k = y_k^o + \eta_k \ , \ \text{var}(\eta_k) = \mu\sigma_y^2 .$$

The identification problem is to estimate the unknown a_i and b_i coefficients based on the noisy input-output records. It is well known that using N pair of input-output samples the EIV method finds the unknown parameters by minimizing the loss function given by

$$J_E = \frac{1}{2} \frac{\mathbf{g}^T \left(\sum_{k=m+1}^{N} \mathbf{x}_k \mathbf{x}_k^T \right) \mathbf{g}}{\mathbf{g}^T \mathbf{C}_E \mathbf{g}},$$

where

$$\mathbf{C}_E = \begin{bmatrix} \sigma_u^2 \mathbf{I}_{m+1} & \mathbf{0}_{m+1,m+1} \\ \mathbf{0}_{m+1,m+1} & \sigma_y^2 \mathbf{I}_{m+1} \end{bmatrix}$$

The loss function can easily be reformulated using matrix notation as follows:

$$J_E = \frac{1}{2} \frac{\mathbf{g}^T \mathbf{X}_E^T \mathbf{X}_E \mathbf{g}}{\mathbf{g}^T \mathbf{C}_E \mathbf{g}} \tag{1}$$

where

$$\mathbf{X}_E = [\mathbf{x}_{m+1} \ ... \ \mathbf{x}_N]^T$$

One way to solve the above minimization problem is to consider it as the following generalized eigenvalue-eigenvector problem:

$$(\mathbf{X}_E^T \mathbf{X}_E - \lambda \mathbf{C}_E) \mathbf{g} = \mathbf{0} .$$

The optimal value of the parameter estimation will be equal to the eigenvector related to the smallest eigenvalue found. The advantage of the EIV method is clearly seen from the above equation, namely the consistent estimation is based on the relatively easy calculation of eigenvectors and eigenvalues of *symmetrical* matrices. Factorizing $\mathbf{C}_E$ according to $\mathbf{C}_E = \tilde{\mathbf{C}}_E^T \tilde{\mathbf{C}}_E$ the generalized eigenvalue calculation for the $(\mathbf{X}_E^T \mathbf{X}_E, \tilde{\mathbf{C}}_E^T \tilde{\mathbf{C}}_E)$ pair can be traced back to a generalized singular value decomposition for $(\mathbf{X}_E, \tilde{\mathbf{C}}_E)$ representing a far more robust numerical algorithm. The EIV can be qualified as a direct strategy in the sense that no iteration is required to obtain the parameter estimation.

3. MAXIMUM LIKELIHOOD ESTIMATION (ML)

To derive a compact form for the Maximum Likelihood estimation let us put all the noisy and noise free observations available into a long observation vector:

$$\mathbf{X}_M = [\mathbf{u}^T \ \mathbf{y}^T] \quad \mathbf{X}^o = [\mathbf{u}_o^T \ \mathbf{y}_o^T]$$

with

$$\mathbf{u} = [u_1, u_2, ..., u_N]^T \quad \mathbf{u}_o = [u_1^o, u_2^o, ..., u_N^o]^T$$

and

$$\mathbf{y} = [y_1, y_2, ..., y_N]^T \quad \mathbf{y}_o = [y_1^o, y_2^o, ..., y_N^o]^T$$

then introduce

$$\mathbf{G}_a = \begin{bmatrix} a_n & a_{n-1} & ... & 1 & & & \\ & a_n & a_{n-1} & ... & 1 & & \\ & & ... & ... & ... & ... & \\ & & & a_n & a_{n-1} & ... & 1 \end{bmatrix}^T$$

and

$$\mathbf{G}_b = \begin{bmatrix} b_n & b_{n-1} & ... & 0 & & & \\ & b_n & b_{n-1} & ... & 0 & & \\ & & ... & ... & ... & ... & \\ & & & b_n & b_{n-1} & ... & 0 \end{bmatrix}^T$$

To set up the likelihood function recall the constraint by $\mathbf{g}^T \mathbf{x}_k^o = 0$ for $k = m+1, m+2, ..., N$:

$$\mathbf{X}^o \mathbf{G} = \mathbf{0}$$

where

$$\mathbf{G} = [\mathbf{G}_b^T \ \ -\mathbf{G}_a^T]^T$$

If the noise components shown in Fig.1 are of Gaussian distribution then the conditional distribution of the measurements is

$$prob\left(\mathbf{X}_M \middle| \mathbf{g}, \mathbf{X}^o\mathbf{G} = \mathbf{0}\right) =$$

$$const.\exp\left(-\frac{1}{2}(\mathbf{X}_M - \mathbf{X}^o)\ \mathbf{C}_M^{\ -1}(\mathbf{X}_M - \mathbf{X}^o)^T\right)$$

where

$$\mathbf{C}_M = \begin{bmatrix} \sigma_u^2\mathbf{I}_N & \mathbf{0}_{N,N} \\ \mathbf{0}_{N,N} & \sigma_y^2\mathbf{I}_N \end{bmatrix}$$

Finding the maximum of the likelihood function leads to minimize the following loss function:

$$J_M = \frac{1}{2}\mathbf{X}_M\mathbf{G}\left[\mathbf{G}^T\mathbf{C}_M\mathbf{G}\right]^{-1}\mathbf{G}^T\mathbf{X}_M^T \quad . \quad (2)$$

The minimization is to be performed by $\mathbf{g}$. In a more detailed fashion the above loss function can be written as

$$J_M = \frac{1}{2}(\mathbf{G}_b^T u - \mathbf{G}_a^T y)^T[\sigma_y^2\mathbf{G}_a^T\mathbf{G}_a + \sigma_u^2\mathbf{G}_b^T\mathbf{G}_b]^{-1}(\mathbf{G}_b^T u - \mathbf{G}_a^T y)$$

or as

$$J_M = \frac{1}{2}\mathbf{g}^T\mathbf{X}_E^T\left[\mathbf{G}^T\mathbf{C}_M\mathbf{G}\right]^{-1}\mathbf{X}_E\mathbf{g}$$

4. DEVELOPING A UNIFIED FRAMEWORK FOR EIV AND ML

In the previous sections it has been shown that both the EIV and the ML identification methods are based on the minimization of loss functions defined by Eqs.(1) and (2), respectively. In this section we are going to develop a unified framework to further discuss these loss functions.

4.1 Derivation of a common form of the loss function for EIV and ML

Considering J_E first introduce

$$\mathbf{D}_E = \mathbf{X}_E^T\mathbf{X}_E$$

which allows to reformulate J_E as follows:

$$J_E = \frac{1}{2}tr\left\{\left(\mathbf{g}^T\mathbf{C}_E\mathbf{g}\right)^{-1}\left(\mathbf{g}^T\mathbf{D}_E\mathbf{g}\right)\right\} \qquad (3)$$

Observe further on that

$$\mathbf{g}^T\mathbf{D}_E\mathbf{g} = \mathbf{g}^T\mathbf{X}_E^T\mathbf{X}_E\mathbf{g} = \mathbf{E}_E^T\mathbf{E}_E$$

where

$$\mathbf{E}_E = \mathbf{X}_E\mathbf{g} = \mathbf{U}_E\mathbf{b} - \mathbf{Y}_E\mathbf{a}$$

represents an error vector expressed by the following parameter vectors and observation matrices

$$\begin{bmatrix} e_{m+1} \\ e_{m+2} \\ \cdots \\ e_N \end{bmatrix} = \begin{bmatrix} u_1 & u_2 & \cdots & u_{m+1} \\ u_2 & u_3 & & u_{m+2} \\ \cdots & & & \\ \cdots & \cdots & \cdots & u_N \end{bmatrix}\begin{bmatrix} b_m \\ b_{m-1} \\ \cdots \\ 0 \end{bmatrix} - \begin{bmatrix} y_1 & y_2 & \cdots & y_{m+1} \\ y_2 & y_3 & & y_{m+2} \\ \cdots & & & \cdots \\ \cdots & \cdots & \cdots & y_N \end{bmatrix}\begin{bmatrix} a_m \\ a_{m-1} \\ \cdots \\ 1 \end{bmatrix}$$

The loss function J_E can than be expressed as follows:

$$J_E = \frac{1}{2}tr\left\{\left(\mathbf{g}^T\mathbf{C}_E\mathbf{g}\right)^{-1}\left(\mathbf{E}_E^T\mathbf{E}_E\right)\right\}$$

On the other hand consider J_M now. Applying the same treatment for J_M by Eq.(2) we have

$$J_M = \frac{1}{2}tr\left\{\left(\mathbf{G}^T\mathbf{C}_M\mathbf{G}\right)^{-1}\left(\mathbf{G}^T\mathbf{D}_M\mathbf{G}\right)\right\} \qquad (4)$$

where

$$\mathbf{D}_M = \mathbf{X}_M^T\mathbf{X}_M$$

or using the $\mathbf{E}_M$ error vector

$$\mathbf{E}_M = \mathbf{X}_M\mathbf{G} = \mathbf{U}_M\mathbf{G}_b - \mathbf{Y}_M\mathbf{G}_a$$

where

$$[e_{m+1} \ e_{m+2} \ \cdots \ e_N] = [u_1 \ u_2 \ \cdots \ u_N]\begin{bmatrix} b_m & 0 & \cdots & 0 \\ b_{m-1} & b_m & & 0 \\ \cdots & & & \cdots \\ 0 & 0 & \cdots & b_1 \\ 0 & 0 & 0 & 0 \end{bmatrix} - [y_1 \ y_2 \ \cdots \ y_N]\begin{bmatrix} a_m & 0 & \cdots & 0 \\ a_{m-1} & a_m & & 0 \\ \cdots & & & \cdots \\ 0 & 0 & 1 & a_1 \\ 0 & 0 & 0 & 1 \end{bmatrix}$$

Based on Eqs. (3) and (4) it can be stated that using the trace operation a rather similar form has been derived for the EIV and the ML method. Beyond the trace operation these forms are based on the error vectors $\mathbf{E}_E$ and $\mathbf{E}_M$, respectively. However, the error vectors have been derived in such a way that $\mathbf{E}_E$ is a column vector, while $\mathbf{E}_M$ is a row vector.

To elaborate a unified framework for the joint discussion of the EIV and the ML techniques needs to introduce an error matrix as follows:

$$\mathbf{E}_q = \mathbf{X}_q\mathbf{G}_q = \mathbf{U}_q\mathbf{G}_q^b - \mathbf{Y}_q\mathbf{G}_q^a$$

where

$$\begin{bmatrix} e_{m+1} & e_{m+2} & \cdots & \cdots \\ e_{m+2} & e_{m+3} & \cdots & \cdots \\ \cdots & \cdots & \cdots & \cdots \\ \cdots & \cdots & \cdots & e_N \end{bmatrix} = \begin{bmatrix} u_1 & u_2 & \cdots & u_q \\ u_2 & u_3 & \cdots & \cdots \\ \cdots & \cdots & \cdots & \cdots \\ \cdots & \cdots & \cdots & u_N \end{bmatrix}\begin{bmatrix} b_m & 0 & \cdots & 0 \\ b_{m-1} & b_m & \cdots & 0 \\ \cdots & \cdots & \cdots & \cdots \\ 0 & 0 & 0 & b_1 \\ 0 & 0 & 0 & 0 \end{bmatrix} - \begin{bmatrix} y_1 & y_2 & \cdots & y_q \\ y_2 & y_3 & \cdots & \cdots \\ \cdots & \cdots & \cdots & \cdots \\ \cdots & \cdots & \cdots & y_N \end{bmatrix}\begin{bmatrix} a_m & 0 & \cdots & 0 \\ a_{m-1} & a_m & \cdots & 0 \\ \cdots & \cdots & \cdots & \cdots \\ 0 & 0 & 1 & a_1 \\ 0 & 0 & 0 & 1 \end{bmatrix}$$

Observe that all the matrices in the above expression are of Hankel and Toeplitz type. Using the error matrix $\mathbf{E}_q$ we have

$$J_q = \frac{1}{2}tr\left\{(\mathbf{G}_q^T\mathbf{C}_q\mathbf{G}_q)^{-1}(\mathbf{G}_q^T\mathbf{X}_q^T\mathbf{X}_q\mathbf{G}_q)\right\}$$

or introducing

$$\mathbf{D}_q = \mathbf{X}_q^T\mathbf{X}_q$$

the general form of the loss function can be obtained:

$$J_q = \frac{1}{2}tr\left\{(\mathbf{G}_q^T\mathbf{C}_q\mathbf{G}_q)^{-1}(\mathbf{G}_q^T\mathbf{D}_q\mathbf{G}_q)\right\} \qquad (5)$$

From Eq. (5) the loss functions defined by Eqs. (1) and (2) can be derived by setting q=$m+1$ (to get the

EIV case) and $q=N$ (to get the ML case). More than just to be able to discuss the EIV and the ML methods in a joint framework, Eq. (5) exhibit a new, highly flexible estimation strategy, due to the fact that q is a free parameter under the constraint of $m+1 \le q \le N$ for the identification. It should be mentioned that the general form of J_q by Eq. (5) is capable to handle the case of noiseless input or output records, as well.

As far as the solution of the minimization procedure is concerned, for $q = m+1$ (EIV case), no iteration is necessary to obtain the parameter estimation. However, for all $m+1 < q \le N$ the solution is iterative in nature. Also, increasing the value of q the estimation will be closer and closer to the ML estimation.

Because the derived method extends the concept used by the Errors-in-Variables method for $q>m+1$, the algorithm will be referred to as EEIV (Extended Errors-in-Variables Method) in the sequel.

4.2 Generalization of the common loss function

In the sequel we are going to show a few variants of the EEIV method just derived in order to elaborate non-iterative parameter estimation algorithms for $m+1 < q \le N$. Consider $\mathbf{G}_q^T \mathbf{C}_q \mathbf{G}_q$ in Eq.5 as one possible setting for a symmetrical weighting matrix in

$$J_q = \frac{1}{2} tr\{\mathbf{W}(\mathbf{G}_q^T \mathbf{D}_q \mathbf{G}_q)\} \qquad (6)$$

where $\mathbf{W}$ has been introduced as a general weighting matrix for the elements of the error matrix, namely

$$J_q = \frac{1}{2} tr\{\mathbf{W}(\mathbf{E}_q^T \mathbf{E}_q)\}$$

Just to show a few special options:

- $\mathbf{W} = \left(\mathbf{G}_q^{oT} \mathbf{C}_q \mathbf{G}_q^o\right)^{-1}$, where $\mathbf{G}_q^o$ is formed based on the a'priori information available.
- In some cases $\mathbf{W}$ can be precalculated. E.g. in case of selecting small sampling time we have $A(q^{-1}) \approx (1-q^{-1})^m$ or in case of no input noise $\left(\mathbf{G}_q^{oT} \mathbf{C}_q \mathbf{G}_q^o\right)^{-1}$ is independent of $B(q^{-1})$. Should all these conditions met $\mathbf{W} = \left(\mathbf{G}_q^{oT} \mathbf{C}_q \mathbf{G}_q^o\right)^{-1}$ will not depend on the parameters to be estimated.
- Another possibility is to choose a weighting matrix of averaging type according to $w_{i,j} = 1$.

Also, note that the minimization of the loss function by Eq.(6) results in $\mathbf{g} = \mathbf{0}$ if no constraints are taken into account. Considering the assumptions made for the system and noise model a reasonable choice for the constraint is $\mathbf{g}^T \mathbf{C}_E \mathbf{g} = 1$. Then the solution of the identification problem requires the minimalization of the loss function by Eq.(6) under the constraint of $\mathbf{g}^T \mathbf{C}_E \mathbf{g} = 1$.

As far as the actual minimization concerned consider the structure of the loss function by Eq.(6). Observe the special structure of the $\mathbf{G}_q$ and $\mathbf{D}_q$ matrices:

$$J_q = \frac{1}{2} \mathbf{g}^T \overline{\mathbf{D}}_E \mathbf{g}$$

where

$$\overline{\mathbf{D}}_E = \sum_{i=1}^{q-m} \sum_{j=1}^{q-m} w_{i,j} \mathbf{X}_{qi}^T \mathbf{X}_{qj}$$

Here

$$\mathbf{X}_{qi} = \left[\mathbf{U}_q(:, i : i+m) \quad \mathbf{Y}_q(:, i : i+m)\right].$$

Consequently, the required calculations can well be kept under control in numerical sense using RQ factorization. The related generalized eigenvalue-eigenvector problem to be solved can be written as

$$\left(\overline{\mathbf{D}}_E - \lambda \mathbf{C}_E\right)\mathbf{g} = \mathbf{0}$$

while the estimated parameter vector will be obtained as the eigenvector belonging to the smallest eigenvalue. It can be seen further on, that in case of sufficient excitation the matrix $\mathbf{D}_q$ is symmetrical and positive definite, while $\mathbf{C}_E$ is symmetrical and positive semi-definite. These conditions guarantee that all the eigenvalues will be non-negative.

Remark: All the relations derived in this paper are valid even for $b_0 \ne 0$. In case of having $b_0 = 0$ by structure the size of the related matrices, e.g. $\mathbf{X}_E, \mathbf{D}_E, \mathbf{C}_E ...$ can be reduced by one.

4.3 Robust numerical solution

Considering Eq. (6) again, as both $\mathbf{D}_q$ and $\mathbf{C}_E$ are symmetrical matrices, $\mathbf{C}_E$ can be factorized by $\mathbf{C}_E = \widetilde{\mathbf{C}}^T \widetilde{\mathbf{C}}$ and this holds for $\mathbf{W}$, as well:

$$\mathbf{W} = \mathbf{V}\mathbf{V}^T$$

Using this decompositions in Eq.(6) we have

$$tr\left((\mathbf{V}\mathbf{V}^T)\mathbf{G}_q^T \mathbf{X}_q^T \mathbf{X}_q \mathbf{G}_q\right) = tr\left(\mathbf{V}^T \mathbf{G}_q^T \mathbf{X}_q^T \mathbf{X}_q \mathbf{G}_q \mathbf{V}\right)$$

to further optimize in the sense of numerical calculations. Applying the RQ factorization of matrix $\mathbf{X}_q$

$$\mathbf{X}_q = \mathbf{Q}_x \mathbf{Z}_q$$

where $\mathbf{Z}_q$ is a matrix of size q by q, the following relation holds

$$\mathbf{X}_q^T \mathbf{X}_q = \mathbf{Z}_q^T \mathbf{Z}_q$$

which in turn allows to reformulate the argument of the trace operation:

$$tr\left(\mathbf{V}^T \mathbf{G}_q^T \mathbf{X}_q^T \mathbf{X}_q \mathbf{G}_q \mathbf{V}\right) = tr\left(\mathbf{V}^T \mathbf{G}_q^T \mathbf{Z}_q^T \mathbf{Z}_q \mathbf{G}_q \mathbf{V}\right)$$

Introducing the following notation:

$$\overline{\mathbf{Z}}_j = \sum_{i=1}^{q-m} v_{ij} \mathbf{Z}_i$$

where $\mathbf{Z}_i = \mathbf{Z}(:, i : i+m)$ and v_{ij} is the *i,j* element of matrix $\mathbf{V}$

$$tr\left(\mathbf{V}^T\mathbf{G}_q{}^T\mathbf{Z}_q^T\mathbf{Z}_q\mathbf{G}_q\mathbf{V}\right)=\mathbf{g}^T\left(\sum_{i=1}^{q-m}\overline{\mathbf{Z}}_i{}^T\overline{\mathbf{Z}}_i\right)\mathbf{g}$$

Finally, the summation can be avoided by introducing the super-matrix $\mathbf{\Phi}$ according to

$$\mathbf{g}^T\left(\sum_{i=1}^{q-m}\overline{\mathbf{Z}}_i{}^T\overline{\mathbf{Z}}_i\right)\mathbf{g}=\mathbf{g}^T\mathbf{\Phi}^T\mathbf{\Phi}\mathbf{g}$$

The above form is a valuable result, it shows that the minimization problem by Eq. (6) can be solved in a numerically robust way, namely as the generalized singular value decomposition (GSVD) problem set up for the matrix pair $(\mathbf{\Phi}, \widetilde{\mathbf{C}}_E)$.

5. A STRAIGHTFORWARD DERIVATION OF THE 4SID ALGORITHM

The basic 4SID estimation is based on the parameter estimation using $\mathbf{D}_q$. A fundamental assumption here is that noiseless input samples are assumed to be available for the identification. The classical derivation of the estimation algorithm is based on the RQ factorization of Hankel matrices constructed by the input-output records:

$$\frac{1}{\sqrt{N}}\begin{bmatrix}\mathbf{U}_q\\ \mathbf{Y}_q\end{bmatrix}^T=\begin{bmatrix}\mathbf{R}_{11} & \mathbf{0}\\ \mathbf{R}_{21} & \mathbf{R}_{22}\end{bmatrix}\begin{bmatrix}\mathbf{Q}_1\\ \mathbf{Q}_2\end{bmatrix}$$

The key point is to derive the observability matrix of the system by reducing the rank of $\mathbf{R}_{22}$ via SVD decomposition. Then the system parameters can be estimated by using the observability matrix, which allows to find the system matrices.

Following the terminology of the EIV now we are going to show another approach to derive the 4SID method. Based on the assumptions made earlier the EEIV technique leads to the following over-determined equation:

$$(\mathbf{D}_q-\lambda\mathbf{C}_q)\mathbf{G}_q\approx\mathbf{0}$$

With no noise acting on the input the covariance matrix partitioning of the above expression leads to

$$\left(\begin{bmatrix}\mathbf{U}_q^T\mathbf{U}_q & \mathbf{U}_q^T\mathbf{Y}_q\\ \mathbf{Y}_q^T\mathbf{U}_q & \mathbf{Y}_q^T\mathbf{Y}_q\end{bmatrix}-\lambda\begin{bmatrix}\mathbf{0} & \mathbf{0}\\ \mathbf{0} & \mathbf{C}_{yq}\end{bmatrix}\right)\begin{bmatrix}\mathbf{G}_q^b\\ -\mathbf{G}_q^a\end{bmatrix}\approx\begin{bmatrix}\mathbf{0}\\ \mathbf{0}\end{bmatrix}.$$

Decomposing this equation results in

$$\mathbf{U}_q^T\mathbf{U}_q\mathbf{G}_q^b-\mathbf{U}_q^T\mathbf{Y}_q\mathbf{G}_q^a\approx\mathbf{0}$$

and

$$\mathbf{Y}_q^T\mathbf{U}_q\mathbf{G}_q^b-\mathbf{Y}_q^T\mathbf{Y}_q\mathbf{G}_q^a\approx\lambda\mathbf{C}_{yq}\mathbf{G}_q^a$$

From the first equation express $\mathbf{G}_q^b$ (the 4SID terminology calls this step as projection) then substitute it to the second equation:

$$(\mathbf{Y}_q^T(\mathbf{I}-\mathbf{U}_q(\mathbf{U}_q^T\mathbf{U}_q)^{-1}\mathbf{U}_q^T)\mathbf{Y}_q-\lambda\mathbf{C}_{yq})\mathbf{G}_q^a\approx\mathbf{0}$$

Assuming that $\mathbf{C}_{yq}$ is a unity matrix the 4SID technique solves the above over-determined equation via SVD decomposition. Since

$$\frac{1}{N}\mathbf{Y}_q^T(\mathbf{I}-\mathbf{U}_q(\mathbf{U}_q^T\mathbf{U}_q)^{-1}\mathbf{U}_q^T)\mathbf{Y}=\mathbf{R}_{22}\mathbf{R}_{22}^T$$

by factorization of $\mathbf{R}_{22}$ the coefficients of the denominator polynomial (a_i) of the system transfer function can be obtained.

6. SIMULATION RESULTS

Simulation results to identify the parameters of a sampled data model of a second order continuous system using two different methods (4SID and EEIV) will be shown. To be able to compare results obtained by using 4SID with those obtained by using EEIV noiseless input records will be assumed. Also, for a clear comparison the loss function

$$J=\frac{1}{2}tr\left(\mathbf{W}\left(\mathbf{G}_q^T\mathbf{D}_q\mathbf{G}_q\right)\right)$$

used by the EEIV method will be replaced by the following one

$$J=\frac{1}{2}tr\left(\mathbf{W}\left(\mathbf{G}_q^{aT}\mathbf{D}_{yyq}\mathbf{G}_q^a\right)\right)$$

where

$$\mathbf{D}_{yyq}=\mathbf{Y}_q^T(\mathbf{I}-\mathbf{U}_q(\mathbf{U}_q^T\mathbf{U}_q)^{-1}\mathbf{U}_q^T)\mathbf{Y}_q$$

is the projected output matrix used by the 4SID terminology. In this way the two different methods will result in identical results for $q=m+1$. As far as the selection of the elements of the weighting matrix $\mathbf{W}$ is concerned, all entries have been selected to unity (1).

To run simulation studies consider a linear, continuous-time, second order process:

$$y_o=\frac{1}{0.1s^2+0.2s+1}u$$

The above process is sampled by $T_s = 0.01$ sec. Also, assume additive white noise with variance of 0.1 acting on the output. Number of samples processed by the identification algorithms is 1100. As far as the excitation is concerned, the input is raised from 0 to 1 at step 100, than set back to zero at step 600. Fig. 2 shows the typical records.

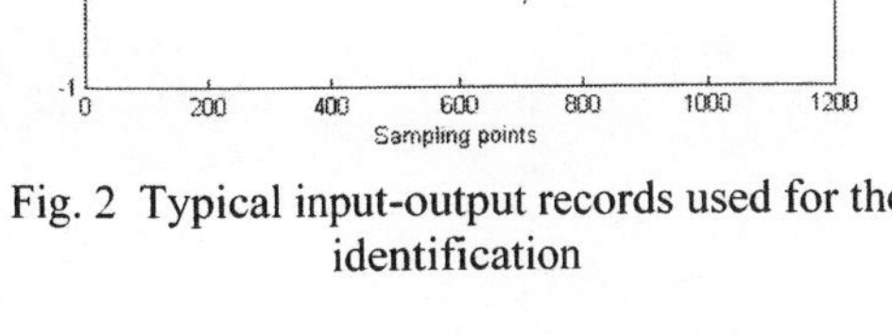

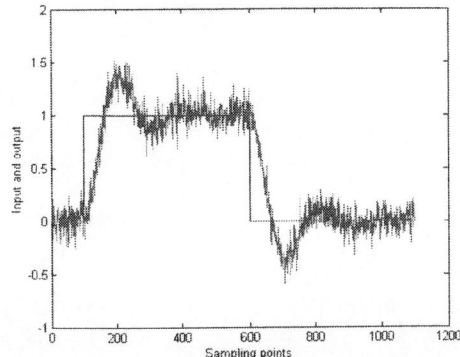

Fig. 2 Typical input-output records used for the identification

Using the same simulation environment several identification runs over 1100 samples will be evaluated. For each identification run the q scaling factor (essentially the size of the Hankel matrices used) will sweep the range from $q=m+1=3$ to $q=20$. As an overall measure for the effectiveness of the parameter estimation the empirical standard deviation will be evaluated:

$$s_k(q) = \sqrt{\frac{1}{L}\sum_{j=1}^{L}(\hat{a}_{k,q,j} - a_k)^2}$$

where j denotes the index of the simulation run, $L=100$ stands for the number of simulation runs performed, while a_k $(k=1,2)$ are the discrete time system coefficients in

$$A(q^{-1}) = 1 + a_1 q^{-1} + a_2 q^{-2} = 1 - 1.9792...q^{-1} + 0.9802...q^{-2}$$

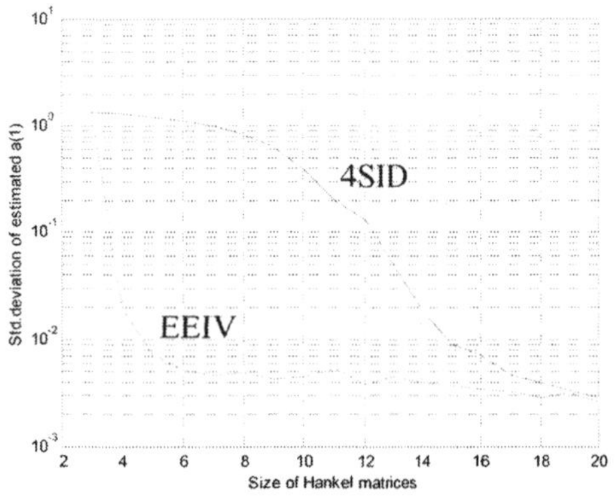

Fig. 3 The standard deviation of the estimated a_1 parameter

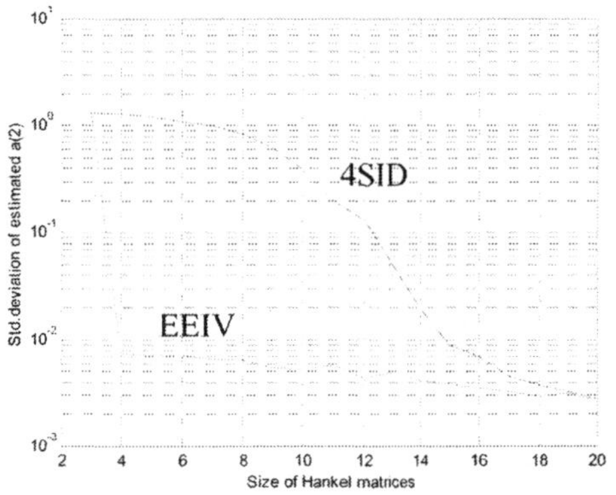

Fig. 4 The standard deviation of the estimated a_2 parameter

Analysing the results shown in Figs. 3 and 4 it is seen how EEIV superiors the results obtained using the 4SID technique. Also, it is quite surprising how quickly the variance of the parameter estimation drops when increasing the size of the Hankel matrices involved in the estimation algorithm. The general experience from a number of simulation studies says that in case of low signal/noise ratio or small sampling time EEIV gives far better results than 4SID.

CONCLUSION

The paper derives a framework suitable to discuss the errors-in-variables (EIV) and the maximum likelihood (ML) estimation algorithms to estimate linear system parameters in a unified way. More than that, it has been shown that the size of the Hankel matrices introduced by this unified framework is not only limited to be set to match the EIV or the ML algorithm, but it can be left as a free design factor to extend the original EIV concept. Utilizing the flexibility while selecting the size of the Hankel matrices results in a very efficient parameter estimation algorithm. The EEIV concept presented in the paper can be generalized for Multiple Input Multiple Output linear systems in a straightforward way.

ACKOWLEGDEMENT

This work has been supported by the fund of the Hungarian Academy of Sciences for control research and the Hungarian National Research Fund (grant number T42741).

REFERENCES

Castaldi, P. and U. Soverini (1996). Identification of dynamic errors-in-variables models. *Automatica,* Vol.32, No.4, pp. 631-636.

Chou, C.T. and M. Verhaegen (1997). Subspace algorithms for the identification of multivariable dynamic errors-in-variables models. *Automatica,* Vol.33, No.10, pp. 1857-1869.

Fernando, K.V. and H. Nicholson (1985). Identification of linear systems with input and output noise: the Koopmans-Levin method. *IEE Proceedings*, Vol.132.Pt.D, No.1, pp. 30-36.

Heij, C. and W. Scherrer (1999). Consistency of system identification by global total least squares. *Automatica* Vol.35, pp. 993-1008.

Koopmans, T. (1937). Linear regression analysis of economic time series. DeErven F.Bohn, N.V. Haarlem, The Netherlands.

Levin, M,J. (1964). Estimation of a system pulse transfer function in the presence of noise. *IEEE Trans. on Automatic Control,* pp. 229-235.

Mahata, K. and T. Söderström (2002): Identification of dynamic errors-in-variables model using prefiltered data. *Proc. 15th IFAC World Congress*, Barcelona, Spain, July 21-26.

Vajk, I. and J. Hetthéssy (2001). An improved version of the Koopmans-Levin algorithm. *ECC'01 European Control Conference*, Porto, Portugal, pp. 1864-1869.

AUTHOR INDEX

Abd-Elrady, E. 1543
Abel, D. 151
Adachi, S. 1423
Adel, M. El 1107, 1339
Agüero, J.C. 771
Ahmed, J. 939
Aihara, S. 1681
Akçay, H. 1843
Akizuki, K. 181
Albertos, P. 1203, 1897
Alcorta-Garcia, M.A. 447
Amaral, W.C. 1807
Andreff, N. 945
Andrieu, C. 1275
Aoun, M. 1333, 1663
Arahal, M.R. 205
Arruda, G.H.M. de 399
Astrid, P. 1393
Ataei, M. 169
Auweraer, H. van der 669, 735

Babuška, V. 699
Bagchi, A. 1681
Bai, E.-W. 819
Bányász, C. 1167
Barabanov, A.E. 1909
Bardow, A. 1185
Barker, H.A. 633, 663
Barreras, M. 429
Barros, P.R. 393, 399
Bars, R. 477
Bartlett, P.L. 1465
Basin, M. 447, 1005
Basseville, M. 693, 1363
Bastogne, T. 1239, 1927
Battaglia, J.-L. 1663
Battipede, M. 279
Bauer, D. 993, 1741, 1753
Becerra, V.M. 363
Beheshti, S. 765
Bekara, M. 777
Belder, K. de 375
Belkoura, L. 705
Bemporad, A. 1789
Benítez, I. 1203
Benveniste, A. 693, 1363
Berendrecht, W.L. 1933
Bergboer, N. 711
Bernstein, D.S. 939, 1537
Besançon, G. 1699
Besançon-Voda, A. 1861
Bhikkaji, B. 1669
Bianchi, M. 975
Bibes, G. 1227
Bissacco, A. 1375
Bitmead, R.R. 591, 651, 1561, 1831
Bittanti, S. 211, 267, 381, 1101
Bloch, G. 453
Blomqvist, A. 1327
Bodson, P. 285
Bogacz, M. 687
Bogaerts, P. 145
Bohlin, T. 1477
Bøjstrup, K. 1633
Bokor, J. 1309, 1657
Bombois, X. 21, 27
Bombois, X.J.A. 1825
Bonné, D. 1615
Bonvin, D. 1137
Börner, M. 261
Bos, R. 1125
Brabanter, J. de 79
Brabec, B. 199
Braems, I. 297, 1819
Brasseur, C. 555
Braun, M.W. 891
Brendel, M. 1627
Bretthauer, G. 1285
Brie, D. 627
Broersen, P.M.T. 1125, 1435
Bukkems, B. 921

Cadic, M. 39
Callafon, R.A. de 507, 519, 1561, 1573
Camacho, E.F. 205
Campello, R.J.G.B. 1807
Candau, Y. 1675, 1687
Carmona, J.C. 1107, 1339
Castelli, S. 1915
Cauberghe, B. 681, 1609
Cerone, V. 849
Chamaillard, Y. 453
Chapuis, R. 1945
Chayanan, S. 1197
Chellappa, M. 939
Chen, H.-F. 1963
Chessari, C. 609
Chiras, N. 1161
Chiuso, A. 237, 855, 1747
Coelho, F.S. 393
Coirault, P. 1227
Corani, G. 1915
Craig, I.K. 411
Crama, P. 61, 831
Crowder, M. 519
Crowe, J. 1621
Csató, L. 789

Dahleh, M.A. 765
Dancre, M. 1399
Danesin, D. 279
Dasgupta, S. 579
Datcu, S. 1675
Date, P. 21, 27
Deistler, M. 1191, 1885
Dekker, A.J. den 127
Deller Jr., J.R. 1969
Delourme, B. 1399
Denis-Vidal, L. 657
Dietz, F. 151
Díez, J.L. 1203
Dijk, D. van 221
Diversi, R. 1555, 1957
Dobrowiecki, T. 1723
Dochain, D. 1945
Dooren, P. van 1369
Doucet, A. 1275
Douma, S.G. 33, 1825
Driessen, P.F. 109
Dunstan, W.J. 651, 1831
Durieu, C. 1017
Dyomin, N.S. 1011

Edelmayer, A. 1657
Eitzinger, B. 717, 723
Enqvist, M. 1873
Erwin, R.S. 699
Espinoza, M. 555

Fanizza, G. 1327
Farachi, F. 267
Fassois, S.D. 157, 933
Favier, G. 1077, 1807
Fernando, P.P. 205
Filter, R. 1215
Fischer, D. 273
Fleury, B.H. 97
Fraanje, R. 867
Friehmelt, H. 387
Fuchs, J.J. 1119, 1315
Fujihira, T. 1423
Fujiwara, Y. 1423
Fukuda, J. 1263

Galdo, G. del 85
Galic, J. 1417
Gan, Q. 1801
Garatti, S. 211
García-Sanz, M. 429
Garnier, H. 405, 597, 969
Garulli, A. 1789
Gasso, K. 1089
Gautier, M. 285, 927
Gawthrop, P.J. 609
Geer, F.C. van 1933
Geering, H.P. 975
Gehrels, J.C. 1933
Georgakis, C. 909
Gerdin, M. 1489

Gestel, T. van 555, 1369
Gevers, M. 489, 645, 747, 1813
Giarré, L. 843
Gilles, F. 777
Gilson, M. 405, 513, 969
Giranzani, M. 381
Girard, A. 1155
Glad, T. 1489
Gland, F. Le 1269
Gleiß, A. 1191
Godfrey, K.R. 633, 663, 1297
Goethals, I. 675, 1369
Gogu, G. 945
Gómez, J.C. 1507
Gomm, J.B. 1083
Goodwin, G.C. 771
Gorlier, G. 567
Goursat, M. 1363
Greblicki, W. 825
Grimble, M.J. 1621
Gröll, L. 303
Grosfils, V. 441
Gruber, K. 1191
Guariso, G. 1915
Guidorzi, R. 1555, 1957
Guillaume, P. 375, 681, 1609
Gunn, S.R. 795
Guo, F. 1285
Guo, X. 1173
Gustafsson, F. 1053, 1245, 1251
Guyader, A. 1269, 1705

Haag, J.E. 145
Haardt, M. 85
Haber, R. 477
Hacioğlu, R. 351
Hagenblad, A. 1053
Han, S.H. 1029
Hanebeck, U.D. 315
Hanner, C. 57
Hanus, R. 501
Hanzon, B. 217, 233, 729, 1885
Hassaine, Y. 1399
Hassani, M.M. 1077
Hassibi, B. 1717
Hatanaka, T. 1903
Heath, W.P. 1951
Heemink, A.W. 1381, 1933
Hennhöfer, M. 85
Henriksen, S. 759
Herrero, J.L.N. 1233
Higelin, P. 453
Higuchi, T. 1263
Hildebrand, R. 489, 645, 1813
Hillström, L. 1651
Hirasawa, K. 1411
Hjalmarsson, H. 1, 45, 495
Hoegaerts, L. 813
Hof, P. van den 513
Hof, P.M.J. van den 33, 1825
Holst, J. 1495
Horváth, G. 801
Hosoyamada, Y. 1047
Hriljac, P. 1245
Hu, J. 1411
Huisman, L. 1645
Hung, P.C.F. 321
Huselstein, E. 969
Hussein, A. 1411

Ibos, L. 1675
Ikeda, K. 309
Inoue, K. 139, 1047
Irwin, G. 321, 1095
Irwin, G.W. 435
Isaksson, A.J. 1477, 1501
Isermann, R. 261, 273

Jager, B. de 921
Jansson, H. 45
Jansson, M. 1585, 1891
Jarvis, A. 597
Jauberthie, C. 657
Jaulin, L. 1819
Jia, L.-J. 1113
Jibetean, D. 1879
Joachim, D. 1969
Johansson, K.H. 291
Johnson Jr., C.R. 585
Johnson, M.A. 1143, 1621
Joly-Blanchard, G. 657
Jørgensen, S.B. 615, 1615, 1633
Jourdan, P. 97
Jutan, A. 1507
Juvva, N.K. 1197

Kaczmarek, P. 1065
Kadirkamanathan, V. 1257
Kameyama, K. 885
Kanae, S. 879, 1113
Kaneko, M. 483
Kaneko, O. 1855
Karimi, A. 1137
Karlström, A. 1795
Kasprzyk, J. 1441
Katayama, T. 873
Katebi, M.R. 1143
Kecman, V. 465
Kee, R. 321
Keesman, K.J. 1921
Keviczky, L. 1167
Khaki-Sedigh, A. 169
Khalil, W. 285
Kibangou, A.Y. 1077
Kieffer, M. 249, 297, 1819
Kinnaert, M. 441, 501
Kivinen, J. 1717
Klein, A.G. 585
Klein, I. 1053
Klerk, E. de 411
Kobayashi, T. 181
Kollár, I. 1459
Kostić, D. 921
Kosut, R.L. 121
Kowalczuk, Z. 1059
Královec, J. 1693
Krief, P. 279
Kristensen, N.R. 615
Kruger, U. 1603
Krutzen, M.P.M. 1939
Küchler, U. 1179
Kuh, A. 807
Kukreja, S.L. 1525
Kumamaru, K. 139, 1047
Kwon, W.H. 1029

Lacort, J.A. 1203
Lacy, S.L. 699
Larimore, W.E. 1453, 1771
Larsson, E.K. 621
Latawiec, K.J. 345
Lecchini, A. 489
Lecoeuche, S. 1597
Lee, H. 891, 915
Lee, J.H. 903, 1639
Leith, D.J. 1281
Leithead, W.E. 1281
Lemaire, C.-E. 927
Lemos, J.M. 951
Levrie, C. 441
Levron, F. 1333
Li, D. 1173
Li, H. 1173
Li, K. 1095
Li, P. 1257
Li, T. 909
Libaux, A. 1239, 1927
Lin, T. 585
Lind, I. 51
Lindkvist, R. 1501
Lindström, E. 1495
Ljung, L. 51, 861, 957, 1483, 1489, 1513, 1549, 1591, 1765, 1873
Lohmann, B. 169
Longo, D. 163, 369
Loon, E.E. van 1921
López-Valcarce, R. 573
Lovera, M. 381, 1345, 1531, 1579
Lübke-Ossenbeck, B. 381
Lundgren, A. 67

MacGregor, J.F. 981
Madani, K. 1663
Madsen, H. 615, 1495
Mahata, K. 1651, 1669
Mäkilä, P.M. 1867
Malti, R. 1333, 1663
Marchi, E. de 381
Marciak, C. 345
Markovsky, I. 1711
Marquardt, W. 1185, 1627
Marques, J.S. 951
Mårtensson, J. 495
Martinet, P. 945
Martinez-Zuniga, R. 1005
Martin, R.K. 585
Matsuüra, Y. 885
Matteï, S. 1675
Matthes, J. 303
Matyus, T. 1191
Maulana, F. 471
McKelvey, T. 57, 1351
McLoone, S. 321
Megretski, A. 1387
Melin, A.M. 699
Mercère, G. 1597
Mevel, L. 693, 1363

Milanese, M. 849, 1519
Milek, J. 199
Millerioux, G. 453
Mittelmann, H.D. 891, 915
Mišković, L. 1137
Moer, W. van 327
Mogami, Y. 309
Moor, B. de 79, 555, 675, 813, 1369, 1711
Moor, B.L.R. de 1759
Morelli, E.A. 639
Mossberg, M. 621
Mosskull, H. 1417
Mourot, G. 1089
Moussaoui, S. 627
M'Saad, M. 423
Murray-Smith, R. 1155, 1281
Muscato, G. 369

Nádai, L. 1657
Nakano, K. 333, 1429
Namvar, M. 1861
Nanto, H. 879
Narasimhan, S. 897
Nazaruddin, Y.Y. 471
Nazin, A. 1513
Nazin, S.A. 1017
Nemeth, J. 61
Németh, J.G. 339
Neuper, C. 139
Niedźwiecki, M. 1065
Nielsen, H.A. 1495
Niemann, H.H. 1633
Ninness, B. 759
Nordin, M. 1501
Nørgaard, M. 1447
Novara, C. 1519
Nunnari, G. 163, 369

Oh, J. 1537
Ohmori, H. 187, 1221
Ohsumi, A. 885
Ohta, Y. 187
Oku, H. 1357
Oliveira, G.H.C. 345
Ooi, S.K. 1939
Opper, M. 789
Orlicki, C.E. 585
Oudjane, N. 1269
Ouladsine, M. 1107, 1339
Oustaloup, A. 1333, 1663
Ouvrard, R. 1227
Ozaki, T. 333, 1429

Pagani, I. 1555
Panciatici, P. 1399
Pan, Y. 1639
Paoletti, S. 1789
Parloir, C. 501
Parloo, E. 1609
Parrilo, P.A. 1483
Peeters, B. 669, 729
Pekpe, K.M. 1089
Pelckmans, K. 79
Peñarrocha, I. 1897
Peng, H. 333, 1429
Pérez, J.A.R. 1233
Perreau, S. 567
Perrier, M. 1945
Pešek, L. 459
Pettersson, N. 291
Pfurtscheller, G. 139
Picci, G. 855, 1747
Pintelon, R. 61, 375, 831, 867, 1459, 1723, 1837
Piroddi, L. 357, 1071
Podsiadly, T. 609
Polderman, J.W. 39
Polyak, B.T. 1017
Pontil, M. 783
Pötscher, B.M. 1735
Poulimenos, A.G. 157, 933
Pouliquen, M. 423
Poulsen, N.K. 1447, 1633
Previdi, F. 1531
Pronzato, L. 537
Půst, L. 459

Qin, S.J. 861, 1591, 1603

Rabitz, H. 121
Ragot, J. 1089
Rahbek, A. 227
Raïssi, T. 1687
Ramdani, N. 1675, 1687, 1819
Rapisarda, P. 1855
Rasmussen, C.E. 1155
Ravn, O. 1447
Rees, D. 1161
Regalia, P.A. 561
Regruto, D. 849
Re, L. Del 1405
Renaud, P. 945
Rengaswamy, R. 897
Ren, L. 435
Ribarits, T. 1885
Richard, A. 627, 1927
Rivera, D.E. 891, 915, 1567
Rohlf, D. 387
Rojek, R. 345
Rolain, Y. 61, 327, 375, 1459, 1723, 1837
Roll, J. 1513
Rosales, E. 1801
Rosenqvist, F. 1795
Rozhkova, S.V. 1011
Ruotolo, R. 669
Rylander, T. 1351

Saccomani, M.P. 1209
Safronova, I.E. 1011
Saisan, P. 1375
Saito, K. 1903
Samson, R. 1945
Sanchez, A. 1143
Sanchis, R. 1897
Sankowski, M. 1059
Sano, A. 187, 193, 1041
Sanyal, A.K. 939
Sassi, G. 279
Sato, J. 1221
Savaresi, S.M. 211, 267, 1101
Sayed, A.H. 1023
Sbarbaro, D. 1155
Schaback, R. 1777
Schaedel, H.M. 963
Scherrer, W. 1849
Schipp, F. 1309
Schlacher, K. 717, 723
Schloßer, A. 151
Schmitz, U. 477
Schoefs, O. 1945
Schön, T. 1251
Schoukens, J. 61, 327, 339, 831, 1459, 1723, 1837
Schuppen, J.H. van 1879
Segouane, A.-K. 777
Sembiring, J. 181
Shafai, E. 975
Shah, J. 261
Shankar, V.N. 1197
Shen, J. 939
Shimomura, T. 309
Shin, S. 417
Shioya, H. 1429
Sijbers, J. 127
Silani, E. 267, 381
Silva, J.G. 951
Šimandl, M. 175, 1693
Sima, V. 1303
Simon, G. 1459
Sitter, G. de 1609
Sjöberg, J. 67
Skelton, R.E. 771
Śliwiński, P. 825
Smola, A.J. 549
Soatto, S. 237, 1375
Söderström, T. 1047, 1543, 1651, 1669
Soemintapoera, K. 181
So, H.C. 1101
Solari, G. 489
Solomou, M. 1161
Song, X. 579
Soverini, U. 1957
Spinelli, W. 357, 1071
Spreitzer, K. 465
Srinivasan, R. 897
Srinivasa, Y.G. 261
Steinbuch, M. 921
Steyer, J.P. 1227
Stoica, P. 1891
Straka, O. 175
Stroet, C.B.M Te 1381
Stucki, A. 97
Subramanian, A. 1023
Suchomski, P. 1035
Sugiyama, M. 73
Sun, L. 193
Suykens, J. 1369
Suykens, J.A.K. 79, 555, 813
Svantesson, T. 91

Tabaru, T. 417
Tadić, V.B. 1275
Takei, Y. 879
Tallfors, M. 1501
Tan, A.H. 633, 663, 1297
Tanaka, H. 873
Tapio, M. 103
Terashima, K. 483

Thierry, É. 537
Thomassin, M. 1927
Tjärnström, F. 1849
Tong, L. 115
Torres, S. 363
Toyoda, Y. 333, 1429
Twerda, A. 1393

Uhl, T. 687
Uosaki, K. 1903

Vajk, I. 1975
Valk, J.L. 939
Valyon, J. 801
Vandanjon, P.-O. 927
Vandewalle, J. 79, 813
Vaněk, F. 459
Vanlanduit, S. 1609
Vasil'iev, V. 1179
Vasseur, C. 1597
Vazquez, E. 1783
Vecchio, A. 669
Veen, A.-J. van der 115
Velardocchia, M. 279
Venture, G. 285
Verboven, P. 681, 1609
Verdult, V. 711, 867
Veres, S.M. 15
Verhaegen, M. 711, 867
Vermeulen, P.T.M. 1381
Viberg, M. 103, 1351
Vicino, A. 1789
Vishwanathan, S.V.N. 549
Vogt, M. 465

Wackernagel, H. 543
Wada, K. 879, 1113
Waele, S. de 1125, 1753
Wahba, G. 525, 531
Wahlberg, B. 1321, 1417
Wallace, J.W. 91
Walmsley, I.A. 121
Walsh, J.M. 585
Walter, E. 249, 297, 1017, 1399, 1783, 1819
Wang, H. 133, 1131
Wang, L. 609
Wang, L.Y. 133
Wang, Y. 1131
Warmuth, M.K. 1717
Weiland, S. 39, 1393, 1645
Wei, X. 1405
Weyer, E. 1191, 1939
Wigren, T. 837, 1543
Willcox, K. 1387
Williams, D. 1083
Williamson, G.A. 351
Wouwer, A.V. 145, 441

Xia, X. 1215
Xin, J. 1041
Xu, A. 1699
Xu, H. 579

Yamamoto, T. 1149
Yang, Z.-J. 879, 1113
Yano, K. 483
Yin, G.G. 133
Yin, X. 97
Young, P.C. 597
Yu, D.L. 1083
Yu, D.W. 1083

Zappa, G. 843
Zeng, J. 507
Zhang, C. 591
Zhang, Q. 1699, 1705
Zhang, X. 1501
Zhu, Y. 1291
Zimmer, M. 273